Miscellaneous

Recreation, Entertainment, and Travel

Science, Technology, and Engineering

Sports

REFERENCE CARD

INTERMEDIATE ALGEBRA Fifth Edition
R. David Gustafson and Peter D. Frisk
ISBN 0-534-36049-1

Brooks/Cole Publishing Company
I(T)P™ An International Thomson Publishing Company
Visit Brooks/Cole on the Internet
http://www.brookscole.com

CHAPTER 1 BASIC CONCEPTS

Properties of exponents:

If there are no divisions by 0, then

$$x^m x^n = x^{m+n} \qquad (x^m)^n = x^{mn}$$

$$(xy)^n = x^n y^n \qquad \left(\frac{x}{y}\right)^n = \frac{x^n}{y^n}$$

$$x^0 = 1 \qquad x^{-n} = \frac{1}{x^n}$$

$$\frac{x^m}{x^n} = x^{m-n} \qquad \left(\frac{x}{y}\right)^{-n} = \left(\frac{y}{x}\right)^n$$

CHAPTER 2 GRAPHS, EQUATIONS OF LINES, AND FUNCTIONS

Midpoint formula: If $P(x_1, y_1)$ and $Q(x_2, y_2)$, the midpoint of segment PQ is

$$M\left(\frac{x_1 + x_2}{2}, \frac{y_1 + y_2}{2}\right)$$

Slope of a nonvertical line: If $x_1 \neq x_2$, then

$$m = \frac{\Delta y}{\Delta x} = \frac{y_2 - y_1}{x_2 - x_1}$$

Equations of a line:

Point–slope form: $y - y_1 = m(x - x_1)$
Slope–intercept form: $y = mx + b$
General form: $Ax + By = C$
Horizontal line: $y = b$
Vertical line: $x = a$

CHAPTER 3 SYSTEMS OF EQUATIONS

$$\begin{vmatrix} a & b \\ c & d \end{vmatrix} = ad - bc$$

CHAPTER 4 INEQUALITIES

If $k > 0$, then

$|x| = k$ is equivalent to $x = k$ or $x = -k$.
$|a| = |b|$ is equivalent to $a = b$ or $a = -b$.
$|x| < k$ is equivalent to $-k < x < k$.
$|x| > k$ is equivalent to $x < -k$ or $x > k$.

CHAPTER 5 POLYNOMIALS AND POLYNOMIAL FUNCTIONS

Factoring formulas:
Difference of two squares:

$$x^2 - y^2 = (x + y)(x - y)$$

Sum and difference of two cubes:

$$x^3 + y^3 = (x + y)(x^2 - xy + y^2)$$
$$x^3 - y^3 = (x - y)(x^2 + xy + y^2)$$

CHAPTER 6 RATIONAL EXPRESSIONS

Variation: If k is a constant, then

$y = kx$ represents direct variation.

$y = \frac{k}{x}$ represents inverse variation.

$y = kxz$ represents joint variation.

$y = \frac{kx}{z}$ represents combined variation.

CHAPTER 7 RATIONAL EXPONENTS AND RADICALS

The distance formula:

$$d(PQ) = \sqrt{(x_2 - x_1)^2 + (y_2 - y_1)^2}$$

If x, y, and n are real numbers and $x = y$, then $x^n = y^n$.

Fractional exponents: If m and n are positive integers and $x > 0$, then

$$x^{m/n} = \sqrt[n]{x^m} = \left(\sqrt[n]{x}\right)^m \qquad x^{-m/n} = \frac{1}{x^{m/n}}$$

If at least one of a or b is positive, then

$$\sqrt[n]{ab} = \sqrt[n]{a}\sqrt[n]{b} \qquad \sqrt[n]{\frac{a}{b}} = \frac{\sqrt[n]{a}}{\sqrt[n]{b}} \; (b \neq 0)$$

CHAPTER 8 QUADRATIC FUNCTIONS, INEQUALITIES, AND ALGEBRA OF FUNCTIONS

The quadratic formula:

$$x = \frac{-b \pm \sqrt{b^2 - 4ac}}{2a} \; (a \neq 0)$$

Complex numbers: If a, b, c, and d are real numbers and $i^2 = -1$, then

$$(a + bi) + (c + di) = (a + c) + (b + d)i$$

$$(a + bi)(c + di) = (ac - bd) + (ad + bc)i$$

$$|a + bi| = \sqrt{a^2 + b^2}$$

CHAPTER 9 EXPONENTIAL AND LOGARITHMIC FUNCTIONS

Compound interest: $A = P\left(1 + \dfrac{r}{k}\right)^{kt}$

Continuous compound interest: $A = Pe^{rt}$

Malthusian population growth: $P = P_0e^{kt}$

Decibel voltage gain: db gain $= 20 \log \dfrac{E_o}{E_I}$

Richter scale: $R = \log \dfrac{A}{P}$

Doubling time: $t = \dfrac{\ln 2}{r}$

Properties of logarithms: If M, N, and b are positive numbers and $b \neq 1$, then

$\log_b 1 = 0 \qquad \log_b b = 1$

$\log_b b^x = x \qquad b^{\log_b x} = x$

$\log_b MN = \log_b M + \log_b N$

$\log_b \dfrac{M}{N} = \log_b M - \log_b N$

$\log_b M^P = p \log_b M \qquad$ If $\log_b x = \log_b y$, then $x = y$.

Change-of-base formula: $\log_b y = \dfrac{\log_a y}{\log_a b}$

pH scale: $\text{pH} = -\log[\text{H}^+]$

Weber-Fechner law: $L = k \ln I$

Carbon dating: $A = A_0 2^{-t/h}$

CHAPTER 10 MORE GRAPHING AND CONIC SECTIONS

Equations of a circle with radius *r*:

$(x - h)^2 + (y - k)^2 = r^2$ center at (h, k)

$x^2 + y^2 = r^2$ center at $(0, 0)$

Equations of a parabola: If $a > 0$, then

Parabola opening	Vertex at origin	Vertex at (h, k)
Up	$y = ax^2$	$y = a(x - h)^2 + k$
Down	$y = -ax^2$	$y = -a(x - h)^2 + k$
Right	$x = ay^2$	$x = a(y - k)^2 + h$
Left	$x = -ay^2$	$x = -a(y - k)^2 + h$

Equations of an ellipse:

Center at $(0, 0)$ if $a > b > 0$

$$\frac{x^2}{a^2} + \frac{y^2}{b^2} = 1 \qquad \frac{x^2}{b^2} + \frac{y^2}{a^2} = 1$$

Center at (h, k)

$$\frac{(x - h)^2}{a^2} + \frac{(y - k)^2}{b^2} = 1 \ (a > b > 0)$$

$$\frac{(x - h)^2}{b^2} + \frac{(y - k)^2}{a^2} = 1 \ (a > b > 0)$$

Equations of a hyperbola:

Center at $(0, 0)$

$$\frac{x^2}{a^2} - \frac{y^2}{b^2} = 1 \qquad \frac{y^2}{a^2} - \frac{x^2}{b^2} = 1$$

Center at (h, k)

$$\frac{(x - h)^2}{a^2} - \frac{(y - k)^2}{b^2} = 1$$

$$\frac{(y - k)^2}{a^2} - \frac{(x - h)^2}{b^2} = 1$$

CHAPTER 11 MISCELLANEOUS TOPICS

The binomial theorem:

$$(a + b)^n = a^n + \frac{n!}{1!(n - 1)!}a^{n-1}b + \frac{n!}{2!(n - 2)!}a^{n-2}b^2 + \cdots + b^n$$

Arithmetic sequence:

$$S_n = \frac{n(a + l)}{2} \quad \text{(sum of the first } n \text{ terms)}$$

Geometric sequence:

$$S_n = \frac{a - ar^n}{1 - r} \ (r \neq 1) \quad \text{(sum of the first } n \text{ terms)}$$

The sum of all the terms of the sequence is

$$S = \frac{a}{1 - r} \quad (|r| < 1)$$

Formulas for permutations and combinations:

$$P(n, r) = \frac{n!}{(n - r)!} \qquad P(n, n) = n!$$

$$P(n, 0) = 1 \qquad C(n, r) = \binom{n}{r} = \frac{n!}{r!(n - r)!}$$

$$C(n, n) = \binom{n}{n} = 1 \qquad C(n, 0) = \binom{n}{0} = 1$$

FIFTH EDITION

Intermediate Algebra

To
Harold and Monie,
Harder and Evelyn,
with love and affection

Books in the Gustafson/Frisk Series

Essential Mathematics with Geometry Third Edition
Beginning Algebra Fifth Edition
Beginning and Intermediate Algebra: An Integrated Approach Second Edition
Intermediate Algebra Fifth Edition
Concepts of Intermediate Algebra
Algebra for College Students Fifth Edition
College Algebra Sixth Edition

FIFTH EDITION

Intermediate Algebra

R. David Gustafson

Rock Valley College

Peter D. Frisk

Rock Valley College

Brooks/Cole Publishing Company

I(T)P™ An International Thomson Publishing Company

Pacific Grove • Albany • Bonn • Boston • Cincinnati • Detroit • London • Madrid • Melbourne
Mexico City • New York • Paris • San Francisco • Singapore • Tokyo • Toronto • Washington

Publisher: *Robert W. Pirtle*
Marketing Team: *Jennifer Huber, Christine Davis, Debra Johnston*
Editorial Assistant: *Peggi Rodgers*
Production Editor: *Ellen Brownstein*
Production Service: *Hoyt Publishing Services*
Manuscript Editor: *David Hoyt*
Permissions Editor: *Carline Haga*
Interior Design: *E. Kelly Shoemaker, Vernon Boes*
Interior Illustration: *Lori Heckelman*
Photo Research: *Terry Powell*
Cover Design: *Roy Neuhaus*
Cover Photo: *Spencer Grant/FPG International*
Art Coordinator: *David Hoyt*
Typesetting: *The Clarinda Company*
Cover Printing: *Phoenix Color Corp.*
Printing and Binding: *World Color Book Services (Taunton)*

For more information, contact:

BROOKS/COLE PUBLISHING COMPANY
511 Forest Lodge Road
Pacific Grove, CA 93950
USA

International Thomson Publishing Europe
Berkshire House 168-173
High Holborn
London WC1V 7AA
England

Thomas Nelson Australia
102 Dodds Street
South Melbourne, 3205
Victoria, Australia

Nelson Canada
1120 Birchmount Road
Scarborough, Ontario
Canada M1K 5G4

International Thomson Editores
Seneca 53
Col. Polanco
11560 México, D. F., México

International Thomson Publishing GmbH
Königswinterer Strasse 418
53227 Bonn
Germany

International Thomson Publishing Asia
221 Henderson Road
#05-10 Henderson Building
Singapore 0315

International Thomson Publishing Japan
Hirakawacho Kyowa Building, 3F
2-2-1 Hirakawacho
Chiyoda-ku, Tokyo 102
Japan

Printed in the United States of America

10 9 8 7 6 5 4 3 2

Library of Congress Cataloging-in-Publication Data

Gustafson, R. David (Roy David), [date]
Intermediate algebra / R. David Gustafson, Peter D. Frisk.—5th ed.
p. cm.
Includes index.
ISBN 0-534-36049-1 (alk. paper)
1. Algebra. I. Frisk, Peter D., [date]. II. Title.
QA154.2.G874 1998
512.9—dc21 98-38389
CIP

THIS BOOK IS PRINTED ON ACID-FREE RECYCLED PAPER

Photo credits: p. 2, Philip Gould/Corbis; **p. 46,** The British Museum; **p. 52 (both),** Frank Rossotto/The Stock Market; **p. 95,** Leif Skoogfors/Corbis; **p. 117,** Courtesy of Texas Instruments; **p. 187,** Roger Ressmeyer/©Corbis; **p. 252,** Ted Streshinsky/Corbis; **p. 316,** Richard T. Nowitz/Corbis; **p. 400,** Rob Rowan; Progressive Image/Corbis; **p. 497,** Roger Ressmeyer/©Corbis; **p. 570,** Charles E. Rotkin/©Corbis; **p. 658,** Ed Young/Corbis; **p. 719,** Archaeological Consulting/Gary Breschini & Trudy Haverstat; **p. 731,** David H. Wells/Corbis; **p. 793,** Liba Taylor/Corbis.

To the Instructor

Intermediate Algebra, Fifth Edition, is the second of a two-volume series designed to prepare students for college mathematics. It presents all of the topics associated with a second course in algebra. We believe that it will hold student attrition to a minimum while preparing students to succeed—whether in college algebra, trigonometry, precalculus, statistics, finite mathematics, liberal arts mathematics, or everyday life.

Our goal has been to write a book that

- Is enjoyable to read.
- Is easy to understand.
- Is relevant.
- Develops the skills necessary for success in future academic courses or on the job.

Although the material has been extensively revised, this fifth edition retains the basic philosophy of the highly successful previous editions. The revisions include several improvements in line with the NCTM standards, the AMATYC Crossroads, and the current trends in mathematics reform. For example, more emphasis has been placed on graphing and problem solving.

GENERAL CHANGES IN THE FIFTH EDITION

The general changes made to the fifth edition are as follows:

- **To modernize the material** with a new table of contents that introduces graphing and functions in Chapter 2.

- **To increase the emphasis on problem solving through realistic applications.** The variety of application problems has been increased significantly, and all application problems are labeled with special titles.

Graphics have been improved. ▶

All application problems have titles. ▶

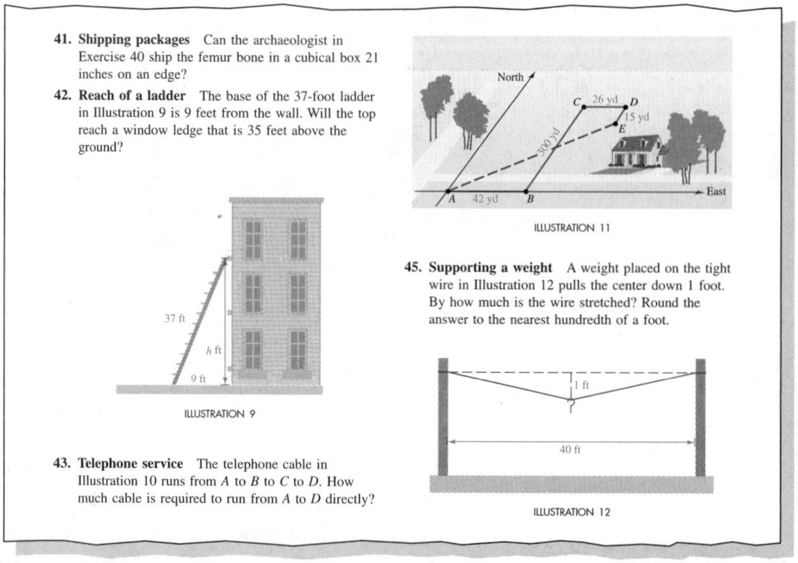

41. Shipping packages Can the archaeologist in Exercise 40 ship the femur bone in a cubical box 21 inches on an edge?

42. Reach of a ladder The base of the 37-foot ladder in Illustration 9 is 9 feet from the wall. Will the top reach a window ledge that is 35 feet above the ground?

ILLUSTRATION 9

43. Telephone service The telephone cable in Illustration 10 runs from A to B to C to D. How much cable is required to run from A to D directly?

ILLUSTRATION 11

45. Supporting a weight A weight placed on the tight wire in Illustration 12 pulls the center down 1 foot. By how much is the wire stretched? Round the answer to the nearest hundredth of a foot.

ILLUSTRATION 12

- **To increase the emphasis on families of functions** and their graphs, including translations of their graphs.
- **To increase the visual interest by the use of color.** We include color not just as a design feature, but to highlight terms that instructors would point to in a classroom discussion.

Definitions are boxed in purple. ▶

Functions are classified into families. ▶

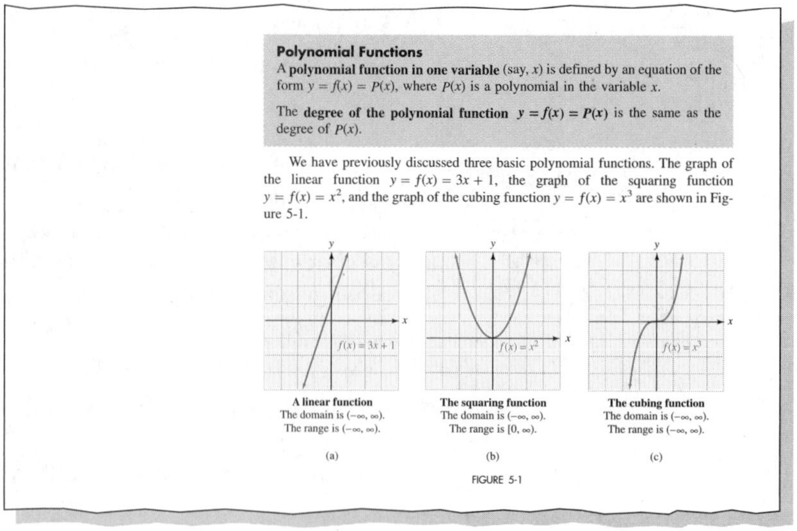

Polynomial Functions

A **polynomial function in one variable** (say, x) is defined by an equation of the form $y = f(x) = P(x)$, where $P(x)$ is a polynomial in the variable x.

The **degree of the polynomial function** $y = f(x) = P(x)$ is the same as the degree of $P(x)$.

We have previously discussed three basic polynomial functions. The graph of the linear function $y = f(x) = 3x + 1$, the graph of the squaring function $y = f(x) = x^2$, and the graph of the cubing function $y = f(x) = x^3$ are shown in Figure 5-1.

A linear function The domain is $(-\infty, \infty)$. The range is $(-\infty, \infty)$. (a)

The squaring function The domain is $(-\infty, \infty)$. The range is $[0, \infty)$. (b)

The cubing function The domain is $(-\infty, \infty)$. The range is $(-\infty, \infty)$. (c)

FIGURE 5-1

- **To fine-tune the presentation of many topics** for better flow of ideas and for clarity.

- **To increase the emphasis on learning mathematics through graphing.** Although graphing calculators are discussed frequently, their use is not mandatory. All of the topics are fully discussed in traditional ways. Of course, we recommend that instructors use the graphing calculator material.

GRAPHING CALCULATORS

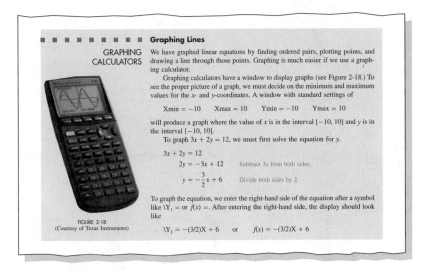

Graphing Lines

We have graphed linear equations by finding ordered pairs, plotting points, and drawing a line through those points. Graphing is much easier if we use a graphing calculator.

Graphing calculators have a window to display graphs (see Figure 2-18.) To see the proper picture of a graph, we must decide on the minimum and maximum values for the x- and y-coordinates. A window with standard settings of

$$\text{Xmin} = -10 \quad \text{Xmax} = 10 \quad \text{Ymin} = -10 \quad \text{Ymax} = 10$$

will produce a graph where the value of x is in the interval $[-10, 10]$ and y is in the interval $[-10, 10]$.

To graph $3x + 2y = 12$, we must first solve the equation for y.

$$3x + 2y = 12$$
$$2y = -3x + 12 \quad \text{Subtract } 3x \text{ from both sides.}$$
$$y = -\frac{3}{2}x + 6 \quad \text{Divide both sides by 2.}$$

To graph the equation, we enter the right-hand side of the equation after a symbol like $\backslash Y_1 =$ or $f(x) =$. After entering the right-hand side, the display should look like

$$\backslash Y_1 = -(3/2)X + 6 \quad \text{or} \quad f(x) = -(3/2)X + 6$$

FIGURE 2-18
(Courtesy of Texas Instruments)

SPECIFIC CHANGES IN THE FIFTH EDITION

To make the book more useful to students, we have:

- **Included Vocabulary and Concepts problems in each exercise set,** to help students clarify basic ideas. The Hints on Studying Algebra that appear in this preface recommend that students begin their study time with review, so Review problems are placed at the beginning of each exercise set. Most exercise sets follow this sequence:

1. Review
2. Vocabulary and Concepts
3. Practice
4. Applications
5. Writing
6. Something to Think About

EXERCISE 9.1

Review problems ▶

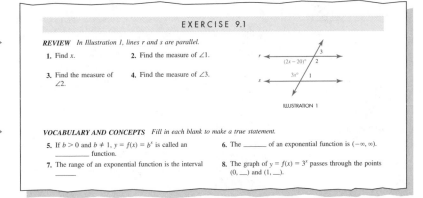

REVIEW *In Illustration 1, lines r and s are parallel.*

1. Find x.
2. Find the measure of $\angle 1$.
3. Find the measure of $\angle 2$.
4. Find the measure of $\angle 3$.

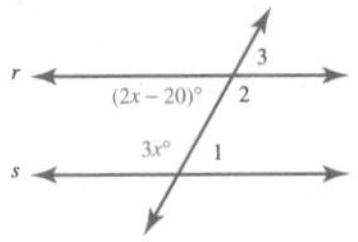

ILLUSTRATION 1

Vocabulary and Concepts problems ▶

VOCABULARY AND CONCEPTS *Fill in each blank to make a true statement.*

5. If $b > 0$ and $b \neq 1$, $y = f(x) = b^x$ is called an ________ function.
6. The ______ of an exponential function is $(-\infty, \infty)$.
7. The range of an exponential function is the interval ______
8. The graph of $y = f(x) = 3^x$ passes through the points (0, __) and (1, __).

Practice problems ▶

PRACTICE *In Exercises 13–16, find each value to four decimal places.*

13. $2^{\sqrt{2}}$ **14.** $7^{\sqrt{2}}$ **15.** $5^{\sqrt{5}}$ **16.** $6^{\sqrt{3}}$

In Exercises 17–20, simplify each expression.

17. $(2^{\sqrt{3}})^{\sqrt{3}}$ **18.** $3^{\sqrt{2}}3^{\sqrt{18}}$ **19.** $7^{\sqrt{3}}7^{\sqrt{12}}$ **20.** $(3^{\sqrt{5}})^{\sqrt{5}}$

In Exercises 21–28, graph each exponential function. Check your work with a graphing calculator.

21. $y = f(x) = 3^x$

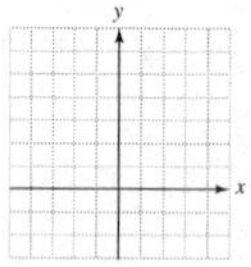

22. $y = f(x) = 5^x$

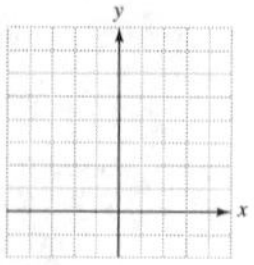

23. $y = f(x) = \left(\frac{1}{3}\right)^x$

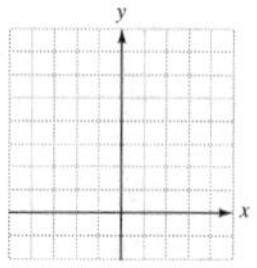

24. $y = f(x) = \left(\frac{1}{5}\right)^x$

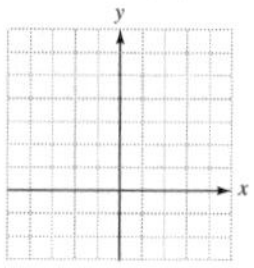

Application problems ▶

APPLICATIONS *In Exercises 41–46, assume that there are no deposits or withdrawals.*

41. Compound interest An initial deposit of \$10,000 earns 8% interest, compounded quarterly. How much will be in the account after 10 years?

42. Compound interest An initial deposit of \$10,000 earns 8% interest, compounded monthly. How much will be in the account after 10 years?

43. Comparing interest rates How much more interest could \$1,000 earn in 5 years, compounded quarterly, if the annual interest rate were $5\frac{1}{2}$% instead of 5%?

44. Comparing savings plans Which institution in Illustration 2 provides the better investment?

Fidelity Savings & Loan
Earn 5.25%
compounded monthly

Union Trust
Money Market Account
paying 5.35%,
compounded annually

interest. In the first account, interest compounds quarterly, and in the second account, interest compounds daily. Find the difference between the accounts after 20 years.

47. Radioactive decay A radioactive material decays according to the formula $A = A_0(\frac{2}{3})^t$, where A_0 is the initial amount present and t is measured in years. Find the amount present in 5 years.

48. Bacteria cultures A colony of 6 million bacteria is growing in a culture medium. (See Illustration 3.) The population P after t hours is given by the formula $P = (6 \times 10^6)(2.3)^t$. Find the population after 4 hours.

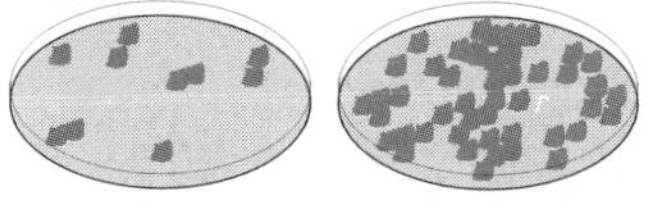

12:00 P.M. 4:00 P.M.

ILLUSTRATION 3

51. Salvage value A small business purchases a computer for \$4,700. It is expected that its value each year will be 75% of its value in the preceding year. If the business disposes of the computer after 5 years, find its salvage value (the value after 5 years).

52. Louisiana Purchase In 1803, the United States acquired territory from France in the Louisiana Purchase. The country doubled its territory by adding 827,000 square miles of land for \$15 million. If the land has appreciated at the rate of 6% each year, what would one square mile of land be worth in 1996?

Writing problems ▶

WRITING

53. If world population is increasing exponentially, why is there cause for concern?

54. How do the graphs of $y = b^x$ differ when $b > 1$ and $0 < b < 1$?

Something to Think About ▶ problems

SOMETHING TO THINK ABOUT

55. In the definition of the exponential function, b could not equal 0. Why not?

56. In the definition of the exponential function, b could not be negative. Why not?

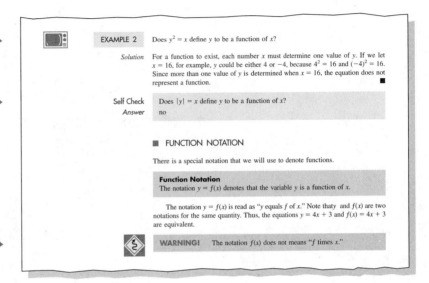

- **Included Self Checks with most examples,** to give students immediate reinforcement.
- **Increased the number of warnings** about common errors.

Examples worked on videotape are marked with a TV icon. ▶

Most examples have Self Checks. ▶

Warnings are marked with a special icon. ▶

EXAMPLE 2 Does $y^2 = x$ define y to be a function of x?

Solution For a function to exist, each number x must determine one value of y. If we let $x = 16$, for example, y could be either 4 or -4, because $4^2 = 16$ and $(-4)^2 = 16$. Since more than one value of y is determined when $x = 16$, the equation does not represent a function. ■

Self Check Does $|y| = x$ define y to be a function of x?

Answer no

■ FUNCTION NOTATION

There is a special notation that we will use to denote functions.

Function Notation
The notation $y = f(x)$ denotes that the variable y is a function of x.

The notation $y = f(x)$ is read as "y equals f of x." Note thaty and $f(x)$ are two notations for the same quantity. Thus, the equations $y = 4x + 3$ and $f(x) = 4x + 3$ are equivalent.

WARNING! The notation $f(x)$ does not means "f times x."

- **Increased the use of tables and graphs,** to help students understand the concept of function.
- **Included more geometric content,** to integrate the subjects of algebra and geometry.

■ GEOMETRIC PROBLEMS

Figure 1-25 illustrates several geometric figures. A **right angle** is an angle whose measure is 90°. A **straight angle** is an angle whose measure is 180°. An **acute angle** is an angle whose measure is greater than 0° but less than 90°.

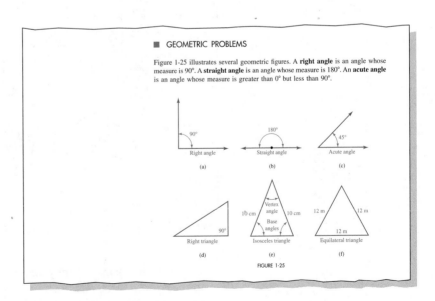

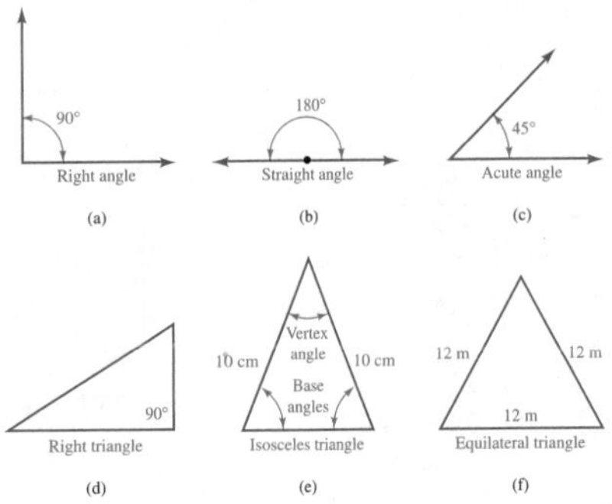

FIGURE 1-25

- **Included topics from statistics** as applications of algebra, to provide background for students, who will encounter these ideas in the future.

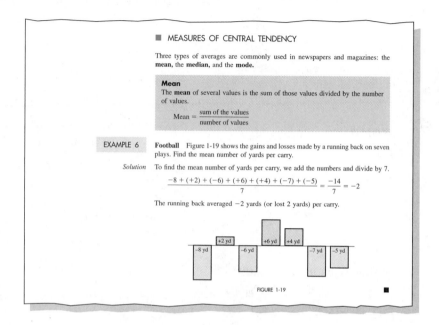

■ MEASURES OF CENTRAL TENDENCY

Three types of averages are commonly used in newspapers and magazines: the **mean,** the **median,** and the **mode.**

Mean
The **mean** of several values is the sum of those values divided by the number of values.

$$\text{Mean} = \frac{\text{sum of the values}}{\text{number of values}}$$

EXAMPLE 6 **Football** Figure 1-19 shows the gains and losses made by a running back on seven plays. Find the mean number of yards per carry.

Solution To find the mean number of yards per carry, we add the numbers and divide by 7.

$$\frac{-8 + (+2) + (-6) + (+6) + (+4) + (-7) + (-5)}{7} = \frac{-14}{7} = -2$$

The running back averaged −2 yards (or lost 2 yards) per carry.

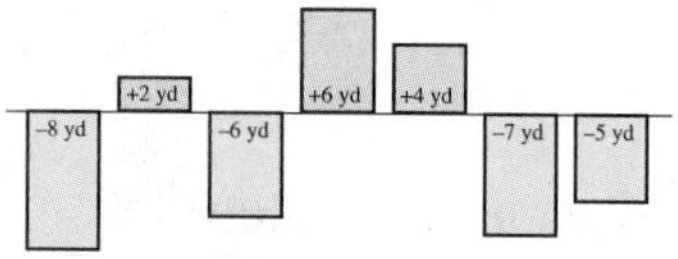

FIGURE 1-19 ■

- **Provided one or two projects** near the end of each chapter—a feature that gives an opportunity for group work or extended projects.

■ ■ ■ ■ ■ ■ ■ ■ ■ **PROJECTS**

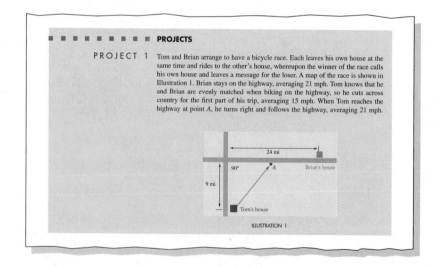

PROJECT 1 Tom and Brian arrange to have a bicycle race. Each leaves his own house at the same time and rides to the other's house, whereupon the winner of the race calls his own house and leaves a message for the loser. A map of the race is shown in Illustration 1. Brian stays on the highway, averaging 21 mph. Tom knows that he and Brian are evenly matched when biking on the highway, so he cuts across country for the first part of his trip, averaging 15 mph. When Tom reaches the highway at point *A*, he turns right and follows the highway, averaging 21 mph.

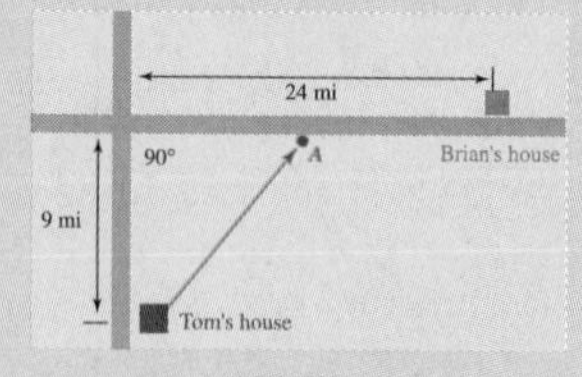

ILLUSTRATION 1

- **Changed the format of the Chapter Summary** to make it more accessible to students. Now the important concepts are listed in a left-hand column, with relevant review exercises beside them in the right-hand column.

CHAPTER SUMMARY

CONCEPTS	REVIEW EXERCISES
SECTION 1.1	*The Real Number System*
A *set* is a collection of objects. **Natural numbers:** 1, 2, 3, 4, 5, . . . **Whole numbers:** 0, 1, 2, 3, 4, 5, . . . **Integers:** . . . , −3, −2, −1, 0, 1, 2, 3, . . .	**1.** List the numbers in $\{-4, -\frac{2}{3}, 0, 1, 2, \pi, 4\}$ that satisfy the given condition. **a.** whole number **b.** natural number **c.** rational number **d.** integer **e.** irrational number **f.** real number **g.** negative number **h.** positive number **i.** prime number **j.** composite number **k.** even integer **l.** odd integer

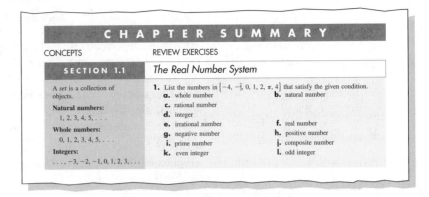

- **Increased the number of Cumulative Review Exercises.** They now occur after chapters 2, 4, 6, 8, 10, and 11.

Specific changes made in the chapters are as follows.

Chapter 1 presents a review of basic topics. More work is done with intervals and interval notation, and more topics from geometry have been included. A problem-solving technique is introduced, using the following steps:

1. Analyze the problem.
2. Form an equation.
3. Solve the equation.
4. State the conclusion.
5. Check the result.

Chapter 2 now includes graphs, equations of lines, and functions. A new introductory section on the rectangular coordinate system introduces the coordinate system with a practical application and illustrates the relationship between tables and graphs. The section concludes by illustrating how to read information given in graphs.

New topics are often introduced with a practical example. ▶

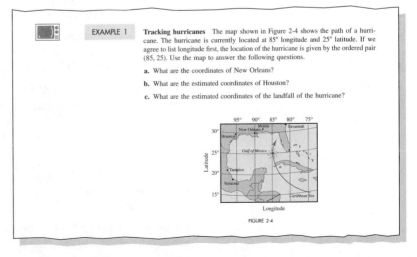

EXAMPLE 1

Tracking hurricanes The map shown in Figure 2-4 shows the path of a hurricane. The hurricane is currently located at 85° longitude and 25° latitude. If we agree to list longitude first, the location of the hurricane is given by the ordered pair (85, 25). Use the map to answer the following questions.

a. What are the coordinates of New Orleans?

b. What are the estimated coordinates of Houston?

c. What are the estimated coordinates of the landfall of the hurricane?

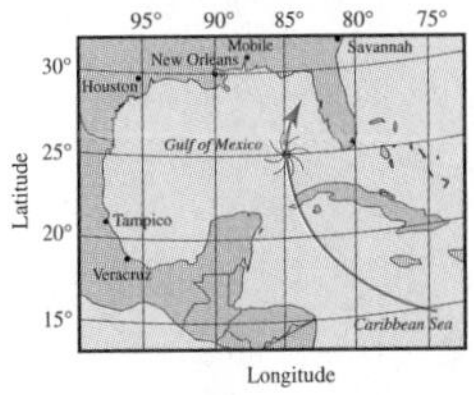

FIGURE 2-4

In Section 2.2, we graph linear equations. Here the relationships between equations, tables, and graphs are emphasized. Graphing calculators are introduced in a special graphing calculator feature. Slopes of lines are introduced as rates of change, with many applications included.

Linear functions and function notation are introduced in Section 2.5. Squaring functions, cubing functions, and absolute value functions are introduced in Section 2.6. In this section, we begin to develop the concepts of translations and reflections of graphs.

Chapter 3 covers systems of equations. Systems are solved by the graphing, elimination, matrix, and determinant methods. The chapter includes many application problems.

Chapter 4 covers linear inequalities, systems of linear inequalities, and linear programming. Equations and inequalities containing absolute values are covered in Section 4.2. Again, many application problems are included.

Chapter 5 covers polynomials, polynomial functions, and factoring. Polynomial functions are introduced in Section 5.1, reviewing linear functions, squaring functions, cubing functions, and translations. In addition, other quadratic polynomial functions are introduced.

Factoring is discussed in Sections 5.4–5.8. All of the traditional topics associated with factoring are included.

Chapter 6 covers rational expressions, begininng with a discussion of simple rational functions, including applications. Section 6.2 covers proportion and variation, including work on similar triangles. Direct variation is related to linear functions, and inverse variation is related to rational functions. Sections 6.3–6.5 cover the standard work on the arithmetic of rational expressions.

Chapter 7 covers rational exponents and radicals. The square root and cube root functions and their translations are introduced in Section 7.1. The Pythagorean theorem and other applications of radicals are discussed in Section 7.2. Equations containing radicals are discussed in Section 7.3, and rational exponents in Section 7.4. Sections 7.5–7.6 discuss the traditional material on manipulation of radical expressions.

Chapter 8 covers quadratic functions, inequalities, and algebra of functions. This chapter builds on the concepts of graphs of functions and their translations, but it is otherwise a fairly standard treatment of the topics.

Chapter 9 covers exponential and logarithmic functions. Their graphs and translations of their graphs are thoroughly discussed, as are base-e exponential functions and base-e logarithmic functions. This chapter includes a wealth of applications.

Chapter 10 covers conic sections, piecewise-defined functions, and step functions. The work includes both conics centered at the origin and conics centered at (h, k). Completing the square is used to write equations of conics in standard form.

Chapter 11 includes a standard treatment of the binomial theorem, sequences, and permutations and combinations.

CALCULATORS

The use of calculators is assumed throughout the book. We believe that students should learn calculator skills in the mathematics classroom. They will then be prepared to use calculators in science and business classes and for nonacademic purposes. The directions within each exercise set indicate which exercises require calculators.

Calculators are used to do mathematics as well as to do computations and graphing.

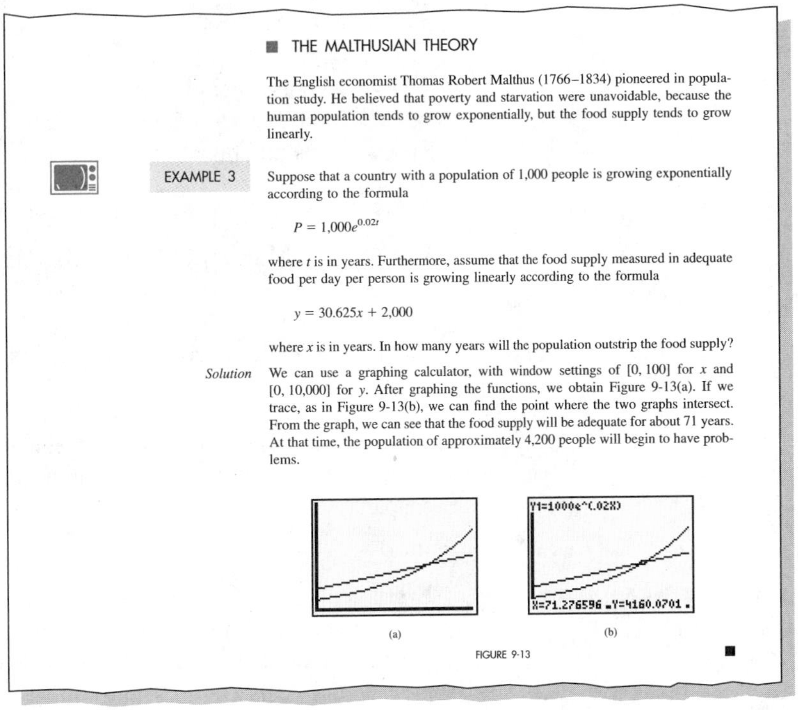

THE MALTHUSIAN THEORY

The English economist Thomas Robert Malthus (1766–1834) pioneered in population study. He believed that poverty and starvation were unavoidable, because the human population tends to grow exponentially, but the food supply tends to grow linearly.

EXAMPLE 3 Suppose that a country with a population of 1,000 people is growing exponentially according to the formula

$$P = 1{,}000e^{0.02t}$$

where t is in years. Furthermore, assume that the food supply measured in adequate food per day per person is growing linearly according to the formula

$$y = 30.625x + 2{,}000$$

where x is in years. In how many years will the population outstrip the food supply?

Solution We can use a graphing calculator, with window settings of [0, 100] for x and [0, 10,000] for y. After graphing the functions, we obtain Figure 9-13(a). If we trace, as in Figure 9-13(b), we can find the point where the two graphs intersect. From the graph, we can see that the food supply will be adequate for about 71 years. At that time, the population of approximately 4,200 people will begin to have problems.

(a) (b)

FIGURE 9-13

ANCILLARIES FOR THE INSTRUCTOR

Annotated Instructor's Edition

Free to professors when the text is adopted, the Instructor's Edition includes the complete text of the student edition, along with answers to all problems printed in blue next to the problems.

Complete Solutions Manual

This manual contains complete step-by-step solutions for all exercises in the text.

Test Manual

Includes printed test forms containing multiple-choice and free-response questions keyed to the text, as well as the answers for instructors.

Thomson World Class Learning™ Testing Tools
This fully integrated suite of programs includes *World Class Test 1.0, World Class Test On-Line 1.0,* and *World Class Manager 1.0.* The program provides text-specific algorithmic testing options designed to offer instructors greater flexibility.

Thomson World Class Learning™ Course
Using *World Class Course,* you can quickly and easily create and update a World Wide Web page specifically for your course or class—including what you plan to cover and when, assignments, grades, and even hot links to other resources on the Internet.

Text-Specific Video Tutorial Series
Free to schools when the text is adopted, this video series features worked-out examples from every section of the text, followed by supplementary examples that give students additional instruction and practice.

ANCILLARIES FOR THE STUDENT

Student Solutions Manual
The manual includes complete solutions to all odd-numbered exercises in the text.

Student Video
This *Greatest Hits* student video features the concepts and skills students traditionally have the most difficulty comprehending. The tape includes examples from each chapter.

Study Guide
Every copy of the text comes shrink-wrapped with a free, two-chapter sample of this text-specific Study Guide. Each chapter of the Study Guide contains chapter objectives, additional explanations of worked examples, exercises involving student participation, cautions, warnings, hints, and end-of-chapter tests.

Interactive Algebra 3.0
This extremely intuitive, text-specific tutorial provides explanations of concepts along with carefully graded, algorithmically generated examples and exercises. Hints are provided when students answer questions incorrectly. A management system provides a report of student progress upon completion of each unit. *Interactive Algebra* may be packaged with the text or sold as a stand-alone supplement.

20 Careers in Mathematics
This booklet, adapted from the Mathematical Association of America's popular *101 Careers in Mathematics*, details rewarding and fascinating careers available and pursued by people with mathematics degrees.

Mastering Mathematics
Richard Manning Smith's *Mastering Mathematics* stresses the importance of a positive attitude and gives students the tools to suceed in any mathematics course.

To the Student

Congratulations. You now own a state-of-the-art textbook that has been written especially for you. We have tried to write a book that you can read and understand. The book includes carefully written narrative and an extensive number of worked examples with Self Checks.

To get the most out of this course, you must read and study the textbook properly. We recommend that you work the examples on paper first and then work the Self Checks. Only after you thoroughly understand the concepts taught in the examples should you attempt to work the exercises. Several ancillary materials may be helpful.

- A *Student Solutions Manual* contains the answers to the odd-numbered exercises.
- *Interactive Algebra 3.0* gives text-specific computer tutorials that cover every section of the text.
- A *Study Guide* contains chapter objectives, additional explanations of worked examples, exercises involving student participation, cautions, warnings, hints, and end-of-chapter tests.
- A complete set of videotapes is available, featuring worked-out examples from every section of the text, followed by supplementary examples.
- A booklet, *20 Careers in Mathematics,* details rewarding and fascinating careers available to people with mathematics degrees.
- The booklet *Mastering Mathematics,* by Richard Manning Smith, discusses the tools necessary to succeed in any mathematics course.

Since the material presented in *Intermediate Algebra*, Fifth Edition, will be of value to you in later years, we suggest that you keep this book. It will be a good source of reference and will keep at your fingertips the material that you have learned here.

We wish you well.

HINTS ON STUDYING ALGEBRA

The phrase "Practice makes perfect" is not quite true. It is *perfect* practice that makes perfect. For this reason, it is important that you learn how to study algebra to get the most out of this course.

Although we all learn differently, there are some hints on how to study algebra that most students find useful. Here are some things you should consider as you work on the material in this course.

Plan a strategy for success. To get where you want to be, you need a goal and a plan. Your goal should be to pass this course with a grade of A or B. To earn one of these grades, you must have a plan to achieve it. A good plan involves several points:

- Getting ready for class
- Attending class
- Doing homework
- Arranging for special help when you need it
- Having a strategy for taking tests

Getting ready for class. To get the most out of every class period, you will need to prepare for class. One of the best things you can do is to preview the material in the text that your instructor will be discussing. Perhaps you will not understand all of what you read, but you will understand it better when the instructor discusses the material in class.

Be sure to do your work every day. If you get behind and attend class without understanding previous material, you will be lost and will become frustrated and discouraged. Make a promise that you will always prepare for class, and then keep that promise.

Attending class. The classroom experience is your opportunity to learn from your instructor. Make the most of it by attending every class. Sit near the front of the room, where you can easily see and hear. It is easy to be distracted and lose interest if you sit in the back of the room. Remember that it is your responsibility to follow the discussion, even though that takes concentration and hard work.

Pay attention to your instructor, and jot down the important things that he or she says. However, do not spend so much time taking notes that you fail to concentrate on what your instructor is explaining. It is much better to listen and understand the big picture than just to copy solutions to problems.

Don't be afraid to ask questions when your instructor asks for them. If something is unclear to you, it is probably unclear to many other students as well. They will appreciate your willingness to ask. Besides, asking questions will make you an active participant in class. This will help you pay attention and keep you alert and involved.

Doing homework. It requires practice to excel at tennis, master a musical instrument, or learn a foreign language. In the same way, it requires practice to learn mathematics. Since practice in mathematics is the homework, homework is your opportunity to practice your skills and experiment with ideas.

It is very important for you to pick a definite time to study and do homework. Set a formal schedule and stick to it. Try to study in a place that is comfortable and quiet. If you can, do some homework shortly after class, or at least before you forget what was discussed in class. This quick follow-up will help you remember the skills and concepts your instructor taught that day.

Each formal study session should include three parts:

1. Begin every study session with a review period. Look over previous chapters and see if you can do a few problems from previous sections, chosen randomly.

Keeping old skills alive will greatly reduce the time you will need to prepare for tests.

2. After reviewing, read the assigned material. Resist the temptation of diving into the exercises without reading and understanding the examples. Instead, work the examples and Self Checks with pencil and paper. Only after you completely understand the principles behind them should you try to work the exercises.

Once you begin to work the exercises, check your answers with those printed in the back of the book. If one of your answers differs from the printed answer, see if the two can be reconciled. Sometimes answers can have more than one form. If you decide that your answer is incorrect, compare your work to the example in the text that most closely resembles the exercise, and try to find your mistake. If you cannot find an error, consult the *Student Solutions Manual*. If nothing works, mark the problem and ask about it in your next class meeting.

3. After completing the written assignment, preview the next section. This preview will be helpful when you hear that material discussed during the next class period.

You probably know the general rule of thumb for college homework: two hours of practice for every hour in class. If mathematics is hard for you, plan on spending even more time on homework.

To make homework more enjoyable, study with one or more friends. The interaction will clarify ideas and help you remember them. If you must study alone, try talking to yourself. A good study technique is to explain the material to yourself out loud.

Arranging for special help. Take advantage of any special help that is available from your instructor. Often, the instructor can clear up difficulties in a very short time.

Find out whether your college has a free tutoring program. Peer tutors can often be of great help.

Taking tests. Students often get nervous before a test, because they are afraid that they will not do well. There are many different reasons for this fear, but the most common one is that students are not confident that they know the material.

To build confidence in your ability to work tests, rework many of the problems in the exercise sets, work the exercises in the Chapter Summaries, and take the Chapter Tests. Check all answers with those printed at the back of the text.

Then guess what the instructor will ask, build your own tests, and work them. Once you know your instructor, you will be surprised at how good you can get at picking test questions. With this preparation, you will have some idea of what will be on the test. You will have more confidence in your ability to do well. You should notice that you are far less nervous before tests, and this will also help your performance.

When you take a test, work slowly and deliberately. Scan the test and work the easy problems first. This will build confidence. Tackle the hardest problems last.

Acknowledgments

We are grateful to the following people, who reviewed the manuscript at various stages of its development. They all had valuable suggestions that have been incorporated into the text.

David Byrd
Enterprise State Junior College

Lee R. Clancy
Golden West College

Linda Crabtree
Longview Community College

Elias Deeba
University of Houston-Downtown

Mary Catherine Dooley
University of New Orleans

Robert B. Eicken
Illinois Central College

Harold Farmer
Wallace Community College-Hanceville

Paul Finster
El Paso Community College

Ruth Flourney
University of Alaska-Anchorage

Mark Foster
Santa Monica College

Lenore Frank
SUNY-Stony Brook

Margaret J. Greene
Florida Community College-Jacksonville

George Grisham
Illinois Central College

Charlotte Grossbeck
SUNY-Cobleskill

David W. Hansen
Monterey Peninsula College

Steven Hatfield
Marshall University

Rose Ann Haw
Mesa College

Denise Hennicke
Collin County College

Dorothy K. Holtgrefe
Seminole Community College

Ingrid Holzner
University of Wisconsin

John Hooker
Southern Illinois University

William A. Hutchings
Diablo Valley College

Mike Judy
Fullerton College

Herbert Kasube
Bradley University

John Robert Kennedy II
Santa Monica College

Diane Koenig
Rock Valley College

Ralph A. Liguori
University of Texas

Thomas McCready
California State University-Chico

Daniel F. Mussa
Southern Illinois University

James W. Newsom
Tidewater Community College

Christine Panoff
University of Michigan-Flint

Kenneth Shabell
Riverside Community College

Pat Stone
Tom Ball College

Salli Takenaka
Santa Monica College

Ray Tebbetts
San Antonio College

Jerry Wilkerson
Missouri Western State College

George J. Witt
Glendale Community College

We are grateful to Diane Koenig and Michael Welden, who read the entire manuscript and worked every problem. We also wish to thank Jerry Frang, Rob Clark, and the mathematics faculty at Rock Valley College for their helpful comments and suggestions.

We are especially grateful to our editor, Robert Pirtle, and our marketing manager, Jennifer Huber, for their encouragement and support. Finally, we express our thanks to Ellen Brownstein, who managed the production process; to David Hoyt, who skillfully edited the manuscript and guided the book to publication; to Lori Heckelman for her magnificent artwork; and to Roy Neuhaus for his creative cover design.

R. David Gustafson
Peter D. Frisk

CONTENTS

1 Basic Concepts

Computer Systems Analyst

MATHEMATICS IN THE WORKPLACE

Computer systems analysts help businesses and scientific research organizations develop computer systems to process and interpret data. Using techniques such as cost accounting, sampling, and mathematical model building, they analyze information and often present the results to management in the form of charts and diagrams.

SAMPLE APPLICATION ■ The process of sorting records into sequential order is a common task in electronic data processing. One sorting technique, called a *selection sort,* requires C comparisons to sort N records, where C and N are related by the formula

$$C = \frac{N(N-1)}{2}$$

How many comparisons are necessary to sort 10,000 records? (See Exercises 83–84 in Exercise 1.2.)

Mathematics is the study of relationships between quantities and magnitudes, as well as the rules that govern these relationships. The language of mathematics is *algebra.* The word *algebra* comes from the title of a book written by the Arabian mathematician Al-Khowarazmi around A.D. 800. Its title, *Ihm al-jabr wa'l muqabalah,* means restoration and reduction, a process then used to solve equations.

In algebra, we will work with expressions that contain numbers and letters. The letters that stand for numbers are called **variables.** Some examples of algebraic expressions are

$$x + 45, \qquad \frac{8x}{3} - \frac{y}{2}, \qquad \frac{1}{2}bh, \qquad \frac{a}{b} + \frac{c}{d}, \qquad \text{and} \qquad 3x + 2y - 12$$

It is the use of variables that distinguishes algebra from arithmetic.

1.1 The Real Number System

■ SETS OF NUMBERS ■ INEQUALITY SYMBOLS ■ INTERVALS ■ ABSOLUTE VALUE OF A NUMBER

Getting Ready

1. Name some different kinds of numbers.
2. Is there a largest number?

■ SETS OF NUMBERS

A **set** is a collection of objects. To denote a set, we often enclose a list of its **elements** with braces. For example,

$\{1, 2, 3, 4\}$ denotes the set with elements 1, 2, 3, and 4.

$\left\{\frac{1}{2}, \frac{2}{3}, \frac{3}{4}\right\}$ denotes the set with elements $\frac{1}{2}$, $\frac{2}{3}$, and $\frac{3}{4}$.

In algebra, we will work with many different sets of numbers. For example,

- To express the number of bedrooms in a house, we use counting numbers: 1, 2, 3, 4, 5, and so on.
- To express temperatures below zero, we use negative numbers: 5° below zero can be denoted as $-5°$, and 12 degrees below zero can be denoted as $-12°$.
- To express interest rates, we use decimals: $5\% = 0.05$, $7\% = 0.07$, and $9.5\% = 0.095$.

In the following definitions, each group of three dots, called an **ellipsis,** shows that the set of numbers continues on forever.

Natural Numbers

The set of **natural numbers** includes the numbers that we use for counting:

$\{1, 2, 3, 4, 5, 6, 7, 8, 9, . . .\}$

Whole Numbers

The set of **whole numbers** includes the natural numbers together with 0:

$\{0, 1, 2, 3, 4, 5, 6, 7, 8, 9, . . .\}$

Integers

The set of **integers** includes the natural numbers, 0, and the negatives of the natural numbers:

$\{. . . , -5, -4, -3, -2, -1, 0, 1, 2, 3, 4, 5, . . .\}$

In Figure 1-1, we graph each of these sets from -6 to 6 on a number line.

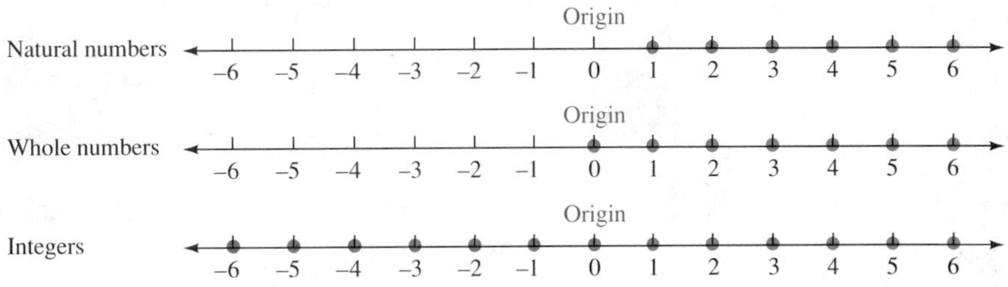

FIGURE 1-1

Since every natural number is also a whole number, we say that the set of natural numbers is a **subset** of the set of whole numbers. Note that the set of natural numbers is also a subset of the integers, and that the set of whole numbers is a subset of the integers.

In Figure 1-1, to each number x there corresponds a point on the number line called its **graph.** Furthermore, to each point there corresponds a number x called its

coordinate. Numbers to the left of 0 are **negative numbers,** and numbers to the right of 0 are **positive numbers.**

WARNING! 0 is neither positive nor negative.

Integers that are exactly divisible by 2 are called **even integers.** Integers that are not exactly divisible by 2 are called **odd integers.**

The set of even integers between -9 and 9 is $\{-8, -6, -4, -2, 0, 2, 4, 6, 8\}$.

The set of odd integers between -8 and 8 is $\{-7, -5, -3, -1, 1, 3, 5, 7\}$.

Two important subsets of the natural numbers are the **prime numbers** and the **composite numbers.**

Prime Numbers
The **prime numbers** are the natural numbers greater than 1 that are divisible only by themselves and 1.

Composite Numbers
The **composite numbers** are the natural numbers greater than 1 that are not prime.

Figure 1-2 shows the graphs of the primes and composites that are less than or equal to 14.

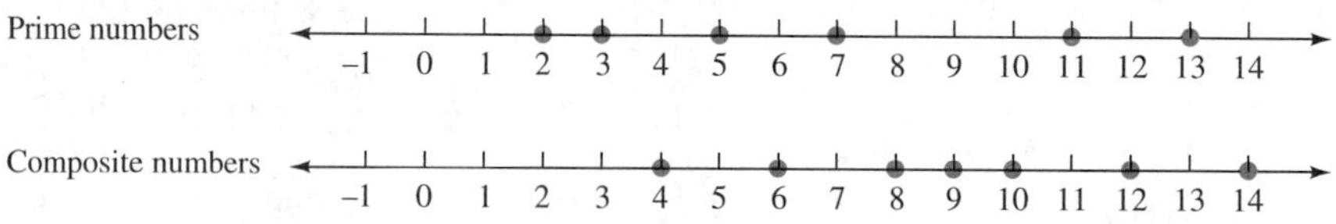

FIGURE 1-2

WARNING! 1 is the only natural number that is neither prime nor composite.

To find the coordinates of more points on the number line, we need the set of **rational numbers.**

Rational Numbers
The set of **rational numbers** includes those numbers that can be written in the form $\frac{a}{b}$ $(b \neq 0)$, where a and b are integers.

Each of the following numbers is an example of a rational number.

$$\frac{8}{5}, \frac{2}{3}, -\frac{44}{23}, \frac{-315}{476}, 0 = \frac{0}{7}, \text{ and } 17 = \frac{17}{1}$$

Each number has an integer numerator and a nonzero integer denominator.

WARNING! Note that $\frac{0}{5} = 0$, because $5 \cdot 0 = 0$. However, the fraction $\frac{5}{0}$ is undefined, because no number multiplied by 0 gives 5.

The fraction $\frac{0}{0}$ is indeterminate, because all numbers multiplied by 0 give 0. Remember that the denominator of a fraction cannot be 0.

EXAMPLE 1 Explain why each number is a rational number: **a.** -7, **b.** 0.125, and **c.** $-0.666. \ldots$

Solution

a. The integer -7 is a rational number, because it can be written as the fraction $\frac{-7}{1}$, where -7 and 1 are integers and the denominator is not 0. Note that all integers are rational numbers.

b. The decimal 0.125 is a rational number, because it can be written as the fraction $\frac{1}{8}$, where 1 and 8 are integers and the denominator is not 0.

c. The decimal $-0.666. \ldots$ is a rational number, because it can be written as the fraction $\frac{-2}{3}$, where -2 and 3 are integers and the denominator is not 0. ■

Self Check Explain why each number is a rational number: **a.** 4 and **b.** 0.5.

Answers **a.** $4 = \frac{4}{1}$, **b.** $0.5 = \frac{1}{2}$

The next example illustrates that every rational number can be written as a decimal that either terminates or repeats a block of digits.

EXAMPLE 2 Change each fraction to decimal form and tell whether the decimal terminates or repeats: **a.** $\frac{3}{4}$ and **b.** $\frac{421}{990}$.

Solution

a. To change $\frac{3}{4}$ to a decimal, we divide 3 by 4 to obtain 0.75. This is a terminating decimal.

b. To change $\frac{421}{990}$ to a decimal, we divide 421 by 990 to obtain $0.4252525 \ldots$. This is a repeating decimal, because the block of digits 25 repeats forever. This decimal can be written as $0.4\overline{25}$, where the overbar indicates the repeating block of digits. ■

Self Check Change each fraction to decimal form and tell whether the decimal terminates or repeats: **a.** $\frac{5}{11}$ and **b.** $\frac{2}{5}$.

Answers **a.** $0.\overline{45}$, repeating decimal, **b.** 0.4, terminating decimal

The rational numbers provide coordinates for many points on the number line that lie between the integers (see Figure 1-3). Note that the integers are a subset of the rational numbers.

Graphs of some rational numbers

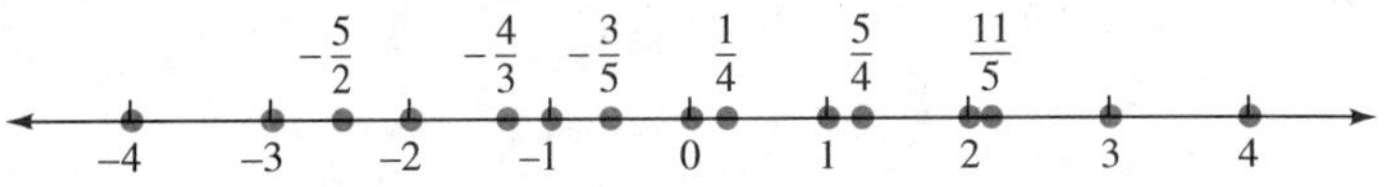

FIGURE 1-3

There are points on the number line whose coordinates are nonterminating, nonrepeating decimals. The coordinates of these points are called **irrational numbers.**

Irrational Numbers

The set of **irrational numbers** includes those numbers whose decimal forms are nonterminating, nonrepeating decimals.

Some examples of irrational numbers are

$$0.313313331\ldots \qquad \sqrt{2} = 1.414213562\ldots \qquad \pi = 3.141592653\ldots$$

If we unite the set of rational numbers (the terminating or repeating decimals) and the set of irrational numbers (the nonterminating, nonrepeating decimals), we obtain the set of **real numbers.**

Real Numbers

The set of **real numbers** includes those numbers that can be expressed as either a terminating, a repeating, or a nonterminating, nonrepeating decimal.

Figure 1-4 shows several points on the number line and their real-number coordinates. The points whose coordinates are real numbers fill up the number line.

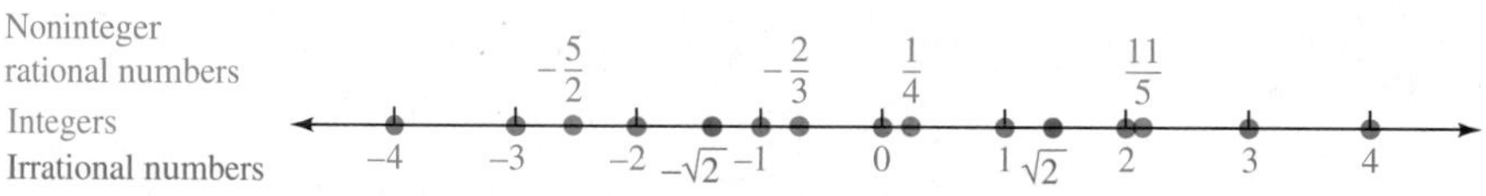

FIGURE 1-4

Figure 1-5 shows how several of the previous sets of numbers are related.

INEQUALITY SYMBOLS

To show that two quantities are not equal, we use one of the following **inequality symbols.**

Symbol	Read as	Examples
$\neq$	"is not equal to"	$6 \neq 9$ and $0.33 \neq \frac{3}{5}$
$<$	"is less than"	$22 < 40$ and $0.25 < 3.1$
$>$	"is greater than"	$19 > 5$ and $\frac{1}{2} > 0.3$
$\leq$	"is less than or equal to"	$35 \leq 35$ and $1.8 \leq 3.5$
$\geq$	"is greater than or equal to"	$29 \geq 29$ and $25.2 \geq 23.7$

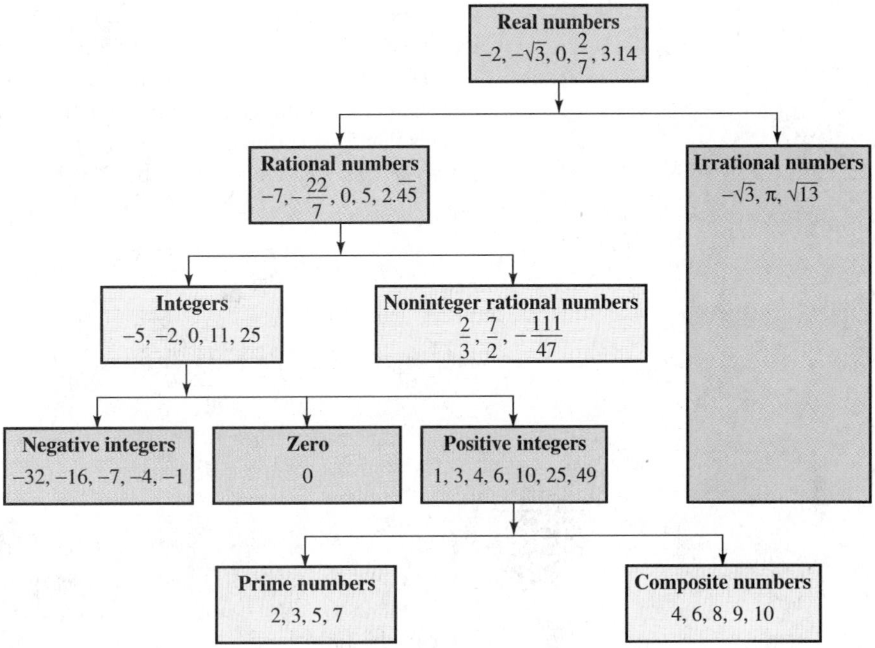

FIGURE 1-5

It is always possible to write an inequality with the inequality symbol pointing in the opposite direction. For example,

$17 < 25$ is equivalent to $25 > 17$.

$5.3 \geq -2.9$ is equivalent to $-2.9 \leq 5.3$.

On the number line, the coordinates of points get larger as we move from left to right, as shown in Figure 1-6. Thus, if a and b are the coordinates of two points, the one to the right is the greater. This suggests the following general principle:

If $a > b$, then point a lies to the right of point b on a number line.

If $a < c$, then point a lies to the left of point c on a number line.

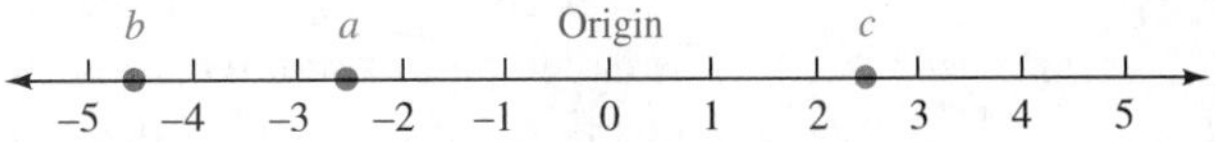

FIGURE 1-6

INTERVALS

The graphs of sets of real numbers are portions of a number line called **intervals.** The graph shown in Figure 1-7(a) includes all real numbers x that are greater than -5. This interval contains the numbers that satisfy the inequality $x > -5$. The parenthesis at -5 indicates that -5 is not included in the interval. We can also express this interval in **interval notation** as $(-5, \infty)$, where the symbol ∞ is read as "infinity." Once again, the parentheses indicate that the endpoints are not included.

The interval shown in Figure 1-7(b) is the graph of the inequality $x \le 7$. This interval contains the numbers that satisfy the inequality $x \le 7$. The bracket at 7 indicates that 7 is included in the interval. To express this interval in interval notation, we write $(-\infty, 7]$. Once again, the bracket indicates that 7 is in the interval.

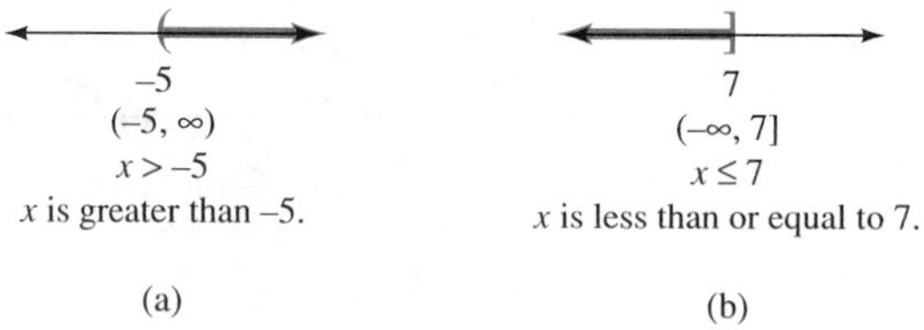

FIGURE 1-7

WARNING! The symbol ∞ (infinity) is not a real number. It is used to indicate that the graph in Figure 1-7(a) extends infinitely far to the right, and the symbol $-\infty$ is used to indicate that the graph in Figure 1-7(b) extends infinitely far to the left.

EXAMPLE 3 Write the inequality $x < 9$ in interval notation and graph it.

Solution The inequality $x < 9$ is satisfied by all real numbers that are less than 9. This is the interval $(-\infty, 9)$. The graph is shown in Figure 1-8.

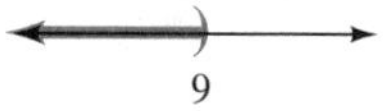

FIGURE 1-8 ■

Self Check Write the inequality $x \ge 5$ in interval notation and graph it.

Answer $[5, \infty)$

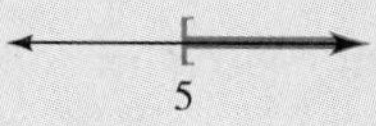

If an interval extends forever in one direction, as in the previous examples, it is called an **unbounded interval.** The following chart lists and illustrates the various types of unbounded intervals

Unbounded Intervals

The interval (a, ∞) includes all real numbers x such that $x > a$.

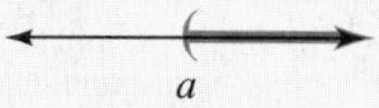

The interval $[a, \infty)$ includes all real numbers x such that $x \ge a$.

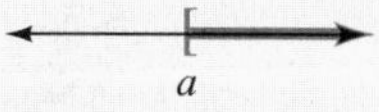

The interval $(-\infty, a)$ includes all real numbers x such that $x < a$.

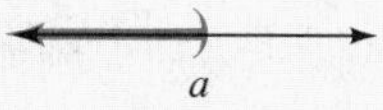

The interval $(-\infty, a]$ includes all real numbers x such that $x \leq a$.

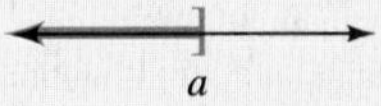

The interval $(-\infty, \infty)$ includes all real numbers x. The graph of this interval is the entire number line.

(2, 15)
$2 < x < 15$
x is between 2 and 15.

FIGURE 1-9

Two inequalities are often written as a single expression to form a **compound inequality.** For example,

$$2 < x < 15$$

is a combination of the inequalities $2 < x$ and $x < 15$. It is read as "2 is less than x, and x is less than 15," and it means that x is between 2 and 15. Its graph is shown in Figure 1-9.

The interval shown in Figure 1-10(a) is denoted by the inequality $-5 \leq x \leq 6$ or as $[-5, 6]$. The brackets indicate that the endpoints are included. The interval shown in Figure 1-10(b) is denoted by the inequality $2 < x \leq 8$ or as $(2, 8]$.

[–5, 6]
$-5 \leq x \leq 6$
x is a number from –5 to 6.

(a)

(2, 8]
$2 < x \leq 8$
x is between 2 and 8, including 8.

(b)

FIGURE 1-10

EXAMPLE 4 Write the inequality $-4 \leq x < 3$ in interval notation and graph it.

Solution This is the interval $[-4, 3)$. Its graph includes all real numbers between -4 and 3, including -4, as shown in Figure 1-11.

FIGURE 1-11 ■

Self Check Write the inequality $-6 < x \leq 10$ in interval notation and graph it.

Answer $(-6, 10]$

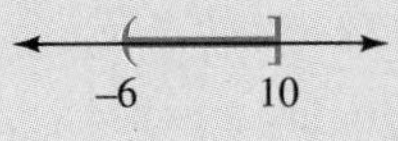

A bounded interval with no endpoints is called an **open interval.** A bounded interval with one endpoint is called a **half-open interval.** Figure 1-12(a) shows the

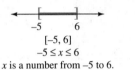

(–4, 3)
$-4 < x < 3$
x is between –4 and 3.

(a)

[–5, 2)
$-5 \leq x < 2$
x is between –5 and 2, including –5.

(b)

FIGURE 1-12

graph of the open interval $(-4, 3)$. Figure 1-12(b) shows the graph of the half-open interval $[-5, 2)$.

Intervals that contain both endpoints are called **closed intervals.** Figure 1-13 shows the graph of the closed interval $[-3, 3]$.

The following chart lists and illustrates the various types of bounded intervals.

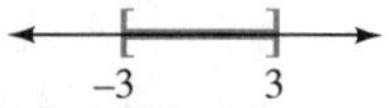

$[-3, 3]$
$-3 \le x \le 3$
x is a number from –3 to 3.

FIGURE 1-13

Open Intervals

The interval (a, b) includes all real numbers x such that $a < x < b$.

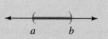

Half-Open Intervals

The interval $[a, b)$ includes all real numbers x such that $a \le x < b$.

The interval $(a, b]$ includes all real numbers x such that $a < x \le b$.

Closed Intervals

The interval $[a, b]$ includes all real numbers x such that $a \le x \le b$.

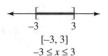

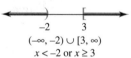

Another type of compound inequality is

$x < -2$ or $x \ge 3$ Read as "x is less than -2 or x is greater than or equal to 3."

This inequality is called the **union** of two intervals. In interval notation, it is written as

$(-\infty, -2) \cup [3, \infty)$ Read the symbol $\cup$ as "union."

Its graph is shown in Figure 1-14.

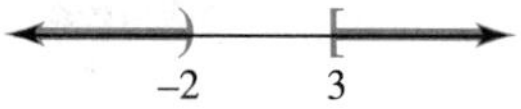

$(-\infty, -2) \cup [3, \infty)$
$x < -2$ or $x \ge 3$
x is less than –2 or x is greater than or equal to 3.

FIGURE 1-14

EXAMPLE 5 Write the inequality $x \le -4$ or $x > 5$ as an interval and graph it.

Solution This is the interval $(-\infty, -4] \cup (5, \infty)$. The graph appears in Figure 1-15.

FIGURE 1-15 ■

Self Check Write the inequality $x < -1$ or $x > 5$ as an interval and graph it.

Answer $(-\infty, -1) \cup (5, \infty)$

ABSOLUTE VALUE OF A NUMBER

The **absolute value** of any real number a, denoted as $|a|$, is the distance on a number line between 0 and the point with coordinate a. For example, the points shown in Figure 1-16 with coordinates of 3 and -3 both lie 3 units from 0. Thus, $|3| = |-3| = 3$.

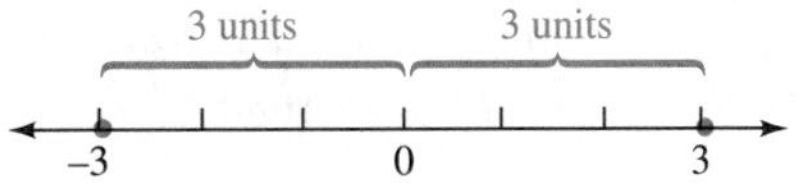

FIGURE 1-16

In general, for any real number a, $|a| = |-a|$.
The absolute value of a number can be defined more formally.

Absolute Value

For any real number x, $\begin{cases} \text{If } x \geq 0, \text{ then } |x| = x. \\ \text{If } x < 0, \text{ then } |x| = -x. \end{cases}$

If x is positive or 0, then x is its own absolute value. However, if x is negative, then $-x$ (which is a positive number) is the absolute value of x. Thus, $|x| \geq 0$ for all real numbers x.

EXAMPLE 6 Find each absolute value.

a. $|3| = 3$ **b.** $|-4| = 4$

c. $|0| = 0$ **d.** $-|-8| = -(8) = -8$ Note that $|-8| = 8$. ■

Self Check Find each absolute value: **a.** $|-9|$ and **b.** $-|-12|$.

Answers **a.** 9, **b.** -12

Orals

1. Define a natural number.
2. Define a whole number.
3. Define an integer.
4. Define a rational number.
5. List the first four prime numbers.
6. List the first five positive even integers.
7. Find $|-6|$.
8. Find $|10|$.

EXERCISE 1.1

REVIEW *To simplify a fraction, factor the numerator and the denominator and divide out common factors. For example,* $\frac{12}{18} = \frac{6 \cdot 2}{6 \cdot 3} = \frac{\not{6} \cdot 2}{\not{6} \cdot 3} = \frac{2}{3}$. *Simplify each fraction.*

1. $\frac{6}{8}$ **2.** $\frac{15}{20}$ **3.** $\frac{32}{40}$ **4.** $\frac{56}{72}$

To multiply fractions, multiply the numerators and multiply the denominators. To divide fractions, invert the divisor and multiply. Always simplify the result if possible.

5. $\frac{1}{4} \cdot \frac{3}{5}$ **6.** $\frac{3}{5} \cdot \frac{20}{27}$ **7.** $\frac{2}{3} \div \frac{3}{7}$ **8.** $\frac{3}{5} \div \frac{9}{15}$

To add (or subtract) fractions, write each fraction with a common denominator and add (or subtract) the numerators and keep the same denominator. Always simplify the result if possible.

9. $\frac{5}{9} + \frac{4}{9}$ **10.** $\frac{16}{7} - \frac{2}{7}$ **11.** $\frac{2}{3} + \frac{4}{5}$ **12.** $\frac{7}{9} - \frac{2}{5}$

VOCABULARY AND CONCEPTS *Fill in each blank to make a true statement.*

13. A ____ is a collection of objects.

14. The numbers 1, 2, 3, 4, 5, 6, . . . form the set of ________ numbers.

15. An ______ integer can be divided exactly by 2.

16. An ______ integer cannot be divided exactly by 2.

17. A prime number is a ________ number that is larger than ___ and can only be divided exactly by ______ and 1.

18. A __________ number is a natural number greater than ___ that is not ________.

19. ___ is neither positive nor negative.

20. The denominator of a fraction can never be ___.

21. A repeating decimal represents a ________ number.

22. A nonrepeating, nonterminating decimal represents an __________ number.

23. The symbol ____ means "is less than."

24. If x is negative, $|x| =$ ____.

In Exercises 25–36, list the elements in the set $\left\{-3, 0, \frac{2}{3}, 1, \sqrt{3}, 2, 9\right\}$ *that satisfy the given condition.*

25. natural number **26.** whole number **27.** integer

28. rational number **29.** irrational number **30.** real number

31. even natural number **32.** odd integer **33.** prime number

34. composite number **35.** odd composite number **36.** even prime number

PRACTICE *In Exercises 37–40, graph each set on the number line.*

37. The set of prime numbers less than 8

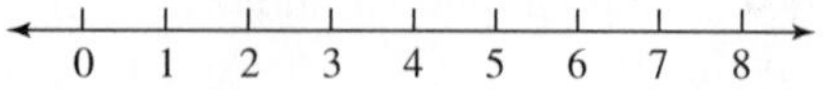

38. The set of integers between −7 and 0

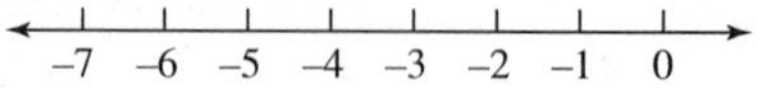

39. The set of odd integers between 10 and 18

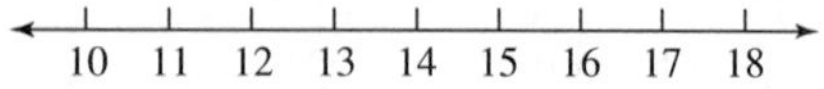

40. The set of composite numbers less than 10

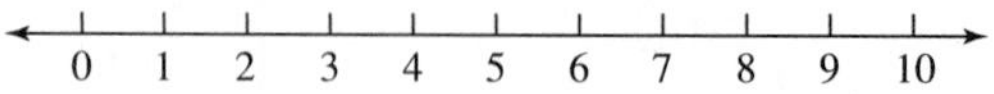

In Exercises 41–44, change each fraction into a decimal and classify the result as a terminating or a repeating decimal.

41. $\frac{7}{8}$

42. $\frac{7}{3}$

43. $-\frac{11}{15}$

44. $-\frac{19}{16}$

In Exercises 45–52, insert either a $<$ or a $>$ symbol to make a true statement.

45. $5 \_\_ 9$

46. $9 \_\_ 0$

47. $-5 \_\_ -10$

48. $-3 \_\_ 10$

49. $-7 \_\_ 7$

50. $0 \_\_ -5$

51. $6 \_\_ -6$

52. $-6 \_\_ -2$

In Exercises 53–60, write each statement with the inequality symbol pointing in the opposite direction.

53. $19 > 12$

54. $-3 \geq -5$

55. $-6 \leq -5$

56. $-10 < 13$

57. $5 \geq -3$

58. $0 \leq 12$

59. $-10 < 0$

60. $-4 > -8$

In Exercises 61–72, graph each interval on the number line.

61. $x > 3$

62. $x < 0$

63. $-3 < x < 2$

64. $-5 \leq x < 2$

65. $0 < x \leq 5$

66. $-4 \leq x \leq -2$

67. $(-2, \infty)$

68. $(-\infty, 4]$

69. $[-6, 9]$

70. $(-1, 3)$

71. $(-2, 4]$

72. $[-5, 2]$

In Exercises 73–80, write each expression without using absolute value symbols. Simplify the result when possible.

73. $|20|$

74. $|-20|$

75. $-|-6|$

76. $-|8|$

77. $|-5| + |-2|$

78. $|12| + |-4|$

79. $|-5| \cdot |4|$

80. $|-6| \cdot |-3|$

81. Find x if $|x| = 3$.

82. Find x if $|x| = 7$.

83. What numbers x are equal to their own absolute values?

84. What numbers x when added to their own absolute values give a sum of 0?

WRITING

85. Explain why the integers are a subset of the rational numbers.

86. Explain why every integer is a rational number, but not every rational number is an integer.

87. Explain why the set of primes together with the set of composites is not the set of natural numbers.

88. Is the absolute value of a number always positive? Explain.

SOMETHING TO THINK ABOUT

89. How many integers have an absolute value that is less than 50?

90. How many odd integers have an absolute value between 20 and 40?

91. The **trichotomy property** of real numbers states that

If a and b are two real numbers, then

$a < b$ or $a = b$ or $a > b$

Explain why this is true.

92. Which of the following statements are always true?

a. $|a + b| = |a| + |b|$

b. $|a \cdot b| = |a| \cdot |b|$

c. $|a + b| \leq |a| + |b|$

1.2 Arithmetic and Properties of Real Numbers

■ ADDING REAL NUMBERS ■ SUBTRACTING REAL NUMBERS ■ MULTIPLYING REAL NUMBERS ■ DIVIDING REAL NUMBERS ■ ORDER OF OPERATIONS ■ MEASURES OF CENTRAL TENDENCY ■ EVALUATING ALGEBRAIC EXPRESSIONS ■ PROPERTIES OF REAL NUMBERS

Getting Ready *Do each operation.*

1. $5 + 4$ **2.** $4 + 5$ **3.** $3 \cdot 4$ **4.** $4 \cdot 3$

5. $12 - 7$ **6.** $15 \div 3$ **7.** $21 \div 7$ **8.** $25 - 19$

In this section, we will show how to add, subtract, multiply, and divide real numbers. We will then discuss several properties of real numbers.

■ ADDING REAL NUMBERS

When two numbers are added, we call the result their **sum.** To find the sum of $+2$ and $+3$, we can use a number line and represent the numbers with arrows, as shown in Figure 1-17(a). Since the endpoint of the second arrow is at $+5$, we have $+2 + (+3) = +5$.

To add -2 and -3, we can draw arrows as shown in Figure 1-17(b). Since the endpoint of the second arrow is at -5, we have $(-2) + (-3) = -5$.

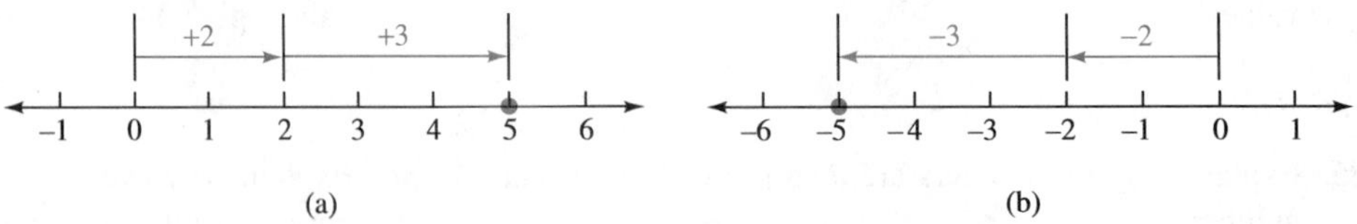

FIGURE 1-17

To add -6 and $+2$, we can draw arrows as shown in Figure 1-18(a). Since the endpoint of the second arrow is at -4, we have $(-6) + (+2) = -4$.

To add $+7$ and -4, we can draw arrows as shown in Figure 1-18(b). Since the endpoint of the final arrow is at $+3$, we have $(+7) + (-4) = +3$.

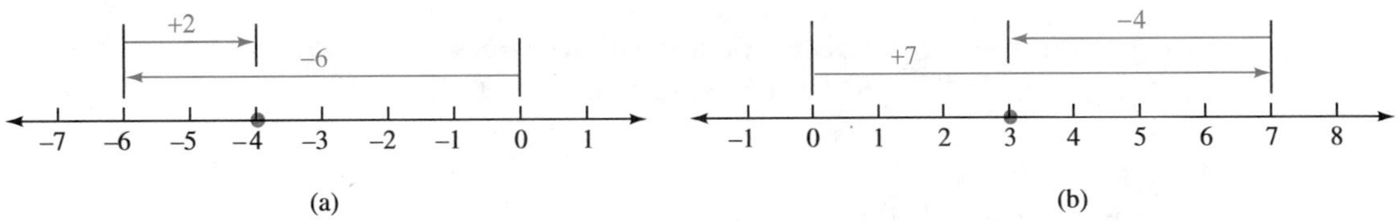

FIGURE 1-18

These examples suggest the following rules.

Adding Real Numbers

With like signs: Add the absolute values of the numbers and keep the common sign.

With unlike signs: Subtract the absolute values of the numbers (the smaller from the larger) and keep the sign of the number with the larger absolute value.

EXAMPLE 1 Add the numbers.

a. $+4 + (+6) = +10$ Add the absolute values and use the common sign: $4 + 6 = +10$.

b. $-5 + (-3) = -8$ Add the absolute values and use the common sign: $-(5 + 3) = -8$.

c. $+9 + (-5) = +4$ Subtract the absolute values and use a + sign: $+(9 - 5) = +4$.

d. $-12 + (+5) = -7$ Subtract the absolute values and use a − sign: $-(12 - 5) = -7$. ■

Self Check Add the numbers: **a.** $-7 + (-2)$, **b.** $-7 + 2$, **c.** $7 + 2$, and **d.** $7 + (-2)$.

Answers **a.** -9, **b.** -5, **c.** 9, **d.** 5

SUBTRACTING REAL NUMBERS

When one number is subtracted from another number, we call the result their **difference.** To find a difference, we can change the subtraction into an equivalent addition. For example, the subtraction $7 - 4$ is equivalent to the addition $7 + (-4)$, because they have the same result:

$$7 - 4 = 3 \quad \text{and} \quad 7 + (-4) = 3$$

This suggests that to subtract two numbers, we change the sign of the number being subtracted and add.

Subtracting Real Numbers
If a and b are real numbers, then $a - b = a + (-b)$.

EXAMPLE 2 Subtract.

a. $12 - 4 = 12 + (-4)$ Change the sign of 4 and add.
$= 8$

b. $-13 - 5 = -13 + (-5)$ Change the sign of 5 and add.
$= -18$

c. $-14 - (-6) = -14 + (+6)$ Change the sign of -6 and add.
$= -8$ ■

Self Check Subtract **a.** $-15 - 4$, **b.** $8 - 5$, and **c.** $-12 - (-7)$.

Answers **a.** -19, **b.** 3, **c.** -5

MULTIPLYING REAL NUMBERS

When two numbers are multiplied, we call the result their **product.** We can find the product of 5 and 4 by using 4 in an addition five times:

$$5(4) = 4 + 4 + 4 + 4 + 4 = 20$$

We can find the product of 5 and -4 by using -4 in an addition five times:

$$5(-4) = (-4) + (-4) + (-4) + (-4) + (-4) = -20$$

Since multiplication by a negative number can be defined as repeated subtraction, we can find the product of -5 and 4 by using 4 in a subtraction five times:

$$-5(4) = -4 - 4 - 4 - 4 - 4$$
$$= -4 + (-4) + (-4) + (-4) + (-4) \quad \text{Change the sign of each 4 and add.}$$
$$= -20$$

We can find the product of -5 and -4 by using -4 in a subtraction five times:

$$-5(-4) = -(-4) - (-4) - (-4) - (-4) - (-4)$$
$$= 4 + 4 + 4 + 4 + 4 \quad \text{Change the sign of each } -4 \text{ and add.}$$
$$= 20$$

The products 5(4) and $-5(-4)$ both equal $+20$, and the products $5(-4)$ and $-5(4)$ both equal -20. These results suggest the first two of the following rules.

Multiplying Real Numbers

With like signs: Multiply their absolute values. The product is positive.
With unlike signs: Multiply their absolute values. The product is negative.
Multiplication by 0: If x is any real number, then $x \cdot 0 = 0 \cdot x = 0$.

EXAMPLE 3 Multiply.

a. $4(-7) = -28$ — Multiply the absolute values: $4 \cdot 7 = 28$. Since the signs are unlike, the product is negative.

b. $-5(-6) = +30$ — Multiply the absolute values: $5 \cdot 6 = 30$. Since the signs are alike, the product is positive.

c. $-7(6) = -42$ — Multiply the absolute values: $7 \cdot 6 = 42$. Since the signs are unlike, the product is negative.

d. $8(6) = +48$ — Multiply the absolute values: $8 \cdot 6 = 48$. Since the signs are alike, the product is positive. ■

Self Check Multiply **a.** $(-6)(5)$, **b.** $(-4)(-8)$, **c.** $(17)(-2)$, and **d.** $(12)(6)$.

Answers **a.** -30, **b.** 32, **c.** -34, **d.** 72

DIVIDING REAL NUMBERS

When two numbers are divided, we call the result their **quotient.** In the division $\frac{x}{y} = q \quad (y \neq 0)$, the quotient q is a number such that $y \cdot q = x$. We can use this relationship to find rules for dividing real numbers. We consider four divisions:

$$\frac{+10}{+2} = +5, \text{ because } +2(+5) = +10 \qquad \frac{-10}{-2} = +5, \text{ because } -2(+5) = -10$$

$$\frac{-10}{+2} = -5, \text{ because } +2(-5) = -10 \qquad \frac{+10}{-2} = -5, \text{ because } -2(-5) = +10$$

These results suggest the first two rules for dividing real numbers.

Dividing Real Numbers

With like signs: Divide their absolute values. The quotient is positive.
With unlike signs: Divide their absolute values. The quotient is negative.
Division by 0: Division by 0 is undefined.

WARNING! If $x \neq 0$, then $\frac{0}{x} = 0$. However, $\frac{x}{0}$ is undefined for any value of x.

EXAMPLE 4 Divide.

a. $\frac{36}{18} = +2$ Divide the absolute values: $\frac{36}{18} = 2$. Since the signs are alike, the quotient is positive.

b. $\frac{-44}{11} = -4$ Divide the absolute values: $\frac{44}{11} = 4$. Since the signs are unlike, the quotient is negative.

c. $\frac{27}{-9} = -3$ Divide the absolute values: $\frac{27}{9} = 3$. Since the signs are unlike, the quotient is negative.

d. $\frac{-64}{-8} = +8$ Divide the absolute values: $\frac{64}{8} = 8$. Since the signs are alike, the quotient is positive. ■

Self Check Divide **a.** $\frac{55}{-5}$, **b.** $\frac{-72}{-6}$, **c.** $\frac{-100}{10}$, **d.** $\frac{50}{25}$.

Answers **a.** -11, **b.** 12, **c.** -10, **d.** 2

ORDER OF OPERATIONS

Suppose that you are asked to contact a friend if you see a rug for sale while traveling in Turkey. After locating a nice one, you send the following message to your friend.

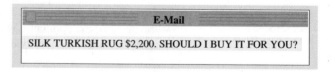

> **E-Mail**
>
> SILK TURKISH RUG $2,200. SHOULD I BUY IT FOR YOU?

The next day, you receive this response.

> **E-Mail**
>
> NO PRICE TOO HIGH! REPEAT...NO! PRICE TOO HIGH.

The first statement says to buy the rug at any price. The second says not to buy it, because it is too expensive. The placement of the exclamation point makes these statements read differently, resulting in different interpretations.

When reading mathematical statements, the same kind of confusion is possible. To illustrate, we consider the expression $5 + 3 \cdot 7$, which contains the operations of addition and multiplication. We can calculate this expression in two different ways. We can do the addition first and then do the multiplication. Or we can do the multiplication first and then do the addition. However, we will get different results.

Method 1: Add First		***Method 2: Multiply First***	
$5 + 3 \cdot 7 = 8 \cdot 7$	Add 5 and 3.	$5 + 3 \cdot 7 = 5 + 21$	Multiply 3 and 7.
$= 56$	Multiply 8 and 7.	$= 26$	Add 5 and 21.

Different results

To eliminate the possibility of getting different answers, we will agree to do multiplications before additions. So Method 1 above is incorrect. The correct calculation of $5 + 3 \cdot 7$ is

$$\begin{aligned} 5 + 3 \cdot 7 &= 5 + 21 \\ &= 26 \end{aligned}$$

To indicate that additions should be done before multiplications, we must use **grouping symbols** such as parentheses ( ), brackets [ ], or braces { }. In the expression $(5 + 3)7$, the parentheses indicate that the addition is to be done first:

$$\begin{aligned} (5 + 3)7 &= 8 \cdot 7 \\ &= 56 \end{aligned}$$

To guarantee that calculations will have one correct result, we will always do calculations in the following order.

Rules for the Order of Operations for Expressions without Exponents

Use the following steps to do all calculations within each pair of grouping symbols, working from the innermost pair to the outermost pair.

1. Do all multiplications and divisions, working from left to right.

2. Do all additions and subtractions, working from left to right.

When all grouping symbols have been removed, repeat the rules above to finish the calculation.

In a fraction, simplify the numerator and the denominator separately. Then simplify the fraction, whenever possible.

d

EXAMPLE 5 Evaluate each expression.

a. $4 + 2 \cdot 3 = 4 + 6$ — Do the multiplication first.

$= 10$ — Then do the addition.

b. $2(3 + 4) = 2 \cdot 7$ — Because of the parentheses, do the addition first.

$= 14$ — Then do the multiplication.

c. $5(3 - 6) \div 3 + 1 = 5(-3) \div 3 + 1$ — Do the subtraction within parentheses.

$= -15 \div 3 + 1$ — Then do the multiplication: $5(-3) = -15$.

$= -5 + 1$ — Then do the division: $-15 \div 3 = -5$.

$= -4$ — Finally, do the addition.

d. $5[3 - 2(6 \div 3 + 1)] = 5[3 - 2(2 + 1)]$ — Do the division within parentheses: $6 \div 3 = 2$.

$= 5[3 - 2(3)]$ — Do the addition: $2 + 1 = 3$.

$= 5(3 - 6)$ — Do the multiplication: $2(3) = 6$.

$= 5(-3)$ — Do the subtraction: $3 - 6 = -3$.

$= -15$ — Do the multiplication.

e. $\frac{4 + 8(3 - 4)}{6 - 2(2)} = \frac{-4}{2}$ — Simplify the numerator and denominator separately.

$= -2$

■

Self Check Evaluate each expression: **a.** $5 + 3 \cdot 4$, **b.** $(5 + 3) \cdot 4$, **c.** $3(5 - 7) \div 6 + 3$, and **d.** $\frac{5 - 2(4 - 6)}{9 - 2 \cdot 3}$.

Answers **a.** 17, **b.** 32, **c.** 2, **d.** 3

MEASURES OF CENTRAL TENDENCY

Three types of averages are commonly used in newspapers and magazines: the **mean,** the **median,** and the **mode.**

Mean

The **mean** of several values is the sum of those values divided by the number of values.

$$\text{Mean} = \frac{\text{sum of the values}}{\text{number of values}}$$

EXAMPLE 6 **Football** Figure 1-19 shows the gains and losses made by a running back on seven plays. Find the mean number of yards per carry.

Solution To find the mean number of yards per carry, we add the numbers and divide by 7.

$$\frac{-8 + (+2) + (-6) + (+6) + (+4) + (-7) + (-5)}{7} = \frac{-14}{7} = -2$$

The running back averaged -2 yards (or lost 2 yards) per carry.

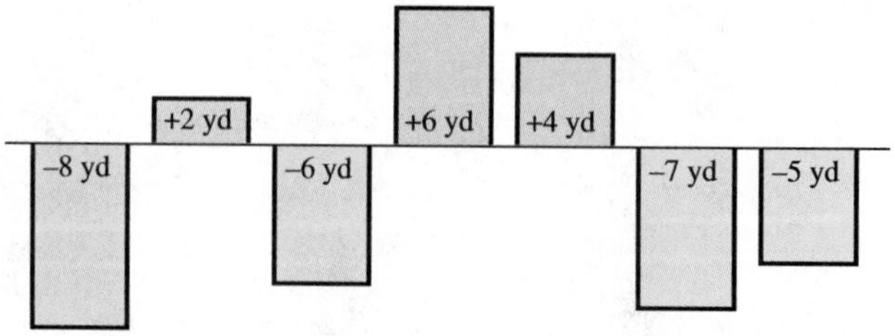

FIGURE 1-19

■

Median

The **median** of several values is the middle value. To find the median,

1. Arrange the values in increasing order.
2. If there is an odd number of values, choose the middle value.
3. If there is an even number of values, find the mean of the middle two values.

Mode

The **mode** of several values is the value that occurs most often.

EXAMPLE 7 Ten workers in a small business have monthly salaries of

\$2,500, \$1,750, \$2,415, \$3,240, \$2,790, \$3,240, \$2,650, \$2,415, \$2,415, \$2,650

Find **a.** the median and **b.** the mode of the distribution.

Solution **a.** To find the median, we first arrange the salaries in increasing order:

\$1,750, \$2,415, \$2,415, \$2,415, \$2,500, \$2,650, \$2,650, \$2,790, \$3,240, \$3,240

Because there is an even number of salaries, the median will be the mean of the middle two scores, \$2,500 and \$2,650.

$$\text{Median} = \frac{\$2{,}500 + \$2{,}650}{2} = \$2{,}575$$

b. Since the salary \$2,415 occurs most often, it is the mode. ■

If two different numbers in a distribution tie for occurring most often, the distribution is called *bimodal.*

Although the mean is probably the most common measure of average, the median and the mode are frequently used. For example, workers' salaries are often compared to the median (average) salary. To say that the modal (average) shoe size is 9 means that a shoe size of 9 occurs more often than any other size.

EVALUATING ALGEBRAIC EXPRESSIONS

Variables and numbers can be combined with the operations of arithmetic to produce **algebraic expressions.** To evaluate algebraic expressions, we substitute numbers for the variables and carry out the arithmetic.

EXAMPLE 8 If $a = 2$, $b = -3$, and $c = -5$, evaluate **a.** $a + bc$ and **b.** $\dfrac{ab + 3c}{b(c - a)}$.

Solution We substitute 2 for a, -3 for b, and -5 for c and simplify.

a. $a + bc = 2 + (-3)(-5)$
$= 2 + (15)$
$= 17$

b. $\dfrac{ab + 3c}{b(c - a)} = \dfrac{2(-3) + 3(-5)}{-3(-5 - 2)}$
$= \dfrac{-6 + (-15)}{-3(-7)}$
$= \dfrac{-21}{21}$
$= -1$ ■

Self Check If $a = 2$, $b = -5$, and $c = 3$, evaluate **a.** $b - ac$ and **b.** $\frac{ab - 2c}{ac + 2b}$.

Answers **a.** -11, **b.** 4

Table 1-1 shows the formulas for the perimeters of several geometric figures.

Figure	Name	Perimeter/ circumference
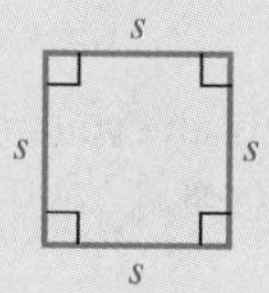	Square	$P = 4s$
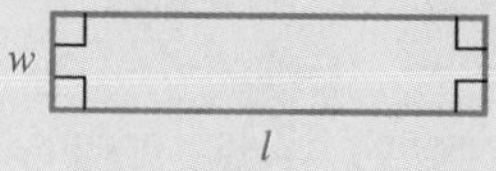	Rectangle	$P = 2l + 2w$
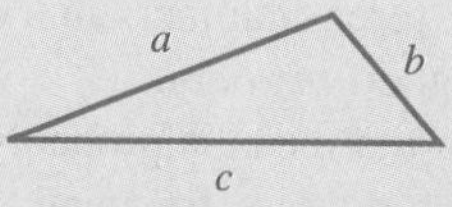	Triangle	$P = a + b + c$
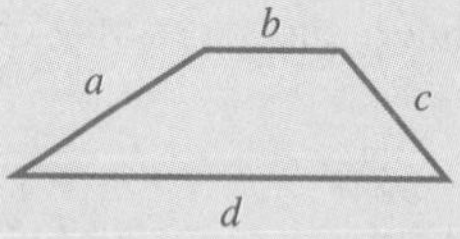	Trapezoid	$P = a + b + c + d$
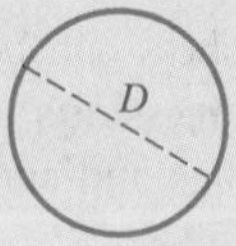	Circle	$C = \pi D$ (π is approximately 3.1416)

TABLE 1-1

EXAMPLE 9 Find the perimeter of the rectangle shown in Figure 1-20.

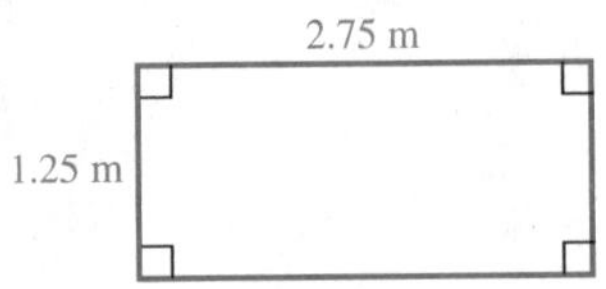

FIGURE 1-20

Solution We substitute 2.75 for l and 1.25 for w into the formula $P = 2l + 2w$ and simplify.

$$P = 2l + 2w$$
$$P = 2(2.75) + 2(1.25)$$
$$= 8.00$$

The perimeter is 8 meters. ■

Self Check Find the perimeter of a rectangle with a length of 8 meters and a width of 5 meters.

Answer 26 m

PROPERTIES OF REAL NUMBERS

When we work with real numbers, we will use certain properties.

Properties of Real Numbers
If a, b, and c are real numbers, the following properties apply.

The associative properties for addition and multiplication

$$(a + b) + c = a + (b + c) \qquad (ab)c = a(bc)$$

The commutative properties for addition and multiplication

$$a + b = b + a \qquad ab = ba$$

The distributive property of multiplication over addition

$$a(b + c) = ab + ac$$

The associative properties enable us to group the numbers in a sum or a product in any way that we wish and still get the same result.

EXAMPLE 10 Simplify each expression.

a. $(2 + 3) + 4 = 5 + 4$
$= 9$

b. $2 + (3 + 4) = 2 + 7$
$= 9$

c. $(2 \cdot 3) \cdot 4 = 6 \cdot 4$
$= 24$

d. $2 \cdot (3 \cdot 4) = 2 \cdot 12$
$= 24$ ■

Self Check Simplify each expression in two ways: **a.** $5 + 4 + (-7)$ and **b.** $(-3)(-5)(6)$.

Answers **a.** 2, **b.** 90

WARNING! Subtraction and division are not associative, because different groupings give different results. For example,

$(8 - 4) - 2 = 4 - 2 = 2$ but $8 - (4 - 2) = 8 - 2 = 6$

$(8 \div 4) \div 2 = 2 \div 2 = 1$ but $8 \div (4 \div 2) = 8 \div 2 = 4$

The commutative properties enable us to add or multiply two numbers in either order and obtain the same result.

EXAMPLE 11 Evaluate each expression.

a. $2 + 3 = 5$ **b.** $3 + 2 = 5$

c. $7 \cdot 9 = 63$ **d.** $9 \cdot 7 = 63$ ■

Self Check Evaluate each expression in two ways: **a.** $7 + (-3)$ and **b.** $7(-3)$.

Answers **a.** 4, **b.** -21

WARNING! Subtraction and division are not commutative, because doing these operations in different orders will give different results. For example,

$8 - 4 = 4$ but $4 - 8 = -4$

$8 \div 4 = 2$ but $4 \div 8 = \dfrac{1}{2}$

The distributive property enables us to evaluate many expressions involving a multiplication over an addition. We can add first inside the parentheses and then multiply, or multiply over the addition first and then add.

EXAMPLE 12 Evaluate each expression.

a. $2(3 + 7) = 2 \cdot 10$
$= 20$

b. $2(3 + 7) = 2 \cdot 3 + 2 \cdot 7$
$= 6 + 14$
$= 20$ ■

Self Check Evaluate $-3(5 + 8)$ in two ways.

Answer -39

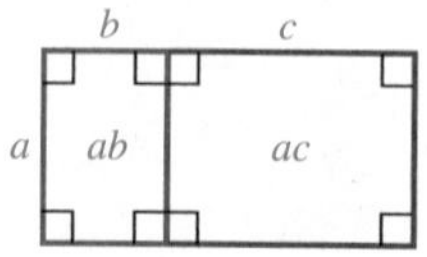

FIGURE 1-21

We can interpret the distributive property geometrically. Since the area of the largest rectangle in Figure 1-21 is the product of its width a and its length $b + c$, its area is $a(b + c)$. The areas of the two smaller rectangles are ab and ac. Since the area of the largest rectangle is equal to the sum of the areas of the smaller rectangles, we have $a(b + c) = ab + ac$.

A more general form of the distributive property is called the **extended distributive property.**

$$a(b + c + d + e + \ldots) = ab + ac + ad + ae + \ldots$$

EXAMPLE 13 Use the distributive property to write each expression without parentheses: **a.** $2(x + 3)$ and **b.** $2(x + y - 7)$.

Solution **a.** $2(x + 3) = 2x + 2 \cdot 3$
$= 2x + 6$

b. $2(x + y - 7) = 2x + 2y - 2 \cdot 7$
$= 2x + 2y - 14$ ■

Self Check Remove parentheses: $-5(a - 2b + 3c)$.

Answer $-5a + 10b - 15c$

The real numbers 0 and 1 have important special properties.

Properties of 0 and 1

Additive identity: The sum of 0 and any number is the number itself.

$$0 + a = a + 0 = a$$

Multiplication property of 0: The product of any number and 0 is 0.

$$a \cdot 0 = 0 \cdot a = 0$$

Multiplicative identity: The product of 1 and any number is the number itself.

$$1 \cdot a = a \cdot 1 = a$$

For example,

$$7 + 0 = 7, \qquad 7(0) = 0, \qquad 1(5) = 5, \qquad \text{and} \qquad (-7)1 = -7$$

If the sum of two numbers is 0, the numbers are called **additive inverses, negatives,** or **opposites** of each other. For example, 6 and -6 are negatives, because $6 + (-6) = 0$.

The Additive Inverse Property

For every real number a, there is a real number $-a$ such that

$$a + (-a) = -a + a = 0$$

The symbol $-(-6)$ means "the negative of negative 6." Because the sum of two numbers that are negatives is 0, we have

$$-6 + [-(-6)] = 0 \qquad \text{and} \qquad -6 + 6 = 0$$

Because -6 has only one additive inverse, it follows that $-(-6) = 6$. In general, the following rule applies.

The Double Negative Rule

If a represents any real number, then $-(-a) = a$.

If the product of two numbers is 1, the numbers are called **multiplicative inverses** or **reciprocals** of each other.

The Multiplicative Inverse Property

For every nonzero real number a, there exists a real number $\frac{1}{a}$ such that

$$a \cdot \frac{1}{a} = \frac{1}{a} \cdot a = 1 \quad (a \neq 0)$$

Some examples of reciprocals are

- 5 and $\frac{1}{5}$ are reciprocals, because $5\left(\frac{1}{5}\right) = 1$.
- $\frac{3}{2}$ and $\frac{2}{3}$ are reciprocals, because $\frac{3}{2}\left(\frac{2}{3}\right) = 1$.
- -0.25 and -4 are reciprocals, because $-0.25(-4) = 1$.

The reciprocal of 0 does not exist, because $\frac{1}{0}$ is undefined.

Orals *Do each operation.*

1. $+2 + (-4)$ **2.** $-5 - 2$ **3.** $7(-4)$

4. $(-7)(-4)$ **5.** $3 + (-3)(2)$ **6.** $\dfrac{-4 + (-2)}{-3 + 5}$

EXERCISE 1.2

REVIEW *Graph each interval on a number line.*

1. $x > 4$ **2.** $x \leq -5$ **3.** $(2, 10]$ **4.** $[-4, 4]$

5. A man bought 32 gallons of gasoline at \$1.29 per gallon and 3 quarts of oil at \$1.35 per quart. The sales tax was included in the price of the gasoline, but 5% sales tax was added to the cost of the oil. Find the total cost.

6. On an adjusted income of \$57,760, a woman must pay taxes according to the schedule shown in Illustration 1. Compute the tax bill.

Over	But not over	Tax	Of the amount over
\$ 0	\$ 22,750	. . . . . 15%	\$ 0
22,750	55,100	\$3,412.50 + 28%	22,750
55,100	115,000	12,470.50 + 31%	55,100

ILLUSTRATION 1

VOCABULARY AND CONCEPTS *Fill in each blank to make a true statement.*

7. To add two numbers with like signs, we add their ________ values and keep the ________ sign.

8. To add two numbers with unlike signs, we ________ their absolute values and keep the sign of the number with the larger absolute value.

9. To subtract one number from another, we ________ the sign of the number that is being subtracted and _____.

10. The product of two real numbers with like signs is ________.

11. The quotient of two real numbers with unlike signs is ________.

12. The denominator of a fraction can never be ___.

13. The three measures of central tendency are the ______, ________, and ______.

14. Write the formula for the circumference of a circle. ________

15. Write the associative property of multiplication. ________________

16. Write the commutative property of addition. ____________

17. Write the distributive property of multiplication over addition. ________________

18. What is the additive additity? ___

19. What is the multiplicative identity? ___

20. $-(-a) =$ ___

PRACTICE *In Exercises 21–48, do the operations.*

21. $-3 + (-5)$ **22.** $2 + (+8)$ **23.** $-7 + 2$ **24.** $3 + (-5)$

25. $-3 - 4$ **26.** $-11 - (-17)$ **27.** $-33 - (-33)$ **28.** $14 - (-13)$

29. $-2(6)$ **30.** $3(-5)$ **31.** $-3(-7)$ **32.** $-2(-5)$

33. $\frac{-8}{4}$ **34.** $\frac{25}{-5}$ **35.** $\frac{-16}{-4}$ **36.** $\frac{-5}{-25}$

37. $\frac{1}{2} + \left(-\frac{1}{3}\right)$ **38.** $-\frac{3}{4} + \left(-\frac{1}{5}\right)$ **39.** $\frac{1}{2} - \left(-\frac{3}{5}\right)$ **40.** $\frac{1}{26} - \frac{11}{13}$

41. $\frac{1}{3} - \frac{1}{2}$ **42.** $\frac{7}{8} - \left(-\frac{3}{4}\right)$ **43.** $\left(-\frac{3}{5}\right)\left(\frac{10}{7}\right)$ **44.** $\left(-\frac{6}{7}\right)\left(-\frac{5}{12}\right)$

45. $\frac{3}{4} \div \left(-\frac{3}{8}\right)$ **46.** $-\frac{3}{5} \div \frac{7}{10}$ **47.** $-\frac{16}{5} \div \left(-\frac{10}{3}\right)$ **48.** $-\frac{5}{24} \div \frac{10}{3}$

In Exercises 49–68, do the operations.

49. $3 + 4 \cdot 5$ **50.** $5 \cdot 3 - 6 \cdot 4$ **51.** $3 - 2 - 1$ **52.** $5 - 3 - 1$

53. $3 - (2 - 1)$ **54.** $5 - (3 - 1)$ **55.** $2 - 3 \cdot 5$ **56.** $6 + 4 \cdot 7$

57. $8 \div 4 \div 2$ **58.** $100 \div 10 \div 5$ **59.** $8 \div (4 \div 2)$ **60.** $100 \div (10 \div 5)$

61. $2 + 6 \div 3 - 5$ **62.** $6 - 8 \div 4 - 2$ **63.** $(2 + 6) \div (3 - 5)$ **64.** $(6 - 8) \div (4 - 2)$

65. $\frac{3(8 + 4)}{2 \cdot 3 - 9}$ **66.** $\frac{5(4 - 1)}{3 \cdot 2 + 5 \cdot 3}$ **67.** $\frac{100(2 - 4)}{1{,}000 \div 10 \div 10}$ **68.** $\frac{8(3) - 4(6)}{5(3) + 3(-7)}$

In Exercises 69–71, use the distribution: 7, 5, 9, 10, 8, 6, 6, 7, 9, 12, 9.

69. Find the mean. **70.** Find the median. **71.** Find the mode.

In Exercises 72–74, use the distribution: 8, 12, 23, 12, 10, 16, 26, 12, 14, 8, 16, 23.

72. Find the median.

73. Find the mode.

74. Find the mean.

In Exercises 75–82, $a = 3$, $b = -2$, $c = -1$, and $d = 2$. Evaluate each expression.

75. $ab + cd$

76. $ad + bc$

77. $a(b + c)$

78. $d(b + a)$

79. $\dfrac{ad + c}{cd + b}$

80. $\dfrac{ab + d}{bd + a}$

81. $\dfrac{ac - bd}{cd - ad}$

82. $\dfrac{bc - ad}{bd + ac}$

In Exercises 83–84, use this information. Sorting records is a common task in data processing. A selection sort requires C comparisons to sort N records, where C and N are related by the formula $C = \frac{N(N-1)}{2}$.

83. How many comparisons are needed to sort 200 records?

84. How many comparisons are needed to sort 10,000 records?

85. Perimeter of a triangle Find the perimeter of a triangle with sides that are 23.5, 37.2, and 39.7 feet long.

86. Perimeter of a trapezoid Find the perimeter of a trapezoid with sides that are 43.27, 47.37, 50.21, and 52.93 centimeters long.

In Exercises 87–98, tell which property of the real numbers justifies each statement.

87. $3 + 7 = 7 + 3$

88. $2 \cdot (9 \cdot 13) = (2 \cdot 9) \cdot 13$

89. $3(2 + 5) = 3 \cdot 2 + 3 \cdot 5$

90. $4 \cdot 3 = 3 \cdot 4$

91. $81 + 0 = 81$

92. $3(9 + 2) = 3 \cdot 9 + 3 \cdot 2$

93. $5 \cdot \dfrac{1}{5} = 1$

94. $3 + (9 + 0) = (9 + 0) + 3$

95. $a + (7 + 8) = (a + 7) + 8$

96. $1 \cdot 3 = 3$

97. $(2 \cdot 3) \cdot 4 = 4 \cdot (2 \cdot 3)$

98. $8 + (-8) = 0$

In Exercises 99–102, use a calculator to verify each statement. Identify the property of real numbers that is being illustrated.

99. $(37.9 + 25.2) + 14.3 = 37.9 + (25.2 + 14.3)$

100. $7.1(3.9 + 8.8) = 7.1 \cdot 3.9 + 7.1 \cdot 8.8$

101. $2.73(4.534 + 57.12) = 2.73 \cdot 4.534 + 2.73 \cdot 57.12$

102. $(6.789 + 345.1) + 27.347 = (345.1 + 6.789) + 27.347$

APPLICATIONS *In Exercises 103–110, use signed numbers to solve each problem.*

103. Earning money One day Scott earned \$22.25 tutoring mathematics and \$39.75 tutoring physics. How much did he earn that day?

104. Losing weight During an illness, Wendy lost 13.5 pounds. She then dieted and lost another 11.5 pounds. How much did she lose?

105. Changing temperatures The temperature rose 17° in 1 hour and then dropped 13° in the next hour. Find the overall change in temperature.

106. Displaying the flag Before the American flag is displayed at half-mast, it should first be raised to the top of the flagpole. How far has the flag in Illustration 2 traveled?

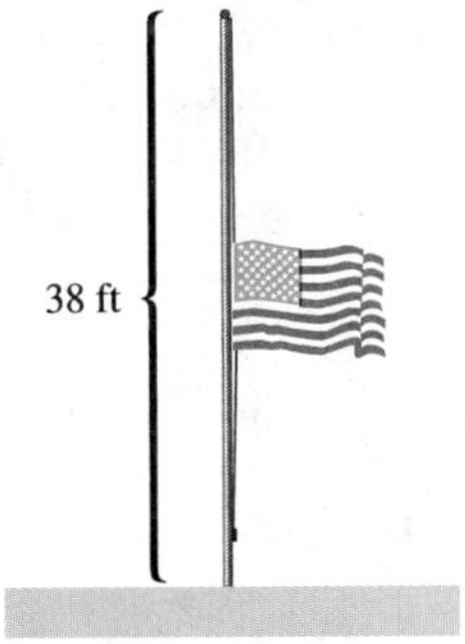

ILLUSTRATION 2

107. Changing temperatures If the temperature has been dropping 4° each hour, how much warmer was it 3 hours ago?

108. Playing slot machines In Las Vegas, Harry lost \$30 per hour playing the slot machines. How much did he lose after gambling for 15 hours?

109. Filling a pool The flow of water from a pipe is filling a pool at the rate of 23 gallons per minute. How much less water was in the pool 5 hours ago?

110. Draining a pool If a drain is emptying a pool at the rate of 12 gallons per minute, how much more water was in the pool 2 hours ago?

In Exercises 111–122, use a calculator to help solve the following problems.

111. Military science An army retreated 2,300 meters. After regrouping, it moved forward 1,750 meters. The next day, it gained another 1,875 meters. Find the army's net gain (or loss).

112. Grooming horses John earned \$8 an hour for grooming horses. After working for 8 hours, he had \$94. How much did he have before he started work?

113. Managing a checkbook Sally started with \$437.37 in a checking account. One month, she had deposits of \$125.18, \$137.26, and \$145.56. That same month, she had withdrawals of \$117.11, \$183.49, and \$122.89. Find her ending balance.

114. Stock averages Illustration 3 shows the daily advances and declines of the Dow Jones average for one week. Find the total gain or loss for the week.

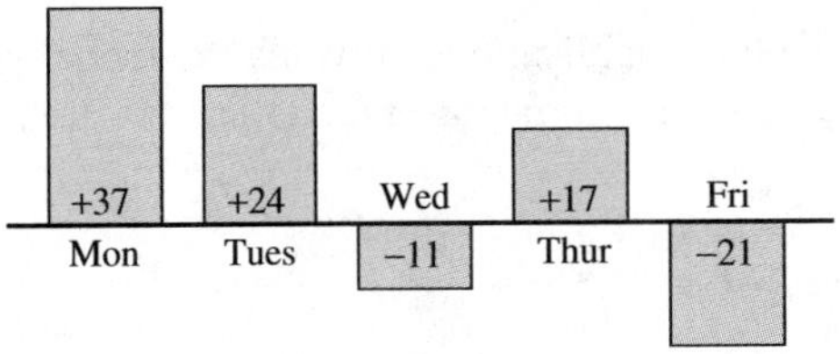

ILLUSTRATION 3

115. Selling clothes If a clerk had the sales shown in Illustration 4 for one week, find the mean of daily sales.

Monday	\$1,525
Tuesday	\$ 785
Wednesday	\$1,628
Thursday	\$1,214
Friday	\$ 917
Saturday	\$1,197

ILLUSTRATION 4

116. Size of viruses Illustration 5 gives the approximate lengths (in centimicrons) of the viruses that cause five common diseases. Find the mean length of the viruses.

Polio	2.5
Influenza	105.1
Pharyngitis	74.9
Chicken pox	137.4
Yellow fever	52.6

ILLUSTRATION 5

117. Calculating grades A student has scores of 75, 82, 87, 80, and 76 on five exams. Find his average (mean) score.

118. Averaging weights The offensive line of a football team has two guards, two tackles, and a center. If the guards weigh 298 and 287 pounds, the tackles 310 and 302 pounds, and the center 303 pounds, find the average (mean) weight of the offensive line.

119. Analyzing ads The businessman who ran the ad that is shown in Illustration 6 earns \$100,000 and employs four students who earn \$10,000 each. Is the ad honest?

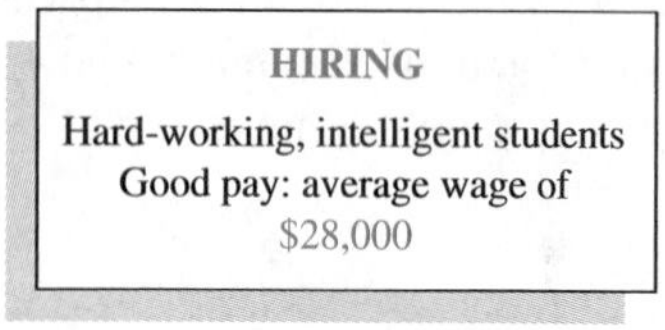
HIRING
Hard-working, intelligent students
Good pay: average wage of
$28,000

ILLUSTRATION 6

120. Averaging grades A student has grades of 78%, 85%, 88%, and 96%. There is one test left, and the student needs to average 90% to earn an A. Does he have a chance?

121. Perimeter of a square Find the perimeter of the square shown in Illustration 7.

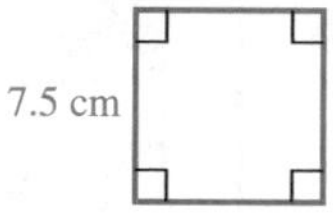
7.5 cm

ILLUSTRATION 7

122. Circumference of a circle To the nearest hundredth, find the circumference of the circle shown in Illustration 8.

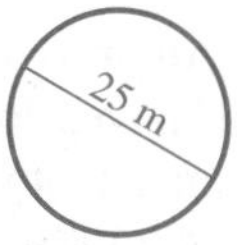
25 m

ILLUSTRATION 8

WRITING

123. The symmetric property of equality states that if $a = b$, then $b = a$. Explain why this property is often confused with the commutative properties. Why do you think this is so?

124. Explain why the mean of two numbers is halfway between the two numbers.

SOMETHING TO THINK ABOUT

125. Pick five numbers and find their mean. Add 7 to each of the numbers to get five new numbers and find their mean. What do you discover? Is this property always true?

126. Take the original five numbers in Exercise 125 and multiply each one by 7 to get five new numbers and find their mean. What do you discover? Is this property always true?

127. Give three applications in which the median would be the most appropriate average to use.

128. Give three applications in which the mode would be the most appropriate average to use.

1.3 Exponents

■ EXPONENTS ■ PROPERTIES OF EXPONENTS ■ ZERO EXPONENTS ■ NEGATIVE EXPONENTS ■ ORDER OF OPERATIONS ■ EVALUATING FORMULAS

Getting Ready *Find each product.*

1. $2 \cdot 2$

2. $3 \cdot 3 \cdot 3$

3. $(-4)(-4)(-4)$

4. $(-3)(-3)(-3)(-3)$ **5.** $\frac{1}{3} \cdot \frac{1}{3} \cdot \frac{1}{3}$ **6.** $-\left(\frac{2}{5} \cdot \frac{2}{5} \cdot \frac{2}{5} \cdot \frac{2}{5}\right)$

EXPONENTS

Exponents indicate repeated multiplication. For example,

$y^2 = y \cdot y$ Read y^2 as "y to the second power" or "y squared."

$z^3 = z \cdot z \cdot z$ Read z^3 as "z to the third power" or "z cubed."

$x^4 = x \cdot x \cdot x \cdot x$ Read x^4 as "x to the fourth power."

These examples suggest the following definition.

Natural-Number Exponents

If n is a natural number, then

$$x^n = \underbrace{x \cdot x \cdot x \cdot \cdots \cdot x}_{n \text{ factors of } x}$$

The exponential expression x^n is called a **power of x,** and we read it as "x to the nth power." In this expression, x is called the **base,** and n is called the **exponent.**

Base ⟶ x^n ⟵ Exponent

A natural-number exponent tells how many times the base of an exponential expression is to be used as a factor in a product.

EXAMPLE 1 Write each number without using exponents.

a. $2^5 = 2 \cdot 2 \cdot 2 \cdot 2 \cdot 2$
$= 32$

b. $(-2)^5 = (-2)(-2)(-2)(-2)(-2)$
$= -32$

c. $-4^4 = -(4^4)$
$= -(4 \cdot 4 \cdot 4 \cdot 4)$
$= -256$

d. $(-4)^4 = (-4)(-4)(-4)(-4)$
$= 256$

e. $\left(\frac{1}{2}a\right)^3 = \left(\frac{1}{2}a\right)\left(\frac{1}{2}a\right)\left(\frac{1}{2}a\right)$
$= \frac{1}{8}a^3$

f. $\left(-\frac{1}{5}b\right)^2 = \left(-\frac{1}{5}b\right)\left(-\frac{1}{5}b\right)$
$= \frac{1}{25}b^2$ ■

Self Check Write each number without using exponents: **a.** 3^4, **b.** $(-5)^3$, and **c.** $\left(-\frac{3}{4}a\right)^2$.

Answers **a.** 81, **b.** -125, **c.** $\frac{9}{16}a^2$

WARNING! Note the difference between $-x^n$ and $(-x)^n$.

$$-x^n = -(\overbrace{x \cdot x \cdot x \cdot \cdots \cdot x}^{n \text{ factors of } x}) \quad \text{and} \quad (-x)^n = \overbrace{(-x)(-x)(-x) \cdot \cdots \cdot (-x)}^{n \text{ factors of } -x}$$

Also, note the difference between ax^n and $(ax)^n$.

$$ax^n = a \cdot \overbrace{x \cdot x \cdot x \cdot \cdots \cdot x}^{n \text{ factors of } x} \quad \text{and} \quad (ax)^n = \overbrace{(ax)(ax)(ax) \cdot \cdots \cdot (ax)}^{n \text{ factors of } ax}$$

PROPERTIES OF EXPONENTS

Since x^5 means that x is to be used as a factor fives times, and since x^3 means that x is to be used as a factor three times, $x^5 \cdot x^3$ means that x will be used as a factor eight times.

$$x^5x^3 = \overbrace{x \cdot x \cdot x \cdot x \cdot x}^{5 \text{ factors of } x} \cdot \overbrace{x \cdot x \cdot x}^{3 \text{ factors of } x} = \overbrace{x \cdot x \cdot x \cdot x \cdot x \cdot x \cdot x \cdot x}^{8 \text{ factors of } x}$$

In general,

$$x^mx^n = \overbrace{x \cdot x \cdot x \cdot \cdots \cdot x}^{m \text{ factors of } x} \cdot \overbrace{x \cdot x \cdot \cdots \cdot x}^{n \text{ factors of } x} = \overbrace{x \cdot x \cdot x \cdot x \cdot \cdots \cdot x}^{m+n \text{ factors of } x}$$

Thus, *to multiply exponential expressions with the same base, we keep the same base and add the exponents.*

The Product Rule of Exponents

If m and n are natural numbers, then

$$x^mx^n = x^{m+n}$$

WARNING! The product rule of exponents applies only to exponential expressions with the same base. The expression x^5y^3, for example, cannot be simplified, because the bases of the exponential expressions are different.

EXAMPLE 2 Simplify each expression.

a. $x^{11}x^5 = x^{11+5}$
$= x^{16}$

b. $a^5a^4a^3 = (a^5a^4)a^3$
$= a^9a^3$
$= a^{12}$

c. $a^2b^3a^3b^2 = a^2a^3b^3b^2$
$= a^5b^5$

d. $-8x^4\left(\frac{1}{4}x^3\right) = -8\left(\frac{1}{4}x^4x^3\right)$
$= -2x^7$

■

Self Check Simplify each expression: **a.** a^3a^5, **b.** $a^2b^3a^3b^4$, and **c.** $-8a^4\left(-\frac{1}{2}a^2b\right)$.

Answers **a.** a^8, **b.** a^5b^7, **c.** $4a^6b$

To find another property of exponents, we simplify $(x^4)^3$, which means x^4 cubed or $x^4 \cdot x^4 \cdot x^4$.

$$(x^4)^3 = x^4 \cdot x^4 \cdot x^4 = \overbrace{x \cdot x \cdot x \cdot x}^{x^4} \cdot \overbrace{x \cdot x \cdot x \cdot x}^{x^4} \cdot \overbrace{x \cdot x \cdot x \cdot x}^{x^4} = x^{12}$$

In general, we have

$$(x^m)^n = \overbrace{x^m \cdot x^m \cdot x^m \cdot \cdots \cdot x^m}^{n \text{ factors of } x^m} = \overbrace{x \cdot x \cdot x \cdot x \cdot x \cdot \cdots \cdot x}^{mn \text{ factors of } x} = x^{mn}$$

Thus, *to raise an exponential expression to a power, we keep the same base and multiply the exponents.*

To find a third property of exponents, we square $3x$ and get

$$(3x)^2 = (3x)(3x) = 3 \cdot 3 \cdot x \cdot x = 3^2x^2 = 9x^2$$

In general, we have

$$(xy)^n = \overbrace{(xy)(xy)(xy) \cdot \cdots \cdot (xy)}^{n \text{ factors of } xy} = \overbrace{xxx \cdot \cdots \cdot x}^{n \text{ factors of } x} \cdot \overbrace{yyy \cdot \cdots \cdot y}^{n \text{ factors of } y} = x^ny^n$$

To find a fourth property of exponents, we cube $\frac{x}{3}$ to get

$$\left(\frac{x}{3}\right)^3 = \frac{x}{3} \cdot \frac{x}{3} \cdot \frac{x}{3} = \frac{x \cdot x \cdot x}{3 \cdot 3 \cdot 3} = \frac{x^3}{3^3} = \frac{x^3}{27}$$

In general, we have

$$\left(\frac{x}{y}\right)^n = \overbrace{\left(\frac{x}{y}\right)\left(\frac{x}{y}\right)\left(\frac{x}{y}\right) \cdot \cdots \cdot \left(\frac{x}{y}\right)}^{n \text{ factors of } x/y} \quad (y \neq 0)$$

$$= \frac{\overbrace{xxx \cdot \cdots \cdot x}^{n \text{ factors of } x}}{\underbrace{yyy \cdot \cdots \cdot y}_{n \text{ factors of } y}}$$ Multiply the numerators and multiply the denominators.

$$= \frac{x^n}{y^n}$$

The previous results are called the **power rules of exponents.**

The Power Rules of Exponents
If m and n are natural numbers, then

$$(x^m)^n = x^{mn} \qquad (xy)^n = x^n y^n \qquad \left(\frac{x}{y}\right)^n = \frac{x^n}{y^n} \quad (y \neq 0)$$

EXAMPLE 3 Simplify each expression.

a. $(3^2)^3 = 3^{2 \cdot 3} = 3^6 = 729$

b. $(x^{11})^5 = x^{11 \cdot 5} = x^{55}$

c. $(x^2x^3)^6 = (x^5)^6 = x^{30}$

d. $(x^2)^4(x^3)^2 = x^8x^6 = x^{14}$ ■

Self Check Simplify each expression: **a.** $(a^5)^8$, **b.** $(a^4a^3)^3$, and **c.** $(a^3)^3(a^2)^3$.

Answers **a.** a^{40}, **b.** a^{21}, **c.** a^{15}

EXAMPLE 4 Simplify each expression. Assume that no denominators are zero.

a. $(x^2y)^3 = (x^2)^3y^3 = x^6y^3$

b. $(x^3y^4)^4 = (x^3)^4(y^4)^4 = x^{12}y^{16}$

c. $\left(\frac{x}{y^2}\right)^4 = \frac{x^4}{(y^2)^4} = \frac{x^4}{y^8}$

d. $\left(\frac{x^3}{y^4}\right)^2 = \frac{(x^3)^2}{(y^4)^2} = \frac{x^6}{y^8}$ ■

Self Check Simplify each expression: **a.** $(a^4b^5)^2$ and **b.** $\left(\frac{a^5}{b^7}\right)^3$ $(b \neq 0)$.

Answers **a.** a^8b^{10}, **b.** $\frac{a^{15}}{b^{21}}$

ZERO EXPONENTS

Since the rules for exponents hold for exponents of 0, we have

$$x^0x^n = x^{0+n} = x^n = 1x^n$$

Because $x^0x^n = 1x^n$, it follows that $x^0 = 1$ $(x \neq 0)$.

Zero Exponents
If $x \neq 0$, then $x^0 = 1$.

WARNING! 0^0 is undefined.

Because of the previous definition, any nonzero base raised to the 0th power is 1. For example, if no variables are zero, then

$$5^0 = 1, \qquad (-7)^0 = 1, \qquad (3ax^3)^0 = 1, \qquad \left(\frac{1}{2}x^5y^7z^9\right)^0 = 1$$

NEGATIVE EXPONENTS

Since the rules for exponents are true for negative integer exponents, we have

$$x^{-n}x^n = x^{-n+n} = x^0 = 1 \quad (x \neq 0)$$

Because $x^{-n} \cdot x^n = 1$ and $\frac{1}{x^n} \cdot x^n = 1$, we define x^{-n} to be the reciprocal of x^n.

Negative Exponents
If n is an integer and $x \neq 0$, then

$$x^{-n} = \frac{1}{x^n} \quad \text{and} \quad \frac{1}{x^{-n}} = x^n$$

WARNING! By the definition of negative exponents, the base cannot be 0. Thus, an expression such as 0^{-5} is undefined.

Because of this definition, we can write expressions containing negative exponents as expressions without negative exponents. For example,

$$5^{-2} = \frac{1}{5^2} = \frac{1}{25} \qquad 10^{-3} = \frac{1}{10^3} = \frac{1}{1{,}000}$$

and if $x \neq 0$, we have

$$(2x)^{-3} = \frac{1}{(2x)^3} = \frac{1}{8x^3} \qquad 3x^{-1} = 3 \cdot \frac{1}{x} = \frac{3}{x}$$

EXAMPLE 5 Write each expression without negative exponents.

a. $x^{-5}x^3 = x^{-5+3}$
$= x^{-2}$
$= \frac{1}{x^2}$

b. $(x^{-3})^{-2} = x^{(-3)(-2)}$
$= x^6$

■

Self Check Write each expression without negative exponents: **a.** $a^{-7}a^3$ and **b.** $(a^{-5})^{-3}$.

Answers **a.** $\frac{1}{a^4}$, **b.** a^{15}

To develop a rule for dividing exponential expressions, we proceed as follows:

$$\frac{x^m}{x^n} = x^m\left(\frac{1}{x^n}\right) = x^m x^{-n} = x^{m+(-n)} = x^{m-n}$$

Thus, *to divide exponential expressions with the same nonzero base, we keep the same base and subtract the exponent in the denominator from the exponent in the numerator.*

The Quotient Rule

If m and n are integers, then

$$\frac{x^m}{x^n} = x^{m-n} \quad (x \neq 0)$$

EXAMPLE 6 Simplify each expression.

a. $\frac{a^5}{a^3} = a^{5-3}$
$= a^2$

b. $\frac{x^{-5}}{x^{11}} = x^{-5-11}$
$= x^{-16}$
$= \frac{1}{x^{16}}$

c. $\frac{x^4x^3}{x^{-5}} = \frac{x^7}{x^{-5}}$
$= x^{7-(-5)}$
$= x^{12}$

d. $\frac{(x^2)^3}{(x^3)^2} = \frac{x^6}{x^6}$
$= x^{6-6}$
$= x^0$
$= 1$

e. $\frac{x^2y^3}{xy^4} = x^{2-1}y^{3-4}$
$= xy^{-1}$
$= \frac{x}{y}$

f. $\left(\frac{a^{-2}b^3}{a^2a^3b^4}\right)^3 = \left(\frac{a^{-2}b^3}{a^5b^4}\right)^3$
$= (a^{-2-5}b^{3-4})^3$
$= (a^{-7}b^{-1})^3$
$= \left(\frac{1}{a^7b}\right)^3$
$= \frac{1}{a^{21}b^3}$ ■

Self Check Simplify each expression: **a.** $\frac{(a^{-2})^3}{(a^2)^{-3}}$ and **b.** $\left(\frac{a^{-2}b^5}{b^8}\right)^{-3}$.

Answers **a.** 1, **b.** a^6b^9

To illustrate one more property of exponents, we consider the following simplification of $\left(\frac{2}{3}\right)^{-4}$.

$$\left(\frac{2}{3}\right)^{-4} = \frac{1}{\left(\frac{2}{3}\right)^{4}} = \frac{1}{\frac{2^4}{3^4}} = 1 \div \frac{2^4}{3^4} = 1 \cdot \frac{3^4}{2^4} = \frac{3^4}{2^4} = \left(\frac{3}{2}\right)^{4}$$

The example suggests that to raise a fraction to a negative power, we can invert the fractional base and then raise it to a positive power.

Fractions to Negative Powers

If n is an integer, then

$$\left(\frac{x}{y}\right)^{-n} = \left(\frac{y}{x}\right)^{n} \quad (x \neq 0,\ y \neq 0)$$

EXAMPLE 7

Write each expression without using parentheses.

a. $$\left(\frac{2}{3}\right)^{-4} = \left(\frac{3}{2}\right)^{4} = \frac{81}{16}$$

b. $$\left(\frac{y^2}{x^3}\right)^{-3} = \left(\frac{x^3}{y^2}\right)^{3} = \frac{x^9}{y^6}$$

c. $$\left(\frac{2x^2}{3y^{-3}}\right)^{-4} = \left(\frac{3y^{-3}}{2x^2}\right)^{4} = \frac{81y^{-12}}{16x^8} = \frac{81}{16x^8} \cdot y^{-12} = \frac{81}{16x^8} \cdot \frac{1}{y^{12}} = \frac{81}{16x^8y^{12}}$$

d. $$\left(\frac{a^{-2}b^3}{a^2a^3b^4}\right)^{-3} = \left(\frac{a^2a^3b^4}{a^{-2}b^3}\right)^{3} = \left(\frac{a^5b^4}{a^{-2}b^3}\right)^{3} = (a^{5-(-2)}b^{4-3})^3 = (a^7b)^3 = a^{21}b^3$$

■

Self Check

Write $\left(\frac{3a^3}{2b^{-2}}\right)^{-5}$ without using parentheses.

Answer

$\frac{32}{243a^{15}b^{10}}$

We summarize the rules of exponents as follows.

Properties of Exponents
If there are no divisions by 0, then for all integers m and n,

$$x^m x^n = x^{m+n} \qquad (x^m)^n = x^{mn} \qquad (xy)^n = x^n y^n \qquad \left(\frac{x}{y}\right)^n = \frac{x^n}{y^n}$$

$$x^0 = 1 \quad (x \neq 0) \qquad x^{-n} = \frac{1}{x^n} \qquad \frac{x^m}{x^n} = x^{m-n} \qquad \left(\frac{x}{y}\right)^{-n} = \left(\frac{y}{x}\right)^n$$

The same rules apply to exponents that are variables.

EXAMPLE 8 Simplify each expression. Assume that $a \neq 0$ and $x \neq 0$.

a. $\dfrac{a^n a}{a^2} = a^{n+1-2}$
$= a^{n-1}$

b. $\dfrac{x^3x^2}{x^n} = x^{3+2-n}$
$= x^{5-n}$

c. $\left(\dfrac{x^n}{x^2}\right)^2 = \dfrac{x^{2n}}{x^4}$
$= x^{2n-4}$

d. $\dfrac{a^n a^{-3}}{a^{-1}} = a^{n+(-3)-(-1)}$
$= a^{n-3+1}$
$= a^{n-2}$ ■

Self Check Simplify each expression (assume that $t \neq 0$): **a.** $\dfrac{t^n t^2}{t^3}$ and **b.** $\left(\dfrac{2t^n}{3t^3}\right)^3$.

Answers **a.** t^{n-1}, **b.** $\dfrac{8t^{3n-9}}{27}$

CALCULATORS

Finding Powers

To find powers of numbers with a scientific calculator, we use the y^x key. For example, to find 5.37^4, we enter these numbers and press these keys:

5.37 y^x 4 = Some calculators have an x^y key.

The display will read 831.5668016 .

To use a graphing calculator, we enter these numbers and press these keys:

5.37 ∧ 4 ENTER

The display will read 5.37^4
831.5668016

If neither of these methods works, consult the owner's manual.

ORDER OF OPERATIONS

When simplifying expressions containing exponents, we find powers before performing additions and multiplications.

EXAMPLE 9 If $x = 2$ and $y = -3$, find the value of $3x + 2y^3$.

Solution

$3x + 2y^3 = 3(2) + 2(-3)^3$	Substitute 2 for x and -3 for y.
$= 3(2) + 2(-27)$	First find the power: $(-3)^3 = -27$.
$= 6 - 54$	Then do the multiplications.
$= -48$	Then do the subtraction.

■

Self Check Evaluate $-2a^2 - 3a$ if $a = -4$.

Answer -20

EVALUATING FORMULAS

Table 1-2 (on page 40) shows the formulas used to compute the areas and volumes of many geometric figures.

EXAMPLE 10 Find the volume of the sphere shown in Figure 1-22.

Solution The formula for the volume of a sphere is $V = \frac{4}{3}\pi r^3$. Since a radius is half as long as a diameter, the radius of the sphere is half of 20 centimeters, or 10 centimeters.

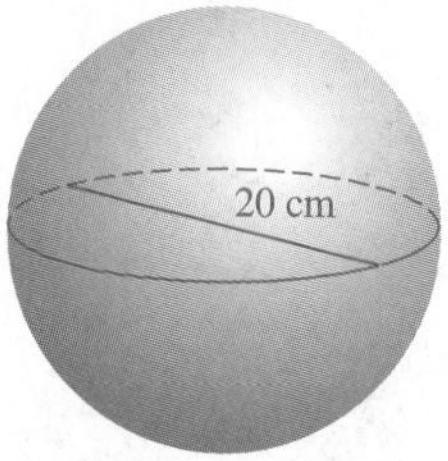

FIGURE 1-22

$$V = \frac{4}{3}\pi r^3$$

$$V = \frac{4}{3}\pi(10)^3 \qquad \text{Substitute 10 for } r.$$

$$\approx 4188.790205$$

Press these keys on a scientific calculator:
10 y^x 3 = × π × 4 ÷ 3 =

To two decimal places, the volume is 4,188.79 cm^3. ■

Self Check Find the volume of a pyramid with a square base, 20 meters on each side, and a height of 21 meters.

Answer 2,800 m^3

Figure	Name	Area	Figure	Name	Volume
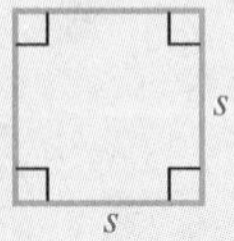	Square	$A = s^2$	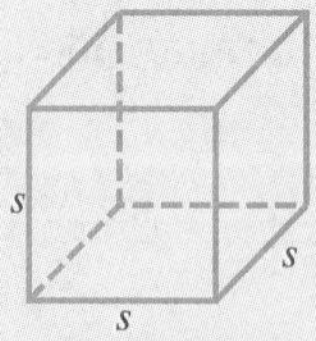	Cube	$V = s^3$
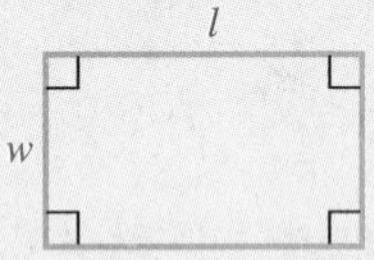	Rectangle	$A = lw$	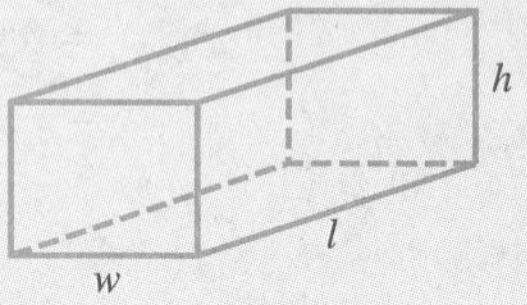	Rectangular solid	$V = lwh$
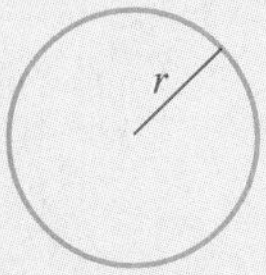	Circle	$A = \pi r^2$	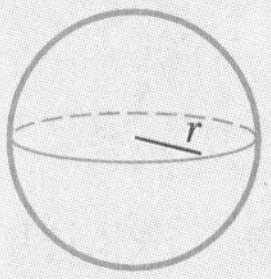	Sphere	$V = \frac{4}{3}\pi r^3$
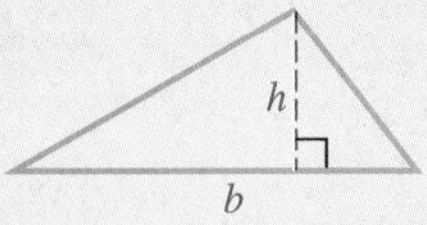	Triangle	$A = \frac{1}{2}bh$	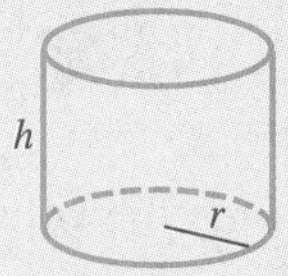	Cylinder	$V = Bh$*
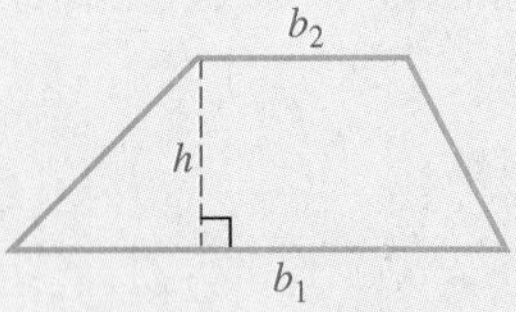	Trapezoid	$A = \frac{1}{2}h(b_1 + b_2)$	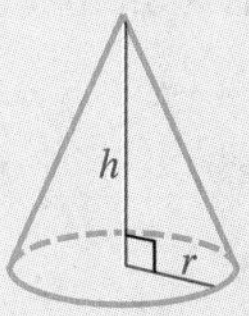	Cone	$V = \frac{1}{3}Bh$*
			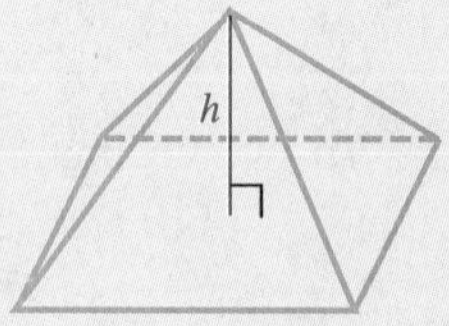	Pyramid	$V = \frac{1}{3}Bh$*

*B represents the area of the base.

TABLE 1-2

Orals *Simplify each expression.*

1. 4^2
2. 3^3
3. x^2x^3
4. y^3y^4
5. 17^0
6. $(x^2)^3$
7. $(a^2b)^3$
8. $\left(\frac{b}{a^2}\right)^2$
9. 5^{-2}
10. $(x^{-2})^{-1}$
11. $\frac{x^5}{x^2}$
12. $\frac{x^2}{x^5}$

EXERCISE 1.3

REVIEW *If $a = 4$, $b = -2$, and $c = 5$, find each value.*

1. $a + b + c$
2. $a - 2b - c$
3. $\frac{ab + 2c}{a + b}$
4. $\frac{ac - bc}{6ab + b}$

VOCABULARY AND CONCEPTS *Fill in each blank to make a true statement.*

5. In the exponential expression x^n, x is called the ____, and n is called the ____.
6. A natural-number exponent tells how many times the base is used as a ____.
7. $x^mx^n =$ ____
8. $(x^m)^n =$ ____
9. $(xy)^n =$ ____
10. $\left(\frac{x}{y}\right)^n = \frac{\_\_}{\_\_}$ $(y \neq 0)$
11. If $a \neq 0$, then $a^0 =$ __.
12. If $a \neq 0$, then $a^{-1} = \frac{\_\_}{\_\_}$.
13. If $x \neq 0$, then $\frac{x^m}{x^n} =$ ____.
14. $\left(\frac{4}{5}\right)^{-3} = (\_\_)^3$.

Write the formula to find each quantity.

15. Area of a square: ____
16. Area of a rectangle: ____
17. Area of a triangle: ____
18. Area of a trapezoid: ____
19. Area of a circle: ____
20. Volume of a cube: ____
21. Volume of a rectangular solid: ____
22. Volume of a sphere: ____
23. Volume of a cylinder: ____
24. Volume of a cone: ____
25. Volume of a pyramid: ____
26. In Exercises 23–25, B represents the area of the ____ of a solid.

PRACTICE *In Exercises 27–34, identify the base and the exponent.*

27. 5^3
28. -7^2
29. $-x^5$
30. $(-t)^4$
31. $2b^6$
32. $(3xy)^5$
33. $(-mn^2)^3$
34. $(-p^2q)^2$

In Exercises 35–106, simplify each expression. Assume that no denominators are zero.

35. 3^2
36. 3^4
37. -3^2
38. -3^4
39. $(-3)^2$
40. $(-3)^3$
41. 5^{-2}
42. 5^{-4}
43. -5^{-2}
44. -5^{-4}
45. $(-5)^{-2}$
46. $(-5)^{-4}$
47. 8^0
48. -9^0
49. $(-8)^0$
50. $(-9)^0$
51. $(-2x)^5$
52. $(-3a)^3$
53. $(-2x)^6$
54. $(-3y)^5$
55. x^2x^3
56. y^3y^4
57. k^0k^7
58. x^8x^{11}
59. $x^2x^3x^5$
60. $y^3y^7y^2$
61. p^9pp^0
62. z^7z^0z
63. aba^3b^4
64. $x^2y^3x^3y^2$
65. $(-x)^2y^4x^3$
66. $-x^2y^7y^3x^{-2}$
67. $(x^4)^7$
68. $(y^7)^5$
69. $(b^{-8})^9$
70. $(z^{12})^2$
71. $(x^3y^2)^4$
72. $(x^2y^5)^2$
73. $(r^{-3}s)^3$
74. $(m^5n^2)^{-3}$
75. $(a^2a^3)^4$
76. $(bb^2b^3)^4$
77. $(-d^2)^3(d^{-3})^3$
78. $(c^3)^2(c^4)^{-2}$
79. $(3x^3y^4)^3$
80. $\left(\frac{1}{2}a^2b^5\right)^4$
81. $\left(-\frac{1}{3}mn^2\right)^6$
82. $(-3p^2q^3)^5$
83. $\left(\frac{a^3}{b^2}\right)^5$
84. $\left(\frac{a^2}{b^3}\right)^4$
85. $\left(\frac{a^{-3}}{b^{-2}}\right)^{-2}$
86. $\left(\frac{k^{-3}}{k^{-4}}\right)^{-1}$
87. $\frac{a^8}{a^3}$
88. $\frac{c^7}{c^2}$
89. $\frac{c^{12}c^5}{c^{10}}$
90. $\frac{a^{33}}{a^2a^3}$
91. $\frac{m^9m^{-2}}{(m^2)^3}$
92. $\frac{a^{10}a^{-3}}{a^5a^{-2}}$
93. $\frac{1}{a^{-4}}$
94. $\frac{3}{b^{-5}}$
95. $\frac{3m^5m^{-7}}{m^2m^{-5}}$
96. $\frac{(2a^{-2})^3}{a^3a^{-4}}$
97. $\left(\frac{4a^{-2}b}{3ab^{-3}}\right)^3$
98. $\left(\frac{2ab^{-3}}{3a^{-2}b^2}\right)^2$
99. $\left(\frac{3a^{-2}b^2}{17a^2b^3}\right)^0$
100. $\frac{a^0 + b^0}{2(a + b)^0}$
101. $\left(\frac{-2a^4b}{a^{-3}b^2}\right)^{-3}$
102. $\left(\frac{-3x^4y^2}{-9x^5y^{-2}}\right)^{-2}$
103. $\left(\frac{2a^3b^2}{3a^{-3}b^2}\right)^{-3}$
104. $\left(\frac{3x^5y^2}{6x^5y^{-2}}\right)^{-4}$
105. $\frac{(3x^2)^{-2}}{x^3x^{-4}x^0}$
106. $\frac{y^{-3}y^{-4}y^0}{(2y^{-2})^3}$

In Exercises 107–114, simplify each expression.

107. $\frac{a^na^3}{a^4}$
108. $\frac{b^9b^7}{b^n}$
109. $\left(\frac{b^n}{b^3}\right)^3$
110. $\left(\frac{a^2}{a^n}\right)^4$
111. $\frac{a^{-n}a^2}{a^3}$
112. $\frac{a^na^{-2}}{a^4}$
113. $\frac{a^{-n}a^{-2}}{a^{-4}}$
114. $\frac{a^n}{a^{-3}a^5}$

In Exercises 115–118, use a calculator to find each value.

115. 1.23^6

116. 0.0537^4

117. -6.25^3

118. $(-25.1)^5$

In Exercises 119–126, use a calculator to verify that each statement is true.

119. $(3.68)^0 = 1$

120. $(2.1)^4(2.1)^3 = (2.1)^7$

121. $(7.2)^2(2.7)^2 = [(7.2)(2.7)]^2$

122. $(3.7)^2 + (4.8)^2 \neq (3.7 + 4.8)^2$

123. $(3.2)^2(3.2)^{-2} = 1$

124. $[(5.9)^3]^2 = (5.9)^6$

125. $(7.23)^{-3} = \dfrac{1}{(7.23)^3}$

126. $\left(\dfrac{5.4}{2.7}\right)^{-4} = \left(\dfrac{2.7}{5.4}\right)^4$

In Exercises 127–134, evaluate each expression when $x = -2$ and $y = 3$.

127. x^2y^3

128. x^3y^2

129. $\dfrac{x^{-3}}{y^3}$

130. $\dfrac{x^2}{y^{-3}}$

131. $(xy^2)^{-2}$

132. $-y^3x^{-2}$

133. $(-yx^{-1})^3$

134. $(-y)^3x^{-2}$

In Exercises 135–142, find the area of each figure. Round all answers to the nearest unit.

135.

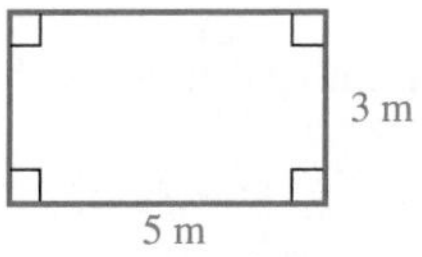

136.

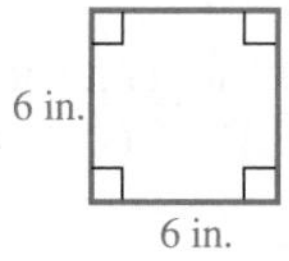

137.

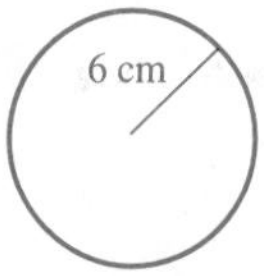

138.

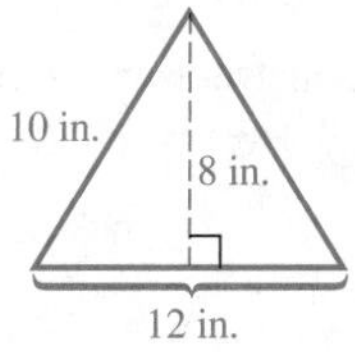

139.

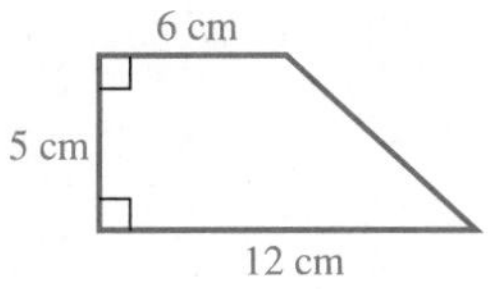

140.

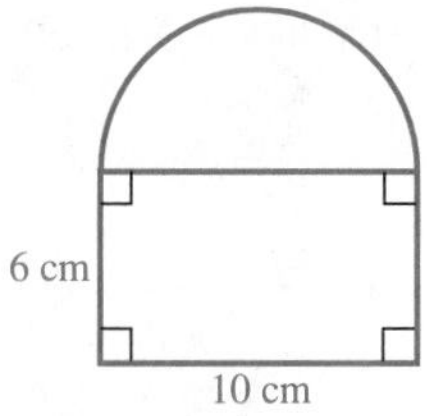

141.

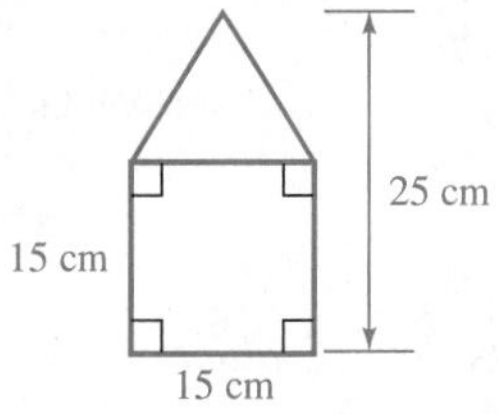

142.

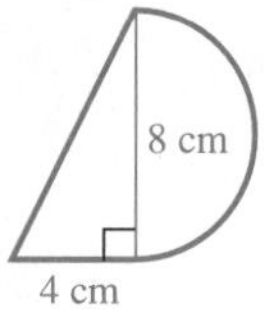

In Exercises 143–150, find the volume of each figure. Round all answers to the nearest unit.

143.

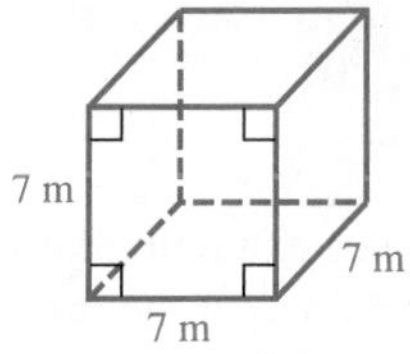

144.

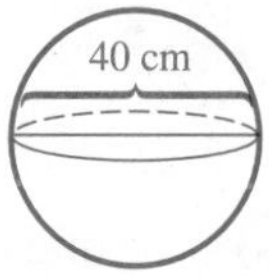

145.

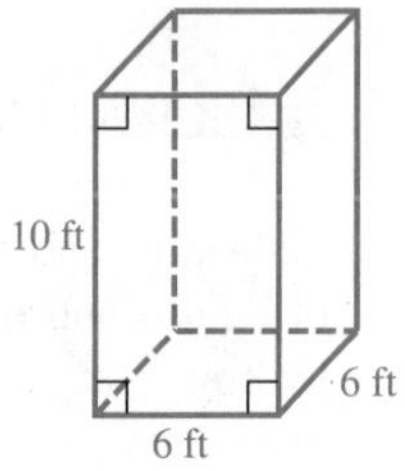

146.

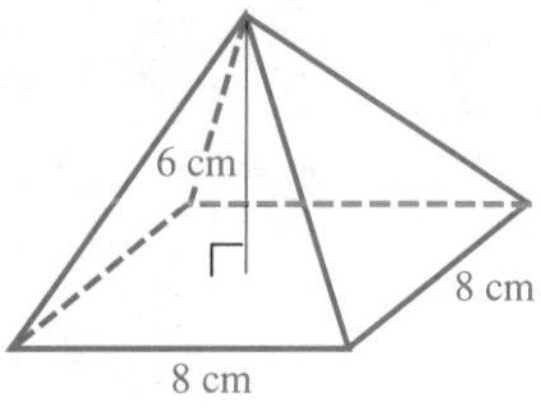

147.

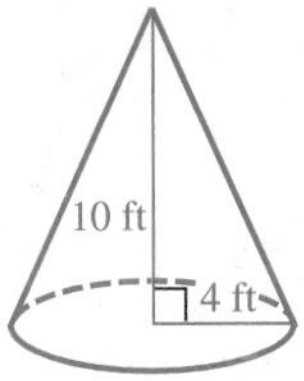

148.

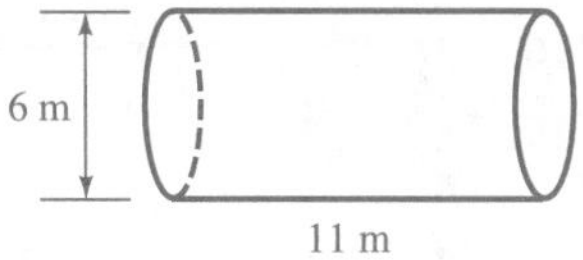

149.

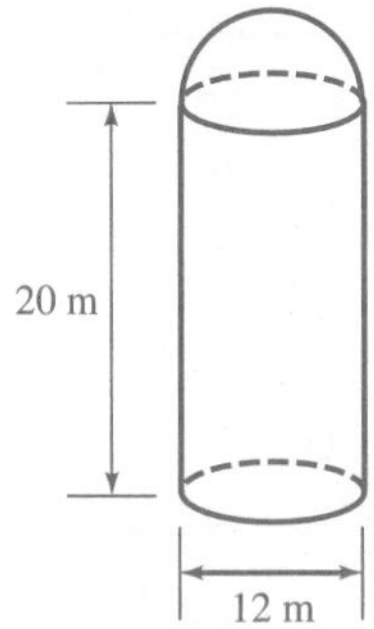

150.

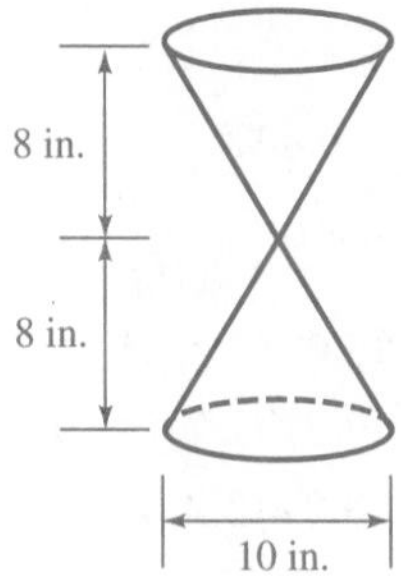

WRITING

151. Explain why a positive number raised to a negative power is positive.

152. Explain the rules that determine the order in which operations are performed.

153. In the definition of x^{-1}, x cannot be 0. Why not?

154. Explain why $(xyz)^2 = x^2y^2z^2$.

SOMETHING TO THINK ABOUT

155. Find the sum: $2^{-1} + 3^{-1} - 4^{-1}$.

156. Simplify $(3^{-1} + 4^{-1})^{-2}$.

157. Construct an example using numbers to show that $x^m + x^n \neq x^{m+n}$.

158. Construct an example using numbers to show that $x^m + y^m \neq (x + y)^m$.

1.4 Scientific Notation

■ SCIENTIFIC NOTATION ■ USING SCIENTIFIC NOTATION TO SIMPLIFY COMPUTATIONS
■ SIGNIFICANT DIGITS ■ PROBLEM SOLVING

Getting Ready *Evaluate each expression.*

1. 10^1 **2.** 10^2 **3.** 10^3 **4.** 10^4

5. 10^{-2} **6.** 10^{-4} **7.** $4(10^3)$ **8.** $7(10^{-4})$

■ SCIENTIFIC NOTATION

Very large and very small numbers occur often in science. For example, the speed of light is approximately 29,980,000,000 centimeters per second, and the mass of a hydrogen atom is approximately 0.000000000000000000000001673 gram. With exponents, we can write these numbers more compactly by using **scientific notation.**

Scientific Notation

A number is written in **scientific notation** when it is written in the form $N \times 10^n$, where $1 \leq |N| < 10$ and n is an integer.

EXAMPLE 1 Change 29,980,000,000 to scientific notation.

Solution The number 2.998 is between 1 and 10. To get 29,980,000,000, the decimal point in 2.998 must be moved ten places to the right. This can be done by multiplying 2.998 by 10^{10}.

$$29{,}980{,}000{,}000 = 2.998 \times 10^{10}$$ ■

Self Check Change 150,000,000 to scientific notation.

Answer 1.5×10^8

EXAMPLE 2 Write 0.000000000000000000000001673 in scientific notation.

Solution The number 1.673 is between 1 and 10. To get 0.000000000000000000000001673, the decimal point in 1.673 must be moved twenty-four places to the left. This can be done by multiplying 1.673 by 10^{-24}.

$$0.000000000000000000000001673 = 1.673 \times 10^{-24}$$ ■

Self Check Change 0.000025 to scientific notation.

Answer 2.5×10^{-5}

EXAMPLE 3 Change -0.0013 to scientific notation.

Solution The absolute value of -1.3 is between 1 and 10. To get -0.0013, we move the decimal point in -1.3 three places to the left by multiplying by 10^{-3}.

$$-0.0013 = -1.3 \times 10^{-3}$$ ■

Self Check Change $-45{,}700$ to scientific notation.

Answer -4.57×10^4

We can change a number written in scientific notation to **standard notation.** For example, to write 9.3×10^7 in standard notation, we multiply 9.3×10^7.

$$9.3 \times 10^7 = 9.3 \times 10{,}000{,}000 = 93{,}000{,}000$$

EXAMPLE 4 Change **a.** 3.7×10^5 and **b.** -1.1×10^{-3} to standard notation.

Solution **a.** Since multiplication by 10^5 moves the decimal point 5 places to the right,

$$3.7 \times 10^5 = 370{,}000$$

b. Since multiplication by 10^{-3} moves the decimal point 3 places to the left,

$$-1.1 \times 10^{-3} = -0.0011$$

■

Self Check Change **a.** -9.6×10^4 and **b.** 5.62×10^{-3} to standard notation.

Answers **a.** $-96{,}000$, **b.** 0.00562

Each of the following numbers is written in both scientific and standard notation. In each case, the exponent gives the number of places that the decimal point moves, and the sign of the exponent indicates the direction that it moves:

$5.32 \times 10^4 = 53200.$ — 4 places to the right

$6.45 \times 10^7 = 64500000.$ — 7 places to the right

$2.37 \times 10^{-4} = 0.000237$ — 4 places to the left

$9.234 \times 10^{-2} = 0.09234$ — 2 places to the left

$4.89 \times 10^0 = 4.89$ — No movement of the decimal point

PERSPECTIVE

The ancient Egyptians developed two systems of writing. In hieroglyphics, each symbol was a picture of an object. Because hieroglyphic writing was usually inscribed in stone, many examples still survive today. For daily life, Egyptians used hieratic writing. Similar to hieroglyphics, hieratic writing was done with ink on papyrus sheets. Papyrus, made from plants, is very delicate and quickly dries and crumbles. Few Egyptian papyri survive today.

Those that do survive provide important clues to the content of ancient mathematics. One papyrus, the Rhind Papyrus, was discovered in 1858 by a British archaeologist, Henry Rhind. Also known as the Ahmes Papyrus after its ancient author, it begins with a description of its contents: *Directions for Obtaining the Knowledge of All Dark Things.*

The Ahmes Papyrus and another, the Moscow Papyrus, together contain 110 mathematical problems and their solutions. Many of these were probably for education, because they represented situations that scribes, priests, and other government and temple administration workers were expected to be able to solve.

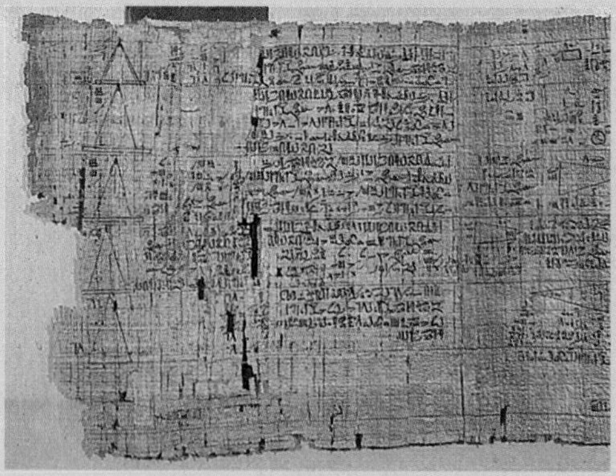

The Ahmes Papyrus
(the British Museum)

Numbers such as 47.2×10^3 and 0.063×10^{-2} appear to be written in scientific notation, because they are the product of a number and a power of 10. However, they are not in scientific notation, because 47.2 and 0.063 are not between 1 and 10.

EXAMPLE 5 Change **a.** 47.2×10^3 and **b.** 0.063×10^{-2} to scientific notation.

Solution Since the first factors are not between 1 and 10, neither number is in scientific notation. However, we can change them to scientific notation as follows:

a.
$$\begin{aligned} \mathbf{47.2} \times 10^3 &= (\mathbf{4.72 \times 10^1}) \times 10^3 && \text{Write 47.2 in scientific notation.} \\ &= 4.72 \times (10^1 \times 10^3) \\ &= 4.72 \times 10^4 \end{aligned}$$

b.
$$\begin{aligned} \mathbf{0.063} \times 10^{-2} &= (\mathbf{6.3 \times 10^{-2}}) \times 10^{-2} && \text{Write 0.063 in scientific notation.} \\ &= 6.3 \times (10^{-2} \times 10^{-2}) \\ &= 6.3 \times 10^{-4} \end{aligned}$$

■

Self Check Change **a.** 27.3×10^2 and **b.** 0.0025×10^{-3} to scientific notation.

Answers **a.** 2.73×10^3, **b.** 2.5×10^{-6}

■ USING SCIENTIFIC NOTATION TO SIMPLIFY COMPUTATIONS

Scientific notation is useful when multiplying and dividing very large or very small numbers.

EXAMPLE 6 Use scientific notation to simplify $\dfrac{(0.00000064)(24{,}000{,}000{,}000)}{(400{,}000{,}000)(0.0000000012)}$.

Solution After changing each number into scientific notation, we can do the arithmetic on the numbers and the exponential expressions separately.

$$\begin{aligned} \frac{(0.00000064)(24{,}000{,}000{,}000)}{(400{,}000{,}000)(0.0000000012)} &= \frac{(6.4 \times 10^{-7})(2.4 \times 10^{10})}{(4 \times 10^8)(1.2 \times 10^{-9})} \\ &= \frac{\mathbf{(6.4)(2.4)}}{\mathbf{(4)(1.2)}} \cdot \frac{\mathbf{10^{-7}10^{10}}}{\mathbf{10^{8}10^{-9}}} \\ &= \mathbf{3.2} \times 10^4 \end{aligned}$$

In standard notation, the result is 32,000. ■

Self Check Simplify $\frac{(320)(25{,}000)}{0.00004}$.

Answer 200,000,000,000

Using Scientific Notation

CALCULATORS

Scientific and graphing calculators often give answers in scientific notation. For example, if we use a calculator to find 301.2^8, the display will read

6.77391496 19 — On a scientific calculator.

301.2 ∧ 8
6.77391496 1E19 — On a graphing calculator.

In either case, the answer is given in scientific notation and is to be interpreted as

$$6.77391496 \times 10^{19}$$

Numbers can also be entered into a calculator in scientific notation. For example, to enter 24,000,000,000 (which is 2.4×10^{10} in scientific notation), we enter these numbers and press these keys:

2.4 EXP 10
2.4 EE 10 } Whichever of these keys is on your calculator.

To use a calculator to simplify

$$\frac{(24{,}000{,}000{,}000)(0.00000006495)}{0.00000004824}$$

we must enter each number in scientific notation, because each number has too many digits to be entered directly. In scientific notation, the three numbers are

$$2.4 \times 10^{10} \qquad 6.495 \times 10^{-8} \qquad 4.824 \times 10^{-8}$$

To use a scientific calculator to simplify the fraction, we enter these numbers and press these keys:

2.4 EXP 10 × 6.495 EXP 8 +/− ÷ 4.824 EXP 8 +/− =

The display will read 3.231343284 10 . In standard notation, the answer is 32,313,432,840.

The steps are similar on a graphing calculator.

SIGNIFICANT DIGITS

If we measure the length of a rectangle and report the length to be 45 centimeters, we have rounded to the nearest centimeter. If we measure more carefully and find the length to be 45.2 centimeters, we have rounded to the nearest tenth of a centimeter. We say that the second measurement is more accurate than the first, because 45.2 has three *significant digits* but 45 has only two.

It is not always easy to know how many significant digits a number has. For example, 270 might be accurate to two or three significant digits. If 270 is rounded to the nearest ten, the number has two significant digits. If 270 is rounded to the nearest unit, it has three significant digits. This ambiguity does not occur when a number is written in scientific notation.

Finding Significant Digits

If a number M is written in scientific notation as $N \times 10^n$, where $1 \le |N| < 10$ and n is an integer, the number of significant digits in M is the same as the number of digits in N.

In a problem where measurements are multiplied or divided, the final result should be rounded so that the answer has the same number of significant digits as the least accurate measurement.

PROBLEM SOLVING

EXAMPLE 7 The earth is approximately 93,000,000 miles from the sun, and Jupiter is approximately 484,000,000 miles from the sun. Assuming the alignment shown in Figure 1-23, how long would it take a spaceship traveling at 7,500 mph to fly from the earth to Jupiter?

Solution When the planets are aligned as shown in the figure, the distance between earth and Jupiter is (484,000,000 − 93,000,000) miles or 391,000,000 miles. To find the length of time in hours for the trip, we divide the distance by the rate.

$$\frac{391{,}000{,}000 \text{ mi}}{7{,}500 \frac{\text{mi}}{\text{hr}}} = \frac{3.91 \times 10^8 \text{ mi}}{7.5 \times 10^3 \ \frac{\text{mi}}{\text{hr}}}$$

There are three significant digits in the numerator and two significant digits in the denominator.

$$\approx 0.5213333 \times 10^5 \text{ mi} \cdot \frac{\text{hr}}{\text{mi}}$$

$$\approx 52{,}133.33 \text{ hr}$$

Since there are 24×365 hours in a year, we can change this result from hours to years by dividing 52,133.33 by (24×365).

$$\frac{52{,}133.33 \text{ hr}}{(24 \times 365) \frac{\text{hr}}{\text{yr}}} \approx 5.9512934 \text{ hr} \cdot \frac{\text{yr}}{\text{hr}} \approx 5.9512934 \text{ yr}$$

Rounding to two significant digits, the trip will take about 6.0 years.

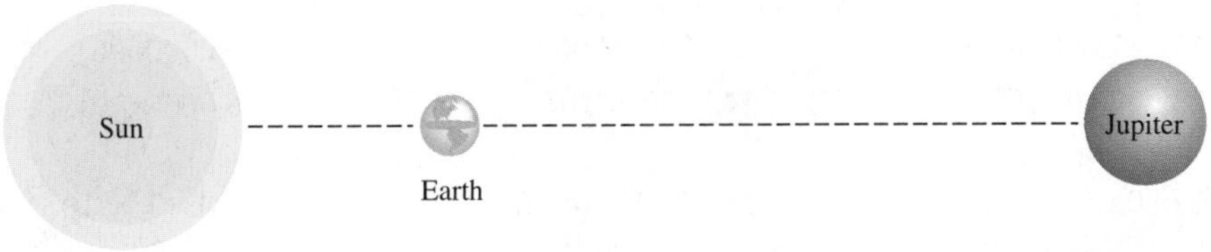

FIGURE 1-23

Self Check In Example 7, how long would it take if the spaceship could travel at 12,000 mph?

Answer about 3.7 years

Orals *Give each numeral in scientific notation.*

1. 352
2. 5,130
3. 0.002
4. 0.00025

Give each numeral in standard notation.

5. 3.5×10^2
6. 4.3×10^3
7. 2.7×10^{-1}
8. 8.5×10^{-2}

EXERCISE 1.4

REVIEW *Write each fraction as a terminating or a repeating decimal.*

1. $\frac{3}{4}$
2. $\frac{4}{5}$
3. $\frac{13}{9}$
4. $\frac{14}{11}$
5. A man raises 3 to the second power, 4 to the third power, and 2 to the fourth power and then finds their sum. What number does he obtain?
6. If $a = -2$, $b = -3$, and $c = 4$, find the value of
$$\frac{5ab - 4ac - 2}{3bc + abc}$$

VOCABULARY AND CONCEPTS *Fill in each blank to make a true statement.*

7. A number is written in scientific notation when it is written in the form $N \times$ ____, where $1 \leq |N| < 10$ and n is an integer.
8. To change 6.31×10^4 to standard notation, we move the decimal point in 6.31 ____ places to the right.
9. To change 6.31×10^{-4} to standard notation, we move the decimal point four places to the ____.
10. The number 6.7×10^3 ____ (> or <) the number $6{,}700{,}000 \times 10^{-4}$.

PRACTICE *In Exercises 11–30, write each numeral in scientific notation.*

11. 3,900
12. 1,700
13. 0.0078
14. 0.068
15. −45,000
16. −547,000
17. −0.00021
18. −0.00078
19. 17,600,000
20. 89,800,000
21. 0.0000096
22. 0.000046
23. 323×10^5
24. 689×10^9
25. $6{,}000 \times 10^{-7}$
26. 765×10^{-5}
27. 0.0527×10^5
28. 0.0298×10^3
29. 0.0317×10^{-2}
30. 0.0012×10^{-3}

In Exercises 31–42, write each numeral in standard notation.

31. 2.7×10^2 **32.** 7.2×10^3 **33.** 3.23×10^{-3} **34.** 6.48×10^{-2}

35. 7.96×10^5 **36.** 9.67×10^6 **37.** 3.7×10^{-4} **38.** 4.12×10^{-5}

39. 5.23×10^0 **40.** 8.67×10^0 **41.** 23.65×10^6 **42.** 75.6×10^{-5}

In Exercises 43–46, write each numeral in scientific notation and do the operations. Give all answers in scientific notation.

43. $\dfrac{(4{,}000)(30{,}000)}{0.0006}$

44. $\dfrac{(0.0006)(0.00007)}{21{,}000}$

45. $\dfrac{(640{,}000)(2{,}700{,}000)}{120{,}000}$

46. $\dfrac{(0.0000013)(0.000090)}{0.00039}$

In Exercises 47–52, write each numeral in scientific notation and do the operations. Give all answers in standard notation.

47. $\dfrac{(0.006)(0.008)}{0.0012}$

48. $\dfrac{(600)(80{,}000)}{120{,}000}$

49. $\dfrac{(220{,}000)(0.000009)}{0.00033}$

50. $\dfrac{(0.00024)(96{,}000{,}000)}{640{,}000{,}000}$

51. $\dfrac{(320{,}000)^2(0.0009)}{12{,}000^2}$

52. $\dfrac{(0.000012)^2(49{,}000)^2}{0.021}$

In Exercises 53–58, use a scientific calculator to evaluate each expression. Round each answer to the appropriate number of significant digits.

53. $23{,}437^3$

54. 0.00034^4

55. $(63{,}480)(893{,}322)$

56. $(0.0000413)(0.0000049)^2$

57. $\dfrac{(69.4)^8(73.1)^2}{(0.0043)^3}$

58. $\dfrac{(0.0031)^4(0.0012)^5}{(0.0456)^{-7}}$

APPLICATIONS *Use scientific notation to find each answer. Round all answers to the proper number of significant digits.*

59. Speed of sound The speed of sound in air is 3.31×10^4 centimeters per second. Find the speed of sound in centimeters per hour.

60. Volume of a tank Find the volume of the tank shown in Illustration 1.

61. Mass of protons If the mass of 1 proton is 0.00000000000000000000000167248 gram, find the mass of 1 million protons.

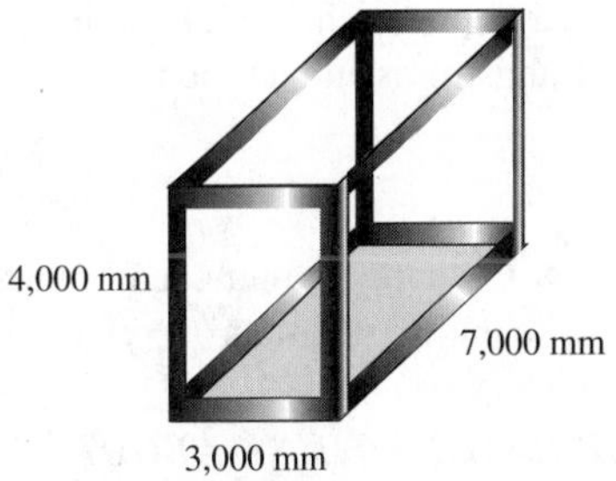

ILLUSTRATION 1

62. Speed of light The speed of light in a vacuum is about 30,000,000,000 centimeters per second. Find the speed of light in mph. (*Hint:* 160,000 cm $\approx$ 1 mile. Read $\approx$ as "is approximately equal to.")

63. Distance to the moon The moon is about 235,000 miles from the earth. Find this distance in inches.

64. Distance to the sun The sun is about 149,700,000 kilometers from the earth. Find this distance in miles. (*Hint:* 1 km $\approx$ 0.6214 mile.)

65. Solar flares Solar flares often produce immense loops of glowing gas ejected from the sun's surface. The flare in Illustration 2 extends about 95,000 kilometers into space. Express this distance in miles. (*Hint:* 1 km $\approx$ 0.6214 mile.)

Frank Rossotto/The Stock Market

ILLUSTRATION 2

66. Distance to the moon The moon is about 378,196 kilometers from the earth. Express this distance in inches. (*Hint:* 1 km $\approx$ 0.6214 mile.)

67. Angstroms per inch One **angstrom** is 0.0000001 millimeter, and one inch is 25.4 millimeters. Find the number of angstroms in one inch.

68. Range of a comet One **astronomical unit** (AU) is the distance from the earth to the sun—about 9.3×10^7 miles. Halley's comet ranges from 0.6 to 18 AU from the sun. Express this range in miles.

69. Flight to Pluto The planet Pluto is approximately 3,574,000,000 miles from the earth. If a spaceship can travel 18,000 mph, how long will it take to reach Pluto?

70. Light year Light travels about 300,000,000 meters per second. A **light year** is the distance that light can travel in one year. How many meters are in one light year?

71. Distance to Alpha Centauri Light travels about 186,000 miles per second. A **parsec** is 3.26 light years. The star Alpha Centauri is 1.3 parsecs from the earth. Express this distance in miles.

72. Life of a comet The mass of the comet shown in Illustration 3 is about 10^{16} grams. When the comet is close to the sun, matter evaporates at the rate of 10^7 grams per second. Calculate the life of the comet if it appears every 50 years and spends ten days close to the sun.

Frank Rossotto/The Stock Market

ILLUSTRATION 3

WRITING

73. Explain how to change a number from standard notation to scientific notation.

74. Explain how to change a number from scientific notation to standard notation.

SOMETHING TO THINK ABOUT

75. Find the highest power of 2 that can be evaluated with a scientific calculator.

76. Find the highest power of 7 that can be evaluated with a scientific calculator.

1.5 Solving Equations

■ EQUATIONS ■ PROPERTIES OF EQUALITY ■ SOLVING LINEAR EQUATIONS ■ COMBINING LIKE TERMS ■ IDENTITIES AND IMPOSSIBLE EQUATIONS ■ FORMULAS

Getting Ready *Fill in each blank to make a true statement.*

1. $\square + 3 = 5$ **2.** $8 - \square = 4$ **3.** $\frac{12}{\square} = 4$ **4.** $\square \cdot 5 = 30$

■ EQUATIONS

An **equation** is a statement indicating that two quantities are equal. The equation $2 + 4 = 6$ is true, and the equation $2 + 4 = 7$ is false. If an equation has a variable (say, x) it can be either true or false, depending on the value of x. For example, if $x = 1$, the equation $7x - 3 = 4$ is true.

$$7(\mathbf{1}) - 3 = 4 \quad \text{Substitute 1 for } x.$$
$$7 - 3 = 4$$
$$4 = 4$$

However, the equation is false for all other values of x. Since 1 makes the equation true, we say that 1 *satisfies* the equation.

The set of numbers that satisfies an equation is called its **solution set.** The elements of the solution set are called **solutions** or **roots** of the equation. Finding the solution set of an equation is called *solving the equation.*

EXAMPLE 1 Determine whether 3 is a solution of $2x + 4 = 10$.

Solution We substitute 3 for x and see whether it satisfies the equation.

$$2x + 4 = 10$$
$$2(\mathbf{3}) + 4 \stackrel{?}{=} 10 \quad \text{Substitute 3 for } x.$$
$$6 + 4 \stackrel{?}{=} 10 \quad \text{First do the multiplication on the left-hand side.}$$
$$10 = 10 \quad \text{Then do the addition.}$$

Since $10 = 10$, the number 3 satisfies the equation. It is a solution. ■

Self Check Is -5 a solution of $2x - 3 = -13$?

Answer yes

■ PROPERTIES OF EQUALITY

To solve an equation, we replace the equation with simpler ones, all having the same solution set. Such equations are called **equivalent equations.**

Equivalent Equations
Equations with the same solution set are called **equivalent equations.**

We continue to replace each resulting equation with an equivalent one until we have isolated the variable on one side of an equation. To isolate the variable, we can use the following properties:

Properties of Equality
If a, b, and c are real numbers and $a = b$, then

$$a + c = b + c \quad \text{and} \quad a - c = b - c$$

and if $c \neq 0$, then

$$ac = bc \quad \text{and} \quad \frac{a}{c} = \frac{b}{c}$$

In words, we can say: *If any quantity is added to (or subtracted from) both sides of an equation, a new equation is formed that is equivalent to the original equation.*

If both sides of an equation are multiplied (or divided) by the same nonzero quantity, a new equation is formed that is equivalent to the original equation.

SOLVING LINEAR EQUATIONS

The easiest equations to solve are **linear equations.**

Linear Equations
A **linear equation in one variable** (say, x) is any equation that can be written in the form

$$ax + c = 0 \quad (a \text{ and } c \text{ are real numbers and } a \neq 0)$$

EXAMPLE 2 Solve $2x + 8 = 0$.

Solution To solve the equation, we will isolate x on the left-hand side.

$$2x + 8 = 0$$

$$2x + 8 - 8 = 0 - 8$$ To eliminate 8 from the left-hand side, subtract 8 from both sides.

$$2x = -8$$ Simplify.

$$\frac{2x}{2} = \frac{-8}{2}$$ To eliminate 2 from the left-hand side, divide both sides by 2.

$$x = -4$$ Simplify.

Check: We substitute -4 for x to verify that it satisfies the original equation.

$$2x + 8 = 0$$
$$2(-4) + 8 \stackrel{?}{=} 0 \qquad \text{Substitute } -4 \text{ for } x.$$
$$-8 + 8 \stackrel{?}{=} 0$$
$$0 = 0$$

Since -4 satisfies the original equation, it is the solution. The solution set is $\{-4\}$. ■

Self Check Solve $-3a + 15 = 0$ and give the solution set.

Answer $\{5\}$

EXAMPLE 3 Solve $3(x - 2) = 20$.

Solution We isolate x on the left-hand side.

$$3(x - 2) = 20$$
$$3x - 6 = 20 \qquad \text{Use the distributive property to remove parentheses.}$$
$$3x - 6 + 6 = 20 + 6 \qquad \text{To eliminate } -6 \text{ from the left-hand side, add 6 to both sides.}$$
$$3x = 26 \qquad \text{Simplify.}$$
$$\frac{3x}{3} = \frac{26}{3} \qquad \text{To eliminate 3 from the left-hand side, divide both sides by 3.}$$
$$x = \frac{26}{3} \qquad \text{Simplify.}$$

Check: $3(x - 2) = 20$

$$3\left(\frac{26}{3} - 2\right) \stackrel{?}{=} 20 \qquad \text{Substitute } \tfrac{26}{3} \text{ for } x.$$
$$3\left(\frac{26}{3} - \frac{6}{3}\right) \stackrel{?}{=} 20 \qquad \text{Get a common denominator: } 2 = \tfrac{6}{3}.$$
$$3\left(\frac{20}{3}\right) \stackrel{?}{=} 20 \qquad \text{Combine the fractions: } \tfrac{26}{3} - \tfrac{6}{3} = \tfrac{20}{3}.$$
$$20 = 20 \qquad \text{Simplify.}$$

Since $\frac{26}{3}$ satisfies the equation, it is the solution. The solution set is $\left\{\frac{26}{3}\right\}$. ■

Self Check Solve $-2(a + 3) = 18$ and give the solution set.

Answer $\{-12\}$

COMBINING LIKE TERMS

To solve more complicated equations, we will need to combine like terms. An **algebraic term** is either a number or the product of numbers (called **constants**) and variables. Some examples of terms are $3x$, $-7y$, y^2, and 8. The **numerical coefficients** of these terms are 3, -7, 1, and 8 (8 can be written as $8x^0$), respectively.

In algebraic expressions, terms are separated by + and − signs. For example, the expression $3x^2 + 2x - 4$ has three terms, and the expression $3x + 7y$ has two terms.

Terms with the same variables with the same exponents are called **like terms** or **similar terms:**

$5x$ and $6x$ are like terms.	$27x^2y^3$ and $-326x^2y^3$ are like terms.
$4x$ and $-17y$ are unlike terms.	Because they have different variables.
$15x^2y$ and $6xy^2$ are unlike terms.	Because the variables have different exponents.

By using the distributive law, we can combine like terms. For example,

$$5x + 6x = (5 + 6)x = 11x \quad \text{and} \quad 32y - 16y = (32 - 16)y = 16y$$

This suggests that to combine like terms, *we add or subtract their numerical coefficients and keep the same variables with the same exponents.*

EXAMPLE 4 Solve $3(2x - 1) = 2x + 9$.

Solution

$3(2x - 1) = 2x + 9$	
$6x - 3 = 2x + 9$	Use the distributive property to remove parentheses.
$6x - 3 + 3 = 2x + 9 + 3$	To eliminate -3 from the left-hand side, add 3 to both sides.
$6x = 2x + 12$	Combine like terms.
$6x - 2x = 2x - 2x + 12$	To eliminate $2x$ from the right-hand side, subtract $2x$ from both sides.
$4x = 12$	Combine like terms.
$x = 3$	To eliminate 4 from the left-hand side, divide both sides by 4.

Check: $3(2x - 1) = 2x + 9$

$3(2 \cdot 3 - 1) \stackrel{?}{=} 2 \cdot 3 + 9$	Substitute 3 for x.
$3(5) \stackrel{?}{=} 6 + 9$	
$15 = 15$	

Since 3 satisfies the equation, it is the solution. The solution set is $\{3\}$. ■

Self Check Solve $-4(3a + 4) = 2a - 4$ and give the solution set.

Answer $\left\{-\frac{6}{7}\right\}$

To solve linear equations, we will follow these steps.

Solving Linear Equations

1. If the equation contains fractions, multiply both sides of the equation by a number that will eliminate the denominators.
2. Use the distributive property to remove all sets of parentheses and combine like terms.
3. Use the addition and subtraction properties to get all variables on one side of the equation and all numbers on the other side. Combine like terms, if necessary.
4. Use the multiplication and division properties to make the coefficient of the variable equal to 1.
5. Check the result by replacing the variable with the possible solution and verifying that the number satisfies the equation.

EXAMPLE 5 Solve the equation $\frac{5}{3}(x-3) = \frac{3}{2}(x-2) + 2$.

Solution *Step 1:* Since 6 is the smallest number that can be divided by both 2 and 3, we multiply both sides of the equation by 6 to eliminate the fractions:

$$\frac{5}{3}(x-3) = \frac{3}{2}(x-2) + 2$$

$$6\left[\frac{5}{3}(x-3)\right] = 6\left[\frac{3}{2}(x-2) + 2\right] \quad \text{To eliminate the fractions, multiply both sides by 6.}$$

$$6 \cdot \frac{5}{3}(x-3) = 6 \cdot \frac{3}{2}(x-2) + 6 \cdot 2 \quad \text{Use the distributive property on the right-hand side.}$$

$$10(x-3) = 9(x-2) + 12 \quad \text{Simplify.}$$

Step 2: We use the distributive property to remove parentheses and then combine like terms.

$$10x - 30 = 9x - 18 + 12$$

$$10x - 30 = 9x - 6$$

Step 3: We use the addition and subtraction properties by adding 30 to both sides and subtracting $9x$ from both sides.

$$10x - 30 - 9x + 30 = 9x - 6 - 9x + 30$$

$$x = 24 \quad \text{Combine like terms.}$$

Since the coefficient of x in the above equation is 1, Step 4 is unnecessary.

Step 5: We check by substituting 24 for x in the original equation and simplifying:

$$\frac{5}{3}(x-3)=\frac{3}{2}(x-2)+2$$
$$\frac{5}{3}(24-3)\stackrel{?}{=}\frac{3}{2}(24-2)+2$$
$$\frac{5}{3}(21)\stackrel{?}{=}\frac{3}{2}(22)+2$$
$$5(7)\stackrel{?}{=}3(11)+2$$
$$35=35$$

Since 24 satisfies the equation, it is the solution. The solution set is $\{24\}$. ■

EXAMPLE 6 Solve $\dfrac{x+2}{5}-4x=\dfrac{8}{5}-\dfrac{x+9}{2}$.

Solution

$$\frac{x+2}{5}-4x=\frac{8}{5}-\frac{x+9}{2}$$

$$10\left(\frac{x+2}{5}-4x\right)=10\left(\frac{8}{5}-\frac{x+9}{2}\right)$$ To eliminate the fractions, multiply both sides by 10.

$$2(x+2)-40x=2(8)-5(x+9)$$ Remove parentheses.

$$2x+4-40x=16-5x-45$$ Remove parentheses.

$$-38x+4=-5x-29$$ Combine like terms.

$$-33x=-33$$ Add $5x$ and -4 to both sides.

$$\frac{-33x}{-33}=\frac{-33}{-33}$$ Divide both sides by -33.

$$x=1$$ Simplify.

Check:
$$\frac{x+2}{5}-4x=\frac{8}{5}-\frac{x+9}{2}$$

$$\frac{1+2}{5}-4(1)\stackrel{?}{=}\frac{8}{5}-\frac{1+9}{2}$$ Substitute 1 for x.

$$\frac{3}{5}-4\stackrel{?}{=}\frac{8}{5}-5$$
$$\frac{3}{5}-\frac{20}{5}\stackrel{?}{=}\frac{8}{5}-\frac{25}{5}$$
$$-\frac{17}{5}=-\frac{17}{5}$$

Since 1 satisfies the equation, it is the solution. The solution set is $\{1\}$. ■

Self Check Solve $\frac{a+5}{5} + 2a = \frac{a}{2} - \frac{a+14}{5}$ and give the solution set.

Answer $\{-2\}$

IDENTITIES AND IMPOSSIBLE EQUATIONS

The equations discussed so far are called **conditional equations.** For these equations, some numbers x satisfy the equation and others do not. An **identity** is an equation that is satisfied by every number x for which both sides of the equation are defined.

EXAMPLE 7 Solve $2(x - 1) + 4 = 4(1 + x) - (2x + 2)$.

Solution

$$2(x - 1) + 4 = 4(1 + x) - (2x + 2)$$

$$2x - 2 + 4 = 4 + 4x - 2x - 2$$

Use the distributive property to remove parentheses.

$$2x + 2 = 2x + 2$$

Combine like terms.

The result $2x + 2 = 2x + 2$ is true for every value of x. Since every number x satisfies the equation, it is an identity. ■

Self Check Solve $3(a + 4) + 5 = 2(a - 1) + a + 19$.

Answer all real numbers

An **impossible equation** or a **contradiction** is an equation that has no solution.

EXAMPLE 8 Solve $\dfrac{x-1}{3} + 4x = \dfrac{3}{2} + \dfrac{13x-2}{3}$.

Solution

$$\frac{x-1}{3} + 4x = \frac{3}{2} + \frac{13x-2}{3}$$

$$6\left(\frac{x-1}{3} + 4x\right) = 6\left(\frac{3}{2} + \frac{13x-2}{3}\right)$$

To eliminate the fractions, multiply both sides by 6.

$$2(x - 1) + 6(4x) = 3(3) + 2(13x - 2)$$

Use the distributive property to remove parentheses.

$$2x - 2 + 24x = 9 + 26x - 4$$

Remove parentheses.

$$26x - 2 = 26x + 5$$

Combine like terms.

$$-2 = 5$$

Subtract $26x$ from both sides.

Since $-2 = 5$ is false, no number x can satisfy the equation. The solution set is the **empty set,** denoted as $\varnothing$. ■

Self Check Solve $\frac{x+5}{5} = \frac{1}{5} + \frac{x}{5}$.

Answer $\varnothing$

FORMULAS

Suppose we want to find the heights of several triangles whose areas and bases are known. It would be tedious to substitute values of A and b into the formula $A = \frac{1}{2}bh$ and then repeatedly solve the formula for h. It is easier to solve for h first and then substitute values for A and b and compute h directly.

To solve a formula for a variable means to isolate that variable on one side of the equation and isolate all other quantities on the other side.

EXAMPLE 9 Solve $A = \frac{1}{2}bh$ for h.

Solution

$$A = \frac{1}{2}bh$$

$$2A = bh \quad \text{To eliminate the fraction, multiply both sides by 2.}$$

$$\frac{2A}{b} = h \quad \text{To isolate } h\text{, divide both sides by } b.$$

$$h = \frac{2A}{b} \quad \text{Write } h \text{ on the left-hand side.}$$

■

Self Check Solve $A = \frac{1}{2}bh$ for b.

Answer $b = \frac{2A}{h}$

EXAMPLE 10 For simple interest, the formula $A = p + prt$ gives the amount of money in an account at the end of a specific time. A represents the amount, p the principal, r the rate of interest, and t the time. We can solve the formula for t as follows:

Solution

$$A = p + prt$$

$$A - p = prt \quad \text{To isolate the term involving } t\text{, subtract } p \text{ from both sides.}$$

$$\frac{A - p}{pr} = t \quad \text{To isolate } t\text{, divide both sides by } pr.$$

$$t = \frac{A - p}{pr} \quad \text{Write } t \text{ on the left-hand side.}$$

■

Self Check Solve $A = p + prt$ for r.

Answer $r = \frac{A - p}{pt}$

EXAMPLE 11 The formula $F = \frac{9}{5}C + 32$ converts degrees Celsius to degrees Fahrenheit. Solve the formula for C.

Solution

$$F = \frac{9}{5}C + 32$$

$$F - 32 = \frac{9}{5}C$$ To isolate the term involving C, subtract 32 from both sides.

$$\frac{5}{9}(F - 32) = \frac{5}{9}\left(\frac{9}{5}C\right)$$ To isolate C, multiply both sides by $\frac{5}{9}$.

$$\frac{5}{9}(F - 32) = C$$ $\frac{5}{9} \cdot \frac{9}{5} = 1$.

$$C = \frac{5}{9}(F - 32)$$

To convert degrees Fahrenheit to degrees Celsius, we can use the formula $C = \frac{5}{9}(F - 32)$. ■

Self Check Solve $S = \frac{180(n-2)}{5}$ for n.

Answer $n = \frac{5S}{180} + 2$ or $n = \frac{5S + 360}{180}$

Orals *Combine like terms.*

1. $5x + 4x$ **2.** $7s^2 - 5s^2$

Tell whether each number is a solution of $2x + 5 = 13$.

3. 3 **4.** 4 **5.** 5 **6.** 6

Solve each equation.

7. $3x - 2 = 7$ **8.** $\frac{1}{2}x - 1 = 5$ **9.** $\frac{x-2}{3} = 1$ **10.** $\frac{x+3}{2} = 3$

EXERCISE 1.5

REVIEW *Simplify each expression.*

1. $(-4)^3$ **2.** -3^3 **3.** $\left(\frac{x+y}{x-y}\right)^0$ **4.** $(x^2x^3)^4$

5. $\left(\frac{x^2x^5}{x^3}\right)^2$ **6.** $\left(\frac{x^4y^3}{x^5y}\right)^3$ **7.** $(2x)^{-3}$ **8.** $\left(\frac{x^2}{y^5}\right)^{-4}$

VOCABULARY AND CONCEPTS *Fill in each blank to make a true statement.*

9. An ______ is a statement that two quantities are equal.

10. If a number is substituted for a variable in an equation and the equation is true, we say that the number ______ the equation.

11. If two equations have the same solution set, they are called ______ equations.

12. If a, b, and c are real numbers and $a = b$, then $a + c = b +$ __ and $a - c = b -$ __.

13. If a, b, and c are real numbers and $a = b$, then $a \cdot c = b \cdot$ __ and $\frac{a}{c} = \frac{}{\_\_}$ $(c \neq 0)$.

14. A number or the product of numbers and variables is called an algebraic ______.

15. ______ terms are terms with the same variables with the same exponents.

16. To combine like terms, add their ______ and keep the same ______ and exponents.

17. An ______ is an equation that is true for all values of its variable.

18. An impossible equation is true for __ values of its variable.

PRACTICE *In Exercises 19–22, tell whether 5 is a solution of each equation.*

19. $3x + 2 = 17$

20. $7x - 2 = 33$

21. $\frac{3}{5}x - 5 = -2$

22. $\frac{2}{5}x + 12 = 8$

In Exercises 23–38, solve each equation.

23. $x + 6 = 8$

24. $y - 7 = 3$

25. $a - 5 = 20$

26. $b + 4 = 18$

27. $2u = 6$

28. $3v = 12$

29. $\frac{x}{4} = 7$

30. $\frac{x}{6} = 8$

31. $3x + 1 = 3$

32. $8x - 2 = 13$

33. $2x + 1 = 13$

34. $2x - 4 = 16$

35. $3(x - 4) = -36$

36. $4(x + 6) = 84$

37. $3(r - 4) = -4$

38. $4(s - 5) = -3$

In Exercises 39–46, tell whether the terms are like terms. If they are, combine them.

39. $2x, 6x$

40. $-3x, 5y$

41. $-5xy, -7yz$

42. $-3t^2, 12t^2$

43. $3x^2, -5x^2$

44. $5y^2, 7xy$

45. $xy, 3xt$

46. $-4x, -5x$

In Exercises 47–76, solve each equation.

47. $3a - 22 = -2a - 7$

48. $a + 18 = 6a - 3$

49. $2(2x + 1) = 15 + 3x$

50. $-2(x + 5) = 30 - x$

51. $3(y - 4) - 6 = y$

52. $2x + (2x - 3) = 5$

53. $5(5 - a) = 37 - 2a$

54. $4a + 17 = 7(a + 2)$

55. $4(y + 1) = -2(4 - y)$

56. $5(r + 4) = -2(r - 3)$

57. $2(a - 5) - (3a + 1) = 0$

58. $8(3a - 5) - 4(2a + 3) = 12$

59. $3(y - 5) + 10 = 2(y + 4)$

60. $2(5x + 2) = 3(3x - 2)$

61. $9(x + 2) = -6(4 - x) + 18$

62. $3(x + 2) - 2 = -(5 + x) + x$

63. $\frac{1}{2}x - 4 = -1 + 2x$

64. $2x + 3 = \frac{2}{3}x - 1$

65. $\frac{x}{2} - \frac{x}{3} = 4$

66. $\frac{x}{2} + \frac{x}{3} = 10$

67. $\frac{x}{6} + 1 = \frac{x}{3}$

68. $\frac{3}{2}(y + 4) = \frac{20 - y}{2}$

69. $5 - \frac{x + 2}{3} = 7 - x$

70. $3x - \frac{2(x + 3)}{3} = 16 - \frac{x + 2}{2}$

71. $\frac{4x - 2}{2} = \frac{3x + 6}{3}$

72. $\frac{t + 4}{2} = \frac{2t - 3}{3}$

73. $\frac{a + 1}{3} + \frac{a - 1}{5} = \frac{2}{15}$

74. $\frac{2z + 3}{3} + \frac{3z - 4}{6} = \frac{z - 2}{2}$

75. $\frac{5a}{2} - 12 = \frac{a}{3} + 1$

76. $\frac{5a}{6} - \frac{5}{2} = -\frac{1}{2} - \frac{a}{6}$

In Exercises 77–84, solve each equation. If the equation is an identity or an impossible equation, so indicate.

77. $4(2 - 3t) + 6t = -6t + 8$

78. $2x - 6 = -2x + 4(x - 2)$

79. $\frac{a + 1}{4} + \frac{2a - 3}{4} = \frac{a}{2} - 2$

80. $\frac{y - 8}{5} + 2 = \frac{2}{5} - \frac{y}{3}$

81. $3(x - 4) + 6 = -2(x + 4) + 5x$

82. $2(x - 3) = \frac{3}{2}(x - 4) + \frac{x}{2}$

83. $y(y + 2) + 1 = y^2 + 2y + 1$

84. $x(x - 3) = x^2 - 2x + 1 - (5 + x)$

In Exercises 85–102, solve each formula for the indicated variable.

85. $A = lw$ for w

86. $p = 4s$ for s

87. $V = \frac{1}{3}Bh$ for B

88. $b = \frac{2A}{h}$ for A

89. $I = prt$ for t

90. $I = prt$ for r

91. $p = 2l + 2w$ for w

92. $p = 2l + 2w$ for l

93. $A = \frac{1}{2}h(B + b)$ for B

94. $A = \frac{1}{2}h(B + b)$ for b

95. $y = mx + b$ for x

96. $y = mx + b$ for m

97. $l = a + (n - 1)d$ for n

98. $l = a + (n - 1)d$ for d

99. $S = \frac{a - lr}{1 - r}$ for l

100. $C = \frac{5}{9}(F - 32)$ for F

101. $S = \frac{n(a + l)}{2}$ for l

102. $S = \frac{n(a + l)}{2}$ for n

APPLICATIONS

103. Force of gravity The masses of the two objects in Illustration 1 are m and M. The force of gravitation F between the masses is

$$F = \frac{GmM}{d^2}$$

where G is a constant and d is the distance between them. Solve for m.

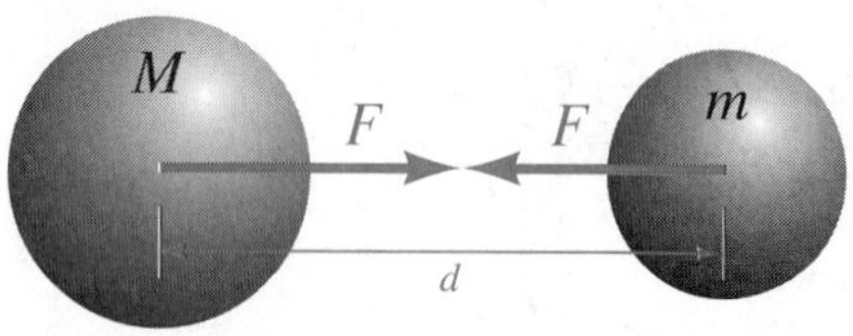

ILLUSTRATION 1

104. Thermodynamics The Gibbs free-energy formula is $G = U - TS + pV$. Solve for S.

105. Converting temperatures Solve the formula $F = \frac{9}{5}C + 32$ for C and find the Celsius temperatures that correspond to Fahrenheit temperatures of 32°, 70°, and 212°.

106. Doubling money A man intends to invest $1,000 at simple interest. Solve the formula $A = p + prt$ for t and find how long it will take to double his money at the rates of 5%, 7%, and 10%.

107. Cost of electricity The cost of electricity in a certain city is given by the formula $C = 0.07n + 6.50$, where C is the cost and n is the number of kilowatt hours used. Solve for n and find the number of kwh used for costs of $49.97, $76.50, and $125.

108. Cost of water A monthly water bill in a certain city is calculated by using the formula $n = \frac{5{,}000C - 17{,}500}{6}$, where n is the number of gallons used and C is the monthly cost. Solve for C and compute the bill for quantities used of 500, 1,200, and 2,500 gallons.

109. Ohm's law The formula $E = IR$, called **Ohm's law,** is used in electronics. Solve for R and then calculate the resistance R if the voltage E is 56 volts and the current I is 7 amperes. (Resistance has units of **ohms.**)

110. Earning interest An amount P, invested at a simple interest rate r, will grow to an amount A in t years, according to the formula $A = P(1 + rt)$. Solve for P. Suppose a man invested some money at 5.5%. If after 5 years, he had $6,693.75 on deposit, what amount did he originally invest?

111. Angles of a polygon A regular polygon has n equal sides and n equal angles. The measure a of an interior angle is given by $a = 180°\left(1 - \frac{2}{n}\right)$. Solve for n. Find the number of sides of the regular polygon in Illustration 2 if an interior angle is 135°.

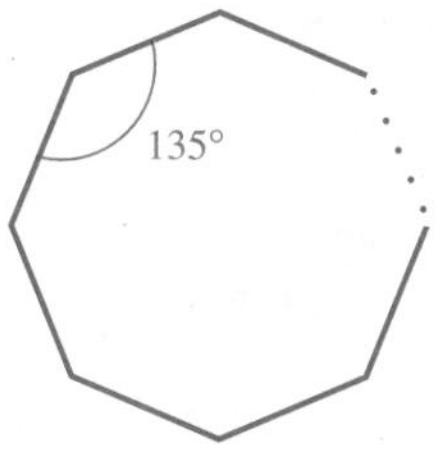

ILLUSTRATION 2

112. Power loss Illustration 3 is the schematic diagram of a resistor connected to a voltage source of 60 volts. As a result, the resistor dissipates power in the form of heat. The power P lost when a voltage V is placed across a resistance R is given by the formula

$$P = \frac{E^2}{R}$$

Solve for R. If P is 4.8 watts and E is 60 volts, find R.

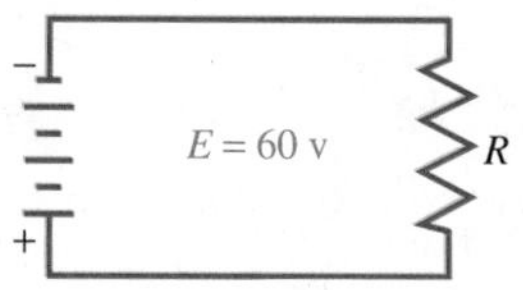

ILLUSTRATION 3

WRITING

113. Explain the difference between a conditional equation, an identity, and an impossible equation.

114. Explain how you would solve an equation.

SOMETHING TO THINK ABOUT *Find the mistake in each solution.*

115.
$$3(x - 2) + 4 = 14$$
$$3x - 2 + 4 = 14$$
$$3x + 2 = 14$$
$$3x = 12$$
$$x = 4$$

116.
$$A = p + prt$$
$$A - p = prt$$
$$A - p - pt = t$$
$$t = A - p - pt$$

1.6 Using Equations to Solve Problems

■ RECREATION PROBLEMS ■ BUSINESS PROBLEMS ■ GEOMETRIC PROBLEMS ■ LEVER PROBLEMS

Getting Ready *Let* $x = 18$.

1. What number is 2 more than x?

2. What number is 4 times x?

3. What number is 5 more than twice x?

4. What number is 2 less than one-half of x?

In this section, we will solve problems. When we translate the words of a problem into mathematics, we are creating a *mathematical model* of the problem. To create these models, we can use Table 1-3 to translate certain words into mathematical operations.

Addition (+)	Subtraction (−)	Multiplication (·)	Division (÷)
added to	subtracted from	multiplied by	divided by
plus	difference	product	quotient
the sum of	less than	times	ratio
more than	less	of	half
increased by	decreased by	twice	

TABLE 1-3

We can then change English phrases into algebraic expressions, as in Table 1-4.

English phrase	Algebraic expression
2 added to some number	$x + 2$
the difference between two numbers	$x - y$
5 times some number	$5x$
the product of 925 and some number	$925x$
5% of some number	$0.05x$
the sum of twice a number and 10	$2x + 10$
the quotient (or ratio) of two numbers	$\frac{x}{y}$
half of a number	$\frac{x}{2}$

TABLE 1-4

Once we know how to change phrases into algebraic expressions, we can solve many problems. The following list of steps provides an excellent strategy for solving problems.

Problem Solving

1. ***Analyze the problem*** by reading it carefully to understand the given facts. What information is given? What vocabulary is given? What are you asked to find? Often a diagram will help you visualize the facts of the problem.
2. ***Form an equation*** by picking a variable to represent the quantity to be found. Then express all other quantities mentioned as expressions involving that variable. Finally, write an equation expressing a quantity in two different ways.
3. ***Solve the equation.***
4. ***State the conclusion.***
5. ***Check the result*** in the words of the problem.

RECREATION PROBLEMS

EXAMPLE 1

Cutting a rope A mountain climber wants to cut a rope 213 feet long into three pieces. If each piece is to be 2 feet longer than the previous one, where should he make the cuts?

Analyze the problem If x represents the length of the shortest piece, the climber wants the lengths of the three pieces to be

$$x, \qquad x + 2, \qquad \text{and} \qquad x + 4$$

feet long. He knows that the sum of these three lengths can be expressed in two ways: as $x + (x + 2) + (x + 4)$ and as 213.

Form an equation Let x represent the length of the first piece of rope. Then $x + 2$ represents the length of the second piece, and $x + 4$ represents the length of the third piece. (See Figure 1-24.)

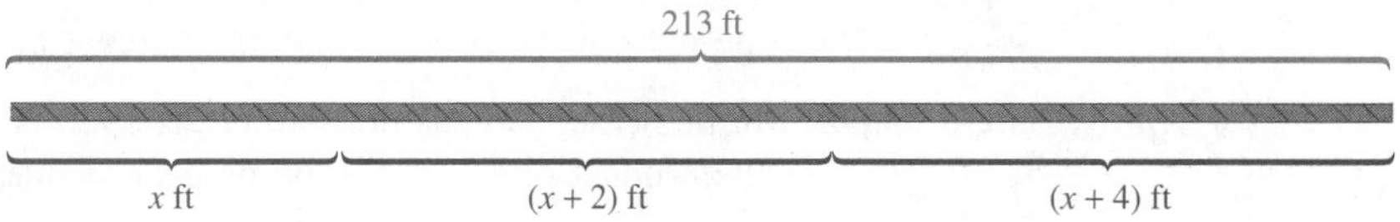

FIGURE 1-24

From Figure 1-24, we see that the sum of the individual pieces must equal the total length of the rope.

The length of the first piece	plus	the length of the second piece	plus	the length of the third piece	equals	the total length of the rope.
x	$+$	$x + 2$	$+$	$x + 4$	$=$	213

Solve the equation We now solve the equation.

$$x + x + 2 + x + 4 = 213$$

$$3x + 6 = 213 \quad \text{Combine like terms.}$$

$$3x = 207 \quad \text{Subtract 6 from both sides.}$$

$$x = 69 \quad \text{Divide both sides by 3.}$$

$$x + 2 = 71$$

$$x + 4 = 73$$

State the conclusion He should make cuts 69 feet from one end and 73 feet from the other end, to get lengths of 69 feet, 71 feet, and 73 feet.

Check the result Each length is 2 feet longer than the previous length, and the sum of the lengths is 213 feet. ■

BUSINESS PROBLEMS

When the regular price of merchandise is reduced, the amount of reduction is called the **markdown** (or discount).

Sale price	=	regular price	−	markdown

Usually, the markdown is expressed as a percent of the regular price.

Markdown	=	percent of markdown	·	regular price

EXAMPLE 2

Finding the percent of markdown A home theater system is on sale for \$777. If the list price was \$925, find the percent of markdown.

Analyze the problem In this case, \$777 is the sale price, \$925 is the regular price, and the markdown is the product of \$925 and the percent of markdown.

Form an equation We can let r represent the percent of markdown, expressed as a decimal. We then substitute \$777 for the sale price and \$925 for the regular price in the formula

Sale price	equals	regular price	minus	markdown.
777	=	925	−	$r \cdot 925$

Solve the equation We now solve the equation.

$$777 = 925 - r \cdot 925$$
$$777 = 925 - 925r$$
$$-148 = -925r \quad \text{Subtract 925 from both sides.}$$
$$0.16 = r \quad \text{Divide both sides by } -925.$$

State the conclusion The percent of markdown is 16%.

Check the result Since the markdown is 16% of \$925, or \$148, the sale price is \$925 − \$148, or \$777. ■

EXAMPLE 3

Portfolio analysis A college foundation owns stock in IBC (selling at \$54 per share), GS (selling at \$65 per share), and ATB (selling at \$105 per share). The foundation owns equal shares of GS and IBC, but five times as many shares of ATB.

If this portfolio is worth \$450,800, how many shares of each type does the foundation own?

Analyze the problem The value of the IBC stock plus the value of the GS stock plus the value of the ATB stock must equal \$450,800.

- If x represents the number of shares of IBC, then \$54x is the value of that stock.
- Since the foundation has equal numbers of shares of GS and of IBC, x also represents the number of shares of GS. The value of this stock is \$65x.
- Since the foundation owns five times as many shares of ATB, it owns $5x$ shares of ATB. The value of this stock is \$105($5x$).

We set the sum of these values equal to \$450,800.

Form an equation We let x represent the number of shares of IBC. Then x also represents the number of shares of GS, and $5x$ represents the number of shares of ATB.

The value of IBC stock	plus	the value of GS stock	plus	the value of ATB stock	equals	the total value of the portfolio.
$54x$	$+$	$65x$	$+$	$105(5x)$	$=$	$450{,}800$

Solve the equation We now solve the equation

$$54x + 65x + 105(5x) = 450{,}800$$

$$54x + 65x + 525x = 450{,}800 \qquad 105(5x) = 525x.$$

$$644x = 450{,}800 \qquad \text{Combine like terms.}$$

$$x = 700 \qquad \text{Divide both sides by 644.}$$

State the conclusion The foundation owns 700 shares of IBC, 700 shares of GS, and 5(700), or 3,500, shares of ABT.

Check the result The value of 700 shares of IBC at $54 per share is $37,800. The value of 700 shares of GS at $65 per share is $45,500. The value of 3,500 shares of ATB at $105 per share is $367,500. The sum of these values is $450,800. ■

GEOMETRIC PROBLEMS

Figure 1-25 illustrates several geometric figures. A **right angle** is an angle whose measure is 90°. A **straight angle** is an angle whose measure is 180°. An **acute angle** is an angle whose measure is greater than 0° but less than 90°.

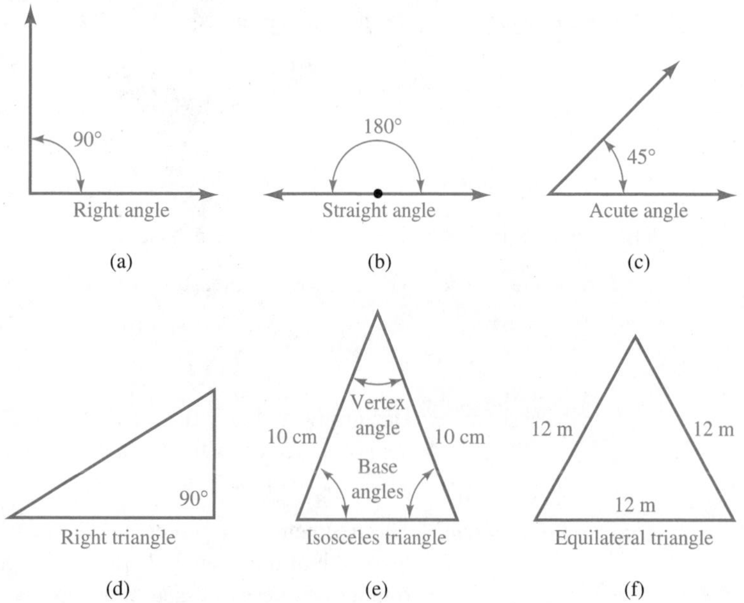

FIGURE 1-25

If the sum of two angles equals 90°, the angles are called **complementary,** and each angle is called the **complement** of the other. If the sum of two angles equals 180°, the angles are called **supplementary,** and each angle is called the **supplement** of the other.

A **right triangle** is a triangle with one right angle. An **isosceles triangle** is a triangle with two sides of equal measure that meet to form the **vertex angle.** The angles opposite the equal sides, called the **base angles,** are also equal. An **equilateral triangle** is a triangle with three equal sides and three equal angles.

EXAMPLE 4 **Angles in a triangle** If the vertex angle of the isosceles triangle shown in Figure 1-25(e) measures 64°, find the measure of each base angle.

Analyze the problem We are given that the vertex angle measures 64°. If we let $x°$ represent the measure of one base angle, the measure of the other base angle is also $x°$. Thus, the sum of the angles in the triangle is $x° + x° + 64°$. Because the sum of the measures of the angles of any triangle is 180°, we know that $x° + x° + 64°$ is equal to 180°.

Form an equation We can form the equation

The measure of one base angle	plus	the measure of the other base angle	plus	the measure of the vertex angle	equals	180°.
x	$+$	x	$+$	64	$=$	180

Solve the equation We now solve the equation.

$$x + x + 64 = 180$$

$$2x + 64 = 180 \quad \text{Combine like terms.}$$

$$2x = 116 \quad \text{Subtract 64 from both sides.}$$

$$x = 58 \quad \text{Divide both sides by 2.}$$

State the conclusion The measure of each base angle is 58°.

Check the result The sum of the measures of each base angle and the vertex angle is 180°:

$$58° + 58° + 64° = 180°$$

■

EXAMPLE 5 **Designing a dog run** A man has 28 meters of fencing to make a rectangular kennel. If he wants the kennel to be 6 meters longer than it is wide, find its dimensions.

Analyze the problem The perimeter P of a rectangle is the distance around it. If w is chosen to represent the width of the kennel, then $w + 6$ represents its length. (See Figure 1-26.) The perimeter can be expressed either as $2w + 2(w + 6)$ or as 28.

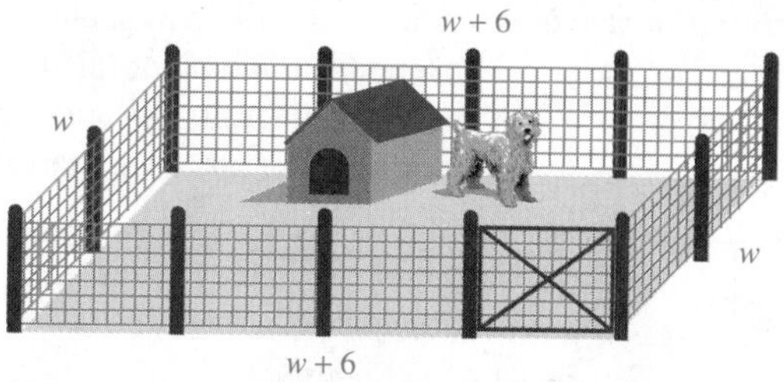

FIGURE 1-26

Form an equation We let w represent the width of the kennel. Then $w + 6$ represents its length.

Two widths	plus	two lengths	equals	the perimeter.
$2 \cdot w$	$+$	$2 \cdot (w + 6)$	$=$	28

Solve the equation We now solve the equation.

$$2 \cdot w + 2 \cdot (w + 6) = 28$$

$$2w + 2w + 12 = 28 \quad \text{Use the distributive property to remove parentheses.}$$

$$4w + 12 = 28 \quad \text{Combine like terms.}$$

$$4w = 16 \quad \text{Subtract 12 from both sides.}$$

$$w = 4 \quad \text{Divide both sides by 4.}$$

$$w + 6 = 10$$

State the conclusion The dimensions of the kennel are 4 meters by 10 meters.

Check the result If a kennel has a width of 4 meters and a length of 10 meters, its length is 6 meters longer than its width, and the perimeter is 2(4) meters + 2(10) meters = 28 meters. ■

■ LEVER PROBLEMS

EXAMPLE 6

Engineering Design engineers must position two hydraulic cylinders as in Figure 1-27 to balance a 9,500-pound force at point A. The first cylinder at the end of the lever exerts a 3,500-pound force. Where should the design engineers position the second cylinder, which is capable of exerting a 5,500-pound force?

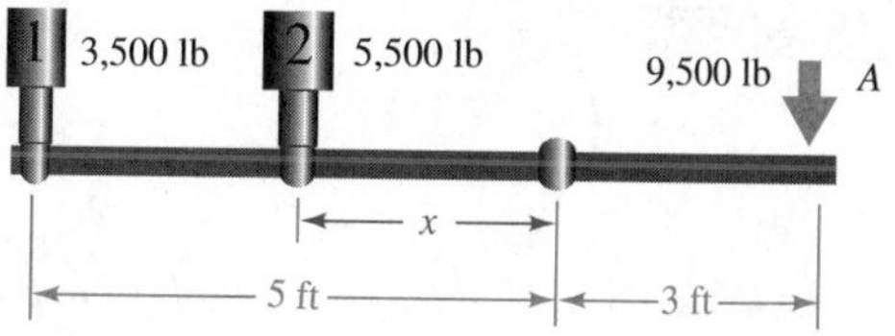

FIGURE 1-27

Analyze the problem The lever will be in balance when the force of the first cylinder multiplied by its distance from the pivot (also called the **fulcrum**) added to the second cylinder's force multiplied by its distance from the fulcrum, is equal to the product of the 9,500-pound force and its distance from the fulcrum.

Form an equation We let x represent the distance from the larger cylinder to the fulcrum.

Force of cylinder 1, times its distance	plus	force of cylinder 2, times its distance	equals	force to be balanced, times its distance.
$3{,}500 \cdot 5$	$+$	$5{,}500x$	$=$	$9{,}500 \cdot 3$

Solve the equation We can now solve the equation.

$$3{,}500 \cdot 5 + 5{,}500x = 9{,}500 \cdot 3$$
$$17{,}500 + 5{,}500x = 28{,}500$$
$$5{,}500x = 11{,}000 \quad \text{Subtract 17,500 from both sides.}$$
$$x = 2 \quad \text{Divide both sides by 5,500.}$$

State the conclusion The design must specify that the second cylinder be positioned 2 feet from the fulcrum.

Check the result $3{,}500 \cdot 5 + 5{,}500 \cdot 2 = 17{,}500 + 11{,}000 = 28{,}500$
$9{,}500 \cdot 3 = 28{,}500$ ■

Orals *Find each value.*

1. 20% of 500 **2.** $33\frac{1}{3}$% of 600

If a stock costs $54, find the cost of

3. 5 shares **4.** x shares

Find the area of the rectangle with the given dimensions.

5. 6 meters long, 4 meters wide **6.** l meters long, $l - 5$ meters wide

EXERCISE 1.6

REVIEW *Simplify each expression.*

1. $\left(\dfrac{3x^{-3}}{4x^2}\right)^{-4}$ **2.** $\left(\dfrac{r^{-3}s^2}{r^2r^3s^{-4}}\right)^{-5}$ **3.** $\dfrac{a^m a^3}{a^2}$ **4.** $\left(\dfrac{b^n}{b^3}\right)^3$

VOCABULARY AND CONCEPTS *Fill in each blank to make a true statement.*

5. The expression ______ represents the phrase "4 more than 5 times x."

6. The expression ______ represents the phrase "6 percent of x."

7. The expression ____ represents the phrase "the value of x shares priced at $40 per share."

8. A right angle is an angle whose measure is ____.

9. A straight angle measures ____.

10. An ____ angle measures greater than 0° but less than 90°.

11. If the sum of the measures of two angles equals 90°, the angles are called ____________ angles.

12. If the sum of the measures of two angles equals ____, the angles are called supplementary angles.

13. If a triangle has a right angle, it is called a ____ triangle.

14. If a triangle has two sides with equal measure, it is called an _______ triangle.

15. The equal sides of an isosceles triangle form the _____ angle.

16. An equilateral triangle has three _____ sides and three _____ angles.

APPLICATIONS

17. **Cutting a rope** A 60-foot rope is cut into four pieces, with each successive piece twice as long as the previous one. Find the length of the longest piece.

18. **Cutting a cable** A 186-foot cable is to be cut into four pieces. Find the length of each piece if each successive piece is 3 feet longer than the previous one.

19. **Cutting a board** The carpenter in Illustration 1 saws a board into two pieces. He wants one piece to be 1 foot longer than twice the length of the shorter piece. Find the length of each piece.

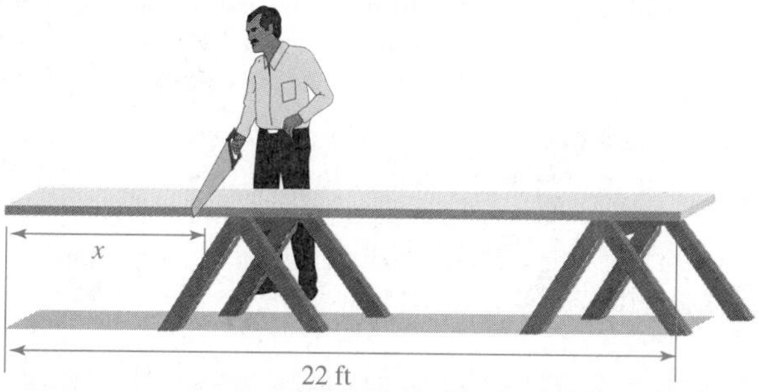

ILLUSTRATION 1

20. **Cutting a beam** A 30-foot steel beam is to be cut into two pieces. The longer piece is to be 2 feet more than three times as long as the shorter piece. Find the length of each piece.

21. **Buying a TV and a VCR** See Illustration 2. If the TV costs $55 more than the VCR, how much does the TV cost?

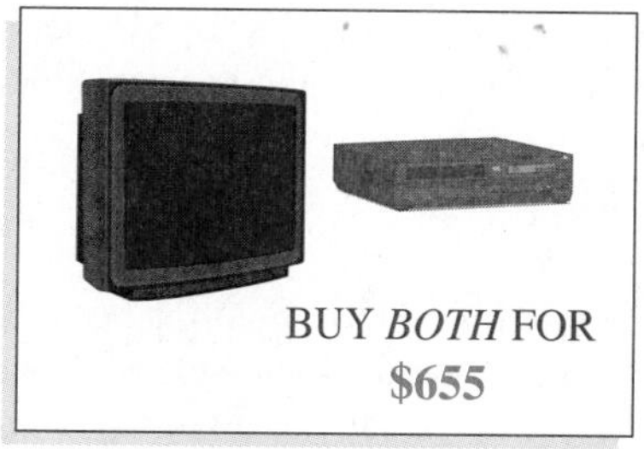

ILLUSTRATION 2

22. **Buying golf clubs** The cost of a set of golf clubs is $590. If the irons cost $40 more than the woods, find the cost of the irons.

23. **Buying a washer and dryer** Find the percent of markdown of the sale in Illustration 3.

ILLUSTRATION 3

24. **Buying furniture** A bedroom set regularly sells for $983. If it is on sale for $737.25, what is the percent of markdown?

25. **Buying books** A bookstore buys a used calculus book for $12 and sells it for $40. Find the percent of markup.

26. **Selling toys** The owner of a gift shop buys stuffed animals for \$18 and sells them for \$30. Find the percent of markup.

27. **Value of an IRA** In an Individual Retirement Account (IRA) valued at \$53,900, a student has 500 shares of stock, some in Big Bank Corporation and some in Safe Savings and Loan. If Big Bank sells for \$115 per share and Safe Savings sells for \$97 per share, how many shares of each does the student own?

28. **Assets of a pension fund** A pension fund owns 12,000 shares in a stock mutual fund and a mutual bond fund. Currently, the stock fund sells for \$12 per share, and the bond fund sells for \$15 per share. How many shares of each does the pension fund own if the value of the securities is \$165,000?

29. **Selling calculators** Last month, a bookstore ran the ad shown in Illustration 4 and sold 85 calculators, generating \$3,875 in sales. How many of each type of calculator did the bookstore sell?

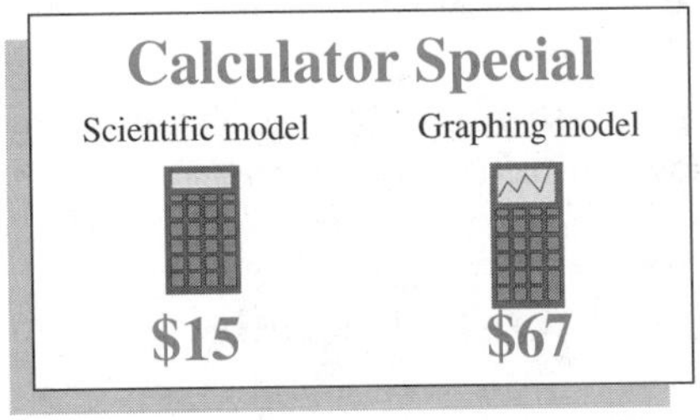

ILLUSTRATION 4

30. **Selling grass seed** A seed company sells two grades of grass seed. A 100-pound bag of a mixture of rye and bluegrass sells for \$245, and a 100-pound bag of bluegrass sells for \$347. How many bags of each are sold in a week when the receipts for 19 bags are \$5,369?

31. **Buying roses** A man with \$21.25 stops after work to order some roses for his wife's birthday. If each rose costs \$1.25 and there is a delivery charge of \$5, how many roses can he buy?

32. **Renting a truck** To move to Wisconsin, a man can rent a truck for \$29.95 per day plus 19¢ per mile. If he keeps the truck for one day, how many miles can he drive for a cost of \$77.45?

33. **Car rental** While waiting for his car to be repaired, a man rents a car for \$12 per day plus 10¢ per mile. If he keeps the car for 2 days, how many miles can he drive for a total cost of \$30? How many miles can he drive for a total cost of \$36?

34. **Computing salaries** A student earns \$17 per day for delivering overnight packages. She is paid \$5 per day plus 60¢ for each package delivered. How many more deliveries must she make each day to increase her daily earnings to \$23?

35. **Finding dimensions** The rectangular garden shown in Illustration 5 is twice as long as it is wide. Find its dimensions.

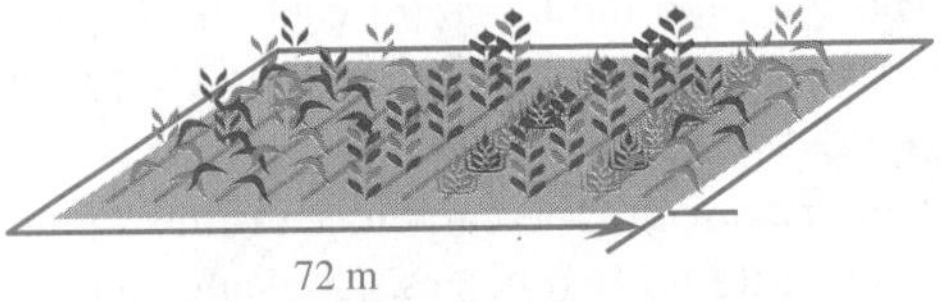

ILLUSTRATION 5

36. **Finding dimensions** The width of a rectangular swimming pool is one-third its length. If its perimeter is 96 meters, find the dimensions of the pool.

37. **Fencing a pasture** A farmer has 624 feet of fencing to enclose the pasture shown in Illustration 6. Because a river runs along one side, fencing will be needed on only three sides. Find the dimensions of the pasture if its length is double its width.

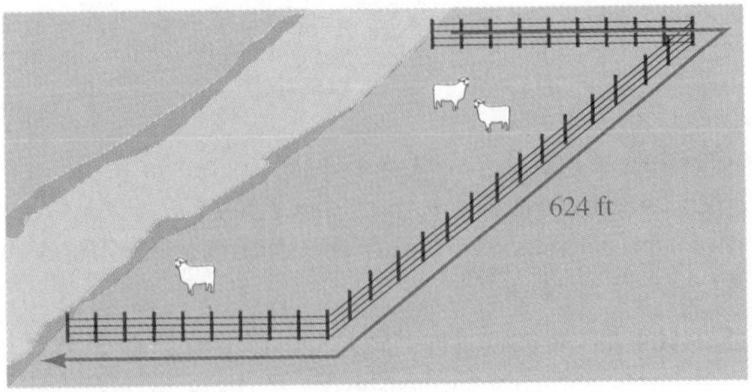

ILLUSTRATION 6

38. Fencing a pen A man has 150 feet of fencing to build the pen shown in Illustration 7. If one end is a square, find the outside dimensions.

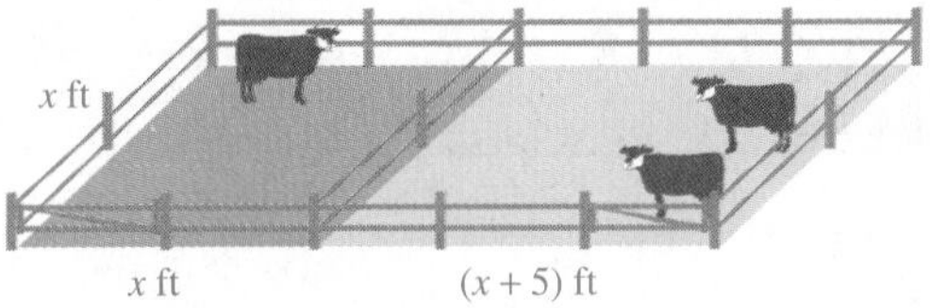

ILLUSTRATION 7

39. Enclosing a swimming pool A woman wants to enclose the swimming pool shown in Illustration 8 and have a walkway of uniform width all the way around. How wide will the walkway be if the woman uses 180 feet of fencing?

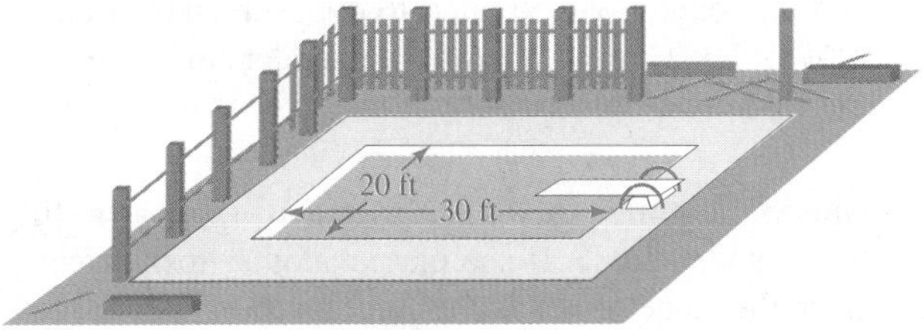

ILLUSTRATION 8

40. Framing a picture An artist wants to frame the picture shown in Illustration 9 with a frame 2 inches wide. How wide will the framed picture be if the artist uses 70 inches of framing material?

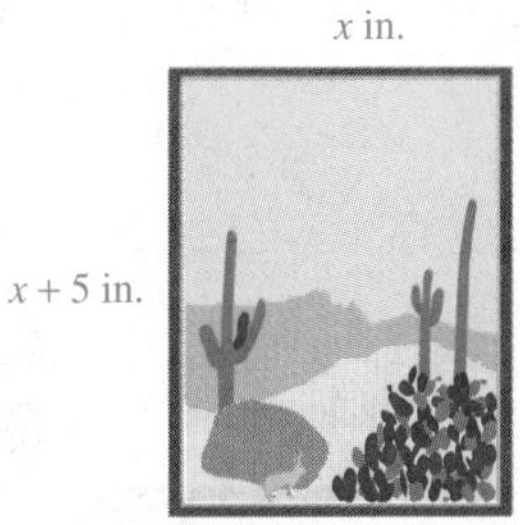

ILLUSTRATION 9

41. Supplementary angles If one of two supplementary angles is 35° larger than the other, find the measure of the smaller angle.

42. Supplementary angles Refer to Illustration 10 and find x.

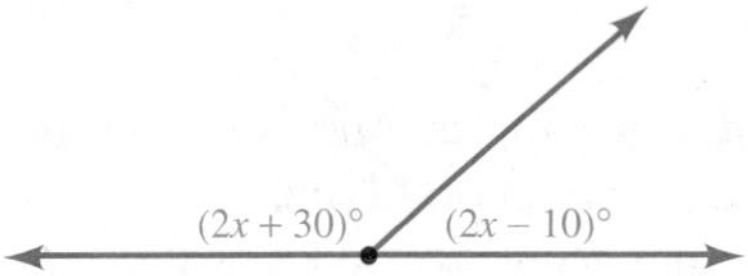

ILLUSTRATION 10

43. Complementary angles If one of two complementary angles is 22° greater than the other, find the measure of the larger angle.

44. Complementary angles in a right triangle Explain why the acute angles in the right triangle shown in Illustration 11 are complementary. Then find the measure of angle A.

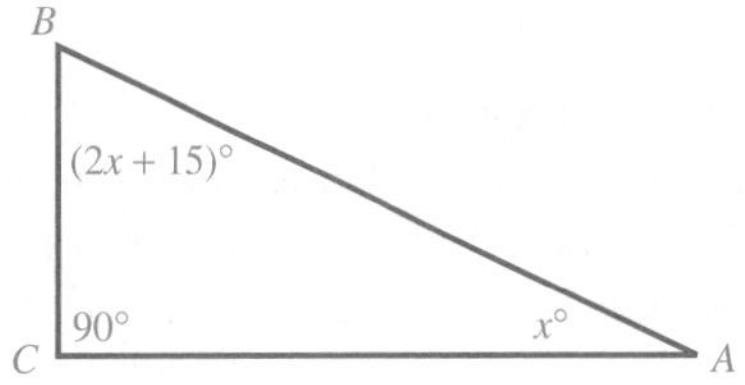

ILLUSTRATION 11

45. Supplementary angles and parallel lines In Illustration 12, lines r and s are cut by a third line l to form angles 1 and 2. When lines r and s are parallel, angles 1 and 2 are supplementary. If angle $1 = (x + 50)°$ and angle $2 = (2x - 20)°$ and lines r and s are parallel, find x.

46. In Illustration 12, find the measure of angle 3. (*Hint:* See Exercise 45.)

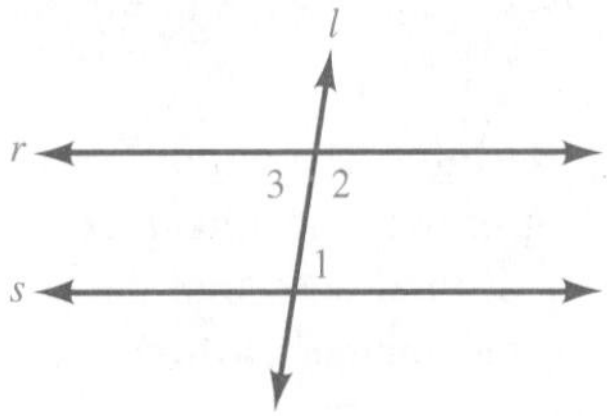

ILLUSTRATION 12

47. Angles of an equilateral triangle Find the measure of each angle in an equilateral triangle.

48. Vertical angles When two lines intersect as in Illustration 13, four angles are formed. Angles that are side-by-side, such as $\angle 1$ (angle 1) and $\angle 2$, are called **adjacent angles.** Angles that are nonadjacent, such as $\angle 1$ and $\angle 3$ or $\angle 2$ and $\angle 4$, are called **vertical angles.** From geometry, we know that if two lines intersect, vertical angles have the same measure. If $m(\angle 1) = (3x + 10)°$ and $m(\angle 3) = (5x - 10)°$, find x. Read $m(\angle 1)$ as "the measure of $\angle 1$."

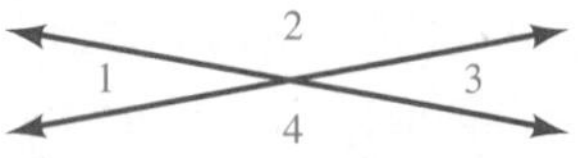

ILLUSTRATION 13

49. If $m(\angle 2) = (6x + 20)°$ in Illustration 13 and $m(\angle 4) = (8x - 20)°$, find $m(\angle 1)$. (See Exercise 48.)

50. Angles of a quadrilateral The sum of the angles of any four-sided figure (called a *quadrilateral*) is 360°. The quadrilateral shown in Illustration 14 has two equal base angles. Find x.

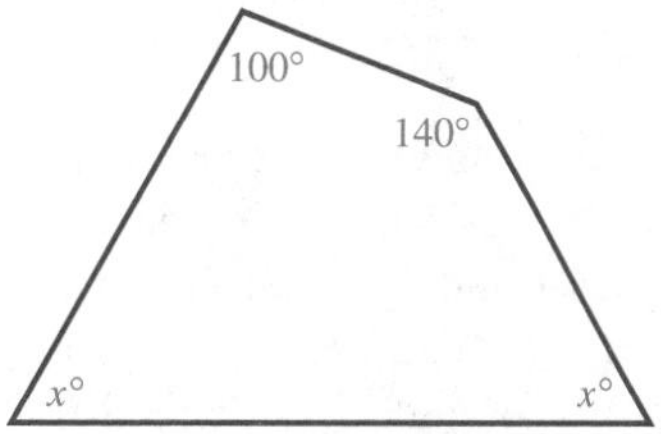

ILLUSTRATION 14

51. Height of a triangle If the height of a triangle with a base of 8 inches is tripled, its area is increased by 96 square inches. Find the height of the triangle.

52. Engineering design The width, w, of the flange in the engineering drawing in Illustration 15 has not yet been determined. Find w so that the area of the 7-in.-by-12-in. rectangular portion is exactly one-half of the total area.

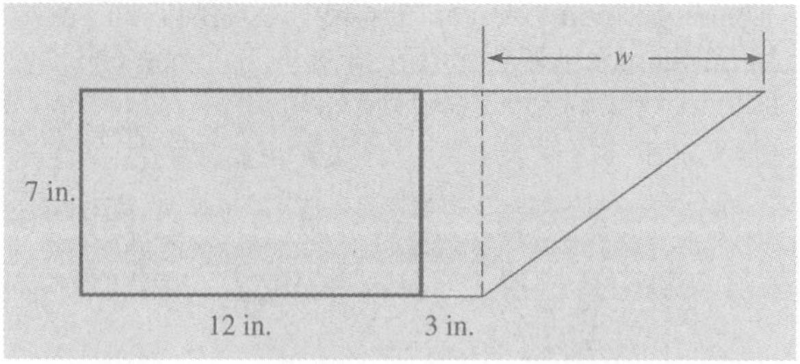

ILLUSTRATION 15

53. Balancing a seesaw A seesaw is 20 feet long, and the fulcrum is in the center. If an 80-pound-boy sits at one end, how far will the boy's 160-pound father have to sit from the fulcrum to balance the seesaw?

54. Establishing equilibrium Two forces—110 pounds and 88 pounds—are applied to opposite ends of an 18-foot lever. How far from the greater force must the fulcrum be placed so that the lever is balanced?

55. Moving a stone A woman uses a 10-foot bar to lift a 210-pound stone. If she places another rock 3 feet from the stone to act as the fulcrum, how much force must she exert to move the stone?

56. Lifting a car A 350-pound football player brags that he can lift a 2,500-pound car. If he uses a 12-foot bar with the fulcrum placed 3 feet from the car, will he be able to lift the car?

57. Balancing a lever Forces are applied to a lever as indicated in Illustration 16. Find x, the distance of the smallest force from the fulcrum.

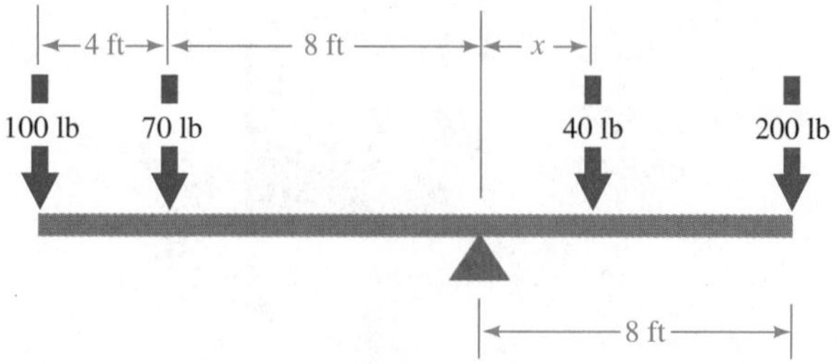

ILLUSTRATION 16

58. Balancing a seesaw Jim and Bob sit at opposite ends of an 18-foot seesaw, with the fulcrum at its center. Jim weighs 160 pounds, and Bob weighs 200 pounds. Kim sits 4 feet in front of Jim, and the seesaw balances. How much does Kim weigh?

59. Temperature scales The Celsius and Fahrenheit temperature scales are related by the equation $C = \frac{5}{9}(F - 32)$. At what temperature will a Fahrenheit and a Celsius thermometer give the same reading?

60. Installing solar heating One solar panel in Illustration 17 is to be 3 feet wider than the other, but to be equally efficient, they must have the same area. Find the width of each.

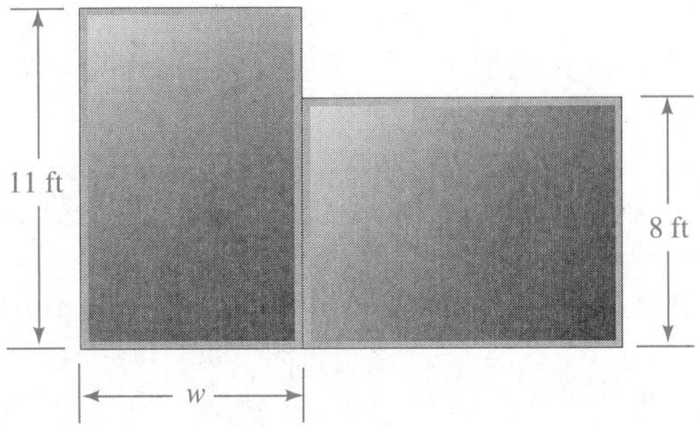

ILLUSTRATION 17

WRITING

61. Explain the steps for solving an applied problem.

62. Explain how to check the solution of an applied problem.

SOMETHING TO THINK ABOUT

63. Find the distance x required to balance the lever in Illustration 18.

64. Interpret the answer to Exercise 63.

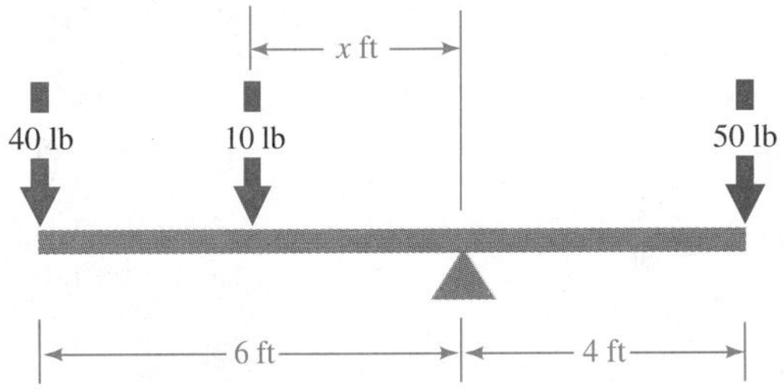

ILLUSTRATION 18

1.7 More Applications of Equations

■ INVESTMENT PROBLEMS ■ UNIFORM MOTION PROBLEMS ■ MIXTURE PROBLEMS

Getting Ready

1. Find 8% of \$500.

2. Write an expression for 8% of $\$x$.

3. If \$9,000 of \$15,000 is invested at 7%, how much is left to invest at another rate?

4. If $\$x$ of \$15,000 is invested at 7%, write an expression for how much is left to be invested at another rate.

5. If a car travels 50 mph for 4 hours, how far will it go?

6. If coffee sells for \$4 per pound, how much will 5 pounds cost?

INVESTMENT PROBLEMS

EXAMPLE 1

Financial planning A professor has \$15,000 to invest for one year, some at 8% and the rest at 7%. If she wants to earn \$1,110 from these investments, how much should she invest at each rate?

Analyze the problem We will add the interest from the 8% investment to the interest from the 7% investment and set the sum equal to the total interest earned.

Simple interest is computed by the formula $i = prt$, where i is the interest earned, p is the principal, r is the annual interest rate, and t is the length of time the principal is invested. In this problem, $t = 1$ year. Thus, if \$$x$ is invested at 8% for one year, the interest earned is $i = \$x(8\%)(1) = \$0.08x$. If the remaining \$$(15{,}000 - x)$ is invested at 7%, the amount earned on that investment is \$$0.07(15{,}000 - x)$. See Figure 1-28. The sum of these amounts should equal \$1,110.

	i	= p	· r	· t
8% investment	$0.08x$	x	0.08	1
7% investment	$0.07(15{,}000 - x)$	$15{,}000 - x$	0.07	1

FIGURE 1-28

Form an equation We can let x represent the number of dollars invested at 8%. Then $15{,}000 - x$ represents the number of dollars invested at 7%.

The interest earned at 8%	plus	the interest earned at 7%	equals	the total interest.
$0.08x$	$+$	$0.07(15{,}000 - x)$	$=$	1,110

Solve the equation We now solve the equation.

$$0.08x + 0.07(15{,}000 - x) = 1{,}110$$

$$8x + 7(15{,}000 - x) = 111{,}000$$ To eliminate the decimals, multiply both sides by 100.

$$8x + 105{,}000 - 7x = 111{,}000$$ Use the distributive property to remove parentheses.

$$x + 105{,}000 = 111{,}000$$ Combine like terms.

$$x = 6{,}000$$ Subtract 105,000 from both sides.

$$15{,}000 - x = 9{,}000$$

State the conclusion She should invest \$6,000 at 8% and \$9,000 at 7%.

Check the result The interest on \$6,000 is 0.08(\$6,000) = \$480. The interest earned on \$9,000 is 0.07(\$9,000) = \$630. The total interest is \$1,110. ■

■ UNIFORM MOTION PROBLEMS

EXAMPLE 2

Travel time A car leaves Rockford traveling toward Wausau at the rate of 55 mph. At the same time, another car leaves Wausau traveling toward Rockford at the rate of 50 mph. How long will it take them to meet if the cities are 157.5 miles apart?

Analyze the problem In this case, the cars are traveling toward each other as shown in Figure 1-29(a). Uniform motion problems are based on the formula $d = rt$, where d is distance, r is rate, and t is time. We can organize the given information in the chart shown in Figure 1-29(b).

We know that one car is traveling at 55 mph and that other is going 50 mph. We also know that they travel for the same amount of time—say, t hours. Thus, the distance that the faster car travels is $55t$ miles, and the distance that the slower car travels is $50t$ miles. The sum of these distances equals 157.5 miles, the distance between the cities.

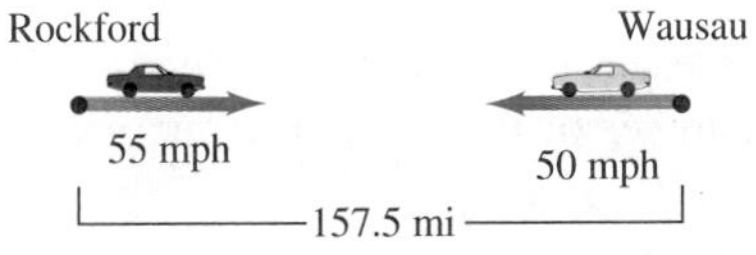

(a)

	Rate	· Time	= Distance
Faster car	55	t	$55t$
Slower car	50	t	$50t$

(b)

FIGURE 1-29

Form an equation We can let t represent the time that each car travels. Then $55t$ represents the distance traveled by the faster car, and $50t$ represents the distance traveled by the slower car.

The distance the faster car goes	plus	the distance the slower car goes	equals	the distance between cities.
$55t$	$+$	$50t$	$=$	157.5

Solve the equation We now solve the equation.

$$55t + 50t = 157.5$$

$$105t = 157.5 \quad \text{Combine like terms.}$$

$$t = 1.5 \quad \text{Divide both sides by 105.}$$

State the conclusion The two cars will meet in $1\frac{1}{2}$ hours.

Check the result The faster car travels 1.5(55) = 82.5 miles. The slower car travels 1.5(50) = 75 miles. The total distance traveled is 157.5 miles. ■

MIXTURE PROBLEMS

EXAMPLE 3

Mixing nuts The owner of a candy store notices that 20 pounds of gourmet cashews are getting stale. They did not sell because of their high price of \$12 per pound. The store owner decides to mix peanuts with the cashews to lower the price per pound. If peanuts sell for \$3 per pound, how many pounds of peanuts must be mixed with the cashews to make a mixture that could be sold for \$6 per pound?

Analyze the problem This problem is based on the formula $V = pn$, where V represents the value, p represents the price per pound, and n represents the number of pounds.

We can let x represent the number of pounds of peanuts to be used and enter the known information in the chart shown in Figure 1-30. The value of the cashews plus the value of the peanuts will be equal to the value of the mixture.

	Price ·	Number of pounds =	Value
Cashews	12	20	240
Peanuts	3	x	$3x$
Mixture	6	$20 + x$	$6(20 + x)$

FIGURE 1-30

Form an equation We can let x represent the number of pounds of peanuts to be used. Then $20 + x$ represents the number of pounds in the mixture.

The value of the cashews	plus	the value of the peanuts	equals	the value of the mixture.
240	$+$	$3x$	$=$	$6(20 + x)$

Solve the equation We now solve the equation.

$240 + 3x = 6(20 + x)$

$240 + 3x = 120 + 6x$ Use the distributive property to remove parentheses.

$120 = 3x$ Subtract $3x$ and 120 from both sides.

$40 = x$ Divide both sides by 3.

State the conclusion The store owner should mix 40 pounds of peanuts with the 20 pounds of cashews.

Check the result The cashews are valued at \$12(20) = \$240, and the peanuts are valued at \$3(40) = \$120, so the mixture is valued at \$6(60) = \$360.

The value of the cashews plus the value of the peanuts equals the value of the mixture. ■

EXAMPLE 4

Milk production A container is partially filled with 12 liters of whole milk containing 4% butterfat. How much 1% milk must be added to get a mixture that is 2% butterfat?

Analyze the problem Since the first container shown in Figure 1-31(a) contains 12 liters of 4% milk, it contains 0.04(12) liters of butterfat. To this container we will add the contents of the second container, which holds $0.01l$ liters of butterfat.

The sum of these two amounts of butterfat $(0.04(12) + 0.01l)$ will equal the amount of butterfat in the third container, which is $0.02(12 + l)$ liters of butterfat. This information is presented in table form in Figure 1-31(b).

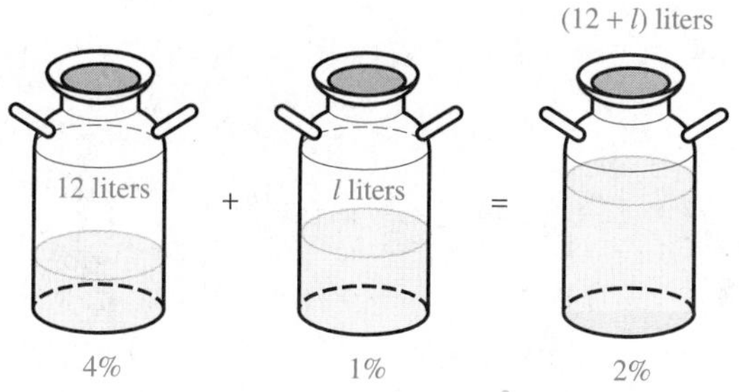

(a)

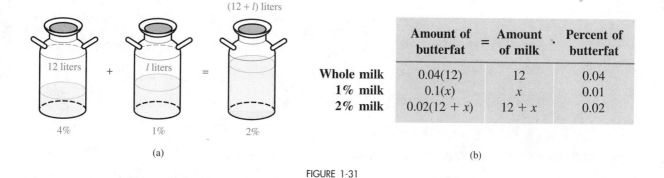

	Amount of butterfat	= Amount of milk	· Percent of butterfat
Whole milk	0.04(12)	12	0.04
1% milk	$0.1(x)$	x	0.01
2% milk	$0.02(12 + x)$	$12 + x$	0.02

(b)

FIGURE 1-31

Form an equation We can let l represent the number of liters of 1% milk to be added. Then

The amount of butterfat in 12 liters of 4% milk	plus	the amount of butterfat in l liters of 1% milk	equals	the amount of butterfat in $(12 + l)$ liters of mixture.
0.04(12)	+	$0.01l$	=	$0.02(12 + l)$

Solve the equation We now solve the equation.

$$0.04(12) + 0.01l = 0.02(12 + l)$$

$$4(12) + 1l = 2(12 + l) \quad \text{Multiply both sides by 100.}$$

$$48 + l = 24 + 2l \quad \text{Use the distributive property to remove parentheses.}$$

$$24 = l \quad \text{Subtract 24 and } l \text{ from both sides.}$$

State the conclusion Thus, 24 liters of 1% milk should be added to get a mixture that is 2% butterfat.

Check the result 12 liters of 4% milk contains 0.48 liters of butterfat. 24 liters of 1% milk contains 0.24 liters of butterfat. This gives a total of 36 liters of a mixture that contains 0.72 liters of butterfat. This is a 2% solution. ■

Orals

Assume that all investments are for one year.

1. How much interest will \$1,500 earn if invested at 6%?

2. Express the amount of interest $\$x$ will earn if invested at 5%.

3. If $\$x$ of \$30,000 is invested at 5%, how would you express the amount left to be invested at 6%?

4. If Brazil nuts are worth $\$x$ per pound, express how much 20 pounds will be worth.

5. If whole milk is 4% butterfat, how much butterfat is in 2 gallons?

6. If whole milk is 4% butterfat, express how much butterfat is in x gallons.

EXERCISE 1.7

REVIEW *Solve each equation.*

1. $9x - 3 = 6x$

2. $7a + 2 = 12 - 4(a - 3)$

3. $\frac{8(y - 5)}{3} = 2(y - 4)$

4. $\frac{t - 1}{3} = \frac{t + 2}{6} + 2$

VOCABULARY AND CONCEPTS *Fill in the blanks to make a true statement.*

5. The formula for simple interest i is $i = prt$, where p is the ________, r is the _____, and t is the _____.

6. Uniform motion problems are based on the formula ______, where d is the ________, r is the _____ of speed, and t is the _____.

7. Dry mixture problems are based on the formula $V = pn$, where V is the ______, p is the ______ per unit, and n is the _______ of units.

8. The liquid mixture problem in Example 4 is based on the idea that the amount of ________ in the first container plus the amount of ________ in the second container is equal to the amount of ________ in the mixture.

APPLICATIONS

9. Investing money Lured by the ad in Illustration 1, a woman invested \$12,000, some in a money market account and the rest in a 5-year CD. How much was invested in each account if the income from both investments is \$1,060 per year?

First Republic Savings and Loan

Account	Rate
NOW	5.5%
Savings	7.5%
Money market	8.0%
Checking	4.0%
5-year CD	9.0%

ILLUSTRATION 1

10. Investing money A man invested \$14,000, some at 7% and some at 10% annual interest. The annual income from these investments was \$1,280. How much did he invest at each rate?

11. Supplemental income A teacher wants to earn \$1,500 per year in supplemental income from a cash gift of \$16,000. She puts \$6,000 in a credit union that pays 7% annual interest. What rate must she earn on the remainder to achieve her goal?

12. Inheriting money Paul split an inheritance between two investments, one paying 7% annual interest and the other 10%. He invested twice as much in the 10% investment as he did in the 7% investment. If his combined annual income from the two investments was \$4,050, how much did he inherit?

13. **Investing money** Maria has some money to invest. If she could invest \$3,000 more, she could qualify for an 11% investment. Otherwise, she could invest the money at 7.5% annual interest. If the 11% investment would yield twice as much annual income as the 7.5% investment, how much does she have on hand to invest?

14. **Supplemental income** A bus driver wants to earn \$3,500 per year in supplemental income from an inheritance of \$40,000. If the driver invests \$10,000 in a mutual fund paying 8%, what rate must he earn on the remainder to achieve his goal?

15. **Concert receipts** For a jazz concert, student tickets were \$2 each, and adult tickets were \$4 each. If 200 tickets were sold and the total receipts were \$750, how many student tickets were sold?

16. **School play** At a school play, 140 tickets were sold, with total receipts of \$290. If adult tickets cost \$2.50 each and student tickets cost \$1.50 each, how many adult tickets were sold?

17. **Computing time** One car leaves Chicago headed for Cleveland, a distance of 343 miles. At the same time, a second car leaves Cleveland headed toward Chicago. If the first car averages 50 mph and the second car averages 48 mph, how long will it take the cars to meet?

18. **Cycling** A cyclist leaves Las Vegas riding at the rate of 18 mph. One hour later, a car leaves Las Vegas going 45 mph in the same direction. How long will it take the car to overtake the cyclist?

19. **Computing distance** At 2 P.M., two cars leave Eagle River, WI, one headed north and one headed south. If the car headed north averages 50 mph and the car headed south averages 60 mph, when will the cars be 165 miles apart?

20. **Running a marathon race** Two marathon runners leave the starting gate, one running 12 mph and the other 10 mph. If they maintain the pace, how long will it take for them to be one-quarter of a mile apart?

21. **Riding a jet ski** A jet ski can go 12 mph in still water. If a rider goes upstream for 3 hours against a current of 4 mph, how long will it take the rider to return? (*Hint:* Upstream speed is $(12 - 4)$ mph; how far can the rider go in 3 hours?)

22. **Taking a walk** Sarah walked north at the rate of 3 mph and returned at the rate of 4 mph. How many miles did she walk if the round trip took 3.5 hours?

23. **Computing travel time** Grant traveled a distance of 400 miles in 8 hours. Part of the time, his rate of speed was 45 mph. The rest of the time, his rate of speed was 55 mph. How long did Grant travel at each rate?

24. **Riding a motorboat** The motorboat in Illustration 2 can go 18 mph in still water. If it can make a trip downstream in 4 hours and make the return trip in 5 hours, find the speed of the current.

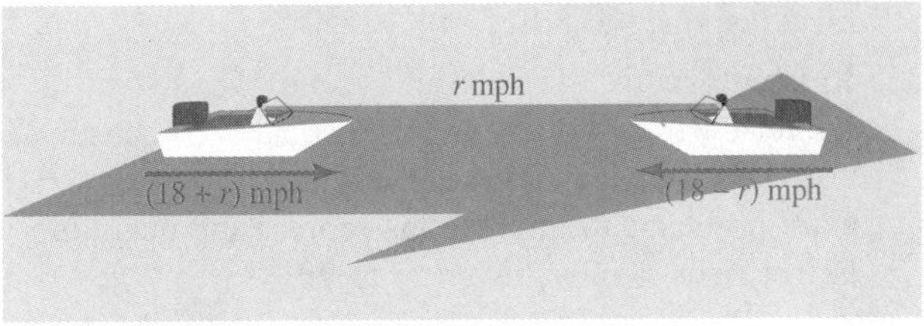

ILLUSTRATION 2

25. **Mixing candies** The owner of a candy store wants to make a 30-pound mixture of two candies to sell for \$1 per pound. If one candy sells for 95¢ per pound and the other for \$1.10 per pound, how many pounds of each should be used?

26. **Computing selling price** A mixture of candy is made to sell for 89¢ per pound. If 32 pounds of candy, selling for 80¢ per pound, are used along with 12 pounds of a more expensive candy, find the price per pound of the better candy.

27. **Diluting solutions** In Illustration 3, how much water should be added to 20 ounces of a 15% solution of alcohol to dilute it to a 10% solution?

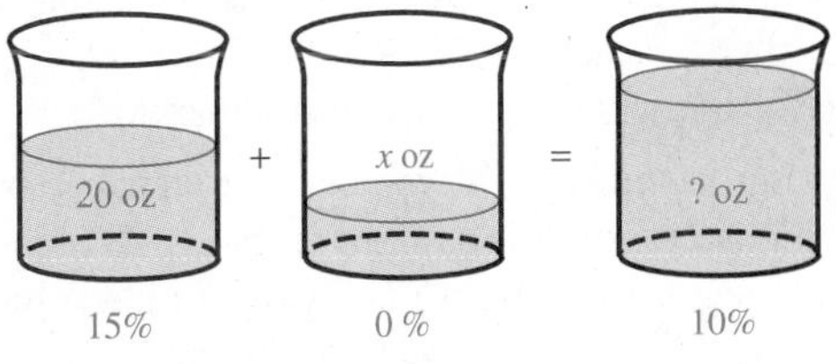

ILLUSTRATION 3

28. Increasing concentration How much water must be boiled away to increase the concentration of 300 gallons of a salt solution from 2% to 3%?

29. Making whole milk Cream is approximately 22% butterfat. How many gallons of cream must be mixed with milk testing at 2% butterfat to get 20 gallons of milk containing 4% butterfat?

30. Mixing solutions How much acid must be added to 60 grams of a solution that is 65% acid to obtain a new solution that is 75% acid?

31. Raising grades A student had a score of 70% on a test that contained 30 questions. To improve his score, the instructor agreed to let him work 15 additional questions. How many must he get right to raise his grade to 80%?

32. Raising grades On a second exam, the student in Exercise 31 earned a score of 60% on a 20-question test. This time, the instructor allowed him to work 20 extra problems to improve his score. How many must he get right to raise his grade to 70%?

33. Computing grades Before the final, Maria had earned a total of 375 points on four tests. To receive an A in the course, she must have 90% of a possible total of 450 points. Find the lowest number of points that she can earn on the final exam and still receive an A.

34. Computing grades A student has earned a total of 435 points on five tests. To receive a B in the course, he must have 80% of a possible total of 600 points. Find the lowest number of points that the student can make on the final exam and still receive a B.

35. Managing a bookstore A bookstore sells a calculus book for \$65. If the bookstore makes a profit of 30% on each book, what does the bookstore pay the publisher for each book? (*Hint:* The retail price = the wholesale price + the markup.)

36. Managing a bookstore A bookstore sells a textbook for \$39.20. If the bookstore makes a profit of 40% on each sale, what does the bookstore pay the publisher for each book? (*Hint:* The retail price = the wholesale price + the markup.)

37. Making furniture A woodworker wants to put two partitions crosswise in a drawer that is 28 inches deep. (See Illustration 4.) He wants to place the partitions so that the spaces created increase by 3 inches from the front to back. If the thickness of each partition is $\frac{1}{2}$ inch, how far from the front end should he place the first partition?

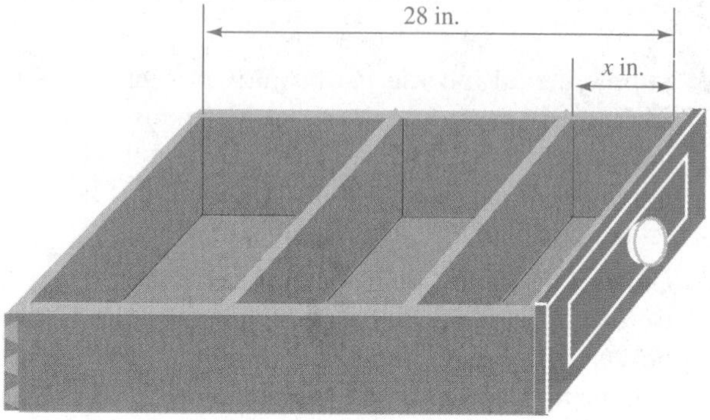

ILLUSTRATION 4

38. Building shelves A carpenter wants to put four shelves on an 8-foot wall so that the five spaces created decrease by 6 inches as we move up the wall. (See Illustration 5.) If the thickness of each shelf is $\frac{3}{4}$ inch, how far will the bottom shelf be from the floor?

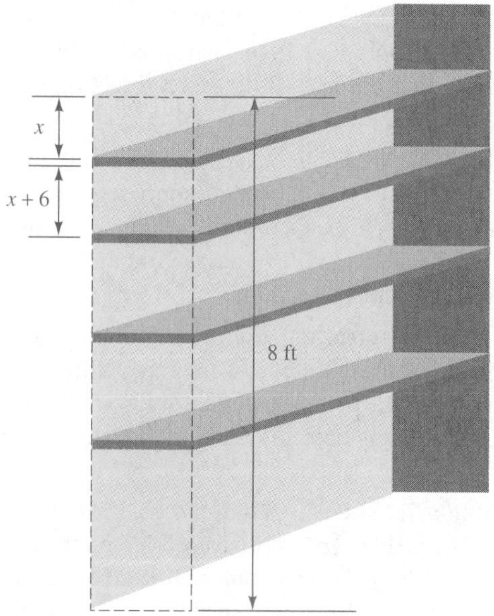

ILLUSTRATION 5

WRITING

39. What do you find most difficult in solving application problems, and why?

40. Which type of application problem do you find easiest, and why?

SOMETHING TO THINK ABOUT

41. Discuss the difficulties in solving this problem:
A man drives 100 miles at 30 mph. How fast should he drive on the return trip to average 60 mph for the entire trip?

42. What difficulties do you encounter when solving this problem?
Adult tickets cost \$4, and student tickets cost \$2. Sales of 71 tickets brought in \$245. How many of each were sold?

PROJECTS

PROJECT 1 You are a member of the Parchdale City Planning Council. The biggest debate in years is taking place over whether or not the town should build an emergency reservoir for use in especially dry periods. The proposed site has room enough for a conical reservoir of diameter 141 feet and depth 85.3 feet.

As long as Parchdale's main water supply is working, the reservoir will be kept full. When a water emergency occurs, however, the reservoir will lose water to evaporation as well as supplying the town with water. The company that has designed the reservoir has given the town the following information.

If D equals the number of consecutive days the reservoir of volume V is used as Parchdale's water supply, then the total amount of water lost to evaporation after D days will be

$$0.1V \cdot \left(\frac{D - 0.7}{D}\right)^2$$

So the volume of water left in the reservoir after it has been used for D days is

$$\text{Volume} = V - (\text{usage per day}) \cdot D - (\text{evaporation})$$

Parchdale uses about 57,500 cubic feet of water per day under emergency conditions. The majority of the council members feel that building the reservoir is a good idea if it will supply the town with water for a week. Otherwise, they will vote against building the reservoir.

a. Calculate the volume of water the reservoir will hold. Remember to use significant digits correctly. Express the answer in scientific notation.

b. Show that the reservoir will supply Parchdale with water for a full week.

c. How much water per day could Parchdale use if the proposed reservoir had to supply the city with water for ten days?

(continued)

PROJECTS *(continued)*

PROJECT 2

Theresa manages the local outlet of Hook n'Slice, a nationwide sporting goods retailer. She recently hired you to help solve some problems. Be sure to provide explanations of how you arrive at your solutions.

a. The company has recently directed that all shipments of golf clubs be sold at a retail price of three times what they cost the company. A shipment of golf clubs arrived yesterday, and they have not yet been labeled. This morning, the main office of Hook n' Slice called to say that the clubs should be sold at 35% off their retail prices. Rather than label each club with its retail price, calculate the sale price, and then relabel the club, Theresa wonders how to go directly from the cost of the club to the sale price. That is, if C is the cost to the company of a certain club, what is the sale price for that club?

1. Develop a formula and answer this question for her.
2. Now try out your formula on a club that cost the company $26. Compare your answer with what you find by following the company procedure of first computing the retail price and then reducing that by 35%.

b. All golf bags have been sale-priced at 20% off for the past few weeks. Theresa has all of them in a large rack, marked with a sign that reads "20% off retail price." The company has now directed that this sale price be reduced by 35%. Rather than relabel every bag with its new sale price, Theresa would like to simply change the sign to say "________% off retail price."

1. Determine what number Theresa should put in the blank.
2. Check your answer by computing the *final* sale price for a golf bag with a $100 retail price in two ways:
   - by using the percentage you told Theresa should go on the sign, and
   - by doing the two separate price reductions.
3. Would the sign read the same if the retail price were first reduced by 35%, and then the sale price reduced by 20%? Explain.

CHAPTER SUMMARY

CONCEPTS | REVIEW EXERCISES

SECTION 1.1 The Real Number System

A *set* is a collection of objects.

Natural numbers:

1, 2, 3, 4, 5, . . .

Whole numbers:

0, 1, 2, 3, 4, 5, . . .

Integers:

. . . , −3, −2, −1, 0, 1, 2, 3, . . .

Even integers:

. . . , −6, −4, −2, 0, 2, 4, 6, . . .

Odd integers:

. . . , −5, −3, −1, 1, 3, 5, . . .

1. List the numbers in $\{-4, -\frac{2}{3}, 0, 1, 2, \pi, 4\}$ that satisfy the given condition.

a. whole number **b.** natural number

c. rational number

d. integer

e. irrational number **f.** real number

g. negative number **h.** positive number

i. prime number **j.** composite number

k. even integer **l.** odd integer

Prime numbers:

2, 3, 5, 7, 11, 13, . . .

Composite numbers:

4, 6, 8, 9, 10, 12, . . .

2. Graph the set of prime numbers between 20 and 30.

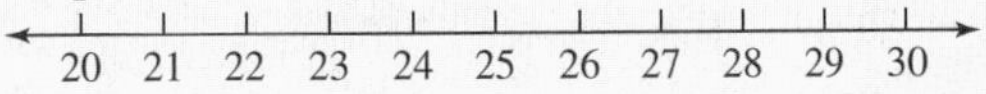

Rational numbers: numbers that can be written as $\frac{a}{b}$ $(b \neq 0)$, where a and b are integers

3. Graph the set of composite numbers between 5 and 13.

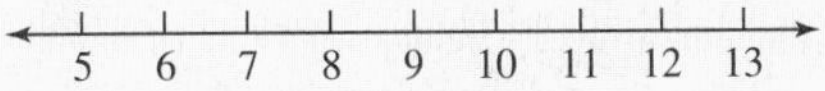

Irrational numbers: numbers that can be written as nonterminating, nonrepeating decimals

Real numbers: numbers that can be written as decimals

4. Graph each interval on the number line.

a. $x \geq -4$ **b.** $-2 < x \leq 6$

c. $(-2, 3)$ **d.** $[2, 6]$

e. $(2, \infty)$ **f.** $(-\infty, -1)$

g. $(-2, 0] \cup (2, 6)$

Absolute value:

$$\begin{cases} \text{If } x \geq 0, \text{ then } |x| = x. \\ \text{If } x < 0, \text{ then } |x| = -x. \end{cases}$$

5. Write each expression without absolute value symbols.

a. $|0|$ **b.** $|-1|$

c. $|8|$ **d.** $-|8|$

SECTION 1.2 Arithmetic and Properties of Real Numbers

Adding and subtracting real numbers:
With like signs:
Add their absolute values and keep the same sign.

With unlike signs:
Subtract their absolute values and keep the sign of the number with the greater absolute value.

$x - y$ is equivalent to $x + (-y)$.

Multiplying and dividing real numbers:
With like signs: Multiply (or divide) their absolute values. The sign is positive.

With unlike signs: Multiply (or divide) their absolute values. The sign is negative.

6. Do the operations and simplify when possible.

a. $3 + (+5)$ **b.** $-6 + (-3)$

c. $-15 + (-13)$ **d.** $25 + 32$

e. $-2 + 5$ **f.** $3 + (-12)$

g. $8 + (-3)$ **h.** $7 + (-9)$

i. $-25 + 12$ **j.** $-30 + 35$

k. $-3 - 10$ **l.** $-8 - (-3)$

m. $27 - (-12)$ **n.** $38 - (-15)$

o. $(+5)(+7)$ **p.** $(-6)(-7)$

q. $\dfrac{-16}{-4}$ **r.** $\dfrac{-25}{-5}$

s. $4(-3)$ **t.** $-3(8)$

u. $\dfrac{-8}{2}$ **v.** $\dfrac{8}{-4}$

Order of operations without exponents:
Do all calculations within grouping symbols, working from the innermost pair to the outermost pair.
1. Do all multiplications and divisions, working from left to right.
2. Do all additions and subtractions, working from left to right.

In a fraction, simplify the numerator and denominator separately, and then simplify the fraction.

7. Simplify each expression.

a. $-4(3 - 6)$ **b.** $3[8 - (-1)]$

c. $-[4 - 2(6 - 4)]$ **d.** $3[-5 + 3(2 - 7)]$

e. $\dfrac{3 - 8}{10 - 5}$ **f.** $\dfrac{-32 - 8}{6 - 16}$

8. Consider the numbers 14, 12, 13, 14, 15, 20, 15, 17, 19, 15.

a. Find the mean. **b.** Find the median.

c. Find the mode. **d.** Can the mean, median, and mode of a group of numbers be the same?

9. Evaluate when $a = 5$, $b = -2$, $c = -3$, and $d = 2$.

a. $\dfrac{3a - 2b}{cd}$ **b.** $\dfrac{3b + 2d}{ac}$

c. $\dfrac{ab + cd}{c(b - d)}$ **d.** $\dfrac{ac - bd}{a(d + c)}$

Associative properties:

$(a + b) + c = a + (b + c)$

$(ab)c = a(bc)$

Commutative properties:

$a + b = b + a$

$ab = ba$

Distributive property:

$a(b + c) = ab + ac$

0 is the **additive identity:**

$a + 0 = 0 + a = a$

1 is the **multiplicative identity:**

$a \cdot 1 = 1 \cdot a = a$

$-a$ is the **negative** (or **additive inverse**) of a:

$a + (-a) = 0$

Double negative rule:

$-(-a) = a$

If $a \neq 0$, then $\frac{1}{a}$ is the **reciprocal** (or **multiplicative inverse**) of a:

$$a\left(\frac{1}{a}\right) = \frac{1}{a} \cdot 1 = 1$$

10. Tell which property justifies each statement.

a. $3(4 + 2) = 3 \cdot 4 + 3 \cdot 2$
b. $3 + (x + 7) = (x + 7) + 3$
c. $3 + (x + 7) = (3 + x) + 7$
d. $3 + 0 = 3$
e. $3 + (-3) = 0$
f. $5(3) = 3(5)$
g. $3(xy) = (3x)y$
h. $3x \cdot 1 = 3x$
i. $a\left(\frac{1}{a}\right) = 1 \quad (a \neq 0)$
j. $-(-x) = x$

SECTION 1.3 Exponents

Properties of exponents:
If there are no divisions by 0,

$$x^n = \underbrace{x \cdot x \cdot x \cdot \cdots \cdot x}_{n \text{ factors of } x}$$

$x^m x^n = x^{m+n} \qquad (x^m)^n = x^{mn}$

$(xy)^n = x^n y^n \qquad \left(\frac{x}{y}\right)^n = \frac{x^n}{y^n}$

$x^0 = 1 \qquad x^{-n} = \frac{1}{x^n}$

$\frac{x^m}{x^n} = x^{m-n} \qquad \left(\frac{x}{y}\right)^{-n} = \left(\frac{y}{x}\right)^n$

11. Simplify each expression and write all answers without negative exponents.

a. 3^6
b. -2^6
c. $(-4)^3$
d. -5^{-4}
e. $(3x^4)(-2x^2)$
f. $(-x^5)(3x^3)$
g. $x^{-4}x^3$
h. $x^{-10}x^{12}$
i. $(3x^2)^3$
j. $(4x^4)^4$
k. $(-2x^2)^5$
l. $-(-3x^3)^5$
m. $(x^2)^{-5}$
n. $(x^{-4})^{-5}$
o. $(3x^{-3})^{-2}$
p. $(2x^{-4})^4$
q. $\frac{x^6}{x^4}$
r. $\frac{x^{12}}{x^7}$

s. $\dfrac{a^7}{a^{12}}$ **t.** $\dfrac{a^4}{a^7}$

u. $\dfrac{y^{-3}}{y^4}$ **v.** $\dfrac{y^5}{y^{-4}}$

w. $\dfrac{x^{-5}}{x^{-4}}$ **x.** $\dfrac{x^{-6}}{x^{-9}}$

12. Simplify each expression and write all answers without negative exponents.

a. $(3x^2y^3)^2$ **b.** $(-3a^3b^2)^{-4}$

c. $\left(\dfrac{3x^2}{4y^3}\right)^{-3}$ **d.** $\left(\dfrac{4y^{-2}}{5y^{-3}}\right)^3$

SECTION 1.4 Scientific Notation

Scientific notation:
$N \times 10^n$, where $1 \le |N| < 10$ and n is an integer.

13. Write each numeral in scientific notation.

a. 19,300,000,000
b. 0.0000000273

14. Write each numeral in standard notation.

a. 7.2×10^7
b. 8.3×10^{-9}

SECTION 1.5 Solving Equations

If a and b are real numbers and $a = b$, then

$a + c = b + c$

$a - c = b - c$

$ac = bc$

$\dfrac{a}{c} = \dfrac{b}{c} \quad (c \neq 0)$

15. Solve each equation.

a. $5x + 12 = 37$ **b.** $-3x - 7 = 20$

c. $4(y - 1) = 28$ **d.** $3(x + 7) = 42$

e. $13(x - 9) - 2 = 7x - 5$ **f.** $\dfrac{8(x - 5)}{3} = 2(x - 4)$

g. $\dfrac{3y}{4} - 13 = -\dfrac{y}{3}$ **h.** $\dfrac{2y}{5} + 5 = \dfrac{14y}{10}$

16. Solve for the indicated quantity.

a. $V = \dfrac{4}{3}\pi r^3$ for r^3

b. $V = \dfrac{1}{3}\pi r^2 h$ for h

c. $v = \dfrac{1}{6}ab(x + y)$ for x

d. $V = \pi h^2\left(r - \dfrac{h}{3}\right)$ for r

SECTION 1.6 Using Equations to Solve Problems

To solve problems, use the following strategy:
1. Analyze the problem.
2. Form an equation.
3. Solve the equation.
4. State the conclusion.
5. Check the result.

17. Carpentry A carpenter wants to cut a 20-foot rafter so that one piece is three times as long as the other. Where should he cut the board?

18. Geometry A rectangle is 4 meters longer than it is wide. If the perimeter of the rectangle is 28 meters, find its area.

19. Balancing a seesaw Sue weighs 48 pounds, and her father weighs 180 pounds. If Sue sits on one end of a 20-foot long seesaw with the fulcrum in the middle, how far from the fulcrum should her father sit to balance the seesaw?

SECTION 1.7 More Applications of Equations

20. Investment problem Sally has \$25,000 to invest. She invests some money at 10% interest and the rest at 9%. If her total annual income from these two investments is \$2,430, how much does she invest at each rate?

21. Mixing solutions How much water must be added to 20 liters of a 12% alcohol solution to dilute it to an 8% solution?

22. Motion problem A car and a motorcycle both leave from the same point and travel in the same direction. (See Illustration 1.) The car travels at an average rate of 55 mph and the motorcycle at an average rate of 40 mph. How long will it take before the vehicles are 5 miles apart?

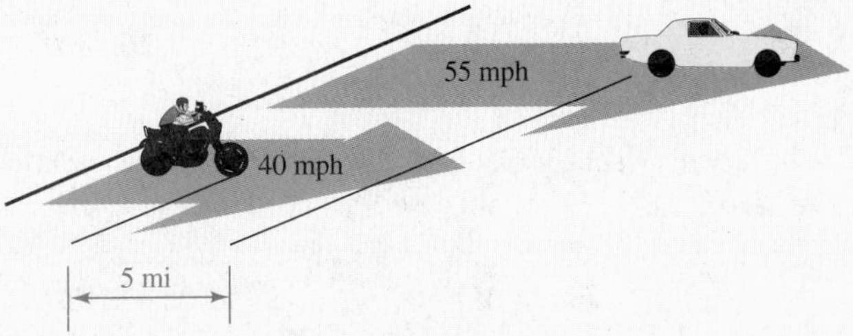

ILLUSTRATION 1

Chapter Test

In Problems 1–2, let $A = \left\{-2, 0, 1, \frac{6}{5}, 2, \sqrt{7}, 5\right\}$.

1. What numbers in A are natural numbers?

2. What numbers in A are irrational numbers?

In Problems 3–4, graph each set on the number line.

3. The set of odd integers between -4 and 6

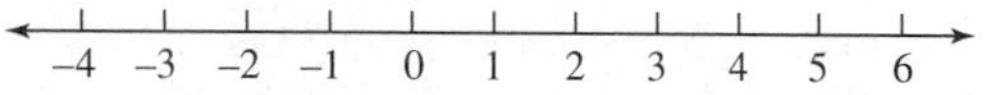

4. The set of prime numbers less than 12

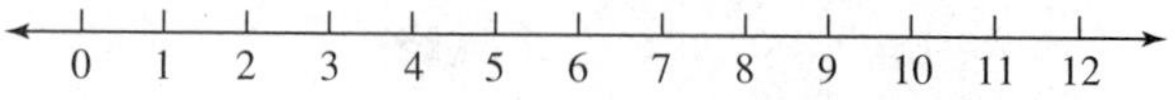

In Problems 5–6, graph each interval on the number line.

5. $[-2, 4)$

6. $(-\infty, -1] \cup [2, \infty)$

In Problems 7–8, write each expression without using absolute value symbols.

7. $-|8|$

8. $|-5|$

In Problems 9–12, do the operations.

9. $7 + (-5)$

10. $-5(-4)$

11. $\dfrac{12}{-3}$

12. $-4 - \dfrac{-15}{3}$

In Problems 13–14, consider the numbers $-2, 0, 2, -2, 3, -1, -1, 1, 1, 2$.

13. Find the mean.

14. Find the median.

In Problems 15–18, $a = 2$, $b = -3$, *and* $c = 4$. *Evaluate each expression.*

15. ab

16. $a + bc$

17. $ab - bc$

18. $\dfrac{-3b + a}{ac - b}$

In Problems 19–20, tell which property of real numbers justifies each statement.

19. $3 + 5 = 5 + 3$

20. $a(b + c) = ab + ac$

In Problems 21–24, simplify each expression. Write all answers without using negative exponents. Assume that no denominators are zero.

21. x^3x^5

22. $(x^2y^3)^3$

23. $(m^{-4})^2$

24. $\left(\dfrac{m^2n^3}{m^4n^{-2}}\right)^{-2}$

In Problems 25–26, write each numeral in scientific notation.

25. 4,700,000

26. 0.00000023

In Problems 27–28, write each numeral in standard notation.

27. 6.53×10^5

28. 24.5×10^{-3}

In Problems 29–30, solve each equation.

29. $9(x + 4) + 4 = 4(x - 5)$

30. $\frac{y - 1}{5} + 2 = \frac{2y - 3}{3}$

31. Solve $P = L + \frac{s}{f}i$ for i.

32. A rectangle has a perimeter of 26 centimeters and is 5 centimeters longer than it is wide. Find its area.

33. Bob invests part of \$10,000 at 9% annual interest and the rest at 8%. If his annual income from these investments is \$860, how much does he invest at 8%?

34. How many liters of water are needed to dilute 20 liters of a 5% salt solution to a 1% solution?

2 Graphs, Equations of Lines, and Functions

MATHEMATICS IN THE WORKPLACE

Economist

Economists study the way a society uses resources such as land, labor, raw materials, and machinery to provide goods and services. Some economists are theoreticians who use mathematical models to explain the causes of recession and inflation. Most economists, however, are concerned with practical applications of economic policy in a particular area.

SAMPLE APPLICATION ■ An electronics firm manufactures tape recorders, receiving \$120 for each recorder it makes. If x represents the number of recorders produced, the income received is determined by the *revenue function*

$$R(x) = 120x$$

The manufacturer has fixed costs of \$12,000 per month and variable costs of \$57.50 for each tape recorder manufactured. Thus, the *cost function* is

$$\begin{aligned} C(x) &= \textit{variable costs} + \textit{fixed costs} \\ &= 57.50x + 12{,}000 \end{aligned}$$

How many recorders must the company sell for revenue to equal cost? (See Exercise 81 in Exercise 2.5.)

It is often said, "A picture is worth a thousand words." In this chapter, we will show how numerical relationships can be described by using mathematical pictures called **graphs.** We will also show how graphs are constructed and how we can obtain important information by reading graphs.

2.1 The Rectangular Coordinate System

■ THE RECTANGULAR COORDINATE SYSTEM ■ GRAPHING MATHEMATICAL RELATIONSHIPS ■ READING GRAPHS ■ FREQUENCY DISTRIBUTIONS ■ STEP GRAPHS

Getting Ready

Graph each set of numbers on the number line.

1. $\{-2, 1, 3\}$

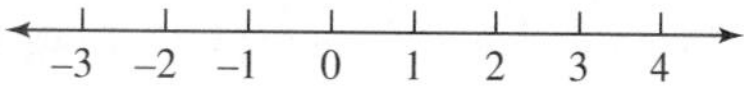

2. All numbers greater than -2

3. All numbers less than or equal to 3

4. All numbers between -3 and 2

■ THE RECTANGULAR COORDINATE SYSTEM

Many cities are laid out on a rectangular grid. For example, on the east side of Rockford, IL, all streets run north and south and all avenues run east and west. (See Figure 2-1.) If we agree to list the street numbers first, every address can be identified by using an ordered pair of numbers. If Jose lives on the corner of Third Street and Sixth Avenue, his address is given by the ordered pair (3, 6).

This is the street. ↓

(3, 6)

↑ This is the avenue.

If Lisa has an address of (6, 3), we know that she lives on the corner of Sixth Street and Third Avenue. From the figure, we can see that

- Bob Anderson's address is (4, 1).
- Rosa Vang's address is (7, 5).
- The address of the store is (8, 2).

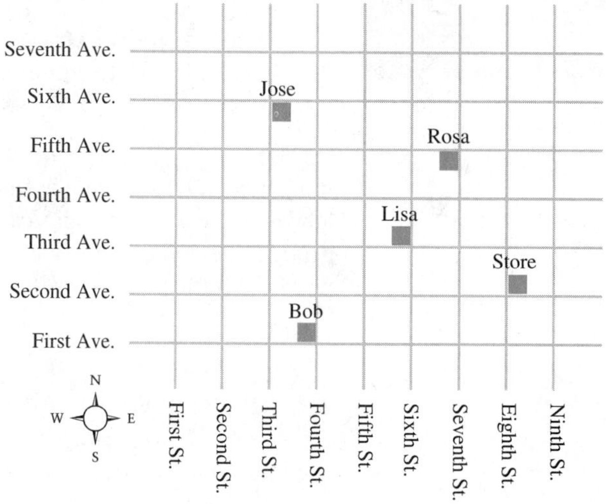

FIGURE 2-1

The idea of associating an ordered pair of numbers with points on a grid is attributed to the 17th-century French mathematician René Descartes. The grid is often called a **rectangular coordinate system,** or **Cartesian coordinate system** after its inventor.

In general, a rectangular coordinate system (see Figure 2-2) is formed by two intersecting perpendicular number lines.

- The horizontal number line is usually called the ***x*-axis.**
- The vertical number line is usually called the ***y*-axis.**

The positive direction on the x-axis is to the right, and the positive direction on the y-axis is upward. If no scale is indicated on the axes, we assume that the axes are scaled in units of 1.

The point where the axes cross is called the **origin.** This is the 0 point on each axis. The two axes form a **coordinate plane** and divide it into four regions called **quadrants,** which are numbered as shown in Figure 2-2.

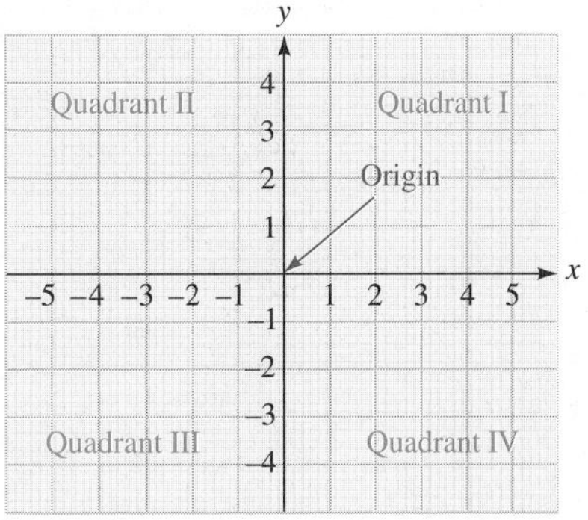

FIGURE 2-2

As with the city streets example, every point in a coordinate plane can be identified by a pair of real numbers x and y, written as (x, y). The first number in the pair is the ***x*-coordinate,** and the second number is the ***y*-coordinate.** The numbers are called the **coordinates** of the point. Some examples of ordered pairs are $(-4, 6)$, $(2, 3)$, and $(6, -4)$.

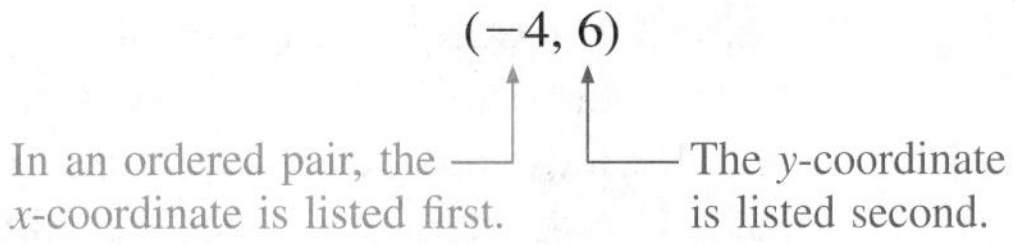

The process of locating a point in the coordinate plane is called **graphing** or **plotting** the point. In Figure 2-3(a), we show how to graph the point Q with coordinates of $(-4, 6)$. Since the ***x*-coordinate** is negative, we start at the origin and move 4 units to the left along the x-axis. Since the ***y*-coordinate** is positive, we then move up 6 units to locate point Q. Point Q is the **graph** of the ordered pair $(-4, 6)$ and lies in quadrant II.

To plot the point $P(2, 3)$, we start at the origin, move 2 units to the right along the x-axis, and then move up 3 units to locate point P. Point P lies in quadrant I. To plot point $R(6, -4)$, we start at the origin and move 6 units to the right and then 4 units down. Point R lies in quadrant IV.

Remember that the x-coordinate of a point tells how far the point is from the y-axis, and its sign tells the direction. The y-coordinate tells how far the point is from the x-axis, and its sign tells the direction.

WARNING! Note that point Q with coordinates of $(-4, 6)$ is not the same as point R with coordinates $(6, -4)$. This illustrates that the order of the coordinates of a point is important. This is why we call the pairs **ordered pairs.**

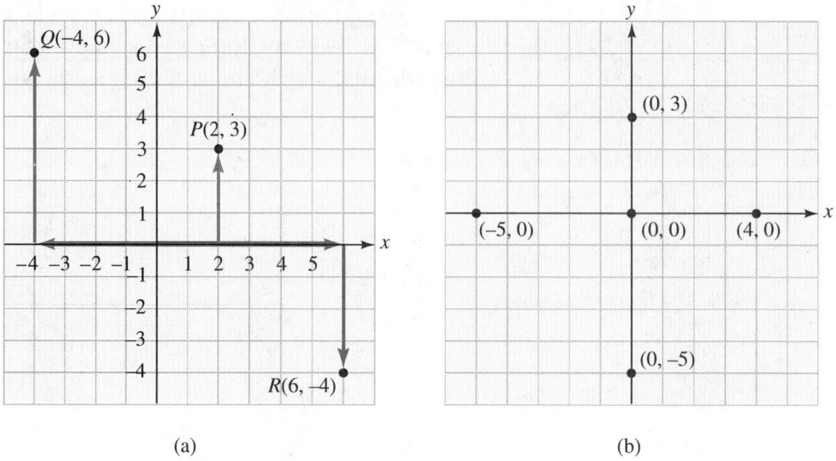

FIGURE 2-3

In Figure 2-3(b), we see that the points $(-5, 0)$, $(0, 0)$, and $(4, 0)$ all lie on the x-axis. In fact, every point with a y-coordinate of 0 will lie on the x-axis. Note that the coordinates of the origin are $(0, 0)$.

In Figure 2-3(b), we also see that the points $(0, -5)$, $(0, 0)$, and $(0, 3)$ all lie on the y-axis. In fact, every point with an x-coordinate of 0 will lie on the y-axis.

EXAMPLE 1

Tracking hurricanes The map shown in Figure 2-4 shows the path of a hurricane. The hurricane is currently located at 85° longitude and 25° latitude. If we agree to list longitude first, the location of the hurricane is given by the ordered pair (85, 25). Use the map to answer the following questions.

a. What are the coordinates of New Orleans?

b. What are the estimated coordinates of Houston?

c. What are the estimated coordinates of the landfall of the hurricane?

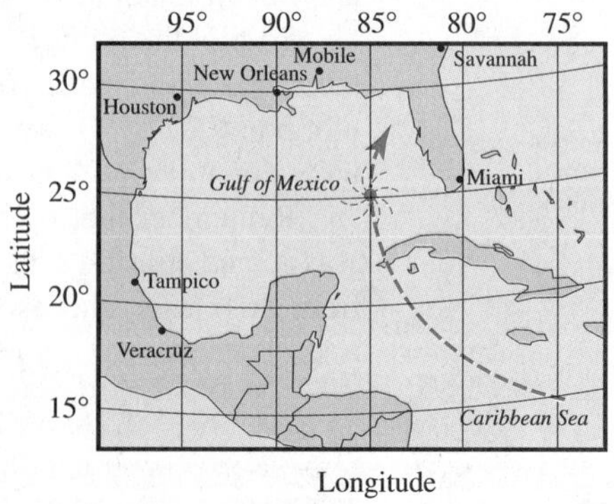

FIGURE 2-4

Solution **a.** Since New Orleans is at 90° longitude and 30° latitude, its coordinates are (90, 30).

b. Since Houston has a longitude that is a little more than 95° and a latitude that is a little less than 30°, its estimated coordinates are (96, 29).

c. The hurricane appears to be headed for the point with coordinates of (83, 29).

■

GRAPHING MATHEMATICAL RELATIONSHIPS

Every day we deal with quantities that are related.

- The distance that we travel depends on how fast we are going.
- Your test score depends on the amount of time you study.
- The height of a toy rocket depends on the time since it was launched.

We can often use graphs to visualize relationships between two quantities. For example, suppose that we know the height of a toy rocket at 1-second intervals from 0 second to 6 seconds. We can list this information in the **table of values** shown in Figure 2-5.

Time (in seconds)	*Height of rocket (in feet)*	
0	0	→ (0, 0)
1	80	→ (1, 80)
2	128	→ (2, 128)
3	144	→ (3, 144)
4	128	→ (4, 128)
5	80	→ (5, 80)
6	0	→ (6, 0)
↑ x-coordinate	↑ y-coordinate	↑ The data in the table can be expressed as ordered pairs (x, y).

FIGURE 2-5

This information can also be used to construct a graph that shows the relationship between the height of the rocket and the time since it was launched. Because the height depends on time, we will associate *time* with the x-axis and the *height* with the y-axis.

To get the graph in Figure 2-6, we plot the seven ordered pairs shown in the table and draw a smooth curve through the resulting data points.

From the graph, we can see that the height of the rocket increases as the time increases from 0 second to 3 seconds. Then the height decreases until the rocket hits the ground in 6 seconds. We can also use the graph to make observations about the height of the rocket at other times. For example, the blue lines on the graph show that in 1.5 seconds, the height of the rocket will be approximately 108 feet.

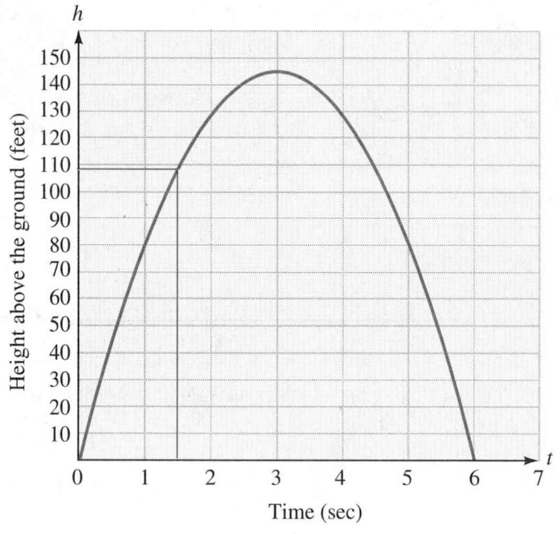

FIGURE 2-6

WARNING! The graph in Figure 2-6 does not show the path of the rocket. It shows the height of the rocket in relation to time since launch.

READING GRAPHS

As we have seen, graphs can be used to describe how quantities change with time. From such a graph, we can determine when a quantity is increasing and when it is decreasing.

a,b

EXAMPLE 2 The graph in Figure 2-7 shows how automobile production has varied in the years since 1900. Refer to the graph and answer the following questions.

a. How many cars were made in 1940?

b. Over which 20-year span did car production increase most rapidly?

c. Why is a dashed line used for the portion of the graph between 1980 and 2000?

Solution **a.** To find the number of cars made in 1940, we find the height of the graph at the point directly above the year 1940. We follow the dashed line from the year 1940 straight up to the graph and then over to the vertical scale. There, we read that about 3.7 million cars were made in 1940.

b. Between 1940 and 1960, the rise of the graph is the greatest. During those years, the production of cars increased most rapidly.

c. Because the results have not been tabulated for the year 2000, the production levels shown on the graph are projections, and the dashed line indicates that the

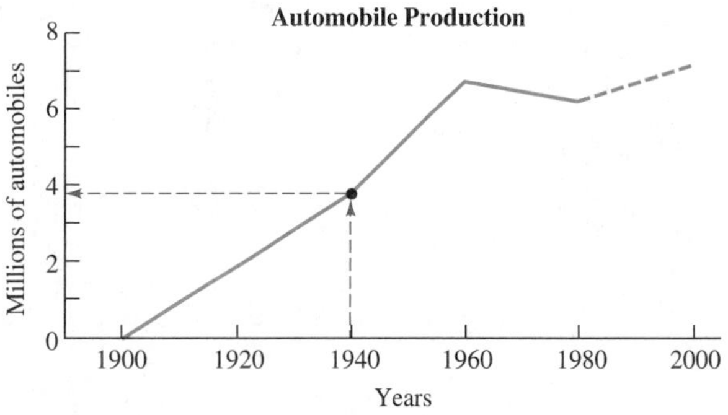

FIGURE 2-7

numbers are not certain. The projected manufacturing level in the year 2000 is about 7.2 million cars, but that number may turn out to be wrong. ■

Self Check Refer to the graph and answer the following questions.

a. How many cars were made in 1950?

b. When did production decrease?

Answers **a.** about 5.2 million, **b.** between 1960 and 1980

C

EXAMPLE 3 The graph in Figure 2-8 shows the annual earnings of General Motors, Ford, and Chrysler from 1980 to 1991. From the graph, answer the following questions.

a. How much money did General Motors make in 1982?

b. Which company lost the most money in 1990?

c. In what years did the earnings of Ford Motor Corporation surpass those of General Motors?

d. Which company's earnings increased the most during the years 1984 to 1988?

Solution **a.** From the key, we see that the earnings of GM are represented by the red line. To find GM's earnings in 1982, we move vertically upward from the year 1982 until we reach the red line. We then move to the left until we reach the number 1 on the vertical scale. Because that scale indicates *billions* of dollars, we see that GM earned approximately $1 billion in 1982.

b. To find which company lost the most money in 1990, we move upward from the year 1990 until we reach the lowest of the three graphs. That graph is a red line, which represents the earnings of GM. So GM lost the most money—about $2 billion.

c. Ford's earnings are represented by the blue line, and GM's earnings are represented by the red line. We then refer to the graph to find the years in which the blue line is above the red line. This is true in 1986, 1987, and 1988, and in 1990 and 1991. In those years, Ford's earnings were higher than GM's.

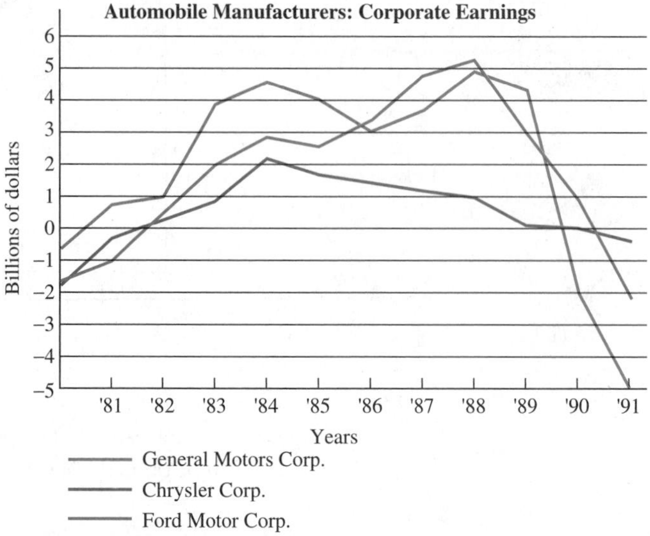

FIGURE 2-8

d. We find the years 1984 and 1988 on the horizontal axis and see which graph rises the most during that time. Chrysler's earnings steadily decreased during that time, and GM's increased slightly. Ford's earnings increased the most, from approximately \$3 billion in 1984 to more than \$5 billion in 1988. ■

Self Check **a.** How much money did Ford lose in 1981? **b.** How much did Ford make in 1988?

Answers **a.** \$1 billion, **b.** over \$5 billion

■ FREQUENCY DISTRIBUTIONS

In statistics, a *distribution* is any set of values, and the *frequency* of each value is the number of times it occurs. For example, suppose we wish to learn about the course loads taken by students at Rock Valley College during the spring semester. To estimate the course loads, we survey 20 students and ask how many courses they are taking, with the following results:

2 courses, 5 courses, 3 courses, 4 courses, 4 courses,
1 course, 4 courses, 3 courses, 2 courses, 5 courses,
3 courses, 4 courses, 3 courses, 1 course, 2 courses,
2 courses, 4 courses, 4 courses, 5 courses, 3 courses

To make the information easier to read at a glance, we can organize the information in the frequency table shown in Figure 2-9(a) and use the information in the table to construct the bar graph shown in Figure 2-9(b). Notice how much easier it is to read the information from the table or bar graph than from the previous list.

Course load	Frequency
One course	2
Two courses	4
Three courses	5
Four courses	6
Five courses	3

(a)

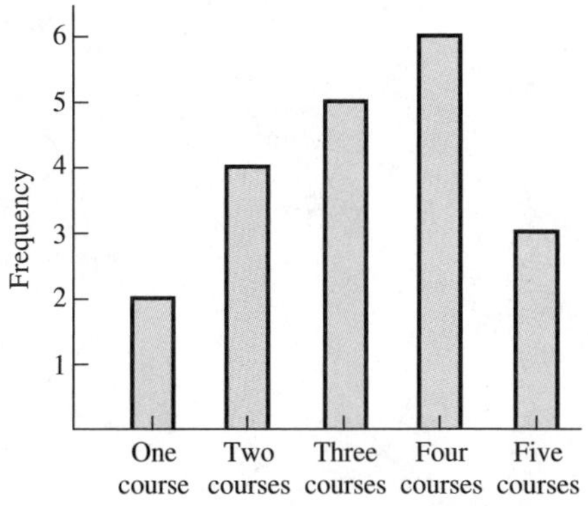

(b)

FIGURE 2-9

In statistics, a *discrete variable* is a variable that has only whole numbers for its values. If x represents the number of courses taken by a student during the spring semester, x can only be a whole number such as 0, 1, 2, and so on. Thus, x is a discrete variable. A bar graph is the most convenient graph to depict the distribution of values of a discrete variable.

STEP GRAPHS

The graph in Figure 2-10 shows the cost of renting a rototiller for different periods of time. For example, the cost of renting a rototiller for 3 days is \$50, which is the y-coordinate of the point with coordinates of (3, 50). The cost of renting the rototiller for more than 4 days up to 5 days is \$70. Since the jumps in cost form steps in the graph, we call the graph a **step graph.**

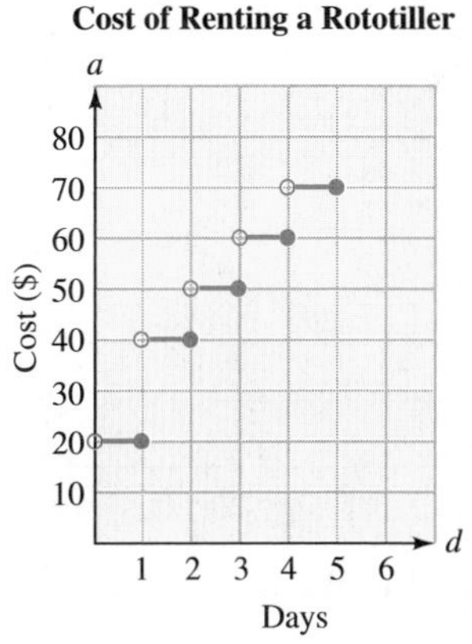

FIGURE 2-10

EXAMPLE 4 Use the information in Figure 2-10 to answer the following questions. Write the results in a table of values.

a. Find the cost of renting the rototiller for 2 days.

b. Find the cost of renting the rototiller for $3\frac{1}{2}$ days.

c. How long can you rent the rototiller if you have \$40?

d. Is the cost per day to rent the rototiller the same?

Solution **a.** We locate 2 on the x-axis and move up to locate the point on the graph directly above the 2. Since the point has coordinates (2, 40), a two-day rental costs \$40. We enter this ordered pair in the table after part **c.**

b. We locate $3\frac{1}{2}$ on the x-axis and move straight up to locate the point on the graph with coordinates $\left(3\frac{1}{2}, 60\right)$, which indicates that a $3\frac{1}{2}$-day rental would cost \$60. We enter this ordered pair in the table after part **c.**

c. We draw a horizontal line through the point labeled 40 on the y-axis. Since this line intersects one of the steps of the graph, we can look down to the x-axis to find the x-values that correspond to a y-value of 40. From the graph, we see that the rototiller can be rented for more than 1 and up to 2 days for \$40. We write (2, 40) in the table below.

Length of rental (days)	*Cost (dollars)*
2	40
$3\frac{1}{2}$	60
2	40

d. No, the cost per day is not the same. If we look at the y-coordinates, we see that for the first day the rental fee is \$20. The second day the cost jumps another \$20. The third day, and all subsequent days, the cost jumps \$10. ■

Self Check In Example 4, find the cost to rent the rototiller for 3 days.

Answer \$50

Orals

1. Explain why the pair $(-2, 4)$ is called an ordered pair.
2. At what point do the coordinate axes intersect?
3. In which quadrant does the graph of $(3, -5)$ lie?
4. On which axis does the point $(0, 5)$ lie?

EXERCISE 2.1

REVIEW

1. Evaluate $-3 - 3(-5)$.
2. Evaluate $(-5)^2 + (-5)$.
3. Simplify $\dfrac{-3 + 5(2)}{9 + 5}$.
4. Simplify $|-1 - 9|$.
5. Solve $-4x + 7 = -21$.
6. Solve $P = 2l + 2w$ for w.

VOCABULARY AND CONCEPTS *Fill in each blank to make a true statement.*

7. The pair of numbers $(6, -2)$ is called an ________.

8. In the ordered pair $(-2, -9)$, -9 is called the ___ coordinate.

9. The point with coordinates $(0, 0)$ is the ______.

10. The x- and y-axes divide the coordinate plane into four regions called ________.

11. Ordered pairs of numbers can be graphed on a ________________ system.

12. The process of locating the position of a point on a coordinate plane is called ________________ the point.

In Exercises 13–20, answer each question or fill in each blank to make a true statement.

13. Do $(3, -2)$ and $(-2, 3)$ represent the same point?

14. In the ordered pair $(-8, 6)$, is -8 associated with the horizontal or the vertical axis?

15. To plot the point with coordinates $(6, -3.5)$, we start at the ______ and move 6 units to the _____ and then 3.5 units ______.

16. To plot the point with coordinates $\left(-6, \frac{3}{2}\right)$, we start at the ______ and move 6 units to the ____ and then $\frac{3}{2}$ units ____.

17. In which quadrant do points with a negative x-coordinate and a positive y-coordinate lie?

18. In which quadrant do points with a positive x-coordinate and a negative y-coordinate lie?

19. Use the graph to complete the table.

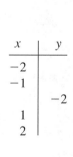

x	y
0	
1	
2	
	3

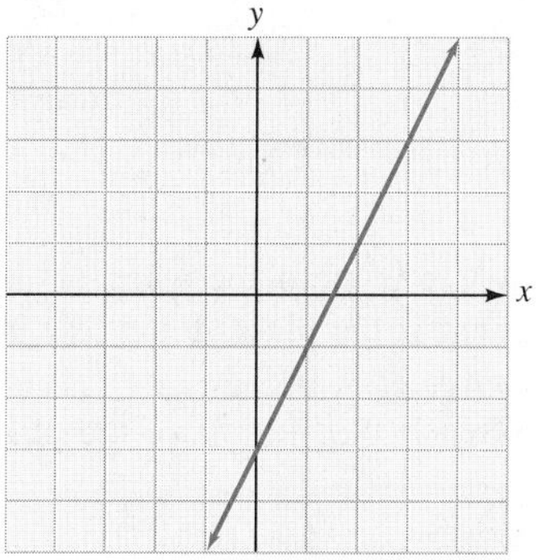

20. Use the graph to complete the table.

x	y
-2	
-1	
	-2
1	
2	

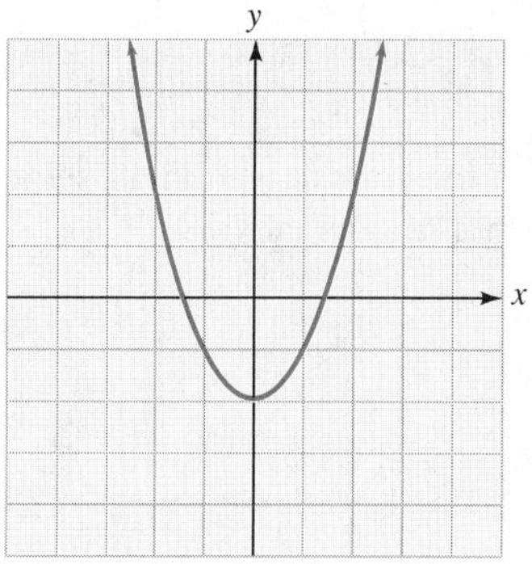

PRACTICE *In Exercises 21–28, plot each point on the rectangular coordinate system shown in Illustration 1.*

21. $A(4, 3)$

22. $B(-2, 1)$

23. $C(3, -2)$

24. $D(-2, -3)$

25. $E(0, 5)$

26. $F(-4, 0)$

27. $G(2, 0)$

28. $H(0, 3)$

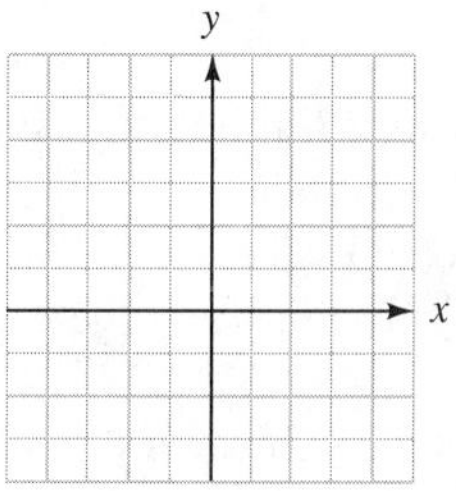

ILLUSTRATION 1

In Exercises 29–36, give the coordinates of each point shown in Illustration 2.

29. A **30.** B

31. C **32.** D

33. E **34.** F

35. G **36.** H

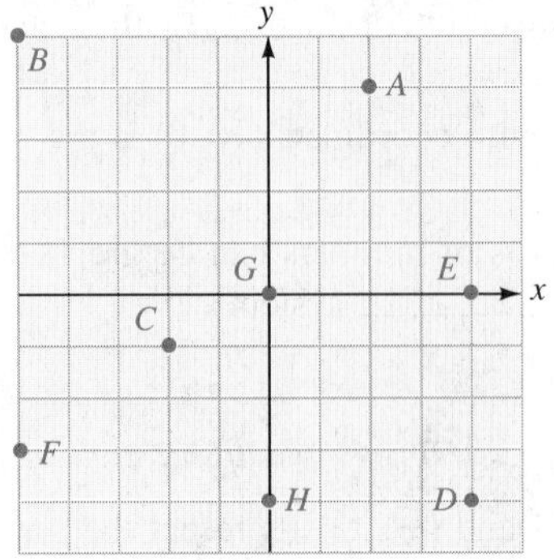

ILLUSTRATION 2

37. Plot the points given in the table and connect them with a line.

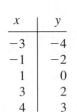

x	y
−3	−4
−1	−2
1	0
3	2
4	3

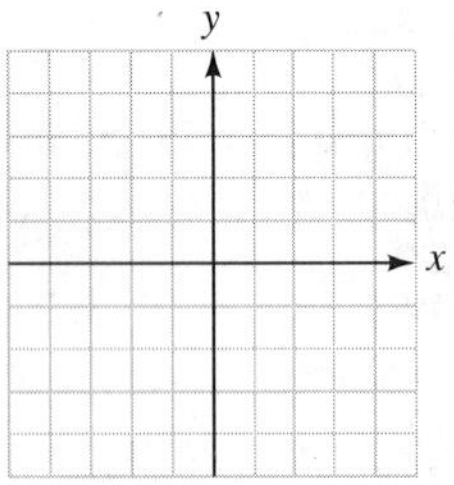

38. Plot the points given in the table and connect them with a smooth curve.

x	y
−2	3
−1	1
0	0
1	1
2	3

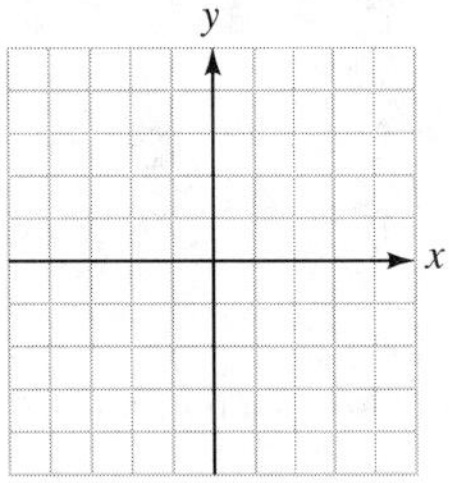

39. The graph shown in Illustration 3 shows the depths of a submarine at certain times.

a. Where is the sub when $t = 2$?

b. What is the sub doing as t increases from $t = 2$ to $t = 3$?

c. How deep is the sub when $t = 4$?

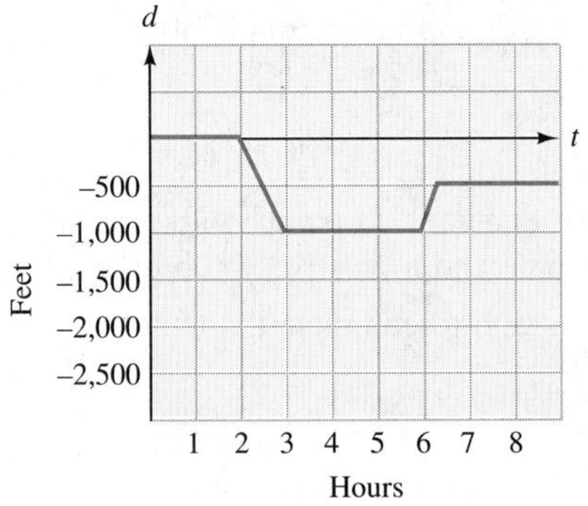

ILLUSTRATION 3

40. The graph shown in Illustration 4 shows the altitudes of a plane at certain times.

a. Where is the plane when $t = 0$?

b. What is the plane doing as t increases from $t = 1$ to $t = 2$?

c. What is the altitude of the plane when $t = 2$?

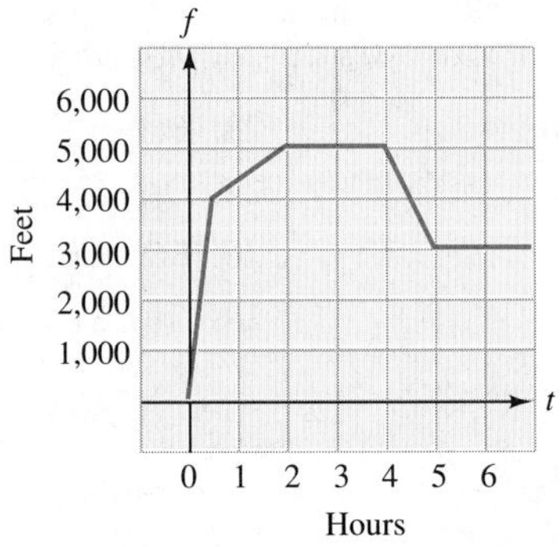

ILLUSTRATION 4

In Exercises 41–46, refer to Illustration 5.

41. Which runner ran faster at the start of the race?

42. Which runner stopped to rest first?

43. Which runner dropped the baton and had to go back and get it?

44. At what times is runner 1 stopped and runner 2 running?

45. Describe what is happening at time D.

46. Which runner won the race?

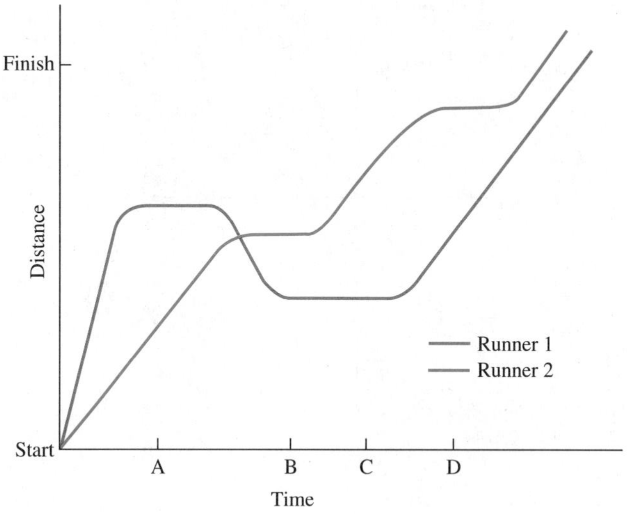

ILLUSTRATION 5

In Exercises 47–48, the marital status of ten faculty members is as follows: single, married, married, widowed, married, divorced, married, divorced, single, divorced.

47. Complete the following frequency table.

Marital status	Frequency
Single	
Married	
Divorced	
Widowed	

48. Use the information in Exercise 47 to construct a bar graph.

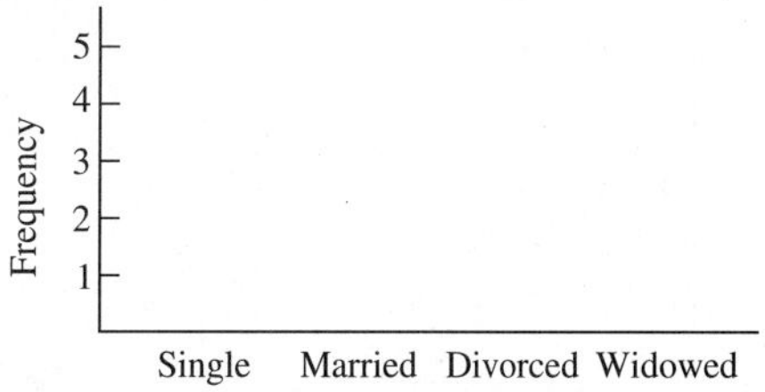

APPLICATIONS

49. Road maps Road maps have a built-in coordinate system to help locate cities. Use the map in Illustration 6 to find the coordinates of Carbondale, Chicago, and Rockford. Express each answer in the form (number, letter).

50. Earthquake damage The map shown in Illustration 7 shows the area where damage was caused by an earthquake.

a. Find the coordinates of the epicenter (the source of the quake).

b. Was damage done at the point (4, 5)?

c. Was damage done at the point $(-1, -4)$?

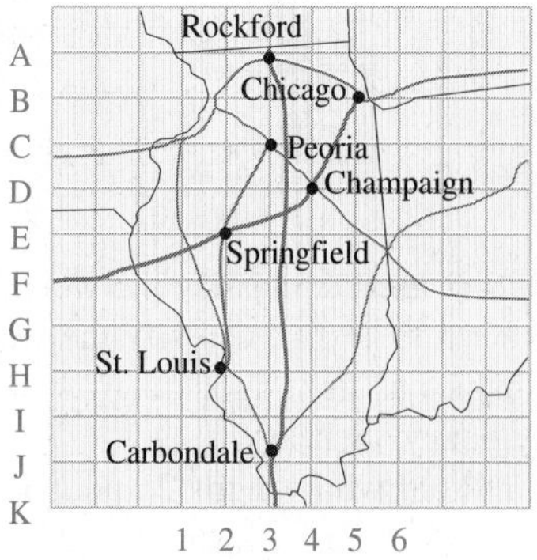

ILLUSTRATION 6

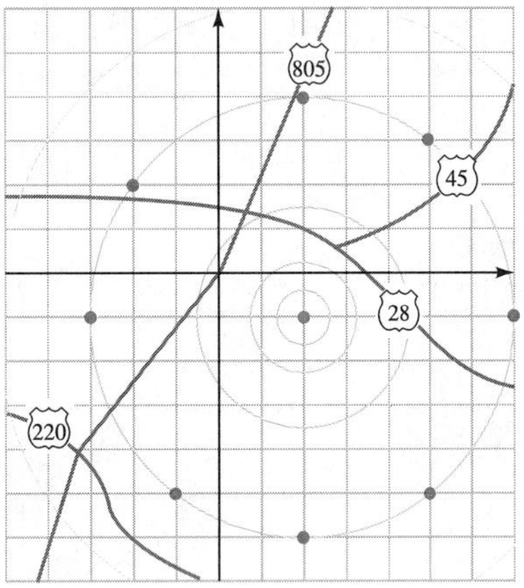

ILLUSTRATION 7

51. Robotics At a factory, a robot can be programmed to make welds on a van body. To do this, an imaginary coordinate system is superimposed on the side of the van, as shown in Illustration 8. Find the coordinates of the welds at points A, B, C, and D.

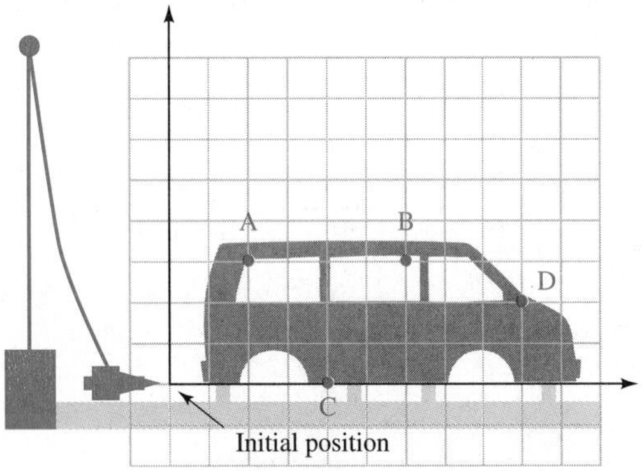

ILLUSTRATION 8

52. Illustration 9 is used to classify the weight of baby boys from birth to 1 year. Classify the weights of the following babies as light, moderately light, average, moderately heavy, or heavy.

a. A 6-month-old who weighs 21 pounds

b. A 10-month-old who weighs 29 pounds

c. A 2-month-old who weighs 9 pounds

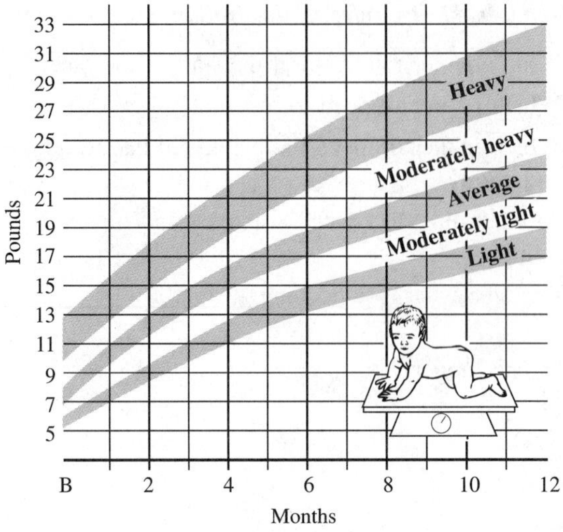

ILLUSTRATION 9

53. Video rental The charges for renting a video are shown in the graph in Illustration 10.

a. Find the charge for a 1-day rental.

b. Find the charge for a 2-day rental.

c. Find the charge if the video is kept for 5 days.

d. Find the charge if the video is kept for a week.

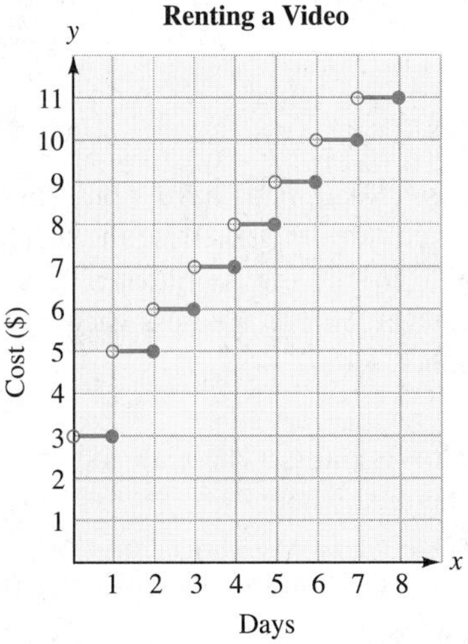

ILLUSTRATION 10

54. Postage rates The graph shown in Illustration 11 gives the first-class postage rates in 1998 for mailing parcels weighing up to 5 ounces.
  a. Find the cost of postage to mail a 3-oz letter.
  b. Find the difference in cost for a 2.75-oz letter and a 3.75-oz letter.
  c. What is the heaviest letter that can be mailed for \$1.01 first class?

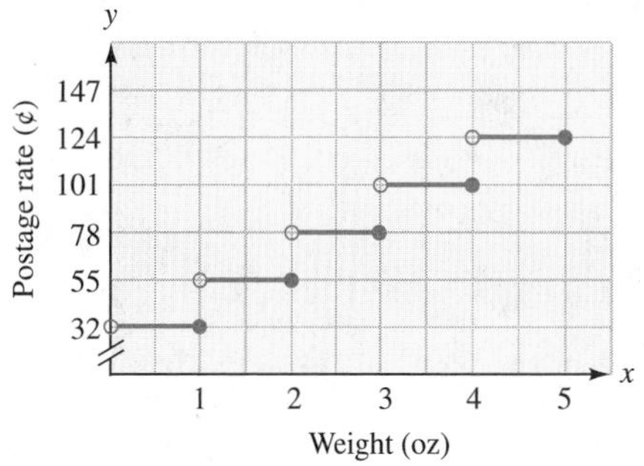

ILLUSTRATION 11

WRITING

55. Explain why the point with coordinates $(-3, 3)$ is not the same as the point with coordinates $(3, -3)$.

56. Define a discrete variable.

57. Explain how to plot the point with coordinates of $(-2, 5)$.

58. Explain why the coordinates of the origin are $(0, 0)$.

SOMETHING TO THINK ABOUT

59. Could you have a coordinate system where the coordinate axes were not perpendicular? How would it be different?

60. René Descartes is famous for saying the words, "I think, therefore I am." What do you think he meant by that?

2.2 Graphing Linear Equations

■ GRAPHING LINEAR EQUATIONS ■ HORIZONTAL AND VERTICAL LINES ■ APPLICATIONS ■ THE MIDPOINT FORMULA

Getting Ready *Find the value of* $y = 2x - 3$ *for the following values of* x.

1. $x = 0$ **2.** $x = 3$ **3.** $x = -2$ **4.** $x = -\frac{5}{2}$

■ GRAPHING LINEAR EQUATIONS

The equation $y = -\frac{1}{2}x + 4$ contains the two variables x and y. The solutions of this equation form a set of ordered pairs of real numbers. For example, the ordered pair $(-4, 6)$ is a solution, because the equation is satisfied when $x = -4$ and $y = 6$.

$$y = -\frac{1}{2}x + 4$$

$$\mathbf{6} = -\frac{1}{2}(\mathbf{-4}) + 4 \qquad \text{Substitute } -4 \text{ for } x \text{ and } 6 \text{ for } y.$$

$$6 = 2 + 4$$

$$6 = 6$$

This pair and others that satisfy the equation are listed in the table of values shown in Figure 2-11.

$y = -\frac{1}{2}x + 4$

x	y
-4	6
-2	5
0	4
2	3
4	2

FIGURE 2-11

The **graph of the equation** $y = -\frac{1}{2}x + 4$ is the graph of all points (x, y) on the rectangular coordinate system whose coordinates satisfy the equation.

EXAMPLE 1 Graph $y = -\frac{1}{2}x + 4$.

Solution To graph the equation, we plot the five ordered pairs listed in the table shown in Figure 2-11. These points appear to lie on a line, as shown in Figure 2-12. In fact, if we were to plot many more pairs that satisfied the equation, it would become obvious that the resulting points will all lie on the line.

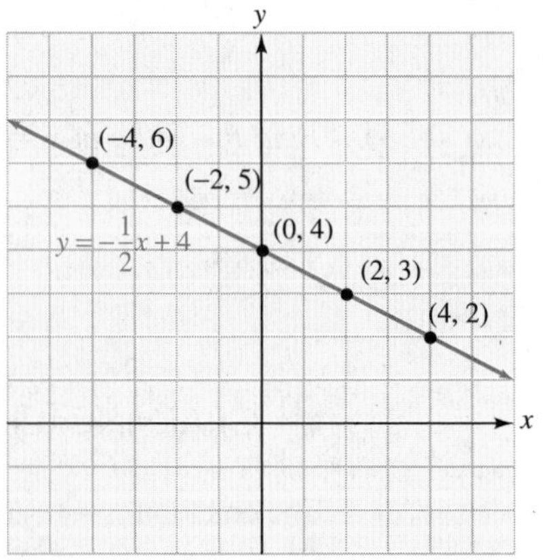

FIGURE 2-12

When we say that the graph of an equation is a line, we imply two things:

1. Every point with coordinates that satisfy the equation will lie on the line.
2. Any point on the line will have coordinates that satisfy the equation. ■

Self Check Graph $y = 2x - 3$.

Answer

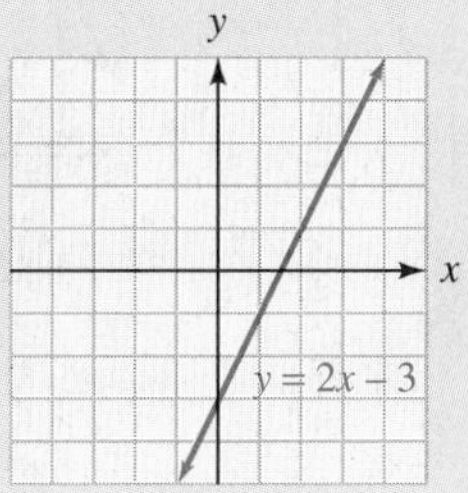

When the graph of an equation is a line, we call the equation a **linear equation.** Linear equations are often written in the form $Ax + By = C$, where A, B, and C stand for specific numbers (called **constants**) and x and y are variables.

EXAMPLE 2 Graph $3x + 2y = 12$.

Solution We can pick values for either x or y, substitute them into the equation, and solve for the other variable. For example, if $x = 2$,

$3x + 2y = 12$	
$3(2) + 2y = 12$	Substitute 2 for x.
$6 + 2y = 12$	Simplify.
$2y = 6$	Subtract 6 from both sides.
$y = 3$	Divide both sides by 2.

The ordered pair (2, 3) satisfies the equation. If $y = 6$,

$3x + 2y = 12$	
$3x + 2(6) = 12$	Substitute 6 for y.
$3x + 12 = 12$	Simplify.
$3x = 0$	Subtract 12 from both sides.
$x = 0$	Divide both sides by 3.

A second ordered pair that satisfies the equation is (0, 6).

These pairs and others that satisfy the equation are shown in Figure 2-13. After we plot each pair, we see that they lie on a line. The graph of the equation is the line shown in the figure.

$3x + 2y = 12$

x	y	(x, y)
-2	9	$(-2, 9)$
0	6	$(0, 6)$
2	3	$(2, 3)$
4	0	$(4, 0)$
6	-3	$(6, -3)$

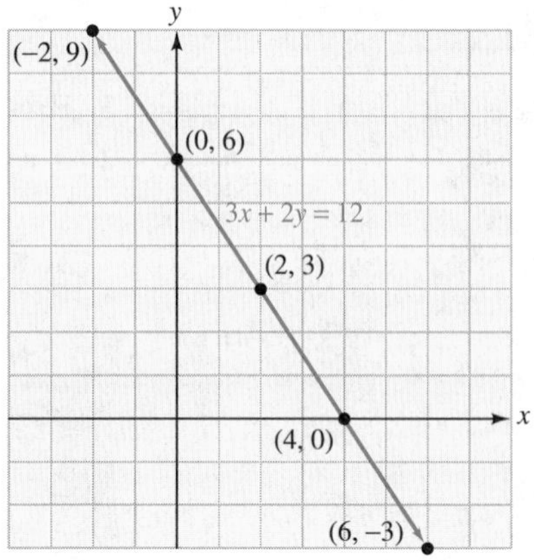

FIGURE 2-13

Self Check Graph $3x - 2y = 6$.

Answer

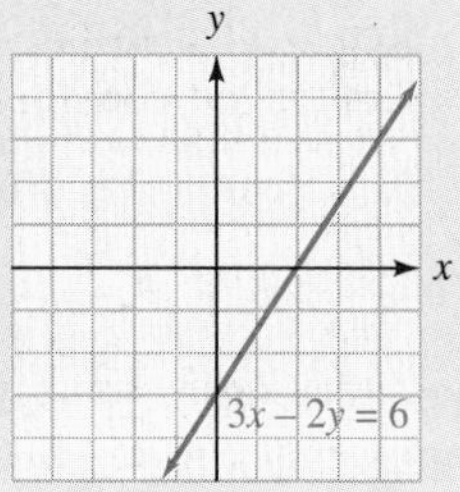

In Example 2, the graph crossed the y-axis at the point with coordinates (0, 6), called the **y-intercept,** and crossed the x-axis at the point with coordinates (4, 0), called the **x-intercept.** In general, we have the following definitions.

Intercepts of a Line

The **y-intercept** of a line is the point $(0, b)$ where the line intersects the y-axis. To find b, substitute 0 for x in the equation of the line and solve for y.

The **x-intercept** of a line is the point $(a, 0)$ where the line intersects the x-axis. To find a, substitute 0 for y in the equation of the line and solve for x.

EXAMPLE 3 Use the x- and y-intercepts to graph $2x + 5y = 10$.

Solution To find the y-intercept, we substitute 0 for x and solve for y:

$$2x + 5y = 10$$
$$2(\mathbf{0}) + 5y = 10 \quad \text{Substitute 0 for } x.$$
$$5y = 10 \quad \text{Simplify.}$$
$$y = 2 \quad \text{Divide both sides by 5.}$$

The y-intercept is the point (0, 2). To find the x-intercept, we substitute 0 for y and solve for x:

$$2x + 5y = 10$$
$$2x + 5(0) = 10 \quad \text{Substitute 0 for } y.$$
$$2x = 10 \quad \text{Simplify.}$$
$$x = 5 \quad \text{Divide both sides by 2.}$$

The x-intercept is the point (5, 0).

Although two points are enough to draw a line, it is a good idea to find and plot a third point as a check. To find the coordinates of a third point, we can substitute any convenient number (such as -5) for x and solve for y:

$$2x + 5y = 10$$
$$2(-5) + 5y = 10 \quad \text{Substitute } -5 \text{ for } x.$$
$$-10 + 5y = 10 \quad \text{Simplify.}$$
$$5y = 20 \quad \text{Add 10 to both sides.}$$
$$y = 4 \quad \text{Divide both sides by 5.}$$

The line will also pass through the point $(-5, 4)$.

A table of ordered pairs and the graph of $2x + 5y = 10$ are shown in Figure 2-14.

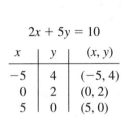

$2x + 5y = 10$

x	y	(x, y)
−5	4	(−5, 4)
0	2	(0, 2)
5	0	(5, 0)

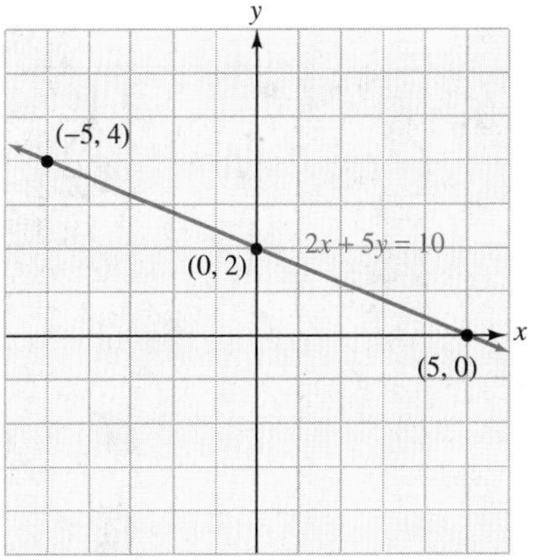

FIGURE 2-14

Self Check Find the x- and y-intercepts and graph $y = 3x + 4$.

Answer

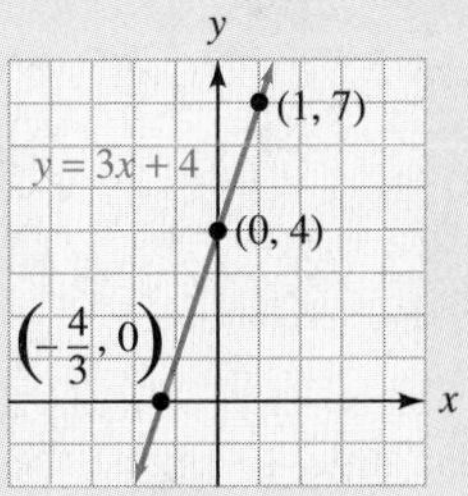

The method of graphing a line used in Example 3 is called the **intercept method.**

HORIZONTAL AND VERTICAL LINES

EXAMPLE 4 Graph **a.** $y = 3$ and **b.** $x = -2$.

Solution **a.** Since the equation $y = 3$ does not contain x, the numbers chosen for x have no effect on y. The value of y is always 3.

After plotting the pairs (x, y) shown in Figure 2-15, we see that the graph is a horizontal line with a y-intercept of $(0, 3)$. The line has no x-intercept.

b. Since the equation $x = -2$ does not contain y, the value of y can be any number.

After plotting the pairs (x, y) shown in Figure 2-15, we see that the graph is a vertical line with an x-intercept of $(-2, 0)$. The line has no y-intercept.

$y = 3$

x	y	(x, y)
−3	3	(−3, 3)
0	3	(0, 3)
2	3	(2, 3)
4	3	(4, 3)

$x = -2$

x	y	(x, y)
−2	−2	(−2, −2)
−2	0	(−2, 0)
−2	2	(−2, 2)
−2	6	(−2, 6)

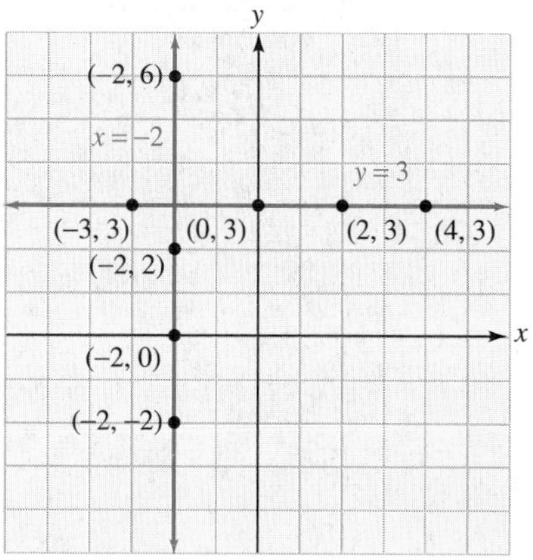

FIGURE 2-15

Self Check Graph $x = 4$ and $y = -3$ on one set of coordinate axes.

Answer

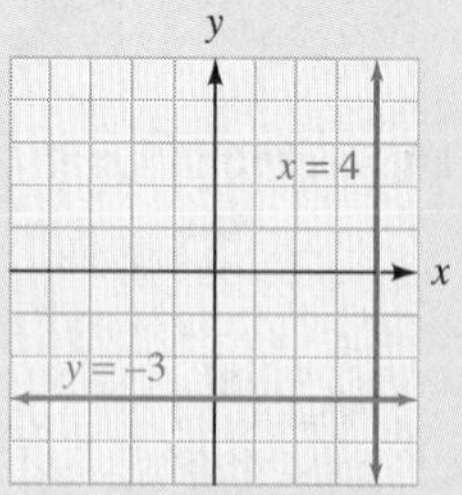

The results of Example 4 suggest the following facts.

Horizontal and Vertical Lines

If a and b are real numbers, then

The graph of $x = a$ is a vertical line with x-intercept at $(a, 0)$. If $a = 0$, the line is the y-axis.

The graph of $y = b$ is a horizontal line with y-intercept at $(0, b)$. If $b = 0$, the line is the x-axis.

APPLICATIONS

EXAMPLE 5 The following table gives the number of miles (y) that a bus can be driven on x gallons of gas. Plot the ordered pairs and estimate how far the bus can be driven on 9 gallons.

x	2	3	4	5	6
y	12	18	24	30	36

Solution Since the distances driven are rather large numbers, we plot the points on a coordinate system where the unit distance on the x-axis is larger than the unit distance on the y-axis. After plotting each ordered pair as in Figure 2-16, we see that the points lie on a line.

To estimate how far the bus can go on 9 gallons, we find 9 on the x-axis, move up to the graph, and then move to the left to locate a y-value of 54. Thus, the bus can be driven approximately 54 miles on 9 gallons of gas.

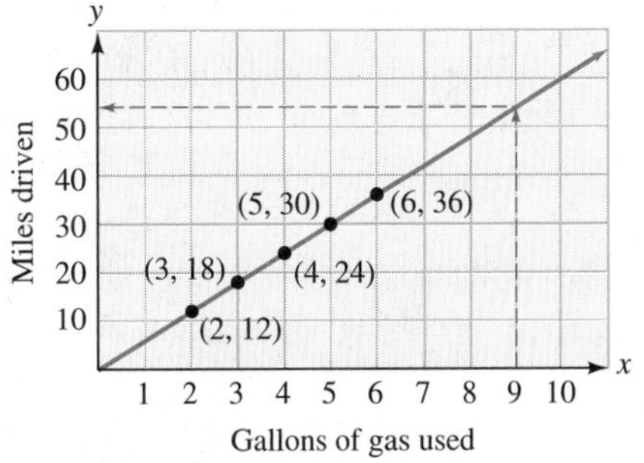

FIGURE 2-16

Self Check In Example 5, how far can the bus go on 10 gallons?

Answer 60 miles

EXAMPLE 6

Depreciation A copy machine purchased for \$6,750 is expected to depreciate according to the formula $y = -950x + 6{,}750$, where y is the value of the copier after x years. When will the copier have no value?

Solution The copier will have no value when its value (y) is 0. To find x when $y = 0$, we substitute 0 for y and solve for x.

$$y = -950x + 6{,}750$$
$$0 = -950x + 6{,}750$$
$$-6{,}750 = -950x \quad \text{Subtract 6,750 from both sides.}$$
$$7.105263158 = x \quad \text{Divide both sides by } -950.$$

The copier will be worthless in about 7.1 years. ■

Self Check In Example 6, when will the copier be worth \$2,000?

Answer 5 years

THE MIDPOINT FORMULA

To distinguish between the coordinates of two points on a line, we often use subscript notation. Point $P(x_1, y_1)$ is read as "point P with coordinates of x sub 1 and y sub 1." Point $Q(x_2, y_2)$ is read as "point Q with coordinates of x sub 2 and y sub 2."

If point M in Figure 2-17 lies midway between points $P(x_1, y_1)$ and $Q(x_2, y_2)$, point M is called the **midpoint** of segment PQ. To find the coordinates of M, we find the mean of the x-coordinates and the mean of the y-coordinates of P and Q.

The Midpoint Formula

The **midpoint** of the line segment with endpoints at $P(x_1, y_1)$ and $Q(x_2, y_2)$ is the point M with coordinates of

$$\left(\frac{x_1 + x_2}{2}, \frac{y_1 + y_2}{2}\right)$$

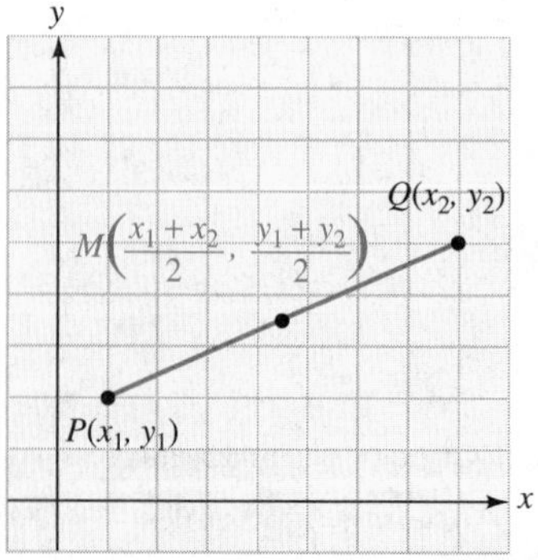

FIGURE 2-17

EXAMPLE 7 Find the midpoint of the line segment joining $P(-2, 3)$ and $Q(3, -5)$.

Solution To find the midpoint, we find the mean of the x-coordinates and the mean of the y-coordinates to get

$$\frac{x_1 + x_2}{2} = \frac{-2 + 3}{2} = \frac{1}{2} \qquad \text{and} \qquad \frac{y_1 + y_2}{2} = \frac{3 + (-5)}{2} = -1$$

The midpoint of segment PQ is the point $M\left(\frac{1}{2}, -1\right)$. ■

Self Check Find the midpoint of the segment joining $P(5, -4)$ and $Q(-3, 5)$.

Answer $\left(1, \frac{1}{2}\right)$

Graphing Lines

GRAPHING CALCULATORS

We have graphed linear equations by finding ordered pairs, plotting points, and drawing a line through those points. Graphing is much easier if we use a graphing calculator.

Graphing calculators have a window to display graphs (see Figure 2-18.) To see the proper picture of a graph, we must decide on the minimum and maximum values for the x- and y-coordinates. A window with standard settings of

$$\text{Xmin} = -10 \qquad \text{Xmax} = 10 \qquad \text{Ymin} = -10 \qquad \text{Ymax} = 10$$

will produce a graph where the value of x is in the interval $[-10, 10]$ and y is in the interval $[-10, 10]$.

To graph $3x + 2y = 12$, we must first solve the equation for y.

$$3x + 2y = 12$$

$$2y = -3x + 12 \qquad \text{Subtract } 3x \text{ from both sides.}$$

$$y = -\frac{3}{2}x + 6 \qquad \text{Divide both sides by 2.}$$

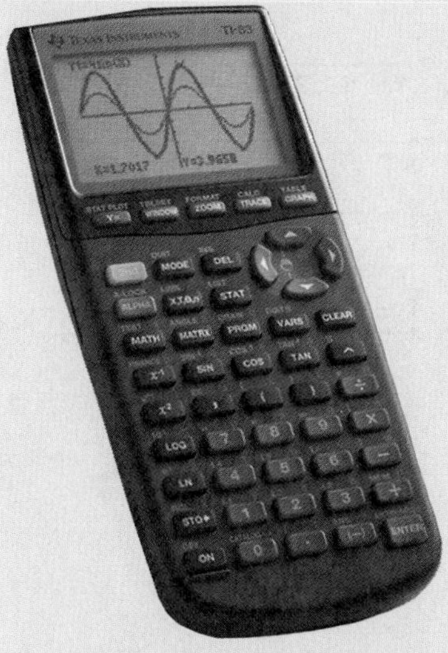

FIGURE 2-18
(Courtesy of Texas Instruments)

To graph the equation, we enter the right-hand side of the equation after a symbol like $\backslash Y_1 =$ or $f(x) =$. After entering the right-hand side, the display should look like

$$\backslash Y_1 = -(3/2)X + 6 \qquad \text{or} \qquad f(x) = -(3/2)X + 6$$

We then press the GRAPH key to get the graph shown in Figure 2-19(a). To show more detail, we can draw the graph in a different window. A window with settings of $[-1, 5]$ for x and $[-2, 7]$ for y will give the graph shown in Figure 2-19(b).

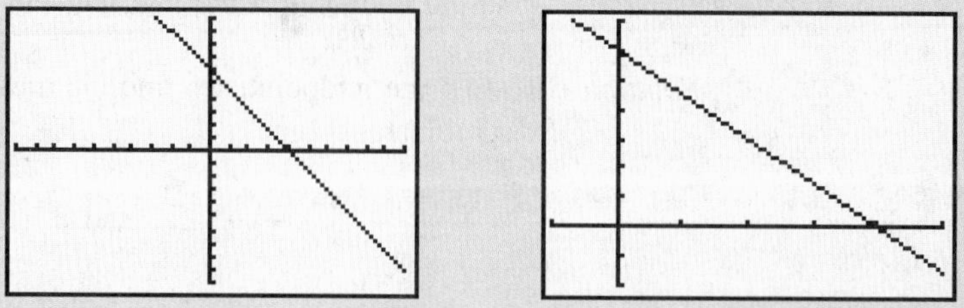

(a) (b)

FIGURE 2-19

We can use a trace command to find the coordinates of any point on a graph. For example, to find the x-intercept of the graph of $2y = -5x - 7$, we solve the equation for y and graph it using the standard window settings of $[-10, 10]$ for x and $[-10, 10]$ for y. We then press the TRACE key to get Figure 2-20(a). We can then use the $>$ and $<$ keys to move the cursor along the line toward the x-intercept until we arrive at a point with the coordinates shown in Figure 2-20(b).

To get better results, we can zoom in to get a magnified picture, trace again, and move the cursor to the point with coordinates shown in Figure 2-20(c). Since the y-coordinate is nearly 0, this point is nearly the x-intercept. We can achieve better results with more zooms.

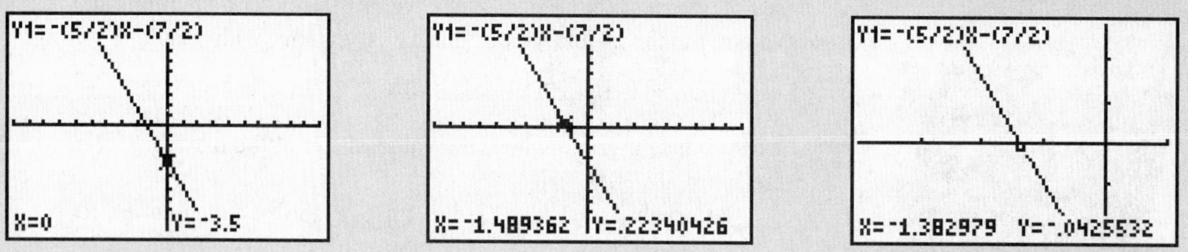

(a) (b) (c)

FIGURE 2-20

Orals *Find the x- and y-intercepts of each line.*

1. $x + y = 3$

2. $3x + y = 6$

3. $x + 4y = 8$

4. $3x - 4y = 12$

Tell whether the graphs of the equations are horizontal or vertical.

5. $x = -6$

6. $y = 8$

Find the midpoint of a line segment with endpoints at

7. $(2, 4), (6, 8)$

8. $(-4, 6), (4, -8)$

EXERCISE 2.2

REVIEW *Do the operations*

1. $4 + 2 \cdot 3^2$

2. $2^2 + 3 \cdot 4$

3. $\dfrac{4(3 + 2) - 5}{12 - 2(3)}$

4. $\dfrac{(4^2 - 3) + 7}{5(2 + 3) - 1}$

5. Write the prime numbers between 10 and 30.

6. Write the first ten composite numbers.

VOCABULARY AND CONCEPTS *Fill in each blank to make a true statement.*

7. The graph of an equation is the graph of all points (x, y) on the rectangular coordinate system whose coordinates ______ the equation.

8. Any equation whose graph is a line is called a ______ equation.

9. The point where a graph intersects the y-axis is called ____________.

10. The point where a graph intersects the ______ is called the x-intercept.

11. The graph of any equation of the form $x = a$ is a ______ line.

12. The graph of any equation of the form $y = b$ is a ________ line.

13. The symbol x_1 is read as "x ______."

14. The midpoint of a line segment joining (a, b) and (c, d) is given by the formula ________.

PRACTICE *In Exercises 15–18, complete the table of solutions for each equation.*

15. $y = -x + 4$

x	y
-1	
0	
2	

16. $y = x - 2$

x	y
-2	
0	
4	

17. $y = 2x - 3$

x	y
-1	
0	
3	

18. $y = -\dfrac{1}{2}x + \dfrac{5}{2}$

x	y
-3	
-1	
3	

In Exercises 19–22, graph each equation. See Exercises 15–18.

19. $y = -x + 4$

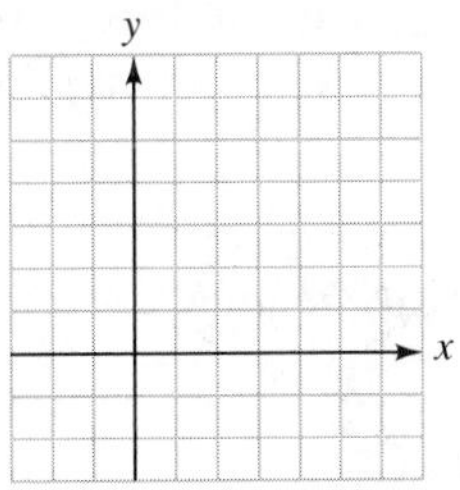

20. $y = x - 2$

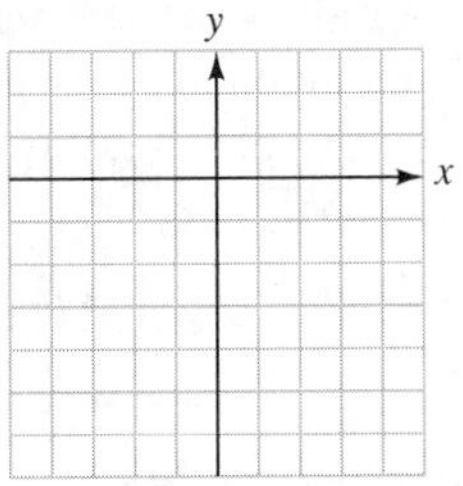

21. $y = 2x - 3$

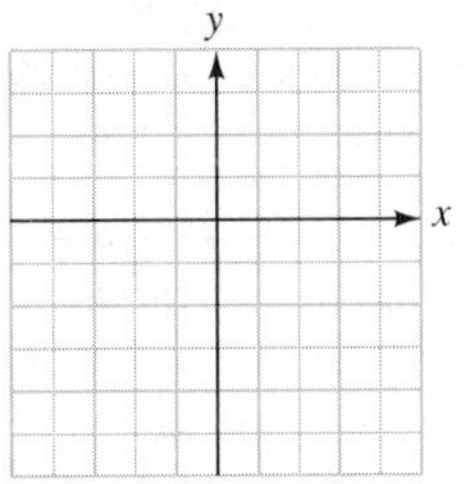

22. $y = -\dfrac{1}{2}x + \dfrac{5}{2}$

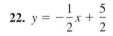

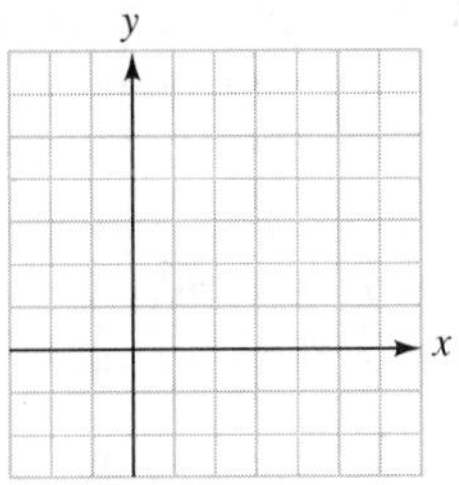

In Exercises 23–34, graph each equation.

23. $3x + 4y = 12$

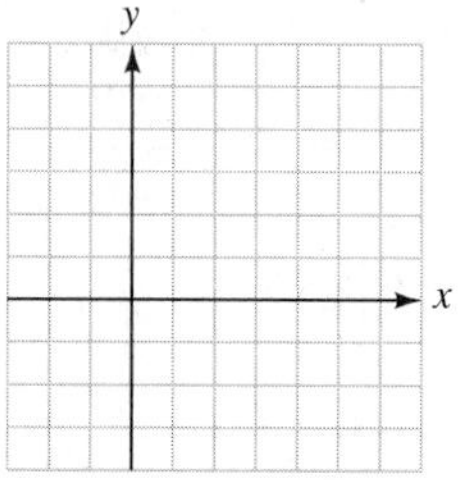

24. $4x - 3y = 12$

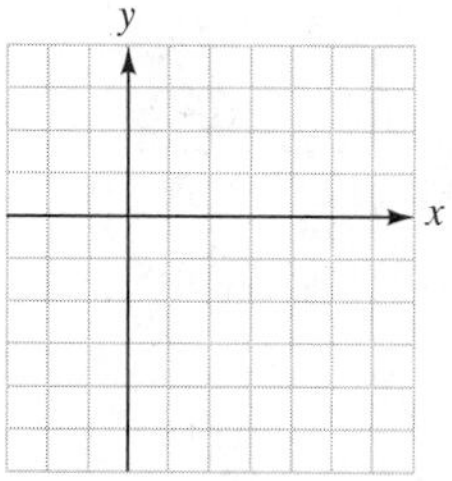

25. $y = -3x + 2$

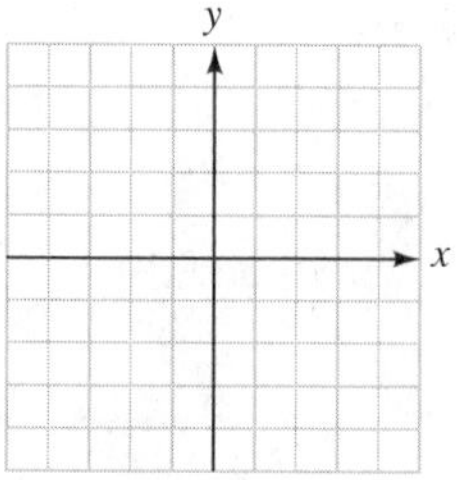

26. $y = 2x - 3$

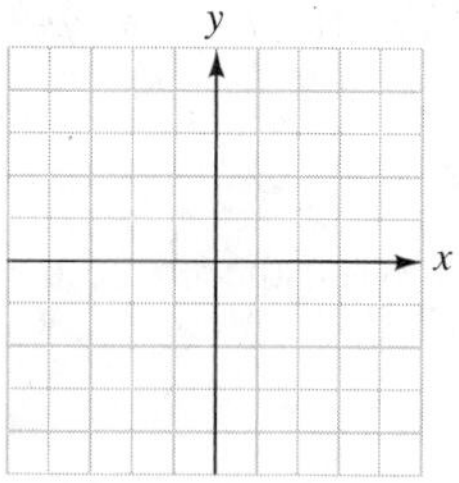

27. $3y = 6x - 9$

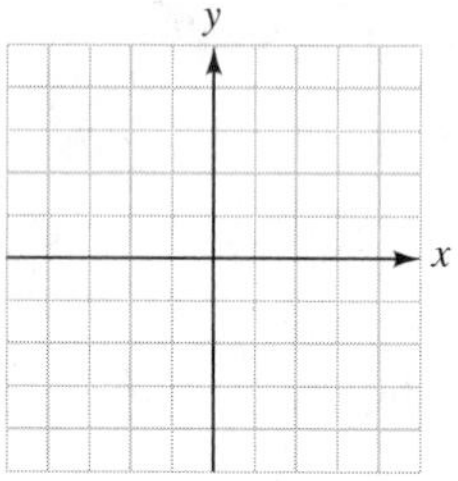

28. $2x = 4y - 10$

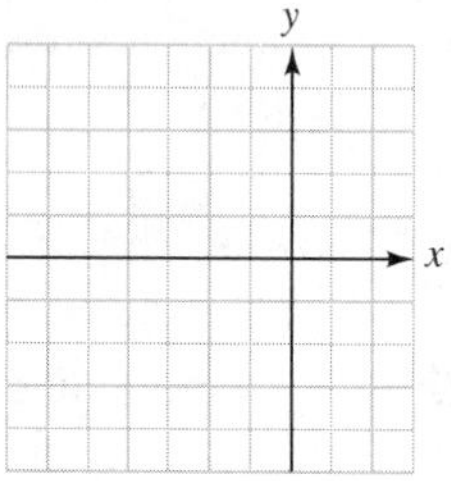

29. $3x + 4y - 8 = 0$

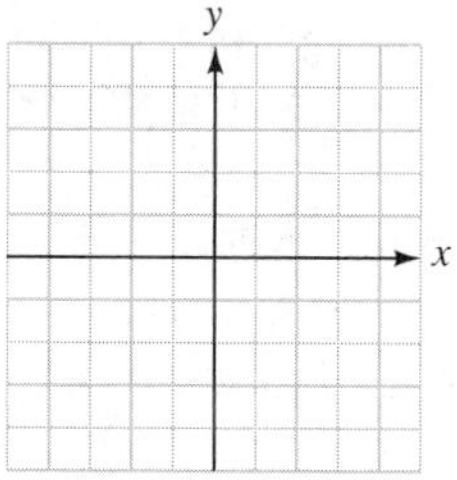

30. $-2y - 3x + 9 = 0$

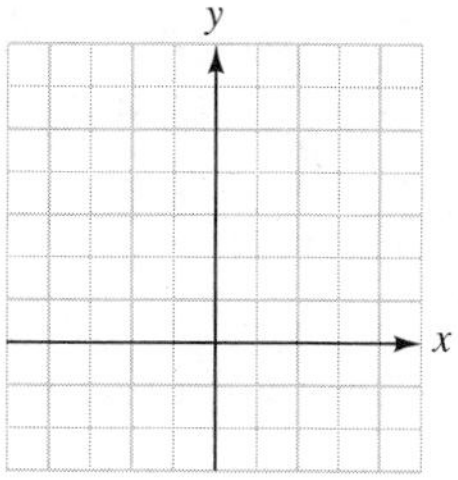

31. $x = 3$

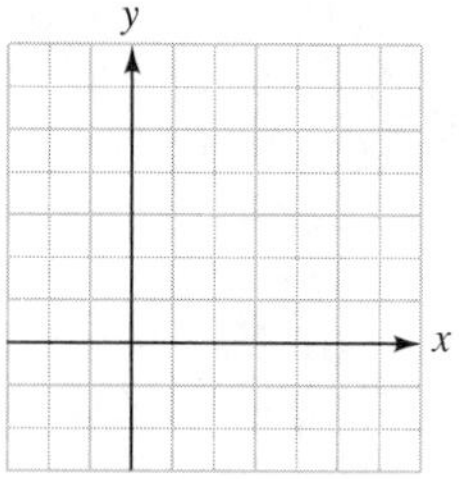

32. $y = -4$

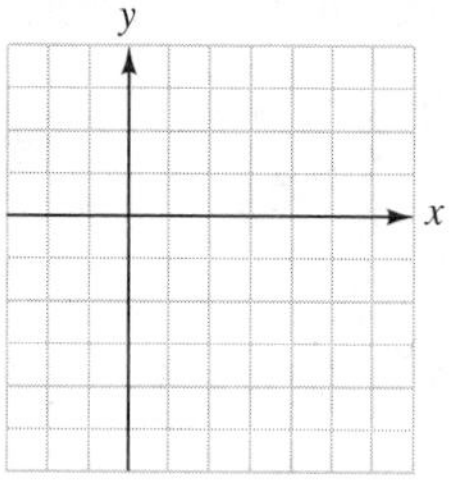

33. $-3y + 2 = 5$

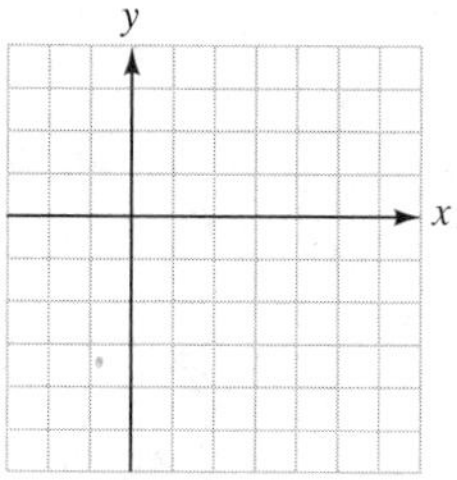

34. $-2x + 3 = 11$

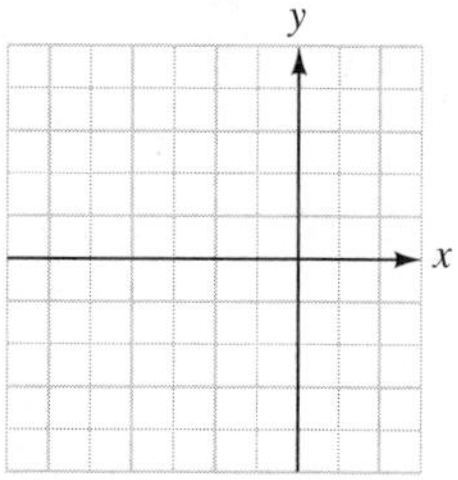

In Exercises 35–44, find the midpoint of segment PQ.

35. $P(0, 0)$, $Q(6, 8)$

36. $P(10, 12)$, $Q(0, 0)$

37. $P(6, 8)$, $Q(12, 16)$

38. $P(10, 4)$, $Q(2, -2)$

39. $P(2, 4)$, $Q(5, 8)$

40. $P(5, 9)$, $Q(8, 13)$

41. $P(-2, -8)$, $Q(3, 4)$

42. $P(-5, -2)$, $Q(7, 3)$

43. $Q(-3, 5)$, $P(-5, -5)$

44. $Q(2, -3)$, $P(4, -8)$

45. Finding the endpoint of a segment If $M(-2, 3)$ is the midpoint of segment PQ and the coordinates of P are $(-8, 5)$, find the coordinates of Q.

46. Finding the endpoint of a segment If $M(6, -5)$ is the midpoint of segment PQ and the coordinates of Q are $(-5, -8)$, find the coordinates of P.

In Exercises 47–50, use a graphing calculator to graph each equation, and then find the x-coordinate of the x-intercept to the nearest hundredth.

47. $y = 3.7x - 4.5$

48. $y = \dfrac{3}{5}x + \dfrac{5}{4}$

49. $1.5x - 3y = 7$

50. $0.3x + y = 7.5$

APPLICATIONS

51. Hourly wages The following table gives amount y (in dollars) that a student can earn for working x hours. Plot the ordered pairs and estimate how much the student will earn for working 8 hours.

x	2	4	5	6
y	12	24	30	36

52. Distance traveled The following table shows how far y (in miles) a biker can travel in x hours. Plot the ordered pairs and estimate how far the biker can go in 8 hours.

x	2	4	5	6
y	30	60	75	90

53. Value of a car The following table shows the value y (in dollars) of a car that is x years old. Plot the ordered pairs and estimate the value of the car when it is 4 years old.

x	0	1	3
y	15,000	12,000	6,000

54. Earning interest The following table shows the amount y (in dollars) in a bank account drawing simple interest left on deposit for x years. Plot the ordered pairs and estimate the value of the account in 6 years.

x	0	1	4
y	1,000	1,050	1,200

55. House appreciation A house purchased for \$125,000 is expected to appreciate according to the formula $y = 7{,}500x + 125{,}000$, where y is the value of the house after x years. Find the value of the house 5 years later.

56. Car depreciation A car purchased for \$17,000 is expected to depreciate according to the formula $y = -1{,}360x + 17{,}000$, where y is the value of the car after x years. When will the car be worthless?

57. Demand equation The number of television sets that consumers buy depends on price. The higher the price, the fewer TVs people will buy. The equation that relates price to the number of TVs sold at that price is called a **demand equation.** If the demand equation for a 13-inch TV is $p = -\frac{1}{10}q + 170$, where p is the price and q is the number of TVs sold at that price, how many TVs will be sold at a price of \$150?

58. Supply equation The number of television sets that manufacturers produce depends on price. The higher the price, the more TVs manufacturers will produce. The equation that relates price to the number of TVs produced at that price is called a **supply equation.** If the supply equation for a 13-inch TV is $p = \frac{1}{10}q + 130$, where p is the price and q is the number of TVs produced for sale at that price, how many TVs will be produced if the price is \$150?

59. TV coverage In Illustration 1, a TV camera is located at point $(-3, 0)$. The camera is to follow the launch of a space shuttle. As the shuttle rises vertically, the camera can tilt back to a line of sight given by $y = 2x + 6$. Estimate how many miles the shuttle will remain in the camera's view.

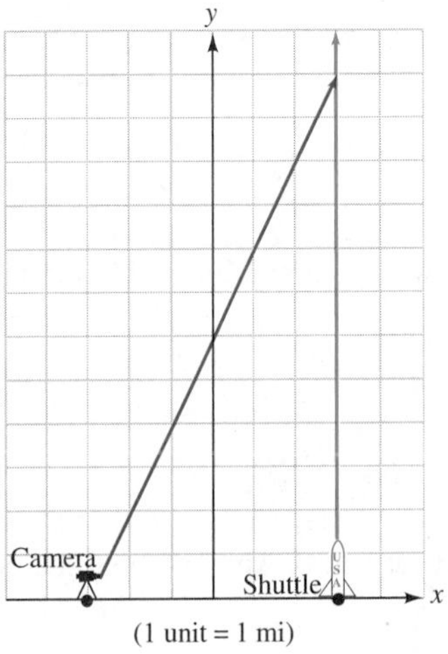

ILLUSTRATION 1

60. Buying tickets Tickets to the circus cost \$5 each from Ticketron plus a \$2 service fee for each block of tickets.

a. Write a linear equation that gives the cost y for a student buying x tickets.

b. Complete the table in Illustration 2 and graph the equation.

c. Use the graph to estimate the cost of buying 5 tickets.

x	y
1	
2	
3	
4	

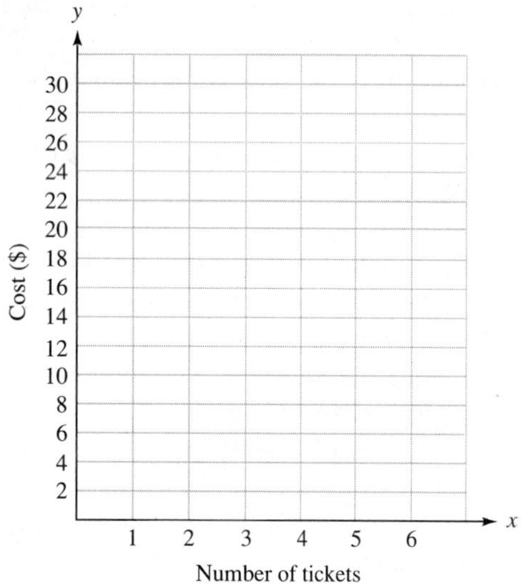

ILLUSTRATION 2

61. Telephone costs In one community, the monthly cost of local telephone service is \$5 per month, plus 25¢ per call.

a. Write a linear equation that gives the cost y for a person making x calls.

b. Complete the table in Illustration 3 and graph the equation.

c. Use the graph to estimate the cost of service in a month when 20 calls were made.

x	y
4	
8	
12	
16	

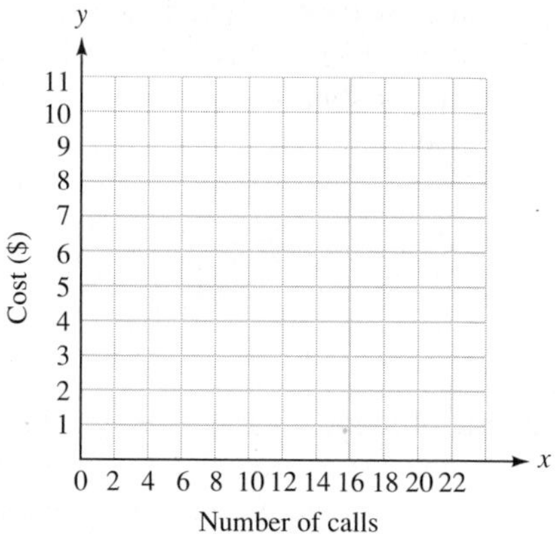

ILLUSTRATION 3

62. Crime prevention The number n of incidents of family violence requiring police response appears to be related to d, the money spent on crisis intervention, by the equation $n = 430 - 0.005d$. What expenditure would reduce the number of incidents to 350?

WRITING

63. Explain how to graph a line using the intercept method.

64. Explain how to determine the quadrant in which the point $P(a, b)$ lies.

SOMETHING TO THINK ABOUT

65. If the line $y = ax + b$ passes only through quadrants I and II, what can be known about the constants a and b?

66. What are the coordinates of the three points that divide the line segment joining $P(a, b)$ and $Q(c, d)$ into four equal parts?

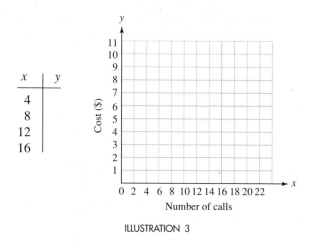

2.3 Slope of a Nonvertical Line

■ SLOPE OF A LINE ■ INTERPRETATION OF SLOPE ■ HORIZONTAL AND VERTICAL LINES ■ SLOPES OF PARALLEL LINES ■ SLOPES OF PERPENDICULAR LINES

Getting Ready *Evaluate each expression.*

1. $\dfrac{6-3}{8-5}$
2. $\dfrac{10-4}{2-8}$
3. $\dfrac{25-12}{9-(-5)}$
4. $\dfrac{-9-(-6)}{-4-10}$

■ SLOPE OF A LINE

A service offered by a computer online company costs \$2 per month plus \$3 for each hour of connect time. The table shown in Figure 2-21(a) gives the cost y for certain numbers of hours x of connect time. If we construct a graph from this data, we get the line shown in Figure 2-21(b).

Hours of connect time						
x	0	1	2	3	4	5
y	2	5	8	11	14	17
Cost						

(a)

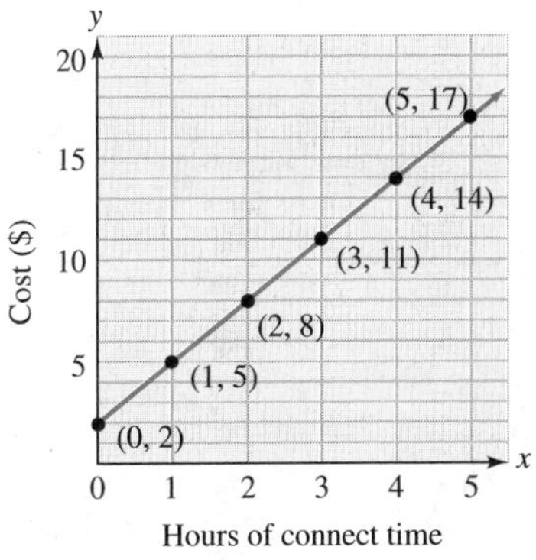

(b)

FIGURE 2-21

From the graph, we can see that if x changes from 0 to 1, y changes from 2 to 5. As x changes from 1 to 2, y changes from 5 to 8, and so on. The ratio of the change in y divided by the change in x is the constant 3.

$$\frac{\text{Change in } y}{\text{Change in } x} = \frac{5-2}{1-0} = \frac{8-5}{2-1} = \frac{11-8}{3-2} = \frac{14-11}{4-3} = \frac{17-14}{5-4} = \frac{3}{1} = 3$$

The ratio of the change in y divided by the change in x between any two points on any line is always a constant. This constant rate of change is called the **slope** of the line.

Slope of a Nonvertical Line

The **slope of the nonvertical line** passing through points $P(x_1, y_1)$ and $Q(x_2, y_2)$ is

$$m = \frac{\text{change in } y}{\text{change in } x} = \frac{y_2 - y_1}{x_2 - x_1} \quad (x_2 \neq x_1)$$

EXAMPLE 1 Find the slope of the line shown in Figure 2-22.

Solution We can let $P(x_1, y_1) = P(-2, 4)$ and $Q(x_2, y_2) = Q(3, -4)$. Then

$$m = \frac{\text{change in } y}{\text{change in } x}$$

$$= \frac{y_2 - y_1}{x_2 - x_1}$$

$$= \frac{\mathbf{-4} - \mathbf{4}}{\mathbf{3} - (\mathbf{-2})} \quad \text{Substitute } -4 \text{ for } y_2,\ 4 \text{ for } y_1,\ 3 \text{ for } x_2, \text{ and } -2 \text{ for } x_1.$$

$$= \frac{-8}{5}$$

$$= -\frac{8}{5}$$

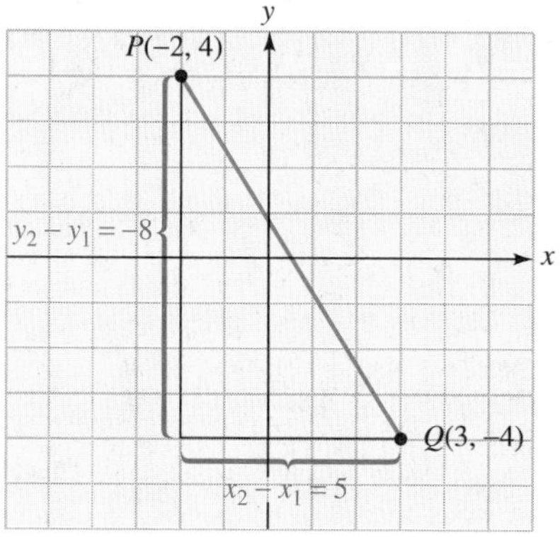

FIGURE 2-22

The slope of the line is $-\frac{8}{5}$. We would obtain the same result if we let $P(x_1, y_1) = P(3, -4)$ and $Q(x_2, y_2) = Q(-2, 4)$. ■

Self Check Find the slope of the line joining the points $(-3, 6)$ and $(4, -8)$.

Answer -2

WARNING! When calculating slope, always subtract the y values and the x values in the same order.

$$m = \frac{y_2 - y_1}{x_2 - x_1} \quad \text{or} \quad m = \frac{y_1 - y_2}{x_1 - x_2}$$

However,

$$m \neq \frac{y_2 - y_1}{x_1 - x_2} \quad \text{and} \quad m \neq \frac{y_1 - y_2}{x_2 - x_1}$$

The change in y (often denoted as Δy) is the **rise** of the line between points P and Q. The change in x (often denoted as Δx) is the **run.** Using this terminology, we can define slope to be the ratio of the rise to the run:

$$m = \frac{\Delta y}{\Delta x} = \frac{\text{rise}}{\text{run}} \quad (\Delta x \neq 0)$$

EXAMPLE 2 Find the slope of the line determined by $3x - 4y = 12$.

Solution We first find the coordinates of two points on the line.

- If $x = 0$, then $y = -3$. The point $(0, -3)$ is on the line.
- If $y = 0$, then $x = 4$. The point $(4, 0)$ is on the line.

We then refer to Figure 2-23 and find the slope of the line between $P(0, -3)$ and $Q(4, 0)$ by substituting 0 for y_2, -3 for y_1, 4 for x_2, and 0 for x_1 in the formula for slope.

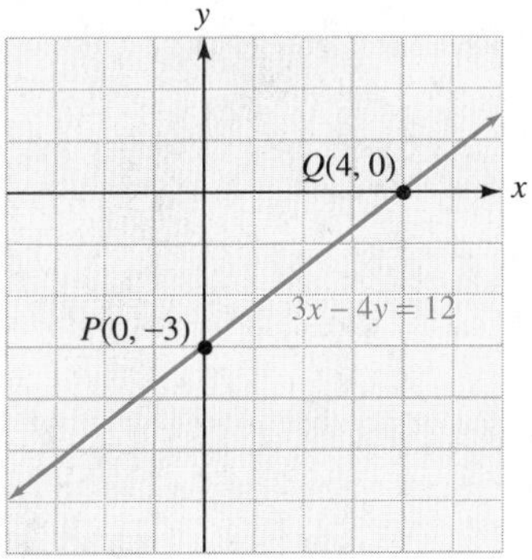

FIGURE 2-23

$$m = \frac{\text{change in } y}{\text{change in } x}$$

$$= \frac{y_2 - y_1}{x_2 - x_1}$$

$$= \frac{0 - (-3)}{4 - 0}$$

$$= \frac{3}{4}$$

The slope of the line is $\frac{3}{4}$. ■

Self Check Find the slope of the line determined by $2x + 5y = 12$.

Answer $-\frac{2}{5}$

■ INTERPRETATION OF SLOPE

Many applied problems involve equations of lines and their slopes.

EXAMPLE 3 **Cost of carpet** A store sells a high-quality carpet for \$25 per square yard, plus a \$20 delivery charge. The total cost c of n square yards is given by the following formula.

$c =$	cost per square yard	$\cdot$	the number of square yards	$+$	the delivery charge.
$c =$	25	$\cdot$	n	$+$	20

Graph this equation and interpret the slope of the line.

Solution We can graph the equation on a coordinate system with a vertical c-axis and a horizontal n-axis. Figure 2-24 shows a table of ordered pairs and the graph.

$c = 25n + 20$

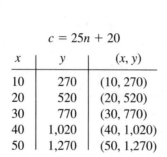

x	y	(x, y)
10	270	(10, 270)
20	520	(20, 520)
30	770	(30, 770)
40	1,020	(40, 1,020)
50	1,270	(50, 1,270)

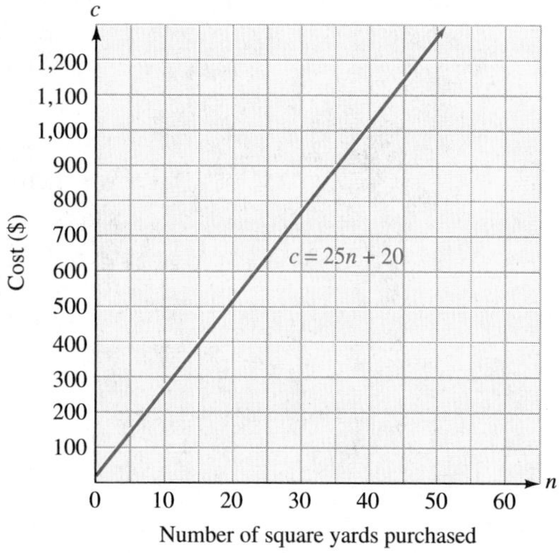

FIGURE 2-24

If we pick the points (30, 770) and (50, 1,270) to find the slope, we have

$$m = \frac{\Delta c}{\Delta n}$$

$$= \frac{c_2 - c_1}{n_2 - n_1}$$

$$= \frac{1{,}270 - 770}{50 - 30} \quad \text{Substitute 1,270 for } c_2\text{, 770 for } c_1\text{, 50 for } n_2\text{, and 30 for } n_1.$$

$$= \frac{500}{20}$$

$$= 25$$

The slope of 25 is the cost of the carpet in dollars per square yard. ■

Self Check Interpret the y-intercept of the graph in Figure 2-24.

Answer The y-coordinate of the y-intercept is the delivery charge.

EXAMPLE 4

Rate of descent It takes a skier 25 minutes to complete the course shown in Figure 2-25. Find his average rate of descent in feet per minute.

Solution To find the average rate of descent, we must find the ratio of the change in altitude to the change in time. To find this ratio, we calculate the slope of the line passing through the points (0, 12,000) and (25, 8,500).

$$\begin{aligned}\text{Average rate of descent} &= \frac{12{,}000 - 8{,}500}{0 - 25} \\ &= \frac{3{,}500}{-25} \\ &= -140\end{aligned}$$

The average rate of descent is -140 ft/min.

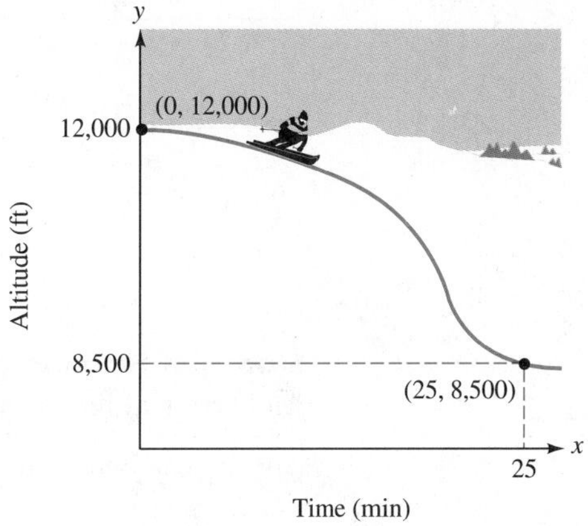

FIGURE 2-25

Self Check In Example 4, find the average rate of descent if the skier completes the course in 20 minutes.

Answer -175 ft/min

HORIZONTAL AND VERTICAL LINES

If $P(x_1, y_1)$ and $Q(x_2, y_2)$ are points on the horizontal line shown in Figure 2-26(a), then $y_1 = y_2$, and the numerator of the fraction

$$\frac{y_2 - y_1}{x_2 - x_1} \qquad \text{On a horizontal line, } x_2 \neq x_1.$$

is 0. Thus, the value of the fraction is 0, and the slope of the horizontal line is 0.

If $P(x_1, y_1)$ and $Q(x_2, y_2)$ are two points on the vertical line shown in Figure 2-26(b), then $x_1 = x_2$, and the denominator of the fraction

$$\frac{y_2 - y_1}{x_2 - x_1} \qquad \text{On a vertical line, } y_2 \neq y_1.$$

is 0. Since the denominator of a fraction cannot be 0, a vertical line has no defined slope.

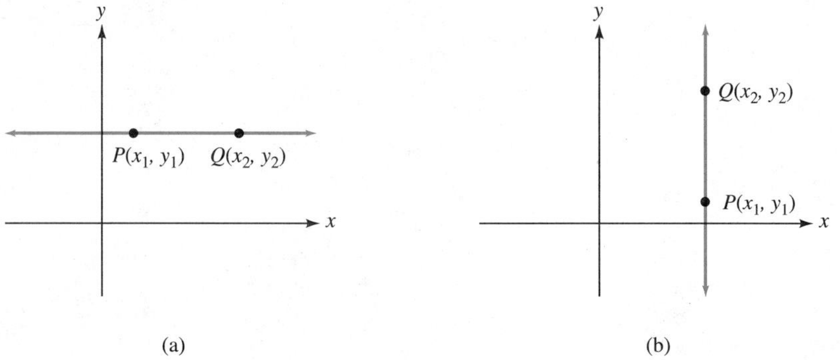

FIGURE 2-26

Slopes of Horizontal and Vertical Lines

All horizontal lines (lines with equations of the form $y = b$) have a slope of 0.

All vertical lines (lines with equations of the form $x = a$) have no defined slope.

If a line rises as we follow it from left to right, as in Figure 2-27(a), its slope is positive. If a line drops as we follow it from left to right, as in Figure 2-27(b), its slope is negative. If a line is horizontal, as in Figure 2-27(c), its slope is 0. If a line is vertical, as in Figure 2-27(d), it has no defined slope.

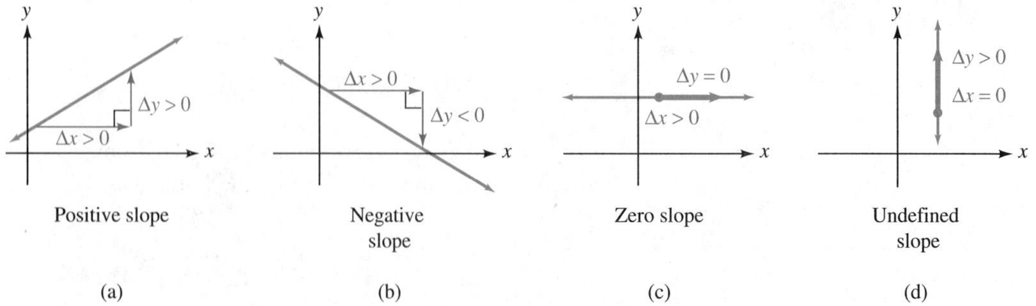

FIGURE 2-27

SLOPES OF PARALLEL LINES

To see a relationship between parallel lines and their slopes, we refer to the parallel lines l_1 and l_2 shown in Figure 2-28, with slopes of m_1 and m_2, respectively. Because right triangles ABC and DEF are similar, it follows that

$$\begin{aligned} m_1 &= \frac{\Delta y \text{ of } l_1}{\Delta x \text{ of } l_1} \\ &= \frac{\Delta y \text{ of } l_2}{\Delta x \text{ of } l_2} \\ &= m_2 \end{aligned}$$

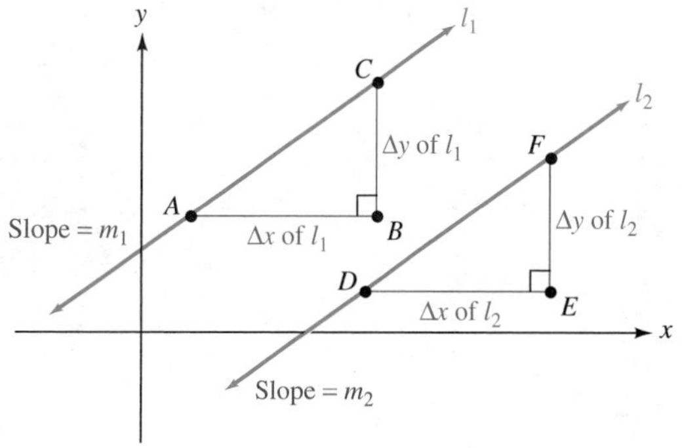

FIGURE 2-28

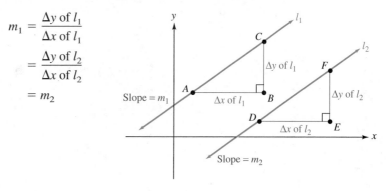

Thus, if two nonvertical lines are parallel, they have the same slope. It is also true that when two lines have the same slope, they are parallel.

Slopes of Parallel Lines
Nonvertical parallel lines have the same slope, and lines having the same slope are parallel.

Since vertical lines are parallel, lines with no defined slope are parallel.

EXAMPLE 5 The lines in Figure 2-29 are parallel. Find y.

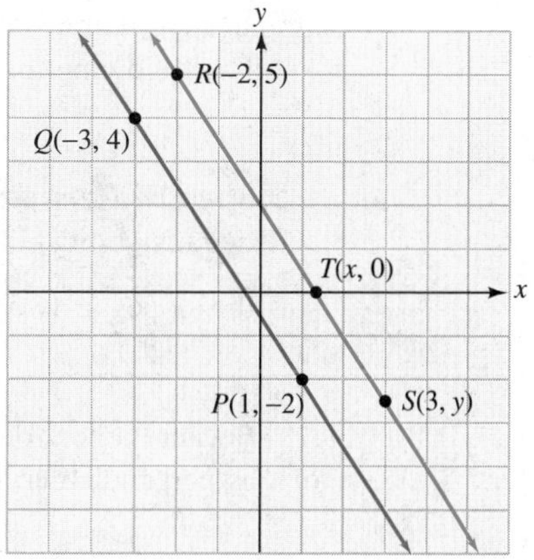

FIGURE 2-29

Solution Since the lines are parallel, they have equal slopes. To find y, we find the slope of each line, set them equal, and solve the resulting equation.

Slope of PQ		***Slope of RS***	
$\frac{-2-4}{1-(-3)}$	$=$	$\frac{y-5}{3-(-2)}$	
$\frac{-6}{4}$	$=$	$\frac{y-5}{5}$	
-30	$=$	$4(y-5)$	Multiply both sides by 20.
-30	$=$	$4y-20$	Use the distributive property.
-10	$=$	$4y$	Add 20 to both sides.
$-\frac{5}{2}$	$=$	y	Divide both sides by 4 and simplify.

Thus, $y = -\frac{5}{2}$. ■

Self Check In Figure 2-29, find x.

Answer $\frac{4}{3}$

■ SLOPES OF PERPENDICULAR LINES

Two real numbers a and b are called **negative reciprocals** if $ab = -1$. For example,

$$-\frac{4}{3} \quad \text{and} \quad \frac{3}{4}$$

are negative reciprocals, because $-\frac{4}{3}\left(\frac{3}{4}\right) = -1$.

The following theorem relates perpendicular lines and their slopes.

Slopes of Perpendicular Lines
If two nonvertical lines are perpendicular, their slopes are negative reciprocals.

If the slopes of two lines are negative reciprocals, the lines are perpendicular.

Because a horizontal line is perpendicular to a vertical line, a line with a slope of 0 is perpendicular to a line with no defined slope.

EXAMPLE 6 Are the lines shown in Figure 2-30 perpendicular?

Solution We find the slopes of the lines and see whether they are negative reciprocals.

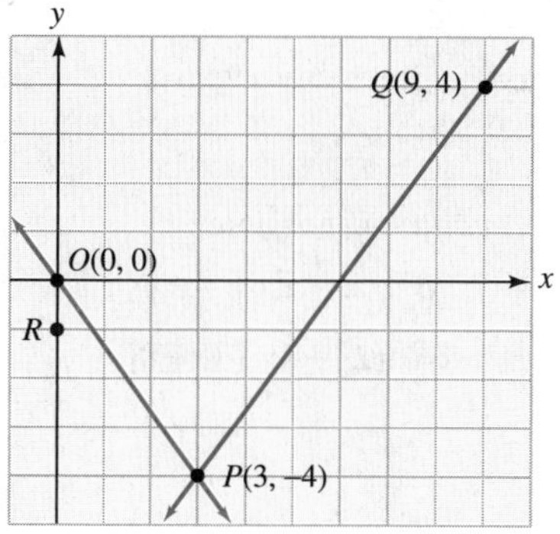

FIGURE 2-30

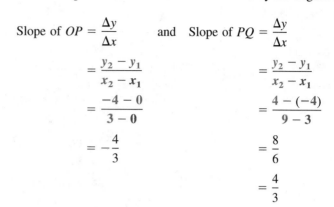

$$\text{Slope of } OP = \frac{\Delta y}{\Delta x} = \frac{y_2 - y_1}{x_2 - x_1} = \frac{-4 - 0}{3 - 0} = -\frac{4}{3}$$

and

$$\text{Slope of } PQ = \frac{\Delta y}{\Delta x} = \frac{y_2 - y_1}{x_2 - x_1} = \frac{4 - (-4)}{9 - 3} = \frac{8}{6} = \frac{4}{3}$$

Since their slopes are not negative reciprocals, the lines are not perpendicular. ■

Self Check In Figure 2-30, is *PR* perpendicular to *PQ*?

Answer no

Orals *Find the slope of the line passing through*

1. (0, 0), (1, 3)
2. (0, 0), (3, 6)
3. Are lines with slopes of −2 and $\frac{8}{-4}$ parallel?
4. Find the negative reciprocal of −0.2.
5. Are lines with slopes of −2 and $\frac{1}{2}$ perpendicular?

EXERCISE 2.3

REVIEW *Simplify each expression. Write all answers without negative exponents.*

1. $(x^3y^2)^3$
2. $\left(\frac{x^5}{x^3}\right)^3$
3. $(x^{-3}y^2)^{-4}$
4. $\left(\frac{x^{-6}}{y^3}\right)^{-4}$
5. $\left(\frac{3x^2y^3}{8}\right)^0$
6. $\left(\frac{x^3x^{-7}y^{-6}}{x^4y^{-3}y^{-2}}\right)^{-2}$

VOCABULARY AND CONCEPTS *Fill in each blank to make a true statement.*

7. Slope is defined as the change in ___ divided by the change in ___.
8. A slope is a rate of _______.
9. The formula to compute slope is $m =$ ______.
10. The change in y (denoted as Δy) is the _____ of the line between points P and Q.

11. The change in x (denoted as Δx) is the ______ of the line between points P and Q.

12. The slope of a __________ line is 0.

13. The slope of a ________ line is undefined.

14. If a line rises as x increases, its slope is ________.

15. ________ lines have the same slope.

16. The slopes of ____________ lines are negative ________.

PRACTICE *In Exercises 17–28, find the slope of the line that passes through the given points, if possible.*

17. (0, 0), (3, 9)

18. (9, 6), (0, 0)

19. (−1, 8), (6, 1)

20. (−5, −8), (3, 8)

21. (3, −1), (−6, 2)

22. (0, −8), (−5, 0)

23. (7, 5), (−9, 5)

24. (2, −8), (3, −8)

25. (−7, −5), (−7, −2)

26. (3, −5), (3, 14)

27. $(a, b), (b, a)$

28. $(a, b), (-b, -a)$

In Exercises 29–36, find the slope of the line determined by each equation.

29. $3x + 2y = 12$

30. $2x - y = 6$

31. $3x = 4y - 2$

32. $x = y$

33. $y = \frac{x - 4}{2}$

34. $x = \frac{3 - y}{4}$

35. $4y = 3(y + 2)$

36. $x + y = \frac{2 - 3y}{3}$

In Exercises 37–42, tell whether the slope of the line in each graph is positive, negative, 0, or undefined.

37.

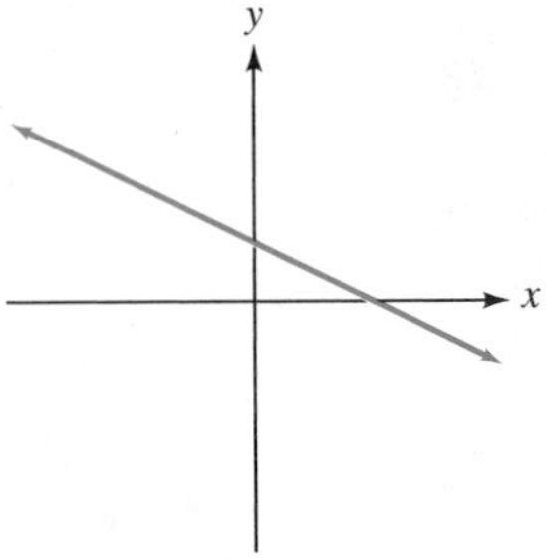

38.

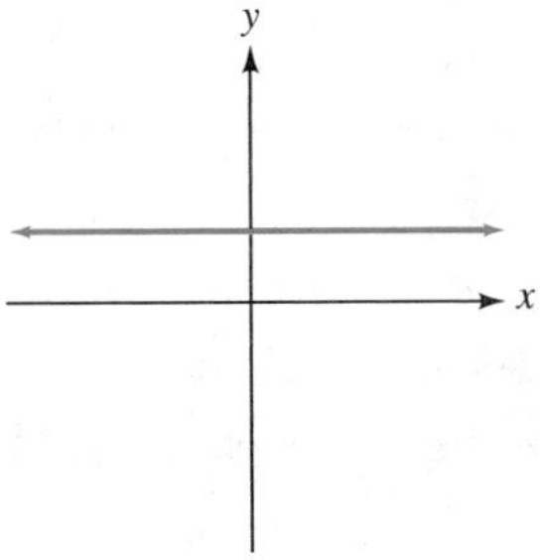

39.

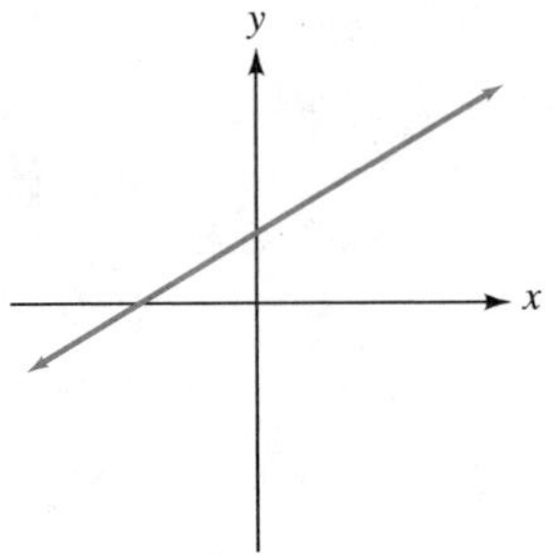

40.

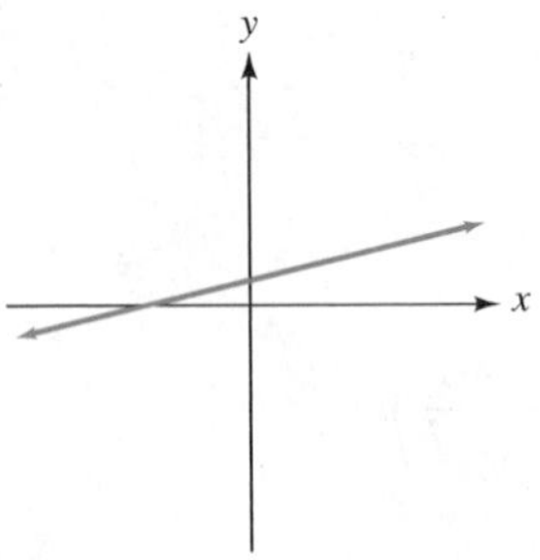

41.

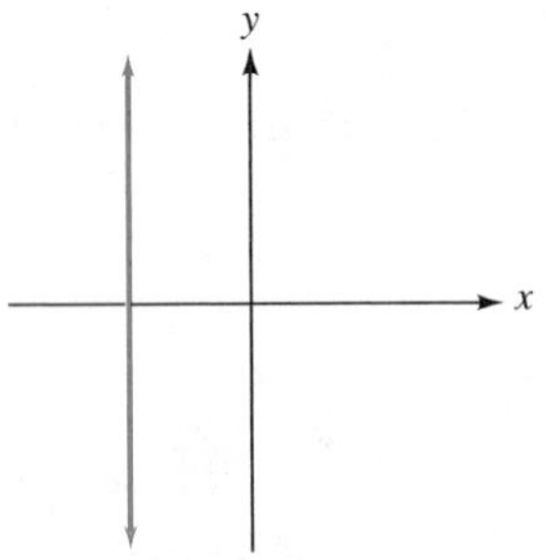

42.

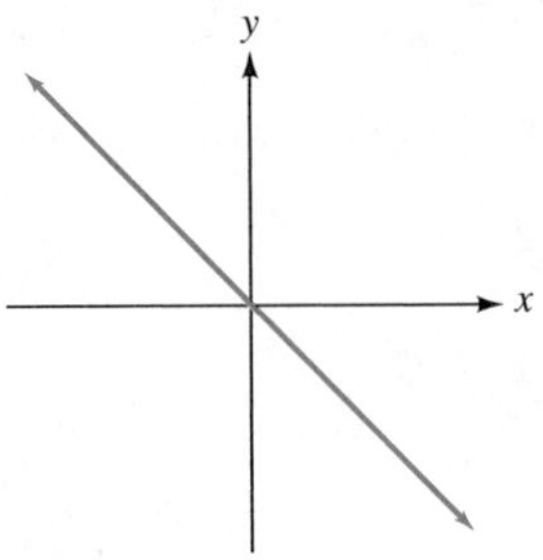

In Exercises 43–48, tell whether the lines with the given slopes are parallel, perpendicular, or neither.

43. $m_1 = 3,\ m_2 = -\frac{1}{3}$

44. $m_1 = \frac{1}{4},\ m_2 = 4$

45. $m_1 = 4,\ m_2 = 0.25$

46. $m_1 = -5,\ m_2 = \dfrac{1}{-0.2}$

47. $m_1 = \dfrac{a}{b},\ m_2 = \left(\dfrac{b}{a}\right)^{-1}$

48. $m_1 = \dfrac{c}{d},\ m_2 = \dfrac{d}{c}$

In Exercises 49–54, tell whether the line PQ is parallel or perpendicular (or neither) to a line with a slope of −2.

49. $P(3, 4),\ Q(4, 2)$

50. $P(6, 4),\ Q(8, 5)$

51. $P(-2, 1),\ Q(6, 5)$

52. $P(3, 4),\ Q(-3, -5)$

53. $P(5, 4),\ Q(6, 6)$

54. $P(-2, 3),\ Q(4, -9)$

In Exercises 55–60, find the slopes of lines PQ and PR and tell whether the points P, Q, and R lie on the same line. (Hint: Two lines with the same slope and a point in common must be the same line.)

55. $P(-2, 4),\ Q(4, 8),\ R(8, 12)$

56. $P(6, 10),\ Q(0, 6),\ R(3, 8)$

57. $P(-4, 10),\ Q(-6, 0),\ R(-1, 5)$

58. $P(-10, -13),\ Q(-8, -10),\ R(-12, -16)$

59. $P(-2, 4),\ Q(0, 8),\ R(2, 12)$

60. $P(8, -4),\ Q(0, -12),\ R(8, -20)$

61. Find the equation of the x-axis and its slope.

62. Find the equation of the y-axis and its slope, if any.

In Exercises 63–68, work each geometry problem.

63. Show that points with coordinates of $(-3, 4)$, $(4, 1)$, and $(-1, -1)$ are the vertices of a right triangle.

64. Show that a triangle with vertices at $(0, 0)$, $(12, 0)$, and $(13, 12)$ is not a right triangle.

65. A square has vertices at points $(a, 0)$, $(0, a)$, $(-a, 0)$, and $(0, -a)$, where $a \neq 0$. Show that its adjacent sides are perpendicular.

66. If a and b are not both 0, show that the points $(2b, a)$, (b, b), and $(a, 0)$ are the vertices of a right triangle.

67. Show that the points $(0, 0)$, $(0, a)$, (b, c), and $(b, a + c)$ are the vertices of a parallelogram. (*Hint:* Opposite sides of a parallelogram are parallel.)

68. If $b \neq 0$, show that the points $(0, 0)$, $(0, b)$, $(8, b + 2)$, and $(12, 3)$ are the vertices of a trapezoid. (*Hint:* A **trapezoid** is a four-sided figure with exactly two sides parallel.)

APPLICATIONS

69. Grade of a road Find the slope of the road shown in Illustration 1. (*Hint:* 1 mi = 5,280 ft.)

70. Pitch of a roof Find the pitch of the roof shown in Illustration 2.

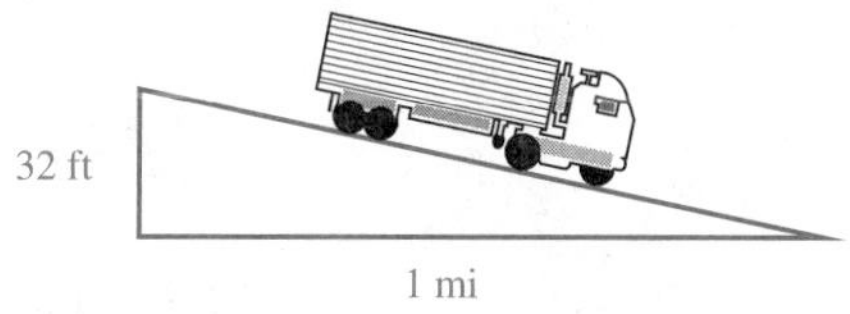

ILLUSTRATION 1

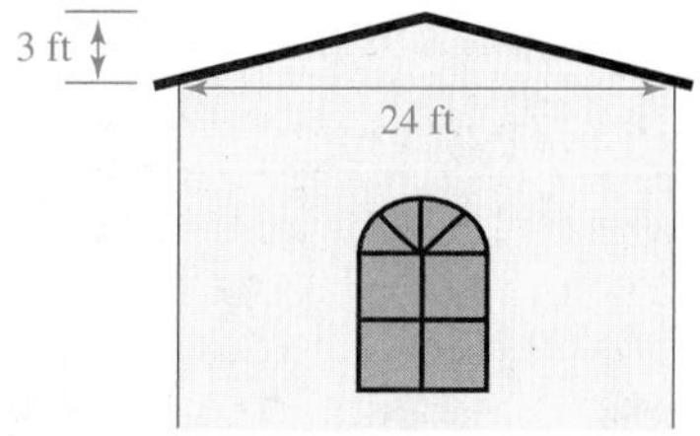

ILLUSTRATION 2

71. **Slope of a ladder** A ladder reaches 18 feet up the side of a building with its base 5 feet from the building. Find the slope of the ladder.

72. **Rate of growth** When a college started an aviation program, the administration agreed to predict enrollments using a straight-line method. If the enrollment during the first year was 8, and the enrollment during the fifth year was 20, find the rate of growth per year (the slope of the line). (See Illustration 3.)

73. **Rate of growth** A small business predicts sales according to a straight-line method. If sales were \$85,000 in the first year and \$125,000 in the third year, find the rate of growth in sales per year (the slope of the line).

74. **Rate of decrease** The price of computer technology has been dropping steadily for the past ten years. If a desktop PC cost \$5,700 ten years ago, and the same computing power cost \$1,499 two years ago, find the rate of decrease per year. (Assume a straight-line model.)

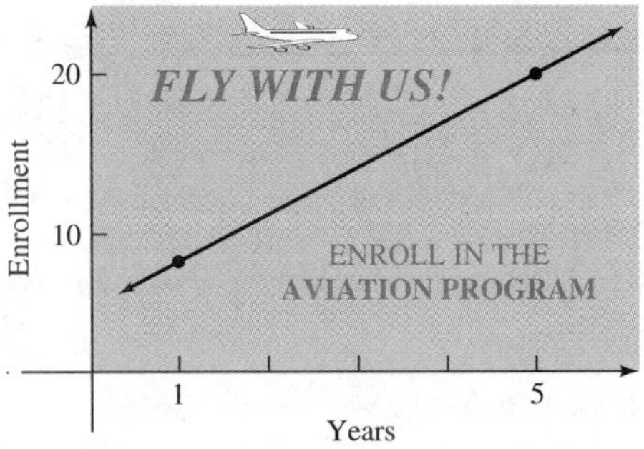

ILLUSTRATION 3

WRITING

75. Explain why a vertical line has no defined slope.

76. Explain how to determine from their slopes whether two lines are parallel, perpendicular, or neither.

SOMETHING TO THINK ABOUT

77. Find the slope of the line $Ax + By = C$. Follow the procedure of Example 2.

78. Follow Example 2 to find the slope of the line $y = mx + b$.

79. The points $(3, a)$, $(5, 7)$, and $(7, 10)$ lie on a line. Find a.

80. The line passing through points $A(1, 3)$ and $B(-2, 7)$ is perpendicular to the line passing through points $C(4, b)$ and $D(8, -1)$. Find b.

2.4 Writing Equations of Lines

■ POINT–SLOPE FORM OF THE EQUATION OF A LINE ■ SLOPE–INTERCEPT FORM OF THE EQUATION OF A LINE ■ USING SLOPE AS AN AID IN GRAPHING ■ GENERAL FORM OF THE EQUATION OF A LINE ■ STRAIGHT-LINE DEPRECIATION ■ CURVE FITTING

Getting Ready *Solve each equation.*

1. $3 = \dfrac{x - 2}{4}$

2. $-2 = 3(x + 1)$

3. Solve $y - 2 = 3(x - 2)$ for y.

4. Solve $Ax + By + 3 = 0$ for x.

POINT–SLOPE FORM OF THE EQUATION OF A LINE

Suppose that line l shown in Figure 2-31 has a slope of m and passes through the point $P(x_1, y_1)$. If $Q(x, y)$ is a second point on line l, we have

$$m = \frac{y - y_1}{x - x_1}$$

or if we multiply both sides by $x - x_1$, we have

1. $y - y_1 = m(x - x_1)$

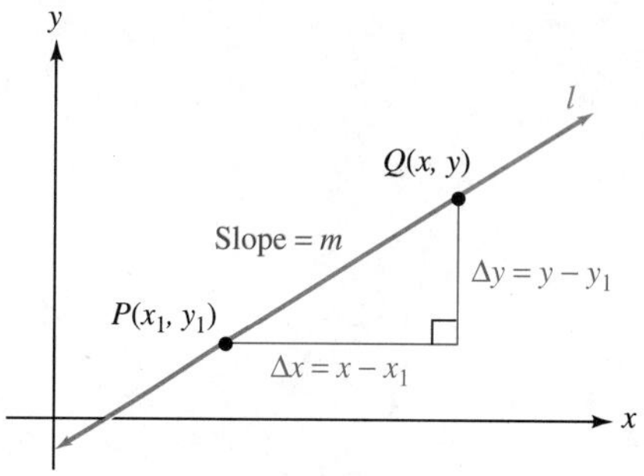

FIGURE 2-31

Because Equation 1 displays the coordinates of the point (x_1, y_1) on the line and the slope m of the line, it is called the **point–slope form** of the equation of a line.

Point–Slope Form

The equation of the line passing through $P(x_1, y_1)$ and with slope m is

$$y - y_1 = m(x - x_1)$$

EXAMPLE 1 Write the equation of the line with a slope of $-\frac{2}{3}$ and passing through $P(-4, 5)$.

Solution We substitute $-\frac{2}{3}$ for m, -4 for x_1, and 5 for y_1 into the point–slope form and simplify.

$$y - y_1 = m(x - x_1)$$

$$y - 5 = -\frac{2}{3}[x - (-4)] \quad \text{Substitute } -\tfrac{2}{3} \text{ for } m, -4 \text{ for } x_1, \text{ and 5 for } y_1.$$

$$y - 5 = -\frac{2}{3}(x + 4) \quad -(-4) = 4.$$

$$y - 5 = -\frac{2}{3}x - \frac{8}{3} \quad \text{Use the distributive property to remove parentheses.}$$

$$y = -\frac{2}{3}x + \frac{7}{3} \quad \text{Add 5 to both sides and simplify.}$$

The equation of the line is $y = -\frac{2}{3}x + \frac{7}{3}$. ■

Self Check Write the equation of the line with slope of $\frac{5}{4}$ and passing through $Q(0, 5)$.

Answer $y = \frac{5}{4}x + 5$

EXAMPLE 2 Write the equation of the line passing through $P(-5, 4)$ and $Q(8, -6)$.

Solution First we find the slope of the line.

$$
\begin{aligned}
m &= \frac{y_2 - y_1}{x_2 - x_1} \\
&= \frac{-6 - 4}{8 - (-5)} && \text{Substitute } -6 \text{ for } y_2,\ 4 \text{ for } y_1,\ 8 \text{ for } x_2, \text{ and } -5 \text{ for } x_1. \\
&= -\frac{10}{13}
\end{aligned}
$$

Because the line passes through both P and Q, we can choose either point and substitute its coordinates into the point–slope form. If we choose $P(-5, 4)$, we substitute -5 for x_1, 4 for y_1, and $-\frac{10}{13}$ for m and proceed as follows.

$$
\begin{aligned}
y - y_1 &= m(x - x_1) \\
y - 4 &= -\frac{10}{13}[x - (-5)] && \text{Substitute } -\tfrac{10}{13} \text{ for } m,\ -5 \text{ for } x_1, \text{ and } 4 \text{ for } y_1. \\
y - 4 &= -\frac{10}{13}(x + 5) && -(-5) = 5. \\
y - 4 &= -\frac{10}{13}x - \frac{50}{13} && \text{Remove parentheses.} \\
y &= -\frac{10}{13}x + \frac{2}{13} && \text{Add 4 to both sides and simplify.}
\end{aligned}
$$

The equation of the line is $y = -\frac{10}{13}x + \frac{2}{13}$. ■

Self Check Write the equation of the line passing through $R(-2, 5)$ and $S(4, -3)$.

Answer $y = -\frac{4}{3}x + \frac{7}{3}$

SLOPE–INTERCEPT FORM OF THE EQUATION OF A LINE

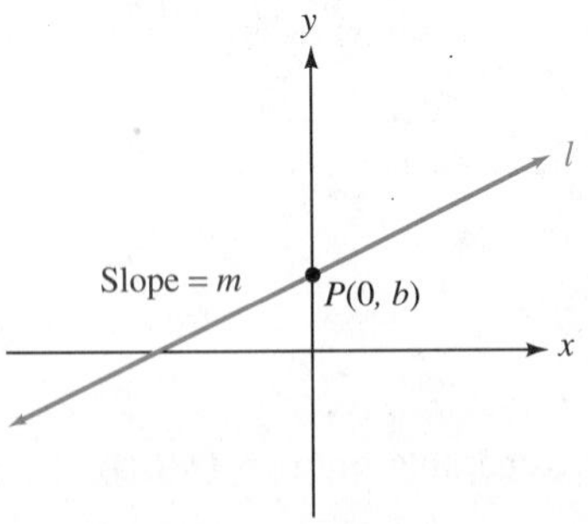

FIGURE 2-32

Since the y-intercept of the line shown in Figure 2-32 is the point $(0, b)$, we can write the equation of the line by substituting 0 for x_1 and b for y_1 in the point–slope form and simplifying.

$$
\begin{aligned}
y - y_1 &= m(x - x_1) \\
y - b &= m(x - 0) \\
y - b &= mx
\end{aligned}
$$

2. $y = mx + b$

Because Equation 2 displays the slope m and the y-coordinate b of the y-intercept, it is called the **slope–intercept form** of the equation of a line.

Slope-Intercept Form

The equation of the line with slope m and y-intercept $(0, b)$ is

$$y = mx + b$$

EXAMPLE 3 Use the slope–intercept form to write the equation of the line with slope 4 that passes through the point $P(5, 9)$.

Solution Since we are given that $m = 4$ and that the ordered pair (5, 9) satisfies the equation, we can substitute 5 for x, 9 for y, and 4 for m in the equation $y = mx + b$ and solve for b.

$$y = mx + b$$

$$9 = 4(5) + b \quad \text{Substitute 9 for } y\text{, 4 for } m\text{, and 5 for } x.$$

$$9 = 20 + b \quad \text{Simplify.}$$

$$-11 = b \quad \text{Subtract 20 from both sides.}$$

Because $m = 4$ and $b = -11$, the equation is $y = 4x - 11$. ■

Self Check Write the equation of the line with slope -2 that passes through the point $Q(-2, 8)$.

Answer $y = -2x + 4$

■ USING SLOPE AS AN AID IN GRAPHING

It is easy to graph a linear equation when it is written in slope–intercept form. For example, to graph $y = \frac{4}{3}x - 2$, we note that $b = -2$ and that the y-intercept is $(0, b) = (0, -2)$. (See Figure 2-33.)

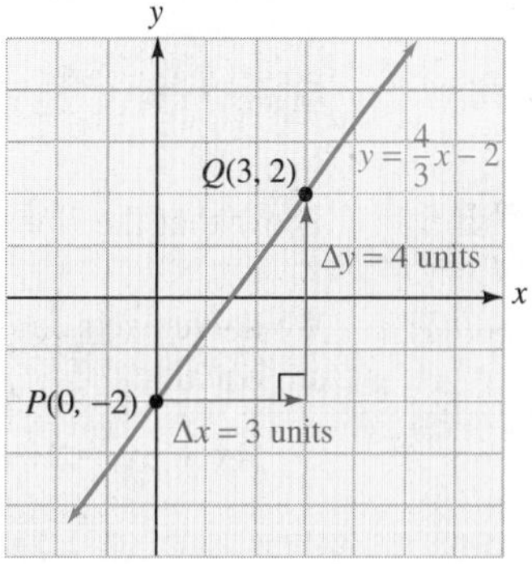

FIGURE 2-33

Because the slope of the line is $\frac{\Delta y}{\Delta x} = \frac{4}{3}$, we can locate another point Q on the line by starting at point P and counting 3 units to the right and 4 units up. The change in x from point P to point Q is $\Delta x = 3$, and the corresponding change in y is $\Delta y = 4$. The line joining points P and Q is the graph of the equation.

EXAMPLE 4 Find the slope and the y-intercept of the line with the equation $2(x - 3) = -3(y + 5)$. Then graph the line.

Solution We write the equation in the form $y = mx + b$ to find the slope m and the y-intercept $(0, b)$.

$$2(x - 3) = -3(y + 5)$$

$$2x - 6 = -3y - 15 \quad \text{Use the distributive property to remove parentheses.}$$

$$2x + 3y - 6 = -15 \quad \text{Add } 3y \text{ to both sides.}$$

$$3y - 6 = -2x - 15 \quad \text{Subtract } 2x \text{ from both sides.}$$

$$3y = -2x - 9 \quad \text{Add 6 to both sides.}$$

$$y = -\frac{2}{3}x - 3 \quad \text{Divide both sides by 3.}$$

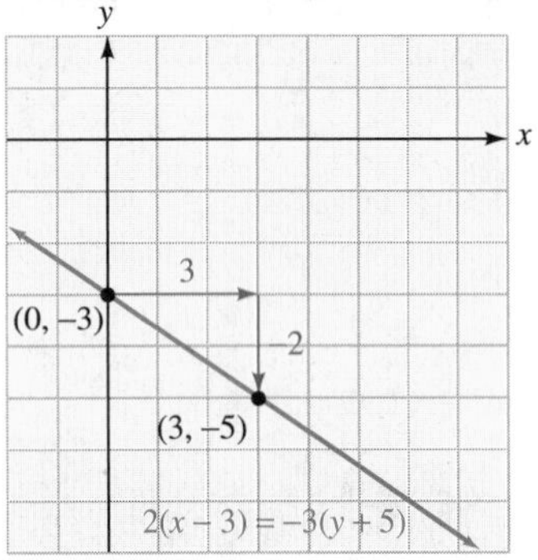

FIGURE 2-34

The slope is $-\frac{2}{3}$, and the y-intercept is $(0, -3)$. To draw the graph, we plot the y-intercept $(0, -3)$ and then locate a second point on the line by moving 3 units to the right and 2 units down. We draw a line through the two points to obtain the graph shown in Figure 2-34. ■

Self Check Find the slope and the y-intercept of the line with the equation $2(y - 1) = 3x + 2$, and graph the line.

Answer $m = \frac{3}{2}$, $(0, 2)$

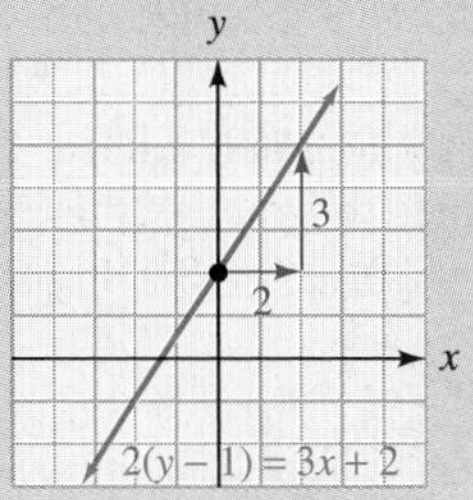

EXAMPLE 5 Show that the lines represented by $4x + 8y = 10$ and $2x = 12 - 4y$ are parallel.

Solution We solve each equation for y to see that the lines are distinct and that their slopes are equal.

$$4x + 8y = 10 \qquad 2x = 12 - 4y$$

$$8y = -4x + 10 \qquad 4y = -2x + 12$$

$$y = -\frac{1}{2}x + \frac{5}{4} \qquad y = -\frac{1}{2}x + 3$$

Since the values of b in these equations are different, the lines are distinct. Since the slope of each line is $-\frac{1}{2}$, they are parallel. ■

Self Check Are lines represented by $3x - 2y = 4$ and $6x = 4(y + 1)$ parallel?

Answer yes

EXAMPLE 6 Show that the lines represented by $4x + 8y = 10$ and $4x - 2y = 21$ are perpendicular.

Solution We solve each equation for y to see that the slopes of their straight-line graphs are negative reciprocals.

$$\begin{aligned} 4x + 8y &= 10 \\ 8y &= -4x + 10 \\ y &= -\frac{1}{2}x + \frac{5}{4} \end{aligned} \qquad \begin{aligned} 4x - 2y &= 21 \\ -2y &= -4x + 21 \\ y &= 2x - \frac{21}{2} \end{aligned}$$

Since the slopes are $-\frac{1}{2}$ and 2 (which are negative reciprocals), the lines are perpendicular. ■

Self Check Are lines represented by $3x + 2y = 6$ and $2x - 3y = 6$ perpendicular?

Answer yes

EXAMPLE 7 Write the equation of the line passing through $P(-2, 5)$ and parallel to the line $y = 8x - 3$.

Solution Since the equation is solved for y, the slope of the line given by $y = 8x - 3$ is the coefficient of x, which is 8. Since the desired equation is to have a graph that is parallel to the graph of $y = 8x - 3$, its slope must also be 8.

We substitute -2 for x_1, 5 for y_1, and 8 for m in the point–slope form and simplify.

$$y - y_1 = m(x - x_1)$$

$$y - 5 = 8[x - (-2)] \quad \text{Substitute 5 for } y_1\text{, 8 for } m\text{, and } -2 \text{ for } x_1.$$

$$y - 5 = 8(x + 2) \quad -(-2) = 2.$$

$$y - 5 = 8x + 16 \quad \text{Use the distributive property to remove parentheses.}$$

$$y = 8x + 21 \quad \text{Add 5 to both sides.}$$

The equation is $y = 8x + 21$. ■

Self Check Write the equation of the line that is parallel to the line $y = 8x - 3$ and passes through the origin.

Answer $y = 8x$

EXAMPLE 8 Write the equation of the line passing through $P(-2, 5)$ and perpendicular to the line $y = 8x - 3$.

Solution The slope of the given line is 8. Thus, the slope of the desired line must be $-\frac{1}{8}$, which is the negative reciprocal of 8.

We substitute -2 for x_1, 5 for y_1, and $-\frac{1}{8}$ for m into the point–slope form and simplify:

$$y - y_1 = m(x - x_1)$$

$$y - 5 = -\frac{1}{8}[x - (-2)] \qquad \text{Substitute 5 for } y_1,\ -\tfrac{1}{8} \text{ for } m, \text{ and } -2 \text{ for } x_1.$$

$$y - 5 = -\frac{1}{8}(x + 2) \qquad -(-2) = 2.$$

$$y - 5 = -\frac{1}{8}x - \frac{1}{4} \qquad \text{Remove parentheses.}$$

$$y = -\frac{1}{8}x - \frac{1}{4} + 5 \qquad \text{Add 5 to both sides.}$$

$$y = -\frac{1}{8}x + \frac{19}{4} \qquad \text{Combine terms: } -\tfrac{1}{4} + \tfrac{20}{4} = \tfrac{19}{4}.$$

The equation is $y = -\frac{1}{8}x + \frac{19}{4}$. ■

Self Check Write the equation of the line that is perpendicular to the line $y = 8x - 3$ and passes through $Q(2, 4)$.

Answer $y = -\frac{1}{8}x + \frac{17}{4}$

GENERAL FORM OF THE EQUATION OF A LINE

Any linear equation that is written in the form $Ax + By = C$, where A, B, and C are constants, is said to be written in **general form.**

WARNING! When writing equations in general form, we usually clear the equation of fractions and make A positive. We will also make A, B, and C as small as possible. For example, the equation $6x + 12y = 24$ can be changed to $x + 2y = 4$ by dividing both sides by 6.

Finding the Slope and *y*-Intercept from the General Form

If A, B, and C are real numbers and $B \neq 0$, the graph of the equation

$$Ax + By = C$$

is a nonvertical line with slope of $-\frac{A}{B}$ and a y-intercept of $\left(0, \frac{C}{B}\right)$.

You will be asked to justify the previous results in the exercises. You will also be asked to show that if $B = 0$, the equation $Ax + By = C$ represents a vertical line with x-intercept of $\left(\frac{C}{A}, 0\right)$.

EXAMPLE 9 Show that the lines represented by $4x + 3y = 7$ and $3x - 4y = 12$ are perpendicular.

Solution To show that the lines are perpendicular, we will show that their slopes are negative reciprocals. The first equation, $4x + 3y = 7$, is written in general form, with $A = 4$, $B = 3$, and $C = 7$. By the previous result, the slope of the line is

$$m_1 = -\frac{A}{B} = -\frac{4}{3}$$

The second equation, $3x - 4y = 12$, is also written in general form, with $A = 3$, $B = -4$, and $C = 12$. The slope of this line is

$$m_2 = -\frac{A}{B} = -\frac{3}{-4} = \frac{3}{4}$$

Since the slopes are negative reciprocals, the lines are perpendicular. ■

Self Check Are the lines $4x + 3y = 7$ and $y = -\frac{4}{3}x + 2$ parallel?

Answer yes

We summarize the various forms for the equation of a line in Table 2-1.

General form of a linear equation	$Ax + By = C$ A and B cannot both be 0.
Slope–intercept form of a linear equation	$y = mx + b$ The slope is m, and the y-intercept is $(0, b)$.
Point–slope form of a linear equation	$y - y_1 = m(x - x_1)$ The slope is m, and the line passes through (x_1, y_1).
A **horizontal line**	$y = b$ The slope is 0, and the y-intercept is $(0, b)$.
A **vertical line**	$x = a$ There is no defined slope, and the x-intercept is $(a, 0)$.

TABLE 2-1

STRAIGHT-LINE DEPRECIATION

For tax purposes, many businesses use *straight-line depreciation* to find the declining value of aging equipment.

EXAMPLE 10

Value of a lathe The owner of a machine shop buys a lathe for \$1,970 and expects it to last for ten years. It can then be sold as scrap for an estimated *salvage value* of \$270. If y represents the value of the lathe after x years of use, and y and x are related by the equation of a line,

a. Find the equation of the line.

b. Find the value of the lathe after $2\frac{1}{2}$ years.

c. Find the economic meaning of the y-intercept of the line.

d. Find the economic meaning of the slope of the line.

Solution

a. To find the equation of the line, we find its slope and use point–slope form to find its equation.

When the lathe is new, its age x is 0, and its value y is \$1,970. When the lathe is 10 years old, $x = 10$ and its value is $y = \$270$. Since the line passes through the points (0, 1,970) and (10, 270), as shown in Figure 2-35, the slope of the line is

$$\begin{aligned} m &= \frac{y_2 - y_1}{x_2 - x_1} \\ &= \frac{\mathbf{270} - \mathbf{1{,}970}}{\mathbf{10} - \mathbf{0}} \\ &= \frac{-1{,}700}{10} \\ &= -170 \end{aligned}$$

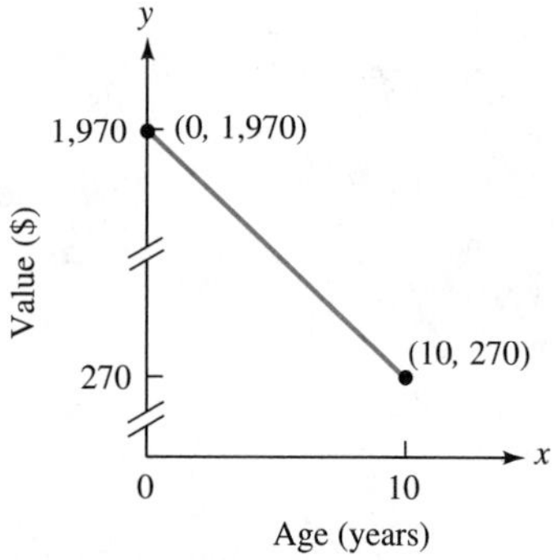

FIGURE 2-35

To find the equation of the line, we substitute -170 for m, 0 for x_1, and 1,970 for y_1 into the point–slope form and simplify.

$$\begin{aligned} y - y_1 &= m(x - x_1) \\ y - \mathbf{1{,}970} &= \mathbf{-170}(x - \mathbf{0}) \end{aligned}$$

3. $$y = -170x + 1{,}970$$

The current value y of the lathe is related to its age x by the equation $y = -170x + 1{,}970$.

b. To find the value of the lathe after $2\frac{1}{2}$ years, we substitute 2.5 for x in Equation 3 and solve for y.

$$\begin{aligned} y &= -170x + 1{,}970 \\ &= -170(\mathbf{2.5}) + 1{,}970 \\ &= -425 + 1{,}970 \\ &= 1{,}545 \end{aligned}$$

In $2\frac{1}{2}$ years, the lathe will be worth \$1,545.

c. The y-intercept of the graph is $(0, b)$, where b is the value of y when $x = 0$.

$$y = -170x + 1{,}970$$
$$y = -170(0) + 1{,}970$$
$$y = 1{,}970$$

Thus, b is the value of a 0-year-old lathe, which is the lathe's original cost, \$1,970.

d. Each year, the value of the lathe decreases by \$170, because the slope of the line is -170. The slope of the depreciation line is the **annual depreciation rate.** ■

CURVE FITTING

In statistics, the process of using one variable to predict another is called **regression.** For example, if we know a man's height, we can make a good prediction about his weight, because taller men usually weigh more than shorter men.

Figure 2-36 shows the results of sampling ten men at random and finding their heights and weights. The graph of the ordered pairs (h, w) is called a **scattergram.**

Man	Height (h) in inches	Weight (w) in pounds
1	66	140
2	68	150
3	68	165
4	70	180
5	70	165
6	71	175
7	72	200
8	74	190
9	75	210
10	75	215

(a)

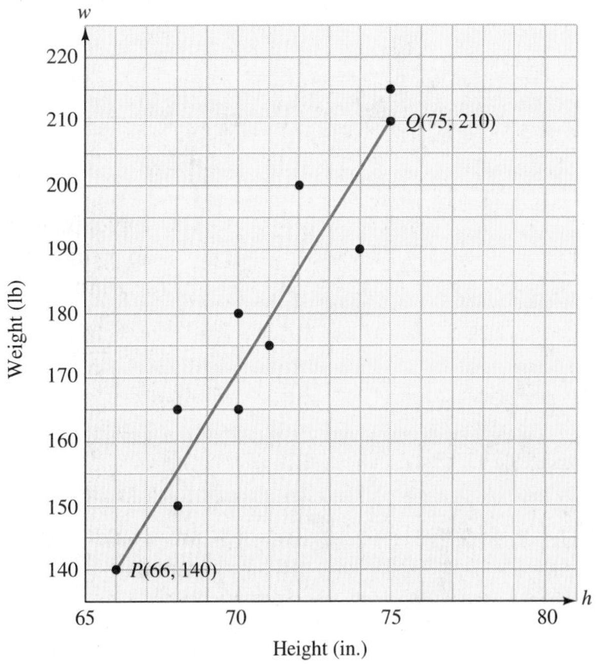

(b)

FIGURE 2-36

To write a **prediction equation** (sometimes called a **regression equation**), we must find the equation of the line that comes closer to all of the points in the scattergram than any other possible line. There are exact methods to find this equation, but we can only approximate it here.

To write an approximation of the regression equation, we place a straightedge on the scattergram shown in Figure 2-36 and draw the line joining two points that seems to best fit all of the points. In the figure, line PQ is drawn, where point P has coordinates of (66, 140) and point Q has coordinates of (75, 210).

Our approximation of the regression equation will be the equation of the line passing through points P and Q. To find the equation of this line, we first find its slope.

$$m = \frac{y_2 - y_1}{x_2 - x_1}$$

$$= \frac{210 - 140}{75 - 66}$$

$$= \frac{70}{9}$$

We can then use point–slope form to find the equation of the line.

$$y - y_1 = m(x - x_1)$$

$$y - 140 = \frac{70}{9}(x - 66) \qquad \text{Choose (66, 140) for } (x_1, y_1).$$

$$y = \frac{70}{9}x - \frac{4{,}620}{9} + 140 \qquad \text{Remove parentheses and add 140 to both sides.}$$

4. $$y = \frac{70}{9}x - \frac{1{,}120}{3} \qquad -\tfrac{4{,}620}{9} + 140 = \tfrac{-1{,}120}{3}.$$

Our approximation of the regression equation is $y = \frac{70}{9}x - \frac{1{,}120}{3}$.

To predict the weight of a man who is 73 inches tall, for example, we substitute 73 for x in Equation 4 and simplify.

$$y = \frac{70}{9}x - \frac{1{,}120}{3}$$

$$y = \frac{70}{9}(73) - \frac{1{,}120}{3}$$

$$y \approx 194.4$$

We would predict that a 73-inch-tall man chosen at random will weigh about 194 pounds.

Orals

Write the point–slope form of the equation of a line with $m = 2$, passing through the given point.

1. (2, 3)

2. (−3, 8)

Write the equation of a line with $m = -3$ and y-intercept of

3. (0, 5)

4. (0, −7)

Tell whether the lines are parallel, perpendicular, or neither.

5. $y = 3x - 4$, $y = 3x + 5$

6. $y = -3x + 7$, $x = 3y - 1$

EXERCISE 2.4

REVIEW EXERCISES *Solve each equation.*

1. $3(x + 2) + x = 5x$

2. $12b + 6(3 - b) = b + 3$

3. $\dfrac{5(2 - x)}{3} - 1 = x + 5$

4. $\dfrac{r - 1}{3} = \dfrac{r + 2}{6} + 2$

5. Mixing alloys In 60 ounces of alloy for watch cases, there are 20 ounces of gold. How much copper must be added to the alloy so that a watch case weighing 4 ounces, made from the new alloy, will contain exactly 1 ounce of gold?

6. Mixing coffee To make a mixture of 80 pounds of coffee worth \$272, a grocer mixes coffee worth \$3.25 a pound with coffee worth \$3.85 a pound. How many pounds of the cheaper coffee should the grocer use?

VOCABULARY AND CONCEPTS *Fill in each blank to make a true statement.*

7. The point–slope form of the equation of a line is ________________.

8. The slope–intercept form of the equation of a line is __________.

9. The general form of the equation of a line is ____________.

10. Two lines are parallel when they have the _____ slope.

11. Two lines are ____________ when their slopes are negative reciprocals.

12. The process that recognizes that equipment loses value with age is called ___________.

PRACTICE *In Exercises 13–16, use point–slope form to write the equation of the line with the given properties. Write each equation in general form.*

13. $m = 5$, passing through $P(0, 7)$

14. $m = -8$, passing through $P(0, -2)$

15. $m = -3$, passing through $P(2, 0)$

16. $m = 4$, passing through $P(-5, 0)$

In Exercises 17–18, use point–slope form to write the equation of each line. Write the equation in general form.

17.

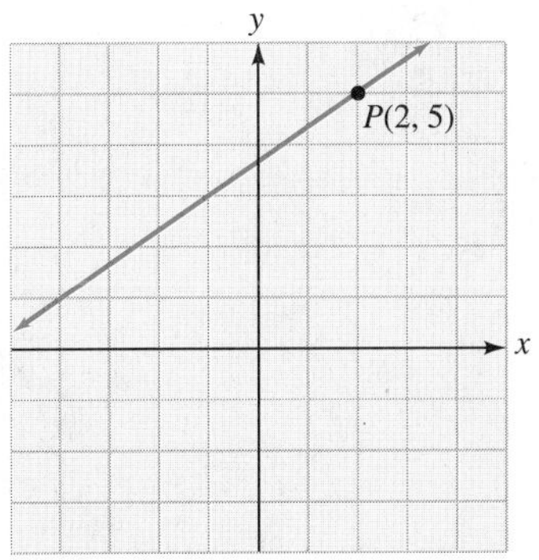

18.

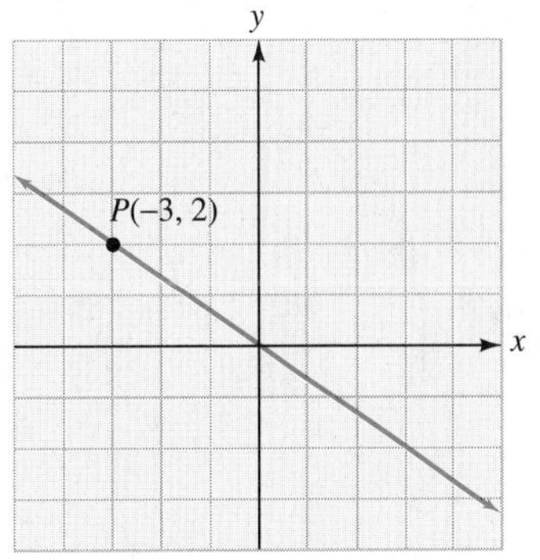

In Exercises 19–22, use point–slope form to write the equation of the line passing through the two given points. Write each equation in slope–intercept form.

19. $P(0, 0)$, $Q(4, 4)$

20. $P(-5, -5)$, $Q(0, 0)$

21. $P(3, 4)$, $Q(0, -3)$

22. $P(4, 0)$, $Q(6, -8)$

In Exercises 23–24, use point–slope form to write the equation of each line. Write each answer in slope–intercept form.

23.

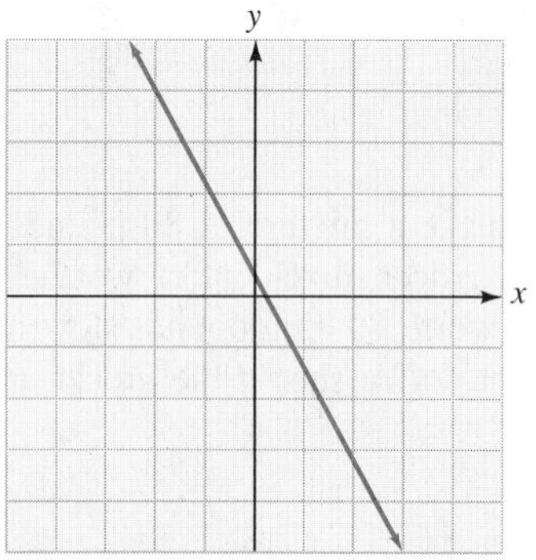

24.

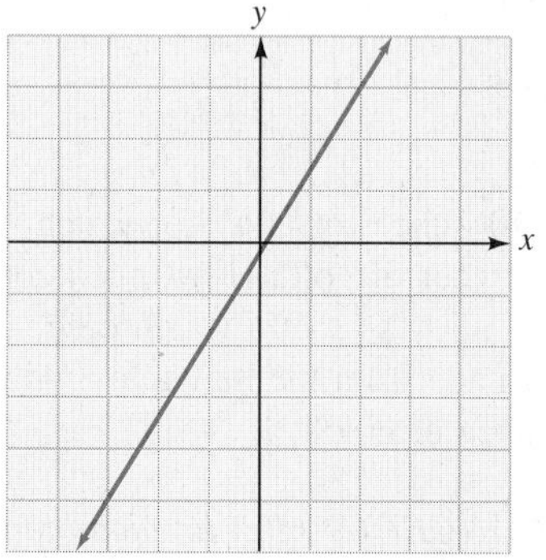

In Exercises 25–32, use the slope–intercept form to write the equation of the line with the given properties. Write each equation in slope–intercept form.

25. $m = 3$, $b = 17$

26. $m = -2$, $b = 11$

27. $m = -7$, passing through $P(7, 5)$

28. $m = 3$, passing through $P(-2, -5)$

29. $m = 0$, passing through $P(2, -4)$

30. $m = -7$, passing through the origin

31. Passing through $P(6, 8)$ and $Q(2, 10)$

32. Passing through $P(-4, 5)$ and $Q(2, -6)$

In Exercises 33–38, write each equation in slope–intercept form to find the slope and the y-intercept. Then use the slope and y-intercept to draw the line.

33. $y + 1 = x$

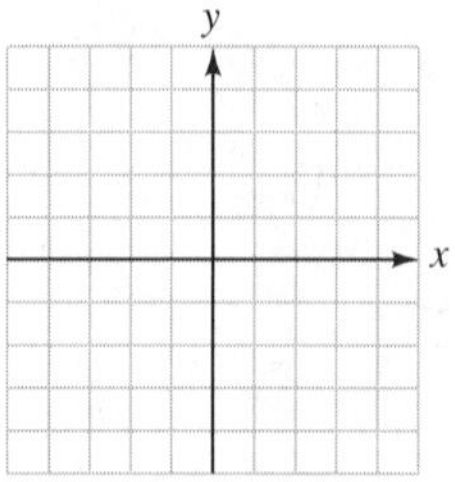

34. $x + y = 2$

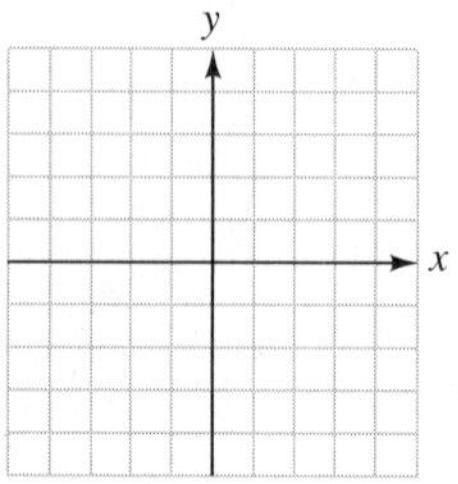

35. $x = \frac{3}{2}y - 3$

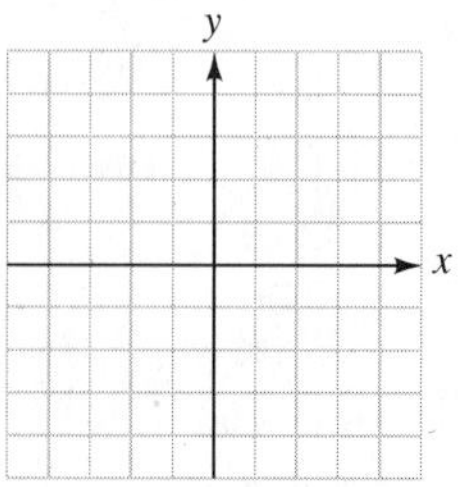

36. $x = -\frac{4}{5}y + 2$

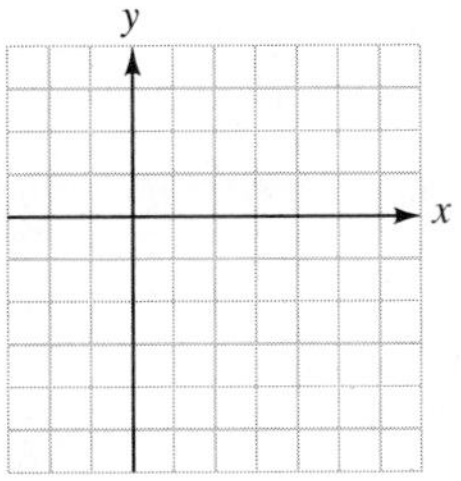

37. $3(y - 4) = -2(x - 3)$

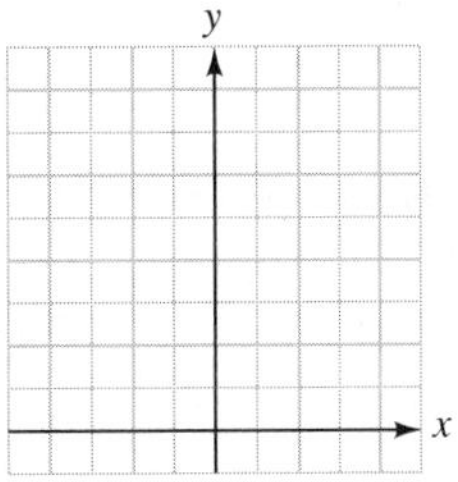

38. $-4(2x + 3) = 3(3y + 8)$

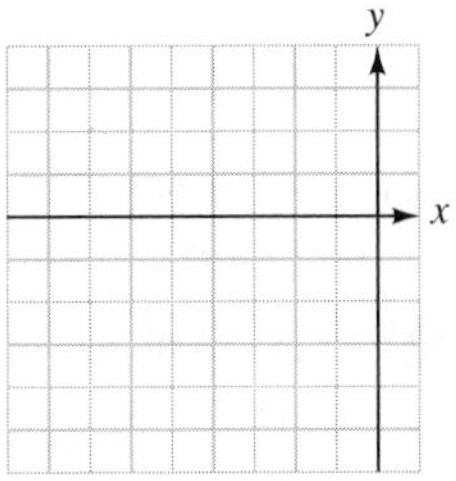

In Exercises 39–44, find the slope and the y-intercept of the line determined by the given equation.

39. $3x - 2y = 8$

40. $-2x + 4y = 12$

41. $-2(x + 3y) = 5$

42. $5(2x - 3y) = 4$

43. $x = \frac{2y - 4}{7}$

44. $3x + 4 = -\frac{2(y - 3)}{5}$

In Exercises 45–56, tell whether the graphs of each pair of equations are parallel, perpendicular, or neither.

45. $y = 3x + 4,\ y = 3x - 7$

46. $y = 4x - 13,\ y = \frac{1}{4}x + 13$

47. $x + y = 2,\ y = x + 5$

48. $x = y + 2,\ y = x + 3$

49. $y = 3x + 7,\ 2y = 6x - 9$

50. $2x + 3y = 9,\ 3x - 2y = 5$

51. $x = 3y + 4,\ y = -3x + 7$

52. $3x + 6y = 1,\ y = \frac{1}{2}x$

53. $y = 3,\ x = 4$

54. $y = -3,\ y = -7$

55. $x = \frac{y - 2}{3},\ 3(y - 3) + x = 0$

56. $2y = 8,\ 3(2 + x) = 2(x + 2)$

In Exercises 57–62, write the equation of the line that passes through the given point and is parallel to the given line. Write the answer in slope–intercept form.

57. $P(0, 0),\ y = 4x - 7$

58. $P(0, 0),\ x = -3y - 12$

59. $P(2, 5),\ 4x - y = 7$

60. $P(-6, 3),\ y + 3x = -12$

61. $P(4, -2),\ x = \frac{5}{4}y - 2$

62. $P(1, -5),\ x = -\frac{3}{4}y + 5$

In Exercises 63–68, write the equation of the line that passes through the given point and is perpendicular to the given line. Write the answer in slope–intercept form.

63. $P(0, 0),\ y = 4x - 7$

64. $P(0, 0),\ x = -3y - 12$

65. $P(2, 5)$, $4x - y = 7$

66. $P(-6, 3)$, $y + 3x = -12$

67. $P(4, -2)$, $x = \frac{5}{4}y - 2$

68. $P(1, -5)$, $x = -\frac{3}{4}y + 5$

In Exercises 69–72, use the method of Example 9 to find whether the graphs determined by each pair of equations are parallel, perpendicular, or neither.

69. $4x + 5y = 20$, $5x - 4y = 20$

70. $9x - 12y = 17$, $3x - 4y = 17$

71. $2x + 3y = 12$, $6x + 9y = 32$

72. $5x + 6y = 30$, $6x + 5y = 24$

73. Find the equation of the line perpendicular to the line $y = 3$ and passing through the midpoint of the segment joining $(2, 4)$ and $(-6, 10)$.

74. Find the equation of the line parallel to the line $y = -8$ and passing through the midpoint of the segment joining $(-4, 2)$ and $(-2, 8)$.

75. Find the equation of the line parallel to the line $x = 3$ and passing through the midpoint of the segment joining $(2, -4)$ and $(8, 12)$.

76. Find the equation of the line perpendicular to the line $x = 3$ and passing through the midpoint of the segment joining $(-2, 2)$ and $(4, -8)$.

77. Solve $Ax + By = C$ for y and thereby show that the slope of its graph is $-\frac{A}{B}$ and its y-intercept is $\left(0, \frac{C}{B}\right)$.

78. Show that the x-intercept of the graph of $Ax + By = C$ is $\left(\frac{C}{A}, 0\right)$.

APPLICATIONS *Assume straight-line depreciation or straight-line appreciation.*

79. Finding a depreciation equation A truck was purchased for \$19,984. Its salvage value at the end of 8 years is expected to be \$1,600. Find the depreciation equation.

80. Finding a depreciation equation A business purchased the computer shown in Illustration 1. It will be depreciated over a 5-year period, when it will probably be worth \$200. Find the depreciation equation.

81. Finding an appreciation equation A famous oil painting was purchased for \$250,000 and is expected to double in value in 5 years. Find the appreciation equation.

82. Finding an appreciation equation A house purchased for \$142,000 is expected to double in value in 8 years. Find its appreciation equation.

83. Finding a depreciation equation Find the depreciation equation for the TV in the want ad in Illustration 2.

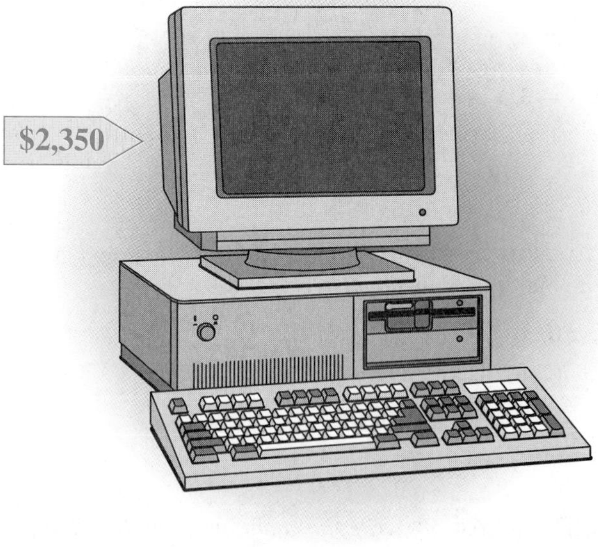

ILLUSTRATION 1

For Sale: 3-year-old 54-inch TV,
$1,750 new. Asking $800.
Call 875-5555. Ask for Mike.

ILLUSTRATION 2

84. Depreciating a lawn mower A lawn mower cost $450 when new and is expected to last 10 years. What will it be worth in $6\frac{1}{2}$ years?

85. Salvage value A copy machine that cost $1,750 when new will be depreciated at the rate of $180 per year. If the useful life of the copier is 7 years, find its salvage value.

86. Annual rate of depreciation A machine that cost $47,600 when new will have a salvage value of $500 after its useful life of 15 years. Find its annual rate of depreciation.

87. Real estate A vacation home is expected to appreciate about $4,000 per year. If the home will be worth $122,000 in 2 years, what will it be worth in 10 years?

88. Car repair A garage charges a fixed amount, plus an hourly rate, to service a car. Use the information in Illustration 3 to find the hourly rate.

A1 Car Repair	
Typical Charges	
2 hours	$143
5 hours	$320

ILLUSTRATION 3

89. Printer charges A printer charges a fixed setup cost, plus $15 for every 100 copies. If 300 copies cost $75, how much will 1,000 copies cost?

90. Predicting burglaries A police department knows that city growth and the number of burglaries are related by a linear equation. City records show that 575 burglaries were reported in a year when the local population was 77,000, and 675 were reported in a year when the population was 87,000. How many burglaries can be expected when the population reaches 110,000?

WRITING

91. Explain how to find the equation of a line passing through two given points.

92. In straight-line depreciation, explain why the slope of the line is called the *rate of depreciation.*

SOMETHING TO THINK ABOUT

In Exercises 93–98, investigate the properties of the slope and the y-intercept by experimenting with the following problems.

93. Graph $y = mx + 2$ for several positive values of m. What do you notice?

94. Graph $y = mx + 2$ for several negative values of m. What do you notice?

95. Graph $y = 2x + b$ for several increasing positive values of b. What do you notice?

96. Graph $y = 2x + b$ for several decreasing negative values of b. What do you notice?

97. How will the graph of $y = \frac{1}{2}x + 5$ compare to the graph of $y = \frac{1}{2}x - 5$?

98. How will the graph of $y = \frac{1}{2}x - 5$ compare to the graph of $y = \frac{1}{2}x$?

99. If the graph of $y = ax + b$ passes through quadrants I, II, and IV, what can be known about the constants a and b?

100. The graph of $Ax + By = C$ passes only through the quadrants I and IV. What is known about the constants A, B, and C?

2.5 Introduction to Functions

■ FUNCTIONS ■ FUNCTION NOTATION ■ FINDING DOMAINS AND RANGES OF FUNCTIONS
■ THE VERTICAL LINE TEST ■ LINEAR FUNCTIONS

Getting Ready *If $y = \frac{3}{2}x - 2$, find the value of y for each value of x.*

1. $x = 2$ **2.** $x = 6$ **3.** $x = -12$ **4.** $x = -\frac{1}{2}$

■ FUNCTIONS

If x and y are real numbers, an equation in x and y determines a correspondence between the values of x and y. To see how, we consider the equation $y = \frac{1}{2}x + 3$. To find the value of y (called an **output value**) that corresponds to $x = 4$ (called an **input value**), we substitute 4 for x and simplify.

$$y = \frac{1}{2}x + 3$$

$$y = \frac{1}{2}(4) + 3 \quad \text{Substitute the input value of 4 for } x.$$

$$= 2 + 3$$

$$= 5$$

The ordered pair (4, 5) satisfies the equation and shows that a y-value of 5 corresponds to an x-value of 4. This ordered pair and others that satisfy the equation appear in the table shown in Figure 2-37. The graph of the equation also appears in the figure.

To see how the table determines the correspondence, we simply find an input in the x-column and then read across to find the corresponding output in the y-column. For example, if we select 2 as an input value, we get 4 as an output value. Thus, a y value of 4 corresponds to an x value of 2.

To see how the graph determines the correspondence, we draw a vertical and a horizontal line through any point (say, point P) on the graph, as shown in Figure 2-37. Because these lines intersect the x-axis at 4 and the y-axis at 5, the point $P(4, 5)$ associates 5 on the y-axis with 4 on the x-axis. This shows that a y value of 5 corresponds to an x value of 4.

In this example, the set of all inputs x is the set of real numbers. The set of all outputs y is also the set of real numbers.

When a correspondence is set up by an equation, a table, or a graph, in which only one y value corresponds to each x value, we call the correspondence a **function.** The set of input values x is called the **domain** of the function, and the set of output values y is called the **range.** Since the value of y usually depends on the number x, we call y the **dependent variable** and x the **independent variable.**

$y = \frac{1}{2}x + 3$

x	y	
−2	2	A y value of 2 corresponds to an x value of −2.
0	3	A y value of 3 corresponds to an x value of 0.
2	4	A y value of 4 corresponds to an x value of 2.
4	5	A y value of 5 corresponds to an x value of 4.
6	6	A y value of 6 corresponds to an x value of 6.
↑ Inputs	↑ Outputs	

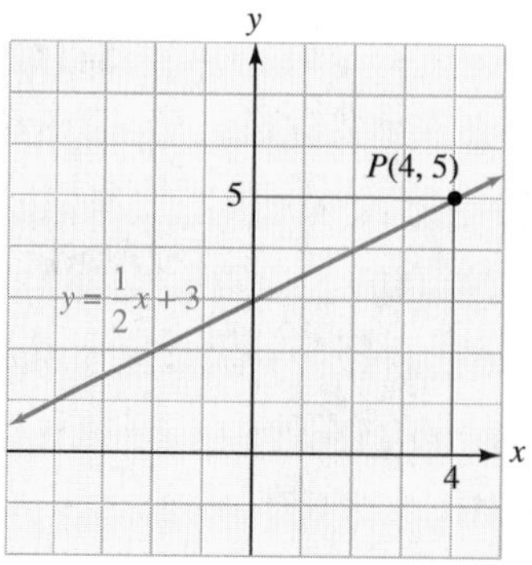

FIGURE 2-37

> **Functions**
> A **function** is a correspondence between a set of input values x (called the **domain**) and a set of output values y (called the **range**), where exactly one y value in the range corresponds to each number x in the domain.

EXAMPLE 1 Does $y = 2x - 3$ define y to be a function of x? If so, find its domain and range and illustrate the function with a table and graph.

Solution For a function to exist, every number x must determine one value of y. To find y in the equation $y = 2x - 3$, we multiply x by 2 and then subtract 3. Since this arithmetic gives one result, each choice of x determines one value of y. Thus, the equation does define y to be a function of x.

Since the input x can be any real number, the domain of the function is the set of real numbers, denoted by the interval $(-\infty, \infty)$. Since the output y can be any real number, the range is also the set of real numbers, denoted as $(-\infty, \infty)$.

A table of values and the graph appear in Figure 2-38.

$y = 2x - 3$

x	y
−4	−11
−2	−7
0	−3
2	1
4	5
6	9
The inputs can be any real number.	The outputs can be any real number.

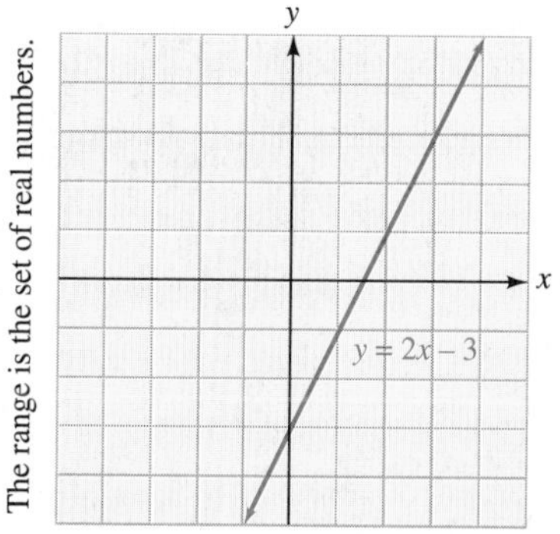

The domain is the set of real numbers.

FIGURE 2-38

■

Self Check Does $y = -2x + 3$ define y to be a function of x?

Answer yes

EXAMPLE 2 Does $y^2 = x$ define y to be a function of x?

Solution For a function to exist, each number x must determine one value of y. If we let $x = 16$, for example, y could be either 4 or -4, because $4^2 = 16$ and $(-4)^2 = 16$. Since more than one value of y is determined when $x = 16$, the equation does not represent a function. ■

Self Check Does $|y| = x$ define y to be a function of x?

Answer no

FUNCTION NOTATION

There is a special notation that we will use to denote functions.

Function Notation
The notation $y = f(x)$ denotes that the variable y is a function of x.

The notation $y = f(x)$ is read as "y equals f of x." Note that y and $f(x)$ are two notations for the same quantity. Thus, the equations $y = 4x + 3$ and $f(x) = 4x + 3$ are equivalent.

WARNING! The notation $f(x)$ does not means "f times x."

The notation $y = f(x)$ provides a way of denoting the value of y (the dependent variable) that corresponds to some number x (the independent variable). For example, if $y = f(x)$, the value of y that is determined by $x = 3$ is denoted by $f(3)$.

EXAMPLE 3 Let $f(x) = 4x + 3$. Find **a.** $f(3)$, **b.** $f(-1)$, **c.** $f(0)$, and **d.** $f(r)$.

Solution

a. We replace x with 3:

$$\begin{aligned} f(x) &= 4x + 3 \\ f(3) &= 4(3) + 3 \\ &= 12 + 3 \\ &= 15 \end{aligned}$$

b. We replace x with -1:

$$\begin{aligned} f(x) &= 4x + 3 \\ f(-1) &= 4(-1) + 3 \\ &= -4 + 3 \\ &= -1 \end{aligned}$$

c. We replace x with 0:

$$\begin{aligned} f(x) &= 4x + 3 \\ f(0) &= 4(0) + 3 \\ &= 3 \end{aligned}$$

d. We replace x with r:

$$\begin{aligned} f(x) &= 4x + 3 \\ f(r) &= 4r + 3 \end{aligned}$$

■

Self Check If $f(x) = -2x - 1$, find **a.** $f(2)$ and **b.** $f(-3)$.

Answer **a.** -5, **b.** 5

To see why function notation is helpful, consider the following sentences:

1. In the equation $y = 4x + 3$, find the value of y when x is 3.
2. In the equation $f(x) = 4x + 3$, find $f(3)$.

Statement 2, which uses $f(x)$ notation, is much more concise.

We can think of a function as a machine that takes some input x and turns it into some output $f(x)$, as shown in Figure 2-39(a). The machine shown in Figure 2-39(b) turns the input number 2 into the output value -3 and turns the input number 6 into the output value -11. The set of numbers that we can put into the machine is the domain of the function, and the set of numbers that comes out is the range.

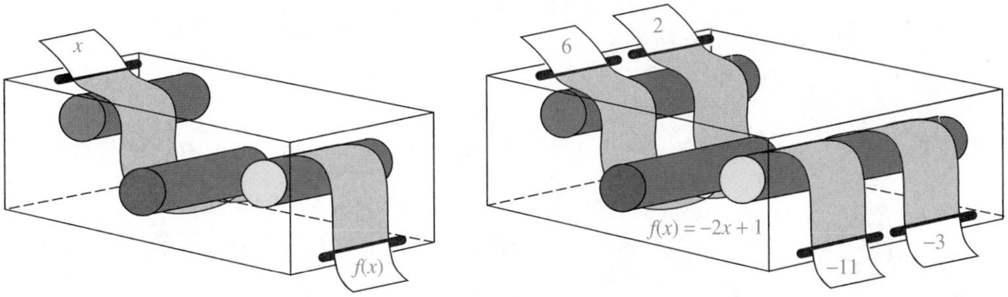

(a) (b)

FIGURE 2-39

The letter f used in the notation $y = f(x)$ represents the word *function.* However, other letters can be used to represent functions. The notations $y = g(x)$ and $y = h(x)$ also denote functions involving the independent variable x.

In Example 4, the equation $y = g(x) = x^2 - 2x$ determines a function, because every possible value of x gives a single value of $g(x)$.

a,d **EXAMPLE 4** Let $g(x) = x^2 - 2x$. Find **a.** $g(\frac{2}{5})$, **b.** $g(s)$, **c.** $g(s^2)$, and **d.** $g(-t)$.

Solution **a.** We replace x with $\frac{2}{5}$:

$$g(x) = x^2 - 2x$$
$$g\left(\frac{2}{5}\right) = \left(\frac{2}{5}\right)^2 - 2\left(\frac{2}{5}\right)$$
$$= \frac{4}{25} - \frac{4}{5}$$
$$= -\frac{16}{25}$$

b. We replace x with s:

$$g(x) = x^2 - 2x$$
$$g(s) = s^2 - 2s$$

c. We replace x with s^2:

$$g(x) = x^2 - 2x$$
$$g(s^2) = (s^2)^2 - 2s^2$$
$$= s^4 - 2s^2$$

d. We replace x with $-t$:

$$g(x) = x^2 - 2x$$
$$g(-t) = (-t)^2 - 2(-t)$$
$$= t^2 + 2t$$

■

Self Check Let $h(x) = -x^2 + 3$. Find **a.** $h(2)$ and **b.** $h(-a)$.

Answers **a.** -1, **b.** $-a^2 + 3$

EXAMPLE 5 Let $f(x) = 4x - 1$. Find **a.** $f(3) + f(2)$ and **b.** $f(a) - f(b)$.

Solution **a.** We find $f(3)$ and $f(2)$ separately.

$$f(x) = 4x - 1 \qquad f(x) = 4x - 1$$
$$f(3) = 4(3) - 1 \qquad f(2) = 4(2) - 1$$
$$= 12 - 1 \qquad = 8 - 1$$
$$= 11 \qquad = 7$$

We then add the results to obtain $f(3) + f(2) = 11 + 7 = 18$.

b. We find $f(a)$ and $f(b)$ separately.

$$f(x) = 4x - 1 \qquad f(x) = 4x - 1$$
$$f(a) = 4a - 1 \qquad f(b) = 4b - 1$$

We then subtract the results to obtain

$$f(a) - f(b) = (4a - 1) - (4b - 1)$$
$$= 4a - 1 - 4b + 1$$
$$= 4a - 4b$$

■

Self Check Let $g(x) = -2x + 3$. Find **a.** $g(-2) + g(3)$ and **b.** $g(\frac{1}{2}) - g(2)$.

Answers **a.** 4, **b.** 3

FINDING DOMAINS AND RANGES OF FUNCTIONS

EXAMPLE 6 Find the domains and ranges of the functions defined by **a.** the ordered pairs $(-2, 4)$, $(0, 6)$, $(2, 8)$ and **b.** the equation $y = \frac{1}{x-2}$.

Solution **a.** The ordered pairs can be placed in a table to show a correspondence between x and y where a single value of y corresponds to each x.

x	y	
-2	4	4 corresponds to -2.
0	6	6 corresponds to 0.
2	8	8 corresponds to 2.

The domain is the set of numbers x: $\{-2, 0, 2\}$. The range is the set of values y: $\{4, 6, 8\}$.

b. The number 2 cannot be substituted for x, because that would make the denominator equal to zero. Since any real number except 2 can be substituted for x in the equation $y = \frac{1}{x-2}$, the domain is the set of all real numbers but 2. This is the interval $(-\infty, 2) \cup (2, \infty)$.

Since a fraction with a numerator of 1 cannot be 0, the range is the set of all real numbers but 0. This is the interval $(-\infty, 0) \cup (0, \infty)$. ■

Self Check Find the domains and ranges of the functions defined by **a.** the ordered pairs $(-3, 5)$, $(-2, 7)$, and $(1, 11)$ and **b.** the equation $y = \frac{2}{x+3}$.

Answers **a.** $\{-3, -2, 1\}$, $\{5, 7, 11\}$, **b.** $(-\infty, -3) \cup (-3, \infty)$, $(-\infty, 0) \cup (0, \infty)$

The *graph of a function* is the graph of the ordered pairs $(x, f(x))$ that define the function. For the graph in Figure 2-40, the domain is shown on the x-axis, and the range is shown on the y-axis. For any x in the domain, there corresponds one value $y = f(x)$ in the range.

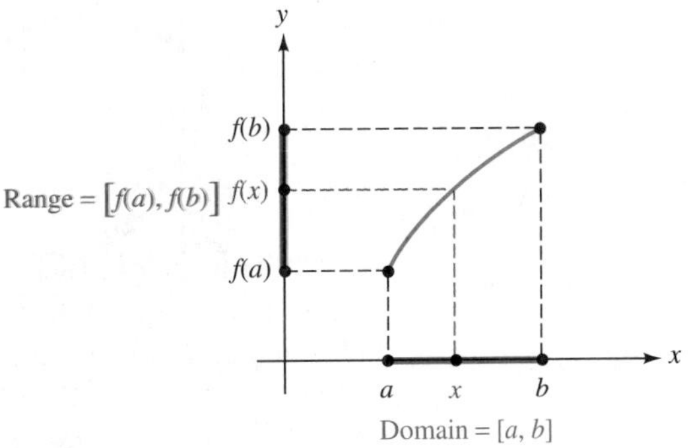

FIGURE 2-40

EXAMPLE 7 Find the domain and range of the function defined by $y = -2x + 1$.

Solution We graph the equation as in Figure 2-41. Since every real number x on the x-axis determines a corresponding value of y, the domain is the interval $(-\infty, \infty)$ shown on the x-axis. Since the values of y can be any real number, the range is the interval $(-\infty, \infty)$ shown on the y-axis.

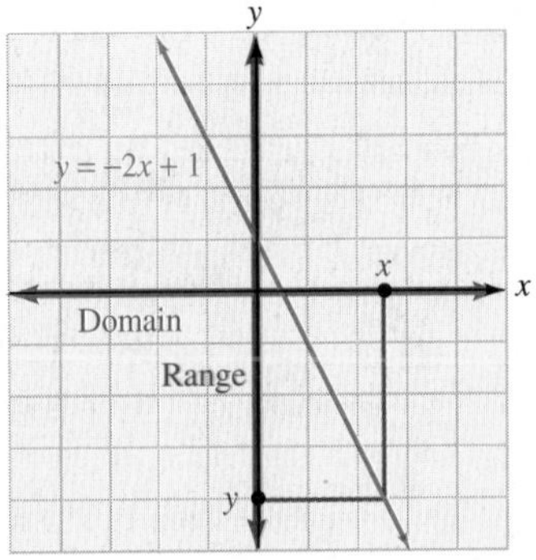

FIGURE 2-41 ■

THE VERTICAL LINE TEST

The **vertical line test** can be used to determine whether the graph of an equation represents a function. If any vertical line intersects a graph more than once, the graph cannot represent a function, because to one number x there would correspond more than one value of y.

The graph in Figure 2-42(a) represents a function, because every vertical line that intersects the graph does so exactly once. The graph in Figure 2-42(b) does not represent a function, because some vertical lines intersect the graph more than once.

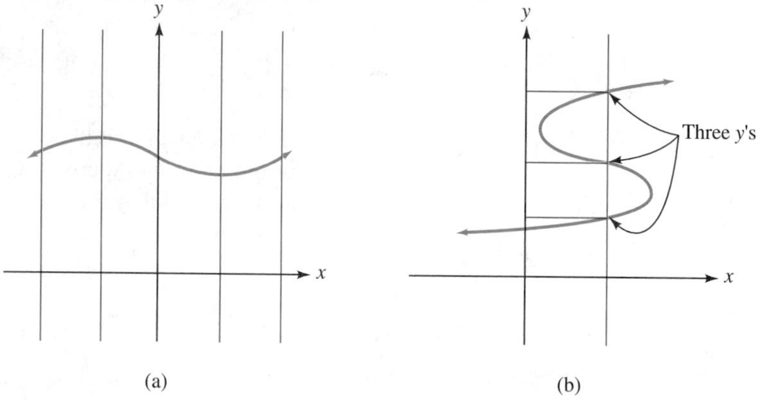

FIGURE 2-42

LINEAR FUNCTIONS

In Section 2.2, we graphed equations whose graphs were lines. These equations define basic functions, called **linear functions.**

> **Linear Functions**
> A **linear function** is a function defined by an equation that can be written in the form
>
> $$f(x) = mx + b \qquad \text{or} \qquad y = mx + b$$
>
> where m is the slope of the line graph and $(0, b)$ is the y-intercept.

EXAMPLE 8 Solve the equation $3x + 2y = 10$ for y to show that it defines a linear function. Then graph it to find its domain and range.

Solution We solve the equation for y as follows:

$$3x + 2y = 10$$

$$2y = -3x + 10 \qquad \text{Subtract } 3x \text{ from both sides.}$$

$$y = -\frac{3}{2}x + 5 \qquad \text{Divide both sides by 2.}$$

Because the given equation is written in the form $y = mx + b$, it defines a linear function. The slope of its line graph is $-\frac{3}{2}$, and the y-intercept is (0, 5). The graph appears in Figure 2-43. From the graph, we can see that both the domain and the range are the interval $(-\infty, \infty)$.

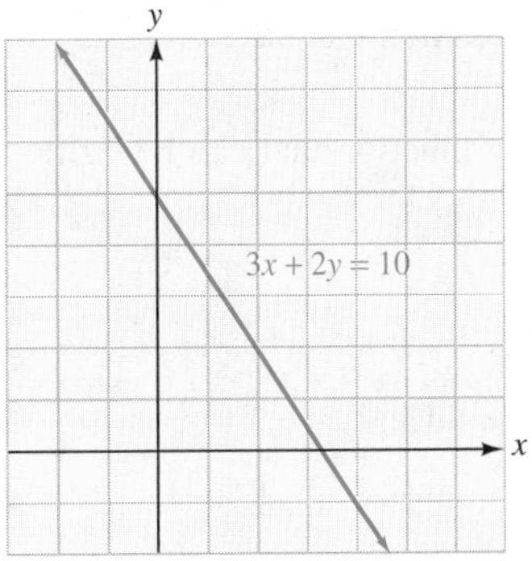

FIGURE 2-43

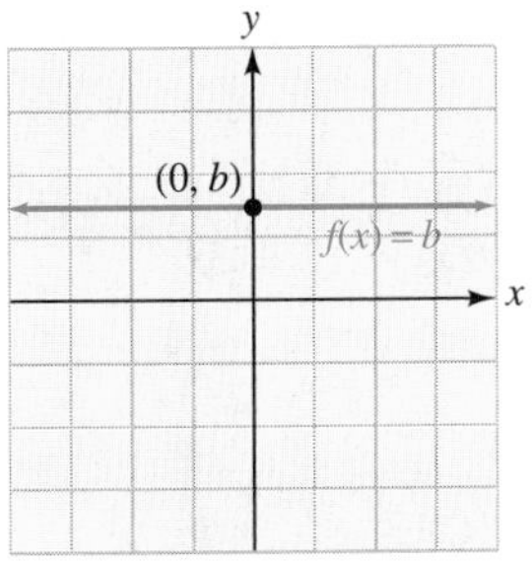

Constant function
Domain: $(-\infty, \infty)$
Range: $\{b\}$

FIGURE 2-44

A special case of a linear function is the **constant function,** defined by the equation $f(x) = b$, where b is a constant. Its graph, domain, and range are shown in Figure 2-44.

Orals *Tell whether each equation or inequality determines y to be a function of x.*

1. $y = 2x + 1$ **2.** $y \geq 2x$ **3.** $y^2 = x$

If $f(x) = 2x + 1$, *find*

4. $f(0)$ **5.** $f(1)$ **6.** $f(-2)$

EXERCISE 2.5

REVIEW *Solve each equation.*

1. $\dfrac{y+2}{2} = 4(y+2)$

2. $\dfrac{3z-1}{6} - \dfrac{3z+4}{3} = \dfrac{z+3}{2}$

3. $\dfrac{2a}{3} + \dfrac{1}{2} = \dfrac{6a-1}{6}$

4. $\dfrac{2x+3}{5} - \dfrac{3x-1}{3} = \dfrac{x-1}{15}$

VOCABULARY AND CONCEPTS *In Exercises 5–12, consider the function* $y = f(x) = 5x - 4$. *Fill in each blank to make a true statement.*

5. Any substitution for x is called an _____ value.

6. The value of __ is called the output value.

7. The independent variable is __.

8. The dependent variable is __.

9. A ______ is a correspondence between a set of input values and a set of output values, where each ______ value determines one ______ value.

10. In a function, the set of all inputs is called the ______ of the function.

11. In a function, the set of all output values is called the ______ of the function.

12. The notation $f(3)$ is the value of __ when $x = 3$.

13. The denominator of a fraction can never be __.

14. If a vertical line intersects a graph more than once, the graph ______ represent a function.

15. A linear function is any function that can be written in the form $y =$ ______.

16. In the function $f(x) = mx + b$, m is the ______ of its graph, and b is the y-coordinate of the ______.

PRACTICE *In Exercises 17–24, tell whether the equation determines y to be a function of x.*

17. $y = 2x + 3$ **18.** $y = -1$ **19.** $y = 2x^2$ **20.** $y^2 = x + 1$

21. $y = 3 + 7x^2$ **22.** $y^2 = 3 - 2x$ **23.** $x = |y|$ **24.** $y = |x|$

In Exercises 25–32, find f(3) and f(−1).

25. $f(x) = 3x$ **26.** $f(x) = -4x$ **27.** $f(x) = 2x - 3$ **28.** $f(x) = 3x - 5$

29. $f(x) = 7 + 5x$ **30.** $f(x) = 3 + 3x$ **31.** $f(x) = 9 - 2x$ **32.** $f(x) = 12 + 3x$

In Exercises 33–40, find f(2) and f(3).

33. $f(x) = x^2$ **34.** $f(x) = x^2 - 2$ **35.** $f(x) = x^3 - 1$ **36.** $f(x) = x^3$

37. $f(x) = (x + 1)^2$ **38.** $f(x) = (x - 3)^2$

39. $f(x) = 2x^2 - x$ **40.** $f(x) = 5x^2 + 2x$

In Exercises 41–48, find f(2) and f(−2).

41. $f(x) = |x| + 2$ **42.** $f(x) = |x| - 5$ **43.** $f(x) = x^2 - 2$ **44.** $f(x) = x^2 + 3$

45. $f(x) = \dfrac{1}{x + 3}$ **46.** $f(x) = \dfrac{3}{x - 4}$ **47.** $f(x) = \dfrac{x}{x - 3}$ **48.** $f(x) = \dfrac{x}{x^2 + 2}$

In Exercises 49–52, find g(w) and g(w + 1).

49. $g(x) = 2x$ **50.** $g(x) = -3x$ **51.** $g(x) = 3x - 5$ **52.** $g(x) = 2x - 7$

In Exercises 53–60, $f(x) = 2x + 1$. Find each value.

53. $f(3) + f(2)$ **54.** $f(1) - f(-1)$ **55.** $f(b) - f(a)$ **56.** $f(b) + f(a)$

57. $f(b) - 1$ **58.** $f(b) - f(1)$ **59.** $f(0) + f\left(-\frac{1}{2}\right)$ **60.** $f(a) + f(2a)$

In Exercises 61–64, find the domain and range of each function.

61. $\{(-2, 3), (4, 5), (6, 7)\}$ **62.** $\{(0, 2), (1, 2), (3, 4)\}$ **63.** $f(x) = \dfrac{1}{x - 4}$ **64.** $f(x) = \dfrac{5}{x + 1}$

In Exercises 65–68, each graph represents a correspondence between x and y. Tell whether the correspondence is a function. If it is, give its domain and range.

65.

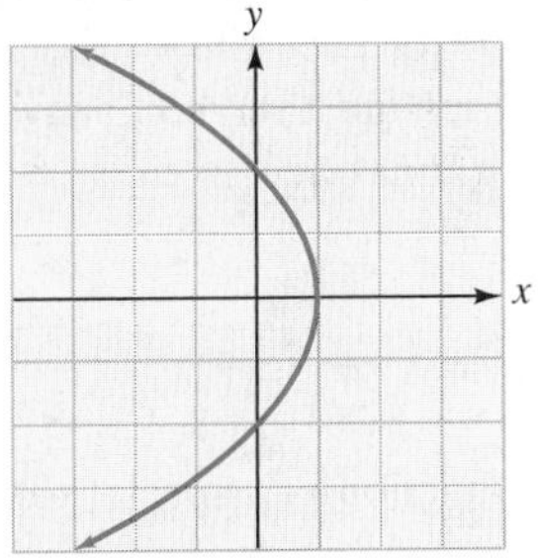

66.

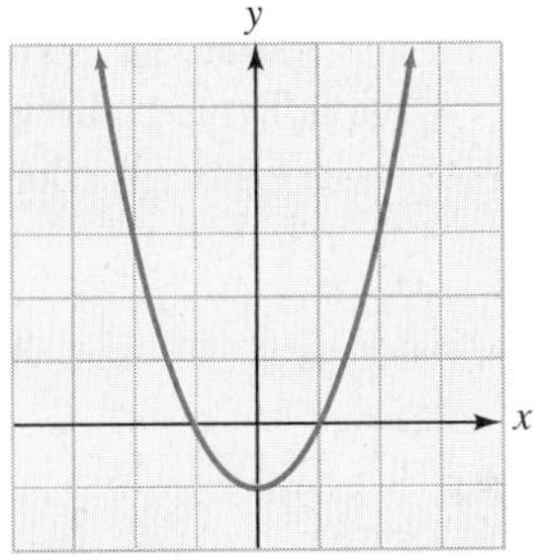

67.

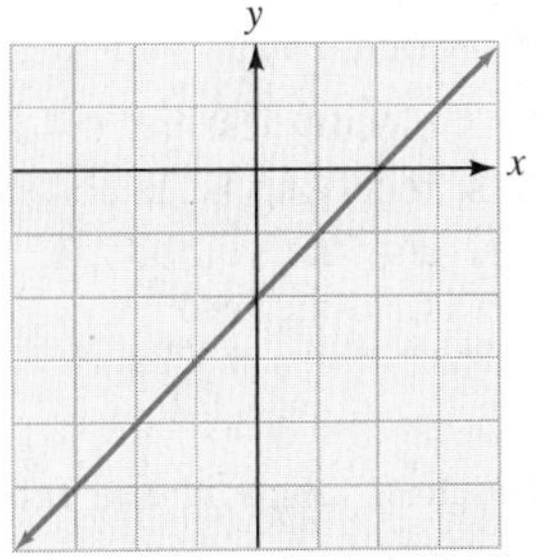

68.

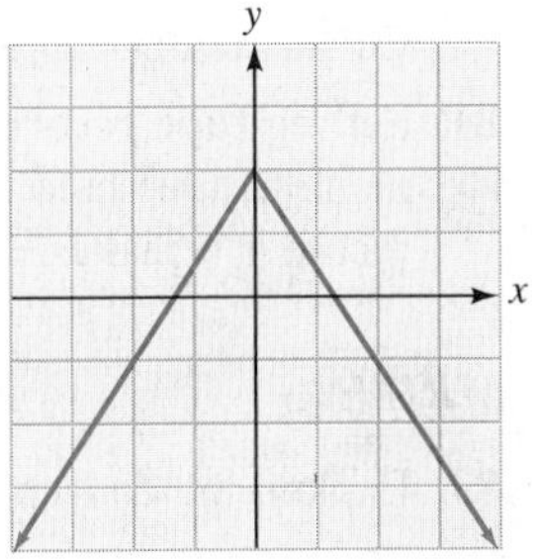

In Exercises 69–72, draw the graph of each linear function. Give the domain and range.

69. $f(x) = 2x - 1$

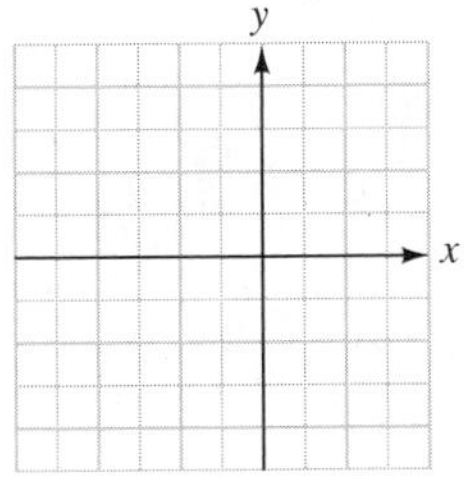

70. $f(x) = -x + 2$

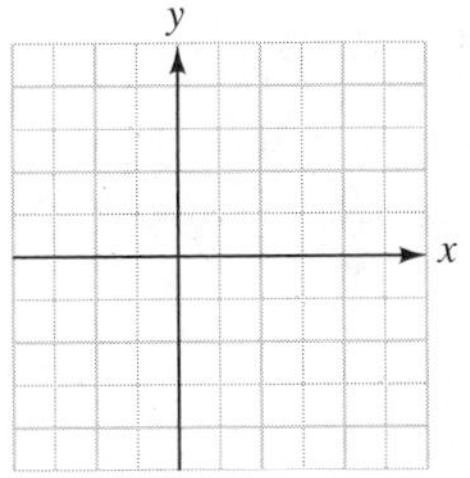

71. $2x - 3y = 6$

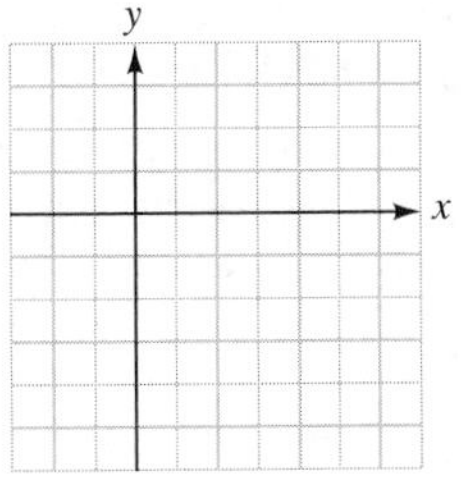

72. $3x + 2y = -6$

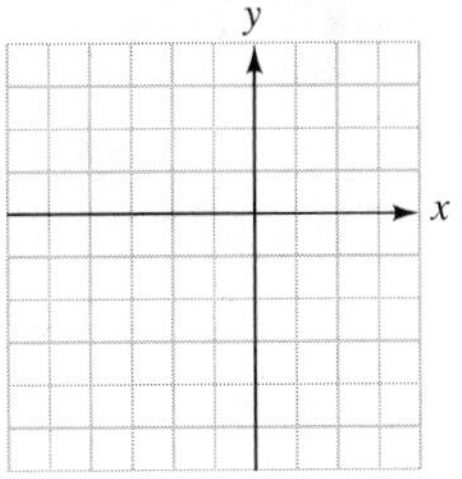

In Exercises 73–76, tell whether each equation defines a linear function.

73. $y = 3x^2 + 2$ **74.** $y = \dfrac{x - 3}{2}$ **75.** $x = 3y - 4$ **76.** $x = \dfrac{8}{y}$

APPLICATIONS

77. Ballistics A bullet shot straight upward is s feet above the ground after t seconds, where $s = f(t) = -16t^2 + 256t$. Find the height of the bullet 3 seconds after it is shot.

78. Artillery fire A mortar shell is s feet above the ground after t seconds, where $s = f(t) = -16t^2 + 512t + 64$. Find the height of the shell 20 seconds after it is fired.

79. Conversion from degrees Celsius to degrees Fahrenheit The temperature in degrees Fahrenheit that is equivalent to a temperature in degrees Celsius is given by the function $F(C) = \frac{9}{5}C + 32$. Find the Fahrenheit temperature that is equivalent to 25° C.

80. Conversion from degrees Fahrenheit to degrees Celsius The temperature in degrees Celsius that is equivalent to a temperature in degrees Fahrenheit is given by the function $C(F) = \frac{5}{9}F - \frac{160}{9}$. Find the Celsius temperature that is equivalent to 14°*F*.

81. Selling tape recorders An electronics firm manufactures tape recorders, receiving $120 for each recorder it makes. If x represents the number of recorders produced, the income received is determined by the *revenue function* $R(x) = 120x$. The manufacturer has fixed costs of $12,000 per month and variable costs of $57.50 for each recorder manufactured. Thus, the *cost function* is $C(x) = 57.50x + 12{,}000$. How many recorders must the company sell for revenue to equal cost?

82. Selling tires A tire company manufactures premium tires, receiving $130 for each tire it makes. If the manufacturer has fixed costs of $15,512.50 per month and variable costs of $93.50 for each tire manufactured, how many tires must the company sell for revenue to equal cost? (*Hint:* See Exercise 81.)

WRITING

83. Explain the concepts of function, domain, and range.

84. Explain why the constant function is a special case of a linear function.

SOMETHING TO THINK ABOUT *Let $f(x) = 2x + 1$ and $g(x) = x^2$. Assume that $f(x) \neq 0$ and $g(x) \neq 0$.*

85. Is $f(x) + g(x)$ equal to $g(x) + f(x)$?

86. Is $f(x) - g(x)$ equal to $g(x) - f(x)$?

2.6 Graphs of Other Functions

■ GRAPHS OF NONLINEAR FUNCTIONS ■ TRANSLATIONS OF GRAPHS ■ REFLECTIONS OF GRAPHS ■ SOLVING EQUATIONS WITH GRAPHING CALCULATORS

Getting Ready *Give the slope and the y-intercept of each linear function.*

1. $f(x) = 2x - 3$

2. $f(x) = -3x + 4$

Find the value of $f(x)$ when $x = 2$ and $x = -1$.

3. $f(x) = 5x - 4$

4. $f(x) = \frac{1}{2}x + 3$

■ GRAPHS OF NONLINEAR FUNCTIONS

If f is a function whose domain and range are sets of real numbers, its graph is the set of all points $(x, f(x))$ in the xy-plane. In other words, the graph of f is the graph of the equation $y = f(x)$. In this section, we will draw the graphs of many basic functions whose graphs are not straight lines.

The first basic function is $f(x) = x^2$ (or $y = x^2$), often called the **squaring function.**

EXAMPLE 1 Graph the function $f(x) = x^2$.

Solution We substitute values for x in the equation and compute the corresponding values of $f(x)$. For example, if $x = -3$, we have

$$\begin{aligned} f(x) &= x^2 \\ f(-3) &= (-3)^2 \quad \text{Substitute } -3 \text{ for } x. \\ &= 9 \end{aligned}$$

The ordered pair $(-3, 9)$ satisfies the equation and will lie on the graph. We list this pair and the others that satisfy the equation in the table shown in Figure 2-45. We plot the points and draw a smooth curve through them to get the graph, called a **parabola.**

$f(x) = x^2$

x	y	$(x, f(x))$
-3	9	$(-3, 9)$
-2	4	$(-2, 4)$
-1	1	$(-1, 1)$
0	0	$(0, 0)$
1	1	$(1, 1)$
2	4	$(2, 4)$
3	9	$(3, 9)$

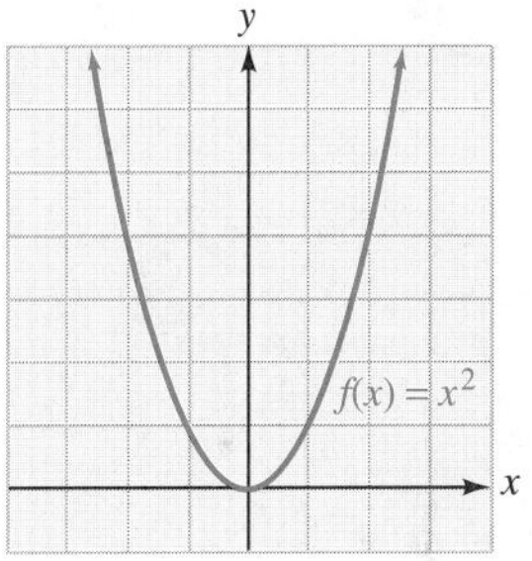

FIGURE 2-45

From the graph, we see that x can be any real number. This indicates that the domain of the squaring function is the set of real numbers, which is the interval $(-\infty, \infty)$. We can also see that y is always positive or zero. This indicates that the range is the set of nonnegative real numbers, which is the interval $[0, \infty)$. ■

Self Check Graph $f(x) = x^2 - 2$ and compare the graph to the graph of $f(x) = x^2$.

Answer The graph has the same shape but is 2 units lower.

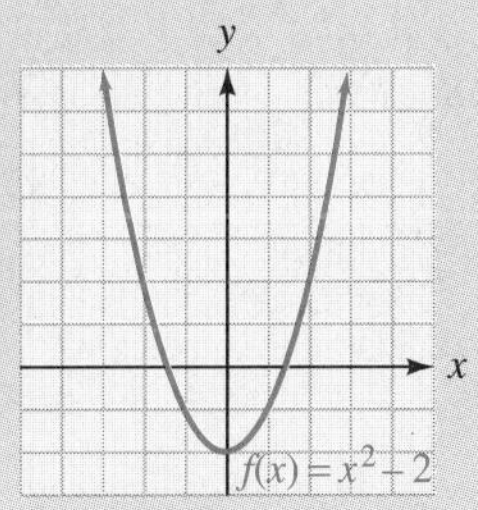

The second basic function is $f(x) = x^3$ (or $y = x^3$), often called the **cubing function.**

EXAMPLE 2 Graph the function $f(x) = x^3$.

Solution We substitute values for x in the equation and compute the corresponding values of $f(x)$. For example, if $x = -2$, we have

$$
\begin{aligned}
f(x) &= x^3 \\
f(-2) &= (-2)^3 && \text{Substitute } -2 \text{ for } x. \\
&= -8
\end{aligned}
$$

The ordered pair $(-2, -8)$ satisfies the equation and will lie on the graph. We list this pair and others that satisfy the equation in the table shown in Figure 2-46. We plot the points and draw a smooth curve through them to get the graph.

$f(x) = x^3$

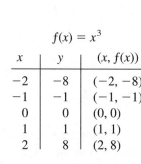

x	y	$(x, f(x))$
-2	-8	$(-2, -8)$
-1	-1	$(-1, -1)$
0	0	(0, 0)
1	1	(1, 1)
2	8	(2, 8)

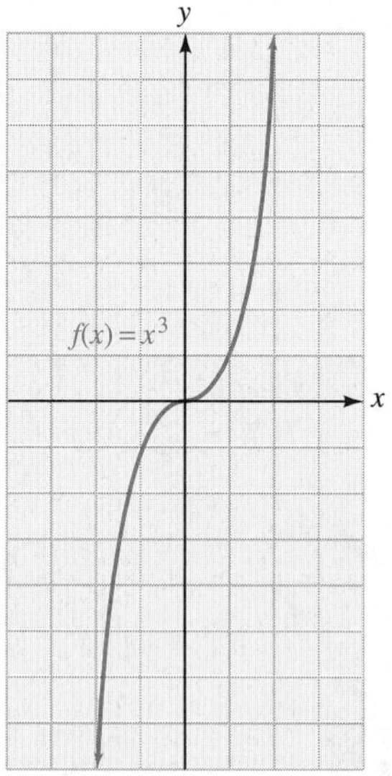

FIGURE 2-46

From the graph, we can see that x can be any real number. This indicates that the domain of the cubing function is the set of real numbers, which is the interval $(-\infty, \infty)$. We can also see that y can be any real number. This indicates that the range is the set of real numbers, which is the interval $(-\infty, \infty)$. ■

Self Check Graph $f(x) = x^3 + 1$ and compare the graph to the graph of $f(x) = x^3$.

Answer The graph has the same shape but is 1 unit higher.

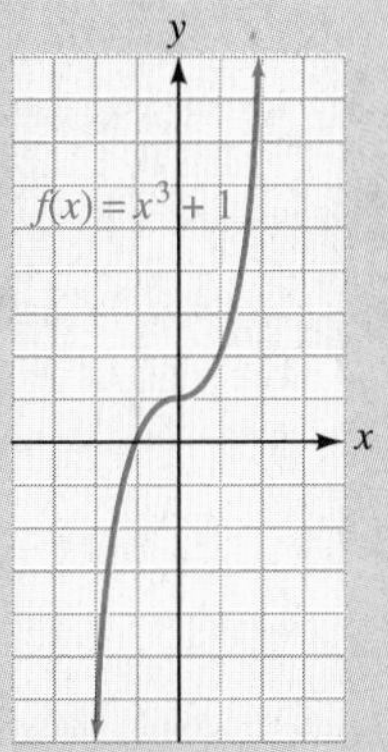

The third basic function is $f(x) = |x|$ (or $y = |x|$), often called the **absolute value function.**

EXAMPLE 3 Graph the function $f(x) = |x|$.

Solution We substitute values for x in the equation and compute the corresponding values of $f(x)$. For example, if $x = -3$, we have

$$\begin{aligned} f(x) &= |x| \\ f(-3) &= |-3| \quad \text{Substitute } -3 \text{ for } x. \\ &= 3 \end{aligned}$$

The ordered pair $(-3, 3)$ satisfies the equation and will lie on the graph. We list this pair and the others that satisfy the equation in the table shown in Figure 2-47. We plot the points and draw a V-shaped line through them to get the graph.

$f(x) = |x|$

x	y	$(x, f(x))$
−3	3	(−3, 3)
−2	2	(−2, 2)
−1	1	(−1, 1)
0	0	(0, 0)
1	1	(1, 1)
2	2	(2, 2)
3	3	(3, 3)

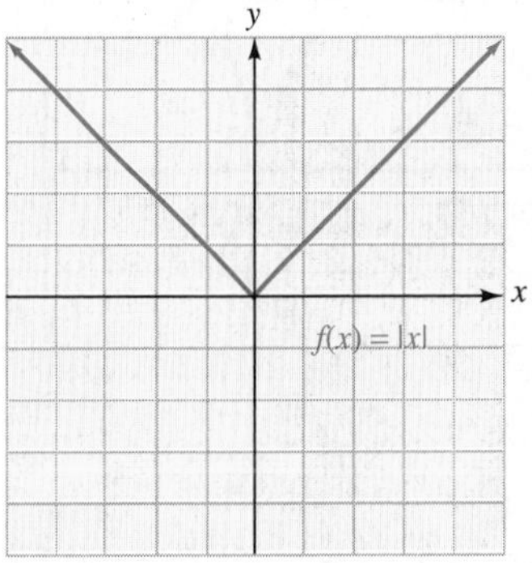

FIGURE 2-47

From the graph, we see that x can be any real number. This indicates that the domain of the absolute value function is the set of real numbers, which is the interval $(-\infty, \infty)$. We can also see that y is always positive or zero. This indicates that the range is the set of nonnegative real numbers, which is the interval $[0, \infty)$. ■

Self Check Graph $f(x) = |x - 2|$ and compare the graph to the graph of $f(x) = |x|$.

Answer The graph has the same shape but is 2 units to the right.

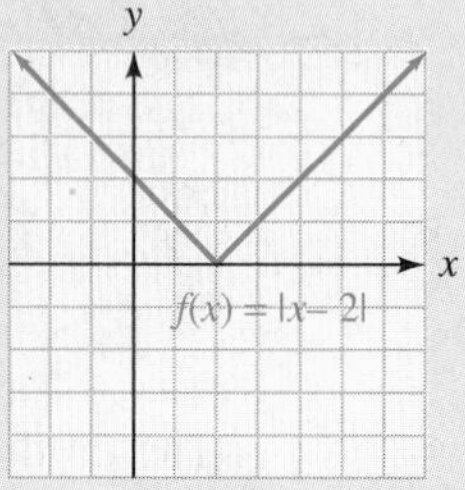

Graphing Functions

GRAPHING CALCULATORS

We can graph nonlinear functions with a graphing calculator. For example, to graph $f(x) = x^2$ in a standard window of $[-10, 10]$ for x and $[-10, 10]$ for y, we enter the function by typing x ^ 2 and press the GRAPH key. We will obtain the graph shown in Figure 2-48(a).

To graph $f(x) = x^3$, we enter the function by typing x ^ 3 and press the GRAPH key to obtain the graph in Figure 2-48(b). To graph $f(x) = |x|$, we enter the function by pressing the ABS key (or selecting "abs" from a menu), typing x, and pressing the GRAPH key to obtain the graph in Figure 2-48(c).

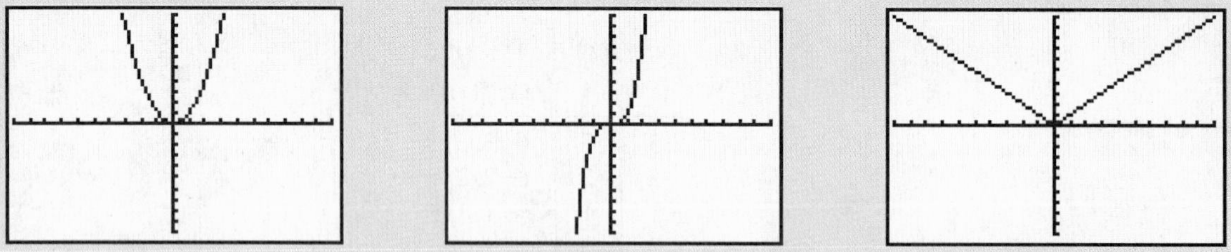

(a) (b) (c)

FIGURE 2-48

When using a graphing calculator, we must be sure that the viewing window does not show a misleading graph. For example, if we graph $f(x) = |x|$ in the window [0, 10] for x and [0, 10] for y, we will obtain a misleading graph that looks like a line. (See Figure 2-49.) This is not true. The proper graph is the V-shaped graph shown in Figure 2-48(c).

One of the challenges of using graphing calculators is finding an appropriate viewing window.

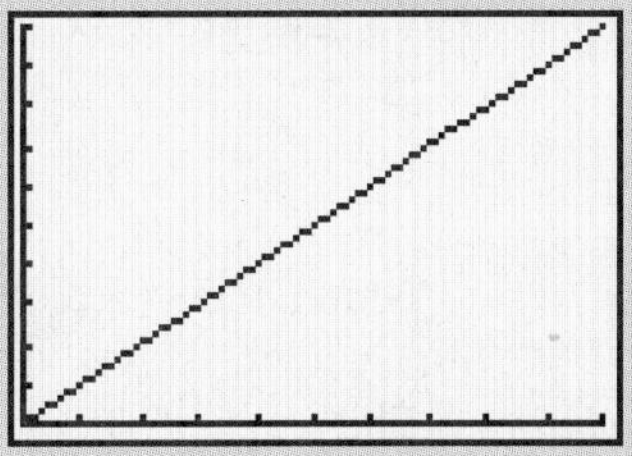

FIGURE 2-49

TRANSLATIONS OF GRAPHS

Examples 1–3 and their Self Checks suggest that the graphs of different functions may be identical except for their positions in the xy-plane. For example, Figure 2-50 shows the graph of $f(x) = x^2 + k$ for three different values of k. If $k = 0$, we get the graph of $f(x) = x^2$. If $k = 3$, we get the graph of $f(x) = x^2 + 3$, which is identical to the graph of $f(x) = x^2$ except that it is shifted 3 units upward. If $k = -4$, we get the graph of $f(x) = x^2 - 4$, which is identical to the graph of $f(x) = x^2$ except that it is shifted 4 units downward. These shifts are called **vertical translations.**

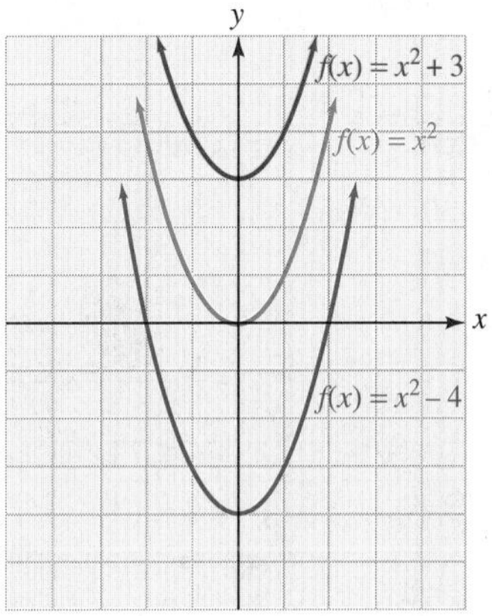

FIGURE 2-50

In general, we can make these observations.

Vertical Translations

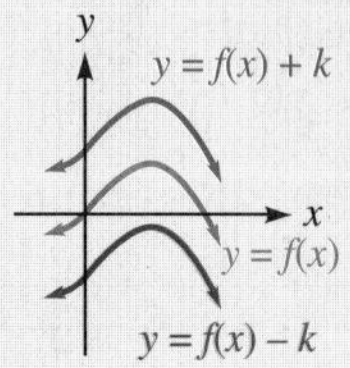

If f is a function and k is a positive number, then

- The graph of $y = f(x) + k$ is identical to the graph of $y = f(x)$ except that it is translated k units upward.
- The graph of $y = f(x) - k$ is identical to the graph of $y = f(x)$ except that it is translated k units downward.

EXAMPLE 4 Graph $f(x) = |x| + 2$.

Solution The graph of $f(x) = |x| + 2$ will be the same V-shaped graph as $f(x) = |x|$, except that it is shifted 2 units up. The graph appears in Figure 2-51.

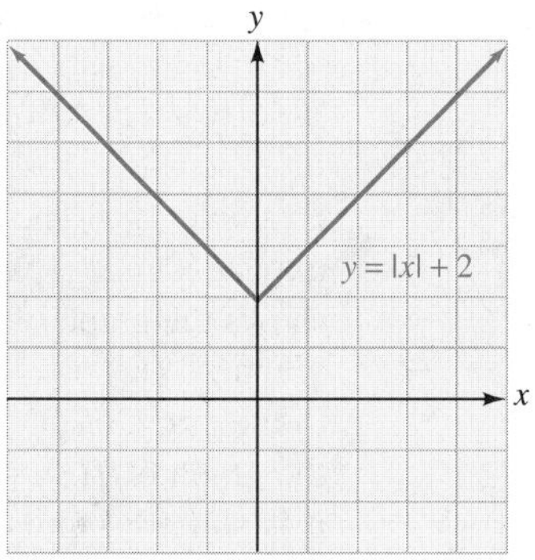

FIGURE 2-51

Self Check Graph $f(x) = |x| - 3$.

Answer

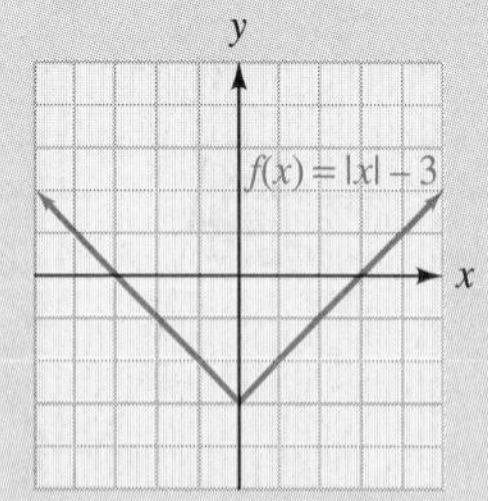

Figure 2-52 shows the graph of $f(x) = (x + h)^2$ for three different values of h. If $h = 0$, we get the graph of $f(x) = x^2$. The graph of $f(x) = (x - 3)^2$ is identical to the graph of $f(x) = x^2$, except that it is shifted 3 units to the right. The graph of

$f(x) = (x + 2)^2$ is identical to the graph of $f(x) = x^2$, except that it is shifted 2 units to the left. These shifts are called **horizontal translations.**

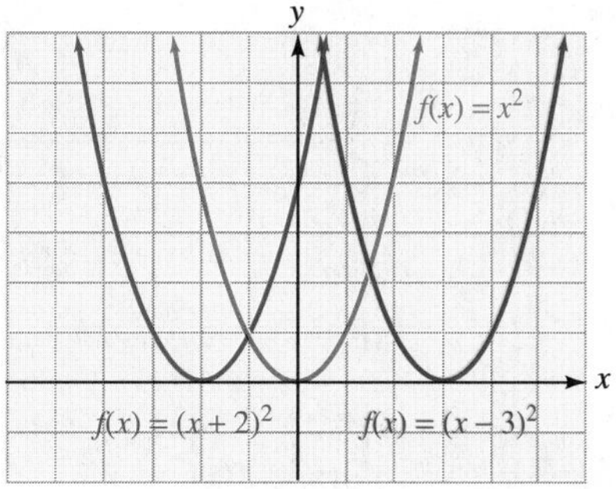

FIGURE 2-52

In general, we can make these observations.

Horizontal Translations

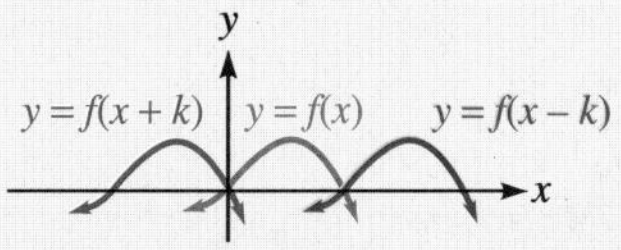

If f is a function and k is a positive number, then

- The graph of $y = f(x - k)$ is identical to the graph of $y = f(x)$ except that it is translated k units to the right.
- The graph of $y = f(x + k)$ is identical to the graph of $y = f(x)$ except that it is translated k units to the left.

EXAMPLE 5 Graph $f(x) = (x - 2)^2$.

Solution The graph of $f(x) = (x - 2)^2$ will be the same shape as the graph of $f(x) = x^2$ except that it is shifted 2 units to the right. The graph appears in Figure 2-53.

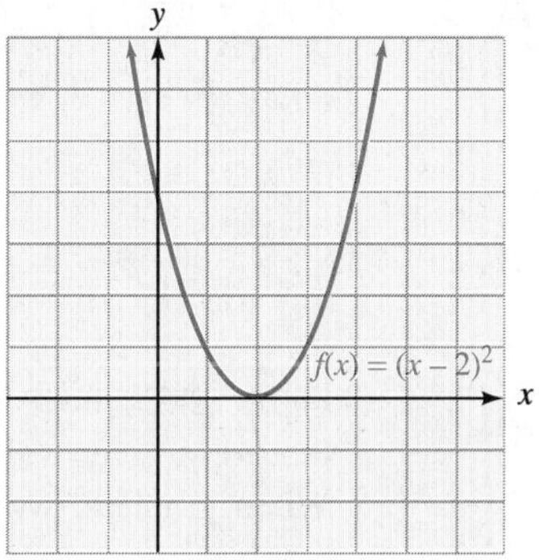

FIGURE 2-53

■

Self Check Graph $f(x) = (x + 3)^3$.

Answer

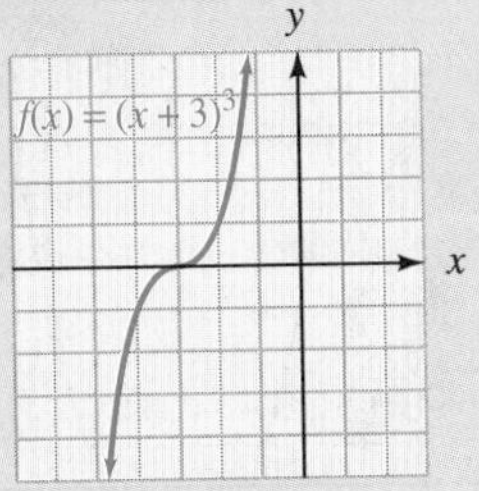

EXAMPLE 6 Graph $f(x) = (x - 3)^2 + 2$.

Solution We can graph this equation by translating the graph of $f(x) = x^2$ to the right 3 units and then up 2 units, as shown in Figure 2-54.

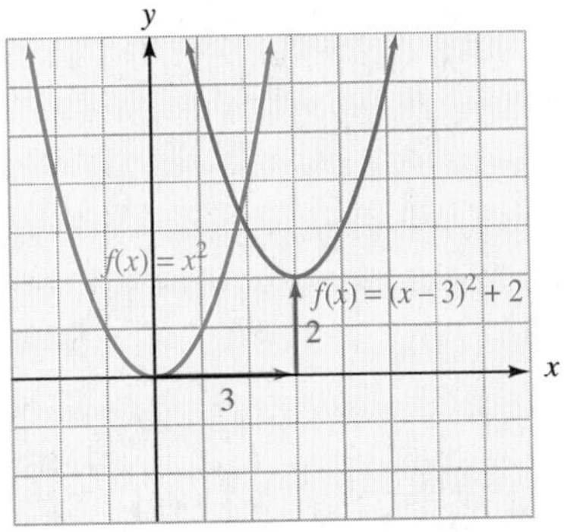

FIGURE 2-54

Self Check Graph $f(x) = |x + 2| - 3$.

Answer

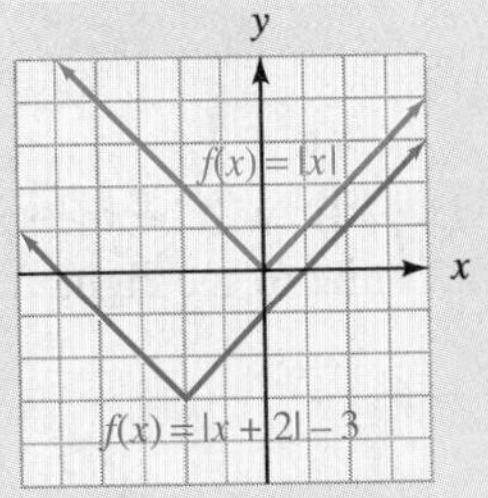

REFLECTIONS OF GRAPHS

Figure 2-55 shows tables of solutions for $f(x) = x^2$ and for $f(x) = -x^2$. We note that for a given value of x, the corresponding y values in the tables are *opposites*. When graphed, we see that the $-$ in $f(x) = -x^2$ has the effect of "flipping" the graph of $f(x) = x^2$ over the x-axis, so that the parabola opens downward. We say that the graph of $f(x) = -x^2$ is a **reflection** of the graph of $f(x) = x^2$ in the x-axis.

$f(x) = x^2$

x	y	$(x, f(x))$
-2	4	$(-2, 4)$
-1	1	$(-1, 1)$
0	0	$(0, 0)$
1	1	$(1, 1)$
2	4	$(2, 4)$

$f(x) = -x^2$

x	y	$(x, f(x))$
-2	-4	$(-2, -4)$
-1	-1	$(-1, -1)$
0	0	$(0, 0)$
1	-1	$(1, -1)$
2	-4	$(2, -4)$

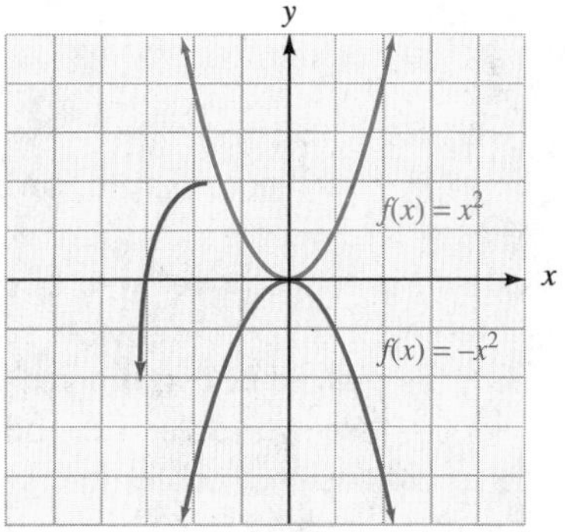

FIGURE 2-55

EXAMPLE 7 **Reflection of a graph.** Graph $f(x) = -x^3$.

Solution To graph $f(x) = -x^3$, we use the graph of $f(x) = x^3$ from Example 2. First, we reflect the portion of the graph of $f(x) = x^3$ in quadrant I to quadrant IV, as shown in Figure 2-56. Then we reflect the portion of the graph of $f(x) = x^3$ in quadrant III to quadrant II.

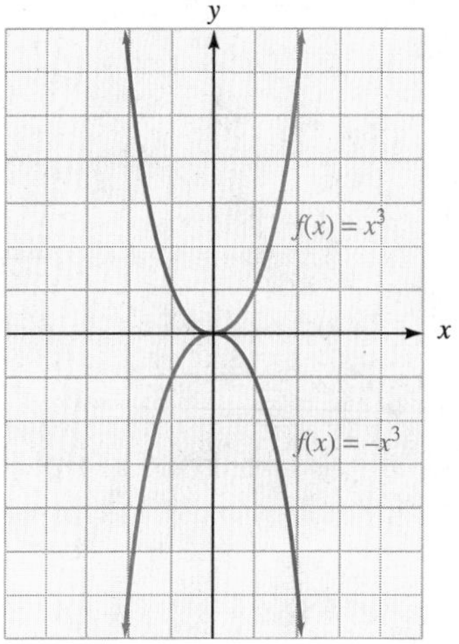

FIGURE 2-56 ■

Self Check Graph $f(x) = -|x|$.

Answer

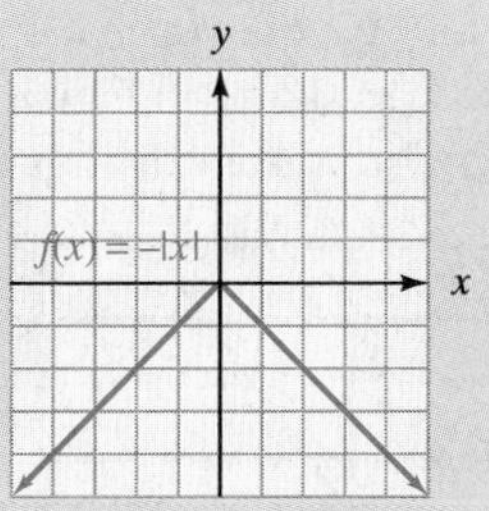

Reflection of a Graph

The graph of $y = -f(x)$ is the graph of $f(x)$ reflected about the x-axis.

PERSPECTIVE

Graphs in Space

In an *xy*-coordinate system, graphs of equations containing the two variables *x* and *y* are lines or curves. Other equations have more than two variables, and graphing them often requires some ingenuity and perhaps the aid of a computer. Graphs of equations with the three variables *x*, *y*, and *z* are viewed in a three-dimensional coordinate system with three axes. The coordinates of points in a three-dimensional coordinate system are ordered triples (x, y, z). For example, the points $P(2, 3, 4)$ and $Q(-1, 2, 3)$ are plotted in Illustration 1.

Graphs of equations in three variables are not lines or curves, but flat planes or curved surfaces. Only the simplest of these equations can be conveniently graphed by hand; a computer provides the best images of others. The graph in Illustration 2 is called a paraboloid; it is the three-dimensional version of a parabola. Illustration 3 models a portion of the vibrating surface of a drum head.

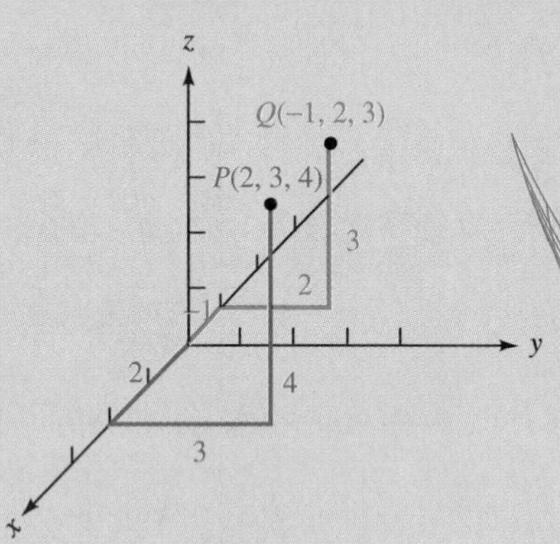

ILLUSTRATION 1

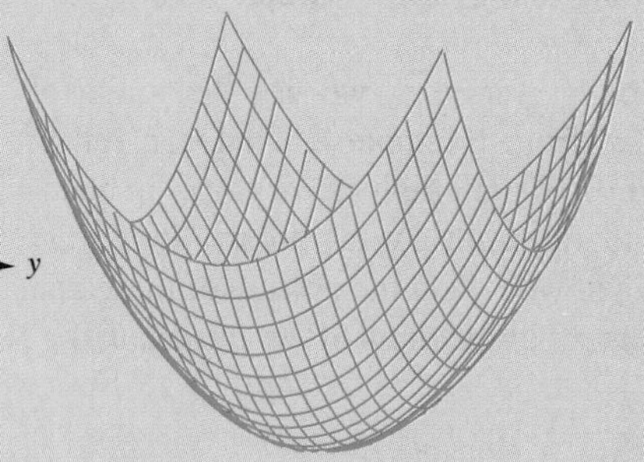

ILLUSTRATION 2

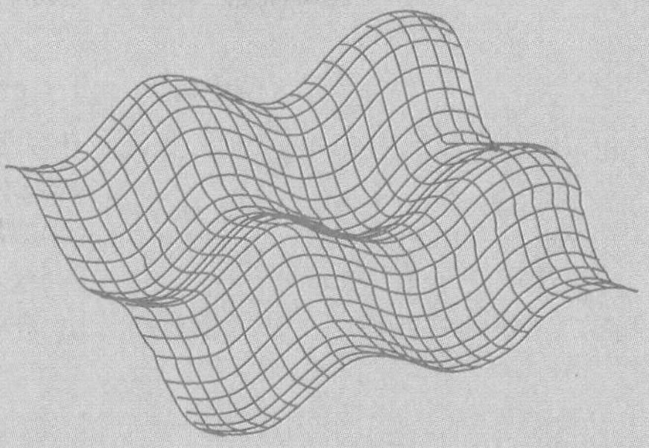

ILLUSTRATION 3

SOLVING EQUATIONS WITH GRAPHING CALCULATORS

Solving Equations

GRAPHING CALCULATORS

To solve the equation $2(x - 3) + 3 = 7$ with a graphing calculator, we can graph the left-hand side and the right-hand side of the equation in the same window, as shown in Figure 2-57(a). We then trace to find the coordinates of the point where the two graphs intersect, as shown in Figure 2-57(b). We can then zoom and trace again to get Figure 2-57(c). From the figure, we see that $x = 5$.

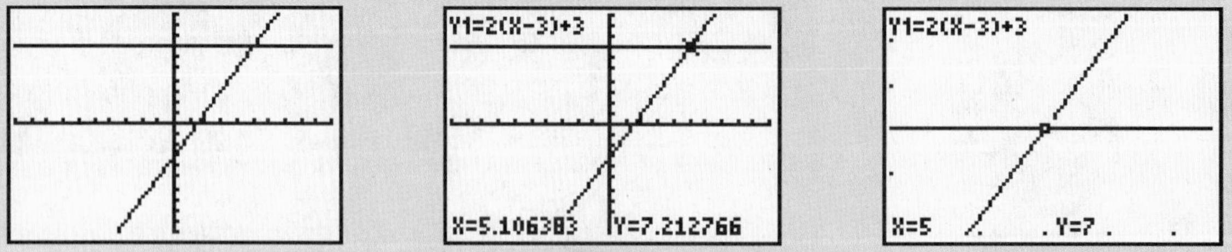

(a) (b) (c)

FIGURE 2-57

Orals

1. Describe a parabola.
2. Describe the graph of $f(x) = |x| + 3$.
3. Describe the graph of $f(x) = x^3 - 4$.
4. Tell why the choice of a viewing window is important.

EXERCISE 2.6

REVIEW

1. List the prime numbers between 40 and 50.
2. State the associative property of addition.
3. State the commutative property of multiplication.
4. What is the additive identity element?
5. What is the multiplicative identity element?
6. Find the multiplicative inverse of $\frac{5}{3}$.

VOCABULARY AND CONCEPTS *Fill in each blank to make a true statement.*

7. The function $f(x) = x^2$ is called the ______ function.
8. The function $f(x) = x^3$ is called the ______ function.
9. The function $f(x) = |x|$ is called the ______ function.
10. Shifting the graph of an equation up or down is called a ______ translation.
11. Shifting the graph of an equation to the left or to the right is called a ______ translation.
12. The graph of $f(x) = x^2 + 5$ is the same as the graph of $f(x) = x^2$ except that it is shifted __ units ___.
13. The graph of $f(x) = x^3 - 2$ is the same as the graph of $f(x) = x^3$ except that it is shifted __ units ______.
14. The graph of $f(x) = (x - 5)^3$ is the same as the graph of $f(x) = x^3$ except that it is shifted __ units ______.
15. The graph of $f(x) = (x + 4)^3$ is the same as the graph of $f(x) = x^3$ except that it is shifted __ units ______.
16. To solve an equation with a graphing calculator, graph ____ sides of the equation and find the ______ of the point where the graphs intersect.

PRACTICE *In Exercises 17–24, graph each function by plotting points. Check your work with a graphing calculator.*

17. $f(x) = x^2 - 3$

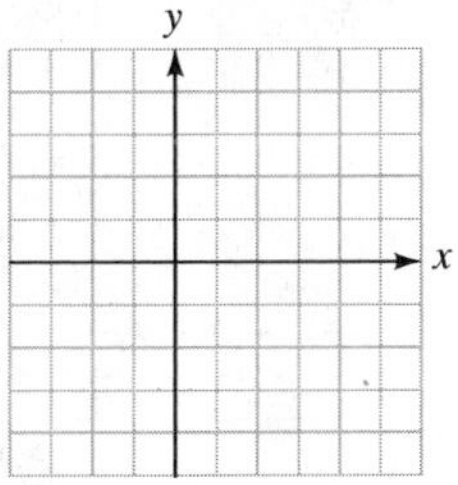

18. $f(x) = x^2 + 2$

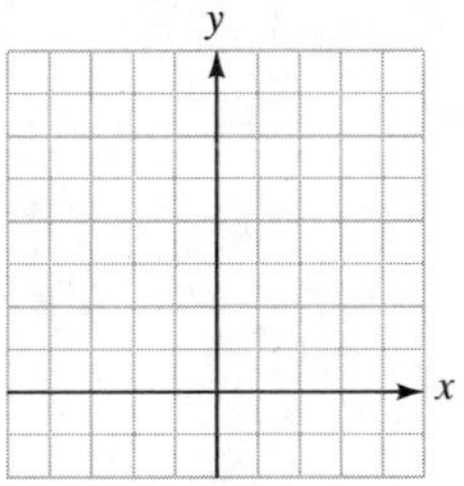

19. $f(x) = (x - 1)^3$

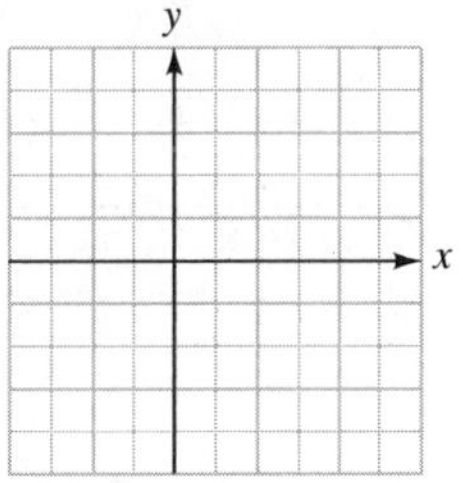

20. $f(x) = (x + 1)^3$

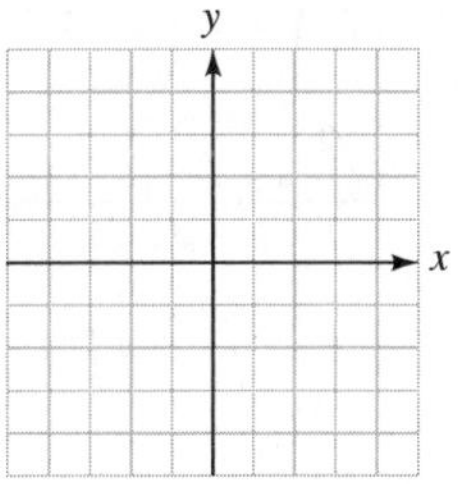

21. $f(x) = |x| - 2$

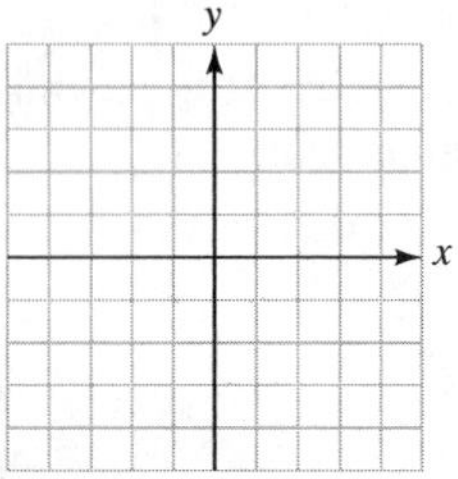

22. $f(x) = |x| + 1$

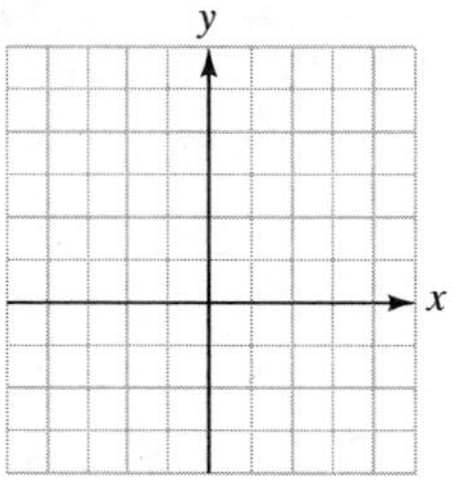

23. $f(x) = |x - 1|$

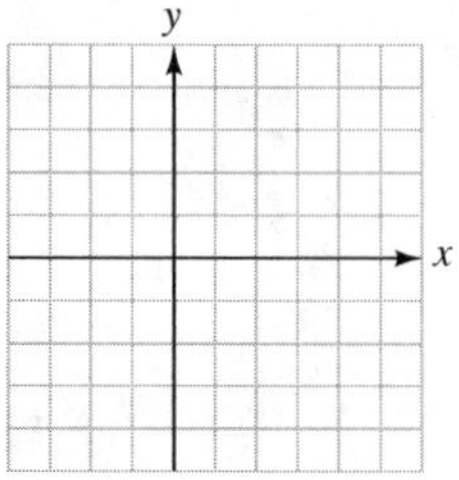

24. $f(x) = |x + 2|$

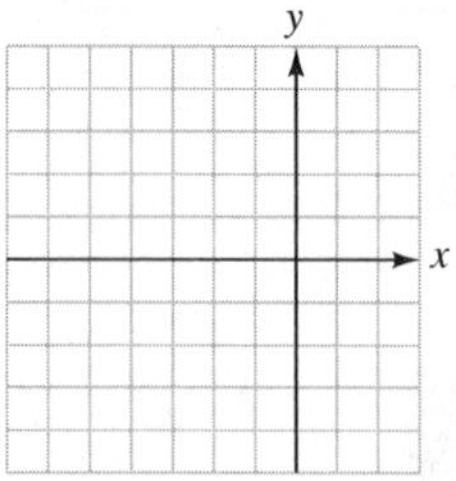

In Exercises 25–32, use a graphing calculator to graph each function, using values of $[-4, 4]$ *for x and* $[-4, 4]$ *for y. The graph is not what it appears to be. Pick a better viewing window and find the true graph.*

25. $f(x) = x^2 + 8$

26. $f(x) = x^3 - 8$

27. $f(x) = |x + 5|$

28. $f(x) = |x - 5|$

29. $f(x) = (x - 6)^2$

30. $f(x) = (x + 9)^2$

31. $f(x) = x^3 + 8$

32. $f(x) = x^3 - 12$

In Exercises 33–40, draw each graph using a translation of the graph of $f(x) = x^2$, $f(x) = x^3$, *or* $f(x) = |x|$.

33. $f(x) = x^2 - 5$

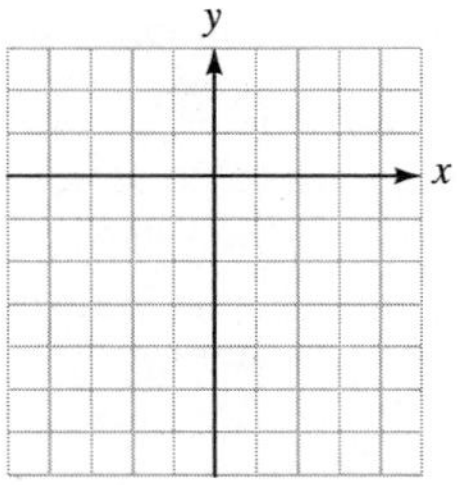

34. $f(x) = x^3 + 4$

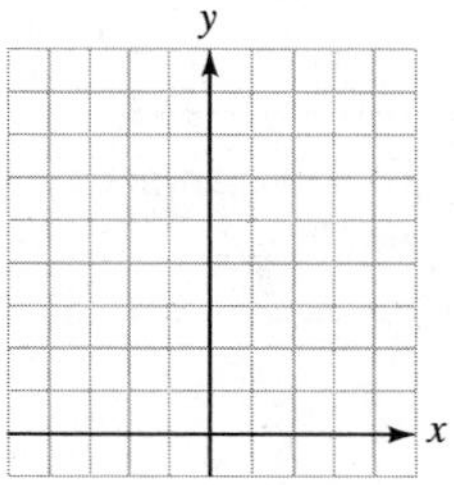

35. $f(x) = (x - 1)^3$

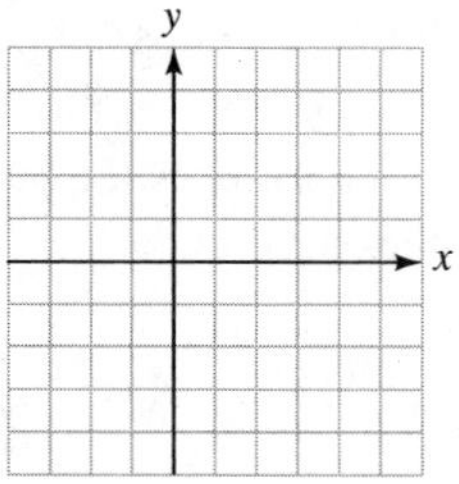

36. $f(x) = (x + 4)^2$

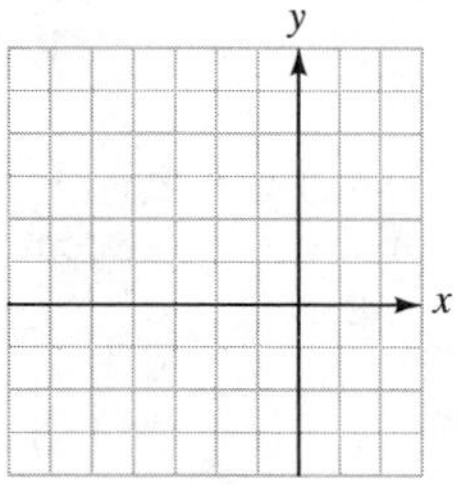

37. $f(x) = |x - 2| - 1$

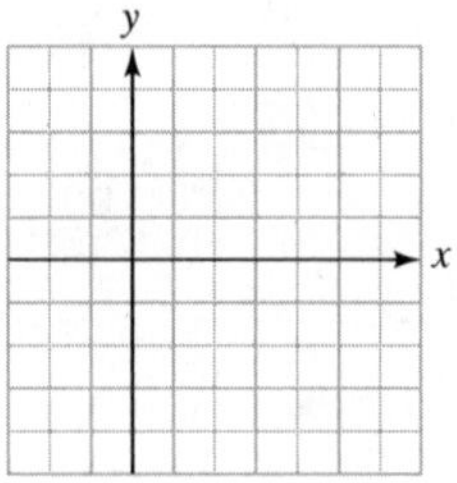

38. $f(x) = (x + 2)^2 - 1$

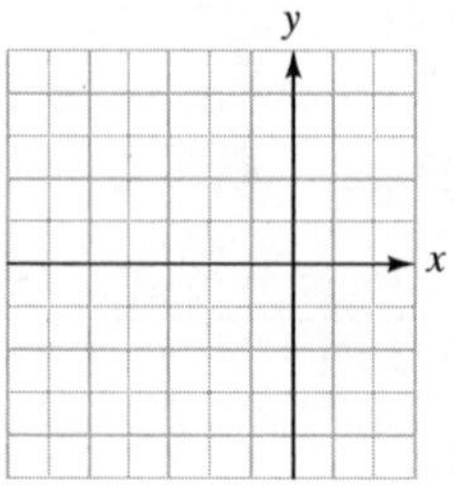

39. $f(x) = (x + 1)^3 - 2$

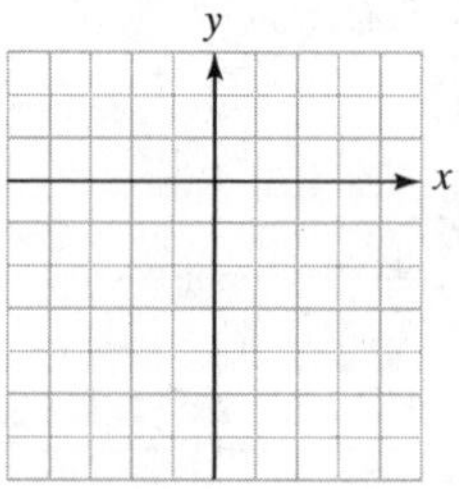

40. $f(x) = |x + 4| + 3$

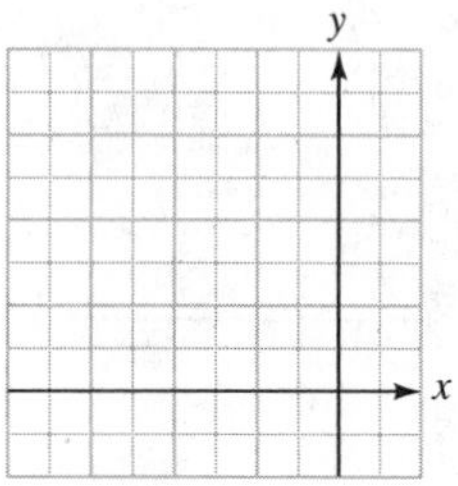

In Exercises 41–44, draw each graph using a translation of the graph of $f(x) = -x^2$, $f(x) = -x^3$, or $f(x) = -|x|$.

41. $f(x) = -|x| + 1$

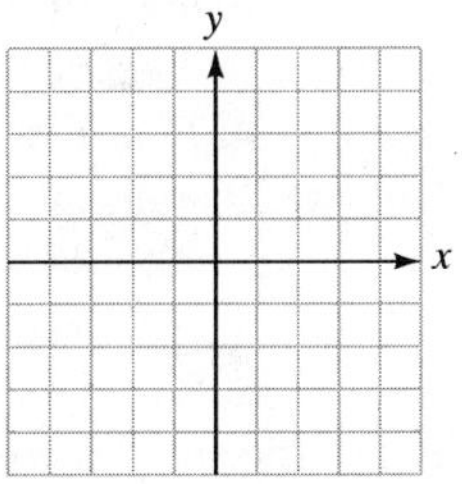

42. $f(x) = -x^3 - 2$

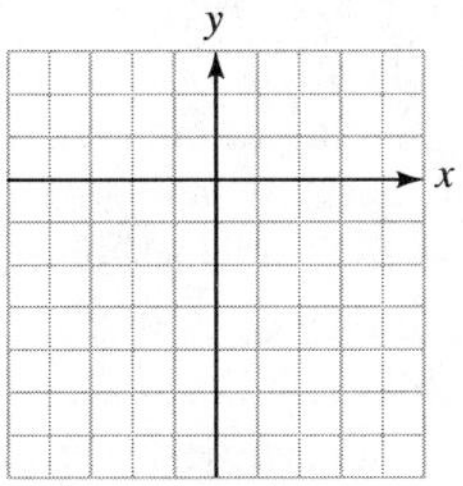

43. $f(x) = -(x - 1)^2$

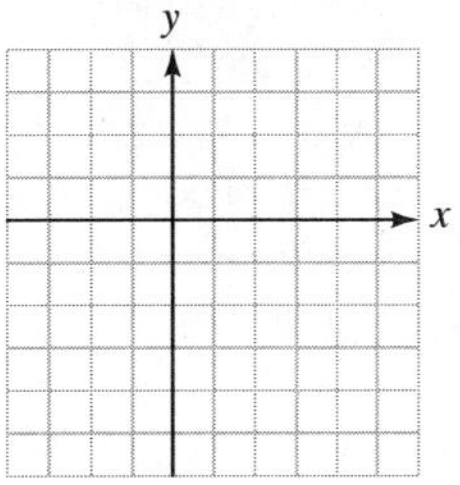

44. $f(x) = -(x + 2)^2$

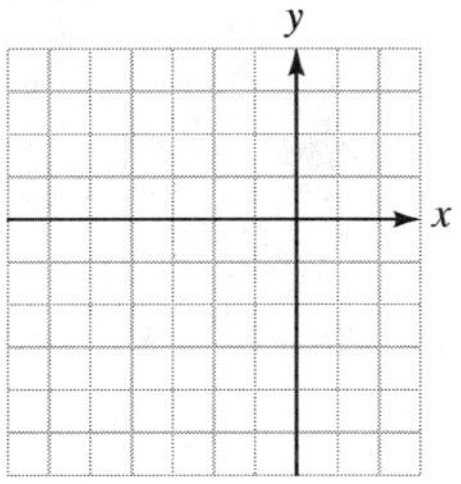

In Exercises 45–50, use a graphing calculator to solve each equation.

45. $3x + 6 = 0$

46. $7x - 21 = 0$

47. $4(x - 1) = 3x$

48. $4(x - 3) - x = x - 6$

49. $11x + 6(3 - x) = 3$

50. $2(x + 2) = 2(1 - x) + 10$

WRITING

51. Explain how to graph an equation by plotting points.

52. Explain why the correct choice of window settings is important when using a graphing calculator.

SOMETHING TO THINK ABOUT

In Exercises 53–60, use a graphing calculator.

53. Use a graphing calculator with settings of $[-10, 10]$ for x and $[-10, 10]$ for y to graph **a.** $y = -x^2$, **b.** $y = -x^2 + 1$, and **c.** $y = -x^2 + 2$. What do you notice?

54. Use a graphing calculator with settings of $[-10, 10]$ for x and $[-10, 10]$ for y to graph **a.** $y = -|x|$, **b.** $y = -|x| - 1$, and **c.** $y = -|x| - 2$. What do you notice?

55. Graph $y = (x + k)^2 + 1$ for several positive values of k. What do you notice?

56. Graph $y = (x + k)^2 + 1$ for several negative values of k. What do you notice?

57. Graph $y = (kx)^2 + 1$ for several positive values of k, where $k > 1$. What do you notice?

58. Graph $y = (kx)^2 + 1$ for several positive values of k, where $0 < k < 1$. What do you notice?

59. Graph $y = (kx)^2 + 1$ for several negative values of k, where $k < -1$. What do you notice?

60. Graph $y = (kx)^2 + 1$ for several negative values of k, where $-1 < k < 0$. What do you notice?

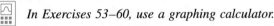

PROJECTS

PROJECT 1

The Board of Administrators of Boondocks County has hired your consulting firm to plan a highway. A new highway is to be built in the outback section of the county, an area with a rugged terrain where road building is hard and expensive. The board has hired you because they want to get everything right the first time.

(continued)

PROJECTS *(continued)*

The two main roads in the outback section are Highway N, running in a straight line north and south, and Highway E, running in a straight line east and west. These two highways meet at an intersection that the locals call Four Corners. The only other county road in the area is Slant Road, which runs in a straight line from northwest to southeast, cutting across Highway N north of Four Corners and Highway E east of Four Corners.

The county clerk is unable to find an official map of the area, but there is an old sketch made by the original road designer. It shows that if a rectangular coordinate system is set up using Highways N and E as the axes and Four Corners as the origin, then the equation of the line representing Slant Road is

$$2x + 3y = 12 \quad \text{(where the unit length is 1 mile)}$$

Given this information, the county wants you to do the following:

a. Update the current information by giving the coordinates of the intersections of Slant Road with Highway N and Highway E.

b. Plan a new highway, Country Drive, that will begin 1 mile north of Four Corners and run in a straight line in a generally northeasterly direction, intersecting Slant Road at right angles. The county wants to know the equation of the line representing Country Drive. You should also state the domain on which this equation is valid as a representation of Country Drive.

PROJECT 2

You are representing your branch of the large Buy-from-Us Corporation at the company's regional meeting, and you are looking forward to presenting your revenue and cost reports to the other branch representatives. But now disaster strikes! The graphs you had planned to present, containing cost and revenue information for this year and last year, are unlabeled! You cannot immediately recognize which graphs represent costs, which represent revenues, and which represent which year. Without these graphs, your presentation will not be effective.

The only other information you have with you is in the notes you made for your talk. From these you are able to glean the following financial data about your branch.

1. All cost and revenue figures on the graphs are rounded to the nearest \$50,000.

2. Costs for the fourth quarter of last year were \$400,000.

3. Revenue was not above \$400,000 for any quarter last year.

4. Last year, your branch lost money during the first quarter.

5. This year, your branch made money during three of the four quarters.
6. Profit during the second quarter of this year was $150,000.

And, of course, you know that profit = revenue − cost.

With this information, you must match each of the graphs (Illustrations 1–4) with one of the following titles:

Costs, This Year	Costs, Last Year
Revenues, This Year	Revenues, Last Year

You should be sure to have sound reasons for your choices—reasons ensuring that no other arrangement of the titles will fit the data. The *last* thing you want to do is present incorrect information to the company bigwigs!

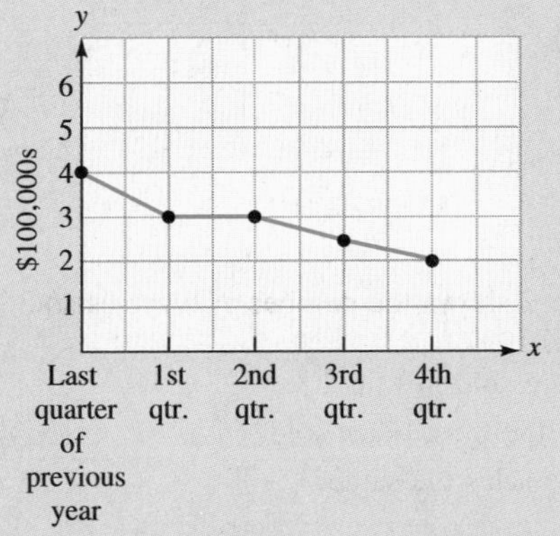

ILLUSTRATION 1

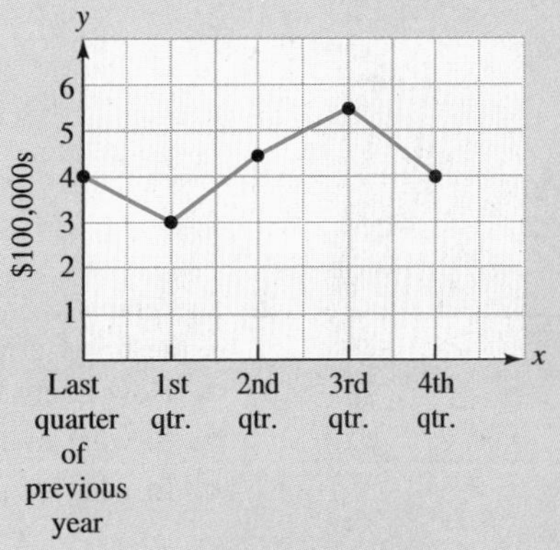

ILLUSTRATION 2

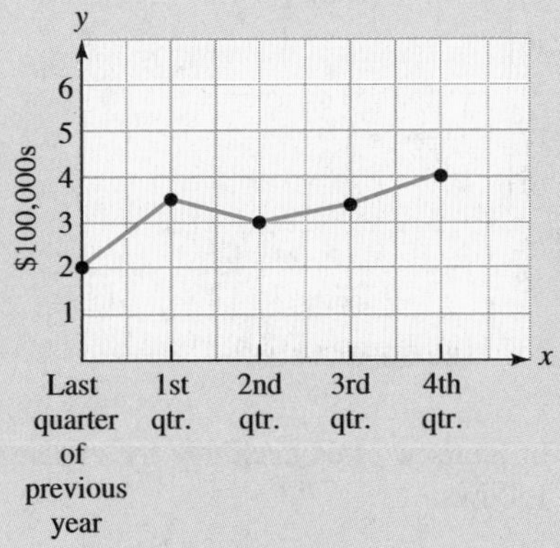

ILLUSTRATION 3

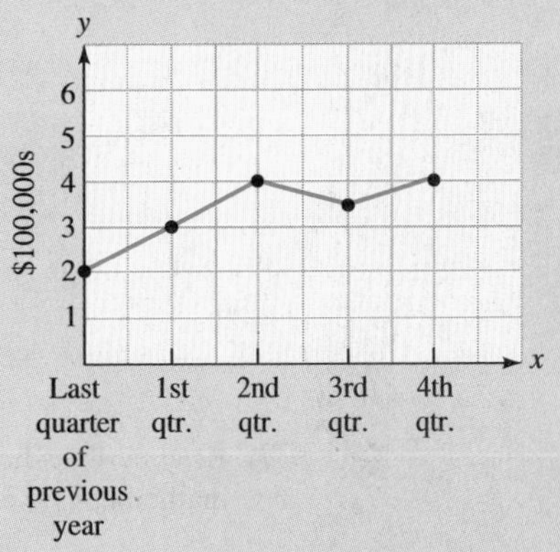

ILLUSTRATION 4

CHAPTER SUMMARY

CONCEPTS | REVIEW EXERCISES

SECTION 2.1 The Rectangular Coordinate System

Information can be presented in tables and graphs.

1. Use the graph in Illustration 1 to complete the table.

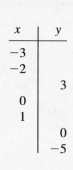

x	y
-3	
-2	
	3
0	
1	
	0
	-5

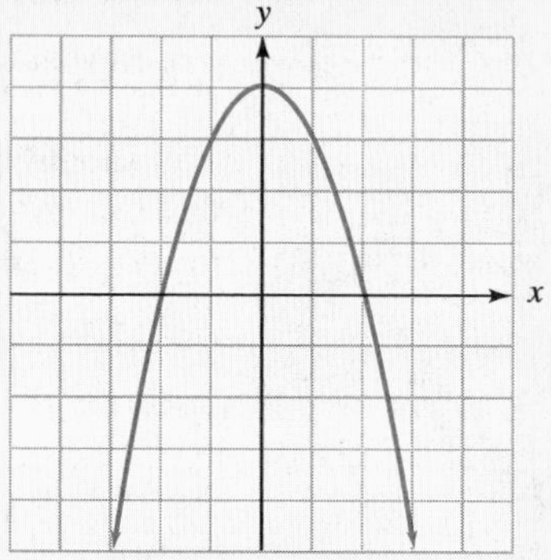

ILLUSTRATION 1

2. The graph in Illustration 2 shows the number of bicycles sold by a bike shop in the first six months of 1997.

a. How many bikes were sold in March?

b. In what month were the most bikes sold?

c. In what months were sales the same?

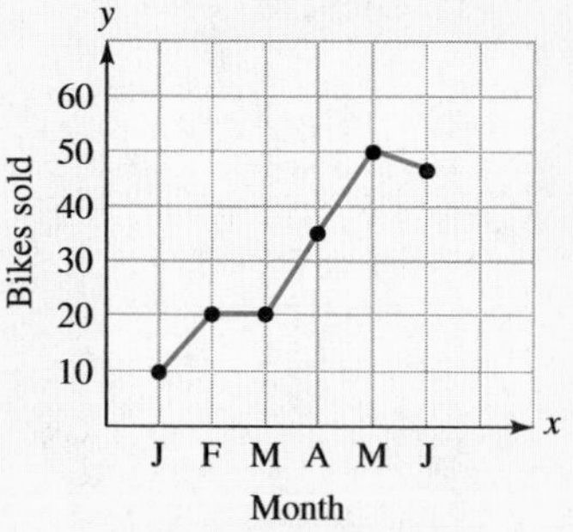

ILLUSTRATION 2

3. Assume that the number of hours worked each day by students enrolled in a mathematics class are as follows:

5, 4, 5, 6, 4, 3, 0, 3, 4, 5, 3, 2, 3, 2, 0,
3, 3, 4, 5, 3, 3, 4, 3, 5, 4, 3, 4, 3, 4, 3

a. Complete the frequency table in Illustration 3.
b. Construct a bar graph for the frequency distribution.

Hours	*Frequency*
0	
1	
	2
3	
4	
	5
6	

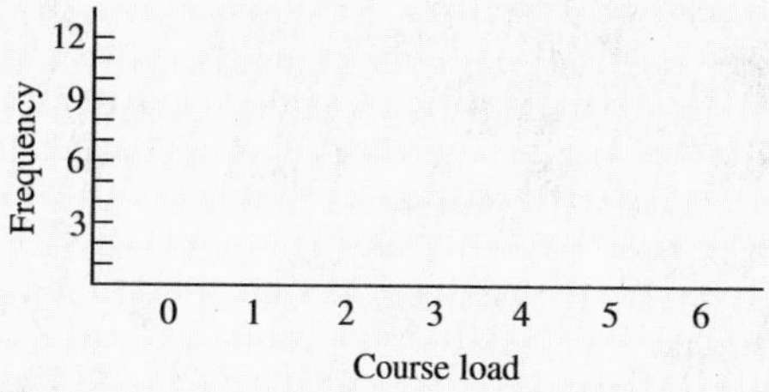

ILLUSTRATION 3

4. The step graph shown in Illustration 4 shows the cost of renting a canoe for different periods of time.
a. What is the cost of a 1-day rental?
b. What is the cost of a 4-day rental?

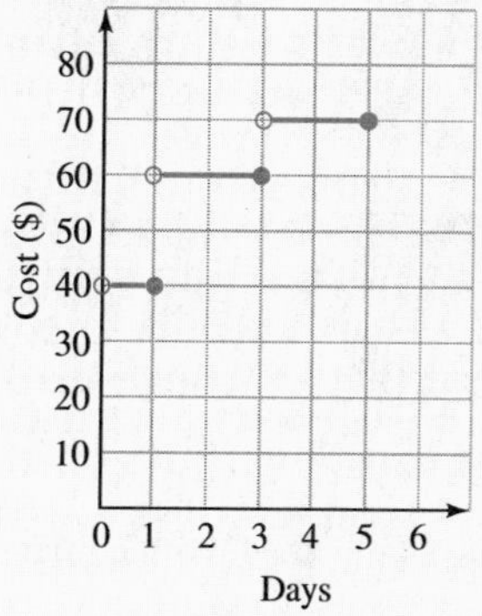

ILLUSTRATION 4

SECTION 2.2 *Graphing Linear Equations*

Graph of a vertical line:

$x = a$

x-intercept at $(a, 0)$

Graph of a horizontal line:

$y = b$

y-intercept at $(0, b)$

5. Graph each equation.
a. $x + y = 4$

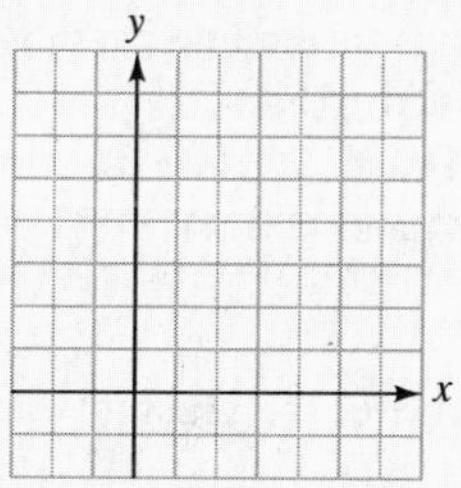

b. $2x - y = 8$

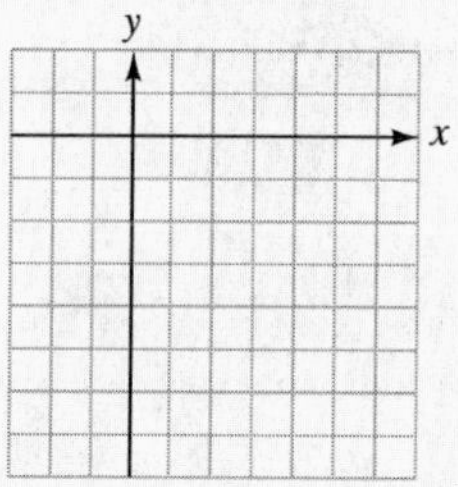

c. $y = 3x + 4$

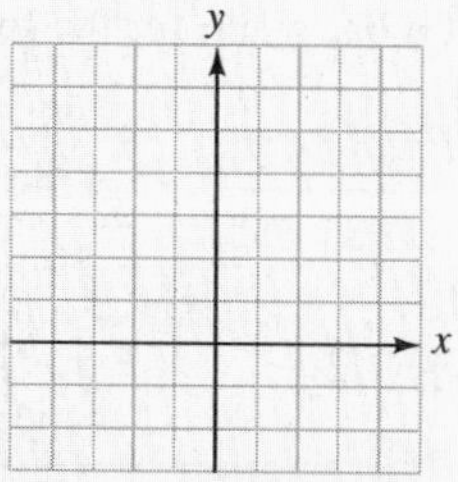

d. $x = 4 - 2y$

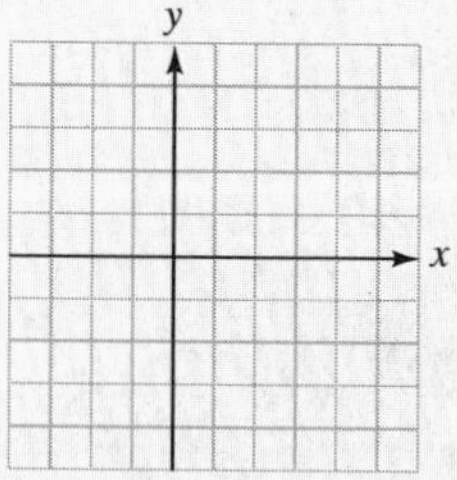

e. $y = 4$

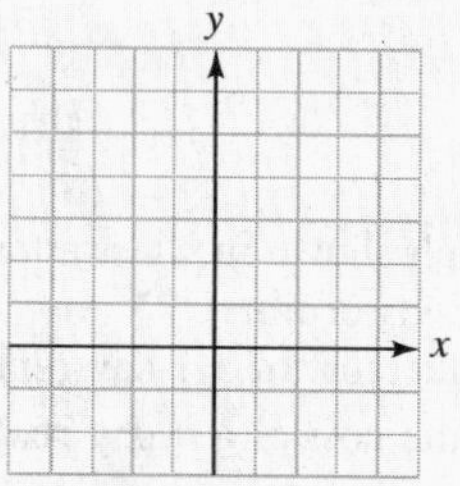

f. $x = -2$

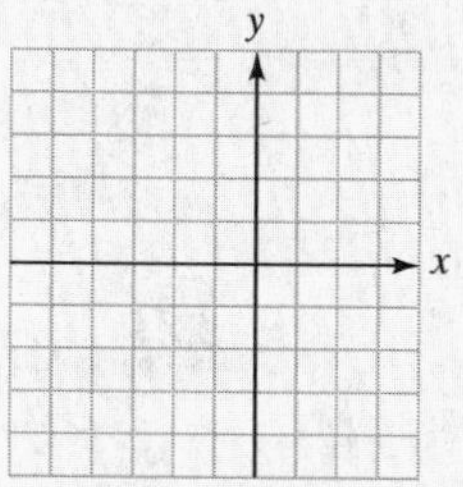

g. $2(x + 3) = x + 2$

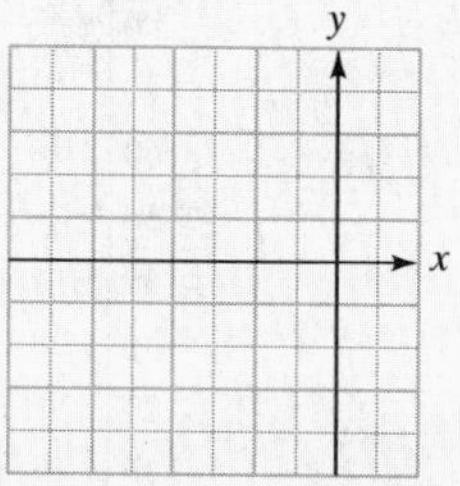

h. $3y = 2(y - 1)$

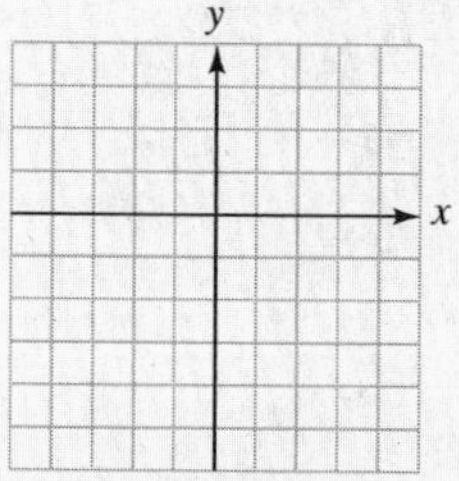

Midpoint formula:
If $P(x_1, y_1)$ and $Q(x_2, y_2)$, the midpoint of PQ is

$$M\left(\frac{x_1 + x_2}{2}, \frac{y_1 + y_2}{2}\right)$$

6. Find the midpoint of the line segment joining $P(-3, 5)$ and $Q(6, 11)$.

SECTION 2.3 Slope of a Nonvertical Line

Slope of a nonvertical line:
If $x_2 \neq x_1$,

$$m = \frac{\Delta y}{\Delta x} = \frac{y_2 - y_1}{x_2 - x_1}$$

Horizontal lines have a slope of 0. Vertical lines have no defined slope.

7. Find the slope of the line passing through points P and Q, if possible.

a. $P(2, 5)$ and $Q(5, 8)$
b. $P(-3, -2)$ and $Q(6, 12)$
c. $P(-3, 4)$ and $Q(-5, -6)$
d. $P(5, -4)$ and $Q(-6, -9)$
e. $P(-2, 4)$ and $Q(8, 4)$
f. $P(-5, -4)$ and $Q(-5, 8)$

8. Find the slope of the graph of each equation, if one exists.

a. $2x - 3y = 18$
b. $2x + y = 8$
c. $-2(x - 3) = 10$
d. $3y + 1 = 7$

Parallel lines have the same slope. The slopes of two nonvertical perpendicular lines are negative reciprocals.

9. Tell whether the lines with the given slopes are parallel, perpendicular, or neither.

a. $m_1 = 4, m_2 = -\frac{1}{4}$

b. $m_1 = 0.5, m_2 = \frac{1}{2}$

c. $m_1 = 0.5, m_2 = -\frac{1}{2}$

d. $m_1 = 5, m_2 = -0.2$

10. If the sales of a new business were \$65,000 in its first year and \$130,000 in its fourth year, find the rate of growth in sales per year.

SECTION 2.4 Writing Equations of Lines

Equations of a line:

Point–slope form:

$y - y_1 = m(x - x_1)$

Slope–intercept form:

$y = mx + b$

General form:

$Ax + By = C$

11. Write the equation of the line with the given properties. Write the equation in general form.

a. Slope of 3; passing through $P(-8, 5)$

b. Passing through $(-2, 4)$ and $(6, -9)$

c. Passing through $(-3, -5)$; parallel to the graph of $3x - 2y = 7$

d. Passing through $(-3, -5)$; perpendicular to the graph of $3x - 2y = 7$

12. A business purchased a copy machine for \$8,700 and will depreciate it on a straight-line basis over the next 5 years. At the end of its useful life, it will be sold as scrap for \$100. Find its depreciation equation.

SECTION 2.5 Introduction to Functions

A **function** is a correspondence between a set of input values x and a set of output values y, where exactly one value of y in the range corresponds to each number x in the domain.

13. Tell whether each equation determines y to be a function of x.

a. $y = 6x - 4$

b. $y = 4 - x$

c. $y^2 = x$

d. $|y| = x^2$

$f(k)$ represents the value of $f(x)$ when $x = k$.

14. Assume that $f(x) = 3x + 2$ and $g(x) = x^2 - 4$ and find each value.

a. $f(-3)$

b. $g(8)$

c. $g(-2)$

d. $f(5)$

The **domain** of a function is the set of input values. The **range** is the set of output values.

15. Find the domain and range of each function.

a. $f(x) = 4x - 1$

b. $f(x) = 3x - 10$

c. $f(x) = x^2 + 1$

d. $f(x) = \dfrac{4}{2 - x}$

e. $f(x) = \dfrac{7}{x - 3}$

f. $y = 7$

The **vertical line test** can be used to determine whether a graph represents a function.

16. Use the vertical line test to determine whether each graph represents a function.

a.

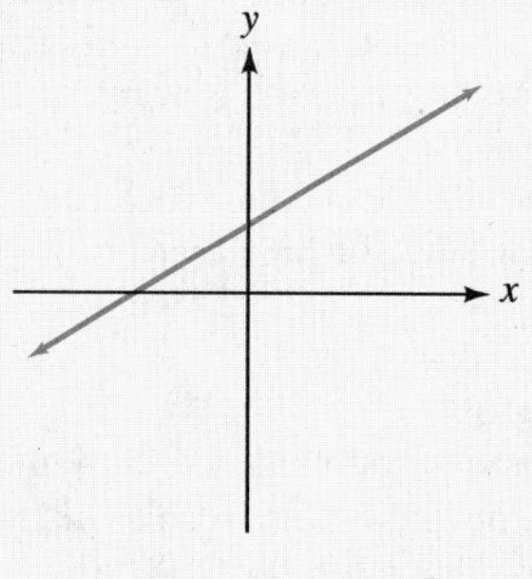

b.

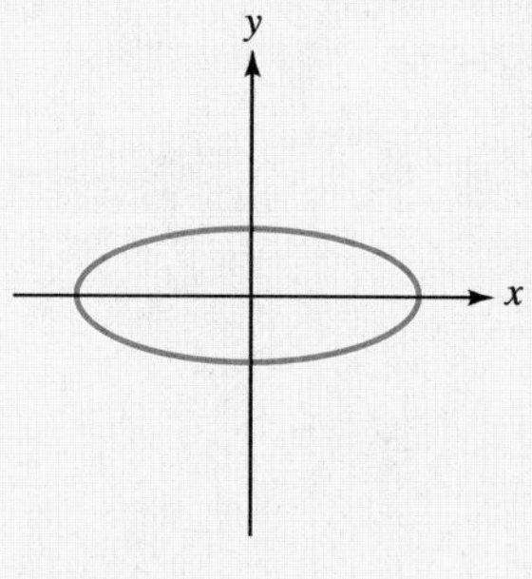

c.

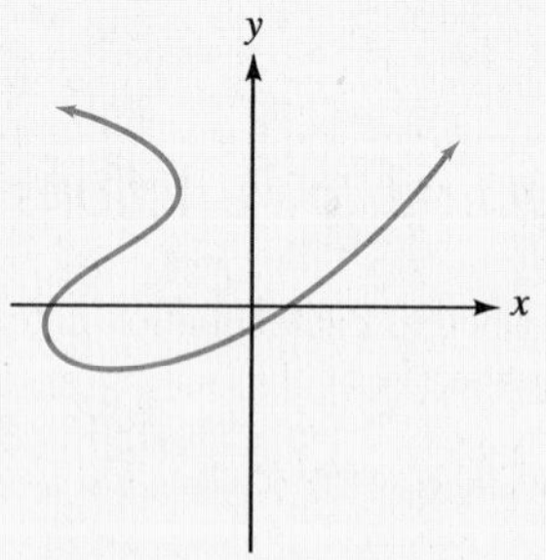

d.

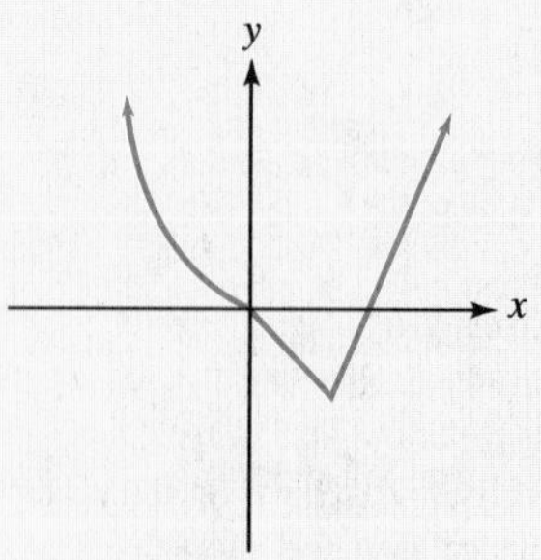

SECTION 2.6 Graphs of Other Functions

Graphs of nonlinear equations are not lines.

17. Graph each function.

a. $f(x) = x^2 - 3$

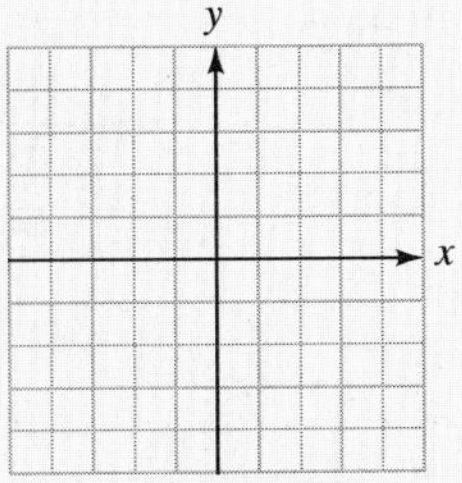

b. $f(x) = |x| - 4$

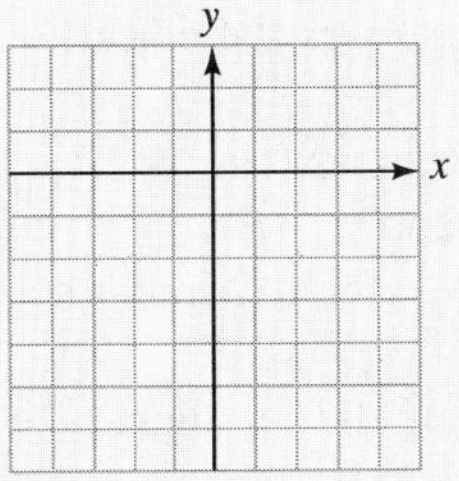

c. $f(x) = (x - 2)^3$

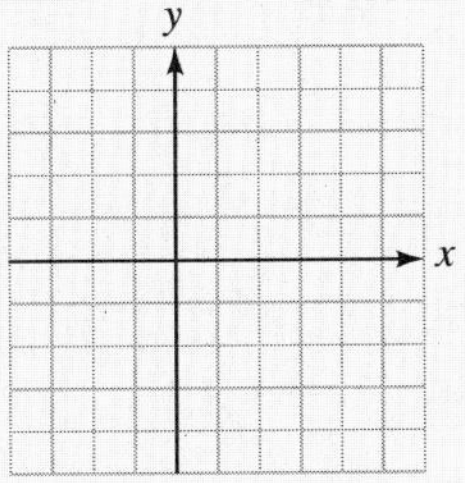

d. $f(x) = (x + 4)^2 - 3$

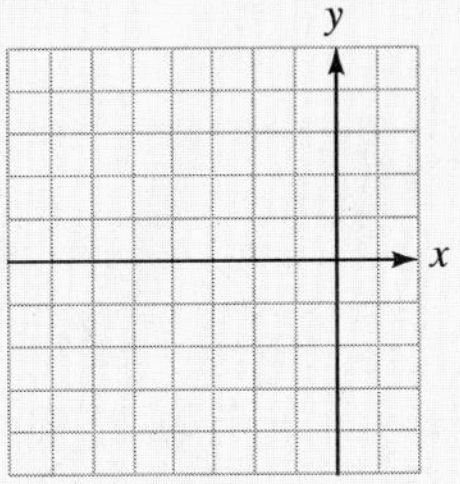

18. Use a graphing calculator to graph each function. Compare the results in Problem 17.

a. $f(x) = x^2 - 3$ **b.** $f(x) = |x| - 4$

c. $f(x) = (x - 2)^3$ **d.** $f(x) = (x + 4)^2 - 3$

19. Tell whether each equation defines a linear function.

a. $y = 3x + 2$ **b.** $y = \dfrac{x + 5}{4}$

c. $4x - 3y = 12$ **d.** $y = x^2 - 25$

20. Graph $f(x) = -|x - 3|$.

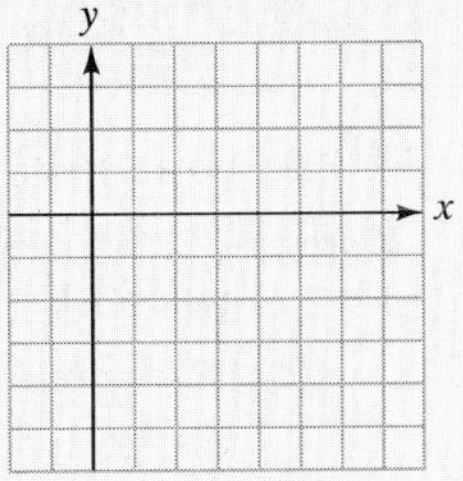

Chapter Test

1. The graph in Illustration 1 shows the height at different times of an object fired straight up into the air.
   a. How high was the object 3 seconds into the flight?
   b. At what times was the object about 110 feet above the ground?
   c. What was the maximum height reached by the object?
   d. How long did the flight take?

2. Graph the equation $2x - 5y = 10$.

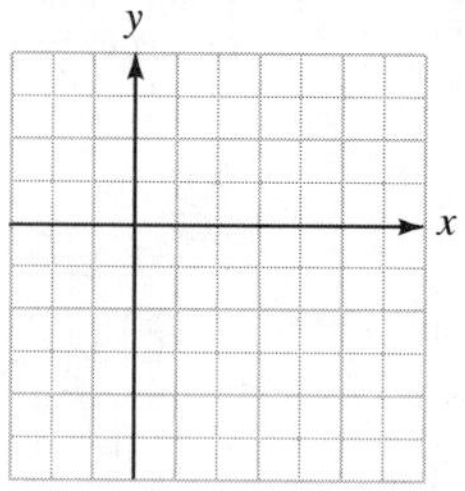

3. Find the x- and y-intercepts of the graph of $y = \frac{x-3}{5}$.

4. Find the midpoint of the line segment shown in Illustration 2.

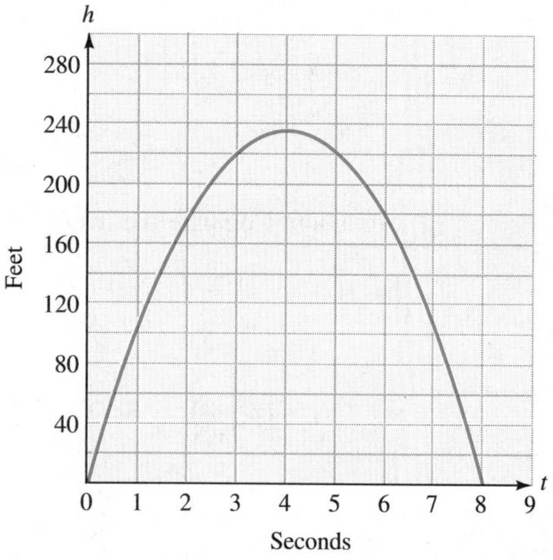

ILLUSTRATION 1

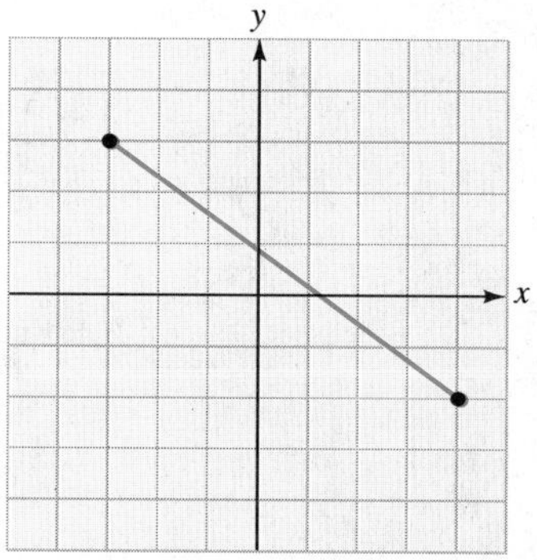

ILLUSTRATION 2

In Problems 5–8, find the slope of each line, if possible.

5. The line through $P(-2, 4)$ and $Q(6, 8)$

6. The graph of $2x - 3y = 8$

7. The graph of $x = 12$.

8. The graph of $y = 12$.

9. Write the equation of the line with slope of $\frac{2}{3}$ that passes through $P(4, -5)$. Give the answer in slope–intercept form.

10. Write the equation of the line that passes through $P(-2, 6)$ and $Q(-4, -10)$. Give the answer in general form.

11. Find the slope and the y-intercept of the graph of $-2(x - 3) = 3(2y + 5)$.

12. Determine whether the graphs of $4x - y = 12$ and $y = \frac{1}{4}x + 3$ are parallel, perpendicular, or neither.

13. Determine whether the graphs of $y = -\frac{2}{3}x + 4$ and $2y = 3x - 3$ are parallel, perpendicular, or neither.

14. Write the equation of the line that passes through the origin and is parallel to the graph of $y = \frac{3}{2}x - 7$.

15. Write the equation of the line that passes through $P(-3, 6)$ and is perpendicular to the graph of $y = -\frac{2}{3}x - 7$.

16. Does $|y| = x$ define y to be a function of x?

17. Find the domain and range of the function $f(x) = |x|$.

18. Find the domain and range of the function $f(x) = x^3$.

In Problems 19–22, $f(x) = 3x + 1$ and $g(x) = x^2 - 2$. Find each value.

19. $f(3)$

20. $g(0)$

21. $f(a)$

22. $g(-x)$

In Problems 23–24, tell whether each graph represents a function.

23.

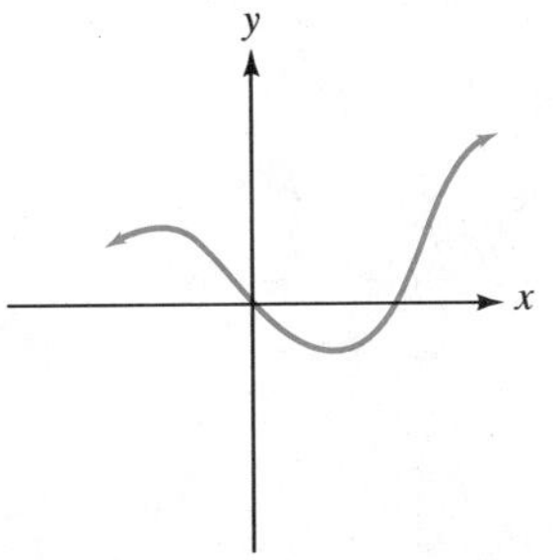

24.

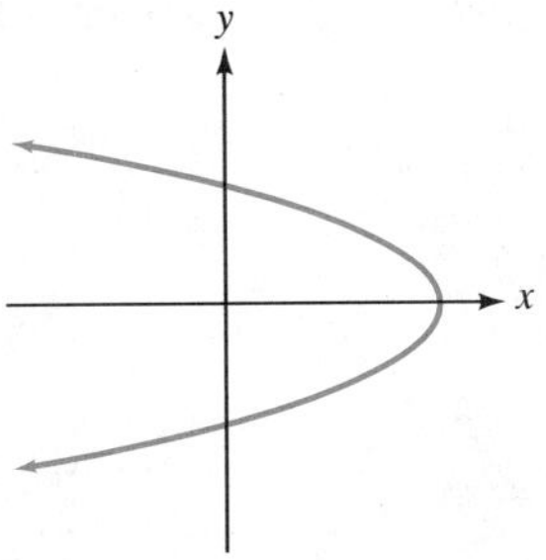

25. Graph $f(x) = x^2 - 1$.

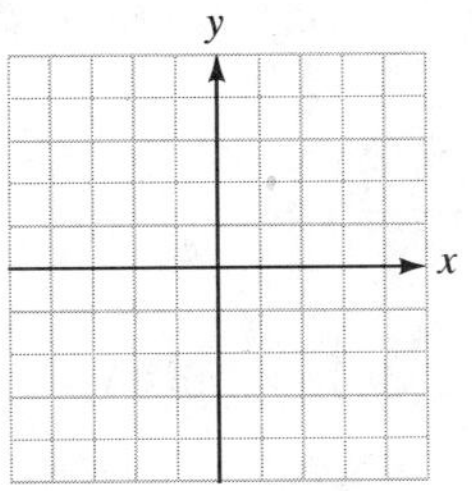

26. Graph $f(x) = -|x + 2|$.

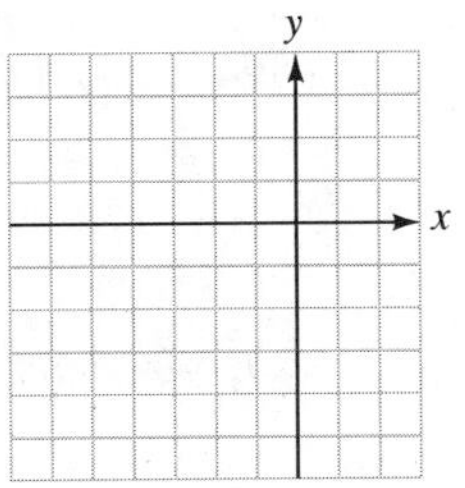

Cumulative Review Exercises

In Exercises 1–10, tell which numbers in the set $\left\{-2, 0, 1, 2, \frac{13}{12}, 6, 7, \sqrt{5}, \pi\right\}$ *are in each category.*

1. Natural numbers
2. Whole numbers
3. Rational numbers
4. Irrational numbers
5. Negative numbers
6. Real numbers
7. Prime numbers
8. Composite numbers
9. Even numbers
10. Odd numbers

In Exercises 11–12, graph each interval on the number line.

11. $-2 < x \leq 5$

12. $[-5, 0) \cup [3, 6]$

In Exercises 13–14, simplify each expression.

13. $-|5| + |-3|$

14. $\dfrac{|-5| + |-3|}{-|4|}$

In Exercises 15–18, do the operations.

15. $2 + 4 \cdot 5$

16. $\dfrac{8 - 4}{2 - 4}$

17. $20 \div (-10 \div 2)$

18. $\dfrac{6 + 3(6 + 4)}{2(3 - 9)}$

In Exercises 19–20, $x = 2$ and $y = -3$. Evaluate each expression.

19. $-x - 2y$

20. $\dfrac{x^2 - y^2}{2x + y}$

In Exercises 21–24, tell which property of real numbers justifies each statement.

21. $(a + b) + c = a + (b + c)$

22. $3(x + y) = 3x + 3y$

23. $(a + b) + c = c + (a + b)$

24. $(ab)c = a(bc)$

In Exercises 25–28, simplify each expression. Assume that all variables are positive numbers and write all answers without negative exponents.

25. $(x^2y^3)^4$

26. $\dfrac{c^4c^8}{(c^5)^2}$

27. $\left(-\dfrac{a^3b^{-2}}{ab}\right)^{-1}$

28. $\left(\dfrac{-3a^3b^{-2}}{6a^{-2}b^3}\right)^0$

29. Change 0.00000497 to scientific notation.

30. Change 9.32×10^8 to standard notation.

In Exercises 31–34, solve each equation.

31. $2x - 5 = 11$

32. $\dfrac{2x - 6}{3} = x + 7$

33. $4(y - 3) + 4 = -3(y + 5)$

34. $2x - \dfrac{3(x - 2)}{2} = 7 - \dfrac{x - 3}{3}$

In Exercises 35–36, solve each formula for the indicated variable.

35. $S = \dfrac{n(a + l)}{2}$ for a

36. $A = \dfrac{1}{2}h(b_1 + b_2)$ for h

37. The sum of three consecutive even integers is 90. Find the integers.

38. A rectangle is three times as long as it is wide. If its perimeter is 112 centimeters, find its dimensions.

39. Tell whether the graph of $2x - 3y = 6$ defines a function.

40. Find the slope of a line passing through $P(-2, 5)$ and $Q(8, -9)$.

41. Write the equation of the line passing through $P(-2, 5)$ and $Q(8, -9)$.

42. Write the equation of the line passing through $P(-2, 3)$ and parallel to the graph of $3x + y = 8$.

In Exercises 43–46, $f(x) = 3x^2 + 2$ and $g(x) = 2x - 1$. Evaluate each expression.

43. $f(-1)$

44. $g(0)$

45. $g(t)$

46. $f(-r)$

In Exercises 47–48, graph each equation and tell whether it is a function. If it is a function, give the domain and range.

47. $y = -x^2 + 1$

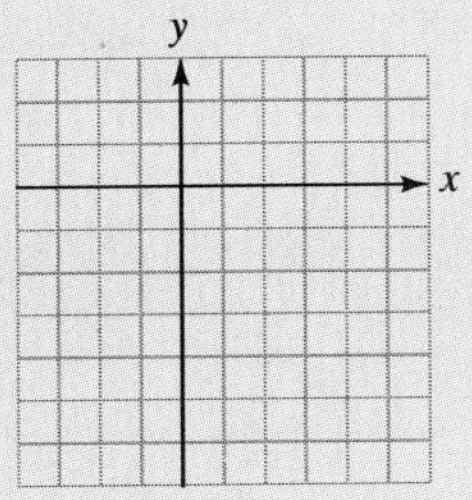

48. $y = \left|\frac{1}{2}x - 3\right|$

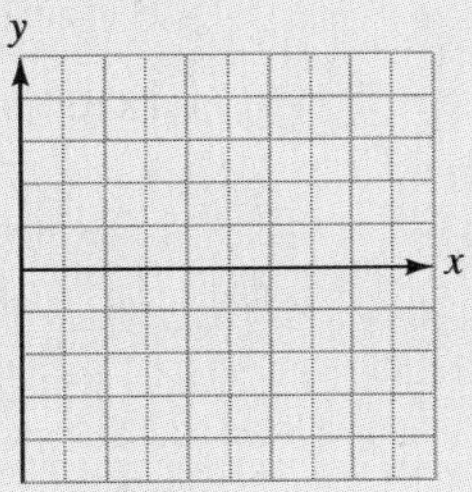

3 Systems of Equations

■ ■ ■ ■ ■ ■ ■ ■ ■ **Electrical/Electronic Engineer**

MATHEMATICS IN THE WORKPLACE

Electrical engineers design, develop, test, and supervise the manufacture of electronic equipment. Electrical engineers who work with electronic equipment are often called electronic engineers.

SAMPLE APPLICATION ■ In a radio, an inductor and a capacitor are used in a resonant circuit to select a wanted radio station at a frequency f and reject all others. The inductance L and the capacitance C determine the inductance reactance X_L and the capacitive reactance X_C of that circuit, where

$$X_L = 2\pi fL \qquad \text{and} \qquad X_C = \frac{1}{2\pi fC}$$

The radio station selected will be at the frequency f, where $X_L = X_C$. Write a formula for f^2 in terms of L and C.

(See Exercise 85 in Exercise 3.2.)

We have considered linear equations with the variables x and y. We found that each equation had infinitely many solutions (x, y), and that we could graph each equation on the rectangular coordinate system. In this chapter, we will discuss many **systems of linear equations** involving two or three equations.

3.1 Solution by Graphing

■ THE GRAPHING METHOD ■ CONSISTENT SYSTEMS ■ INCONSISTENT SYSTEMS ■ DEPENDENT EQUATIONS

Getting Ready

In Exercises 1–4, let $y = -3x + 2$.

1. Find y when $x = 0$.

2. Find y when $x = 3$.

3. Find y when $x = -3$.

4. Find y when $x = -\frac{1}{3}$.

5. Find five pairs of numbers with a sum of 12.

6. Find five pairs of numbers with a difference of 3.

■ THE GRAPHING METHOD

In the pair of equations

$$\begin{cases} x + 2y = 4 \\ 2x - y = 3 \end{cases} \qquad \text{(called a system of equations)}$$

there are infinitely many ordered pairs (x, y) that satisfy the first equation and infinitely many ordered pairs (x, y) that satisfy the second equation. However, there is only one ordered pair (x, y) that satisfies both equations at the same time. The process of finding this ordered pair is called *solving the system.*

We follow these steps to solve a system of two equations in two variables by graphing.

The Graphing Method

1. On a single set of coordinate axes, graph each equation.
2. Find the coordinates of the point (or points) where the graphs intersect. These coordinates give the solution of the system.
3. If the graphs have no point in common, the system has no solution.
4. If the graphs of the equations coincide, the system has infinitely many solutions.
5. Check the solution in both of the original equations.

CONSISTENT SYSTEMS

When a system of equations has a solution (as in Example 1), the system is called a **consistent system.**

EXAMPLE 1 Solve the system $\begin{cases} x + 2y = 4 \\ 2x - y = 3 \end{cases}$.

Solution We graph both equations on one set of coordinate axes, as shown in Figure 3-1.

$x + 2y = 4$

x	y	(x, y)
4	0	(4, 0)
0	2	(0, 2)
−2	3	(−2, 3)

$2x - y = 3$

x	y	(x, y)
1	−1	(1, −1)
0	−3	(0, −3)
−1	−5	(−1, −5)

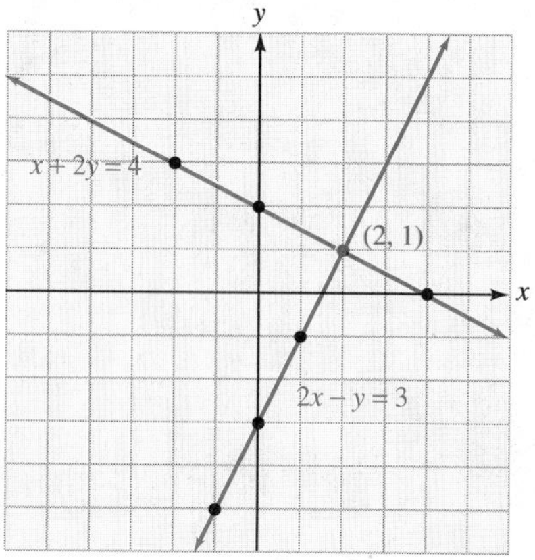

FIGURE 3-1

Although infinitely many ordered pairs (x, y) satisfy $x + 2y = 4$, and infinitely many ordered pairs (x, y) satisfy $2x - y = 3$, only the coordinates of the point where the graphs intersect satisfy both equations. Since the intersection point has coordinates of (2, 1), the solution is the ordered pair (2, 1) or $x = 2$ and $y = 1$.

When we check the solution, we substitute 2 for x and 1 for y in both equations and verify that (2, 1) satisfies each one.

Self Check Solve $\begin{cases} 2x + y = 4 \\ x - 3y = -5 \end{cases}$.

Answer (1, 2)

INCONSISTENT SYSTEMS

When a system has no solution (as in Example 2), it is called an **inconsistent system.**

EXAMPLE 2 Solve the system $\begin{cases} 2x + 3y = 6 \\ 4x + 6y = 24 \end{cases}$, if possible.

Solution We graph both equations on one set of coordinate axes, as shown in Figure 3-2. In this example, the graphs are parallel, because the slopes of the two lines are equal and their y-intercepts are different. We can see that the slope of each line is $-\frac{2}{3}$ by writing each equation in slope–intercept form.

Since the graphs are parallel lines, the lines do not intersect, and the system does not have a solution. It is an inconsistent system.

$2x + 3y = 6$

x	y	(x, y)
3	0	(3, 0)
0	2	(0, 2)
−3	4	(−3, 4)

$4x + 6y = 24$

x	y	(x, y)
6	0	(6, 0)
0	4	(0, 4)
−3	6	(−3, 6)

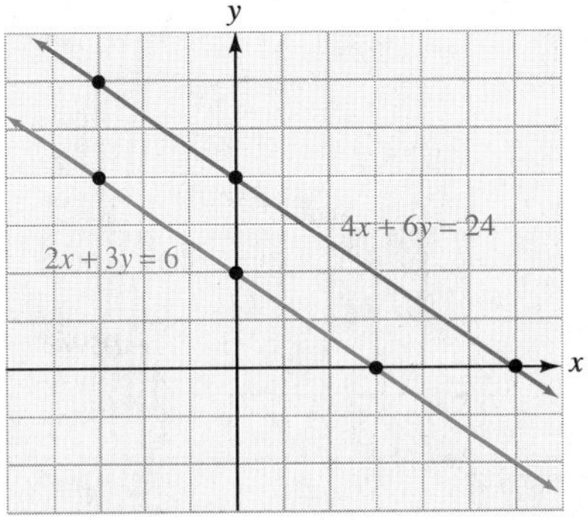

FIGURE 3-2

Self Check Solve $\begin{cases} 2x - 3y = 6 \\ y = \frac{2}{3}x - 3 \end{cases}$.

Answer no solutions

DEPENDENT EQUATIONS

When the equations of a system have different graphs (as in Examples 1 and 2), the equations are called **independent equations.** Two equations with the same graph are called **dependent equations.**

EXAMPLE 3 Solve the system $\begin{cases} 2y - x = 4 \\ 2x + 8 = 4y \end{cases}$.

Solution We graph each equation on one set of coordinate axes, as shown in Figure 3-3. Since the graphs coincide, the system has infinitely many solutions. Any ordered pair (x, y) that satisfies one equation satisfies the other also.

From the tables of ordered pairs shown in Figure 3-3, we see that $(-4, 0)$ and $(0, 2)$ are solutions. We can find infinitely many more solutions by finding additional ordered pairs (x, y) that satisfy either equation.

Because the two equations have the same graph, they are dependent equations.

$2y - x = 4$

x	y	(x, y)
-4	0	$(-4, 0)$
0	2	$(0, 2)$

$2x + 8 = 4y$

x	y	(x, y)
-4	0	$(-4, 0)$
0	2	$(0, 2)$

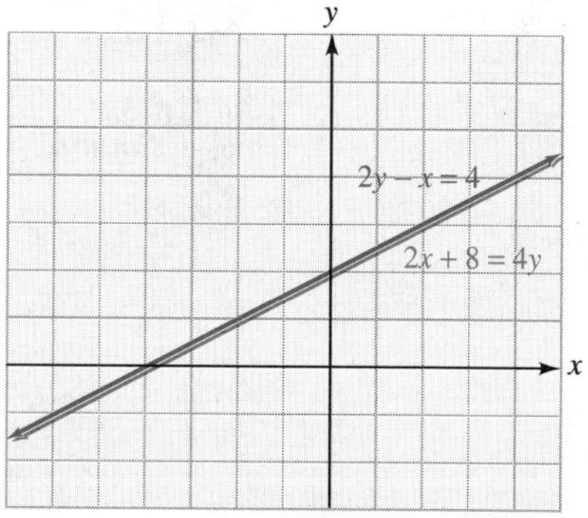

FIGURE 3-3

Self Check Solve $\begin{cases} 2x - y = 4 \\ x = \frac{1}{2}y + 2 \end{cases}$.

Answer infinitely many solutions (three of them are $(0, -4)$, $(2, 0)$, and $(-3, -10)$)

We summarize the possibilities that can occur when two equations, each with two variables, are graphed.

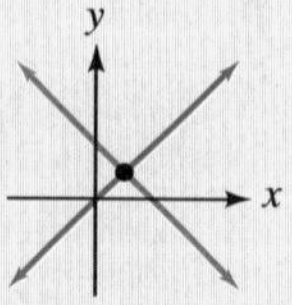

If the lines are different and intersect, the equations are independent and the system is consistent. **One solution exists.**

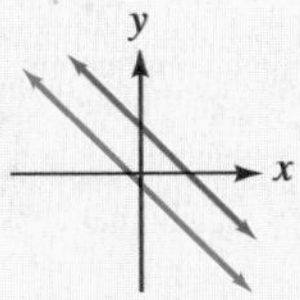

If the lines are different and parallel, the equations are independent and the system is inconsistent. **No solution exists.**

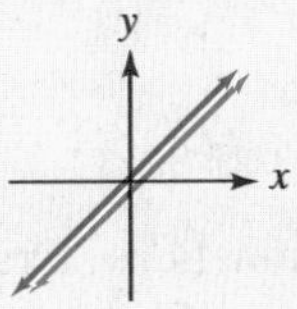

If the lines coincide, the equations are dependent and the system is consistent. **Infinitely many solutions exist.**

If the equations in two systems are equivalent, then the systems are called **equivalent**. In Example 4, we solve a more difficult system by changing it into a simpler equivalent system.

EXAMPLE 4 Solve the system $\begin{cases} \frac{3}{2}x - y = \frac{5}{2} \\ x + \frac{1}{2}y = 4 \end{cases}$.

Solution We multiply both sides of $\frac{3}{2}x - y = \frac{5}{2}$ by 2 to eliminate the fractions and obtain the equation $3x - 2y = 5$. We multiply both sides of $x + \frac{1}{2}y = 4$ by 2 to eliminate the fractions and obtain the equation $2x + y = 8$.

The new system

$$\begin{cases} 3x - 2y = 5 \\ 2x + y = 8 \end{cases}$$

is equivalent to the original system and is easier to solve, since it has no fractions. If we graph each equation in the new system, as in Figure 3-4, we see that the coordinates of the point where the two lines intersect are (3, 2). Verify that $x = 3$ and $y = 2$ satisfy each equation in the original system.

$3x - 2y = 5$

x	y	(x, y)
0	$-\frac{5}{2}$	$\left(0, -\frac{5}{2}\right)$
$\frac{5}{3}$	0	$\left(\frac{5}{3}, 0\right)$

$2x + y = 8$

x	y	(x, y)
4	0	(4, 0)
1	6	(1, 6)

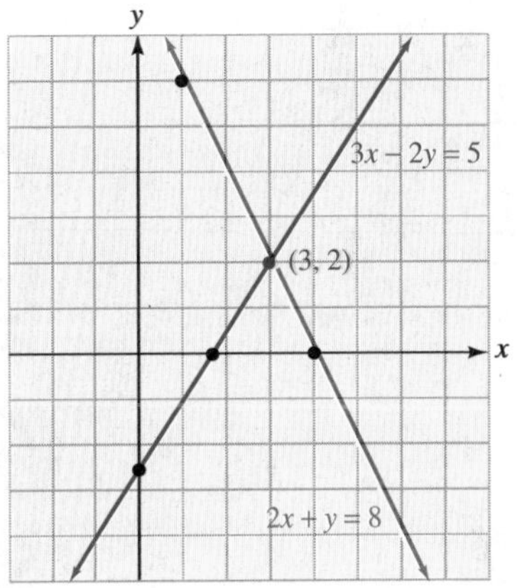

FIGURE 3-4

■

Self Check Solve $\begin{cases} \frac{5}{2}x - y = 2 \\ x + \frac{1}{3}y = 3 \end{cases}$.

Answer (2, 3)

Solving Systems by Graphing

GRAPHING CALCULATORS

The graphing method has limitations. First, the method is limited to equations with two variables. Systems with three or more variables cannot be solved graphically. Second, it is often difficult to find exact solutions graphically. However, the trace and zoom capabilities of graphing calculators enable us to get very good approximations of such solutions.

To solve the system

$$\begin{cases} 3x + 2y = 12 \\ 2x - 3y = 12 \end{cases}$$

with a graphing calculator, we must first solve each equation for y so that we can enter them into a graphing calculator. After solving for y, we obtain the following equivalent system.

$$\begin{cases} y = -\dfrac{3}{2}x + 6 \\ y = \dfrac{2}{3}x - 4 \end{cases}$$

If we use window settings of $[-10, 10]$ for x and $[-10, 10]$ for y, the graphs of the equations will look like those in Figure 3-5(a). If we zoom in on the intersection point of the two lines and trace, we will get an approximate solution like the one shown in Figure 3-5(b). To get better results, we can do more zooms. Verify that the exact solution is $x = \frac{60}{13}$ and $y = -\frac{12}{13}$.

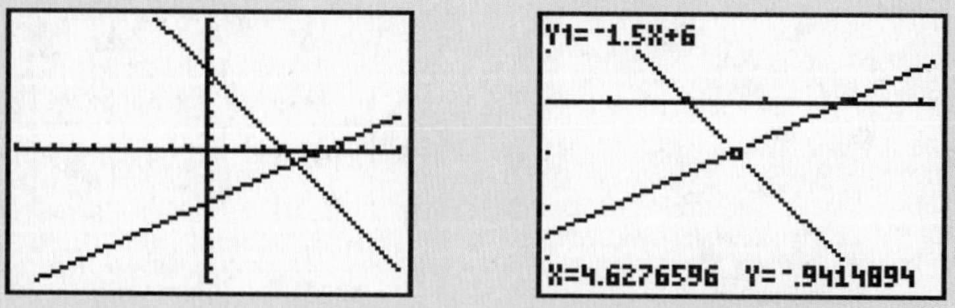

(a) (b)

FIGURE 3-5

Orals *Tell whether the following systems will have one solution, no solutions, or infinitely many solutions.*

1. $\begin{cases} y = 2x \\ y = 2x + 5 \end{cases}$

2. $\begin{cases} y = 2x \\ y = x + x \end{cases}$

3. $\begin{cases} y = 2x \\ y = -2x \end{cases}$

4. $\begin{cases} y = 2x + 1 \\ 2x = y \end{cases}$

EXERCISE 3.1

REVIEW *Write each number in scientific notation.*

1. 93,000,000

2. 0.0000000236

3. 345×10^2

4. 752×10^{-5}

VOCABULARY AND CONCEPTS *Fill in each blank to make a true statement.*

5. If two or more equations are considered at the same time, they are called a _______ of equations.

6. When a system of equations has one or more solutions, it is called a _________ system.

7. If a system has no solutions, it is called an __________ system.

8. If two equations have different graphs, they are called __________ equations.

9. Two equations with the same graph are called _________ equations.

10. If the equations in two systems are equivalent, the systems are called _________ systems.

PRACTICE *In Exercises 11–14, tell whether the ordered pair is a solution of the system of equations.*

11. $(1, 2)$; $\begin{cases} y = 2x \\ y = \frac{1}{2}x + \frac{3}{2} \end{cases}$

12. $(-1, 2)$; $\begin{cases} y = 3x + 5 \\ y = x + 4 \end{cases}$

13. $(2, -3)$; $\begin{cases} y = \frac{1}{2}x - 2 \\ 3x + 2y = 0 \end{cases}$

14. $(-4, 3)$; $\begin{cases} 4x - y = -19 \\ 3x + 2y = -6 \end{cases}$

In Exercises 15–34, solve each system by graphing, if possible.

15. $\begin{cases} x + y = 6 \\ x - y = 2 \end{cases}$

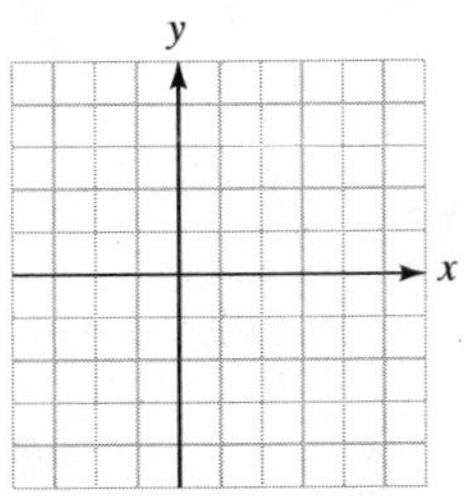

16. $\begin{cases} x - y = 4 \\ 2x + y = 5 \end{cases}$

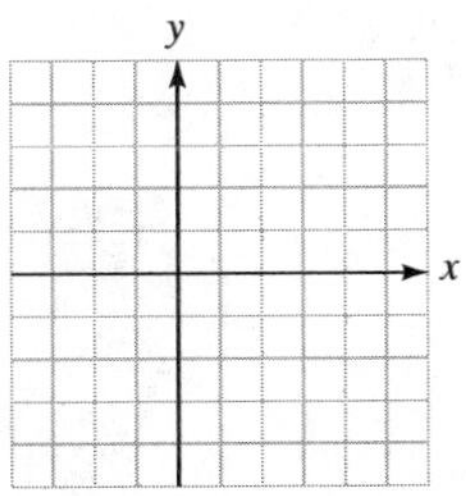

17. $\begin{cases} 2x + y = 1 \\ x - 2y = -7 \end{cases}$

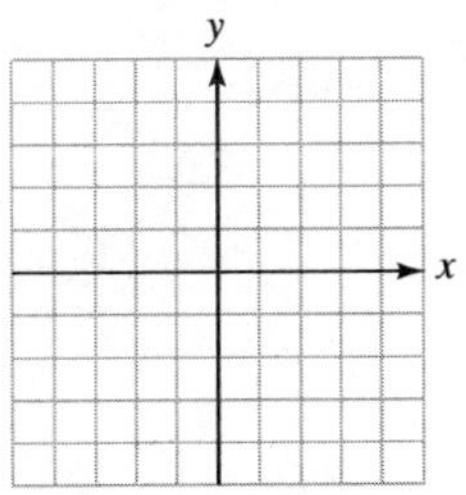

18. $\begin{cases} 3x - y = -3 \\ 2x + y = -7 \end{cases}$

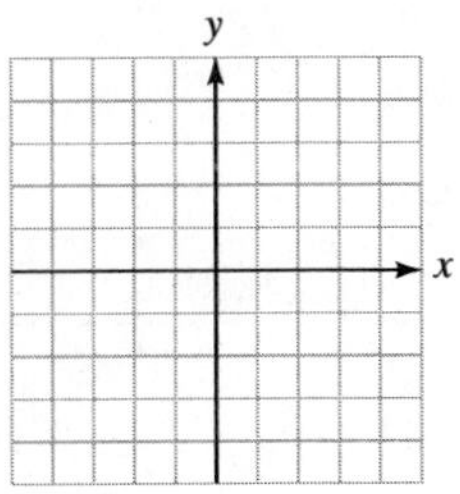

19. $\begin{cases} x = 13 - 4y \\ 3x = 4 + 2y \end{cases}$

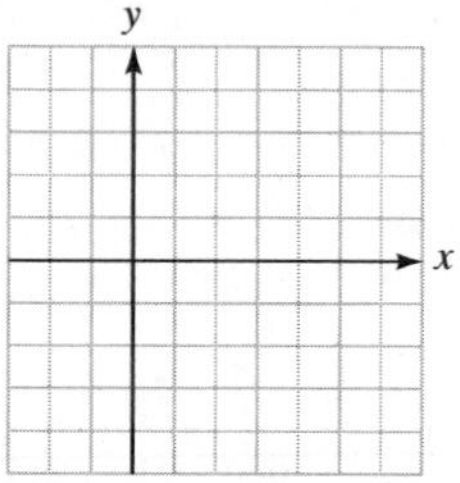

20. $\begin{cases} 3x = 7 - 2y \\ 2x = 2 + 4y \end{cases}$

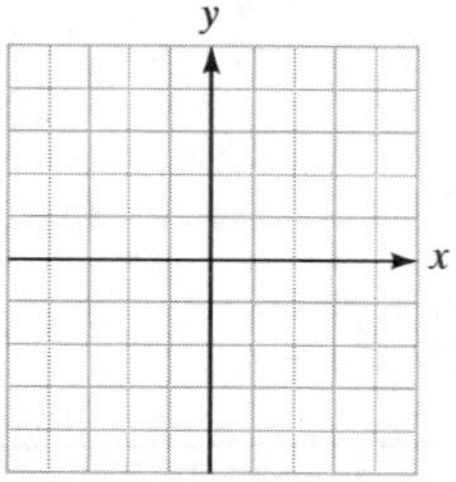

21. $\begin{cases} x = 3 - 2y \\ 2x + 4y = 6 \end{cases}$

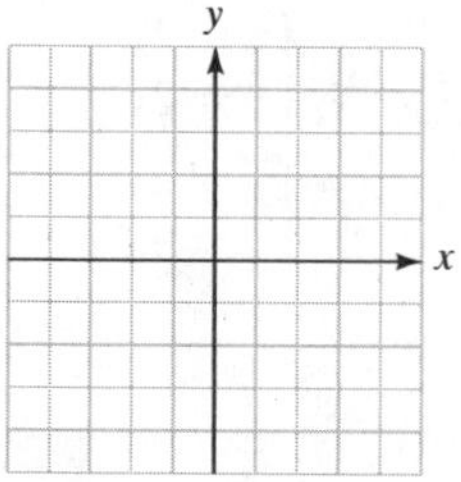

22. $\begin{cases} 3x = 5 - 2y \\ 3x + 2y = 7 \end{cases}$

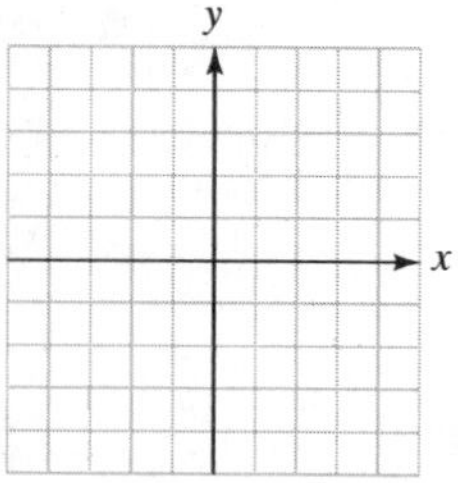

23. $\begin{cases} x = 2 \\ y = \dfrac{4 - x}{2} \end{cases}$

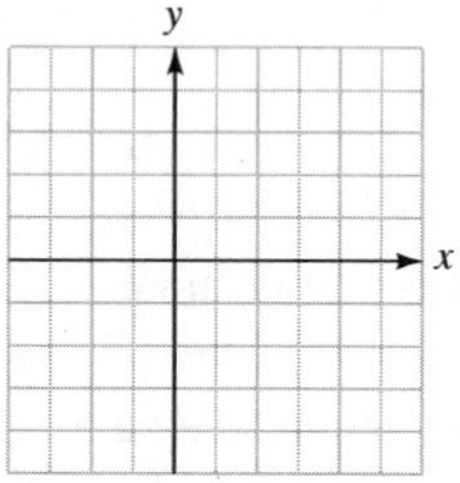

24. $\begin{cases} y = -2 \\ x = \dfrac{4 + 3y}{2} \end{cases}$

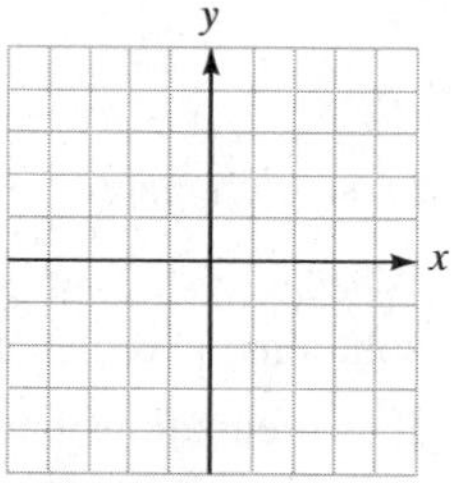

25. $\begin{cases} y = 3 \\ x = 2 \end{cases}$

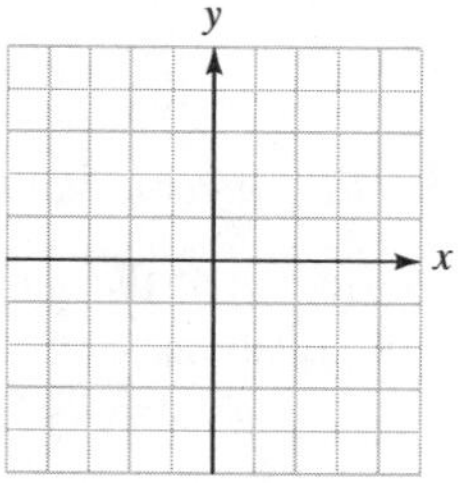

26. $\begin{cases} 2x + 3y = -15 \\ 2x + y = -9 \end{cases}$

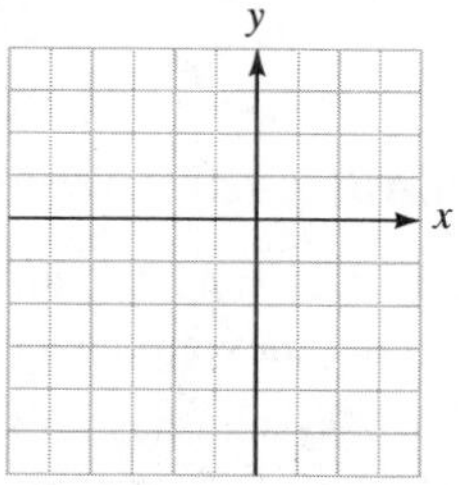

27. $\begin{cases} x = \dfrac{11 - 2y}{3} \\ y = \dfrac{11 - 6x}{4} \end{cases}$

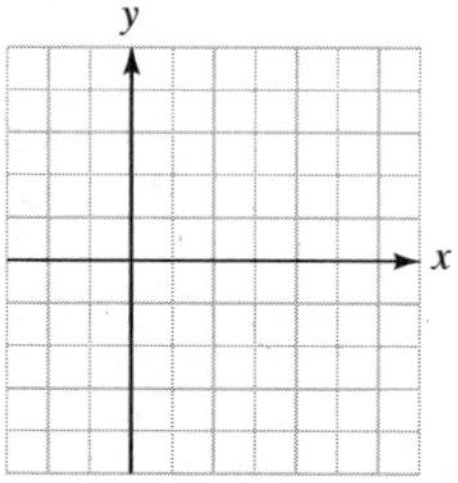

28. $\begin{cases} x = \dfrac{1 - 3y}{4} \\ y = \dfrac{12 + 3x}{2} \end{cases}$

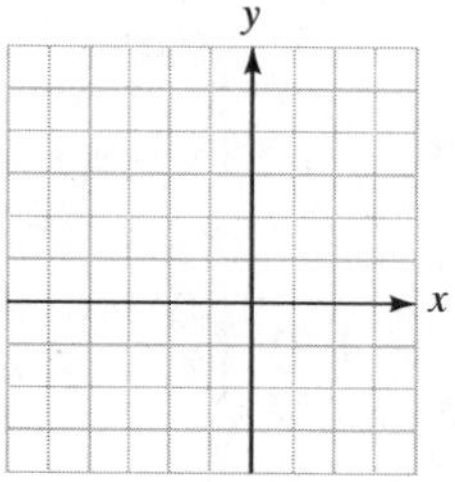

29. $\begin{cases} \dfrac{5}{2}x + y = \dfrac{1}{2} \\ 2x - \dfrac{3}{2}y = 5 \end{cases}$

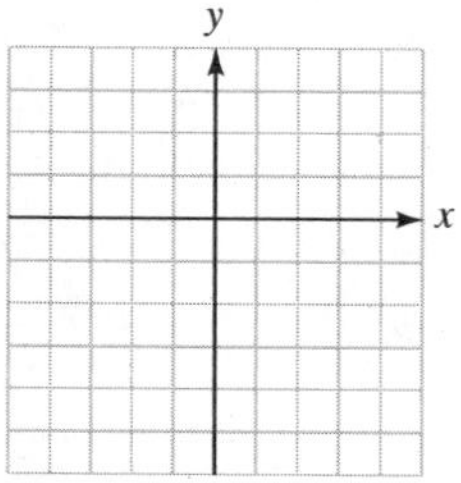

30. $\begin{cases} \dfrac{5}{2}x + 3y = 6 \\ y = \dfrac{24 - 10x}{12} \end{cases}$

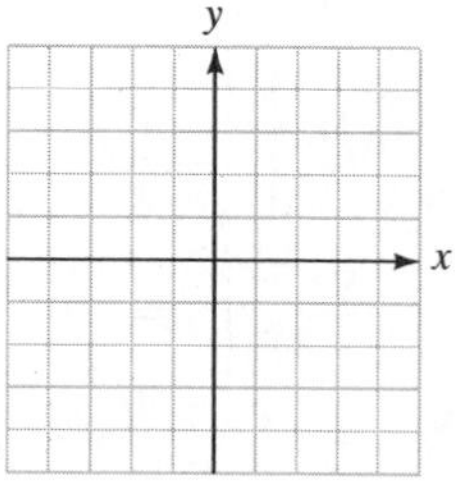

31. $\begin{cases} x = \dfrac{5y - 4}{2} \\ x - \dfrac{5}{3}y + \dfrac{1}{3} = 0 \end{cases}$

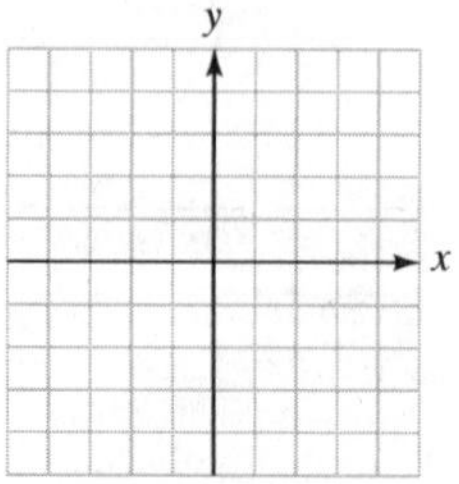

32. $\begin{cases} 2x = 5y - 11 \\ 3x = 2y \end{cases}$

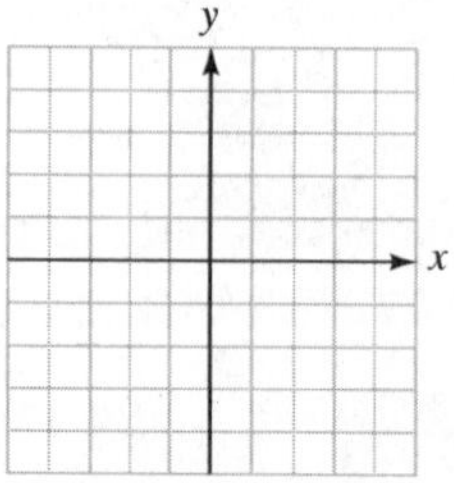

33. $\begin{cases} x = -\dfrac{3}{2}y \\ x = \dfrac{3}{2}y - 2 \end{cases}$

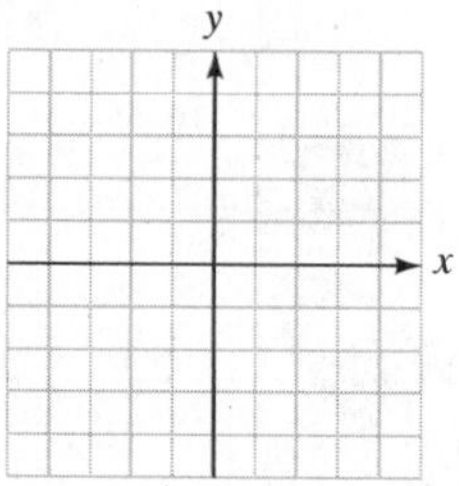

34. $\begin{cases} x = \dfrac{3y - 1}{4} \\ y = \dfrac{4 - 8x}{3} \end{cases}$

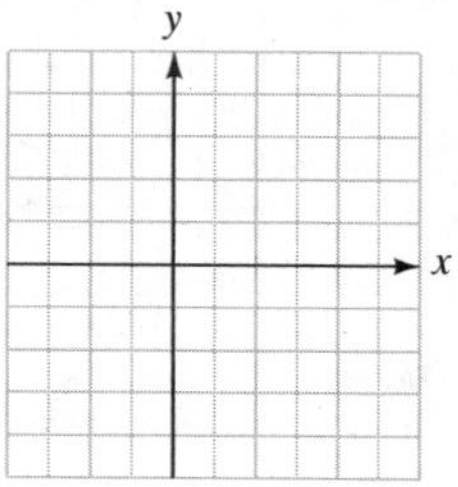

In Exercises 35–38, use a graphing calculator to solve each system. Give all answers to the nearest hundredth.

35. $\begin{cases} y = 3.2x - 1.5 \\ y = -2.7x - 3.7 \end{cases}$

36. $\begin{cases} y = -0.45x + 5 \\ y = 5.55x - 13.7 \end{cases}$

37. $\begin{cases} 1.7x + 2.3y = 3.2 \\ y = 0.25x + 8.95 \end{cases}$

38. $\begin{cases} 2.75x = 12.9y - 3.79 \\ 7.1x - y = 35.76 \end{cases}$

APPLICATIONS

39. Navigation Two ships are sailing on the same coordinate system. One ship is following a path described by the equation $2x + 3y = 6$, and the other is following a path described by the equation $2x - 3y = 9$.

a. Is there a possibility of a collision?

b. Find the coordinates of the danger point.

c. Is a collision a certainty?

40. Navigation Two airplanes are flying at the same altitude and in the same coordinate system. One plane is following a path described by the equation $y = \frac{2}{5}x - 2$, and the other is following a path described by the equation $x = \frac{5y + 7}{2}$. Is there a possibility of a collision?

WRITING

41. Explain how to solve a system of two equations in two variables.

42. Can a system of two equations in two variables have exactly two solutions? Why or why not?

SOMETHING TO THINK ABOUT

43. Form an independent system of equations with a solution of $(-5, 2)$.

44. Form a dependent system of equations with a solution of $(-5, 2)$.

3.2 Solution by Elimination

■ THE SUBSTITUTION METHOD ■ THE ADDITION METHOD ■ AN INCONSISTENT SYSTEM ■ A SYSTEM WITH INFINITELY MANY SOLUTIONS ■ REPEATING DECIMALS ■ PROBLEM SOLVING ■ BREAK-POINT ANALYSIS ■ PARALLELOGRAMS

Getting Ready *Remove parentheses.*

1. $3(2x - 7)$

2. $-4(3x + 5)$

Substitute $x - 3$ for y and remove parentheses.

3. $3y$

4. $-2(y + 2)$

Add the left-hand sides and the right-hand sides of the equations in each system.

5. $\begin{cases} 2x + 5y = 7 \\ 5x - 5y = 8 \end{cases}$

6. $\begin{cases} 3a - 4b = 12 \\ -3a - 5b = 15 \end{cases}$

The graphing method provides a way to visualize the process of solving systems of equations. However, it cannot be used to solve systems of higher order, such as three equations, each with three variables. In this section, we will discuss algebraic methods that will enable us to solve such systems.

THE SUBSTITUTION METHOD

To solve a system of two equations (each with two variables) by substitution, we use the following steps.

The Substitution Method

1. If necessary, solve one equation for one of its variables, preferably a variable with a coefficient of 1.
2. Substitute the resulting expression for the variable obtained in Step 1 into the other equation and solve that equation.
3. Find the value of the other variable by substituting the value of the variable found in Step 2 into any equation containing both variables.
4. State the solution.
5. Check the solution in both of the original equations.

EXAMPLE 1 Solve the system $\begin{cases} 4x + y = 13 \\ -2x + 3y = -17 \end{cases}$.

Solution *Step 1:* We solve the first equation for y, because y has a coefficient of 1 and no fractions are introduced.

$$4x + y = 13$$

1. $y = -4x + 13$ Subtract $4x$ from both sides.

Step 2: We then substitute $-4x + 13$ for y in the second equation of the system and solve for x.

$$-2x + 3y = -17$$

$$-2x + 3(-4x + 13) = -17$$ Substitute $-4x + 13$ for y.

$$-2x - 12x + 39 = -17$$ Use the distributive property to remove parentheses.

$$-14x = -56$$ Combine like terms and subtract 39 from both sides.

$$x = 4$$ Divide both sides by -14.

Step 3: To find y, we substitute 4 for x in Equation 1 and simplify:

$$y = -4x + 13$$

$$= -4(4) + 13$$ Substitute 4 for x.

$$= -3$$

Step 4: The solution is $x = 4$ and $y = -3$, or just $(4, -3)$. The graphs of these two equations would intersect at the point $(4, -3)$.

Step 5: To verify that this solution satisfies both equations, we substitute $x = 4$ and $y = -3$ into each equation in the system and simplify.

$$\begin{aligned} 4x + y &= 13 \\ 4(4) + (-3) &\stackrel{?}{=} 13 \\ 16 - 3 &\stackrel{?}{=} 13 \\ 13 &= 13 \end{aligned} \qquad \begin{aligned} -2x + 3y &= -17 \\ -2(4) + 3(-3) &\stackrel{?}{=} -17 \\ -8 - 9 &\stackrel{?}{=} -17 \\ -17 &= -17 \end{aligned}$$

Since the ordered pair $(4, -3)$ satisfies both equations of the system, the solution $(4, -3)$ checks. ■

Self Check Solve $\begin{cases} x + 3y = 9 \\ 2x - y = -10 \end{cases}$.

Answer $(-3, 4)$

EXAMPLE 2 Solve the system $\begin{cases} \frac{4}{3}x + \frac{1}{2}y = -\frac{2}{3} \\ \frac{1}{2}x + \frac{2}{3}y = \frac{5}{3} \end{cases}$.

Solution First we find an equivalent system without fractions by multiplying both sides of each equation by 6.

2. $8x + 3y = -4$
3. $3x + 4y = 10$

Because no variable in either equation has a coefficient of 1, it is impossible to avoid fractions when solving for a variable. We solve Equation 3 for x.

$$3x + 4y = 10$$

$$3x = -4y + 10 \qquad \text{Subtract } 4y \text{ from both sides.}$$

4. $$x = -\frac{4}{3}y + \frac{10}{3} \qquad \text{Divide both sides by 3.}$$

We then substitute $-\frac{4}{3}y + \frac{10}{3}$ for x in Equation 2 and solve for y.

$$8x + 3y = -4$$

$$8\left(-\frac{4}{3}y + \frac{10}{3}\right) + 3y = -4 \qquad \text{Substitute } -\tfrac{4}{3}y + \tfrac{10}{3} \text{ for } x.$$

$$-\frac{32}{3}y + \frac{80}{3} + 3y = -4 \qquad \text{Use the distributive property to remove parentheses.}$$

$$-32y + 80 + 9y = -12 \qquad \text{Multiply both sides by 3.}$$

$$-23y = -92 \qquad \text{Combine like terms and subtract 80 from both sides.}$$

$$y = 4 \qquad \text{Divide both sides by } -23.$$

We can find x by substituting 4 for y in Equation 4 and simplifying:

$$\begin{aligned} x &= -\frac{4}{3}y + \frac{10}{3} \\ &= -\frac{4}{3}(4) + \frac{10}{3} && \text{Substitute 4 for } y. \\ &= -\frac{6}{3} && -\tfrac{16}{3} + \tfrac{10}{3} = -\tfrac{6}{3}. \\ &= -2 \end{aligned}$$

The solution is the ordered pair $(-2, 4)$. Verify that this solution satisfies both equations in the original system. ■

Self Check Solve $\begin{cases} \frac{2}{3}x + \frac{1}{2}y = 1 \\ \frac{1}{3}x - \frac{3}{2}y = 4 \end{cases}$.

Answer $(3, -2)$

■ THE ADDITION METHOD

In the addition method, we combine the equations of the system in a way that will eliminate the terms involving one of the variables.

The Addition Method

1. Write both equations of the system in general form.
2. Multiply the terms of one or both of the equations by constants chosen to make the coefficients of x (or y) differ only in sign.
3. Add the equations and solve the resulting equation, if possible.
4. Substitute the value obtained in Step 3 into either of the original equations and solve for the remaining variable.
5. State the solution obtained in Steps 3 and 4.
6. Check the solution in both of the original equations.

EXAMPLE 3 Solve the system $\begin{cases} 4x + y = 13 \\ -2x + 3y = -17 \end{cases}$.

Solution *Step 1:* This is the system discussed in Example 1. Since both equations are already written in general form, Step 1 is unnecessary.

Step 2: To solve the system by addition, we multiply the second equation by 2 to make the coefficients of x differ only in sign.

$$\textbf{5.}\quad \begin{cases} 4x + y = 13 \\ -4x + 6y = -34 \end{cases}$$

Step 3: When these equations are added, the terms involving x drop out, and we get

$$7y = -21$$
$$y = -3 \qquad \text{Divide both sides by 7.}$$

Step 4: To find x, we substitute -3 for y in either of the original equations and solve for x. If we use Equation 5, we have

$$4x + y = 13$$
$$4x + (-3) = 13 \qquad \text{Substitute } -3 \text{ for } y.$$
$$4x = 16 \qquad \text{Add 3 to both sides.}$$
$$x = 4 \qquad \text{Divide both sides by 4.}$$

Step 5: The solution is $x = 4$ and $y = -3$, or just $(4, -3)$.

Step 6: The check was completed in Example 1. ■

Self Check Solve $\begin{cases} 3x + 2y = 0 \\ 2x - y = -7 \end{cases}$.

Answer $(-2, 3)$

EXAMPLE 4 Use the addition method to solve $\begin{cases} \frac{4}{3}x + \frac{1}{2}y = -\frac{2}{3} \\ \frac{1}{2}x + \frac{2}{3}y = \frac{5}{3} \end{cases}$.

Solution This is the system discussed in Example 2. To solve it by addition, we find an equivalent system with no fractions by multiplying both sides of each equation by 6 to obtain

$$\begin{array}{ll} \textbf{6.} & 8x + 3y = -4 \\ \textbf{7.} & 3x + 4y = 10 \end{array}$$

To make the y-terms drop out when we add the equations, we multiply both sides of Equation 6 by 4 and both sides of Equation 7 by -3 to get

$$\begin{cases} 32x + 12y = -16 \\ -9x - 12y = -30 \end{cases}$$

When these equations are added, the y-terms drop out, and we get

$$23x = -46$$
$$x = -2 \qquad \text{Divide both sides by 23.}$$

To find y, we substitute -2 for x in either Equation 6 or Equation 7. If we substitute -2 for x in Equation 7, we get

$$3x + 4y = 10$$
$$3(-2) + 4y = 10 \quad \text{Substitute } -2 \text{ for } x.$$
$$-6 + 4y = 10 \quad \text{Simplify.}$$
$$4y = 16 \quad \text{Add 6 to both sides.}$$
$$y = 4 \quad \text{Divide both sides by 4.}$$

The solution is the ordered pair $(-2, 4)$. ■

Self Check Solve $\begin{cases} \frac{4}{3}x + \frac{1}{2}y = -3 \\ \frac{1}{2}x + \frac{2}{3}y = -\frac{1}{6} \end{cases}$.

Answer $(-3, 2)$

■ AN INCONSISTENT SYSTEM

EXAMPLE 5 Solve $\begin{cases} y = 2x + 4 \\ 8x - 4y = 7 \end{cases}$, if possible.

Solution Because the first equation is already solved for y, we use the substitution method.

$$8x - 4y = 7$$
$$8x - 4(2x + 4) = 7 \quad \text{Substitute } 2x + 4 \text{ for } y.$$

We then solve this equation for x:

$$8x - 8x - 16 = 7 \quad \text{Use the distributive property to remove parentheses.}$$
$$-16 = 7 \quad \text{Combine like terms.}$$

This impossible result shows that the equations in the system are independent and that the system is inconsistent. Since the system has no solution, the graphs of the equations in the system are parallel. ■

Self Check Solve $\begin{cases} x = -\frac{5}{2}y + 5 \\ y = -\frac{2}{5}x + 5 \end{cases}$.

Answer no solution

A SYSTEM WITH INFINITELY MANY SOLUTIONS

EXAMPLE 6 Solve $\begin{cases} 4x + 6y = 12 \\ -2x - 3y = -6 \end{cases}$.

Solution Since the equations are written in general form, we use the addition method. To make the x-terms drop out when we add the equations, we multiply both sides of the second equation by 2 to get

$$\begin{cases} 4x + 6y = 12 \\ -4x - 6y = -12 \end{cases}$$

After adding the left-hand sides and the right-hand sides, we get

$$\begin{aligned} 0x + 0y &= 0 \\ 0 &= 0 \end{aligned}$$

Here, both the $x-$ and y-terms drop out. The true statement $0 = 0$ shows that the equations in this system are dependent and that the system is consistent.

Note that the equations of the system are equivalent, because when the equation is multiplied by -2, it becomes the first equation. The line graphs of these equations would coincide. Since any ordered pair that satisfies one of the equations also satisfies the other, there are infinitely many solutions. Verify that three of them are $(0, 2)$, $(3, 0)$, and $(6, -2)$. ■

Self Check Solve $\begin{cases} x = -\frac{5}{2}y + 5 \\ y = -\frac{2}{5}x + 2 \end{cases}$.

Answer infinitely many solutions (three of them are $(0, 2)$, $(5, 0)$, and $(10, -2)$)

REPEATING DECIMALS

We have seen how to change fractions into decimal form. By using systems of equations, we can change repeating decimals into fractional form. For example, to write $0.2\overline{54}$ as a fraction, we note that the decimal has a repeating block of two digits and then form an equation by setting x equal to the decimal.

8. $x = 0.2\ 54\ 54\ 54\ .\ .\ .$

We then form another equation by multiplying both sides of Equation 8 by 10^2.

9. $100x = 25.4\ 54\ 54\ 54\ .\ .\ .$ $\quad 10^2 = 100.$

We can subtract each side of Equation 8 from the corresponding side of Equation 9 to obtain

$$\begin{array}{r} 100x = 25.4\ 54\ 54\ 54\ldots \\ \underline{x = \ \ 0.2\ 54\ 54\ 54\ldots} \\ 99x = 25.2 \end{array}$$

Finally, we solve $99x = 25.2$ for x and simplify the fraction.

$$x = \frac{25.2}{99} = \frac{25.2 \cdot \mathbf{10}}{99 \cdot \mathbf{10}} = \frac{252}{990} = \frac{\cancel{18} \cdot 14}{\cancel{18} \cdot 55} = \frac{14}{55}$$

We can use a calculator to verify that the decimal representation of $\frac{14}{55}$ is $0.2\overline{54}$.

The key step in the solution was multiplying both sides of Equation 8 by 10^2. If there had been n digits in the repeating block of the decimal, we would have multiplied both sides of Equation 8 by 10^n.

PROBLEM SOLVING

To solve problems using two variables, we follow the same problem-solving strategy discussed in Chapter 1, except that we use two variables and form two equations instead of one.

EXAMPLE 7 **Retail sales** A store advertises two types of cordless telephones, one selling for \$67 and the other for \$100. If the receipts from the sale of 36 phones totaled \$2,940, how many of each type were sold?

Analyze the problem We can let x represent the number of phones sold for \$67 and let y represent the number of phones sold for \$100. Then the receipts for the sale of the lower-priced phones are \$67x, and the receipts for the sale of the higher-priced phones are \$100y.

Form two equations The information of the problem gives the following two equations:

The number of lower-priced phones	+	the number of higher-priced phones	=	the total number of phones.
x	+	y	=	36
The value of the lower-priced phones	+	the value of the higher-priced phones	=	the total receipts.
$67x$	+	$100y$	=	2,940

Solve the system To find out how many of each type of phone were sold, we must solve the following system:

$$\begin{cases} x + y = 36 & \textbf{(10)} \\ 67x + 100y = 2{,}940 & \textbf{(11)} \end{cases}$$

We multiply both sides of Equation 10 by -100, add the resulting equation to Equation 11, and solve for x:

$$\begin{aligned} -100x - 100y &= -3{,}600 \\ 67x + 100y &= 2{,}940 \\ \hline -33x &= -660 \\ x &= 20 \qquad \text{Divide both sides by } -33. \end{aligned}$$

To find y, we substitute 20 for x in Equation 10 and solve for y:

$$\begin{aligned} x + y &= 36 \\ 20 + y &= 36 \qquad \text{Substitute 20 for } x. \\ y &= 16 \qquad \text{Subtract 20 from both sides.} \end{aligned}$$

State the conclusion The store sold 20 of the lower-priced phones and 16 of the higher-priced phones.

Check the result If 20 of one type were sold and 16 of the other type were sold, a total of 36 phones were sold.

Since the value of the lower-priced phones is 20($67) = $1,340 and the value of the higher-priced phones is 16($100) = $1,600, the total receipts are $2,940. ■

EXAMPLE 8

Mixing solutions How many ounces of a 5% saline solution and how many ounces of a 20% saline solution must be mixed together to obtain 50 ounces of a 15% saline solution?

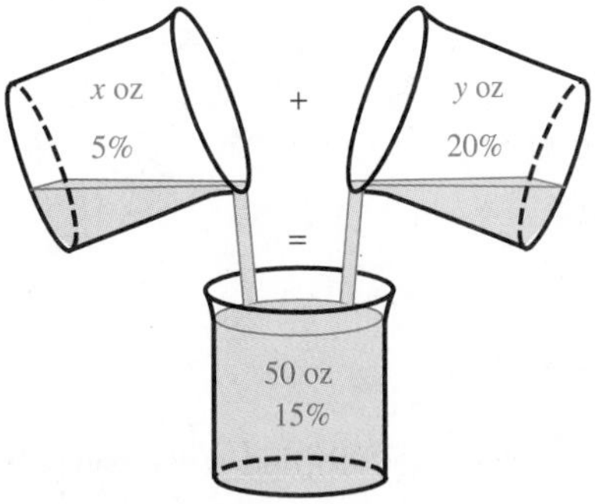

FIGURE 3-6

Analyze the problem We can let x represent the number of ounces of the 5% solution and let y represent the number of ounces of the 20% solution that are to be mixed. Then the amount of salt in the 5% solution is $0.05x$, and the amount of salt in the 20% solution is $0.20y$. (See Figure 3-6.)

Form two equations The information of the problem gives the following two equations:

The number of ounces of 5% solution	+	the number of ounces of 20% solution	=	the total number of ounces in the mixture.
x	+	y	=	50

The salt in the 5% solution	+	the salt in the 20% solution	=	the salt in the mixture.
$0.05x$	+	$0.20y$	=	$0.15(50)$

Solve the system To find out how many ounces of each are needed, we solve the following system:

$$\begin{cases} x + y = 50 & (12) \\ 0.05x + 0.20y = 7.5 & (13) \end{cases} \qquad 0.15(50) = 7.5.$$

To solve this system by substitution, we can solve Equation 12 for y

$$x + y = 50$$

14. $y = 50 - x$ Subtract x from both sides.

and then substitute $50 - x$ for y in Equation 13.

$$0.05x + 0.20y = 7.5$$

$$0.05x + 0.20(\mathbf{50 - x}) = 7.5 \quad \text{Substitute } 50 - x \text{ for } y.$$

$$5x + 20(50 - x) = 750 \quad \text{Multiply both sides by 100.}$$

$$5x + 1{,}000 - 20x = 750 \quad \text{Use the distributive property to remove parentheses.}$$

$$-15x = -250 \quad \text{Combine like terms and subtract 1,000 from both sides.}$$

$$x = \frac{-250}{-15} \quad \text{Divide both sides by } -15.$$

$$x = \frac{50}{3} \quad \text{Simplify } \tfrac{-250}{-15}.$$

To find y, we can substitute $\frac{50}{3}$ for x in Equation 14:

$$y = 50 - x$$

$$= 50 - \mathbf{\frac{50}{3}} \quad \text{Substitute } \tfrac{50}{3} \text{ for } x.$$

$$= \frac{100}{3}$$

State the conclusion To obtain 50 ounces of a 15% solution, we must mix $16\frac{2}{3}$ ounces of the 5% solution with $33\frac{1}{3}$ ounces of the 20% solution.

Check the result We note that $16\frac{2}{3}$ ounces of solution plus $33\frac{1}{3}$ ounces of solution equals the required 50 ounces of solution. We also note that 5% of $16\frac{2}{3} \approx 0.83$, and 20% of $33\frac{1}{3} \approx 6.67$, giving a total of 7.5, which is 15% of 50. ■

BREAK-POINT ANALYSIS

Running a machine involves both *setup costs* and *unit costs*. Setup costs include the cost of preparing a machine to do a certain job. Unit costs depend on the number of items to be manufactured, including costs of raw materials and labor.

Suppose that a certain machine has a setup cost of \$600 and a unit cost of \$3. If x items will be manufactured using this machine, the cost will be

$$\text{Cost} = 600 + 3x \qquad \text{Cost = setup cost + unit cost} \times \text{the number of items.}$$

Furthermore, suppose that a larger and more efficient machine has a setup cost of \$800 and a unit cost of \$2. The cost of manufacturing x items using this machine is

$$\text{Cost on larger machine} = 800 + 2x$$

The *break point* is the number of units x that need to be manufactured to make the cost the same using either machine. It can be found by setting the two costs equal to each other and solving for x.

$$600 + 3x = 800 + 2x$$
$$x = 200 \qquad \text{Subtract 600 and } 2x \text{ from both sides.}$$

The break point is 200 units, because the cost using either machine is \$1,200 when $x = 200$.

$$\begin{aligned}\text{Cost on small machine} &= 600 + 3x \\ &= 600 + 3(200) \\ &= 600 + 600 \\ &= 1{,}200\end{aligned} \qquad \begin{aligned}\text{Cost on large machine} &= 800 + 2x \\ &= 800 + 2(200) \\ &= 800 + 400 \\ &= 1{,}200\end{aligned}$$

EXAMPLE 9 One machine has a setup cost of \$400 and a unit cost of \$1.50, and another machine has a setup cost of \$500 and a unit cost of \$1.25. Find the break point.

Analyze the problem The cost C_1 of manufacturing x units on machine 1 is $\$1.50x + \$400$ (the number of units manufactured times \$1.50, plus the setup cost of \$400). The cost C_2 of manufacturing the same number of units on machine 2 is $\$1.25x + \$500$ (the number of units manufactured times \$1.25, plus the setup cost of \$500). The break point occurs when the costs are equal ($C_1 = C_2$).

Form two equations If x represents the number of items to be manufactured, the cost C_1 using machine 1 is

The cost of using machine 1	=	the cost of manufacturing x units	+	the setup cost.
C_1	=	$1.5x$	+	400

The cost C_2 using machine 2 is

The cost of using machine 2	=	the cost of manufacturing x units	+	the setup cost.
C_2	=	$1.25x$	+	500

Solve the system To find the break point, we must solve the system $\begin{cases} C_1 = 1.5x + 400 \\ C_2 = 1.25x + 500 \end{cases}$. Since the break point occurs when $C_1 = C_2$, we can substitute $1.5x + 400$ for C_2 to get

$$\begin{aligned} 1.5x + 400 &= 1.25x + 500 \\ 1.5x &= 1.25x + 100 && \text{Subtract 400 from both sides.} \\ 0.25x &= 100 && \text{Subtract } 1.25x \text{ from both sides.} \\ x &= 400 && \text{Divide both sides by 0.25.} \end{aligned}$$

State the conclusion The break point is 400 units.

Check the result For 400 units, the cost using machine 1 is $400 + 1.5(400) = 400 + 600 = 1{,}000$. The cost using machine 2 is $500 + 1.25(400) = 500 + 500 = 1{,}000$. Since the costs are equal, the break point is 400. ■

PARALLELOGRAMS

A **parallelogram** is a four-sided figure with its opposite sides parallel. (See Figure 3-7(a).) Here are some important facts about parallelograms.

1. Opposite sides of a parallelogram have the same length.
2. Opposite angles of a parallelogram have the same measure.
3. Consecutive angles of a parallelogram are supplementary.
4. A diagonal of a parallelogram (see Figure 3-7(b)) divides the parallelogram into two *congruent triangles*—triangles with the same shape and same area.
5. In Figure 3-7(b), angles 1 and 2, and angles 3 and 4, are called pairs of *alternate interior angles.* When a diagonal intersects two parallel sides of a parallelogram, all pairs of alternate interior angles have the same measure.

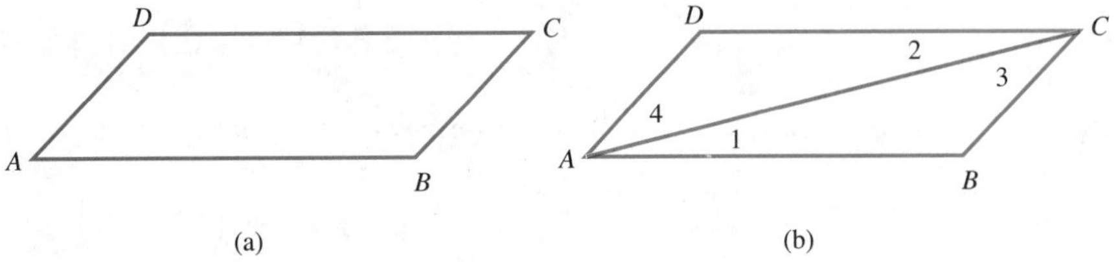

FIGURE 3-7

EXAMPLE 10 Refer to the parallelogram shown in Figure 3-8 and find the values of x and y.

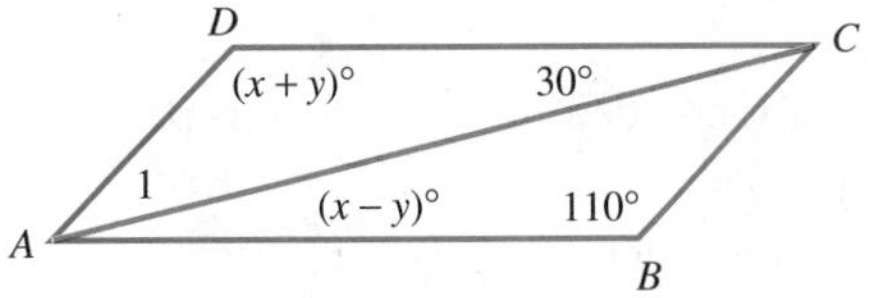

FIGURE 3-8

Solution Since diagonal AC intersects two parallel sides, the alternate interior angles that are formed have the same measure. Thus, $(x - y)° = 30°$. Since opposite angles of a parallelogram have the same measure, we know that $(x + y)° = 110°$. We can form the following system of equations and solve it by addition.

$$\begin{aligned} &\textbf{15.}\quad \begin{cases} x - y = 30 \\ x + y = 110 \end{cases} \\ &\textbf{16.} \end{aligned}$$

$2x = 140$	Add Equations 15 and 16.
$x = 70$	Divide both sides by 2.

We can substitute 70 for x in Equation 16 and solve for y.

$x + y = 110$	
$\mathbf{70} + y = 110$	Substitute 70 for x.
$y = 40$	Subtract 70 from both sides.

Thus, $x = 70$ and $y = 40$. ■

Self Check Find the measure of angle 1. (*Hint:* The sum of the angles of a triangle equals 180°.)

Answer 40°

Orals *Solve each system for x.*

1. $\begin{cases} y = 2x \\ x + y = 6 \end{cases}$

2. $\begin{cases} y = -x \\ 2x + y = 4 \end{cases}$

3. $\begin{cases} x - y = 6 \\ x + y = 2 \end{cases}$

4. $\begin{cases} x + y = 4 \\ 2x - y = 5 \end{cases}$

EXERCISE 3.2

REVIEW *Simplify each expression. Write all answers without using negative exponents.*

1. $(a^2a^3)^2(a^4a^2)^2$
2. $\left(\dfrac{a^2b^3c^4d}{ab^2c^3d^4}\right)^{-3}$
3. $\left(\dfrac{-3x^3y^4}{x^{-5}y^3}\right)^{-4}$
4. $\dfrac{3t^0 - 4t^0 + 5}{5t^0 + 2t^0}$

VOCABULARY AND CONCEPTS *Fill in each blank to make a true statement.*

5. Running a machine involves both ______ costs and ______ costs.
6. The ______ point is the number of units that need to be manufactured to make the cost the same on either of two machines.
7. A ____________ is a four-sided figure with both pairs of opposite sides parallel.
8. ________ sides of a parallelogram have the same length.
9. ________ angles of a parallelogram have the same measure.
10. __________ angles of a parallelogram are supplementary.

PRACTICE *In Exercises 11–22, solve each system by substitution, if possible.*

11. $\begin{cases} y = x \\ x + y = 4 \end{cases}$
12. $\begin{cases} y = x + 2 \\ x + 2y = 16 \end{cases}$
13. $\begin{cases} x - y = 2 \\ 2x + y = 13 \end{cases}$
14. $\begin{cases} x - y = -4 \\ 3x - 2y = -5 \end{cases}$
15. $\begin{cases} x + 2y = 6 \\ 3x - y = -10 \end{cases}$
16. $\begin{cases} 2x - y = -21 \\ 4x + 5y = 7 \end{cases}$
17. $\begin{cases} 3x = 2y - 4 \\ 6x - 4y = -4 \end{cases}$
18. $\begin{cases} 8x = 4y + 10 \\ 4x - 2y = 5 \end{cases}$
19. $\begin{cases} 3x - 4y = 9 \\ x + 2y = 8 \end{cases}$
20. $\begin{cases} 3x - 2y = -10 \\ 6x + 5y = 25 \end{cases}$
21. $\begin{cases} 2x + 2y = -1 \\ 3x + 4y = 0 \end{cases}$
22. $\begin{cases} 5x + 3y = -7 \\ 3x - 3y = 7 \end{cases}$

In Exercises 23–34, solve each system by addition, if possible.

23. $\begin{cases} x - y = 3 \\ x + y = 7 \end{cases}$
24. $\begin{cases} x + y = 1 \\ x - y = 7 \end{cases}$
25. $\begin{cases} 2x + y = -10 \\ 2x - y = -6 \end{cases}$
26. $\begin{cases} x + 2y = -9 \\ x - 2y = -1 \end{cases}$
27. $\begin{cases} 2x + 3y = 8 \\ 3x - 2y = -1 \end{cases}$
28. $\begin{cases} 5x - 2y = 19 \\ 3x + 4y = 1 \end{cases}$
29. $\begin{cases} 4x + 9y = 8 \\ 2x - 6y = -3 \end{cases}$
30. $\begin{cases} 4x + 6y = 5 \\ 8x - 9y = 3 \end{cases}$
31. $\begin{cases} 8x - 4y = 16 \\ 2x - 4 = y \end{cases}$
32. $\begin{cases} 2y - 3x = -13 \\ 3x - 17 = 4y \end{cases}$
33. $\begin{cases} x = \dfrac{3}{2}y + 5 \\ 2x - 3y = 8 \end{cases}$
34. $\begin{cases} x = \dfrac{2}{3}y \\ y = 4x + 5 \end{cases}$

In Exercises 35–42, solve each system by any method.

35. $\begin{cases} \dfrac{x}{2} + \dfrac{y}{2} = 6 \\ \dfrac{x}{2} - \dfrac{y}{2} = -2 \end{cases}$
36. $\begin{cases} \dfrac{x}{2} - \dfrac{y}{3} = -4 \\ \dfrac{x}{2} + \dfrac{y}{9} = 0 \end{cases}$
37. $\begin{cases} \dfrac{3}{4}x + \dfrac{2}{3}y = 7 \\ \dfrac{3}{5}x - \dfrac{1}{2}y = 18 \end{cases}$
38. $\begin{cases} \dfrac{2}{3}x - \dfrac{1}{4}y = -8 \\ \dfrac{1}{2}x - \dfrac{3}{8}y = -9 \end{cases}$

39. $\begin{cases} \dfrac{3x}{2} - \dfrac{2y}{3} = 0 \\ \dfrac{3x}{4} + \dfrac{4y}{3} = \dfrac{5}{2} \end{cases}$

40. $\begin{cases} \dfrac{3x}{5} + \dfrac{5y}{3} = 2 \\ \dfrac{6x}{5} - \dfrac{5y}{3} = 1 \end{cases}$

41. $\begin{cases} \dfrac{2}{5}x - \dfrac{1}{6}y = \dfrac{7}{10} \\ \dfrac{3}{4}x - \dfrac{2}{3}y = \dfrac{19}{8} \end{cases}$

42. $\begin{cases} \dfrac{5}{6}x + \dfrac{2}{3}y = \dfrac{7}{6} \\ \dfrac{10}{7}x - \dfrac{4}{9}y = \dfrac{17}{21} \end{cases}$

In Exercises 43–46, write each repeating decimal as a fraction. Simplify the answer when possible.

43. $0.\overline{3}$

44. $0.\overline{29}$

45. $-0.34\overline{89}$

46. $-2.3\overline{47}$

In Exercises 47–50, solve each system for x and y. Solve for $\frac{1}{x}$ and $\frac{1}{y}$ first.

47. $\begin{cases} \dfrac{1}{x} + \dfrac{1}{y} = \dfrac{5}{6} \\ \dfrac{1}{x} - \dfrac{1}{y} = \dfrac{1}{6} \end{cases}$

48. $\begin{cases} \dfrac{1}{x} + \dfrac{1}{y} = \dfrac{9}{20} \\ \dfrac{1}{x} - \dfrac{1}{y} = \dfrac{1}{20} \end{cases}$

49. $\begin{cases} \dfrac{1}{x} + \dfrac{2}{y} = -1 \\ \dfrac{2}{x} - \dfrac{1}{y} = -7 \end{cases}$

50. $\begin{cases} \dfrac{3}{x} - \dfrac{2}{y} = -30 \\ \dfrac{2}{x} - \dfrac{3}{y} = -30 \end{cases}$

APPLICATIONS *Use two variables and two equations to solve each problem.*

51. Merchandising A pair of shoes and a sweater cost \$98. If the sweater cost \$16 more than the shoes, how much did the sweater cost?

52. Merchandising A sporting goods salesperson sells 2 fishing reels and 5 rods for \$270. The next day, the salesperson sells 4 reels and 2 rods for \$220. How much does each cost?

53. Electronics Two resistors in the voltage divider circuit in Illustration 1 have a total resistance of 1,375 ohms. To provide the required voltage, R_1 must be 125 ohms greater than R_2. Find both resistances.

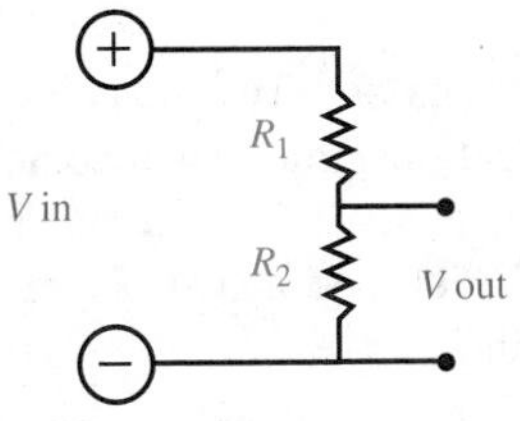

ILLUSTRATION 1

54. Stowing baggage A small aircraft can carry 950 pounds of baggage, distributed between two storage compartments. On one flight, the plane is fully loaded, with 150 pounds more baggage in one compartment than the other. How much is stowed in each compartment?

55. Geometry problem The rectangular field in Illustration 2 is surrounded by 72 meters of fencing. If the field is partitioned as shown, a total of 88 meters of fencing is required. Find the dimensions of the field.

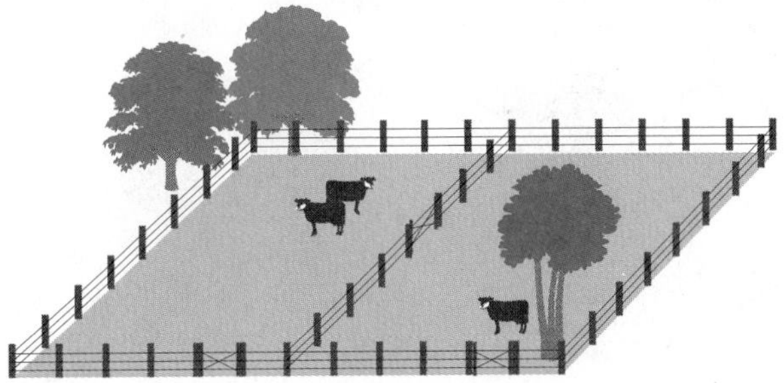

ILLUSTRATION 2

56. Geometry In a right triangle, one acute angle is 15° greater than two times the other acute angle. Find the difference between the angles.

57. Investment income Part of \$8,000 was invested at 10% interest and the rest at 12%. If the annual income from these investments was \$900, how much was invested at each rate?

58. Investment income Part of \$12,000 was invested at 6% interest and the rest at 7.5%. If the annual income from these investments was \$810, how much was invested at each rate?

59. Mixing a solution How many ounces of the two alcohol solutions in Illustration 3 must be mixed to obtain 100 ounces of a 12.2% solution?

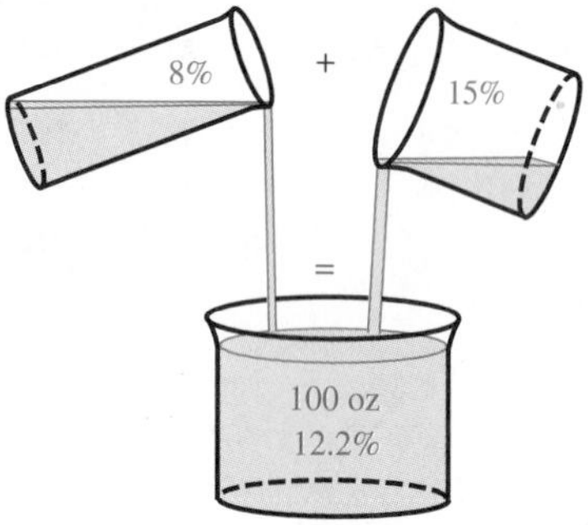

ILLUSTRATION 3

60. Mixing candy How many pounds each of candy shown in Illustration 4 must be mixed to obtain 60 pounds of candy that is worth \$3 per pound?

ILLUSTRATION 4

61. Travel A car travels 50 miles in the same time that a plane travels 180 miles. The speed of the plane is 143 mph faster than the speed of the car. Find the speed of the car.

62. Travel A car and a truck leave Rockford at the same time, heading in opposite directions. When they are 350 miles apart, the car has gone 70 miles farther than the truck. How far has the car traveled?

63. **Making bicycles** A bicycle manufacturer builds racing bikes and mountain bikes, with the per-unit manufacturing costs shown in Illustration 5. The company has budgeted \$15,900 for labor and \$13,075 for materials. How many bicycles of each type can be built?

Model	Cost of materials	Cost of labor
Racing	\$55	\$60
Mountain	\$70	\$90

ILLUSTRATION 5

64. Farming A farmer keeps some animals on a strict diet. Each animal is to receive 15 grams of protein and 7.5 grams of carbohydrates. The farmer uses two food mixes with nutrients shown in Illustration 6. How many grams of each mix should be used to provide the correct nutrients for each animal?

Mix	Protein	Carbohydrates
Mix A	12%	9%
Mix B	15%	5%

ILLUSTRATION 6

65. Milling brass plates Two machines can mill a brass plate. One machine has a setup cost of \$300 and a cost per plate of \$2. The other machine has a setup cost of \$500 and a cost per plate of \$1. Find the break point.

66. Printing books A printer has two presses. One has a setup cost of \$210 and can print the pages of a certain book for \$5.98. The other press has a setup cost of \$350 and can print the pages of the same book for \$5.95. Find the break point.

67. Managing a computer store The manager of a computer store knows that his fixed costs are $8,925 per month and that his unit cost is $850 for every computer sold. If he can sell all the computers he can get for $1,275 each, how many computers must he sell each month to break even?

68. Managing a beauty shop A beauty shop specializing in permanents has fixed costs of $2,101.20 per month. The owner estimates that the cost for each permanent is $23.60. This cost covers labor, chemicals, and electricity. If her shop can give as many permanents as she wants at a price of $44 each, how many must be given each month to break even?

69. Running a small business A person invests $18,375 to set up a small business that produces a piece of computer software that will sell for $29.95. If each piece can be produced for $5.45, how many pieces must be sold to break even?

70. Running a record company Three people invest $35,000 each to start a record company that will produce reissues of classic jazz. Each release will be a set of 3 CDs that will retail for $15 per disc. If each set can be produced for $18.95, how many sets must be sold for the investors to make a profit?

In Exercises 71–74, a paint manufacturer can choose between two processes for manufacturing house paint, with monthly costs shown in Illustration 7. Assume that the paint sells for $18 per gallon.

71. For process A, how many gallons must be sold for the manufacturer to break even?

72. For process B, how many gallons must be sold for the manufacturer to break even?

73. If expected sales are 6,000 gallons per month, which process should the company use?

74. If expected sales are 7,000 gallons per month, which process should the company use?

Process	Fixed costs	Unit cost (per gallon)
A	$32,500	$13
B	$80,600	$ 5

ILLUSTRATION 7

In Exercises 75–80, a manufacturer of automobile water pumps is considering retooling for one of two manufacturing processes, with monthly fixed costs and unit costs as indicated in Illustration 8. Each water pump can be sold for $50.

75. For process A, how many water pumps must be sold for the manufacturer to break even?

76. For process B, how many water pumps must be sold for the manufacturer to break even?

77. If expected sales are 550 per month, which process should be used?

78. If expected sales are 600 per month, which process should be used?

79. If expected sales are 650 per month, which process should be used?

80. At what monthly sales level is process B better?

Process	Fixed costs	Unit cost
A	$12,390	$29
B	$20,460	$17

ILLUSTRATION 8

81. Geometry If two angles are supplementary, their sum is 180°. If the difference between two supplementary angles is 110°, find the measure of each angle.

82. Geometry If two angles are complementary, their sum is 90°. If one of two complementary angles is 16° greater than the other, find the measure of each angle.

In Exercises 83–84, Illustrations 9 and 10 are parallelograms.

83. Find x and y in Illustration 9.

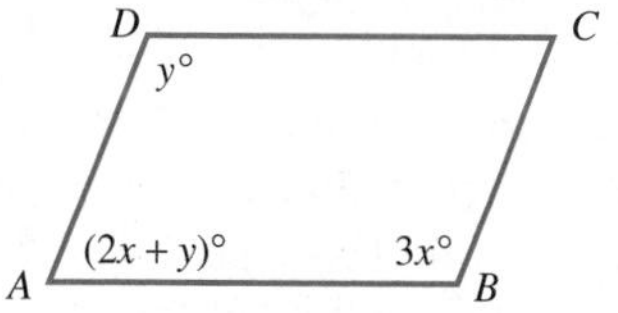

ILLUSTRATION 9

84. Find x and y in Illustration 10.

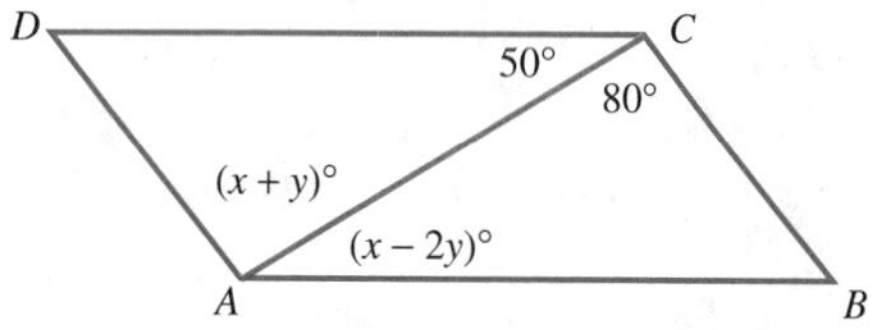

ILLUSTRATION 10

85. Selecting radio frequencies In a radio, an inductor and a capacitor are used in a resonant circuit to select a wanted radio station at a frequency f and reject all others. The inductance L and the capacitance C determine the inductance reactance X_L and the capacitive reactance X_C of that circuit, where

$$X_L = 2\pi f L \quad \text{and} \quad X_C = \frac{1}{2\pi f C}$$

The radio station selected will be at the frequency f where $X_L = X_C$. Write a formula for f^2 in terms of L and C.

86. Choosing salary plans A sales clerk can choose from two salary options: **1.** a straight 7% commission and **2.** \$150 + 2% commission. How much would the clerk have to sell for each plan to produce the same monthly paycheck?

WRITING

87. Tell which method you would use to solve the following system. Why?

$$\begin{cases} y = 3x + 1 \\ 3x + 2y = 12 \end{cases}$$

88. Tell which method you would use to solve the following system. Why?

$$\begin{cases} 2x + 4y = 9 \\ 3x - 5y = 20 \end{cases}$$

SOMETHING TO THINK ABOUT

89. Under what conditions will a system of two equations in two variables be inconsistent?

90. Under what conditions will the equations of a system of two equations in two variables be dependent?

3.3 Solutions of Three Equations in Three Variables

■ SOLVING THREE EQUATIONS IN THREE VARIABLES ■ A CONSISTENT SYSTEM ■ AN INCONSISTENT SYSTEM ■ SYSTEMS WITH DEPENDENT EQUATIONS ■ PROBLEM SOLVING ■ CURVE FITTING

Getting Ready *Tell whether the equation $x + 2y + 3z = 6$ is satisfied by the following values.*

1. (1, 1, 1)

2. (−2, 1, 2)

3. (2, −2, −1)

4. (2, 2, 0)

SOLVING THREE EQUATIONS IN THREE VARIABLES

We now extend the definition of a linear equation to include equations of the form $ax + by + cz = d$. The solution of a system of three linear equations with three variables is an ordered triple of numbers. For example, the solution of the system

$$\begin{cases} 2x + 3y + 4z = 20 \\ 3x + 4y + 2z = 17 \\ 3x + 2y + 3z = 16 \end{cases}$$

is the triple (1, 2, 3), since each equation is satisfied if $x = 1$, $y = 2$, and $z = 3$.

$$\begin{aligned} 2x + 3y + 4z &= 20 \\ 2(1) + 3(2) + 4(3) &= 20 \\ 2 + 6 + 12 &= 20 \\ 20 &= 20 \end{aligned} \qquad \begin{aligned} 3x + 4y + 2z &= 17 \\ 3(1) + 4(2) + 2(3) &= 17 \\ 3 + 8 + 6 &= 17 \\ 17 &= 17 \end{aligned} \qquad \begin{aligned} 3x + 2y + 3z &= 16 \\ 3(1) + 2(2) + 3(3) &= 16 \\ 3 + 4 + 9 &= 16 \\ 16 &= 16 \end{aligned}$$

The graph of an equation of the form $ax + by + cz = d$ is a flat surface called a **plane**. A system of three linear equations in three variables is consistent or inconsistent, depending on how the three planes corresponding to the three equations intersect. Figure 3-9 illustrates some of the possibilities.

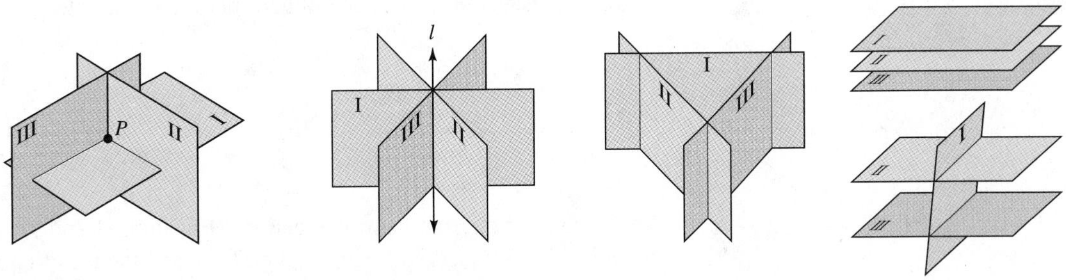

The three planes intersect at a single point P: One solution

(a)

The three planes have a line l in common: Infinitely many solutions

(b)

The three planes have no point in common: No solutions

(c)

FIGURE 3-9

To solve a system of three linear equations in three variables, we follow these steps.

Solving Three Equations in Three Variables

1. Pick any two equations and eliminate a variable.
2. Pick a different pair of equations and eliminate the same variable.
3. Solve the resulting pair of two equations in two variables.
4. To find the value of the third variable, substitute the values of the two variables found in Step 3 into any equation containing all three variables and solve the equation.
5. Check the solution in all three of the original equations.

A CONSISTENT SYSTEM

EXAMPLE 1 Solve the system $\begin{cases} 2x + y + 4z = 12 \\ x + 2y + 2z = 9 \\ 3x - 3y - 2z = 1 \end{cases}$.

Solution We are given the system

$$\begin{array}{ll} \textbf{1.} & \\ \textbf{2.} & \\ \textbf{3.} & \end{array}\begin{cases} 2x + y + 4z = 12 \\ x + 2y + 2z = 9 \\ 3x - 3y - 2z = 1 \end{cases}$$

If we pick Equations 2 and 3 and add them, the variable z is eliminated:

$$\begin{array}{lrcl} \textbf{2.} & x + 2y + 2z &=& 9 \\ \textbf{3.} & 3x - 3y - 2z &=& 1 \\ \hline \textbf{4.} & 4x - y &=& 10 \end{array}$$

We now pick a different pair of equations (Equations 1 and 3) and eliminate z again. If each side of Equation 3 is multiplied by 2 and the resulting equation is added to Equation 1, z is again eliminated:

$$\begin{array}{lrcl} \textbf{1.} & 2x + y + 4z &=& 12 \\ & 6x - 6y - 4z &=& 2 \\ \hline \textbf{5.} & 8x - 5y &=& 14 \end{array}$$

Equations 4 and 5 form a system of two equations in two variables:

$$\begin{array}{l} \textbf{4.} \\ \textbf{5.} \end{array}\begin{cases} 4x - y = 10 \\ 8x - 5y = 14 \end{cases}$$

To solve this system, we multiply Equation 4 by -5 and add the resulting equation to Equation 5 to eliminate y:

$$\begin{array}{lrcl} & -20x + 5y &=& -50 \\ \textbf{5.} & 8x - 5y &=& 14 \\ \hline & -12x \quad &=& -36 \\ & x &=& 3 \end{array} \qquad \text{Divide both sides by } -12.$$

To find y, we substitute 3 for x in any equation containing only x and y (such as Equation 5) and solve for y:

$$\begin{array}{lrcll} \textbf{5.} & 8x - 5y &=& 14 & \\ & 8(3) - 5y &=& 14 & \text{Substitute 3 for } x. \\ & 24 - 5y &=& 14 & \text{Simplify.} \\ & -5y &=& -10 & \text{Subtract 24 from both sides.} \\ & y &=& 2 & \text{Divide both sides by } -5. \end{array}$$

To find z, we substitute 3 for x and 2 for y in an equation containing x, y, and z (such as Equation 1) and solve for z:

$$\begin{array}{lrcll} \textbf{1.} & 2x + y + 4z &=& 12 & \\ & 2(3) + 2 + 4z &=& 12 & \text{Substitute 3 for } x \text{ and 2 for } y. \\ & 8 + 4z &=& 12 & \text{Simplify.} \\ & 4z &=& 4 & \text{Subtract 8 from both sides.} \\ & z &=& 1 & \text{Divide both sides by 4.} \end{array}$$

The solution of the system is $(x, y, z) = (3, 2, 1)$. Verify that these values satisfy each equation in the original system. ■

Self Check Solve $\begin{cases} 2x + y + 4z = 16 \\ x + 2y + 2z = 11 \\ 3x - 3y - 2z = -9 \end{cases}$.

Answer (1, 2, 3)

■ AN INCONSISTENT SYSTEM

EXAMPLE 2 Solve the system $\begin{cases} 2x + y - 3z = -3 \\ 3x - 2y + 4z = 2 \\ 4x + 2y - 6z = -7 \end{cases}$.

Solution We are given the system of equations

$$\begin{array}{l} \textbf{1.} \\ \textbf{2.} \\ \textbf{3.} \end{array} \begin{cases} 2x + y - 3z = -3 \\ 3x - 2y + 4z = 2 \\ 4x + 2y - 6z = -7 \end{cases}$$

We can multiply Equation 1 by 2 and add the resulting equation to Equation 2 to eliminate y:

$$\begin{array}{ll} & 4x + 2y - 6z = -6 \\ \textbf{2.} & \underline{3x - 2y + 4z = 2} \\ \textbf{4.} & 7x - 2z = -4 \end{array}$$

We now add Equations 2 and 3 to eliminate y again:

$$\begin{array}{ll} \textbf{2.} & 3x - 2y + 4z = 2 \\ \textbf{3.} & \underline{4x + 2y - 6z = -7} \\ \textbf{5.} & 7x - 2z = -5 \end{array}$$

Equations 4 and 5 form the system

$$\begin{array}{l} \textbf{4.} \\ \textbf{5.} \end{array} \begin{cases} 7x - 2z = -4 \\ 7x - 2z = -5 \end{cases}$$

Since $7x - 2z$ cannot equal both -4 and -5, this system is inconsistent. Thus, it has no solution. ■

Self Check Solve $\begin{cases} 2x + y - 3z = 8 \\ 3x - 2y + 4z = 10 \\ 4x + 2y - 6z = -5 \end{cases}$.

Answer no solution

■ SYSTEMS WITH DEPENDENT EQUATIONS

When the equations in a system of two equations in two variables were dependent, the system had infinitely many solutions. This is not always true for systems of three equations in three variables. In fact, a system can have dependent equations and still be inconsistent. Figure 3-10 illustrates the different possibilities.

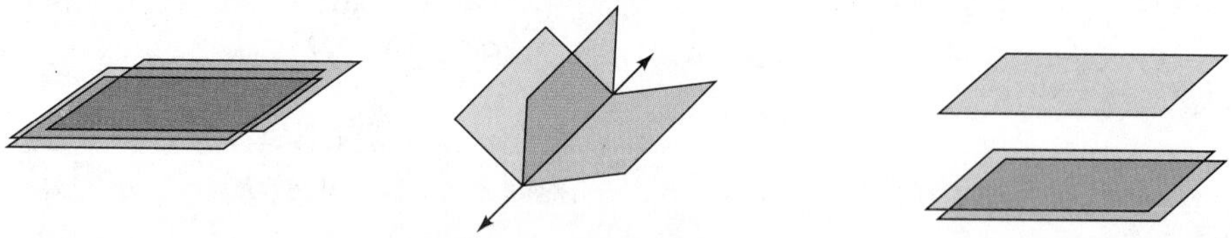

When three planes coincide, the equations are dependent, and there are infinitely many solutions.

(a)

When three planes intersect in a common line, the equations are dependent, and there are infinitely many solutions.

(b)

When two planes coincide and are parallel to a third plane, the system is inconsistent, and there are no solutions.

(c)

FIGURE 3-10

EXAMPLE 3 Solve the system $\begin{cases} 3x - 2y + z = -1 \\ 2x + y - z = 5 \\ 5x - y = 4 \end{cases}$.

Solution We can add the first two equations to get

$$\begin{aligned} 3x - 2y + z &= -1 \\ 2x + y - z &= 5 \\ \hline \textbf{1.}\quad 5x - y \phantom{{}+z} &= 4 \end{aligned}$$

Since Equation 1 is the same as the third equation of the system, the equations of the system are dependent, and there will be infinitely many solutions. From a graphical perspective, the equations represent three planes that intersect in a common line, as shown in Figure 3-10(b).

To write the general solution to this system, we can solve Equation 1 for y to get

$$\begin{aligned} 5x - y &= 4 \\ -y &= -5x + 4 && \text{Subtract } 5x \text{ from both sides.} \\ y &= 5x - 4 && \text{Multiply both sides by } -1. \end{aligned}$$

We can then substitute $5x - 4$ for y in the first equation of the system and solve for z to get

$$\begin{aligned} 3x - 2y + z &= -1 \\ 3x - 2(5x - 4) + z &= -1 && \text{Substitute } 5x - 4 \text{ for } y. \\ 3x - 10x + 8 + z &= -1 && \text{Use the distributive property to remove parentheses.} \\ -7x + 8 + z &= -1 && \text{Combine like terms} \\ z &= 7x - 9 && \text{Add } 7x \text{ and } -8 \text{ to both sides.} \end{aligned}$$

Since we have found the values of y and z in terms of x, every solution to the system has the form $(x, 5x - 4, 7x - 9)$, where x can be any real number. For example,

If $x = 1$, a solution is $(1, 1, -2)$. $5(1) - 4 = 1$, and $7(1) - 9 = -2$.

If $x = 2$, a solution is $(2, 6, 5)$. $5(2) - 4 = 6$, and $7(2) - 9 = 5$.

If $x = 3$, a solution is $(3, 11, 12)$. $5(3) - 4 = 11$, and $7(3) - 9 = 12$. ■

Self Check Solve $\begin{cases} 3x + 2y + z = -1 \\ 2x - y - z = 5 \\ 5x + y = 4 \end{cases}$.

Answer infinitely many solutions; a general solution is $(x, 4 - 5x, -9 + 7x)$; three solutions are $(1, -1, -2)$, $(2, -6, 5)$, and $(3, -11, 12)$

PROBLEM SOLVING

EXAMPLE 4

Manufacturing hammers A company manufactures three types of hammers—good, better, and best. The cost of manufacturing each type of hammer is \$4, \$6, and \$7, respectively, and the hammers sell for \$6, \$9, and \$12. Each day, the cost of manufacturing 100 hammers is \$520, and the daily revenue from their sale is \$810. How many of each type are manufactured?

Analyze the problem If we let x represent the number of good hammers, y represent the number of better hammers, and z represent the number of best hammers, we know that

The total number of hammers is $x + y + z$.

The cost of manufacturing the good hammers will be \$$4x$ (\$4 times x hammers).

The cost of manufacturing the better hammers will be \$$6y$ (\$6 times y hammers).

The cost of manufacturing the best hammers will be \$$7z$ (\$7 times z hammers).

The revenue received by selling the good hammers is \$$6x$ (\$6 times x hammers).

The revenue received by selling the better hammers is \$$9y$ (\$9 times y hammers).

The revenue received by selling the best hammers is \$$12z$ (\$12 times z hammers).

Form three equations Since x represents the number of good hammers manufactured, y represents the number of better hammers manufactured, and z represents the number of best hammers manufactured, we have

The number of good hammers	+	the number of better hammers	+	the number of best hammers	=	the total number of hammers.
x	+	y	+	z	=	100
The cost of good hammers	+	the cost of better hammers	+	the cost of best hammers	=	the total cost.
$4x$	+	$6y$	+	$7z$	=	520
The revenue from good hammers	+	the revenue from better hammers	+	the revenue from best hammers	=	the total revenue.
$6x$	+	$9y$	+	$12z$	=	810

Solve the system We must now solve the system

$$\begin{array}{ll} \textbf{1.} & \\ \textbf{2.} & \\ \textbf{3.} & \end{array} \begin{cases} x + y + z = 100 \\ 4x + 6y + 7z = 520 \\ 6x + 9y + 12z = 810 \end{cases}$$

If we multiply Equation 1 by -7 and add the result to Equation 2, we get

$$\begin{array}{rl} & -7x - 7y - 7z = -700 \\ & \underline{\;\;4x + 6y + 7z = \;\;\;520} \\ \textbf{4.} & -3x - y \qquad\;\; = -180 \end{array}$$

If we multiply Equation 1 by -12 and add the result to Equation 3, we get

$$\begin{array}{rl} & -12x - 12y - 12z = -1{,}200 \\ & \underline{\;\;6x + \;\;9y + 12z = \qquad 810} \\ \textbf{5.} & -6x - \;\;3y \qquad\;\; = \;\; -390 \end{array}$$

If we multiply Equation 4 by -3 and add it to Equation 5, we get

$$\begin{aligned} 9x + 3y &= \;\;540 \\ \underline{-6x - 3y} &\underline{= -390} \\ 3x \qquad &= 150 \\ x &= 50 \qquad \text{Divide both sides by 3.} \end{aligned}$$

To find y, we substitute 50 for x in Equation 4:

$$\begin{aligned} -3x - y &= -180 \\ -3(50) - y &= -180 && \text{Substitute 50 for } x. \\ -y &= -30 && \text{Add 150 to both sides.} \\ y &= 30 && \text{Divide both sides by } -1. \end{aligned}$$

To find z, we substitute 50 for x and 30 for y in Equation 1:

$$\begin{aligned} x + y + z &= 100 \\ 50 + 30 + z &= 100 \\ z &= 20 && \text{Subtract 80 from both sides.} \end{aligned}$$

State the conclusion The company manufactures 50 good hammers, 30 better hammers, and 20 best hammers each day.

Check the result Check the solution in each equation in the original system. ■

CURVE FITTING

EXAMPLE 5 **Curve fitting** The equation of the parabola shown in Figure 3-11 is of the form $y = ax^2 + bx + c$. Find the equation of the parabola.

Solution Since the parabola passes through the points shown in the figure, each pair of coordinates satisfies the equation $y = ax^2 + bx + c$. If we substitute the x- and y-values of each point into the equation and simplify, we obtain the following system of equations.

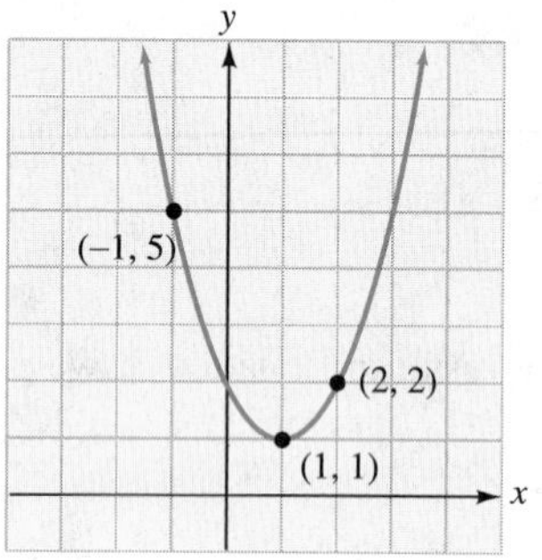

FIGURE 3-11

1. $a - b + c = 5$

2. $a + b + c = 1$

3. $4a + 2b + c = 2$

If we add Equations 1 and 2, we obtain $2a + 2c = 6$. If we multiply Equation 1 by 2 and add the result to Equation 3, we get $6a + 3c = 12$. We can then divide both sides of $2a + 2c = 6$ by 2 and divide both sides of $6a + 3c = 12$ by 3 to get the system

$$\begin{cases} a + c = 3 & \textbf{4.} \\ 2a + c = 4 & \textbf{5.} \end{cases}$$

If we multiply Equation 4 by -1 and add the result to Equation 5, we get $a = 1$. To find c, we can substitute 1 for a in Equation 4 and find that $c = 2$. To find b, we can substitute 1 for a and 2 for c in Equation 2 and find that $b = -2$.

After we substitute these values of a, b, and c into the equation $y = ax^2 + bx + c$, we have the equation of the parabola.

$$y = ax^2 + bx + c$$
$$y = 1x^2 + (-2)x + 2$$
$$y = x^2 - 2x + 2$$

■

Orals *Is the triple a solution of the system?*

1. $(1, 1, 1)$, $\begin{cases} 2x + y - 3z = 0 \\ 3x - 2y + 4z = 5 \\ 4x + 2y - 6z = 0 \end{cases}$

2. $(2, 0, 1)$, $\begin{cases} 3x + 2y - z = 5 \\ 2x - 3y + 2z = 4 \\ 4x - 2y + 3z = 10 \end{cases}$

EXERCISE 3.3

REVIEW *Consider the line passing through* $P(-2, -4)$ *and* $Q(3, 5)$.

1. Find the slope of line PQ.

2. Write the equation of line PQ in general form

Let $f(x) = 2x^2 + 1$. *Find each value.*

3. $f(0)$

4. $f(-2)$

5. $f(s)$

6. $f(2t)$

VOCABULARY AND CONCEPTS *Fill in each blank to make a true statement.*

7. The graph of the equation $2x + 3y + 4z = 5$ is a flat surface called a ____.

8. When three planes coincide, the equations of the system are ________, and there are ________ many solutions.

9. When three planes intersect in a line, the system will have ________ many solutions.

10. When three planes are parallel, the system will have ___ solutions.

PRACTICE *In Exercises 11–12, tell whether the given triple is a solution of the given system.*

11. (2, 1, 1), $\begin{cases} x - y + z = 2 \\ 2x + y - z = 4 \\ 2x - 3y + z = 2 \end{cases}$

12. (−3, 2, −1), $\begin{cases} 2x + 2y + 3z = -1 \\ 3x + y - z = -6 \\ x + y + 2z = 1 \end{cases}$

In Exercises 13–24, solve each system.

13. $\begin{cases} x + y + z = 4 \\ 2x + y - z = 1 \\ 2x - 3y + z = 1 \end{cases}$

14. $\begin{cases} x + y + z = 4 \\ x - y + z = 2 \\ x - y - z = 0 \end{cases}$

15. $\begin{cases} 2x + 2y + 3z = 10 \\ 3x + y - z = 0 \\ x + y + 2z = 6 \end{cases}$

16. $\begin{cases} x - y + z = 4 \\ x + 2y - z = -1 \\ x + y - 3z = -2 \end{cases}$

17. $\begin{cases} a + b + 2c = 7 \\ a + 2b + c = 8 \\ 2a + b + c = 9 \end{cases}$

18. $\begin{cases} 2a + 3b + c = 2 \\ 4a + 6b + 2c = 5 \\ a - 2b + c = 3 \end{cases}$

19. $\begin{cases} 2x + y - z = 1 \\ x + 2y + 2z = 2 \\ 4x + 5y + 3z = 3 \end{cases}$

20. $\begin{cases} 4x + 3z = 4 \\ 2y - 6z = -1 \\ 8x + 4y + 3z = 9 \end{cases}$

21. $\begin{cases} 2x + 3y + 4z = 6 \\ 2x - 3y - 4z = -4 \\ 4x + 6y + 8z = 12 \end{cases}$

22. $\begin{cases} x - 3y + 4z = 2 \\ 2x + y + 2z = 3 \\ 4x - 5y + 10z = 7 \end{cases}$

23. $\begin{cases} x + \frac{1}{3}y + z = 13 \\ \frac{1}{2}x - y + \frac{1}{3}z = -2 \\ x + \frac{1}{2}y - \frac{1}{3}z = 2 \end{cases}$

24. $\begin{cases} x - \frac{1}{5}y - z = 9 \\ \frac{1}{4}x + \frac{1}{5}y - \frac{1}{2}z = 5 \\ 2x + y + \frac{1}{6}z = 12 \end{cases}$

APPLICATIONS

25. Integer problem The sum of three integers is 18. The third integer is four times the second, and the second integer is 6 more than the first. Find the integers.

26. Integer problem The sum of three integers is 48. If the first integer is doubled, the sum is 60. If the second integer is doubled, the sum is 63. Find the integers.

27. Geometry problem The sum of the angles in any triangle is 180°. In triangle *ABC*, angle *A* is 100° less than the sum of angles *B* and *C*, and angle *C* is 40° less than twice angle *B*. Find each angle.

28. Geometry problem The sum of the angles of any four-sided figure is 360°. In the quadrilateral shown in Illustration 1, angle *A* = angle *B*, angle *C* is 20° greater than angle *A*, and angle *D* = 40°. Find the angles.

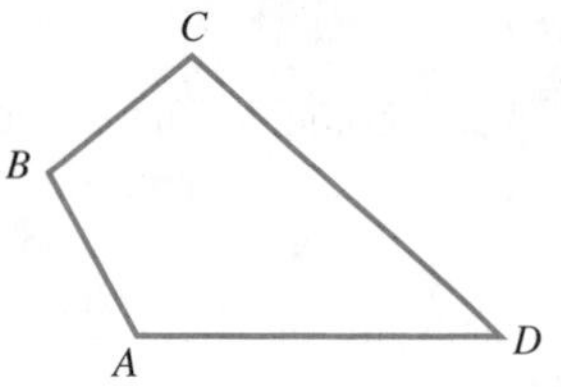

ILLUSTRATION 1

29. Nutritional planning One unit of each of three foods contains the nutrients shown in Illustration 2. How many units of each must be used to provide exactly 11 grams of fat, 6 grams of carbohydrates, and 10 grams of protein?

Food	Fat	Carbohydrates	Protein
A	1	1	2
B	2	1	1
C	2	1	2

ILLUSTRATION 2

30. Nutritional planning One unit of each of three foods contains the nutrients shown in Illustration 3. How many units of each must be used to provide exactly 14 grams of fat, 9 grams of carbohydrates, and 9 grams of protein?

Food	Fat	Carbohydrates	Protein
A	2	1	2
B	3	2	1
C	1	1	2

ILLUSTRATION 3

31. Making statues An artist makes three types of ceramic statues at a monthly cost of $650 for 180 statues. The manufacturing costs for the three types are $5, $4, and $3. If the statues sell for $20, $12, and $9, respectively, how many of each type should be made to produce $2,100 in monthly revenue?

32. Manufacturing footballs A factory manufactures three types of footballs at a monthly cost of $2,425 for 1,125 footballs. The manufacturing costs for the three types of footballs are $4, $3, and $2. These footballs sell for $16, $12, and $10, respectively. How many of each type are manufactured if the monthly profit is $9,275? (*Hint:* Profit = income − cost.)

33. Concert tickets Tickets for a concert cost $5, $3, and $2. Twice as many $5 tickets were sold as $2 tickets. The receipts for 750 tickets were $2,625. How many of each price ticket were sold?

34. Mixing nuts The owner of a candy store mixed some peanuts worth $3 per pound, some cashews worth $9 per pound, and some Brazil nuts worth $9 per pound to get 50 pounds of a mixture that would sell for $6 per pound. She used 15 fewer pounds of cashews than peanuts. How many pounds of each did she use?

35. Chainsaw sculpting A north woods sculptor carves three types of statues with a chainsaw. The time required for carving, sanding, and painting a totem pole, a bear, and a deer are shown in Illustration 4. How many of each should be produced to use all available labor hours?

	Totem pole	Bear	Deer	Time available
Carving	2 hours	2 hours	1 hour	14 hours
Sanding	1 hour	2 hours	2 hours	15 hours
Painting	3 hours	2 hours	2 hours	21 hours

ILLUSTRATION 4

36. Making clothing A clothing manufacturer makes coats, shirts, and slacks. The time required for cutting, sewing, and packaging each item are shown in Illustration 5. How many of each should be made to use all available labor hours?

	Coats	Shirts	Slacks	Time available
Cutting	20 min	15 min	10 min	115 hr
Sewing	60 min	30 min	24 min	280 hr
Packaging	5 min	12 min	6 min	65 hr

ILLUSTRATION 5

37. Curve fitting Find the equation of the parabola shown in Illustration 6.

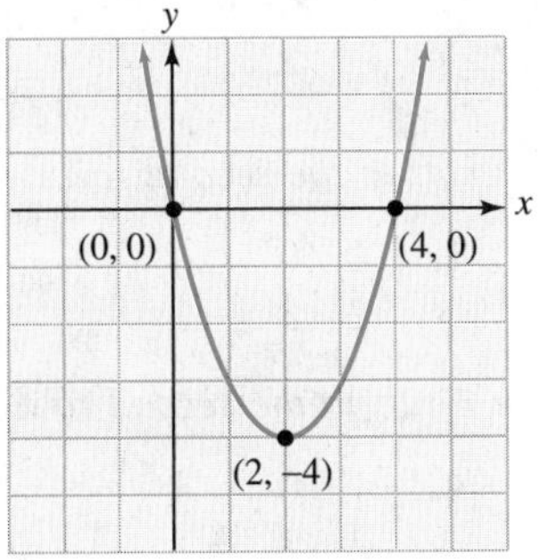

ILLUSTRATION 6

38. Curve fitting Find the equation of the parabola shown in Illustration 7.

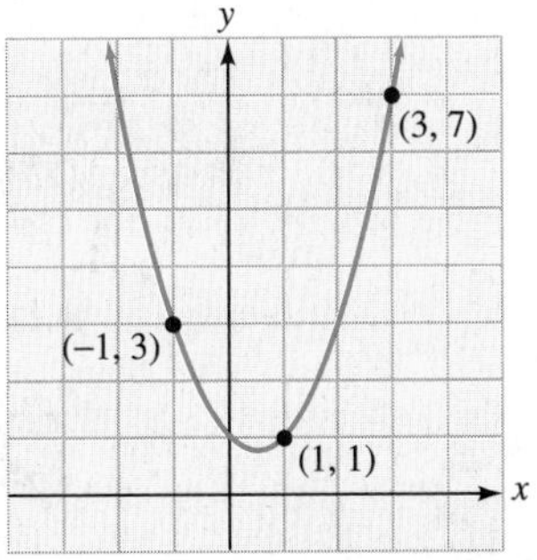

ILLUSTRATION 7

In Exercises 39–40, the equation of a circle is of the form $x^2 + y^2 + cx + dy + e = 0$.

39. Curve fitting Find the equation of the circle shown in Illustration 8.

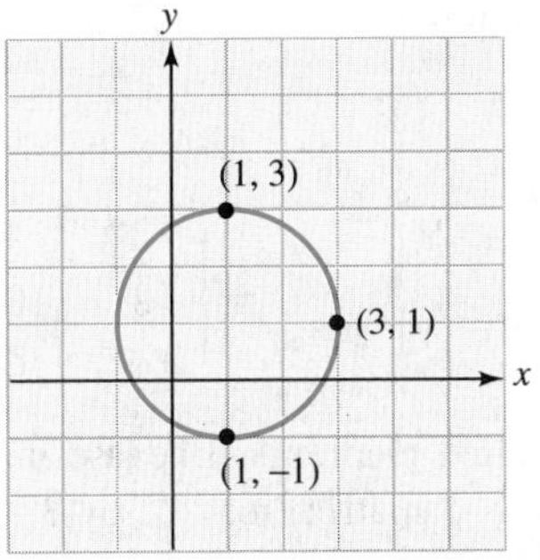

ILLUSTRATION 8

40. Curve fitting Find the equation of the circle shown in Illustration 9.

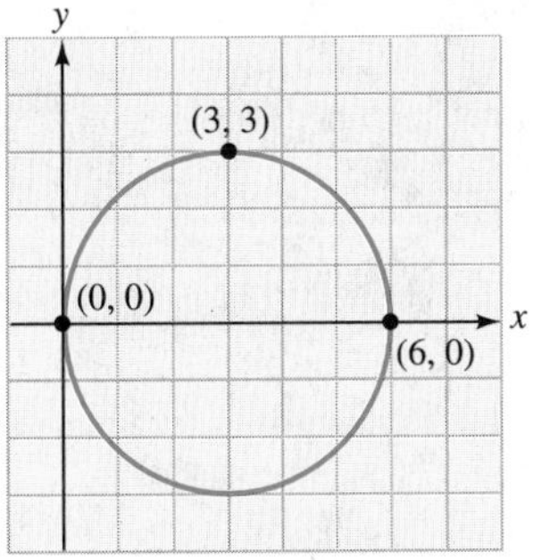

ILLUSTRATION 9

WRITING

41. What makes a system of three equations in three variables inconsistent?

42. What makes the equations of a system of three equations in three variables dependent?

SOMETHING TO THINK ABOUT

43. Solve the system

$$\begin{cases} x + y + z + w = 3 \\ x - y - z - w = -1 \\ x + y - z - w = 1 \\ x + y - z + w = 3 \end{cases}$$

44. Solve the system

$$\begin{cases} 2x + y + z + w = 3 \\ x - 2y - z + w = -3 \\ x - y - 2z - w = -3 \\ x + y - z + 2w = 4 \end{cases}$$

3.4 Solution by Matrices

■ MATRICES ■ GAUSSIAN ELIMINATION ■ SYSTEMS WITH MORE EQUATIONS THAN VARIABLES
■ SYSTEMS WITH MORE VARIABLES THAN EQUATIONS

Getting Ready *Multiply the first row by 2 and add the result to the second row.*

1.
$$\begin{matrix} 2 & 3 & 5 \\ 1 & 2 & 3 \end{matrix}$$

2.
$$\begin{matrix} -1 & 0 & 4 \\ 2 & 3 & -1 \end{matrix}$$

Multiply the first row by −1 and add the result to the second row.

3.
$$\begin{matrix} 2 & 3 & 5 \\ 1 & 2 & 3 \end{matrix}$$

4.
$$\begin{matrix} -1 & 0 & 4 \\ 2 & 3 & -1 \end{matrix}$$

■ MATRICES

Another method of solving systems of equations involves rectangular arrays of numbers called *matrices.*

Matrix
A **matrix** is any rectangular array of numbers.

Some examples of matrices are

$$A = \begin{bmatrix} 1 & 2 & 3 \\ 4 & 5 & 6 \end{bmatrix} \qquad B = \begin{bmatrix} 1 & 2 \\ 3 & 4 \\ 5 & 6 \end{bmatrix} \qquad C = \begin{bmatrix} 2 & 4 & 6 \\ 8 & 10 & 12 \\ 14 & 16 & 18 \end{bmatrix}$$

The numbers in each matrix are called **elements.** Because matrix A has two rows and three columns, it is called a 2×3 matrix (read "2 by 3" matrix). Matrix B is a 3×2 matrix, because the matrix has three rows and two columns. Matrix C is a 3×3 matrix (three rows and three columns).

Any matrix with the same number of rows and columns, like matrix C, is called a **square matrix.**

To show how to use matrices to solve systems of linear equations, we consider the system

$$\begin{cases} x - 2y - z = 6 \\ 2x + 2y - z = 1 \\ -x - y + 2z = 1 \end{cases}$$

which can be represented by the following matrix, called an **augmented matrix:**

$$\begin{bmatrix} 1 & -2 & -1 & 6 \\ 2 & 2 & -1 & 1 \\ -1 & -1 & 2 & 1 \end{bmatrix}$$

The first three columns of the augmented matrix form a 3×3 matrix called a **coefficient matrix**. It is determined by the coefficients of x, y, and z in the equations of the system. The 3×1 matrix in the last column is determined by the constants in the equations.

Coefficient matrix ***Column of constants***

$$\begin{bmatrix} 1 & -2 & -1 \\ 2 & 2 & -1 \\ -1 & -1 & 2 \end{bmatrix} \qquad \begin{bmatrix} 6 \\ 1 \\ 1 \end{bmatrix}$$

Each row of the augmented matrix represents one equation of the system:

$$\begin{bmatrix} 1 & -2 & -1 & 6 \\ 2 & 2 & -1 & 1 \\ -1 & -1 & 2 & 1 \end{bmatrix} \begin{matrix} \longleftrightarrow \\ \longleftrightarrow \\ \longleftrightarrow \end{matrix} \begin{cases} x - 2y - z = 6 \\ 2x + 2y - z = 1 \\ -x - y + 2z = 1 \end{cases}$$

GAUSSIAN ELIMINATION

To solve a 3×3 system of equations by **Gaussian elimination**, we transform an augmented matrix into the following matrix that has all 0's below its main diagonal, which is formed by the elements a, e, and h.

$$\begin{bmatrix} a & b & c & d \\ 0 & e & f & g \\ 0 & 0 & h & i \end{bmatrix} \quad (a, b, c, \ldots, i \text{ are real numbers})$$

We can often write a matrix in this form, called **triangular form**, by using three operations called **elementary row operations**.

Elementary Row Operations

1. Any two rows of a matrix can be interchanged.
2. Any row of a matrix can be multiplied by a nonzero constant.
3. Any row of a matrix can be changed by adding a nonzero constant multiple of another row to it.

- A type 1 row operation corresponds to interchanging two equations of a system.
- A type 2 row operation corresponds to multiplying both sides of an equation by a nonzero constant.
- A type 3 row operation corresponds to adding a nonzero multiple of one equation to another.

None of these operations will change the solution of the given system of equations.

After we have written the matrix in triangular form, we can solve the corresponding system of equations by a back substitution process, as shown in Example 1.

EXAMPLE 1 Solve the system $\begin{cases} x - 2y - z = 6 \\ 2x + 2y - z = 1 \\ -x - y + 2z = 1 \end{cases}$.

Solution We can represent the system with the following augmented matrix:

$$\begin{bmatrix} 1 & -2 & -1 & 6 \\ 2 & 2 & -1 & 1 \\ -1 & -1 & 2 & 1 \end{bmatrix}$$

To get 0's under the 1 in the first column, we use a type 3 row operation twice:

Multiply row 1 by -2 and add to row 2. Multiply row 1 by 1 and add to row 3.

$$\begin{bmatrix} 1 & -2 & -1 & 6 \\ 2 & 2 & -1 & 1 \\ -1 & -1 & 2 & 1 \end{bmatrix} \approx \begin{bmatrix} 1 & -2 & -1 & 6 \\ 0 & 6 & 1 & -11 \\ -1 & -1 & 2 & 1 \end{bmatrix} \approx \begin{bmatrix} 1 & -2 & -1 & 6 \\ 0 & 6 & 1 & -11 \\ 0 & -3 & 1 & 7 \end{bmatrix}$$

The symbol "$\approx$" is read as "is row equivalent to." Each of the above matrices represents an equivalent system of equations.

To get a 0 under the 6 in the second column of the last matrix, we use another type 3 row operation:

Multiply row 2 by $\frac{1}{2}$ and add to row 3.

$$\begin{bmatrix} 1 & -2 & -1 & 6 \\ 0 & 6 & 1 & -11 \\ 0 & -3 & 1 & 7 \end{bmatrix} \approx \begin{bmatrix} 1 & -2 & -1 & 6 \\ 0 & 6 & 1 & -11 \\ 0 & 0 & \frac{3}{2} & \frac{3}{2} \end{bmatrix}$$

We can use a type 2 row operation to simplify further.

Multiply row 3 by $\frac{2}{3}$, which is the reciprocal of $\frac{3}{2}$.

$$\begin{bmatrix} 1 & -2 & -1 & 6 \\ 0 & 6 & 1 & -11 \\ 0 & 0 & \frac{3}{2} & \frac{3}{2} \end{bmatrix} \approx \begin{bmatrix} 1 & -2 & -1 & 6 \\ 0 & 6 & 1 & -11 \\ 0 & 0 & 1 & 1 \end{bmatrix}$$

The final matrix represents the system of equations

$$\begin{cases} \textbf{1.} \quad x - 2y - z = 6 \\ \textbf{2.} \quad 0x + 6y + z = -11 \\ \textbf{3.} \quad 0x + 0y + z = 1 \end{cases}$$

From Equation 3, we can read that $z = 1$. To find y, we substitute 1 for z in Equation 2 and solve for y:

2. $6y + z = -11$

$6y + 1 = -11$ Substitute 1 for z.

$6y = -12$ Subtract 1 from both sides.

$y = -2$ Divide both sides by 6.

Thus, $y = -2$. To find x, we substitute 1 for z and -2 for y in Equation 1 and solve for x:

1.	$x - 2y - z = 6$	
	$x - 2(\mathbf{-2}) - \mathbf{1} = 6$	Substitute 1 for z and -2 for y.
	$x + 3 = 6$	Simplify.
	$x = 3$	Subtract 3 from both sides.

Thus, $x = 3$. The solution of the given system is $(3, -2, 1)$. Verify that this triple satisfies each equation of the original system. ■

Self Check Solve $\begin{cases} x - 2y - z = 2 \\ 2x + 2y - z = -5. \\ -x - y + 2z = 7 \end{cases}$

Answer $(1, -2, 3)$

■ SYSTEMS WITH MORE EQUATIONS THAN VARIABLES

We can use matrices to solve systems that have more equations than variables.

EXAMPLE 2 Solve the system $\begin{cases} x + y = -1 \\ 2x - y = 7 \\ -x + 2y = -8 \end{cases}$.

Solution This system can be represented by a 3×3 augmented matrix:

$$\begin{bmatrix} 1 & 1 & -1 \\ 2 & -1 & 7 \\ -1 & 2 & -8 \end{bmatrix}$$

To get 0's under the 1 in the first column, we do a type 3 row operation twice:

Multiply row 1 by -2 and add to row 2. Multiply row 1 by 1 and add to row 3.

$$\begin{bmatrix} 1 & 1 & -1 \\ 2 & -1 & 7 \\ -1 & 2 & -8 \end{bmatrix} \approx \begin{bmatrix} 1 & 1 & -1 \\ 0 & -3 & 9 \\ -1 & 2 & -8 \end{bmatrix} \approx \begin{bmatrix} 1 & 1 & -1 \\ 0 & -3 & 9 \\ 0 & 3 & -9 \end{bmatrix}$$

We can do other row operations to get a 0 under the -3 in the second column and then 1 in the second row, second column.

Multiply row 2 by 1 and add to row 3. Multiply row 2 by $-\frac{1}{3}$.

$$\begin{bmatrix} 1 & 1 & -1 \\ 0 & -3 & 9 \\ 0 & 3 & -9 \end{bmatrix} \approx \begin{bmatrix} 1 & 1 & -1 \\ 0 & -3 & 9 \\ 0 & 0 & 0 \end{bmatrix} \approx \begin{bmatrix} 1 & 1 & -1 \\ 0 & 1 & -3 \\ 0 & 0 & 0 \end{bmatrix}$$

The final matrix represents the system

$$\begin{cases} x + y = -1 \\ 0x + y = -3 \\ 0x + 0y = 0 \end{cases}$$

The third equation can be discarded, because $0x + 0y = 0$ for all x and y. From the second equation, we can read that $y = -3$. To find x, we substitute -3 for y in the first equation and solve for x:

$$\begin{aligned} x + y &= -1 \\ x + (-3) &= -1 && \text{Substitute } -3 \text{ for } y. \\ x &= 2 && \text{Add 3 to both sides.} \end{aligned}$$

The solution is $(2, -3)$. Verify that this solution satisfies all three equations of the original system. ■

Self Check Solve $\begin{cases} x + y = 1 \\ 2x - y = 8 \\ -x + 2y = -7 \end{cases}$.

Answer $(3, -2)$

If the last row of the final matrix in Example 2 had been of the form $0x + 0y = k$, where $k \neq 0$, the system would not have a solution. No values of x and y could make the expression $0x + 0y$ equal to a nonzero constant k.

SYSTEMS WITH MORE VARIABLES THAN EQUATIONS

We can also solve many systems that have more variables than equations.

EXAMPLE 3 Solve the system $\begin{cases} x + y - 2z = -1 \\ 2x - y + z = -3 \end{cases}$.

Solution We start by doing a type 3 row operation to get a 0 under the 1 in the first column.

$$\begin{bmatrix} 1 & 1 & -2 & -1 \\ 2 & -1 & 1 & -3 \end{bmatrix} \approx \begin{bmatrix} 1 & 1 & -2 & -1 \\ 0 & -3 & 5 & -1 \end{bmatrix} \quad \text{Multiply row 1 by } -2 \text{ and add to row 2.}$$

We then do a type 2 row operation:

Multiply row 2 by $-\frac{1}{3}$.

$$\begin{bmatrix} 1 & 1 & -2 & -1 \\ 0 & -3 & 5 & -1 \end{bmatrix} \approx \begin{bmatrix} 1 & 1 & -2 & -1 \\ 0 & 1 & -\frac{5}{3} & \frac{1}{3} \end{bmatrix}$$

The final matrix represents the system

$$\begin{cases} x + y - 2z = -1 \\ y - \frac{5}{3}z = \frac{1}{3} \end{cases}$$

We add $\frac{5}{3}z$ to both sides of the second equation to obtain

$$y = \frac{1}{3} + \frac{5}{3}z$$

We have not found a specific value for y. However, we have found y in terms of z. To find a value of x in terms of z, we substitute $\frac{1}{3} + \frac{5}{3}z$ for y in the first equation and simplify to get

$$x + y - 2z = -1$$

$$x + \frac{1}{3} + \frac{5}{3}z - 2z = -1 \quad \text{Substitute } \tfrac{1}{3} + \tfrac{5}{3}z \text{ for } y.$$

$$x + \frac{1}{3} - \frac{1}{3}z = -1 \quad \text{Combine like terms.}$$

$$x - \frac{1}{3}z = -\frac{4}{3} \quad \text{Subtract } \tfrac{1}{3} \text{ from both sides.}$$

$$x = -\frac{4}{3} + \frac{1}{3}z \quad \text{Add } \tfrac{1}{3}z \text{ to both sides.}$$

A solution of this system must have the form

$$\left(-\frac{4}{3} + \frac{1}{3}z,\ \frac{1}{3} + \frac{5}{3}z,\ z\right)$$

This solution is called "a general solution" of the system.

for all values of z. This system has infinitely many solutions, a different one for each value of z. For example,

- If $z = 0$, the corresponding solution is $\left(-\frac{4}{3}, \frac{1}{3}, 0\right)$.
- If $z = 1$, the corresponding solution is $(-1, 2, 1)$.

Verify that both of these solutions satisfy each equation of the given system. ■

Self Check Solve $\begin{cases} x + y - 2z = 11 \\ 2x - y + z = -2 \end{cases}$.

Answer infinitely many solutions; a general solution is $\left(3 + \frac{1}{3}z, 8 + \frac{5}{3}z, z\right)$; two solutions are $(2, 3, -3)$ and $(3, 8, 0)$

PERSPECTIVE

Matrices with the same number of rows and columns can be added. We simply add their corresponding elements. For example,

$$\begin{bmatrix} 2 & 3 & -4 \\ -1 & 2 & 5 \end{bmatrix} + \begin{bmatrix} 3 & -1 & 0 \\ 4 & 3 & 2 \end{bmatrix}$$
$$= \begin{bmatrix} 2+3 & 3+(-1) & -4+0 \\ -1+4 & 2+3 & 5+2 \end{bmatrix}$$
$$= \begin{bmatrix} 5 & 2 & -4 \\ 3 & 5 & 7 \end{bmatrix}$$

To multiply a matrix by a constant, we multiply each element of the matrix by the constant. For example,

$$5 \cdot \begin{bmatrix} 2 & 3 & -4 \\ -1 & 2 & 5 \end{bmatrix}$$
$$= \begin{bmatrix} 5 \cdot 2 & 5 \cdot 3 & 5 \cdot (-4) \\ 5 \cdot (-1) & 5 \cdot 2 & 5 \cdot 5 \end{bmatrix}$$
$$= \begin{bmatrix} 10 & 15 & -20 \\ -5 & 10 & 25 \end{bmatrix}$$

Since matrices provide a good way to store information in computers, they are often used in applied problems. For example, suppose there are 66 security officers employed at either the downtown office or the suburban office:

Downtown Office

	Male	Female
Day shift	12	18
Night shift	3	0

Suburban Office

	Male	Female
Day shift	14	12
Night shift	5	2

The information about the employees is contained in the following matrices.

$$D = \begin{bmatrix} 12 & 18 \\ 3 & 0 \end{bmatrix} \quad \text{and} \quad S = \begin{bmatrix} 14 & 12 \\ 5 & 2 \end{bmatrix}$$

The entry in the first row-first column in matrix D gives the information that 12 males work the day shift at the downtown office. Company management can add the matrices D and S to find corporate-wide totals:

$$D + S = \begin{bmatrix} 12 & 18 \\ 3 & 0 \end{bmatrix} + \begin{bmatrix} 14 & 12 \\ 5 & 2 \end{bmatrix}$$
$$= \begin{bmatrix} 26 & 30 \\ 8 & 2 \end{bmatrix}$$

We interpret the total to mean:

	Male	Female
Day shift	26	30
Night shift	8	2

If one-third of the force in each category at the downtown location retires, the downtown staff would be reduced to $\frac{2}{3}D$ people. We can compute $\frac{2}{3}D$ by multiplying each entry by $\frac{2}{3}$.

$$\frac{2}{3}D = \frac{2}{3}\begin{bmatrix} 12 & 18 \\ 3 & 0 \end{bmatrix}$$
$$= \begin{bmatrix} 8 & 12 \\ 2 & 0 \end{bmatrix}$$

After retirements, downtown staff would be

	Male	Female
Day shift	8	12
Night shift	2	0

Orals *Consider the system* $\begin{cases} 3x + 2y = 8 \\ 4x - 3y = 6 \end{cases}$.

1. Find the coefficient matrix.

2. Find the augmented matrix.

Tell whether each matrix is in triangular form.

3. $\begin{bmatrix} 4 & 1 & 5 \\ 0 & 2 & 7 \\ 0 & 0 & 4 \end{bmatrix}$

4. $\begin{bmatrix} 8 & 5 & 2 \\ 0 & 4 & 5 \\ 0 & 7 & 0 \end{bmatrix}$

EXERCISE 3.4

REVIEW *Write each number in scientific notation.*

1. 93,000,000

2. 0.00045

3. 63×10^3

4. 0.33×10^3

VOCABULARY AND CONCEPTS *Fill in each blank to make a true statement.*

5. A ______ is a rectangular array of numbers.

6. The numbers in a matrix are called its ________.

7. A 3×4 matrix has __ rows and 4 ________.

8. A ______ matrix has the same number of rows as columns.

9. An _________ matrix of a system of equations includes the coefficient matrix and the column of constants.

10. If a matrix has all 0's below its main diagonal, it is written in ________ form.

11. A ______ row operation corresponds to interchanging two equations in a system of equations.

12. A type 2 row operation corresponds to _________ both sides of an equation by a nonzero constant.

13. A type 3 row operation corresponds to adding a _______ multiple of one equation to another.

14. In the Gaussian method of solving systems of equations, we transform the _________ matrix into triangular form and finish the solution by using _____ substitution.

PRACTICE *In Exercises 15–18, use a row operation to find the missing number in the second matrix.*

15. $\begin{bmatrix} 2 & 1 & 1 \\ 5 & 4 & 1 \end{bmatrix}$ $\begin{bmatrix} 2 & 1 & 1 \\ 3 & 3 & \square \end{bmatrix}$

16. $\begin{bmatrix} -1 & 3 & 2 \\ 1 & -2 & 3 \end{bmatrix}$ $\begin{bmatrix} -1 & 3 & 2 \\ \square & 1 & 5 \end{bmatrix}$

17. $\begin{bmatrix} 3 & -2 & 1 \\ -1 & 2 & 4 \end{bmatrix}$ $\begin{bmatrix} 3 & -2 & 1 \\ -2 & 4 & \square \end{bmatrix}$

18. $\begin{bmatrix} 2 & 1 & -3 \\ 2 & 6 & 1 \end{bmatrix}$ $\begin{bmatrix} 6 & 3 & \square \\ 2 & 6 & 1 \end{bmatrix}$

In Exercises 19–30, use matrices to solve each system of equations. Each system has one solution.

19. $\begin{cases} x + y = 2 \\ x - y = 0 \end{cases}$

20. $\begin{cases} x + y = 3 \\ x - y = -1 \end{cases}$

21. $\begin{cases} x + 2y = -4 \\ 2x + y = 1 \end{cases}$

22. $\begin{cases} 2x - 3y = 16 \\ -4x + y = -22 \end{cases}$

23. $\begin{cases} 3x + 4y = -12 \\ 9x - 2y = 6 \end{cases}$

24. $\begin{cases} 5x - 4y = 10 \\ x - 7y = 2 \end{cases}$

25. $\begin{cases} x + y + z = 6 \\ x + 2y + z = 8 \\ x + y + 2z = 9 \end{cases}$

26. $\begin{cases} x - y + z = 2 \\ x + 2y - z = 6 \\ 2x - y - z = 3 \end{cases}$

27. $\begin{cases} 2x + y + 3z = 3 \\ -2x - y + z = 5 \\ 4x - 2y + 2z = 2 \end{cases}$

28. $\begin{cases} 3x + 2y + z = 8 \\ 6x - y + 2z = 16 \\ -9x + y - z = -20 \end{cases}$

29. $\begin{cases} 3x - 2y + 4z = 4 \\ x + y + z = 3 \\ 6x - 2y - 3z = 10 \end{cases}$

30. $\begin{cases} 2x + 3y - z = -8 \\ x - y - z = -2 \\ -4x + 3y + z = 6 \end{cases}$

In Exercises 31–38, use matrices to solve each system of equations. If a system has no solution, so indicate.

31. $\begin{cases} x + y = 3 \\ 3x - y = 1 \\ 2x + y = 4 \end{cases}$

32. $\begin{cases} x - y = -5 \\ 2x + 3y = 5 \\ x + y = 1 \end{cases}$

33. $\begin{cases} 2x - y = 4 \\ x + 3y = 2 \\ -x - 4y = -2 \end{cases}$

34. $\begin{cases} 3x - 2y = 5 \\ x + 2y = 7 \\ -3x - y = -11 \end{cases}$

35. $\begin{cases} 2x + y = 7 \\ x - y = 2 \\ -x + 3y = -2 \end{cases}$

36. $\begin{cases} 3x - y = 2 \\ -6x + 3y = 0 \\ -x + 2y = -4 \end{cases}$

37. $\begin{cases} x + 3y = 7 \\ x + y = 3 \\ 3x + y = 5 \end{cases}$

38. $\begin{cases} x + y = 3 \\ x - 2y = -3 \\ x - y = 1 \end{cases}$

In Exercises 39–42, use matrices to help find a general solution of each system of equations.

39. $\begin{cases} x + 2y + 3z = -2 \\ -x - y - 2z = 4 \end{cases}$

40. $\begin{cases} 2x - 4y + 3z = 6 \\ -4x + 6y + 4z = -6 \end{cases}$

41. $\begin{cases} x - y = 1 \\ y + z = 1 \\ x + z = 2 \end{cases}$

42. $\begin{cases} x + z = 1 \\ x + y = 2 \\ 2x + y + z = 3 \end{cases}$

In Exercises 43–46, remember these facts from geometry.

Two angles whose measures add up to 90° are complementary.

Two angles whose measures add up to 180° are supplementary.

The sum of the measures of the interior angles in a triangle is 180°.

43. Geometry One angle is 46° larger than its complement. Find the angles.

44. Geometry One angle is 28° larger than its supplement. Find the angles.

45. Geometry In Illustration 1, angle B is 25° more than angle A, and angle C is 5° less than twice angle A. Find each angle in the triangle.

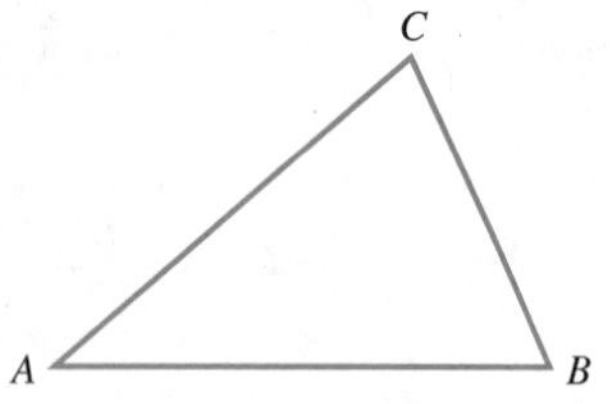

ILLUSTRATION 1

46. Geometry In Illustration 2, angle A is 10° less than angle B, and angle B is 10° less than angle C. Find each angle in the triangle.

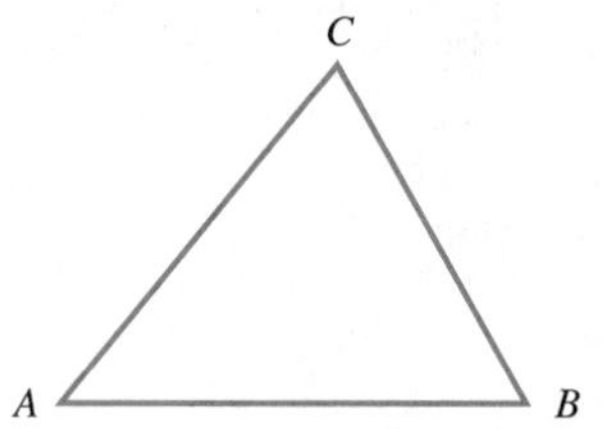

ILLUSTRATION 2

In Exercises 47–48, remember that the equation of a parabola is of the form $y = ax^2 + bx + c$.

47. Curve fitting Find the equation of the parabola passing through the points (0, 1), (1, 2), and (−1, 4).

48. Curve fitting Find the equation of the parabola passing through the points (0, 1), (1, 1), and (−1, −1).

WRITING

49. Explain how to check the solution of a system of equations.

50. Explain how to perform a type 3 row operation.

SOMETHING TO THINK ABOUT

51. If the system represented by

$$\begin{bmatrix} 1 & 1 & 0 & 1 \\ 0 & 0 & 1 & 2 \\ 0 & 0 & 0 & k \end{bmatrix}$$

has no solution, what do you know about k?

52. Is it possible for a system with fewer equations than variables to have no solution? Illustrate.

3.5 Solution by Determinants

■ DETERMINANTS ■ CRAMER'S RULE

Getting Ready *Find each value.*

1. 3(−4) − 2(5)

2. 5(2) − 3(−4)

3. 2(2 − 5) − 3(5 − 2) + 2(4 − 3)

4. −3(5 − 2) + 2(3 + 1) − 2(5 + 1)

■ DETERMINANTS

An idea closely related to the concept of matrix is the **determinant.** A determinant is a number that is associated with a square matrix. For any square matrix A, the symbol $|A|$ represents the determinant of A.

Value of a 2 × 2 Determinant

If a, b, c, and d are real numbers, the **determinant** of the matrix $\begin{bmatrix} a & b \\ c & d \end{bmatrix}$ is

$$\begin{vmatrix} a & b \\ c & d \end{vmatrix} = ad - bc$$

The determinant of a 2×2 matrix is the number that is equal to the product of the numbers on the major diagonal

$$\begin{vmatrix} a & b \\ c & d \end{vmatrix}$$

minus the product of the numbers on the other diagonal

$$\begin{vmatrix} a & b \\ c & d \end{vmatrix}$$

b

EXAMPLE 1 Evaluate the determinants **a.** $\begin{vmatrix} 3 & 2 \\ 6 & 9 \end{vmatrix}$ and **b.** $\begin{vmatrix} -5 & \frac{1}{2} \\ -1 & 0 \end{vmatrix}$.

Solution **a.** $\begin{vmatrix} 3 & 2 \\ 6 & 9 \end{vmatrix} = 3(9) - 2(6)$

$= 27 - 12$

$= 15$

b. $\begin{vmatrix} -5 & \frac{1}{2} \\ -1 & 0 \end{vmatrix} = -5(0) - \frac{1}{2}(-1)$

$= 0 + \frac{1}{2}$

$= \frac{1}{2}$ ■

Self Check Evaluate $\begin{vmatrix} 4 & -3 \\ 2 & 1 \end{vmatrix}$.

Answer 10

A 3×3 determinant is evaluated by expanding by **minors.**

Value of a 3 × 3 Determinant

Minor of a_1 ↓ Minor of b_1 ↓ Minor of c_1 ↓

$$\begin{vmatrix} a_1 & b_1 & c_1 \\ a_2 & b_2 & c_2 \\ a_3 & b_3 & c_3 \end{vmatrix} = a_1\begin{vmatrix} b_2 & c_2 \\ b_3 & c_3 \end{vmatrix} - b_1\begin{vmatrix} a_2 & c_2 \\ a_3 & c_3 \end{vmatrix} + c_1\begin{vmatrix} a_2 & b_2 \\ a_3 & b_3 \end{vmatrix}$$

To find the minor of a_1, we find the determinant formed by crossing out the elements of the matrix that are in the same row and column as a_1:

$$\begin{vmatrix} a_1 & b_1 & c_1 \\ a_2 & b_2 & c_2 \\ a_3 & b_3 & c_3 \end{vmatrix}$$

The minor of a_1 is $\begin{vmatrix} b_2 & c_2 \\ b_3 & c_3 \end{vmatrix}$.

To find the minor of b_1, we cross out the elements of the matrix that are in the same row and column as b_1:

$$\begin{vmatrix} \not{a_1} & \not{b_1} & \not{c_1} \\ a_2 & \not{b_2} & c_2 \\ a_3 & \not{b_3} & c_3 \end{vmatrix} \qquad \text{The minor of } b_1 \text{ is } \begin{vmatrix} a_2 & c_2 \\ a_3 & c_3 \end{vmatrix}.$$

To find the minor of c_1, we cross out the elements of the matrix that are in the same row and column as c_1:

$$\begin{vmatrix} \not{a_1} & \not{b_1} & \not{c_1} \\ a_2 & b_2 & \not{c_2} \\ a_3 & b_3 & \not{c_3} \end{vmatrix} \qquad \text{The minor of } c_1 \text{ is } \begin{vmatrix} a_2 & b_2 \\ a_3 & b_3 \end{vmatrix}.$$

EXAMPLE 2 Evaluate the determinant $\begin{vmatrix} 1 & 3 & -2 \\ 2 & 1 & 3 \\ 1 & 2 & 3 \end{vmatrix}$.

Solution

Minor of 1 ↓   Minor of 3 ↓   Minor of −2 ↓

$$\begin{vmatrix} \mathbf{1} & \mathbf{3} & \mathbf{-2} \\ 2 & 1 & 3 \\ 1 & 2 & 3 \end{vmatrix} = \mathbf{1}\begin{vmatrix} 1 & 3 \\ 2 & 3 \end{vmatrix} - \mathbf{3}\begin{vmatrix} 2 & 3 \\ 1 & 3 \end{vmatrix} + \mathbf{(-2)}\begin{vmatrix} 2 & 1 \\ 1 & 2 \end{vmatrix}$$

$$= 1(3 - 6) - 3(6 - 3) - 2(4 - 1)$$

$$= -3 - 9 - 6$$

$$= -18$$

■

Self Check Evaluate $\begin{vmatrix} 2 & -1 & 3 \\ 1 & 2 & -2 \\ 3 & 1 & 1 \end{vmatrix}$.

Answer 0

We can evaluate a 3×3 determinant by expanding it along any row or column. To determine the signs between the terms of the expansion of a 3×3 determinant, we use the following array of signs.

Array of Signs for a 3 × 3 Determinant

$$\begin{matrix} + & - & + \\ - & + & - \\ + & - & + \end{matrix}$$

EXAMPLE 3 Evaluate $\begin{vmatrix} 1 & 3 & -2 \\ 2 & 1 & 3 \\ 1 & 2 & 3 \end{vmatrix}$ by expanding on the middle column.

Solution This is the determinant of Example 2. To expand it along the middle column, we use the signs of the middle column of the array of signs:

$$\begin{vmatrix} 1 & \mathbf{3} & -2 \\ 2 & \mathbf{1} & 3 \\ 1 & \mathbf{2} & 3 \end{vmatrix} = -\mathbf{3}\underbrace{\begin{vmatrix} 2 & 3 \\ 1 & 3 \end{vmatrix}}_{\text{Minor of 3}} + \mathbf{1}\underbrace{\begin{vmatrix} 1 & -2 \\ 1 & 3 \end{vmatrix}}_{\text{Minor of 1}} - \mathbf{2}\underbrace{\begin{vmatrix} 1 & -2 \\ 2 & 3 \end{vmatrix}}_{\text{Minor of 2}}$$

$$= -3(6 - 3) + 1[3 - (-2)] - 2[3 - (-4)]$$

$$= -3(3) + 1(5) - 2(7)$$

$$= -9 + 5 - 14$$

$$= -18$$

As expected, we get the same value as in Example 2. ■

Self Check Evaluate $\begin{vmatrix} 2 & -1 & 3 \\ 1 & 2 & 2 \\ 3 & 1 & 1 \end{vmatrix}$.

Answer -20

Evaluating Determinants

GRAPHING CALCULATORS

It is possible to use a graphing calculator to evaluate determinants. For example, to evaluate the determinant in Example 3, we first enter the matrix by pressing the MATRIX key, selecting EDIT, and pressing the ENTER key. We then enter the dimensions and the elements of the matrix to get Figure 3-12(a). We then press 2nd QUIT to clear the screen. We then press MATRIX, select MATH, and press 1 to get Figure 3-12(b). We then press MATRIX, select NAMES, and press 1 to get Figure 3-12(c). To get the value of the determinant, we now press ENTER to get Figure 3-12(d), which shows that the value of the determinant is -18.

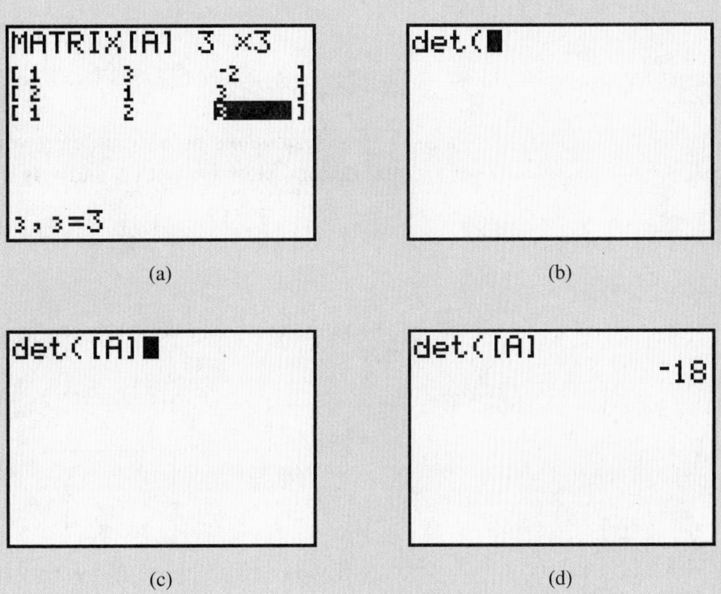

FIGURE 3-12

CRAMER'S RULE

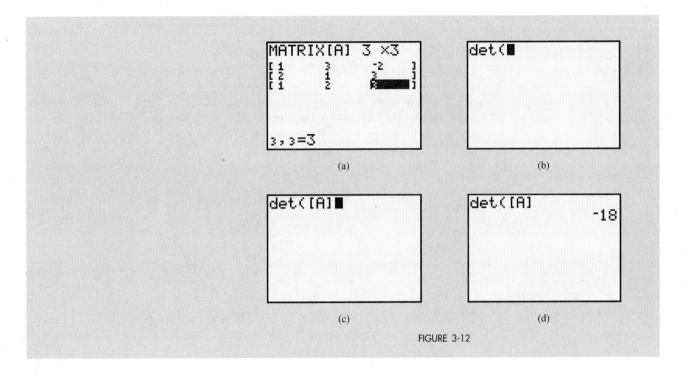

The method of using determinants to solve systems of equations is called **Cramer's rule,** named after the 18th-century mathematician Gabriel Cramer. To develop Cramer's rule, we consider the system

$$\begin{cases} ax + by = e \\ cx + dy = f \end{cases}$$

where x and y are variables and a, b, c, d, e, and f are constants.

If we multiply both sides of the first equation by d and multiply both sides of the second equation by $-b$, we can add the equations and eliminate y:

$$\begin{array}{rcl} adx + bdy &=& ed \\ -bcx - bdy &=& -bf \\ \hline adx - bcx \qquad &=& ed - bf \end{array}$$

To solve for x, we use the distributive property to write $adx - bcx$ as $(ad - bc)x$ on the left-hand side and divide each side by $ad - bc$:

$$(ad - bc)x = ed - bf$$

$$x = \frac{ed - bf}{ad - bc} \quad (ad - bc \neq 0)$$

We can find y in a similar manner. After eliminating the variable x, we get

$$y = \frac{af - ec}{ad - bc} \quad (ad - bc \neq 0)$$

Determinants provide an easy way of remembering these formulas. Note that the denominator for both x and y is

$$\begin{vmatrix} a & b \\ c & d \end{vmatrix} = ad - bc$$

The numerators can be expressed as determinants also:

$$x = \frac{ed - bf}{ad - bc} = \frac{\begin{vmatrix} e & b \\ f & d \end{vmatrix}}{\begin{vmatrix} a & b \\ c & d \end{vmatrix}} \quad \text{and} \quad y = \frac{af - ec}{ad - bc} = \frac{\begin{vmatrix} a & e \\ c & f \end{vmatrix}}{\begin{vmatrix} a & b \\ c & d \end{vmatrix}}$$

If we compare these formulas with the original system

$$\begin{cases} ax + by = e \\ cx + dy = f \end{cases}$$

we note that in the expressions for x and y above, the denominator determinant is formed by using the coefficients a, b, c, and d of the variables in the equations. The numerator determinants are the same as the denominator determinant, except that the column of coefficients of the variable for which we are solving is replaced with the column of constants e and f.

Cramer's Rule for Two Equations in Two Variables

The solution of the system $\begin{cases} ax + by = e \\ cx + dy = f \end{cases}$ is given by

$$x = \frac{D_x}{D} = \frac{\begin{vmatrix} e & b \\ f & d \end{vmatrix}}{\begin{vmatrix} a & b \\ c & d \end{vmatrix}} \quad \text{and} \quad y = \frac{D_y}{D} = \frac{\begin{vmatrix} a & e \\ c & f \end{vmatrix}}{\begin{vmatrix} a & b \\ c & d \end{vmatrix}}$$

If every determinant is 0, the system is consistent but the equations are dependent.

If $D = 0$ and D_x or D_y is nonzero, the system is inconsistent.

If $D \neq 0$, the system is consistent and the equations are independent.

EXAMPLE 4 Use Cramer's rule to solve $\begin{cases} 4x - 3y = 6 \\ -2x + 5y = 4 \end{cases}$.

Solution The value of x is the quotient of two determinants. The denominator determinant is made up of the coefficients of x and y:

$$D = \begin{vmatrix} 4 & -3 \\ -2 & 5 \end{vmatrix}$$

To solve for x, we form the numerator determinant from the denominator determinant by replacing its first column (the coefficients of x) with the column of constants (6 and 4).

To solve for y, we form the numerator determinant from the denominator determinant by replacing the second column (the coefficients of y) with the column of constants (6 and 4).

To find the values of x and y, we evaluate each determinant:

$$x = \frac{\begin{vmatrix} 6 & -3 \\ 4 & 5 \end{vmatrix}}{\begin{vmatrix} 4 & -3 \\ -2 & 5 \end{vmatrix}} = \frac{6(5) - (-3)(4)}{4(5) - (-3)(-2)} = \frac{30 + 12}{20 - 6} = \frac{42}{14} = 3$$

$$y = \frac{\begin{vmatrix} 4 & 6 \\ -2 & 4 \end{vmatrix}}{\begin{vmatrix} 4 & -3 \\ -2 & 5 \end{vmatrix}} = \frac{4(4) - 6(-2)}{4(5) - (-3)(-2)} = \frac{16 + 12}{20 - 6} = \frac{28}{14} = 2$$

The solution of this system is (3, 2). Verify that $x = 3$ and $y = 2$ satisfy each equation in the given system. ■

Self Check Solve $\begin{cases} 2x - 3y = -16 \\ 3x + 5y = 14 \end{cases}$.

Answer (−2, 4)

EXAMPLE 5 Use Cramer's rule to solve $\begin{cases} 7x = 8 - 4y \\ 2y = 3 - \frac{7}{2}x \end{cases}$.

Solution We multiply both sides of the second equation by 2 to eliminate the fraction and write the system in the form

$$\begin{cases} 7x + 4y = 8 \\ 7x + 4y = 6 \end{cases}$$

When we attempt to use Cramer's rule to solve this system for x, we obtain

$$x = \frac{\begin{vmatrix} 8 & 4 \\ 6 & 4 \end{vmatrix}}{\begin{vmatrix} 7 & 4 \\ 7 & 4 \end{vmatrix}} = \frac{8}{0} \quad \text{which is undefined}$$

Since the denominator determinant is 0 and the numerator determinant is not 0, the system is inconsistent. It has no solutions.

We can see directly from the system that it is inconsistent. For any values of x and y, it is impossible that 7 times x plus 4 times y could be both 8 and 6. ■

Self Check Solve $\begin{cases} 3x = 8 - 4y \\ y = \frac{5}{2} - \frac{3}{4}x \end{cases}$.

Answer no solutions

Cramer's Rule for Three Equations in Three Variables

The solution of the system $\begin{cases} ax + by + cz = j \\ dx + ey + fz = k \\ gx + hy + iz = l \end{cases}$ is given by

$$x = \frac{D_x}{D}, \quad y = \frac{D_y}{D}, \quad \text{and} \quad z = \frac{D_z}{D}$$

where

$$D = \begin{vmatrix} a & b & c \\ d & e & f \\ g & h & i \end{vmatrix} \qquad D_x = \begin{vmatrix} j & b & c \\ k & e & f \\ l & h & i \end{vmatrix}$$

$$D_y = \begin{vmatrix} a & j & c \\ d & k & f \\ g & l & i \end{vmatrix} \qquad D_z = \begin{vmatrix} a & b & j \\ d & e & k \\ g & h & l \end{vmatrix}$$

If every determinant is 0, the system is consistent but the equations are dependent.

If $D = 0$ and D_x or D_y or D_z is nonzero, the system is inconsistent.

If $D \neq 0$, the system is consistent and the equations are independent.

EXAMPLE 6 Use Cramer's rule to solve $\begin{cases} 2x + y + 4z = 12 \\ x + 2y + 2z = 9 \\ 3x - 3y - 2z = 1 \end{cases}$.

Solution The denominator determinant is the determinant formed by the coefficients of the variables. The numerator determinants are formed by replacing the coefficients of the variable being solved for by the column of constants. We form the quotients for x, y, and z and evaluate the determinants:

$$x = \frac{\begin{vmatrix} 12 & 1 & 4 \\ 9 & 2 & 2 \\ 1 & -3 & -2 \end{vmatrix}}{\begin{vmatrix} 2 & 1 & 4 \\ 1 & 2 & 2 \\ 3 & -3 & -2 \end{vmatrix}} = \frac{12\begin{vmatrix} 2 & 2 \\ -3 & -2 \end{vmatrix} - 1\begin{vmatrix} 9 & 2 \\ 1 & -2 \end{vmatrix} + 4\begin{vmatrix} 9 & 2 \\ 1 & -3 \end{vmatrix}}{2\begin{vmatrix} 2 & 2 \\ -3 & -2 \end{vmatrix} - 1\begin{vmatrix} 1 & 2 \\ 3 & -2 \end{vmatrix} + 4\begin{vmatrix} 1 & 2 \\ 3 & -3 \end{vmatrix}} = \frac{12(2) - (-20) + 4(-29)}{2(2) - (-8) + 4(-9)} = \frac{-72}{-24} = 3$$

$$y = \frac{\begin{vmatrix} 2 & 12 & 4 \\ 1 & 9 & 2 \\ 3 & 1 & -2 \end{vmatrix}}{\begin{vmatrix} 2 & 1 & 4 \\ 1 & 2 & 2 \\ 3 & -3 & -2 \end{vmatrix}} = \frac{2\begin{vmatrix} 9 & 2 \\ 1 & -2 \end{vmatrix} - 12\begin{vmatrix} 1 & 2 \\ 3 & -2 \end{vmatrix} + 4\begin{vmatrix} 1 & 9 \\ 3 & 1 \end{vmatrix}}{-24} = \frac{2(-20) - 12(-8) + 4(-26)}{-24} = \frac{-48}{-24} = 2$$

$$z = \frac{\begin{vmatrix} 2 & 1 & 12 \\ 1 & 2 & 9 \\ 3 & -3 & 1 \end{vmatrix}}{\begin{vmatrix} 2 & 1 & 4 \\ 1 & 2 & 2 \\ 3 & -3 & -2 \end{vmatrix}} = \frac{2\begin{vmatrix} 2 & 9 \\ -3 & 1 \end{vmatrix} - 1\begin{vmatrix} 1 & 9 \\ 3 & 1 \end{vmatrix} + 12\begin{vmatrix} 1 & 2 \\ 3 & -3 \end{vmatrix}}{-24} = \frac{2(29) - 1(-26) + 12(-9)}{-24} = \frac{-24}{-24} = 1$$

The solution of this system is (3, 2, 1). ■

Self Check Solve $\begin{cases} x + y + 2z = 6 \\ 2x - y + z = 9 \\ x + y - 2z = -6 \end{cases}$.

Answer (2, −2, 3)

Orals *Evaluate each determinant.*

1. $\begin{vmatrix} 2 & 1 \\ 1 & 1 \end{vmatrix}$

2. $\begin{vmatrix} 0 & 2 \\ 1 & 1 \end{vmatrix}$

3. $\begin{vmatrix} 0 & 1 \\ 0 & 1 \end{vmatrix}$

When using Cramer's rule to solve the system $\begin{cases} x + 2y = 5 \\ 2x - y = 4 \end{cases}$,

4. Find the denominator determinant for x.

5. Find the numerator determinant for x.

6. Find the numerator determinant for y.

EXERCISE 3.5

REVIEW *Solve each equation.*

1. $3(x + 2) - (2 - x) = x - 5$

2. $\frac{3}{7}x = 2(x + 11)$

3. $\frac{5}{3}(5x + 6) - 10 = 0$

4. $5 - 3(2x - 1) = 2(4 + 3x) - 24$

VOCABULARY AND CONCEPTS *Fill in each blank to make a true statement.*

5. A determinant is a ______ that is associated with a square matrix.

6. The value of $\begin{vmatrix} a & b \\ c & d \end{vmatrix}$ is ______.

7. The minor of b_1 in $\begin{vmatrix} a_1 & b_1 & c_1 \\ a_2 & b_2 & c_2 \\ a_3 & b_3 & c_3 \end{vmatrix}$ is ______.

8. We can evaluate a determinant by expanding it along any ____ or ______.

9. The denominator determinant for the value of x in the system $\begin{cases} 3x + 4y = 7 \\ 2x - 3y = 5 \end{cases}$ is ______.

10. If the denominator determinant for y in a system of equations is zero, the equations of the system are ______ or the system is ______.

PRACTICE *In Exercises 11–28, evaluate each determinant.*

11. $\begin{vmatrix} 2 & 3 \\ -2 & 1 \end{vmatrix}$

12. $\begin{vmatrix} 3 & -2 \\ -2 & 4 \end{vmatrix}$

13. $\begin{vmatrix} -1 & 2 \\ 3 & -4 \end{vmatrix}$

14. $\begin{vmatrix} -1 & -2 \\ -3 & -4 \end{vmatrix}$

15. $\begin{vmatrix} x & y \\ y & x \end{vmatrix}$

16. $\begin{vmatrix} x + y & y - x \\ x & y \end{vmatrix}$

17. $\begin{vmatrix} 1 & 0 & 1 \\ 0 & 1 & 0 \\ 1 & 1 & 1 \end{vmatrix}$

18. $\begin{vmatrix} 1 & 2 & 0 \\ 0 & 1 & 2 \\ 0 & 0 & 1 \end{vmatrix}$

19. $\begin{vmatrix} -1 & 2 & 1 \\ 2 & 1 & -3 \\ 1 & 1 & 1 \end{vmatrix}$

20. $\begin{vmatrix} 1 & 2 & 3 \\ 1 & 2 & 3 \\ 1 & 2 & 3 \end{vmatrix}$ **21.** $\begin{vmatrix} 1 & -2 & 3 \\ -2 & 1 & 1 \\ -3 & -2 & 1 \end{vmatrix}$ **22.** $\begin{vmatrix} 1 & 1 & 2 \\ 2 & 1 & -2 \\ 3 & 1 & 3 \end{vmatrix}$

23. $\begin{vmatrix} 1 & 2 & 3 \\ 4 & 5 & 6 \\ 7 & 8 & 9 \end{vmatrix}$ **24.** $\begin{vmatrix} 1 & 4 & 7 \\ 2 & 5 & 8 \\ 3 & 6 & 9 \end{vmatrix}$ **25.** $\begin{vmatrix} a & 2a & -a \\ 2 & -1 & 3 \\ 1 & 2 & -3 \end{vmatrix}$

26. $\begin{vmatrix} 1 & 2b & -3 \\ 2 & -b & 2 \\ 1 & 3b & 1 \end{vmatrix}$ **27.** $\begin{vmatrix} 1 & a & b \\ 1 & 2a & 2b \\ 1 & 3a & 3b \end{vmatrix}$ **28.** $\begin{vmatrix} a & b & c \\ 0 & b & c \\ 0 & 0 & c \end{vmatrix}$

In Exercises 29–54, use Cramer's rule to solve each system of equations, if possible.

29. $\begin{cases} x + y = 6 \\ x - y = 2 \end{cases}$ **30.** $\begin{cases} x - y = 4 \\ 2x + y = 5 \end{cases}$ **31.** $\begin{cases} 2x + y = 1 \\ x - 2y = -7 \end{cases}$ **32.** $\begin{cases} 3x - y = -3 \\ 2x + y = -7 \end{cases}$

33. $\begin{cases} 2x + 3y = 0 \\ 4x - 6y = -4 \end{cases}$ **34.** $\begin{cases} 4x - 3y = -1 \\ 8x + 3y = 4 \end{cases}$ **35.** $\begin{cases} y = \dfrac{-2x + 1}{3} \\ 3x - 2y = 8 \end{cases}$ **36.** $\begin{cases} 2x + 3y = -1 \\ x = \dfrac{y - 9}{4} \end{cases}$

37. $\begin{cases} y = \dfrac{11 - 3x}{2} \\ x = \dfrac{11 - 4y}{6} \end{cases}$ **38.** $\begin{cases} x = \dfrac{12 - 6y}{5} \\ y = \dfrac{24 - 10x}{12} \end{cases}$ **39.** $\begin{cases} x = \dfrac{5y - 4}{2} \\ y = \dfrac{3x - 1}{5} \end{cases}$ **40.** $\begin{cases} y = \dfrac{1 - 5x}{2} \\ x = \dfrac{3y + 10}{4} \end{cases}$

41. $\begin{cases} x + y + z = 4 \\ x + y - z = 0 \\ x - y + z = 2 \end{cases}$ **42.** $\begin{cases} x + y + z = 4 \\ x - y + z = 2 \\ x - y - z = 0 \end{cases}$ **43.** $\begin{cases} x + y + 2z = 7 \\ x + 2y + z = 8 \\ 2x + y + z = 9 \end{cases}$ **44.** $\begin{cases} x + 2y + 2z = 10 \\ 2x + y + 2z = 9 \\ 2x + 2y + z = 1 \end{cases}$

45. $\begin{cases} 2x + y - z = 1 \\ x + 2y + 2z = 2 \\ 4x + 5y + 3z = 3 \end{cases}$ **46.** $\begin{cases} 4x + 3z = 4 \\ 2y - 6z = -1 \\ 8x + 4y + 3z = 9 \end{cases}$ **47.** $\begin{cases} 2x + y + z = 5 \\ x - 2y + 3z = 10 \\ x + y - 4z = -3 \end{cases}$ **48.** $\begin{cases} 3x + 2y - z = -8 \\ 2x - y + 7z = 10 \\ 2x + 2y - 3z = -10 \end{cases}$

49. $\begin{cases} 2x + 3y + 4z = 6 \\ 2x - 3y - 4z = -4 \\ 4x + 6y + 8z = 12 \end{cases}$ **50.** $\begin{cases} x - 3y + 4z - 2 = 0 \\ 2x + y + 2z - 3 = 0 \\ 4x - 5y + 10z - 7 = 0 \end{cases}$

51. $\begin{cases} x + y = 1 \\ \dfrac{1}{2}y + z = \dfrac{5}{2} \\ x - z = -3 \end{cases}$ **52.** $\begin{cases} 3x + 4y + 14z = 7 \\ -\dfrac{1}{2}x - y + 2z = \dfrac{3}{2} \\ x + \dfrac{3}{2}y + \dfrac{5}{2}z = 1 \end{cases}$ **53.** $\begin{cases} 2x - y + 4z + 2 = 0 \\ 5x + 8y + 7z = -8 \\ x + 3y + z + 3 = 0 \end{cases}$ **54.** $\begin{cases} \dfrac{1}{2}x + y + z + \dfrac{3}{2} = 0 \\ x + \dfrac{1}{2}y + z - \dfrac{1}{2} = 0 \\ x + y + \dfrac{1}{2}z + \dfrac{1}{2} = 0 \end{cases}$

In Exercises 55–58, evaluate each determinant and solve the resulting equation.

55. $\begin{vmatrix} x & 1 \\ 3 & 2 \end{vmatrix} = 1$

56. $\begin{vmatrix} x & -x \\ 2 & -3 \end{vmatrix} = -5$

57. $\begin{vmatrix} x & -2 \\ 3 & 1 \end{vmatrix} = \begin{vmatrix} 4 & 2 \\ x & 3 \end{vmatrix}$

58. $\begin{vmatrix} x & 3 \\ x & 2 \end{vmatrix} = \begin{vmatrix} 3 & 2 \\ 1 & 1 \end{vmatrix}$

APPLICATIONS

59. Making investments A student wants to average a 6.6% return by investing $20,000 in the three stocks listed in Illustration 1. Because HiTech is considered to be a high-risk investment, he wants to invest three times as much in SaveTel and HiGas combined as he invests in HiTech. How much should he invest in each stock?

Stock	Rate of return
HiTech	10%
SaveTel	5%
HiGas	6%

ILLUSTRATION 1

60. Making investments See Illustration 2. A woman wants to average a $7\frac{1}{3}\%$ return by investing $30,000 in three certificates of deposit. She wants to invest five times as much in the 8% CD as in the 6% CD. How much should she invest in each CD?

Type of CD	Rate of return
12 month	6%
24 month	7%
36 month	8%

ILLUSTRATION 2

In Exercises 61–64, use a graphing calculator to evaluate each determinant.

61. $\begin{vmatrix} 2 & -3 & 4 \\ -1 & 2 & 4 \\ 3 & -3 & 1 \end{vmatrix}$

62. $\begin{vmatrix} -3 & 2 & -5 \\ 3 & -2 & 6 \\ 1 & -3 & 4 \end{vmatrix}$

63. $\begin{vmatrix} 2 & 1 & -3 \\ -2 & 2 & 4 \\ 1 & -2 & 2 \end{vmatrix}$

64. $\begin{vmatrix} 4 & 2 & -3 \\ 2 & -5 & 6 \\ 2 & 5 & -2 \end{vmatrix}$

WRITING

65. Tell how to find the minor of an element of a determinant.

66. Tell how to find x when solving a system of linear equations by Cramer's rule.

SOMETHING TO THINK ABOUT

67. Show that

$$\begin{vmatrix} x & y & 1 \\ -2 & 3 & 1 \\ 3 & 5 & 1 \end{vmatrix} = 0$$

is the equation of the line passing through $(-2, 3)$ and $(3, 5)$.

68. Show that

$$\frac{1}{2}\begin{vmatrix} 0 & 0 & 1 \\ 3 & 0 & 1 \\ 0 & 4 & 1 \end{vmatrix}$$

is the area of the triangle with vertices at $(0, 0)$, $(3, 0)$, and $(0, 4)$.

Determinants with more than 3 rows and 3 columns can be evaluated by expanding them by minors. The sign array for a 4 × 4 determinant is

$$\begin{matrix} + & - & + & - \\ - & + & - & + \\ + & - & + & - \\ - & + & - & + \end{matrix}$$

Evaluate each determinant.

69. $\begin{vmatrix} 1 & 0 & 2 & 1 \\ 2 & 1 & 1 & 3 \\ 1 & 1 & 1 & 1 \\ 2 & 1 & 1 & 1 \end{vmatrix}$

70. $\begin{vmatrix} 1 & 2 & -1 & 1 \\ -2 & 1 & 3 & -1 \\ 0 & 1 & 1 & 2 \\ 2 & 0 & 3 & 1 \end{vmatrix}$

PROJECTS

PROJECT 1

The number of units of a product that will be produced, and the number of units that will be sold, depends on the unit price of the product. As the unit price gets higher, the product will be produced in greater quantity, because the producer will make more money on each item. The *supply* of the product will grow, and we say that supply *is a function of* (or *depends on*) the unit price. As the price rises, fewer consumers will buy the product, and the *demand* will decrease. The demand for the product is also a function of the unit price.

In this project, we will assume that both supply and demand are *linear* functions of the unit price. Thus, the graph of supply (the y-coordinate) versus price (the x-coordinate) is a line with positive slope. The graph of the demand function is a line with negative slope. Because these two lines cannot be parallel, they must intersect. The price at which supply equals demand is called the *market price*: At this price, the same number of units of the product will be sold as are manufactured.

You work for Soda Pop Inc. Your boss has given you the task of analyzing the sales figures for the past year for the soda market in a small city. You have been provided with the following supply and demand functions. (Supply and demand are measured in cases per week; p, the price per case, is measured in dollars.)

The demand for soda is $D(p) = 19{,}000 - 2{,}200p$.

The supply of soda is $S(p) = 3{,}000 + 1{,}080p$.

Both functions are true for values of p from \$3.50 to \$5.75.

(continued)

PROJECTS *(continued)*

Graph both functions on the same set of coordinate axes, being sure to label each graph and include any other important information. Then write a report for your supervisor that answers the following questions. Include any supporting work.

a. Explain why producers will be able to sell all of the soda they make when the price is \$3.50 per case. How much money will the producers take in from these sales?

b. How much money will producers take in from sales when the price is \$5.75 per case? How much soda will have to be warehoused? (That is, how much extra soda will have been made?)

c. Find the market price for soda (to the nearest cent). How many cases per week will be sold at this price? How much money will the producers take in from sales at the market price?

d. Explain why prices always tend toward the market price. That is, explain why the unit price will rise if the demand is greater than the supply, and why the unit price will fall if supply is greater than demand.

PROJECT 2

Goodstuff Produce Company has two large water canals that feed the irrigation ditches on its fruit farm. One of these canals runs directly north and south, and the other runs directly east and west. The canals cross at the center of the farm property (the origin) and divide the farm into four quadrants. The company is interested in digging some new irrigation ditches in a portion of the northeast quadrant. You have been hired to plan the layout of the new system.

Your design is to make use of ditch Z, which is already present. This ditch runs from a point 300 meters north of the origin to a point 400 meters east of the origin. The owners of Goodstuff want two new ditches dug.

- Ditch A is to begin at a point 100 meters north of the origin and follow a line that travels 3 meters north for every 7 meters it travels east until it intersects ditch Z.
- Ditch B is to run from the origin to ditch Z in such a way that it exactly bisects the area in the northeast quadrant that is south of both ditch Z and ditch A.

You are to provide the equations of the lines that the three ditches follow, as well as the exact location of the gates that will be installed where the ditches intersect one another. Be sure to provide explanations and organized work that will clearly display the desired information and assure the owners of Goodstuff that they will get exactly what they want.

CHAPTER SUMMARY

CONCEPTS | REVIEW EXERCISES

SECTION 3.1 Solution by Graphing

If a system of equations has at least one solution, the system is a **consistent system**. Otherwise, the system is an **inconsistent system**.

If the graphs of the equations of a system are distinct, the equations are **independent equations**. Otherwise, the equations are **dependent equations**.

1. Solve each system by the graphing method.

a. $\begin{cases} 2x + y = 11 \\ -x + 2y = 7 \end{cases}$

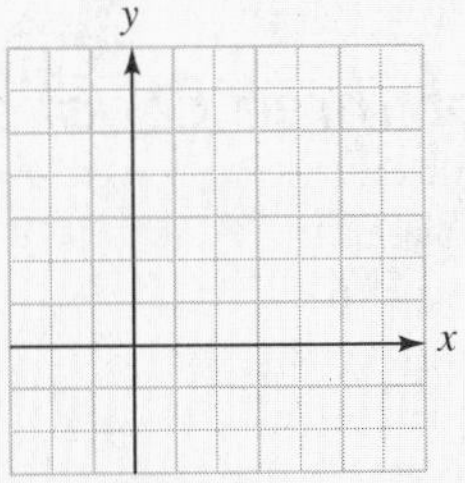

b. $\begin{cases} 3x + 2y = 0 \\ 2x - 3y = -13 \end{cases}$

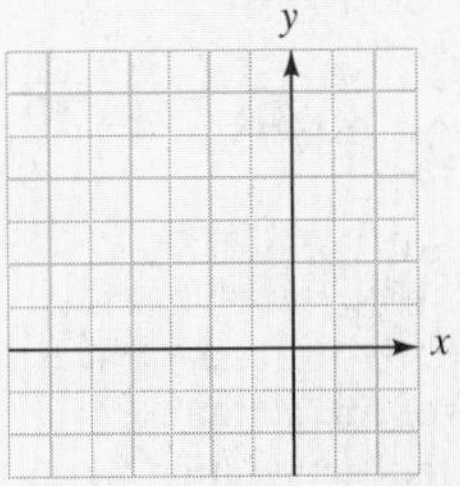

c. $\begin{cases} \frac{1}{2}x + \frac{1}{3}y = 2 \\ y = 6 - \frac{3}{2}x \end{cases}$

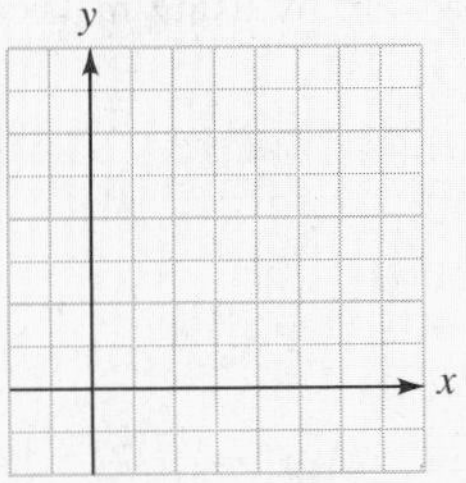

d. $\begin{cases} \frac{1}{3}x - \frac{1}{2}y = 1 \\ 6x - 9y = 2 \end{cases}$

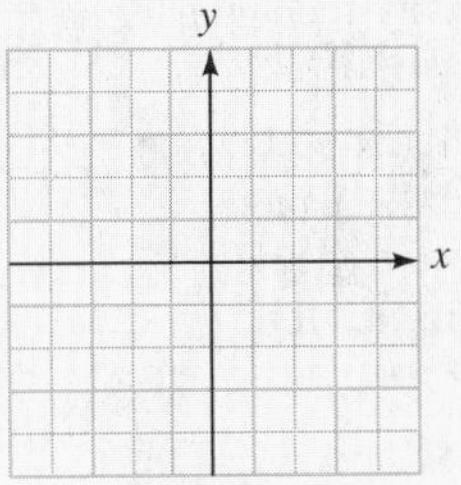

SECTION 3.2 Solution by Elimination

To solve a system by *substitution*, solve one equation for a variable. Then substitute the expression found for that variable into the other equation. Then solve for the other variable.

2. Solve each system by substitution.

a. $\begin{cases} y = x + 4 \\ 2x + 3y = 7 \end{cases}$

b. $\begin{cases} y = 2x + 5 \\ 3x - 5y = -4 \end{cases}$

c. $\begin{cases} x + 2y = 11 \\ 2x - y = 2 \end{cases}$

d. $\begin{cases} 2x + 3y = -2 \\ 3x + 5y = -2 \end{cases}$

To solve a system by *addition*, combine the equations of the system in a way that will eliminate one of the variables.

3. Solve each system by addition.

a. $\begin{cases} x + y = -2 \\ 2x + 3y = -3 \end{cases}$

b. $\begin{cases} 3x + 2y = 1 \\ 2x - 3y = 5 \end{cases}$

c. $\begin{cases} x + \frac{1}{2}y = 7 \\ -2x = 3y - 6 \end{cases}$

d. $\begin{cases} y = \dfrac{x - 3}{2} \\ x = \dfrac{2y + 7}{2} \end{cases}$

SECTION 3.3 Solutions of Three Equations in Three Variables

A system of three linear equations in three variables can be solved by using a combination of the addition method and the substitution method.

4. Solve each system.

a. $\begin{cases} x + y + z = 6 \\ x - y - z = -4 \\ -x + y - z = -2 \end{cases}$

b. $\begin{cases} 2x + 3y + z = -5 \\ -x + 2y - z = -6 \\ 3x + y + 2z = 4 \end{cases}$

SECTION 3.4 Solution by Matrices

A **matrix** is any rectangular array of numbers.

Systems of linear equations can be solved by using matrices and the method of **Gaussian elimination**.

5. Solve each system by using matrices.

a. $\begin{cases} x + 2y = 4 \\ 2x - y = 3 \end{cases}$

b. $\begin{cases} x + y + z = 6 \\ 2x - y + z = 1 \\ 4x + y - z = 5 \end{cases}$

c. $\begin{cases} x + y = 3 \\ x - 2y = -3 \\ 2x + y = 4 \end{cases}$

d. $\begin{cases} x + 2y + z = 2 \\ 2x + 5y + 4z = 5 \end{cases}$

SECTION 3.5 Solution by Determinants

A **determinant of a square matrix** is a real number.

$$\begin{vmatrix} a & b \\ c & d \end{vmatrix} = ad - bc$$

$$\begin{vmatrix} a_1 & b_1 & c_1 \\ a_2 & b_2 & c_2 \\ a_3 & b_3 & c_3 \end{vmatrix} = a_1 \begin{vmatrix} b_2 & c_2 \\ b_3 & c_3 \end{vmatrix} - b_1 \begin{vmatrix} a_2 & c_2 \\ a_3 & c_3 \end{vmatrix} + c_1 \begin{vmatrix} a_2 & b_2 \\ a_3 & b_3 \end{vmatrix}$$

6. Evaluate each determinant.

a. $\begin{vmatrix} 2 & 3 \\ -4 & 3 \end{vmatrix}$

b. $\begin{vmatrix} -3 & -4 \\ 5 & -6 \end{vmatrix}$

c. $\begin{vmatrix} -1 & 2 & -1 \\ 2 & -1 & 3 \\ 1 & -2 & 2 \end{vmatrix}$

d. $\begin{vmatrix} 3 & -2 & 2 \\ 1 & -2 & -2 \\ 2 & 1 & -1 \end{vmatrix}$

7. Use Cramer's rule to solve each system.

a. $\begin{cases} 3x + 4y = 10 \\ 2x - 3y = 1 \end{cases}$ **b.** $\begin{cases} 2x - 5y = -17 \\ 3x + 2y = 3 \end{cases}$

c. $\begin{cases} x + 2y + z = 0 \\ 2x + y + z = 3 \\ x + y + 2z = 5 \end{cases}$ **d.** $\begin{cases} 2x + 3y + z = 2 \\ x + 3y + 2z = 7 \\ x - y - z = -7 \end{cases}$

Chapter Test

1. Solve $\begin{cases} 2x + y = 5 \\ y = 2x - 3 \end{cases}$ by graphing.

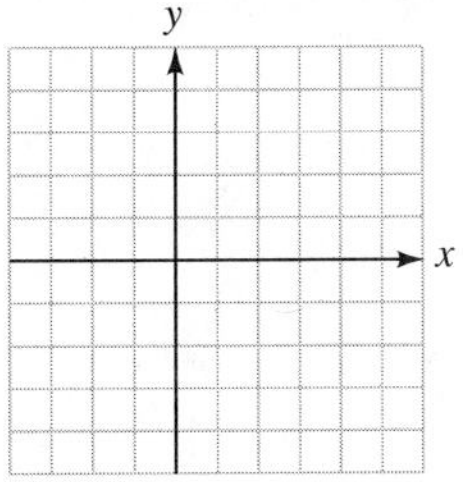

2. Use substitution to solve $\begin{cases} 2x - 4y = 14 \\ x = -2y + 7 \end{cases}$.

3. Use addition to solve $\begin{cases} 2x + 3y = -5 \\ 3x - 2y = 12 \end{cases}$.

4. Use any method to solve $\begin{cases} \dfrac{x}{2} - \dfrac{y}{4} = -4 \\ x + y = -2 \end{cases}$.

In Problems 5–6, consider the system $\begin{cases} 3(x + y) = x - 3 \\ -y = \dfrac{2x + 3}{3} \end{cases}$.

5. Are the equations of the system dependent or independent?

6. Is the system consistent or inconsistent?

In Problems 7–8, use an elementary row operation to find the missing number in the second matrix.

7. $\begin{bmatrix} 1 & 2 & -1 \\ 2 & -2 & 3 \end{bmatrix}, \begin{bmatrix} 1 & 2 & -1 \\ -1 & -8 & \square \end{bmatrix}$

8. $\begin{bmatrix} -1 & 3 & 6 \\ 3 & -2 & 4 \end{bmatrix}, \begin{bmatrix} -1 & 3 & 6 \\ 5 & \square & -8 \end{bmatrix}$

In Problems 9–10, consider the system $\begin{cases} x + y + z = 4 \\ x + y - z = 6 \\ 2x - 3y + z = -1 \end{cases}$.

9. Write the augmented matrix that represents the system.

10. Write the coefficient matrix that represents the system.

In Problems 11–12, use matrices to solve each system.

11. $\begin{cases} x + y = 4 \\ 2x - y = 2 \end{cases}$

12. $\begin{cases} x + y = 2 \\ x - y = -4 \\ 2x + y = 1 \end{cases}$

In Problems 13–16, evaluate each determinant.

13. $\begin{vmatrix} 2 & -3 \\ 4 & 5 \end{vmatrix}$

14. $\begin{vmatrix} -3 & -4 \\ -2 & 3 \end{vmatrix}$

15. $\begin{vmatrix} 1 & 2 & 0 \\ 2 & 0 & 3 \\ 1 & -2 & 2 \end{vmatrix}$

16. $\begin{vmatrix} 2 & -1 & 1 \\ 3 & 1 & 0 \\ 0 & 1 & 2 \end{vmatrix}$

In Problems 17–20, consider the system $\begin{cases} x - y = -6 \\ 3x + y = -6 \end{cases}$, *which is to be solved with Cramer's rule.*

17. When solve for x, what is the numerator determinant? **(Don't evaluate it.)**

18. When solving for y, what is the denominator determinant? **(Don't evaluate it.)**

19. Solve the system for x.

20. Solve the system for y.

In Problems 21–22, consider the system $\begin{cases} x + y + z = 4 \\ x + y - z = 6 \\ 2x - 3y + z = -1 \end{cases}$.

21. Solve for x.

22. Solve for z.

4 Inequalities

MATHEMATICS IN THE WORKPLACE

Statistician

Statisticians devise, carry out, and interpret the numerical results of surveys and experiments. They apply their knowledge to subject areas such as economics, human behavior, natural science, and engineering. There are many career opportunities for people trained in statistics who also have knowledge in a field of application such as manufacturing, engineering, scientific research, or business.

SAMPLE APPLICATION ■ To estimate the mean property tax paid by homeowners living in Rockford, Illinois, a researcher selects a random sample of homeowners and computes the mean tax paid by those homeowners. If the researcher knows that the standard deviation σ of all tax bills is \$120, how large must the sample size be before the researcher can be 95% certain that the sample mean will be within \$35 of the population mean?
(See Example 8 in Section 4.1.)

4.1 LINEAR INEQUALITIES

■ INEQUALITIES ■ PROPERTIES OF INEQUALITIES ■ LINEAR INEQUALITIES ■ COMPOUND INEQUALITIES ■ PROBLEM SOLVING ■ FINDING A SAMPLE SIZE

Getting Ready *Graph each set on the number line.*

1. All real numbers greater than -3

2. All real numbers less than 4

3. All real numbers less than or equal to 5

4. All real numbers greater than or equal to -1

■ INEQUALITIES

Inequalities are statements indicating that two quantities are unequal. Inequalities can be recognized because they contain one or more of the following symbols.

Inequality Symbols

$a \neq b$ means "*a* is not equal to *b*."
$a < b$ means "*a* is less than *b*."
$a > b$ means "*a* is greater than *b*."
$a \leq b$ means "*a* is less than or equal to *b*."
$a \geq b$ means "*a* is greater than or equal to *b*."

To show that 2 is less than 3, we can write $2 < 3$. To show that x is greater than 4 or equal to 4, we can write $x \geq 4$.

By definition, $a < b$ means that "a is less than b," but it also means that $b > a$. If a is to the left of b on the number line, then $a < b$.

In Chapter 1, we saw that many inequalities can be graphed as intervals on a number line. Table 4-1 shows the possibilities that exist when a and b are real numbers.

Kind of interval	Inequality	Graph	Interval
Unbounded intervals	$x > a$	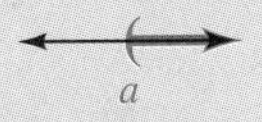	(a, ∞)
	$x < a$	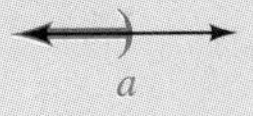	$(-\infty, a)$
	$x \geq a$	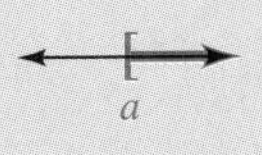	$[a, \infty)$
	$x \leq a$	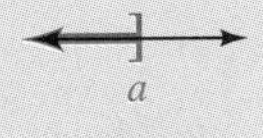	$(-\infty, a]$
Open intervals	$a < x < b$		(a, b)
Half-open intervals	$a \leq x < b$	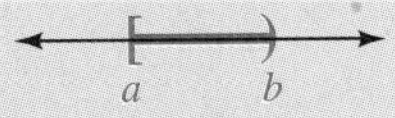	$[a, b)$
	$a < x \leq b$		$(a, b]$
Closed intervals	$a \leq x \leq b$	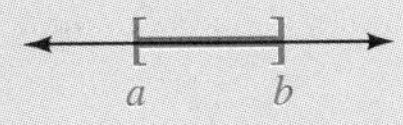	$[a, b]$

TABLE 4-1

WARNING! Remember that in interval notation, the notation $(-3, 4)$ represents the set of real numbers between -3 and 4, and not the coordinates of a point on a graph.

Inequalities like $x + 1 > x$, which are true for all x, are called **absolute inequalities.** Inequalities like $3x + 2 < 8$, which are true for some x but not others, are called **conditional inequalities.**

PROPERTIES OF INEQUALITIES

There are several basic properties of inequalities.

Trichotomy Property
For any real numbers a and b, exactly one of the following statements is true:

$$a < b, \qquad a = b, \qquad \text{or} \qquad a > b$$

The trichotomy property indicates that one and only one of the following statements is true about any two real numbers. Either

- the first is less than the second,
- the first is equal to the second, or
- the first is greater than the second.

Transitive Property
If a, b, and c are real numbers with $a < b$ and $b < c$, then $a < c$.

If a, b, and c are real numbers with $a > b$ and $b > c$, then $a > c$.

The first part of the transitive property indicates that

- If a first number is less than a second number and
- the second number is less than a third number, then
- the first number is less than the third.

The second part of the transitive property is similar, with the words "is greater than" substituted for "is less than."

Property 1 of Inequalities
Any real number can be added to (or subtracted from) both sides of an inequality to produce another inequality with the same direction.

Property 1 indicates that any number can be added to both sides of a true inequality to get another true inequality with the same direction (sometimes called the *sense* or *order* of an inequality). For example, if we add 4 to both sides of the inequality $3 < 12$, we get

$$3 + 4 < 12 + 4$$
$$7 < 16$$

and the $<$ symbol is unchanged.

Subtracting 4 from both sides of $3 < 12$ does not change the direction of the inequality either.

$$3 - 4 < 12 - 4$$
$$-1 < 8$$

Property 2 of Inequalities
If both sides of an inequality are multiplied (or divided) by a positive number, another inequality results with the same direction as the original one.

Property 2 indicates that both sides of a true inequality can be multiplied by any positive number to get another true inequality with the same direction. For example, if both sides of the inequality $-4 < 6$ are multiplied by $+2$, we get

$$2(-4) < 2(6)$$
$$-8 < 12$$

and the $<$ symbol is unchanged.

Dividing both sides by $+2$ does not change the direction of the inequality either.

$$\frac{-4}{2} < \frac{6}{2}$$
$$-2 < 3$$

Property 3 of Inequalities
If both sides of an inequality are multiplied (or divided) by a negative number, another inequality results, but with the opposite direction from the original inequality.

Property 3 indicates that if both sides of a true inequality are multiplied by a negative number, another true inequality results, but with the opposite direction. For example, if both sides of the inequality $-4 < 6$ are multiplied by -2, we get

$$-4 < 6$$
$$-2(-4) > -2(6) \qquad \text{Change } < \text{ to } >.$$
$$8 > -12$$

and the $<$ symbol changes to a $>$ symbol.

Dividing both sides by -2 also changes the direction of the inequality.

$$-4 < 6$$
$$\frac{-4}{-2} > \frac{6}{-2}$$
$$2 > -3$$

WARNING! We must remember to change the direction of an inequality symbol every time we multiply or divide both sides by a negative number.

LINEAR INEQUALITIES

Linear Inequalities
A **linear inequality** in x is any inequality that can be expressed in one of the following forms (with $a \neq 0$).

$$ax + c < 0 \qquad ax + c > 0 \qquad ax + c \leq 0 \qquad \text{or} \qquad ax + c \geq 0$$

We can solve linear inequalities just as we solve linear equations, with one exception. If we multiply or divide both sides by a negative number, we must change the direction of the inequality.

EXAMPLE 1 Solve the linear inequality $3(2x - 9) < 9$. Give the result in interval notation and graph the interval.

Solution We use the same steps as for solving equations.

$$3(2x - 9) < 9$$

$$6x - 27 < 9 \qquad \text{Use the distributive property to remove parentheses.}$$

$$6x < 36 \qquad \text{Add 27 to both sides.}$$

$$x < 6 \qquad \text{Divide both sides by 6.}$$

FIGURE 4-1

The solution set is the interval $(-\infty, 6)$, whose graph is shown in Figure 4-1. The parenthesis at 6 indicates that 6 is not included in the solution set. ■

Self Check Solve $2(3x + 2) < 16$.

Answer $(-\infty, 2)$

EXAMPLE 2 Solve the linear inequality $-4(3x + 2) \leq 16$.

Solution We use the same steps as for solving equations.

$$-4(3x + 2) \leq 16$$

$$-12x - 8 \leq 16 \qquad \text{Use the distributive property to remove parentheses.}$$

$$-12x \leq 24 \qquad \text{Add 8 to both sides.}$$

$$x \geq -2 \qquad \text{Divide both sides by } -12 \text{ and reverse the } \leq \text{ symbol.}$$

FIGURE 4-2

The solution set is the interval $[-2, \infty)$, whose graph is shown in Figure 4-2. The bracket at -2 indicates that -2 is included in the solution set. ■

Self Check Solve $-3(2x - 4) \leq 24$.

Answer $[-2, \infty)$

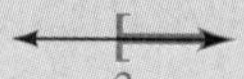

EXAMPLE 3 Solve the linear inequality $\frac{2}{3}(x + 2) > \frac{4}{5}(x - 3)$.

Solution We use the same steps as for solving equations.

$$\frac{2}{3}(x + 2) > \frac{4}{5}(x - 3)$$

$$\mathbf{15} \cdot \frac{2}{3}(x + 2) > \mathbf{15} \cdot \frac{4}{5}(x - 3) \qquad \text{Multiply both sides by 15, the LCD of 3 and 5.}$$

$$10(x + 2) > 12(x - 3) \qquad \text{Simplify.}$$

$$10x + 20 > 12x - 36 \qquad \text{Use the distributive property to remove parentheses.}$$

$$-2x + 20 > -36 \qquad \text{Subtract } 12x \text{ from both sides.}$$

$$-2x > -56 \qquad \text{Subtract 20 from both sides.}$$

$$x < 28 \qquad \text{Divide both sides by } -2 \text{ and reverse the } > \text{ symbol.}$$

FIGURE 4-3

The solution set is the interval $(-\infty, 28)$, whose graph is shown in Figure 4-3. ■

Self Check Solve $\frac{3}{2}(x + 2) < \frac{3}{5}(x - 3)$.

Answer $\left(-\infty, -\frac{16}{3}\right)$

■ COMPOUND INEQUALITIES

To say that x is between -3 and 8, we write the inequality

$$-3 < x < 8 \qquad \text{Read as “}-3 \text{ is less than } x \text{ and } x \text{ is less than 8.”}$$

This inequality is called a **compound inequality,** because it is a combination of two inequalities:

$$-3 < x \qquad \text{and} \qquad x < 8$$

The word *and* indicates that both inequalities are true at the same time.

Compound Inequalities
The compound inequality $c < x < d$ is equivalent to $c < x$ and $x < d$.

WARNING! The inequality $c < x < d$ means $c < x$ and $x < d$. It does not mean $c < x$ or $x < d$.

EXAMPLE 4 Solve the compound inequality $-3 \le 2x + 5 < 7$.

Solution This inequality means that $2x + 5$ is between -3 and 7. We can solve it by isolating x between the inequality symbols.

$$-3 \le 2x + 5 < 7$$

$$-8 \le 2x < 2 \qquad \text{Subtract 5 from all three parts.}$$

$$-4 \le x < 1 \qquad \text{Divide all three parts by 2.}$$

FIGURE 4-4

The solution set is the interval $[-4, 1)$, whose graph is shown in Figure 4-4. ■

Self Check Solve $-5 \le 3x - 8 \le 7$.

Answer $[1, 5]$

EXAMPLE 5 Solve the inequality $x + 3 < 2x - 1 < 4x - 3$.

Solution Here it is impossible to isolate x between the inequality symbols, so we solve each of the linear inequalities separately.

$$x + 3 < 2x - 1 \quad \text{and} \quad 2x - 1 < 4x - 3$$

$$4 < x \qquad\qquad 2 < 2x$$

$$x > 4 \qquad\qquad 1 < x$$

$$x > 1$$

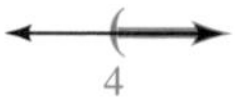

FIGURE 4-5

Only those x where $x > 4$ and $x > 1$ are in the solution set. Since all numbers greater than 4 are also greater than 1, the solutions are the numbers x where $x > 4$. Thus, the solution set is the interval $(4, \infty)$, whose graph is shown in Figure 4-5. ■

Self Check Solve $x + 2 < 3x + 1 < 5x + 3$.

Answer $\left(\frac{1}{2}, \infty\right)$

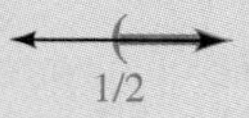

To say that x is less than -5 or greater than 10, we write the inequality

$$x < -5 \quad \text{or} \quad x > 10 \qquad \text{Read as “}x\text{ is less than }-5\text{ or }x\text{ is greater than 10.”}$$

This inequality is also a **compound inequality.** The word *or* indicates that only one of the inequalities needs to be true to make the entire statement true.

EXAMPLE 6 Solve the compound inequality $x \le -3$ or $x \ge 8$.

Solution The word *or* in the statement $x \le -3$ or $x \ge 8$ indicates that only one of the inequalities needs to be true to make the statement true. The graph of the inequality is shown in Figure 4-6. The solution set is $(-\infty, -3] \cup [8, \infty)$. ■

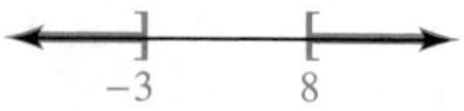

FIGURE 4-6

Self Check Solve $x < -5$ or $x \geq 3$.

Answer $(-\infty, -5) \cup [3, \infty)$

WARNING! In the statement $x \leq -3$ or $x \geq 8$, it is incorrect to string the inequalities together as $8 \leq x \leq -3$, because that would imply that $8 \leq -3$, which is false.

PROBLEM SOLVING

EXAMPLE 7 Suppose that a long-distance telephone call costs 36¢ for the first three minutes and 11¢ for each additional minute. For how many minutes can a person talk for less than \$2?

Analyze the problem We can let x represent the total number of minutes that the call can last. Then the cost of the call will be 36¢ for the first three minutes plus 11¢ times the number of additional minutes, where the number of additional minutes is $x - 3$ (the total number of minutes minus 3 minutes). The cost of the call is to be less than \$2.

Form an inequality With this information, we can form the inequality

The cost of the first three minutes	+	the cost of the additional minutes	<	\$2.
0.36	+	$0.11(x - 3)$	<	2

Solve the inequality We can solve the inequality as follows:

$$0.36 + 0.11(x - 3) < 2$$

$$36 + 11(x - 3) < 200$$ To eliminate the decimal points, multiply both sides by 100.

$$36 + 11x - 33 < 200$$ Use the distributive property to remove parentheses.

$$11x + 3 < 200$$ Combine like terms.

$$11x < 197$$ Subtract 3 from both sides.

$$x < 17.\overline{90}$$ Divide both sides by 11.

State the conclusion Since the phone company doesn't bill for part of a minute, the longest that the person can talk is 17 minutes.

Check the result If the call lasts 17 minutes, the customer will be billed \$0.36 + \$0.11(14) = \$1.90. If the call lasts 18 minutes, the customer will be billed \$0.36 + \$0.11(15) = \$2.01. ■

Solving Linear Inequalities

GRAPHING CALCULATORS

We can solve linear inequalities with a graphing approach. For example, to solve the inequality $3(2x - 9) < 9$, we can graph $y = 3(2x - 9)$ and $y = 9$ using window settings of $[-10, 10]$ for x and $[-10, 10]$ for y to get Figure 4-7(a). We can trace to see that the graph of $y = 3(2x - 9)$ is below the graph of $y = 9$ for x values in the interval $(-\infty, 6)$. See Figure 4-7(b). This interval is the solution, because in this interval, $3(2x - 9) < 9$.

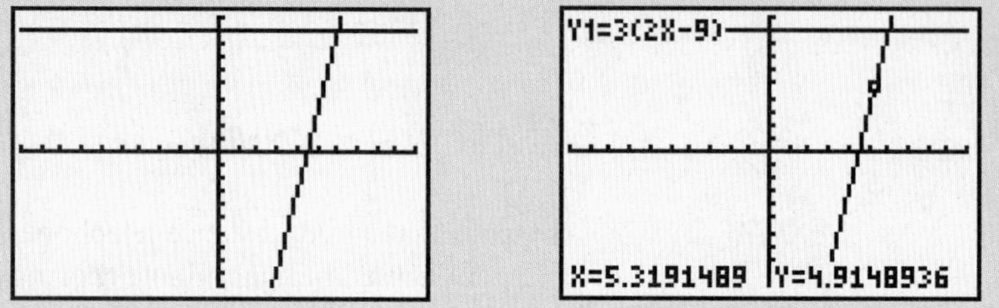

(a) (b)

FIGURE 4-7

FINDING A SAMPLE SIZE

In statistics, researchers often estimate the mean of a population from the results of a random sample taken from the population.

EXAMPLE 8 A researcher wants to estimate the mean (average) real estate tax paid by homeowners living in Rockford, Illinois. To do so, he decides to select a *random sample* of homeowners and compute the mean tax paid by the homeowners in that sample. How large must the sample be for the researcher to be 95% certain that his computed sample mean will be within \$35 of the true population mean—that is, within \$35 of the mean tax paid by all homeowners in the city? Assume that the standard deviation σ of all tax bills in the city is known to be \$120.

Solution From elementary statistics, the researcher has the formula

$$\frac{3.84\sigma^2}{N} < E^2$$

where σ^2, called the *population variance,* is the square of the standard deviation, E is the maximum acceptable error, and N is the sample size. The researcher substitutes 120 for σ and 35 for E in the previous formula and solves for N.

$$\frac{3.84(120)^2}{N} < 35^2$$

$$\frac{55{,}296}{N} < 1{,}225 \quad \text{Simplify.}$$

$$55{,}296 < 1{,}225N \quad \text{Multiply both sides by } N.$$

$$45.13959184 < N \quad \text{Divide both sides by 1,225.}$$

To be 95% certain that the sample mean will be within \$35 of the true population mean, the researcher must sample more than 45.13959184 homeowners. Thus, the sample must contain at least 46 homeowners. ■

Orals *Solve each inequality.*

1. $2x < 4$

2. $3x + 1 \geq 10$

3. $-3x > 12$

4. $-\frac{x}{2} \leq 4$

5. $-2 < 2x < 8$

6. $3 \leq \frac{x}{3} \leq 4$

EXERCISE 4.1

REVIEW *Simplify each expression.*

1. $\left(\frac{t^3t^5t^{-6}}{t^2t^{-4}}\right)^{-3}$

2. $\left(\frac{a^{-2}b^3a^5b^{-2}}{a^6b^{-5}}\right)^{-4}$

3. A man invests \$1,200 in baking equipment to make pies. Each pie requires \$3.40 in ingredients. If the man can sell all the pies he can make for \$5.95 each, how many pies will he have to make to earn a profit?

4. A woman invested \$15,000, part at 7% annual interest and the rest at 8%. If she earned \$2,200 in income over a two-year period, how much did she invest at 7%?

VOCABULARY AND CONCEPTS *Fill in each blank to make a true statement.*

5. The symbol for "is not equal to" is ___.

6. The symbol for "is greater than" is ___.

7. The symbol for "is less than" is ___.

8. The symbol for "is less than or equal to" is ___.

9. The symbol for "is greater than or equal to" is ___.

10. If a and b are two numbers, then $a < b$, ______, or ______.

11. If $a < b$ and $b < c$, then ______.

12. If both sides of an inequality are multiplied by a _______ number, the direction of the inequality remains the same.

13. If both sides of an inequality are divided by a negative number, the direction of the inequality symbol must be _______.

14. $3x + 2 > 7$ is an example of a ______ inequality.

15. The inequality $c < x < d$ is equivalent to ______ and ______.

16. The word "or" between two inequality statements indicates that only ____ of the inequalities needs to be true for the entire statement to be true.

PRACTICE *In Exercises 17–30, solve each inequality. Give the result in interval notation and graph the solution set.*

17. $x + 4 < 5$

18. $x - 5 > 2$

19. $-3x - 1 \le 5$

20. $-2x + 6 \ge 16$

21. $5x - 3 > 7$

22. $7x - 9 < 5$

23. $3(z - 2) \le 2(z + 7)$

24. $5(3 + z) > -3(z + 3)$

25. $-11(2 - b) < 4(2b + 2)$

26. $-9(h - 3) + 2h \le 8(4 - h)$

27. $\frac{1}{2}y + 2 \ge \frac{1}{3}y - 4$

28. $\frac{1}{4}x - \frac{1}{3} \le x + 2$

29. $\frac{2}{3}x + \frac{3}{2}(x - 5) \le x$

30. $\frac{5}{9}(x + 3) - \frac{4}{3}(x - 3) \ge x - 1$

In Exercises 31–52, solve each compound inequality. Give the result in interval notation and graph the solution set.

31. $-2 < -b + 3 < 5$

32. $2 < -t - 2 < 9$

33. $15 > 2x - 7 > 9$

34. $25 > 3x - 2 > 7$

35. $-6 < -3(x - 4) \le 24$

36. $-4 \le -2(x + 8) < 8$

37. $0 \ge \frac{1}{2}x - 4 > 6$

38. $-6 \le \frac{1}{3}a + 1 < 0$

39. $0 \le \frac{4 - x}{3} \le 2$

40. $-2 \le \frac{5 - 3x}{2} \le 2$

41. $x + 3 < 3x - 1 < 2x + 2$

42. $x - 1 \le 2x + 4 \le 3x - 1$

43. $4x \ge -x + 5 \ge 3x - 4$

44. $x + 2 < -\frac{1}{3}x < \frac{1}{2}x$

45. $5(x + 1) \le 4(x + 3) < 3(x - 1)$

46. $-5(2 + x) < 4x + 1 < 3x$

47. $3x + 2 < 8$ or $2x - 3 > 11$

48. $3x + 4 < -2$ or $3x + 4 > 10$

49. $-4(x + 2) \geq 12$ or $3x + 8 < 11$

50. $5(x - 2) \geq 0$ and $-3x < 9$

51. $x < -3$ and $x > 3$

52. $x < 3$ or $x > -3$

53. If $x > -3$, must it be true that $x^2 > 9$?

54. If $x > 2$, must it be true that $x^2 > 4$?

APPLICATIONS *Use a calculator to help solve each problem.*

55. Renting a rototiller The cost of renting a rototiller is \$15.50 for the first hour and \$7.95 for each additional hour. How long can a person have the rototiller if the cost is to be less than \$50?

56. Renting a truck The cost of renting a truck is \$29.95 for the first hour and \$8.95 for each additional hour. How long can a person have the truck if the cost is to be less than \$110?

57. Investing money If a woman invests \$10,000 at 8% annual interest, how much more must she invest at 9% so that her annual income will exceed \$1,250?

58. Investing money If a man invests \$8,900 at 5.5% annual interest, how much more must he invest at 8.75% so that his annual income will be more than \$1,500?

59. Buying compact discs A student can afford to spend up to \$330 on a stereo system and some compact discs. If the stereo costs \$175 and the discs are \$8.50 each, find the greatest number of discs he can buy.

60. Buying a computer A student who can afford to spend up to \$2,000 sees the ad shown in Illustration 1. If she buys a computer, find the greatest number of CD-ROMs that she can buy.

ILLUSTRATION 1

61. Averaging grades A student has scores of 70, 77, and 85 on three exams. What score is needed on a fourth exam to make an average of 80 or better?

62. Averaging grades A student has scores of 70, 79, 85, and 88 on four exams. What score does she need on the fifth exam to keep her average above 80?

63. Planning a work schedule Tom can earn \$5 an hour for working at the college library and \$9 an hour for construction work. To save time for study, he wants to limit his work to 20 hours a week but still earn more than \$125. How many hours can he work at the library?

64. Scheduling equipment An excavating company charges \$300 an hour for the use of a backhoe and \$500 an hour for the use of a bulldozer. (Part of an hour counts as a full hour.) The company employs one operator for 40 hours per week. If the company wants to take in at least \$18,500 each week, how many hours per week can it schedule the operator to use a backhoe?

65. Choosing a medical plan A college provides its employees with a choice of the two medical plans shown in Illustration 2. For what size hospital bills is Plan 2 better than Plan 1? (*Hint:* The cost to the employee includes both the deductible payment and the employee's insurance copayment.)

Plan 1	Plan 2
Employee pays \$100 Plan pays 70% of the rest	Employee pays \$200 Plan pays 80% of the rest

ILLUSTRATION 2

66. Choosing a medical plan To save costs, the college in Exercise 65 raised the employee deductible, as shown in Illustration 3. For what size hospital bills is Plan 2 better than Plan 1? (*Hint:* The cost to the employee includes both the deductible payment and the employee's insurance copayment.)

Plan 1	Plan 2
Employee pays \$200 Plan pays 70% of the rest	Employee pays \$400 Plan pays 80% of the rest

ILLUSTRATION 3

In Exercises 67–70, use a graphing calculator to solve each inequality.

67. $2x + 3 < 5$

68. $3x - 2 > 4$

69. $5x + 2 \geq -18$

70. $3x - 4 \leq 20$

In Exercises 71–72, solve each problem.

71. Choosing sample size How large would the sample have to be for the researcher in Example 8 to be 95% certain that the true population mean would be within \$20 of the sample mean?

72. Choosing sample size How large would the sample have to be for the researcher in Example 8 to be 95% certain that the true population mean would be within \$10 of the sample mean?

WRITING

73. The techniques for solving linear equations and linear inequalities are similar, yet different. Explain.

74. Explain the concepts of *absolute* inequality and *conditional* inequality.

SOMETHING TO THINK ABOUT

75. Which of these relations is transitive?
a. $=$ **b.** $\leq$ **c.** $\not\geq$ **d.** $\neq$

76. Solve $\frac{1}{3} > \frac{1}{x}$. The following solution is not correct. Why?

$$\frac{1}{3} > \frac{1}{x}$$

$$3x\left(\frac{1}{3}\right) > 3x\left(\frac{1}{x}\right) \qquad \text{Multiply both sides by } 3x.$$

$$x > 3 \qquad \text{Simplify.}$$

4.2 Equations and Inequalities with Absolute Values

■ ABSOLUTE VALUE ■ ABSOLUTE VALUE FUNCTIONS ■ EQUATIONS OF THE FORM $|x| = k$
■ EQUATIONS WITH TWO ABSOLUTE VALUES ■ INEQUALITIES OF THE FORM $|x| < k$
■ INEQUALITIES OF THE FORM $|x| > k$

Getting Ready *Find each value. Assume that $x = -5$.*

1. $-(-5)$ **2.** -0 **3.** $-(x - 5)$ **4.** $-(-x)$

Solve each inequality and give the result in interval notation.

5. $-4 < x + 1 < 5$

6. $x < -3$ or $x > 3$

ABSOLUTE VALUE

Recall the following definition of the absolute value of x.

> **Absolute Value**
> If $x \geq 0$, then $|x| = x$.
> If $x < 0$, then $|x| = -x$.

This definition gives a way to associate a nonnegative real number with any real number.

- If $x \geq 0$, then x (which is positive or 0) is its own absolute value.
- If $x < 0$, then $-x$ (which is positive) is the absolute value.

Either way, $|x|$ is positive or 0:

$$|x| \geq 0 \quad \text{for all real numbers } x$$

EXAMPLE 1 Find **a.** $|9|$, **b.** $|-5|$, **c.** $|0|$, and **d.** $|2 - \pi|$.

Solution **a.** Since $9 \geq 0$, 9 is its own absolute value: $|9| = 9$.

b. Since $-5 < 0$, the negative of -5 is the absolute value:

$$|-5| = -(-5) = 5$$

c. Since $0 \geq 0$, 0 is its own absolute value: $|0| = 0$.

d. Since $\pi \approx 3.14$, it follows that $2 - \pi < 0$. Thus,

$$|2 - \pi| = -(2 - \pi) = \pi - 2$$

■

Self Check Find **a.** $|-3|$ and **b.** $|\pi - 2|$.

Answers **a.** 3, **b.** $\pi - 2$

WARNING! The placement of a $-$ sign in an expression containing an absolute value symbol is important. For example, $|-19| = 19$, but $-|19| = -19$.

ABSOLUTE VALUE FUNCTIONS

In Section 2.6, we graphed the absolute value function $y = f(x) = |x|$ and considered many translations of its graph. The work in Example 2 reviews these concepts.

EXAMPLE 2 Graph the absolute value function $y = f(x) = |x - 1| + 3$.

Solution We graph this function by translating the graph of $y = f(x) = |x|$ 1 unit to the right and 3 units up, as shown in Figure 4-8.

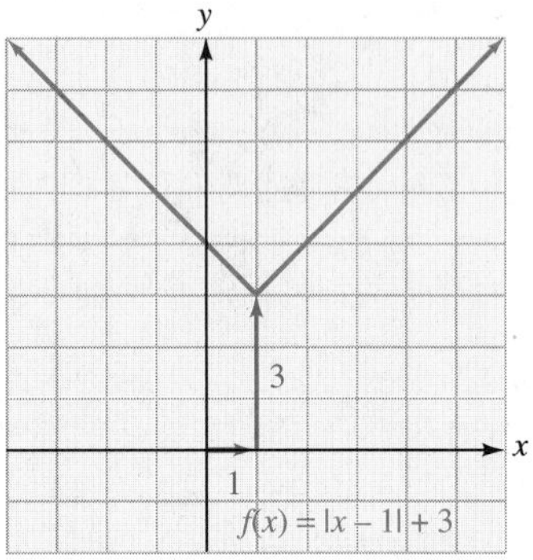

FIGURE 4-8 ■

Self Check Graph $y = f(x) = |x + 2| - 3$.

Answer

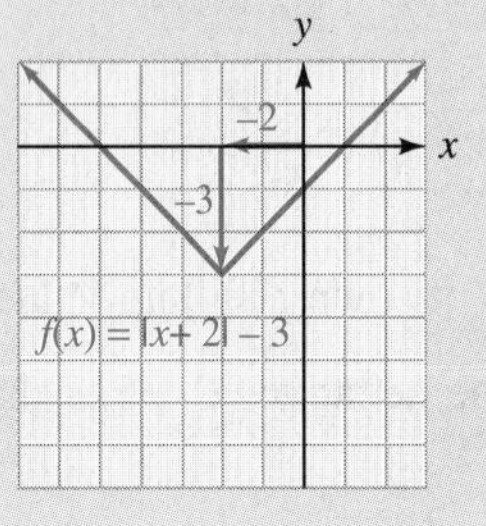

We also saw that the graph of $y = f(x) = -|x|$ is the same as the graph of $y = f(x) = |x|$ reflected about the x-axis, as shown in Figure 4-9.

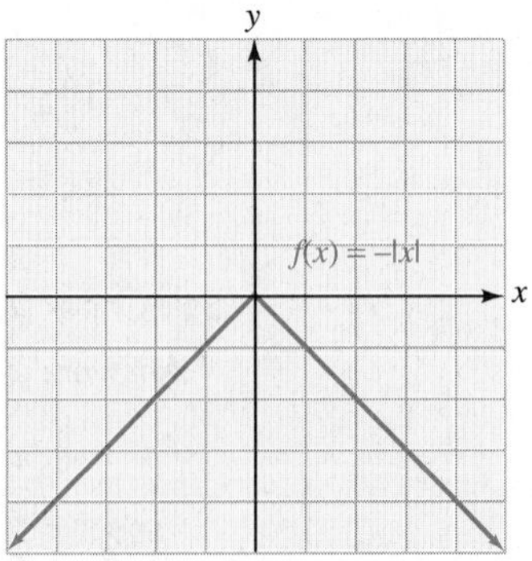

FIGURE 4-9

EXAMPLE 3 Graph the absolute value function $y = f(x) = -|x - 1| + 3$.

Solution We graph this function by translating the graph of $y = f(x) = -|x|$ 1 unit to the right and 3 units up, as shown in Figure 4-10.

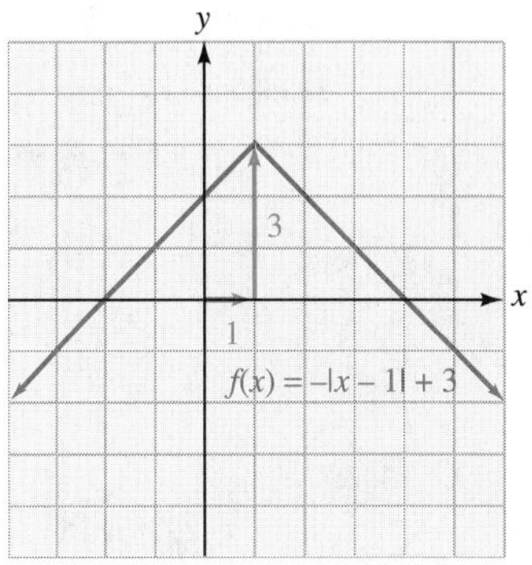

FIGURE 4-10

Self Check Graph $y = f(x) = -|x + 2| - 3$.

Answer

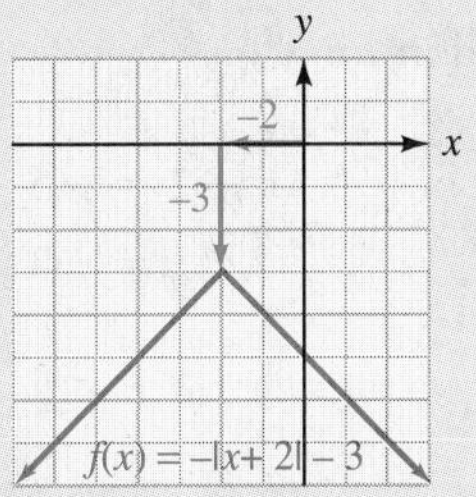

EQUATIONS OF THE FORM $|x| = k$

In the equation $|x| = 5$, x can be either 5 or -5, because $|5| = 5$ and $|-5| = 5$. In the equation $|x| = 8$, x can be either 8 or -8. In general, the following is true.

Absolute Value Equations

If $k > 0$, then

$|x| = k$ is equivalent to $x = k$ or $x = -k$

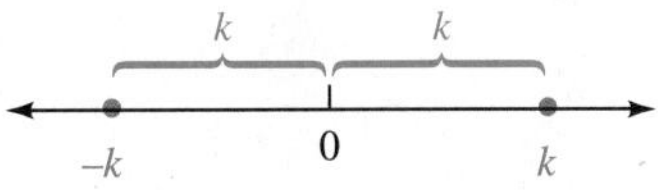

FIGURE 4-11

The absolute value of a number represents the distance on the number line from a point to the origin. The solutions of $|x| = k$ are the coordinates of the two points that lie exactly k units from the origin. See Figure 4-11.

The equation $|x - 3| = 7$ indicates that a point on the number line with a coordinate of $x - 3$ is 7 units from the origin. Thus, $x - 3$ can be either 7 or -7.

$$x - 3 = 7 \quad \text{or} \quad x - 3 = -7$$
$$x = 10 \qquad\qquad x = -4$$

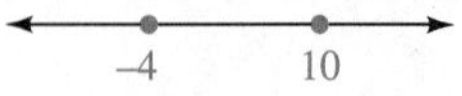

FIGURE 4-12

The solutions of the equation $|x - 3| = 7$ are 10 and -4, as shown in Figure 4-12. If either of these numbers is substituted for x in $|x - 3| = 7$, the equation is satisfied:

$$\begin{aligned} |x - 3| &= 7 \\ |\mathbf{10} - 3| &\stackrel{?}{=} 7 \\ |7| &\stackrel{?}{=} 7 \\ 7 &= 7 \end{aligned} \qquad \begin{aligned} |x - 3| &= 7 \\ |\mathbf{-4} - 3| &\stackrel{?}{=} 7 \\ |-7| &\stackrel{?}{=} 7 \\ 7 &= 7 \end{aligned}$$

EXAMPLE 4 Solve the equation $|3x - 2| = 5$.

Solution We can write $|3x - 2| = 5$ as

$$3x - 2 = 5 \quad \text{or} \quad 3x - 2 = -5$$

and solve each equation for x:

$$\begin{aligned} 3x - 2 &= 5 \\ 3x &= 7 \\ x &= \frac{7}{3} \end{aligned} \quad \text{or} \quad \begin{aligned} 3x - 2 &= -5 \\ 3x &= -3 \\ x &= -1 \end{aligned}$$

Verify that both solutions check. ■

Self Check Solve $|2x - 3| = 7$.

Answer 5, -2

EXAMPLE 5 Solve the equation $\left|\frac{2}{3}x + 3\right| + 4 = 10$.

Solution We first subtract 4 from both sides to isolate the absolute value on the left-hand side.

$$\left|\frac{2}{3}x + 3\right| + 4 = 10$$

$$\textbf{1.} \quad \left|\frac{2}{3}x + 3\right| = 6 \qquad \text{Subtract 4 from both sides.}$$

We can now write Equation 1 as

$$\frac{2}{3}x + 3 = 6 \quad \text{or} \quad \frac{2}{3}x + 3 = -6$$

and solve each equation for x:

$$\frac{2}{3}x + 3 = 6 \quad \text{or} \quad \frac{2}{3}x + 3 = -6$$

$$\frac{2}{3}x = 3 \qquad\qquad \frac{2}{3}x = -9$$

$$2x = 9 \qquad\qquad 2x = -27$$

$$x = \frac{9}{2} \qquad\qquad x = -\frac{27}{2}$$

Verify that both solutions check. ■

Self Check Solve $\left|\frac{3}{5}x - 2\right| - 3 = 4$.

Answer $15, -\frac{25}{3}$

WARNING! Since the absolute value of a quantity cannot be negative, equations such as $|7x + \frac{1}{2}| = -4$ have no solution. Since there are no solutions, their solution sets are empty. Recall that an **empty set** is denoted by the symbol $\varnothing$.

EXAMPLE 6 Solve the equation $|\frac{1}{2}x - 5| - 4 = -4$.

Solution We first isolate the absolute value on the left-hand side.

$$\left|\frac{1}{2}x - 5\right| - 4 = -4$$

$$\left|\frac{1}{2}x - 5\right| = 0 \qquad \text{Add 4 to both sides.}$$

Since 0 is the only number whose absolute value is 0, the binomial $\frac{1}{2}x - 5$ must be 0, and we have

$$\frac{1}{2}x - 5 = 0$$

$$\frac{1}{2}x = 5 \qquad \text{Add 5 to both sides.}$$

$$x = 10 \qquad \text{Multiply both sides by 2.}$$

Verify that 10 satisfies the original equation. ■

Self Check Solve $|\frac{2}{3}x + 4| + 4 = 5$.

Answer $-\frac{15}{2}, -\frac{9}{2}$

EQUATIONS WITH TWO ABSOLUTE VALUES

The equation $|a| = |b|$ is true when $a = b$ or when $a = -b$. For example,

$$|3| = |3| \quad \text{or} \quad |3| = |-3|$$
$$3 = 3 \qquad\qquad 3 = 3$$

In general, the following statement is true.

Equations with Two Absolute Values
If a and b represent algebraic expressions, the equation $|a| = |b|$ is equivalent to the pair of equations

$$a = b \quad \text{or} \quad a = -b$$

EXAMPLE 7 Solve the equation $|5x + 3| = |3x + 25|$.

Solution This equation is true when $5x + 3 = 3x + 25$, or when $5x + 3 = -(3x + 25)$. We solve each equation for x.

$$\begin{aligned} 5x + 3 &= 3x + 25 \\ 2x &= 22 \\ x &= 11 \end{aligned} \quad \text{or} \quad \begin{aligned} 5x + 3 &= -(3x + 25) \\ 5x + 3 &= -3x - 25 \\ 8x &= -28 \\ x &= -\frac{28}{8} \\ x &= -\frac{7}{2} \end{aligned}$$

Verify that both solutions check. ■

Self Check Solve $|2x - 3| = |4x + 9|$.

Answer $-1, -6$

INEQUALITIES OF THE FORM $|x| < k$

FIGURE 4-13

The inequality $|x| < 5$ indicates that a point with a coordinate of x is less than 5 units from the origin. (See Figure 4-13.) Thus, x is between -5 and 5, and

$$|x| < 5 \quad \text{is equivalent to} \quad -5 < x < 5$$

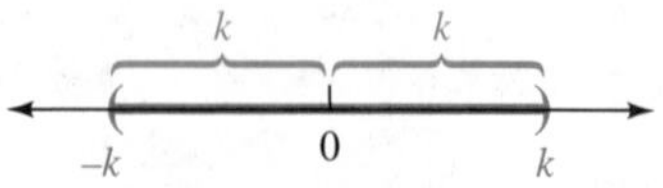

FIGURE 4-14

The solution set of the inequality $|x| < k$ $(k > 0)$ includes the coordinates of the points on the number line that are less than k units from the origin. (See Figure 4-14.)

Solving $|x| < k$ and $|x| \le k$

If $k > 0$, then

$|x| < k$ is equivalent to $-k < x < k$

$|x| \le k$ is equivalent to $-k \le x \le k \quad (k \ge 0)$

EXAMPLE 8 Solve the inequality $|2x - 3| < 9$.

Solution We write the inequality as a compound inequality and solve for x.

$|2x - 3| < 9$ is equivalent to $-9 < 2x - 3 < 9$

$-9 < 2x - 3 < 9$

$-6 < 2x < 12$ Add 3 to all three parts.

$-3 < x < 6$ Divide all parts by 2.

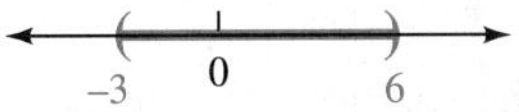

FIGURE 4-15

Any number between -3 and 6, not including either -3 or 6, is in the solution set. This is the interval $(-3, 6)$, whose graph is shown in Figure 4-15. ■

Self Check Solve $|3x + 2| < 4$.

Answer $\left(-2, \frac{2}{3}\right)$

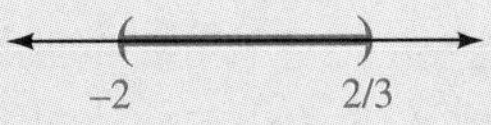

EXAMPLE 9 Solve the inequality $|2 - 3x| \le 5$.

Solution We write the expression as a compound inequality and solve for x:

$|2 - 3x| \le 5$ is equivalent to $-5 \le 2 - 3x \le 5$

$-5 \le 2 - 3x \le 5$

$-7 \le -3x \le 3$ Subtract 2 from all parts.

$\frac{7}{3} \ge x \ge -1$ Divide all three parts by -3 and reverse the directions of the inequality symbols.

$-1 \le x \le \frac{7}{3}$ Write the inequality with the inequality symbols pointing in the opposite direction.

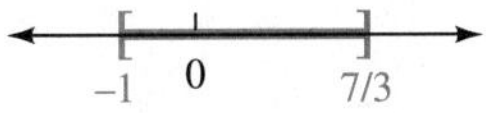

FIGURE 4-16

The solution set is the interval $\left[-1, \frac{7}{3}\right]$, whose graph is shown in Figure 4-16. ■

Self Check Solution $|3 - 2x| \le 5$.

Answer $[-1, 4]$

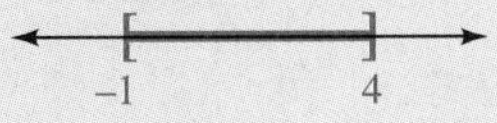

INEQUALITIES OF THE FORM $|x| > k$

The inequality $|x| > 5$ indicates that a point with a coordinate of x is more than 5 units from the origin. See Figure 4-17. Thus, $x < -5$ or $x > 5$.

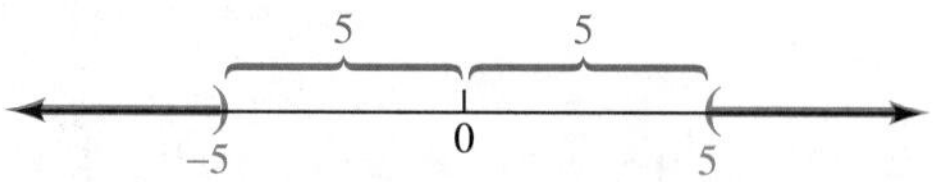

FIGURE 4-17

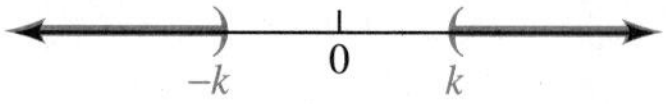

FIGURE 4-18

In general, the inequality $|x| > k$ can be interpreted to mean that a point with coordinate x is more than k units from the origin. (See Figure 4-18.)

Thus,

$|x| > k$ is equivalent to $x < -k$ or $x > k$

The *or* indicates an either/or situation. To be in the solution set, x needs to satisfy only one of the two conditions.

Solving $|x| > k$ and $|x| \geq k$

If $k \geq 0$, then

$|x| > k$ is equivalent to $x < -k$ or $x > k$

$|x| \geq k$ is equivalent to $x \leq -k$ or $x \geq k$

EXAMPLE 10 Solve the inequality $|5x - 10| > 20$.

Solution We write the inequality as two separate inequalities and solve each one for x.

$|5x - 10| > 20$ is equivalent to $5x - 10 < -20$ or $5x - 10 > 20$

$5x - 10 < -20$	or	$5x - 10 > 20$	
$5x < -10$		$5x > 30$	Add 10 to both sides.
$x < -2$		$x > 6$	Divide both sides by 5.

Thus, x is either less than -2 or greater than 6:

$x < -2$ or $x > 6$

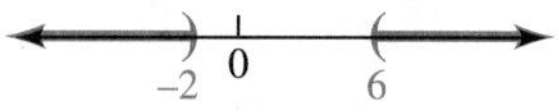

FIGURE 4-19

This is the interval $(-\infty, -2) \cup (6, \infty)$, whose graph appears in Figure 4-19. ■

Self Check Solve $|3x + 4| > 13$.

Answer $\left(-\infty, -\frac{17}{3}\right) \cup (3, \infty)$

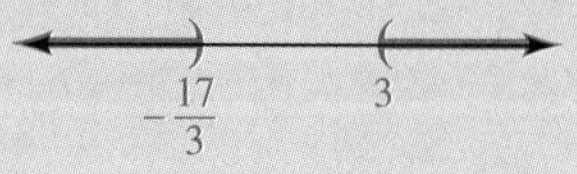

EXAMPLE 11 Solve the inequality $\left|\frac{3-x}{5}\right| \geq 6$.

Solution We write the inequality as two separate inequalities:

$$\left|\frac{3-x}{5}\right| \geq 6 \quad \text{is equivalent to} \quad \frac{3-x}{5} \leq -6 \quad \text{or} \quad \frac{3-x}{5} \geq 6$$

Then we solve each one for x:

$$\frac{3-x}{5} \leq -6 \qquad \text{or} \qquad \frac{3-x}{5} \geq 6$$

$$3 - x \leq -30 \qquad\qquad 3 - x \geq 30 \qquad \text{Multiply both sides by 5.}$$

$$-x \leq -33 \qquad\qquad -x \geq 27 \qquad \text{Subtract 3 from both sides.}$$

$$x \geq 33 \qquad\qquad x \leq -27 \qquad \text{Divide both sides by } -1 \text{ and reverse the direction of the inequality symbol.}$$

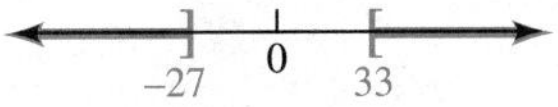

FIGURE 4-20

The solution set is the interval $(-\infty, -27] \cup [33, \infty)$, whose graph appears in Figure 4-20. ■

Self Check Solve $\left|\frac{3-x}{6}\right| \geq 5$.

Answer $(-\infty, -27] \cup [33, \infty)$

EXAMPLE 12 Solve the inequality $\left|\frac{2}{3}x - 2\right| - 3 > 6$.

Solution We begin by adding 3 to both sides to isolate the absolute value on the left-hand side. We then proceed as follows:

$$\left|\frac{2}{3}x - 2\right| - 3 > 6$$

$$\left|\frac{2}{3}x - 2\right| > 9 \qquad \text{Add 3 to both sides to isolate the absolute value.}$$

$$\frac{2}{3}x - 2 < -9 \qquad \text{or} \qquad \frac{2}{3}x - 2 > 9$$

$$\frac{2}{3}x < -7 \qquad\qquad \frac{2}{3}x > 11 \qquad \text{Add 2 to both sides.}$$

$$2x < -21 \qquad\qquad 2x > 33 \qquad \text{Multiply both sides by 3.}$$

$$x < -\frac{21}{2} \qquad\qquad x > \frac{33}{2} \qquad \text{Divide both sides by 2.}$$

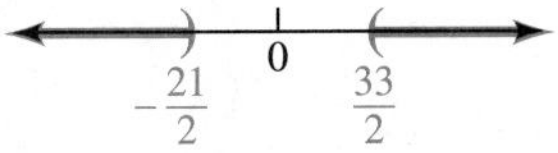

FIGURE 4-21

The solution set is $\left(-\infty, -\frac{21}{2}\right) \cup \left(\frac{33}{2}, \infty\right)$, whose graph appears in Figure 4-21. ■

Self Check Solve $\left|\frac{3}{4}x + 2\right| - 1 > 3.$

Answer $(-\infty, -8) \cup \left(\frac{8}{3}, \infty\right)$

EXAMPLE 13 Solve the inequality $|3x - 5| \geq -2$.

Solution Since the absolute value of any number is nonnegative, and since any nonnegative number is greater than -2, the inequality is true for all x. The solution set is the interval $(-\infty, \infty)$, whose graph appears in Figure 4-22. ■

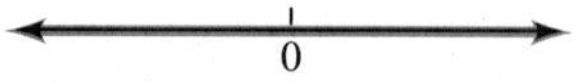

FIGURE 4-22

Self Check Solve $|3x - 5| \leq -2$, if possible.

Answer no solution

Solving Absolute Value Inequalities

GRAPHING CALCULATORS

We can also solve absolute value inequalities by a graphing method. For example, to solve the inequality $|2x - 3| < 9$, we graph the equations $y = |2x - 3|$ and $y = 9$ on the same coordinate system. If we use window settings of $[-5, 15]$ for x and $[-5, 15]$ for y, we get the graph shown in Figure 4-23.

The inequality $|2x - 3| < 9$ will be true for all x-coordinates of points that lie on the graph of $y = |2x - 3|$ and below the graph of $y = 9$. By using the trace feature, we can see that these values of x are in the interval $(-3, 6)$.

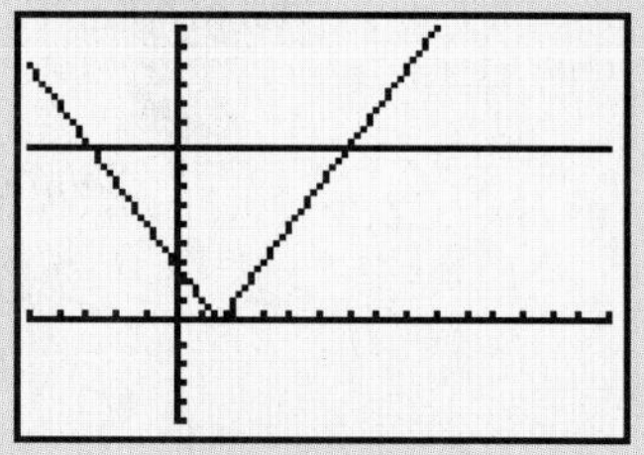
FIGURE 4-23

Orals *Find each absolute value.*

1. $|-5|$ **2.** $-|5|$ **3.** $-|-6|$ **4.** $-|4|$

Solve each equation or inequality.

5. $|x| = 8$ **6.** $|x| = -5$

7. $|x| < 8$ **8.** $|x| > 8$

9. $|x| \geq 4$ **10.** $|x| \leq 7$

EXERCISE 4.2

REVIEW *Solve each equation or formula.*

1. $3(2a - 1) = 2a$

2. $\frac{t}{6} - \frac{t}{3} = -1$

3. $\frac{5x}{2} - 1 = \frac{x}{3} + 12$

4. $4b - \frac{b + 9}{2} = \frac{b + 2}{5} - \frac{8}{5}$

5. $A = p + prt$ for t

6. $P = 2w + 2l$ for l

VOCABULARY AND CONCEPTS *Fill in each blank to make a true statement.*

7. If $x \geq 0$, $|x| =$ ___.

8. If $x < 0$, $|x| =$ ____.

9. $|x| \geq$ ___ for all real numbers x.

10. The graph of the function $y = f(x) = |x - 2| - 4$ is the same as the graph of $y = f(x) = |x|$ except that it has been translated 2 units to the ______ and 4 units ______.

11. The graph of $y = f(x) = -|x|$ is the same as the graph of $y = f(x) = |x|$ except that it has been ________ about the x-axis.

12. If $k > 0$, then $|x| = k$ is equivalent to ______________.

13. $|a| = |b|$ is equivalent to ______________.

14. If $k > 0$, then $|x| < k$ is equivalent to ___________.

15. If $k \geq 0$, then $|x| \geq k$ is equivalent to ______________.

16. The equation $|x - 4| < -5$ has ___ solutions.

PRACTICE *In Exercises 17–24, find the value of each expression.*

17. $|8|$

18. $|-18|$

19. $-|2|$

20. $-|-20|$

21. $-|-30|$

22. $-|25|$

23. $|\pi - 4|$

24. $|2\pi - 4|$

In Exercises 25–28, graph each absolute value function.

25. $f(x) = |x| - 2$

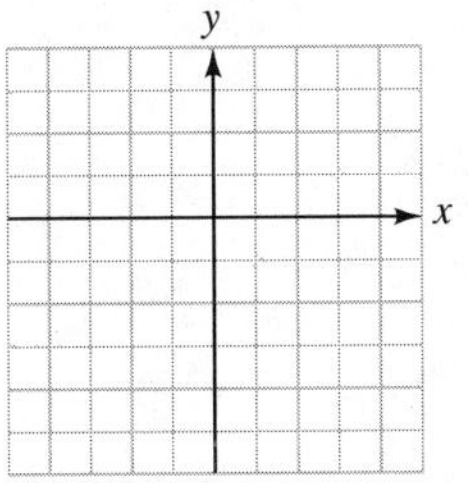

26. $f(x) = -|x| + 1$

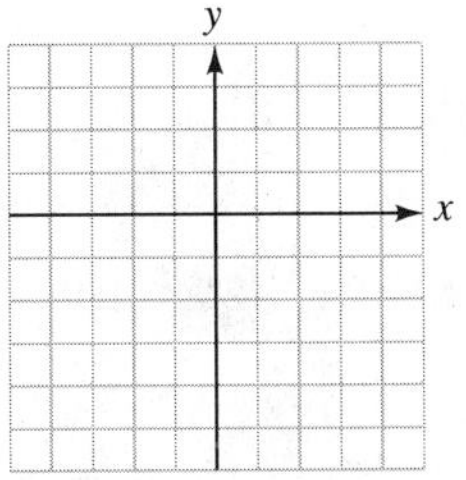

27. $f(x) = -|x + 4|$

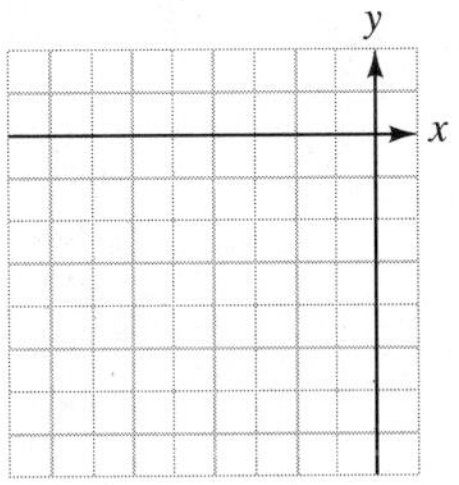

28. $f(x) = |x - 1| + 2$

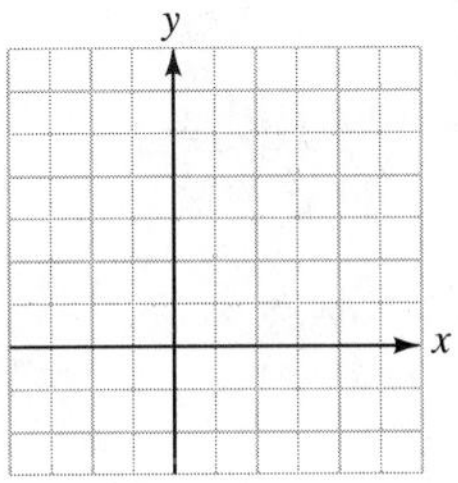

In Exercises 29–50, solve each equation, if possible.

29. $|x| = 4$

30. $|x| = 9$

31. $|x - 3| = 6$

32. $|x + 4| = 8$

33. $|2x - 3| = 5$

34. $|4x - 4| = 20$

35. $|3x + 2| = 16$

36. $|5x - 3| = 22$

37. $\left|\frac{7}{2}x + 3\right| = -5$

38. $|2x + 10| = 0$

39. $\left|\frac{x}{2} - 1\right| = 3$

40. $\left|\frac{4x - 64}{4}\right| = 32$

41. $|3 - 4x| = 5$

42. $|8 - 5x| = 18$

43. $|3x + 24| = 0$

44. $|x - 21| = -8$

45. $\left|\frac{3x + 48}{3}\right| = 12$

46. $\left|\frac{x}{2} + 2\right| = 4$

47. $|x + 3| + 7 = 10$

48. $|2 - x| + 3 = 5$

49. $\left|\frac{3}{5}x - 4\right| - 2 = -2$

50. $\left|\frac{3}{4}x + 2\right| + 4 = 4$

In Exercises 51–62, solve each equation, if possible.

51. $|2x + 1| = |3x + 3|$

52. $|5x - 7| = |4x + 1|$

53. $|3x - 1| = |x + 5|$

54. $|3x + 1| = |x - 5|$

55. $|2 - x| = |3x + 2|$

56. $|4x + 3| = |9 - 2x|$

57. $\left|\frac{x}{2} + 2\right| = \left|\frac{x}{2} - 2\right|$

58. $|7x + 12| = |x - 6|$

59. $\left|x + \frac{1}{3}\right| = |x - 3|$

60. $\left|x - \frac{1}{4}\right| = |x + 4|$

61. $|3x + 7| = -|8x - 2|$

62. $-|17x + 13| = |3x - 14|$

In Exercises 63–102, solve each inequality. Write the solution set in interval notation and graph it.

63. $|2x| < 8$

64. $|3x| < 27$

65. $|x + 9| \le 12$

66. $|x - 8| \le 12$

67. $|3x + 2| \le -3$

68. $|3x - 2| < 10$

69. $|4x - 1| \le 7$

70. $|5x - 12| < -5$

71. $|3 - 2x| < 7$

72. $|4 - 3x| \le 13$

73. $|5x| > 5$

74. $|7x| > 7$

75. $|x - 12| > 24$

76. $|x + 5| \geq 7$

77. $|3x + 2| > 14$

78. $|2x - 5| > 25$

79. $|4x + 3| > -5$

80. $|4x + 3| > 0$

81. $|2 - 3x| \geq 8$

82. $|-1 - 2x| > 5$

83. $-|2x - 3| < -7$

84. $-|3x + 1| < -8$

85. $|8x - 3| > 0$

86. $|7x + 2| > -8$

87. $\left|\dfrac{x - 2}{3}\right| \leq 4$

88. $\left|\dfrac{x - 2}{3}\right| > 4$

89. $|3x + 1| + 2 < 6$

90. $|3x - 2| + 2 \geq 0$

91. $3|2x + 5| \geq 9$

92. $-2|3x - 4| < 16$

93. $|5x - 1| + 4 \leq 0$

94. $-|5x - 1| + 2 < 0$

95. $\left|\dfrac{1}{3}x + 7\right| + 5 > 6$

96. $\left|\dfrac{1}{2}x - 3\right| - 4 < 2$

97. $\left|\dfrac{1}{5}x - 5\right| + 4 > 4$

98. $\left|\dfrac{1}{6}x + 6\right| + 2 < 2$

99. $\left|\dfrac{1}{7}x + 1\right| \leq 0$

100. $|2x + 1| + 2 \leq 2$

101. $\left|\dfrac{x - 5}{10}\right| \leq 0$

102. $\left|\dfrac{3}{5}x - 2\right| + 3 \leq 3$

In Exercises 103–106, write each compound inequality as an inequality using absolute values.

103. $-4 < x < 4$

104. $x < -4$ or $x > 4$

105. $x + 3 < -6$ or $x + 3 > 6$

106. $-5 \leq x - 3 \leq 5$

APPLICATIONS

107. Finding temperature ranges The temperatures on a summer day satisfied the inequality $|t - 78°| \le 8°$, where t is a temperature in degrees Fahrenheit. Express the range of temperatures as a compound inequality.

108. Finding operating temperatures A car CD player has an operating temperature of $|t - 40°| < 80°$, where t is a temperature in degrees Fahrenheit. Express this range of temperatures as a compound inequality.

109. Range of camber angles The specifications for a car state that the camber angle c of its wheels should be $0.6° \pm 0.5°$. Express this range with an inequality containing absolute value symbols.

110. Tolerance of sheet steel A sheet of steel is to be 0.25 inch thick with a tolerance of 0.015 inch. Express this specification with an inequality containing absolute value symbols.

WRITING

111. Explain how to find the absolute value of a given number.

112. Explain why the equation $|x| + 5 = 0$ has no solution.

113. Explain the use of parentheses and brackets when graphing inequalities.

114. If $k > 0$, explain the differences between the solution sets of $|x| < k$ and $|x| > k$.

SOMETHING TO THINK ABOUT

115. For what values of k does $|x| + k = 0$ have exactly two solutions?

116. For what value of k does $|x| + k = 0$ have exactly one solution?

117. Under what conditions is $|x| + |y| > |x + y|$?

118. Under what conditions is $|x| + |y| = |x + y|$?

4.3 Linear Inequalities in Two Variables

■ GRAPHING LINEAR INEQUALITIES ■ GRAPHING COMPOUND INEQUALITIES ■ PROBLEM SOLVING

Getting Ready *Do the coordinates of the points satisfy the equation* $y = 5x + 2$?

1. (0, 2) **2.** $\left(-\frac{2}{5}, 0\right)$ **3.** (3, 18) **4.** (−3, −13)

Do the coordinates of the points satisfy the equation $3x - 2y = 12$?

5. (0, −6) **6.** (−4, 0) **7.** (−2, 9) **8.** (−4, −12)

The **graph of a linear inequality** in x and y is the graph of all ordered pairs (x, y) that satisfy the inequality.

Linear Inequalities

A **linear inequality** in x and y is any inequality that can be written in the form

$$Ax + By < C \quad \text{or} \quad Ax + By > C \quad \text{or} \quad Ax + By \leq C \quad \text{or} \quad Ax + By \geq C$$

where A, B, and C are real numbers and A and B are not both 0.

Because the inequality $y > 3x + 2$ can be written in the form $-3x + y > 2$, it is an example of a linear inequality in x and y.

GRAPHING LINEAR INEQUALITIES

To graph $y > 3x + 2$, we first note that exactly one of the following statements is true:

$$y < 3x + 2, \qquad y = 3x + 2, \qquad \text{or} \qquad y > 3x + 2$$

The graph of the equation $y = 3x + 2$ is the line shown in Figure 4-24(a). The graphs of $y < 3x + 2$ and $y > 3x + 2$ are half-planes, one on each side of that line. The graph of $y = 3x + 2$ is a boundary line separating the two half-planes. It is drawn with a broken line to show that it is not part of the graph of $y > 3x + 2$.

To find which half-plane is the graph of $y > 3x + 2$, we can substitute the coordinates of any point from either half-plane. Since an easy point to use is the origin, we substitute its coordinates (0, 0) into the inequality and simplify.

$y > 3x + 2$

$\mathbf{0} > 3(\mathbf{0}) + 2$ Substitute 0 for x and 0 for y.

$0 \not> 2$

Since the coordinates of the origin don't satisfy $y > 3x + 2$, the origin is not part of the graph. Thus, the half-plane on the other side of the broken line is the graph, which is shown in Figure 4-24(b).

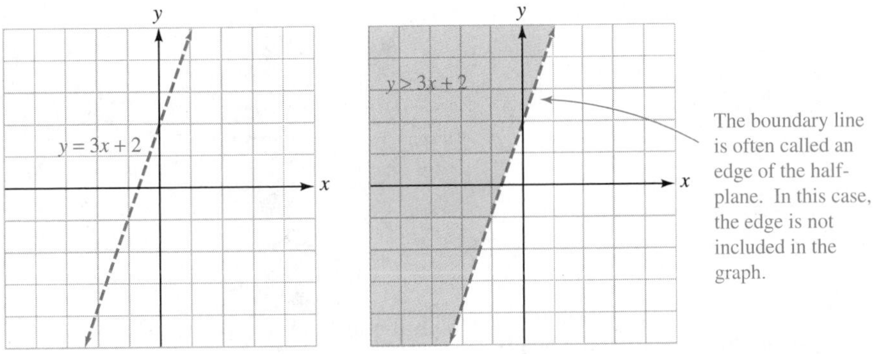
y
x
y = 3x + 2
y
x
y > 3x + 2
The boundary line is often called an edge of the half-plane. In this case, the edge is not included in the graph.

(a) (b)

FIGURE 4-24

EXAMPLE 1 Graph the inequality $2x - 3y \le 6$.

Solution This inequality is the combination of the inequality $2x - 3y < 6$ and the equation $2x - 3y = 6$.

We start by graphing $2x - 3y = 6$ to find the boundary line that separates the two half-planes. This time, we draw the solid line shown in Figure 4-25(a), because equality is permitted. To decide which half-plane represents $2x - 3y < 6$, we check to see whether the coordinates of the origin satisfy the inequality.

$$2x - 3y < 6$$
$$2(\mathbf{0}) - 3(\mathbf{0}) < 6 \qquad \text{Substitute 0 for } x \text{ and 0 for } y.$$
$$0 < 6$$

Since the coordinates of the origin satisfy the inequality, the origin is in the half-plane that is the graph of $2x - 3y < 6$. The graph is shown in Figure 4-25(b).

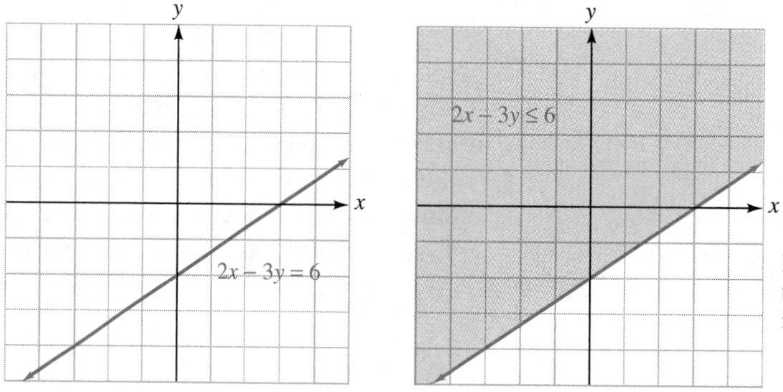

In this case, the edge is included.

(a) (b)

FIGURE 4-25

Self Check Graph $3x - 2y \ge 6$.

Answer

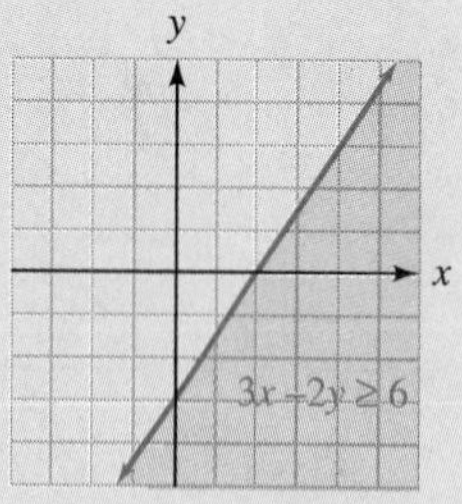

EXAMPLE 2 Graph the inequality $y < 2x$.

Solution We graph $y = 2x$, as shown in Figure 4-26(a). Because it is not part of the inequality, we draw the edge as a broken line.

To decide which half-plane is the graph of $y < 2x$, we check to see whether the coordinates of some fixed point satisfy the inequality. We cannot use the origin as a test point, because the edge passes through the origin. However, we can choose a different point—say, (3, 1).

$$y < 2x$$

$$1 < 2(3) \quad \text{Substitute 1 for } y \text{ and 3 for } x.$$

$$1 < 6$$

Since $1 < 6$ is a true inequality, the point (3, 1) satisfies the inequality and is in the graph, which is shown in Figure 4-26(b).

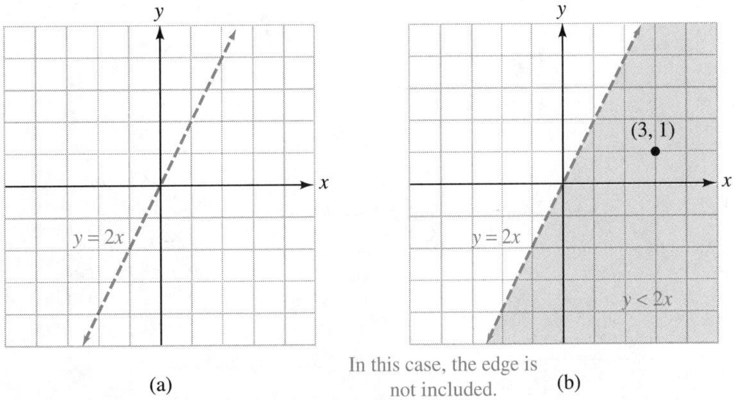

FIGURE 4-26

Self Check Graph $y > 2x$.

Answer

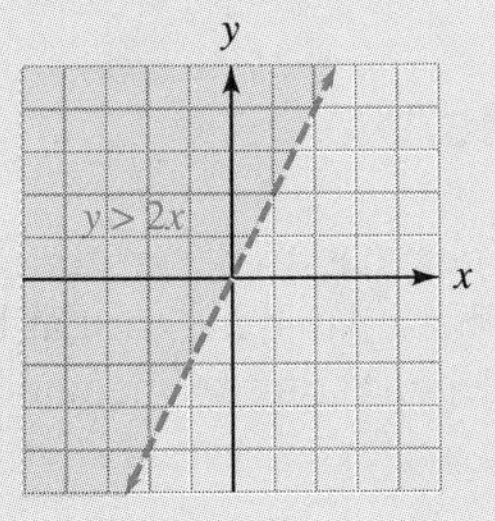

GRAPHING COMPOUND INEQUALITIES

EXAMPLE 3 Graph the inequality $2 < x \le 5$.

Solution The inequality $2 < x \le 5$ is equivalent to the following two inequalities:

$$2 < x \quad \text{and} \quad x \le 5$$

Its graph will contain all points in the plane that satisfy the inequalities $2 < x$ and $x \leq 5$ simultaneously. These points are in the shaded region of Figure 4-27.

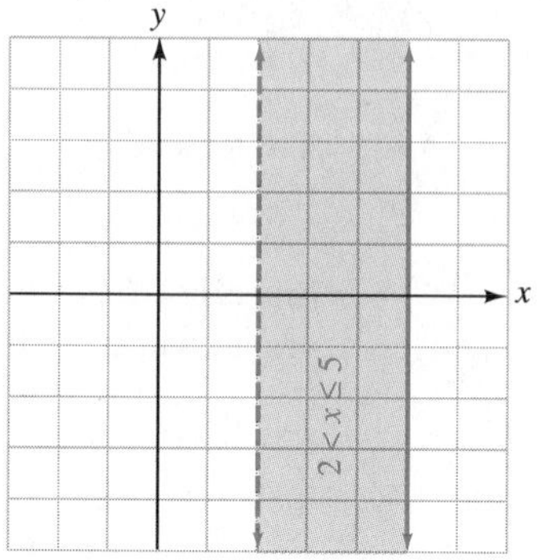

FIGURE 4-27

■

Self Check Graph $-2 \leq x < 3$.

Answer

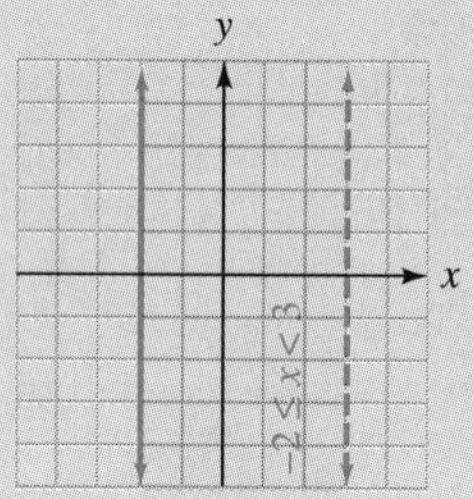

PROBLEM SOLVING

EXAMPLE 4 **Earning money** Rick has two part-time jobs, one paying \$7 per hour and the other paying \$5 per hour. He must earn at least \$140 per week to pay his expenses while attending college. Write an inequality that shows the various ways he can schedule his time to achieve his goal.

Analyze the problem If we let x represent the number of hours per week he works on the first job, he will earn $\$7x$ per week on the first job. If we let y represent the number of hours per week he works on the second job, he will earn $\$5y$ per week on the second job. To achieve his goal, the sum of these two incomes must be at least \$140.

Form an inequality Since x represents the number of hours per week he works on the first job and y represents the number of hours he works per week on the second job, we have

The hourly rate on the first job	·	the hours worked on the first job	+	The hourly rate on the second job	·	the hours worked on the second job	≥	\$140.
\$7	·	x	+	\$5	·	y	≥	\$140

Solve the inequality The graph of $7x + 5y \geq 140$ is shown in Figure 4-28. Any point in the shaded region indicates a way that he can schedule his time and earn \$140 or more per week. For example, if he works 10 hours on the first job and 15 hours on the second job, he will earn

$$\begin{aligned}\$7(10) + \$5(15) &= \$70 + \$75 \\ &= \$145\end{aligned}$$

If he works 5 hours on the first job and 25 hours on the second job, he will earn

$$\begin{aligned}\$7(5) + \$5(25) &= \$35 + \$125 \\ &= \$160\end{aligned}$$

Since Rick cannot work a negative number of hours, the graph has no meaning when x or y is negative, so only the first quadrant of the graph is shown.

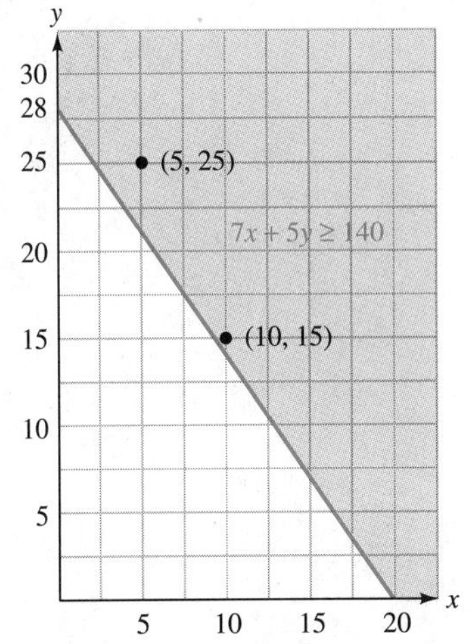

FIGURE 4-28

Graphing Inequalities

GRAPHING CALCULATORS

Some calculators (such as the TI-83) have a graphing-style icon in the $y =$ editor. See Figure 4-29(a). Some of the different graphing styles are as follows.

\	line	A straight line or curved graph is shown.
◥	above	Shading covers the area above a graph.
◣	below	Shading covers the area below a graph.

We can change the icon by placing the cursor on it and pressing the ENTER key.

(continued)

Graphing Inequalities *(continued)*

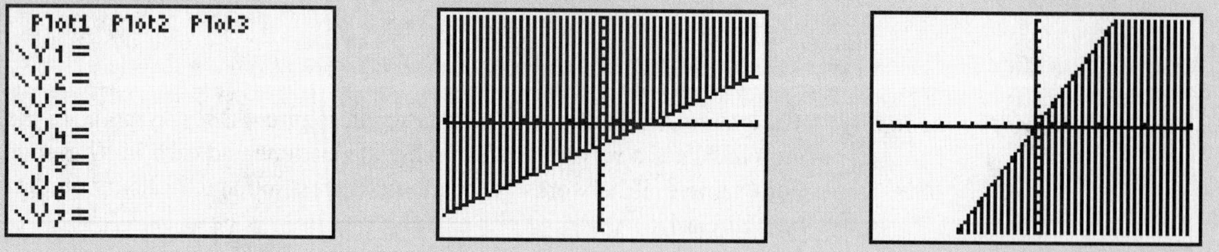

(a) (b) (c)

FIGURE 4-29

To graph the inequality of Example 1 using window settings of $x = [-10, 10]$ and $y = [-10, 10]$, we change the graphing-style icon to "above" (◥), enter the equation $2x - 3y = 6$ as $y = \frac{2}{3}x - 2$, and press the GRAPH key to get Figure 4-29(b).

To graph the inequality of Example 2 using window settings of $x = [-10, 10]$ and $y = [-10, 10]$, we change the graphing-style icon to "below" (◣), enter the equation $y = 2x$, and press the GRAPH key to get Figure 4-29(c).

If your calculator does not have a graphing-style icon, you can graph linear inequalities with a shade feature. To do so, consult your owner's manual.

It is important to note that graphing calculators do not distinguish between solid and dashed lines to show whether or not the edge of a region is included within the graph.

Orals *Do the points satisfy* $2x + 3y < 12$?

1. (0, 0) **2.** (3, 2) **3.** (2, 3) **4.** (−1, 4)

Do the points satisfy $3x - 2y \geq 12$?

5. (0, 0) **6.** (3, 2) **7.** (2, −3) **8.** (5, 1)

EXERCISE 4.3

REVIEW *Solve each system of equations.*

1. $\begin{cases} x + y = 4 \\ x - y = 2 \end{cases}$

2. $\begin{cases} 2x - y = -4 \\ x + 2y = 3 \end{cases}$

3. $\begin{cases} 3x + y = 3 \\ 2x - 3y = 13 \end{cases}$

4. $\begin{cases} 2x - 5y = 8 \\ 5x + 2y = -9 \end{cases}$

VOCABULARY AND CONCEPTS *Fill in each blank to make a true statement.*

5. $3x + 2y < 12$ is an example of a ______ inequality.

6. Graphs of linear inequalities in two variables are ______.

7. The boundary line of a half-plane is called an ______.

8. If $y < \frac{1}{2}x - 2$ and $y > \frac{1}{2}x - 2$ are false, then ______.

PRACTICE *In Exercises 9–24, graph each inequality.*

9. $y > x + 1$

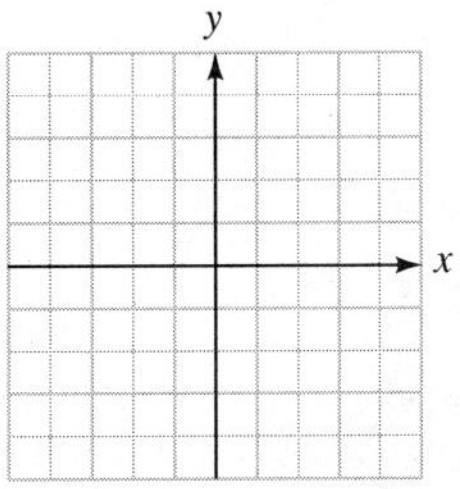

10. $y < 2x - 1$

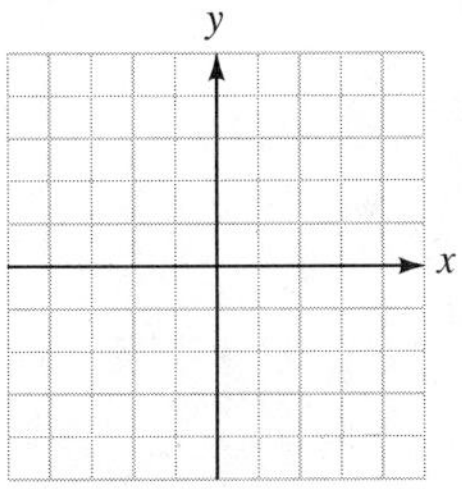

11. $y \geq x$

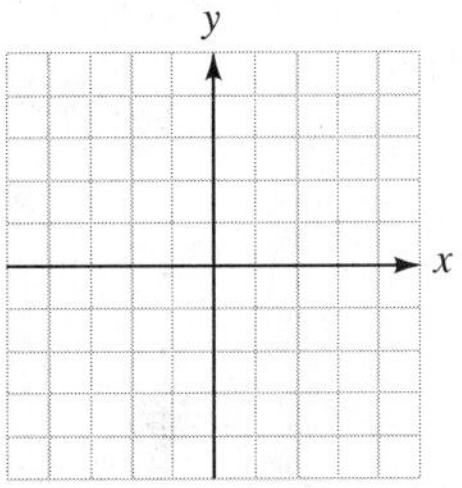

12. $y \leq 2x$

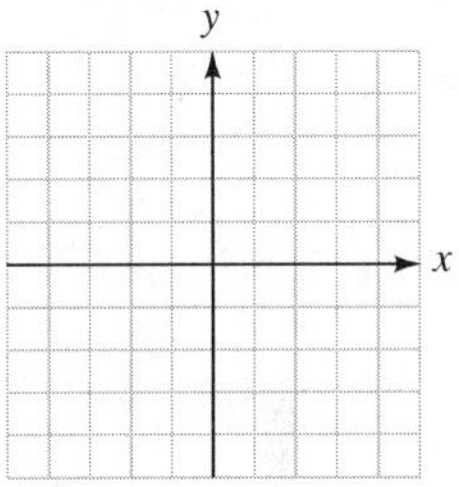

13. $2x + y \leq 6$

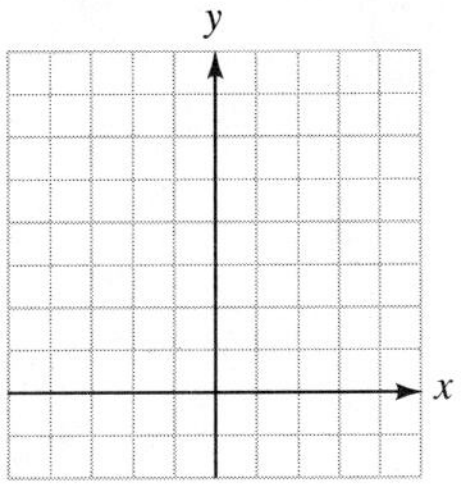

14. $x - 2y \geq 4$

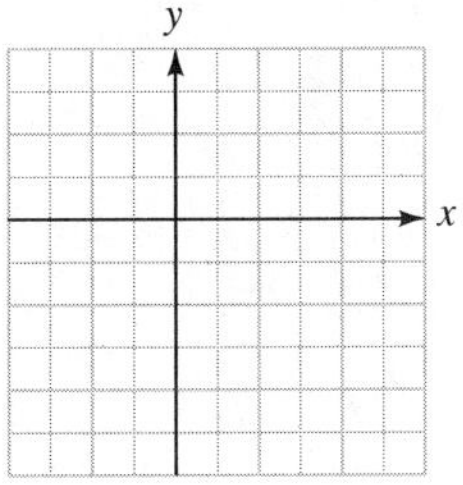

15. $3x \geq -y + 3$

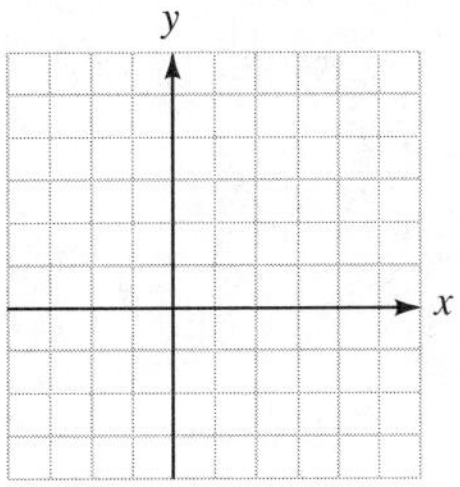

16. $2x \leq -3y - 12$

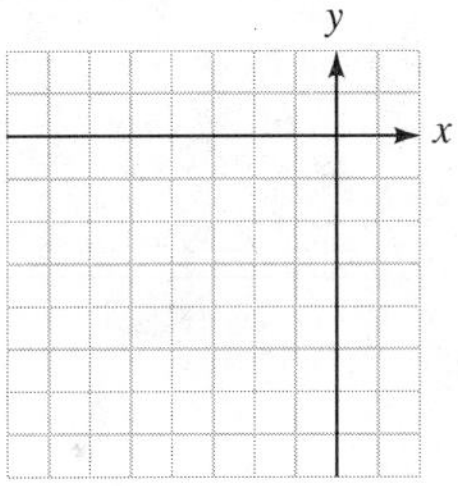

17. $y \geq 1 - \dfrac{3}{2}x$

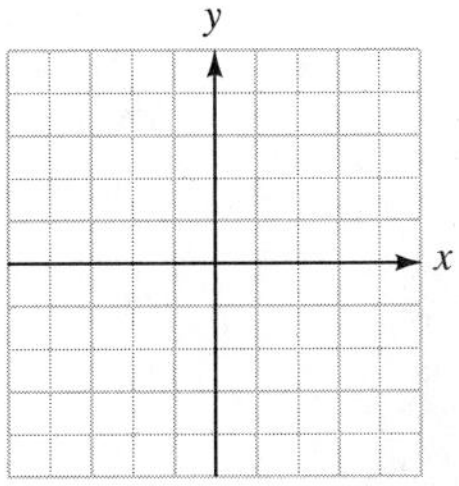

18. $y < \dfrac{1}{3}x - 1$

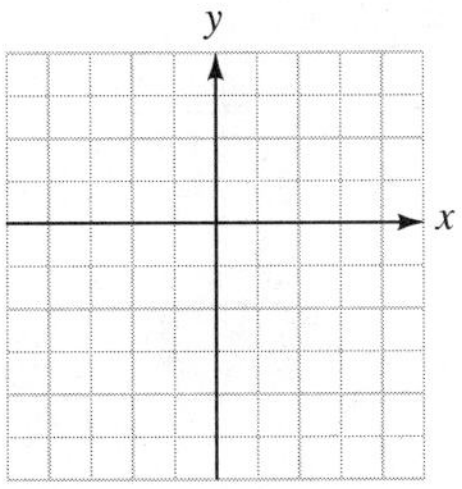

19. $x < 4$

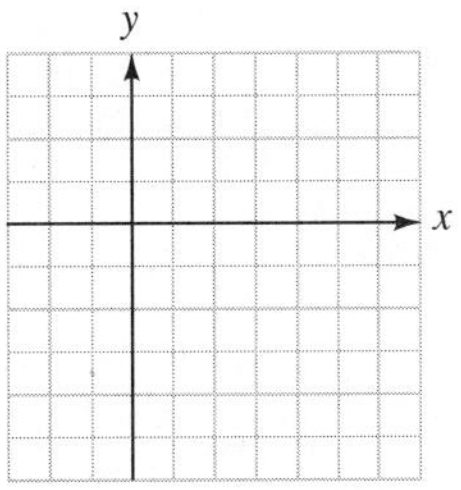

20. $y \geq -2$

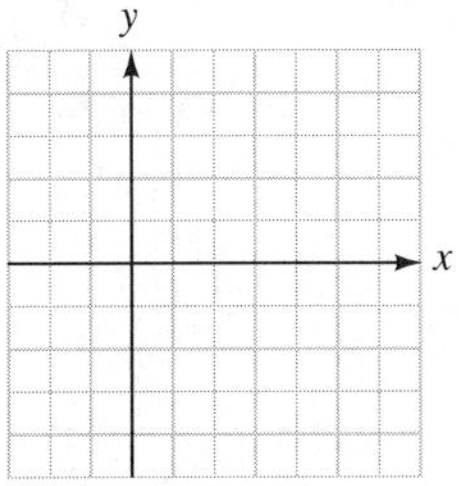

21. $-2 \leq x < 0$

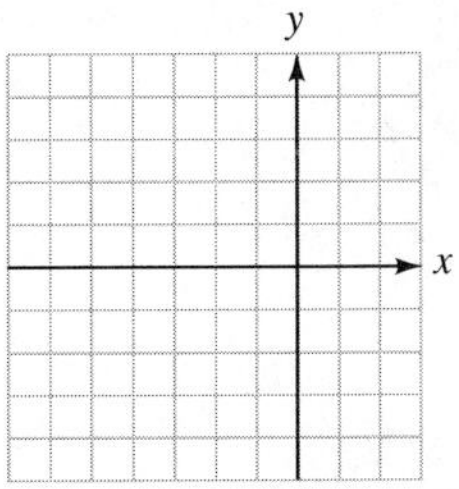

22. $-3 < y \leq -1$

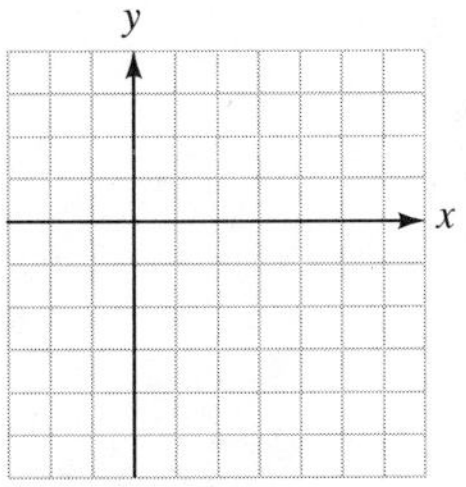

23. $y < -2$ or $y > 3$

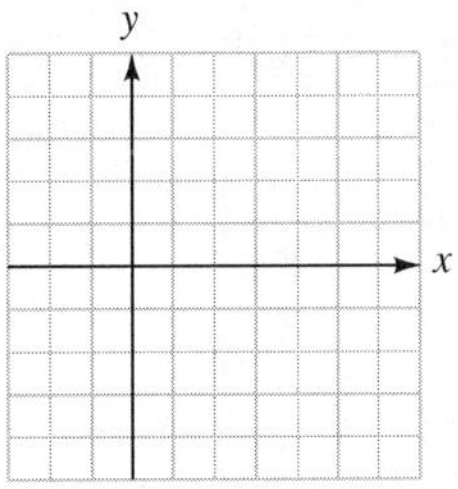

24. $-x \leq 1$ or $x \geq 2$

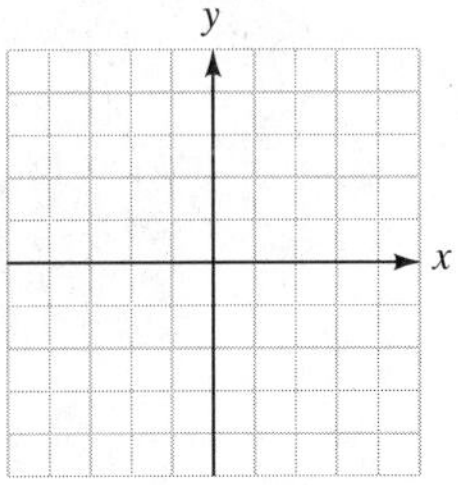

In Exercises 25–34, find the equation of the boundary line or lines. Then give the inequality whose graph is shown.

25.

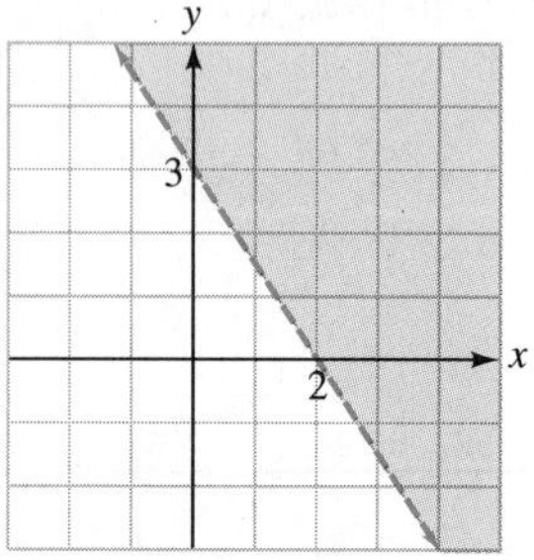

26.

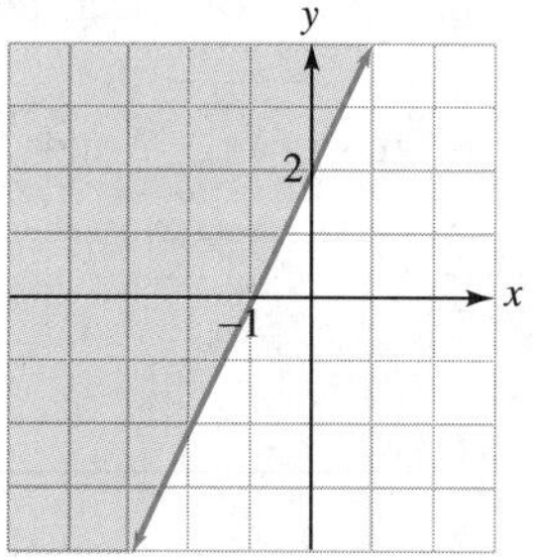

27.

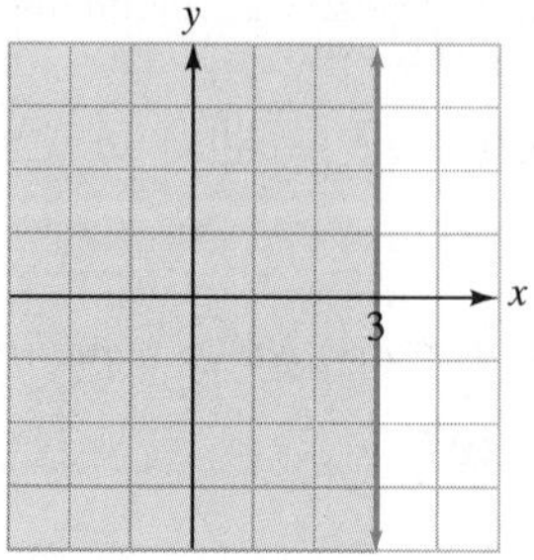

28.

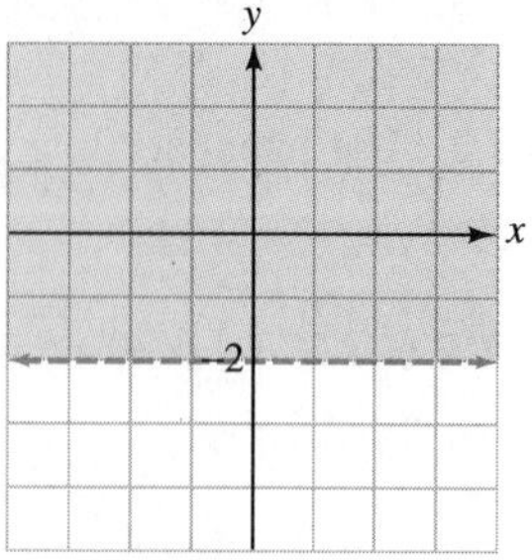

29.

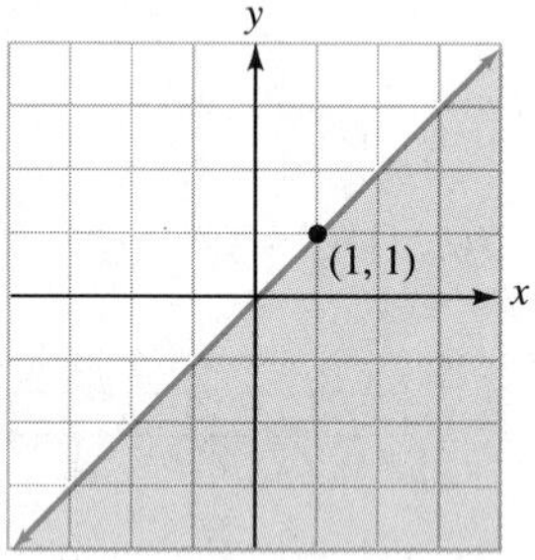

30.

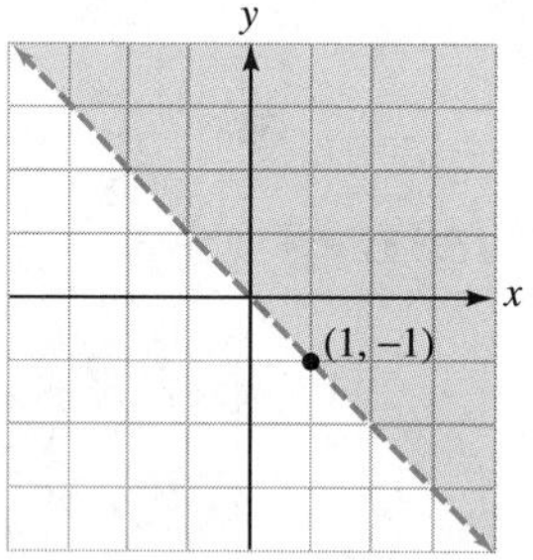

31.

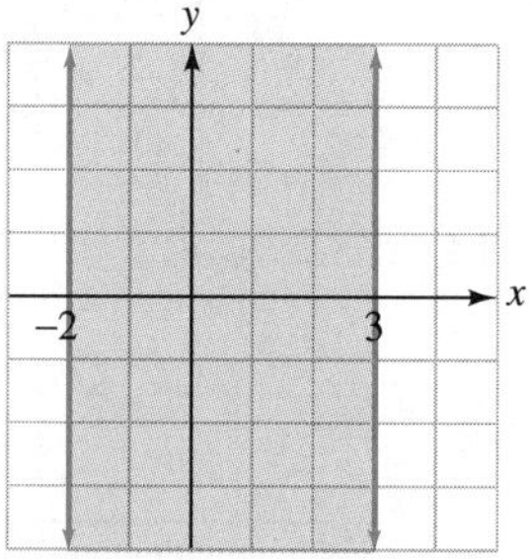

32.

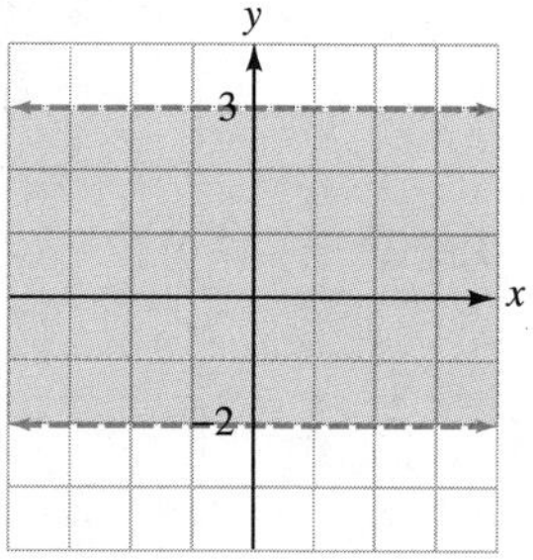

33.

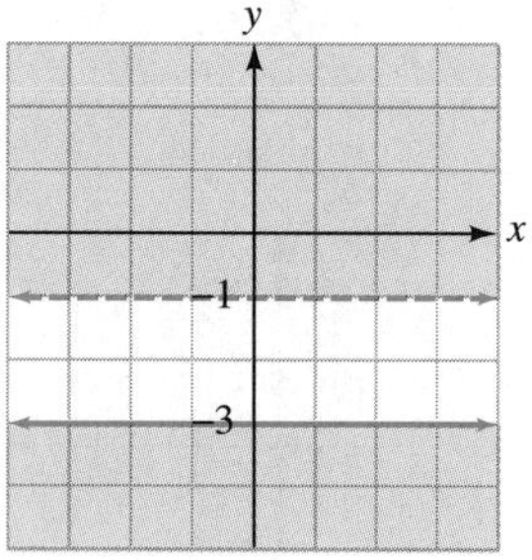

34.

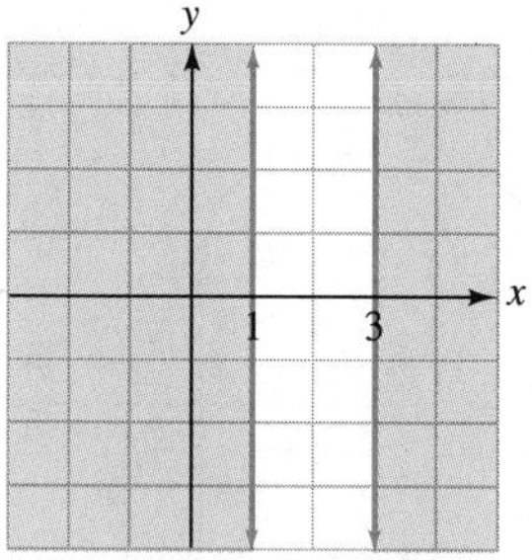

In Exercises 35–38, use a graphing calculator to graph each inequality.

35. $y < 0.27x - 1$

36. $y > -3.5x + 2.7$

37. $y \geq -2.37x + 1.5$

38. $y \leq 3.37x - 1.7$

APPLICATIONS *In Exercises 39–44, graph each inequality for nonnegative values of x and y. Then give some ordered pairs that satisfy the inequality.*

39. Figuring taxes On average, it takes an accountant 1 hour to complete a simple tax return and 3 hours to complete a complicated return. If the accountant wants to work no more than 9 hours per day, use Illustration 1 to graph an inequality that shows the possible ways that simple returns (x) and complicated returns (y) can be completed each day.

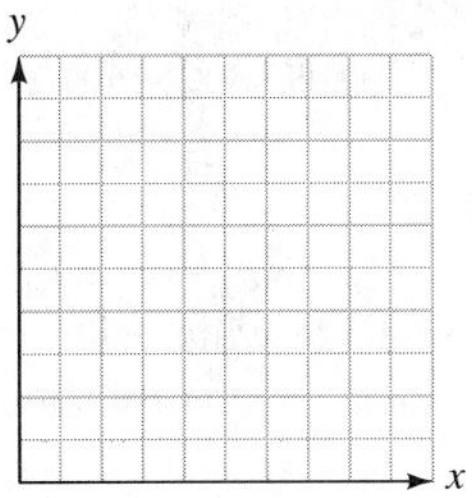

ILLUSTRATION 1

40. Selling trees During a sale, a garden store sold more than $2,000 worth of trees. If a 6-foot maple costs $100 and a 5-foot pine costs $125, use Illustration 2 to graph an inequality that shows the possible ways that maple trees (x) and pine trees (y) were sold.

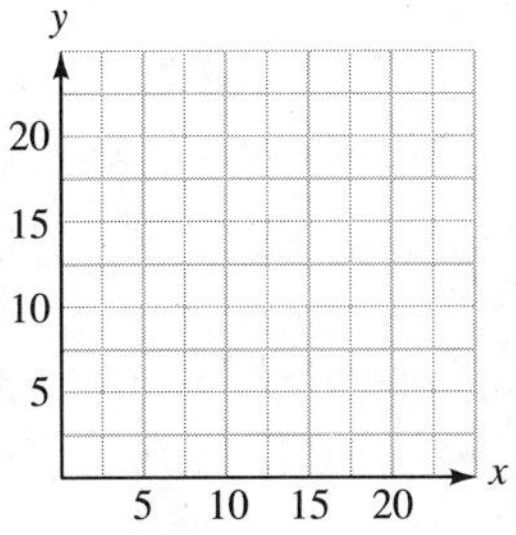

ILLUSTRATION 2

41. Choosing housekeepers One housekeeper charges $6 per hour, and another charges $7 per hour. If Sarah can afford no more than $42 per week to clean her house, use Illustration 3 to graph an inequality that shows the possible ways that she can hire the first housekeeper (x) and the second housekeeper (y).

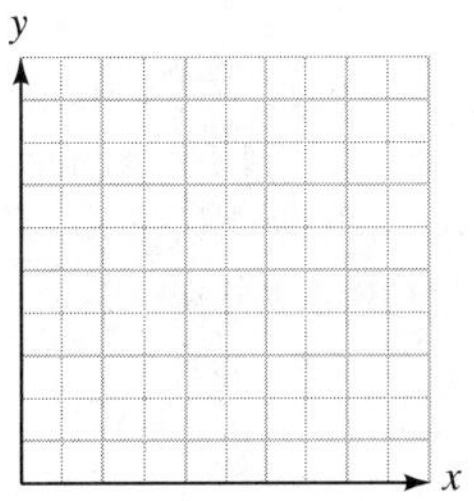

ILLUSTRATION 3

42. Making sporting goods A sporting goods manufacturer allocates at least 1,200 units of time per day to make fishing rods and reels. If it takes 10 units of time to make a rod and 15 units of time to make a reel, use Illustration 4 to graph an inequality that shows the possible ways to schedule the time to make rods (x) and reels (y).

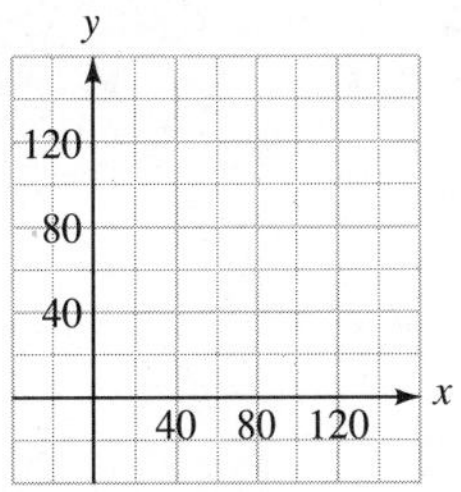

ILLUSTRATION 4

43. Investing A woman has up to $6,000 to invest. If stock in Traffico sells for $50 per share and stock in Cleanco sells for $60 per share, use Illustration 5 to graph an inequality that shows the possible ways that she can buy shares of Traffico (x) and Cleanco (y).

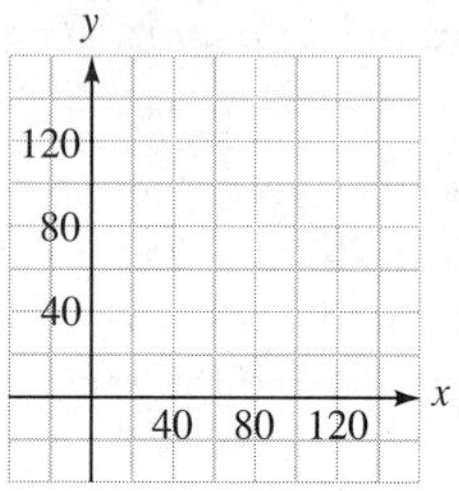

ILLUSTRATION 5

44. **Buying concert tickets** Tickets to a concert cost \$6 for reserved seats and \$4 for general admission. If receipts must be at least \$10,200 to meet expenses, use Illustration 6 to graph an inequality that shows the possible ways that the box office can sell reserved seats (x) and general admission tickets (y).

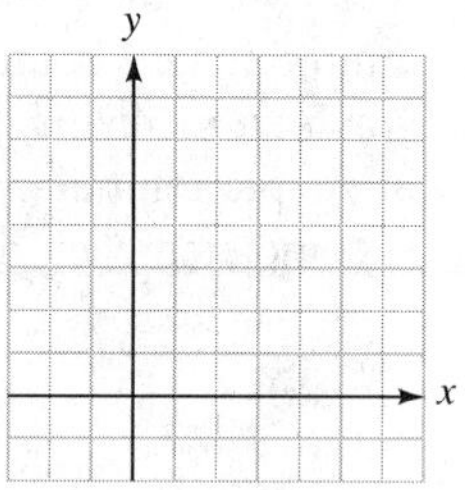

ILLUSTRATION 6

WRITING

45. Explain how to decide where to draw the boundary of the graph of a linear inequality, and whether to draw it as a solid or a broken line.

46. Explain how to decide which side of the boundary of the graph of a linear inequality should be shaded.

SOMETHING TO THINK ABOUT

47. Can an inequality be an identity, one that is satisfied by all (x, y) pairs? Illustrate.

48. Can an inequality have no solutions? Illustrate.

4.4 Systems of Inequalities

■ SYSTEMS OF INEQUALITIES ■ PROBLEM SOLVING

Getting Ready *Do the coordinates of the points satisfy the inequality* $y > 2x - 3$ *and* $y < x + 1$?

1. (0, 0) **2.** (1, 1) **3.** (−2, 0) **4.** (0, 4)

Do the coordinates of the points satisfy the inequality $2x + 3y \leq 6$ *and* $x - y \geq 4$?

5. (0, 0) **6.** (−2, 2) **7.** (1, −3) **8.** (0, 2)

■ SYSTEMS OF INEQUALITIES

We now consider the graphs of systems of inequalities in the variables x and y. These graphs will usually be the intersection of half-planes.

EXAMPLE 1 Graph the solution set of $\begin{cases} x + y \leq 1 \\ 2x - y > 2 \end{cases}$.

Solution On one set of coordinate axes, we graph each inequality as shown in Figure 4-30.

The graph of $x + y \leq 1$ includes the line graph of the equation $x + y = 1$ and all points below it. Since the edge is included, we draw it as a solid line.

The graph of $2x - y > 2$ contains the points below the graph of the equation $2x - y = 2$. Since the edge is not included, we draw it as a broken line.

The area where the half-planes intersect represents the solution of the system of inequalities, because any point in that region has coordinates that will satisfy both inequalities.

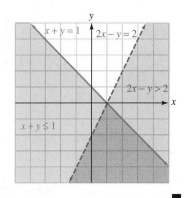

$x + y = 1$

x	y	(x, y)
0	1	(0, 1)
1	0	(1, 0)

$2x - y = 2$

x	y	(x, y)
0	−2	(0, −2)
1	0	(1, 0)

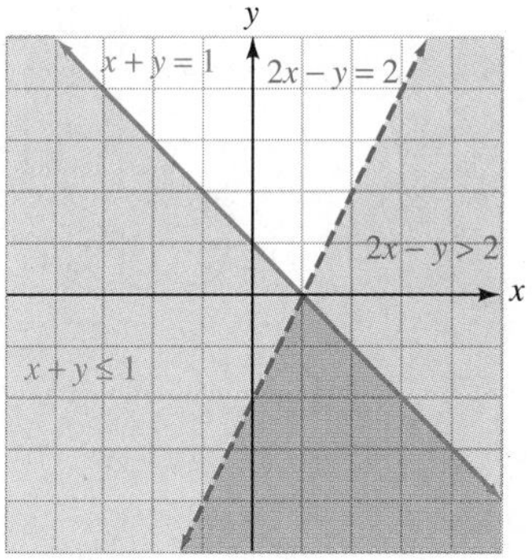

FIGURE 4-30

Self Check Graph $\begin{cases} x + y \geq 1 \\ 2x - y < 2 \end{cases}$.

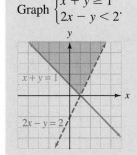

Answer

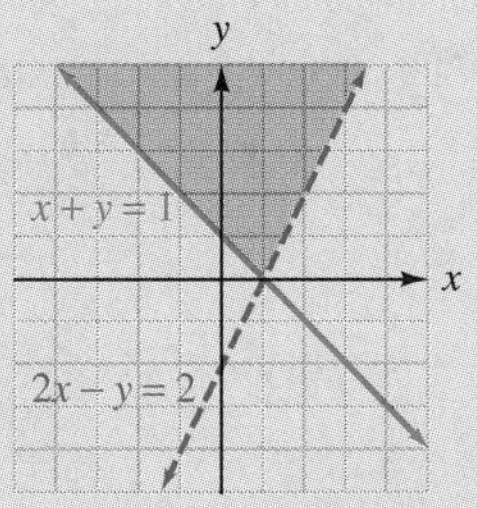

EXAMPLE 2 Graph the solution set of $\begin{cases} y < x^2 \\ y > \dfrac{x^2}{4} - 2 \end{cases}$.

Solution The graph of $y = x^2$ is the parabola shown in Figure 4-31, which opens upward and has its vertex at the origin. The points with coordinates that satisfy $y < x^2$ are the points that lie below the parabola.

The graph of $y = \frac{x^2}{4} - 2$ is a parabola opening upward, with vertex at $(0, -2)$. This time, the points that lie above the parabola satisfy the inequality. Thus, the solution of the system is the area between the parabolas.

$y = x^2$

x	y	(x, y)
0	0	(0, 0)
1	1	(1, 1)
−1	1	(−1, 1)
2	4	(2, 4)
−2	4	(−2, 4)

$y = \dfrac{x^2}{4} - 2$

x	y	(x, y)
0	−2	(0, −2)
2	−1	(2, −1)
−2	−1	(−2, −1)
4	2	(4, 2)
−4	2	(−4, 2)

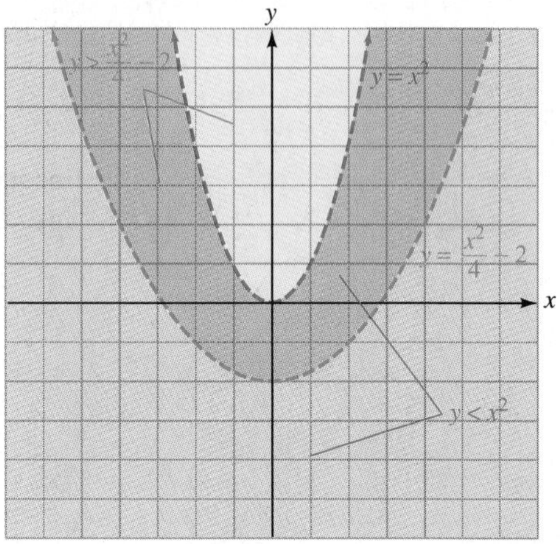

FIGURE 4-31

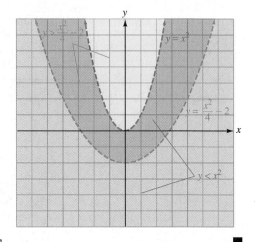

Self Check Graph $\begin{cases} y \geq x^2 \\ y \leq x + 2 \end{cases}$.

Answer

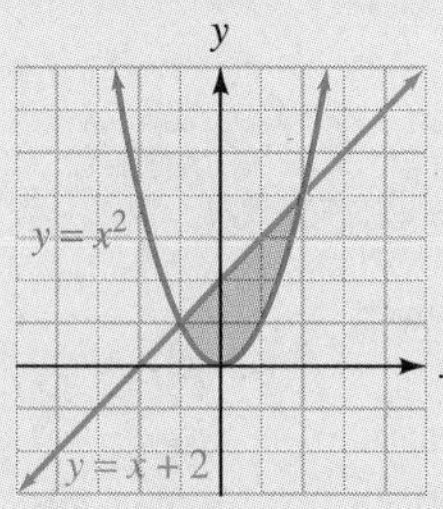

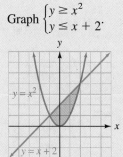

Solving Systems of Inequalities

GRAPHING CALCULATORS

To solve the system of Example 1, we use window settings of $x = [-10, 10]$ and $y = [-10, 10]$. To graph $x + y \leq 1$, we enter the equation $x + y = 1$ $(y = -x + 1)$ and change the graphing-style icon to below (◣). To graph $2x - y > 2$, we enter the equation $2x - y = 2$ $(y = 2x - 2)$ and change the graphing-style icon to below (◣). Finally, we press the GRAPH key to obtain Figure 4-32(a).

To solve the system of Example 2, we enter the equation $y = x^2$ and change the graphing-style icon to below (◣). We then enter the equation $y = \frac{x^2}{4} - 2$ and change the graphing-style icon to above (◥). Finally, we press the GRAPH key to obtain Figure 4-32(b).

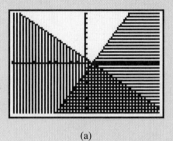

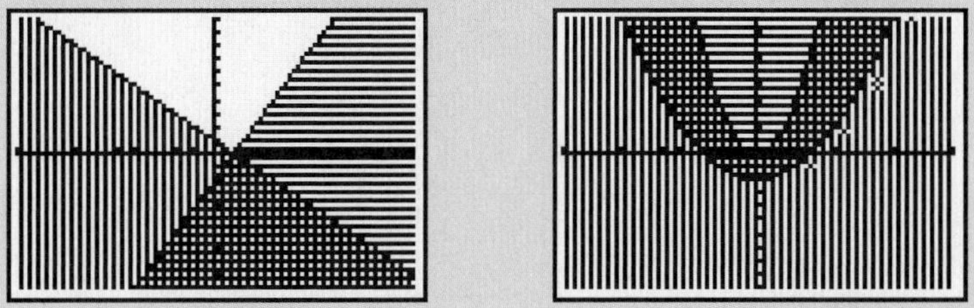

(a) (b)

FIGURE 4-32

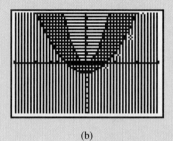

EXAMPLE 3 Graph the solution set of $\begin{cases} x \geq 1 \\ y \geq x \\ 4x + 5y < 20 \end{cases}$.

Solution The graph of $x \geq 1$ includes the points that lie on the graph of $x = 1$ and to the right, as shown in Figure 4-33(a).

The graph of $y \geq x$ includes the points that lie on the graph of $y = x$ and above it, as shown in Figure 4-33(b).

The graph of $4x + 5y < 20$ includes the points that lie below the graph of $4x + 5y = 20$, as shown in Figure 4-33(c).

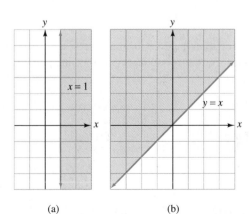

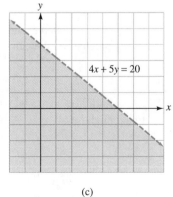

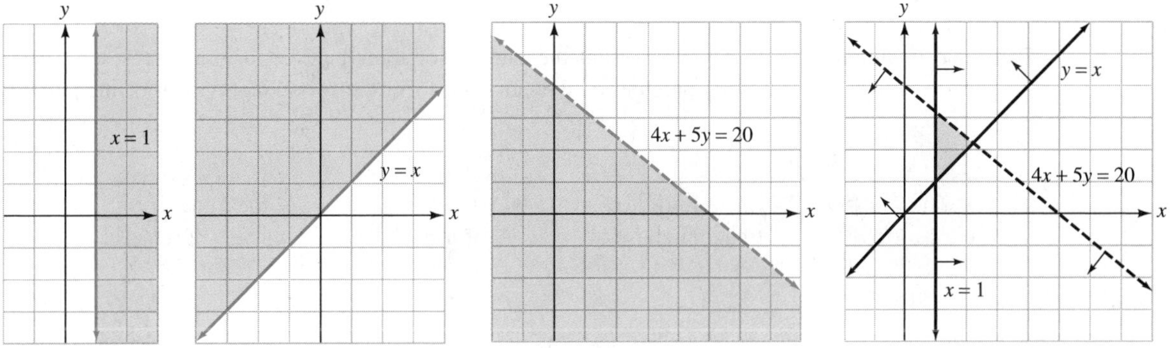

(a) (b) (c) (d)

FIGURE 4-33

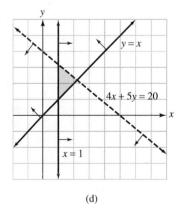

If we merge these graphs onto one set of coordinate axes, we see that the graph of the system includes the points that lie within a shaded triangle together with the points on two of the three sides of the triangle, as shown in Figure 4-33(d). ■

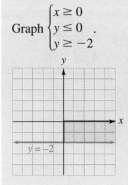

Self Check Graph $\begin{cases} x \geq 0 \\ y \leq 0 \\ y \geq -2 \end{cases}$.

Answer

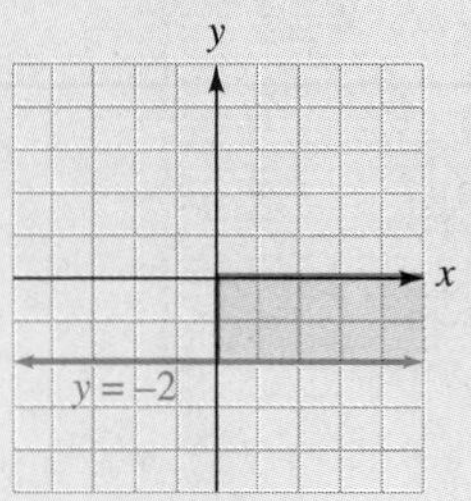

■ PROBLEM SOLVING

EXAMPLE 4 **Landscaping** A homeowner budgets from \$300 to \$600 for trees and bushes to landscape his yard. After shopping around, he finds that good trees cost \$150 and mature bushes cost \$75. What combinations of trees and bushes can he afford to buy?

Analyze the problem If x represents the number of trees purchased, then \$150x will be the cost of the trees. If y represents the number of bushes purchased, then \$75y will be the cost of the bushes. We know that the homeowner wants the sum of these costs to be from \$300 to \$600.

Form two inequalities We let x represent the number of trees purchased and y represent the number of bushes purchased. We can then form the following system of inequalities:

The cost of a tree	·	the number of trees purchased	+	the cost of a bush	·	the number of bushes purchased	≥	\$300.
\$150	·	x	+	\$75	·	y	≥	\$300

The cost of a tree	·	the number of trees purchased	+	the cost of a bush	·	the number of bushes purchased	≤	\$600.
\$150	·	x	+	\$75	·	y	≤	\$600

Solve the system We graph the system

$$\begin{cases} 150x + 75y \geq 300 \\ 150x + 75y \leq 600 \end{cases}$$

as in Figure 4-34. The coordinates of each point shown in the graph give a possible combination of trees (x) and bushes (y) that can be purchased.

State the conclusion These possibilities are

(0, 4), (0, 5), (0, 6), (0, 7), (0, 8)
(1, 2), (1, 3), (1, 4), (1, 5), (1, 6)
(2, 0), (2, 1), (2, 2), (2, 3), (2, 4)
(3, 0), (3, 1), (3, 2), (4, 0)

Only these points can be used, because the homeowner cannot buy a portion of a tree.

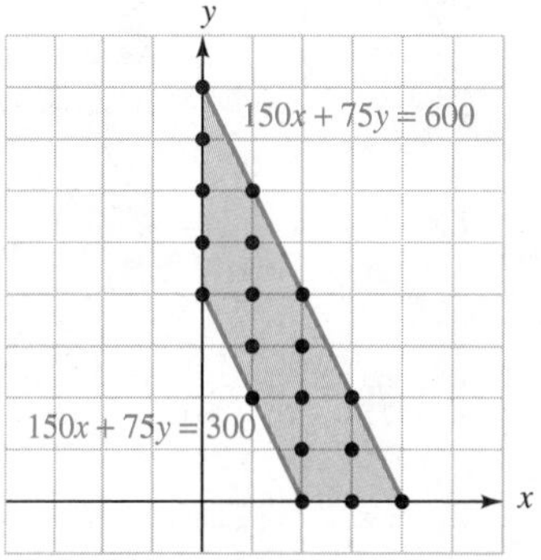

FIGURE 4-34

Check the result Check some of the ordered pairs to verify that they satisfy both inequalities.

Orals *Do the coordinates* (1, 1) *satisfy both inequalities?*

1. $\begin{cases} y < x + 1 \\ y > x - 1 \end{cases}$

2. $\begin{cases} y > 2x - 3 \\ y < -x + 3 \end{cases}$

EXERCISE 4.4

REVIEW *Solve each formula for the given variable.*

1. $A = p + prt$ for r

2. $C = \frac{5}{9}(F - 32)$ for F

3. $z = \frac{x - \mu}{\sigma}$ for x

4. $P = 2l + 2w$ for w

5. $l = a + (n - 1)d$ for d

6. $z = \frac{x - \mu}{\sigma}$ for μ

VOCABULARY AND CONCEPTS *Fill in each blank to make a true statement.*

7. To solve a system of inequalities by graphing, we graph each inequality. The solution is the region where the graphs ________.

8. If an edge is included in the graph of an inequality, we draw it as a _____ graph.

PRACTICE *In Exercises 9–24, graph the solution set of each system of inequalities.*

9. $\begin{cases} y < 3x + 2 \\ y < -2x + 3 \end{cases}$

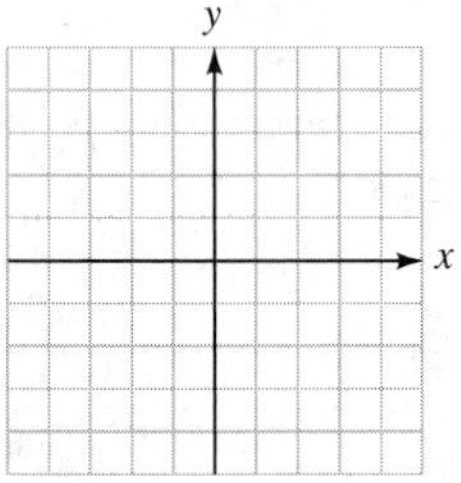

10. $\begin{cases} y \leq x - 2 \\ y \geq 2x + 1 \end{cases}$

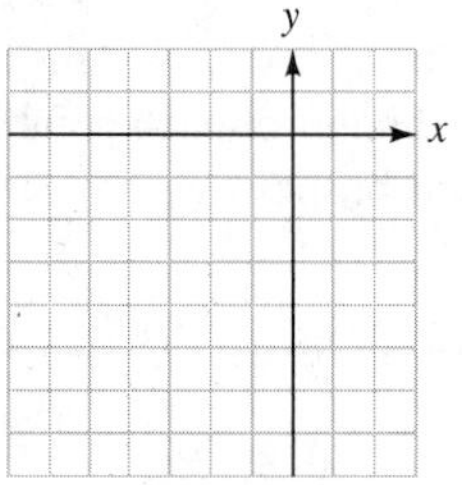

11. $\begin{cases} 3x + 2y > 6 \\ x + 3y \leq 2 \end{cases}$

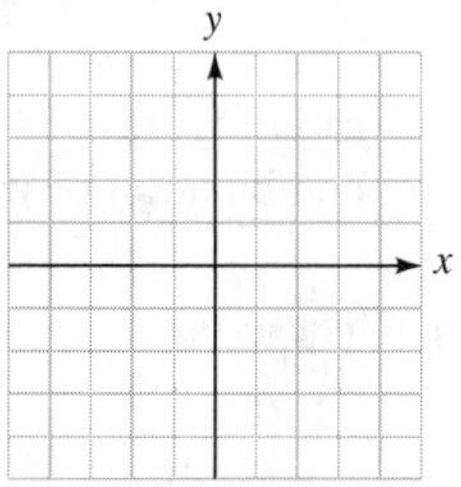

12. $\begin{cases} x + y < 2 \\ x + y \leq 1 \end{cases}$

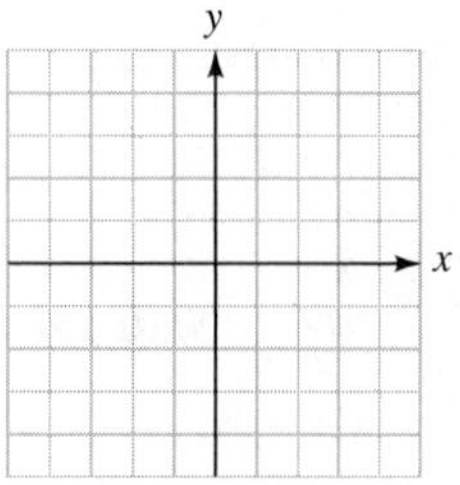

13. $\begin{cases} 3x + y \leq 1 \\ -x + 2y \geq 6 \end{cases}$

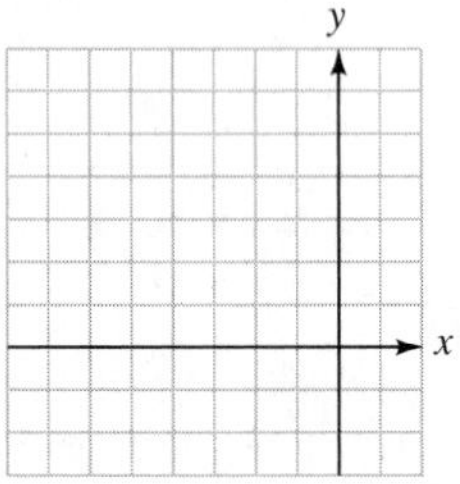

14. $\begin{cases} x + 2y < 3 \\ 2x + 4y < 8 \end{cases}$

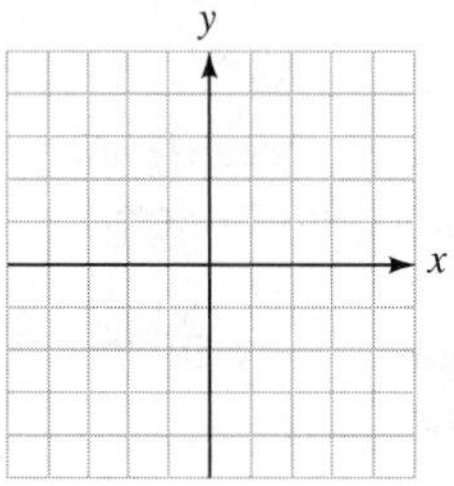

15. $\begin{cases} 2x - y > 4 \\ y < -x^2 + 2 \end{cases}$

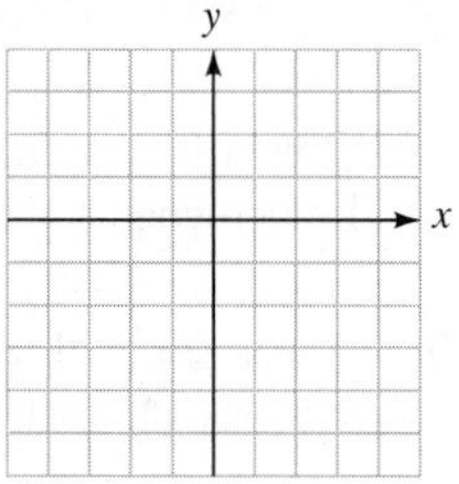

16. $\begin{cases} x \leq y^2 \\ y \geq x \end{cases}$

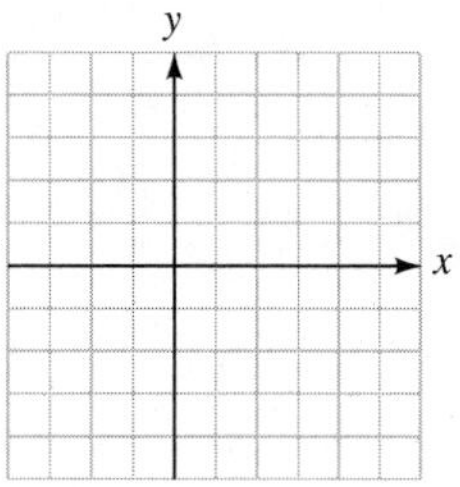

17. $\begin{cases} y > x^2 - 4 \\ y < -x^2 + 4 \end{cases}$

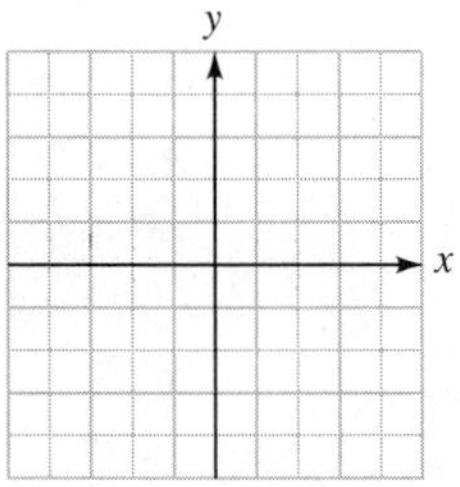

18. $\begin{cases} x \geq y^2 \\ y \geq x^2 \end{cases}$

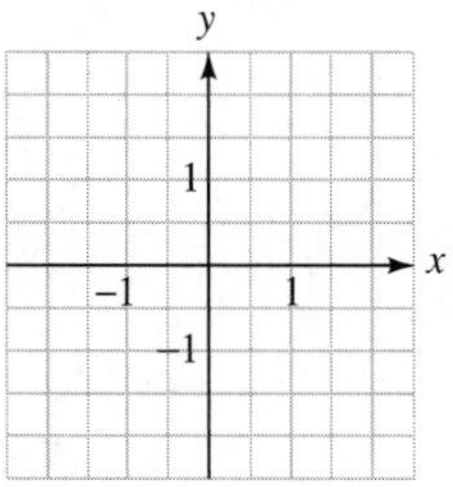

19. $\begin{cases} 2x + 3y \leq 6 \\ 3x + y \leq 1 \\ x \leq 0 \end{cases}$

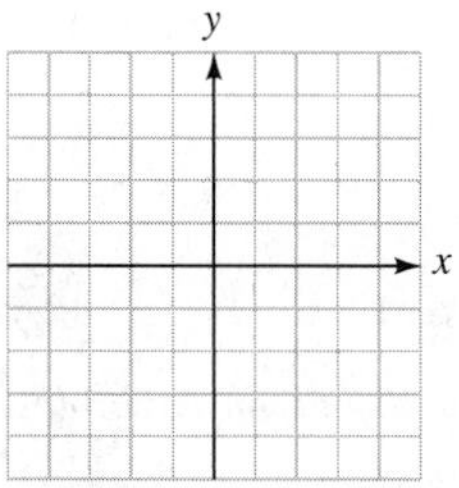

20. $\begin{cases} 2x + y \leq 2 \\ y \geq x \\ x \geq 0 \end{cases}$

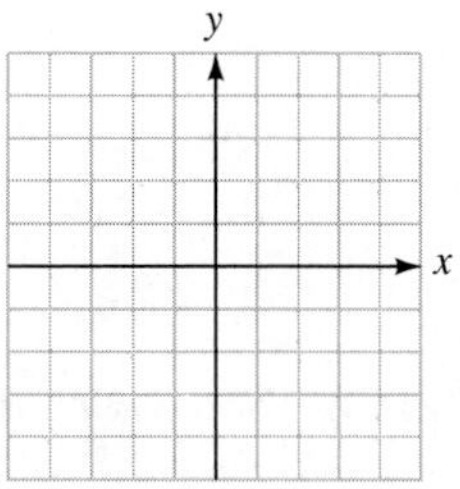

21. $\begin{cases} x - y < 4 \\ y \leq 0 \\ x \geq 0 \end{cases}$

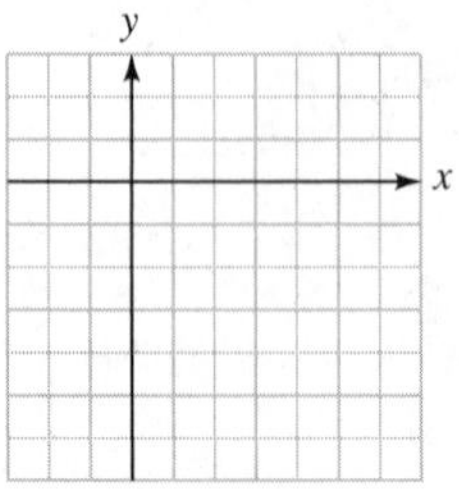

22. $\begin{cases} x + y \leq 4 \\ x \geq 0 \\ y \geq 0 \end{cases}$

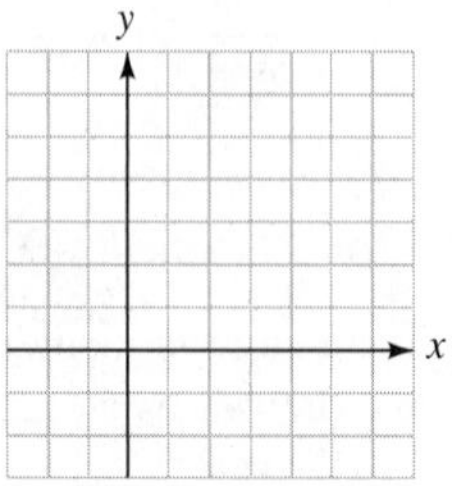

23. $\begin{cases} x \geq 0 \\ y \geq 0 \\ 9x + 3y \leq 18 \\ 3x + 6y \leq 18 \end{cases}$

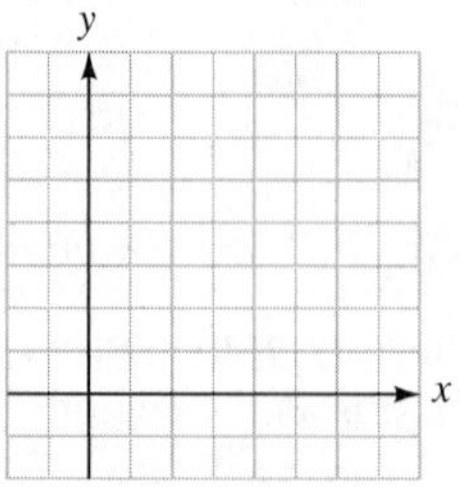

24. $\begin{cases} x + y \geq 1 \\ x - y \leq 1 \\ x - y \geq 0 \\ x \leq 2 \end{cases}$

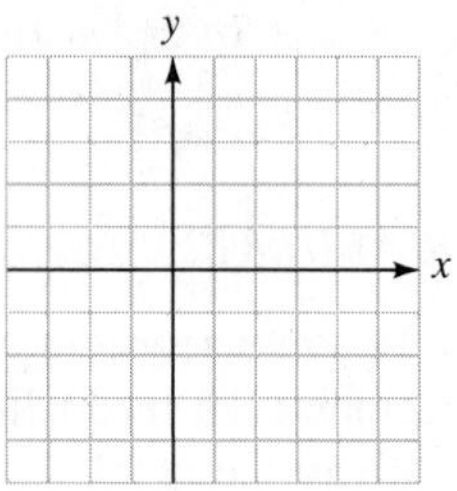

In Exercises 25–26, use a graphing calculator to solve each system.

25. $\begin{cases} y < 3x + 2 \\ y < -2x + 3 \end{cases}$ (See Exercise 9.)

26. $\begin{cases} y > x^2 - 4 \\ y < -x^2 + 4 \end{cases}$ (See Exercise 17.)

APPLICATIONS *In Exercises 27–30, graph each system of inequalities and give two possible solutions to each problem.*

27. Buying compact discs Melodic Music has compact discs on sale for either $10 or $15. If a customer wants to spend at least $30 but no more than $60 on CDs, use Illustration 1 to graph a system of inequalities that will show the possible ways a customer can buy $10 CDs ($x$) and $15 CDs ($y$).

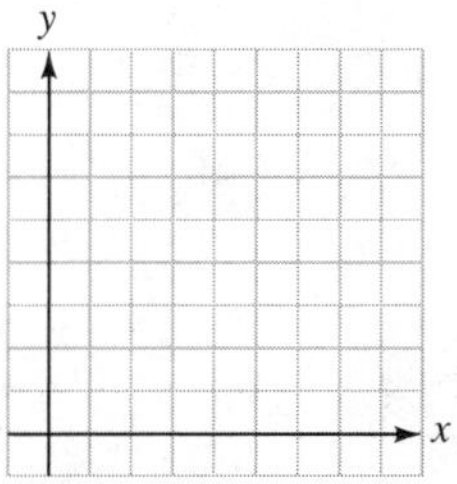

ILLUSTRATION 1

28. Buying boats Dry Boat Works wholesales aluminum boats for $800 and fiberglass boats for $600. Northland Marina wants to order at least $2,400 worth, but no more than $4,800 worth of boats. Use Illustration 2 to graph a system of inequalities that will show the possible combinations of aluminum boats (x) and fiberglass boats (y) that can be ordered.

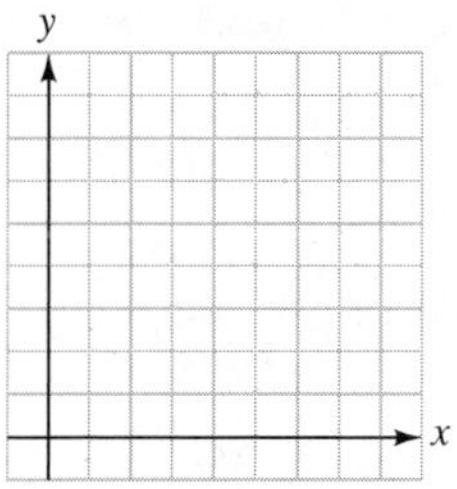

ILLUSTRATION 2

29. Buying furniture A distributor wholesales desk chairs for $150 and side chairs for $100. Best Furniture wants to order no more than $900 worth of chairs, including more side chairs than desk chairs. Use Illustration 3 to graph a system of inequalities that will show the possible combinations of desk chairs (x) and side chairs (y) that can be ordered.

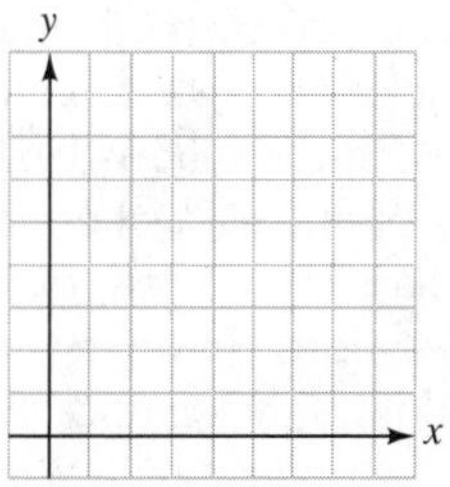

ILLUSTRATION 3

30. Ordering furnace equipment Bolden Heating Company wants to order no more than $2,000 worth of electronic air cleaners and humidifiers from a wholesaler that charges $500 for air cleaners and $200 for humidifiers. If Bolden wants more humidifiers than air cleaners, use Illustration 4 to graph a system of inequalities that will show the possible combinations of air cleaners (x) and humidifiers (y) that can be ordered.

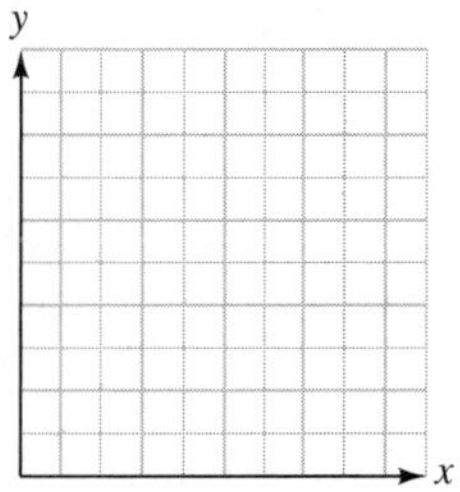

ILLUSTRATION 4

WRITING

31. When graphing a system of linear inequalities, explain how to decide which region to shade.

32. Explain how a system of two linear inequalities might have no solution.

SOMETHING TO THINK ABOUT

33. The solution of a system of inequalities in two variables is *bounded* if it is possible to draw a circle around it. Can the solution of two linear inequalities be bounded?

34. The solution of $\begin{cases} y \geq |x| \\ y \leq k \end{cases}$ has an area of 25. Find k.

PERSPECTIVE

How to Solve It

As a young student, George Polya (1888–1985) enjoyed his studies of mathematics and understood the solutions presented by his teachers. However, Polya had questions still asked by mathematics students today: "Yes, the solution works, but how is it possible to come up with such a solution? How could I discover such things by myself?" These questions still concerned him years later when, as Professor of Mathematics at Stanford University, he developed an approach to teaching mathematics that was very popular with faculty and students. His book, *How to Solve It,* became a bestseller.

Polya's problem-solving approach involves four steps.

George Polya
(1888–1985)

- *Understand the problem.* What is the unknown? What information is known? What are the conditions?
- *Devise a plan.* Have you seen anything like it before? Do you know any related problems you have solved before? If you can't solve the proposed problem, can you solve a similar but easier problem?
- *Carry out the plan.* Check each step. Can you explain why each step is correct?
- *Look back.* Examine the solution. Can you check the result? Can you use the result, or the method, to solve any other problem?

4.5 Linear Programming

■ LINEAR PROGRAMMING ■ APPLICATIONS OF LINEAR PROGRAMMING

Getting Ready *Evaluate* $2x + 3y$ *for each pair of coordinates.*

1. (0, 0) **2.** (3, 0) **3.** (2, 2) **4.** (0, 4)

LINEAR PROGRAMMING

Systems of inequalities provide the basis for an area of applied mathematics known as **linear programming.** Linear programming helps answer questions such as *"How can a business make as much money as possible?"* or *"How can I plan a nutritious meal at a school cafeteria for the least cost?"* Linear programming was developed during World War II when it became necessary to move huge quantities of people, materials, and supplies as efficiently and economically as possible.

The solutions to such problems depend on certain **constraints:** A business has limited resources, and a meal must contain sufficient vitamins and minerals. In linear programming, the constraints are expressed as a system of linear inequalities. Any solution that satisfies the system of inequalities is called a **feasible solution.** The quantity to be maximized (or minimized) is expressed as a linear function of several variables.

To solve a linear program, we will maximize or minimize a function subject to given constraints. For example, suppose that we wish to maximize the profit P to be earned by a company, where the profit P is given by the equation $P = y + 2x$. Also suppose that P is subject to the following constraints:

$$\begin{cases} x + y \geq 1 \\ x - y \leq 1 \\ x - y \geq 0 \\ x \leq 2 \end{cases}$$

We can find the solution of this system and then find the coordinates of each corner of the region R shown in Figure 4-35(a). We can then rewrite the equation

$$P = y + 2x \qquad \text{in the equivalent form} \qquad y = -2x + P$$

This is the equation for a set of parallel lines, each with a slope of -2 and a y-intercept of P. Many such lines pass through the region R. To decide which of these provides the maximum value of P, we refer to Figure 4-35(b) and locate the line with the greatest y-intercept. Since line l has the greatest y-intercept and crosses region R at the corner (2, 2), the maximum value of P (subject to the given constraints) is

$$\begin{aligned} P &= y + 2x \\ &= 2 + 2(2) \\ &= 6 \end{aligned}$$

Thus, the profit P has a maximum value of 6, subject to the given constraints. This profit occurs when $x = 2$ and $y = 2$.

In a linear program, the function to be maximized (or minimized) is called the **objective function.** The solution region R for the set of constraints is called a **feasibility region.**

The preceding discussion illustrates the following important fact.

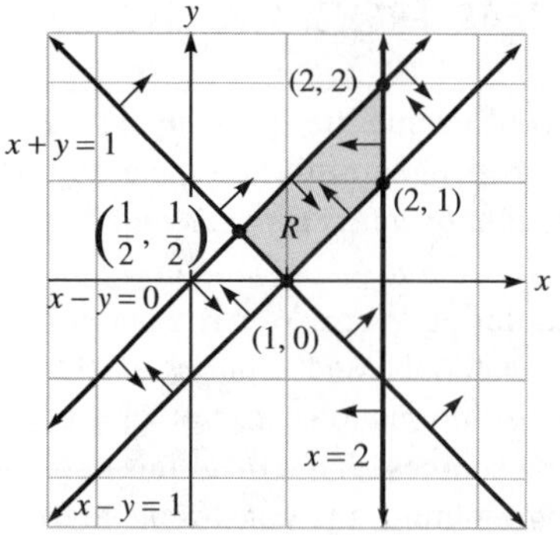

(a)

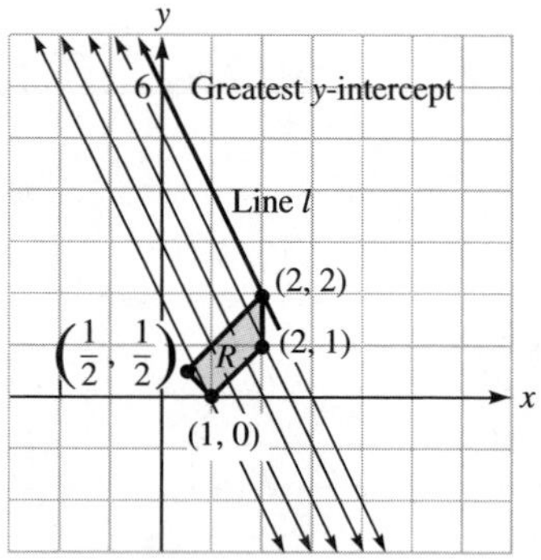

(b)

FIGURE 4-35

Maximum or Minimum of an Objective Function

If a linear function, subject to the constraints of a system of linear inequalities in two variables, attains a maximum or a minimum value, that value will occur at a corner or along an entire edge of the region R that represents the solution of the system.

EXAMPLE 1 If $P = 2x + 3y$, find the maximum value of P subject to the given constraints:

$$\begin{cases} x \geq 0 \\ y \geq 0 \\ x + y \leq 4 \\ 2x + y \leq 6 \end{cases}$$

Solution We solve the system of inequalities to find the feasibility region R shown in Figure 4-36. The coordinates of its corners are (0, 0), (3, 0), (0, 4), and (2, 2).

Because the maximum value of P will occur at a corner of R, we substitute the coordinates of each corner point into the objective function

$$P = 2x + 3y$$

We then find the one that gives the maximum value of P.

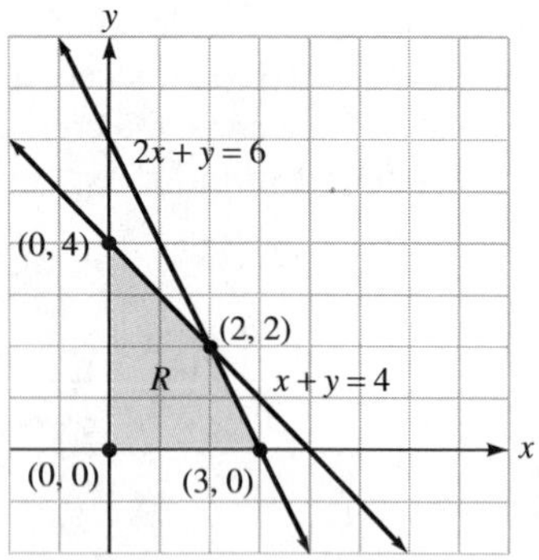

FIGURE 4-36

Point	$P = 2x + 3y$
(0, 0)	$P = 2(0) + 3(0) = 0$
(3, 0)	$P = 2(3) + 3(0) = 6$
(2, 2)	$P = 2(2) + 3(2) = 10$
(0, 4)	$P = 2(0) + 3(4) = 12$

The maximum value $P = 12$ occurs when $x = 0$ and $y = 4$. ■

Self Check Find the maximum value of $P = x + 2y$ subject to the constraints of Example 1.

Answer 8

EXAMPLE 2 If $P = 3x + 2y$, find the minimum value of P subject to the given constraints:

$$\begin{cases} x + y \geq 1 \\ x - y \leq 1 \\ x - y \geq 0 \\ x \leq 2 \end{cases}$$

Solution We refer to the feasibility region shown in Figure 4-37 with corners at $\left(\frac{1}{2}, \frac{1}{2}\right)$, (2, 2), (2, 1), and (1, 0). Because the minimum value of P occurs at a corner point of R, we substitute the coordinates of each corner point into the objective function

$$P = 3x + 2y$$

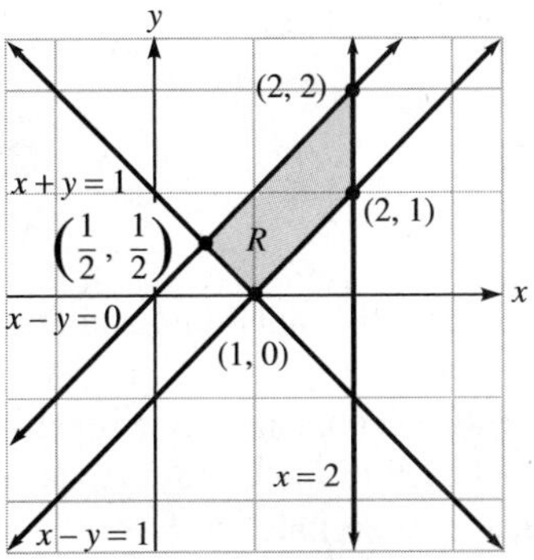

FIGURE 4-37

Then we find which one gives the minimum value of P.

Point	$P = 3x + 2y$
$\left(\frac{1}{2}, \frac{1}{2}\right)$	$P = 3\left(\frac{1}{2}\right) + 2\left(\frac{1}{2}\right) = \frac{5}{2}$
(2, 2)	$P = 3(2) + 2(2) = 10$
(2, 1)	$P = 3(2) + 2(1) = 8$
(1, 0)	$P = 3(1) + 2(0) = 3$

The minimum value $P = \frac{5}{2}$ occurs when $x = \frac{1}{2}$ and $y = \frac{1}{2}$. ■

Self Check Find the minimum value of $P = 2x + y$ subject to the constraints of Example 2.

Answer $\frac{3}{2}$

APPLICATIONS OF LINEAR PROGRAMMING

EXAMPLE 3 **Maximizing income** An accountant earns \$80 for preparing an individual tax return and \$150 for preparing a commercial tax return. On average, an individual return requires 3 hours of time, of which 1 hour is spent working on a computer. Each commercial return requires 4 hours of time, of which 2 hours are spent working on a computer. If the accountant's time is limited to 240 hours and the available computer time is limited to 100 hours, how should he divide his time between individual and commercial returns to maximize his income?

Solution Suppose that x represents the number of individual returns completed and y represents the number of commercial returns completed. Because each of the x individual returns brings in \$80 and each of the y commercial returns brings in \$150, the profit function P is given by the equation

$$P = 80x + 150y$$

The following table provides the information about time requirements:

	Individual return	Commercial return	Time available
Accountant's time	3	4	240
Computer time	1	2	100

The profit is subject to the following constraints:

$$\begin{cases} x \geq 0 \\ y \geq 0 \\ 3x + 4y \leq 240 \\ x + 2y \leq 100 \end{cases}$$

The inequalities $x \geq 0$ and $y \geq 0$ indicate that the number of individual and commercial returns cannot be negative.

The inequality $3x + 4y \leq 240$ is a constraint on the accountant's time. Each of the x individual returns takes 3 hours of his time, and each of the y commercial returns takes 4 hours. The sum of these two amounts of time must be less than or equal to his available time, which is 240 hours.

The inequality $x + 2y \leq 100$ is a constraint on computer time. Each of the x individual returns takes 1 hour of computer time, and each of the y commercial returns takes 2 hours of computer time. The sum of x and $2y$ must be less than or equal to the available time, which is 100 hours.

We graph each of the constraints to find the feasibility region R, as in Figure 4-38. The four corners of region R have coordinates of (0, 0), (80, 0), (40, 30), and (0, 50).

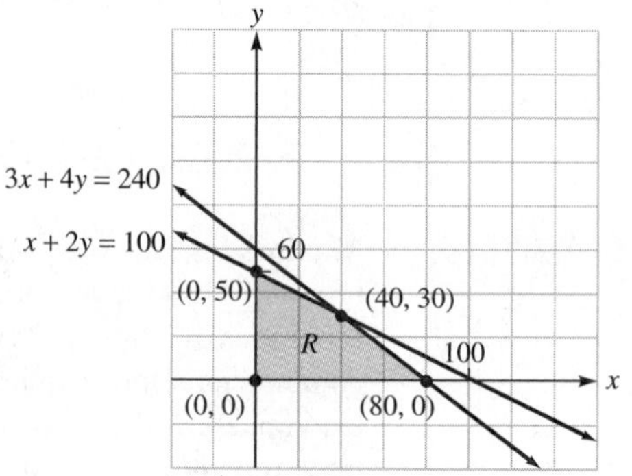

FIGURE 4-38

We substitute the coordinates of each corner point into the profit function

$$P = 80x + 150y$$

to find the maximum profit P.

Point	$P = 80x + 150y$
(0, 0)	$P = 80(0) + 150(0) = 0$
(80, 0)	$P = 80(80) + 150(0) = 6{,}400$
(40, 30)	$P = 80(40) + 150(30) = 7{,}700$
(0, 50)	$P = 80(0) + 150(50) = 7{,}500$

The accountant will maximize his income at $7,700 if he prepares 40 individual returns and 30 commercial returns. ■

EXAMPLE 4

Diet problem Two diet supplements are Vigortab and Robust. Each Vigortab tablet contains 3 units of calcium, 20 units of vitamin C, and 40 units of iron and costs $0.50. Each Robust tablet contains 4 units of calcium, 40 units of vitamin C, and 30 units of iron and costs $0.60. At least 24 units of calcium, 200 units of vitamin C, and 120 units of iron are required for the daily needs of one patient. How many tablets of each supplement should be taken daily for a minimum cost? Find the daily minimum cost.

Solution We can let x represent the number of Vigortab tablets to be taken daily and y represent the number of Robust tablets to be taken daily. We then construct a table displaying the calcium, vitamin C, iron, and cost information.

	Vigortab	Robust	Amount required
Calcium	3	4	24
Vitamin C	20	40	200
Iron	40	30	120
Cost	$0.50	$0.60	

Since cost is to be minimized, the cost function is the objective function

$$C = 0.50x + 0.60y$$

We then write the constraints. Since there are requirements for calcium, vitamin C, and iron, there is a constraint for each. Note that neither x nor y can be negative.

Calcium	$3x + 4y \geq 24$
Vitamin C	$20x + 40y \geq 200$
Iron	$40x + 30y \geq 120$
Nonnegative constraints	$x \geq 0,\ y \geq 0$

We graph the inequalities to find the feasibility region and the coordinates of its corners, as in Figure 4-39.

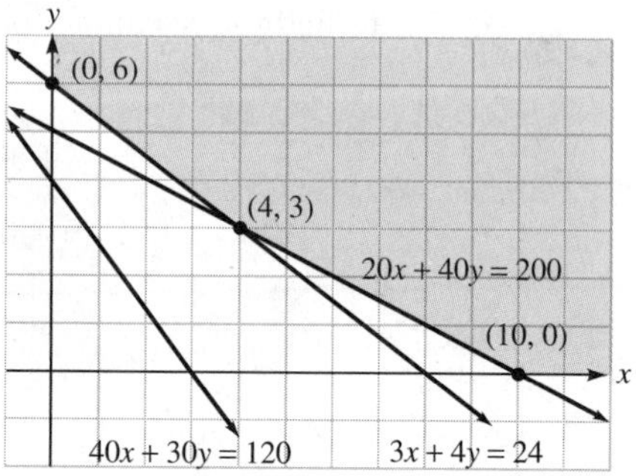

FIGURE 4-39

In this case, the feasibility region is not bounded on all sides. The coordinates of the corner points are (0, 6), (4, 3), and (10, 0). To find the minimum cost, we substitute each pair of coordinates into the objective function:

Point	$C = 0.50x + 0.60y$
(0, 6)	$C = 0.50(0) + 0.60(6) = 3.60$
(4, 3)	$C = 0.50(4) + 0.60(3) = 3.80$
(10, 0)	$C = 0.50(10) + 0.60(0) = 5.00$

A minimum cost will occur if no Vigortab and 6 Robust tablets are taken daily. The minimum daily cost is $3.60. ■

EXAMPLE 5

Production-schedule problem A television program director must schedule comedy skits and musical numbers for prime-time variety shows. Each comedy skit requires 2 hours of rehearsal time, costs $3,000, and brings in $20,000 from the show's sponsors. Each musical number requires 1 hour of rehearsal time, costs $6,000, and generates $12,000. If 250 hours are available for rehearsal and $600,000 is budgeted for comedy and music, how many segments of each type should be produced to maximize income? Find the maximum income.

Solution We can let x represent the number of comedy skits and y the number of musical numbers to be scheduled. We then construct a table with information about rehearsal time, production cost, and income generated.

	Comedy	Musical	Available
Rehearsal time (hr)	2	1	250
Cost ($1,000s)	3	6	600
Generated income ($1,000s)	20	12	

Since each of the x comedy skits generates \$20 thousand, the income generated by the comedy skits is \$$20x$ thousand. The musical numbers produce \$$12y$ thousand. The objective function to be maximized is

$$V = 20x + 12y$$

Then we write the constraints. Since there are limits on rehearsal time and budget, there is a constraint for each. Note that neither x nor y can be negative.

Constraint on rehearsal time	$2x + y \leq 250$
Constraint on cost	$3x + 6y \leq 600$
Nonnegative constraints	$x \geq 0,\ y \geq 0$

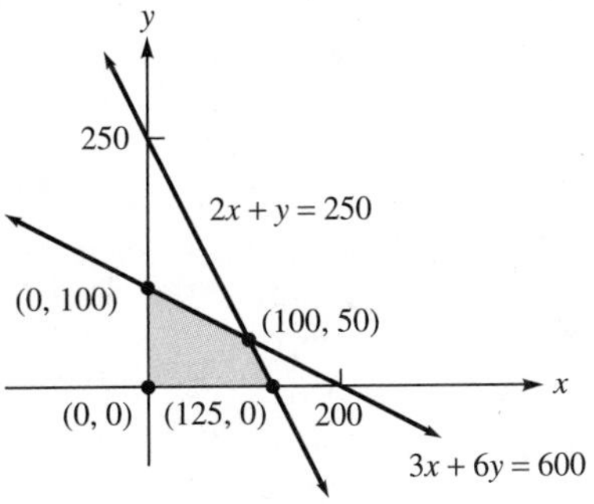

FIGURE 4-40

We graph the inequalities to find the feasibility region shown in Figure 4-40 and find the coordinates of each corner point.

The coordinates of the corner points of the feasibility region are (0, 0), (0, 100), (100, 50), and (125, 0). To find the maximum income, we substitute each pair of coordinates into the objective function:

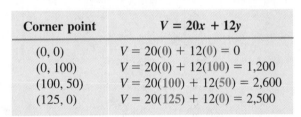

Corner point	$V = 20x + 12y$
(0, 0)	$V = 20(0) + 12(0) = 0$
(0, 100)	$V = 20(0) + 12(100) = 1{,}200$
(100, 50)	$V = 20(100) + 12(50) = 2{,}600$
(125, 0)	$V = 20(125) + 12(0) = 2{,}500$

Maximum income occurs if 100 comedy skits and 50 musical numbers are scheduled. The maximum income will be 2,600 thousand dollars, or \$2,600,000. ■

Orals *Evaluate $P = 2x + 5y$ when*

1. $x = 0,\ y = 5$
2. $x = 2,\ y = 1$

Find the corner points of the region determined by

3. $\begin{cases} x \geq 0 \\ y \geq 0 \\ x + y \leq 3 \end{cases}$
4. $\begin{cases} x \geq 0 \\ y \geq 0 \\ x + 2y \leq 4 \end{cases}$

EXERCISE 4.5

REVIEW *Consider the line passing through $P(-2, 4)$ and $Q(5, 7)$.*

1. Find the slope of line PQ.
2. Write the equation of line PQ in general form.
3. Write the equation of line PQ in slope–intercept form.
4. Write the equation of the line that passes through the origin and is parallel to line PQ.

VOCABULARY AND CONCEPTS *Fill in each blank to make a true statement.*

5. In a linear program, the inequalities are called ________.

6. Ordered pairs that satisfy the constraints of a linear program are called ________ solutions.

7. The function to be maximized (or minimized) in a linear program is called the ________ function.

8. The objective function of a linear program attains a maximum (or minimum), subject to the constraints, at a ______ or along an ______ of the feasibility region.

PRACTICE *In Exercises 9–16, maximize P subject to the following constraints.*

9. $P = 2x + 3y$
$$\begin{cases} x \geq 0 \\ y \geq 0 \\ x + y \leq 4 \end{cases}$$

10. $P = 3x + 2y$
$$\begin{cases} x \geq 0 \\ y \geq 0 \\ x + y \leq 4 \end{cases}$$

11. $P = y + \frac{1}{2}x$
$$\begin{cases} x \geq 0 \\ y \geq 0 \\ 2y - x \leq 1 \\ y - 2x \geq -2 \end{cases}$$

12. $P = 4y - x$
$$\begin{cases} x \leq 2 \\ y \geq 0 \\ x + y \geq 1 \\ 2y - x \leq 1 \end{cases}$$

13. $P = 2x + y$
$$\begin{cases} y \geq 0 \\ y - x \leq 2 \\ 2x + 3y \leq 6 \\ 3x + y \leq 3 \end{cases}$$

14. $P = x - 2y$
$$\begin{cases} x + y \leq 5 \\ y \leq 3 \\ x \leq 2 \\ x \geq 0 \\ y \geq 0 \end{cases}$$

15. $P = 3x - 2y$
$$\begin{cases} x \leq 1 \\ x \geq -1 \\ y - x \leq 1 \\ x - y \leq 1 \end{cases}$$

16. $P = x - y$
$$\begin{cases} 5x + 4y \leq 20 \\ y \leq 5 \\ x \geq 0 \\ y \geq 0 \end{cases}$$

In Exercises 17–24, minimize P subject to the following constraints.

17. $P = 5x + 12y$
$$\begin{cases} x \geq 0 \\ y \geq 0 \\ x + y \leq 4 \end{cases}$$

18. $P = 3x + 6y$
$$\begin{cases} x \geq 0 \\ y \geq 0 \\ x + y \leq 4 \end{cases}$$

19. $P = 3y + x$
$$\begin{cases} x \geq 0 \\ y \geq 0 \\ 2y - x \leq 1 \\ y - 2x \geq -2 \end{cases}$$

20. $P = 5y + x$
$$\begin{cases} x \leq 2 \\ y \geq 0 \\ x + y \geq 1 \\ 2y - x \leq 1 \end{cases}$$

21. $P = 6x + 2y$
$$\begin{cases} y \geq 0 \\ y - x \leq 2 \\ 2x + 3y \leq 6 \\ 3x + y \leq 3 \end{cases}$$

22. $P = 2y - x$
$$\begin{cases} x \geq 0 \\ y \geq 0 \\ x + y \leq 5 \\ x + 2y \geq 2 \end{cases}$$

23. $P = 2x - 2y$
$$\begin{cases} x \leq 1 \\ x \geq -1 \\ y - x \leq 1 \\ x - y \leq 1 \end{cases}$$

24. $P = y - 2x$
$$\begin{cases} x + 2y \leq 4 \\ 2x + y \leq 4 \\ x + 2y \geq 2 \\ 2x + y \geq 2 \end{cases}$$

APPLICATIONS *In Exercises 25–32, write the objective function and the inequalities that describe the constraints in each problem. Graph the feasibility region, showing the corner points. Then find the maximum or minimum value of the objective function.*

25. Making furniture Two woodworkers, Tom and Carlos, bring in \$100 for making a table and \$80 for making a chair. On average, Tom must work 3 hours and Carlos 2 hours to make a chair. Tom must work 2 hours and Carlos 6 hours to make a table. If neither wishes to work more than 42 hours per week, how many tables and how many chairs should they make each week to maximize their income? Find the maximum income.

	Table	Chair	Time available
Income ($)	100	80	
Tom's time (hr)	2	3	42
Carlos's time (hr)	6	2	42

26. Making crafts Two artists, Nina and Rob, make yard ornaments. They bring in $80 for each wooden snowman they make and $64 for each wooden Santa Claus. On average, Nina must work 4 hours and Rob 2 hours to make a snowman. Nina must work 3 hours and Rob 4 hours to make a Santa Claus. If neither wishes to work more than 20 hours per week, how many of each ornament should they make each week to maximize their income? Find the maximum income.

	Snowman	Santa Claus	Time available
Income ($)	80	64	
Nina's time (hr)	4	3	20
Rob's time (hr)	2	4	20

27. Inventories An electronics store manager stocks from 20 to 30 IBM-compatible computers and from 30 to 50 Macintosh computers. There is room in the store to stock up to 60 computers. The manager receives a commission of $50 on the sale of each IBM-compatible computer and $40 on the sale of each Macintosh computer. If the manager can sell all of the computers, how many should she stock to maximize her commissions? Find the maximum commission.

Inventory	IBM	Macintosh
Minimum	20	30
Maximum	30	50
Commission	$50	$40

28. Diet problem A diet requires at least 16 units of vitamin C and at least 34 units of vitamin B complex. Two food supplements are available that provide these nutrients in the amounts and costs shown in the table. How much of each should be used to minimize the cost?

Supplement	Vitamin C	Vitamin B	Cost
A	3 units/g	2 units/g	3¢/g
B	2 units/g	6 units/g	4¢/g

29. Production Manufacturing VCRs and TVs requires the use of the electronics, assembly, and finishing departments of a factory, according to the following schedule:

	Hours for VCR	Hours for TV	Hours available per week
Electronics	3	4	180
Assembly	2	3	120
Finishing	2	1	60

Each VCR has a profit of $40, and each TV has a profit of $32. How many VCRs and TVs should be manufactured weekly to maximize profit? Find the maximum profit.

30. Production problem A company manufactures one type of computer chip that runs at 233 MHz and another that runs at 400 MHz. The company can make a maximum of 50 fast chips per day and a maximum of 100 slow chips per day. It takes 6 hours to make a fast chip and 3 hours to make a slow chip, and the company's employees can provide up to 360 hours of labor per day. If the company makes a profit of $20 on each 400-MHz chip and $27 on each 233-MHz chip, how many of each type should be manufactured to earn the maximum profit?

31. Financial planning A stockbroker has $200,000 to invest in stocks and bonds. She wants to invest at least $100,000 in stocks and at least $50,000 in bonds. If stocks have an annual yield of 9% and bonds have an annual yield of 7%, how much should she invest in each to maximize her income? Find the maximum return.

32. **Production** A small country exports soybeans and flowers. Soybeans require 8 workers per acre, flowers require 12 workers per acre, and 100,000 workers are available. Government contracts require that there be at least 3 times as many acres of soybeans as flowers planted. It costs \$250 per acre to plant soybeans and \$300 per acre to plant flowers, and there is a budget of \$3 million. If the profit from soybeans is \$1,600 per acre and the profit from flowers is \$2,000 per acre, how many acres of each crop should be planted to maximize profit? Find the maximum profit.

WRITING

33. What is meant by the constraints of a linear program?

34. What is meant by a feasible solution of a linear program?

SOMETHING TO THINK ABOUT

35. Try to construct a linear programming problem. What difficulties do you encounter?

36. Try to construct a linear programming problem that will have a maximum at every point along an edge of the feasibility region.

PROJECTS

PROJECT 1 A farmer is building a machine shed onto his barn, as shown in Illustration 1. The shed is to be 12 feet wide, and of course h_2 must be no more than 20 feet. In order for all of the shed to be useful for storing machinery, h_1 must be at least 6 feet. For the roof to shed rain and melting snow adequately, the slope of the roof must be at least $\frac{1}{2}$, but to be easily shingled, it must have a slope that is no greater than 1.

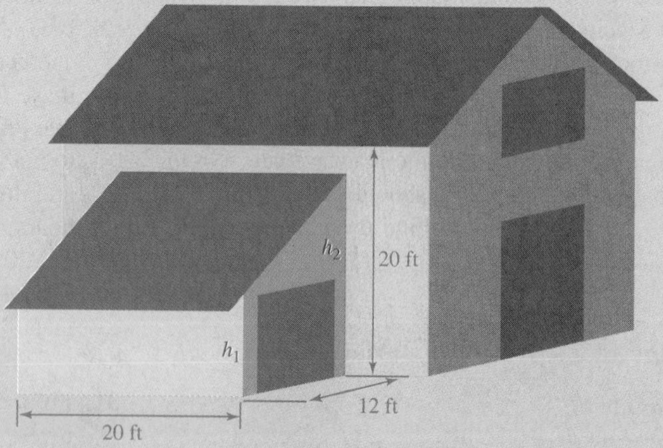

ILLUSTRATION 1

a. Represent on a graph all of the possible values for h_1 and h_2, subject to the constraints listed above.

b. The farmer wishes to minimize the construction costs while still making sure that the shed is large enough for his purposes. He does this by setting a lower bound on the volume of the shed (3,000 cubic feet) and then minimizing the surface area of the walls that must be built. The volume of the shed can be expressed in a formula that contains h_1 and h_2. Derive this formula, and include the volume restriction in your design constraints. Then find the dimensions that will minimize the total area of the two ends of the shed and the outside wall. (The inner wall is already present as a wall of the barn and therefore involves no new cost.)

PROJECT 2 Knowing any three points on the graph of a parabolic function is enough to determine the equation of that parabola. It follows that for any three points that could possibly lie on a parabola, there is exactly one parabola that passes through those points, and this parabola will have the equation $y = ax^2 + bx + c$ for appropriate a, b, and c.

a. In order for a set of three points to lie on the graph of a parabolic function, no two of the points can have the same x-coordinate, and not all three can have the same y-coordinate. Explain why we need these restrictions.

b. Suppose that the points (1, 3) and (2, 6) are on the graph of a parabola. What restrictions would have to be placed on a and b to guarantee that the y-intercept of the parabola has an absolute value of 4 or less? Can $(-1, 8)$ be a third point on such a parabola?

CHAPTER SUMMARY

CONCEPTS	REVIEW EXERCISES
SECTION 4.1	*Linear Inequalities*
Trichotomy property: $a < b$, $a = b$, or $a > b$	**1.** Solve each inequality. Give each solution set in interval notation and graph it. **a.** $5(x - 2) \leq 5$
Transitive properties: If $a < b$ and $b < c$, then $a < c$. If $a > b$ and $b > c$, then $a > c$.	**b.** $3x + 4 > 10$

Properties of inequality:
If a and b are real numbers and $a < b$, then

$a + c < b + c$

$a - c < b - c$

$ac < bc \quad (c > 0)$

$ac > bc \quad (c < 0)$

$\frac{a}{c} < \frac{b}{c} \quad (c > 0)$

$\frac{a}{c} > \frac{b}{c} \quad (c < 0)$

$c < x < d$ is equivalent to $c < x$ and $x < d$.

c. $\frac{1}{3}y - 2 \geq \frac{1}{2}y + 2$

d. $\frac{7}{4}(x + 3) < \frac{3}{8}(x - 3)$

e. $3 < 3x + 4 < 10$

f. $4x > 3x + 2 > x - 3$

g. $-5 \leq 2x - 3 < 5$

2. A woman invests \$10,000 at 6% annual interest. How much more must she invest at 7% so that her annual income is at least \$2,000?

SECTION 4.2 Equations and Inequalities with Absolute Values

If $x \geq 0$, $|x| = x$.
If $x < 0$, $|x| = -x$.

3. Find each absolute value.

a. $|-7|$ **b.** $|8|$

c. $-|7|$ **d.** $-|-12|$

4. Graph each function.

a. $f(x) = |x + 1| - 3$

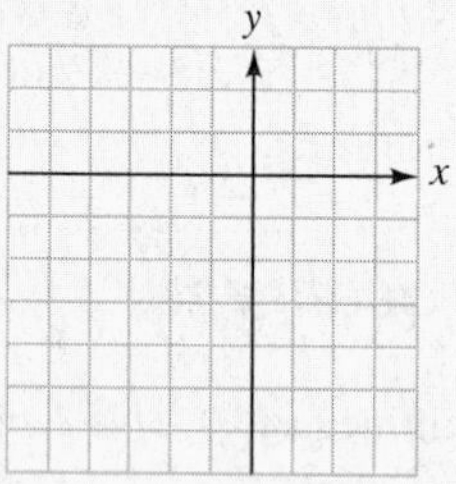

b. $f(x) = |x - 2| + 1$

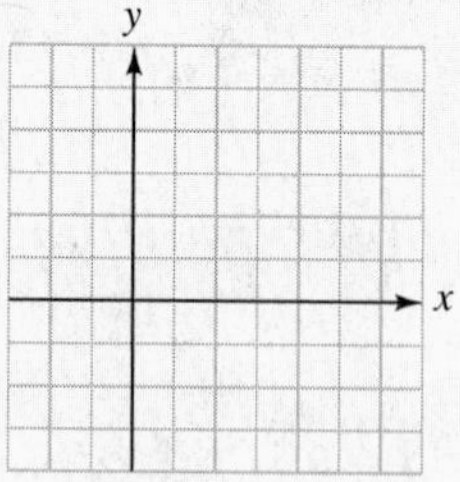

If $k > 0$, $|x| = k$ is equivalent to $x = k$ or $x = -k$.

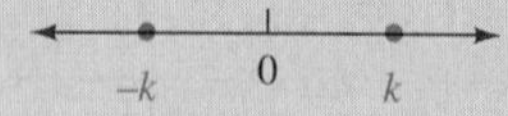

5. Solve and check each equation.

a. $|3x + 1| = 10$ **b.** $\left|\frac{3}{2}x - 4\right| = 9$

c. $\left|\frac{2 - x}{3}\right| = 4$ **d.** $|3x + 2| = |2x - 3|$

$|a| = |b|$ is equivalent to $a = b$ or $a = -b$.

e. $|5x - 4| = |4x - 5|$ **f.** $\left|\frac{3 - 2x}{2}\right| = \left|\frac{3x - 2}{3}\right|$

If $k > 0$, $|x| < k$ is equivalent to $-k < x < k$.

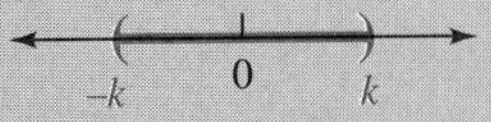

$|x| > k$ is equivalent to $x < -k$ or $x > k$.

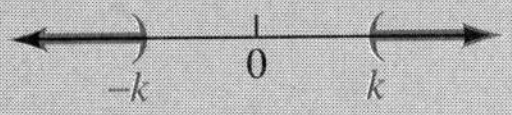

6. Solve each inequality. Give the solution in interval notation and graph it.

a. $|2x + 7| < 3$

b. $|5 - 3x| \leq 14$

c. $\left| \frac{2}{3}x + 14 \right| < 0$

d. $\left| \frac{1 - 5x}{3} \right| > 7$

e. $|3x \geq 8| \geq 4$

f. $\left| \frac{3}{2}x - 14 \right| \geq 0$

SECTION 4.3 Linear Inequalities in Two Variables

To graph a linear inequality in x and y, graph the boundary line, and then use a test point to decide which side of the boundary should be shaded.

7. Graph each inequality on the coordinate plane.

a. $2x + 3y > 6$

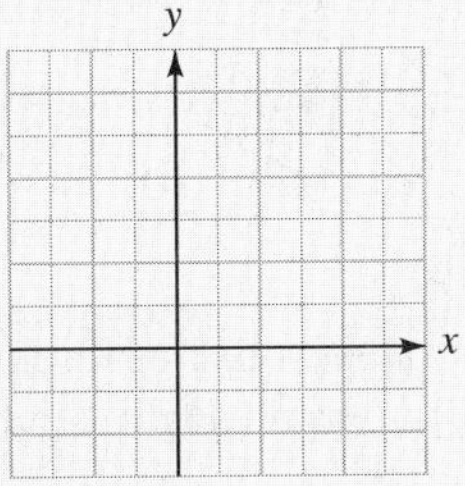

b. $y \leq 4 - x$

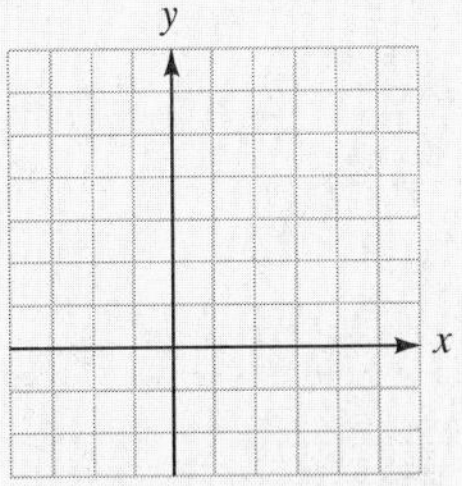

c. $-2 < x < 4$

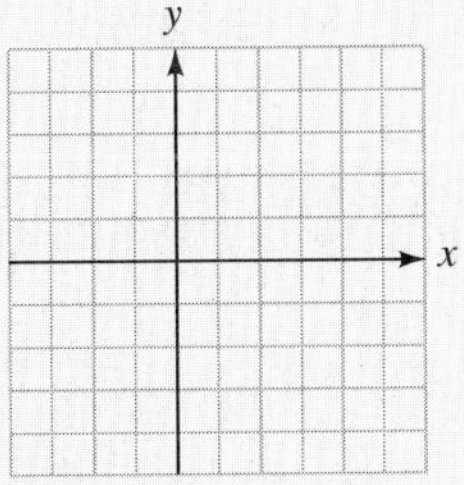

d. $y \leq -2$ or $y > 1$

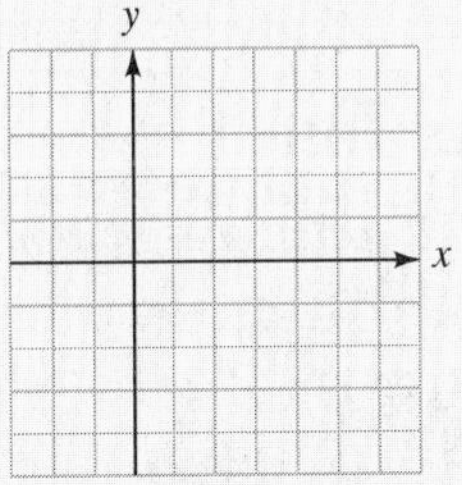

SECTION 4.4 Systems of Inequalities

Systems of inequalities can be solved by graphing.

8. Graph the solution set of each system of inequalities.

a. $\begin{cases} y \geq x + 1 \\ 3x + 2y < 6 \end{cases}$

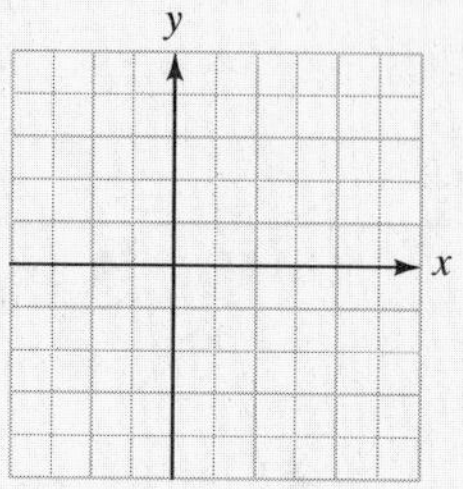

b. $\begin{cases} y \geq x^2 - 4 \\ y < x + 3 \end{cases}$

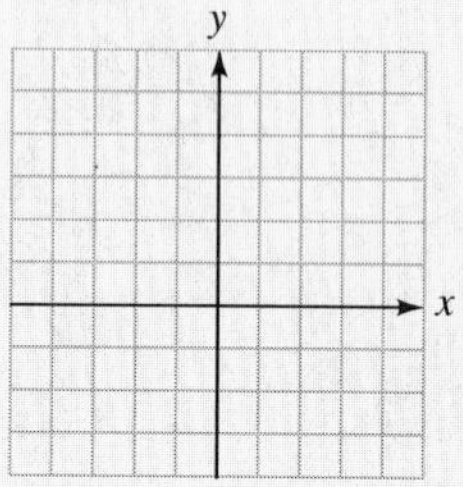

SECTION 4.5 Linear Programming

If a linear function, subject to the constraints of a system of linear inequalities in two variables, attains a maximum or a minimum value, that value will occur at a corner or along an entire edge of the region R that represents the solution of the system.

9. Maximize $P = 2x + y$ subject to $\begin{cases} x \geq 0 \\ y \geq 0 \\ x + y \leq 3 \end{cases}$.

10. A company manufactures fertilizers X and Y. Each 50-pound bag requires three ingredients, which are available in the limited quantities shown below.

Ingredient	Number of pounds in fertilizer X	Number of pounds in fertilizer Y	Total number of pounds available
Nitrogen	6	10	20,000
Phosphorus	8	6	16,400
Potash	6	4	12,000

The profit on each bag of fertilizer X is \$6, and on each bag of Y, \$5. How many bags of each should be produced to maximize profit?

Chapter Test

In Problems 1–2, graph the solution of each inequality. Also give the solution in interval notation.

1. $-2(2x + 3) \geq 14$

2. $-2 < \dfrac{x - 4}{3} < 4$

In Problems 3–4, write each expression without absolute value symbols.

3. $|5 - 8|$

4. $|4\pi - 4|$

In Problems 5–6, graph each function.

5. $f(x) = |x + 1| - 4$

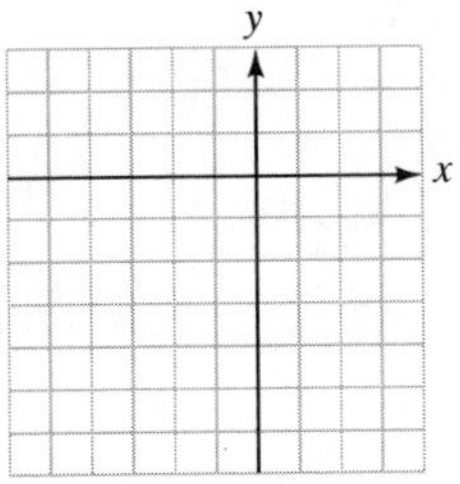

6. $f(x) = |x - 2| + 3$

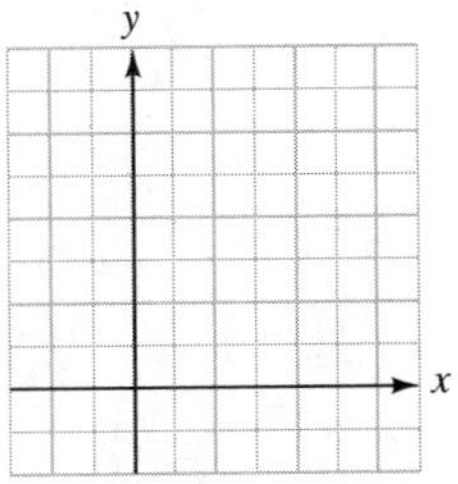

In Problems 7–10, solve each equation.

7. $|2x + 3| = 11$

8. $|4 - 3x| = 19$

9. $|3x + 4| = |x + 12|$

10. $|3 - 2x| = |2x + 3|$

In Problems 11–14, graph the solution of each inequality. Also give the solution in interval notation.

11. $|x + 3| \le 4$

12. $|2x - 4| > 22$

13. $|4 - 2x| > 2$

14. $|2x - 4| \le 2$

In Problems 15–16, graph each inequality.

15. $3x + 2y \ge 6$

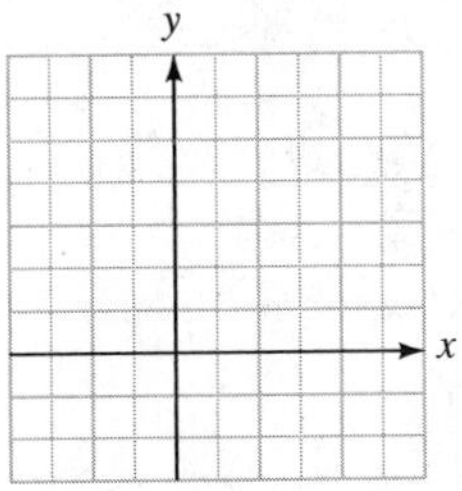

16. $-2 \le y < 5$

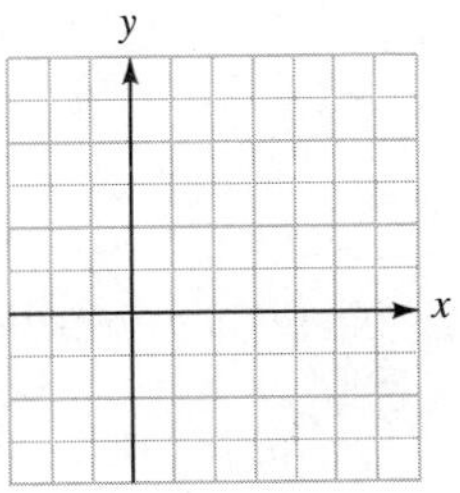

In Problems 17–18, use graphing to solve each system.

17. $\begin{cases} 2x - 3y \geq 6 \\ y \leq -x + 1 \end{cases}$

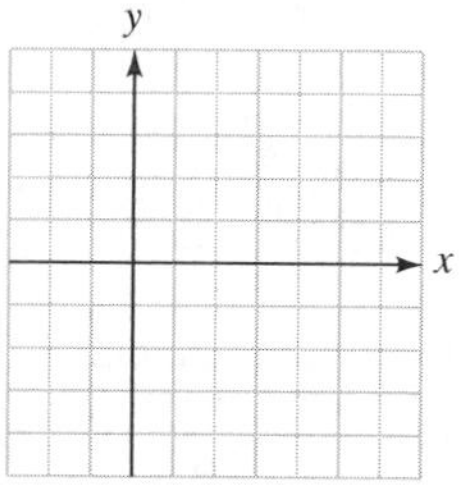

18. $\begin{cases} y \geq x^2 \\ y < x + 3 \end{cases}$

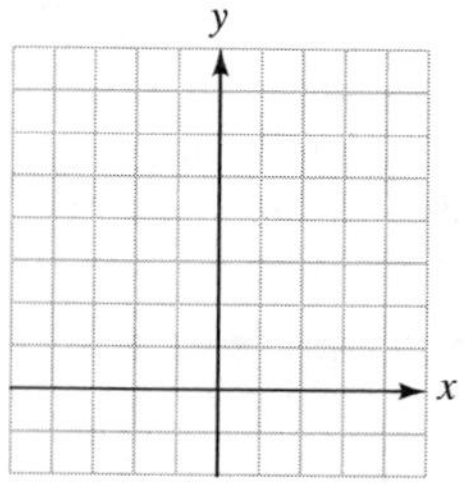

19. Maximize $P = 3x - y$ subject to $\begin{cases} y \geq 1 \\ y \leq 2 \\ y \leq 3x + 1 \\ x \leq 1 \end{cases}$.

Cumulative Review Exercises

1. Draw a number line and graph the prime numbers from 50 to 60.

2. Find the additive inverse of -5.

In Exercises 3–4, evaluate each expression when $x = 2$ and $y = -4$.

3. $x - xy$

4. $\dfrac{x^2 - y^2}{3x + y}$

In Exercises 5–8, simplify each expression.

5. $(x^2x^3)^2$

6. $(x^2)^3(x^4)^2$

7. $\left(\dfrac{x^3}{x^5}\right)^{-2}$

8. $\dfrac{a^2b^n}{a^nb^2}$

9. Write 32,600,000 in scientific notation.

10. Write 0.000012 in scientific notation.

In Exercises 11–14, solve each equation, if possible.

11. $3x - 6 = 20$

12. $6(x - 1) = 2(x + 3)$

13. $\dfrac{5b}{2} - 10 = \dfrac{b}{3} + 3$

14. $2a - 5 = -2a + 4(a - 2) + 1$

In Exercises 15–16, tell whether the lines represented by the equations are parallel, perpendicular, or neither.

15. $3x + 2y = 12$, $2x - 3y = 5$

16. $3x = y + 4$, $y = 3(x - 4) - 1$

17. Write the equation of the line passing through $P(-2, 3)$ and perpendicular to the graph of $3x + y = 8$.

18. Solve the formula $A = \frac{1}{2}h(b_1 + b_2)$ for h.

In Exercises 19–20, $f(x) = 3x^2 - x$. Find each value.

19. $f(2)$

20. $f(-2)$

21. Use graphing to solve $\begin{cases} 2x + y = 5 \\ x - 2y = 0 \end{cases}$.

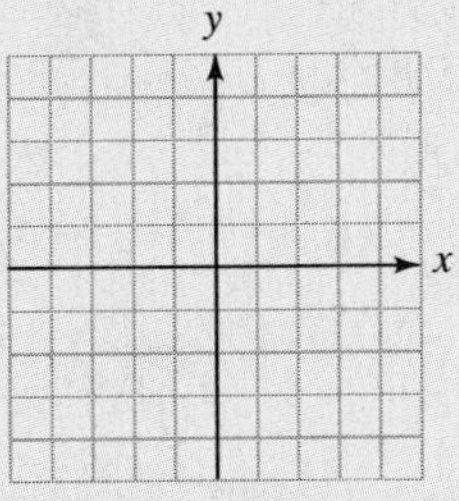

22. Use substitution to solve $\begin{cases} 3x + y = 4 \\ 2x - 3y = -1 \end{cases}$.

23. Use addition to solve $\begin{cases} x + 2y = -2 \\ 2x - y = 6 \end{cases}$.

24. Solve $\begin{cases} \dfrac{x}{10} + \dfrac{y}{5} = \dfrac{1}{2} \\ \dfrac{x}{2} - \dfrac{y}{5} = \dfrac{13}{10} \end{cases}$.

25. Solve $\begin{cases} x + y + z = 1 \\ 2x - y - z = -4 \\ x - 2y + z = 4 \end{cases}$.

26. Solve $\begin{cases} x + 2y + 3z = 1 \\ 3x + 2y + z = -1 \\ 2x + 3y + z = -2 \end{cases}$.

27. Evaluate $\begin{vmatrix} 3 & -2 \\ 1 & -1 \end{vmatrix}$.

28. Evaluate $\begin{vmatrix} 2 & 3 & -1 \\ -1 & -1 & 2 \\ 4 & 1 & -1 \end{vmatrix}$.

In Exercises 29–30, use Cramer's rule to solve each system.

29. $\begin{cases} 4x - 3y = -1 \\ 3x + 4y = -7 \end{cases}$

30. $\begin{cases} x - 2y - z = -2 \\ 3x + y - z = 6 \\ 2x - y + z = -1 \end{cases}$

In Exercises 31–32, solve each inequality.

31. $-3(x - 4) \geq x - 32$

32. $-8 < -3x + 1 < 10$

In Exercises 33–34, solve each equation.

33. $|4x - 3| = 9$

34. $|2x - 1| = |3x + 4|$

In Exercises 35–36, solve each inequality.

35. $|3x - 2| \leq 4$

36. $|2x + 3| - 1 > 4$

In Exercises 37–38, use graphing to solve each inequality.

37. $2x - 3y \leq 12$

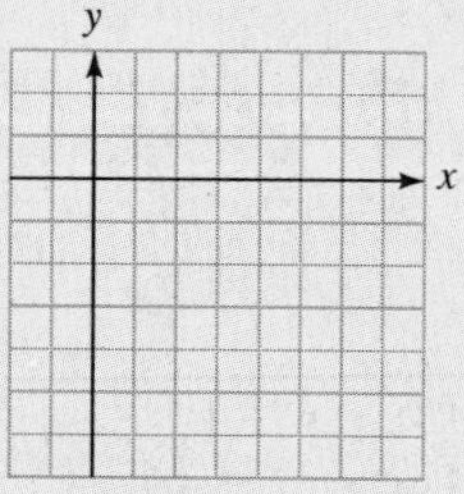

38. $3 > x \geq -2$

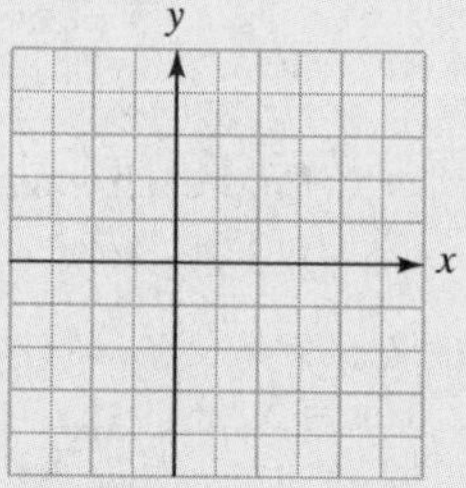

In Exercises 39–40, use graphing to solve each system of inequalities.

39. $\begin{cases} 3x - 2y < 6 \\ y < -x + 2 \end{cases}$

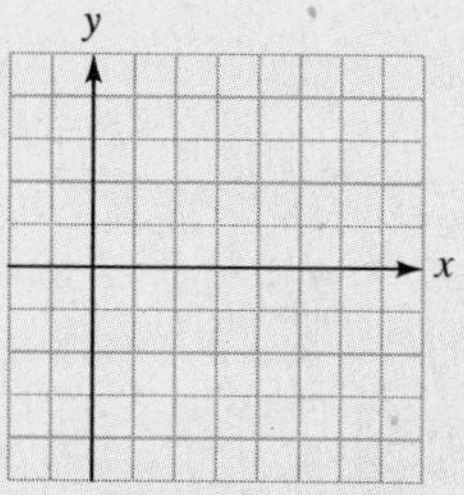

40. $\begin{cases} y < x + 2 \\ 3x + y \leq 6 \end{cases}$

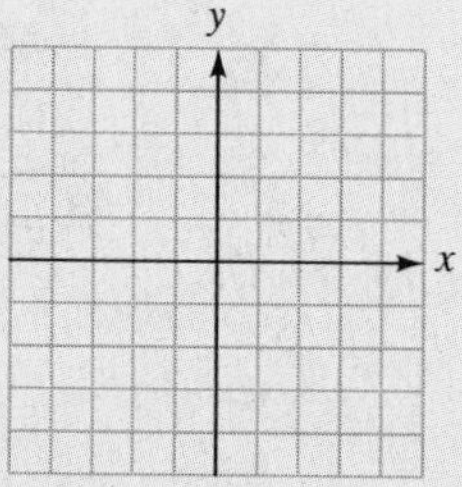

41. To conduct an experiment with mice and rats, a researcher will place the animals into one of two mazes for the number of minutes shown in Illustration 1. Find the greatest number of animals that can be used in this experiment.

	Time per mouse	Time per rat	Time available
Maze 1	12 min	8 min	240 min
Maze 2	10 min	15 min	300 min

ILLUSTRATION 1

5 Polynomials and Polynomial Functions

MATHEMATICS IN THE WORKPLACE

Computer Programmer

Computers process vast quantities of information rapidly and accurately when they are given programs to follow. Computer programmers write those programs, which logically list the steps the machine must follow to organize data, solve a problem, or do other tasks. Applications programmers are usually oriented toward business, engineering, or science. System programmers maintain the software that controls a computer system.

SAMPLE APPLICATION ■ Computers take more time to do multiplications than additions. To make a program run as quickly as possible, a computer programmer wants to write the polynomial $3x^4 + 2x^3 + 5x^2 + 7x + 1$ in a form that requires fewer multiplications. Write the polynomial so that it contains only four multiplications.
(See Example 10 in Section 5.4.)

5.1 Polynomials and Polynomial Functions

■ POLYNOMIALS ■ DEGREE OF A POLYNOMIAL ■ EVALUATING POLYNOMIALS ■ POLYNOMIAL FUNCTIONS IN ONE VARIABLE

Getting Ready *Write each expression using exponents.*

1. $3aabb$ **2.** $-5xxxy$

3. $4pp + 7qq$ **4.** $aaa - bb$

■ POLYNOMIALS

Algebraic terms are expressions that contain constants and/or variables. Some examples are

$$17, \quad 9x, \quad 15y^2, \quad \text{and} \quad -24x^4y^5$$

The *numerical coefficient* of 17 is 17. The numerical coefficients of the remaining terms are 9, 15, and -24, respectively.

A **polynomial** is the sum of one or more algebraic terms whose variables have whole-number exponents.

Polynomial in One Variable
A **polynomial in one variable** (say, x) is the sum of one or more terms of the form ax^n, where a is a real number and n is a whole number.

The following expressions are polynomials in x. Note that $17 = 17x^0$.

$$3x^2 + 2x, \qquad \frac{3}{2}x^5 - \frac{7}{3}x^4 - \frac{8}{3}x^3, \qquad \text{and} \qquad 19x^{20} + \sqrt{3}x^{14} + 4.5x^{11} - 17$$

The following expressions are not polynomials.

$$\frac{2x}{x^2 + 1}, \qquad x^{1/2} - 1, \qquad \text{and} \qquad x^{-3} + 2x$$

The first expression is the quotient of two polynomials, and the last two have exponents that are not whole numbers.

Polynomial in More than One Variable

A **polynomial in several variables** (say, x, y, and z) is the sum of one or more terms of the form $ax^m y^n z^p$, where a is a real number and m, n, and p are whole numbers.

The following expressions are polynomials in more than one variable.

$$3xy, \qquad 5x^2y + 2yz^3 - 3xz, \qquad \text{and} \qquad u^2v^2w^2 + x^3y^3 + 1$$

A polynomial with one term is called a **monomial,** a polynomial with two terms is called a **binomial,** and a polynomial with three terms is called a **trinomial:**

Monomials	*Binomials*	*Trinomials*
$2x^3$	$2x^4 + 5$	$2x^3 + 4x^2 + 3$
a^2b	$-17t^{45} - 3xy$	$3mn^3 - m^2n^3 + 7n$
$3x^3y^5z^2$	$32x^{13}y^5 + 47x^3yz$	$-12x^5y^2 + 13x^4y^3 - 7x^3y^3$

DEGREE OF A POLYNOMIAL

Because the variable x occurs three times as a factor in the monomial $2x^3$, the monomial is called a *third-degree monomial* or a *monomial of degree 3.* The monomial $3x^3y^5z^2$ is called a *monomial of degree 10,* because the variables x, y, and z occur as factors a total of ten times. These examples illustrate the following definition.

Degree of a Monomial

If $a \neq 0$, the **degree of ax^n** is n. The degree of a monomial containing several variables is the sum of the exponents on those variables.

EXAMPLE 1 Find the degree of **a.** $3x^4$, **b.** $-18x^3y^2z^{12}$, **c.** $4^7x^2y^3$, and **d.** 3.

Solution **a.** $3x^4$ is a monomial of degree 4, because the variable x occurs as a factor four times.

b. $-18x^3y^2z^{12}$ is a monomial of degree 17, because the sum of the exponents on the variables is 17.

c. $4^7x^2y^3$ is a monomial of degree 5, because the sum of the exponents on the variables is 5.

d. 3 is a monomial of degree 0, because $3 = 3x^0$. ■

Self Check Find the degree of **a.** $-12a^5$ and **b.** $8a^3b^2$.

Answers **a.** 5, **b.** 5

WARNING! Since $a \neq 0$ in the previous definition, 0 has no defined degree.

We define the degree of a polynomial by considering the degrees of each of its terms.

Degree of a Polynomial

The **degree of a polynomial** is the same as the degree of the term in the polynomial with largest degree.

EXAMPLE 2 Find the degree of each polynomial: **a.** $3x^5 + 4x^2 + 7$, **b.** $7x^2y^8 - 3xy$, **c.** $3x + 2y - xy$, and **d.** $18x^2y^3 - 12x^7y^2 + 3x^9y^3 - 3$.

Solution **a.** $3x^5 + 4x^2 + 7$ is a trinomial of degree 5, because the largest degree of the three monomials is 5.

b. $7x^2y^8 - 3xy$ is a binomial of degree 10.

c. $3x + 2y - xy$ is a trinomial of degree 2.

d. $18x^2y^3 - 12x^7y^2 + 3x^9y^3 - 3$ is a polynomial of degree 12. ■

Self Check Find the degree of $7a^3b^2 - 14a^2b^4$.

Answer 6

If the terms of a polynomial in one variable are written so that the exponents decrease as we move from left to right, we say that the terms are written with their exponents in descending order. If the terms are written so that the exponents increase as we move from left to right, we say that the terms are written with their exponents in ascending order.

EXAMPLE 3 Write the exponents of $7x^2 - 5x^4 + 3x + 2x^3 - 1$ in **a.** descending order and **b.** ascending order.

Solution **a.** $-5x^4 + 2x^3 + 7x^2 + 3x - 1$

b. $-1 + 3x + 7x^2 + 2x^3 - 5x^4$ ■

Self Check Write the exponents of $3x^2 + 2x^4 - 5x - 3$ in descending order.

Answer $2x^4 + 3x^2 - 5x - 3$

In the following polynomial, the exponents on x are in descending order, and the exponents on y are in ascending order.

$$7x^4 - 2x^3y + 4x^2y^2 - 8xy^3 + 12y^4$$

EVALUATING POLYNOMIALS

Polynomials in one variable are often denoted by expressions such as

$P(x)$ and $Q(t)$ Read $P(x)$ as "P of x" and $Q(t)$ as "Q of t."

where the letter within the parentheses represents the variable of the polynomial.

The symbols $P(x)$ and $Q(t)$ could represent the polynomials

$$P(x) = x^6 + 4x^5 - 3x^2 + x - 2 \quad \text{and} \quad Q(t) = t^2 - 3t + 5$$

To evaluate a polynomial at a specific value of its variable, we substitute the value of the variable and simplify. For example, to evaluate $P(x)$ at $x = 1$, we substitute 1 for x and simplify.

$$\begin{aligned} P(x) &= x^6 + 4x^5 - 3x^2 + x - 2 \\ P(1) &= (1)^6 + 4(1)^5 - 3(1)^2 + 1 - 2 \quad \text{Substitute 1 for } x. \\ &= 1 + 4 - 3 + 1 - 2 \\ &= 1 \end{aligned}$$

Note that to each number x there will correspond a single value $P(x)$.

EXAMPLE 4 **Height of a rocket** If a toy rocket is launched straight up with an initial velocity of 128 feet per second, its height h (in feet) above the ground after t seconds is given by the polynomial

$$P(t) = -16t^2 + 128t \quad \text{The height } h \text{ is the value } P(t).$$

Find the height of the rocket at **a.** 0 second, **b.** 3 seconds, and **c.** 7.9 seconds.

Solution **a.** To find the height at 0 second, we substitute 0 for t and simplify.

$$\begin{aligned} P(t) &= -16t^2 + 128t \\ P(0) &= -16(0)^2 + 128(0) \\ &= 0 \end{aligned}$$

At 0 second, the rocket is on the ground waiting to be launched.

b. To find the height at 3 seconds, we substitute 3 for t and simplify.

$$\begin{aligned} P(3) &= -16(3)^2 + 128(3) \\ &= -16(9) + 384 \\ &= -144 + 384 \\ &= 240 \end{aligned}$$

At 3 seconds, the height of the rocket is 240 feet.

c. To find the height at 7.9 seconds, we substitute 7.9 for t and simplify.

$$\begin{aligned} P(7.9) &= -16(7.9)^2 + 128(7.9) \\ &= -16(62.41) + 1{,}011.2 \\ &= -998.56 + 1{,}011.2 \\ &= 12.64 \end{aligned}$$

At 7.9 seconds, the height is 12.64 feet. It has fallen nearly all the way back to earth. ■

Self Check In Example 4, find the height of the rocket at 4 seconds.

Answer 256 feet

EXAMPLE 5 For the polynomial function $P(x) = 3x^2 - 2x + 7$, find **a.** $P(a)$ and **b.** $P(-2t)$.

Solution **a.** $P(a) = 3(a)^2 - 2(a) + 7$
$= 3a^2 - 2a + 7$

b. $P(-2t) = 3(-2t)^2 - 2(-2t) + 7$
$= 12t^2 + 4t + 7$ ■

Self Check In Example 5, find $P(2t)$.

Answer $12t^2 - 4t + 7$

To evaluate polynomials with more than one variable, we substitute values for the variables into the polynomial and simplify.

EXAMPLE 6 Evaluate $4x^2y - 5xy^3$ at $x = 3$ and $y = -2$.

Solution We substitute 3 for x and -2 for y and simplify.

$$\begin{aligned} 4x^2y - 5xy^3 &= 4(3)^2(-2) - 5(3)(-2)^3 \\ &= 4(9)(-2) - 5(3)(-8) \\ &= -72 + 120 \\ &= 48 \end{aligned}$$

■

Self Check Evaluate $3ab^2 - 2a^2b$ at $a = 2$ and $b = -3$.

Answer 78

POLYNOMIAL FUNCTIONS IN ONE VARIABLE

Since any value we substitute for x in the polynomial $P(x)$ gives a single result, the equation $y = P(x)$ defines a function.

Polynomial Functions

A **polynomial function in one variable** (say, x) is defined by an equation of the form $y = f(x) = P(x)$, where $P(x)$ is a polynomial in the variable x.

The **degree of the polynomial function $y = f(x) = P(x)$** is the same as the degree of $P(x)$.

We have previously discussed three basic polynomial functions. The graph of the linear function $y = f(x) = 3x + 1$, the graph of the squaring function $y = f(x) = x^2$, and the graph of the cubing function $y = f(x) = x^3$ are shown in Figure 5-1.

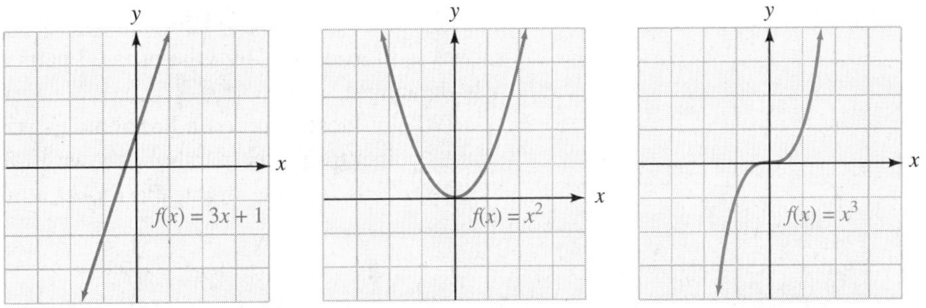

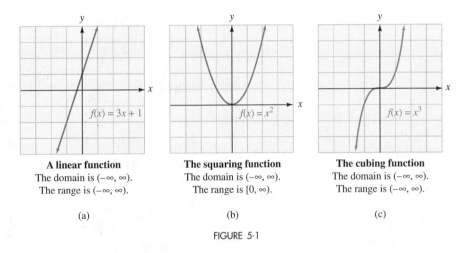

A linear function
The domain is $(-\infty, \infty)$.
The range is $(-\infty, \infty)$.

(a)

The squaring function
The domain is $(-\infty, \infty)$.
The range is $[0, \infty)$.

(b)

The cubing function
The domain is $(-\infty, \infty)$.
The range is $(-\infty, \infty)$.

(c)

FIGURE 5-1

We have seen that the graphs of many polynomial functions are translations or reflections of these basic graphs. For example, the graph of the function $f(x) = x^2 - 2$ is the graph of $f(x) = x^2$ translated 2 units downward, as shown in Figure 5-2(a). The graph of $f(x) = -x^2$ is the graph of $f(x) = x^2$ reflected about the x-axis, as shown in Figure 5-2(b).

In Example 4, we saw that a polynomial function can describe (or model) the flight of a rocket. Since the height (h) of the rocket depends on time (t), we say that the height is a function of time, and we can write $h = f(t)$. To graph this function, we can make a table of values, plot the points, and join them with a smooth curve, as shown in Example 7.

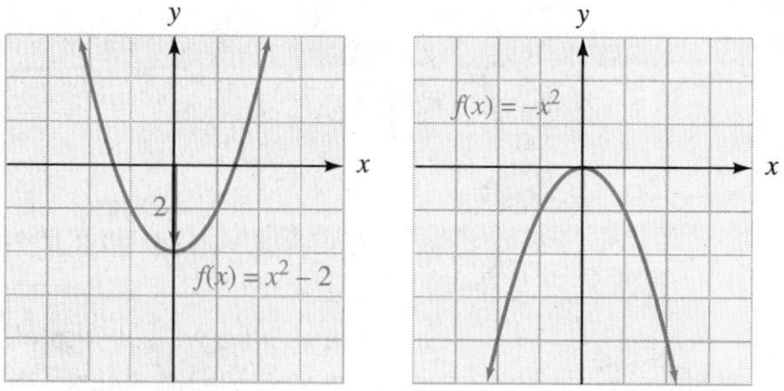

The domain is $(-\infty, \infty)$.
The range is $[-2, \infty)$.

(a)

The domain is $(-\infty, \infty)$.
The range is $(-\infty, 0]$.

(b)

FIGURE 5-2

EXAMPLE 7 Graph $h = f(t) = -16t^2 + 128t$.

Solution From Example 4, we have seen that

When $t = 0$, then $h = 0$.
When $t = 3$, then $h = 240$.

These ordered pairs and others that satisfy $h = -16t^2 + 128t$ are given in the table shown in Figure 5-3. Here the ordered pairs are pairs (t, h), where t is the time and h is the height.

We plot these pairs on a horizontal t-axis and a vertical h-axis and join the resulting points to get the parabola shown in the figure. From the graph, we can see that 4 seconds into the flight, the rocket attains a maximum height of 256 feet.

$h = f(t) = -16t^2 + 128t$

t	$h = f(t)$	$(t, f(t))$
0	0	(0, 0)
1	112	(1, 112)
2	192	(2, 192)
3	240	(3, 240)
4	256	(4, 256)
5	240	(5, 240)
6	192	(6, 192)
7	112	(7, 112)
8	0	(8, 0)

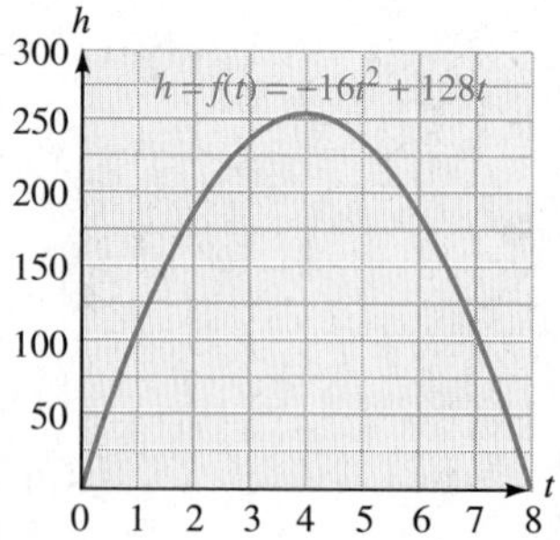

FIGURE 5-3 ■

Self Check Use the graph in Figure 5-3 to estimate the height of the rocket at 6.5 seconds.

Answer 156 feet

WARNING! The parabola shown in Figure 5-3 describes the height of the rocket in relation to time. It does not show the path of the rocket. The rocket goes straight up and then comes straight down.

Graphing Polynomial Functions

GRAPHING CALCULATORS

We can graph polynomial functions with a graphing calculator. For example, to graph $y = P(t) = -16t^2 + 128t$, we can use window settings of [0, 8] for x and [0, 260] for y to get the parabola shown in Figure 5-4(a).

We can trace to estimate the height of the rocket for any number of seconds into the flight. Figure 5-4(b) shows that the height of the rocket 1.6 seconds into the flight is approximately 165 feet.

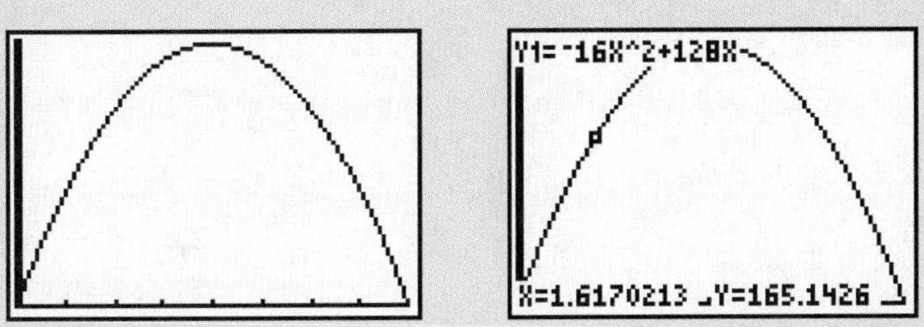

(a) (b)

FIGURE 5-4

Orals *Give the degree of each polynomial.*

1. $8x^3$ **2.** $-4x^3y$ **3.** $4x^2y + 2x$ **4.** $3x + 4xyz$

If $P(x) = 2x + 1$, *find each value.*

5. $P(0)$ **6.** $P(2)$ **7.** $P(-1)$ **8.** $P(-2)$

EXERCISE 5.1

REVIEW *Write each expression with a single exponent.*

1. a^3a^2 **2.** $\frac{b^3b^3}{b^4}$ **3.** $\frac{3(y^3)^{10}}{y^3y^4}$ **4.** $\frac{4x^{-4}x^5}{2x^{-6}}$

5. The distance from Mars to the sun is about 114,000,000 miles. Express this number in scientific notation.

6. One angstrom is about 0.0000001 millimeter. Express this number in scientific notation.

VOCABULARY AND CONCEPTS *In Exercises 7–12, fill in each blank to make a true statement.*

7. A polynomial is the ____ of one or more algebraic terms whose variables have ______-number exponents.

8. A monomial is a polynomial with ____ term.

9. A ______ is a polynomial with two terms.

10. A ______ is a polynomial with three terms.

11. The equation $y = P(x)$, where $P(x)$ is a polynomial, defines a function, because each input value x determines ____ output value y.

12. The degree of the polynomial function $y = P(x)$ is the same as the degree of ____.

In Exercises 13–20, classify each polynomial as a monomial, a binomial, a trinomial, or none of these.

13. $3x^2$

14. $2y^3 + 4y^2$

15. $3x^2y - 2x + 3y$

16. $a^2 + b^2$

17. $x^2 - y^2$

18. $\frac{17}{2}x^3 + 3x^2 - x - 4$

19. 5

20. $8x^3y^5$

In Exercises 21–28, find the degree of each polynomial.

21. $3x^2 + 2$

22. x^{17}

23. $4x^8 + 3x^2y^4$

24. $19x^2y^4 - y^{10}$

25. $4x^2 - 5y^3z^3t^4$

26. $7x$

27. 121

28. $x^2y^3z^4 + z^{12}$

In Exercises 29–32, write each polynomial with the exponents on x in descending order.

29. $3x - 2x^4 + 7 - 5x^2$

30. $-x^2 + 3x^5 - 7x + 3x^3$

31. $a^2x - ax^3 + 7a^3x^5 - 5a^3x^2$

32. $4x^2y^7 - 3x^5y^2 + 4x^3y^3 - 2x^4y^6 + 5x^6$

In Exercises 33–36, write each polynomial with the exponents on y in ascending order.

33. $4y^2 - 2y^5 + 7y - 5y^3$

34. $y^3 + 3y^2 + 8y^4 - 2$

35. $5x^3y^6 + 2x^4y - 5x^3y^3 + x^5y^7 - 2y^4$

36. $-x^3y^2 + x^2y^3 - 2x^3y + x^7y^6 - 3x^6$

PRACTICE *In Exercises 37–40, consider the polynomial $P(x) = 2x^2 + x + 2$. Find each value.*

37. $P(0)$

38. $P(1)$

39. $P(-2)$

40. $P(-3)$

In Exercises 41–44, the height h, in feet, of a ball shot straight up with an initial velocity of 64 feet per second is given by the polynomial function $h = f(t) = -16t^2 + 64t$. Find the height of the ball after the given number of seconds.

41. 0 second

42. 1 second

43. 2 seconds

44. 4 seconds

In Exercises 45–52, find each value when $x = 2$ and $y = -3$.

45. $x^2 + y^2$ **46.** $x^3 + y^3$ **47.** $x^3 - y^3$ **48.** $x^2 - y^2$

49. $3x^2y + xy^3$ **50.** $8xy - xy^2$ **51.** $-2xy^2 + x^2y$ **52.** $-x^3y - x^2y^2$

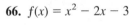

In Exercises 53–58, use a calculator to find each value when $x = 3.7$, $y = -2.5$, and $z = 8.9$.

53. x^2y **54.** xyz^2 **55.** $\frac{x^2}{z^2}$ **56.** $\frac{z^3}{y^2}$

57. $\frac{x + y + z}{xyz}$ **58.** $\frac{x + yz}{xy + z}$

In Exercises 59–66, graph each polynomial function. Check your work with a graphing calculator.

59. $f(x) = x^2 + 2$

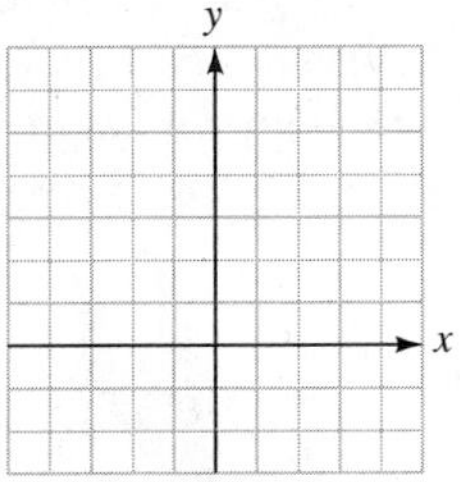

60. $f(x) = x^3 - 2$

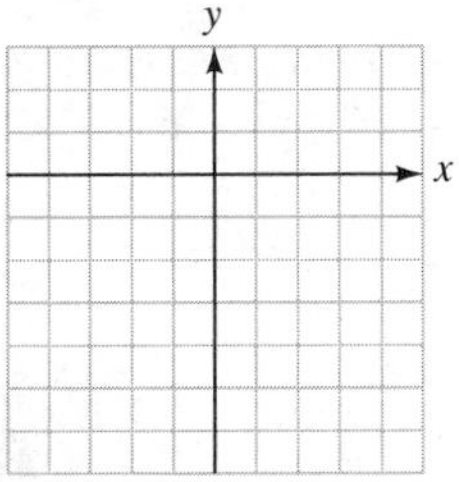

61. $f(x) = -x^3$

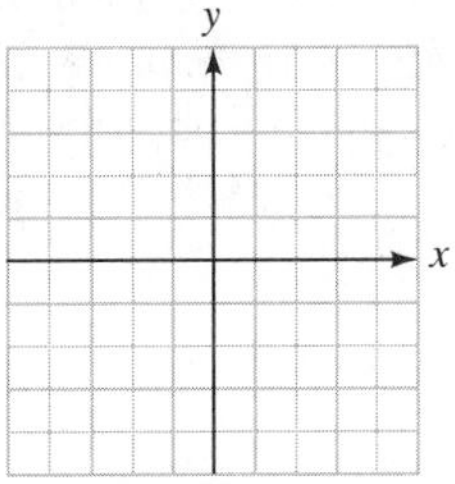

62. $f(x) = -x^2 + 1$

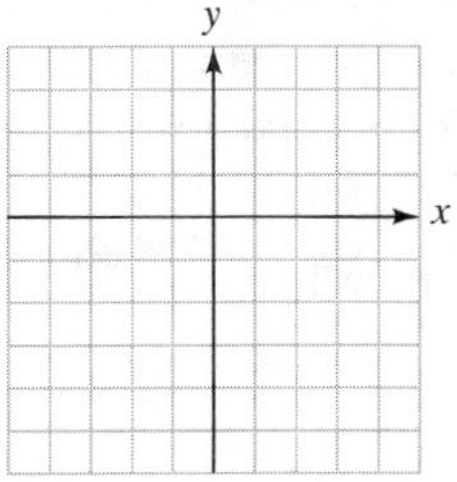

63. $f(x) = -x^3 + x$

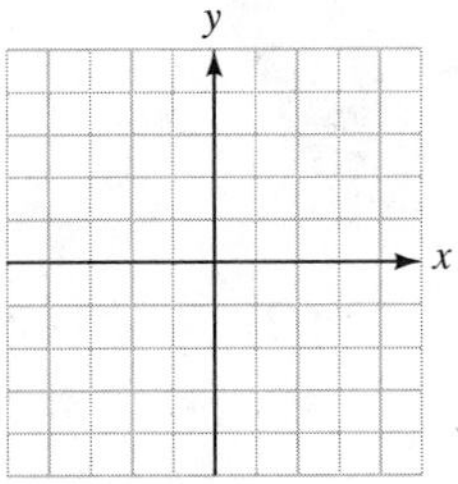

64. $f(x) = x^3 - x$

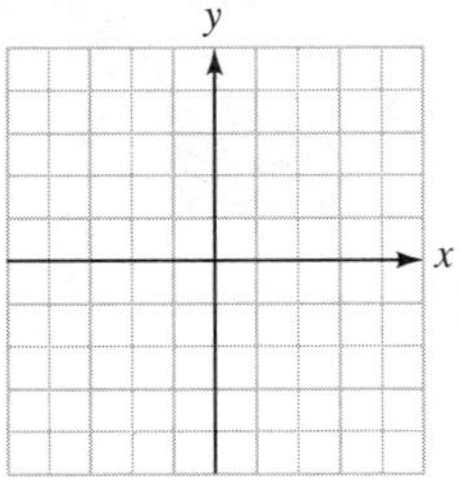

65. $f(x) = x^2 - 2x + 1$

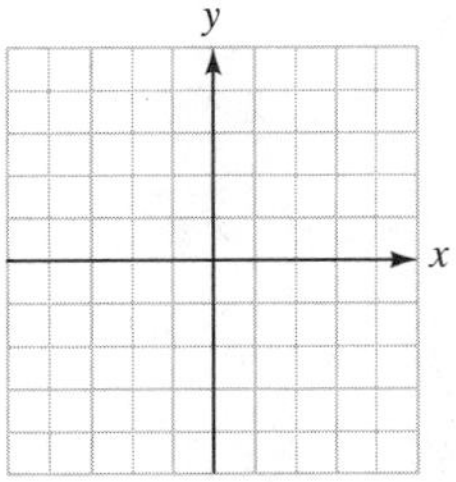

66. $f(x) = x^2 - 2x - 3$

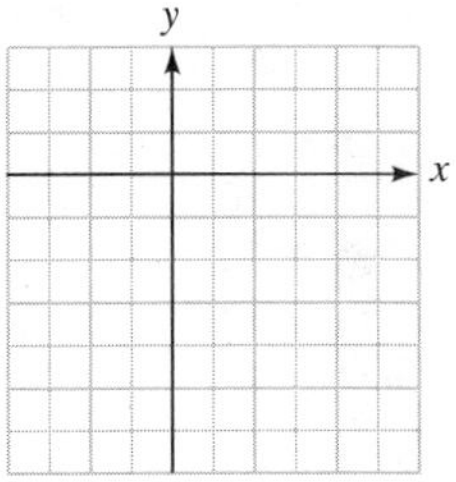

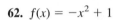

In Exercises 67–70, use a graphing calculator to graph each polynomial function. Use window settings of $[-4, 6]$ for x and $[-5, 5]$ for y.

67. $f(x) = 2.75x^2 - 4.7x + 1.5$ **68.** $f(x) = -2.5x^2 + 1.7x + 3.2$

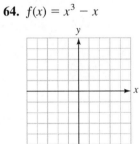

69. $f(x) = 0.25x^2 - 0.5x - 2.5$ **70.** $f(x) = 0.37x^2 - 1.4x - 1.5$

APPLICATIONS *In Exercises 71–74, the number of feet that a car travels before stopping depends on the driver's reaction time and the braking distance. See Illustration 1. For one driver, the stopping distance d is given by the polynomial function $d = f(v) = 0.04v^2 + 0.9v$, where v is the speed of the car. Find the stopping distance for each of the following speeds.*

71. 30 mph

72. 50 mph

73. 60 mph

74. 70 mph

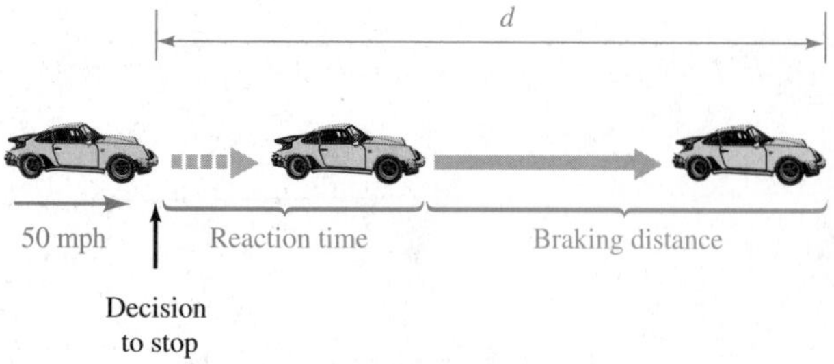

ILLUSTRATION 1

In Exercises 75–78, a rectangular sheet of metal will be used to make a rain gutter by bending up its sides, as shown in Illustration 2. Since a cross section is a rectangle, the cross-sectional area is the product of its length and width, and the capacity c of the gutter is a polynomial function of x: $c = f(x) = -2x^2 + 12x$. Find the capacity for each value of x.

75. 1 inch

76. 2 inches

77. 3 inches

78. 4 inches

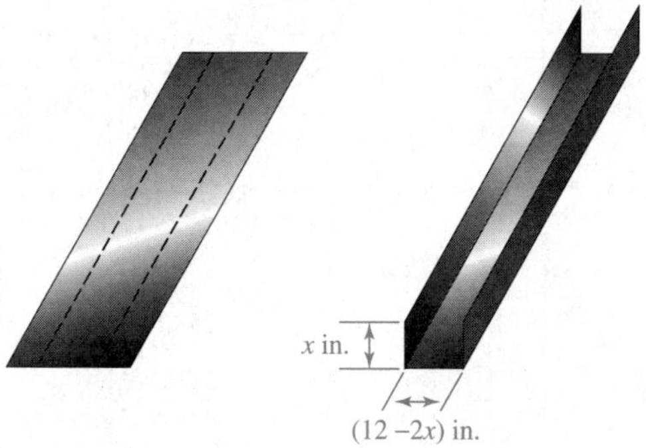

ILLUSTRATION 2

WRITING

79. Explain how to find the degree of a polynomial.

80. Explain why $P(x) = x^6 + 4x^5 - 3x^2 + x - 2$ defines a function.

SOMETHING TO THINK ABOUT

81. If $P(x) = x^2 - 5x$, is $P(2) + P(3) = P(2 + 3)$?

82. If $P(x) = x^3 - 3x$, is $P(2) - P(3) = P(2 - 3)$?

83. If $P(x) = x^2 - 2x - 3$, find $P(P(0))$.

84. If $P(x) = 2x^2 - x - 5$, find $P(P(-1))$.

85. Graph $f(x) = x^2$, $f(x) = 2x^2$, and $f(x) = 4x^2$. What do you discover?

86. Graph $f(x) = x^2$, $f(x) = \frac{1}{2}x^2$, and $f(x) = \frac{1}{4}x^2$. What do you discover?

5.2 Adding and Subtracting Polynomials

■ COMBINING LIKE TERMS ■ ADDING POLYNOMIALS ■ SUBTRACTING POLYNOMIALS ■ ADDING AND SUBTRACTING MULTIPLES OF POLYNOMIALS

Getting Ready *Use the distributive property to remove parentheses.*

1. $2(x + 3)$

2. $-4(2x - 5)$

3. $\frac{5}{2}(2x + 6)$

4. $-\frac{2}{3}(3x - 9)$

■ COMBINING LIKE TERMS

Recall that when terms have the same variables with the same exponents, they are called **like** or **similar terms.**

- $3x^2$, $5x^2$, and $7x^2$ are like terms, because they have the same variables with the same exponents.
- $5x^3y^2$, $17x^3y^2$, and $103x^3y^2$ are like terms, because they have the same variables with the same exponents.
- $4x^4y^2$, $12xy^5$, and $98x^7y^9$ are unlike terms. They have the same variables, but with different exponents.
- $3x^4y$ and $5x^4z^2$ are unlike terms. They have different variables.

The distributive property enables us to combine like terms. For example,

$3x + 7x = (3 + 7)x$	Use the distributive property.
$= 10x$	$3 + 7 = 10.$
$5x^2y^3 + 22x^2y^3 = (5 + 22)x^2y^3$	Use the distributive property.
$= 27x^2y^3$	$5 + 22 = 27.$
$9xy^4 + 6xy^4 + xy^4 = 9xy^4 + 6xy^4 + 1xy^4$	$xy^4 = 1xy^4.$
$= (9 + 6 + 1)xy^4$	Use the distributive property.
$= 16xy^4$	$9 + 6 + 1 = 16.$

The results of the previous example suggest that to add like terms, *we add their numerical coefficients and keep the same variables with the same exponents.*

WARNING! The terms in the following binomials cannot be combined, because they are not like terms.

$$3x^2 - 5y^2, \qquad -2a^2 + 3a^3, \qquad \text{and} \qquad 5y^2 + 17xy$$

ADDING POLYNOMIALS

To add polynomials, we use the distributive property to remove parentheses and combine like terms, whenever possible.

EXAMPLE 1 Add $3x^2 - 2x + 4$ and $2x^2 + 4x - 3$.

Solution

$(3x^2 - 2x + 4) + (2x^2 + 4x - 3)$

$= 1(3x^2 - 2x + 4) + 1(2x^2 + 4x - 3)$ — Each polynomial has an understood coefficient of 1.

$= 3x^2 - 2x + 4 + 2x^2 + 4x - 3$ — Use the distributive property to remove parentheses.

$= 3x^2 + 2x^2 - 2x + 4x + 4 - 3$ — Use the commutative property of addition to get the terms involving x^2 together and the terms involving x together.

$= 5x^2 + 2x + 1$ — Combine like terms. ■

Self Check Add $2a^2 - 3a + 5$ and $5a^2 - 4a - 2$.

Answer $7a^2 - 7a + 3$

EXAMPLE 2 Add $5x^3y^2 + 4x^2y^3$ and $-2x^3y^2 + 5x^2y^3$.

Solution

$(5x^3y^2 + 4x^2y^3) + (-2x^3y^2 + 5x^2y^3)$

$= 1(5x^3y^2 + 4x^2y^3) + 1(-2x^3y^2 + 5x^2y^3)$

$= 5x^3y^2 + 4x^2y^3 - 2x^3y^2 + 5x^2y^3$

$= 5x^3y^2 - 2x^3y^2 + 4x^2y^3 + 5x^2y^3$

$= 3x^3y^2 + 9x^2y^3$ ■

Self Check Add $6a^2b^3 + 5a^3b^2$ and $-3a^2b^3 - 2a^3b^2$.

Answer $3a^2b^3 + 3a^3b^2$

The additions in Examples 1 and 2 can be done by aligning the terms vertically.

$$\begin{array}{r} 3x^2 - 2x + 4 \\ \underline{2x^2 + 4x - 3} \\ 5x^2 + 2x + 1 \end{array} \qquad \begin{array}{r} 5x^3y^2 + 4x^2y^3 \\ \underline{-2x^3y^2 + 5x^2y^3} \\ 3x^3y^2 + 9x^2y^3 \end{array}$$

SUBTRACTING POLYNOMIALS

To subtract one monomial from another, we add the negative (or opposite) of the monomial that is to be subtracted.

EXAMPLE 3 Do each subtraction.

a. $8x^2 - 3x^2 = 8x^2 + (-3x^2)$
$= 5x^2$

b. $3x^2y - 9x^2y = 3x^2y + (-9x^2y)$
$= -6x^2y$

c. $-5x^5y^3z^2 - 3x^5y^3z^2 = -5x^5y^3z^2 + (-3x^5y^3z^2)$
$= -8x^5y^3z^2$ ■

Self Check Subtract **a.** $-2a^2b^3 - 5a^2b^3$ and **b.** $-2a^2b^3 - (-5a^2b^3)$.

Answers **a.** $-7a^2b^3$, **b.** $3a^2b^3$

To subtract polynomials, we use the distributive property to remove parentheses and combine like terms, whenever possible.

EXAMPLE 4 Do each subtraction.

a. $(8x^3y + 2x^2y) - (2x^3y - 3x^2y)$

$= 1(8x^3y + 2x^2y) - 1(2x^3y - 3x^2y)$ Insert the understood coefficients of 1.

$= 8x^3y + 2x^2y - 2x^3y + 3x^2y$ Use the distributive property to remove parentheses.

$= 6x^3y + 5x^2y$ Combine like terms.

b. $(3rt^2 + 4r^2t^2) - (8rt^2 - 4r^2t^2 + r^3t^2)$

$= 1(3rt^2 + 4r^2t^2) - 1(8rt^2 - 4r^2t^2 + r^3t^2)$ Insert the understood coefficients of 1.

$= 3rt^2 + 4r^2t^2 - 8rt^2 + 4r^2t^2 - r^3t^2$ Use the distributive property to remove parentheses.

$= -5rt^2 + 8r^2t^2 - r^3t^2$ Combine like terms. ■

Self Check Subtract $(6a^2b^3 - 2a^2b^2) - (-2a^2b^3 + a^2b^2)$.

Answer $8a^2b^3 - 3a^2b^2$

To subtract polynomials in vertical form, we add the negative (or opposite) of the polynomial that is being subtracted.

$$- \begin{array}{r} 8x^3y + 2x^2y \\ \underline{2x^3y - 3x^2y} \end{array} \Rightarrow + \begin{array}{r} 8x^3y + 2x^2y \\ \underline{-2x^3y + 3x^2y} \\ 6x^3y + 5x^2y \end{array}$$

■ ADDING AND SUBTRACTING MULTIPLES OF POLYNOMIALS

To add multiples of one polynomial to another, or subtract multiples of one polynomial from another, we use the distributive property to remove parentheses and combine like terms.

EXAMPLE 5 Simplify $3(2x^2 + 4x - 7) - 2(3x^2 - 4x - 5)$.

Solution

$$3(2x^2 + 4x - 7) - 2(3x^2 - 4x - 5)$$
$$= 6x^2 + 12x - 21 - 6x^2 + 8x + 10$$
$$= 20x - 11 \quad \text{Combine like terms.}$$

■

Self Check Simplify $-2(3a^3 + a^2) - 5(2a^3 - a^2 + a)$.

Answer $-16a^3 + 3a^2 - 5a$

Orals *Combine like terms.*

1. $4x^2 + 5x^2$
2. $3y^2 - 5y^2$

Do the operations.

3. $(x^2 + 2x + 1) + (2x^2 - 2x + 1)$
4. $(x^2 + 2x + 1) - (2x^2 - 2x + 1)$
5. $(2x^2 - x - 3) + (x^2 - 3x - 1)$
6. $(2x^2 - x - 3) - (x^2 - 3x - 1)$

EXERCISE 5.2

REVIEW *Solve each inequality. Give the result in interval notation.*

1. $2x + 3 \leq 11$
2. $\frac{2}{3}x + 5 > 11$
3. $|x - 4| < 5$
4. $|2x + 1| \geq 7$

VOCABULARY AND CONCEPTS *In Exercises 5–8, fill in each blank to make a true statement.*

5. If two algebraic terms have the same variables with the same ________, they are called like terms.
6. $-6a^3b^2$ and $7a^2b^3$ are ______ terms.
7. To add two like monomials, we add their numerical ________ and keep the same variables with the same exponents.
8. To subtract one like monomial from another, we add the ________________ of the monomial that is to be subtracted.

In Exercises 9–16, tell whether the terms are like or unlike terms. If they are like terms, combine them.

9. $3x$, $7x$
10. $-8x$, $3y$
11. $7x$, $7y$
12. $3mn$, $5mn$
13. $3r^2t^3$, $-8r^2t^3$
14. $9u^2v$, $10u^2v$
15. $9x^2y^3$, $3x^2y^2$
16. $27x^6y^4z$, $8x^6y^4z^2$

PRACTICE *In Exercises 17–24, simplify each expression.*

17. $8x + 4x$

18. $-2y + 16y$

19. $5x^3y^2z - 3x^3y^2z$

20. $8wxy - 12wxy$

21. $-2x^2y^3 + 3xy^4 - 5x^2y^3$

22. $3ab^4 - 4a^2b^2 - 2ab^4 + 2a^2b^2$

23. $(3x^2y)^2 + 2x^4y^2 - x^4y^2$

24. $(5x^2y^4)^3 - (5x^3y^6)^2$

In Exercises 25–36, do each operation.

25. $(3x^2 + 2x + 1) + (-2x^2 - 7x + 5)$

26. $(-2a^2 - 5a - 7) + (-3a^2 + 7a + 1)$

27. $(-a^2 + 2a + 3) - (4a^2 - 2a - 1)$

28. $(x^2 - 3x + 8) - (3x^2 + x + 3)$

29. $(7y^3 + 4y^2 + y + 3) + (-8y^3 - y + 3)$

30. $(6x^3 + 3x - 2) - (2x^3 + 3x^2 + 5)$

31. $(3x^2 + 4x - 3) + (2x^2 - 3x - 1) - (x^2 + x + 7)$

32. $(-2x^2 + 6x + 5) - (-4x^2 - 7x + 2) - (4x^2 + 10x + 5)$

33. $(3x^3 - 2x + 3) + (4x^3 + 3x^2 - 2) + (-4x^3 - 3x^2 + x + 12)$

34. $(x^4 - 3x^2 + 4) + (-2x^4 - x^3 + 3x^2) + (3x^2 + 2x + 1)$

35. $(3y^2 - 2y + 4) + [(2y^2 - 3y + 2) - (y^2 + 4y + 3)]$

36. $(-t^2 - t - 1) - [(t^2 + 3t - 1) - (-2t^2 + 4)]$

In Exercises 37–40, add the polynomials.

37.
$$\begin{array}{r} 3x^3 - 2x^2 + 4x - 3 \\ -2x^3 + 3x^2 + 3x - 2 \\ 5x^3 - 7x^2 + 7x - 12 \\ \hline \end{array}$$

38.
$$\begin{array}{r} 7a^3 \qquad\quad + 3a + 7 \\ -2a^3 + 4a^2 \qquad\quad - 13 \\ 3a^3 - 3a^2 + 4a + 5 \\ \hline \end{array}$$

39.
$$\begin{array}{r} -2y^4 - 2y^3 + 4y^2 - 3y + 10 \\ -3y^4 + 7y^3 - y^2 + 14y - 3 \\ -3y^3 - 5y^2 - 5y + 7 \\ -4y^4 + y^3 - 13y^2 + 14y - 2 \\ \hline \end{array}$$

40.
$$\begin{array}{r} 17t^4 + 3t^3 - 2t^2 - 3t + 4 \\ -12t^4 - 2t^3 + 3t^2 - 5t - 17 \\ -2t^4 - 7t^3 + 4t^2 + 12t - 5 \\ 5t^4 + t^3 + 5t^2 - 13t + 12 \\ \hline \end{array}$$

In Exercises 41–44, subtract the bottom polynomial from the top polynomial.

41.
$$\begin{array}{r} 3x^2 - 4x + 17 \\ 2x^2 + 4x - 5 \\ \hline \end{array}$$

42.
$$\begin{array}{r} -2y^2 - 4y + 3 \\ 3y^2 + 10y - 5 \\ \hline \end{array}$$

43.
$$\begin{array}{r} -5y^3 + 4y^2 - 11y + 3 \\ -2y^3 - 14y^2 + 17y - 32 \\ \hline \end{array}$$

44.
$$\begin{array}{r} 17x^4 - 3x^2 - 65x - 12 \\ 23x^4 + 14x^2 + 3x - 23 \\ \hline \end{array}$$

In Exercises 45–58, simplify each expression.

45. $3(x + 2) + 2(x - 5)$

46. $-2(x - 4) + 5(x + 1)$

47. $-6(t - 4) - 5(t - 1)$

48. $4(a + 5) - 3(a - 1)$

49. $2(x^3 + x^2) + 3(2x^3 - x^2)$

50. $3(y^2 + 2y) - 4(y^2 - 4)$

51. $-3(2m - n) + 2(m - 3n)$

52. $5(p - 2q) - 4(2p + q)$

53. $-5(2x^3 + 7x^2 + 4x) - 2(3x^3 - 4x^2 - 4x)$

54. $-3(3a^2 + 4b^3 + 7) + 4(5a^2 - 2b^3 + 3)$

55. $4(3z^2 - 4z + 5) + 6(-2z^2 - 3z + 4) - 2(4z^2 + 3z - 5)$

56. $-3(4x^3 - 2x^2 + 4) - 4(3x^3 + 4x^2 + 3x) + 5(3x - 4)$

57. $5(2a^2 + 4a - 2) - 2(-3a^2 - a + 12) - 2(a^2 + 3a - 5)$

58. $-2(2b^2 - 3b + 3) + 3(3b^2 + 2b - 8) - (3b^2 - b + 4)$

APPLICATIONS

59. Write a polynomial that represents the perimeter of the triangle shown in Illustration 1.

60. Write a polynomial that represents the perimeter of the rectangle shown in Illustration 2.

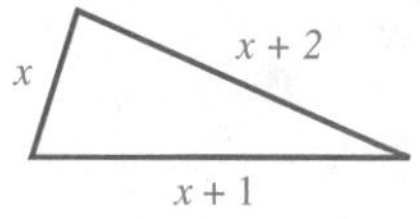

ILLUSTRATION 1

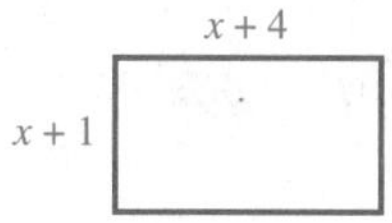

ILLUSTRATION 2

In Exercises 61–64, consider the following information. If a house is purchased for \$125,000 *and is expected to appreciate* \$1,100 *per year, its value y after x years is given by the polynomial function* $y = f(x) = 1{,}100x + 125{,}000$.

61. Value of a house Find the expected value of the house in 10 years.

62. Value of a house A second house is purchased for \$150,000 and is expected to appreciate \$1,400 per year.

a. Find a polynomial function that will give the value y of the house in x years.

b. Find the value of the second house after 12 years.

63. Value of two houses Find one polynomial function that will give the combined value y of both houses after x years.

64. Value of two houses Find the value of the two houses after 20 years by

a. substituting 20 into the polynomial functions

$$y = f(x) = 1{,}100x + 125{,}000 \quad \text{and}$$
$$y = f(x) = 1{,}400x + 150{,}000$$

and adding.

b. substituting 20 into the result of Exercise 63.

In Exercises 65–68, consider the following information. A business bought two cars, one for \$16,600 *and the other for* \$19,200. *The first car is expected to depreciate* \$2,100 *per year and the second* \$2,700 *per year.*

65. Value of a car Find a polynomial function that will give the value of the first car after x years.

66. Value of a car Find a polynomial function that will give the value of the second car after x years.

67. Value of two cars Find one polynomial function that will give the value of both cars after x years.

68. Value of two cars In two ways, find the total value of the cars after 3 years.

WRITING

69. Explain why the terms x^2y and xy^2 are not like terms.

70. Explain how to recognize like terms, and how to add them.

SOMETHING TO THINK ABOUT

71. Find the difference when $3x^2 + 4x - 3$ is subtracted from the sum of $-2x^2 - x + 7$ and $5x^2 + 3x - 1$.

72. Find the difference when $8x^3 + 2x^2 - 1$ is subtracted from the sum of $x^2 + x + 2$ and $2x^3 - x + 9$.

73. Find the sum when $2x^2 - 4x + 3$ minus $8x^2 + 5x - 3$ is added to $-2x^2 + 7x - 4$.

74. Find the sum when $7x^3 - 4x$ minus $x^2 + 2$ is added to $5 + 3x$.

5.3 Multiplying Polynomials

■ MULTIPLYING MONOMIALS ■ MULTIPLYING A POLYNOMIAL BY A MONOMIAL ■ MULTIPLYING A POLYNOMIAL BY A POLYNOMIAL ■ THE FOIL METHOD ■ SPECIAL PRODUCTS ■ AN APPLICATION OF MULTIPLYING POLYNOMIALS ■ MULTIPLYING EXPRESSIONS THAT ARE NOT POLYNOMIALS

Getting Ready *Simplify.*

1. $(3a)(4)$ **2.** $4aaa(a)$ **3.** $-7a^3 \cdot a$ **4.** $12a^3a^2$

Use the distributive property to remove parentheses.

5. $2(a + 4)$ **6.** $a(a - 3)$

7. $-4(a - 3)$ **8.** $-2b(b + 2)$

■ MULTIPLYING MONOMIALS

In Section 1.3, we saw that to multiply one monomial by another, *we multiply the numerical factors and then multiply the variable factors.*

EXAMPLE 1 Do each multiplication: **a.** $(3x^2)(6x^3)$, **b.** $(-8x)(2y)(xy)$, and **c.** $(2a^3b)(-7b^2c)(-12ac^4)$.

Solution We can use the commutative and associative properties of multiplication to rearrange the factors and regroup the numbers.

a. $$\begin{aligned}(3x^2)(6x^3) &= 3 \cdot x^2 \cdot 6 \cdot x^3 \\ &= (3 \cdot 6)(x^2 \cdot x^3) \\ &= 18x^5\end{aligned}$$

b. $$\begin{aligned}(-8x)(2y)(xy) &= -8 \cdot x \cdot 2 \cdot y \cdot x \cdot y \\ &= (-8 \cdot 2) \cdot x \cdot x \cdot y \cdot y \\ &= -16x^2y^2\end{aligned}$$

c. $$\begin{aligned}(2a^3b)(-7b^2c)(-12ac^4) &= 2 \cdot a^3 \cdot b \cdot (-7) \cdot b^2 \cdot c \cdot (-12) \cdot a \cdot c^4 \\ &= 2(-7)(-12) \cdot a^3 \cdot a \cdot b \cdot b^2 \cdot c \cdot c^4 \\ &= 168a^4b^3c^5\end{aligned}$$

■

Self Check Multiply **a.** $(-2a^3)(4a^2)$ and **b.** $(-5b^3)(-3a)(a^2b)$.

Answers **a.** $-8a^5$, **b.** $15a^3b^4$

■ MULTIPLYING A POLYNOMIAL BY A MONOMIAL

To multiply a polynomial by a monomial, *we use the distributive property and multiply each term of the polynomial by the monomial.*

EXAMPLE 2 Do each multiplication: **a.** $3x^2(6xy + 3y^2)$, **b.** $5x^3y^2(xy^3 - 2x^2y)$, and **c.** $-2ab^2(3bz - 2az + 4z^3)$

Solution We can use the distributive property to remove parentheses.

a. $$\begin{aligned}3x^2(6xy + 3y^2) &= 3x^2 \cdot 6xy + 3x^2 \cdot 3y^2 \\ &= 18x^3y + 9x^2y^2\end{aligned}$$

b. $$\begin{aligned}5x^3y^2(xy^3 - 2x^2y) &= 5x^3y^2 \cdot xy^3 - 5x^3y^2 \cdot 2x^2y \\ &= 5x^4y^5 - 10x^5y^3\end{aligned}$$

c. $$\begin{aligned}-2ab^2(3bz - 2az + 4z^3) &= -2ab^2 \cdot 3bz - (-2ab^2) \cdot 2az + (-2ab^2) \cdot 4z^3 \\ &= -6ab^3z + 4a^2b^2z - 8ab^2z^3\end{aligned}$$

■

Self Check Multiply $-2a^2(a^2 - a + 3)$.

Answer $-4a^4 + 2a^3 - 6a^2$

MULTIPLYING A POLYNOMIAL BY A POLYNOMIAL

To multiply a polynomial by a polynomial, we use the distributive property repeatedly.

b

EXAMPLE 3 Do each multiplication: **a.** $(3x + 2)(4x + 9)$ and **b.** $(2a - b)(3a^2 - 4ab + b^2)$.

Solution We can use the distributive property twice to remove parentheses.

$$\textbf{a.}\quad (3x + 2)(4x + 9) = (3x + 2) \cdot 4x + (3x + 2) \cdot 9$$
$$= 12x^2 + 8x + 27x + 18$$
$$= 12x^2 + 35x + 18 \qquad \text{Combine like terms.}$$

$$\textbf{b.}\quad (2a - b)(3a^2 - 4ab + b^2)$$
$$= (2a - b)3a^2 - (2a - b)4ab + (2a - b)b^2$$
$$= 6a^3 - 3a^2b - 8a^2b + 4ab^2 + 2ab^2 - b^3$$
$$= 6a^3 - 11a^2b + 6ab^2 - b^3$$

■

Self Check Multiply $(2a + b)(3a - 2b)$.

Answer $6a^2 - ab - 2b^2$

The results of Example 3 suggest that to multiply one polynomial by another, we *multiply each term of one polynomial by each term of the other polynomial and combine like terms, when possible.*

In the next example, we organize the work done in Example 3 vertically.

EXAMPLE 4 Do each multiplication.

a.

$$\begin{array}{rrrl}
 & 3x & +\ 2 & \\
 & 4x & +\ 9 & \\ \hline
12x^2 & +\ 8x & & \longleftarrow 4x(3x + 2) \\
 & +\ 27x & +\ 18 & \longleftarrow 9(3x + 2) \\ \hline
12x^2 & +\ 35x & +\ 18 &
\end{array}$$

b.

$$\begin{array}{rrrrl}
3a^2 & -\ 4ab & +\ b^2 & & \\
 & 2a & -\ b & & \\ \hline
6a^3 & -\ 8a^2b & +\ 2ab^2 & & \longleftarrow 2a(3a^2 - 4ab + b^2) \\
 & -\ 3a^2b & +\ 4ab^2 & -\ b^3 & \longleftarrow -b(3a^2 - 4ab + b^2) \\ \hline
6a^3 & -\ 11a^2b & +\ 6ab^2 & -\ b^3 &
\end{array}$$

■

Self Check Multiply: $\begin{array}{r} 3x^2 + 2x - 5 \\ \underline{2x + 1} \end{array}$

Answer $6x^3 + 7x^2 - 8x - 5$

THE FOIL METHOD

When multiplying two binomials, the distributive property requires that each term of one binomial be multiplied by each term of the other binomial. This fact can be emphasized by drawing lines to show the indicated products. For example, to multiply $3x + 2$ and $x + 4$, we can write

First terms   Last terms

$$\begin{aligned}(3x + 2)(x + 4) &= 3x \cdot x + 3x \cdot 4 + 2 \cdot x + 2 \cdot 4 \\ &= 3x^2 + 12x + 2x + 8 \\ &= 3x^2 + 14x + 8 \quad \text{Combine like terms.}\end{aligned}$$

Inner terms   Outer terms

We note that

- the product of the **F**irst terms is $3x^2$,
- the product of the **O**uter terms is $12x$,
- the product of the **I**nner terms is $2x$, and
- the product of the **L**ast terms is 8.

This scheme is called the **FOIL** method of multiplying two binomials. FOIL is an acronym for **F**irst terms, **O**uter terms, **I**nner terms, and **L**ast terms. Of course, the resulting terms of the products must be combined, if possible.

It is easy to multiply binomials by sight if we use the FOIL method. We find the product of the first terms, then find the products of the outer terms and the inner terms and add them (when possible), and then find the product of the last terms.

EXAMPLE 5 Find each product.

a. $(2x - 3)(3x + 2) = 6x^2 - 5x - 6$

The middle term of $-5x$ in the result comes from combining the outer and inner products of $+4x$ and $-9x$:

$$4x + (-9x) = -5x$$

b. $(3x + 1)(3x + 4) = 9x^2 + 15x + 4$

The middle term of $+15x$ in the result comes from combining the products $+12x$ and $+3x$:

$$12x + 3x = 15x$$

c. $(4x - y)(2x + 3y) = 8x^2 + 10xy - 3y^2$

The middle term of $+10xy$ in the result comes from combining the products $+12xy$ and $-2xy$:

$$12xy - 2xy = 10xy$$

■

Self Check Multiply $(3a + 4b)(2a - b)$.

Answer $6a^2 + 5ab - 4b^2$

SPECIAL PRODUCTS

It is easy to square a binomial by using the FOIL method.

b

EXAMPLE 6 Find each square: **a.** $(x + y)^2$ and **b.** $(x - y)^2$.

Solution We multiply each term of one binomial by each term of the other binomial, and then we combine like terms.

a. $(x + y)^2 = (x + y)(x + y)$

$= x^2 + xy + xy + y^2$ Use the FOIL method.

$= x^2 + 2xy + y^2$ Combine like terms.

We see that the square of the binomial is the square of the first term, plus twice the product of the terms, plus the square of the last term. This product can be illustrated graphically as shown in Figure 5-5.

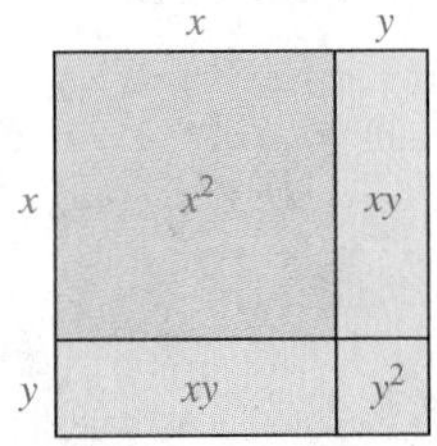

FIGURE 5-5

The area of the large square is the product of its length and width: $(x + y)(x + y) = (x + y)^2$.

The area of the large square is also the sum of its four pieces: $x^2 + xy + xy + y^2 = x^2 + 2xy + y^2$.

Thus, $(x + y)^2 = x^2 + 2xy + y^2$.

b. $(x - y)^2 = (x - y)(x - y)$

$= x^2 - xy - xy + y^2$ Use the FOIL method.

$= x^2 - 2xy + y^2$ Combine like terms.

We see that the square of the binomial is the square of the first term, minus twice the product of the terms, plus the square of the last term. For a geometric interpretation, see Exercise 115. ■

Self Check Find the squares: **a.** $(a + 2)^2$ and **b.** $(a - 4)^2$.

Answers **a.** $a^2 + 4a + 4$, **b.** $a^2 - 8a + 16$

EXAMPLE 7 Multiply $(x + y)(x - y)$.

Solution

$$\begin{aligned}(x + y)(x - y) &= x^2 - xy + xy - y^2 && \text{Use the FOIL method.}\\ &= x^2 - y^2 && \text{Combine like terms.}\end{aligned}$$

From this example, we see that the product of the sum of two quantities and the difference of two quantities is the square of the first quantity minus the square of the second quantity.

For a geometric interpretation, see Exercise 116. ■

Self Check Multiply $(a + 3)(a - 3)$.

Answer $a^2 - 9$

The products discussed in Examples 6 and 7 are called **special products.** Because they occur so often, it is useful to learn their forms.

Special Product Formulas

$$(x + y)^2 = (x + y)(x + y) = x^2 + 2xy + y^2$$
$$(x - y)^2 = (x - y)(x - y) = x^2 - 2xy + y^2$$
$$(x + y)(x - y) = x^2 - y^2$$

Because $x^2 + 2xy + y^2 = (x + y)^2$ and $x^2 - 2xy + y^2 = (x - y)^2$, the two trinomials are called **perfect square trinomials.**

WARNING! The squares $(x + y)^2$ and $(x - y)^2$ have trinomials for their products. Don't forget to write the middle terms in these products. Remember that

$$(x + y)^2 \neq x^2 + y^2 \quad \text{and that} \quad (x - y)^2 \neq x^2 - y^2$$

Also remember that the product $(x + y)(x - y)$ is the binomial $x^2 - y^2$.

At first, the expression $3[x^2 - 2(x + 3)]$ doesn't look like a polynomial. However, if we remove the parentheses and the brackets, it takes on the form of a polynomial.

$$\begin{aligned}3[x^2 - 2(x + 3)] &= 3[x^2 - 2x - 6]\\ &= 3x^2 - 6x - 18\end{aligned}$$

If an expression has one set of grouping symbols that is enclosed within another set, we always eliminate the inner set first.

EXAMPLE 8 Find the product of $-2[y^3 + 3(y^2 - 2)]$ and $5[y^2 - 2(y + 1)]$.

Solution We change each expression into polynomial form:

$$\begin{array}{ll} -2[y^3 + 3(y^2 - 2)] & 5[y^2 - 2(y + 1)] \\ -2(y^3 + 3y^2 - 6) & 5(y^2 - 2y - 2) \\ -2y^3 - 6y^2 + 12 & 5y^2 - 10y - 10 \end{array}$$

Then we do the multiplication:

$$\begin{array}{rrrrrr} & & -2y^3 & -6y^2 & & +12 \\ & & & 5y^2 & -10y & -10 \\ \hline -10y^5 & -30y^4 & & +60y^2 & & \\ & +20y^4 & +60y^3 & & -120y & \\ & & +20y^3 & +60y^2 & & -120 \\ \hline -10y^5 & -10y^4 & +80y^3 & +120y^2 & -120y & -120 \end{array}$$

■

Self Check Find the product of $2[a^2 + 3(a - 2)]$ and $3[a^2 + 3(a - 1)]$.

Answer $6a^4 + 36a^3 - 162a + 108$

AN APPLICATION OF MULTIPLYING POLYNOMIALS

The profit p earned on the sale of one or more items is given by the formula

$$p = r - c$$

where r is the revenue taken in and c is the wholesale cost.

If a salesperson has 12 vacuum cleaners and sells them for \$225 each, the revenue will be $r = \$(12 \cdot 225) = \$2{,}700$. This illustrates the following formula for finding the revenue r:

$$r = \boxed{\text{number of items sold } (x)} \cdot \boxed{\text{selling price of each item } (p)} = xp = px$$

EXAMPLE 9 **Selling vacuum cleaners** Over the years, a saleswoman has found that the number of vacuum cleaners she can sell depends on price. The lower the price, the more she can sell. She has determined that the number of vacuums (x) that she can sell at a price (p) is related by the equation $x = -\frac{2}{25}p + 28$.

a. Find a formula for the revenue r.

b. How much revenue will be taken in if the vacuums are priced at \$250?

Solution **a.** To find a formula for revenue, we substitute $-\frac{2}{25}p + 28$ for x in the formula $r = px$ and solve for r.

$$r = px \qquad \text{The formula for revenue.}$$

$$r = p\left(-\frac{2}{25}p + 28\right) \qquad \text{Substitute } -\tfrac{2}{25}p + 28 \text{ for } x.$$

$$= -\frac{2}{25}p^2 + 28p \qquad \text{Multiply the polynomials.}$$

b. To find how much revenue will be taken in if the vacuums are priced at \$250, we substitute 250 for p in the formula for revenue.

$$r = -\frac{2}{25}p^2 + 28p \qquad \text{The formula for revenue.}$$

$$r = -\frac{2}{25}(250)^2 + 28(250) \qquad \text{Substitute 250 for } p.$$

$$= -5{,}000 + 7{,}000$$

$$= 2{,}000$$

The revenue will be \$2,000. ■

MULTIPLYING EXPRESSIONS THAT ARE NOT POLYNOMIALS

The following examples show how to use the FOIL method to multiply expressions that are not polynomials.

EXAMPLE 10 Find the product of $x^{-2} + y$ and $x^2 - y^{-2}$.

Solution We multiply each term of the second expression by each term of the first expression and then simplify.

$$(x^{-2} + y)(x^2 - y^{-2}) = x^{-2}x^2 - x^{-2}y^{-2} + yx^2 - yy^{-2}$$

$$= x^{-2+2} - x^{-2}y^{-2} + yx^2 - y^{1+(-2)}$$

$$= x^0 - \frac{1}{x^2y^2} + x^2y - y^{-1}$$

$$= 1 - \frac{1}{x^2y^2} + x^2y - \frac{1}{y}$$ ■

Self Check Multiply $(a^{-3} + b)(a^2 - b^{-1})$.

Answer $\frac{1}{a} - \frac{1}{a^3b} + a^2b - 1$

EXAMPLE 11 Find the product of $x^n + 2x$ and $x^n + 3x^{-n}$.

Solution We multiply each term of the second expression by each term of the first expression and then simplify:

$$\begin{aligned}(x^n + 2x)(x^n + 3x^{-n}) &= x^n x^n + x^n(3x^{-n}) + 2x(x^n) + 2x(3x^{-n})\\ &= x^{n+n} + 3x^{n+(-n)} + 2x^{1+n} + 6xx^{-n}\\ &= x^{2n} + 3x^0 + 2x^{n+1} + 6x^{1+(-n)}\\ &= x^{2n} + 3 + 2x^{n+1} + 6x^{1-n}\end{aligned}$$

Self Check Multiply $(a^n + b)(a^n + b^n)$.

Answer $a^{2n} + a^n b^n + a^n b + b^{n+1}$

PERSPECTIVE

An important figure in the history of mathematics, François Vieta (1540–1603) was one of the first to use notation close to that which we use today. Trained as a lawyer, Vieta served in the parliament of Brittany and as the personal lawyer of Henry of Navarre. If he had continued as a successful lawyer, Vieta might now be forgotten. However, he lost his job.

When political opposition forced him out of office in 1584, Vieta had time to devote himself entirely to his hobby, mathematics. He studied the writings of earlier mathematicians and adapted and improved their ideas. Vieta was the first to use letters to represent unknown numbers, but he did not use our modern notation for exponents. To us, his notation seems awkward. For example, what we would write as

$$(x + 1)^3 = x^3 + 3x^2 + 3x + 1$$

Vieta would have written as

$\overline{x + 1}$ cubus aequalis x cubus + x quad. 3 + x in 3 + 1

Orals *Find each product.*

1. $(-2a^2b)(3ab^2)$
2. $(4xy^2)(-2xy)$
3. $3a^2(2a - 1)$
4. $-4n^2(4m - n)$
5. $(x + 1)(2x + 1)$
6. $(3y - 2)(2y + 1)$

EXERCISE 5.3

REVIEW *Let $a = -2$ and $b = 4$ and find the absolute value of each expression.*

1. $|3a - b|$
2. $|ab - b^2|$
3. $-|a^2b - b^0|$
4. $\left|\dfrac{a^3b^2 + ab}{2(ab)^2 - a^3}\right|$

5. A woman owns 200 shares of ABC Company, valued at \$125 per share, and 350 shares of WD Company, valued at \$75 per share. One day, ABC rose $1\frac{1}{2}$ points and WD fell $1\frac{1}{2}$ points. Find the current value of her portfolio.

6. One light year is approximately 5,870,000,000,000 miles. Write this number in scientific notation.

VOCABULARY AND CONCEPTS *Fill in each blank to make a true statement.*

7. To multiply a monomial by a monomial, we multiply the numerical factors and then multiply the ______ factors.

8. To multiply a polynomial by a monomial, we multiply each term of the polynomial by the ______.

9. To multiply a polynomial by a polynomial, we multiply each ____ of one polynomial by each term of the other polynomial.

10. FOIL is an acronym for ____ terms, ____ terms, ____ terms, and ____ terms.

11. $(x + y)^2 = (x + y)(x + y) =$ ________

12. $(x - y)^2 = (x - y)(x - y) =$ ________

13. $(x + y)(x - y) =$ ______

14. $x^2 + 2xy + y^2$ and $x^2 - 2xy + y^2$ are called ________ trinomials.

PRACTICE *In Exercises 15–38, find each product.*

15. $(2a^2)(-3ab)$

16. $(-3x^2y)(3xy)$

17. $(-3ab^2c)(5ac^2)$

18. $(-2m^2n)(-4mn^3)$

19. $(4a^2b)(-5a^3b^2)(6a^4)$

20. $(2x^2y^3)(4xy^5)(-5y^6)$

21. $(5x^3y^2)^4\left(\frac{1}{5}x^{-2}\right)^2$

22. $(4a^{-2}b^{-1})^2(2a^3b^4)^4$

23. $(-5xx^2)(-3xy)^4$

24. $(-2a^2ab^2)^3(-3ab^2b^2)$

25. $3(x + 2)$

26. $-5(a + b)$

27. $-a(a - b)$

28. $y^2(y - 1)$

29. $3x(x^2 + 3x)$

30. $-2x(3x^2 - 2)$

31. $-2x(3x^2 - 3x + 2)$

32. $3a(4a^2 + 3a - 4)$

33. $5a^2b^3(2a^4b - 5a^0b^3)$

34. $-2a^3b(3a^0b^4 - 2a^2b^3)$

35. $7rst(r^2 + s^2 - t^2)$

36. $3x^2yz(x^2 - 2y + 3z^2)$

37. $4m^2n(-3mn)(m + n)$

38. $-3a^2b^3(2b)(3a + b)$

In Exercises 39–68, find each product. If possible, find the product by sight.

39. $(x + 2)(x + 3)$

40. $(y - 3)(y + 4)$

41. $(z - 7)(z - 2)$

42. $(x + 3)(x - 5)$

43. $(2a + 1)(a - 2)$

44. $(3b - 1)(2b - 1)$

45. $(3t - 2)(2t + 3)$

46. $(p + 3)(3p - 4)$

47. $(3y - z)(2y - z)$

48. $(2m + n)(3m + n)$

49. $(2x - 3y)(x + 2y)$

50. $(3y + 2z)(y - 3z)$

51. $(3x + y)(3x - 3y)$

52. $(2x - y)(3x + 2y)$

53. $(4a - 3b)(2a + 5b)$

54. $(3a + 2b)(2a - 7b)$

55. $(x + 2)^2$

56. $(x - 3)^2$

57. $(a - 4)^2$

58. $(y + 5)^2$

59. $(2a + b)^2$

60. $(a - 2b)^2$

61. $(2x - y)^2$

62. $(3m + 4n)(3m + 4n)$

63. $(x + 2)(x - 2)$

64. $(z + 3)(z - 3)$

65. $(a + b)(a - b)$

66. $(p + q)(p - q)$

67. $(2x + 3y)(2x - 3y)$

68. $(3a + 4b)(3a - 4b)$

In Exercises 69–78, find each product.

69. $(x - y)(x^2 + xy + y^2)$

70. $(x + y)(x^2 - xy + y^2)$

71. $(3y + 1)(2y^2 + 3y + 2)$

72. $(a + 2)(3a^2 + 4a - 2)$

73. $(2a - b)(4a^2 + 2ab + b^2)$

74. $(x - 3y)(x^2 + 3xy + 9y^2)$

75. $(2x - 1)[2x^2 - 3(x + 2)]$

76. $(x + 1)^2[x^2 - 2(x + 2)]$

77. $(a + b)(a - b)(a - 3b)$

78. $(x - y)(x + 2y)(x - 2y)$

In Exercises 79–86, find each product. Write all answers without negative exponents.

79. $x^3(2x^2 + x^{-2})$

80. $x^{-4}(2x^{-3} - 5x^2)$

81. $x^3y^{-6}z^{-2}(3x^{-2}y^2z - x^3y^{-4})$

82. $ab^{-2}c^{-3}(a^{-4}bc^3 + a^{-3}b^4c^3)$

83. $(x^{-1} + y)(x^{-1} - y)$

84. $(x^{-1} - y)(x^{-1} - y)$

85. $(2x^{-3} + y^3)(2x^3 - y^{-3})$

86. $(5x^{-4} - 4y^2)(5x^2 - 4y^{-4})$

In Exercises 87–96, find each product. Consider n to be a whole number.

87. $x^n(x^{2n} - x^n)$

88. $a^{2n}(a^n + a^{2n})$

89. $(x^n + 1)(x^n - 1)$

90. $(x^n - a^n)(x^n + a^n)$

91. $(x^n - y^n)(x^n - y^{-n})$

92. $(x^n + y^n)(x^n + y^{-n})$

93. $(x^{2n} + y^{2n})(x^{2n} - y^{2n})$

94. $(a^{3n} - b^{3n})(a^{3n} + b^{3n})$

95. $(x^n + y^n)(x^n + 1)$

96. $(1 - x^n)(x^{-n} - 1)$

In Exercises 97–110, simplify each expression.

97. $3x(2x + 4) - 3x^2$

98. $2y - 3y(y^2 + 4)$

99. $3pq - p(p - q)$

100. $-4rs(r - 2) + 4rs$

101. $2m(m - n) - (m + n)(m - 2n)$

102. $-3y(2y + z) + (2y - z)(3y + 2z)$

103. $(x + 3)(x - 3) + (2x - 1)(x + 2)$

104. $(2b + 3)(b - 1) - (b + 2)(3b - 1)$

105. $(3x - 4)^2 - (2x + 3)^2$

106. $(3y + 1)^2 + (2y - 4)^2$

107. $3(x - 3y)^2 + 2(3x + y)^2$

108. $2(x - y^2)^2 - 3(y^2 + 2x)^2$

109. $5(2y - z)^2 + 4(y + 2z)^2$

110. $3(x + 2z)^2 - 2(2x - z)^2$

In Exercises 111–114, use a calculator to find each product.

111. $(3.21x - 7.85)(2.87x + 4.59)$

112. $(7.44y + 56.7)(-2.1y - 67.3)$

113. $(-17.3y + 4.35)^2$

114. $(-0.31x + 29.3)(-81x - 0.2)$

115. Refer to Illustration 1 and answer the following questions.
a. Find the area of the large square.
b. Find the area of square I.
c. Find the area of rectangle II.
d. Find the area of rectangle III.
e. Find the area of square IV.
Use the answers to the preceding questions to show that $(x - y)^2 = x^2 - 2xy + y^2$.

116. Refer to Illustration 2 and answer the following questions.
a. Find the area of rectangle $ABCD$.
b. Find the area of rectangle I.
c. Find the area of rectangle II.
Use the answers to the preceding questions to show that $(x + y)(x - y) = x^2 - y^2$.

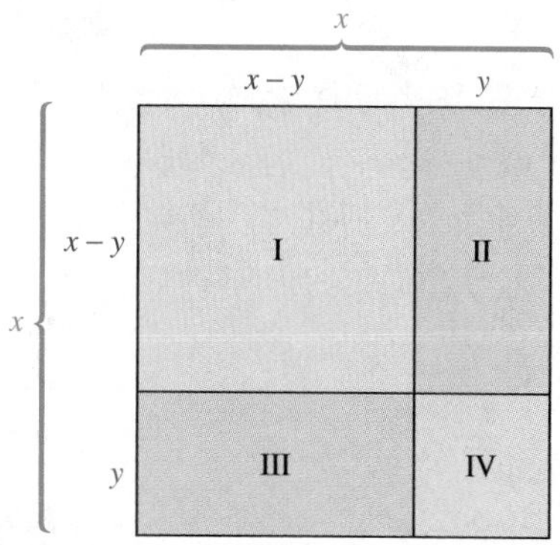

ILLUSTRATION 1

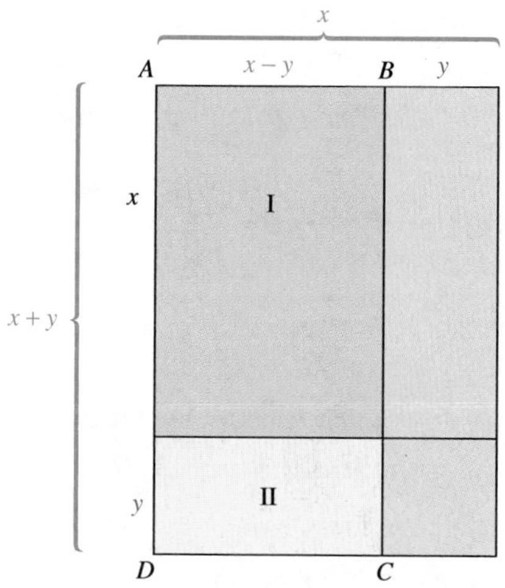

ILLUSTRATION 2

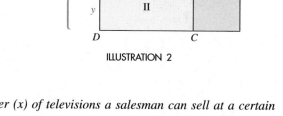

APPLICATIONS *In Exercises 117–120, assume that the number (x) of televisions a salesman can sell at a certain price (p) is given by the equation* $x = -\frac{1}{5}p + 90$.

117. Find the number of TVs he will sell if the price is $375.

118. If he sold 20 TVs, what was the price?

119. Write a formula for the revenue when x TVs are sold.

120. Find the revenue generated by the sales of the TVs if they are priced at $400 each.

121. Write a polynomial that represents the area of the rectangle shown in Illustration 3.

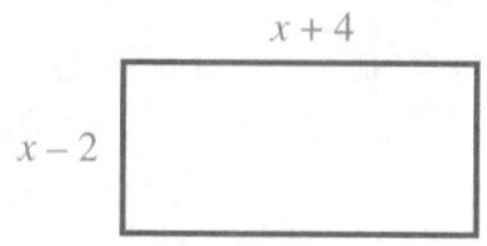

ILLUSTRATION 3

122. Write a polynomial that represents the area of the triangle shown in Illustration 4.

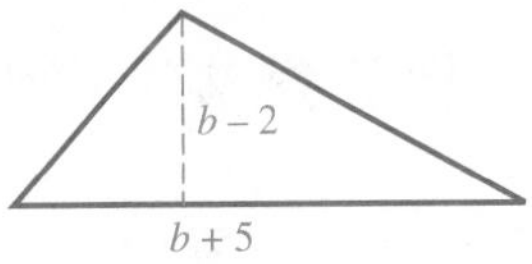

ILLUSTRATION 4

WRITING

123. Explain how to use the FOIL method.

124. Explain how to multiply two trinomials.

SOMETHING TO THINK ABOUT

125. The numbers 0.35×10^7 and 1.96×10^7 both involve the same power of 10. Find their sum.

126. Without converting to standard notation, find the sum: $1.435 \times 10^8 + 2.11 \times 10^7$. (*Hint:* The first number in the previous exercise is not in scientific notation.)

5.4 The Greatest Common Factor and Factoring by Grouping

■ PRIME-FACTORED FORM OF A NATURAL NUMBER ■ FACTORING OUT THE GREATEST COMMON FACTOR ■ AN APPLICATION OF FACTORING ■ FACTORING BY GROUPING ■ FORMULAS

Getting Ready *Simplify each expression by removing parentheses.*

1. $4(a + 2)$ **2.** $-5(b - 5)$ **3.** $a(a + 5)$

4. $-b(b - 3)$ **5.** $6a(a + 2b)$ **6.** $-2p^2(3p + 5)$

■ PRIME-FACTORED FORM OF A NATURAL NUMBER

In this section, we will reverse the operation of multiplication and show how to find the factors of a known product. The process of finding the individual factors of a product is called **factoring.**

If one number a divides a second number b, then a is called a **factor** of b. For example, because 3 divides 24, it is a factor of 24. Each number in the following list is a factor of 24, because each number divides 24.

1, 2, 3, 4, 6, 8, 12, and 24

To factor a natural number means to write it as a product of other natural numbers. If each factor is a prime number, the natural number is said to be written in **prime-factored form.** Example 1 shows how to find the prime-factored forms of 60, 84, and 180, respectively.

EXAMPLE 1 Find the prime factorization of each number.

a. $60 = 6 \cdot 10$
$= 2 \cdot 3 \cdot 2 \cdot 5$
$= 2^2 \cdot 3 \cdot 5$

b. $84 = 4 \cdot 21$
$= 2 \cdot 2 \cdot 3 \cdot 7$
$= 2^2 \cdot 3 \cdot 7$

c. $180 = 10 \cdot 18$
$= 2 \cdot 5 \cdot 3 \cdot 6$
$= 2 \cdot 5 \cdot 3 \cdot 3 \cdot 2$
$= 2^2 \cdot 3^2 \cdot 5$ ■

Self Check Find the prime factorization of 120.
Answer $2^3 \cdot 3 \cdot 5$

The largest natural number that divides 60, 84, and 180 is called the **greatest common factor (GCF)** of the numbers. Because 60, 84, and 180 all have two factors of 2 and one factor of 3, the GCF of these three numbers is $2^2 \cdot 3 = 12$. We note that

$$\frac{60}{12} = 5, \qquad \frac{84}{12} = 7, \qquad \text{and} \qquad \frac{180}{12} = 15$$

There is no natural number greater than 12 that divides 60, 84, and 180.

Algebraic monomials can also have greatest common factors.

EXAMPLE 2 Find the GCF of $6a^2b^3c$, $9a^3b^2c$, and $18a^4c^3$.

Solution We begin by factoring each monomial:

$$6a^2b^3c = 3 \cdot 2 \cdot a \cdot a \cdot b \cdot b \cdot b \cdot c$$
$$9a^3b^2c = 3 \cdot 3 \cdot a \cdot a \cdot a \cdot b \cdot b \cdot c$$
$$18a^4c^3 = 2 \cdot 3 \cdot 3 \cdot a \cdot a \cdot a \cdot a \cdot c \cdot c \cdot c$$

Since each monomial has one factor of 3, two factors of a, and one factor of c in common, their GCF is

$$3^1 \cdot a^2 \cdot c^1 = 3a^2c$$ ■

Self Check Find the GCF of $-24x^3y^3$, $3x^3y$, and $18x^2y^2$.
Answer $3x^2y$

To find the GCF of several monomials, we follow these steps.

Steps for Finding the GCF

1. Find the prime-factored form of each monomial.
2. Identify the prime factors and variable factors that are common to each monomial.
3. Find the product of the factors found in Step 2 with each factor raised to the smallest power that occurs in any one monomial.

FACTORING OUT THE GREATEST COMMON FACTOR

We have seen that the distributive property provides a method for multiplying a polynomial by a monomial. For example,

$$\begin{aligned} 2x^3y^3(3x^2 - 4y^3) &= 2x^3y^3 \cdot 3x^2 - 2x^3y^3 \cdot 4y^3 \\ &= 6x^5y^3 - 8x^3y^6 \end{aligned}$$

If the product of a multiplication is $6x^5y^3 - 8x^3y^6$, we can use the distributive property to find the individual factors.

$$\begin{aligned} 6x^5y^3 - 8x^3y^6 &= 2x^3y^3 \cdot 3x^2 - 2x^3y^3 \cdot 4y^3 \\ &= 2x^3y^3(3x^2 - 4y^3) \end{aligned}$$

Since $2x^3y^3$ is the GCF of the terms of $6x^5y^3 - 8x^3y^6$, this process is called **factoring out the greatest common factor.**

EXAMPLE 3 Factor $25a^3b + 15ab^3$.

Solution We begin by factoring each monomial:

$$\begin{aligned} 25a^3b &= 5 \cdot 5 \cdot a \cdot a \cdot a \cdot b \\ 15ab^3 &= 5 \cdot 3 \cdot a \cdot b \cdot b \cdot b \end{aligned}$$

Since each term has at least one factor of 5, one factor of a, and one factor of b in common and there are no other common factors, $5ab$ is the GCF of the two terms. We can use the distributive property to factor it out.

$$\begin{aligned} 25a^3b + 15ab^3 &= 5ab \cdot 5a^2 + 5ab \cdot 3b^2 \\ &= 5ab(5a^2 + 3b^2) \end{aligned}$$

■

Self Check Factor $9x^4y^2 - 12x^3y^3$.

Answer $3x^3y^2(3x - 4y)$

EXAMPLE 4 Factor $3xy^2z^3 + 6xyz^3 - 3xz^2$.

Solution We begin by factoring each monomial:

$$\begin{aligned} 3xy^2z^3 &= 3 \cdot x \cdot y \cdot y \cdot z \cdot z \cdot z \\ 6xyz^3 &= 3 \cdot 2 \cdot x \cdot y \cdot z \cdot z \cdot z \\ -3xz^2 &= -3 \cdot x \cdot z \cdot z \end{aligned}$$

Since each term has one factor of 3, one factor of x, and two factors of z in common, and because there are no other common factors, $3xz^2$ is the GCF of the three terms. We can use the distributive property to factor it out.

$$\begin{aligned} 3xy^2z^3 + 6xyz^3 - 3xz^2 &= 3xz^2 \cdot y^2z + 3xz^2 \cdot 2yz - 3xz^2 \cdot 1 \\ &= 3xz^2(y^2z + 2yz - 1) \end{aligned}$$

WARNING! The last term $-3xz^2$ of the given trinomial has an understood coefficient of -1. When the $3xz^2$ is factored out, remember to write the -1. ■

Self Check Factor $2a^4b^2 + 6a^3b^2 - 4a^2b$.

Answer $2a^2b(a^2b + 3ab - 2)$

A polynomial that cannot be factored is called a **prime polynomial** or an **irreducible polynomial.**

EXAMPLE 5

Factor $3x^2 + 4y + 7$, if possible.

Solution We factor each monomial:

$$3x^2 = 3 \cdot x \cdot x \qquad 4y = 2 \cdot 2 \cdot y \qquad 7 = 7$$

There are no common factors other than 1. This polynomial is an example of a prime polynomial. ■

Self Check Factor $6a^3 + 7b^2 + 5$.

Answer a prime polynomial

EXAMPLE 6

Factor the negative of the GCF from $-6u^2v^3 + 8u^3v^2$.

Solution Because the GCF of the two terms is $2u^2v^2$, the negative of the GCF is $-2u^2v^2$. To factor out $-2u^2v^2$, we proceed as follows:

$$\begin{aligned}-6u^2v^3 + 8u^3v^2 &= -2u^2v^2 \cdot 3v + 2u^2v^2 \cdot 4u \\ &= -2u^2v^2 \cdot 3v - (-2u^2v^2)4u \\ &= -2u^2v^2(3v - 4u)\end{aligned}$$

■

Self Check Factor out the negative of the GCF from $-8a^2b^2 - 12ab^3$.

Answer $-4ab^2(2a + 3b)$

We can also factor out common factors with variable exponents.

EXAMPLE 7

Factor x^{2n} from $x^{4n} + x^{3n} + x^{2n}$.

Solution We write the trinomial as $x^{2n} \cdot x^{2n} + x^{2n} \cdot x^n + x^{2n} \cdot 1$ and factor out x^{2n}.

$$\begin{aligned}x^{4n} + x^{3n} + x^{2n} &= x^{2n} \cdot x^{2n} + x^{2n} \cdot x^n + x^{2n} \cdot 1 \\ &= x^{2n}(x^{2n} + x^n + 1)\end{aligned}$$

■

Self Check Factor a^n from $a^{6n} + a^{3n} - a^n$.

Answer $a^n(a^{5n} + a^{2n} - 1)$

EXAMPLE 8 Factor $a^{-2}b^{-2}$ from $a^{-2}b - a^3b^{-2}$.

Solution We write $a^{-2}b - a^3b^{-2}$ as $a^{-2}b^{-2} \cdot b^3 - a^{-2}b^{-2} \cdot a^5$ and factor out $a^{-2}b^{-2}$.

$$\begin{aligned} a^{-2}b - a^3b^{-2} &= a^{-2}b^{-2} \cdot b^3 - a^{-2}b^{-2} \cdot a^5 \\ &= a^{-2}b^{-2}(b^3 - a^5) \end{aligned}$$

■

Self Check Factor $x^{-3}y^{-2}$ from $x^{-3}y^{-2} - x^6y^{-2}$.

Answer $x^{-3}y^{-2}(1 - x^9)$

A common factor can have more than one term. For example, in the expression

$$x(a + b) + y(a + b)$$

the binomial $a + b$ is a factor of both terms. We can factor it out to get

$$\begin{aligned} x(a + b) + y(a + b) &= (a + b)x + (a + b)y && \text{Use the commutative property of multiplication.} \\ &= (a + b)(x + y) \end{aligned}$$

EXAMPLE 9 Factor $a(x - y + z) - b(x - y + z) + 3(x - y + z)$.

Solution We can factor out the GCF of $(x - y + z)$.

$$\begin{aligned} &a(x - y + z) - b(x - y + z) + 3(x - y + z) \\ &\quad = (x - y + z)a - (x - y + z)b + (x - y + z)3 \\ &\quad = (x - y + z)(a - b + 3) \end{aligned}$$

■

Self Check Factor $x(a + b - c) - y(a + b - c)$.

Answer $(a + b - c)(x - y)$

AN APPLICATION OF FACTORING

EXAMPLE 10 The polynomial $3x^4 + 2x^3 + 5x^2 + 7x + 1$ can be written as

$$3 \cdot x \cdot x \cdot x \cdot x + 2 \cdot x \cdot x \cdot x + 5 \cdot x \cdot x + 7 \cdot x + 1$$

to illustrate that it involves 10 multiplications and 4 additions. Since computers take more time to do multiplications than additions, a programmer wants to write the polynomial in a way that will require fewer multiplications. Write the polynomial so that it contains only four multiplications.

Solution Factor the common x from the first four terms of the polynomial to get

$$3x^4 + 2x^3 + 5x^2 + 7x + 1 = x(3x^3 + 2x^2 + 5x + 7) + 1$$

Now factor the common x from the first three terms of the polynomial in the parentheses to get

$$3x^4 + 2x^3 + 5x^2 + 7x + 1 = x[x(3x^2 + 2x + 5) + 7] + 1$$

Again, factor an x from the first two terms of the polynomial within the set of parentheses to get

1. $$3x^4 + 2x^3 + 5x^2 + 7x + 1 = x\{x[x(3x + 2) + 5] + 7\} + 1$$

To emphasize that there are now only four multiplications, we rewrite the right-hand side of Equation 1 as

$$x \cdot \{x \cdot [x \cdot (3 \cdot x + 2) + 5] + 7\} + 1$$

The right-hand side of Equation 1 is called the *nested form* of the polynomial. ■

FACTORING BY GROUPING

Suppose that we wish to factor

$$ac + ad + bc + bd$$

Although there is no factor common to all four terms, there is a common factor of a in the first two terms and a common factor of b in the last two terms. We can factor out these common factors to get

$$ac + ad + bc + bd = a(c + d) + b(c + d)$$

We can now factor out the common factor of $c + d$ on the right-hand side:

$$ac + ad + bc + bd = (c + d)(a + b)$$

The grouping in this type of problem is not always unique, but the factorization is. For example, if we write the expression $ac + ad + bc + bd$ in the form

$$ac + bc + ad + bd$$

and factor c from the first two terms and d from the last two terms, we obtain

$$\begin{aligned} ac + bc + ad + bd &= c(a + b) + d(a + b) \\ &= (a + b)(c + d) \end{aligned}$$

The method used in the previous examples is called **factoring by grouping.**

EXAMPLE 11 Factor $3ax^2 + 3bx^2 + a + 5bx + 5ax + b$.

Solution Although there is no factor common to all six terms, $3x^2$ can be factored out of the first two terms, and $5x$ can be factored out of the fourth and fifth terms to get

$$3ax^2 + 3bx^2 + a + 5bx + 5ax + b = 3x^2(a + b) + a + 5x(b + a) + b$$

This result can be written in the form

$$3ax^2 + 3bx^2 + a + 5bx + 5ax + b = 3x^2(a + b) + 5x(a + b) + (a + b)$$

Since $a + b$ is common to all three terms, it can be factored out to get

$$3ax^2 + 3bx^2 + a + 5bx + 5ax + b = (a + b)(3x^2 + 5x + 1)$$ ■

Self Check Factor $4x^3 + 3x^2 + x + 4x^2y + 3xy + y$.

Answer $(x + y)(4x^2 + 3x + 1)$

To factor an expression, it is often necessary to factor more than once, as the following example illustrates.

EXAMPLE 12 Factor $3x^3y - 4x^2y^2 - 6x^2y + 8xy^2$.

Solution We begin by factoring out the common factor of xy.

$$3x^3y - 4x^2y^2 - 6x^2y + 8xy^2 = xy(3x^2 - 4xy - 6x + 8y)$$

We can now factor $3x^2 - 4xy - 6x + 8y$ by grouping:

$$\begin{aligned} &3x^3y - 4x^2y^2 - 6x^2y + 8xy^2 \\ &\quad = xy(3x^2 - 4xy - 6x + 8y) \\ &\quad = xy[x(3x - 4y) - 2(3x - 4y)] && \text{Factor } x \text{ from } 3x^2 - 4xy \text{ and } -2 \text{ from } -6x + 8y. \\ &\quad = xy(3x - 4y)(x - 2) && \text{Factor out } 3x - 4y. \end{aligned}$$

Because no more factoring can be done, the factorization is complete. ■

Self Check Factor $3a^3b + 3a^2b - 2a^2b^2 - 2ab^2$.

Answer $ab(3a - 2b)(a + 1)$

WARNING! Whenever you factor an expression, always factor it completely. Each factor of a completely factored expression will be prime.

FORMULAS

Factoring is often required to solve a literal equation for one of its variables.

EXAMPLE 13 **Electronics** The formula $r_1r_2 = rr_2 + rr_1$ is used in electronics to relate the combined resistance r of two resistors wired in parallel. The variable r_1 represents the resistance of the first resistor, and the variable r_2 represents the resistance of the second. Solve for r_2.

Solution To isolate r_2 on one side of the equation, we get all terms involving r_2 on the left-hand side and all terms not involving r_2 on the right-hand side. We then proceed as follows:

$$r_1r_2 = rr_2 + rr_1$$

$$r_1r_2 - rr_2 = rr_1 \quad \text{Subtract } rr_2 \text{ from both sides.}$$

$$r_2(r_1 - r) = rr_1 \quad \text{Factor out } r_2 \text{ on the left-hand side.}$$

$$r_2 = \frac{rr_1}{r_1 - r} \quad \text{Divide both sides by } r_1 - r.$$

■

Self Check Solve $A = p + prt$ for p.

Answer $p = \dfrac{A}{1 + rt}$

Orals *Factor each expression.*

1. $3x^2 - x$

2. $7t^3 + 14t^2$

3. $-3a^2 - 6a$

4. $-4x^2 + 12x$

5. $3(a + b) + x(a + b)$

6. $a(m - n) - b(m - n)$

EXERCISE 5.4

REVIEW *Do each multiplication.*

1. $(a + 4)(a - 4)$

2. $(2b + 3)(2b - 3)$

3. $(4r^2 + 3s)(4r^2 - 3s)$

4. $(5a + 2b^3)(5a - 2b^3)$

5. $(m + 4)(m^2 - 4m + 16)$

6. $(p - q)(p^2 + pq + q^2)$

VOCABULARY AND CONCEPTS *Fill in each blank to make a true statement.*

7. The process of finding the individual factors of a known product is called ________.

8. If a natural number is written as the product of prime factors, we say that it is written in ____________ form.

9. The abbreviation GCF stands for ____________________.

10. If a polynomial cannot be factored, it is called a ______ polynomial or an _________ polynomial.

PRACTICE *In Exercises 11–18, find the prime-factored form of each number.*

11. 6

12. 10

13. 135

14. 98

15. 128

16. 357

17. 325

18. 288

In Exercises 19–26, find the GCF of each set of monomials.

19. 36, 48
20. 45, 75
21. 42, 36, 98
22. 16, 40, 60
23. $4a^2b$, $8a^3c$
24. $6x^3y^2z$, $9xyz^2$
25. $18x^4y^3z^2$, $-12xy^2z^3$
26. $6x^2y^3$, $24xy^3$, $40x^2y^2z^3$

In Exercises 27–30, complete each factorization.

27. $3a - 12 = 3(a - \quad)$
28. $5t + 25 = 5(t + \quad)$
29. $8z^2 + 2z = 2z(4z + \quad)$
30. $9t^3 - 3t^2 = 3t^2(3t - \quad)$

In Exercises 31–50, factor each expression, if possible.

31. $2x + 8$
32. $3y - 9$
33. $2x^2 - 6x$
34. $3y^3 + 3y^2$
35. $5xy + 12ab^2$
36. $7x^2 + 14x$
37. $15x^2y - 10x^2y^2$
38. $11m^3n^2 - 12x^2y$
39. $63x^3y^2 + 81x^2y^4$
40. $33a^3b^4c - 16xyz$
41. $14r^2s^3 + 15t^6$
42. $13ab^2c^3 - 26a^3b^2c$
43. $27z^3 + 12z^2 + 3z$
44. $25t^6 - 10t^3 + 5t^2$
45. $24s^3 - 12s^2t + 6st^2$
46. $18y^2z^2 + 12y^2z^3 - 24y^4z^3$
47. $45x^{10}y^3 - 63x^7y^7 + 81x^{10}y^{10}$
48. $48u^6v^6 - 16u^4v^4 - 3u^6v^3$
49. $25x^3 - 14y^3 + 36x^3y^3$
50. $9m^4n^3p^2 + 18m^2n^3p^4 - 27m^3n^4p$

In Exercises 51–60, factor out the negative of the greatest common factor.

51. $-3a - 6$
52. $-6b + 12$
53. $-3x^2 - x$
54. $-4a^3 + a^2$
55. $-6x^2 - 3xy$
56. $-15y^3 + 25y^2$
57. $-18a^2b - 12ab^2$
58. $-21t^5 + 28t^3$
59. $-63u^3v^6z^9 + 28u^2v^7z^2 - 21u^3v^3z^4$
60. $-56x^4y^3z^2 - 72x^3y^4z^5 + 80xy^2z^3$

In Exercises 61–70, factor out the designated factor.

61. x^2 from $x^{n+2} + x^{n+3}$
62. y^3 from $y^{n+3} + y^{n+5}$
63. y^n from $2y^{n+2} - 3y^{n+3}$
64. x^n from $4x^{n+3} - 5x^{n+5}$
65. x^{-2} from $x^4 - 5x^6$
66. y^{-4} from $7y^4 + y$
67. t^{-3} from $t^5 + 4t^{-6}$
68. p^{-5} from $6p^3 - p^{-2}$
69. $4y^{-2n}$ from $8y^{2n} + 12 + 16y^{-2n}$
70. $7x^{-3n}$ from $21x^{6n} + 7x^{3n} + 14$

In Exercises 71–82, factor each expression.

71. $4(x + y) + t(x + y)$

72. $5(a - b) - t(a - b)$

73. $(a - b)r - (a - b)s$

74. $(x + y)u + (x + y)v$

75. $3(m + n + p) + x(m + n + p)$

76. $x(x - y - z) + y(x - y - z)$

77. $(x + y)(x + y) + z(x + y)$

78. $(a - b)^2 + (a - b)$

79. $(u + v)^2 - (u + v)$

80. $a(x - y) - (x - y)^2$

81. $-a(x + y) + b(x + y)$

82. $-bx(a - b) - cx(a - b)$

In Exercises 83–94, factor by grouping.

83. $ax + bx + ay + by$

84. $ar - br + as - bs$

85. $x^2 + yx + 2x + 2y$

86. $2c + 2d - cd - d^2$

87. $3c - cd + 3d - c^2$

88. $x^2 + 4y - xy - 4x$

89. $a^2 - 4b + ab - 4a$

90. $7u + v^2 - 7v - uv$

91. $ax + bx - a - b$

92. $x^2y - ax - xy + a$

93. $x^2 + xy + xz + xy + y^2 + zy$

94. $ab - b^2 - bc + ac - bc - c^2$

In Exercises 95–100, factor by grouping. Factor out all common monomials first.

95. $mpx + mqx + npx + nqx$

96. $abd - abe + acd - ace$

97. $x^2y + xy^2 + 2xyz + xy^2 + y^3 + 2y^2z$

98. $a^3 - 2a^2b + a^2c - a^2b + 2ab^2 - abc$

99. $2n^4p - 2n^2 - n^3p^2 + np + 2mn^3p - 2mn$

100. $a^2c^3 + ac^2 + a^3c^2 - 2a^2bc^2 - 2bc^2 + c^3$

In Exercises 101–112, solve for the indicated variable.

101. $r_1r_2 = rr_2 + rr_1$ for r_1

102. $r_1r_2 = rr_2 + rr_1$ for r

103. $d_1d_2 = fd_2 + fd_1$ for f

104. $d_1d_2 = fd_2 + fd_1$ for d_1

105. $b^2x^2 + a^2y^2 = a^2b^2$ for a^2

106. $b^2x^2 + a^2y^2 = a^2b^2$ for b^2

107. $S(1 - r) = a - lr$ for r

108. $Sn = (n - 2)180°$ for n

109. $H(a + b) = 2ab$ for a

110. $H(a + b) = 2ab$ for b

111. $3xy - x = 2y + 3$ for y

112. $x(5y + 3) = y - 1$ for y

APPLICATIONS *In Exercises 113–114, write each polynomial in nested form.*

113. $2x^3 + 5x^2 - 2x + 8$

114. $4x^4 + 5x^2 - 3x + 9$

WRITING

115. Explain how to find the greatest common factor of two natural numbers.

116. Explain how to recognize when a number is prime.

SOMETHING TO THINK ABOUT

117. Pick two natural numbers. Divide their product by their greatest common factor. The result is called the **least common multiple** of the two numbers you picked. Why?

118. The number 6 is called a **perfect number,** because the sum of all the divisors of 6 is twice 6: $1 + 2 + 3 + 6 = 12$. Verify that 28 is also a perfect number.

If the greatest common factor of several terms is 1*, the terms are called* ***relatively prime.*** *In Exercises 119–124, tell whether the terms in each set are relatively prime.*

119. 14, 45

120. 24, 63, 112

121. 60, 28, 36

122. 55, 49, 78

123. $12x^2y$, $5ab^3$, $35x^2b^3$

124. $18uv$, $25rs$, $12rsuv$

5.5 The Difference of Two Squares; the Sum and Difference of Two Cubes

■ PERFECT SQUARES ■ THE DIFFERENCE OF TWO SQUARES ■ THE SUM AND DIFFERENCE OF TWO CUBES ■ FACTORING BY GROUPING

Getting Ready *Multiply.*

1. $(a + b)(a - b)$

2. $(5p + q)(5p - q)$

3. $(3m + 2n)(3m - 2n)$

4. $(2a^2 + b^2)(2a^2 - b^2)$

5. $(a - 3)(a^2 + 3a + 9)$

6. $(p + 2)(p^2 - 2p + 4)$

■ PERFECT SQUARES

To factor the difference of two squares, it is helpful to know the first 20 integers that are **perfect squares.**

1, 4, 9, 16, 25, 36, 49, 64, 81, 100, 121, 144, 169, 196, 225, 256, 289, 324, 361, 400

Expressions like $x^6y^4z^2$ are also perfect squares, because they can be written as the square of another quantity:

$$x^6y^4z^2 = (x^3y^2z)^2$$

All exponential expressions that have even-numbered exponents are perfect squares.

THE DIFFERENCE OF TWO SQUARES

In Section 5.3, we developed the special product formula

1. $(x + y)(x - y) = x^2 - y^2$

The binomial $x^2 - y^2$ is called the **difference of two squares,** because x^2 represents the square of x, y^2 represents the square of y, and $x^2 - y^2$ represents the difference of these squares.

Equation 1 can be written in reverse order to give a formula for factoring the difference of two squares.

Factoring the Difference of Two Squares

$$x^2 - y^2 = (x + y)(x - y)$$

If we think of the difference of two squares as the square of a **F**irst quantity minus the square of a **L**ast quantity, we have the formula

$$F^2 - L^2 = (F + L)(F - L)$$

and we say: *To factor the square of a* **F**irst *quantity minus the square of a* **L**ast *quantity, we multiply the* **F**irst *plus the* **L**ast *by the* **F**irst *minus the* **L**ast.

EXAMPLE 1 Factor $49x^2 - 16$.

Solution We can write $49x^2 - 16$ in the form $(7x)^2 - (4)^2$ and use the formula for factoring the difference of two squares:

$$\begin{array}{ccccc} F^2 & - & L^2 & = & (F + L)(F - L) \\ \downarrow & & \downarrow & & \downarrow \quad \downarrow \quad \downarrow \quad \downarrow \\ (7x)^2 & - & 4^2 & = & (7x + 4)(7x - 4) \end{array}$$

Substitute $7x$ for F and 4 for L.

We can verify this result by multiplication.

$$(7x + 4)(7x - 4) = 49x^2 - 28x + 28x - 16$$
$$= 49x^2 - 16$$

■

Self Check Factor $81p^2 - 25$.

Answer $(9p + 5)(9p - 5)$

WARNING! Expressions such as $(7x)^2 + 4^2$ that represent the sum of two squares cannot be factored in the real-number system. Thus, the binomial $49x^2 + 16$ is a prime binomial.

EXAMPLE 2 Factor $64a^4 - 25b^2$.

Solution We can write $64a^4 - 25b^2$ in the form $(8a^2)^2 - (5b)^2$ and use the formula for factoring the difference of two squares.

$$\begin{array}{ccccccc} F^2 & - & L^2 & = & (F + L) & (F & - L) \\ \downarrow & & \downarrow & & \downarrow \quad \downarrow & \downarrow & \downarrow \\ (8a^2)^2 & - & (5b)^2 & = & (8a^2 + 5b) & (8a^2 & - 5b) \end{array}$$

Verify by multiplication. ■

Self Check Factor $36r^4 - s^2$.

Answer $(6r^2 + s)(6r^2 - s)$

EXAMPLE 3 Factor $x^4 - 1$.

Solution Because the binomial is the difference of the squares of x^2 and 1, it factors into the sum of x^2 and 1, and the difference of x^2 and 1.

$$\begin{aligned} x^4 - 1 &= (x^2)^2 - (1)^2 \\ &= (x^2 + 1)(x^2 - 1) \end{aligned}$$

The factor $x^2 + 1$ is the sum of two squares and is prime. However, the factor $x^2 - 1$ is the difference of two squares and can be factored as $(x + 1)(x - 1)$. Thus,

$$\begin{aligned} x^4 - 1 &= (x^2 + 1)(x^2 - 1) \\ &= (x^2 + 1)(x + 1)(x - 1) \end{aligned}$$ ■

Self Check Factor $a^4 - 16$.

Answer $(a^2 + 4)(a + 2)(a - 2)$

EXAMPLE 4 Factor $(x + y)^4 - z^4$.

Solution This expression is the difference of two squares that can be factored:

$$\begin{aligned} (x + y)^4 - z^4 &= [(x + y)^2]^2 - (z^2)^2 \\ &= [(x + y)^2 + z^2][(x + y)^2 - z^2] \end{aligned}$$

The factor $(x + y)^2 + z^2$ is the sum of two squares and is prime. However, the factor $(x + y)^2 - z^2$ is the difference of two squares and can be factored as $(x + y + z)(x + y - z)$. Thus,

$$\begin{aligned} (x + y)^4 - z^4 &= [(x + y)^2 + z^2][(x + y)^2 - z^2] \\ &= [(x + y)^2 + z^2](x + y + z)(x + y - z) \end{aligned}$$ ■

Self Check Factor $(a - b)^2 - c^2$.

Answer $(a - b + c)(a - b - c)$

When possible, we will always factor out a common factor before factoring the difference of two squares. The factoring process is easier when all common factors are factored out first.

EXAMPLE 5 Factor $2x^4y - 32y$.

Solution

$$\begin{aligned} 2x^4y - 32y &= 2y(x^4 - 16) && \text{Factor out } 2y. \\ &= 2y(x^2 + 4)(x^2 - 4) && \text{Factor } x^4 - 16. \\ &= 2y(x^2 + 4)(x + 2)(x - 2) && \text{Factor } x^2 - 4. \end{aligned}$$

■

Self Check Factor $3a^4 - 3$.

Answer $3(a^2 + 1)(a + 1)(a - 1)$

THE SUM AND DIFFERENCE OF TWO CUBES

The number 64 is called a **perfect cube,** because $4^3 = 64$. To factor the sum or difference of two cubes, it is helpful to know the first ten positive perfect cubes:

1, 8, 27, 64, 125, 216, 343, 512, 729, 1,000

Expressions like $x^9y^6z^3$ are also perfect cubes, because they can be written as the cube of another quantity:

$$x^9y^6z^3 = (x^3y^2z)^3$$

To find formulas for factoring the sum or difference of two cubes, we use the following product formulas:

2. $(x + y)(x^2 - xy + y^2) = x^3 + y^3$

3. $(x - y)(x^2 + xy + y^2) = x^3 - y^3$

To verify Equation 2, we multiply $x^2 - xy + y^2$ by $x + y$ and see that the product is $x^3 + y^3$.

$$\begin{aligned} (x + y)(x^2 - xy + y^2) &= (x + y)x^2 - (x + y)xy + (x + y)y^2 \\ &= x \cdot x^2 + y \cdot x^2 - x \cdot xy - y \cdot xy + x \cdot y^2 + y \cdot y^2 \\ &= x^3 + x^2y - x^2y - xy^2 + xy^2 + y^3 \\ &= x^3 + y^3 && \text{Combine like terms.} \end{aligned}$$

Equation 3 can also be verified by multiplication.

If we write Equations 2 and 3 in reverse order, we have the formulas for factoring the sum and difference of two cubes.

Sum and Difference of Two Cubes

$$x^3 + y^3 = (x + y)(x^2 - xy + y^2)$$
$$x^3 - y^3 = (x - y)(x^2 + xy + y^2)$$

If we think of the sum of two cubes as the sum of the cube of a **F**irst quantity plus the cube of a **L**ast quantity, we have the formula

$$\text{F}^3 + \text{L}^3 = (\text{F} + \text{L})(\text{F}^2 - \text{FL} + \text{L}^2)$$

To factor the cube of a First quantity plus the cube of a Last quantity, we multiply the sum of the First and Last by

- *the First squared*
- *minus the First times the Last*
- *plus the Last squared.*

The formula for the difference of two cubes is

$$\text{F}^3 - \text{L}^3 = (\text{F} - \text{L})(\text{F}^2 + \text{FL} + \text{L}^2)$$

To factor the cube of a First quantity minus the cube of a Last quantity, we multiply the difference of the First and Last by

- *the First squared*
- *plus the First times the Last*
- *plus the Last squared.*

EXAMPLE 6 Factor $a^3 + 8$.

Solution Since $a^3 + 8$ can be written as $a^3 + 2^3$, we have the sum of two cubes, which factors as follows:

$$\begin{aligned}\text{F}^3 + \text{L}^3 &= (\text{F} + \text{L})(\text{F}^2 - \text{FL} + \text{L}^2)\\ &\downarrow \\ a^3 + 2^3 &= (a + 2)(a^2 - a\,2 + 2^2)\\ &= (a + 2)(a^2 - 2a + 4)\end{aligned}$$

Thus, $a^3 + 8 = (a + 2)(a^2 - 2a + 4)$. Check by multiplication. ■

Self Check Factor $p^3 + 27$.

Answer $(p + 3)(p^2 - 3p + 9)$

EXAMPLE 7 Factor $27a^3 - 64b^3$.

Solution Since $27a^3 - 64b^3$ can be written as $(3a)^3 - (4b)^3$, we have the difference of two cubes, which factors as follows:

$$\begin{array}{l} \mathrm{F}^3 - \mathrm{L}^3 = (\mathrm{F} - \mathrm{L})(\mathrm{F}^2 + \mathrm{F}\ \mathrm{L} + \mathrm{L}^2) \\ \quad\downarrow \qquad \downarrow \qquad\quad \downarrow \quad \downarrow \qquad \downarrow \qquad \downarrow \ \downarrow \qquad \downarrow \\ (3a)^3 - (4b)^3 = (3a - 4b)[(3a)^2 + (3a)(4b) + (4b)^2] \\ \qquad\qquad\qquad = (3a - 4b)(9a^2 + 12ab + 16b^2) \end{array}$$

Thus, $27a^3 - 64b^3 = (3a - 4b)(9a^2 + 12ab + 16b^2)$. Check by multiplication. ■

Self Check Factor $8p^3 - 27q^3$.

Answer $(2p - 3q)(4p^2 + 6pq + 9q^2)$

EXAMPLE 8 Factor $a^3 - (c + d)^3$.

Solution

$$\begin{aligned} a^3 - (c + d)^3 &= [a - (c + d)][a^2 + a(c + d) + (c + d)^2] \\ &= (a - c - d)(a^2 + ac + ad + c^2 + 2cd + d^2) \end{aligned}$$ ■

Self Check Factor $(p + q)^3 - r^3$.

Answer $(p + q - r)(p^2 + 2pq + q^2 + pr + qr + r^2)$

EXAMPLE 9 Factor $x^6 - 64$.

Solution This expression can be considered as the difference of two squares or the difference of two cubes. If we treat it as the difference of two squares, it factors into the product of a sum and a difference.

$$\begin{aligned} x^6 - 64 &= (x^3)^2 - 8^2 \\ &= (x^3 + 8)(x^3 - 8) \end{aligned}$$

Each of these factors further, however, for one is the sum of two cubes and the other is the difference of two cubes:

$$x^6 - 64 = (x + 2)(x^2 - 2x + 4)(x - 2)(x^2 + 2x + 4)$$ ■

Self Check Factor $x^6 - 1$.

Answer $(x + 1)(x^2 - x + 1)(x - 1)(x^2 + x + 1)$

EXAMPLE 10 Factor $2a^5 + 128a^2$.

Solution We first factor out the common factor of $2a^2$ to obtain

$$2a^5 + 128a^2 = 2a^2(a^3 + 64)$$

Then we factor $a^3 + 64$ as the sum of two cubes to obtain

$$2a^5 + 128a^2 = 2a^2(a + 4)(a^2 - 4a + 16)$$

■

Self Check Factor $3x^5 + 24x^2$.

Answer $3x^2(x + 2)(x^2 - 2x + 4)$

EXAMPLE 11 Factor $16r^{6m} - 54t^{3n}$.

Solution

$16r^{6m} - 54t^{3n} = 2(8r^{6m} - 27t^{3n})$	Factor out 2.
$= 2[(2r^{2m})^3 - (3t^n)^3]$	Write $8r^{6m}$ as $(2r^{2m})^3$ and $27t^{3n}$ as $(3t^n)^3$.
$= 2[(2r^{2m} - 3t^n)(4r^{4m} + 6r^{2m}t^n + 9t^{2n})]$	Factor $(2r^{2m})^3 - (3t^n)^3$.

■

Self Check Factor $2a^{3m} - 16b^{3n}$.

Answer $2(a^m - 2b^n)(a^{2m} + 2a^mb^n + 4b^{2n})$

FACTORING BY GROUPING

EXAMPLE 12 Factor $x^2 - y^2 + x - y$.

Solution If we group the first two terms and factor the difference of two squares, we have

$x^2 - y^2 + x - y = (x + y)(x - y) + (x - y)$	Factor $x^2 - y^2$.
$= (x - y)(x + y + 1)$	Factor out $x - y$.

■

Self Check Factor $a^2 - b^2 + a + b$.

Answer $(a + b)(a - b + 1)$

Orals *Factor each expression, if possible.*

1. $x^2 - 1$

2. $a^4 - 16$

3. $x^3 + 1$

4. $a^3 - 8$

5. $2x^2 - 8$

6. $x^4 + 25$

EXERCISE 5.5

REVIEW *Do each multiplication.*

1. $(x + 1)(x + 1)$
2. $(2m - 3)(m - 2)$
3. $(2m + n)(2m + n)$
4. $(3m - 2n)(3m - 2n)$
5. $(a + 4)(a + 3)$
6. $(3b + 2)(2b - 5)$
7. $(4r - 3s)(2r - s)$
8. $(5a - 2b)(3a + 4b)$

VOCABULARY AND CONCEPTS *Fill in each blank to make a true statement.*

9. Write the first ten perfect square natural numbers. ______________________
10. $p^2 - q^2 = (p + q)$______
11. The sum of two squares ______ be factored.
12. Write the first ten perfect cube natural numbers. ______________________
13. $p^3 + q^3 = (p + q)$__________
14. $p^3 - q^3 = (p - q)$__________

PRACTICE *In Exercises 15–34, factor each expression, if possible.*

15. $x^2 - 4$
16. $y^2 - 9$
17. $9y^2 - 64$
18. $16x^4 - 81y^2$
19. $x^2 + 25$
20. $144a^2 - b^4$
21. $625a^2 - 169b^4$
22. $4y^2 + 9z^4$
23. $81a^4 - 49b^2$
24. $64r^6 - 121s^2$
25. $36x^4y^2 - 49z^4$
26. $4a^2b^4c^6 - 9d^8$
27. $(x + y)^2 - z^2$
28. $a^2 - (b - c)^2$
29. $(a - b)^2 - c^2$
30. $(m + n)^2 - p^4$
31. $x^4 - y^4$
32. $16a^4 - 81b^4$
33. $256x^4y^4 - z^8$
34. $225a^4 - 16b^8c^{12}$

In Exercises 35–42, factor each expression.

35. $2x^2 - 288$
36. $8x^2 - 72$
37. $2x^3 - 32x$
38. $3x^3 - 243x$
39. $5x^3 - 125x$
40. $6x^4 - 216x^2$
41. $r^2s^2t^2 - t^2x^4y^2$
42. $16a^4b^3c^4 - 64a^2bc^6$

In Exercises 43–52, factor each expression.

43. $r^3 + s^3$
44. $t^3 - v^3$
45. $x^3 - 8y^3$
46. $27a^3 + b^3$
47. $64a^3 - 125b^6$
48. $8x^6 + 125y^3$
49. $125x^3y^6 + 216z^9$
50. $1{,}000a^6 - 343b^3c^6$
51. $x^6 + y^6$
52. $x^9 + y^9$

In Exercises 53–60, factor each expression.

53. $5x^3 + 625$

54. $2x^3 - 128$

55. $4x^5 - 256x^2$

56. $2x^6 + 54x^3$

57. $128u^2v^3 - 2t^3u^2$

58. $56rs^2t^3 + 7rs^2v^6$

59. $(a + b)x^3 + 27(a + b)$

60. $(c - d)r^3 - (c - d)s^3$

In Exercises 61–70, factor each expression. Assume that m and n are natural numbers.

61. $x^{2m} - y^{4n}$

62. $a^{4m} - b^{8n}$

63. $100a^{4m} - 81b^{2n}$

64. $25x^{8m} - 36y^{4n}$

65. $x^{3n} - 8$

66. $a^{3m} + 64$

67. $a^{3m} + b^{3n}$

68. $x^{6m} - y^{3n}$

69. $2x^{6m} + 16y^{3m}$

70. $24 + 3c^{3m}$

In Exercises 71–76, factor each expression by grouping.

71. $a^2 - b^2 + a + b$

72. $x^2 - y^2 - x - y$

73. $a^2 - b^2 + 2a - 2b$

74. $m^2 - n^2 + 3m + 3n$

75. $2x + y + 4x^2 - y^2$

76. $m - 2n + m^2 - 4n^2$

WRITING

77. Describe the pattern used to factor the difference of two squares.

78. Describe the patterns used to factor the sum and the difference of two cubes.

SOMETHING TO THINK ABOUT

79. Factor $x^{32} - y^{32}$.

80. Find the error in this proof that 2 = 1.

$$x = y$$
$$x^2 = xy$$
$$x^2 - y^2 = xy - y^2$$
$$(x + y)(x - y) = y(x - y)$$
$$\frac{(x + y)\cancel{(x - y)}}{\cancel{x - y}} = \frac{y\cancel{(x - y)}}{\cancel{x - y}}$$
$$x + y = y$$
$$y + y = y$$
$$2y = y$$
$$\frac{2y}{y} = \frac{y}{y}$$
$$2 = 1$$

5.6 Factoring Trinomials

■ SPECIAL PRODUCT FORMULAS ■ FACTORING TRINOMIALS WITH LEAD COEFFICIENTS OF 1 ■ FACTORING TRINOMIALS WITH LEAD COEFFICIENTS OTHER THAN 1 ■ TEST FOR FACTORABILITY ■ USING SUBSTITUTION TO FACTOR TRINOMIALS ■ FACTORING BY GROUPING ■ USING GROUPING TO FACTOR TRINOMIALS

Getting Ready *Multiply.*

1. $(a + 3)(a - 4)$ **2.** $(3a + 5)(2a - 1)$

3. $(a + b)(a + 2b)$ **4.** $(2a + b)(a - b)$

5. $(3a - 2b)(2a - 3b)$ **6.** $(4a - 3b)^2$

■ SPECIAL PRODUCT FORMULAS

Many trinomials can be factored by using the following special product formulas.

1. $(x + y)(x + y) = x^2 + 2xy + y^2$

2. $(x - y)(x - y) = x^2 - 2xy + y^2$

To factor the perfect square trinomial $x^2 + 6x + 9$, we note that it can be written in the form $x^2 + 2(3)x + 3^2$. If $y = 3$, this form matches the right-hand side of Equation 1. Thus, $x^2 + 6x + 9$ factors as

$$\begin{aligned} x^2 + 6x + 9 &= x^2 + 2(3)x + 3^2 \\ &= (x + 3)(x + 3) \end{aligned}$$

This result can be verified by multiplication:

$$\begin{aligned} (x + 3)(x + 3) &= x^2 + 3x + 3x + 9 \\ &= x^2 + 6x + 9 \end{aligned}$$

To factor the perfect square trinomial $x^2 - 4xz + 4z^2$, we note that it can be written in the form $x^2 - 2x(2z) + (2z)^2$. If $y = 2z$, this form matches the right-hand side of Equation 2. Thus, $x^2 - 4xz + 4z^2$ factors as

$$\begin{aligned} x^2 - 4xz + 4z^2 &= x^2 - 2x(2z) + (2z)^2 \\ &= (x - 2z)(x - 2z) \end{aligned}$$

This result can also be verified by multiplication.

We begin our discussion of *general trinomials* by considering trinomials with lead coefficients (the coefficient of the squared term) of 1.

FACTORING TRINOMIALS WITH LEAD COEFFICIENTS OF 1

Since the product of two binomials is often a trinomial, we expect that many trinomials will factor as two binomials. For example, to factor $x^2 + 7x + 12$, we must find two binomials $x + a$ and $x + b$ such that

$$x^2 + 7x + 12 = (x + a)(x + b)$$

where $ab = 12$ and $ax + bx = 7x$.

To find the numbers a and b, we list the possible factorizations of 12 and find the one where the sum of the factors is 7.

The one to choose ↓ (points to 4(3))

$$12(1) \quad 6(2) \quad 4(3) \quad -12(-1) \quad -6(-2) \quad -4(-3)$$

Thus, $a = 4$, $b = 3$, and

$$x^2 + 7x + 12 = (x + a)(x + b)$$

3. $x^2 + 7x + 12 = (x + 4)(x + 3)$ Substitute 4 for a and 3 for b.

This factorization can be verified by multiplying $x + 4$ and $x + 3$ and observing that the product is $x^2 + 7x + 12$.

Because of the commutative property of multiplication, the order of the factors in Equation 3 is not important.

To factor trinomials with lead coefficients of 1, we follow these steps.

Factoring Trinomials

1. Write the trinomial in descending powers of one variable.
2. List the factorizations of the third term of the trinomial.
3. Pick the factorization where the sum of the factors is the coefficient of the middle term.

EXAMPLE 1 Factor $x^2 - 6x + 8$.

Solution Since the trinomial is written in descending powers of x, we can move to Step 2 and list the possible factorizations of the third term, which is 8.

The one to choose ↓ (points to −4(−2))

$$8(1) \quad 4(2) \quad -8(-1) \quad -4(-2)$$

In the trinomial, the coefficient of the middle term is -6. The only factorization where the sum of the factors is -6 is $-4(-2)$. Thus, $a = -4$, $b = -2$, and

$$x^2 - 6x + 8 = (x + a)(x + b)$$
$$= (x - 4)(x - 2) \quad \text{Substitute } -4 \text{ for } a \text{ and } -2 \text{ for } b.$$

We can verify this result by multiplication:

$$(x - 4)(x - 2) = x^2 - 2x - 4x + 8$$
$$= x^2 - 6x + 8$$

■

Self Check Factor $a^2 - 7a + 12$.

Answer $(a - 4)(a - 3)$

EXAMPLE 2 Factor $-x + x^2 - 12$.

Solution We begin by writing the trinomial in descending powers of x:

$$-x + x^2 - 12 = x^2 - x - 12$$

The possible factorizations of the third term are

$$12(-1) \quad 6(-2) \quad 4(-3) \quad 1(-12) \quad 2(-6) \quad 3(-4)$$

(The one to choose: $3(-4)$)

In the trinomial, the coefficient of the middle term is -1. The only factorization where the sum of the factors is -1 is $3(-4)$. Thus, $a = 3$, $b = -4$, and

$$\begin{aligned} -x + x^2 - 12 &= x^2 - x - 12 \\ &= (x + a)(x + b) \\ &= (x + 3)(x - 4) \end{aligned}$$ Substitute 3 for a and -4 for b. ■

Self Check Factor $-3a + a^2 - 10$.

Answer $(a + 2)(a - 5)$

EXAMPLE 3 Factor $30x - 4xy - 2xy^2$.

Solution We begin by writing the trinomial in descending powers of y:

$$30x - 4xy - 2xy^2 = -2xy^2 - 4xy + 30x$$

Each term in this trinomial has a common monomial factor of $-2x$, which we will factor out.

$$30x - 4xy - 2xy^2 = -2x(y^2 + 2y - 15)$$

To factor $y^2 + 2y - 15$, we list the factors of -15 and find the pair whose sum is 2.

$$15(-1) \quad 5(-3) \quad 1(-15) \quad 3(-5)$$

(The one to choose: $5(-3)$)

The only factorization where the sum of the factors is 2 (the coefficient of the middle term of $y^2 + 2y - 15$) is $5(-3)$. Thus, $a = 5$, $b = -3$, and

$$\begin{aligned} 30x - 4xy - 2xy^2 &= -2x(y^2 + 2y - 15) \\ &= -2x(y + 5)(y - 3) \end{aligned}$$ ■

Self Check Factor $18a + 3ab - 3ab^2$.

Answer $-3a(b + 2)(b - 3)$

WARNING! In Example 3, be sure to include all factors in the final result. It is a common error to forget to write the $-2x$.

FACTORING TRINOMIALS WITH LEAD COEFFICIENTS OTHER THAN 1

There are more combinations of factors to consider when factoring trinomials with lead coefficients other than 1. To factor $5x^2 + 7x + 2$, for example, we must find two binomials of the form $ax + b$ and $cx + d$ such that

$$5x^2 + 7x + 2 = (ax + b)(cx + d)$$

Since the first term of the trinomial $5x^2 + 7x + 2$ is $5x^2$, the first terms of the binomial factors must be $5x$ and x.

$$5x^2 + 7x + 2 = (\overset{5x^2}{5x + b)(x} + d)$$

Since the product of the last terms must be 2, and the sum of the products of the outer and inner terms must be $7x$, we must find two numbers whose product is 2 that will give a middle term of $7x$.

$$5x^2 + 7x + 2 = (5x + \overset{2}{b)(x + d})$$

$$\text{O} + \text{I} = 7x$$

Since 2(1) and $(-2)(-1)$ give a product of 2, there are four possible combinations to consider:

$$(5x + 2)(x + 1) \qquad (5x - 2)(x - 1)$$
$$(5x + 1)(x + 2) \qquad (5x - 1)(x - 2)$$

Of these possibilities, only the one printed in color gives the correct middle term of $7x$. Thus,

4. $5x^2 + 7x + 2 = (5x + 2)(x + 1)$

We can verify this result by multiplication:

$$\begin{aligned}(5x + 2)(x + 1) &= 5x^2 + 5x + 2x + 2 \\ &= 5x^2 + 7x + 2\end{aligned}$$

TEST FOR FACTORABILITY

If a trinomial has the form $ax^2 + bx + c$, with integer coefficients and $a \neq 0$, we can test to see whether it is factorable.

- If the value of $b^2 - 4ac$ is a perfect square, the trinomial can be factored using only integers.
- If the value is not a perfect square, the trinomial cannot be factored using only integers.

For example, $5x^2 + 7x + 2$ is a trinomial in the form $ax^2 + bx + c$ with

$$a = 5, \qquad b = 7, \qquad \text{and} \qquad c = 2$$

For this trinomial, the value of $b^2 - 4ac$ is

$$\begin{aligned} b^2 - 4ac &= 7^2 - 4(5)(2) && \text{Substitute 7 for } b\text{, 5 for } a\text{, and 2 for } c. \\ &= 49 - 40 \\ &= 9 \end{aligned}$$

Since 9 is a perfect square, the trinomial is factorable. Its factorization is shown in Equation 4 on the previous page.

Test for Factorability

A trinomial of the form $ax^2 + bx + c$, with integer coefficients and $a \neq 0$, will factor into two binomials with integer coefficients if the value of $b^2 - 4ac$ is a perfect square. If $b^2 - 4ac = 0$, the factors will be the same.

EXAMPLE 4 Factor $3p^2 - 4p - 4$.

Solution In the trinomial, $a = 3$, $b = -4$, and $c = -4$. To see whether it factors, we evaluate $b^2 - 4ac$.

$$\begin{aligned} b^2 - 4ac &= (-4)^2 - 4(3)(-4) && \text{Substitute } -4 \text{ for } b\text{, 3 for } a\text{, and } -4 \text{ for } c. \\ &= 16 + 48 \\ &= 64 \end{aligned}$$

Since 64 is a perfect square, the trinomial is factorable.

To factor the trinomial, we note that the first terms of the binomial factors must be $3p$ and p to give the first term of $3p^2$.

$$3p^2 - 4p - 4 = \overbrace{(3p + ?)(p} ^{3p^2} + ?)$$

The product of the last terms must be -4, and the sum of the products of the outer terms and the inner terms must be $-4p$.

$$3p^2 - 4p - 4 = (3p + ?)(p + ?) \qquad (-4;\ \text{O} + \text{I} = -4p)$$

Because $1(-4)$, $-1(4)$ and $-2(2)$ all give a product of -4, there are six possible combinations to consider:

$$(3p+1)(p-4) \qquad (3p-4)(p+1)$$
$$(3p-1)(p+4) \qquad (3p+4)(p-1)$$
$$(3p-2)(p+2) \qquad (3p+2)(p-2)$$

Of these possibilities, only the one printed in color gives the correct middle term of $-4p$. Thus,

$$3p^2 - 4p - 4 = (3p+2)(p-2)$$

Self Check Factor $4q^2 - 9q - 9$.

Answer $(4q+3)(q-3)$

EXAMPLE 5 Factor $4t^2 - 3t - 5$, if possible.

Solution In the trinomial, $a = 4$, $b = -3$, and $c = -5$. To see whether the trinomial is factorable, we evaluate $b^2 - 4ac$ by substituting the values of a, b, and c.

$$\begin{aligned} b^2 - 4ac &= (-3)^2 - 4(4)(-5) \\ &= 9 + 80 \\ &= 89 \end{aligned}$$

Since 89 is not a perfect square, the trinomial is not factorable using only integer coefficients. The polynomial is prime.

Self Check Factor $5a^2 - 8a + 2$, if possible.

Answer a prime polynomial

To factor general trinomials, the following hints are helpful.

Factoring a General Trinomial

1. Write the trinomial in descending powers of one variable.
2. Factor out any greatest common factor (including -1 if that is necessary to make the coefficient of the first term positive).
3. Test the trinomial for factorability.
4. When the sign of the first term of a trinomial is + and the sign of the third term is +, the signs between the terms of each binomial factor are the same as the sign of the middle term of the trinomial.
   When the sign of the first term is + and the sign of the third term is −, the signs between the terms of the binomials are opposites.
5. Try various combinations of first terms and last terms until you find the one that works.
6. Check the factorization by multiplication.

EXAMPLE 6 Factor $24y + 10xy - 6x^2y$.

Solution We write the trinomial in descending powers of x and factor out the common factor of $-2y$:

$$\begin{aligned} 24y + 10xy - 6x^2y &= -6x^2y + 10xy + 24y \\ &= -2y(3x^2 - 5x - 12) \end{aligned}$$

In the trinomial $3x^2 - 5x - 12$, $a = 3$, $b = -5$, and $c = -12$.

$$\begin{aligned} b^2 - 4ac &= (-5)^2 - 4(3)(-12) \\ &= 25 + 144 \\ &= 169 \end{aligned}$$

Since 169 is a perfect square, the trinomial will factor.

Since the sign of the third term of $3x^2 - 5x - 12$ is $-$, the signs between the binomial factors will be opposite. Because the first term is $3x^2$, the first terms of the binomial factors must be $3x$ and x.

$$-2y(3x^2 - 5x - 12) = -2y(\overset{3x^2}{3x \qquad)(x} \qquad)$$

The product of the last terms must be -12, and the sum of the product of the outer terms and the product of the inner terms must be $-5x$.

$$-2y(3x^2 - 5x - 12) = -2y(3x \quad \overset{-12}{?)(x \quad ?)}$$

$$\text{O} + \text{I} = -5x$$

Since $1(-12)$, $2(-6)$, $3(-4)$, $12(-1)$, $6(-2)$, and $4(-3)$ all give a product of -12, there are 12 possible combinations to consider.

	$(3x + 1)(x - 12)$	$(3x - 12)(x + 1)$
	$(3x + 2)(x - 6)$	$(3x - 6)(x + 2)$
	$(3x + 3)(x - 4)$	$(3x - 4)(x + 3)$
	$(3x + 12)(x - 1)$	$(3x - 1)(x + 12)$
	$(3x + 6)(x - 2)$	$(3x - 2)(x + 6)$
The one to choose →	$(3x + 4)(x - 3)$	$(3x - 3)(x + 4)$

The combinations marked in blue cannot work, because one of the factors has a common factor. This implies that $3x^2 - 5x - 12$ would have a common factor, which it doesn't.

After mentally trying the remaining combinations, we find that only $(3x + 4)(x - 3)$ gives the correct middle term of $-5x$.

$$\begin{aligned} 24y + 10xy - 6x^2y &= -2y(3x^2 - 5x - 12) \\ &= -2y(3x + 4)(x - 3) \end{aligned}$$

Verify this result by multiplication. ■

Self Check Factor $9b - 6a^2b - 3ab$.

Answer $-3b(2a + 3)(a - 1)$

EXAMPLE 7 Factor $6y + 13x^2y + 6x^4y$.

Solution We write the trinomial in descending powers of x and factor out the common factor of y to obtain

$$\begin{aligned} 6y + 13x^2y + 6x^4y &= 6x^4y + 13x^2y + 6y \\ &= y(6x^4 + 13x^2 + 6) \end{aligned}$$

A test of the trinomial $6x^4 + 13x^2 + 6$ will show that it will factor. Since the coefficients of the first and last terms are positive, the signs between the terms in each binomial will be +.

Since the first term of the trinomial is $6x^4$, the first terms of the binomial factors must be either $2x^2$ and $3x^2$ or x^2 and $6x^2$.

Since the product of the last terms of the binomial factors must be 6, we must find two numbers whose product is 6 that will lead to a middle term of $13x^2$. After trying some combinations, we find the one that works.

$$\begin{aligned} 6y + 13x^2y + 6x^4y &= y(6x^4 + 13x^2 + 6) \\ &= y(2x^2 + 3)(3x^2 + 2) \end{aligned}$$

Verify this result by multiplication. ■

Self Check Factor $4b + 11a^2b + 6a^4b$.

Answer $b(2a^2 + 1)(3a^2 + 4)$

EXAMPLE 8 Factor $x^{2n} + x^n - 2$.

Solution Since the first term is x^{2n}, the first terms of the binomial factors must be x^n and x^n.

$$x^{2n} + x^n - 2 = (\overbrace{x^n \qquad)(x^n}^{x^{2n}} \qquad)$$

Since the third term of the trinomial is -2, the last terms of the binomial factors must have opposite signs, have a product of -2, and lead to a middle term of x^n. The only combination that works is

$$x^{2n} + x^n - 2 = (x^n + 2)(x^n - 1)$$

Verify this result by multiplication. ■

Self Check Factor $a^{2n} + a^n - 6$.

Answer $(a^n + 3)(a^n - 2)$

USING SUBSTITUTION TO FACTOR TRINOMIALS

EXAMPLE 9 Factor $(x + y)^2 + 7(x + y) + 12$.

Solution We rewrite the trinomial $(x + y)^2 + 7(x + y) + 12$ as $z^2 + 7z + 12$, where $z = (x + y)$. The trinomial $z^2 + 7z + 12$ factors as $(z + 4)(z + 3)$.

To find the factorization of $(x + y)^2 + 7(x + y) + 12$, we substitute $x + y$ for z in the expression $(z + 4)(z + 3)$ to obtain

$$z^2 + 7z + 12 = (z + 4)(z + 3)$$
$$(x + y)^2 + 7(x + y) + 12 = (x + y + 4)(x + y + 3)$$

■

Self Check Factor $(a + b)^2 - 3(a + b) - 10$.

Answer $(a + b + 2)(a + b - 5)$

FACTORING BY GROUPING

EXAMPLE 10 Factor $x^2 + 6x + 9 - z^2$.

Solution We group the first three terms together and factor the trinomial to get

$$x^2 + 6x + 9 - z^2 = (x + 3)(x + 3) - z^2$$
$$= (x + 3)^2 - z^2$$

We can now factor the difference of two squares to get

$$x^2 + 6x + 9 - z^2 = (x + 3 + z)(x + 3 - z)$$

■

Self Check Factor $a^2 + 4a + 4 - b^2$.

Answer $(a + 2 + b)(a + 2 - b)$

USING GROUPING TO FACTOR TRINOMIALS

The method of factoring by grouping can be used to help factor trinomials of the form $ax^2 + bx + c$. For example, to factor the trinomial $6x^2 + 7x - 3$, we proceed as follows:

1. First find the product ac: $6(-3) = -18$. This number is called the **key number.**
2. Find two factors of the key number -18 whose sum is $b = 7$:

$$9(-2) = -18 \qquad \text{and} \qquad 9 + (-2) = 7$$

3. Use the factors 9 and -2 as coefficients of two terms to be placed between $6x^2$ and -3:

$$6x^2 + 7x - 3 = 6x^2 + 9x - 2x - 3$$

4. Factor by grouping:

$$6x^2 + 9x - 2x - 3 = 3x(2x + 3) - (2x + 3)$$
$$= (2x + 3)(3x - 1) \quad \text{Factor out } 2x + 3.$$

We can verify this factorization by multiplication.

EXAMPLE 11 Factor $10x^2 + 13x - 3$.

Solution Since $a = 10$ and $c = -3$ in the trinomial, $ac = -30$. We now find two factors of -30 whose sum is $+13$. Two such factors are 15 and -2. We use these factors as coefficients of two terms to be placed between $10x^2$ and -3:

$$10x^2 + 15x - 2x - 3$$

Finally, we factor by grouping.

$$5x(2x + 3) - (2x + 3) = (2x + 3)(5x - 1)$$

Thus, $10x^2 + 13x - 3 = (2x + 3)(5x - 1)$. ■

Self Check Factor $15a^2 + 17a - 4$.

Answer $(3a + 4)(5a - 1)$

Orals *Factor each expression.*

1. $x^2 + 3x + 2$
2. $x^2 + 5x + 4$
3. $x^2 - 5x + 6$
4. $x^2 - 3x - 4$
5. $2x^2 + 3x + 1$
6. $3x^2 + 4x + 1$

EXERCISE 5.6

REVIEW *Solve each equation.*

1. $\dfrac{2 + x}{11} = 3$
2. $\dfrac{3y - 12}{2} = 9$
3. $\dfrac{2}{3}(5t - 3) = 38$
4. $3(p + 2) = 4p$
5. $11r + 6(3 - r) = 3$
6. $2q^2 - 9 = q(q + 3) + q^2$

VOCABULARY AND CONCEPTS *Fill in each blank to make a true statement.*

7. $(x + y)(x + y) = x^2 +$ ______
8. $(x - y)(x - y) = x^2 -$ ______
9. $(x + y)(x - y) =$ ______
10. If ______ is a perfect square, the trinomial $ax^2 + bx + c$ will factor using ______ coefficients.

PRACTICE *In Exercises 11–18, complete each factorization.*

11. $x^2 + 5x + 6 = (x + 3)(\_\_\_\_\_)$

12. $x^2 - 6x + 8 = (x - 4)(\_\_\_\_\_)$

13. $x^2 + 2x - 15 = (x + 5)(\_\_\_\_\_)$

14. $x^2 - 3x - 18 = (x - 6)(\_\_\_\_\_)$

15. $2a^2 + 9a + 4 = (\_\_\_\_\_)(a + 4)$

16. $6p^2 - 5p - 4 = (\_\_\_\_\_)(2p + 1)$

17. $4m^2 + 8mn + 3n^2 = (\_\_\_\_\_)(2m + n)$

18. $12r^2 + 5rs - 2s^2 = (\_\_\_\_\_)(3r + 2s)$

In Exercises 19–28, use a special product formula to factor each perfect square trinomial.

19. $x^2 + 2x + 1$

20. $y^2 - 2y + 1$

21. $a^2 - 18a + 18$

22. $b^2 + 12b + 36$

23. $4y^2 + 4y + 1$

24. $9x^2 + 6x + 1$

25. $9b^2 - 12b + 4$

26. $4a^2 - 12a + 9$

27. $9z^2 + 24z + 16$

28. $16z^2 - 24z + 9$

In Exercises 29–40, test each trinomial for factorability and factor it, if possible.

29. $x^2 + 9x + 8$

30. $y^2 + 7y + 6$

31. $x^2 - 7x + 10$

32. $c^2 - 7c + 12$

33. $b^2 + 8b + 18$

34. $x^2 - 12x + 35$

35. $x^2 - x - 30$

36. $a^2 + 4a - 45$

37. $a^2 + 5a - 50$

38. $b^2 + 9b - 36$

39. $y^2 - 4y - 21$

40. $x^2 + 4x - 28$

In Exercises 41–52, factor each trinomial. If the lead coefficient is negative, begin by factoring out −1.

41. $3x^2 + 12x - 63$

42. $2y^2 + 4y - 48$

43. $a^2b^2 - 13ab^2 + 22b^2$

44. $a^2b^2x^2 - 18a^2b^2x + 81a^2b^2$

45. $b^2x^2 - 12bx^2 + 35x^2$

46. $c^3x^2 + 11c^3x - 42c^3$

47. $-a^2 + 4a + 32$

48. $-x^2 - 2x + 15$

49. $-3x^2 + 15x - 18$

50. $-2y^2 - 16y + 40$

51. $-4x^2 + 4x + 80$

52. $-5a^2 + 40a - 75$

In Exercises 53–78, factor each trinomial. Factor out all common monomials first (including −1 if the lead coefficient is negative). If a trinomial is prime, so indicate.

53. $6y^2 + 7y + 2$

54. $6x^2 - 11x + 3$

55. $8a^2 + 6a - 9$

56. $15b^2 + 4b - 4$

57. $6x^2 - 5x - 4$

58. $18y^2 - 3y - 10$

59. $5x^2 + 4x + 1$

60. $6z^2 + 17z + 12$

61. $8x^2 - 10x + 3$

62. $4a^2 + 20a + 3$

63. $a^2 - 3ab - 4b^2$

64. $b^2 + 2bc - 80c^2$

65. $2y^2 + yt - 6t^2$

66. $3x^2 - 10xy - 8y^2$

67. $3x^3 - 10x^2 + 3x$

68. $3t^3 - 3t^2 + t$

69. $-3a^2 + ab + 2b^2$

70. $-2x^2 + 3xy + 5y^2$

71. $-4x^2 - 9 + 12x$

72. $6x + 4 + 9x^2$

73. $5a^2 + 45b^2 - 30ab$

74. $-90x^2 + 2 - 8x$

75. $8x^2z + 6xyz + 9y^2z$

76. $x^3 - 60xy^2 + 7x^2y$

77. $21x^4 - 10x^3 - 16x^2$

78. $16x^3 - 50x^2 + 36x$

In Exercises 79–88, factor each trinomial.

79. $x^4 + 8x^2 + 15$

80. $x^4 + 11x^2 + 24$

81. $y^4 - 13y^2 + 30$

82. $y^4 - 13y^2 + 42$

83. $a^4 - 13a^2 + 36$

84. $b^4 - 17b^2 + 16$

85. $z^4 - z^2 - 12$

86. $c^4 - 8c^2 - 9$

87. $4x^3 + x^6 + 3$

88. $a^6 - 2 + a^3$

In Exercises 89–96, factor each expression. Assume that n is a natural number.

89. $x^{2n} + 2x^n + 1$

90. $x^{4n} - 2x^{2n} + 1$

91. $2a^{6n} - 3a^{3n} - 2$

92. $b^{2n} - b^n - 6$

93. $x^{4n} + 2x^{2n}y^{2n} + y^{4n}$

94. $y^{6n} + 2y^{3n}z + z^2$

95. $6x^{2n} + 7x^n - 3$

96. $12y^{4n} + 10y^{2n} + 2$

In Exercises 97–102, factor each expression.

97. $(x + 1)^2 + 2(x + 1) + 1$

98. $(a + b)^2 - 2(a + b) + 1$

99. $(a + b)^2 - 2(a + b) - 24$

100. $(x - y)^2 + 3(x - y) - 10$

101. $6(x + y)^2 - 7(x + y) - 20$

102. $2(x - z)^2 + 9(x - z) + 4$

In Exercises 103–112, factor each expression.

103. $x^2 + 4x + 4 - y^2$

104. $x^2 - 6x + 9 - 4y^2$

105. $x^2 + 2x + 1 - 9z^2$

106. $x^2 + 10x + 25 - 16z^2$

107. $c^2 - 4a^2 + 4ab - b^2$

108. $4c^2 - a^2 - 6ab - 9b^2$

109. $a^2 - b^2 + 8a + 16$

110. $a^2 + 14a - 25b^2 + 49$

111. $4x^2 - z^2 + 4xy + y^2$

112. $x^2 - 4xy - 4z^2 + 4y^2$

In Exercises 113–120, use grouping to help factor each trinomial.

113. $a^2 - 17a + 16$

114. $b^2 - 4b - 21$

115. $2u^2 + 5u + 3$

116. $6y^2 + 5y - 6$

117. $20r^2 - 7rs - 6s^2$

118. $6s^2 + st - 12t^2$

119. $20u^2 + 19uv + 3v^2$

120. $12m^2 + mn - 6n^2$

WRITING

121. Explain how you would factor -1 from a trinomial.

122. Explain how you would test the polynomial $ax^2 + bx + c$ for factorability.

SOMETHING TO THINK ABOUT

123. Because it is the difference of two squares, $x^2 - q^2$ always factors. Does the test for factorability predict this?

124. The polynomial $ax^2 + ax + a$ factors, because a is a common factor. Does the test for factorability predict this? If not, is there something wrong with the test? Explain.

5.7 Summary of Factoring Techniques

■ FACTORING RANDOM POLYNOMIALS

Getting Ready *Factor each polynomial.*

1. $3p^2q - 3pq^2$

2. $4p^2 - 9q^2$

3. $p^2 + 5p - 6$

4. $6p^2 - 13pq + 6q^2$

■ FACTORING RANDOM POLYNOMIALS

In this section, we will discuss ways to approach a randomly chosen factoring problem. For example, suppose we wish to factor the trinomial

$$x^2y^2z^3 + 7xy^2z^3 + 6y^2z^3$$

We begin by attempting to identify the problem type. The first type to look for is **factoring out a common monomial.** Because the trinomial has a common monomial factor of y^2z^3, we factor it out:

$$x^2y^2z^3 + 7xy^2z^3 + 6y^2z^3 = y^2z^3(x^2 + 7x + 6)$$

We note that $x^2 + 7x + 6$ is a trinomial that can be factored as $(x + 6)(x + 1)$. Thus,

$$\begin{aligned} x^2y^2z^3 + 7xy^2z^3 + 6y^2z^3 &= y^2z^3(x^2 + 7x + 6) \\ &= y^2z^3(x + 6)(x + 1) \end{aligned}$$

To identify the type of factoring problem, we follow these steps.

Identifying Factoring Problem Types

1. Factor out all common monomial factors.
2. If an expression has two terms, check to see if the problem type is
   a. **The difference of two squares:** $(x^2 - y^2) = (x + y)(x - y)$
   b. **The sum of two cubes:** $(x^3 + y^3) = (x + y)(x^2 - xy + y^2)$
   c. **The difference of two cubes:** $(x^3 - y^3) = (x - y)(x^2 + xy + y^2)$
3. If an expression has three terms, attempt to factor it as a **trinomial.**
4. If an expression has four or more terms, try factoring by **grouping.**
5. Continue until each individual factor is prime.
6. Check the results by multiplying.

EXAMPLE 1 Factor $48a^4c^3 - 3b^4c^3$.

Solution We begin by factoring out the common monomial factor of $3c^3$:

$$48a^4c^3 - 3b^4c^3 = 3c^3(16a^4 - b^4)$$

Since the expression $16a^4 - b^4$ has two terms, we check to see whether it is the difference of two squares, which it is. As the difference of two squares, it factors as $(4a^2 + b^2)(4a^2 - b^2)$.

$$\begin{aligned} 48a^4c^3 - 3b^4c^3 &= 3c^3(16a^4 - b^4) \\ &= 3c^3(4a^2 + b^2)(4a^2 - b^2) \end{aligned}$$

The binomial $4a^2 + b^2$ is the sum of two squares and is prime. However, $4a^2 - b^2$ is the difference of two squares and factors as $(2a + b)(2a - b)$.

$$\begin{aligned} 48a^4c^3 - 3b^4c^3 &= 3c^3(16a^4 - b^4) \\ &= 3c^3(4a^2 + b^2)(4a^2 - b^2) \\ &= 3c^3(4a^2 + b^2)(2a + b)(2a - b) \end{aligned}$$

Since each of the individual factors is prime, the factorization is complete. ■

Self Check Factor $3p^4r^3 - 3q^4r^3$.
Answer $3r^3(p^2 + q^2)(p + q)(p - q)$

EXAMPLE 2 Factor $x^5y + x^2y^4 - x^3y^3 - y^6$.

Solution We begin by factoring out the common monomial factor of y:

$$x^5y + x^2y^4 - x^3y^3 - y^6 = y(x^5 + x^2y^3 - x^3y^2 - y^5)$$

Because the expression $x^5 + x^2y^3 - x^3y^2 - y^5$ has four terms, we try factoring by grouping to obtain

$$\begin{aligned} x^5y + x^2y^4 - x^3y^3 - y^6 &= y(x^5 + x^2y^3 - x^3y^2 - y^5) && \text{Factor out } y. \\ &= y[x^2(x^3 + y^3) - y^2(x^3 + y^3)] && \text{Factor by grouping.} \\ &= y(x^3 + y^3)(x^2 - y^2) && \text{Factor out } x^3 + y^3. \end{aligned}$$

Finally, we factor $x^3 + y^3$ (the sum of two cubes) and $x^2 - y^2$ (the difference of two squares) to obtain

$$x^5y + x^2y^4 - x^3y^3 - y^6 = y(x + y)(x^2 - xy + y^2)(x + y)(x - y)$$

Because each of the individual factors is prime, the factorization is complete. ■

Self Check Factor $a^5p - a^3b^2p + a^2b^3p - b^5p$.

Answer $p(a + b)(a^2 - ab + b^2)(a + b)(a - b)$

EXAMPLE 3 Factor $p^{-4} - p^{-2} - 6$ and write the result without using negative exponents.

Solution We factor this expression as if it were a trinomial:

$$\begin{aligned} p^{-4} - p^{-2} - 6 &= (p^{-2} - 3)(p^{-2} + 2) \\ &= \left(\frac{1}{p^2} - 3\right)\left(\frac{1}{p^2} + 2\right) \end{aligned}$$

■

Self Check Factor $x^{-4} - x^{-2} - 12$.

Answer $\left(\frac{1}{x^2} + 3\right)\left(\frac{1}{x^2} - 4\right)$

EXAMPLE 4 Factor $x^3 + 5x^2 + 6x + x^2y + 5xy + 6y$.

Solution Since there are more than three terms, we try factoring by grouping. We can factor x from the first three terms and y from the last three terms and proceed as follows:

$$\begin{aligned} &x^3 + 5x^2 + 6x + x^2y + 5xy + 6y \\ &\quad = x(x^2 + 5x + 6) + y(x^2 + 5x + 6) \\ &\quad = (x^2 + 5x + 6)(x + y) && \text{Factor out } x^2 + 5x + 6. \\ &\quad = (x + 3)(x + 2)(x + y) && \text{Factor } x^2 + 5x + 6. \end{aligned}$$

■

Self Check Factor $a^3 - 5a^2 + 6a + a^2b - 5ab + 6b$.

Answer $(a - 2)(a - 3)(a + b)$

EXAMPLE 5 Factor $x^4 + 2x^3 + x^2 + x + 1$.

Solution Since there are more than three terms, we try factoring by grouping. We can factor x^2 from the first three terms and proceed as follows:

$$\begin{aligned} x^4 + 2x^3 + x^2 + x + 1 &= x^2(x^2 + 2x + 1) + (x + 1) \\ &= x^2(x + 1)(x + 1) + (x + 1) && \text{Factor } x^2 + 2x + 1. \\ &= (x + 1)[x^2(x + 1) + 1] && \text{Factor out } x + 1. \\ &= (x + 1)(x^3 + x^2 + 1) \end{aligned}$$

■

Self Check Factor $a^4 - a^3 - 2a^2 + a - 2$.

Answer $(a - 2)(a^3 + a^2 + 1)$

Orals *Factor each expression.*

1. $x^2 - y^2$
2. $2x^3 - 4x^4$
3. $x^2 + 4x + 4$
4. $x^2 - 5x + 6$
5. $x^3 - 8$
6. $x^3 + 8$

EXERCISE 5.7

REVIEW *Do the operations.*

1. $(3a^2 + 4a - 2) + (4a^2 - 3a - 5)$
2. $(-4b^2 - 3b - 2) - (3b^2 - 2b + 5)$
3. $5(2y^2 - 3y + 3) - 2(3y^2 - 2y + 6)$
4. $4(3x^2 + 3x + 3) + 3(x^2 - 3x - 4)$
5. $(m + 4)(m - 2)$
6. $(3p + 4q)(2p - 3q)$

VOCABULARY AND CONCEPTS *Fill in each blank to make a true statement.*

7. In any factoring problem, always factor out any __________ first.
8. If an expression has two terms, check to see whether the problem type is the ________ of two squares, the sum of two ______, or the ________ of two cubes.
9. If an expression has three terms, try to factor it as a ________.
10. If an expression has four or more terms, try factoring it by ________.

PRACTICE *In Exercises 11–62, factor each polynomial, if possible.*

11. $x^2 + 8x + 16$
12. $20 + 11x - 3x^2$
13. $8x^3y^3 - 27$
14. $3x^2y + 6xy^2 - 12xy$
15. $xy - ty + xs - ts$
16. $bc + b + cd + d$

17. $25x^2 - 16y^2$

18. $27x^9 - y^3$

19. $12x^2 + 52x + 35$

20. $12x^2 + 14x - 6$

21. $6x^2 - 14x + 8$

22. $12x^2 - 12$

23. $56x^2 - 15x + 1$

24. $7x^2 - 57x + 8$

25. $4x^2y^2 + 4xy^2 + y^2$

26. $100z^2 - 81t^2$

27. $x^3 + (a^2y)^3$

28. $4x^2y^2z^2 - 26x^2y^2z^3$

29. $2x^3 - 54$

30. $4(xy)^3 + 256$

31. $ae + bf + af + be$

32. $a^2x^2 + b^2y^2 + b^2x^2 + a^2y^2$

33. $2(x + y)^2 + (x + y) - 3$

34. $(x - y)^3 + 125$

35. $625x^4 - 256y^4$

36. $2(a - b)^2 + 5(a - b) + 3$

37. $36x^4 - 36$

38. $6x^2 - 63 - 13x$

39. $2x^6 + 2y^6$

40. $x^4 - x^4y^4$

41. $a^4 - 13a^2 + 36$

42. $x^4 - 17x^2 + 16$

43. $x^2 + 6x + 9 - y^2$

44. $x^2 + 10x + 25 - y^8$

45. $4x^2 + 4x + 1 - 4y^2$

46. $9x^2 - 6x + 1 - 25y^2$

47. $x^2 - y^2 - 2y - 1$

48. $a^2 - b^2 + 4b - 4$

49. $x^5 + x^2 - x^3 - 1$

50. $x^5 - x^2 - 4x^3 + 4$

51. $x^5 - 9x^3 + 8x^2 - 72$

52. $x^5 - 4x^3 - 8x^2 + 32$

53. $2x^5z - 2x^2y^3z - 2x^3y^2z + 2y^5z$

54. $x^2y^3 - 4x^2y - 9y^3 + 36y$

55. $x^{2m} - x^m - 6$

56. $a^{2n} - b^{2n}$

57. $a^{3n} - b^{3n}$

58. $x^{3m} + y^{3m}$

59. $x^{-2} + 2x^{-1} + 1$

60. $4a^{-2} - 12a^{-1} + 9$

61. $6x^{-2} - 5x^{-1} - 6$

62. $x^{-4} - y^{-4}$

WRITING

63. What is your strategy for factoring a polynomial?

64. Explain how you can know that your factorization is correct.

SOMETHING TO THINK ABOUT

65. If you have the choice of factoring a polynomial as the difference of two squares or as the difference of two cubes, which do you do first? Why?

66. Can several polynomials have a greatest common factor? Find the GCF of $2x^2 + 7x + 3$ and $x^2 - 2x - 15$.

67. Factor $x^4 + x^2 + 1$. (*Hint:* Add and subtract x^2.)

68. Factor $x^4 + 7x^2 + 16$. (*Hint:* Add and subtract x^2.)

5.8 Solving Equations by Factoring

■ SOLVING QUADRATIC EQUATIONS ■ SOLVING HIGHER-DEGREE POLYNOMIAL EQUATIONS
■ PROBLEM SOLVING

Getting Ready *Factor each polynomial.*

1. $2a^2 - 4a$ **2.** $a^2 - 25$

3. $6a^2 + 5a - 6$ **4.** $6a^3 - a^2 - 2a$

■ SOLVING QUADRATIC EQUATIONS

An equation such as $3x^2 + 4x - 7 = 0$ or $-5y^2 + 3y + 8 = 0$ is called a **quadratic** (or **second-degree**) equation.

Quadratic Equations
A **quadratic equation** is any equation that can be written in the form

$$ax^2 + bx + c = 0$$

where a, b, and c are real numbers and $a \neq 0$.

Many quadratic equations can be solved by factoring and then using the **zero-factor property.**

Zero-Factor Property
If a and b are real numbers, then

If $ab = 0$, then $a = 0$ or $b = 0$.

The zero-factor property states that *if the product of two or more numbers is 0, then at least one of the numbers must be 0.*

To solve the quadratic equation $x^2 + 5x + 6 = 0$, we factor its left-hand side to obtain

$$(x + 3)(x + 2) = 0$$

Since the product of $x + 3$ and $x + 2$ is 0, then at least one of the factors is 0. Thus, we can set each factor equal to 0 and solve each resulting linear equation for x:

$$x + 3 = 0 \quad \text{or} \quad x + 2 = 0$$
$$x = -3 \quad | \quad x = -2$$

To check these solutions, we first substitute -3 and then -2 for x in the equation and verify that each number satisfies the equation.

$$\begin{array}{rcl|rcl} x^2 + 5x + 6 &=& 0 & x^2 + 5x + 6 &=& 0 \\ (-3)^2 + 5(-3) + 6 &\stackrel{?}{=}& 0 & (-2)^2 + 5(-2) + 6 &\stackrel{?}{=}& 0 \\ 9 - 15 + 6 &\stackrel{?}{=}& 0 & 4 - 10 + 6 &\stackrel{?}{=}& 0 \\ 0 &=& 0 & 0 &=& 0 \end{array}$$

(or)

Both -3 and -2 are solutions, because both satisfy the equation.

EXAMPLE 1 Solve $3x^2 + 6x = 0$.

Solution We factor the left-hand side, set each factor equal to 0, and solve each resulting equation for x.

$$3x^2 + 6x = 0$$

$$3x(x + 2) = 0 \qquad \text{Factor out the common factor of } 3x.$$

$$3x = 0 \quad \text{or} \quad x + 2 = 0$$

$$x = 0 \qquad\qquad x = -2$$

Verify that both solutions check. ■

Self Check Solve $4p^2 - 12p = 0$.

Answer 0, 3

WARNING! In Example 1, do not divide both sides by $3x$, or you will lose the solution $x = 0$.

EXAMPLE 2 Solve $x^2 - 16 = 0$.

Solution We factor the difference of two squares on the left-hand side, set each factor equal to 0, and solve each resulting equation.

$$x^2 - 16 = 0$$

$$(x + 4)(x - 4) = 0$$

$$x + 4 = 0 \quad \text{or} \quad x - 4 = 0$$

$$x = -4 \quad | \quad x = 4$$

Verify that both solutions check. ■

Self Check Solve $a^2 - 81 = 0$.

Answer 9, −9

Many equations that do not appear to be quadratic can be put into quadratic form ($ax^2 + bx + c = 0$) and then solved by factoring.

EXAMPLE 3 Solve $x = \frac{6}{5} - \frac{6}{5}x^2$.

Solution We write the equation in quadratic form and then solve by factoring.

$$x = \frac{6}{5} - \frac{6}{5}x^2$$

$$5x = 6 - 6x^2 \quad \text{Multiply both sides by 5.}$$

$$6x^2 + 5x - 6 = 0 \quad \text{Add } 6x^2 \text{ to both sides and subtract 6 from both sides.}$$

$$(3x - 2)(2x + 3) = 0 \quad \text{Factor the trinomial.}$$

$$3x - 2 = 0 \quad \text{or} \quad 2x + 3 = 0 \quad \text{Set each factor equal to 0.}$$

$$3x = 2 \quad \bigg| \quad 2x = -3$$

$$x = \frac{2}{3} \quad \bigg| \quad x = -\frac{3}{2}$$

Verify that both solutions check. ■

Self Check Solve $x = \frac{6}{7}x^2 - \frac{3}{7}$.

Answer $\frac{3}{2}, -\frac{1}{3}$

WARNING! To solve a quadratic equation by factoring, be sure to set the quadratic polynomial equal to 0 before factoring and using the zero-factor property. Do not make the following error:

$$6x^2 + 5x = 6$$

$$x(6x + 5) = 6 \quad \text{If the product of two numbers is 6, neither number need be 6. For example, } 2 \cdot 3 = 6.$$

$$x = 6 \quad \text{or} \quad 6x + 5 = 6$$

$$\bigg| \quad x = \frac{1}{6}$$

Neither solution checks.

GRAPHING CALCULATORS

Solving Quadratic Equations

To solve a quadratic equation such as $x^2 + 4x - 5 = 0$ with a graphing calculator, we can use standard window settings of $[-10, 10]$ for x and $[-10, 10]$ for y and graph the quadratic function $y = x^2 + 4x - 5$, as shown in Figure 5-6(a). We can then trace to find the x-coordinates of the x-intercepts of the parabola. See Figures 5-6(b) and 5-6(c). For better results, we can zoom in. Since these are the numbers x that make $y = 0$, they are the solutions of the equation.

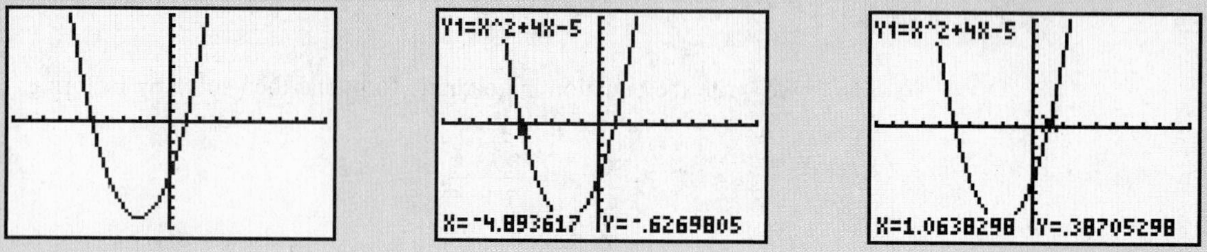

(a) (b) (c)

FIGURE 5-6

SOLVING HIGHER-DEGREE POLYNOMIAL EQUATIONS

We can solve many polynomial equations with degree greater than 2 by factoring.

EXAMPLE 4 Solve $6x^3 - x^2 - 2x = 0$.

Solution We factor x from the third-degree polynomial on the left-hand side and proceed as follows:

$$
\begin{aligned}
6x^3 - x^2 - 2x &= 0 \\
x(6x^2 - x - 2) &= 0 && \text{Factor out } x. \\
x(3x - 2)(2x + 1) &= 0 && \text{Factor } 6x^2 - x - 2. \\
x = 0 \quad \text{or} \quad 3x - 2 &= 0 \quad \text{or} \quad 2x + 1 = 0 && \text{Set each factor equal to 0.} \\
x &= \frac{2}{3} \qquad\quad x = -\frac{1}{2}
\end{aligned}
$$

Verify that the solutions check. ■

Self Check Solve $5x^3 + 13x^2 - 6x = 0$.

Answer $0, \frac{2}{5}, -3$

EXAMPLE 5 Solve $x^4 - 5x^2 + 4 = 0$.

Solution We factor the trinomial on the left-hand side and proceed as follows:

$$x^4 - 5x^2 + 4 = 0$$
$$(x^2 - 1)(x^2 - 4) = 0$$
$$(x + 1)(x - 1)(x + 2)(x - 2) = 0 \qquad \text{Factor } x^2 - 1 \text{ and } x^2 - 4.$$
$$x + 1 = 0 \quad \text{or} \quad x - 1 = 0 \quad \text{or} \quad x + 2 = 0 \quad \text{or} \quad x - 2 = 0$$
$$x = -1 \quad | \quad x = 1 \quad | \quad x = -2 \quad | \quad x = 2$$

Verify that each solution checks. ■

Self Check Solve $a^4 - 13a + 36 = 0$.

Answer 2, −2, 3, −3

Solving Equations

GRAPHING CALCULATORS

To solve the equation $x^4 - 5x^2 + 4 = 0$ with a graphing calculator, we can use window settings of $[-6, 6]$ for x and $[-5, 10]$ for y and graph the polynomial function $y = x^4 - 5x^2 + 4$ as shown in Figure 5-7. We can then read the values of x that make $y = 0$. They are $x = -2, -1, 1$, and 2. If the x-coordinates of the x-intercepts were not obvious, we could get their values by using the trace and zoom features.

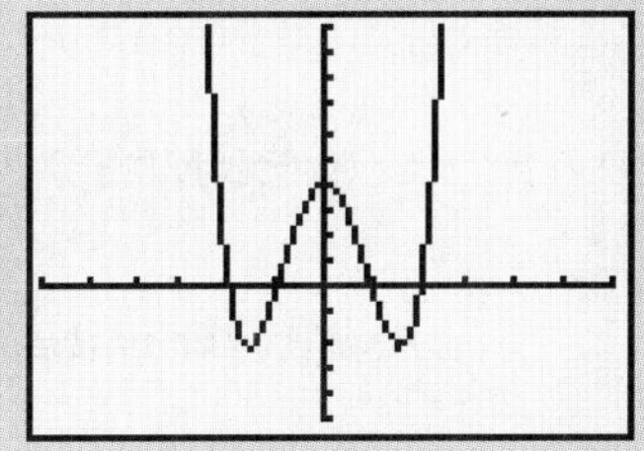

FIGURE 5-7

■ PROBLEM SOLVING

EXAMPLE 6 **Finding dimensions of a truss** The width of the triangular truss shown in Figure 5-8 is 3 times its height. The area of the triangle is 96 square feet. Find its width and height.

Analyze the problem We can let x be the positive number that represents the height of the truss. Then $3x$ represents its width. We can substitute x for h, $3x$ for b, and 96 for A in the formula for the area of a triangle and solve for x.

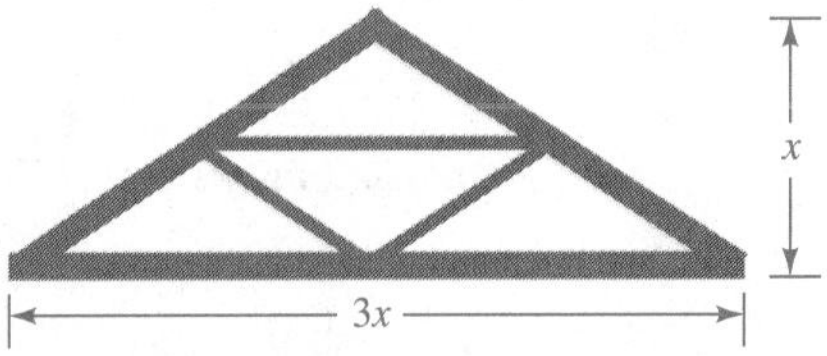

FIGURE 5-8

Form and solve an equation

$$A = \frac{1}{2}bh$$

$$96 = \frac{1}{2}(3x)x$$

$$192 = 3x^2 \quad \text{Multiply both sides by 2.}$$

$$64 = x^2 \quad \text{Divide both sides by 3.}$$

$$0 = x^2 - 64 \quad \text{Subtract 64 from both sides.}$$

$$0 = (x + 8)(x - 8) \quad \text{Factor the difference of two squares.}$$

$$x + 8 = 0 \quad \text{or} \quad x - 8 = 0$$

$$x = -8 \quad | \quad x = 8$$

State the conclusion Since the height of a triangle cannot be negative, we must discard the negative solution. Thus, the height of the truss is 8 feet, and its width is 3(8), or 24 feet.

Check the result The area of a triangle with a width (base) of 24 feet and a height of 8 feet is 96 square feet:

$$A = \frac{1}{2}bh = \frac{1}{2}(24)(8) = 12(8) = 96$$

The solution checks. ■

EXAMPLE 7

Ballistics If the initial velocity of an object thrown straight up into the air is 176 feet per second, when will it hit the ground?

Analyze the problem The height of an object thrown straight up into the air with an initial velocity of v feet per second is given by the formula

$$h = vt - 16t^2$$

The height h is in feet, and t represents the number of seconds since the object was released.

Form and solve an equation When the object hits the ground, its height will be 0. Thus, we set h equal to 0, set v equal to 176, and solve for t.

$$h = vt - 16t^2$$

$$0 = 176t - 16t^2$$

$$0 = 16t(11 - t) \quad \text{Factor out } 16t.$$

$$16t = 0 \quad \text{or} \quad 11 - t = 0 \quad \text{Set each factor equal to 0.}$$

$$t = 0 \quad | \quad t = 11$$

State the conclusion When $t = 0$, the object's height above the ground is 0 feet, because it has just been released. When $t = 11$, the height is again 0 feet, and the object has hit the ground. The solution is 11 seconds.

Check the result Verify that $h = 0$ when $t = 11$. ■

Orals *Solve each equation.*

1. $(x - 2)(x - 3) = 0$
2. $(x + 4)(x - 2) = 0$
3. $(x - 2)(x - 3)(x + 1) = 0$
4. $(x + 3)(x + 2)(x - 5)(x - 6) = 0$

EXERCISE 5.8

REVIEW

1. List the prime numbers less than 10.
2. List the composite numbers between 7 and 17.
3. The formula for the volume of a sphere is $V = \frac{4}{3}\pi r^3$. Find the volume when $r = 21.23$ centimeters. Round the answer to the nearest hundredth.
4. The formula for the volume of a cone is $V = \frac{1}{3}\pi r^2 h$. Find the volume when $r = 12.33$ meters and $h = 14.7$ meters. Round the answer to the nearest hundredth.

VOCABULARY AND CONCEPTS *Fill in the blank to make a true statement.*

5. A quadratic equation is any equation that can be written in the form ______________.
6. If a and b are real numbers, then if ______, $a = 0$ or $b = 0$.

PRACTICE *In Exercises 7–30, solve each equation.*

7. $4x^2 + 8x = 0$
8. $x^2 - 9 = 0$
9. $y^2 - 16 = 0$
10. $5y^2 - 10y = 0$
11. $x^2 + x = 0$
12. $x^2 - 3x = 0$
13. $5y^2 - 25y = 0$
14. $y^2 - 36 = 0$
15. $z^2 + 8z + 15 = 0$
16. $w^2 + 7w + 12 = 0$
17. $y^2 - 7y + 6 = 0$
18. $n^2 - 5n + 6 = 0$
19. $y^2 - 7y + 12 = 0$
20. $x^2 - 3x + 2 = 0$
21. $x^2 + 6x + 8 = 0$
22. $x^2 + 9x + 20 = 0$
23. $3m^2 + 10m + 3 = 0$
24. $2r^2 + 5r + 3 = 0$
25. $2y^2 - 5y + 2 = 0$
26. $2x^2 - 3x + 1 = 0$
27. $2x^2 - x - 1 = 0$
28. $2x^2 - 3x - 5 = 0$
29. $3s^2 - 5s - 2 = 0$
30. $8t^2 + 10t - 3 = 0$

In Exercises 31–42, write each equation in quadratic form and solve it by factoring.

31. $x(x - 6) + 9 = 0$
32. $x^2 + 8(x + 2) = 0$
33. $8a^2 = 3 - 10a$
34. $5z^2 = 6 - 13z$
35. $b(6b - 7) = 10$
36. $2y(4y + 3) = 9$

37. $\frac{3a^2}{2} = \frac{1}{2} - a$

38. $x^2 = \frac{1}{2}(x + 1)$

39. $\frac{1}{2}x^2 - \frac{5}{4}x = -\frac{1}{2}$

40. $\frac{1}{4}x^2 + \frac{3}{4}x = 1$

41. $x\left(3x + \frac{22}{5}\right) = 1$

42. $x\left(\frac{x}{11} - \frac{1}{7}\right) = \frac{6}{77}$

In Exercises 43–54, solve each equation.

43. $x^3 + x^2 = 0$

44. $2x^4 + 8x^3 = 0$

45. $y^3 - 49y = 0$

46. $2z^3 - 200z = 0$

47. $x^3 - 4x^2 - 21x = 0$

48. $x^3 + 8x^2 - 9x = 0$

49. $z^4 - 13z^2 + 36 = 0$

50. $y^4 - 10y^2 + 9 = 0$

51. $3a(a^2 + 5a) = -18a$

52. $7t^3 = 2t\left(t + \frac{5}{2}\right)$

53. $\frac{x^2(6x + 37)}{35} = x$

54. $x^2 = -\frac{4x^3(3x + 5)}{3}$

In Exercises 55–60, use grouping to help solve each equation.

55. $x^3 + 3x^2 - x - 3 = 0$

56. $x^3 - x^2 - 4x + 4 = 0$

57. $2r^3 + 3r^2 - 18r - 27 = 0$

58. $3s^3 - 2s^2 - 3s + 2 = 0$

59. $3y^3 + y^2 = 4(3y + 1)$

60. $w^3 + 16 = w(w + 16)$

61. **Integer problem** The product of two consecutive even integers is 288. Find the integers.

62. **Integer problem** The product of two consecutive odd integers is 143. Find the integers.

63. **Integer problem** The sum of the squares of two consecutive positive integers is 85. Find the integers.

64. **Integer problem** The sum of the squares of three consecutive positive integers is 77. Find the integers.

APPLICATIONS

65. **Geometry** Find the perimeter of the rectangle in Illustration 1.

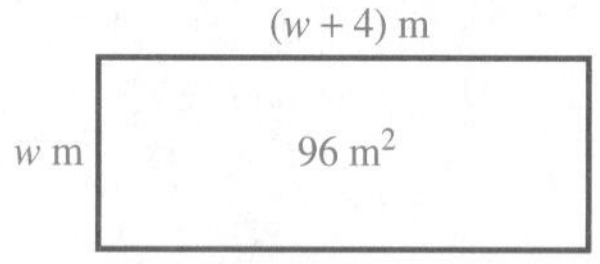

ILLUSTRATION 1

66. **Geometry** One side of a rectangle is three times longer than another. If its area is 147 square centimeters, find its dimensions.

67. **Geometry** Find the dimensions of the rectangle shown in Illustration 2, given that its area is 375 square feet.

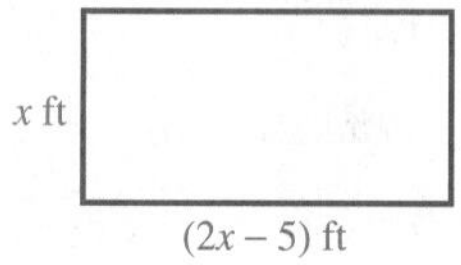

ILLUSTRATION 2

68. Geometry Find the height of the triangle shown in Illustration 3, given that its area is 162 square centimeters.

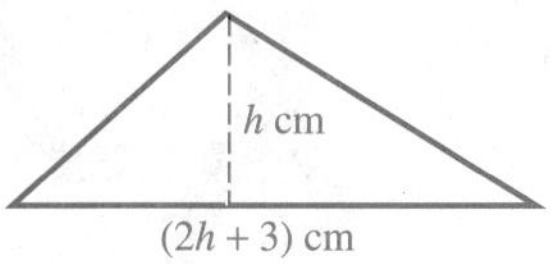

ILLUSTRATION 3

69. Fine arts An artist intends to paint a 60-square-foot mural on the large wall shown in Illustration 4. Find the dimensions of the mural if the artist leaves a border of uniform width around it.

ILLUSTRATION 4

70. Gardening A woman plans to use one-fourth of her 48-foot-by-100-foot rectangular backyard to plant a garden. Find the perimeter of the garden if the length is to be 40 feet greater than the width.

71. Architecture The rectangular room shown in Illustration 5 is twice as long as it is wide. It is divided into two rectangular parts by a partition, positioned as shown. If the larger part of the room contains 560 square feet, find the dimensions of the entire room.

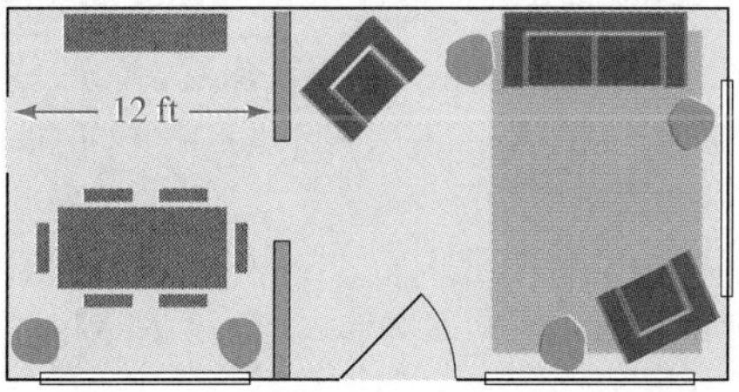

ILLUSTRATION 5

72. Perimeter of a square If the length of one side of a square is increased by 4 inches, the area of the square becomes 9 times greater. Find the perimeter of the original square.

73. Time of flight After how many seconds will an object hit the ground if it is thrown upward with an initial velocity of 160 feet per second?

74. Time of flight After how many seconds will an object hit the ground if it is thrown upward with an initial velocity of 208 feet per second?

75. Ballistics The muzzle velocity of a cannon is 480 feet per second. If a cannonball is fired vertically, at what times will it be at a height of 3,344 feet?

76. Ballistics A slingshot can provide an initial velocity of 128 feet per second. At what times will a stone, shot vertically upward, be 192 feet above the ground?

77. Winter recreation The length of the rectangular ice-skating rink in Illustration 6 is 20 meters greater than twice its width. Find the width.

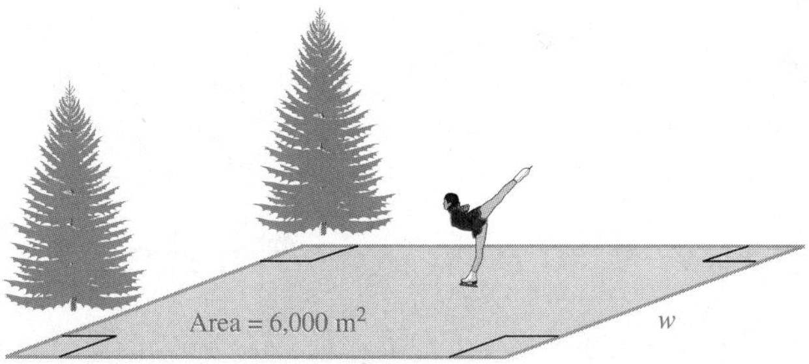

ILLUSTRATION 6

78. Carpentry A 285-square-foot room is 4 feet longer than it is wide. What length of crown molding is needed to trim the perimeter of the ceiling?

79. Designing a swimming pool Building codes require that the rectangular swimming pool in Illustration 7 be surrounded by a uniform-width walkway of at least 516 square feet. If the length of the pool is 10 feet less than twice the width, how wide should the border be?

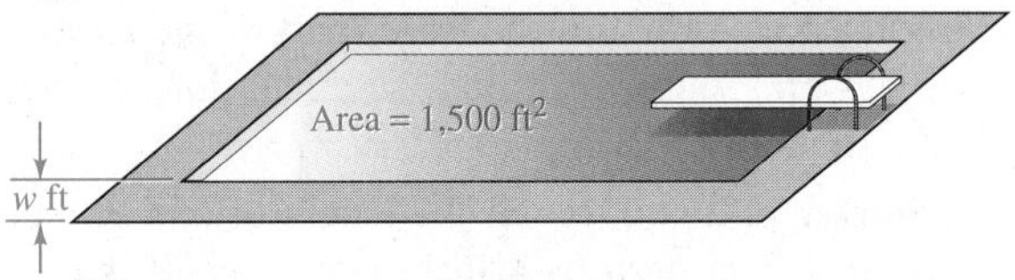

ILLUSTRATION 7

80. House construction The formula for the area of a trapezoid is $A = \frac{h(B + b)}{2}$. The area of the trapezoidal truss in Illustration 8 is 44 square feet. Find the height of the truss if the shorter base is the same as the height.

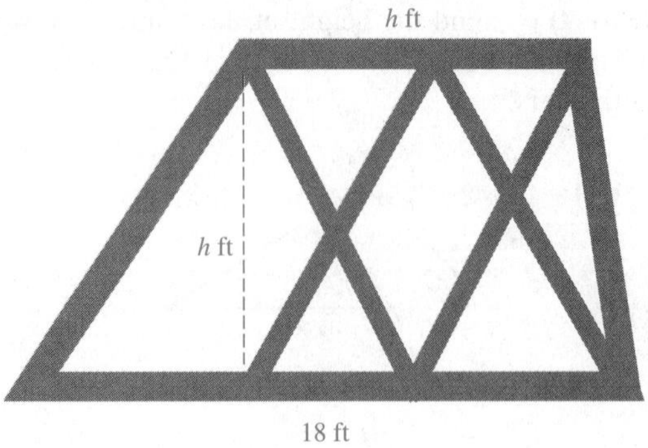

ILLUSTRATION 8

In Exercises 81–84, use a graphing calculator to find the solutions of each equation, if any exist. If an answer is not exact, give the answer to the nearest hundredth.

81. $x^2 - 4x + 7 = 0$

82. $2x^2 - 7x + 4 = 0$

83. $-3x^3 - 2x^2 + 5 = 0$

84. $-2x^3 - 3x - 5 = 0$

WRITING

85. Describe the steps for solving an application problem.

86. Explain how to check the solution of an application problem.

SOMETHING TO THINK ABOUT *Find a quadratic equation with the given roots.*

87. 3, 5

88. −2, 6

89. 0, −5

90. $\frac{1}{2}, \frac{1}{3}$

PROJECTS

PROJECT 1

Mac operates a bait shop on the Snarly River. The shop is located between two sharp bends in the river, and the area around Mac's shop has recently become a popular hiking and camping area. Mac has decided to produce some maps of the area for the use of the visitors. Although he knows the region well, he has very little idea of the actual distances from one location to another. What he knows is:

1. Big Falls, a beautiful waterfall on Snarly River, is due east of Mac's.
2. Grandview Heights, a fabulous rock climbing site, is due west of Mac's, right on the river.
3. Foster's General Store, the only sizable camping and climbing outfitter in the area, is located on the river some distance west and north of Mac's.

Mac hires an aerial photographer to take pictures of the area, with some surprising results. If Mac's bait shop is treated as the origin of a coordinate system, with the y-axis running north–south and the x-axis running east–west, then on the domain $-4 \leq x \leq 4$ (where the units are miles), the river follows the curve

$$P(x) = \frac{1}{4}(x^3 - x^2 - 6x)$$

The aerial photograph is shown in Illustration 1.

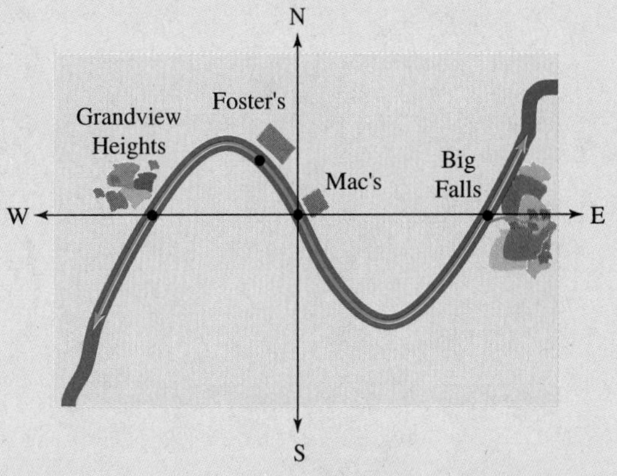

ILLUSTRATION 1

a. Mac would like to include on his maps the exact locations (relative to Mac's) of Big Falls and Grandview Heights. Find these for him, and explain to Mac why your answers must be correct.

b. Mac and Foster have determined that Foster's General Store is 0.7 miles west of the bait shop. Since it is on the river, it is a bit north as well. They decide that to promote business, they will join together to clear a few campsites in the region bordered by the straight-line paths that run between Mac's and Foster's, Mac's and Grandview Heights, and Foster's and Grandview Heights. If they clear 1 campsite for each 40 full acres of area, how many campsites can they put in? (*Hint:* A square mile contains 640 acres.)

c. A path runs in a straight line directly southeast (along the line $y = -x$) from Mac's to the river. How far east and how far south has a hiker on this trail traveled when he or she reaches the river?

PROJECT 2

The rate at which fluid flows through a cylindrical pipe, or any cylinder-shaped tube (an artery, for instance), is

$$\text{Velocity of fluid flow} = V = \frac{p}{nL}(R^2 - r^2)$$

(continued)

PROJECTS *(continued)*

where p is the difference in pressure between the two ends of the tube, L is the length of the tube, R is the radius of the tube, and n is the *viscosity constant,* a measure of the thickness of a fluid. Since the variable r represents the distance from the center of the tube, $0 \le r \le R$. (See Illustration 2.) In most situations, p, L, R, and n are constants, so V is a function of r.

$$V(r) = \frac{p}{nL}(R^2 - r^2)$$

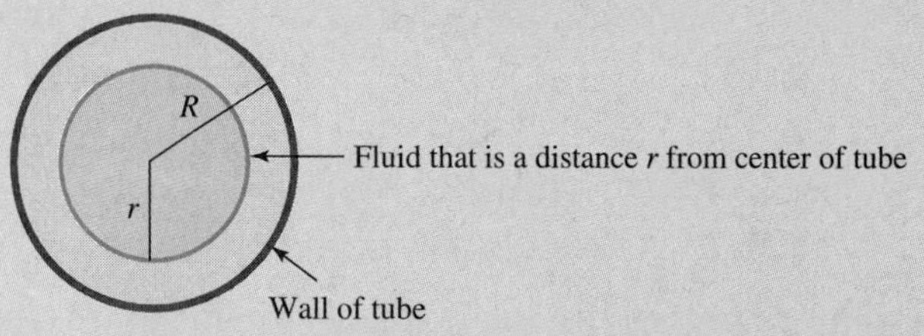

ILLUSTRATION 2

It can be shown that the velocity of a fluid moving in the tube depends on how far from the center (or how close to the wall of the tube) the fluid is.

a. Consider a pipe with radius 5 centimeters and length 60 centimeters. Suppose that $p = 15$ and $n = 0.001$ (water has a viscosity of approximately 0.001). Find the velocity of the fluid at the center of the pipe. The units of measurement for V will be centimeters per second.
Find the velocity of the fluid when it is halfway between the center and the wall of the pipe. What percent of the velocity at the center of the pipe does this represent? Where in the pipe is the fluid flowing with a velocity of 4,000 centimeters per second?

b. Suppose that the situation is the same, but the fluid is now machinery oil, with a viscosity of 0.15. Answer the same questions as in part **a**, except find where in the pipe the oil flows with a velocity of 15 cm per second. Note that the oil travels at a much slower speed than water.

c. Medical doctors use various methods to increase the rate of blood flow through an artery. The patient may take a drug that "thins the blood" (lowers its viscosity) or a drug that dilates the artery, or the patient may undergo angioplasty, a surgical procedure that widens the canal through which the blood passes. Explain why each of these increases the velocity V at a given distance r from the center of the artery.

CHAPTER SUMMARY

CONCEPTS

REVIEW EXERCISES

SECTION 5.1

Polynomial and Polynomial Functions

The **degree of a polynomial** is the degree of the term with highest degree contained within the polynomial.

If $P(x)$ is a polynomial in x, then $P(r)$ is the value of the polynomial at $x = r$.

1. Find the degree of $P(x) = 3x^5 + 4x^3 + 2$.

2. Find the degree of $9x^2y + 13x^3y^2 + 8x^4y^4$.

3. Find each value when $P(x) = -x^2 + 4x + 6$.

a. $P(0)$ **b.** $P(1)$

c. $P(-t)$ **d.** $P(z)$

4. Graph each function.

a. $f(x) = x^3 - 1$

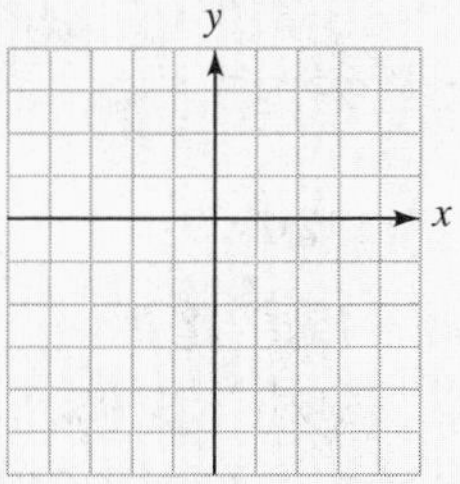

b. $f(x) = x^2 - 2x$

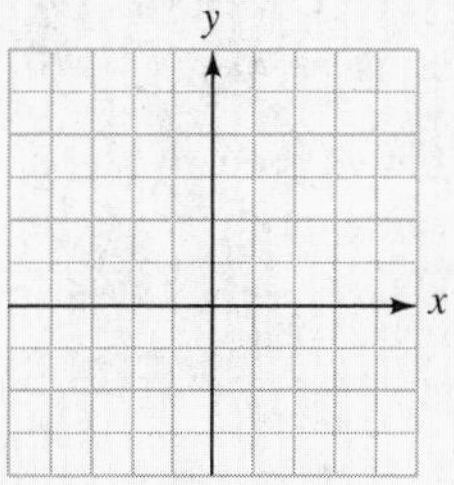

SECTION 5.2

Adding and Subtracting Polynomials

To add polynomials, add their like terms.

To subtract polynomials, add the negative of the polynomial that is to be subtracted from the other polynomial.

5. Simplify each expression.

a. $(3x^2 + 4x + 9) + (2x^2 - 2x + 7)$

b. $(4x^3 + 4x^2 + 7) - (-2x^3 - x - 2)$

c. $(2x^2 - 5x + 9) - (x^2 - 3) - (-3x^2 + 4x - 7)$

d. $2(7x^3 - 6x^2 + 4x - 3) - 3(7x^3 + 6x^2 + 4x - 3)$

SECTION 5.3

Multiplying Polynomials

To multiply monomials, multiply their numerical factors and multiply their variable factors.

6. Find each product.

a. $(8a^2b^2)(-2abc)$

b. $(-3xy^2z)(2xz^3)$

To multiply a polynomial by a monomial, multiply each term of the polynomial by the monomial.

7. Find each product.

a. $2xy^2(x^3y - 4xy^5)$

b. $a^2b(a^2 + 2ab + b^2)$

To multiply polynomials, multiply each term of one polynomial by each term of the other polynomial.

8. Find each product.

a. $(8x - 5)(2x + 3)$

b. $(3x + 2)(2x - 4)$

c. $(5x - 4)(3x - 2)$

d. $(3x^2 - 2)(x^2 - x + 2)$

SECTION 5.4 The Greatest Common Factor and Factoring by Grouping

Always factor out common monomial factors as the first step in a factoring problem.

Use the distributive property to factor out common monomial factors.

9. Factor each expression.

a. $4x + 8$

b. $3x^2 - 6x$

c. $5x^2y^3 - 10xy^2$

d. $7a^4b^2 + 49a^3b$

e. $-8x^2y^3z^4 - 12x^4y^3z^2$

f. $12a^6b^4c^2 + 15a^2b^4c^6$

g. $27x^3y^3z^3 + 81x^4y^5z^2 - 90x^2y^3z^7$

h. $-36a^5b^4c^2 + 60a^7b^5c^3 - 24a^2b^3c^7$

i. Factor x^n from $x^{2n} + x^n$.

j. Factor y^{2n} from $y^{2n} - y^{4n}$.

k. Factor x^{-2} from $x^{-4} - x^{-2}$.

l. Factor a^{-3} from $a^6 + 1$.

m. $5x^2(x + y)^3 - 15x^3(x + y)^4$

n. $-49a^3b^2(a - b)^4 + 63a^2b^4(a - b)^3$

If an expression has four or more terms, try to factor the expression by grouping.

10. Factor each expression.

a. $xy + 2y + 4x + 8$

b. $ac + bc + 3a + 3b$

c. $x^4 + 4y + 4x^2 + x^2y$

d. $a^5 + b^2c + a^2c + a^3b^2$

11. Solve for the indicated variable.

a. $S = 2wh + 2wl + 2lh$ for h

b. $S = 2wh + 2wl + 2lh$ for l

SECTION 5.5 The Difference of Two Squares; the Sum and Difference of Two Cubes

Difference of two squares:

$x^2 - y^2 = (x + y)(x - y)$

12. Factor each expression, if possible.

a. $z^2 - 16$
b. $y^2 - 121$
c. $x^2y^4 - 64z^6$
d. $a^2b^2 + c^2$
e. $(x + z)^2 - t^2$
f. $c^2 - (a + b)^2$
g. $2x^4 - 98$
h. $3x^6 - 300x^2$

Sum of two cubes:

$x^3 + y^3 = (x + y)(x^2 - xy + y^2)$

Difference of two cubes:

$x^3 - y^3 = (x - y)(x^2 + xy + y^2)$

13. Factor each expression, if possible.

a. $x^3 + 343$
b. $a^3 - 125$
c. $8y^3 - 512$
d. $4x^3y + 108yz^3$

SECTION 5.6 Factoring Trinomials

Special product formulas:

$x^2 + 2xy + y^2 = (x + y)(x + y)$

$x^2 - 2xy + y^2 = (x - y)(x - y)$

14. Factor each expression, if possible.

a. $x^2 + 10x + 25$
b. $a^2 - 14a + 49$

Test for factorability:
A trinomial of the form $ax^2 + bx + c$ will factor with integer coefficients if $b^2 - 4ac$ is a perfect square.

15. Factor each expression, if possible.

a. $y^2 + 21y + 20$
b. $z^2 - 11z + 30$
c. $-x^2 - 3x + 28$
d. $y^2 - 5y - 24$
e. $4a^2 - 5a + 1$
f. $3b^2 + 2b + 1$
g. $7x^2 + x + 2$
h. $-15x^2 + 14x + 8$
i. $y^3 + y^2 - 2y$
j. $2a^4 + 4a^3 - 6a^2$
k. $-3x^2 - 9x - 6$
l. $8x^2 - 4x - 24$
m. $15x^2 - 57xy - 12y^2$
n. $30x^2 + 65xy + 10y^2$
o. $24x^2 - 23xy - 12y^2$
p. $14x^2 + 13xy - 12y^2$

SECTION 5.7 Summary of Factoring Techniques

Use these steps to factor a random expression:

1. Factor out all common monomial factors.

2. If an expression has two terms, check to see if it is

a. The difference of two squares:
$(x^2 - y^2) = (x + y)(x - y)$

b. The sum of two cubes:
$(x^3 + y^3) = (x + y)(x^2 - xy + y^2)$

c. The difference of two cubes:
$(x^3 - y^3) = (x - y)(x^2 + xy + y^2)$

3. If an expression has three terms, attempt to factor it as a **trinomial.**

4. If an expression has four or more terms, try factoring by **grouping.**

5. Continue until each individual factor is prime.

6. Check the results by multiplying.

16. Factor each expression, if possible.

a. $x^3 + 5x^2 - 6x$

b. $3x^2y - 12xy - 63y$

c. $z^2 - 4 + zx - 2x$

d. $x^2 + 2x + 1 - p^2$

e. $x^2 + 4x + 4 - 4p^4$

f. $y^2 + 3y + 2 + 2x + xy$

g. $x^{2m} + 2x^m - 3$

h. $x^{-2} - x^{-1} - 2$

SECTION 5.8 Solving Equations by Factoring

Zero-factor property:

If $xy = 0$, then $x = 0$ or $y = 0$.

17. Solve each equation.

a. $4x^2 - 3x = 0$

b. $x^2 - 36 = 0$

c. $12x^2 + 4x - 5 = 0$

d. $7y^2 - 37y + 10 = 0$

e. $t^2(15t - 2) = 8t$

f. $3u^3 = u(19u + 14)$

18. Volume The volume V of the rectangular solid in Illustration 1 is given by the formula $V = lwh$, where l is its length, w is its width, and h is its height. If the volume is 840 cubic centimeters, the length is 12 centimeters, and the width exceeds the height by 3 centimeters, find the height.

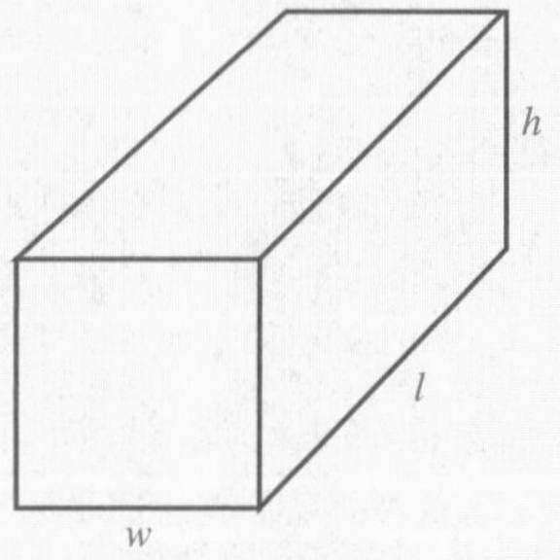

ILLUSTRATION 1

19. Volume of a pyramid
The volume of the pyramid in Illustration 2 is given by the formula $V = \frac{Bh}{3}$, where B is the area of its base and h is its height. The volume of the pyramid is 1,020 cubic meters. Find the dimensions of its rectangular base if one edge of the base is 3 meters longer than the other and the height of the pyramid is 9 meters.

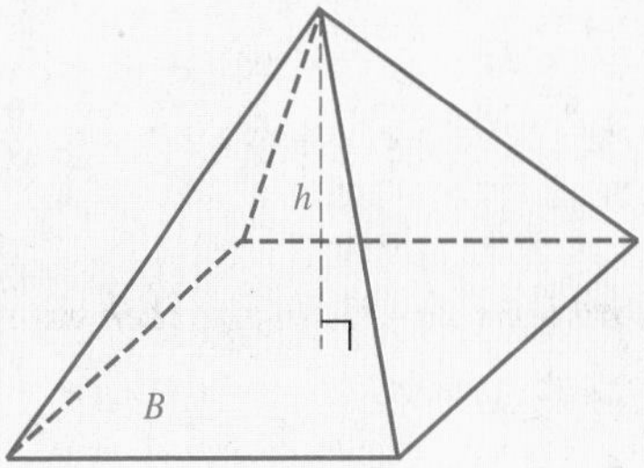

ILLUSTRATION 2

Chapter Test

In Problems 1–2, find the degree of each polynomial.

1. $3x^3 - 4x^5 - 3x^2 - 5$

2. $3x^5y^3 - x^8y^2 + 2x^9y^4 - 3x^2y^5 + 4$

In Problems 3–4, let $P(x) = -3x^2 + 2x - 1$ and find each value.

3. $P(2)$

4. $P(-1)$

5. Graph the function $f(x) = x^2 + 2x$.

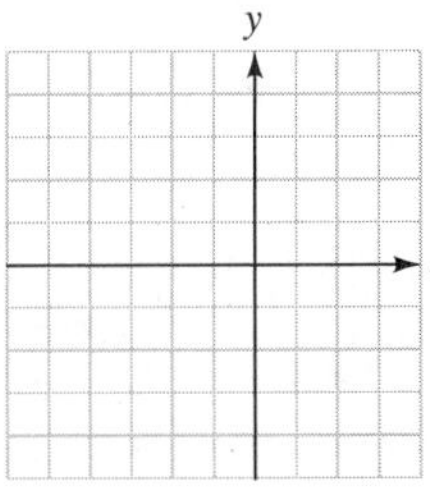

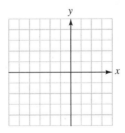

In Problems 6–13, do the operations.

6. $(2y^2 + 4y + 3) + (3y^2 - 3y - 4)$

7. $(-3u^2 + 2u - 7) - (u^2 + 7)$

8. $3(2a^2 - 4a + 2) - 4(-a^2 - 3a - 4)$

9. $-2(2x^2 - 2) + 3(x^2 + 5x - 2)$

10. $(3x^3y^2z)(-2xy^{-1}z^3)$

11. $-5a^2b(3ab^3 - 2ab^4)$

12. $(z + 4)(z - 4)$

13. $(3x - 2)(4x + 3)$

In Problems 14–17, factor each polynomial.

14. $3xy^2 + 6x^2y$

15. $12a^3b^2c - 3a^2b^2c^2 + 6abc^3$

16. Factor y^n from $x^2y^{n+2} + y^n$.

17. Factor b^n from $a^nb^n - ab^{-n}$.

In Problems 18–19, factor each expression.

18. $(u - v)r + (u - v)s$

19. $ax - xy + ay - y^2$

In Problems 20–25, factor each polynomial.

20. $x^2 - 49$

21. $2x^2 - 32$

22. $4y^4 - 64$

23. $b^3 + 125$

24. $b^3 - 27$

25. $3u^3 - 24$

In Problems 26–31, factor each expression.

26. $a^2 - 5a - 6$

27. $6b^2 + b - 2$

28. $6u^2 + 9u - 6$

29. $20r^2 - 15r - 5$

30. $x^{2n} + 2x^n + 1$

31. $x^2 + 6x + 9 - y^2$

32. Solve for r: $r_1r_2 - r_2r = r_1r$

33. Solve for x: $x^2 - 5x - 6 = 0$

34. Integer problem The product of two consecutive positive integers is 156. Find their sum.

35. Preformed concrete The slab of concrete in Illustration 1 is twice as long as it is wide. The area in which it is placed includes a 1-foot-wide border of 70 square feet. Find the dimensions of the slab.

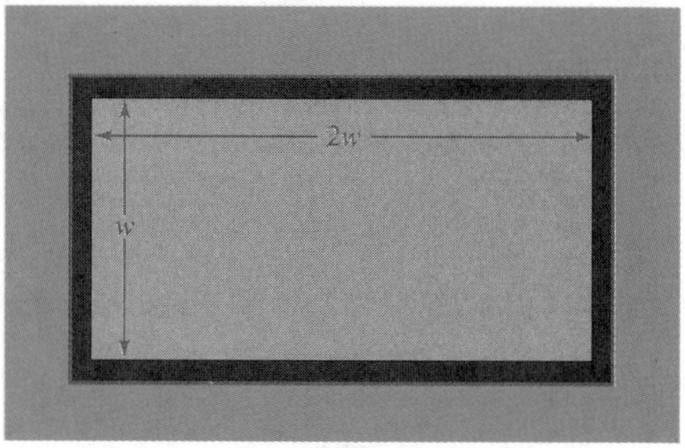

ILLUSTRATION 1

6 Rational Expressions

Mechanical Engineer

MATHEMATICS IN THE WORKPLACE

Mechanical engineers design and develop power-producing machines such as internal combustion engines, steam and gas turbines, and jet rocket engines, as well as power-using machines such as refrigeration and air-conditioning equipment, machine tools, printing presses, and steel rolling mills. Many mechanical engineers do research, test, and design work. Others work in maintenance, technical sales, and production operations. Many teach in colleges and universities or work as consultants.

SAMPLE APPLICATION ■ The stiffness of the shaft shown in Illustration 1 is given by the formula

$$k = \frac{1}{\frac{1}{k_1} + \frac{1}{k_2}}$$

ILLUSTRATION 1

where k_1 and k_2 are the individual stiffnesses of each section. If the stiffness, k_2, of Section 2 is 4,200,000 in. lb/rad and design specifications require that the overall stiffness, k, of the entire shaft be 1,900,000 in. lb/rad, what must the stiffness of Section 1 be?
(See Exercise 64 in Exercise 6.6.)

6.1 Rational Functions and Simplifying Rational Expressions

■ RATIONAL EXPRESSIONS ■ RATIONAL FUNCTIONS ■ FINDING THE DOMAIN AND RANGE OF A RATIONAL FUNCTION ■ SIMPLIFYING FRACTIONS ■ SIMPLIFYING FRACTIONS BY FACTORING OUT −1 ■ PROBABILITY

Getting Ready *Simplify each fraction.*

1. $\frac{6}{8}$ **2.** $\frac{12}{15}$ **3.** $\frac{-25}{65}$ **4.** $\frac{-49}{-63}$

■ RATIONAL EXPRESSIONS

Rational expressions are fractions that indicate the quotient of two polynomials, such as

$$\frac{3x}{x-7}, \quad \frac{5m+n}{8m+16}, \quad \text{and} \quad \frac{a^3+2a^2+7}{2a^2-5a+4}$$

Since division by 0 is undefined, the value of a polynomial occurring in the denominator of a rational expression cannot be 0. For example, x cannot be 7 in the rational expression $\frac{3x}{x-7}$, because the denominator would be 0. In the rational expression $\frac{5m+n}{8m+16}$, m cannot be -2, because that would make the denominator equal to 0.

RATIONAL FUNCTIONS

Rational expressions often define functions. For example, if the cost of subscribing to an on-line information network is \$6 per month plus \$1.50 per hour of access time, the average (mean) hourly cost of the service is the total monthly cost, divided by the number of hours of access time:

$$\bar{c} = \frac{C}{n} = \frac{1.50n + 6}{n}$$

$\bar{c}$ is the mean hourly cost, C is the total monthly cost, and n is the number of hours the service is used.

The function

1. $$\bar{c} = f(n) = \frac{1.50n + 6}{n} \quad (n > 0)$$

gives the mean hourly cost of using the information network for n hours per month.

Figure 6-1 shows the graph of the rational function $\bar{c} = f(n) = \frac{1.50n+6}{n}$ $(n > 0)$. Since $n > 0$, the domain of this function is the interval $(0, \infty)$.

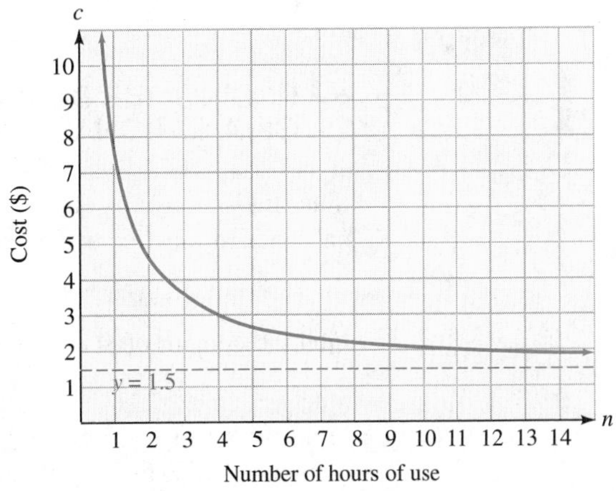

FIGURE 6-1

From the graph, we can see that the mean hourly cost decreases as the number of hours of access time increases. Since the cost of each extra hour of access time is \$1.50, the mean hourly cost can approach \$1.50 but never drop below it. Thus, the graph of the function approaches the line $y = 1.5$ as n increases without bound. When a graph approaches a line as the dependent variable gets large, we call the line an **asymptote.** The line $y = 1.5$ is a **horizontal asymptote** of the graph.

As n gets smaller and approaches 0, the graph approaches the y-axis but never touches it. The y-axis is a **vertical asymptote** of the graph.

EXAMPLE 1 Find the mean hourly cost when the network described above is used for **a.** 3 hours and **b.** 70.4 hours.

Solution **a.** To find the mean hourly cost for 3 hours of access time, we substitute 3 for n in Equation 1 and simplify:

$$\bar{c} = f(3) = \frac{1.50(3) + 6}{3} = 3.5$$

The mean hourly cost for 3 hours of access time is \$3.50.

b. To find the mean hourly cost for 70.4 hours of access time, we substitute 70.4 for n in Equation 1 and simplify:

$$\bar{c} = f(70.4) = \frac{1.50(70.4) + 6}{70.4} = 1.585227273$$

The mean hourly cost for 70.4 hours of access time is approximately \$1.59. ■

Self Check In Example 1, find the mean hourly cost when the network is used for 5 hours.

Answer \$2.70

FINDING THE DOMAIN AND RANGE OF A RATIONAL FUNCTION

Since division by 0 is undefined, any values that make the denominator 0 in a rational function must be excluded from the domain of the function.

EXAMPLE 2 Find the domain of $f(x) = \dfrac{3x + 2}{x^2 + x - 6}$.

Solution From the set of real numbers, we must exclude any values of x that make the denominator 0. To find these values, we set $x^2 + x - 6$ equal to 0 and solve for x.

$$x^2 + x - 6 = 0$$

$$(x + 3)(x - 2) = 0 \qquad \text{Factor.}$$

$$x + 3 = 0 \quad \text{or} \quad x - 2 = 0 \qquad \text{Set each factor equal to 0.}$$

$$x = -3 \qquad\qquad x = 2 \qquad \text{Solve each linear equation.}$$

Thus, the domain of the function is the set of all real numbers except −3 and 2. In interval notation, the domain is $(-\infty, -3) \cup (-3, 2) \cup (2, \infty)$. ■

Self Check Find the domain of $f(x) = \dfrac{x^2 + 1}{x - 2}$.

Answer $(-\infty, 2) \cup (2, \infty)$

Finding the Domain and Range of a Function

GRAPHING CALCULATORS

We can find the domain and range of the function in Example 2 by looking at its graph. If we use window settings of $[-10, 10]$ for x and $[-10, 10]$ for y and graph the function

$$f(x) = \frac{3x + 2}{x^2 + x - 6}$$

we will obtain the graph in Figure 6-2(a).

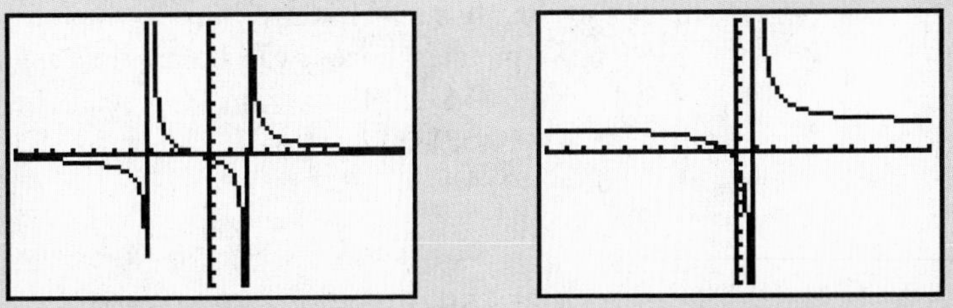

(a) (b)

FIGURE 6-2

From the figure, we can see that

- As x approaches -3 from the left, the values of y decrease, and the graph approaches the vertical line $x = -3$.
- As x approaches -3 from the right, the values of y increase, and the graph approaches the vertical line $x = -3$.

From the figure, we can also see that

- As x approaches 2 from the left, the values of y decrease, and the graph approaches the vertical line $x = 2$.
- As x approaches 2 from the right, the values of y increase, and the graph approaches the vertical line $x = 2$.

The lines $x = -3$ and $x = 2$ are vertical asymptotes. Although the vertical lines in the graph appear to be the graphs of $x = -3$ and $x = 2$, they are not. Graphing calculators draw graphs by connecting dots whose x-coordinates are close together. Often, when two such points straddle a vertical asymptote and their y-coordinates are far apart, the calculator draws a line between them anyway, producing what appears to be a vertical asymptote. If you set your calculator to dot mode instead of connected mode, the vertical lines will not appear.

From Figure 6-2(a), we can also see that

- As x increases to the right of 2, the values of y decrease and approach the value $y = 0$.
- As x decreases to the left of -3, the values of y increase and approach the value $y = 0$.

The line $y = 0$ (the x-axis) is a horizontal asymptote. Graphing calculators do not draw lines that appear to be horizontal asymptotes.

From the graph, we can see that all real numbers x, except -3 and 2, give a value of y. This confirms that the domain of the function is $(-\infty, -3) \cup (-3, 2) \cup (2, \infty)$. We can also see that y can be any value. Thus, the range is $(-\infty, \infty)$.

To find the domain and range of the function $f(x) = \frac{2x+1}{x-1}$, we use a calculator to draw the graph shown in Figure 6-2(b). From this graph, we can see that the line $x = 1$ is a vertical asymptote and that the line $y = 2$ is a horizontal asymptote. Since x can be any real number except 1, the domain is the interval $(-\infty, 1) \cup (1, \infty)$. Since y can be any value except 2, the range is $(-\infty, 2) \cup (2, \infty)$.

SIMPLIFYING FRACTIONS

Since rational expressions are the fractions of algebra, the familiar rules for arithmetic fractions apply.

Properties of Fractions

If there are no divisions by 0, then

1. $\frac{a}{b} = \frac{c}{d}$ if and only if $ad = bc$
2. $\frac{a}{1} = a$ and $\frac{a}{a} = 1$
3. $\frac{ak}{bk} = \frac{a}{b} \cdot \frac{k}{k} = \frac{a}{b}$
4. $-\frac{a}{b} = \frac{-a}{b} = \frac{a}{-b}$

Property 3 of fractions is true because

$$\frac{ak}{bk} = \frac{a}{b} \cdot \frac{k}{k} = \frac{a}{b} \cdot 1 = \frac{a}{b} \quad (k \neq 0)\ (b \neq 0)$$

Any number times 1 is the number itself.

To simplify fractions, we will use Property 3, which enables us to divide out factors that are common to the numerator and the denominator.

EXAMPLE 3 Simplify **a.** $\frac{10k}{25k^2}$ and **b.** $\frac{-8y^3z^5}{6y^4z^3}$.

Solution We factor each numerator and denominator and divide out all common factors:

a. $$\frac{10k}{25k^2} = \frac{5 \cdot 2 \cdot k}{5 \cdot 5 \cdot k \cdot k} = \frac{\overset{1}{\cancel{5}} \cdot 2 \cdot \overset{1}{\cancel{k}}}{\underset{1}{\cancel{5}} \cdot 5 \cdot \underset{1}{\cancel{k}} \cdot k} = \frac{2}{5k}$$

b. $$\frac{-8y^3z^5}{6y^4z^3} = \frac{-2 \cdot 4 \cdot y \cdot y \cdot y \cdot z \cdot z \cdot z \cdot z \cdot z}{2 \cdot 3 \cdot y \cdot y \cdot y \cdot y \cdot z \cdot z \cdot z} = \frac{-\overset{1}{\cancel{2}} \cdot 4 \cdot \overset{1}{\cancel{y}} \cdot \overset{1}{\cancel{y}} \cdot \overset{1}{\cancel{y}} \cdot \overset{1}{\cancel{z}} \cdot \overset{1}{\cancel{z}} \cdot \overset{1}{\cancel{z}} \cdot z \cdot z}{\underset{1}{\cancel{2}} \cdot 3 \cdot \underset{1}{\cancel{y}} \cdot \underset{1}{\cancel{y}} \cdot \underset{1}{\cancel{y}} \cdot y \cdot \underset{1}{\cancel{z}} \cdot \underset{1}{\cancel{z}} \cdot \underset{1}{\cancel{z}}} = -\frac{4z^2}{3y}$$

■

Self Check Simplify $\frac{-12a^4b^2}{-3ab^4}$.

Answer $\frac{4a^3}{b^2}$

The fractions in Example 3 can also be simplified by using the rules of exponents:

$$\frac{10k}{25k^2} = \frac{5 \cdot 2}{5 \cdot 5} k^{1-2} = \frac{2}{5} \cdot k^{-1} = \frac{2}{5} \cdot \frac{1}{k} = \frac{2}{5k}$$

$$\frac{-8y^3z^5}{6y^4z^3} = \frac{-2 \cdot 4}{2 \cdot 3} y^{3-4}z^{5-3} = \frac{-4}{3} \cdot y^{-1}z^2 = -\frac{4}{3} \cdot \frac{1}{y} \cdot \frac{z^2}{1} = -\frac{4z^2}{3y}$$

EXAMPLE 4 Simplify $\frac{x^2 - 16}{x + 4}$.

Solution We factor $x^2 - 16$ and use the fact that $\frac{x+4}{x+4} = 1$.

$$\frac{x^2 - 16}{x + 4} = \frac{\overset{1}{\cancel{(x+4)}}(x-4)}{1\underset{1}{\cancel{(x+4)}}} = \frac{x-4}{1} = x - 4$$

■

Self Check Simplify $\frac{x^2 - 9}{x - 3}$.

Answer $x + 3$

Checking an Algebraic Simplification

GRAPHING CALCULATORS

To show that the simplification in Example 4 is correct, we can graph the functions

$$f(x) = \frac{x^2-16}{x+4} \quad \text{and} \quad g(x) = x - 4$$

(See Figure 6-3.) Except for the point where $x = -4$, the graphs are the same. The point where $x = -4$ is excluded from the graph of $f(x) = \frac{x^2-16}{x+4}$, because -4 is not in the domain of f. However, graphing calculators do not show that this point is excluded. The point where $x = -4$ is included in the graph of $g(x) = x - 4$, because -4 is in the domain of g.

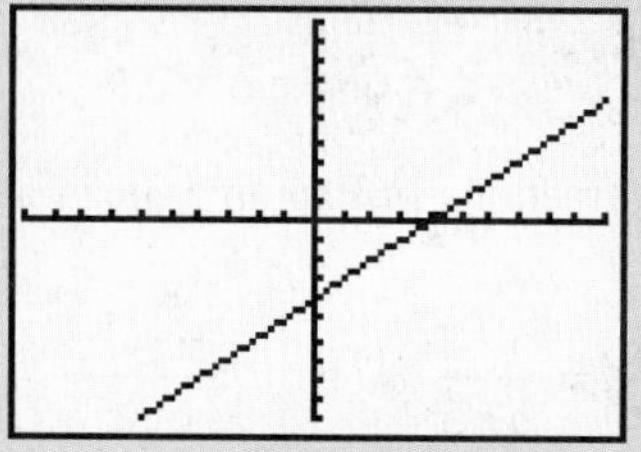

$f(x) = \frac{x^2-16}{x+4}$

(a)

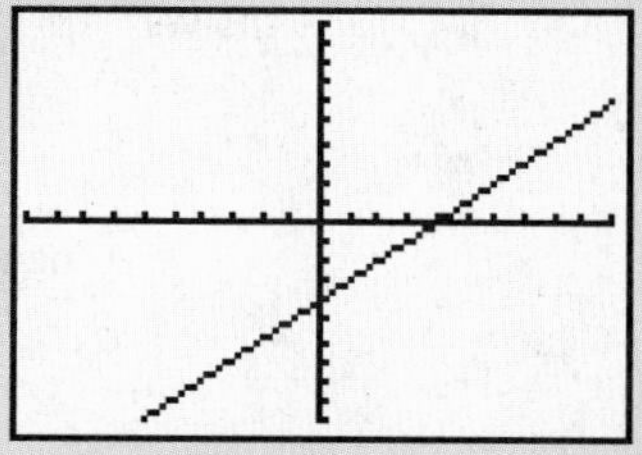

$g(x) = x - 4$

(b)

FIGURE 6-3

EXAMPLE 5 Simplify $\frac{2x^2 + 11x + 12}{3x^2 + 11x - 4}$.

Solution We factor the numerator and denominator and use the fact that $\frac{x+4}{x+4} = 1$.

$$\frac{2x^2 + 11x + 12}{3x^2 + 11x - 4} = \frac{(2x+3)\overset{1}{\cancel{(x+4)}}}{(3x-1)\underset{1}{\cancel{(x+4)}}}$$

$$= \frac{2x+3}{3x-1}$$

WARNING! Do not divide out the x's in $\frac{2x+3}{3x-1}$. The x in the numerator is a factor of the first term only. It is not a factor of the entire numerator. Likewise, the x in the denominator is a factor of the first term only. It is not a factor of the entire denominator.

■

Self Check Simplify $\dfrac{2x^2 + 7x - 15}{2x^2 + 13x + 15}$.

Answer $\frac{2x - 3}{2x + 3}$

SIMPLIFYING FRACTIONS BY FACTORING OUT −1

To simplify $\frac{b-a}{a-b}$ $(a \neq b)$, we factor -1 from the numerator and divide out any factors common to both the numerator and the denominator:

$$\begin{aligned}\frac{b-a}{a-b} &= \frac{-a+b}{a-b} \quad (a \neq b)\\ &= \frac{-\overset{1}{\cancel{(a-b)}}}{\underset{1}{\cancel{(a-b)}}} \qquad \frac{a-b}{a-b} = 1.\\ &= \frac{-1}{1}\\ &= -1\end{aligned}$$

In general, we have the following principle.

Quotient of a Quantity and Its Opposite
The quotient of any nonzero quantity and its negative (or opposite) is -1.

EXAMPLE 6 Simplify $\dfrac{3x^2 - 10xy - 8y^2}{4y^2 - xy}$.

Solution We factor the numerator and denominator and note that because $x - 4y$ and $4y - x$ are negatives, their quotient is -1.

$$\begin{aligned}\frac{3x^2 - 10xy - 8y^2}{4y^2 - xy} &= \frac{(3x+2y)\overset{-1}{\cancel{(x-4y)}}}{y\underset{1}{\cancel{(4y-x)}}}\\ &= \frac{-(3x+2y)}{y}\\ &= \frac{-3x-2y}{y}\end{aligned}$$

■

Self Check Simplify $\dfrac{2a^2 - 5ab - 3b^2}{3b^2 - ab}$.

Answer $-\frac{2a + b}{b}$

Many fractions we shall encounter are already in simplified form. For example, to attempt to simplify the fraction

$$\frac{x^2 + xa + 2x + 2a}{x^2 + x - 6}$$

we factor the numerator and denominator and divide out any common factors:

$$\frac{x^2 + xa + 2x + 2a}{x^2 + x - 6} = \frac{x(x + a) + 2(x + a)}{(x - 2)(x + 3)} = \frac{(x + a)(x + 2)}{(x - 2)(x + 3)}$$

Since there are no common factors in the numerator and denominator, the fraction cannot be simplified.

EXAMPLE 7 Simplify $\dfrac{(x^2 + 2x)(x^2 + 2x - 3)}{(x^2 + x - 2)(x^2 + 3x)}$.

Solution We factor the polynomials in the numerator and denominator and divide out all common factors:

$$\frac{(x^2 + 2x)(x^2 + 2x - 3)}{(x^2 + x - 2)(x^2 + 3x)} = \frac{\overset{1}{\cancel{x}}\overset{1}{\cancel{(x + 2)}}\overset{1}{\cancel{(x + 3)}}\overset{1}{\cancel{(x - 1)}}}{\underset{1}{\cancel{(x + 2)}}\underset{1}{\cancel{(x - 1)}}\underset{1}{\cancel{x}}\underset{1}{\cancel{(x + 3)}}}$$

$$= 1$$

■

Self Check Simplify $\dfrac{(a^2 + 4a)(a^2 - a - 2)}{a(a^2 + 2a - 8)}$.

Answer $a + 1$

WARNING! Only factors that are common to the entire numerator and the entire denominator can be divided out. Terms common to both the numerator and denominator cannot be divided out. It is incorrect to divide out the common term of 3 in the following simplification, because doing so gives a wrong answer.

$$\frac{3 + 7}{3} = \frac{\overset{1}{\cancel{3}} + 7}{\underset{1}{\cancel{3}}} = \frac{1 + 7}{1} = 8$$

The correct simplification is $\frac{3+7}{3} = \frac{10}{3}$.

The 3's in the fraction $\frac{5 + 3(2)}{3(4)}$ cannot be divided out, because the 3 in the numerator is a factor of the second term only. To be divided out, the 3 must be a factor of the entire numerator.

It is not correct to divide out the y in the fraction $\dfrac{x^2y+6x}{y}$, because y is not a factor of the entire numerator.

PROBABILITY

If we toss a coin, it can land in two ways: either heads or tails. Since the outcomes of heads or tails are equally likely, if we were to toss the same coin many times, we would get heads about half of the time. We say that the **probability** of getting heads on a single toss of a coin is $\frac{1}{2}$.

If records show that out of 100 days with today's weather conditions, 30 have received rain, the weather service reports, "Today, there is a $\frac{30}{100}$ or 30% probability of rain."

Activities such as tossing a coin, rolling a die, drawing a card, and predicting rain are called **experiments.** For any experiment, a list of all possible outcomes is called a **sample space.** For example, the sample space, S, for the experiment of tossing two coins is the set

$$S = \{(\text{H, H}), (\text{H, T}), (\text{T, H}), (\text{T, T})\}$$ There are 4 possible outcomes.

where the ordered pair (H, T) represents the outcome "heads on the first coin and tails on the second."

An **event** is a subset of the sample space of an experiment. For example, if E is the event "getting at least one heads" in the experiment of tossing two coins, then

$$E = \{(\text{H, H}), (\text{H, T}), (\text{T, H})\}$$ There are 3 ways of getting at least one heads.

Because the outcome of getting at least one heads can occur in 3 out of 4 possible ways, we say that the probability of E is $\frac{3}{4}$, and we write

$$P(E) = P(\text{at least one heads}) = \frac{3}{4}$$

Probability of an Event

If a sample space of an experiment has n distinct and equally likely outcomes and E is an event that occurs in s of those ways, then the **probability of E** is

$$P(E) = \frac{s}{n}$$

EXAMPLE 8 Find the probability of the event "tossing a sum of 7 on one toss of two dice."

Solution We first find the sample space of the experiment "rolling two dice a single time." If we use ordered pair notation and let the first number in each ordered pair be the result on the first die and the second number be the result on the second die, the sample space contains the following elements:

(1, 1), (1, 2), (1, 3), (1, 4), (1, 5), **(1, 6)**
(2, 1), (2, 2), (2, 3), (2, 4), **(2, 5)**, (2, 6)
(3, 1), (3, 2), (3, 3), **(3, 4)**, (3, 5), (3, 6)
(4, 1), (4, 2), **(4, 3)**, (4, 4), (4, 5), (4, 6)
(5, 1), **(5, 2)**, (5, 3), (5, 4), (5, 5), (5, 6)
(6, 1), (6, 2), (6, 3), (6, 4), (6, 5), (6, 6)

Since there are 6 ordered pairs whose numbers give a sum of 7 out of a total of 36 equally likely outcomes, we have

$$P(E) = P(\text{tossing a 7}) = \frac{s}{n} = \frac{6}{36} = \frac{1}{6}$$

■

Self Check Find the probability of tossing a sum of 11 on one toss of two dice.

Answer $\frac{1}{18}$

Orals *Evaluate the function* $f(x) = \frac{2x-3}{x}$ *when*

1. $x = 1$

2. $x = 3$

Give the domain of each function.

3. $f(x) = \dfrac{x+7}{2x-4}$

4. $f(x) = \dfrac{3x-4}{x^2-9}$

Simplify each fraction.

5. $\dfrac{25}{30}$

6. $\dfrac{x^2}{xy}$

7. $\dfrac{2x-4}{x-2}$

8. $\dfrac{x-2}{2-x}$

EXERCISE 6.1

REVIEW *Factor each expression.*

1. $3x^2 - 9x$

2. $6t^2 - 5t - 6$

3. $27x^6 + 64y^3$

4. $x^2 + ax + 2x + 2a$

VOCABULARY AND CONCEPTS *Fill in each blank to make a true statement.*

5. If a fraction is the quotient of two polynomials, it is called a ________ expression.

6. The denominator of a fraction can never be __.

7. If a graph approaches a line, the line is called an ________.

8. $\dfrac{a}{b} = \dfrac{c}{d}$ if and only if ________

9. $\dfrac{a}{1} =$ __

10. $\dfrac{a}{a} =$ __, provided that $a \neq$ __

11. $\dfrac{ak}{bk} =$ __, provided that $b \neq$ __ and $k \neq 0$.

12. For an ________, a list of all possible outcomes is called a ________ space.

PRACTICE *In Exercises 13–16, the time, t, it takes to travel* 600 *miles is a function of the mean rate of speed,* $r: t = f(r) = \frac{600}{r}$. *Find t for each value of r.*

13. 30 mph

14. 40 mph

15. 50 mph

16. 60 mph

In Exercises 17–20, suppose the cost (in dollars) of removing p% of the pollution in a river is given by the function $c = f(p) = \frac{50{,}000p}{100 - p}$ $(0 \leq p < 100)$. *Find the cost of removing each percent of pollution.*

17. 10%

18. 30%

19. 50%

20. 80%

In Exercises 21–26, a service club wants to publish a directory of its members. Some investigation shows that the cost of typesetting and photography will be \$700 *and the cost of printing each directory will be* \$1.25.

21. Find a function that gives the total cost c of printing x directories.

22. Find a function that gives the mean cost per directory $\overline{c}$ of printing x directories.

23. Find the total cost of printing 500 directories.

24. Find the mean cost per directory if 500 directories are printed.

25. Find the mean cost per directory if 1,000 directories are printed.

26. Find the mean cost per directory if 2,000 directories are printed.

In Exercises 27–32, an electric company charges \$7.50 *per month plus* 9¢ *for each kilowatt hour (kwh) of electricity used.*

27. Find a function that gives the total cost c of n kwh of electricity.

28. Find a function that gives the mean cost per kwh, $\overline{c}$, when using n kwh.

29. Find the total cost for using 775 kwh.

30. Find the mean cost per kwh when 775 kwh are used.

31. Find the mean cost per kwh when 1,000 kwh are used.

32. Find the mean cost per kwh when 1,200 kwh are used.

In Exercises 33–34, the function $f(t) = \frac{t^2+2t}{2t+2}$ *gives the number of days it would take two construction crews, working together, to frame a house that crew* 1 *(working alone) could complete in t days and crew* 2 *(working alone) could complete in* $t + 2$ *days.*

33. If crew 1 could frame a certain house in 15 days, how long would it take both crews working together?

34. If crew 2 could frame a certain house in 20 days, how long would it take both crews working together?

In Exercises 35–36, the function $f(t) = \frac{t^2+3t}{2t+3}$ *gives the number of hours it would take two pipes to fill a pool that the larger pipe (working alone) could fill in t hours and the smaller pipe (working alone) could fill in* $t + 3$ *hours.*

35. If the smaller pipe could fill a pool in 7 hours, how long would it take both pipes to fill the pool?

36. If the larger pipe could fill a pool in 8 hours, how long would it take both pipes to fill the pool?

In Exercises 37–40, use a graphing calculator to graph each rational function. From the graph, determine its domain.

37. $f(x) = \dfrac{x}{x-2}$

38. $f(x) = \dfrac{x+2}{x}$

39. $f(x) = \dfrac{x+1}{x^2-4}$

40. $f(x) = \dfrac{x-2}{x^2-3x-4}$

In Exercises 41–100, simplify each fraction when possible.

41. $\dfrac{12}{18}$

42. $\dfrac{25}{55}$

43. $-\dfrac{112}{36}$

44. $-\dfrac{49}{21}$

45. $\dfrac{288}{312}$

46. $\dfrac{144}{72}$

47. $-\dfrac{244}{74}$

48. $-\dfrac{512}{236}$

49. $\dfrac{12x^3}{3x}$

50. $-\dfrac{15a^2}{25a^3}$

51. $\dfrac{-24x^3y^4}{18x^4y^3}$

52. $\dfrac{15a^5b^4}{21b^3c^2}$

53. $\dfrac{(3x^3)^2}{9x^4}$

54. $\dfrac{8(x^2y^3)^3}{2(xy^2)^2}$

55. $-\dfrac{11x(x-y)}{22(x-y)}$

56. $\dfrac{x(x-2)^2}{(x-2)^3}$

57. $\dfrac{9y^2(y-z)}{21y(y-z)^2}$

58. $\dfrac{-3ab^2(a-b)}{9ab(b-a)}$

59. $\dfrac{(a-b)(c-d)}{(c-d)(a-b)}$

60. $\dfrac{(p+q)(p-r)}{(r-p)(p+q)}$

61. $\dfrac{x+y}{x^2-y^2}$

62. $\dfrac{x-y}{x^2-y^2}$

63. $\dfrac{5x-10}{x^2-4x+4}$

64. $\dfrac{y-xy}{xy-x}$

65. $\dfrac{12-3x^2}{x^2-x-2}$

66. $\dfrac{x^2+2x-15}{x^2-25}$

67. $\dfrac{3x+6y}{x+2y}$

68. $\dfrac{x^2+y^2}{x+y}$

69. $\dfrac{x^3+8}{x^2-2x+4}$

70. $\dfrac{x^2+3x+9}{x^3-27}$

71. $\dfrac{x^2+2x+1}{x^2+4x+3}$

72. $\dfrac{6x^2+x-2}{8x^2+2x-3}$

73. $\dfrac{3m-6n}{3n-6m}$

74. $\dfrac{ax+by+ay+bx}{a^2-b^2}$

75. $\dfrac{4x^2+24x+32}{16x^2+8x-48}$

76. $\dfrac{a^2-4}{a^3-8}$

77. $\dfrac{3x^2-3y^2}{x^2+2y+2x+yx}$

78. $\dfrac{x^2+x-30}{x^2-x-20}$

79. $\dfrac{4x^2+8x+3}{6+x-2x^2}$

80. $\dfrac{6x^2+13x+6}{6-5x-6x^2}$

81. $\dfrac{a^3 + 27}{4a^2 - 36}$

82. $\dfrac{a - b}{b^2 - a^2}$

83. $\dfrac{2x^2 - 3x - 9}{2x^2 + 3x - 9}$

84. $\dfrac{6x^2 - 7x - 5}{2x^2 + 5x + 2}$

85. $\dfrac{(m + n)^3}{m^2 + 2mn + n^2}$

86. $\dfrac{x^3 - 27}{3x^2 - 8x - 3}$

87. $\dfrac{m^3 - mn^2}{mn^2 + m^2n - 2m^3}$

88. $\dfrac{p^3 + p^2q - 2pq^2}{pq^2 + p^2q - 2p^3}$

89. $\dfrac{x^4 - y^4}{(x^2 + 2xy + y^2)(x^2 + y^2)}$

90. $\dfrac{-4x - 4 + 3x^2}{4x^2 - 2 - 7x}$

91. $\dfrac{4a^2 - 9b^2}{2a^2 - ab - 6b^2}$

92. $\dfrac{x^2 + 2xy}{x + 2y + x^2 - 4y^2}$

93. $\dfrac{x - y}{x^3 - y^3 - x + y}$

94. $\dfrac{2x^2 + 2x - 12}{x^3 + 3x^2 - 4x - 12}$

95. $\dfrac{x^6 - y^6}{x^4 + x^2y^2 + y^4}$

96. $\dfrac{6xy - 4x - 9y + 6}{6y^2 - 13y + 6}$

97. $\dfrac{(x^2 - 1)(x + 1)}{(x^2 - 2x + 1)^2}$

98. $\dfrac{(x^2 + 2x + 1)(x^2 - 2x + 1)}{(x^2 - 1)^2}$

99. $\dfrac{(2x^2 + 3xy + y^2)(3a + b)}{(x + y)(2xy + 2bx + y^2 + by)}$

100. $\dfrac{(x - 1)(6ax + 9x + 4a + 6)}{(3x + 2)(2ax - 2a + 3x - 3)}$

In Exercises 101–104, assume that you roll a die one time. Find the probability of each event.

101. Rolling a 2

102. Rolling a 3 or a 4

103. Rolling a 10

104. Rolling an even number

In Exercises 105–108, assume that you draw one card from a standard deck containing 52 cards. Find the probability of each event.

105. Drawing a black card

106. Drawing a jack

107. Drawing an ace

108. Drawing a 4 or a 5

APPLICATIONS

109. Winning a lottery Find the probability of winning a lottery with a single ticket if 6,000,000 tickets are sold and each ticket has the same chance of being the winning number.

110. Winning a raffle Find the probability of winning a raffle if 375 tickets are sold, each with the same chance of being the winning number, and you purchased 3 tickets.

111. Probability of parents having all girls Find the probability that two parents with three children have all girls.

112. Probability of parents having girls Find the probability that two parents with three children have exactly two girls.

WRITING

113. Explain how to simplify a rational expression.

114. Explain how to recognize that a rational expression is in lowest terms.

SOMETHING TO THINK ABOUT

115. A student compares an answer of $\frac{a-3b}{2b-a}$ to an answer of $\frac{3b-a}{a-2b}$. Are the two answers the same?

116. Is this work correct? Explain.

$$\frac{3x^2+6}{3y} = \frac{\cancel{3}x^2+6}{\cancel{3}y} = \frac{x^2+6}{y}$$

117. In which parts can you divide out the 4's?

a. $\frac{4x}{4y}$ **b.** $\frac{4x}{x+4}$

c. $\frac{4+x}{4+y}$ **d.** $\frac{4x}{4+4y}$

118. In which parts can you divide out the 3's?

a. $\frac{3x+3y}{3z}$ **b.** $\frac{3(x+y)}{3x+y}$

c. $\frac{3x+3y}{3a-3b}$ **d.** $\frac{x+3}{3y}$

In Exercises 119–121, interpret each statement.

119. The probability of a man being 20 feet tall is 0.

120. The probability of dying is 1.

121. Explain why all probabilities, p, are in the interval $0 \le p \le 1$.

6.2 Proportion and Variation

■ RATIOS ■ PROPORTIONS ■ SOLVING PROPORTIONS ■ SIMILAR TRIANGLES ■ DIRECT VARIATION ■ INVERSE VARIATION ■ JOINT VARIATION ■ COMBINED VARIATION

Getting Ready *Solve each equation.*

1. $30k = 70$

2. $\frac{k}{4{,}000^2} = 90$

Graph each function and classify it as a linear function or a rational function.

3. $f(x) = 3x$

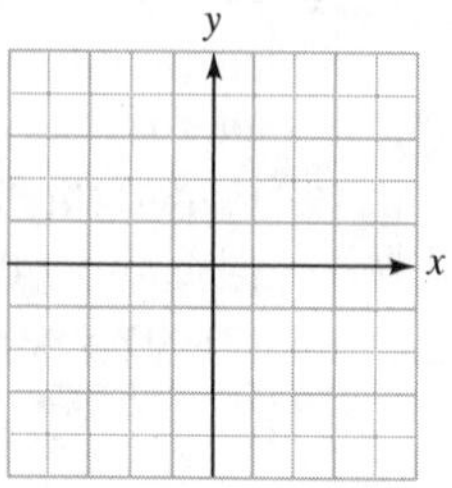

4. $f(x) = \frac{3}{x}\ (x > 0)$

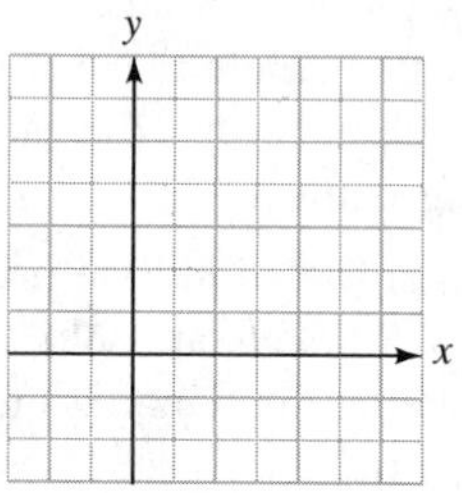

RATIOS

The quotient of two numbers is often called a **ratio.** For example, the fraction $\frac{2}{3}$ can be read as "the ratio of 2 to 3." Some more examples of ratios are

$$\frac{4x}{7y} \quad \text{(the ratio of } 4x \text{ to } 7y) \quad \text{and} \quad \frac{x-2}{3x} \quad \text{(the ratio of } x-2 \text{ to } 3x)$$

Ratios are often used to express **unit costs,** such as the cost per pound of ground round steak.

$$\begin{matrix}\text{The cost of a package of ground round.} \rightarrow \\ \text{The weight of the package.} \rightarrow\end{matrix} \frac{\$18.75}{5 \text{ lb}} = \$3.75 \text{ per lb} \leftarrow \text{The cost per pound.}$$

Ratios are also used to express **rates,** such as an average rate of speed.

$$\begin{matrix}\text{A distance traveled} \rightarrow \\ \text{in a period of time.} \rightarrow\end{matrix} \frac{372 \text{ miles}}{6 \text{ hours}} = 62 \text{ mph} \leftarrow \text{The average rate of speed.}$$

PROPORTIONS

An equation indicating that two ratios are equal is called a **proportion.** Two examples of proportions are

$$\frac{1}{4} = \frac{2}{8} \quad \text{and} \quad \frac{4}{7} = \frac{12}{21}$$

In the proportion $\frac{a}{b} = \frac{c}{d}$, a and d are called the **extremes** of the proportion and b and c are called the **means.**

To develop a fundamental property of proportions, we suppose that

$$\frac{a}{b} = \frac{c}{d}$$

is a proportion and multiply both sides by bd to obtain

$$bd\left(\frac{a}{b}\right) = bd\left(\frac{c}{d}\right)$$

$$\frac{\cancel{b}da}{\cancel{b}} = \frac{b\cancel{d}c}{\cancel{d}}$$

$$ad = bc$$

Thus, if $\frac{a}{b} = \frac{c}{d}$, then $ad = bc$. This shows that in a proportion, the *product of the extremes equals the product of the means.*

SOLVING PROPORTIONS

EXAMPLE 1 Solve $\frac{x+1}{x} = \frac{x}{x+2}$ $(x \neq 0, -2)$.

Solution

$$\frac{x+1}{x} = \frac{x}{x+2}$$

$$(x+1)(x+2) = x \cdot x$$ In a proportion, the product of the extremes equals the product of the means.

$$x^2 + 3x + 2 = x^2$$ Multiply.

$$3x + 2 = 0$$ Subtract x^2 from both sides.

$$x = -\frac{2}{3}$$ Subtract 2 from both sides and divide by 3.

Thus, $x = -\frac{2}{3}$. ■

Self Check Solve $\frac{x-1}{x} = \frac{x}{x+3}$ $(x \neq 0, -3)$.

Answer $\frac{3}{2}$

EXAMPLE 2 Solve $\frac{5a+2}{2a} = \frac{18}{a+4}$ $(a \neq 0, -4)$.

Solution

$$\frac{5a+2}{2a} = \frac{18}{a+4}$$

$$(5a+2)(a+4) = 2a(18)$$ In a proportion, the product of the extremes equals the product of the means.

$$5a^2 + 22a + 8 = 36a$$ Multiply.

$$5a^2 - 14a + 8 = 0$$ Subtract $36a$ from both sides.

$$(5a-4)(a-2) = 0$$ Factor.

$$5a - 4 = 0 \quad \text{or} \quad a - 2 = 0$$ Set each factor equal to 0.

$$5a = 4 \qquad a = 2$$ Solve each linear equation.

$$a = \frac{4}{5}$$

Thus, $a = \frac{4}{5}$ or $a = 2$. ■

Self Check Solve $\frac{3x+1}{12} = \frac{x}{x+2}$ $(x \neq -2)$.

Answer $\frac{2}{3}$, 1

EXAMPLE 3 **Grocery shopping** If 7 pears cost \$2.73, how much will 11 pears cost?

Solution We can let c represent the cost of 11 pears. The price per pound of 7 pears is $\frac{\$2.73}{7}$, and the price per pound of 11 pears is $\frac{\$c}{11}$. Since these ratios are equal, we can set up and solve the following proportion.

$$\frac{2.73}{7} = \frac{c}{11}$$

$11(2.73) = 7c$ In a proportion, the product of the extremes is equal to the product of the means.

$30.03 = 7c$ Multiply.

$\frac{30.03}{7} = c$ Divide both sides by 7.

$c = 4.29$ Simplify.

Eleven pears will cost \$4.29. ■

Self Check In Example 3, how much will 28 pears cost?

Answer \$10.92

SIMILAR TRIANGLES

If two angles of one triangle have the same measure as two angles of a second triangle, the triangles will have the same shape. In this case, we call the triangles **similar triangles.** Here are some facts about similar triangles.

Similar Triangles

If two triangles are similar, then

1. the three angles of the first triangle have the same measure, respectively, as the three angles of the second triangle.
2. the lengths of all corresponding sides are in proportion.

The triangles shown in Figure 6-4 are similar triangles. The corresponding sides are in proportion.

$$\frac{2}{4} = \frac{x}{2x}, \quad \frac{x}{2x} = \frac{1}{2}, \quad \frac{1}{2} = \frac{2}{4}$$

The properties of similar triangles often enable us to find the measures of the sides of a triangle indirectly. For example, on a sunny day we can find the height of a tree and stay safely on the ground.

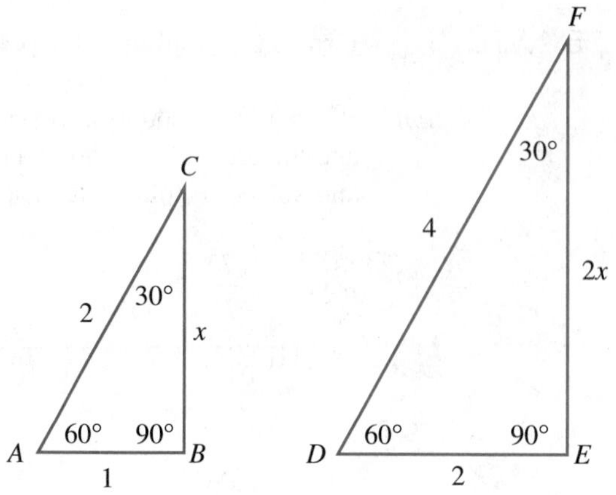

FIGURE 6-4

EXAMPLE 4 A tree casts a shadow of 29 feet at the same time as a vertical yardstick casts a shadow of 2.5 feet. Find the height of the tree.

Solution Refer to Figure 6-5, which shows the triangles determined by the tree and its shadow, and the yardstick and its shadow. Because the triangles have the same shape, they are similar, and the measures of their corresponding sides are in proportion. If we let h represent the height of the tree, we can find h by setting up and solving the following proportion.

$$\frac{h}{3} = \frac{29}{2.5}$$

$$2.5h = 3(29)$$ In a proportion, the product of the extremes is equal to the product of the means.

$$2.5h = 87$$ Simplify.

$$h = 34.8$$ Divide both sides by 2.5.

The tree is about 35 feet tall.

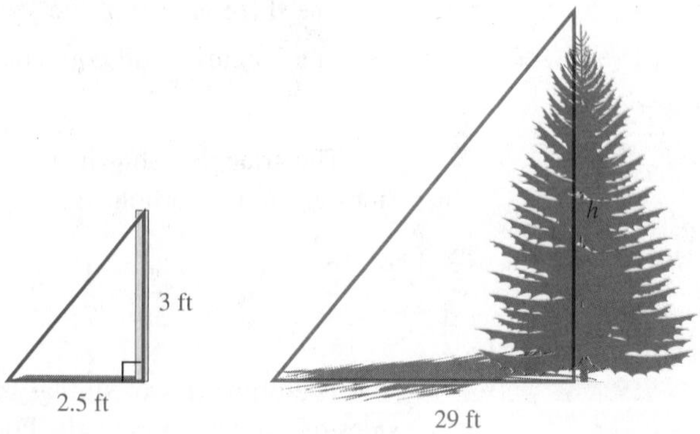

FIGURE 6-5

■

DIRECT VARIATION

To introduce direct variation, we consider the formula

$$C = \pi D$$

for the circumference of a circle, where C is the circumference, D is the diameter, and $\pi \approx 3.14159$. If we double the diameter of a circle, we determine another circle with a larger circumference C_1 such that

$$C_1 = \pi(2D) = 2\pi D = 2C$$

Thus, doubling the diameter results in doubling the circumference. Likewise, if we triple the diameter, we triple the circumference.

In this formula, we say that the variables C and D *vary directly,* or that they are *directly proportional.* This is because as one variable gets larger, so does the other, and in a predictable way. In this example, the constant π is called the *constant of variation* or the *constant of proportionality.*

Direct Variation
The words "y varies directly with x" or "y is directly proportional to x" mean that $y = kx$ for some nonzero constant k. The constant k is called the **constant of variation** or the **constant of proportionality.**

Since the formula for direct variation ($y = kx$) defines a linear function, its graph is always a line with a y-intercept at the origin. The graph of $y = kx$ appears in Figure 6-6 for three positive values of k.

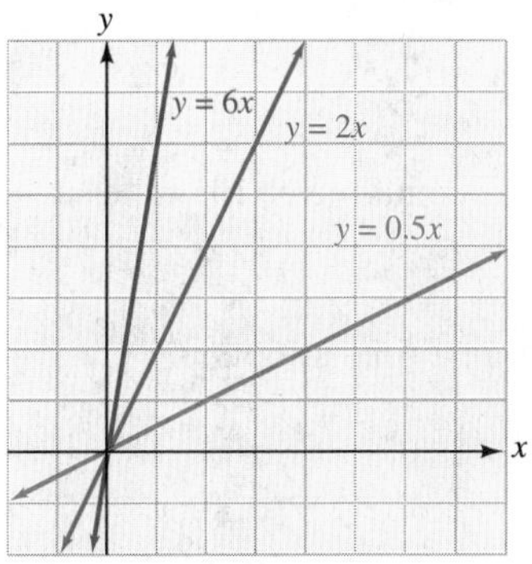

FIGURE 6-6

One example of direct variation is Hooke's law from physics. Hooke's law states that the distance a spring will stretch varies directly with the force that is applied to it.

If d represents a distance and f represents a force, Hooke's law is expressed mathematically as

$$d = kf$$

where k is the constant of variation. If the spring stretches 10 inches when a weight of 6 pounds is attached, k can be found as follows:

$$d = kf$$

$$10 = k(6) \quad \text{Substitute 10 for } d \text{ and 6 for } f.$$

$$\frac{5}{3} = k \quad \text{Divide both sides by 6 and simplify.}$$

To find the force required to stretch the spring a distance of 35 inches, we can solve the equation $d = kf$ for f, with $d = 35$ and $k = \frac{5}{3}$.

$$d = kf$$

$$35 = \frac{5}{3}f \quad \text{Substitute 35 for } d \text{ and } \tfrac{5}{3} \text{ for } k.$$

$$105 = 5f \quad \text{Multiply both sides by 3.}$$

$$21 = f \quad \text{Divide both sides by 5.}$$

The force required to stretch the spring a distance of 35 inches is 21 pounds.

EXAMPLE 5 **Direct variation** The distance traveled in a given time is directly proportional to the speed. If a car travels 70 miles at 30 mph, how far will it travel in the same time at 45 mph?

Solution The words *distance is directly proportional to speed* can be expressed by the equation

1. $d = ks$

where d is distance, k is the constant of variation, and s is the speed. To find k, we substitute 70 for d and 30 for s, and solve for k.

$$d = ks$$

$$70 = k(30)$$

$$k = \frac{7}{3}$$

To find the distance traveled at 45 mph, we substitute $\frac{7}{3}$ for k and 45 for s in Equation 1 and simplify.

$$d = ks$$

$$d = \frac{7}{3}(45)$$

$$= 105$$

In the time it took to go 70 miles at 30 mph, the car could travel 105 miles at 45 mph. ■

Self Check In Example 5, how far will the car travel in the same time at 60 mph?

Answer 140 mi

INVERSE VARIATION

In the formula $w = \frac{12}{l}$, w gets smaller as l gets larger, and w gets larger as l gets smaller. Since these variables vary in opposite directions in a predictable way, we say that the variables *vary inversely,* or that they are *inversely proportional.* The constant 12 is the constant of variation.

Inverse Variation

The words "y varies inversely with x" or "y is inversely proportional to x" mean that $y = \frac{k}{x}$ for some nonzero constant k. The constant k is called the **constant of variation.**

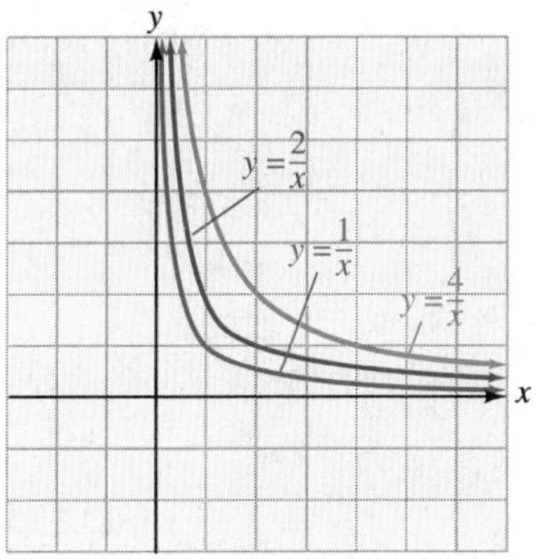

FIGURE 6-7

The formula for inverse variation $\left(y = \frac{k}{x}\right)$ defines a rational function. The graph of $y = \frac{k}{x}$ appears in Figure 6-7 for three positive values of k.

Because of gravity, an object in space is attracted to the earth. The force of this attraction varies inversely with the square of the distance from the object to the center of the earth.

If f represents the force and d represents the distance, this information can be expressed by the equation

$$f = \frac{k}{d^2}$$

If we know that an object 4,000 miles from the center of the earth is attracted to the earth with a force of 90 pounds, we can find k.

$$f = \frac{k}{d^2}$$

$$90 = \frac{k}{4{,}000^2} \quad \text{Substitute 90 for } f \text{ and 4,000 for } d.$$

$$k = 90(4{,}000^2)$$

$$= 1.44 \times 10^9$$

To find the force of attraction when the object is 5,000 miles from the center of the earth, we proceed as follows:

$$f = \frac{k}{d^2}$$

$$f = \frac{1.44 \times 10^9}{5{,}000^2} \quad \text{Substitute } 1.44 \times 10^9 \text{ for } k \text{ and 5,000 for } d.$$

$$= 57.6$$

The object will be attracted to the earth with a force of 57.6 pounds when it is 5,000 miles from the earth's center.

EXAMPLE 6

Light intensity The intensity I of light received from a light source varies inversely with the square of the distance from the light source. If the intensity of a light source 4 feet from an object is 8 candelas, find the intensity at a distance of 2 feet.

Solution The words *intensity varies inversely with the square of the distance d* can be expressed by the equation

$$I = \frac{k}{d^2}$$

To find k, we substitute 8 for I and 4 for d and solve for k.

$$I = \frac{k}{d^2}$$

$$8 = \frac{k}{4^2}$$

$$128 = k$$

To find the intensity when the object is 2 feet from the light source, we substitute 2 for d and 128 for k and simplify.

$$I = \frac{k}{d^2}$$

$$I = \frac{128}{2^2}$$

$$= 32$$

The intensity at 2 feet is 32 candelas. ■

Self Check In Example 6, find the intensity at a distance of 8 feet.

Answer 2 candelas

JOINT VARIATION

There are times when one variable varies with the product of several variables. For example, the area of a triangle varies directly with the product of its base and height:

$$A = \frac{1}{2}bh$$

Such variation is called *joint variation.*

Joint Variation
If one variable varies directly with the product of two or more variables, the relationship is called **joint variation.** If y varies jointly with x and z, then $y = kxz$. The nonzero constant k is called the **constant of variation.**

EXAMPLE 7 The volume V of a cone varies jointly with its height h and the area of its base B. If $V = 6\text{ cm}^3$ when $h = 3$ cm and $B = 6\text{ cm}^2$, find V when $h = 2$ cm and $B = 8\text{ cm}^2$.

Solution The words *V varies jointly with h and B* can be expressed by the equation

$V = khB$ The relationship can also be read as "V is directly proportional to the product of h and B."

We can find k by substituting 6 for V, 3 for h, and 6 for B.

$$V = khB$$
$$6 = k(3)(6)$$
$$6 = k(18)$$
$$\frac{1}{3} = k \qquad \text{Divide both sides by 18; } \tfrac{6}{18} = \tfrac{1}{3}.$$

To find V when $h = 2$ and $B = 8$, we substitute these values into the formula $V = \frac{1}{3}hB$.

$$V = \frac{1}{3}hB$$
$$V = \left(\frac{1}{3}\right)(2)(8)$$
$$= \frac{16}{3}$$

The volume is $5\frac{1}{3}$ cm^3. ■

COMBINED VARIATION

Many applied problems involve a combination of direct and inverse variation. Such variation is called **combined variation.**

EXAMPLE 8 **Building highways** The time it takes to build a highway varies directly with the length of the road, but inversely with the number of workers. If it takes 100 workers 4 weeks to build 2 miles of highway, how long will it take 80 workers to build 10 miles of highway?

Solution We can let t represent the time in weeks, l represent the length in miles, and w represent the number of workers. The relationship between these variables can be expressed by the equation

$$t = \frac{kl}{w}$$

We substitute 4 for t, 100 for w, and 2 for l to find k:

$$4 = \frac{k(2)}{100}$$
$$400 = 2k \qquad \text{Multiply both sides by 100.}$$
$$200 = k \qquad \text{Divide both sides by 2.}$$

We now substitute 80 for w, 10 for l, and 200 for k in the equation $t = \frac{kl}{w}$ and simplify:

$$t = \frac{kl}{w}$$

$$t = \frac{200(10)}{80}$$

$$= 25$$

It will take 25 weeks for 80 workers to build 10 miles of highway. ■

Self Check In Example 8, how long will it take 60 workers to build 6 miles of highway?

Answer 20 weeks

Orals *Solve each proportion.*

1. $\frac{x}{2} = \frac{3}{6}$ **2.** $\frac{3}{x} = \frac{4}{12}$ **3.** $\frac{5}{7} = \frac{2}{x}$

Express each sentence with a formula.

4. a varies directly with b.

5. a varies inversely with b.

6. a varies jointly with b and c.

7. a varies directly with b but inversely with c.

EXERCISE 6.2

REVIEW *Simplify each expression.*

1. $(x^2x^3)^2$

2. $\left(\frac{a^3a^5}{a^{-2}}\right)^3$

3. $\frac{b^0 - 2b^0}{b^0}$

4. $\left(\frac{2r^{-2}r^{-3}}{4r^{-5}}\right)^{-3}$

5. Write 35,000 in scientific notation.

6. Write 0.00035 in scientific notation.

7. Write 2.5×10^{-3} in standard notation.

8. Write 2.5×10^4 in standard notation.

VOCABULARY AND CONCEPTS *Fill in each blank to make a true statement.*

9. Ratios are used to express ________ and _____.

10. An equation stating that two ratios are equal is called a _________.

11. In a proportion, the product of the ________ is equal to the product of the _______.

12. If two angles of one triangle have the same measure as two angles of a second triangle, the triangles are _______.

13. The equation $y = kx$ indicates ______ variation.

14. The equation $y = \frac{k}{x}$ indicates _______ variation.

15. Inverse variation is represented by a _______ function.

16. Direct variation is represented by a ______ function through the origin.

17. The equation $y = kxz$ indicates _____ variation.

18. The equation $y = \frac{kx}{z}$ indicates ________ variation.

Tell whether the graph represents direct variation, inverse variation, or neither.

19.

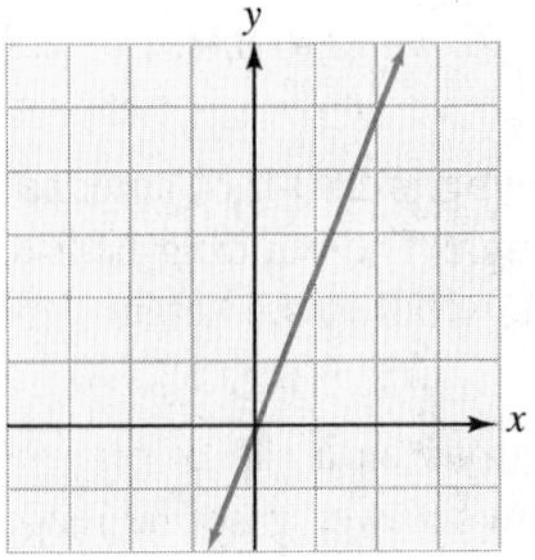

20.

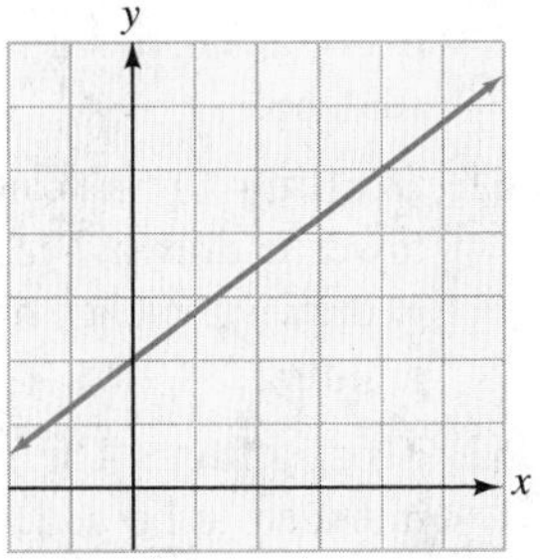

21.

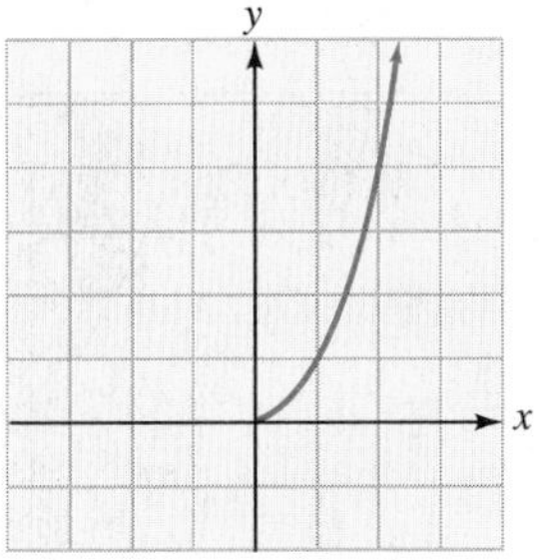

22.

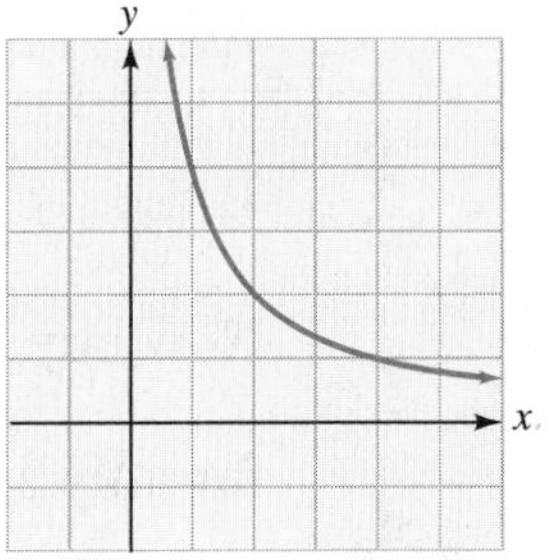

PRACTICE *In Exercises 23–38, solve each proportion for the variable, if possible.*

23. $\frac{x}{5} = \frac{15}{25}$

24. $\frac{4}{y} = \frac{6}{27}$

25. $\frac{r-2}{3} = \frac{r}{5}$

26. $\frac{x+1}{x-1} = \frac{6}{4}$

27. $\frac{3}{n} = \frac{2}{n+1}$

28. $\frac{4}{x+3} = \frac{3}{5}$

29. $\frac{5}{5z+3} = \frac{2z}{2z^2+6}$

30. $\frac{9t+6}{t(t+3)} = \frac{7}{t+3}$

31. $\frac{2}{c} = \frac{c-3}{2}$

32. $\frac{y}{4} = \frac{4}{y}$

33. $\frac{2}{3x} = \frac{6x}{36}$

34. $\frac{2}{x+6} = \frac{-2x}{5}$

35. $\frac{2(x+3)}{3} = \frac{4(x-4)}{5}$

36. $\frac{x+4}{5} = \frac{3(x-2)}{3}$

37. $\frac{1}{x+3} = \frac{-2x}{x+5}$

38. $\frac{x-1}{x+1} = \frac{2}{3x}$

In Exercises 39–46, express each sentence as a formula.

39. A varies directly with the square of p.

40. z varies inversely with the cube of t.

41. v varies inversely with the cube of r.

42. r varies directly with the square of s.

43. B varies jointly with m and n.

44. C varies jointly with x, y, and z.

45. P varies directly with the square of a, and inversely with the cube of j.

46. M varies inversely with the cube of n, and jointly with x and the square of z.

In Exercises 47–54, express each formula in words. In each formula, k is the constant of variation.

47. $L = kmn$

48. $P = \frac{km}{n}$

49. $E = kab^2$

50. $U = krs^2t$

51. $X = \frac{kx^2}{y^2}$

52. $Z = \frac{kw}{xy}$

53. $R = \frac{kL}{d^2}$

54. $e = \frac{kPL}{A}$

APPLICATIONS *In Exercises 55–64, set up and solve the required proportion.*

55. Selling shirts Sport shirts are on sale (see Illustration 1). How much will 5 shirts cost?

ILLUSTRATION 1

56. Cooking A recipe requires four 16-ounce bottles of catsup to make two gallons of spaghetti sauce. How many bottles are needed to make 10 gallons of sauce?

57. Gas consumption A car gets 42 mpg. How much gas will be needed to drive 315 miles?

58. Model railroading An HO-scale model railroad engine is 9 inches long. The HO scale is 87 feet to 1 foot. How long is a real engine?

59. Hobbies Standard dollhouse scale is 1 inch to 1 foot. Heidi's dollhouse is 32 inches wide. How wide would it be if it were a real house?

60. Staffing A school board has determined that there should be 3 teachers for every 50 students. How many teachers are needed for an enrollment of 2,700 students?

61. Drafting In a scale drawing, a 280-foot antenna tower is drawn 7 inches high. The building next to it is drawn 2 inches high. How tall is the actual building?

62. Mixing fuel The instructions on a can of oil intended to be added to lawnmower gasoline read:

Recommended	Gasoline	Oil
50 to 1	6 gal	16 oz

Are these instructions correct? (*Hint:* There are 128 ounces in 1 gallon.)

63. Recommended dosage The recommended child's dose of the sedative hydroxine is 0.006 gram per kilogram of body mass. Find the dosage for a 30-kg child.

64. Body mass The proper dose of the antibiotic cephalexin in children is 0.025 gram per kilogram of body mass. Find the mass of a child receiving a $1\frac{1}{8}$-gram dose.

In Exercises 65–70, use similar triangles to help solve each problem.

65. Height of a tree A tree casts a shadow of 28 feet at the same time as a 6-foot man casts a shadow of 4 feet. (See Illustration 2.) Find the height of the tree.

66. Height of a flagpole A man places a mirror on the ground and sees the reflection of the top of a flagpole, as in Illustration 3. The two triangles in the illustration are similar. Find the height, h, of the flagpole.

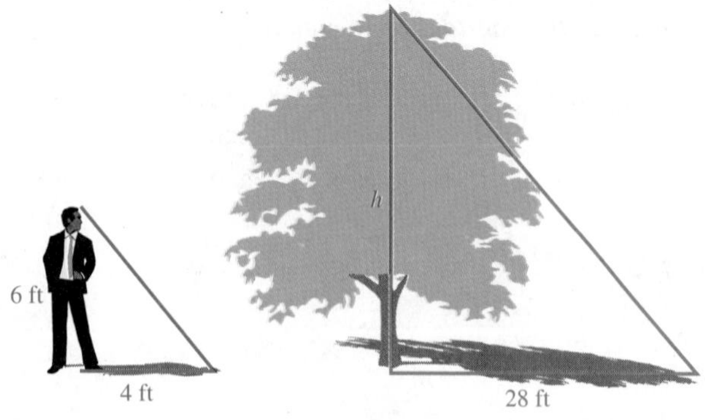

ILLUSTRATION 2

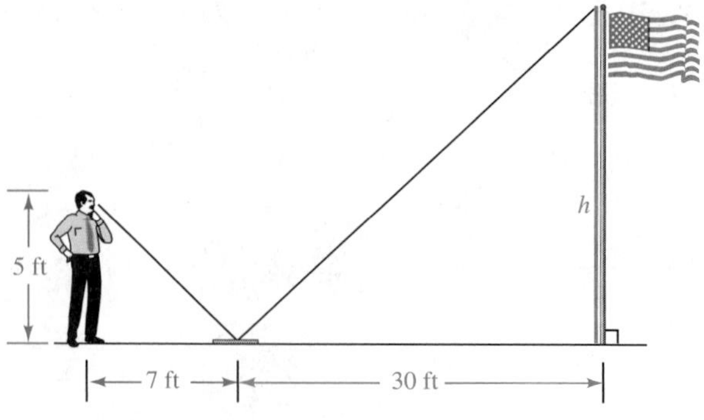

ILLUSTRATION 3

67. **Width of a river** Use the dimensions in Illustration 4 to find w, the width of the river. The two triangles in the illustration are similar.

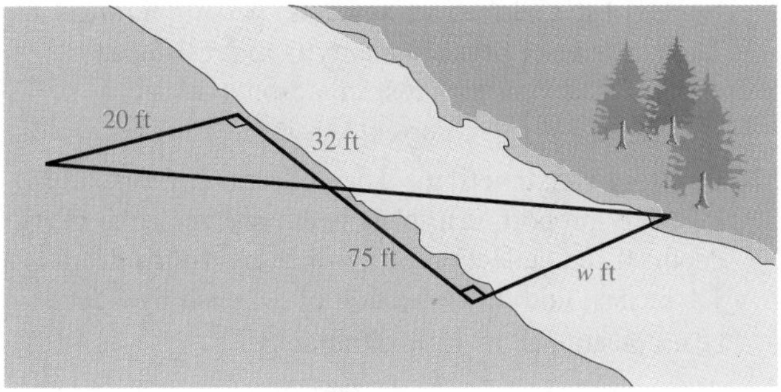

ILLUSTRATION 4

68. **Flight path** An airplane ascends 150 feet as it flies a horizontal distance of 1,000 feet. How much altitude will it gain as it flies a horizontal distance of 1 mile? See Illustration 5. (*Hint:* 5,280 feet = 1 mile.)

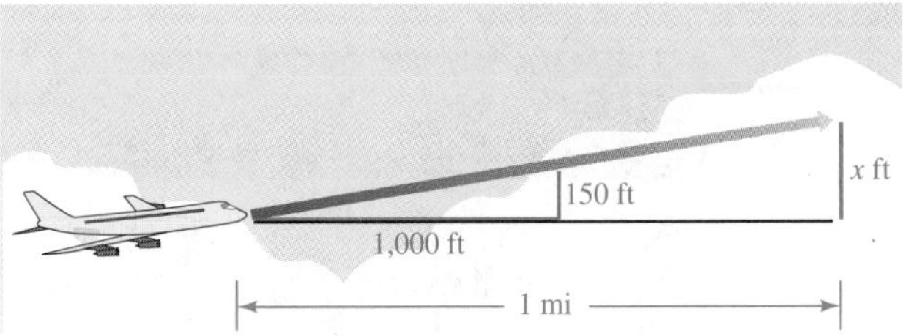

ILLUSTRATION 5

69. **Flight path** An airplane descends 1,350 feet as it flies a horizontal distance of 1 mile. How much altitude is lost as it flies a horizontal distance of 5 miles?

70. **Ski runs** A ski course with $\frac{1}{2}$ mile of horizontal run falls 100 feet in every 300 feet of run. Find the height of the hill.

In Exercises 71–90, solve each problem.

71. **Area of a circle** The area of a circle varies directly with the square of its radius. The constant of variation is π. Find the area of a circle with a radius of 6 inches.

72. **Falling objects** An object in free fall travels a distance s that is directly proportional to the square of the time t. If an object falls 1,024 feet in 8 seconds, how far will it fall in 10 seconds?

73. **Finding distance** The distance that a car can go is directly proportional to the number of gallons of gasoline it consumes. If a car can go 288 miles on 12 gallons of gasoline, how far can it go on a full tank of 18 gallons?

74. **Farming** A farmer's harvest in bushels varies directly with the number of acres planted. If 8 acres can produce 144 bushels, how many acres are required to produce 1,152 bushels?

75. **Farming** The length of time that a given number of bushels of corn will last when feeding cattle varies inversely with the number of animals. If x bushels will feed 25 cows for 10 days, how long will the feed last for 10 cows?

76. **Geometry** For a fixed area, the length of a rectangle is inversely proportional to its width. A rectangle has a width of 18 feet and a length of 12 feet. If the length is increased to 16 feet, find the width.

77. **Gas pressure** Under constant temperature, the volume occupied by a gas is inversely proportional to the pressure applied. If the gas occupies a volume of 20 cubic inches under a pressure of 6 pounds per square inch, find the volume when the gas is subjected to a pressure of 10 pounds per square inch.

78. **Value of a boat** The value of a boat usually varies inversely with its age. If a boat is worth \$7,000 when it is 3 years old, how much will it be worth when it is 7 years old?

79. **Organ pipes** The frequency of vibration of air in an organ pipe is inversely proportional to the length of the pipe. (See Illustration 6.) If a pipe 2 feet long vibrates 256 times per second, how many times per second will a 6-foot pipe vibrate?

80. **Geometry** The area of a rectangle varies jointly with its length and width. If both the length and the width are tripled, by what factor is the area multiplied?

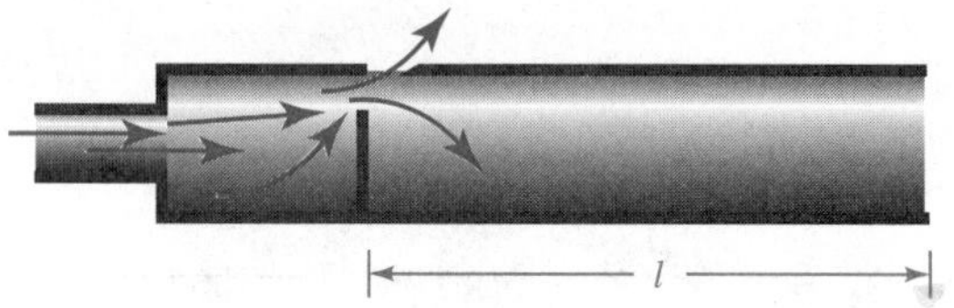

ILLUSTRATION 6

81. **Geometry** The volume of a rectangular solid varies jointly with its length, width, and height. If the length is doubled, the width is tripled, and the height is doubled, by what factor is the volume multiplied?

82. **Costs of a trucking company** The costs incurred by a trucking company vary jointly with the number of trucks in service and the number of hours they are used. When 4 trucks are used for 6 hours each, the costs are $1,800. Find the costs of using 10 trucks, each for 12 hours.

83. **Storing oil** The number of gallons of oil that can be stored in a cylindrical tank varies jointly with the height of the tank and the square of the radius of its base. The constant of proportionality is 23.5. Find the number of gallons that can be stored in the cylindrical tank in Illustration 7.

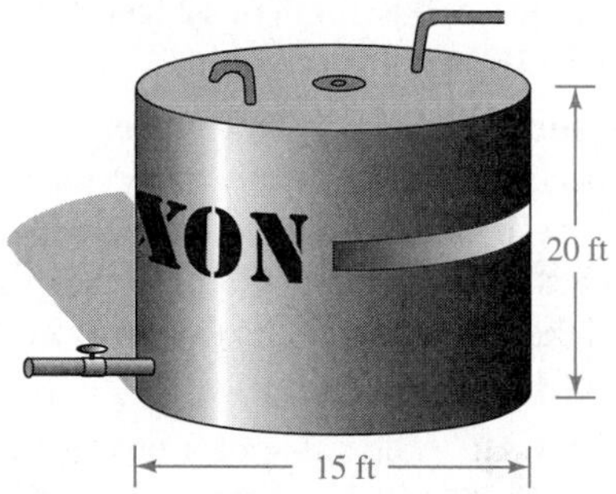

ILLUSTRATION 7

84. **Finding the constant of variation** A quantity l varies jointly with x and y and inversely with z. If the value of l is 30 when $x = 15$, $y = 5$, and $z = 10$, find k.

85. **Electronics** The voltage (in volts) measured across a resistor is directly proportional to the current (in amperes) flowing through the resistor. The constant of variation is the **resistance** (in ohms). If 6 volts is measured across a resistor carrying a current of 2 amperes, find the resistance.

86. **Electronics** The power (in watts) lost in a resistor (in the form of heat) is directly proportional to the square of the current (in amperes) passing through it. The constant of proportionality is the resistance (in ohms). What power is lost in a 5-ohm resistor carrying a 3-ampere current?

87. **Building construction** The deflection of a beam is inversely proportional to its width and the cube of its depth. If the deflection of a 4-inch-by-4-inch beam is 1.1 inches, find the deflection of a 2-inch-by-8-inch beam positioned as in Illustration 8.

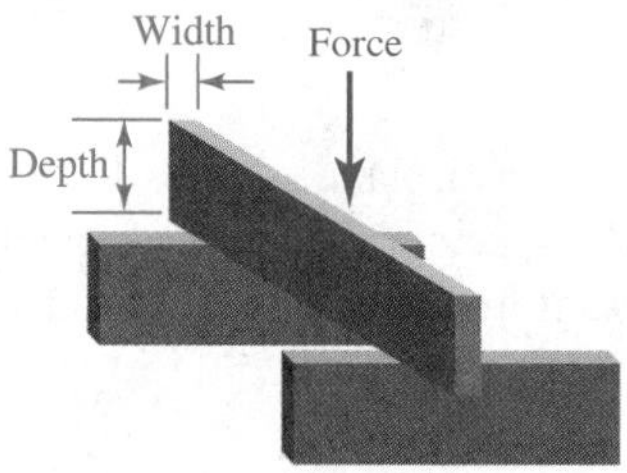

ILLUSTRATION 8

88. **Building construction** Find the deflection of the beam in Exercise 87 when the beam is positioned as in Illustration 9.

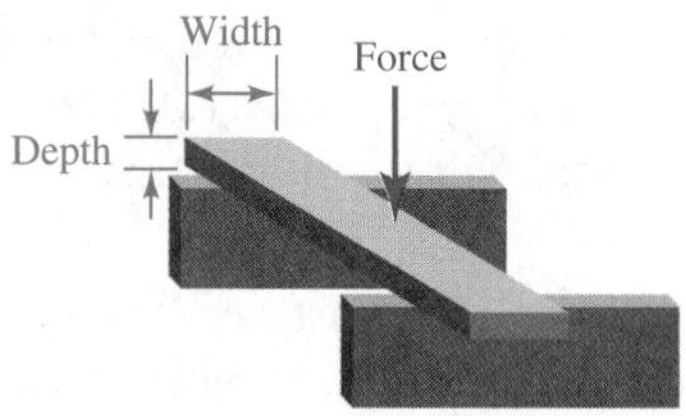

ILLUSTRATION 9

89. **Gas pressure** The pressure of a certain amount of gas is directly proportional to the temperature (measured in degrees Kelvin) and inversely proportional to the volume. A sample of gas at a pressure of 1 atmosphere occupies a volume of 1 cubic meter at a temperature of 273 Kelvin. When heated, the gas expands to twice its volume, but the pressure remains constant. To what temperature is it heated?

90. Tension A yo-yo, twirled at the end of a string, is kept in its circular path by the tension of the string. The tension T is directly proportional to the square of the speed s and inversely proportional to the radius r of the circle. In Illustration 10, the tension is 32 pounds when the speed is 8 feet/second and the radius is 6 feet. Find the tension when the speed is 4 feet/second and the radius is 3 feet.

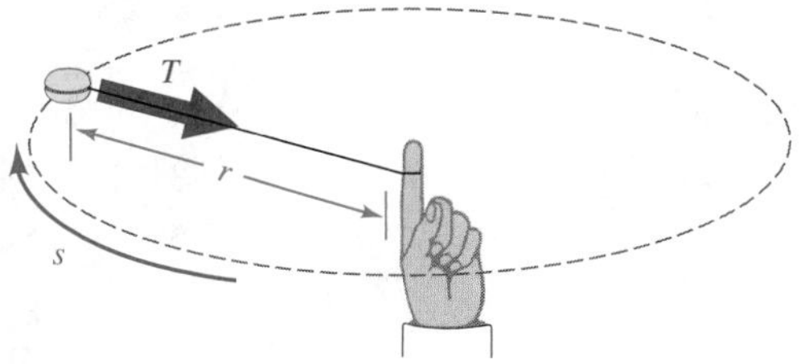

ILLUSTRATION 10

WRITING

91. Explain the terms *means* and *extremes.*

92. Distinguish between a *ratio* and a *proportion.*

93. Explain the term *joint variation.*

94. Explain why the equation $\frac{y}{x} = k$ indicates that y varies directly with x.

SOMETHING TO THINK ABOUT

95. As temperature increases on the Fahrenheit scale, it also increases on the Celsius scale. Is this direct variation? Explain.

96. As the cost of a purchase (less than \$5) increases, the amount of change received from a five-dollar bill decreases. Is this inverse variation? Explain.

6.3 Multiplying and Dividing Rational Expressions

■ MULTIPLYING RATIONAL EXPRESSIONS ■ DIVIDING RATIONAL EXPRESSIONS ■ MIXED OPERATIONS

Getting ready *Simplify each expression.*

1. $\frac{45}{30}$ **2.** $\frac{72}{180}$ **3.** $\frac{600}{450}$ **4.** $\frac{210}{45}$

■ MULTIPLYING RATIONAL EXPRESSIONS

In Section 6.1, we introduced four basic properties of fractions. We now provide the rule for multiplying fractions.

Multiplying Fractions

If no denominators are 0, then

$$\frac{a}{b} \cdot \frac{c}{d} = \frac{a \cdot c}{b \cdot d} = \frac{ac}{bd}$$

Thus, *to multiply two fractions, we multiply the numerators and multiply the denominators.*

$$\frac{3}{5} \cdot \frac{2}{7} = \frac{3 \cdot 2}{5 \cdot 7} = \frac{6}{35}$$

$$\frac{4}{7} \cdot \frac{5}{8} = \frac{4 \cdot 5}{7 \cdot 8} = \frac{\overset{1}{\cancel{2}} \cdot \overset{1}{\cancel{2}} \cdot 5}{7 \cdot \underset{1}{\cancel{2}} \cdot \underset{1}{\cancel{2}} \cdot 2} = \frac{5}{14} \qquad \frac{2}{2} = 1.$$

The same rule applies to algebraic fractions. If $t \neq 0$, then

$$\frac{x^2y}{t} \cdot \frac{xy^3}{t^3} = \frac{x^2y \cdot xy^3}{tt^3} = \frac{x^2x \cdot yy^3}{t^4} = \frac{x^3y^4}{t^4}$$

EXAMPLE 1 Find the product of $\dfrac{x^2 - 6x + 9}{x}$ and $\dfrac{x^2}{x - 3}$.

Solution We multiply the numerators and multiply the denominators and then simplify the resulting fraction.

$$\frac{x^2 - 6x + 9}{x} \cdot \frac{x^2}{x - 3} = \frac{(x^2 - 6x + 9)x^2}{x(x - 3)} \qquad \text{Multiply the numerators and multiply the denominators.}$$

$$= \frac{(x - 3)(x - 3)xx}{x(x - 3)} \qquad \text{Factor the numerator.}$$

$$= \frac{\overset{1}{\cancel{(x - 3)}}(x - 3)\overset{1}{\cancel{x}}x}{\underset{1}{\cancel{x}}\underset{1}{\cancel{(x - 3)}}} \qquad \frac{x - 3}{x - 3} = 1 \text{ and } \frac{x}{x} = 1.$$

$$= x(x - 3)$$

■

Self Check Multiply $\dfrac{a^2 + 6a + 9}{a} \cdot \dfrac{a^3}{a + 3}$.

Answer $a^2(a + 3)$

Checking an Algebraic Simplification

GRAPHING CALCULATORS

We can check the simplification in Example 1 by graphing the rational functions $f(x) = \left(\frac{x^2-6x+9}{x}\right)\left(\frac{x^2}{x-3}\right)$, shown in Figure 6-8(a), and $g(x) = x(x-3)$, shown in Figure 6-8(b), and observing that the graphs are the same, except that 0 and 3 are not included in the domain of the first function.

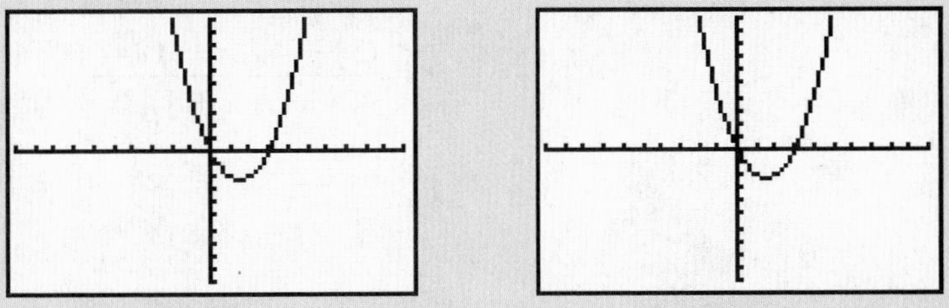

(a) (b)

FIGURE 6-8

EXAMPLE 2 Multiply $\dfrac{x^2-x-6}{x^2-4} \cdot \dfrac{x^2+x-6}{x^2-9}$.

Solution

$$\frac{x^2-x-6}{x^2-4} \cdot \frac{x^2+x-6}{x^2-9}$$

$$= \frac{(x^2-x-6)(x^2+x-6)}{(x^2-4)(x^2-9)}$$ Multiply the numerators and multiply the denominators.

$$= \frac{(x-3)(x+2)(x+3)(x-2)}{(x+2)(x-2)(x+3)(x-3)}$$ Factor the polynomials.

$$= \frac{\overset{1}{\cancel{(x-3)}}\overset{1}{\cancel{(x+2)}}\overset{1}{\cancel{(x+3)}}\overset{1}{\cancel{(x-2)}}}{\underset{1}{\cancel{(x+2)}}\underset{1}{\cancel{(x-2)}}\underset{1}{\cancel{(x+3)}}\underset{1}{\cancel{(x-3)}}}$$ $\frac{x-3}{x-3}=1$, $\frac{x+2}{x+2}=1$, $\frac{x+3}{x+3}=1$, and $\frac{x-2}{x-2}=1$.

$$= 1$$

■

Self Check Multiply $\dfrac{a^2+a-6}{a^2-9} \cdot \dfrac{a^2-a-6}{a^2-4}$.

Answer 1

WARNING! Note that when all factors divide out, the result is 1, not 0.

EXAMPLE 3 Multiply $\dfrac{6x^2+5x-4}{2x^2+5x+3} \cdot \dfrac{8x^2+6x-9}{12x^2+7x-12}$.

Solution

$$\frac{6x^2+5x-4}{2x^2+5x+3}\cdot\frac{8x^2+6x-9}{12x^2+7x-12}$$

$$=\frac{(6x^2+5x-4)(8x^2+6x-9)}{(2x^2+5x+3)(12x^2+7x-12)}$$ Multiply the numerators and multiply the denominators.

$$=\frac{(3x+4)(2x-1)(4x-3)(2x+3)}{(2x+3)(x+1)(3x+4)(4x-3)}$$ Factor the polynomials.

$$=\frac{\overset{1}{\cancel{(3x+4)}}(2x-1)\overset{1}{\cancel{(4x-3)}}\overset{1}{\cancel{(2x+3)}}}{\underset{1}{\cancel{(2x+3)}}(x+1)\underset{1}{\cancel{(3x+4)}}\underset{1}{\cancel{(4x-3)}}}$$ $\frac{3x+4}{3x+4}=1$, $\frac{4x-3}{4x-3}=1$, and $\frac{2x+3}{2x+3}=1$.

$$=\frac{2x-1}{x+1}$$

■

Self Check Multiply $\frac{2a^2+5a-12}{2a^2+11a+12}\cdot\frac{2a^2-3a-9}{2a^2-a-3}$.

Answer $\frac{a-3}{a+1}$

EXAMPLE 4 Multiply $(2x-x^2)\cdot\frac{x}{x^2-5x+6}$.

Solution

$$(2x-x^2)\cdot\frac{x}{x^2-5x+6}$$

$$=\frac{2x-x^2}{1}\cdot\frac{x}{x^2-5x+6}$$ Write $2x-x^2$ as $\frac{2x-x^2}{1}$.

$$=\frac{(2x-x^2)x}{1(x^2-5x+6)}$$ Multiply the fractions.

$$=\frac{x\overset{-1}{\cancel{(2-x)}}x}{\underset{1}{\cancel{(x-2)}}(x-3)}$$ Factor out x and note that the quotient of any nonzero quantity and its negative is -1.

$$=\frac{-x^2}{x-3}$$

Since $\frac{-a}{b}=-\frac{a}{b}$, the $-$ sign can be written in front of the fraction. For this reason, the final result can be written as

$$-\frac{x^2}{x-3}$$

■

Self Check Multiply $\frac{x^2+5x+6}{(x^2+4x)(x+2)}\cdot x^3+4x^2$.

Answer $x(x+3)$

In Examples 1–4, we would obtain the same answers if we had factored first and divided out the common factors before we multiplied.

DIVIDING RATIONAL EXPRESSIONS

Here is the rule for dividing fractions.

Dividing Fractions
If no denominators are 0, then

$$\frac{a}{b} \div \frac{c}{d} = \frac{a}{b} \cdot \frac{d}{c} = \frac{ad}{bc}$$

We can prove this rule as follows:

$$\frac{a}{b} \div \frac{c}{d} = \frac{\frac{a}{b}}{\frac{c}{d}} \cdot 1 = \frac{\frac{a}{b}}{\frac{c}{d}} \cdot \frac{\frac{d}{c}}{\frac{d}{c}} = \frac{\frac{a}{b} \cdot \frac{d}{c}}{\frac{c}{d} \cdot \frac{d}{c}} = \frac{\frac{a}{b} \cdot \frac{d}{c}}{\frac{cd}{cd}} = \frac{\frac{a}{b} \cdot \frac{d}{c}}{1} = \frac{a}{b} \cdot \frac{d}{c}$$

Thus, *to divide two fractions, we can invert the divisor and multiply.*

$$\begin{aligned} \frac{3}{5} \div \frac{2}{7} &= \frac{3}{5} \cdot \frac{7}{2} \\ &= \frac{3 \cdot 7}{5 \cdot 2} \\ &= \frac{21}{10} \end{aligned}$$

$$\begin{aligned} \frac{4}{7} \div \frac{2}{21} &= \frac{4}{7} \cdot \frac{21}{2} \\ &= \frac{4 \cdot 21}{7 \cdot 2} \\ &= \frac{\overset{1}{\cancel{2}} \cdot 2 \cdot 3 \cdot \overset{1}{\cancel{7}}}{\underset{1}{\cancel{7}} \cdot \underset{1}{\cancel{2}}} \\ &= 6 \end{aligned}$$

The same rule applies to algebraic fractions.

$$\frac{x^2}{y^3z^2} \div \frac{x^2}{yz^3} = \frac{x^2}{y^3z^2} \cdot \frac{yz^3}{x^2}$$ Invert the divisor and multiply.

$$= \frac{x^2yz^3}{x^2y^3z^2}$$ Multiply the numerators and the denominators.

$$= x^{2-2}y^{1-3}z^{3-2}$$ To divide exponential expressions with the same base, keep the base and subtract the exponents.

$$= x^0y^{-2}z^1$$

$$= 1 \cdot y^{-2} \cdot z$$ $x^0 = 1$.

$$= \frac{z}{y^2}$$ $y^{-2} = \frac{1}{y^2}$.

EXAMPLE 5 Divide $\dfrac{x^3+8}{x+1} \div \dfrac{x^2-2x+4}{2x^2-2}$.

Solution We invert the divisor and multiply.

$$\begin{aligned}
\frac{x^3+8}{x+1} \div \frac{x^2-2x+4}{2x^2-2} &= \frac{x^3+8}{x+1} \cdot \frac{2x^2-2}{x^2-2x+4} \\
&= \frac{(x^3+8)(2x^2-2)}{(x+1)(x^2-2x+4)} \\
&= \frac{(x+2)\overset{1}{\cancel{(x^2-2x+4)}}2\overset{1}{\cancel{(x+1)}}(x-1)}{\underset{1}{\cancel{(x+1)}}\underset{1}{\cancel{(x^2-2x+4)}}} && \frac{x^2-2x+4}{x^2-2x+4}=1,\ \frac{x+1}{x+1}=1. \\
&= 2(x+2)(x-1)
\end{aligned}$$

■

Self Check Divide $\dfrac{x^3-8}{x-1} \div \dfrac{x^2+2x+4}{3x^2-3x}$.

Answer $3x(x-2)$

EXAMPLE 6 Divide $\dfrac{x^2-4}{x-1} \div (x-2)$.

Solution

$$\begin{aligned}
\frac{x^2-4}{x-1} \div (x-2) &= \frac{x^2-4}{x-1} \div \frac{x-2}{1} && \text{Write } x-2 \text{ as a fraction with a denominator of 1.} \\
&= \frac{x^2-4}{x-1} \cdot \frac{1}{x-2} && \text{Invert the divisor and multiply.} \\
&= \frac{x^2-4}{(x-1)(x-2)} && \text{Multiply the numerators and the denominators.} \\
&= \frac{(x+2)\overset{1}{\cancel{(x-2)}}}{(x-1)\underset{1}{\cancel{(x-2)}}} && \text{Factor } x^2-4. \\
&= \frac{x+2}{x-1} && \frac{x-2}{x-2}=1.
\end{aligned}$$

■

Self Check Divide $\dfrac{a^3-9a}{a-3} \div (a^2+3a)$.

Answer 1

MIXED OPERATIONS

EXAMPLE 7 Simplify $\frac{x^2+2x-3}{6x^2+5x+1} \div \frac{2x^2-2}{2x^2-5x-3} \cdot \frac{6x^2+4x-2}{x^2-2x-3}$.

Solution Since multiplications and divisions are done in order from left to right, we change the division to a multiplication

$$\left(\frac{x^2+2x-3}{6x^2+5x+1} \div \frac{2x^2-2}{2x^2-5x-3}\right)\frac{6x^2+4x-2}{x^2-2x-3} = \left(\frac{x^2+2x-3}{6x^2+5x+1} \cdot \frac{2x^2-5x-3}{2x^2-2}\right)\frac{6x^2+4x-2}{x^2-2x-3}$$

and then multiply the fractions and simplify the result.

$$= \frac{(x^2+2x-3)(2x^2-5x-3)(6x^2+4x-2)}{(6x^2+5x+1)(2x^2-2)(x^2-2x-3)}$$

$$= \frac{(x+3)\overset{1}{\cancel{(x-1)}}\overset{1}{\cancel{(2x+1)}}\overset{1}{\cancel{(x-3)}}\overset{1}{\cancel{2}}(3x-1)\overset{1}{\cancel{(x+1)}}}{(3x+1)\underset{1}{\cancel{(2x+1)}}\underset{1}{\cancel{2}}(x+1)\underset{1}{\cancel{(x-1)}}\underset{1}{\cancel{(x-3)}}\underset{1}{\cancel{(x+1)}}}$$

$$= \frac{(x+3)(3x-1)}{(3x+1)(x+1)}$$

■

Orals *Multiply the fractions and simplify, if possible.*

1. $\frac{3}{2} \cdot \frac{3}{4}$ **2.** $\frac{3x}{7} \cdot \frac{7}{6x}$ **3.** $\frac{x-2}{y} \cdot \frac{y}{x+2}$

Divide the fractions and simplify, if possible.

4. $\frac{3}{4} \div \frac{4}{3}$ **5.** $\frac{5a}{b} \div \frac{a}{b}$ **6.** $\frac{x^2y}{ab} \div \frac{2x^2}{ba}$

EXERCISE 6.3

REVIEW *Do each operation.*

1. $-2a^2(3a^3 - a^2)$

2. $(2t - 1)^2$

3. $(m^n + 2)(m^n - 2)$

4. $(3b^{-n} + c)(b^{-n} - c)$

VOCABULARY AND CONCEPTS *Fill in each blank to make a true statement.*

5. $\frac{a}{b} \cdot \frac{c}{d} =$ ___ $(b \neq 0, d \neq 0)$

6. $\frac{a}{b} \div \frac{c}{d} =$ ___ $(b \neq 0, c \neq 0, d \neq 0)$

7. The denominator of a fraction cannot be ___.

8. $\frac{a+1}{a+1} =$ ___ $(a \neq -1)$

PRACTICE *In Exercises 9–62, do the operations and simplify.*

9. $\frac{3}{4} \cdot \frac{5}{3} \cdot \frac{8}{7}$

10. $-\frac{5}{6} \cdot \frac{3}{7} \cdot \frac{14}{25}$

11. $-\frac{6}{11} \div \frac{36}{55}$

12. $\frac{17}{12} \div \frac{34}{3}$

13. $\frac{x^2y^2}{cd} \cdot \frac{c^{-2}d^2}{x}$

14. $\frac{a^{-2}b^2}{x^{-1}y} \cdot \frac{a^4b^4}{x^2y^3}$

15. $\frac{-x^2y^{-2}}{x^{-1}y^{-3}} \div \frac{x^{-3}y^2}{x^4y^{-1}}$

16. $\frac{(a^3)^2}{b^{-1}} \div \frac{(a^3)^{-2}}{b^{-1}}$

17. $\frac{x^2 + 2x + 1}{x} \cdot \frac{x^2 - x}{x^2 - 1}$

18. $\frac{a + 6}{a^2 - 16} \cdot \frac{3a - 12}{3a + 18}$

19. $\frac{2x^2 - x - 3}{x^2 - 1} \cdot \frac{x^2 + x - 2}{2x^2 + x - 6}$

20. $\frac{9x^2 + 3x - 20}{3x^2 - 7x + 4} \cdot \frac{3x^2 - 5x + 2}{9x^2 + 18x + 5}$

21. $\frac{x^2 - 16}{x^2 - 25} \div \frac{x + 4}{x - 5}$

22. $\frac{a^2 - 9}{a^2 - 49} \div \frac{a + 3}{a + 7}$

23. $\frac{a^2 + 2a - 35}{12x} \div \frac{ax - 3x}{a^2 + 4a - 21}$

24. $\frac{x^2 - 4}{2b - bx} \div \frac{x^2 + 4x + 4}{2b + bx}$

25. $\frac{3t^2 - t - 2}{6t^2 - 5t - 6} \cdot \frac{4t^2 - 9}{2t^2 + 5t + 3}$

26. $\frac{2p^2 - 5p - 3}{p^2 - 9} \cdot \frac{2p^2 + 5p - 3}{2p^2 + 5p + 2}$

27. $\frac{3n^2 + 5n - 2}{12n^2 - 13n + 3} \div \frac{n^2 + 3n + 2}{4n^2 + 5n - 6}$

28. $\frac{8y^2 - 14y - 15}{6y^2 - 11y - 10} \div \frac{4y^2 - 9y - 9}{3y^2 - 7y - 6}$

29. $(x + 1) \cdot \frac{1}{x^2 + 2x + 1}$

30. $\frac{x^2 - 4}{x} \div (x + 2)$

31. $(x^2 - x - 2) \cdot \frac{x^2 + 3x + 2}{x^2 - 4}$

32. $(2x^2 - 9x - 5) \cdot \frac{x}{2x^2 + x}$

33. $(2x^2 - 15x + 25) \div \frac{2x^2 - 3x - 5}{x + 1}$

34. $(x^2 - 6x + 9) \div \frac{x^2 - 9}{x + 3}$

35. $\frac{x^3 + y^3}{x^3 - y^3} \div \frac{x^2 - xy + y^2}{x^2 + xy + y^2}$

36. $\frac{x^2 - 6x + 9}{4 - x^2} \div \frac{x^2 - 9}{x^2 - 8x + 12}$

37. $\frac{m^2 - n^2}{2x^2 + 3x - 2} \cdot \frac{2x^2 + 5x - 3}{n^2 - m^2}$

38. $\frac{x^2 - y^2}{2x^2 + 2xy + x + y} \cdot \frac{2x^2 - 5x - 3}{yx - 3y - x^2 + 3x}$

39. $\frac{ax + ay + bx + by}{x^3 - 27} \cdot \frac{x^2 + 3x + 9}{xc + xd + yc + yd}$

40. $\frac{x^2 + 3x + yx + 3y}{x^2 - 9} \cdot \frac{x - 3}{x + 3}$

41. $\frac{x^2 - x - 6}{x^2 - 4} \cdot \frac{x^2 - x - 2}{9 - x^2}$

42. $\frac{2x^2 - 7x - 4}{20 - x - x^2} \div \frac{2x^2 - 9x - 5}{x^2 - 25}$

43. $\frac{2x^2 + 3xy + y^2}{y^2 - x^2} \div \frac{6x^2 + 5xy + y^2}{2x^2 - xy - y^2}$

44. $\frac{p^3 - q^3}{q^2 - p^2} \cdot \frac{q^2 + pq}{p^3 + p^2q + pq^2}$

45. $\frac{3x^2y^2}{6x^3y} \cdot \frac{-4x^7y^{-2}}{18x^{-2}y} \div \frac{36x}{18y^{-2}}$

46. $\frac{9ab^3}{7xy} \cdot \frac{14xy^2}{27z^3} \div \frac{18a^2b^2x}{3z^2}$

47. $(4x + 12) \cdot \frac{x^2}{2x - 6} \div \frac{2}{x - 3}$

48. $(4x^2 - 9) \div \frac{2x^2 + 5x + 3}{x + 2} \div (2x - 3)$

49. $\frac{2x^2 - 2x - 4}{x^2 + 2x - 8} \cdot \frac{3x^2 + 15x}{x + 1} \div \frac{4x^2 - 100}{x^2 - x - 20}$

50. $\frac{6a^2 - 7a - 3}{a^2 - 1} \div \frac{4a^2 - 12a + 9}{a^2 - 1} \cdot \frac{2a^2 - a - 3}{3a^2 - 2a - 1}$

51. $\frac{2t^2 + 5t + 2}{t^2 - 4t + 16} \div \frac{t + 2}{t^3 + 64} \div \frac{2t^3 + 9t^2 + 4t}{t + 1}$

52. $\frac{a^6 - b^6}{a^4 - a^3b} \cdot \frac{a^3}{a^4 + a^2b^2 + b^4} \div \frac{1}{a}$

53. $\frac{x^4 - 3x^2 - 4}{x^4 - 1} \cdot \frac{x^2 + 3x + 2}{x^2 + 4x + 4}$

54. $\frac{x^3 + 2x^2 + 4x + 8}{y^2 - 1} \cdot \frac{y^2 + 2y + 1}{x^4 - 16}$

55. $(x^2 - x - 6) \div (x - 3) \div (x - 2)$

56. $(x^2 - x - 6) \div [(x - 3) \div (x - 2)]$

57. $\frac{3x^2 - 2x}{3x + 2} \div (3x - 2) \div \frac{3x}{3x - 3}$

58. $(2x^2 - 3x - 2) \div \frac{2x^2 - x - 1}{x - 2} \div (x - 1)$

59. $\frac{2x^2 + 5x - 3}{x^2 + 2x - 3} \div \left(\frac{x^2 + 2x - 35}{x^2 - 6x + 5} \div \frac{x^2 - 9x + 14}{2x^2 - 5x + 2}\right)$

60. $\frac{x^2 - 4}{x^2 - x - 6} \div \left(\frac{x^2 - x - 2}{x^2 - 8x + 15} \cdot \frac{x^2 - 3x - 10}{x^2 + 3x + 2}\right)$

61. $\frac{x^2 - x - 12}{x^2 + x - 2} \div \frac{x^2 - 6x + 8}{x^2 - 3x - 10} \cdot \frac{x^2 - 3x + 2}{x^2 - 2x - 15}$

62. $\frac{4x^2 - 10x + 6}{x^4 - 3x^3} \div \frac{2x - 3}{2x^3} \cdot \frac{x - 3}{2x - 2}$

WRITING

63. Explain how to multiply two rational expressions.

64. Explain how to divide one rational expression by another.

SOMETHING TO THINK ABOUT *Insert either a multiplication or a division symbol in each box to make a true statement.*

65. $\frac{x^2}{y} \square \frac{x}{y^2} \square \frac{x^2}{y^2} = \frac{x^3}{y}$

66. $\frac{x^2}{y} \square \frac{x}{y^2} \square \frac{x^2}{y^2} = \frac{y^3}{x}$

6.4 Adding and Subtracting Rational Expressions

■ ADDING AND SUBTRACTING RATIONAL EXPRESSIONS WITH LIKE DENOMINATORS ■ ADDING AND SUBTRACTING RATIONAL EXPRESSIONS WITH UNLIKE DENOMINATORS ■ FINDING THE LEAST COMMON DENOMINATOR ■ MIXED OPERATIONS

Getting Ready *Tell whether the fractions are equal.*

1. $\frac{3}{5}, \frac{27}{45}$

2. $\frac{3}{7}, \frac{18}{40}$

3. $\frac{8}{13}, \frac{40}{65}$

4. $\frac{7}{25}, \frac{42}{150}$

5. $\frac{23}{32}, \frac{46}{66}$

6. $\frac{8}{9}, \frac{64}{72}$

ADDING AND SUBTRACTING RATIONAL EXPRESSIONS WITH LIKE DENOMINATORS

Fractions with like denominators are added and subtracted according to the following rules.

Adding and Subtracting Fractions

If there are no divisions by 0, then

$$\frac{a}{b} + \frac{c}{b} = \frac{a + c}{b} \quad \text{and} \quad \frac{a}{b} - \frac{c}{b} = \frac{a - c}{b}$$

In words, *we add (or subtract) fractions with like denominators by adding (or subtracting) the numerators and keeping the common denominator.* Whenever possible, we should simplify the result.

EXAMPLE 1 Simplify **a.** $\frac{17}{22} + \frac{13}{22}$, **b.** $\frac{3}{2x} + \frac{7}{2x}$, and **c.** $\frac{4x}{x+2} - \frac{7x}{x+2}$.

Solution

a.
$$\frac{17}{22} + \frac{13}{22} = \frac{17 + 13}{22} = \frac{30}{22} = \frac{15 \cdot \not{2}}{11 \cdot \not{2}} = \frac{15}{11}$$

b.
$$\frac{3}{2x} + \frac{7}{2x} = \frac{3 + 7}{2x} = \frac{10}{2x} = \frac{\not{2} \cdot 5}{\not{2} \cdot x} = \frac{5}{x}$$

c.
$$\frac{4x}{x+2} - \frac{7x}{x+2} = \frac{4x - 7x}{x+2} = \frac{-3x}{x+2}$$

■

Self Check Simplify **a.** $\frac{5}{3a} - \frac{2}{3a}$ and **b.** $\frac{3a}{a-2} + \frac{2a}{a-2}$.

Answers **a.** $\frac{1}{a}$, **b.** $\frac{5a}{a-2}$

GRAPHING CALCULATORS **Checking Algebra**

We can check the subtraction in Part c of Example 1 by graphing the rational functions $f(x) = \frac{4x}{x+2} - \frac{7x}{x+2}$, shown in Figure 6-9(a), and $g(x) = \frac{-3x}{x+2}$, shown in Figure 6-9(b), and observing that the graphs are the same. Note that -2 is not in the domain of either function.

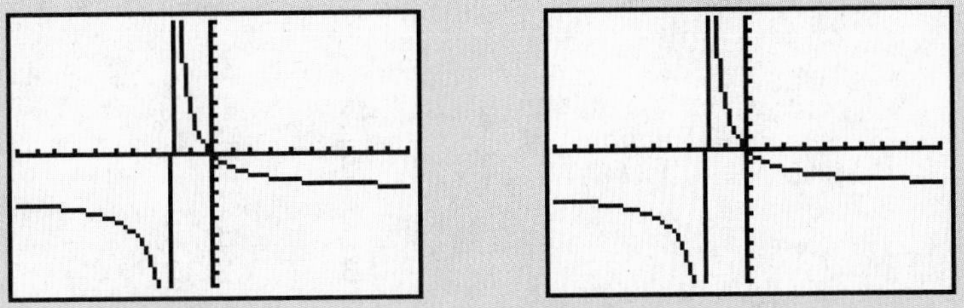

(a) (b)

FIGURE 6-9

ADDING AND SUBTRACTING RATIONAL EXPRESSIONS WITH UNLIKE DENOMINATORS

To add or subtract fractions with unlike denominators, we change them into fractions with a common denominator. When the denominators of the fractions are negatives (opposites), we can multiply one of the fractions by 1, written in the form $\frac{-1}{-1}$, to get a common denominator.

EXAMPLE 2 Add $\frac{x}{x-y}+\frac{y}{y-x}$.

Solution

$$\frac{x}{x-y}+\frac{y}{y-x}=\frac{x}{x-y}+\left(\frac{-1}{-1}\right)\frac{y}{y-x}$$ $\frac{-1}{-1}=1$, and multiplying a fraction by 1 does not change its value.

$$=\frac{x}{x-y}+\frac{-y}{-y+x}$$ Multiply.

$$=\frac{x}{x-y}+\frac{-y}{x-y}$$ $-y+x=x-y$.

$$=\frac{x-y}{x-y}$$ Add the numerators and keep the common denominator.

$$=1$$ Simplify. ■

Self Check Add $\frac{2a}{a-b}+\frac{b}{b-a}$.

Answer $\frac{2a-b}{a-b}$

When the denominators of two or more fractions are different, we often have to multiply one or more fractions by 1, written in some appropriate form, to get a common denominator.

EXAMPLE 3 Simplify **a.** $\frac{2}{3} - \frac{3}{2}$ and **b.** $\frac{3}{x} + \frac{4}{y}$.

Solution **a.**

$$\frac{2}{3} - \frac{3}{2} = \frac{2}{3} \cdot \mathbf{1} - \frac{3}{2} \cdot \mathbf{1}$$ Multiply each fraction by 1.

$$= \frac{2}{3} \cdot \frac{\mathbf{2}}{\mathbf{2}} - \frac{3}{2} \cdot \frac{\mathbf{3}}{\mathbf{3}}$$ $\frac{2}{2} = 1$ and $\frac{3}{3} = 1$.

$$= \frac{2 \cdot 2}{3 \cdot 2} - \frac{3 \cdot 3}{2 \cdot 3}$$ Multiply the fractions by multiplying their numerators and their denominators.

$$= \frac{4}{6} - \frac{9}{6}$$ Simplify.

$$= \frac{-5}{6}$$ Subtract the numerators and keep the common denominator.

$$= -\frac{5}{6}$$

b.

$$\frac{3}{x} + \frac{4}{y} = \frac{3}{x} \cdot \mathbf{1} + \frac{4}{y} \cdot \mathbf{1}$$ Multiply each fraction by 1.

$$= \frac{3}{x} \cdot \frac{y}{y} + \frac{4}{y} \cdot \frac{x}{x}$$ $\frac{y}{y} = 1$ $(y \neq 0)$ and $\frac{x}{x} = 1$ $(x \neq 0)$.

$$= \frac{3y}{xy} + \frac{4x}{xy}$$ Multiply the fractions by multiplying their numerators and their denominators.

$$= \frac{3y + 4x}{xy}$$ Add the numerators and keep the common denominator. ■

Self Check Simplify $\frac{5}{a} - \frac{7}{b}$.

Answer $\frac{5b - 7a}{ab}$

EXAMPLE 4 Simplify **a.** $3 + \frac{7}{x - 2}$ and **b.** $\frac{4x}{x + 2} - \frac{7x}{x - 2}$.

Solution **a.**

$$3 + \frac{7}{x - 2} = \frac{3}{1} + \frac{7}{x - 2}$$ $3 = \frac{3}{1}$.

$$= \frac{3(x - 2)}{1(x - 2)} + \frac{7}{x - 2}$$ $\frac{x-2}{x-2} = 1$, and multiplying 3 by 1 does not change its value.

$$= \frac{3x - 6}{x - 2} + \frac{7}{x - 2}$$ Remove parentheses.

$$= \frac{3x - 6 + 7}{x - 2}$$ Add the numerators and keep the common denominator.

$$= \frac{3x + 1}{x - 2}$$ Simplify.

b. $\dfrac{4x}{x+2} - \dfrac{7x}{x-2}$

$$= \frac{4x(x-2)}{(x+2)(x-2)} - \frac{(x+2)7x}{(x+2)(x-2)}$$

$\frac{x-2}{x-2} = 1$ and $\frac{x+2}{x+2} = 1$.

$$= \frac{(4x^2 - 8x) - (7x^2 + 14x)}{(x+2)(x-2)}$$

Multiply in the numerators, subtract the numerators, and keep the common denominator.

$$= \frac{4x^2 - 8x - 7x^2 - 14x}{(x+2)(x-2)}$$

Use the distributive property to remove parentheses in the numerator.

$$= \frac{-3x^2 - 22x}{(x+2)(x-2)}$$

Simplify the numerator.

$$= \frac{-x(3x+22)}{(x+2)(x-2)}$$

Factor the numerator to see whether the fraction can be simplified.

WARNING! The $-$ sign between the fractions in Step 1 of part **b** applies to both terms of $7x^2 + 14x$.

■

Self Check Simplify $\dfrac{3a}{a+3} - \dfrac{5a}{a-3}$.

Answer $\dfrac{-2a^2 - 24a}{(a+3)(a-3)}$

FINDING THE LEAST COMMON DENOMINATOR

When adding fractions with unlike denominators, it is easiest if we change the fractions into fractions having the smallest common denominator possible, called the **least** (or lowest) **common denominator (LCD).**

Suppose we have the fractions $\frac{1}{12}$, $\frac{1}{20}$, and $\frac{1}{35}$. To find the LCD of these fractions, we first find the prime factorizations of each denominator.

$$12 = 4 \cdot 3 = 2^2 \cdot 3$$
$$20 = 4 \cdot 5 = 2^2 \cdot 5$$
$$35 = 5 \cdot 7$$

Since the LCD is the smallest number that can be divided by 12, 20, and 35, it must contain factors of 2^2, 3, 5, and 7.

$$\text{LCD} = 2^2 \cdot 3 \cdot 5 \cdot 7 = 420$$

To find the least common denominator of several fractions, we follow these steps.

Finding the LCD

1. Factor the denominator of each fraction.
2. List the different factors of each denominator.
3. Write each factor found in Step 2 to the highest power that occurs in any one factorization.
4. The LCD is the product of the factors to their highest powers found in Step 3.

EXAMPLE 5 Find the LCD of $\frac{1}{x^2 + 7x + 6}$, $\frac{3}{x^2 - 36}$, and $\frac{5}{x^2 + 12x + 36}$.

Solution We factor each denominator:

$$x^2 + 7x + 6 = (x + 6)(x + 1)$$
$$x^2 - 36 = (x + 6)(x - 6)$$
$$x^2 + 12x + 36 = (x + 6)(x + 6) = (x + 6)^2$$

and list the individual factors:

$$x + 6, \qquad x + 1, \qquad \text{and} \qquad x - 6$$

To find the LCD, we use the highest power of each of these factors:

$$\text{LCD} = (x + 6)^2(x + 1)(x - 6)$$

■

Self Check Find the LCD of $\frac{1}{a^2 - 25}$ and $\frac{7}{a^2 + 7a + 10}$.

Answer $(a + 5)(a+ 2)(a - 5)$

EXAMPLE 6 Simplify $\frac{x}{x^2 - 2x + 1} + \frac{3}{x^2 - 1}$.

Solution We factor each denominator to find the LCD:

$$x^2 - 2x + 1 = (x - 1)(x - 1) = (x - 1)^2$$
$$x^2 - 1 = (x + 1)(x - 1)$$

The LCD is $(x - 1)^2(x + 1)$.

We now write each fraction with its denominator in factored form and change them into fractions with an LCD of $(x - 1)^2(x + 1)$. Then we add the fractions.

$$\frac{x}{x^2 - 2x + 1} + \frac{3}{x^2 - 1}$$
$$= \frac{x}{(x - 1)(x - 1)} + \frac{3}{(x + 1)(x - 1)}$$
$$= \frac{x(x + 1)}{(x - 1)(x - 1)(x + 1)} + \frac{3(x - 1)}{(x + 1)(x - 1)(x - 1)}$$
$$= \frac{x^2 + x + 3x - 3}{(x - 1)(x - 1)(x + 1)}$$
$$= \frac{x^2 + 4x - 3}{(x - 1)^2(x + 1)}$$

This result does not simplify. ■

Self Check Simplify $\frac{a}{a^2 - 4a + 4} - \frac{2}{a^2 - 4}$.

Answer $\frac{a^2 + 4}{(a - 2)^2(a + 2)}$

EXAMPLE 7 Simplify $\frac{3x}{x - 1} - \frac{2x^2 + 3x - 2}{(x + 1)(x - 1)}$.

Solution We write each fraction in a form having the LCD of $(x + 1)(x - 1)$, remove the resulting parentheses in the first numerator, do the subtraction, and simplify.

1.
$$\frac{3x}{x - 1} - \frac{2x^2 + 3x - 2}{(x + 1)(x - 1)} = \frac{(x + 1)3x}{(x + 1)(x - 1)} - \frac{2x^2 + 3x - 2}{(x + 1)(x - 1)}$$
$$= \frac{3x^2 + 3x}{(x + 1)(x - 1)} - \frac{2x^2 + 3x - 2}{(x + 1)(x - 1)}$$
$$= \frac{3x^2 + 3x - (2x^2 + 3x - 2)}{(x + 1)(x - 1)}$$
$$= \frac{3x^2 + 3x - 2x^2 - 3x + 2}{(x + 1)(x - 1)}$$
$$= \frac{x^2 + 2}{(x + 1)(x - 1)}$$

WARNING! The $-$ sign between the fractions in Equation 1 affects every term of the numerator $2x^2 + 3x - 2$. Whenever we subtract one fraction from another, we must remember to subtract each term of the numerator in the second fraction.

■

Self Check Simplify $\dfrac{2a}{a+2} - \dfrac{a^2 - 4a + 4}{a^2 + a - 2}$.

Answer $\dfrac{a^2 + 2a - 4}{(a+2)(a-1)}$

MIXED OPERATIONS

EXAMPLE 8 Simplify $\dfrac{2x}{x^2 - 4} - \dfrac{1}{x^2 - 3x + 2} + \dfrac{x+1}{x^2 + x - 2}$.

Solution We factor each denominator to find the LCD:

$$\text{LCD} = (x+2)(x-2)(x-1)$$

We then write each fraction as a fraction with the LCD as its denominator and do the subtraction and addition.

$$\frac{2x}{x^2 - 4} - \frac{1}{x^2 - 3x + 2} + \frac{x+1}{x^2 + x - 2}$$

$$= \frac{2x}{(x-2)(x+2)} - \frac{1}{(x-2)(x-1)} + \frac{x+1}{(x-1)(x+2)}$$

$$= \frac{2x(x-1)}{(x-2)(x+2)(x-1)} - \frac{1(x+2)}{(x-2)(x-1)(x+2)} + \frac{(x+1)(x-2)}{(x-1)(x+2)(x-2)}$$

$$= \frac{2x(x-1) - 1(x+2) + (x+1)(x-2)}{(x+2)(x-2)(x-1)}$$

$$= \frac{2x^2 - 2x - x - 2 + x^2 - x - 2}{(x+2)(x-2)(x-1)}$$

$$= \frac{3x^2 - 4x - 4}{(x+2)(x-2)(x-1)}$$

Here, the final result does simplify. Thus,

$$\frac{2x}{x^2 - 4} - \frac{1}{x^2 - 3x + 2} + \frac{x+1}{x^2 + x - 2}$$

$$= \frac{3x^2 - 4x - 4}{(x+2)(x-2)(x-1)}$$

$$= \frac{(3x+2)(x-2)}{(x+2)(x-2)(x-1)} \qquad \text{Factor the numerator.}$$

$$= \frac{3x+2}{(x+2)(x-1)} \qquad \frac{x-2}{x-2} = 1.$$

■

EXAMPLE 9 Simplify $\left(\dfrac{x^2}{x-2}+\dfrac{4}{2-x}\right)^2$.

Solution We do the addition within the parentheses. Since the denominators are negatives (opposites) of each other, we can write the fractions with a common denominator by multiplying both the numerator and denominator of $\frac{4}{2-x}$ by -1. We can then add the fractions, simplify, and square the result.

$$\begin{aligned}\left(\frac{x^2}{x-2}+\frac{4}{2-x}\right)^2 &= \left[\frac{x^2}{x-2}+\frac{(-1)4}{(-1)(2-x)}\right]^2\\ &= \left[\frac{x^2}{x-2}+\frac{-4}{x-2}\right]^2\\ &= \left[\frac{x^2-4}{x-2}\right]^2\\ &= \left[\frac{(x+2)(x-2)}{x-2}\right]^2\\ &= (x+2)^2\\ &= x^2+4x+4\end{aligned}$$

■

Self Check Simplify $\left(\dfrac{a^2}{a-3}-\dfrac{9-6a}{3-a}\right)^2$.

Answer a^2-6a+9

Orals *Add or subtract the fractions and simplify the result, if possible.*

1. $\dfrac{x}{2}+\dfrac{x}{2}$

2. $\dfrac{3a}{4}-\dfrac{a}{4}$

3. $\dfrac{x}{x+2}+\dfrac{2}{x+2}$

4. $\dfrac{2a}{a+4}-\dfrac{a-4}{a+4}$

5. $\dfrac{2x}{3}+\dfrac{x}{2}$

6. $\dfrac{5}{x}-\dfrac{3}{y}$

EXERCISE 6.4

REVIEW *Graph each interval on a number line.*

1. $(-1, 4]$

2. $(-\infty, -5] \cup [4, \infty)$

Solve each formula for the indicated letter.

3. $P = 2l + 2w$; for w

4. $S = \frac{a - lr}{1 - r}$; for a

VOCABULARY AND CONCEPTS *Fill in each blank to make a true statement.*

5. $\frac{a}{b} + \frac{c}{b} =$ ______

6. $\frac{a}{b} - \frac{c}{b} =$ ______

7. To subtract fractions with like denominators, we ______ the numerators and ______ the common denominator.

8. To add fractions with like denominators, we ______ the numerators and keep the ______ denominator.

9. The abbreviation for the least common denominator is ______.

10. To find the LCD of several fractions, we ______ each denominator and use each factor to the ______ power that it appears in any factorization.

PRACTICE *In Exercises 11–26, do the operations and simplify the result when possible.*

11. $\frac{3}{4} + \frac{7}{4}$

12. $\frac{5}{11} + \frac{2}{11}$

13. $\frac{10}{33} - \frac{21}{33}$

14. $\frac{8}{15} - \frac{2}{15}$

15. $\frac{3}{4y} + \frac{8}{4y}$

16. $\frac{5}{3z^2} - \frac{6}{3z^2}$

17. $\frac{3}{a+b} - \frac{a}{a+b}$

18. $\frac{x}{x+4} + \frac{5}{x+4}$

19. $\frac{3x}{2x+2} + \frac{x+4}{2x+2}$

20. $\frac{4y}{y-4} - \frac{16}{y-4}$

21. $\frac{3x}{x-3} - \frac{9}{x-3}$

22. $\frac{9x}{x-y} - \frac{9y}{x-y}$

23. $\frac{5x}{x+1} + \frac{3}{x+1} - \frac{2x}{x+1}$

24. $\frac{4}{a+4} - \frac{2a}{a+4} + \frac{3a}{a+4}$

25. $\frac{3(x^2+x)}{x^2-5x+6} + \frac{-3(x^2-x)}{x^2-5x+6}$

26. $\frac{2x+4}{x^2+13x+12} - \frac{x+3}{x^2+13x+12}$

In Exercises 27–34, the denominators of several fractions are given. Find the LCD.

27. 8, 12, 18

28. 10, 15, 28

29. $x^2 + 3x, x^2 - 9$

30. $3y^2 - 6y, 3y(y - 4)$

31. $x^3 + 27, x^2 + 6x + 9$

32. $x^3 - 8, x^2 - 4x + 4$

33. $2x^2 + 5x + 3, 4x^2 + 12x + 9, x^2 + 2x + 1$

34. $2x^2 + 5x + 3, 4x^2 + 12x + 9, 4x + 6$

In Exercises 35–96, do the operations and simplify the result when possible.

35. $\frac{1}{2} + \frac{1}{3}$

36. $\frac{5}{6} + \frac{2}{7}$

37. $\frac{7}{15} - \frac{17}{25}$

38. $\frac{8}{9} - \frac{5}{12}$

39. $\frac{a}{2} + \frac{2a}{5}$

40. $\frac{b}{6} + \frac{3a}{4}$

41. $\frac{3a}{2} - \frac{4b}{7}$

42. $\frac{2m}{3} - \frac{4n}{5}$

43. $\frac{3}{4x} + \frac{2}{3x}$

44. $\frac{2}{5a} + \frac{3}{2b}$

45. $\frac{3a}{2b} - \frac{2b}{3a}$

46. $\frac{5m}{2n} - \frac{3n}{4m}$

47. $\frac{a + b}{3} + \frac{a - b}{7}$

48. $\frac{x - y}{2} + \frac{x + y}{3}$

49. $\frac{3}{x + 2} + \frac{5}{x - 4}$

50. $\frac{2}{a + 4} - \frac{6}{a + 3}$

51. $\frac{x + 2}{x + 5} - \frac{x - 3}{x + 7}$

52. $\frac{7}{x + 3} + \frac{4x}{x + 6}$

53. $x + \frac{1}{x}$

54. $2 - \frac{1}{x + 1}$

55. $\frac{x + 8}{x - 3} - \frac{x - 14}{3 - x}$

56. $\frac{3 - x}{2 - x} + \frac{x - 1}{x - 2}$

57. $\frac{2a + 1}{3a + 2} - \frac{a - 4}{2 - 3a}$

58. $\frac{4}{x - 2} + \frac{5}{4 - x^2}$

59. $\frac{x}{x^2 + 5x + 6} + \frac{x}{x^2 - 4}$

60. $\frac{x}{3x^2 - 2x - 1} + \frac{4}{3x^2 + 10x + 3}$

61. $\frac{4}{x^2 - 2x - 3} - \frac{x}{3x^2 - 7x - 6}$

62. $\frac{2a}{a^2 - 2a - 8} + \frac{3}{a^2 - 5a + 4}$

63. $\frac{8}{x^2 - 9} + \frac{2}{x - 3} - \frac{6}{x}$

64. $\frac{x}{x^2 - 4} - \frac{x}{x + 2} + \frac{2}{x}$

65. $\frac{x}{x + 1} - \frac{x}{1 - x^2} + \frac{1}{x}$

66. $\frac{y}{y - 2} - \frac{2}{y + 2} - \frac{-8}{4 - y^2}$

67. $2x + 3 + \frac{1}{x + 1}$

68. $x + 1 + \frac{1}{x - 1}$

69. $1 + x - \frac{x}{x - 5}$

70. $2 - x + \frac{3}{x - 9}$

71. $\frac{3x}{x - 1} - 2x - x^2$

72. $\frac{23}{x - 1} + 4x - 5x^2$

73. $\frac{y + 4}{y^2 + 7y + 12} - \frac{y - 4}{y + 3} + \frac{47}{y + 4}$

74. $\frac{x + 3}{2x^2 - 5x + 2} - \frac{3x - 1}{x^2 - x - 2}$

75. $\frac{3}{x + 1} - \frac{2}{x - 1} + \frac{x + 3}{x^2 - 1}$

76. $\frac{2}{x - 2} + \frac{3}{x + 2} - \frac{x - 1}{x^2 - 4}$

77. $\frac{x - 2}{x^2 - 3x} + \frac{2x - 1}{x^2 + 3x} - \frac{2}{x^2 - 9}$

78. $\frac{2}{x - 1} - \frac{2x}{x^2 - 1} - \frac{x}{x^2 + 2x + 1}$

79. $\frac{5}{x^2 - 25} - \frac{3}{2x^2 - 9x - 5} + 1$

80. $\frac{3x}{2x - 1} + \frac{x + 1}{3x + 2} + \frac{2x}{6x^3 + x^2 - 2x}$

81. $\frac{3x}{x-3} + \frac{4}{x-2} - \frac{5x}{x^3 - 5x^2 + 6x}$

82. $\frac{2x-1}{x^2+x-6} - \frac{3x-5}{x^2-2x-15} + \frac{2x-3}{x^2-7x+10}$

83. $2 + \frac{4a}{a^2-1} - \frac{2}{a+1}$

84. $\frac{a}{a-1} - \frac{a+1}{2a-2} + a$

85. $\frac{x+5}{2x^2-2} + \frac{x}{2x+2} - \frac{3}{x-1}$

86. $\frac{a}{2-a} + \frac{3}{a-2} - \frac{3a-2}{a^2-4}$

87. $\frac{a}{a-b} + \frac{b}{a+b} + \frac{a^2+b^2}{b^2-a^2}$

88. $\frac{1}{x+y} - \frac{1}{x-y} - \frac{2y}{y^2-x^2}$

89. $\frac{7n^2}{m-n} + \frac{3m}{n-m} - \frac{3m^2-n}{m^2-2mn+n^2}$

90. $\frac{3b}{2a-b} + \frac{2a-1}{b-2a} - \frac{3a^2+b}{b^2-4ab+4a^2}$

91. $\frac{m+1}{m^2+2m+1} + \frac{m-1}{m^2-2m+1} + \frac{2}{m^2-1}$ (*Hint:* Think about this before finding the LCD.)

92. $\frac{a+2}{a^2+3a+2} + \frac{a-1}{a^2-1} + \frac{3}{a+1}$

93. $\left(\frac{1}{x-1} + \frac{1}{1-x}\right)^2$

94. $\left(\frac{1}{a-1} - \frac{1}{1-a}\right)^2$

95. $\left(\frac{x}{x-3} + \frac{3}{3-x}\right)^3$

96. $\left(\frac{2y}{y+4} + \frac{8}{y+4}\right)^3$

97. Show that $\frac{a}{b} + \frac{c}{d} = \frac{ad+bc}{bd}$.

98. Show that $\frac{a}{b} - \frac{c}{d} = \frac{ad-bc}{bd}$.

WRITING

99. Explain how to find the least common denominator.

100. Explain how to add two fractions.

SOMETHING TO THINK ABOUT

101. Find the error:

$$\frac{8x+2}{5} - \frac{3x+8}{5}$$
$$= \frac{8x+2-3x+8}{5}$$
$$= \frac{5x+10}{5}$$
$$= x+2$$

102. Find the error:

$$\frac{(x+y)^2}{2} + \frac{(x-y)^2}{3}$$
$$= \frac{3 \cdot (x+y)^2}{3 \cdot 2} + \frac{2 \cdot (x-y)^2}{2 \cdot 3}$$
$$= \frac{3x^2+3y^2+2x^2-2y^2}{6}$$
$$= \frac{5x^2+y^2}{6}$$

6.5 Complex Fractions

■ COMPLEX FRACTIONS ■ SIMPLIFYING COMPLEX FRACTIONS

Getting Ready *Use the distributive property to remove parentheses and simplify.*

1. $2\left(1 + \frac{1}{2}\right)$ **2.** $12\left(\frac{1}{6} - 3\right)$ **3.** $a\left(\frac{3}{a} + 2\right)$ **4.** $b\left(\frac{5}{b} - 2\right)$

■ COMPLEX FRACTIONS

A **complex fraction** is a fraction that has a fraction in its numerator or its denominator or both. Examples of complex fractions are

$$\frac{\frac{3a}{b}}{\frac{6ac}{b^2}}, \quad \frac{\frac{2}{x} + 1}{3 + x}, \quad \text{and} \quad \frac{\frac{1}{x} + \frac{1}{y}}{\frac{1}{x} - \frac{1}{y}}$$

■ SIMPLIFYING COMPLEX FRACTIONS

We can use two methods to simplify the complex fraction

$$\frac{\frac{3a}{b}}{\frac{6ac}{b^2}}$$

In one method, we eliminate the fractions in the numerator and denominator by writing the complex fraction as a division and using the division rule for fractions:

$$\frac{\frac{3a}{b}}{\frac{6ac}{b^2}} = \frac{3a}{b} \div \frac{6ac}{b^2}$$

$$= \frac{3a}{b} \cdot \frac{b^2}{6ac} \qquad \text{Invert the divisor and multiply.}$$

$$= \frac{b}{2c} \qquad \text{Multiply the fractions and simplify.}$$

In the other method, we eliminate the fractions in the numerator and denominator by multiplying the fraction by 1, written in the form $\frac{b^2}{b^2}$. We use $\frac{b^2}{b^2}$, because b^2 is the LCD of $\frac{3a}{b}$ and $\frac{6ac}{b^2}$.

$$\frac{\frac{3a}{b}}{\frac{6ac}{b^2}} = \frac{b^2 \cdot \frac{3a}{b}}{b^2 \cdot \frac{6ac}{b^2}}$$

$$= \frac{\frac{3ab^2}{b}}{\frac{6acb^2}{b^2}}$$

$$= \frac{3ab}{6ac} \quad \text{Simplify the fractions in the numerator and denominator.}$$

$$= \frac{b}{2c} \quad \text{Divide out the common factor of } 3a.$$

With either method, the result is the same.

EXAMPLE 1 Simplify $\dfrac{\frac{2}{x} + 1}{3 + x}$.

Method 1 We add the fractions in the numerator and proceed as follows:

$$\frac{\frac{2}{x} + 1}{3 + x} = \frac{\frac{2}{x} + \frac{x}{x}}{\frac{3 + x}{1}} \quad \text{Write 1 as } \frac{x}{x} \text{ and } 3 + x \text{ as } \frac{3 + x}{1}.$$

$$= \frac{\frac{2 + x}{x}}{\frac{3 + x}{1}} \quad \text{Add } \frac{2}{x} \text{ and } \frac{x}{x} \text{ to get } \frac{2 + x}{x}.$$

$$= \frac{2 + x}{x} \div \frac{3 + x}{1} \quad \text{Write the complex fraction as a division.}$$

$$= \frac{2 + x}{x} \cdot \frac{1}{3 + x} \quad \text{Invert the divisor and multiply.}$$

$$= \frac{2 + x}{x^2 + 3x} \quad \text{After noting that there are no common factors, multiply the numerators and multiply the denominators.}$$

Method 2 To eliminate the denominator of x, we multiply the numerator and the denominator by x.

$$\frac{\frac{2}{x}+1}{3+x}=\frac{x\left(\frac{2}{x}+1\right)}{x(3+x)}$$

$$=\frac{2+x}{x^2+3x}$$ Use the distributive property to remove parentheses and simplify. ■

Self Check Simplify $\frac{\frac{3}{a}+2}{2+a}$.

Answer $\frac{2a+3}{a^2+2a}$

Checking Algebra

GRAPHING CALCULATORS

We can check the simplification in Example 1 by graphing the functions

$$f(x)=\frac{\frac{2}{x}+1}{3+x}$$

shown in Figure 6-10(a), and

$$g(x)=\frac{2+x}{x^2+3x}$$

shown in Figure 6-10(b), and observing that the graphs are the same. Each graph has window settings of $[-5, 3]$ for x and $[-6, 6]$ for y.

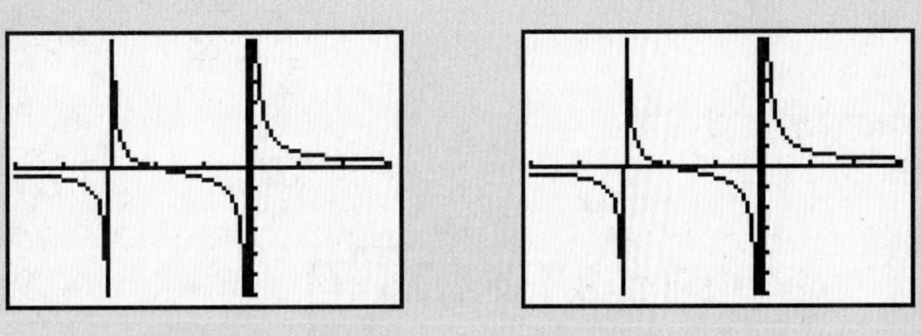

(a) (b)

FIGURE 6-10

EXAMPLE 2 Simplify $\dfrac{\frac{1}{x} + \frac{1}{y}}{\frac{1}{x} - \frac{1}{y}}$.

Method 1 We add the fractions in the numerator and in the denominator and proceed as follows:

$$\frac{\frac{1}{x} + \frac{1}{y}}{\frac{1}{x} - \frac{1}{y}} = \frac{\frac{1y}{xy} + \frac{x1}{xy}}{\frac{1y}{xy} - \frac{x1}{xy}}$$

$$= \frac{\frac{y+x}{xy}}{\frac{y-x}{xy}}$$ Add the fractions in the numerator and subtract the fractions in the denominator.

$$= \frac{y+x}{xy} \div \frac{y-x}{xy}$$ Write the complex fraction as a division.

$$= \frac{y+x}{xy} \cdot \frac{xy}{y-x}$$ Invert the divisor and multiply.

$$= \frac{y+x}{y-x}$$ Multiply the numerators and multiply the denominators and simplify.

Method 2 We multiply the numerator and denominator by xy (the LCD of the fractions appearing in the complex fraction) and simplify.

$$\frac{\frac{1}{x} + \frac{1}{y}}{\frac{1}{x} - \frac{1}{y}} = \frac{xy\left(\frac{1}{x} + \frac{1}{y}\right)}{xy\left(\frac{1}{x} - \frac{1}{y}\right)}$$

$$= \frac{\frac{xy}{x} + \frac{xy}{y}}{\frac{xy}{x} - \frac{xy}{y}}$$ Use the distributive property to remove parentheses.

$$= \frac{y+x}{y-x}$$ Simplify each fraction. ■

Self Check Simplify $\dfrac{\frac{1}{a} - \frac{1}{b}}{\frac{1}{a} + \frac{1}{b}}$.

Answer $\frac{b-a}{b+a}$

EXAMPLE 3 Simplify $\dfrac{x^{-1} + y^{-1}}{x^{-2} - y^{-2}}$.

Method 1 We proceed as follows:

$$\frac{x^{-1}+y^{-1}}{x^{-2}-y^{-2}} = \frac{\dfrac{1}{x}+\dfrac{1}{y}}{\dfrac{1}{x^2}-\dfrac{1}{y^2}}$$ Write the fraction without using negative exponents.

$$= \frac{\dfrac{y}{xy}+\dfrac{x}{xy}}{\dfrac{y^2}{x^2y^2}-\dfrac{x^2}{x^2y^2}}$$ Get a common denominator in the numerator and denominator.

$$= \frac{\dfrac{y+x}{xy}}{\dfrac{y^2-x^2}{x^2y^2}}$$ Add the fractions in the numerator and denominator.

$$= \frac{y+x}{xy} \div \frac{y^2-x^2}{x^2y^2}$$ Write the fraction as a division.

$$= \frac{y+x}{xy} \cdot \frac{xxyy}{(y-x)(y+x)}$$ Invert and factor $y^2 - x^2$.

$$= \frac{(y+x)xxyy}{xy(y-x)(y+x)}$$ Multiply the numerators and the denominators.

$$= \frac{xy}{y-x}$$ Divide out the common factors of x, y, and $y + x$ in the numerator and denominator.

Method 2 We multiply both numerator and denominator by x^2y^2, the LCD of the fractions, and proceed as follows:

$$\frac{x^{-1}+y^{-1}}{x^{-2}-y^{-2}} = \frac{\dfrac{1}{x}+\dfrac{1}{y}}{\dfrac{1}{x^2}-\dfrac{1}{y^2}}$$ Write the fraction without negative exponents.

$$= \frac{x^2y^2\left(\dfrac{1}{x}+\dfrac{1}{y}\right)}{x^2y^2\left(\dfrac{1}{x^2}-\dfrac{1}{y^2}\right)}$$ Multiply numerator and denominator by x^2y^2.

$$= \frac{xy^2+x^2y}{y^2-x^2}$$ Use the distributive property to remove parentheses.

$$= \frac{xy(y+x)}{(y+x)(y-x)}$$ Factor the numerator and denominator.

$$= \frac{xy}{y-x}$$ Divide out $y + x$. ■

Self Check Simplify $\dfrac{a^{-2} + b^{-2}}{a^{-1} - b^{-1}}$.

Answer $\dfrac{b^2 + a^2}{ab(b - a)}$

PERSPECTIVE

Each of the complex fractions in the list

$$1 + \frac{1}{2}, \quad 1 + \cfrac{1}{1 + \cfrac{1}{2}}, \quad 1 + \cfrac{1}{1 + \cfrac{1}{1 + \cfrac{1}{2}}}, \quad 1 + \cfrac{1}{1 + \cfrac{1}{1 + \cfrac{1}{1 + \cfrac{1}{2}}}}, \quad \ldots$$

can be simplified by using the value of the expression preceding it. For example, to simplify the second expression in the list, replace $1 + \frac{1}{2}$ with $\frac{3}{2}$:

$$1 + \cfrac{1}{1 + \cfrac{1}{2}} = 1 + \cfrac{1}{\cfrac{3}{2}} = 1 + \frac{2}{3} = \frac{5}{3}$$

To simplify the third expression, we replace $1 + \cfrac{1}{1 + \cfrac{1}{2}}$ with $\frac{5}{3}$:

$$1 + \cfrac{1}{1 + \cfrac{1}{1 + \cfrac{1}{2}}} = 1 + \cfrac{1}{\cfrac{5}{3}} = 1 + \frac{3}{5} = \frac{8}{5}$$

The complex fractions in the list simplify to the fractions $\frac{3}{2}, \frac{5}{3}, \frac{8}{5}, \frac{13}{8}$. The decimal values of these fractions get closer and closer to the irrational number 1.61803398875 . . . , which is known as the **golden ratio.** This number often appears in the architecture of the ancient Greeks and Egyptians. The width of the stairs in front of the Greek Parthenon (Illustration 1), divided by the building's height, is the golden ratio. The height of the triangular face of the Great Pyramid of Cheops (Illustration 2), divided by the pyramid's width, is also the golden ratio.

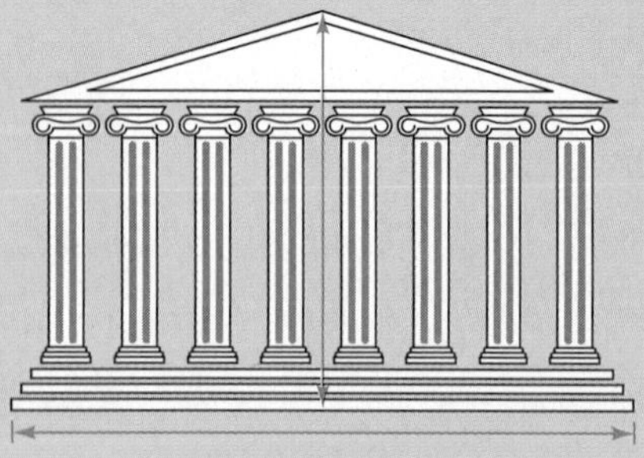

ILLUSTRATION 1

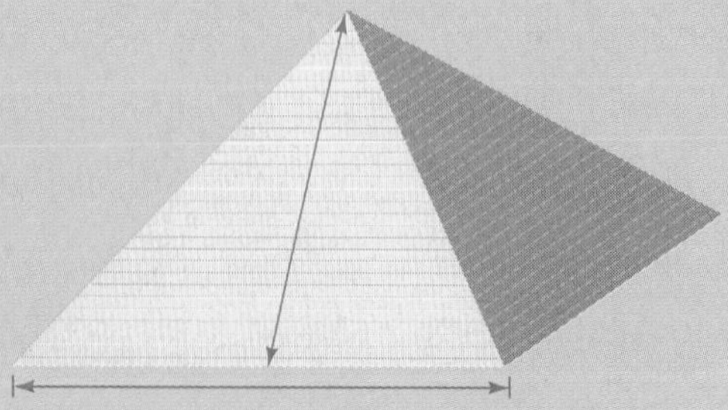

ILLUSTRATION 2

WARNING! $x^{-1} + y^{-1}$ means $\frac{1}{x} + \frac{1}{y}$, and $(x + y)^{-1}$ means $\frac{1}{x+y}$. Thus,

$$x^{-1} + y^{-1} \neq (x + y)^{-1}$$

EXAMPLE 4 Simplify $\dfrac{\dfrac{2x}{1 - \dfrac{1}{x}} + 3}{3 - \dfrac{2}{x}}$.

Solution We begin by multiplying the numerator and denominator of the fraction

$$\frac{2x}{1 - \dfrac{1}{x}}$$

by x to eliminate the complex fraction in the numerator of the given fraction.

$$\frac{\dfrac{2x}{1 - \dfrac{1}{x}} + 3}{3 - \dfrac{2}{x}} = \frac{\dfrac{x2x}{x\left(1 - \dfrac{1}{x}\right)} + 3}{3 - \dfrac{2}{x}}$$

$$= \frac{\dfrac{2x^2}{x - 1} + 3}{3 - \dfrac{2}{x}}$$

We then multiply the numerator and denominator of the previous fraction by $x(x - 1)$, the LCD of $\frac{2x^2}{x-1}$, 3, and $\frac{2}{x}$, and simplify:

$$\frac{\dfrac{2x}{1 - \dfrac{1}{x}} + 3}{3 - \dfrac{2}{x}} = \frac{x(x - 1)\left(\dfrac{2x^2}{x - 1} + 3\right)}{x(x - 1)\left(3 - \dfrac{2}{x}\right)}$$

$$= \frac{2x^3 + 3x(x - 1)}{3x(x - 1) - 2(x - 1)}$$

$$= \frac{2x^3 + 3x^2 - 3x}{3x^2 - 5x + 2}$$

This result does not simplify. ■

Self Check Simplify $\dfrac{3 + \dfrac{2}{a}}{\dfrac{2a}{1 + \dfrac{1}{a}} + 3}$.

Answer $\dfrac{(3a + 2)(a + 1)}{a(2a^2 + 3a + 3)}$

Orals *Simplify each complex fraction.*

1. $\dfrac{\frac{3}{4}}{\frac{5}{4}}$
2. $\dfrac{\frac{a}{b}}{\frac{d}{b}}$
3. $\dfrac{\frac{3}{4}}{\frac{3}{8}}$
4. $\dfrac{\frac{x + y}{x}}{\frac{x - y}{x}}$
5. $\dfrac{\frac{x}{y} - 1}{\frac{x}{y}}$
6. $\dfrac{1 + \frac{a}{b}}{\frac{a}{b}}$

EXERCISE 6.5

REVIEW *Solve each equation.*

1. $\dfrac{8(a - 5)}{3} = 2(a - 4)$
2. $\dfrac{3t^2}{5} + \dfrac{7t}{10} = \dfrac{3t + 6}{5}$
3. $a^4 - 13a^2 + 36 = 0$
4. $|2x - 1| = 9$

VOCABULARY AND CONCEPTS *Fill in each blank to make a true statement.*

5. A ________ fraction is a fraction that has fractions in its numerator or denominator.
6. The fraction $\dfrac{\frac{a}{b}}{\frac{c}{d}}$ is equivalent to $\dfrac{a}{b}$ ___ $\dfrac{c}{d}$.

PRACTICE *In Exercises 7–60, simplify each complex fraction.*

7. $\dfrac{\frac{1}{2}}{\frac{3}{4}}$
8. $-\dfrac{\frac{3}{4}}{\frac{1}{2}}$
9. $\dfrac{-\frac{2}{3}}{\frac{6}{9}}$
10. $\dfrac{\frac{11}{18}}{\frac{22}{27}}$

11. $\dfrac{\frac{1}{2}+\frac{1}{3}}{\frac{1}{4}}$

12. $\dfrac{\frac{1}{4}-\frac{1}{5}}{\frac{1}{3}}$

13. $\dfrac{\frac{1}{2}-\frac{2}{3}}{\frac{2}{3}+\frac{1}{2}}$

14. $\dfrac{\frac{2}{3}+\frac{4}{5}}{\frac{2}{5}-\frac{1}{3}}$

15. $\dfrac{\frac{4x}{y}}{\frac{6xz}{y^2}}$

16. $\dfrac{\frac{5t^4}{9x}}{\frac{2t}{18x}}$

17. $\dfrac{5ab^2}{\frac{ab}{25}}$

18. $\dfrac{\frac{6a^2b}{4t}}{3a^2b^2}$

19. $\dfrac{\frac{x-y}{xy}}{\frac{y-x}{x}}$

20. $\dfrac{\frac{x^2+5x+6}{3xy}}{\frac{x^2-9}{6xy}}$

21. $\dfrac{\frac{1}{x}-\frac{1}{y}}{xy}$

22. $\dfrac{xy}{\frac{1}{x}-\frac{1}{y}}$

23. $\dfrac{\frac{1}{a}+\frac{1}{b}}{\frac{1}{a}}$

24. $\dfrac{\frac{1}{b}}{\frac{1}{a}-\frac{1}{b}}$

25. $\dfrac{1+\frac{x}{y}}{1-\frac{x}{y}}$

26. $\dfrac{\frac{x}{y}+1}{1-\frac{x}{y}}$

27. $\dfrac{\frac{y}{x}-\frac{x}{y}}{\frac{1}{x}+\frac{1}{y}}$

28. $\dfrac{\frac{y}{x}-\frac{x}{y}}{\frac{1}{y}-\frac{1}{x}}$

29. $\dfrac{\frac{1}{a}-\frac{1}{b}}{\frac{a}{b}-\frac{b}{a}}$

30. $\dfrac{\frac{1}{a}+\frac{1}{b}}{\frac{a}{b}-\frac{b}{a}}$

31. $\dfrac{x+1-\frac{6}{x}}{\frac{1}{x}}$

32. $\dfrac{x-1-\frac{2}{x}}{\frac{x}{3}}$

33. $\dfrac{5xy}{1+\frac{1}{xy}}$

34. $\dfrac{3a}{a+\frac{1}{a}}$

35. $\dfrac{1+\frac{6}{x}+\frac{8}{x^2}}{1+\frac{1}{x}-\frac{12}{x^2}}$

36. $\dfrac{1-x-\frac{2}{x}}{\frac{6}{x^2}+\frac{1}{x}-1}$

37. $\dfrac{\frac{1}{a+1}+1}{\frac{3}{a-1}+1}$

38. $\dfrac{2+\frac{3}{x+1}}{\frac{1}{x}+x+x^2}$

39. $\dfrac{x^{-1}+y^{-1}}{x}$

40. $\dfrac{x^{-1}-y^{-1}}{y}$

41. $\dfrac{y}{x^{-1}-y^{-1}}$

42. $\dfrac{x^{-1}+y^{-1}}{(x+y)^{-1}}$

43. $\dfrac{x^{-1}+y^{-1}}{x^{-1}-y^{-1}}$

44. $\dfrac{(x+y)^{-1}}{x^{-1}+y^{-1}}$

45. $\dfrac{x+y}{x^{-1}+y^{-1}}$

46. $\dfrac{x-y}{x^{-1}-y^{-1}}$

47. $\dfrac{x-y^{-2}}{y-x^{-2}}$

48. $\dfrac{x^{-2}-y^{-2}}{x^{-1}-y^{-1}}$

49. $\dfrac{1+\dfrac{a}{b}}{1-\dfrac{a}{1-\dfrac{a}{b}}}$

50. $\dfrac{1+\dfrac{2}{1+\dfrac{a}{b}}}{1-\dfrac{a}{b}}$

51. $\dfrac{x-\dfrac{1}{x}}{1+\dfrac{1}{\dfrac{1}{x}}}$

52. $\dfrac{\dfrac{a^2+3a+4}{ab}}{2+\dfrac{3+a}{\dfrac{2}{a}}}$

53. $\dfrac{b}{b+\dfrac{2}{2+\dfrac{1}{2}}}$

54. $\dfrac{2y}{y-\dfrac{y}{3-\dfrac{1}{2}}}$

55. $a+\dfrac{a}{1+\dfrac{a}{a+1}}$

56. $b+\dfrac{b}{1-\dfrac{b+1}{b}}$

57. $\dfrac{x-\dfrac{1}{1-\dfrac{x}{2}}}{\dfrac{3}{x+\dfrac{2}{3}}-x}$

58. $\dfrac{\dfrac{2x}{x-\dfrac{1}{x}}-\dfrac{1}{x}}{2x+\dfrac{2x}{1-\dfrac{1}{x}}}$

59. $\dfrac{2x+\dfrac{1}{2-\dfrac{x}{2}}}{\dfrac{4}{\dfrac{x}{2}-2}-x}$

60. $\dfrac{3x-\dfrac{1}{3-\dfrac{x}{2}}}{\dfrac{3}{\dfrac{x}{2}-3}+x}$

In Exercises 61–62, factor each denominator and simplify the complex fraction.

61. $\dfrac{\dfrac{1}{x^2+3x+2}+\dfrac{1}{x^2+x-2}}{\dfrac{3x}{x^2-1}-\dfrac{x}{x+2}}$

62. $\dfrac{\dfrac{1}{x^2-1}-\dfrac{2}{x^2+4x+3}}{\dfrac{2}{x^2+2x-3}+\dfrac{1}{x+3}}$

APPLICATIONS

63. Engineering The stiffness k of the shaft shown in Illustration 1 is given by the formula

$$k = \frac{1}{\frac{1}{k_1} + \frac{1}{k_2}}$$

ILLUSTRATION 1

where k_1 and k_2 are the individual stiffnesses of each section. Simplify the complex fraction.

64. Transportation If a car travels a distance d_1 at a speed s_1, and then travels a distance d_2 at a speed s_2, the average (mean) speed is given by the formula

$$\bar{s} = \frac{d_1 + d_2}{\frac{d_1}{s_1} + \frac{d_2}{s_2}}$$

Simplify the complex fraction.

WRITING

65. There are two methods used to simplify a complex fraction. Explain one of them.

66. Explain the other method of simplifying a complex fraction.

SOMETHING TO THINK ABOUT

67. Simplify $(x^{-1}y^{-1})(x^{-1} + y^{-1})^{-1}$.

68. Simplify $[(x^{-1} + 1)^{-1} + 1]^{-1}$.

6.6 Equations Containing Rational Expressions

■ SOLVING RATIONAL EQUATIONS ■ FORMULAS ■ PROBLEM SOLVING

Getting Ready *Solve each proportion.*

1. $\frac{3}{x} = \frac{6}{9}$

2. $\frac{x + 1}{6} = \frac{2}{x}$

■ SOLVING RATIONAL EQUATIONS

If an equation contains one or more rational expressions, it is called a **rational equation.** Some examples of rational equations are

$$\frac{3}{5} + \frac{7}{x + 2} = 2, \quad \frac{x + 3}{x - 3} = \frac{2}{x^2 - 4}, \quad \text{and} \quad \frac{-x^2 + 10}{x^2 - 1} + \frac{3x}{x - 1} = \frac{2x}{x + 1}$$

To solve rational equations, we can multiply both sides of the equation by the LCD of the fractions in the equation to clear it of fractions.

EXAMPLE 1 Solve $\frac{3}{5} + \frac{7}{x+2} = 2$.

Solution We note that x cannot be -2, because this would give a 0 in the denominator of $\frac{7}{x+2}$. If $x \neq -2$, we can multiply both sides of the equation by $5(x + 2)$ and simplify to get

$$5(x+2)\left(\frac{3}{5} + \frac{7}{x+2}\right) = 5(x+2)2$$

$$5(x+2)\left(\frac{3}{5}\right) + 5(x+2)\left(\frac{7}{x+2}\right) = 5(x+2)2$$ Use the distributive property on the left-hand side.

$$3(x+2) + 5(7) = 10(x+2)$$ Simplify.

$$3x + 6 + 35 = 10x + 20$$ Use the distributive property and simplify.

$$3x + 41 = 10x + 20$$ Simplify.

$$-7x = -21$$ Add $-10x$ and -41 to both sides.

$$x = 3$$ Divide both sides by -7.

Check: To check, we substitute 3 for x in the original equation and simplify:

$$\frac{3}{5} + \frac{7}{x+2} = 2$$

$$\frac{3}{5} + \frac{7}{3+2} \stackrel{?}{=} 2$$

$$\frac{3}{5} + \frac{7}{5} \stackrel{?}{=} 2$$

$$2 = 2$$

■

Self Check $\frac{2}{5} + \frac{5}{x-2} = \frac{29}{10}$.

Answer 4

GRAPHING CALCULATORS

Solving Equations

To use a graphing calculator to approximate the solution of $\frac{3}{5} + \frac{7}{x+2} = 2$, we graph the functions $f(x) = \frac{3}{5} + \frac{7}{x+2}$ and $g(x) = 2$. If we use window settings of $[-10, 10]$ for x and $[-10, 10]$ for y, we will obtain the graph shown in Figure 6-11(a).

If we trace and move the cursor close to the intersection point of the two graphs, we will get the approximate value of x shown in Figure 6-11(b). If we zoom twice and trace again, we get the results shown in Figure 6-11(c). Algebra will show that the exact solution is 3, as shown in Example 1.

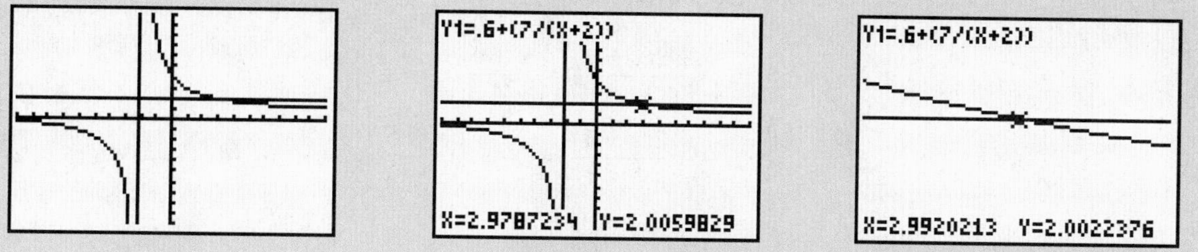

(a) (b) (c)

FIGURE 6-11

EXAMPLE 2 Solve $\dfrac{-x^2+10}{x^2-1} + \dfrac{3x}{x-1} = \dfrac{2x}{x+1}$.

Solution We note that x cannot be 1 or -1, because this would give a 0 in the denominator of a fraction. If $x \neq 1$ and $x \neq -1$, we can clear the equation of fractions by multiplying both sides by the LCD of the three fractions and proceed as follows:

$$\frac{-x^2+10}{x^2-1} + \frac{3x}{x-1} = \frac{2x}{x+1}$$

$$\frac{-x^2+10}{(x+1)(x-1)} + \frac{3x}{x-1} = \frac{2x}{x+1} \quad \text{Factor } x^2-1.$$

$$\frac{(x+1)(x-1)(-x^2+10)}{(x+1)(x-1)} + \frac{3x(x+1)(x-1)}{x-1} = \frac{2x(x+1)(x-1)}{x+1} \quad \text{Multiply both sides by } (x+1)(x-1).$$

$$-x^2+10+3x(x+1) = 2x(x-1) \quad \text{Divide out common factors.}$$

$$-x^2+10+3x^2+3x = 2x^2-2x \quad \text{Remove parentheses.}$$

$$2x^2+10+3x = 2x^2-2x \quad \text{Combine like terms.}$$

$$10+3x = -2x \quad \text{Subtract } 2x^2 \text{ from both sides.}$$

$$10+5x = 0 \quad \text{Add } 2x \text{ to both sides.}$$

$$5x = -10 \quad \text{Subtract 10 from both sides.}$$

$$x = -2 \quad \text{Divide both sides by 5.}$$

Verify that -2 is a solution of the original equation. The solution set is $\{-2\}$. ■

When we multiply both sides of an equation by a quantity that contains a variable, we can get false solutions, called **extraneous solutions.** We must exclude extraneous solutions from the solution set of an equation.

EXAMPLE 3 Solve $\dfrac{2(x+1)}{x-3} = \dfrac{x+5}{x-3}$.

Solution We start by noting that x cannot be 3, because this would give a 0 in the denominator of a fraction. If $x \neq 3$, we can clear the equation of fractions by multiplying both sides by $x - 3$.

$$\frac{2(x+1)}{x-3} = \frac{x+5}{x-3}$$

$$(x-3)\frac{2(x+1)}{x-3} = (x-3)\frac{x+5}{x-3} \qquad \text{Multiply both sides by } x-3.$$

$$2(x+1) = x+5 \qquad \text{Simplify.}$$

$$2x+2 = x+5 \qquad \text{Remove parentheses.}$$

$$x+2 = 5 \qquad \text{Subtract } x \text{ from both sides.}$$

$$x = 3 \qquad \text{Subtract 2 from both sides.}$$

Since x cannot be 3, the 3 must be discarded. This equation has no solutions. Its solution set is the empty set, $\varnothing$. ■

Self Check Solve $\dfrac{6}{x-6} + 3 = \dfrac{x}{x-6}$.

Answer no solutions; 6 is extraneous

EXAMPLE 4 Solve $\dfrac{x+1}{5} - 2 = -\dfrac{4}{x}$.

Solution We note that x cannot be 0, because this would give a 0 in the denominator of a fraction. If $x \neq 0$, we can clear the equation of fractions by multiplying both sides by $5x$.

$$\frac{x+1}{5} - 2 = -\frac{4}{x}$$

$$5x\left(\frac{x+1}{5} - 2\right) = 5x\left(-\frac{4}{x}\right) \qquad \text{Multiply both sides by } 5x.$$

$$x(x+1) - 10x = -20 \qquad \text{Remove parentheses and simplify.}$$

$$x^2 + x - 10x = -20 \qquad \text{Remove parentheses.}$$

$$x^2 - 9x + 20 = 0 \qquad \text{Combine like terms and add 20 to both sides.}$$

$$(x-5)(x-4) = 0 \qquad \text{Factor } x^2 - 9x + 20.$$

$$x - 5 = 0 \quad \text{or} \quad x - 4 = 0 \qquad \text{Set each factor equal to 0.}$$

$$x = 5 \qquad\qquad x = 4$$

Since 4 and 5 both satisfy the original equation, the solution set is $\{4, 5\}$. ■

Self Check Solve $a + \frac{2}{3} = \frac{2a - 12}{3(a - 3)}$.

Answer 1, 2

FORMULAS

Many formulas must be cleared of fractions before we can solve them for a specific variable.

EXAMPLE 5 Solve $\frac{1}{r} = \frac{1}{r_1} + \frac{1}{r_2}$ for r.

Solution

$$\frac{1}{r} = \frac{1}{r_1} + \frac{1}{r_2}$$

$$\frac{rr_1r_2}{r} = \frac{rr_1r_2}{r_1} + \frac{rr_1r_2}{r_2}$$ Multiply both sides by rr_1r_2.

$$r_1r_2 = rr_2 + rr_1$$ Simplify each fraction.

$$r_1r_2 = r(r_2 + r_1)$$ Factor out r on the right-hand side.

$$\frac{r_1r_2}{r_2 + r_1} = r$$ Divide both sides by $r_2 + r_1$.

$$r = \frac{r_1r_2}{r_2 + r_1}$$

■

Self Check Solve $\frac{1}{a} - \frac{1}{b} = 1$ for b.

Answer $b = \frac{a}{1 - a}$

PROBLEM SOLVING

EXAMPLE 6 **Drywalling a house** A contractor knows that one crew can drywall a house in 4 days and that another crew can drywall the same house in 5 days. One day must be allowed for the plaster coat to dry. If the contractor uses both crews, can the house be ready for painting in 4 days?

Analyze the problem Because 1 day is necessary for drying, the drywallers must complete their work in 3 days. Since the first crew can drywall the house in 4 days, it can do $\frac{1}{4}$ of the job in 1 day. Since the second crew can drywall the house in 5 days, it can do $\frac{1}{5}$ of the job in 1 day. If it takes x days for both crews to finish the house, together they can do $\frac{1}{x}$ of the job in 1 day. The amount of work the first crew can do in 1 day plus the amount of work the second crew can do in 1 day equals the amount of work both crews can do in 1 day when working together.

Form an equation If x represents the number of days it takes for both crews to drywall the house, we can form the equation

What crew 1 can do in one day	+	what crew 2 can do in one day	=	what they can do together in one day.
$\frac{1}{4}$	+	$\frac{1}{5}$	=	$\frac{1}{x}$

Solve the equation We can solve this equation as follows.

$$20x\left(\frac{1}{4}+\frac{1}{5}\right) = 20x\left(\frac{1}{x}\right) \qquad \text{Multiply both sides by } 20x.$$

$$5x + 4x = 20 \qquad \text{Remove parentheses and simplify.}$$

$$9x = 20 \qquad \text{Combine like terms.}$$

$$x = \frac{20}{9} \qquad \text{Divide both sides by 9.}$$

State the conclusion Since it will take $2\frac{2}{9}$ days for both crews to drywall the house and it takes 1 day for drying, it will be ready for painting in $3\frac{2}{9}$ days, which is less than 4 days.

Check the result. ■

EXAMPLE 7

Driving to a convention A man drove 200 miles to a convention. Because of road construction, his average speed on the return trip was 10 mph less than his average speed going to the convention. If the return trip took 1 hour longer, how fast did he drive in each direction?

Analyze the problem Because the distance traveled is given by the formula

$d = rt$ (d is distance, r is the rate of speed, and t is time)

the formula for time is

$$t = \frac{d}{r}$$

We can organize the given information in the chart shown in Figure 6-12.

	Rate	· Time	= Distance
Going	r	$\frac{200}{r}$	200
Returning	$r - 10$	$\frac{200}{r-10}$	200

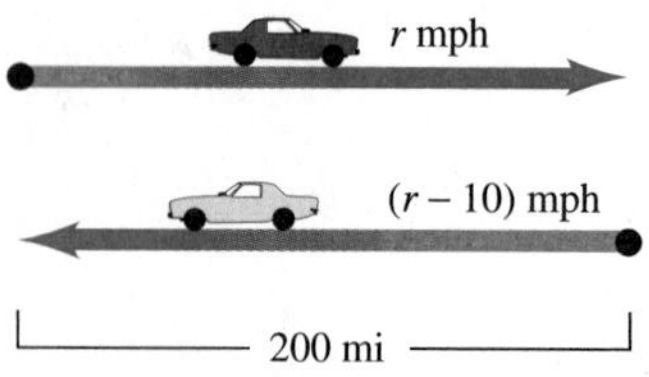

FIGURE 6-12

Form an equation Let r represent the average rate of speed going to the meeting. Then $r - 10$ represents the average rate of speed on the return trip. Because the return trip took 1 hour longer, we can form the following equation:

The time it took to travel to the convention	plus 1 =	the time it took to return.
$\dfrac{200}{r}$	$+ 1 =$	$\dfrac{200}{r - 10}$

Solve the equation We can solve the equation as follows:

$$r(r - 10)\left(\frac{200}{r} + 1\right) = r(r - 10)\left(\frac{200}{r - 10}\right) \quad \text{Multiply both sides by } r(r - 10).$$

$$200(r - 10) + r(r - 10) = 200r \quad \text{Remove parentheses and simplify.}$$

$$200r - 2{,}000 + r^2 - 10r = 200r \quad \text{Remove parentheses.}$$

$$r^2 - 10r - 2{,}000 = 0 \quad \text{Subtract } 200r \text{ from both sides.}$$

$$(r - 50)(r + 40) = 0 \quad \text{Factor } r^2 - 10r - 2{,}000.$$

$$r - 50 = 0 \quad \text{or} \quad r + 40 = 0 \quad \text{Set each factor equal to 0.}$$

$$r = 50 \qquad\qquad r = -40$$

State the conclusion We must exclude the solution of -40, because a speed cannot be negative. Thus, the man averaged 50 mph going to the convention, and he averaged $50 - 10$, or 40 mph, returning.

Check the result At 50 mph, the 200-mile trip took 4 hours. At 40 mph, the return trip took 5 hours, which is 1 hour longer. ■

EXAMPLE 8

A river cruise The Forest City Queen can make a 9-mile trip down the Rock River and return in a total of 1.6 hours. If the riverboat travels 12 mph in still water, find the speed of the current in the Rock River.

Analyze the problem We can let c represent the speed of the current. Since the boat travels 12 mph and a current of c mph pushes the boat while it is going downstream, the speed of the boat going downstream is $(12 + c)$ mph. On the return trip, the current pushes against the boat, and its speed is $(12 - c)$ mph. Since $t = \frac{d}{r}$ $\left(\text{time} = \frac{\text{distance}}{\text{rate}}\right)$, the time required for the downstream leg of the trip is $\frac{9}{12 + c}$ hours, and the time required for the upstream leg of the trip is $\frac{9}{12 - c}$ hours.

We can organize this information in the chart shown in Figure 6-13.

Furthermore, we know that the total time required for the round trip is 1.6 hours.

Form an equation If c represents the speed of the current, then $12 + c$ represents the speed of the boat going downstream, and $12 - c$ represents the speed of the boat going upstream.

	Rate	· Time	= Distance
Going downstream	$12 + c$	$\frac{9}{12+c}$	9
Going upstream	$12 - c$	$\frac{9}{12-c}$	9

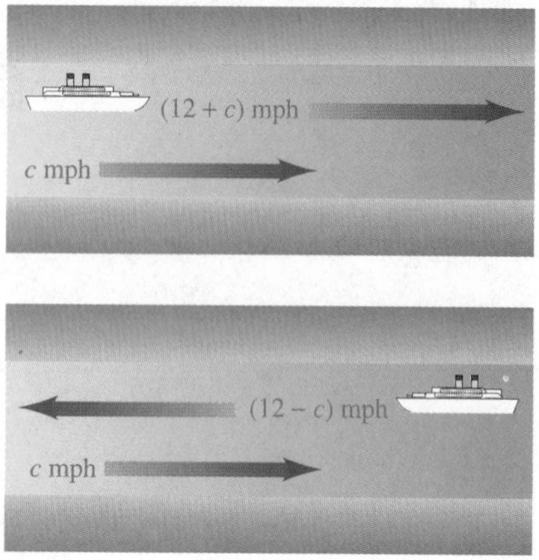

FIGURE 6-13

Since the time going downstream is $\frac{9}{12+c}$ hours, the time going upstream is $\frac{9}{12-c}$ hours, and total time is 1.6 hours $\left(\frac{8}{5}\text{ hours}\right)$, we have

The time it takes to travel downstream	+	The time it takes to travel upstream	=	the total time for the round trip.
$\frac{9}{12+c}$	+	$\frac{9}{12-c}$	=	$\frac{8}{5}$

Solve the equation We can multiply both sides of this equation by $5(12 + c)(12 - c)$ to clear it of fractions and proceed as follows:

$$\frac{5(12+c)(12-c)9}{12+c} + \frac{5(12+c)(12-c)9}{12-c} = \frac{5(12+c)(12-c)8}{5}$$

$$\begin{aligned}
45(12 - c) + 45(12 + c) &= 8(12 + c)(12 - c) && \frac{12+c}{12+c} = 1 \text{ and } \frac{12-c}{12-c} = 1. \\
540 - 45c + 540 + 45c &= 8(144 - c^2) && \text{Multiply.} \\
1{,}080 &= 1{,}152 - 8c^2 && \text{Combine like terms and multiply.} \\
8c^2 - 72 &= 0 && \text{Add } 8c^2 \text{ and } -1{,}152 \text{ to both sides.} \\
c^2 - 9 &= 0 && \text{Divide both sides by 8.} \\
(c + 3)(c - 3) &= 0 && \text{Factor } c^2 - 9. \\
c + 3 = 0 \quad &\text{or} \quad c - 3 = 0 && \text{Set each factor equal to 0.} \\
c = -3 \quad & \quad\quad\;\; c = 3
\end{aligned}$$

State the conclusion Since the current cannot be negative, the apparent solution -3 must be discarded. Thus, the current in the Rock River is 3 mph.

Check the result. ■

Orals *Solve each equation.*

1. $\frac{4}{x} = 2$
2. $\frac{9}{y} = 3$
3. $\frac{4}{p} + \frac{5}{p} = 9$
4. $\frac{5}{r} - \frac{2}{r} = 1$
5. $\frac{4}{y} - \frac{1}{y} = 3$
6. $\frac{8}{t} - \frac{2}{t} = 3$

EXERCISE 6.6

REVIEW *Simplify each expression. Write each answer without using negative exponents.*

1. $(m^2n^{-3})^{-2}$
2. $\frac{a^{-1}}{a^{-1} + 1}$
3. $\frac{a^0 + 2a^0 - 3a^0}{(a - b)^0}$
4. $(4x^{-2} + 3)(2x - 4)$

VOCABULARY AND CONCEPTS *Fill in each blank to make a true statement.*

5. If an equation contains a rational expression, it is called a ______ equation.
6. A false solution to an equation is called an ______ solution.

PRACTICE *In Exercises 7–36, solve each equation. If a solution is extraneous, so indicate.*

7. $\frac{1}{4} + \frac{9}{x} = 1$
8. $\frac{1}{3} - \frac{10}{x} = -3$
9. $\frac{34}{x} - \frac{3}{2} = -\frac{13}{20}$
10. $\frac{1}{2} + \frac{7}{x} = 2 + \frac{1}{x}$
11. $\frac{3}{y} + \frac{7}{2y} = 13$
12. $\frac{2}{x} + \frac{1}{2} = \frac{7}{2x}$
13. $\frac{x+1}{x} - \frac{x-1}{x} = 0$
14. $\frac{2}{x} + \frac{1}{2} = \frac{9}{4x} - \frac{1}{2x}$
15. $\frac{7}{5x} - \frac{1}{2} = \frac{5}{6x} + \frac{1}{3}$
16. $\frac{x-3}{x-1} - \frac{2x-4}{x-1} = 0$
17. $\frac{3-5y}{2+y} = \frac{3+5y}{2-y}$
18. $\frac{x}{x-2} = 1 + \frac{1}{x-3}$
19. $\frac{a+2}{a+1} = \frac{a-4}{a-3}$
20. $\frac{z+2}{z+8} - \frac{z-3}{z-2} = 0$
21. $\frac{x+2}{x+3} - 1 = \frac{1}{3 - 2x - x^2}$
22. $\frac{x-3}{x-2} - \frac{1}{x} = \frac{x-3}{x}$
23. $\frac{x}{x+2} = 1 - \frac{3x+2}{x^2 + 4x + 4}$
24. $\frac{3+2a}{a^2 + 6 + 5a} + \frac{2-5a}{a^2 - 4} = \frac{2-3a}{a^2 - 6 + a}$
25. $\frac{2}{x-2} + \frac{1}{x+1} = \frac{1}{x^2 - x - 2}$
26. $\frac{5}{y-1} + \frac{3}{y-3} = \frac{8}{y-2}$
27. $\frac{a-1}{a+3} - \frac{1-2a}{3-a} = \frac{2-a}{a-3}$
28. $\frac{5}{2z^2 + z - 3} - \frac{2}{2z+3} = \frac{z+1}{z-1} - 1$
29. $\frac{5}{x+4} + \frac{1}{x+4} = x - 1$
30. $\frac{2}{x-1} + \frac{x-2}{3} = \frac{4}{x-1}$

31. $\frac{3}{x+1} - \frac{x-2}{2} = \frac{x-2}{x+1}$

32. $\frac{x-4}{x-3} + \frac{x-2}{x-3} = x-3$

33. $\frac{2}{x-3} + \frac{3}{4} = \frac{17}{2x}$

34. $\frac{30}{y-2} + \frac{24}{y-5} = 13$

35. $\frac{x+4}{x+7} - \frac{x}{x+3} = \frac{3}{8}$

36. $\frac{5}{x+4} - \frac{1}{3} = \frac{x-1}{x}$

In Exercises 37–42, solve each formula for the indicated variable.

37. $\frac{1}{p} + \frac{1}{q} = \frac{1}{f}$ for f

38. $\frac{1}{p} + \frac{1}{q} = \frac{1}{f}$ for p

39. $S = \frac{a - lr}{1 - r}$ for r

40. $H = \frac{2ab}{a+b}$ for a

41. $\frac{1}{R} = \frac{1}{r_1} + \frac{1}{r_2} + \frac{1}{r_3}$ for R

42. $\frac{1}{R} = \frac{1}{r_1} + \frac{1}{r_2} + \frac{1}{r_3}$ for r_1

APPLICATIONS

43. **Focal length** The design of a camera lens uses the equation

$$\frac{1}{f} = \frac{1}{s_1} + \frac{1}{s_2}$$

which relates the focal length f of a lens to the image distance s_1 and the object distance s_2. Find the focal length of the lens in Illustration 1. (*Hint:* Convert feet to inches.)

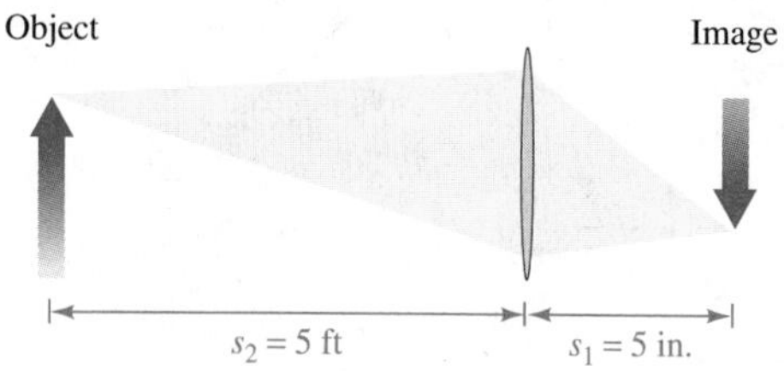

ILLUSTRATION 1

44. **Lensmaker's formula** The focal length f of a lens is given by the lensmakers formula

$$\frac{1}{f} = 0.6\left(\frac{1}{r_1} + \frac{1}{r_2}\right)$$

where f is the focal length of the lens and r_1 and r_2 are the radii of the two circular surfaces. Find the focal length of the lens in Illustration 2.

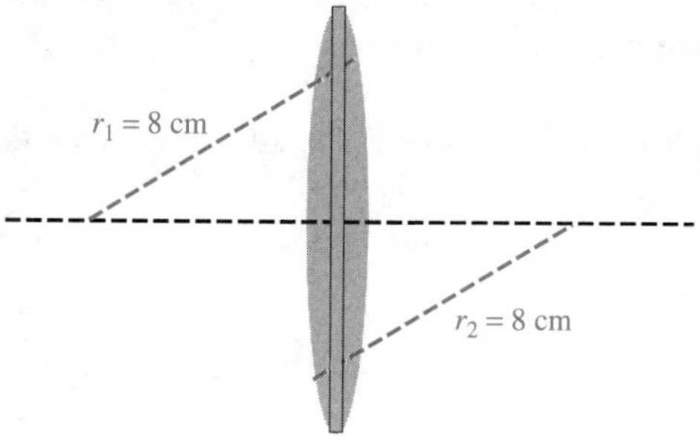

ILLUSTRATION 2

45. **House painting** If one painter can paint a house in 5 days and another painter can paint the same house in 3 days, how long will it take them to paint the house working together?

46. **Reading proof** A proofreader can read 250 pages in 8 hours, and a second proofreader can read 250 pages in 10 hours. If they both work on a 250-page book, can they meet a five-hour deadline?

47. **Storing corn** In 10 minutes, a conveyor belt can move 1,000 bushels of corn into the storage bin shown in Illustration 3. A smaller belt can move 1,000 bushels to the storage bin in 14 minutes. If both belts are used, how long will it take to move 1,000 bushels to the storage bin?

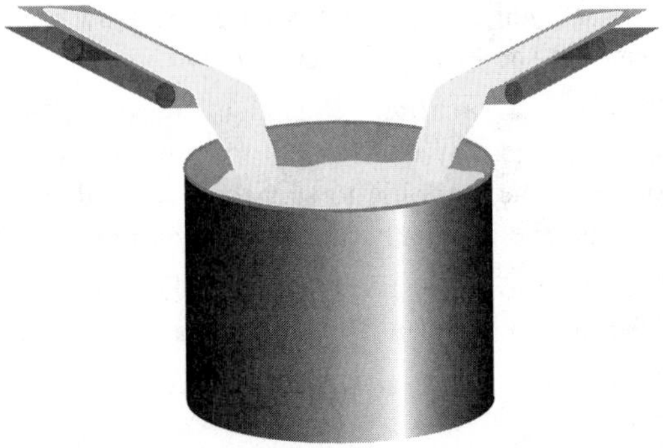

ILLUSTRATION 3

48. **Roofing a house** One roofing crew can finish a 2,800-square-foot roof in 12 hours, and another crew can do the job in 10 hours. If they work together, can they finish before a predicted rain in 5 hours?

49. **Draining a pool** A drain can empty the swimming pool shown in Illustration 4 in 3 days. A second drain can empty the pool in 2 days. How long will it take to empty the pool if both drains are used?

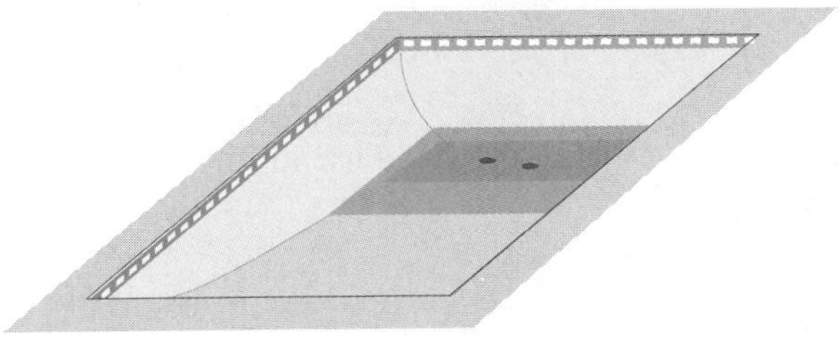

ILLUSTRATION 4

50. **Filling a pool** A pipe can fill a pool in 9 hours. If a second pipe is also used, the pool can be filled in 3 hours. How long would it take the second pipe alone to fill the pool?

51. **Filling a pond** One pipe can fill a pond in 3 weeks, and a second pipe can fill the pond in 5 weeks. However, evaporation and seepage can empty the pond in 10 weeks. If both pipes are used, how long will it take to fill the pond?

52. **Housecleaning** Sally can clean the house in 6 hours, and her father can clean the house in 4 hours. Sally's younger brother, Dennis, can completely mess up the house in 8 hours. If Sally and her father clean and Dennis plays, how long will it take to clean the house?

53. **Touring the countryside** A man bicycles 5 mph faster than he can walk. He bicycles 24 miles and walks back along the same route in 11 hours. How fast does he walk?

54. **Finding rates** Two trains made the same 315-mile run. Since one train traveled 10 mph faster than the other, it arrived 2 hours earlier. Find the speed of each train.

55. **Train travel** A train traveled 120 miles from Freeport to Chicago and returned the same distance in a total time of 5 hours. If the train traveled 20 mph slower on the return trip, how fast did the train travel in each direction?

56. **Time on the road** A car traveled from Rockford to Chicago in 3 hours less time than it took a second car to travel from Rockford to St. Louis. If the cars traveled at the same average speed, how long was the first driver on the road? (See Illustration 5.)

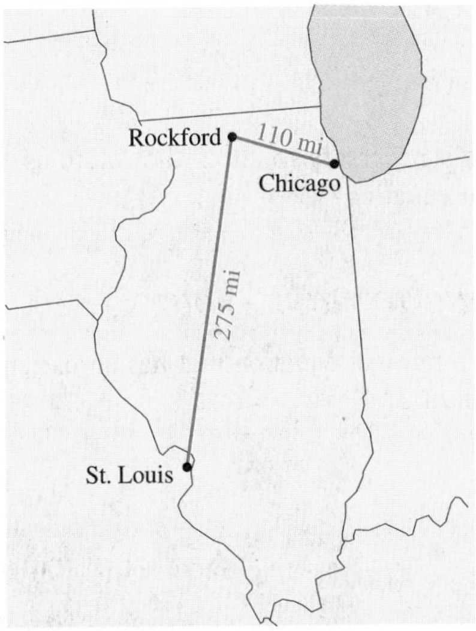

ILLUSTRATION 5

57. **Boating** A man can drive a motorboat 45 miles down the Rock River in the same amount of time that he can drive it 27 miles upstream. Find the speed of the current if the speed of the boat is 12 mph in still water.

58. Rowing a boat A woman who can row 3 mph in still water rows 10 miles downstream on the Eagle River and returns upstream in a total of 12 hours. Find the speed of the current.

59. Aviation A plane that can fly 340 mph in still air can fly 200 miles downwind in the same amount of time that it can fly 140 miles upwind. Find the velocity of the wind.

60. Aviation An airplane can fly 650 miles with the wind in the same amount of time as it can fly 475 miles against the wind. If the wind speed is 40 mph, find the speed of the plane in still air.

61. Discount buying A repairman purchased some washing-machine motors for a total of $224. When the unit cost decreased by $4, he was able to buy one extra motor for the same total price. How many motors did he buy originally?

62. Price increase An appliance store manager bought several microwave ovens for a total of $1,800. When her unit cost increased by $25, she was able to buy one less oven for the same total price. How many ovens did she buy originally?

63. Stretching a vacation A student saved $1,200 for a trip to Europe. By cutting $20 from her daily expenses, she was able to stay three extra days. How long had she originally planned to be gone?

64. Engineering The stiffness of the shaft shown in Illustration 6 is given by the formula

$$k = \frac{1}{\frac{1}{k_1} + \frac{1}{k_2}}$$

ILLUSTRATION 6

where k_1 and k_2 are the individual stiffnesses of each section. If the stiffness, k_2, of Section 2 is 4,200,000 in. lb/rad, and the design specifications require that the stiffness, k, of the entire shaft be 1,900,000 in. lb/rad, what must the stiffness, k_1, of Section 1 be?

WRITING

65. Why is it necessary to check the solutions of a rational equation?

66. Explain the steps you would use to solve a rational equation.

SOMETHING TO THINK ABOUT

67. Invent a rational equation that has an extraneous solution of 3.

68. Solve $(x - 1)^{-1} - x^{-1} = 6^{-1}$.

6.7 Dividing Polynomials

- DIVIDING A MONOMIAL BY A MONOMIAL
- DIVIDING A POLYNOMIAL BY A MONOMIAL
- DIVIDING A POLYNOMIAL BY A POLYNOMIAL
- THE CASE OF THE MISSING TERMS

Getting Ready *Simplify the fraction or do the division.*

1. $\frac{16}{24}$ **2.** $-\frac{45}{81}$ **3.** $23\overline{)115}$ **4.** $32\overline{)512}$

DIVIDING A MONOMIAL BY A MONOMIAL

In Example 1, we review how to divide a monomial by a monomial.

EXAMPLE 1 Simplify $(3a^2b^3) \div (2a^3b)$.

Method 1 We write the expression as a fraction and divide out all common factors:

$$\frac{3a^2b^3}{2a^3b} = \frac{3aabbb}{2aaab}$$
$$= \frac{3aabbb}{2aaab}$$
$$= \frac{3b^2}{2a}$$

Method 2 We write the expression as a fraction and use the rules of exponents:

$$\frac{3a^2b^3}{2a^3b} = \frac{3}{2}a^{2-3}b^{3-1}$$
$$= \frac{3}{2}a^{-1}b^2$$
$$= \frac{3}{2}\left(\frac{1}{a}\right)\frac{b^2}{1}$$
$$= \frac{3b^2}{2a}$$

■

Self Check Simplify $\frac{6x^3y^2}{8x^2y^3}$.

Answer $\frac{3x}{4y}$

DIVIDING A POLYNOMIAL BY A MONOMIAL

In Example 2, we use the fact that $\frac{a}{b} = \frac{1}{b} \cdot a$ $(b \neq 0)$ to divide a polynomial by a monomial.

EXAMPLE 2 Divide $4x^3y^2 + 3xy^5 - 12xy$ by $3x^2y^3$.

Solution We write the division as a product and use the distributive property to remove parentheses.

$$\frac{4x^3y^2 + 3xy^5 - 12xy}{3x^2y^3} = \frac{1}{3x^2y^3}(4x^3y^2 + 3xy^5 - 12xy)$$
$$= \frac{4x^3y^2}{3x^2y^3} + \frac{3xy^5}{3x^2y^3} + \frac{-12xy}{3x^2y^3}$$

We then simplify each of the fractions on the right-hand side of the equal sign to get

$$\begin{aligned}\frac{4x^3y^2 + 3xy^5 - 12xy}{3x^2y^3} &= \frac{4x}{3y} + \frac{y^2}{x} + \frac{-4}{xy^2}\\ &= \frac{4x}{3y} + \frac{y^2}{x} - \frac{4}{xy^2}\end{aligned}$$

■

Self Check Simplify $\dfrac{8a^3b^4 - 4a^4b^2 + a^2b^2}{4a^2b^2}$.

Answer $2ab^2 - a^2 + \frac{1}{4}$

■ DIVIDING A POLYNOMIAL BY A POLYNOMIAL

There is an **algorithm** (a repeating series of steps) to use when the divisor is not a monomial. To use the division algorithm to divide $x^2 + 7x + 12$ by $x + 4$, we write the division in long division form and proceed as follows:

$$\begin{array}{r} x \\ x+4\overline{)x^2+7x+12} \end{array}$$

How many times does x divide x^2? $\frac{x^2}{x} = x$. Place the x in the quotient.

$$\begin{array}{r} x \\ x+4\overline{)x^2+7x+12}\\ \underline{x^2+4x}\\ 3x+12 \end{array}$$

Multiply each term in the divisor by x to get $x^2 + 4x$, subtract $x^2 + 4x$ from $x^2 + 7x$, and bring down the 12.

$$\begin{array}{r} x+3 \\ x+4\overline{)x^2+7x+12}\\ \underline{x^2+4x}\\ 3x+12 \end{array}$$

How many times does x divide $3x$? $\frac{3x}{x} = +3$. Place the $+3$ in the quotient.

$$\begin{array}{r} x+3 \\ x+4\overline{)x^2+7x+12}\\ \underline{x^2+4x}\\ 3x+12\\ \underline{3x+12}\\ 0 \end{array}$$

Multiply each term in the divisor by 3 to get $3x + 12$, and subtract $3x + 12$ from $3x + 12$ to get 0.

The division process stops when the result of the subtraction is a constant or a polynomial with degree less than the degree of the divisor. Here, the quotient is $x + 3$ and the remainder is 0.

We can check the quotient by multiplying the divisor by the quotient. The product should be the dividend.

$$\underbrace{\text{Divisor}}\cdot\underbrace{\text{quotient}} = \underbrace{\text{dividend}}$$
$$(x + 4)\ (x + 3) = x^2 + 7x + 12 \qquad \text{The quotient checks.}$$

EXAMPLE 3 Divide $2a^3 + 9a^2 + 5a - 6$ by $2a + 3$.

Solution

$$\begin{array}{rl} & a^2 \\ 2a + 3 & \overline{)2a^3 + 9a^2 + 5a - 6} \end{array}$$

How many times does $2a$ divide $2a^3$? $\frac{2a^3}{2a} = a^2$. Place a^2 in the quotient.

$$\begin{array}{rl} & a^2 \\ 2a + 3 & \overline{)2a^3 + 9a^2 + 5a - 6} \\ & \underline{\ \ 2a^3 + 3a^2} \\ & \qquad\ \ 6a^2 + 5a \end{array}$$

Multiply each term in the divisor by a^2 to get $2a^3 + 3a^2$, subtract $2a^3 + 3a^2$ from $2a^3 + 9a^2$, and bring down the $5a$.

$$\begin{array}{rl} & a^2 + 3a \\ 2a + 3 & \overline{)2a^3 + 9a^2 + 5a - 6} \\ & \underline{\ \ 2a^3 + 3a^2} \\ & \qquad\ \ 6a^2 + 5a \end{array}$$

How many times does $2a$ divide $6a^2$? $6a^2/2a = 3a$. Place the $+3a$ in the quotient.

$$\begin{array}{rl} & a^2 + 3a \\ 2a + 3 & \overline{)2a^3 + 9a^2 + 5a - 6} \\ & \underline{\ \ 2a^3 + 3a^2} \\ & \qquad\ \ 6a^2 + 5a \\ & \qquad\ \ \underline{6a^2 + 9a} \\ & \qquad\qquad\ \ - 4a - 6 \end{array}$$

Multiply each term in the divisor by $3a$ to get $6a^2 + 9a$, subtract $6a^2 + 9a$ from $6a^2 + 5a$, and bring down -6.

$$\begin{array}{rl} & a^2 + 3a\ - 2 \\ 2a + 3 & \overline{)2a^3 + 9a^2 + 5a - 6} \\ & \underline{\ \ 2a^3 + 3a^2} \\ & \qquad\ \ 6a^2 + 5a \\ & \qquad\ \ \underline{6a^2 + 9a} \\ & \qquad\qquad\ \ - 4a - 6 \end{array}$$

How many times does $2a$ divide $-4a$? $-4a/2a = -2$. Place the -2 in the quotient.

$$\begin{array}{rl} & a^2 + 3a\ - 2 \\ 2a + 3 & \overline{)2a^3 + 9a^2 + 5a - 6} \\ & \underline{\ \ 2a^3 + 3a^2} \\ & \qquad\ \ 6a^2 + 5a \\ & \qquad\ \ \underline{6a^2 + 9a} \\ & \qquad\qquad\ \ - 4a - 6 \\ & \qquad\qquad\ \ \underline{- 4a - 6} \\ & \qquad\qquad\qquad\quad\ \ 0 \end{array}$$

Multiply each term in the divisor by -2 to get $-4a - 6$, and subtract $-4a - 6$ from $-4a - 6$ to get 0.

Since the remainder is 0, the quotient is $a^2 + 3a - 2$. We can check the quotient by verifying that

$$\underbrace{\text{Divisor}}\cdot\underbrace{\text{quotient}} = \underbrace{\text{dividend}}$$
$$(2a + 3)\ (a^2 + 3a - 2) = 2a^3 + 9a^2 + 5a - 6$$

■

EXAMPLE 4 Divide $3x^3 + 2x^2 - 3x + 8$ by $x - 2$.

Solution

$$
\begin{array}{r}
3x^2 + 8x + 13 \\
x - 2 \overline{)\, 3x^3 + 2x^2 - 3x + 8} \\
\underline{3x^3 - 6x^2} \\
8x^2 - 3x \\
\underline{8x^2 - 16x} \\
13x + 8 \\
\underline{13x - 26} \\
34
\end{array}
$$

This division gives a quotient of $3x^2 + 8x + 13$ and a remainder of 34. It is common to form a fraction with the remainder as the numerator and the divisor as denominator and to write the result as

$$3x^2 + 8x + 13 + \frac{34}{x - 2}$$

To check, we verify that

$$(x - 2)\left(3x^2 + 8x + 13 + \frac{34}{x - 2}\right) = 3x^3 + 2x^2 - 3x + 8$$

■

Self Check Divide: $a - 3\overline{)2a^3 + 3a^2 - a + 2}$.

Answer $2a^2 + 9a + 26 + \frac{80}{a - 3}$

EXAMPLE 5 Divide $-9x + 8x^3 + 10x^2 - 9$ by $3 + 2x$.

Solution The division algorithm works best when the polynomials in the dividend and the divisor are written in descending powers of x. We can use the commutative property of addition to rearrange the terms. Then the division is routine:

$$
\begin{array}{r}
4x^2 - x - 3 \\
2x + 3 \overline{)\, 8x^3 + 10x^2 - 9x - 9} \\
\underline{8x^3 + 12x^2} \\
-2x^2 - 9x \\
\underline{-2x^2 - 3x} \\
-6x - 9 \\
\underline{-6x - 9} \\
0
\end{array}
$$

Thus,

$$\frac{-9x + 8x^3 + 10x^2 - 9}{3 + 2x} = 4x^2 - x - 3$$

■

Self Check Divide: $2 + 3a\overline{)-4a + 15a^2 + 18a^3 - 4}$.

Answer $6a^2 + a - 2$

THE CASE OF THE MISSING TERMS

EXAMPLE 6 Divide $8x^3 + 1$ by $2x + 1$.

Solution When we write the terms in the dividend in descending powers of x, we see that the terms involving x^2 and x are missing. We must include the terms $0x^2$ and $0x$ in the dividend or leave spaces for them. Then the division is routine.

$$
\begin{array}{r}
4x^2 - 2x \;\; + 1 \\
2x + 1 \overline{)\,8x^3 + 0x^2 + 0x + 1} \\
\underline{8x^3 + 4x^2} \\
-4x^2 + 0x \\
\underline{-4x^2 - 2x} \\
2x + 1 \\
\underline{2x + 1} \\
0
\end{array}
$$

Thus,

$$\frac{8x^3 + 1}{2x + 1} = 4x^2 - 2x + 1$$

■

Self Check Divide: $3a - 1\overline{)27a^3 - 1}$.

Answer $9a^2 + 3a + 1$

EXAMPLE 7 Divide $-17x^2 + 5x + x^4 + 2$ by $x^2 - 1 + 4x$.

Solution We write the problem with the divisor and the dividend in descending powers of x. After leaving space for the missing term in the dividend, we proceed as follows:

$$
\begin{array}{r}
x^2 - 4x \\
x^2 + 4x - 1 \overline{)\,x^4 - 17x^2 + 5x + 2} \\
\underline{x^4 + 4x^3 - x^2} \\
-4x^3 - 16x^2 + 5x \\
\underline{-4x^3 - 16x^2 + 4x} \\
x + 2
\end{array}
$$

This division gives a quotient of $x^2 - 4x$ and a remainder of $x + 2$.

$$\frac{-17x^2 + 5x + x^4 + 2}{x^2 - 1 + 4x} = x^2 - 4x + \frac{x + 2}{x^2 + 4x - 1}$$

■

Self Check Divide: $\dfrac{2a^2 + 3a^3 + a^4 - 7 + a}{a^2 - 2a + 1}$.

Answer $a^2 + 5a + 11 + \dfrac{18a - 18}{a^2 - 2a + 1}$

Orals *Divide.*

1. $\dfrac{6x^2y^2}{2xy}$
2. $\dfrac{4ab^2 + 8a^2b}{2ab}$
3. $\dfrac{x^2 + 2x + 1}{x + 1}$
4. $\dfrac{x^2 - 4}{x - 2}$

EXERCISE 6.7

REVIEW *Remove parentheses and simplify.*

1. $2(x^2 + 4x - 1) + 3(2x^2 - 2x + 2)$
2. $3(2a^2 - 3a + 2) - 4(2a^2 + 4a - 7)$
3. $-2(3y^3 - 2y + 7) - 3(y^2 + 2y - 4) + 4(y^3 + 2y - 1)$
4. $3(4y^3 + 3y - 2) + 2(3y^2 - y + 3) - 5(2y^3 - y^2 - 2)$

VOCABULARY AND CONCEPTS *Fill in each blank to make a true statement.*

5. $\dfrac{a}{b} = \underline{\quad} \cdot a \ (b \neq 0)$
6. The division ________ is a repeating series of steps used to do a long division.
7. divisor · ________ + remainder = dividend
8. If a polynomial is divided by $3a - 2$ and the quotient is $3a^2 + 5$ with a remainder of 6, we usually write the result as $3a^2 + 5 + \dfrac{\quad}{\underline{\qquad}}$.

PRACTICE *In Exercises 9–26, do each division. Write each answer without using negative exponents.*

9. $\dfrac{4x^2y^3}{8x^5y^2}$
10. $\dfrac{25x^4y^7}{5xy^9}$
11. $\dfrac{33a^{-2}b^2}{44a^2b^{-2}}$
12. $\dfrac{-63a^4b^{-3}}{81a^{-3}b^3}$
13. $\dfrac{45x^{-2}y^{-3}t^0}{-63x^{-1}y^4t^2}$
14. $\dfrac{112a^0b^2c^{-3}}{48a^4b^0c^4}$
15. $\dfrac{-65a^{2n}b^nc^{3n}}{-15a^nb^{-n}c}$
16. $\dfrac{-32x^{-3n}y^{-2n}z}{40x^{-2}y^{-n}z^{n+1}}$
17. $\dfrac{4x^2 - x^3}{6x}$
18. $\dfrac{5y^4 + 45y^3}{15y^2}$

19. $\dfrac{4x^2y^3 + x^3y^2}{6xy}$

20. $\dfrac{3a^3y^2 - 18a^4y^3}{27a^2y^2}$

21. $\dfrac{24x^6y^7 - 12x^5y^{12} + 36xy}{48x^2y^3}$

22. $\dfrac{9x^4y^3 + 18x^2y - 27xy^4}{9x^3y^3}$

23. $\dfrac{3a^{-2}b^3 - 6a^2b^{-3} + 9a^{-2}}{12a^{-1}b}$

24. $\dfrac{4x^3y^{-2} + 8x^{-2}y^2 - 12y^4}{12x^{-1}y^{-1}}$

25. $\dfrac{x^ny^n - 3x^{2n}y^{2n} + 6x^{3n}y^{3n}}{x^ny^n}$

26. $\dfrac{2a^n - 3a^nb^{2n} - 6b^{4n}}{a^nb^{n-1}}$

In Exercises 27–62, do each division.

27. $\dfrac{x^2 + 5x + 6}{x + 3}$

28. $\dfrac{x^2 - 5x + 6}{x - 3}$

29. $\dfrac{x^2 + 10x + 21}{x + 3}$

30. $\dfrac{x^2 + 10x + 21}{x + 7}$

31. $\dfrac{6x^2 - x - 12}{2x + 3}$

32. $\dfrac{6x^2 - x - 12}{2x - 3}$

33. $\dfrac{3x^3 - 2x^2 + x + 6}{x - 1}$

34. $\dfrac{4a^3 + a^2 - 3a + 7}{a + 1}$

35. $\dfrac{6x^3 + 11x^2 - x - 2}{3x - 2}$

36. $\dfrac{6x^3 + 11x^2 - x + 10}{2x + 3}$

37. $\dfrac{6x^3 - x^2 - 6x - 9}{2x - 3}$

38. $\dfrac{16x^3 + 16x^2 - 9x - 5}{4x + 5}$

39. $\dfrac{2a + 1 + a^2}{a + 1}$

40. $\dfrac{a - 15 + 6a^2}{2a - 3}$

41. $\dfrac{6y - 4 + 10y^2}{5y - 2}$

42. $\dfrac{-10xy + x^2 + 16y^2}{x - 2y}$

43. $\dfrac{-18x + 12 + 6x^2}{x - 1}$

44. $\dfrac{27x + 23x^2 + 6x^3}{2x + 3}$

45. $\dfrac{-9x^2 + 8x + 9x^3 - 4}{3x - 2}$

46. $\dfrac{6x^2 + 8x^3 - 13x + 3}{4x - 3}$

47. $\dfrac{13x + 16x^4 + 3x^2 + 3}{4x + 3}$

48. $\dfrac{3x^2 + 9x^3 + 4x + 4}{3x + 2}$

49. $\dfrac{a^3 + 1}{a - 1}$

50. $\dfrac{27a^3 - 8b^3}{3a - 2b}$

51. $\dfrac{15a^3 - 29a^2 + 16}{3a - 4}$

52. $\dfrac{4x^3 - 12x^2 + 17x - 12}{2x - 3}$

53. $y - 2\overline{)-24y + 24 + 6y^2}$

54. $3 - a\overline{)21a - a^2 - 54}$

55. $2x + y\overline{)32x^5 + y^5}$

56. $3x - y\overline{)81x^4 - y^4}$

57. $x^2 - 2\overline{)x^6 - x^4 + 2x^2 - 8}$

58. $x^2 + 3\overline{)x^6 + 2x^4 - 6x^2 - 9}$

59. $\dfrac{x^4 + 2x^3 + 4x^2 + 3x + 2}{x^2 + x + 2}$

60. $\dfrac{2x^4 + 3x^3 + 3x^2 - 5x - 3}{2x^2 - x - 1}$

61. $\dfrac{x^3 + 3x + 5x^2 + 6 + x^4}{x^2 + 3}$

62. $\dfrac{x^5 + 3x + 2}{x^3 + 1 + 2x}$

In Exercises 63–64, use a calculator to help find each quotient.

63. $x - 2\overline{)9.8x^2 - 3.2x - 69.3}$

64. $2.5x - 3.7\overline{)-22.25x^2 - 38.9x - 16.65}$

APPLICATIONS

65. Find an expression for the length of the longer sides of the rectangle shown in Illustration 1.

66. Find an expression for the height of the triangle shown in Illustration 2.

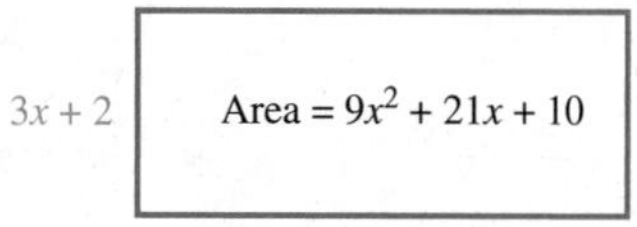

ILLUSTRATION 1

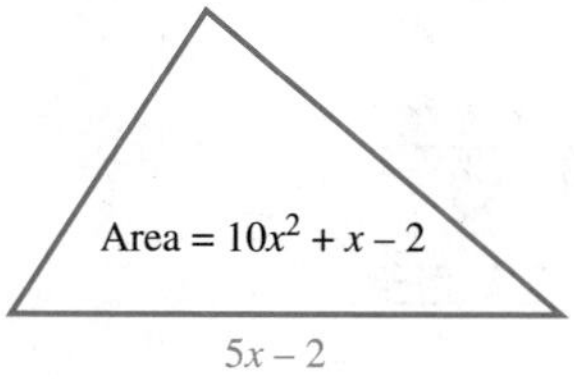

ILLUSTRATION 2

WRITING

67. Explain how to divide a monomial by a monomial.

68. Explain how to check the result of a division problem.

SOMETHING TO THINK ABOUT

69. Since 6 is a factor of 24, 6 divides 24 with no remainder. Decide whether $2x - 3$ is a factor of $10x^2 - x - 21$.

70. Is $x - 1$ a factor of $x^5 - 1$?

6.8 Synthetic Division (Optional)

■ SYNTHETIC DIVISION ■ THE REMAINDER THEOREM ■ THE FACTOR THEOREM

Getting Ready *Do each division and find P(2).*

1. $x - 2\overline{)x^2 - x - 1}$

2. $x - 2\overline{)x^2 + x + 3}$

SYNTHETIC DIVISION

There is a shortcut method, called **synthetic division,** that we can use to divide a polynomial by a binomial of the form $x - r$. To see how it works, we consider the division of $4x^3 - 5x^2 - 11x + 20$ by $x - 2$.

$$
\begin{array}{rl}
 & 4x^2 + 3x \;-\; 5 \\
x - 2\,) & \overline{4x^3 - 5x^2 - 11x + 20} \\
 & \underline{4x^3 - 8x^2} \\
 & \quad 3x^2 - 11x \\
 & \quad \underline{3x^2 - \;\;6x} \\
 & \qquad\quad -5x + 20 \\
 & \qquad\quad \underline{-5x + 10} \\
 & \qquad\qquad\quad 10 \quad \text{(remainder)}
\end{array}
\qquad
\begin{array}{rl}
 & 4 \;\; 3 - 5 \\
1 - 2\,) & \overline{4 - 5 - 11 \;\; 20} \\
 & \underline{4 - 8} \\
 & \quad 3 - 11 \\
 & \quad \underline{3 - \;6} \\
 & \qquad -5 \;\; 20 \\
 & \qquad \underline{-5 \;\; 10} \\
 & \qquad\qquad 10 \quad \text{(remainder)}
\end{array}
$$

On the left is the long division, and on the right is the same division with the variables and their exponents removed. The various powers of x can be remembered without actually writing them, because the exponents of the terms in the divisor, dividend, and quotient were written in descending order.

We can further shorten the version on the right. The numbers printed in color need not be written, because they are duplicates of the numbers above them. Thus, we can write the division in the following form:

$$
\begin{array}{r|rrrr}
 & 4 & 3 & -5 & \\
\hline
1 - 2\,) & 4 & -5 & -11 & 20 \\
 & & -8 & & \\
 & & 3 & & \\
 & & & -6 & \\
 & & & -5 & \\
 & & & & 10 \\
 & & & & 10
\end{array}
$$

We can shorten the process further by compressing the work vertically and eliminating the 1 (the coefficient of x in the divisor):

$$
\begin{array}{r|rrrr}
 & 4 & 3 & -5 & \\
\hline
-2\,) & 4 & -5 & -11 & 20 \\
 & & -8 & -6 & 10 \\
\hline
 & & 3 & -5 & 10
\end{array}
$$

If we write the 4 in the quotient on the bottom line, the bottom line gives the coefficients of the quotient and the remainder. If we eliminate the top line, the division appears as follows:

$$
\begin{array}{r|rrrr}
-2 & 4 & -5 & -11 & 20 \\
 & & -8 & -6 & 10 \\
\hline
 & 4 & 3 & -5 & 10
\end{array}
$$

The bottom line was obtained by subtracting the middle line from the top line. If we replace the -2 in the divisor by $+2$, the division process will reverse the signs of every entry in the middle line, and then the bottom line can be obtained by addition. This gives the final form of the synthetic division.

$$\begin{array}{r|rrrr} +2 & 4 & -5 & -11 & 20 \\ & & 8 & 6 & -10 \\ \hline & 4 & 3 & -5 & 10 \end{array}$$

The coefficients of the dividend.

The coefficients of the quotient and the remainder.

Thus,

$$\frac{4x^3 - 5x^2 - 11x + 20}{x - 2} = 4x^2 + 3x - 5 + \frac{10}{x - 2}$$

EXAMPLE 1 Divide $6x^2 + 5x - 2$ by $x - 5$.

Solution We write the coefficients in the dividend and the 5 in the divisor in the following form:

$$\begin{array}{r|rrr} 5 & 6 & 5 & -2 \\ & & & \\ \hline & & & \end{array}$$

Then we follow these steps:

$$\begin{array}{r|rrr} 5 & 6 & 5 & -2 \\ & & & \\ \hline & 6 & & \end{array}$$

Begin by bringing down the 6.

$$\begin{array}{r|rrr} 5 & 6 & 5 & -2 \\ & & 30 & \\ \hline & 6 & & \end{array}$$

Multiply 5 by 6 to get 30.

$$\begin{array}{r|rrr} 5 & 6 & 5 & -2 \\ & & 30 & \\ \hline & 6 & 35 & \end{array}$$

Add 5 and 30 to get 35.

$$\begin{array}{r|rrr} 5 & 6 & 5 & -2 \\ & & 30 & 175 \\ \hline & 6 & 35 & \end{array}$$

Multiply 35 by 5 to get 175.

$$\begin{array}{r|rrr} 5 & 6 & 5 & -2 \\ & & 30 & 175 \\ \hline & 6 & 35 & 173 \end{array}$$

Add -2 and 175 to get 173.

The numbers 6 and 35 represent the quotient $6x + 35$, and 173 is the remainder. Thus,

$$\frac{6x^2 + 5x - 2}{x - 5} = 6x + 35 + \frac{173}{x - 5}$$

■

EXAMPLE 2 Divide $5x^3 + x^2 - 3$ by $x - 2$.

Solution We begin by writing

$$\begin{array}{r|rrrr} 2 & 5 & 1 & 0 & -3 \end{array}$$

Write 0 for the coefficient of x, the missing term.

and complete the division as follows:

$$\begin{array}{r|rrrr} 2 & 5 & 1 & 0 & -3 \\ & & 10 & & \\ \hline & 5 & 11 & & \end{array} \qquad \begin{array}{r|rrrr} 2 & 5 & 1 & 0 & -3 \\ & & 10 & 22 & \\ \hline & 5 & 11 & 22 & \end{array} \qquad \begin{array}{r|rrrr} 2 & 5 & 1 & 0 & -3 \\ & & 10 & 22 & 44 \\ \hline & 5 & 11 & 22 & 41 \end{array}$$

Thus,

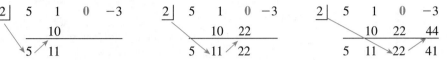

$$\frac{5x^3 + x^2 - 3}{x - 2} = 5x^2 + 11x + 22 + \frac{41}{x - 2}$$

■

EXAMPLE 3 Divide $5x^2 + 6x^3 + 2 - 4x$ by $x + 2$.

Solution First, we write the dividend with the exponents in descending order.

$$6x^3 + 5x^2 - 4x + 2$$

Then we write the divisor in $x - r$ form: $x - (-2)$. Using synthetic division, we begin by writing

$$\begin{array}{r|rrrr} -2 & 6 & 5 & -4 & 2 \\ & & & & \\ \hline & & & & \end{array}$$

and complete the division.

$$\begin{array}{r|rrrr} -2 & 6 & 5 & -4 & 2 \\ & & -12 & 14 & -20 \\ \hline & 6 & -7 & 10 & -18 \end{array}$$

Thus,

$$\frac{5x^2 + 6x^3 + 2 - 4x}{x + 2} = 6x^2 - 7x + 10 + \frac{-18}{x + 2}$$

■

Self Check Divide $2x - 4x^2 + 3x^3 - 3$ by $x - 1$.

Answer $3x^2 - x + 1 + \frac{-2}{x-1}$

THE REMAINDER THEOREM

Synthetic division is important in mathematics because of the **remainder theorem.**

Remainder Theorem
If a polynomial $P(x)$ is divided by $x - r$, the remainder is $P(r)$.

We illustrate the remainder theorem in the next example.

EXAMPLE 4 Let $P(x) = 2x^3 - 3x^2 - 2x + 1$. Find **a.** $P(3)$ and **b.** the remainder when $P(x)$ is divided by $x - 3$.

Solution **a.** $P(3) = 2(3)^3 - 3(3)^2 - 2(3) + 1$ Substitute 3 for x.

$$= 2(27) - 3(9) - 6 + 1$$
$$= 54 - 27 - 6 + 1$$
$$= \mathbf{22}$$

b. We use synthetic division to find the remainder when $P(x) = 2x^3 - 3x^2 - 2x + 1$ is divided by $x - 3$.

$$\begin{array}{r|rrrr} 3 & 2 & -3 & -2 & 1 \\ & & 6 & 9 & 21 \\ \hline & 2 & 3 & 7 & \mathbf{22} \end{array}$$

The remainder is 22.

The results of parts **a** and **b** show that when $P(x)$ is divided by $x - 3$, the remainder is $P(3)$. ■

It is often easier to find $P(r)$ by using synthetic division than by substituting r for x in $P(x)$. This is especially true if r is a decimal.

THE FACTOR THEOREM

Recall that if two quantities are multiplied, each is called a **factor** of the product. Thus, $x - 2$ is one factor of $6x - 12$, because $6(x - 2) = 6x - 12$. A theorem, called the **factor theorem,** tells us how to find one factor of a polynomial if the remainder of a certain division is 0.

Factor Theorem
If $P(x)$ is a polynomial in x, then

$P(r) = 0$ if and only if $x - r$ is a factor of $P(x)$

If $P(x)$ is a polynomial in x and if $P(r) = 0$, r is a **zero** of the polynomial.

EXAMPLE 5 Let $P(x) = 3x^3 - 5x^2 + 3x - 10$. Show that **a.** $P(2) = 0$ and **b.** $x - 2$ is a factor of $P(x)$.

Solution **a.** Use the remainder theorem to evaluate $P(2)$ by dividing $P(x) = 3x^3 - 5x^2 + 3x - 10$ by $x - 2$.

$$\begin{array}{r|rrrr} 2 & 3 & -5 & 3 & -10 \\ & & 6 & 2 & 10 \\ \hline & 3 & 1 & 5 & 0 \end{array}$$

The remainder in this division is 0. By the remainder theorem, the remainder is $P(2)$. Thus, $P(2) = 0$, and 2 is a zero of the polynomial.

b. Because the remainder is 0, the numbers 3, 1, and 5 in the synthetic division in part **a** represent the quotient $3x^2 + x + 5$. Thus,

$$\underbrace{(x-2)}_{\text{Divisor}} \cdot \underbrace{(3x^2 + x + 5)}_{\text{quotient}} + \underbrace{0}_{\text{remainder}} = \underbrace{3x^3 - 5x^2 + 3x - 10}_{\text{the dividend, } P(x)}$$

or

$$(x-2)(3x^2 + x + 5) = 3x^3 - 5x^2 + 3x - 10$$

Thus, $x - 2$ is a factor of $P(x)$. ■

The result in Example 5 is true, because the remainder, $P(2)$, is 0. If the remainder had not been 0, then $x - 2$ would not have been a factor of $P(x)$.

Approximating Zeros of Polynomials

GRAPHING CALCULATORS

We can use a graphing calculator to approximate the real zeros of a polynomial function $f(x)$. For example, to find the real zeros of $f(x) = 2x^3 - 6x^2 + 7x - 21$, we graph the function as in Figure 6-14.

It is clear from the figure that the function f has a zero at $x = 3$.

$$\begin{aligned} f(3) &= 2(3)^3 - 6(3)^2 + 7(3) - 21 && \text{Substitute 3 for } x. \\ &= 2(27) - 6(9) + 21 - 21 \\ &= 0 \end{aligned}$$

From the factor theorem, we know that $x - 3$ is a factor of the polynomial. To find the other factor, we can synthetically divide by 3.

$$\begin{array}{r|rrrr} 3 & 2 & -6 & 7 & -21 \\ & & 6 & 0 & 21 \\ \hline & 2 & 0 & 7 & 0 \end{array}$$

Thus, $f(x) = (x - 3)(2x^2 + 7)$. Since $2x^2 + 7$ cannot be factored over the real numbers, we can conclude that 3 is the only real zero of the polynomial function.

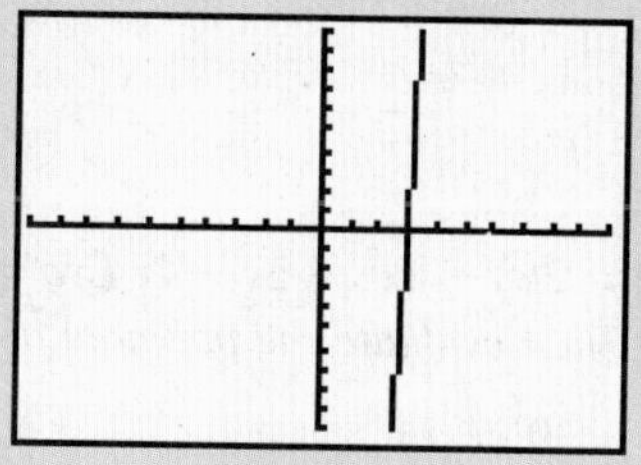

FIGURE 6-14

Orals *Find the remainder in each division.*

1. $(x^2 + 2x + 1) \div (x - 2)$

2. $(x^2 - 4) \div (x + 1)$

Tell whether $x - 2$ is a factor of each polynomial.

3. $x^3 - 2x^2 + x - 2$

4. $x^3 + 4x^2 - 1$

EXERCISE 6.8

REVIEW *Let $f(x) = 3x^2 + 2x - 1$ and find each value.*

1. $f(1)$

2. $f(-2)$

3. $f(2a)$

4. $f(-t)$

Remove parentheses and simplify.

5. $2(x^2 + 4x - 1) + 3(2x^2 - 2x + 2)$

6. $-2(3y^3 - 2y + 7) - 3(y^2 + 2y - 4) + 4(y^3 + 2y - 1)$

VOCABULARY AND CONCEPTS *Fill in each blank to make a true statement.*

7. If a polynomial $P(x)$ is divided by $x - r$, the remainder is ____.

8. If $P(x)$ is a polynomial in x, then $P(r) = 0$ if and only if ____ is a factor of $P(x)$.

PRACTICE *In Exercises 9–22, use synthetic division to do each division.*

9. $(x^2 + x - 2) \div (x - 1)$

10. $(x^2 + x - 6) \div (x - 2)$

11. $(x^2 - 7x + 12) \div (x - 4)$

12. $(x^2 - 6x + 5) \div (x - 5)$

13. $(x^2 + 8 + 6x) \div (x + 4)$

14. $(x^2 - 15 - 2x) \div (x + 3)$

15. $(x^2 - 5x + 14) \div (x + 2)$

16. $(x^2 + 13x + 42) \div (x + 6)$

17. $(3x^3 - 10x^2 + 5x - 6) \div (x - 3)$

18. $(2x^3 - 9x^2 + 10x - 3) \div (x - 3)$

19. $(2x^3 - 5x - 6) \div (x - 2)$

20. $(4x^3 + 5x^2 - 1) \div (x + 2)$

21. $(5x^2 + 6x^3 + 4) \div (x + 1)$

22. $(4 - 3x^2 + x) \div (x - 4)$

In Exercises 23–28, use a calculator and synthetic division to do each division.

23. $(7.2x^2 - 2.1x + 0.5) \div (x - 0.2)$

24. $(8.1x^2 + 3.2x - 5.7) \div (x - 0.4)$

25. $(2.7x^2 + x - 5.2) \div (x + 1.7)$

26. $(1.3x^2 - 0.5x - 2.3) \div (x + 2.5)$

27. $(9x^3 - 25) \div (x + 57)$

28. $(0.5x^3 + x) \div (x - 2.3)$

In Exercises 29–36, let $P(x) = 2x^3 - 4x^2 + 2x - 1$. Evaluate $P(x)$ by substituting the given value of x into the polynomial and simplifying. Then evaluate the polynomial by using the remainder theorem and synthetic division.

29. $P(1)$

30. $P(2)$

31. $P(-2)$

32. $P(-1)$

33. $P(3)$

34. $P(-4)$

35. $P(0)$

36. $P(4)$

In Exercises 37–44, let $Q(x) = x^4 - 3x^3 + 2x^2 + x - 3$. Evaluate $Q(x)$ by substituting the given value of x into the polynomial and simplifying. Then evaluate the polynomial by using the remainder theorem and synthetic division.

37. $Q(-1)$
38. $Q(1)$
39. $Q(2)$
40. $Q(-2)$
41. $Q(3)$
42. $Q(0)$
43. $Q(-3)$
44. $Q(-4)$

In Exercises 45–52, use the remainder theorem and synthetic division to find $P(r)$.

45. $P(x) = x^3 - 4x^2 + x - 2;\ r = 2$
46. $P(x) = x^3 - 3x^2 + x + 1;\ r = 1$
47. $P(x) = 2x^3 + x + 2;\ r = 3$
48. $P(x) = x^3 + x^2 + 1;\ r = -2$
49. $P(x) = x^4 - 2x^3 + x^2 - 3x + 2;\ r = -2$
50. $P(x) = x^5 + 3x^4 - x^2 + 1;\ r = -1$
51. $P(x) = 3x^5 + 1;\ r = -\frac{1}{2}$
52. $P(x) = 5x^7 - 7x^4 + x^2 + 1;\ r = 2$

In Exercises 53–56, use the factor theorem and tell whether the first expression is a factor of $P(x)$.

53. $x - 3;\ P(x) = x^3 - 3x^2 + 5x - 15$
54. $x + 1;\ P(x) = x^3 + 2x^2 - 2x - 3$ (*Hint:* Write $x + 1$ as $x - (-1)$.)
55. $x + 2;\ P(x) = 3x^2 - 7x + 4$ (*Hint:* Write $x + 2$ as $x - (-2)$.)
56. $x;\ P(x) = 7x^3 - 5x^2 - 8x$ (*Hint:* $x = x - 0$.)

In Exercises 57–58, use a calculator to work each problem.

57. Find 2^6 by using synthetic division to evaluate the polynomial $P(x) = x^6$ at $x = 2$. Then check the answer by evaluating 2^6 with a calculator.
58. Find $(-3)^5$ by using synthetic division to evaluate the polynomial $P(x) = x^5$ at $x = -3$. Then check the answer by evaluating $(-3)^5$ with a calculator.

WRITING

59. If you are given $P(x)$, explain how to use synthetic division to calculate $P(a)$.
60. Explain the factor theorem.

SOMETHING TO THINK ABOUT *Suppose that $P(x) = x^{100} - x^{99} + x^{98} - x^{97} + \cdots + x^2 - x + 1$.*

61. Find the remainder when $P(x)$ is divided by $x - 1$.
62. Find the remainder when $P(x)$ is divided by $x + 1$.

PROJECTS

PROJECT 1 Suppose that a motorist usually makes a certain 80-mile trip at an average speed of 50 mph.

a. How much time does the driver save by increasing her rate by 5 mph? By 10 mph? By 15 mph? Give all answers to the nearest minute.

(continued)

PROJECTS (continued)

b. Consider your work in part **a,** and then find a formula that will tell how much time is saved if the motorist travels x mph faster than 50 mph. That is, find an expression involving x that will give the time saved by traveling $(50 + x)$ mph instead of 50 mph.

c. Find a formula that will give the time saved on a trip of d miles by traveling at an average rate that is x mph faster than a usual speed of y mph. When you have this formula, simplify it into a nice, compact form.

d. Test your formula by doing part **a** over again using the formula. The answers you get should agree with those found earlier.

e. Use your formula to solve the following problem. Every holiday season, Kurt and Ellen travel to visit their relatives, a distance of 980 miles. Under normal circumstances, they can average 60 mph during the trip. However, improved roads will enable them to travel 4 mph faster this year. How much time (to the nearest minute) will they save on this year's trip? How much faster than normal (to the nearest tenth of a mph) would Kurt and Ellen have to travel to save two hours?

PROJECT 2

By careful cross-fertilization among a number of species of corn, Professor Greenthumb has succeeded in creating a miracle hybrid corn plant. Under ideal conditions, the plant will produce twice as much corn as an ordinary hybrid. However, the new hybrid is very sensitive to the amount of sun and water it receives. If conditions are not ideal, the corn yield diminishes.

To determine what yield the plants will have, daily measurements record the amount of sun the plants receive (the sun index, S) and the amount of water they receive (W, measured in millimeters). Then the daily yield diminution is calculated using the formula

$$\text{Daily yield diminution} = \frac{S^3 + W^3 - (S + W)^2}{(S + W)^3}$$

At the end of a 100-day growing season, the daily yield diminutions are added together to make the total diminution (D). The yield for the year is then determined using the formula

$$\text{Annual yield} = 1 - 0.01D$$

The resulting decimal is the percent of maximum yield that the corn plants will achieve. Remember that maximum yield for these plants is twice the yield of an ordinary corn plant.

As Greenthumb's research assistant, you have been asked to handle a few questions that have come up regarding her research.

a. First, show that if $S = W = 2$, the daily yield diminution is 0. These would be the optimal conditions for Greenthumb's plants.

b. Now suppose that for each day of the 100-day growing season, the sun index is 8, and 6 millimeters of rain fall on the plants. Find the daily yield diminution. To the nearest tenth of a percent, what percentage of the yield of normal plants will the special hybrid plants produce?

c. Show that when the daily yield diminution is 0.5 for each of the 100 days in the growing season, the annual yield will be 0.5 (exactly the yield of ordinary corn plants). Now suppose that through the use of an irrigation system, you arrange for the corn to receive 10 millimeters of rain each day. What would the sun index have to be each day to give an annual yield of 0.5? To answer this question, first simplify the daily yield diminution formula. Do this *before* you substitute any numbers into the formula.

d. Another assistant is also working with Greenthumb's formula, but he is getting different results. Something is wrong; for the situation given in part **b,** he finds that the daily yield diminution for one day is 34. He simplified Greenthumb's formula first, then substituted in the values for S and W. He shows you the following work. Find all of his mistakes and explain what he did wrong in making each of them.

$$\begin{aligned}
\text{Daily yield diminution} &= \frac{S^3 + W^3 - (S + W)^2}{(S + W)^3} \\
&= \frac{S^3 + W^3 - S^2 + W^2}{(S + W)^3} \\
&= \frac{(S^3 + W^3) - (S^2 - W^2)}{(S + W)^3} \\
&= \frac{(S + W) \cdot (S^2 + SW + W^2) - (S + W) \cdot (S - W)}{(S + W)^3} \\
&= \frac{(S^2 + SW + W^2) - S - W}{(S + W)^2} \\
&= \frac{S^2 + SW + W^2 - S - W}{S^2 + W^2} \\
&= SW - S - W
\end{aligned}$$

So for $S = 8$ and $W = 6$, he gets 34 as the daily yield diminution.

CHAPTER SUMMARY

CONCEPTS | REVIEW EXERCISES

SECTION 6.1 Rational Functions and Simplifying Rational Expressions

1. Graph the rational function $f(x) = \frac{3x+2}{x}$ $(x > 0)$. Find the equation of the horizontal and vertical asymptotes.

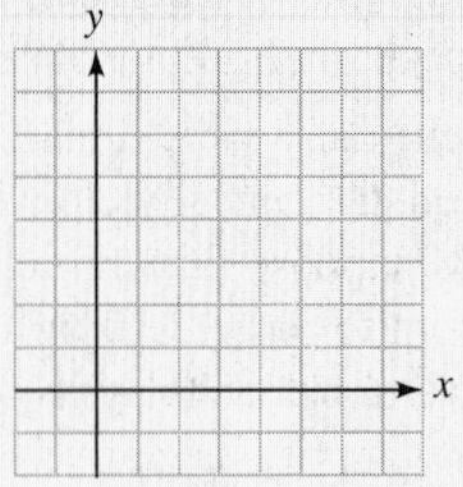

Division by 0 is undefined.

$$\frac{ak}{bk} = \frac{a}{b} \quad (b \neq 0 \text{ and } k \neq 0)$$

To simplify a fraction, factor the numerator and denominator and divide out all factors common to the numerator and denominator.

2. Simplify each fraction.

a. $\dfrac{248x^2y}{576xy^2}$ **b.** $\dfrac{212m^3n}{588m^2n^3}$

c. $\dfrac{x^2-49}{x^2+14x+49}$ **d.** $\dfrac{x^2+6x+36}{x^3-216}$

e. $\dfrac{x^2-2x+4}{2x^3+16}$ **f.** $\dfrac{x-y}{y-x}$

g. $\dfrac{2m-2n}{n-m}$ **h.** $\dfrac{ac-ad+bc-bd}{d^2-c^2}$

3. Find each probability.
a. Rolling an 11 on one roll of two dice
b. Three children, all boys

SECTION 6.2 Proportion and Variation

In a proportion, the product of the extremes is equal to the product of the means.

4. Solve each proportion.

a. $\dfrac{x+1}{8} = \dfrac{4x-2}{24}$ **b.** $\dfrac{1}{x+6} = \dfrac{x+10}{12}$

If two angles of one triangle have the same measure as two angles of a second triangle, the triangles are similar.

5. Find the height of a tree if it casts a 44-foot shadow when a 4-foot tree casts a $2\frac{1}{2}$-foot shadow.

Direct variation:

$y = kx$ (k is a constant)

6. Assume that x varies directly with y. If $x = 12$ when $y = 2$, find x when $y = 12$.

Inverse variation:

$y = \frac{k}{x}$ (k is a constant)

7. Assume that x varies inversely with y. If $x = 24$ when $y = 3$, find y when $x = 12$.

Joint variation:

$y = kxz$ (k is a constant)

8. Assume that x varies jointly with y and z. Find the constant of variation if $x = 24$ when $y = 3$ and $z = 4$.

Combined variation:

$y = \frac{kx}{y}$ (k is a constant)

9. Assume that x varies directly with t and inversely with y. Find the constant of variation if $x = 2$ when $t = 8$ and $y = 64$.

SECTION 6.3 *Multiplying and Dividing Rational Expressions*

$\frac{a}{b} \cdot \frac{c}{d} = \frac{ac}{bd}$ $(b, d \neq 0)$

$\frac{a}{b} \div \frac{c}{d} = \frac{a}{b} \cdot \frac{d}{c}$ $(b, d, c \neq 0)$

10. Do the operations and simplify.

a. $\frac{x^2 + 4x + 4}{x^2 - x - 6} \cdot \frac{x^2 - 9}{x^2 + 5x + 6}$

b. $\frac{x^3 - 64}{x^2 + 4x + 16} \div \frac{x^2 - 16}{x + 4}$

c. $\frac{x^2 + 3x + 2}{x^2 - x - 6} \cdot \frac{3x^2 - 3x}{x^2 - 3x - 4} \div \frac{x^2 + 3x + 2}{x^2 - 2x - 8}$

d. $\frac{x^2 - x - 6}{x^2 - 3x - 10} \div \frac{x^2 - x}{x^2 - 5x} \cdot \frac{x^2 - 4x + 3}{x^2 - 6x + 9}$

SECTION 6.4 *Adding and Subtracting Rational Expressions*

$\frac{a}{b} + \frac{c}{b} = \frac{a + c}{b}$ $(b \neq 0)$

$\frac{a}{b} - \frac{c}{b} = \frac{a - c}{b}$ $(b \neq 0)$

To find the LCD of two fractions, factor each denominator and use each factor the greatest number of times that it appears in any one denominator. The product of these factors is the LCD.

11. Do the operations and simplify.

a. $\frac{5y}{x - y} - \frac{3}{x - y}$

b. $\frac{3x - 1}{x^2 + 2} + \frac{3(x - 2)}{x^2 + 2}$

c. $\frac{3}{x + 2} + \frac{2}{x + 3}$

d. $\frac{4x}{x - 4} - \frac{3}{x + 3}$

e. $\frac{2x}{x + 1} + \frac{3x}{x + 2} + \frac{4x}{x^2 + 3x + 2}$

f. $\frac{5x}{x - 3} + \frac{5}{x^2 - 5x + 6} + \frac{x + 3}{x - 2}$

g. $\frac{3(x + 2)}{x^2 - 1} - \frac{2}{x + 1} + \frac{4(x + 3)}{x^2 - 2x + 1}$

h. $\frac{-2(3 + x)}{x^2 + 6x + 9} + \frac{3(x + 2)}{x^2 - 6x + 9} - \frac{1}{x^2 - 9}$

SECTION 6.5 Complex Fractions

A fraction that has a fraction in its numerator or its denominator is called a **complex fraction.**

12. Simplify each complex fraction.

a. $\dfrac{\frac{3}{x} - \frac{2}{y}}{xy}$

b. $\dfrac{\frac{1}{x} + \frac{2}{y}}{\frac{2}{x} - \frac{1}{y}}$

c. $\dfrac{2x + 3 + \frac{1}{x}}{x + 2 + \frac{1}{x}}$

d. $\dfrac{6x + 13 + \frac{6}{x}}{6x + 5 - \frac{6}{x}}$

e. $\dfrac{1 - \frac{1}{x} - \frac{2}{x^2}}{1 + \frac{4}{x} + \frac{3}{x^2}}$

f. $\dfrac{x^{-1} + 1}{x + 1}$

g. $\dfrac{x^{-1} - y^{-1}}{x^{-1} + y^{-1}}$

h. $\dfrac{(x - y)^{-2}}{x^{-2} - y^{-2}}$

SECTION 6.6 Equations Containing Rational Expressions

Multiplying both sides of an equation by a quantity that contains a variable can lead to false solutions. All possible solutions of a rational equation must be checked.

13. Solve each equation, if possible.

a. $\dfrac{4}{x} - \dfrac{1}{10} = \dfrac{7}{2x}$

b. $\dfrac{2}{x + 5} - \dfrac{1}{6} = \dfrac{1}{x + 4}$

c. $\dfrac{2(x - 5)}{x - 2} = \dfrac{6x + 12}{4 - x^2}$

d. $\dfrac{7}{x + 9} - \dfrac{x + 2}{2} = \dfrac{x + 4}{x + 9}$

14. Solve each formula for the indicated variable.

a. $\dfrac{x^2}{a^2} - \dfrac{y^2}{b^2} = 1$ for y^2

b. $H = \dfrac{2ab}{a + b}$ for b

15. Trip length Traffic reduced Jim's usual speed by 10 mph, which lengthened his 200-mile trip by 1 hour. Find his usual average speed.

16. Flying speed On a 600-mile trip, a pilot can save 30 minutes by increasing her usual speed by 40 mph. Find her usual speed.

17. Draining a tank If one outlet pipe can drain a tank in 24 hours and another pipe can drain the tank in 36 hours, how long will it take for both pipes to drain the tank?

18. Siding a house Two men have estimated that they can side a house in 8 days. If one of them, who could have sided the house alone in 14 days, gets sick, how long will it take the other man to side the house alone?

SECTION 6.7 Dividing Polynomials

To find the quotient of two monomials, express the quotient as a fraction and use the rules of exponents to simplify.

19. Do each division.

a. $\dfrac{-5x^6y^3}{10x^3y^6}$ **b.** $\dfrac{30x^3y^2 - 15x^2y - 10xy^2}{-10xy}$

c. $(3x^2 + 13xy - 10y^2) \div (3x - 2y)$

d. $(2x^3 + 7x^2 + 3 + 4x) \div (2x + 3)$

SECTION 6.8 Synthetic Division (Optional)

Remainder theorem:
If a polynomial $P(x)$ is divided by $x - r$, then the remainder is $P(r)$.

Factor theorem:
If $P(x)$ is divided by $x - r$, then $P(r) = 0$, if and only if $x - r$ is a factor of $P(x)$.

20. Use the factor theorem to decide whether the first expression is a factor of $P(x)$.

a. $x - 5$; $P(x) = x^3 - 3x^2 - 8x - 10$

b. $x + 5$; $P(x) = x^3 + 4x^2 - 5x + 5$ (*Hint:* Write $x + 5$ as $x - (-5)$.)

Chapter Test

In Problems 1–4, simplify each fraction.

1. $\dfrac{-12x^2y^3z^2}{18x^3y^4z^2}$ **2.** $\dfrac{2x + 4}{x^2 - 4}$ **3.** $\dfrac{3y - 6z}{2z - y}$ **4.** $\dfrac{2x^2 + 7x + 3}{4x + 12}$

5. Find the probability of tossing a coin three times and getting no heads.

6. Find the height of a tree that casts a shadow of 12 feet when a vertical yardstick casts a shadow of 2 feet.

7. Solve the proportion $\dfrac{3}{x - 2} = \dfrac{x + 3}{2x}$.

8. V varies inversely with t. If $V = 55$ when $t = 20$, find t when $V = 75$.

In Problems 9–18, do the operations and simplify, if necessary. Write all answers without negative exponents.

9. $\dfrac{x^2y^{-2}}{x^3z^2} \cdot \dfrac{x^2z^4}{y^2z}$

10. $\dfrac{(x+1)(x+2)}{10} \cdot \dfrac{5}{x+2}$

11. $\dfrac{u^2+5u+6}{u^2-4} \cdot \dfrac{u^2-5u+6}{u^2-9}$

12. $\dfrac{x^3+y^3}{4} \div \dfrac{x^2-xy+y^2}{2x+2y}$

13. $\dfrac{xu+2u+3x+6}{u^2-9} \cdot \dfrac{2u-6}{x^2+3x+2}$

14. $\dfrac{a^2+7a+12}{a+3} \div \dfrac{16-a^2}{a-4}$

15. $\dfrac{3t}{t+3} + \dfrac{9}{t+3}$

16. $\dfrac{3w}{w-5} + \dfrac{w+10}{5-w}$

17. $\dfrac{2}{r} + \dfrac{r}{s}$

18. $\dfrac{x+2}{x+1} - \dfrac{x+1}{x+2}$

In Problems 19–20, simplify each complex fraction.

19. $\dfrac{\dfrac{2u^2w^3}{v^2}}{\dfrac{4uw^4}{uv}}$

20. $\dfrac{\dfrac{x}{y}+\dfrac{1}{2}}{\dfrac{x}{2}-\dfrac{1}{y}}$

In Problems 21–22, solve each equation.

21. $\dfrac{2}{x-1} + \dfrac{5}{x+2} = \dfrac{11}{x+2}$

22. $\dfrac{u-2}{u-3} + 3 = u + \dfrac{u-4}{3-u}$

In Problems 23–24, solve each formula for the indicated variable.

23. $\dfrac{x^2}{a^2} + \dfrac{y^2}{b^2} = 1$ for a^2

24. $\dfrac{1}{r} = \dfrac{1}{r_1} + \dfrac{1}{r_2}$ for r_2

25. Sailing time A boat sails a distance of 440 nautical miles. If the boat had averaged 11 nautical miles more each day, the trip would have required 2 fewer days. How long did the trip take?

26. Investing A student can earn $300 interest annually by investing in a certificate of deposit at a certain interest rate. If she were to receive an annual interest rate that is 4% higher, she could receive the same annual interest by investing $2,000 less. How much would she invest at each rate?

27. Divide: $\dfrac{18x^2y^3 - 12x^3y^2 + 9xy}{-3xy^4}$.

28. Divide: $(6x^3 + 5x^2 - 2) \div (2x - 1)$.

29. Find the remainder: $\dfrac{x^3 - 4x^2 + 5x + 3}{x + 1}$.

30. Optional Use synthetic division to find the remainder when $4x^3 + 3x^2 + 2x - 1$ is divided by $x - 2$.

$$2\,\rfloor\;\; 4 \quad 3 \quad 2 \quad -1$$

$$\overline{\qquad\qquad\qquad\qquad}$$

Cumulative Review Exercises

In Exercises 1–4, simplify each expression.

1. $a^3b^2a^5b^2$

2. $\dfrac{a^3b^6}{a^7b^2}$

3. $\left(\dfrac{2a^2}{3b^4}\right)^{-4}$

4. $\left(\dfrac{x^{-2}y^3}{x^2x^3y^4}\right)^{-3}$

In Exercises 5–6, write each number in standard notation.

5. 4.25×10^4

6. 7.12×10^{-4}

In Exercises 7–8, solve each equation.

7. $\dfrac{a+2}{5} - \dfrac{8}{5} = 4a - \dfrac{a+9}{2}$

8. $\dfrac{3x-4}{6} - \dfrac{x-2}{2} = \dfrac{-2x-3}{3}$

In Exercises 9–12, find the slope of the line with the given properties.

9. Passing through $P(-2, 5)$ and $Q(4, 10)$

10. Has an equation of $3x + 4y = 13$

11. Parallel to a line with equation of $y = 3x + 2$

12. Perpendicular to a line with equation of $y = 3x + 2$

In Exercises 13–16, let $f(x) = x^2 - 2x$ and find each value.

13. $f(0)$

14. $f(-2)$

15. $f\left(\dfrac{2}{5}\right)$

16. $f(t-1)$

17. Express as a formula: *y varies directly with the product of x and y, but inversely with r.*

18. Does the graph below represent a function?

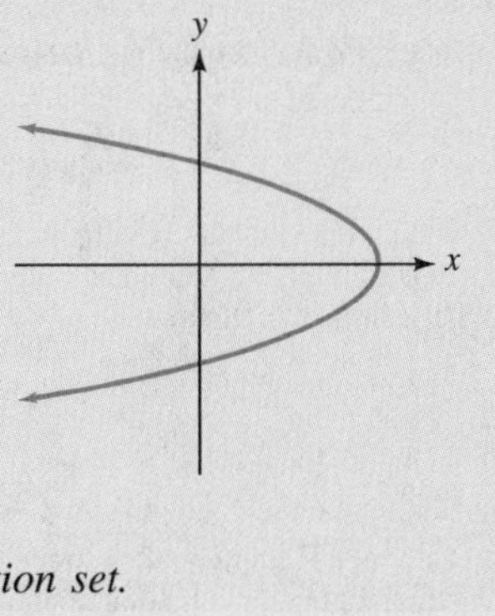

In Exercises 19–20, solve each inequality and graph the solution set.

19. $x - 2 \le 3x + 1 \le 5x - 4$

20. $\left|\dfrac{3a}{5} - 2\right| + 1 \ge \dfrac{6}{5}$

21. Is $3 + x + x^2$ a monomial, a binomial, or a trinomial?

22. Find the degree of $3 + x^2y + 17x^3y^4$.

23. If $f(x) = -3x^3 + x - 4$, find $f(-2)$.

24. Graph $y = f(x) = 2x^2 - 3$.

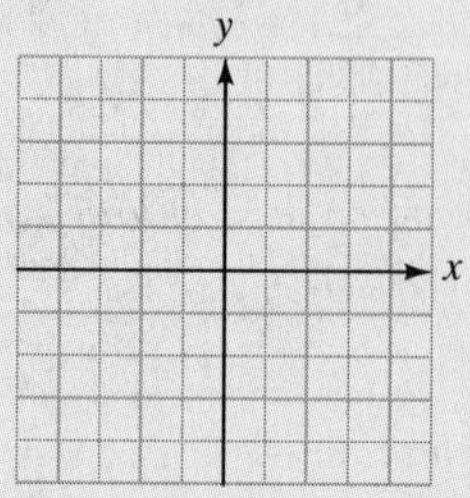

In Exercises 25–28, do the operations and simplify.

25. $(3x^2 - 2x + 7) + (-2x^2 + 2x + 5) + (3x^2 - 4x + 2)$

26. $(-5x^2 + 3x + 4) - (-2x^2 + 3x + 7)$

27. $(3x + 4)(2x - 5)$

28. $(2x^n - 1)(x^n + 2)$

In Exercises 29–40, factor each expression.

29. $3r^2s^3 - 6rs^4$

30. $5(x - y) - a(x - y)$

31. $xu + yv + xv + yu$

32. $81x^4 - 16y^4$

33. $8x^3 - 27y^6$

34. $6x^2 + 5x - 6$

35. $9x^2 - 30x + 25$

36. $15x^2 - x - 6$

37. $27a^3 + 8b^3$

38. $6x^2 + x - 35$

39. $x^2 + 10x + 25 - y^4$

40. $y^2 - x^2 + 4x - 4$

In Exercises 41–42, solve each equation by factoring.

41. $x^3 - 4x = 0$

42. $6x^2 + 7 = -23x$

In Exercises 43–46, simplify each expression.

43. $\dfrac{2x^2y + xy - 6y}{3x^2y + 5xy - 2y}$

44. $\dfrac{x^2 - 4}{x^2 + 9x + 20} \div \dfrac{x^2 + 5x + 6}{x^2 + 4x - 5} \cdot \dfrac{x^2 + 3x - 4}{(x - 1)^2}$

45. $\dfrac{2}{x + y} + \dfrac{3}{x - y} - \dfrac{x - 3y}{x^2 - y^2}$

46. $\dfrac{\dfrac{a}{b} + b}{a - \dfrac{b}{a}}$

In Exercises 47–48, solve each equation.

47. $\dfrac{5x-3}{x+2}=\dfrac{5x+3}{x-2}$

48. $\dfrac{3}{x-2}+\dfrac{x^2}{(x+3)(x-2)}=\dfrac{x+4}{x+3}$

In Exercises 49–50, do the operations.

49. $(x^2+9x+20)\div(x+5)$

50. $(2x^2+4x-x^3+3)\div(x-1)$

7 Rational Exponents and Radicals

MATHEMATICS IN THE WORKPLACE

Photographer

Photographers use cameras and film to portray people, objects, places, and events. Some specialize in scientific, medical, or engineering photography and provide illustrations and documentation for publications and research reports. Others specialize in portrait, fashion, or industrial photography and provide pictures for catalogs and other publications. Photojournalists capture newsworthy events, people, and places, and their work is seen in newspapers and magazines, as well as on television.

SAMPLE APPLICATION ■ Many cameras have an adjustable lens opening called an *aperture.* The aperture controls the amount of light passing through the lens. Various lenses—wide-angle, close-up, and telephoto—are distinguished by their *focal length.* The *f-number* of a lens is its focal length divided by the diameter of its circular aperture:

$$f\text{-number} = \frac{f}{d} \qquad \text{f is the focal length, and d is the diameter of the aperture.}$$

A lens with a focal length of 12 centimeters and an aperture with a diameter of 6 centimeters has an f-number of $\frac{12}{6}$ and is called an $f/2$ lens. If the area of the aperture is reduced to admit half as much light, the f-number of the lens will change. Find the new f-number.
See Section 7.6, Example 5.

7.1 Radical Expressions

■ SQUARE ROOTS ■ SQUARE ROOTS OF EXPRESSIONS WITH VARIABLES ■ THE SQUARE ROOT FUNCTION ■ CUBE ROOTS ■ *n*TH ROOTS ■ STANDARD DEVIATION

Getting Ready

Find each power.

1. 0^2 **2.** 4^2 **3.** $(-4)^2$ **4.** -4^2

5. $\left(\frac{2}{5}\right)^3$ **6.** $\left(-\frac{3}{4}\right)^4$ **7.** $(7xy)^2$ **8.** $(7xy)^3$

In this chapter, we will reverse the squaring process and learn how to find **square roots** of numbers. We will also discuss how to find other roots of numbers.

■ SQUARE ROOTS

When solving problems, we must often find what number must be squared to obtain a second number a. If such a number can be found, it is called a **square root of *a*.** For example,

- 0 is a square root of 0, because $0^2 = 0$.
- 4 is a square root of 16, because $4^2 = 16$.

- -4 is a square root of 16, because $(-4)^2 = 16$.
- $7xy$ is a square root of $49x^2y^2$, because $(7xy)^2 = 49x^2y^2$.
- $-7xy$ is a square root of $49x^2y^2$, because $(-7xy)^2 = 49x^2y^2$.

The preceding examples illustrate the following definition.

Square Root of *a*
The number b is a **square root of *a*** if $b^2 = a$.

All positive numbers have two real number square roots: one that is positive and one that is negative.

EXAMPLE 1 Find the two square roots of 121.

Solution The two square roots of 121 are 11 and -11, because

$$11^2 = 121 \quad \text{and} \quad (-11)^2 = 121$$

■

Self Check Find the square roots of 144.

Answers 12, -12

In the following definition, the symbol $\sqrt{}$ is called a **radical sign,** and the number x within the radical sign is called the **radicand.**

Principal Square Root
If $x > 0$, the **principal square root of *x*** is the positive square root of x, denoted as $\sqrt{x}$.

The principal square root of 0 is 0: $\sqrt{0} = 0$.

By definition, the principal square root of a positive number is always positive. Although 5 and -5 are both square roots of 25, only 5 is the principal square root. The radical $\sqrt{25}$ represents 5. The radical $-\sqrt{25}$ represents -5.

EXAMPLE 2 Simplify each radical.

a. $\sqrt{1} = 1$ **b.** $\sqrt{81} = 9$

c. $-\sqrt{81} = -9$ **d.** $-\sqrt{225} = -15$

e. $\sqrt{\frac{1}{4}} = \frac{1}{2}$ **f.** $-\sqrt{\frac{16}{121}} = -\frac{4}{11}$

g. $\sqrt{0.04} = 0.2$ **h.** $-\sqrt{0.0009} = -0.03$

■

Self Check Simplify **a.** $-\sqrt{49}$ and **b.** $\sqrt{\frac{25}{49}}$.

Answers **a.** -7, **b.** $\frac{5}{7}$

Numbers such as 1, 4, 9, 16, 49, and 1,600 are called **integer squares,** because each one is the square of an integer. The square root of every integer square is a rational number.

$$\sqrt{1} = 1, \quad \sqrt{4} = 2, \quad \sqrt{9} = 3, \quad \sqrt{16} = 4, \quad \sqrt{49} = 7, \quad \sqrt{1{,}600} = 40$$

The square roots of many positive integers are not rational numbers. For example, $\sqrt{11}$ is an **irrational number.** To find an approximate value of $\sqrt{11}$, we enter 11 into a scientific calculator and press the $\sqrt{\ }$ key.

$\sqrt{11} \approx 3.31662479$ Read $\approx$ as "is approximately equal to."

Square roots of negative numbers are not real numbers. For example, $\sqrt{-9}$ is not a real number, because no real number squared equals -9. Square roots of negative numbers come from a set called the **imaginary numbers,** which we will discuss in the next chapter.

SQUARE ROOTS OF EXPRESSIONS WITH VARIABLES

If $x \neq 0$, the positive number x^2 has x and $-x$ for its two square roots. To denote the positive square root of $\sqrt{x^2}$, we must know whether x is positive or negative.

If $x > 0$, we can write

$\sqrt{x^2} = x$ $\sqrt{x^2}$ represents the positive square root of x^2, which is x.

If x is negative, then $-x > 0$, and we can write

$\sqrt{x^2} = -x$ $\sqrt{x^2}$ represents the positive square root of x^2, which is $-x$.

If we don't know whether x is positive or negative, we must use absolute value symbols to guarantee that $\sqrt{x^2}$ is positive.

Definition of $\sqrt{x^2}$
If x can be any real number, then

$$\sqrt{x^2} = |x|$$

EXAMPLE 3 Simplify each expression.

If x can be any real number, we have

a. $\sqrt{16x^2} = \sqrt{(4x)^2}$ — Write $16x^2$ as $(4x)^2$.

$= |4x|$ — Because $(|4x|)^2 = 16x^2$. Since x could be negative, absolute value symbols are needed.

$= 4|x|$ — Since 4 is a positive constant in the product $4x$, we can write it outside the absolute value symbols.

b. $\sqrt{x^2 + 2x + 1}$

$= \sqrt{(x + 1)^2}$ — Factor $x^2 + 2x + 1$.

$= |x + 1|$ — Because $(x + 1)^2 = x^2 + 2x + 1$. Since $x + 1$ can be negative (for example, when $x = -5$), absolute value symbols are needed.

c. $\sqrt{x^4} = x^2$ — Because $(x^2)^2 = x^4$. Since $x^2 \geq 0$, no absolute value symbols are needed. ■

Self Check Simplify **a.** $\sqrt{25a^2}$ and **b.** $\sqrt{16a^4}$.

Answers **a.** $5|a|$, **b.** $4a^2$

■ THE SQUARE ROOT FUNCTION

Since there is one principal square root for every nonnegative real number x, the equation $y = f(x) = \sqrt{x}$ determines a function, called the **square root function.**

EXAMPLE 4 Graph $y = f(x) = \sqrt{x}$ and find its domain and range.

Solution We can make a table of values and plot points to get the graph shown in Figure 7-1(a), or we can use a graphing calculator with window settings of $[-1, 9]$ for x and $[-2, 5]$ for y to get the graph shown in Figure 7-1(b). Since the equation defines a function, its graph passes the vertical line test.

$y = f(x) = \sqrt{x}$

x	$f(x)$	$(x, f(x))$
0	0	(0, 0)
1	1	(1, 1)
4	2	(4, 2)
9	3	(9, 3)

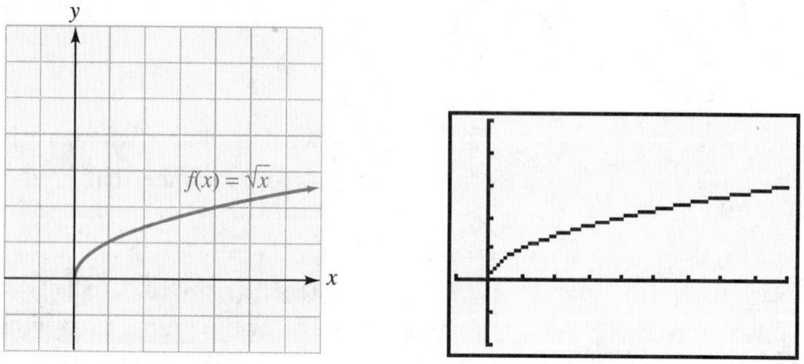

(a) (b)

FIGURE 7-1

From either graph, we can see that the domain and the range are the set of nonnegative real numbers, which is the interval $[0,\infty)$. ■

Self Check Graph $y = f(x) = \sqrt{x} + 2$ and compare the graph to the graph of $y = f(x) = \sqrt{x}$.

Answer It is 2 units higher.

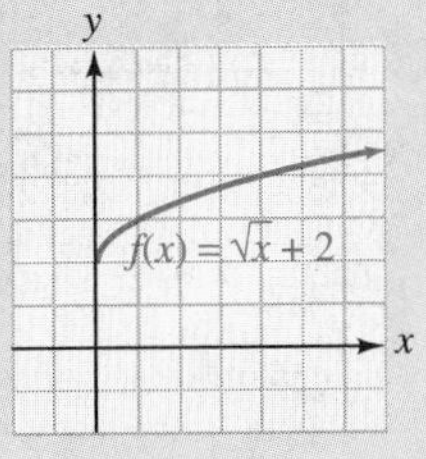

The graphs of many functions are translations or reflections of the square root function. For example, if $k > 0$

- The graph of $f(x) = \sqrt{x} + k$ is the graph of $f(x) = \sqrt{x}$ translated k units up.
- The graph of $f(x) = \sqrt{x} - k$ is the graph of $f(x) = \sqrt{x}$ translated k units down.
- The graph of $f(x) = \sqrt{x + k}$ is the graph of $f(x) = \sqrt{x}$ translated k units to the left.
- The graph of $f(x) = \sqrt{x - k}$ is the graph of $f(x) = \sqrt{x}$ translated k units to the right.
- The graph of $f(x) = -\sqrt{x}$ is the graph of $f(x) = \sqrt{x}$ reflected about the x-axis.

EXAMPLE 5 Graph $y = f(x) = -\sqrt{x + 4} - 2$ and find its domain and range.

Solution This graph will be the reflection of $f(x) = \sqrt{x}$ about the x-axis, translated 4 units to the left and 2 units down. See Figure 7-2(a). We can confirm this graph by using a graphing calculator with window settings of $[-5, 6]$ for x and $[-6, 2]$ for y to get the graph shown in Figure 7-2(b).

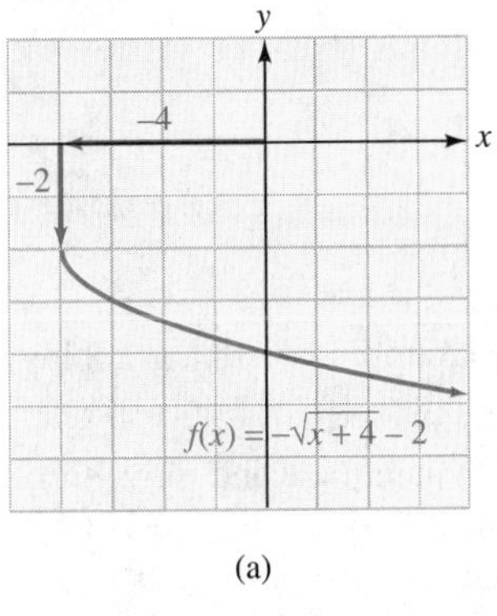

(a)

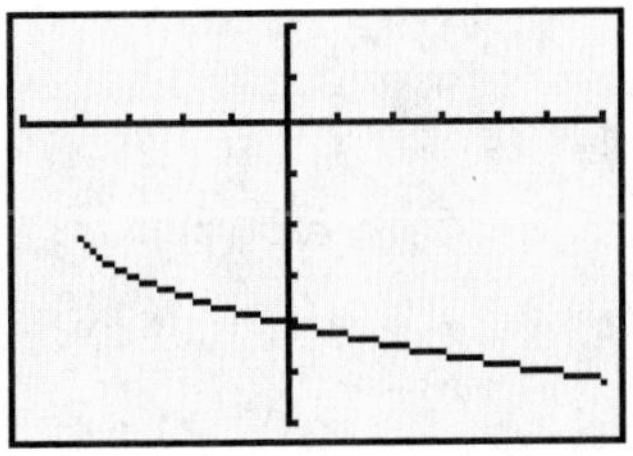

(b)

FIGURE 7-2

From either graph, we can see that the domain is the interval $[-4, \infty)$ and that the range is the interval $(-\infty, -2]$. ■

Self Check Graph $f(x) = \sqrt{x-2} - 4$.

Answer

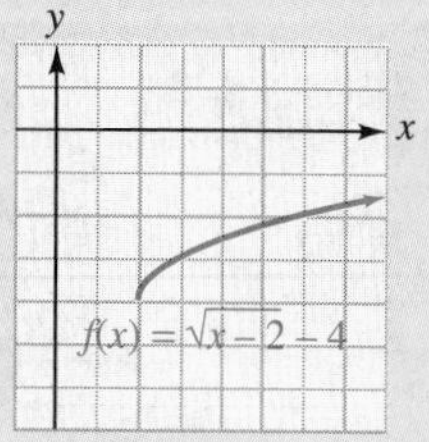

EXAMPLE 6 **Period of a pendulum** The **period of a pendulum** is the time required for the pendulum to swing back and forth to complete one cycle. (See Figure 7-3.) The period t (in seconds) is a function of the pendulum's length l, which is defined by the formula

$$t = f(l) = 2\pi\sqrt{\frac{l}{32}}$$

Find the period of a pendulum that is 5 feet long.

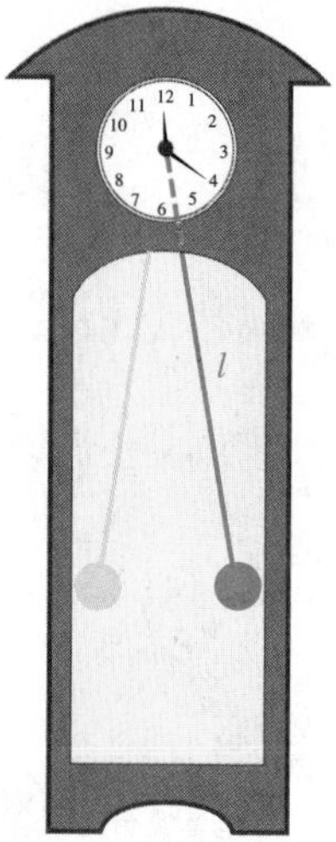

FIGURE 7-3

Solution We substitute 5 for l in the formula and simplify.

$$t = 2\pi\sqrt{\frac{l}{32}}$$

$$t = 2\pi\sqrt{\frac{5}{32}}$$

$$\approx 2.483647066$$

To the nearest tenth, the period is 2.5 seconds. ■

Self Check To the nearest hundredth, find the period of a pendulum that is 3 feet long.

Answer 1.92 sec

GRAPHING CALCULATORS

To solve Example 6 with a graphing calculator with window settings of $[-2, 10]$ for x and $[-2, 10]$ for y, we graph the function $f(x) = 2\pi\sqrt{\frac{x}{32}}$, as in Figure 7-4(a). We then trace and move the cursor toward an x value of 5 until we see the coordinates shown in Figure 7-4(b). The period is given by the y value shown in the screen. By zooming in, we can get better results.

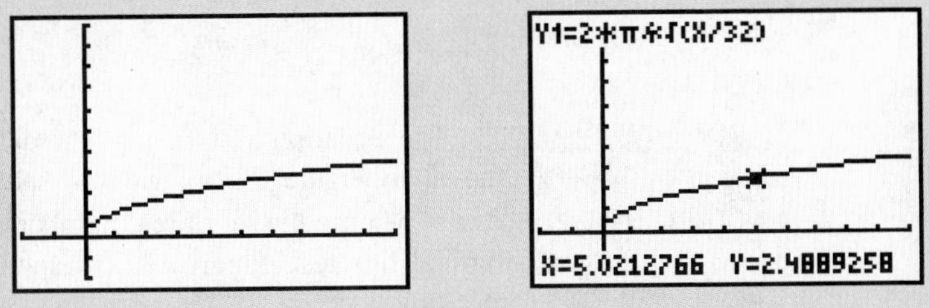

(a) (b)

FIGURE 7-4

CUBE ROOTS

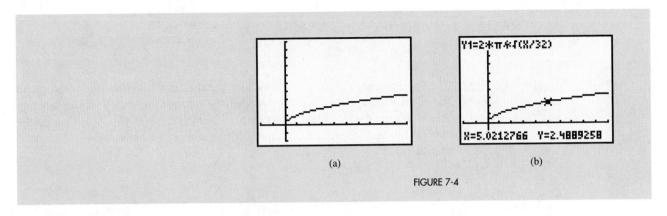

The **cube root of x** is any number whose cube is x. For example,

4 is a cube root of 64, because $4^3 = 64$.

$3x^2y$ is a cube root of $27x^6y^3$, because $(3x^2y)^3 = 27x^6y^3$.

$-2y$ is a cube root of $-8y^3$, because $(-2y)^3 = -8y^3$.

Cube Roots

The **cube root of x** is denoted as $\sqrt[3]{x}$ and is defined by

$$\sqrt[3]{x} = y \quad \text{if} \quad y^3 = x$$

We note that 64 has two real-number square roots, 8 and -8. However, 64 has only one real-number cube root, 4, because 4 is the only real number whose cube is 64. Since every real number has exactly one real cube root, it is unnecessary to use absolute value symbols when simplifying cube roots.

Definition of $\sqrt[3]{x^3}$

If x is any real number, then

$$\sqrt[3]{x^3} = x$$

b,d

EXAMPLE 7 Simplify each radical.

a. $\sqrt[3]{125} = 5$ — Because $5^3 = 5 \cdot 5 \cdot 5 = 125$.

b. $\sqrt[3]{\frac{1}{8}} = \frac{1}{2}$ — Because $\left(\frac{1}{2}\right)^3 = \frac{1}{2} \cdot \frac{1}{2} \cdot \frac{1}{2} = \frac{1}{8}$.

c. $\sqrt[3]{-27x^3} = -3x$ — Because $(-3x)^3 = (-3x)(-3x)(-3x) = -27x^3$.

d. $\sqrt[3]{-\frac{8a^3}{27b^3}} = -\frac{2a}{3b}$ — Because $\left(-\frac{2a}{3b}\right)^3 = \left(-\frac{2a}{3b}\right)\left(-\frac{2a}{3b}\right)\left(-\frac{2a}{3b}\right) = -\frac{8a^3}{27b^3}$.

e. $\sqrt[3]{0.216x^3y^6} = 0.6xy^2$ — Because $(0.6xy^2)^3 = (0.6xy^2)(0.6xy^2)(0.6xy^2) = 0.216x^3y^6$. ■

Self Check Simplify **a.** $\sqrt[3]{1{,}000}$, **b.** $\sqrt[3]{\frac{1}{27}}$, and **c.** $\sqrt[3]{125a^3}$.

Answers **a.** 10, **b.** $\frac{1}{3}$, **c.** $5a$

The equation $y = f(x) = \sqrt[3]{x}$ defines a **cube root function.** From the graph shown in Figure 7-5(a), we can see that the domain and range of the function $f(x) = \sqrt[3]{x}$ are the set of real numbers. Note that the graph of $f(x) = \sqrt[3]{x}$ passes the vertical line test. Figures 7-5(b) and 7-5(c) show several translations of the cube root function.

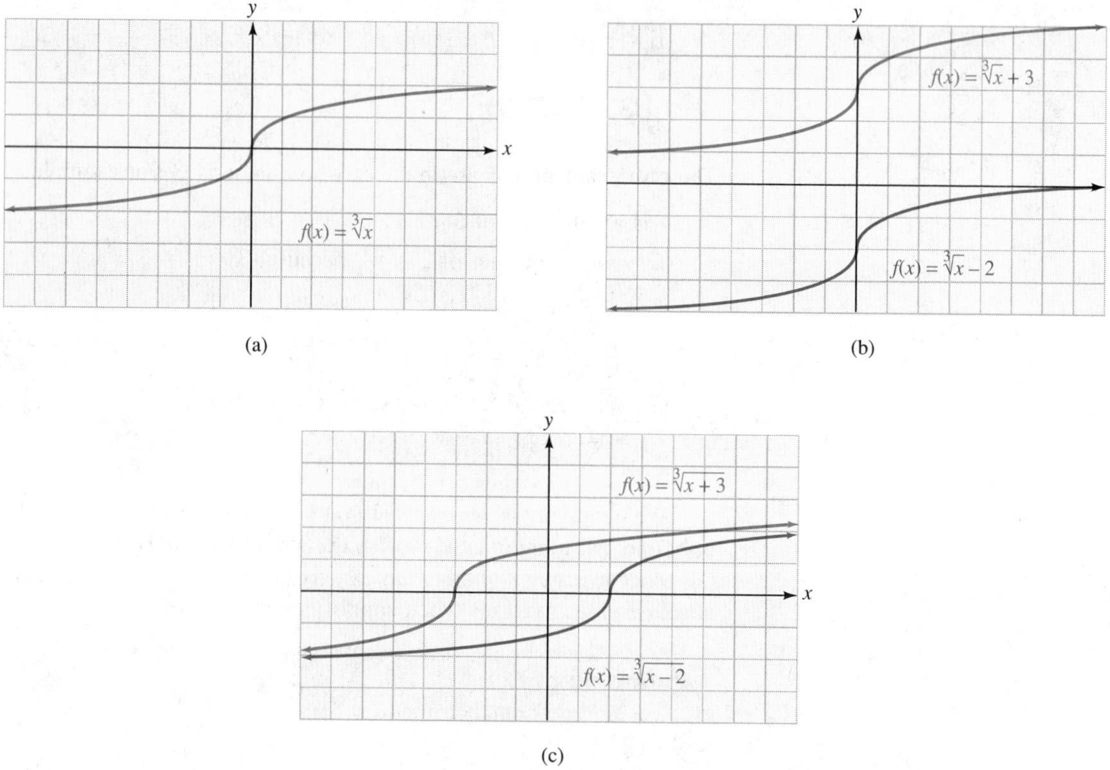

FIGURE 7-5

nTH ROOTS

Just as there are square roots and cube roots, there are fourth roots, fifth roots, sixth roots, and so on.

When n is an odd natural number, the radical $\sqrt[n]{x}$ $(n > 1)$ represents an **odd root.** Since every real number has just one real nth root when n is odd, we don't need to worry about absolute value symbols when finding odd roots. For example,

$$\sqrt[5]{243} = \sqrt[5]{3^5} = 3 \qquad \text{because} \qquad 3^5 = 243$$

$$\sqrt[7]{-128x^7} = \sqrt[7]{(-2x)^7} = -2x \qquad \text{because} \qquad (-2x)^7 = -128x^7$$

When n is an even natural number, the radical $\sqrt[n]{x}$ $(n > 1)$ represents an **even root.** In this case, there will be one positive and one negative real nth root. For example, the two real sixth roots of 729 are 3 and -3, because $3^6 = 729$ and $(-3)^6 = 729$. When finding even roots, we often use absolute value symbols to guarantee that the principal nth root is positive.

$\sqrt[4]{(-3)^4} = |-3| = 3$ — We could also simplify this as follows: $\sqrt[4]{(-3)^4} = \sqrt[4]{81} = 3$.

$\sqrt[6]{729x^6} = \sqrt[6]{(3x)^6} = |3x| = 3|x|$ — The absolute value symbols guarantee that the sixth root is positive.

In general, we have the following rules.

Rules for $\sqrt[n]{x^n}$

If x is a real number and $n > 1$, then

If n is an odd natural number, then $\sqrt[n]{x^n} = x$.

If n is an even natural number, then $\sqrt[n]{x^n} = |x|$.

In the radical $\sqrt[n]{x}$, n is called the **index** (or **order**) of the radical. When the index is 2, the radical is a square root, and we usually do not write the index.

$$\sqrt[2]{x} = \sqrt{x}$$

WARNING! When n is even $(n > 1)$ and $x < 0$, $\sqrt[n]{x}$ is not a real number. For example, $\sqrt[4]{-81}$ is not a real number, because no real number raised to the 4th power is -81.

EXAMPLE 8 Simplify each radical.

a. $\sqrt[4]{625} = 5$, because $5^4 = 625$ — Read $\sqrt[4]{625}$ as "the fourth root of 625."

b. $\sqrt[5]{-32} = -2$, because $(-2)^5 = -32$ — Read $\sqrt[5]{-32}$ as "the fifth root of -32."

c. $\sqrt[6]{\dfrac{1}{64}} = \dfrac{1}{2}$, because $\left(\dfrac{1}{2}\right)^6 = \dfrac{1}{64}$ — Read $\sqrt[6]{\frac{1}{64}}$ as "the sixth root of $\frac{1}{64}$."

d. $\sqrt[7]{10^7} = 10$, because $10^7 = 10^7$ — Read $\sqrt[7]{10^7}$ as "the seventh root of 10^7." ■

Self Check Simplify **a.** $\sqrt[4]{\frac{1}{81}}$ and **b.** $\sqrt[5]{10^5}$.

Answers **a.** $\frac{1}{3}$, **b.** 10

b,d

EXAMPLE 9 Simplify each radical. Assume that x can be any real number.

Solution

a. $\sqrt[5]{x^5} = x$ — Since n is odd, absolute value symbols aren't needed.

b. $\sqrt[4]{16x^4} = |2x| = 2|x|$ — Since n is even and x can be negative, absolute value symbols are needed to guarantee that the result is positive.

c. $\sqrt[6]{(x+4)^6} = |x+4|$ — Absolute value symbols are needed to guarantee that the result is positive.

d. $\sqrt[3]{(x+1)^3} = x + 1$ — Since n is odd, absolute value symbols aren't needed.

e.
$$\begin{aligned}\sqrt{(x^2+4x+4)^2} &= \sqrt{[(x+2)^2]^2} && \text{Factor } x^2+4x+4.\\ &= \sqrt{(x+2)^4}\\ &= (x+2)^2 && \text{Since } (x+2)^2 \text{ is always positive, absolute value symbols aren't needed.}\end{aligned}$$

■

Self Check Simplify **a.** $\sqrt[4]{16a^8}$ and **b.** $\sqrt[5]{(a+5)^5}$.

Answers **a.** $2a^2$, **b.** $a + 5$

We summarize the definitions concerning $\sqrt[n]{x}$ as follows.

Summary of the Definitions of $\sqrt[n]{x}$

If n is a natural number greater than 1 and x is a real number, then

If $x > 0$, then $\sqrt[n]{x}$ is the positive number such that $\left(\sqrt[n]{x}\right)^n = x$.

If $x = 0$, then $\sqrt[n]{x} = 0$.

If $x < 0$ $\begin{cases}\text{and } n \text{ is odd, then } \sqrt[n]{x} \text{ is the real number such that } \left(\sqrt[n]{x}\right)^n = x.\\ \text{and } n \text{ is even, then } \sqrt[n]{x} \text{ is not a real number.}\end{cases}$

■ STANDARD DEVIATION

In statistics, the *standard deviation* is used to tell which of a set of distributions is the most variable. To see how to compute the standard deviation of a distribution, we consider the distribution 4, 5, 5, 8, 13 and construct the following table.

Original terms	Mean of the distribution	Differences (original term minus mean)	Squares of the differences from the mean
4	7	−3	9
5	7	−2	4
5	7	−2	4
8	7	1	1
13	7	6	36

The **standard deviation** of the distribution is the positive square root of the mean of the numbers shown in column 4 of the table.

$$\text{Standard deviation} = \sqrt{\frac{\text{sum of the squares of the differences from the mean}}{\text{number of differences}}}$$

$$= \sqrt{\frac{9 + 4 + 4 + 1 + 36}{5}}$$

$$= \sqrt{\frac{54}{5}}$$

$$\approx 3.286335345$$

To the nearest hundredth, the standard deviation of the given distribution is 3.29.

The symbol for standard deviation is σ, the lowercase Greek letter *sigma.*

EXAMPLE 10 Which of the following distributions has the most variability: **a.** 3, 5, 7, 8, 12 or **b.** 1, 4, 6, 11?

Solution We compute the standard deviation of each distribution.

a.

Original terms	Mean of the distribution	Differences (original term minus mean)	Squares of the differences from the mean
3	7	−4	16
5	7	−2	4
7	7	0	0
8	7	1	1
12	7	5	25

$$\sigma = \sqrt{\frac{16 + 4 + 0 + 1 + 25}{5}} = \sqrt{\frac{46}{5}} \approx 3.03$$

b.

Original terms	Mean of the distribution	Differences (original term minus mean)	Squares of the differences from the mean
1	5.5	−4.5	20.25
4	5.5	−1.5	2.25
6	5.5	0.5	0.25
11	5.5	5.5	30.25

$$\sigma = \sqrt{\frac{20.25 + 2.25 + 0.25 + 30.25}{4}} = \sqrt{\frac{53}{4}} \approx 3.64$$

Since the standard deviation for the second distribution is greater than the standard deviation for the first distribution, the second distribution has the greater variability. ■

Orals *Simplify each radical, if possible.*

1. $\sqrt{9}$
2. $-\sqrt{16}$
3. $\sqrt[3]{-8}$
4. $\sqrt[5]{32}$
5. $\sqrt{64x^2}$
6. $\sqrt[3]{-27x^3}$
7. $\sqrt{-3}$
8. $\sqrt[4]{(x+1)^8}$

EXERCISE 7.1

REVIEW *Simplify each fraction.*

1. $\dfrac{x^2 + 7x + 12}{x^2 - 16}$
2. $\dfrac{a^3 - b^3}{b^2 - a^2}$

Do the operations.

3. $\dfrac{x^2 - x - 6}{x^2 - 2x - 3} \cdot \dfrac{x^2 - 1}{x^2 + x - 2}$
4. $\dfrac{x^2 - 3x - 4}{x^2 - 5x + 6} \div \dfrac{x^2 - 2x - 3}{x^2 - x - 2}$
5. $\dfrac{3}{m+1} + \dfrac{3m}{m-1}$
6. $\dfrac{2x+3}{3x-1} - \dfrac{x-4}{2x+1}$

VOCABULARY AND CONCEPTS *Fill in each blank to make a true statement.*

7. $5x^2$ is the square root of $25x^4$, because _____ $= 25x^4$.
8. b is a square root of a if _____.
9. The principal square root of x ($x > 0$) is the _____ square root of x.
10. $\sqrt{x^2} =$ ____
11. The graph of $f(x) = \sqrt{x} + 3$ is the graph of $f(x) = \sqrt{x}$ translated __ units ___.
12. The graph of $f(x) = \sqrt{x + 5}$ is the graph of $f(x) = \sqrt{x}$ translated __ units to the ___.
13. $\sqrt[3]{x} = y$ if _____.
14. $\sqrt[3]{x^3} =$ __

15. When n is an odd number, $\sqrt[n]{x}$ represents an ____ root.

16. When n is an ____ number, $\sqrt[n]{x}$ represents an even root.

17. $\sqrt{0} =$ __

18. The ________ deviation of a set of numbers is the positive square root of the squares of the differences of the numbers and their mean.

In Exercises 19–22, identify the radicand in each expression.

19. $\sqrt{3x^2}$ **20.** $5\sqrt{x}$ **21.** $ab^2\sqrt{a^2 + b^3}$ **22.** $\frac{1}{2}x\sqrt{\frac{x}{y}}$

PRACTICE *In Exercises 23–38, find each square root, if possible.*

23. $\sqrt{121}$ **24.** $\sqrt{144}$ **25.** $-\sqrt{64}$ **26.** $-\sqrt{1}$

27. $\sqrt{\frac{1}{9}}$ **28.** $-\sqrt{\frac{4}{25}}$ **29.** $-\sqrt{\frac{25}{49}}$ **30.** $\sqrt{\frac{49}{81}}$

31. $\sqrt{-25}$ **32.** $\sqrt{0.25}$ **33.** $\sqrt{0.16}$ **34.** $\sqrt{-49}$

35. $\sqrt{(-4)^2}$ **36.** $\sqrt{(-9)^2}$ **37.** $\sqrt{-36}$ **38.** $-\sqrt{-4}$

In Exercises 39–42, use a calculator to find each square root. Give the answer to four decimal places.

39. $\sqrt{12}$ **40.** $\sqrt{340}$ **41.** $\sqrt{679.25}$ **42.** $\sqrt{0.0063}$

In Exercises 43–50, find each square root. Assume that all variables are unrestricted, and use absolute value symbols when necessary.

43. $\sqrt{4x^2}$ **44.** $\sqrt{16y^4}$ **45.** $\sqrt{(t + 5)^2}$ **46.** $\sqrt{(a + 6)^2}$

47. $\sqrt{(-5b)^2}$ **48.** $\sqrt{(-8c)^2}$ **49.** $\sqrt{a^2 + 6a + 9}$ **50.** $\sqrt{x^2 + 10x + 25}$

In Exercises 51–54, find each value given that $f(x) = \sqrt{x - 4}$.

51. $f(4)$ **52.** $f(8)$ **53.** $f(20)$ **54.** $f(29)$

In Exercises 55–58, find each value given that $f(x) = \sqrt{x^2 + 1}$. *Give each answer to four decimal places.*

55. $f(4)$ **56.** $f(6)$ **57.** $f(2.35)$ **58.** $f(21.57)$

In Exercises 59–62, graph each function and find its domain and range.

59. $f(x) = \sqrt{x + 4}$ **60.** $f(x) = -\sqrt{x - 2}$ **61.** $f(x) = -\sqrt{x} - 3$ **62.** $f(x) = \sqrt[3]{x} - 1$

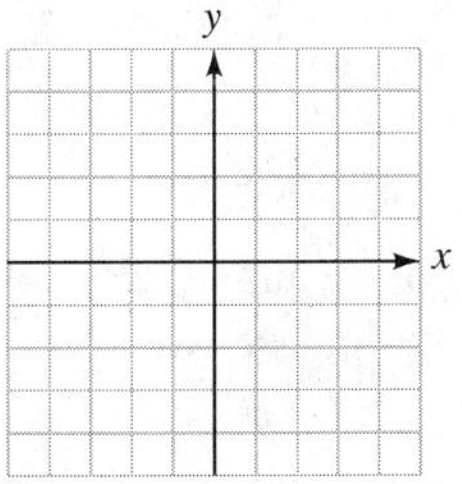

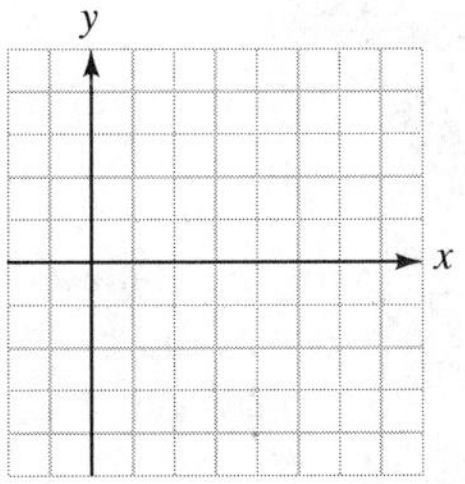

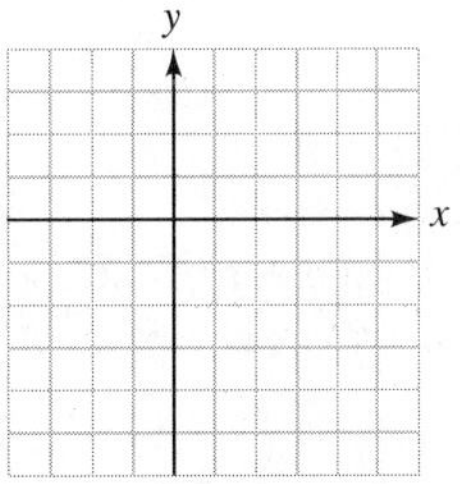

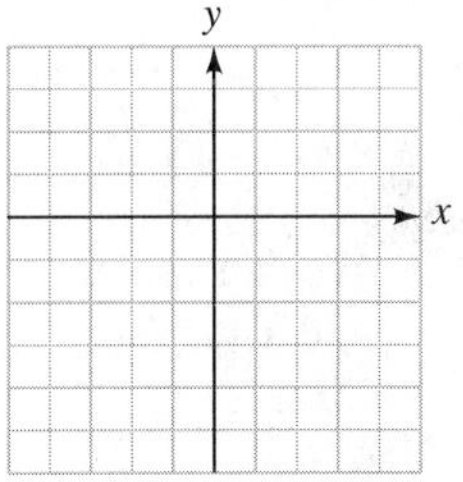

In Exercises 63–78, simplify each cube root.

63. $\sqrt[3]{1}$ **64.** $\sqrt[3]{-8}$ **65.** $\sqrt[3]{-125}$ **66.** $\sqrt[3]{512}$

67. $\sqrt[3]{-\frac{8}{27}}$ **68.** $\sqrt[3]{\frac{125}{216}}$ **69.** $\sqrt[3]{0.064}$ **70.** $\sqrt[3]{0.001}$

71. $\sqrt[3]{8a^3}$ **72.** $\sqrt[3]{-27x^6}$ **73.** $\sqrt[3]{-1{,}000p^3q^3}$ **74.** $\sqrt[3]{343a^6b^3}$

75. $\sqrt[3]{-\frac{1}{8}m^6n^3}$ **76.** $\sqrt[3]{\frac{27}{1{,}000}a^6b^6}$ **77.** $\sqrt[3]{0.008z^9}$ **78.** $\sqrt[3]{0.064s^9t^6}$

In Exercises 79–90, simplify each radical, if possible.

79. $\sqrt[4]{81}$ **80.** $\sqrt[6]{64}$ **81.** $-\sqrt[5]{243}$ **82.** $-\sqrt[4]{625}$

83. $\sqrt[5]{-32}$ **84.** $\sqrt[6]{729}$ **85.** $\sqrt[4]{\frac{16}{625}}$ **86.** $\sqrt[5]{-\frac{243}{32}}$

87. $-\sqrt[5]{-\frac{1}{32}}$ **88.** $\sqrt[6]{-729}$ **89.** $\sqrt[4]{-256}$ **90.** $-\sqrt[4]{\frac{81}{256}}$

In Exercises 91–102, simplify each radical. Assume that all variables are unrestricted, and use absolute value symbols where necessary.

91. $\sqrt[4]{16x^4}$ **92.** $\sqrt[5]{32a^5}$ **93.** $\sqrt[3]{8a^3}$ **94.** $\sqrt[6]{64x^6}$

95. $\sqrt[4]{\frac{1}{16}x^4}$ **96.** $\sqrt[4]{\frac{1}{81}x^8}$ **97.** $\sqrt[4]{x^{12}}$ **98.** $\sqrt[8]{x^{24}}$

99. $\sqrt[5]{-x^5}$ **100.** $\sqrt[3]{-x^6}$ **101.** $\sqrt[3]{-27a^6}$ **102.** $\sqrt[5]{-32x^5}$

In Exercises 103–106, simplify each radical. Assume that all variables are unrestricted, and use absolute value symbols when necessary.

103. $\sqrt[25]{(x+2)^{25}}$ **104.** $\sqrt[44]{(x+4)^{44}}$ **105.** $\sqrt[8]{0.00000001x^{16}y^8}$ **106.** $\sqrt[5]{0.00032x^{10}y^5}$

107. Find the standard deviation of the following distribution to the nearest hundredth: 2, 5, 5, 6, 7.

108. Find the standard deviation of the following distribution to the nearest hundredth: 3, 6, 7, 9, 11, 12.

109. Statistics In statistics, the formula

$$s_{\bar{x}} = \frac{s}{\sqrt{N}}$$

gives an estimate of the standard error of the mean. Find $s_{\bar{x}}$ to four decimal places when $s = 65$ and $N = 30$.

110. Statistics In statistics, the formula

$$\sigma_{\bar{x}} = \frac{\sigma}{\sqrt{N}}$$

gives the standard deviation of means of samples of size N. Find $\sigma_{\bar{x}}$ to four decimal places when $\sigma = 12.7$ and $N = 32$.

APPLICATIONS *Use a calculator to solve each problem.*

111. Radius of a circle The radius r of a circle is given by the formula $r = \sqrt{\frac{A}{\pi}}$, where A is its area. Find the radius of a circle whose area is 9π square units.

112. Diagonal of a baseball diamond The diagonal d of a square is given by the formula $d = \sqrt{2s^2}$, where s is the length of each side. Find the diagonal of the baseball diamond shown in Illustration 1.

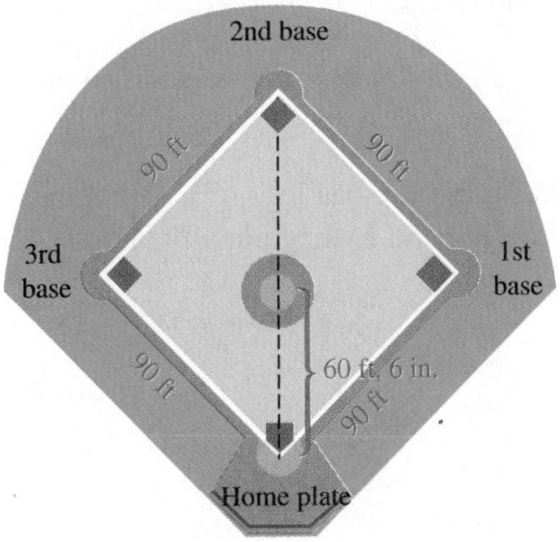

ILLUSTRATION 1

113. Falling objects The time t (in seconds) that it will take for an object to fall a distance of s feet is given by the formula $t = \sqrt{s}/4$. If a stone is dropped down a 256-foot well, how long will it take it to hit bottom?

114. Law enforcement Police sometimes use the formula $s = k\sqrt{l}$ to estimate the speed s (in mph) of a car involved in an accident. In this formula, l is the length of the skid in feet, and k is a constant depending on the condition of the pavement. For wet pavement, $k \approx 3.24$. How fast was a car going if its skid was 400 feet on wet pavement?

115. Electronics When the resistance in a circuit is 18 ohms, the current I (measured in amperes) and the power P (measured in watts) are related by the formula $I = \sqrt{P/18}$. Find the current used by an electrical appliance that is rated at 980 watts.

116. Medicine The approximate pulse rate p (in beats per minute) of an adult who is t inches tall is given by the formula

$$p = \frac{590}{\sqrt{t}}$$

Find the approximate pulse rate of an adult who is 71 inches tall.

WRITING

117. If x is any real number, then $\sqrt{x^2} = x$ is not correct. Explain

118. If x is any real number, then $\sqrt[3]{x^3} = |x|$ is not correct. Explain.

SOMETHING TO THINK ABOUT

119. Is $\sqrt{x^2 - 4x + 4} = x - 2$? What are the exceptions?

120. When is $\sqrt{x^2} \neq x$?

7.2 Applications of Radicals

■ THE PYTHAGOREAN THEOREM ■ THE DISTANCE FORMULA

Getting Ready *Evaluate each expression.*

1. $3^2 + 4^2$

2. $5^2 + 12^2$

3. $(5 - 2)^2 + (2 + 1)^2$

4. $(111 - 21)^2 + (60 - 4)^2$

THE PYTHAGOREAN THEOREM

If we know the lengths of two legs of a right triangle, we can always find the length of the **hypotenuse** (the side opposite the 90° angle) by using the **Pythagorean theorem.**

Pythagorean Theorem
If a and b are the lengths of two legs of a right triangle and c is the length of the hypotenuse, then

$$a^2 + b^2 = c^2$$

In words, the Pythagorean theorem says,

In any right triangle, the square of the hypotenuse is equal to the sum of the squares of the two legs.

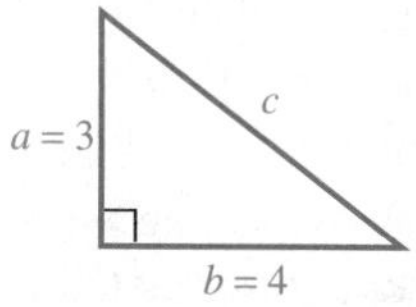

FIGURE 7-6

Suppose the right triangle shown in Figure 7-6 has legs of length 3 and 4 units. To find the length of the hypotenuse, we use the Pythagorean theorem.

$$a^2 + b^2 = c^2$$
$$3^2 + 4^2 = c^2$$
$$9 + 16 = c^2$$
$$25 = c^2$$
$$\sqrt{25} = \sqrt{c^2} \quad \text{Take the positive square root of both sides.}$$
$$5 = c$$

The length of the hypotenuse is 5 units.

EXAMPLE 1 **Fighting fires** To fight a forest fire, the forestry department plans to clear a rectangular fire break around the fire, as shown in Figure 7-7. Crews are equipped with mobile communications with a 3,000-yard range. Can crews at points A and B remain in radio contact?

Solution Points A, B, and C form a right triangle. To find the distance c from point A to point B, we can use the Pythagorean theorem, substituting 2,400 for a and 1,000 for b and solving for c.

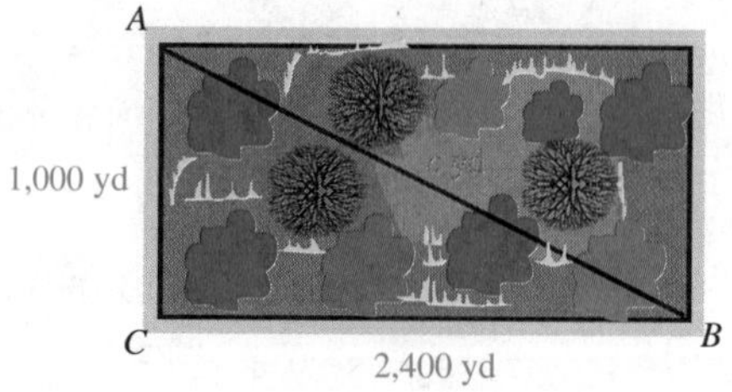

FIGURE 7-7

$$a^2 + b^2 = c^2$$
$$2{,}400^2 + 1{,}000^2 = c^2$$
$$5{,}760{,}000 + 1{,}000{,}000 = c^2$$
$$6{,}760{,}000 = c^2$$
$$\sqrt{6{,}760{,}000} = \sqrt{c^2} \quad \text{Take the positive square root of both sides.}$$
$$2{,}600 = c \quad \text{Use a calculator to find the square root.}$$

The two crews are 2,600 yards apart. Because this distance is less than the range of the radios, they can communicate. ■

Self Check Can the crews communicate if $b = 1{,}500$ yards?

Answer yes

PERSPECTIVE

Pythagoras was a teacher. Although it was unusual for schools at that time, his classes were coeducational. According to some legends, Pythagoras married one of his students. He and his followers formed a secret society with two rules: membership was for life, and members could not reveal the secrets they knew.

Much of their teaching was good mathematics, but some ideas were strange. To them, numbers were sacred. Because beans were used as counters to represent numbers, Pythagoreans refused to eat beans. They also believed that the *only* numbers were the whole numbers. To them, fractions were not numbers; $\frac{2}{3}$ was just a way of comparing the whole numbers 2 and 3. They believed that whole numbers were the building blocks of the universe, just as atoms are to us. The basic Pythagorean doctrine was, "All things are number," and they meant *whole* number.

The Pythagorean theorem was an important discovery of the Pythagorean school, yet it caused some division in the ranks. The right triangle in Illustration 1 has two legs of length 1. By the Pythagorean theorem, the length of the hypotenuse is $\sqrt{2}$. One of their own group, Hippasus of Metapontum, discovered that $\sqrt{2}$ is an irrational number: There are *no* whole numbers a and b that make the fraction $\frac{a}{b}$ exactly equal to $\sqrt{2}$. This discovery was not appreciated by the other Pythagoreans. How could everything in the universe be described with whole numbers, when the side of this simple triangle couldn't? The Pythagoreans had a choice. Either revise and expand their beliefs, or cling to the old. According to legend, the group was at sea at the time of the discovery. Rather than upset the system, they threw Hippasus overboard.

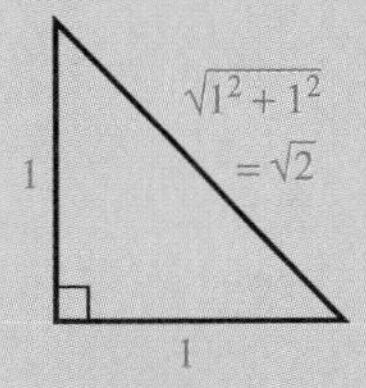

ILLUSTRATION 1

■ THE DISTANCE FORMULA

With the *distance formula,* we can find the distance between any two points that are graphed on a rectangular coordinate system.

To find the distance d between points $P(x_1, y_1)$ and $Q(x_2, y_2)$ shown in Figure 7-8, we construct the right triangle PRQ. The distance between P and R is $|x_2 - x_1|$, and the distance between R and Q is $|y_2 - y_1|$. We apply the Pythagorean theorem to the right triangle PRQ to get

$$[d(PQ)]^2 = |x_2 - x_1|^2 + |y_2 - y_1|^2$$

Read $d(PQ)$ as "the distance between P and Q."

$$= (x_2 - x_1)^2 + (y_2 - y_1)^2$$

Because $|x_2 - x_1|^2 = (x_2 - x_1)^2$ and $|y_2 - y_1|^2 = (y_2 - y_1)^2$.

or

1. $d(PQ) = \sqrt{(x_2 - x_1)^2 + (y_2 - y_1)^2}$

Equation 1 is called the **distance formula.**

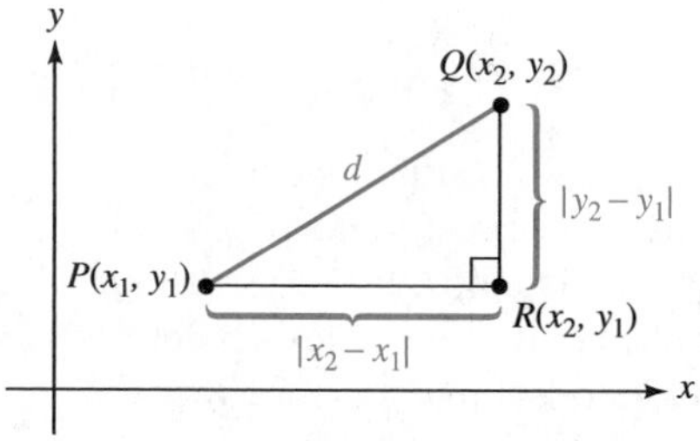

FIGURE 7-8

Distance Formula

The distance between two points $P(x_1, y_1)$ and $Q(x_2, y_2)$ is given by the formula

$$d(PQ) = \sqrt{(x_2 - x_1)^2 + (y_2 - y_1)^2}$$

EXAMPLE 2 Find the distance between points $P(-2, 3)$ and $Q(4, -5)$.

Solution To find the distance, we can use the distance formula by substituting 4 for x_2, -2 for x_1, -5 for y_2, and 3 for y_1.

$$\begin{aligned} d(PQ) &= \sqrt{(x_2 - x_1)^2 + (y_2 - y_1)^2} \\ &= \sqrt{[4 - (-2)]^2 + (-5 - 3)^2} \\ &= \sqrt{(4 + 2)^2 + (-5 - 3)^2} \\ &= \sqrt{6^2 + (-8)^2} \\ &= \sqrt{36 + 64} \\ &= \sqrt{100} \\ &= 10 \end{aligned}$$

The distance between P and Q is 10 units. ■

Self Check Find the distance between $P(-2, -2)$ and $Q(3, 10)$.

Answer 13

EXAMPLE 3 **Building a freeway** In a city, streets run north and south, and avenues run east and west. Streets and avenues are 850 feet apart. The city plans to construct a straight freeway from the intersection of 25th Street and 8th Avenue to the intersection of 115th Street and 64th Avenue. How long will the freeway be?

Solution We can represent the roads of the city by the coordinate system in Figure 7-9, where the units on each axis represent 850 feet. We represent the end of the freeway at 25th Street and 8th Avenue by the point $(x_1, y_1) = (25, 8)$. The other end is $(x_2, y_2) = (115, 64)$.

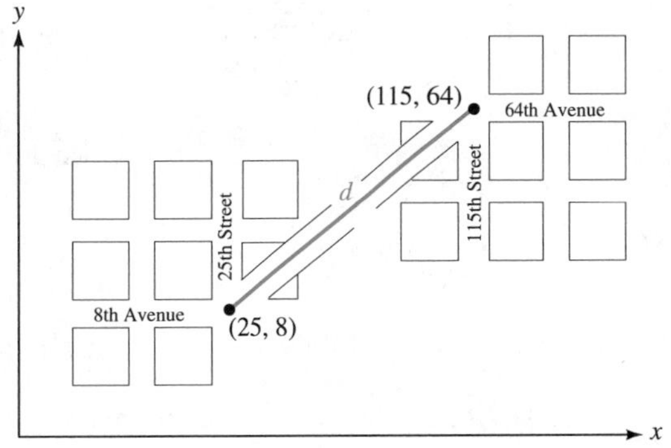

FIGURE 7-9

We can use the distance formula to find the length of the freeway.

$$\begin{aligned} d &= \sqrt{(x_2 - x_1)^2 + (y_2 - y_1)^2} \\ d &= \sqrt{(115 - 25)^2 + (64 - 8)^2} \\ &= \sqrt{90^2 + 56^2} \\ &= \sqrt{8{,}100 + 3{,}136} \\ &= \sqrt{11{,}236} \\ &= 106 \end{aligned}$$

Use a calculator to find the square root.

Because each unit represents 850 feet, the length of the freeway is $106(850) = 90{,}100$ feet. Since 5,280 feet = 1 mile, we can divide 90,100 by 5,280 to convert 90,100 feet to 17.064394 miles. Thus, the freeway will be about 17 miles long. ■

EXAMPLE 4 **Bowling** The velocity, v, of an object after it has fallen d feet is given by the equation $v^2 = 64d$. An inexperienced bowler lofts the ball 4 feet. With what velocity does it strike the alley?

Solution We find the velocity by substituting 4 for d in the equation $v^2 = 64d$ and solving for v.

$$v^2 = 64d$$
$$v^2 = 64(4)$$
$$v^2 = 256$$
$$v = \sqrt{256}$$ Take the square root of both sides. Only the positive square root is meaningful.
$$= 16$$

The ball strikes the alley with a velocity of 16 feet per second. ■

Orals *Evaluate each expression.*

1. $\sqrt{25}$
2. $\sqrt{100}$
3. $\sqrt{169}$
4. $\sqrt{3^2 + 4^2}$
5. $\sqrt{8^2 + 6^2}$
6. $\sqrt{5^2 + 12^2}$
7. $\sqrt{5^2 - 3^2}$
8. $\sqrt{5^2 - 4^2}$
9. $\sqrt{169 - 12^2}$

EXERCISE 7.2

REVIEW *Find each product.*

1. $(4x + 2)(3x - 5)$
2. $(3y - 5)(2y + 3)$
3. $(5t + 4s)(3t - 2s)$
4. $(4r - 3)(2r^2 + 3r - 4)$

VOCABULARY AND CONCEPTS *Fill in each blank to make a true statement.*

5. In a right triangle, the side opposite the 90° angle is called the ________.
6. In a right triangle, the two shorter sides are called _____.
7. If a and b are the lengths of two legs of a right triangle and c is the length of the hypotenuse, then __________.
8. In any right triangle, the square of the hypotenuse is equal to the _____ of the squares of the two _____.
9. With the ________ formula, we can find the distance between two points on a rectangular coordinate system.
10. $d(PQ) =$ ________________

PRACTICE *In Exercises 11–14, the lengths of two legs of the right triangle ABC shown in Illustration 1 are given. Find the length of the missing side.*

11. $a = 6$ ft and $b = 8$ ft
12. $a = 10$ cm and $c = 26$ cm
13. $b = 18$ m and $c = 82$ m
14. $a = 14$ in. and $c = 50$ in.

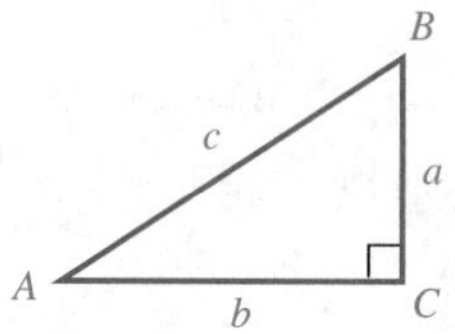

ILLUSTRATION 1

APPLICATIONS *In Exercises 15–16, give each answer to the nearest tenth.*

15. Geometry Find the length of the diagonal of one of the faces of the cube shown in Illustration 2.

16. Geometry Find the length of the diagonal of the cube shown in Illustration 2.

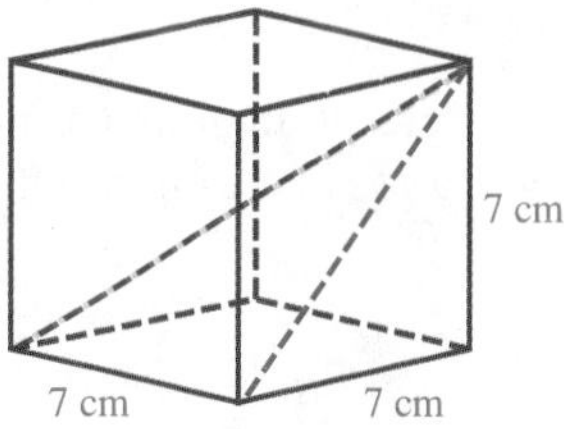

ILLUSTRATION 2

In Exercises 17–26, find the distance between P and Q. If an answer is not exact, give the answer to the nearest tenth.

17. $Q(0, 0)$, $P(3, -4)$

18. $Q(0, 0)$, $P(-6, 8)$

19. $P(2, 4)$, $Q(5, 8)$

20. $P(5, 9)$, $Q(8, 13)$

21. $P(-2, -8)$, $Q(3, 4)$

22. $P(-5, -2)$, $Q(7, 3)$

23. $P(6, 8)$, $Q(12, 16)$

24. $P(10, 4)$, $Q(2, -2)$

25. $Q(-3, 5)$, $P(-5, -5)$

26. $Q(2, -3)$, $P(4, -8)$

27. Geometry Show that a triangle with vertices at $(-2, 4)$, $(2, 8)$, and $(6, 4)$ is isosceles.

28. Geometry Show that a triangle with vertices at $(-2, 13)$, $(-8, 9)$, and $(-2, 5)$ is isosceles.

29. Finding the equation of a line Every point on the line CD in Illustration 3 is equidistant from points A and B. Use the distance formula to find the equation of line CD.

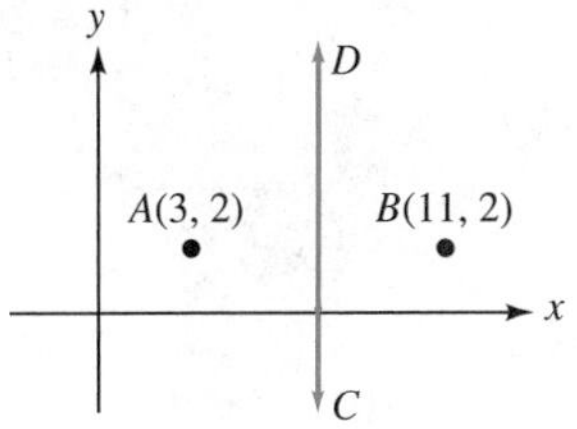

ILLUSTRATION 3

30. Geometry Show that a triangle with vertices at $(2, 3)$, $(-3, 4)$, and $(1, -2)$ is a right triangle. (*Hint:* If the Pythagorean relation holds, the triangle is a right triangle.)

31. Geometry Find the coordinates of the two points on the x-axis that are $\sqrt{5}$ units from the point $(5, 1)$.

32. Geometry The square in Illustration 4 has an area of 18 square units, and its diagonals lie on the x- and y-axes. Find the coordinates of each corner of the square.

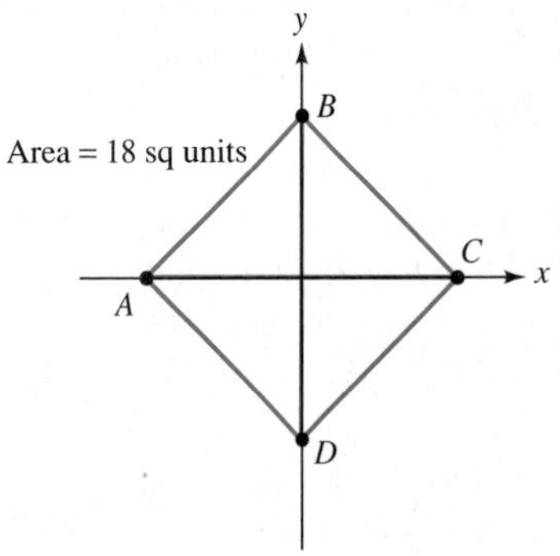

ILLUSTRATION 4

APPLICATIONS

33. Sailing Refer to the sailboat in Illustration 5. How long must a rope be to fasten the top of the mast to the bow?

34. Carpentry The gable end of the roof shown in Illustration 6 is divided in half by a vertical brace. Find the distance from eaves to peak.

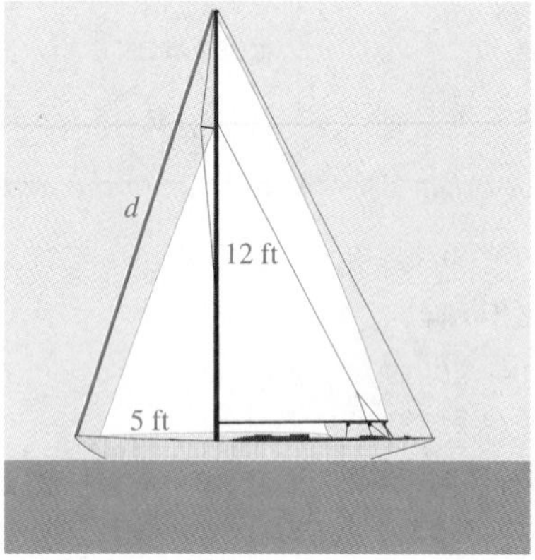

ILLUSTRATION 5

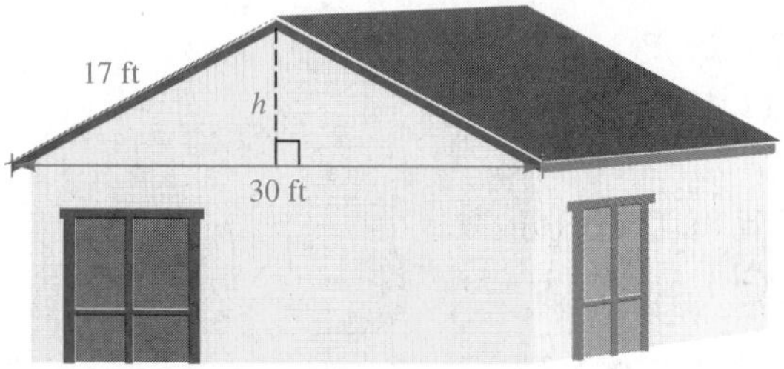

ILLUSTRATION 6

In Exercises 35–38, use a calculator. The baseball diamond shown in Illustration 7 is a square, 90 feet on a side.

35. Baseball How far must a catcher throw the ball to throw out a runner stealing second base?

36. Baseball In baseball, the pitcher's mound is 60 feet, 6 inches from home plate. How far from the mound is second base?

37. Baseball If the third baseman fields a ground ball 10 feet directly behind third base, how far must he throw the ball to throw a runner out at first base?

38. Baseball The shortstop fields a grounder at a point one-third of the way from second base to third base. How far will he have to throw the ball to make an out at first base?

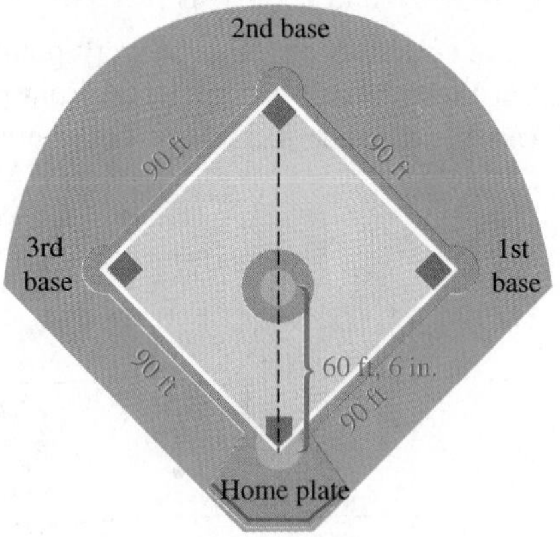

ILLUSTRATION 7

In Exercises 39–48, use a calculator.

39. Packing a tennis racket The diagonal d of a rectangular box with dimensions $a \times b \times c$ is given by

$$d = \sqrt{a^2 + b^2 + c^2}$$

See Illustration 8. Will the racket fit in the shipping carton?

40. Shipping packages A delivery service won't accept a package for shipping if any dimension exceeds 21 inches. An archaeologist wants to ship a 36-inch femur bone. Will it fit in a 3-inch-tall box that has a 21-inch-square base?

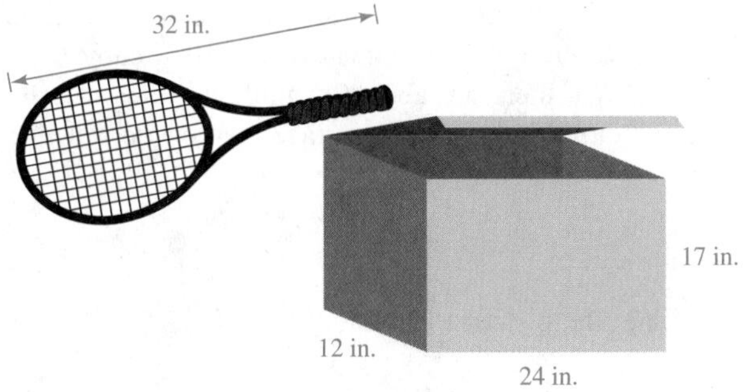

ILLUSTRATION 8

41. **Shipping packages** Can the archaeologist in Exercise 40 ship the femur bone in a cubical box 21 inches on an edge?

42. **Reach of a ladder** The base of the 37-foot ladder in Illustration 9 is 9 feet from the wall. Will the top reach a window ledge that is 35 feet above the ground?

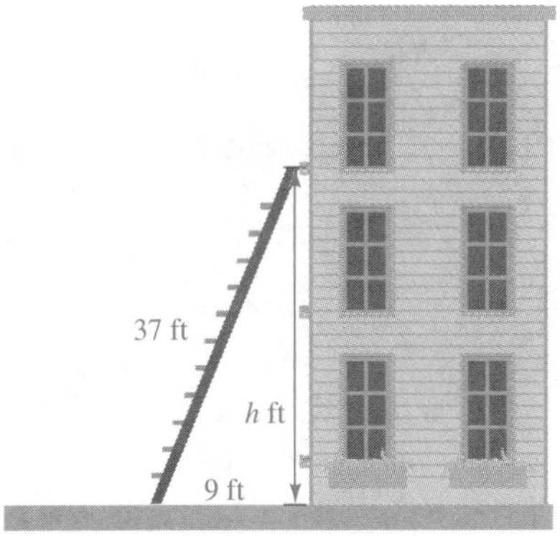

ILLUSTRATION 9

43. **Telephone service** The telephone cable in Illustration 10 runs from A to B to C to D. How much cable is required to run from A to D directly?

44. **Electric service** The power company routes its lines as in Illustration 11. How much wire could be saved by going directly from A to E?

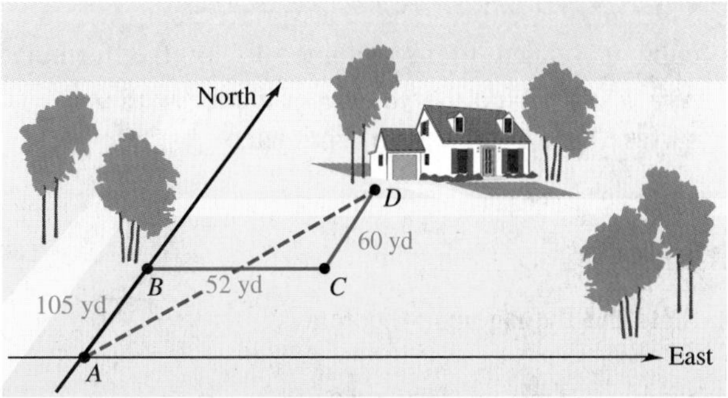

ILLUSTRATION 10

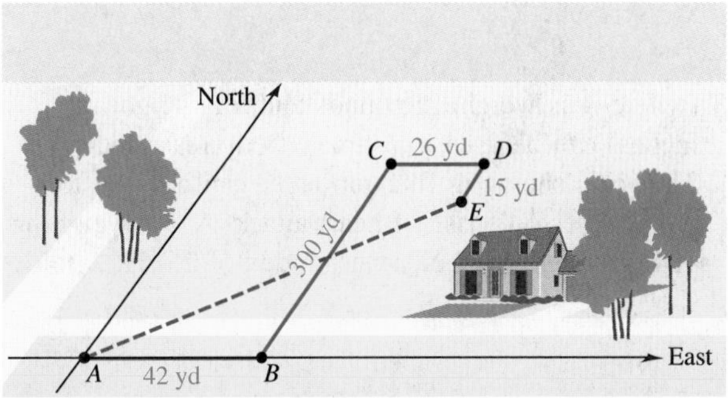

ILLUSTRATION 11

45. **Supporting a weight** A weight placed on the tight wire in Illustration 12 pulls the center down 1 foot. By how much is the wire stretched? Round the answer to the nearest hundredth of a foot.

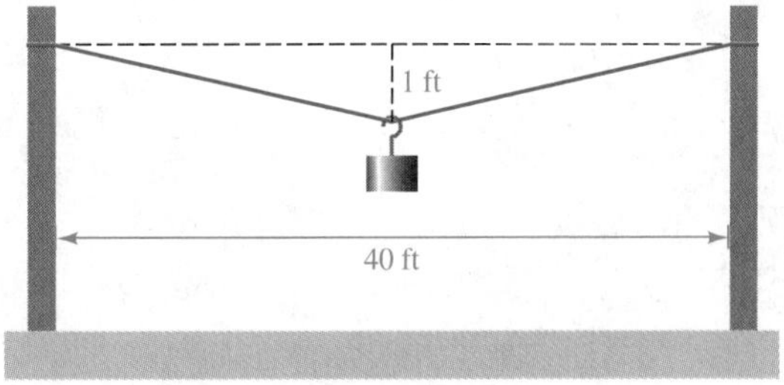

ILLUSTRATION 12

46. **Geometry** The side, s, of a square with area A square feet is given by the formula $s = \sqrt{A}$. Find the perimeter of a square with an area of 49 square feet.

47. Volume of a cube The total surface area, A, of a cube is related to its volume, V, by the formula $A = 6\sqrt[3]{V^2}$. Find the volume of a cube with a surface area of 24 square centimeters.

48. Area of many cubes A grain of table salt is a cube with a volume of approximately 6×10^{-6} cubic inches, and there are about 1.5 million grains of salt in one cup. Find the total surface area of the salt in one cup. (See Exercise 47.)

WRITING

49. State the Pythagorean theorem.

50. Explain the distance formula.

SOMETHING TO THINK ABOUT

51. The formula

$$I = \frac{703w}{h^2}$$

(where w is weight in pounds and h is height in inches) can be used to estimate body mass index, I. The scale shown in Illustration 13 can be used to judge a person's risk of heart attack. A girl weighing 104 pounds has a body mass index of 25. How tall is she?

20–26	normal
27–29	higher risk
30 and above	very high risk

ILLUSTRATION 13

52. What is the risk of a heart attack for a man who is 6 feet tall and weighs 220 pounds?

7.3 Radical Equations

■ EQUATIONS CONTAINING ONE RADICAL ■ EQUATIONS CONTAINING TWO RADICALS
■ EQUATIONS CONTAINING THREE RADICALS

Getting Ready *Find each power.*

1. $(\sqrt{a})^2$ **2.** $(\sqrt{5x})^2$ **3.** $(\sqrt{x+4})^2$ **4.** $(\sqrt[4]{y-3})^4$

As we will see, the solutions of many problems involve equations containing radicals. To solve these equations, we will use the **power rule.**

The Power Rule
If x, y, and n are real numbers and $x = y$, then
$$x^n = y^n$$

If we raise both sides of an equation to the same power, the resulting equation might not be equivalent to the original equation. For example, if we square both sides of the equation

1. $x = 3$ With a solution set of $\{3\}$.

we obtain the equation

2. $x^2 = 9$ With a solution set of $\{3, -3\}$.

Equations 1 and 2 are not equivalent, because they have different solution sets, and the solution -3 of Equation 2 does not satisfy Equation 1. Since raising both sides of an equation to the same power can produce an equation with roots that don't satisfy the original equation, we must always check each suspected solution in the original equation.

EQUATIONS CONTAINING ONE RADICAL

EXAMPLE 1 Solve $\sqrt{x + 3} = 4$.

Solution To eliminate the radical, we apply the power rule by squaring both sides of the equation, and proceed as follows:

$$\sqrt{x+3} = 4$$

$$\left(\sqrt{x+3}\right)^2 = (4)^2 \quad \text{Square both sides.}$$

$$x + 3 = 16$$

$$x = 13 \quad \text{Subtract 3 from both sides.}$$

We must check the apparent solution of 13 to see whether it satisfies the original equation.

Check: $\sqrt{x+3} = 4$

$$\sqrt{13+3} \stackrel{?}{=} 4 \quad \text{Substitute 13 for } x.$$

$$\sqrt{16} \stackrel{?}{=} 4$$

$$4 = 4$$

Since 13 satisfies the original equation, it is a solution. ■

Self Check Solve $\sqrt{a - 2} = 3$.

Answer 11

To solve an equation with radicals, we follow these steps.

Solving an Equation Containing Radicals

1. Isolate one radical expression on one side of the equation.
2. Raise both sides of the equation to the power that is the same as the index of the radical.
3. Solve the resulting equation. If it still contains a radical, go back to Step 1.
4. Check the possible solutions to eliminate the ones that do not satisfy the original equation.

EXAMPLE 2

Height of a bridge The distance d (in feet) that an object will fall in t seconds is given by the formula

$$t = \sqrt{\frac{d}{16}}$$

To find the height of a bridge, a man drops a stone into the water (see Figure 7-10). If it takes the stone 3 seconds to hit the water, how far above the river is the bridge?

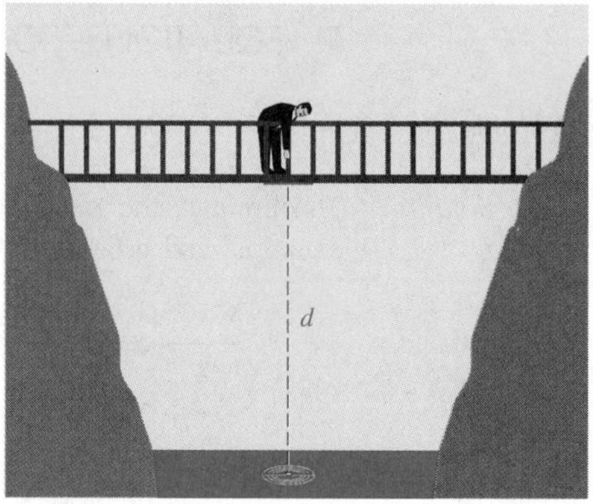

FIGURE 7-10

Solution We substitute 3 for t in the formula and solve for d.

$$t = \sqrt{\frac{d}{16}}$$

$$3 = \sqrt{\frac{d}{16}}$$

$$9 = \frac{d}{16} \quad \text{Square both sides.}$$

$$144 = d \quad \text{Multiply both sides by 16.}$$

The bridge is 144 feet above the river. ■

Self Check In Example 2, how high is the bridge if it takes 4 seconds for the stone to hit the water?

Answer 256 ft

EXAMPLE 3

Solve $\sqrt{3x + 1} + 1 = x$.

Solution We first subtract 1 from both sides to isolate the radical. Then, to eliminate the radical, we square both sides of the equation and proceed as follows:

$$\sqrt{3x+1}+1=x$$

$\sqrt{3x+1}=x-1$	Subtract 1 from both sides.
$\left(\sqrt{3x+1}\right)^2=(x-1)^2$	Square both sides to eliminate the square root.
$3x+1=x^2-2x+1$	$(x-1)^2 \neq x^2-1$. Instead, $(x-1)^2 =$ $(x-1)(x-1)=x^2-x-x+1=x^2-2x+1$.
$0=x^2-5x$	Subtract $3x$ and 1 from both sides.
$0=x(x-5)$	Factor x^2-5x.
$x=0$ or $x-5=0$	Set each factor equal to 0.
$x=0$ $\quad x=5$	

We must check each apparent solution to see whether it satisfies the original equation.

Check:

$\sqrt{3x+1}+1=x$	$\sqrt{3x+1}+1=x$
$\sqrt{3(0)+1}+1 \stackrel{?}{=} 0$	$\sqrt{3(5)+1}+1 \stackrel{?}{=} 5$
$\sqrt{1}+1 \stackrel{?}{=} 0$	$\sqrt{16}+1 \stackrel{?}{=} 5$
$2 \neq 0$	$5=5$

Since 0 does not check, it must be discarded. The only solution of the original equation is 5. ■

Self Check Solve $\sqrt{4x+1}+1=x$.

Answer 6; 0 is extraneous

Solving Equations Containing Radicals

GRAPHING CALCULATORS

To find approximate solutions for $\sqrt{3x+1}+1=x$ with a graphing calculator, we use window settings of $[-5, 10]$ for x and $[-2, 8]$ for y and graph the functions $f(x)=\sqrt{3x+1}+1$ and $g(x)=x$, as in Figure 7-11(a). We then trace to find the approximate x-coordinate of their intersection point, as in Figure 7-11(b). After repeated zooms, we will see that $x=5$.

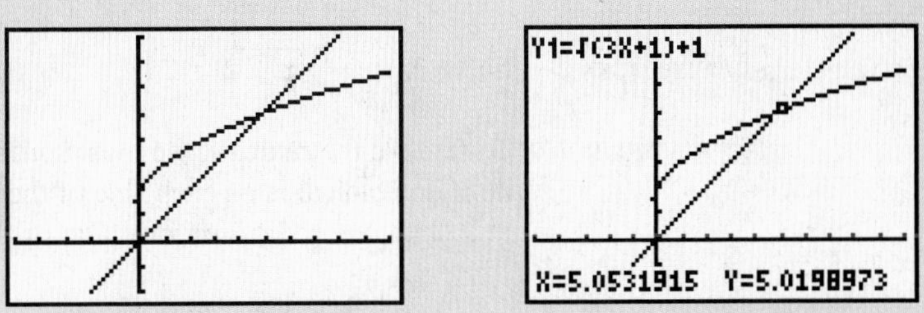

(a) (b)

FIGURE 7-11

EXAMPLE 4 Solve $\sqrt[3]{x^3 + 7} = x + 1$.

Solution To eliminate the radical, we cube both sides of the equation and proceed as follows:

$$\begin{aligned}
\sqrt[3]{x^3 + 7} &= x + 1 \\
\left(\sqrt[3]{x^3 + 7}\right)^3 &= (x + 1)^3 && \text{Cube both sides to eliminate the cube root.} \\
x^3 + 7 &= x^3 + 3x^2 + 3x + 1 \\
0 &= 3x^2 + 3x - 6 && \text{Subtract } x^3 \text{ and 7 from both sides.} \\
0 &= x^2 + x - 2 && \text{Divide both sides by 3.} \\
0 &= (x + 2)(x - 1)
\end{aligned}$$

$$\begin{aligned}
x + 2 &= 0 \quad \text{or} \quad & x - 1 &= 0 \\
x &= -2 & x &= 1
\end{aligned}$$

We check each apparent solution to see whether it satisfies the original equation.

Check:

$$\begin{aligned}
\sqrt[3]{x^3 + 7} &= x + 1 & \qquad \sqrt[3]{x^3 + 7} &= x + 1 \\
\sqrt[3]{(-2)^3 + 7} &\stackrel{?}{=} -2 + 1 & \sqrt[3]{1 + 7} &\stackrel{?}{=} 1 + 1 \\
\sqrt[3]{-8 + 7} &\stackrel{?}{=} -1 & \sqrt[3]{8} &\stackrel{?}{=} 2 \\
\sqrt[3]{-1} &\stackrel{?}{=} -1 & 2 &= 2 \\
-1 &= -1
\end{aligned}$$

Both solutions satisfy the original equation. ■

Self Check Solve $\sqrt[3]{x^3 + 8} = x + 2$.

Answer $0, -2$

■ EQUATIONS CONTAINING TWO RADICALS

When more than one radical appears in an equation, it is often necessary to apply the power rule more than once.

EXAMPLE 5 Solve $\sqrt{x} + \sqrt{x + 2} = 2$.

Solution To remove the radicals, we must square both sides of the equation. This is easier to do if one radical is on each side of the equation. So we subtract $\sqrt{x}$ from both sides to isolate one radical on one side of the equation.

$$\begin{aligned}
\sqrt{x} + \sqrt{x + 2} &= 2 \\
\sqrt{x + 2} &= 2 - \sqrt{x} && \text{Subtract } \sqrt{x} \text{ from both sides.} \\
\left(\sqrt{x + 2}\right)^2 &= \left(2 - \sqrt{x}\right)^2 && \text{Square both sides to eliminate the square root.}
\end{aligned}$$

$$x + 2 = 4 - 4\sqrt{x} + x \qquad (2-\sqrt{x})(2-\sqrt{x}) = 4 - 2\sqrt{x} - 2\sqrt{x} + x = 4 - 4\sqrt{x} + x$$

$$2 = 4 - 4\sqrt{x} \qquad \text{Subtract } x \text{ from both sides.}$$

$$-2 = -4\sqrt{x} \qquad \text{Subtract 4 from both sides.}$$

$$\frac{1}{2} = \sqrt{x} \qquad \text{Divide both sides by } -4.$$

$$\frac{1}{4} = x \qquad \text{Square both sides.}$$

Check:

$$\sqrt{x} + \sqrt{x+2} = 2$$

$$\sqrt{\frac{1}{4}} + \sqrt{\frac{1}{4} + 2} \stackrel{?}{=} 2$$

$$\frac{1}{2} + \sqrt{\frac{9}{4}} \stackrel{?}{=} 2$$

$$\frac{1}{2} + \frac{3}{2} \stackrel{?}{=} 2$$

$$2 = 2$$

The solution checks. ■

Self Check Solve $\sqrt{a} + \sqrt{a+3} = 3$.

Answer 1

Solving Equations Containing Radicals

GRAPHING CALCULATORS

To find approximate solutions for $\sqrt{x} + \sqrt{x+2} = 5$ with a graphing calculator, we use window settings of $[-2, 10]$ for x and $[-2, 8]$ for y and graph the functions $f(x) = \sqrt{x} + \sqrt{x+2}$ and $g(x) = 5$, as in Figure 7-12(a). We then trace to find an approximation of the x-coordinate of their intersection point, as in Figure 7-12(b). From the figure, we can see that $x \approx 5.15$. We can zoom to get better results.

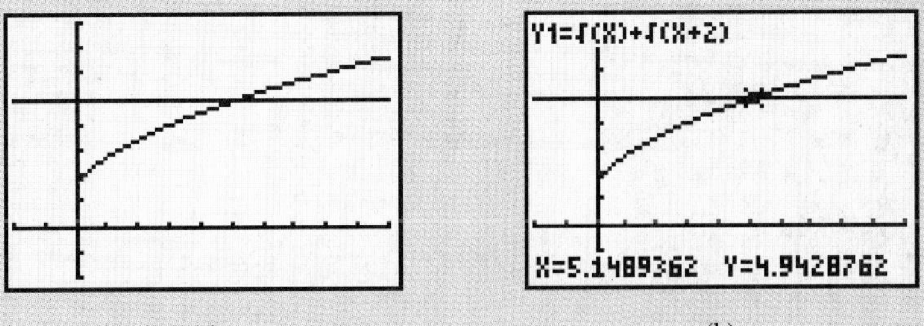

(a) (b)

FIGURE 7-12

EQUATIONS CONTAINING THREE RADICALS

EXAMPLE 6 Solve $\sqrt{x+2} + \sqrt{2x} = \sqrt{18 - x}$.

Solution In this case, it is impossible to isolate one radical on each side of the equation, so we begin by squaring both sides. Then we proceed as follows.

$$\sqrt{x+2} + \sqrt{2x} = \sqrt{18 - x}$$

$$\left(\sqrt{x+2} + \sqrt{2x}\right)^2 = \left(\sqrt{18 - x}\right)^2 \quad \text{Square both sides to eliminate one square root.}$$

$$x + 2 + 2\sqrt{x+2}\sqrt{2x} + 2x = 18 - x$$

$$2\sqrt{x+2}\sqrt{2x} = 16 - 4x \quad \text{Subtract } 3x \text{ and 2 from both sides.}$$

$$\sqrt{x+2}\sqrt{2x} = 8 - 2x \quad \text{Divide both sides by 2.}$$

$$\left(\sqrt{x+2}\sqrt{2x}\right)^2 = (8 - 2x)^2 \quad \text{Square both sides to eliminate the other square roots.}$$

$$(x + 2)2x = 64 - 32x + 4x^2$$

$$2x^2 + 4x = 64 - 32x + 4x^2$$

$$0 = 2x^2 - 36x + 64 \quad \text{Write the equation in quadratic form.}$$

$$0 = x^2 - 18x + 32 \quad \text{Divide both sides by 2.}$$

$$0 = (x - 16)(x - 2) \quad \text{Factor the trinomial.}$$

$$x - 16 = 0 \quad \text{or} \quad x - 2 = 0 \quad \text{Set each factor equal to 0.}$$

$$x = 16 \qquad\qquad x = 2$$

Verify that 2 satisfies the equation, but 16 does not. Thus, the only solution is 2. ■

Self Check Solve $\sqrt{3x + 4} + \sqrt{x + 9} = \sqrt{x + 25}$.

Answer 0

Orals *Solve each equation.*

1. $\sqrt{x+2} = 3$

2. $\sqrt{x-2} = 1$

3. $\sqrt[3]{x+1} = 1$

4. $\sqrt[3]{x-1} = 2$

5. $\sqrt[4]{x-1} = 2$

6. $\sqrt[5]{x+1} = 2$

EXERCISE 7.3

REVIEW *If* $f(x) = 3x^2 - 4x + 2$, *find each quantity.*

1. $f(0)$

2. $f(-3)$

3. $f(2)$

4. $f\left(\frac{1}{2}\right)$

VOCABULARY AND CONCEPTS *Fill in each blank to make a true statement.*

5. If x, y, and n are real numbers and $x = y$, then ______.
6. When solving equations containing radicals, try to ______ one radical expression on one side of the equation.
7. To solve the equation $\sqrt{x+4} = 5$, we first ______ both sides.
8. To solve the equation $\sqrt[3]{x+4} = 2$, we first ______ both sides.
9. Squaring both sides of an equation can introduce ______ solutions.
10. Always remember to ______ the solutions of an equation containing radicals to eliminate any ______ solutions.

PRACTICE *In Exercises 11–64, solve each equation. Write all solutions and cross out those that are extraneous.*

11. $\sqrt{5x-6} = 2$
12. $\sqrt{7x-10} = 12$
13. $\sqrt{6x+1} + 2 = 7$
14. $\sqrt{6x+13} - 2 = 5$
15. $2\sqrt{4x+1} = \sqrt{x+4}$
16. $\sqrt{3(x+4)} = \sqrt{5x-12}$
17. $\sqrt[3]{7n-1} = 3$
18. $\sqrt[3]{12m+4} = 4$
19. $\sqrt[4]{10p+1} = \sqrt[4]{11p-7}$
20. $\sqrt[4]{10y+2} = 2\sqrt[4]{2}$
21. $x = \dfrac{\sqrt{12x-5}}{2}$
22. $x = \dfrac{\sqrt{16x-12}}{2}$
23. $\sqrt{x+2} = \sqrt{4-x}$
24. $\sqrt{6-x} = \sqrt{2x+3}$
25. $2\sqrt{x} = \sqrt{5x-16}$
26. $3\sqrt{x} = \sqrt{3x+12}$
27. $r - 9 = \sqrt{2r-3}$
28. $-s - 3 = 2\sqrt{5-s}$
29. $\sqrt{-5x+24} = 6 - x$
30. $\sqrt{-x+2} = x - 2$
31. $\sqrt{y+2} = 4 - y$
32. $\sqrt{22y+86} = y + 9$
33. $\sqrt{x}\sqrt{x+16} = 15$
34. $\sqrt{x}\sqrt{x+6} = 4$
35. $\sqrt[3]{x^3-7} = x - 1$
36. $\sqrt[3]{x^3+56} - 2 = x$
37. $\sqrt[4]{x^4+4x^2-4} = -x$
38. $\sqrt[4]{8x-8} + 2 = 0$
39. $\sqrt[4]{12t+4} + 2 = 0$
40. $u = \sqrt[4]{u^4-6u^2+24}$
41. $\sqrt{2y+1} = 1 - 2\sqrt{y}$
42. $\sqrt{u} + 3 = \sqrt{u-3}$
43. $\sqrt{y+7} + 3 = \sqrt{y+4}$
44. $1 + \sqrt{z} = \sqrt{z+3}$
45. $\sqrt{v} + \sqrt{3} = \sqrt{v+3}$
46. $\sqrt{x} + 2 = \sqrt{x+4}$
47. $2 + \sqrt{u} = \sqrt{2u+7}$
48. $5r + 4 = \sqrt{5r+20} + 4r$
49. $\sqrt{6t+1} - 3\sqrt{t} = -1$
50. $\sqrt{4s+1} - \sqrt{6s} = -1$
51. $\sqrt{2x+5} + \sqrt{x+2} = 5$
52. $\sqrt{2x+5} + \sqrt{2x+1} + 4 = 0$
53. $\sqrt{z-1} + \sqrt{z+2} = 3$
54. $\sqrt{16v+1} + \sqrt{8v+1} = 12$
55. $\sqrt{x-5} - \sqrt{x+3} = 4$
56. $\sqrt{x+8} - \sqrt{x-4} = -2$
57. $\sqrt{x+1} + \sqrt{3x} = \sqrt{5x+1}$
58. $\sqrt{3x} - \sqrt{x+1} = \sqrt{x-2}$
59. $\sqrt{\sqrt{a} + \sqrt{a+8}} = 2$
60. $\sqrt{\sqrt{2y} - \sqrt{y-1}} = 1$
61. $\dfrac{6}{\sqrt{x+5}} = \sqrt{x}$
62. $\dfrac{\sqrt{2x}}{\sqrt{x+2}} = \sqrt{x-1}$
63. $\sqrt{x+2} + \sqrt{2x-3} = \sqrt{11-x}$
64. $\sqrt{8-x} - \sqrt{3x-8} = \sqrt{x-4}$

APPLICATIONS

65. Highway design A curve banked at 8° will accommodate traffic traveling s mph if the radius of the curve is r feet, according to the formula $s = 1.45\sqrt{r}$. If engineers expect 65-mph traffic, what radius should they specify? (See Illustration 1.)

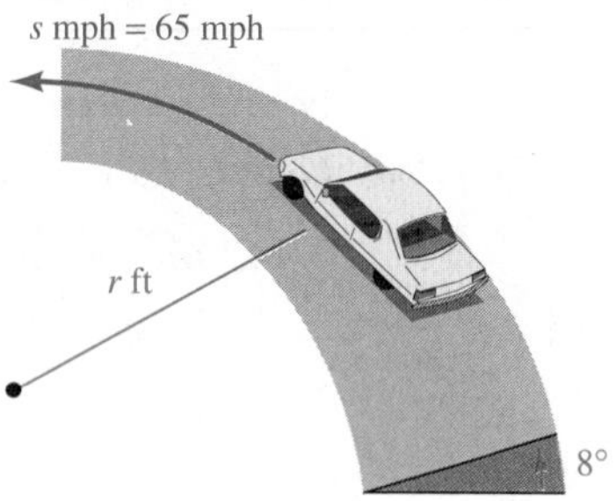

ILLUSTRATION 1

66. Horizon distance The higher a lookout tower is built, the farther an observer can see. (See Illustration 2.) That distance d (called the *horizon distance,* measured in miles) is related to the height h of the observer (measured in feet) by the formula $d = 1.4\sqrt{h}$. How tall must a lookout tower be to see the edge of the forest, 25 miles away?

67. Generating power The power generated by a windmill is related to the velocity of the wind by the formula

$$v = \sqrt[3]{\frac{P}{0.02}}$$

where P is the power (in watts) and v is the velocity of the wind (in mph). Find the speed of the wind when the windmill is generating 500 watts of power.

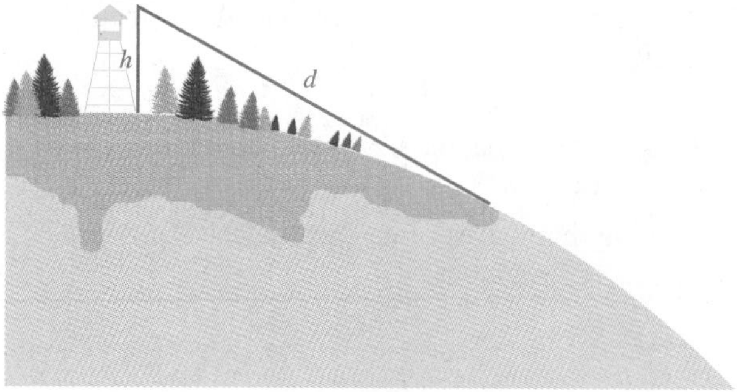

ILLUSTRATION 2

68. Carpentry During construction, carpenters often brace walls as shown in Illustration 3, where the length of the brace is given by the formula

$$l = \sqrt{f^2 + h^2}$$

If a carpenter nails a 10-ft brace to the wall 6 feet above the floor, how far from the base of the wall should he nail the brace to the floor?

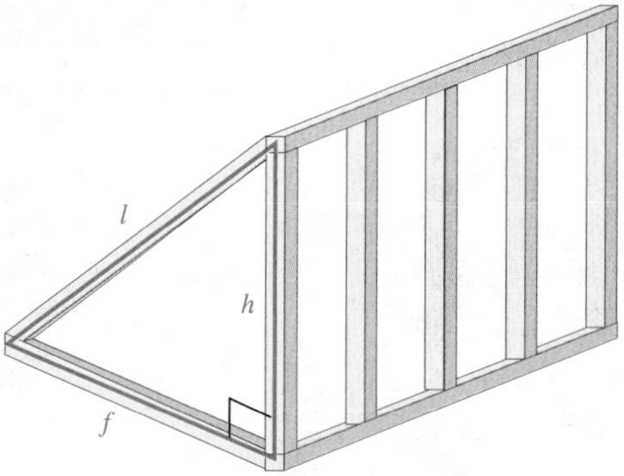

ILLUSTRATION 3

In Exercises 69–70, use a graphing calculator.

69. Marketing The number of wrenches that will be produced at a given price can be predicted by the formula $s = \sqrt{5x}$, where s is the supply (in thousands) and x is the price (in dollars). If the demand, d, for wrenches can be predicted by the formula $d = \sqrt{100 - 3x^2}$, find the equilibrium price.

70. Marketing The number of footballs that will be produced at a given price can be predicted by the formula $s = \sqrt{23x}$, where s is the supply (in thousands) and x is the price (in dollars). If the demand, d, for footballs can be predicted by the formula $d = \sqrt{312 - 2x^2}$, find the equilibrium price.

WRITING

71. If both sides of an equation are raised to the same power, the resulting equation might not be equivalent to the original equation. Explain.

72. Explain why you must check each apparent solution of a radical equation.

SOMETHING TO THINK ABOUT

73. Solve $\sqrt[3]{2x} = \sqrt{x}$. (*Hint:* Square and then cube both sides.)

74. Solve $\sqrt[4]{x} = \sqrt{\frac{x}{4}}$.

7.4 Rational Exponents

■ RATIONAL EXPONENTS ■ EXPONENTIAL EXPRESSIONS WITH VARIABLES IN THEIR BASES ■ FRACTIONAL EXPONENTS WITH NUMERATORS OTHER THAN 1 ■ NEGATIVE FRACTIONAL EXPONENTS ■ SIMPLIFYING RADICAL EXPRESSIONS

Getting Ready *Simplify each expression.*

1. x^3x^4 **2.** $(a^3)^4$ **3.** $\frac{a^8}{a^4}$ **4.** a^0

5. x^{-4} **6.** $(ab^2)^3$ **7.** $\left(\frac{b^2}{c^3}\right)^3$ **8.** $(a^2a^3)^2$

■ RATIONAL EXPONENTS

We have seen that positive integer exponents indicate the number of times that a base is to be used as a factor in a product. For example, x^5 means that x is to be used as a factor five times.

$$x^5 = \overbrace{x \cdot x \cdot x \cdot x \cdot x}^{\text{5 factors of } x}$$

Furthermore, we recall the following properties of exponents.

Rules of Exponents
If there are no divisions by 0, then for all integers m and n,

1. $x^mx^n = x^{m+n}$ **2.** $(x^m)^n = x^{mn}$ **3.** $(xy)^n = x^ny^n$ **4.** $\left(\frac{x}{y}\right)^n = \frac{x^n}{y^n}$

5. $x^0 = 1 \quad (x \neq 0)$ **6.** $x^{-n} = \frac{1}{x^n}$ **7.** $\frac{x^m}{x^n} = x^{m-n}$ **8.** $\left(\frac{x}{y}\right)^{-n} = \left(\frac{y}{x}\right)^n$

It is possible to raise many bases to fractional powers. Since we want fractional exponents to obey the same rules as integer exponents, the square of $10^{1/2}$ must be 10, because

$$\begin{aligned}(10^{1/2})^2 &= 10^{(1/2)2} && \text{Keep the base and multiply the exponents.}\\ &= 10^1 && \tfrac{1}{2}\cdot 2 = 1.\\ &= 10 && 10^1 = 10.\end{aligned}$$

However, we have seen that

$$\left(\sqrt{10}\right)^2 = 10$$

Since $(10^{1/2})^2$ and $\left(\sqrt{10}\right)^2$ both equal 10, we define $10^{1/2}$ to be $\sqrt{10}$. Likewise, we define

$$10^{1/3} \text{ to be } \sqrt[3]{10} \qquad \text{and} \qquad 10^{1/4} \text{ to be } \sqrt[4]{10}$$

Rational Exponents

If n $(n > 1)$ is a natural number and $\sqrt[n]{x}$ is a real number, then

$$x^{1/n} = \sqrt[n]{x}$$

EXAMPLE 1 Simplify each expression.

a. $9^{1/2} = \sqrt{9} = 3$

b. $-\left(\dfrac{16}{9}\right)^{1/2} = -\sqrt{\dfrac{16}{9}} = -\dfrac{4}{3}$

c. $(-64)^{1/3} = \sqrt[3]{-64} = -4$

d. $16^{1/4} = \sqrt[4]{16} = 2$

e. $\left(\dfrac{1}{32}\right)^{1/5} = \sqrt[5]{\dfrac{1}{32}} = \dfrac{1}{2}$

f. $0^{1/8} = \sqrt[8]{0} = 0$

g. $-(32x^5)^{1/5} = -\sqrt[5]{32x^5} = -2x$

h. $(xyz)^{1/4} = \sqrt[4]{xyz}$ ■

Self Check Assume that $x > 0$. Simplify **a.** $16^{1/2}$, **b.** $\left(\frac{27}{8}\right)^{1/3}$, and **c.** $-(16x^4)^{1/4}$.

Answers **a.** 4, **b.** $\frac{3}{2}$, **c.** $-2x$

EXAMPLE 2 Write each radical as an expression with a fractional exponent: **a.** $\sqrt[4]{5xyz}$ and **b.** $\sqrt[5]{\dfrac{xy^2}{15}}$.

Solution **a.** $\sqrt[4]{5xyz} = (5xyz)^{1/4}$ **b.** $\sqrt[5]{\dfrac{xy^2}{15}} = \left(\dfrac{xy^2}{15}\right)^{1/5}$ ■

Self Check Write the radical with a fractional exponent: $\sqrt[6]{4ab}$.

Answer $(4ab)^{1/6}$

EXPONENTIAL EXPRESSIONS WITH VARIABLES IN THEIR BASES

As with radicals, when n is odd in the expression $x^{1/n}$ $(n > 1)$, there is exactly one real nth root, and we don't have to worry about absolute value symbols.

When n is even, there are two nth roots. Since we want the expression $x^{1/n}$ to represent the positive nth root, we must often use absolute value symbols to guarantee that the simplified result is positive. Thus, if n is even,

$$(x^n)^{1/n} = |x|$$

When n is even and x is negative, the expression $x^{1/n}$ is not a real number.

c,e

EXAMPLE 3 Assume that all variables can be any real number, and simplify each expression.

a. $(-27x^3)^{1/3} = -3x$ — Because $(-3x)^3 = -27x^3$. Since n is odd, no absolute value symbols are needed.

b. $(49x^2)^{1/2} = |7x|$
$= 7|x|$ — Because $(|7x|)^2 = 49x^2$. Since $7x$ can be negative, absolute value symbols are needed.

c. $(256a^8)^{1/8} = 2|a|$ — Because $(2|a|)^8 = 256a^8$. Since a can be any real number, $2a$ can be negative. Thus, absolute value symbols are needed.

d. $[(y + 1)^2]^{1/2} = |y + 1|$ — Because $|y + 1|^2 = (y + 1)^2$. Since y can be any real number, $y + 1$ can be negative, and the absolute value symbols are needed.

e. $(25b^4)^{1/2} = 5b^2$ — Because $(5b^2)^2 = 25b^4$. Since $b^2 \geq 0$, no absolute value symbols are needed.

f. $(-256x^4)^{1/4}$ is not a real number. — Because no real number raised to the 4th power is $-256x^4$. ■

Self Check Simplify each expression: **a.** $(625a^4)^{1/4}$ and **b.** $(b^4)^{1/2}$.

Answer **a.** $5|a|$, **b.** b^2

We summarize the cases as follows.

Summary of the Definitions of $x^{1/n}$

If n is a natural number greater than 1 and x is a real number, then

If $x > 0$, then $x^{1/n}$ is the positive number such that $(x^{1/n})^n = x$.

If $x = 0$, then $x^{1/n} = 0$.

If $x < 0$ $\begin{cases} \text{and } n \text{ is odd, then } x^{1/n} \text{ is the real number such that } (x^{1/n})^n = x. \\ \text{and } n \text{ is even, then } x^{1/n} \text{ is not a real number.} \end{cases}$

FRACTIONAL EXPONENTS WITH NUMERATORS OTHER THAN 1

We can extend the definition of $x^{1/n}$ to include fractional exponents with numerators other than 1. For example, since $4^{3/2}$ can be written as $(4^{1/2})^3$, we have

$$4^{3/2} = (4^{1/2})^3 = \left(\sqrt{4}\right)^3 = 2^3 = 8$$

Thus, we can simplify $4^{3/2}$ by cubing the square root of 4. We can also simplify $4^{3/2}$ by taking the square root of 4 cubed.

$$4^{3/2} = (4^3)^{1/2} = 64^{1/2} = \sqrt{64} = 8$$

In general, we have the following rule.

Changing from Rational Exponents to Radicals

If m and n are positive integers, $x \geq 0$, and $\frac{m}{n}$ is in simplified form, then

$$x^{m/n} = \left(\sqrt[n]{x}\right)^m = \sqrt[n]{x^m}$$

Because of the previous definition, we can interpret $x^{m/n}$ in two ways:

1. $x^{m/n}$ means the mth power of the nth root of x.
2. $x^{m/n}$ means the nth root of the mth power of x.

EXAMPLE 4 Simplify each expression.

a. $27^{2/3} = \left(\sqrt[3]{27}\right)^2 = 3^2 = 9$ or $27^{2/3} = \sqrt[3]{27^2} = \sqrt[3]{729} = 9$

b. $\left(\frac{1}{16}\right)^{3/4} = \left(\sqrt[4]{\frac{1}{16}}\right)^3 = \left(\frac{1}{2}\right)^3 = \frac{1}{8}$ or $\left(\frac{1}{16}\right)^{3/4} = \sqrt[4]{\left(\frac{1}{16}\right)^3} = \sqrt[4]{\frac{1}{4{,}096}} = \frac{1}{8}$

c. $(-8x^3)^{4/3} = \left(\sqrt[3]{-8x^3}\right)^4 = (-2x)^4 = 16x^4$ or $(-8x^3)^{4/3} = \sqrt[3]{(-8x^3)^4} = \sqrt[3]{4{,}096x^{12}} = 16x^4$ ■

Self Check Simplify **a.** $16^{3/2}$ and **b.** $(-27x^6)^{2/3}$.

Answers **a.** 64, **b.** $9x^4$

To avoid large numbers, it is usually better to find the root of the base first, as shown in Example 4.

NEGATIVE FRACTIONAL EXPONENTS

To be consistent with the definition of negative integer exponents, we define $x^{-m/n}$ as follows.

Definition of $x^{-m/n}$

If m and n are positive integers, $\frac{m}{n}$ is in simplified form, and $x^{1/n}$ is a real number, then

$$x^{-m/n} = \frac{1}{x^{m/n}} \quad \text{and} \quad \frac{1}{x^{-m/n}} = x^{m/n} \quad (x \neq 0)$$

a

EXAMPLE 5 Write each expression without using negative exponents, if possible.

a. $64^{-1/2} = \frac{1}{64^{1/2}}$

$= \frac{1}{8}$

b. $16^{-3/2} = \frac{1}{16^{3/2}}$

$= \frac{1}{(16^{1/2})^3}$

$= \frac{1}{64}$ $\quad (16^{1/2})^3 = 4^3 = 64.$

c. $(-32x^5)^{-2/5} = \frac{1}{(-32x^5)^{2/5}} \quad (x \neq 0)$

$= \frac{1}{[(-32x^5)^{1/5}]^2}$

$= \frac{1}{(-2x)^2}$

$= \frac{1}{4x^2}$

d. $(-16)^{-3/4}$ is not a real number, because $(-16)^{1/4}$ is not a real number.

■

Self Check Write without using negative exponents: **a.** $25^{-3/2}$ and **b.** $(-27a^3)^{-2/3}$.

Answers **a.** $\frac{1}{125}$, **b.** $\frac{1}{9a^2}$

WARNING! By definition, 0^0 is undefined. A base of 0 raised to a negative power is also undefined, because 0^{-2} would equal $\frac{1}{0^2}$, which is undefined since we cannot divide by 0.

We can use the laws of exponents to simplify many expressions with fractional exponents. If all variables represent positive numbers, no absolute value symbols are necessary.

EXAMPLE 6 Assume that all variables represent positive numbers. Write all answers without using negative exponents.

a. $5^{2/7}5^{3/7} = 5^{2/7+3/7}$	Use the rule $x^m x^n = x^{m+n}$.
$= 5^{5/7}$	Add: $\frac{2}{7} + \frac{3}{7} = \frac{5}{7}$.
b. $(5^{2/7})^3 = 5^{(2/7)(3)}$	Use the rule $(x^m)^n = x^{mn}$.
$= 5^{6/7}$	Multiply: $\frac{2}{7}(3) = \frac{6}{7}$.
c. $(a^{2/3}b^{1/2})^6 = (a^{2/3})^6(b^{1/2})^6$	Use the rule $(xy)^n = x^n y^n$.
$= a^{12/3}b^{6/2}$	Use the rule $(x^m)^n = x^{mn}$ twice.
$= a^4b^3$	Simplify the exponent.
d. $\dfrac{a^{8/3}a^{1/3}}{a^2} = a^{8/3+1/3-2}$	Use the rules $x^m x^n = x^{m+n}$ and $\frac{x^m}{x^n} = x^{m-n}$.
$= a^{8/3+1/3-6/3}$	$2 = \frac{6}{3}$.
$= a^{3/3}$	$\frac{8}{3} + \frac{1}{3} - \frac{6}{3} = \frac{3}{3}$.
$= a$	$\frac{3}{3} = 1$. ■

Self Check Simplify **a.** $(x^{1/3}y^{3/2})^6$ and **b.** $\dfrac{x^{5/3}x^{2/3}}{x^{1/3}}$.

Answers **a.** x^2y^9, **b.** x^2

EXAMPLE 7 Assume that all variables represent positive numbers, and do the operations. Write all answers without using negative exponents.

a. $a^{4/5}(a^{1/5} + a^{3/5}) = a^{4/5}a^{1/5} + a^{4/5}a^{3/5}$	Use the distributive property.
$= a^{4/5+1/5} + a^{4/5+3/5}$	Use the rule $x^m x^n = x^{m+n}$.
$= a^{5/5} + a^{7/5}$	Simplify the exponents.
$= a + a^{7/5}$	

WARNING! Note that $a + a^{7/5} \neq a^{1+7/5}$. The expression $a + a^{7/5}$ cannot be simplified, because a and $a^{7/5}$ are not like terms.

b. $x^{1/2}(x^{-1/2} + x^{1/2}) = x^{1/2}x^{-1/2} + x^{1/2}x^{1/2}$	Use the distributive property.
$= x^{1/2-1/2} + x^{1/2+1/2}$	Use the rule $x^m x^n = x^{m+n}$.
$= x^0 + x^1$	Simplify.
$= 1 + x$	$x^0 = 1$.

c. $(x^{2/3} + 1)(x^{2/3} - 1) = x^{4/3} - x^{2/3} + x^{2/3} - 1$ Use the FOIL method.

$= x^{4/3} - 1$

d. $(x^{1/2} + y^{1/2})^2 = (x^{1/2} + y^{1/2})(x^{1/2} + y^{1/2})$

$= x + 2x^{1/2}y^{1/2} + y$ Use the FOIL method. ■

SIMPLIFYING RADICAL EXPRESSIONS

We can simplify many radical expressions by using the following steps.

Using Fractional Exponents to Simplify Radicals

1. Change the radical expression into an exponential expression with rational exponents.
2. Simplify the rational exponents.
3. Change the exponential expression back into a radical.

C

EXAMPLE 8 Simplify **a.** $\sqrt[4]{3^2}$, **b.** $\sqrt[8]{x^6}$, and **c.** $\sqrt[9]{27x^6y^3}$.

Solution **a.** $\sqrt[4]{3^2} = (3^2)^{1/4}$ Change the radical to an exponential expression.

$= 3^{2/4}$ Use the rule $(x^m)^n = x^{mn}$.

$= 3^{1/2}$ $\frac{2}{4} = \frac{1}{2}$.

$= \sqrt{3}$ Change back to radical notation.

b. $\sqrt[8]{x^6} = (x^6)^{1/8}$ Change the radical to an exponential expression.

$= x^{6/8}$ Use the rule $(x^m)^n = x^{mn}$.

$= x^{3/4}$ $\frac{6}{8} = \frac{3}{4}$.

$= (x^3)^{1/4}$ $\frac{3}{4} = 3\left(\frac{1}{4}\right)$.

$= \sqrt[4]{x^3}$ Change back to radical notation.

c. $\sqrt[9]{27x^6y^3} = (3^3x^6y^3)^{1/9}$ Write 27 as 3^3 and change the radical to an exponential expression.

$= 3^{3/9}x^{6/9}y^{3/9}$ Raise each factor to the $\frac{1}{9}$ power by multiplying the fractional exponents.

$= 3^{1/3}x^{2/3}y^{1/3}$ Simplify each fractional exponent.

$= (3x^2y)^{1/3}$ Use the rule $(xy)^n = x^ny^n$.

$= \sqrt[3]{3x^2y}$ Change back to radical notation. ■

Self Check Simplify **a.** $\sqrt[6]{3^3}$ and **b.** $\sqrt[4]{64x^2y^2}$.

Answers **a.** $\sqrt{3}$, **b.** $\sqrt{8xy}$

Orals *Simplify each expression.*

1. $4^{1/2}$
2. $9^{1/2}$
3. $27^{1/3}$
4. $1^{1/4}$
5. $4^{3/2}$
6. $8^{2/3}$
7. $\left(\frac{1}{4}\right)^{1/2}$
8. $\left(\frac{1}{4}\right)^{-1/2}$
9. $(8x^3)^{1/3}$
10. $(16x^8)^{1/4}$

EXERCISE 7.4

REVIEW *Solve each inequality.*

1. $5x - 4 < 11$
2. $2(3t - 5) \geq 8$
3. $\frac{4}{5}(r - 3) > \frac{2}{3}(r + 2)$
4. $-4 < 2x - 4 \leq 8$
5. How much water must be added to 5 pints of a 20% alcohol solution to dilute it to a 15% solution?
6. A grocer bought some boxes of apples for \$70. However, 4 boxes were spoiled. The grocer sold the remaining boxes at a profit of \$2 each. How many boxes did the grocer sell if she managed to break even?

VOCABULARY AND CONCEPTS *Fill in each blank to make a true statement.*

7. $a^4 =$ ________
8. $a^m a^n =$ _____
9. $(a^m)^n =$ ____
10. $(ab)^n =$ _____
11. $\left(\frac{a}{b}\right)^n =$ ___
12. $a^0 =$ __, provided $a \neq$ __
13. $a^{-n} =$ __, provided $a \neq$ __
14. $\frac{a^m}{a^n} =$ _____
15. $\left(\frac{a}{b}\right)^{-n} =$ ___
16. $x^{1/n} =$ ____
17. $(x^n)^{1/n} =$ ____, provided n is even
18. $x^{m/n} = \sqrt[n]{x^m} =$ ______

PRACTICE *In Exercises 19–26, change each expression into radical notation.*

19. $7^{1/3}$
20. $26^{1/2}$
21. $(3x)^{1/4}$
22. $(4ab)^{1/6}$
23. $\left(\frac{1}{2}x^3y\right)^{1/4}$
24. $\left(\frac{3}{4}a^2b^2\right)^{1/5}$
25. $(x^2 + y^2)^{1/2}$
26. $(x^3 + y^3)^{1/3}$

In Exercises 27–34, change each radical to an exponential expression.

27. $\sqrt{11}$
28. $\sqrt[3]{12}$
29. $\sqrt[4]{3a}$
30. $3\sqrt[5]{a}$
31. $\sqrt[6]{\frac{1}{7}abc}$
32. $\sqrt[7]{\frac{3}{8}p^2q}$
33. $\sqrt[3]{a^2 - b^2}$
34. $\sqrt{x^2 + y^2}$

In Exercises 35–54, simplify each expression, if possible.

35. $4^{1/2}$ **36.** $25^{1/2}$ **37.** $8^{1/3}$ **38.** $125^{1/3}$

39. $16^{1/4}$ **40.** $625^{1/4}$ **41.** $32^{1/5}$ **42.** $0^{1/5}$

43. $\left(\frac{1}{4}\right)^{1/2}$ **44.** $\left(\frac{1}{16}\right)^{1/2}$ **45.** $\left(\frac{1}{8}\right)^{1/3}$ **46.** $\left(\frac{1}{16}\right)^{1/4}$

47. $-16^{1/4}$ **48.** $-125^{1/3}$ **49.** $(-27)^{1/3}$ **50.** $(-125)^{1/3}$

51. $(-64)^{1/2}$ **52.** $(-243)^{1/5}$ **53.** $0^{1/3}$ **54.** $(-216)^{1/2}$

In Exercises 55–62, simplify each expression, if possible. Assume that all variables are unrestricted, and use absolute value symbols when necessary.

55. $(25y^2)^{1/2}$ **56.** $(-27x^3)^{1/3}$ **57.** $(16x^4)^{1/4}$ **58.** $(-16x^4)^{1/2}$

59. $(243x^5)^{1/5}$ **60.** $[(x + 1)^4]^{1/4}$ **61.** $(-64x^8)^{1/4}$ **62.** $[(x + 5)^3]^{1/3}$

In Exercises 63–74, simplify each expression.

63. $36^{3/2}$ **64.** $27^{2/3}$ **65.** $81^{3/4}$ **66.** $100^{3/2}$

67. $144^{3/2}$ **68.** $1{,}000^{2/3}$ **69.** $\left(\frac{1}{8}\right)^{2/3}$ **70.** $\left(\frac{4}{9}\right)^{3/2}$

71. $(25x^4)^{3/2}$ **72.** $(27a^3b^3)^{2/3}$ **73.** $\left(\frac{8x^3}{27}\right)^{2/3}$ **74.** $\left(\frac{27}{64y^6}\right)^{2/3}$

In Exercises 75–90, write each expression without using negative exponents. Assume that all variables represent positive numbers.

75. $4^{-1/2}$ **76.** $8^{-1/3}$ **77.** $(4)^{-3/2}$ **78.** $25^{-5/2}$

79. $(16x^2)^{-3/2}$ **80.** $(81c^4)^{-3/2}$ **81.** $(-27y^3)^{-2/3}$ **82.** $(-8z^9)^{-2/3}$

83. $(-32p^5)^{-2/5}$ **84.** $(16q^6)^{-5/2}$ **85.** $\left(\frac{1}{4}\right)^{-3/2}$ **86.** $\left(\frac{4}{25}\right)^{-3/2}$

87. $\left(\frac{27}{8}\right)^{-4/3}$ **88.** $\left(\frac{25}{49}\right)^{-3/2}$ **89.** $\left(-\frac{8x^3}{27}\right)^{-1/3}$ **90.** $\left(\frac{16}{81y^4}\right)^{-3/4}$

In Exercises 91–102, do the operations. Write answers without negative exponents. Assume that all variables represent positive numbers.

91. $5^{4/9}5^{4/9}$ **92.** $4^{2/5}4^{2/5}$ **93.** $(4^{1/5})^3$ **94.** $(3^{1/3})^5$

95. $\frac{9^{4/5}}{9^{3/5}}$ **96.** $\frac{7^{2/3}}{7^{1/2}}$ **97.** $\frac{7^{1/2}}{7^0}$ **98.** $5^{1/3}5^{-5/3}$

99. $6^{-2/3}6^{-4/3}$ **100.** $\frac{3^{4/3}3^{1/3}}{3^{2/3}}$ **101.** $\frac{2^{5/6}2^{1/3}}{2^{1/2}}$ **102.** $\frac{5^{1/3}5^{1/2}}{5^{1/3}}$

In Exercises 103–114, do the operations. Assume that all variables are positive, and write all answers without using negative exponents.

103. $a^{2/3}a^{1/3}$
104. $b^{3/5}b^{1/5}$
105. $(a^{2/3})^{1/3}$
106. $(t^{4/5})^{10}$
107. $(a^{1/2}b^{1/3})^{3/2}$
108. $(a^{3/5}b^{3/2})^{2/3}$
109. $(mn^{-2/3})^{-3/5}$
110. $(r^{-2}s^{3})^{1/3}$
111. $\dfrac{(4x^3y)^{1/2}}{(9xy)^{1/2}}$
112. $\dfrac{(27x^3y)^{1/3}}{(8xy^2)^{2/3}}$
113. $(27x^{-3})^{-1/3}$
114. $(16a^{-2})^{-1/2}$

In Exercises 115–126, do the multiplications. Assume that all variables represent positive numbers, and write all answers without using negative exponents.

115. $y^{1/3}(y^{2/3} + y^{5/3})$
116. $y^{2/5}(y^{-2/5} + y^{3/5})$
117. $x^{3/5}(x^{7/5} - x^{2/5} + 1)$
118. $x^{4/3}(x^{2/3} + 3x^{5/3} - 4)$
119. $(x^{1/2} + 2)(x^{1/2} - 2)$
120. $(x^{1/2} + y^{1/2})(x^{1/2} - y^{1/2})$
121. $(x^{2/3} - x)(x^{2/3} + x)$
122. $(x^{1/3} + x^2)(x^{1/3} - x^2)$
123. $(x^{2/3} + y^{2/3})^2$
124. $(a^{1/2} - b^{2/3})^2$
125. $(a^{3/2} - b^{3/2})^2$
126. $(x^{-1/2} - x^{1/2})^2$

In Exercises 127–130, use rational exponents to simplify each radical. Assume that all variables represent positive numbers.

127. $\sqrt[6]{p^3}$
128. $\sqrt[8]{q^2}$
129. $\sqrt[4]{25b^2}$
130. $\sqrt[9]{-8x^6}$

WRITING

131. Explain how you would decide whether $a^{1/n}$ is a real number.

132. The expression $(a^{1/2} + b^{1/2})^2$ is not equal to $a + b$. Explain.

SOMETHING TO THINK ABOUT

133. The fraction $\frac{2}{4}$ is equal to $\frac{1}{2}$. Is $16^{2/4}$ equal to $16^{1/2}$? Explain.

134. How would you evaluate an expression with a mixed-number exponent? For example, what is $8^{1\frac{1}{3}}$? What is $25^{2\frac{1}{2}}$? Discuss.

7.5 Simplifying and Combining Radical Expressions

■ PROPERTIES OF RADICALS ■ SIMPLIFYING RADICAL EXPRESSIONS ■ ADDING AND SUBTRACTING RADICAL EXPRESSIONS ■ SOME SPECIAL TRIANGLES

Getting Ready *Simplify each radical. Assume that all variables represent positive numbers.*

1. $\sqrt{225}$
2. $\sqrt{576}$
3. $\sqrt[3]{125}$
4. $\sqrt[3]{343}$
5. $\sqrt{16x^4}$
6. $\sqrt{\dfrac{64}{121}x^6}$
7. $\sqrt[3]{27a^3b^9}$
8. $\sqrt[3]{-8a^{12}}$

PROPERTIES OF RADICALS

Many properties of exponents have counterparts in radical notation. For example, because $a^{1/n}b^{1/n} = (ab)^{1/n}$, we have

1. $\sqrt[n]{a}\sqrt[n]{b} = \sqrt[n]{ab}$

For example,

$$\sqrt{5}\sqrt{5} = \sqrt{5 \cdot 5} = \sqrt{5^2} = 5$$
$$\sqrt[3]{7x}\sqrt[3]{49x^2} = \sqrt[3]{7x \cdot 7^2x^2} = \sqrt[3]{7^3 \cdot x^3} = 7x$$
$$\sqrt[4]{2x^3}\sqrt[4]{8x} = \sqrt[4]{2x^3 \cdot 2^3x} = \sqrt[4]{2^4 \cdot x^4} = 2|x|$$

If we rewrite Equation 1, we have the following rule.

Multiplication Property of Radicals
If $\sqrt[n]{a}$ and $\sqrt[n]{b}$ are real numbers, then

$$\sqrt[n]{ab} = \sqrt[n]{a}\sqrt[n]{b}$$

As long as all radicals represent real numbers, *the nth root of the product of two numbers is equal to the product of their nth roots.*

WARNING! The multiplication property of radicals applies to the *n*th root of the product of two numbers. There is no such property for sums or differences. For example,

$$\sqrt{9+4} \neq \sqrt{9} + \sqrt{4} \qquad \sqrt{9-4} \neq \sqrt{9} - \sqrt{4}$$
$$\sqrt{13} \neq 3 + 2 \qquad \sqrt{5} \neq 3 - 2$$
$$\sqrt{13} \neq 5 \qquad \sqrt{5} \neq 1$$

Thus, $\sqrt{a+b} \neq \sqrt{a} + \sqrt{b}$ and $\sqrt{a-b} \neq \sqrt{a} - \sqrt{b}$.

A second property of radicals involves quotients. Because

$$\frac{a^{1/n}}{b^{1/n}} = \left(\frac{a}{b}\right)^{1/n}$$

it follows that

2. $\dfrac{\sqrt[n]{a}}{\sqrt[n]{b}} = \sqrt[n]{\dfrac{a}{b}} \quad (b \neq 0)$

For example,

$$\frac{\sqrt{8x^3}}{\sqrt{2x}} = \sqrt{\frac{8x^3}{2x}} = \sqrt{4x^2} = 2x \quad (x > 0)$$
$$\frac{\sqrt[3]{54x^5}}{\sqrt[3]{2x^2}} = \sqrt[3]{\frac{54x^5}{2x^2}} = \sqrt[3]{27x^3} = 3x$$

If we rewrite Equation 2, we have the following rule.

Quotient Property of Radicals

If $\sqrt[n]{a}$ and $\sqrt[n]{b}$ are real numbers, then

$$\sqrt[n]{\frac{a}{b}} = \frac{\sqrt[n]{a}}{\sqrt[n]{b}} \quad (b \neq 0)$$

As long as all radicals represent real numbers, *the nth root of the quotient of two numbers is equal to the quotient of their nth roots.*

SIMPLIFYING RADICAL EXPRESSIONS

A radical expression is said to be in simplest form when each of the following statements is true.

Simplified Form of a Radical Expression

A radical expression is in simplest form when

1. No radicals appear in the denominator of a fraction.
2. The radicand contains no fractions or negative numbers.
3. Each prime and variable factor in the radicand appears to a power that is less than the index of the radical.

EXAMPLE 1 Simplify **a.** $\sqrt{12}$, **b.** $\sqrt{98}$, and **c.** $\sqrt[3]{54}$.

Solution **a.** Recall that numbers that are squares of integers, such as 1, 4, 9, 16, 25, and 36, are *perfect squares.* To simplify $\sqrt{12}$, we first factor 12 so that one factor is the largest perfect square that divides 12. Since 4 is the largest perfect square factor of 12, we write 12 as $4 \cdot 3$, use the multiplication property of radicals, and simplify.

$$\begin{aligned} \sqrt{12} &= \sqrt{4 \cdot 3} && \text{Write 12 as } 4 \cdot 3. \\ &= \sqrt{4}\sqrt{3} && \sqrt{4 \cdot 3} = \sqrt{4}\sqrt{3}. \\ &= 2\sqrt{3} && \sqrt{4} = 2. \end{aligned}$$

b. The largest perfect square factor of 98 is 49. Thus,

$$\begin{aligned} \sqrt{98} &= \sqrt{49 \cdot 2} && \text{Write 98 as } 49 \cdot 2. \\ &= \sqrt{49}\sqrt{2} && \sqrt{49 \cdot 2} = \sqrt{49}\sqrt{2}. \\ &= 7\sqrt{2} && \sqrt{49} = 7. \end{aligned}$$

c. Numbers that are cubes of integers, such as 1, 8, 27, 64, 125, and 216, are called *perfect cubes.* Since the largest perfect cube factor of 54 is 27, we have

$$\begin{aligned} \sqrt[3]{54} &= \sqrt[3]{27 \cdot 2} && \text{Write 54 as } 27 \cdot 2. \\ &= \sqrt[3]{27}\sqrt[3]{2} && \sqrt[3]{27 \cdot 2} = \sqrt[3]{27}\sqrt[3]{2}. \\ &= 3\sqrt[3]{2} && \sqrt[3]{27} = 3. \end{aligned}$$

■

Self Check Simplify **a.** $\sqrt{20}$ and **b.** $\sqrt[3]{24}$.

Answers **a.** $2\sqrt{5}$, **b.** $2\sqrt[3]{3}$

EXAMPLE 2 Simplify **a.** $\sqrt{\dfrac{15}{49x^2}}$ $(x > 0)$ and **b.** $\sqrt[3]{\dfrac{10x^2}{27y^6}}$ $(y \neq 0)$.

Solution **a.** We can write the square root of the quotient as the quotient of the square roots and simplify the denominator. Since $x > 0$, we have

$$\sqrt{\frac{15}{49x^2}} = \frac{\sqrt{15}}{\sqrt{49x^2}}$$
$$= \frac{\sqrt{15}}{7x}$$

b. We can write the cube root of the quotient as the quotient of two cube roots. Since $y \neq 0$, we have

$$\sqrt[3]{\frac{10x^2}{27y^6}} = \frac{\sqrt[3]{10x^2}}{\sqrt[3]{27y^6}}$$
$$= \frac{\sqrt[3]{10x^2}}{3y^2}$$

■

Self Check Simplify **a.** $\sqrt{\dfrac{11}{36a^2}}$ $(a > 0)$ and **b.** $\sqrt[3]{\dfrac{8a^2}{125y^3}}$ $(y \neq 0)$.

Answers **a.** $\dfrac{\sqrt{11}}{6a}$, **b.** $\dfrac{2\sqrt[3]{a^2}}{5y}$

EXAMPLE 3 Simplify each expression. Assume that all variables represent positive numbers.

a. $\sqrt{128a^5}$, **b.** $\sqrt[3]{24x^5}$, **c.** $\dfrac{\sqrt{45xy^2}}{\sqrt{5x}}$, and **d.** $\dfrac{\sqrt[3]{-432x^5}}{\sqrt[3]{8x}}$.

Solution **a.** We write $128a^5$ as $64a^4 \cdot 2a$ and use the multiplication property of radicals.

$\sqrt{128a^5} = \sqrt{64a^4 \cdot 2a}$	$64a^4$ is the largest perfect square that divides $128a^5$.
$= \sqrt{64a^4}\sqrt{2a}$	Use the multiplication property of radicals.
$= 8a^2\sqrt{2a}$	$\sqrt{64a^4} = 8a^2$.

b. We write $24x^5$ as $8x^3 \cdot 3x^2$ and use the multiplication property of radicals.

$\sqrt[3]{24x^5} = \sqrt[3]{8x^3 \cdot 3x^2}$	$8x^3$ is the largest perfect cube that divides $24x^5$.
$= \sqrt[3]{8x^3}\sqrt[3]{3x^2}$	Use the multiplication property of radicals.
$= 2x\sqrt[3]{3x^2}$	$\sqrt[3]{8x^3} = 2x$.

c. We can write the quotient of the square roots as the square root of a quotient.

$$\frac{\sqrt{45xy^2}}{\sqrt{5x}} = \sqrt{\frac{45xy^2}{5x}}$$ Use the quotient property of radicals.

$$= \sqrt{9y^2}$$ Simplify the fraction.

$$= 3y$$

d. We can write the quotient of the cube roots as the cube root of a quotient.

$$\frac{\sqrt[3]{-432x^5}}{\sqrt[3]{8x}} = \sqrt[3]{\frac{-432x^5}{8x}}$$ Use the quotient property of radicals.

$$= \sqrt[3]{-54x^4}$$ Simplify the fraction.

$$= \sqrt[3]{-27x^3 \cdot 2x}$$ $-27x^3$ is the largest perfect cube that divides $-54x^4$.

$$= \sqrt[3]{-27x^3}\sqrt[3]{2x}$$ Use the multiplication property of radicals.

$$= -3x\sqrt[3]{2x}$$

■

Self Check Simplify **a.** $\sqrt{98b^3}$, **b.** $\sqrt[3]{54y^5}$, and **c.** $\frac{\sqrt{50ab^2}}{\sqrt{2a}}$. Assume that all variables represent positive numbers.

Answers **a.** $7b\sqrt{2b}$, **b.** $3y\sqrt[3]{2y^2}$, **c.** $5b$

To simplify more complicated radicals, we can use the prime factorization of the radicand to find its perfect square factors. For example, to simplify $\sqrt{3{,}168x^5y^7}$, we first find the prime factorization of $3{,}168x^5y^7$.

$$3{,}168x^5y^7 = 2^5 \cdot 3^2 \cdot 11 \cdot x^5 \cdot y^7$$

Then we have

$$\sqrt{3{,}168x^5y^7} = \sqrt{2^4 \cdot 3^2 \cdot x^4 \cdot y^6 \cdot 2 \cdot 11 \cdot x \cdot y}$$

$$= \sqrt{2^4 \cdot 3^2 \cdot x^4 \cdot y^6}\sqrt{2 \cdot 11 \cdot x \cdot y}$$ Write each perfect square under the left radical and each nonperfect square under the right radical.

$$= 2^2 \cdot 3x^2y^3\sqrt{22xy}$$

$$= 12x^2y^3\sqrt{22xy}$$

■ ADDING AND SUBTRACTING RADICAL EXPRESSIONS

Radical expressions with the same index and the same radicand are called **like** or **similar radicals.** For example, $3\sqrt{2}$ and $2\sqrt{2}$ are like radicals. However,

$3\sqrt{5}$ and $4\sqrt{2}$ are not like radicals, because the radicands are different.

$3\sqrt{5}$ and $2\sqrt[3]{5}$ are not like radicals, because the indexes are different.

We can often combine like terms. For example, to simplify the expression $3\sqrt{2} + 2\sqrt{2}$, we use the distributive property to factor out $\sqrt{2}$ and simplify.

$$\begin{aligned} 3\sqrt{2} + 2\sqrt{2} &= (3 + 2)\sqrt{2} \\ &= 5\sqrt{2} \end{aligned}$$

Radicals with the same index but different radicands can often be written as like radicals. For example, to simplify the expression $\sqrt{27} - \sqrt{12}$, we simplify both radicals and then combine the like radicals.

$$\begin{aligned} \sqrt{27} - \sqrt{12} &= \sqrt{9 \cdot 3} - \sqrt{4 \cdot 3} \\ &= \sqrt{9}\sqrt{3} - \sqrt{4}\sqrt{3} && \sqrt{ab} = \sqrt{a}\sqrt{b} \\ &= 3\sqrt{3} - 2\sqrt{3} && \sqrt{9} = 3 \text{ and } \sqrt{4} = 2. \\ &= (3 - 2)\sqrt{3} && \text{Factor out } \sqrt{3}. \\ &= \sqrt{3} \end{aligned}$$

As the previous examples suggest, we can use the following rule to add or subtract radicals.

Adding and Subtracting Radicals

To add or subtract radicals, simplify each radical and combine all like radicals. To combine like radicals, add the coefficients and keep the common radical.

EXAMPLE 4 Simplify $2\sqrt{12} - 3\sqrt{48} + 3\sqrt{3}$.

Solution We simplify each radical separately and combine like radicals.

$$\begin{aligned} 2\sqrt{12} - 3\sqrt{48} + 3\sqrt{3} &= 2\sqrt{4 \cdot 3} - 3\sqrt{16 \cdot 3} + 3\sqrt{3} \\ &= 2\sqrt{4}\sqrt{3} - 3\sqrt{16}\sqrt{3} + 3\sqrt{3} \\ &= 2(2)\sqrt{3} - 3(4)\sqrt{3} + 3\sqrt{3} \\ &= 4\sqrt{3} - 12\sqrt{3} + 3\sqrt{3} \\ &= (4 - 12 + 3)\sqrt{3} \\ &= -5\sqrt{3} \end{aligned}$$

■

Self Check Simplify $3\sqrt{75} - 2\sqrt{12} + 2\sqrt{48}$.

Answer $19\sqrt{3}$

EXAMPLE 5 Simplify $\sqrt[3]{16} - \sqrt[3]{54} + \sqrt[3]{24}$.

Solution We simplify each radical separately and combine like radicals:

$$\begin{aligned} \sqrt[3]{16} - \sqrt[3]{54} + \sqrt[3]{24} &= \sqrt[3]{8 \cdot 2} - \sqrt[3]{27 \cdot 2} + \sqrt[3]{8 \cdot 3} \\ &= \sqrt[3]{8}\sqrt[3]{2} - \sqrt[3]{27}\sqrt[3]{2} + \sqrt[3]{8}\sqrt[3]{3} \\ &= 2\sqrt[3]{2} - 3\sqrt[3]{2} + 2\sqrt[3]{3} \\ &= -\sqrt[3]{2} + 2\sqrt[3]{3} \end{aligned}$$

WARNING! We cannot combine $-\sqrt[3]{2}$ and $2\sqrt[3]{3}$, because the radicals have different radicands.

■

Self Check Simplify $\sqrt[3]{24} - \sqrt[3]{16} + \sqrt[3]{54}$.

Answer $2\sqrt[3]{3} + \sqrt[3]{2}$

EXAMPLE 6 Simplify $\sqrt[3]{16x^4} + \sqrt[3]{54x^4} - \sqrt[3]{-128x^4}$ $(x > 0)$.

Solution We simplify each radical separately, factor out $\sqrt[3]{2x}$, and simplify.

$$\begin{aligned}
&\sqrt[3]{16x^4} + \sqrt[3]{54x^4} - \sqrt[3]{-128x^4} \\
&\quad = \sqrt[3]{8x^3 \cdot 2x} + \sqrt[3]{27x^3 \cdot 2x} - \sqrt[3]{-64x^3 \cdot 2x} \\
&\quad = \sqrt[3]{8x^3}\sqrt[3]{2x} + \sqrt[3]{27x^3}\sqrt[3]{2x} - \sqrt[3]{-64x^3}\sqrt[3]{2x} \\
&\quad = 2x\sqrt[3]{2x} + 3x\sqrt[3]{2x} + 4x\sqrt[3]{2x} \\
&\quad = (2x + 3x + 4x)\sqrt[3]{2x} \\
&\quad = 9x\sqrt[3]{2x}
\end{aligned}$$

■

Self Check Simplify $\sqrt{32x^3} + \sqrt{50x^3} - \sqrt{18x^3}$ $(x > 0)$.

Answer $6x\sqrt{2x}$

SOME SPECIAL TRIANGLES

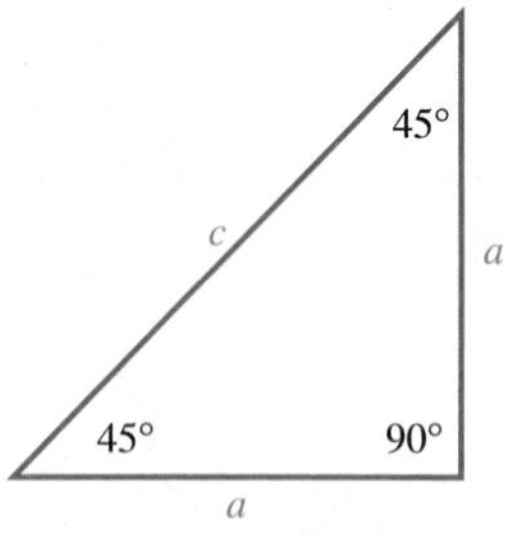

FIGURE 7-13

An **isosceles right triangle** is a right triangle with two legs of equal length. If we know the length of one leg of an isosceles right triangle, we can use the Pythagorean theorem to find the length of the hypotenuse. Since the triangle shown in Figure 7-13 is a right triangle, we have

$c^2 = a^2 + a^2$	Use the Pythagorean theorem.
$c^2 = 2a^2$	Combine like terms.
$\sqrt{c^2} = \sqrt{2a^2}$	Take the positive square root of both sides.
$c = a\sqrt{2}$	$\sqrt{2a^2} = \sqrt{2}\sqrt{a^2} = \sqrt{2}a = a\sqrt{2}$. No absolute value symbols are needed, because a is positive.

Thus, *in an isosceles right triangle, the length of the hypotenuse is the length of one leg times* $\sqrt{2}$.

EXAMPLE 7 If one leg of the isosceles right triangle shown in Figure 7-13 is 10 feet long, find the length of the hypotenuse.

Solution Since the length of the hypotenuse is the length of a leg times $\sqrt{2}$, we have

$$c = 10\sqrt{2}$$

The length of the hypotenuse is $10\sqrt{2}$ feet. To two decimal places, the length is 14.14 feet. ■

Self Check Find the length of the hypotenuse of an isosceles right triangle if one leg is 12 meters long.

Answer $12\sqrt{2}$ m

If the length of the hypotenuse of an isosceles right triangle is known, we can use the Pythagorean theorem to find the length of each leg.

EXAMPLE 8 Find the length of each leg of the isosceles right triangle shown in Figure 7-14.

Solution We use the Pythagorean theorem.

$$c^2 = a^2 + a^2$$

$$25^2 = 2a^2 \quad \text{Substitute 25 for } c \text{ and combine like terms.}$$

$$\frac{625}{2} = a^2 \quad \text{Square 25 and divide both sides by 2.}$$

$$\sqrt{\frac{625}{2}} = a \quad \text{Take the positive square root of both sides.}$$

$$a = 17.67766953 \quad \text{Use a calculator.}$$

To two decimal places, the length is 17.68 units. ■

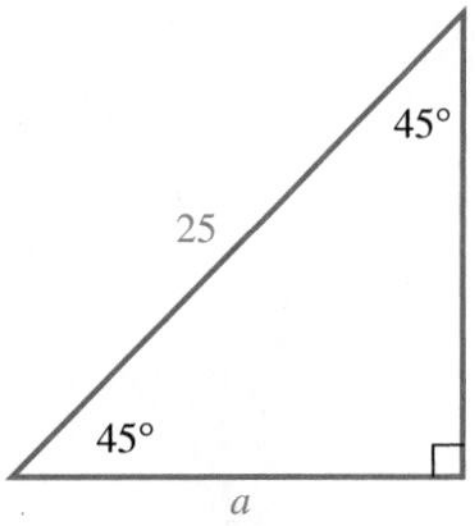

FIGURE 7-14

From geometry, we know that an **equilateral triangle** is a triangle with three sides of equal length and three 60° angles. If an **altitude** is drawn upon the base of an equilateral triangle, as shown in Figure 7-15, it bisects the base and divides the triangle into two 30°–60°–90° triangles. We can see that the shortest leg of each 30°–60°–90° triangle is a units long. Thus,

The shorter leg of a 30°–60°–90° right triangle is half as long as its hypotenuse.

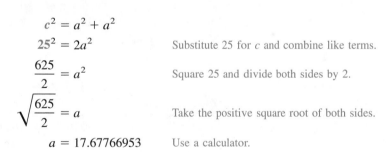

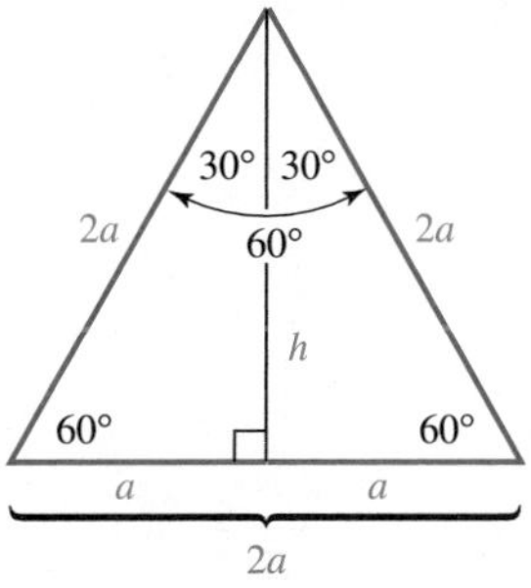

FIGURE 7-15

We can find the length of the altitude, h, by using the Pythagorean theorem.

$a^2 + h^2 = (2a)^2$

$a^2 + h^2 = 4a^2$ — $(2a)^2 = (2a)(2a) = 4a^2$.

$h^2 = 3a^2$ — Subtract a^2 from both sides.

$h = \sqrt{3a^2}$ — Take the positive square root of both sides.

$h = a\sqrt{3}$ — $\sqrt{3a^2} = \sqrt{3}\sqrt{a^2} = a\sqrt{3}$. No absolute value symbols are needed, because a is positive

Thus,

The length of the longer leg is the length of the shorter side times $\sqrt{3}$.

EXAMPLE 9 Find the length of the hypotenuse and the longer leg of the right triangle shown in Figure 7-16.

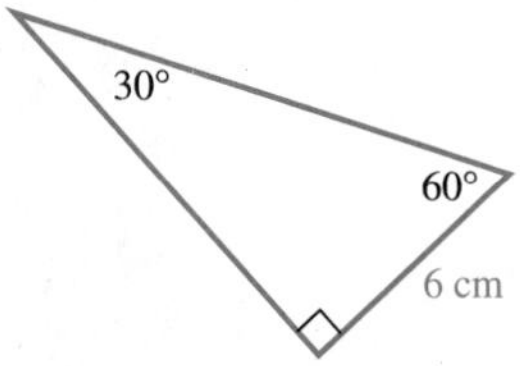

FIGURE 7-16

Solution Since the shorter leg of a 30°–60°–90° right triangle is half as long as its hypotenuse, the hypotenuse is 12 centimeters long.

Since the length of the longer leg is the length of the shorter leg times $\sqrt{3}$, the longer leg is $6\sqrt{3}$ (about 10.39) centimeters long. ■

Self Check Find the length of the hypotenuse and the longer leg of a 30°–60°–90° right triangle if the shorter leg is 8 centimeters long.

Answers 16 cm, $8\sqrt{3}$ cm

EXAMPLE 10 Find the length of each leg of the triangle shown in Figure 7-17.

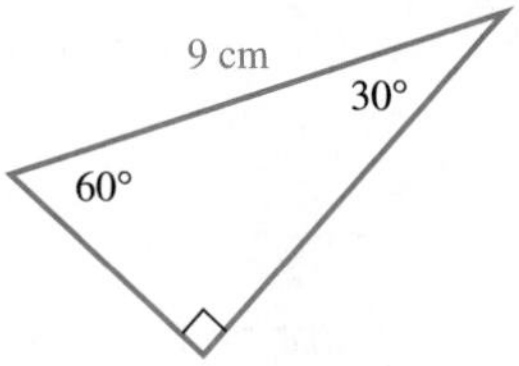

FIGURE 7-17

Solution Since the shorter leg of a 30°–60°–90° right triangle is half as long as its hypotenuse, the shorter leg is $\frac{9}{2}$ centimeters long.

Since the length of the longer leg is the length of the shorter leg times $\sqrt{3}$, the longer leg is $\frac{9}{2}\sqrt{3}$ (or about 7.79) centimeters long. ■

Orals *Simplify.*

1. $\sqrt{7}\sqrt{7}$ **2.** $\sqrt[3]{4^2}\sqrt[3]{4}$ **3.** $\dfrac{\sqrt[3]{54}}{\sqrt[3]{2}}$

Simplify each expression. Assume that $b \neq 0$.

4. $\sqrt{18}$ **5.** $\sqrt[3]{16}$ **6.** $\sqrt[3]{\dfrac{3x^2}{64b^6}}$

Combine like terms.

7. $3\sqrt{3} + 4\sqrt{3}$

8. $5\sqrt{7} - 2\sqrt{7}$

9. $2\sqrt[3]{9} + 3\sqrt[3]{9}$

10. $10\sqrt[5]{4} - 2\sqrt[5]{4}$

EXERCISE 7.5

REVIEW *Do each operation.*

1. $3x^2y^3(-5x^3y^{-4})$

2. $-2a^2b^{-2}(4a^{-2}b^4 - 2a^2b + 3a^3b^2)$

3. $(3t + 2)^2$

4. $(5r - 3s)(5r + 2s)$

5. $2p - 5\overline{)6p^2 - 7p - 25}$

6. $3m + n\overline{)6m^3 - m^2n + 2mn^2 + n^3}$

VOCABULARY AND CONCEPTS *Fill in each blank to make a true statement.*

7. $\sqrt[n]{ab} =$ ______

8. $\sqrt[n]{\dfrac{a}{b}} =$ ___

PRACTICE *In Exercises 9–24, simplify each expression. Assume that all variables represent positive numbers.*

9. $\sqrt{6}\sqrt{6}$

10. $\sqrt{11}\sqrt{11}$

11. $\sqrt{t}\sqrt{t}$

12. $-\sqrt{z}\sqrt{z}$

13. $\sqrt[3]{5x^2}\sqrt[3]{25x}$

14. $\sqrt[4]{25a}\sqrt[4]{25a^3}$

15. $\dfrac{\sqrt{500}}{\sqrt{5}}$

16. $\dfrac{\sqrt{128}}{\sqrt{2}}$

17. $\dfrac{\sqrt{98x^3}}{\sqrt{2x}}$

18. $\dfrac{\sqrt{75y^5}}{\sqrt{3y}}$

19. $\dfrac{\sqrt{180ab^4}}{\sqrt{5ab^2}}$

20. $\dfrac{\sqrt{112ab^3}}{\sqrt{7ab}}$

21. $\dfrac{\sqrt[3]{48}}{\sqrt[3]{6}}$

22. $\dfrac{\sqrt[3]{64}}{\sqrt[3]{8}}$

23. $\dfrac{\sqrt[3]{189a^4}}{\sqrt[3]{7a}}$

24. $\dfrac{\sqrt[3]{243x^7}}{\sqrt[3]{9x}}$

In Exercises 25–44, simplify each radical.

25. $\sqrt{20}$

26. $\sqrt{8}$

27. $-\sqrt{200}$

28. $-\sqrt{250}$

29. $\sqrt[3]{80}$

30. $\sqrt[3]{270}$

31. $\sqrt[3]{-81}$

32. $\sqrt[3]{-72}$

33. $\sqrt[4]{32}$

34. $\sqrt[4]{48}$

35. $\sqrt[5]{96}$

36. $\sqrt[7]{256}$

37. $\sqrt{\dfrac{7}{9}}$

38. $\sqrt{\dfrac{3}{4}}$

39. $\sqrt[3]{\dfrac{7}{64}}$

40. $\sqrt[3]{\dfrac{4}{125}}$

41. $\sqrt[4]{\dfrac{3}{10{,}000}}$

42. $\sqrt[5]{\dfrac{4}{243}}$

43. $\sqrt[5]{\dfrac{3}{32}}$

44. $\sqrt[6]{\dfrac{5}{64}}$

In Exercises 45–64, simplify each radical. Assume that all variables represent positive numbers.

45. $\sqrt{50x^2}$ **46.** $\sqrt{75a^2}$ **47.** $\sqrt{32b}$ **48.** $\sqrt{80c}$

49. $-\sqrt{112a^3}$ **50.** $\sqrt{147a^5}$ **51.** $\sqrt{175a^2b^3}$ **52.** $\sqrt{128a^3b^5}$

53. $-\sqrt{300xy}$ **54.** $\sqrt{200x^2y}$ **55.** $\sqrt[3]{-54x^6}$ **56.** $-\sqrt[3]{-81a^3}$

57. $\sqrt[3]{16x^{12}y^3}$ **58.** $\sqrt[3]{40a^3b^6}$ **59.** $\sqrt[4]{32x^{12}y^4}$ **60.** $\sqrt[5]{64x^{10}y^5}$

61. $\sqrt{\dfrac{z^2}{16x^2}}$ **62.** $\sqrt{\dfrac{b^4}{64a^8}}$ **63.** $\sqrt[4]{\dfrac{5x}{16z^4}}$ **64.** $\sqrt[3]{\dfrac{11a^2}{125b^6}}$

In Exercises 65–104, simplify and combine like radicals. All variables represent positive numbers.

65. $4\sqrt{2x}+6\sqrt{2x}$ **66.** $6\sqrt[3]{5y}+3\sqrt[3]{5y}$

67. $8\sqrt[5]{7a^2}-7\sqrt[5]{7a^2}$ **68.** $10\sqrt[6]{12xyz}-\sqrt[6]{12xyz}$

69. $\sqrt{3}+\sqrt{27}$ **70.** $\sqrt{8}+\sqrt{32}$

71. $\sqrt{2}-\sqrt{8}$ **72.** $\sqrt{20}-\sqrt{125}$

73. $\sqrt{98}-\sqrt{50}$ **74.** $\sqrt{72}-\sqrt{200}$

75. $3\sqrt{24}+\sqrt{54}$ **76.** $\sqrt{18}+2\sqrt{50}$

77. $\sqrt[3]{24}+\sqrt[3]{3}$ **78.** $\sqrt[3]{16}+\sqrt[3]{128}$

79. $\sqrt[3]{32}-\sqrt[3]{108}$ **80.** $\sqrt[3]{80}-\sqrt[3]{10{,}000}$

81. $2\sqrt[3]{125}-5\sqrt[3]{64}$ **82.** $3\sqrt[3]{27}+12\sqrt[3]{216}$

83. $14\sqrt[4]{32}-15\sqrt[4]{162}$ **84.** $23\sqrt[4]{768}+\sqrt[4]{48}$

85. $3\sqrt[4]{512}+2\sqrt[4]{32}$ **86.** $4\sqrt[4]{243}-\sqrt[4]{48}$

87. $\sqrt{98}-\sqrt{50}-\sqrt{72}$ **88.** $\sqrt{20}+\sqrt{125}-\sqrt{80}$

89. $\sqrt{18}+\sqrt{300}-\sqrt{243}$ **90.** $\sqrt{80}-\sqrt{128}+\sqrt{288}$

91. $2\sqrt[3]{16}-\sqrt[3]{54}-3\sqrt[3]{128}$ **92.** $\sqrt[4]{48}-\sqrt[4]{243}-\sqrt[4]{768}$

93. $\sqrt{25y^2z}-\sqrt{16y^2z}$ **94.** $\sqrt{25yz^2}+\sqrt{9yz^2}$

95. $\sqrt{36xy^2}+\sqrt{49xy^2}$ **96.** $3\sqrt{2x}-\sqrt{8x}$

97. $2\sqrt[3]{64a}+2\sqrt[3]{8a}$ **98.** $3\sqrt[4]{x^4y}-2\sqrt[4]{x^4y}$

99. $\sqrt{y^5}-\sqrt{9y^5}-\sqrt{25y^5}$ **100.** $\sqrt{8y^7}+\sqrt{32y^7}-\sqrt{2y^7}$

101. $\sqrt[5]{x^6y^2}+\sqrt[5]{32x^6y^2}+\sqrt[5]{x^6y^2}$ **102.** $\sqrt[3]{xy^4}+\sqrt[3]{8xy^4}-\sqrt[3]{27xy^4}$

103. $\sqrt{x^2+2x+1}+\sqrt{x^2+2x+1}$ **104.** $\sqrt{4x^2+12x+9}+\sqrt{9x^2+6x+1}$

In Exercises 105–112, find the missing lengths in each triangle. Give each answer to two decimal places.

105.

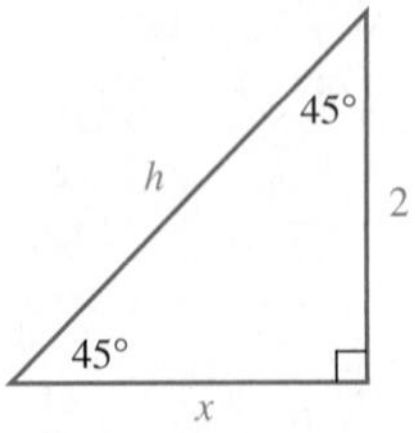

106.

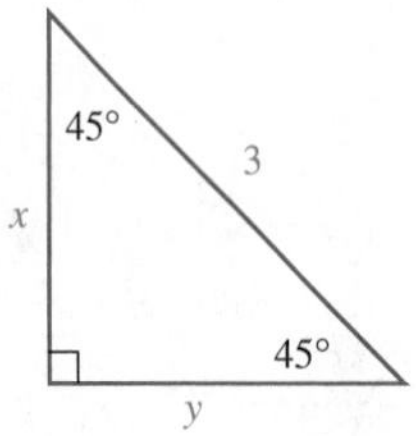

107.

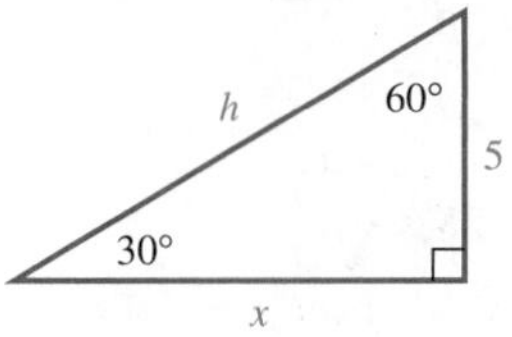

108.

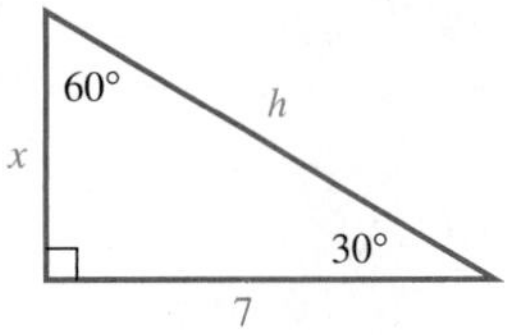

109.

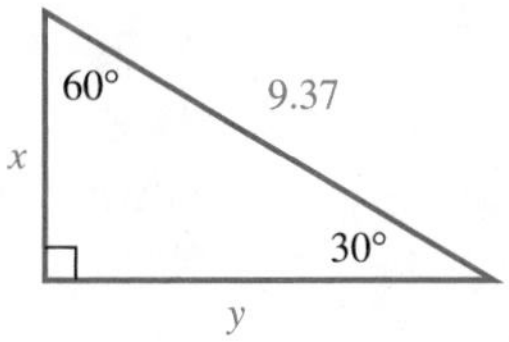

110.

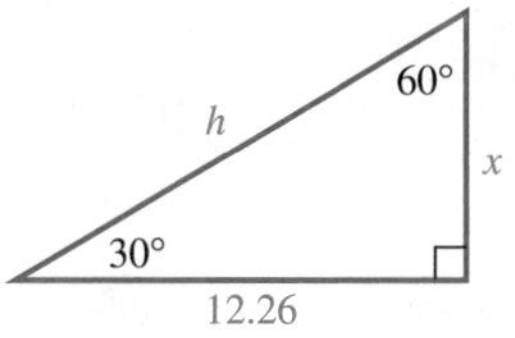

111.

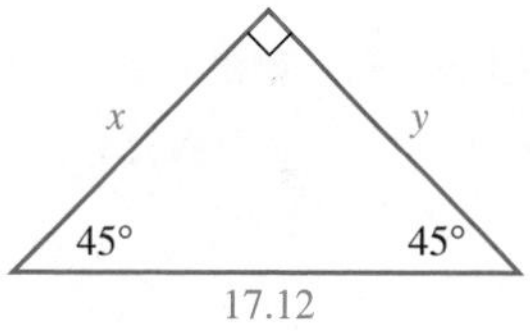

112.

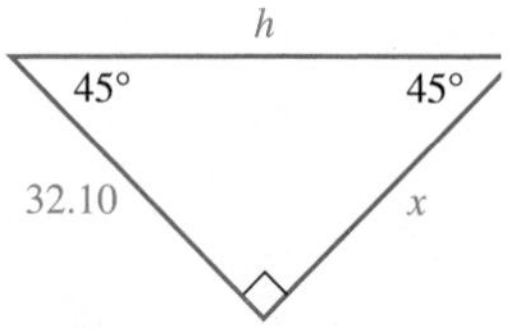

WRITING

113. Explain how to recognize like radicals.

114. Explain how to combine like radicals.

SOMETHING TO THINK ABOUT

115. Can you find any numbers a and b such that $\sqrt{a+b} = \sqrt{a} + \sqrt{b}$?

116. Find the sum: $\sqrt{3} + \sqrt{3^2} + \sqrt{3^3} + \sqrt{3^4} + \sqrt{3^5}$.

7.6 Multiplying and Dividing Radical Expressions

■ MULTIPLYING A MONOMIAL BY A MONOMIAL ■ MULTIPLYING A POLYNOMIAL BY A MONOMIAL ■ MULTIPLYING A POLYNOMIAL BY A POLYNOMIAL ■ PROBLEM SOLVING ■ RATIONALIZING DENOMINATORS ■ RATIONALIZING NUMERATORS

Getting Ready *Do each operation and simplify, if possible.*

1. a^3a^4 **2.** $\dfrac{b^5}{b^2}$ **3.** $a(a-2)$ **4.** $3b^2(2b+3)$

5. $(a+2)(a-5)$ **6.** $(2a+3b)(2a-3b)$

■ MULTIPLYING A MONOMIAL BY A MONOMIAL

Radical expressions with the same index can be multiplied and divided.

EXAMPLE 1 Multiply $3\sqrt{6}$ by $2\sqrt{3}$.

Solution We use the commutative and associative properties of multiplication to multiply the coefficients and the radicals separately. Then we simplify any radicals in the product, if possible.

$$3\sqrt{6} \cdot 2\sqrt{3} = 3(2)\sqrt{6}\sqrt{3} \quad \text{Multiply the coefficients and multiply the radicals.}$$
$$= 6\sqrt{18} \quad 3(2) = 6 \text{ and } \sqrt{6}\sqrt{3} = \sqrt{18}.$$
$$= 6\sqrt{9}\sqrt{2} \quad \sqrt{18} = \sqrt{9 \cdot 2} = \sqrt{9}\sqrt{2}.$$
$$= 6(3)\sqrt{2} \quad \sqrt{9} = 3.$$
$$= 18\sqrt{2}$$

■

Self Check Multiply $-2\sqrt{7}$ by $5\sqrt{2}$.

Answer $-10\sqrt{14}$

MULTIPLYING A POLYNOMIAL BY A MONOMIAL

To multiply a polynomial by a monomial, we use the distributive property to remove parentheses and then simplify each resulting term, if possible.

EXAMPLE 2 Multiply $3\sqrt{3}(4\sqrt{8} - 5\sqrt{10})$.

Solution

$$
\begin{aligned}
&3\sqrt{3}(4\sqrt{8} - 5\sqrt{10}) \\
&= 3\sqrt{3} \cdot 4\sqrt{8} - 3\sqrt{3} \cdot 5\sqrt{10} && \text{Use the distributive property.} \\
&= 12\sqrt{24} - 15\sqrt{30} && \text{Multiply the coefficients and multiply the radicals.} \\
&= 12\sqrt{4}\sqrt{6} - 15\sqrt{30} \\
&= 12(2)\sqrt{6} - 15\sqrt{30} \\
&= 24\sqrt{6} - 15\sqrt{30}
\end{aligned}
$$

■

Self Check Multiply $4\sqrt{2}(3\sqrt{5} - 2\sqrt{8})$.

Answer $12\sqrt{10} - 32$

MULTIPLYING A POLYNOMIAL BY A POLYNOMIAL

To multiply a binomial by a binomial, we use the FOIL method.

EXAMPLE 3 Multiply $(\sqrt{7} + \sqrt{2})(\sqrt{7} - 3\sqrt{2})$.

Solution

$$
\begin{aligned}
&(\sqrt{7} + \sqrt{2})(\sqrt{7} - 3\sqrt{2}) \\
&= (\sqrt{7})^2 - 3\sqrt{7}\sqrt{2} + \sqrt{2}\sqrt{7} - 3\sqrt{2}\sqrt{2} \\
&= 7 - 3\sqrt{14} + \sqrt{14} - 3(2) \\
&= 7 - 2\sqrt{14} - 6 \\
&= 1 - 2\sqrt{14}
\end{aligned}
$$

■

Self Check Multiply $(\sqrt{5} + 2\sqrt{3})(\sqrt{5} - \sqrt{3})$.

Answer $-1 + \sqrt{15}$

Technically, the expression $\sqrt{3x} - \sqrt{5}$ is not a polynomial, because the variable does not have a whole-number exponent $\left(\sqrt{3x} = 3^{1/2}x^{1/2}\right)$. However, we will multiply such expressions as if they were polynomials.

EXAMPLE 4 Multiply $\left(\sqrt{3x} - \sqrt{5}\right)\left(\sqrt{2x} + \sqrt{10}\right)$.

Solution

$$\left(\sqrt{3x} - \sqrt{5}\right)\left(\sqrt{2x} + \sqrt{10}\right)$$

$$\begin{aligned} &= \sqrt{3x}\sqrt{2x} + \sqrt{3x}\sqrt{10} - \sqrt{5}\sqrt{2x} - \sqrt{5}\sqrt{10} \\ &= \sqrt{6x^2} + \sqrt{30x} - \sqrt{10x} - \sqrt{50} \\ &= \sqrt{6}\sqrt{x^2} + \sqrt{30x} - \sqrt{10x} - \sqrt{25}\sqrt{2} \\ &= \sqrt{6}x + \sqrt{30x} - \sqrt{10x} - 5\sqrt{2} \end{aligned}$$

■

Self Check Multiply $\left(\sqrt{x} + 1\right)\left(\sqrt{x} - 3\right)$.

Answer $x - 2\sqrt{x} - 3$

WARNING! It is important to draw radical signs carefully so that they completely cover the radicand, but no more than the radicand. To avoid confusion, we often write an expression such as $\sqrt{6}x$ in the form $x\sqrt{6}$.

PROBLEM SOLVING

EXAMPLE 5 **Photography** Many camera lenses (see Figure 7-18) have an adjustable opening called the *aperture,* which controls the amount of light passing through the lens. The *f-number* of a lens is its *focal length* divided by the diameter of its circular aperture:

$$f\text{-number} = \frac{f}{d}$$

f is the focal length, and d is the diameter of the aperture.

FIGURE 7-18

A lens with a focal length of 12 centimeters and an aperture with a diameter of 6 centimeters has an f-number of $\frac{12}{6}$ and is an $f/2$ lens. If the area of the aperture is reduced to admit half as much light, the f-number of the lens will change. Find the new f-number.

Solution We first find the area of the aperture when its diameter is 6 centimeters.

$A = \pi r^2$ The formula for the area of a circle.

$A = \pi(3)^2$ Since a radius is half the diameter, substitute 3 for r.

$A = 9\pi$

When the size of the aperture is reduced to admit half as much light, the area of the aperture will be $\frac{9\pi}{2}$ square centimeters. To find the diameter of a circle with this area, we proceed as follows:

$A = \pi r^2$	The formula for the area of a circle.
$\frac{9\pi}{2} = \pi\left(\frac{d}{2}\right)^2$	Substitute $\frac{9\pi}{2}$ for A and $\frac{d}{2}$ for r.
$\frac{9\pi}{2} = \frac{\pi d^2}{4}$	$\left(\frac{d}{2}\right)^2 = \frac{d^2}{4}$.
$18 = d^2$	Multiply both sides by 4, and divide both sides by π.
$d = 3\sqrt{2}$	$\sqrt{18} = \sqrt{9}\sqrt{2} = 3\sqrt{2}$.

Since the focal length of the lens is still 12 centimeters and the diameter is now $3\sqrt{2}$ centimeters, the new f-number of the lens is

$f\text{-number} = \frac{f}{d} = \frac{12}{3\sqrt{2}}$	Substitute 12 for f and $3\sqrt{2}$ for d.
≈ 2.828427125	Use a calculator.

The lens is now an $f/2.8$ lens. ■

RATIONALIZING DENOMINATORS

To divide radical expressions, we **rationalize the denominator** of a fraction to replace the denominator with a rational number. For example, to divide $\sqrt{70}$ by $\sqrt{3}$, we write the division as the fraction

$$\frac{\sqrt{70}}{\sqrt{3}}$$

To eliminate the radical in the denominator, we multiply the numerator and the denominator by a number that will give a perfect square under the radical in the denominator. Because $3 \cdot 3 = 9$ and 9 is a perfect square, $\sqrt{3}$ is such a number.

$\frac{\sqrt{70}}{\sqrt{3}} = \frac{\sqrt{70} \cdot \sqrt{3}}{\sqrt{3} \cdot \sqrt{3}}$	Multiply numerator and denominator by $\sqrt{3}$.
$= \frac{\sqrt{210}}{3}$	Multiply the radicals.

Since there is no radical in the denominator and $\sqrt{210}$ cannot be simplified, the expression $\sqrt{210}/3$ is in simplest form, and the division is complete.

EXAMPLE 6 Rationalize the denominator: **a.** $\sqrt{\frac{20}{7}}$ and **b.** $\frac{4}{\sqrt[3]{2}}$.

Solution **a.** We first write the square root of the quotient as the quotient of two square roots:

$$\sqrt{\frac{20}{7}} = \frac{\sqrt{20}}{\sqrt{7}}$$

We then proceed as follows:

$$\frac{\sqrt{20}}{\sqrt{7}} = \frac{\sqrt{20} \cdot \sqrt{7}}{\sqrt{7} \cdot \sqrt{7}} \quad \text{Multiply numerator and denominator by } \sqrt{7}.$$

$$= \frac{\sqrt{140}}{7} \quad \text{Multiply the radicals.}$$

$$= \frac{2\sqrt{35}}{7} \quad \text{Simplify } \sqrt{140}: \sqrt{140} = \sqrt{4 \cdot 35} = \sqrt{4}\sqrt{35} = 2\sqrt{35}.$$

b. Since the denominator is a cube root, we multiply the numerator and the denominator by a number that will give a perfect cube under the radical sign. Since $2 \cdot 4 = 8$ is a perfect cube, $\sqrt[3]{4}$ is such a number.

$$\frac{4}{\sqrt[3]{2}} = \frac{4 \cdot \sqrt[3]{4}}{\sqrt[3]{2} \cdot \sqrt[3]{4}} \quad \text{Multiply numerator and denominator by } \sqrt[3]{4}.$$

$$= \frac{4\sqrt[3]{4}}{\sqrt[3]{8}} \quad \text{Multiply the radicals in the denominator.}$$

$$= \frac{4\sqrt[3]{4}}{2} \quad \sqrt[3]{8} = 2.$$

$$= 2\sqrt[3]{4} \quad \text{Simplify.}$$

■

Self Check Rationalize the denominator: $\frac{5}{\sqrt[4]{3}}$.

Answer $\frac{5\sqrt[4]{27}}{3}$

EXAMPLE 7 Rationalize the denominator of $\frac{\sqrt[3]{5}}{\sqrt[3]{18}}$.

Solution We multiply the numerator and the denominator by a number that will result in a perfect cube under the radical sign in the denominator.

Since 216 is the smallest perfect cube that is divisible by 18 ($216 \div 18 = 12$), multiplying the numerator and the denominator by $\sqrt[3]{12}$ will give the smallest possible perfect cube under the radical in the denominator.

$$\frac{\sqrt[3]{5}}{\sqrt[3]{18}} = \frac{\sqrt[3]{5} \cdot \sqrt[3]{12}}{\sqrt[3]{18} \cdot \sqrt[3]{12}} \qquad \text{Multiply numerator and denominator by } \sqrt[3]{12}.$$

$$= \frac{\sqrt[3]{60}}{\sqrt[3]{216}} \qquad \text{Multiply the radicals.}$$

$$= \frac{\sqrt[3]{60}}{6} \qquad \sqrt[3]{216} = 6.$$

■

Self Check Rationalize the denominator: $\frac{\sqrt{5}}{\sqrt{18}}$.

Answer $\frac{\sqrt{10}}{6}$

EXAMPLE 8 Rationalize the denominator of $\frac{\sqrt{5xy^2}}{\sqrt{xy^3}}$ (x and y are positive numbers).

Solution

Method 1

$$\frac{\sqrt{5xy^2}}{\sqrt{xy^3}} = \sqrt{\frac{5xy^2}{xy^3}} = \sqrt{\frac{5}{y}} = \frac{\sqrt{5}}{\sqrt{y}} = \frac{\sqrt{5}\sqrt{y}}{\sqrt{y}\sqrt{y}} = \frac{\sqrt{5y}}{y}$$

Method 2

$$\frac{\sqrt{5xy^2}}{\sqrt{xy^3}} = \sqrt{\frac{5xy^2}{xy^3}} = \sqrt{\frac{5}{y}} = \sqrt{\frac{5 \cdot y}{y \cdot y}} = \frac{\sqrt{5y}}{\sqrt{y^2}} = \frac{\sqrt{5y}}{y}$$

■

Self Check Rationalize the denominator: $\frac{\sqrt{4ab^3}}{\sqrt{2a^2b^2}}$.

Answer $\frac{\sqrt{2ab}}{a}$

To rationalize the denominator of a fraction with square roots in a binomial denominator, we multiply its numerator and denominator by the *conjugate* of its denominator. Conjugate binomials are binomials with the same terms but with opposite signs between their terms.

Conjugate Binomials

The **conjugate** of the binomial $a + b$ is $a - b$, and the conjugate of $a - b$ is $a + b$.

EXAMPLE 9 Rationalize the denominator of $\dfrac{1}{\sqrt{2} + 1}$.

Solution We multiply the numerator and denominator of the fraction by $\sqrt{2} - 1$, which is the conjugate of the denominator.

$$\frac{1}{\sqrt{2}+1} = \frac{1(\sqrt{2}-1)}{(\sqrt{2}+1)(\sqrt{2}-1)}$$

$$= \frac{\sqrt{2}-1}{(\sqrt{2})^2 - 1} \qquad (\sqrt{2}+1)(\sqrt{2}-1) = (\sqrt{2})^2 - 1.$$

$$= \frac{\sqrt{2}-1}{2-1} \qquad (\sqrt{2})^2 = 2.$$

$$= \sqrt{2} - 1 \qquad \frac{\sqrt{2}-1}{2-1} = \frac{\sqrt{2}-1}{1} = \sqrt{2} - 1.$$

■

Self Check Rationalize the denominator of $\dfrac{2}{\sqrt{3} + 1}$.

Answer $\sqrt{3} - 1$

EXAMPLE 10 Rationalize the denominator of $\dfrac{\sqrt{x} + \sqrt{2}}{\sqrt{x} - \sqrt{2}}$ $(x > 0)$.

Solution We multiply the numerator and denominator by $\sqrt{x} + \sqrt{2}$, which is the conjugate of $\sqrt{x} - \sqrt{2}$, and simplify.

$$\frac{\sqrt{x}+\sqrt{2}}{\sqrt{x}-\sqrt{2}} = \frac{(\sqrt{x}+\sqrt{2})(\sqrt{x}+\sqrt{2})}{(\sqrt{x}-\sqrt{2})(\sqrt{x}+\sqrt{2})}$$

$$= \frac{x + \sqrt{2x} + \sqrt{2x} + 2}{x - 2} \qquad \text{Use the FOIL method.}$$

$$= \frac{x + 2\sqrt{2x} + 2}{x - 2}$$

■

Self Check Rationalize the denominator of $\dfrac{\sqrt{x} - \sqrt{2}}{\sqrt{x} + \sqrt{2}}$.

Answer $\dfrac{x - 2\sqrt{2x} + 2}{x - 2}$

RATIONALIZING NUMERATORS

In calculus, we sometimes have to rationalize a numerator by multiplying the numerator and denominator of the fraction by the conjugate of the numerator.

EXAMPLE 11 Rationalize the numerator of $\dfrac{\sqrt{x}-3}{\sqrt{x}}$ $(x > 0)$.

Solution We multiply the numerator and denominator by $\sqrt{x}+3$, which is the conjugate of the numerator.

$$\begin{aligned}\frac{\sqrt{x}-3}{\sqrt{x}} &= \frac{(\sqrt{x}-3)(\sqrt{x}+3)}{\sqrt{x}(\sqrt{x}+3)}\\ &= \frac{x+3\sqrt{x}-3\sqrt{x}-9}{x+3\sqrt{x}}\\ &= \frac{x-9}{x+3\sqrt{x}}\end{aligned}$$

The final expression is not in simplified form. However, this nonsimplified form is sometimes desirable in calculus. ■

Self Check Rationalize the numerator of $\dfrac{\sqrt{x}+3}{\sqrt{x}}$.

Answer $\dfrac{x-9}{x-3\sqrt{x}}$

Orals *Multiply and simplify.*

1. $\sqrt{3}\sqrt{3}$ **2.** $\sqrt[3]{2}\sqrt[3]{2}\sqrt[3]{2}$ **3.** $\sqrt{3}\sqrt{9}$

4. $\sqrt{a^3b}\sqrt{ab}$ **5.** $3\sqrt{2}(\sqrt{2}+1)$ **6.** $(\sqrt{2}+1)(\sqrt{2}-1)$

7. $\dfrac{1}{\sqrt{2}}$ **8.** $\dfrac{1}{\sqrt{3}-1}$

EXERCISE 7.6

REVIEW *Solve each equation.*

1. $\dfrac{2}{3-a}=1$

2. $5(s-4)=-5(s-4)$

3. $\dfrac{8}{b-2}+\dfrac{3}{2-b}=-\dfrac{1}{b}$

4. $\dfrac{2}{x-2}+\dfrac{1}{x+1}=\dfrac{1}{(x+1)(x-2)}$

VOCABULARY AND CONCEPTS *Fill in each blank to make a true statement.*

5. To multiply $2\sqrt{7}$ by $3\sqrt{5}$, we multiply __ by 3 and then multiply ____ by ____.
6. To multiply $2\sqrt{5}(3\sqrt{8} + \sqrt{3})$, we use the ________ property to remove parentheses and simplify each resulting term.
7. To multiply $(\sqrt{3} + \sqrt{2})(\sqrt{3} - 2\sqrt{2})$, we can use the _____ method.
8. The conjugate of $\sqrt{x} + 1$ is ______.
9. To rationalize the denominator of $\frac{1}{\sqrt{3} - 1}$, multiply both the numerator and denominator by the ________ of the denominator.
10. To rationalize the numerator of $\frac{\sqrt{5} + 2}{\sqrt{5} - 2}$, multiply both the numerator and denominator by ______.

PRACTICE *In Exercises 11–34, do each multiplication and simplify, if possible. All variables represent positive numbers.*

11. $\sqrt{2}\sqrt{8}$
12. $\sqrt{3}\sqrt{27}$
13. $\sqrt{5}\sqrt{10}$
14. $\sqrt{7}\sqrt{35}$
15. $2\sqrt{3}\sqrt{6}$
16. $3\sqrt{11}\sqrt{33}$
17. $\sqrt[3]{5}\sqrt[3]{25}$
18. $\sqrt[3]{7}\sqrt[3]{49}$
19. $(3\sqrt[3]{9})(2\sqrt[3]{3})$
20. $(2\sqrt[3]{16})(-\sqrt[3]{4})$
21. $\sqrt[3]{2}\sqrt[3]{12}$
22. $\sqrt[3]{3}\sqrt[3]{18}$
23. $\sqrt{ab^3}\sqrt{ab}$
24. $\sqrt{8x}\sqrt{2x^3y}$
25. $\sqrt{5ab}\sqrt{5a}$
26. $\sqrt{15rs^2}\sqrt{10r}$
27. $\sqrt[3]{5r^2s}\sqrt[3]{2r}$
28. $\sqrt[3]{3xy^2}\sqrt[3]{9x^3}$
29. $\sqrt[3]{a^5b}\sqrt[3]{16ab^5}$
30. $\sqrt[3]{3x^4y}\sqrt[3]{18x}$
31. $\sqrt{x(x + 3)}\sqrt{x^3(x + 3)}$
32. $\sqrt{y^2(x + y)}\sqrt{(x + y)^3}$
33. $\sqrt[3]{6x^2(y + z)^2}\sqrt[3]{18x(y + z)}$
34. $\sqrt[3]{9x^2y(z + 1)^2}\sqrt[3]{6xy^2(z + 1)}$

In Exercises 35–54, do each multiplication and simplify. All variables represent positive numbers.

35. $3\sqrt{5}(4 - \sqrt{5})$
36. $2\sqrt{7}(3\sqrt{7} - 1)$
37. $3\sqrt{2}(4\sqrt{3} + 2\sqrt{7})$
38. $-\sqrt{3}(\sqrt{7} - \sqrt{5})$
39. $-2\sqrt{5x}(4\sqrt{2x} - 3\sqrt{3})$
40. $3\sqrt{7t}(2\sqrt{7t} + 3\sqrt{3t^2})$
41. $(\sqrt{2} + 1)(\sqrt{2} - 3)$
42. $(2\sqrt{3} + 1)(\sqrt{3} - 1)$
43. $(4\sqrt{x} + 3)(2\sqrt{x} - 5)$
44. $(7\sqrt{y} + 2)(3\sqrt{y} - 5)$
45. $(\sqrt{5z} + \sqrt{3})(\sqrt{5z} + \sqrt{3})$
46. $(\sqrt{3p} - \sqrt{2})(\sqrt{3p} + \sqrt{2})$
47. $(\sqrt{3x} - \sqrt{2y})(\sqrt{3x} + \sqrt{2y})$
48. $(\sqrt{3m} + \sqrt{2n})(\sqrt{3m} + \sqrt{2n})$
49. $(2\sqrt{3a} - \sqrt{b})(\sqrt{3a} + 3\sqrt{b})$
50. $(5\sqrt{p} - \sqrt{3q})(\sqrt{p} + 2\sqrt{3q})$
51. $(3\sqrt{2r} - 2)^2$
52. $(2\sqrt{3t} + 5)^2$
53. $-2(\sqrt{3x} + \sqrt{3})^2$
54. $3(\sqrt{5x} - \sqrt{3})^2$

In Exercises 55–78, rationalize each denominator. All variables represent positive numbers.

55. $\sqrt{\frac{1}{7}}$
56. $\sqrt{\frac{5}{3}}$
57. $\sqrt{\frac{2}{3}}$
58. $\sqrt{\frac{3}{2}}$
59. $\frac{\sqrt{5}}{\sqrt{8}}$
60. $\frac{\sqrt{3}}{\sqrt{50}}$
61. $\frac{\sqrt{8}}{\sqrt{2}}$
62. $\frac{\sqrt{27}}{\sqrt{3}}$

63. $\dfrac{1}{\sqrt[3]{2}}$

64. $\dfrac{2}{\sqrt[3]{6}}$

65. $\dfrac{3}{\sqrt[3]{9}}$

66. $\dfrac{2}{\sqrt[3]{a}}$

67. $\dfrac{\sqrt[3]{2}}{\sqrt[3]{9}}$

68. $\dfrac{\sqrt[3]{9}}{\sqrt[3]{54}}$

69. $\dfrac{\sqrt{8x^2y}}{\sqrt{xy}}$

70. $\dfrac{\sqrt{9xy}}{\sqrt{3x^2y}}$

71. $\dfrac{\sqrt{10xy^2}}{\sqrt{2xy^3}}$

72. $\dfrac{\sqrt{5ab^2c}}{\sqrt{10abc}}$

73. $\dfrac{\sqrt[3]{4a^2}}{\sqrt[3]{2ab}}$

74. $\dfrac{\sqrt[3]{9x}}{\sqrt[3]{3xy}}$

75. $\dfrac{1}{\sqrt[4]{4}}$

76. $\dfrac{1}{\sqrt[5]{2}}$

77. $\dfrac{1}{\sqrt[5]{16}}$

78. $\dfrac{4}{\sqrt[4]{32}}$

In Exercises 79–94, do each division by rationalizing the denominator and simplifying. All variables represent positive numbers.

79. $\dfrac{1}{\sqrt{2} - 1}$

80. $\dfrac{3}{\sqrt{3} - 1}$

81. $\dfrac{\sqrt{2}}{\sqrt{5} + 3}$

82. $\dfrac{\sqrt{3}}{\sqrt{3} - 2}$

83. $\dfrac{\sqrt{3} + 1}{\sqrt{3} - 1}$

84. $\dfrac{\sqrt{2} - 1}{\sqrt{2} + 1}$

85. $\dfrac{\sqrt{7} - \sqrt{2}}{\sqrt{2} + \sqrt{7}}$

86. $\dfrac{\sqrt{3} + \sqrt{2}}{\sqrt{3} - \sqrt{2}}$

87. $\dfrac{2}{\sqrt{x} + 1}$

88. $\dfrac{3}{\sqrt{x} - 2}$

89. $\dfrac{x}{\sqrt{x} - 4}$

90. $\dfrac{2x}{\sqrt{x} + 1}$

91. $\dfrac{2z - 1}{\sqrt{2z} - 1}$

92. $\dfrac{3t - 1}{\sqrt{3t} + 1}$

93. $\dfrac{\sqrt{x} - \sqrt{y}}{\sqrt{x} + \sqrt{y}}$

94. $\dfrac{\sqrt{x} + \sqrt{y}}{\sqrt{x} - \sqrt{y}}$

In Exercises 95–100, rationalize each numerator. All variables represent positive numbers.

95. $\dfrac{\sqrt{3} + 1}{2}$

96. $\dfrac{\sqrt{5} - 1}{2}$

97. $\dfrac{\sqrt{x} + 3}{x}$

98. $\dfrac{2 + \sqrt{x}}{5x}$

99. $\dfrac{\sqrt{x} + \sqrt{y}}{\sqrt{x}}$

100. $\dfrac{\sqrt{x} - \sqrt{y}}{\sqrt{x} + \sqrt{y}}$

APPLICATIONS

101. Photography In Example 5, we saw that a lens with a focal length of 12 centimeters and an aperture $3\sqrt{2}$ centimeters in diameter is an $f/2.8$ lens. Find the f-number if the area of the aperture is again cut in half.

102. Photography In Exercise 101, we saw that a lens with a focal length of 12 centimeters and an aperture 3 centimeters in diameter is an $f/4$ lens. Find the f-number if the area of the aperture is again cut in half.

WRITING

103. Explain how to simplify a fraction with the monomial denominator $\sqrt[3]{3}$.

104. Explain how to simplify a fraction with the monomial denominator $\sqrt[3]{9}$.

SOMETHING TO THINK ABOUT *Assume that x is a rational number.*

105. Change the numerator of $\dfrac{\sqrt{x} - 3}{4}$ to a rational number.

106. Rationalize the numerator of $\dfrac{2\sqrt{3x} + 4}{\sqrt{3x} - 1}$.

PROJECTS

PROJECT 1

Tom and Brian arrange to have a bicycle race. Each leaves his own house at the same time and rides to the other's house, whereupon the winner of the race calls his own house and leaves a message for the loser. A map of the race is shown in Illustration 1. Brian stays on the highway, averaging 21 mph. Tom knows that he and Brian are evenly matched when biking on the highway, so he cuts across country for the first part of his trip, averaging 15 mph. When Tom reaches the highway at point A, he turns right and follows the highway, averaging 21 mph.

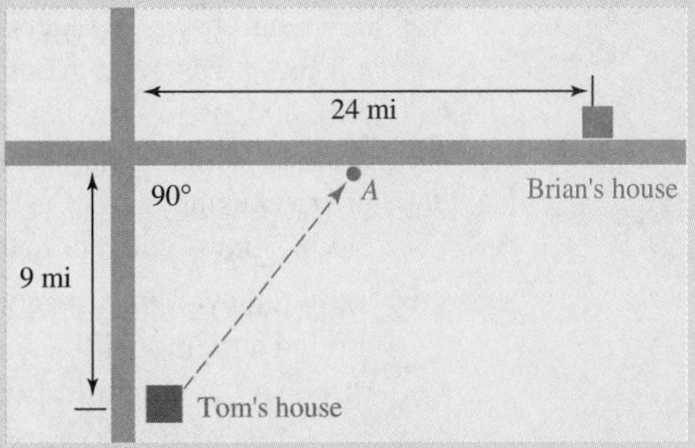

ILLUSTRATION 1

Tom and Brian never meet during the race, and amazingly, the race is a tie. Each of them calls the other at exactly the same moment!

a. How long (to the nearest second) did it take each person to complete the race?

b. How far from the intersection of the two highways is point A? (*Hint:* Set the travel times for Brian and Tom equal to each other. You may find two answers, but only one of them matches all of the information.)

(continued)

PROJECTS *(continued)*

c. Show that if Tom had started straight across country for Brian's house (in order to minimize the distance he had to travel), he would have lost the race. By how much time (to the nearest second) would he have lost? Then show that if Tom had biked across country to a point 9 miles from the intersection of the two highways, he would have won the race. By how much time (to the nearest second) would he have won?

PROJECT 2

You have been hired by the accounts receivable department of Widget Industries. Your boss, I. Tally, the head of accounting, informs you that those who purchase from Widget Industries are billed according to a pricing structure given by the following function:

$$\text{Price per widget in an order of } x \text{ widgets} = p(x) = p_0 - \frac{p_0}{2}\left(1 - \sqrt{1 - \frac{x^2}{k^2}}\right)$$

where x is the number of widgets ordered, p_0 is the standard price of a widget, and k is the maximum number of widgets that can be ordered at one time.

The formula was developed because it allows for changes in the standard price and the maximum order size without requiring that the entire pricing scheme be redone. Note that the total cost of an order of x widgets is $x \cdot p(x)$.

Tally is quite a taskmaster. On your first day, he assigns the following problems to you.

a. The pricing function is designed to benefit those who order the largest number of widgets. To see this, calculate the price (as a percentage of p_0) that a customer would pay for each widget if the customer ordered 100, 500, 800, or 1,000 widgets. Assume that $k = 1{,}000$. For what value of x is the cost per widget a minimum, and what is that minimum? Looking at the graph of $p(x)$ (using, say, 8 as p_0) on a graphing calculator will help you see how buying more widgets means that each widget costs less.

b. Tally tells you that the company lost money in the past because someone sent around a memo with a simplified, but incorrect, pricing function. Find the error(s) in the simplification given below.

$$\begin{aligned} p(x) &= p_0 - \frac{p_0}{2}\left(1 - \sqrt{1 - \frac{x^2}{k^2}}\right) \\ &= p_0 - \frac{p_0}{2}\left[1 - \left(1 - \frac{x}{k}\right)\right] \\ &= p_0 - \frac{p_0}{2}\left(\frac{x}{k}\right) \\ &= p_0\left(1 - \frac{x}{2k}\right) \end{aligned}$$

Show that $x = 0$ and 1,000 are the *only* values of x for which the two formulas agree. (*Hint:* Set the two formulas equal to each other and solve for x.)

Now calculate the amount of money lost by using the incorrect formula, rather than the correct formula, on orders of 100, 500, 800, and 1,000 widgets. For these calculations, assume that $k = 1{,}000$ and $p_0 = \$5.60$.

c. Now is your chance to show Tally a thing or two. The pricing function *can* be simplified. Demonstrate to Tally that his pricing function is equivalent to

$$p(x) = \frac{p_0}{2}\left(1 + \frac{\sqrt{k^2 - x^2}}{k}\right)$$

Use this function to help a customer who wants to know how many widgets she must order to receive a 30% discount off the standard price p_0. Again, assume that $k = 1{,}000$ when doing this calculation.

CHAPTER SUMMARY

CONCEPTS | REVIEW EXERCISES

SECTION 7.1 Radical Expressions

If n is an even natural number,

$\sqrt[n]{a^n} = |a|$

If n is an odd natural number,

$\sqrt[n]{a^n} = a$

If n is a natural number greater than 1 and x is a real number, then

If $x > 0$, then $\sqrt[n]{x}$ is the positive number such that $\left(\sqrt[n]{x}\right)^n = x$.

If $x = 0$, then $\sqrt[n]{x} = 0$.

If $x < 0$, and n is odd, $\sqrt[n]{x}$ is the real number such that $\left(\sqrt[n]{x}\right)^n = x$.

If $x < 0$, and n is even, $\sqrt[n]{x}$ is not a real number.

1. Simplify each radical. Assume that x can be any number.

a. $\sqrt{49}$ **b.** $-\sqrt{121}$

c. $-\sqrt{36}$ **d.** $\sqrt{225}$

e. $\sqrt[3]{-27}$ **f.** $-\sqrt[3]{216}$

g. $\sqrt[4]{625}$ **h.** $\sqrt[5]{-32}$

i. $\sqrt{25x^2}$ **j.** $\sqrt{x^2 + 4x + 4}$

k. $\sqrt[3]{27a^6b^3}$ **l.** $\sqrt[4]{256x^8y^4}$

2. Graph each function.

a. $y = f(x) = \sqrt{x + 2}$

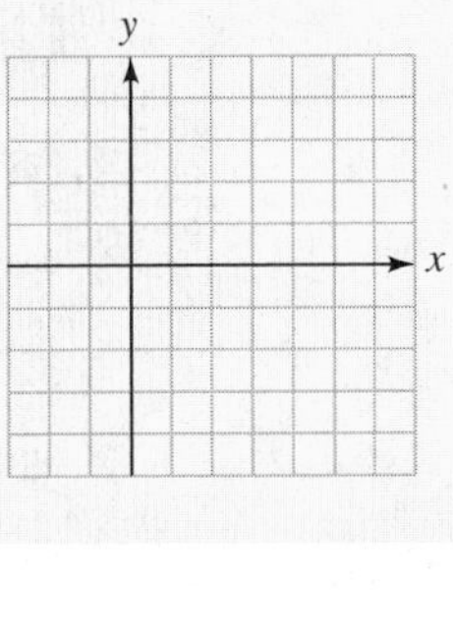

b. $y = f(x) = -\sqrt{x - 1}$

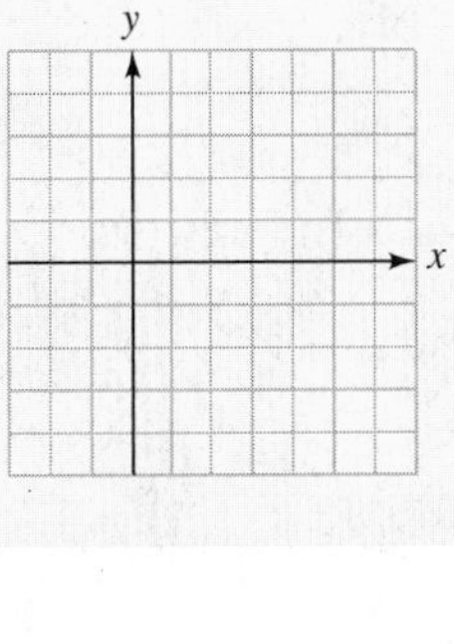

c. $y = f(x) = -\sqrt{x} + 2$

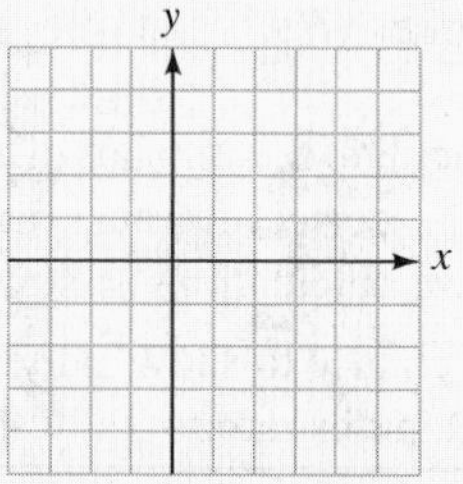

d. $y = f(x) = -\sqrt[3]{x} + 3$

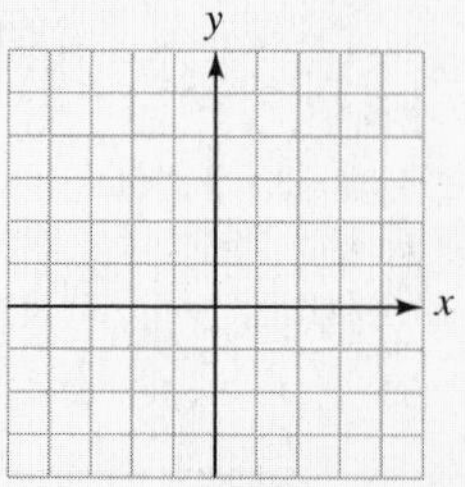

3. Consider the distribution 4, 8, 12, 16, 20.
a. Find the mean of the distribution.
b. Find the standard deviation.

SECTION 7.2 Applications of Radicals

The Pythagorean theorem: If a and b are the lengths of the legs of a right triangle and c is the length of the hypotenuse, then $a^2 + b^2 = c^2$.

In Exercises 4–5, the horizon distance d (measured in miles) is related to the height h (measured in feet) of the observer by the formula $d = 1.4\sqrt{h}$.

4. View from a submarine A submarine's periscope extends 4.7 feet above the surface. How far away is the horizon?

5. View from a submarine How far out of the water must a submarine periscope extend to provide a 4-mile horizon?

6. Sailing A technique called *tacking* allows a sailboat to make progress into the wind. A sailboat follows the course in Illustration 1. Find d, the distance the boat advances into the wind.

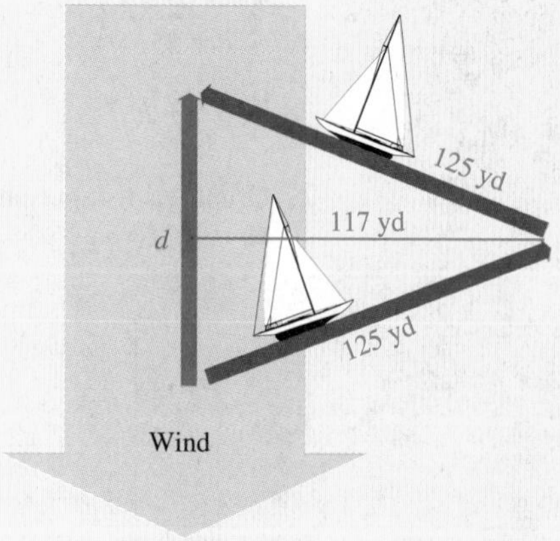

ILLUSTRATION 1

7. Communications Some campers 3,900 yards from a highway are chatting with truckers on a citizen's band radio with an 8,900-yard range. Over what length of highway can these conversations take place? (See Illustration 2.)

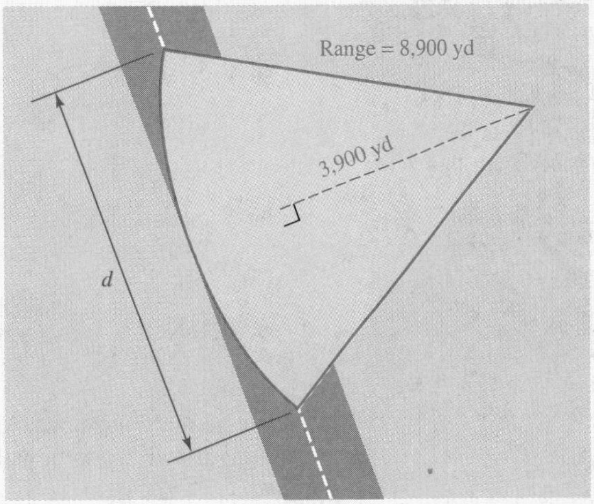

ILLUSTRATION 2

The distance formula:

$d(PQ) =$
$\sqrt{(x_2 - x_1)^2 + (y_2 - y_1)^2}$

8. Find the distance between points $P(0, 0)$ and $Q(5, -12)$.

9. Find the distance between points $P(-4, 6)$ and $Q(-2, 8)$. Give the result to the nearest hundredth.

SECTION 7.3 *Radical Equations*

The power rule:

If $x = y$, then $x^n = y^n$.

Raising both sides of an equation to the same power can lead to extraneous solutions. Be sure to check all suspected solutions.

10. Solve each equation.

a. $\sqrt{y+3} = \sqrt{2y-19}$

b. $u = \sqrt{25u - 144}$

c. $r = \sqrt{12r - 27}$

d. $\sqrt{z+1} + \sqrt{z} = 2$

e. $\sqrt{2x+5} - \sqrt{2x} = 1$

f. $\sqrt[3]{x^3 + 8} = x + 2$

SECTION 7.4 Rational Exponents

If n ($n > 1$) is a natural number and $\sqrt[n]{x}$ is a real number, then $x^{1/n} = \sqrt[n]{x}$.

If n is even, $(x^n)^{1/n} = |x|$.

If n is a natural number greater than 1 and x is a real number, then

- If $x > 0$, then $x^{1/n}$ is the positive number such that $(x^{1/n})^n = x$.
- If $x = 0$, then $x^{1/n} = 0$.
- If $x < 0$, and n is odd, then $x^{1/n}$ is the real number such that $(x^{1/n})^n = x$.
- If $x < 0$ and n is even, then $x^{1/n}$ is not a real number.

If m and n are positive integers and $x > 0$,

$$x^{m/n} = \sqrt[n]{x^m} = \left(\sqrt[n]{x}\right)^m$$

$$x^{-m/n} = \frac{1}{x^{m/n}}$$

$$\frac{1}{x^{-m/n}} = x^{m/n} \quad (x \neq 0)$$

11. Simplify each expression, if possible. Assume that all variables represent positive numbers.

a. $25^{1/2}$ **b.** $-36^{1/2}$
c. $9^{3/2}$ **d.** $16^{3/2}$
e. $(-8)^{1/3}$ **f.** $-8^{2/3}$
g. $8^{-2/3}$ **h.** $8^{-1/3}$
i. $-49^{5/2}$ **j.** $\frac{1}{25^{5/2}}$
k. $\left(\frac{1}{4}\right)^{-3/2}$ **l.** $\left(\frac{4}{9}\right)^{-3/2}$
m. $(27x^3y)^{1/3}$ **n.** $(81x^4y^2)^{1/4}$
o. $(25x^3y^4)^{3/2}$ **p.** $(8u^2v^3)^{-2/3}$

12. Do the multiplications. Assume that all variables represent positive numbers and write all answers without negative exponents.

a. $5^{1/4}5^{1/2}$ **b.** $a^{3/7}a^{2/7}$
c. $u^{1/2}(u^{1/2} - u^{-1/2})$ **d.** $v^{2/3}(v^{1/3} + v^{4/3})$
e. $(x^{1/2} + y^{1/2})^2$
f. $(a^{2/3} + b^{2/3})(a^{2/3} - b^{2/3})$

13. Simplify each expression. Assume that all variables are positive.

a. $\sqrt[6]{5^2}$ **b.** $\sqrt[8]{x^4}$
c. $\sqrt[9]{27a^3b^6}$ **d.** $\sqrt[4]{25a^2b^2}$

SECTION 7.5 Simplifying and Combining Radical Expressions

Properties of radicals:

$$\sqrt[n]{ab} = \sqrt[n]{a}\sqrt[n]{b}$$

$$\sqrt[n]{\frac{a}{b}} = \frac{\sqrt[n]{a}}{\sqrt[n]{b}} \quad (b \neq 0)$$

14. Simplify each expression.

a. $\sqrt{240}$ **b.** $\sqrt[3]{54}$
c. $\sqrt[4]{32}$ **d.** $\sqrt[5]{96}$
e. $\sqrt{8x^3}$ **f.** $\sqrt{18x^4y^3}$
g. $\sqrt[3]{16x^5y^4}$ **h.** $\sqrt[3]{54x^7y^3}$
i. $\frac{\sqrt{32x^3}}{\sqrt{2x}}$ **j.** $\frac{\sqrt[3]{16x^5}}{\sqrt[3]{2x^2}}$
k. $\sqrt[3]{\frac{2a^2b}{27x^3}}$ **l.** $\sqrt{\frac{17xy}{64a^4}}$

Like radicals can be combined by addition and subtraction:

$3\sqrt{2} + 5\sqrt{2} = 8\sqrt{2}$

Radicals that are not similar can often be simplified to radicals that are similar and then combined:

$$\begin{aligned}\sqrt{2} + \sqrt{8} &= \sqrt{2} + \sqrt{4}\sqrt{2} \\ &= \sqrt{2} + 2\sqrt{2} \\ &= 3\sqrt{2}\end{aligned}$$

15. Simplify and combine like radicals. Assume that all variables represent positive numbers.

a. $\sqrt{2} + \sqrt{8}$

b. $\sqrt{20} - \sqrt{5}$

c. $2\sqrt[3]{3} - \sqrt[3]{24}$

d. $\sqrt[4]{32} + 2\sqrt[4]{162}$

e. $2x\sqrt{8} + 2\sqrt{200x^2} + \sqrt{50x^2}$

f. $3\sqrt{27a^3} - 2a\sqrt{3a} + 5\sqrt{75a^3}$

g. $\sqrt[3]{54} - 3\sqrt[3]{16} + 4\sqrt[3]{128}$

h. $2\sqrt[4]{32x^5} + 4\sqrt[4]{162x^5} - 5x\sqrt[4]{512x}$

In an isosceles right triangle, the length of the hypotenuse is the length of one leg times $\sqrt{2}$.

16. Find the length of the hypotenuse of an isosceles right triangle whose legs measure 7 meters.

The shorter leg of a 30°–60°–90° triangle is half as long as the hypotenuse. The longer leg is the length of the shorter leg times $\sqrt{3}$.

17. The hypotenuse of a 30°–60°–90° triangle measures $12\sqrt{3}$ centimeters. Find the length of each leg.

18. Find x to two decimal places.

a.

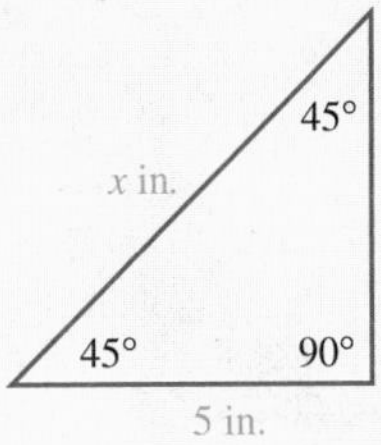

b.

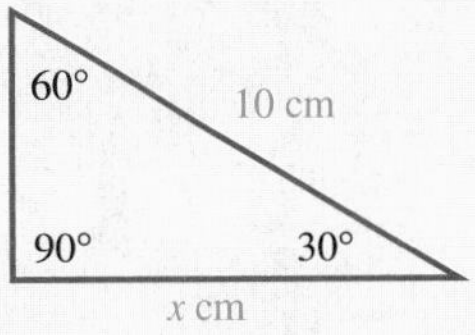

SECTION 7.6 Multiplying and Dividing Radical Expressions

If two radicals have the same index, they can be multiplied:

$$\begin{aligned}\sqrt{3x}\sqrt{6x} &= \sqrt{18x^2}\ (x > 0) \\ &= 3x\sqrt{2}\end{aligned}$$

19. Simplify each expression. Assume that all variables represent positive numbers.

a. $(2\sqrt{5})(3\sqrt{2})$

b. $2\sqrt{6}\sqrt{216}$

c. $\sqrt{9x}\sqrt{x}$

d. $\sqrt[3]{3}\sqrt[3]{9}$

e. $-\sqrt[3]{2x^2}\sqrt[3]{4x}$

f. $-\sqrt[4]{256x^5y^{11}}\sqrt[4]{625x^9y^3}$

g. $\sqrt{2}(\sqrt{8} - 3)$

h. $\sqrt{2}(\sqrt{2} + 3)$

i. $\sqrt{5}(\sqrt{2}-1)$
j. $\sqrt{3}(\sqrt{3}+\sqrt{2})$
k. $(\sqrt{2}+1)(\sqrt{2}-1)$
l. $(\sqrt{3}+\sqrt{2})(\sqrt{3}+\sqrt{2})$
m. $(\sqrt{x}+\sqrt{y})(\sqrt{x}-\sqrt{y})$
n. $(2\sqrt{u}+3)(3\sqrt{u}-4)$

To rationalize the binomial denominator of a fraction, multiply the numerator and the denominator by the conjugate of the binomial in the denominator.

20. Rationalize each denominator.

a. $\dfrac{1}{\sqrt{3}}$ **b.** $\dfrac{\sqrt{3}}{\sqrt{5}}$

c. $\dfrac{x}{\sqrt{xy}}$ **d.** $\dfrac{\sqrt[3]{uv}}{\sqrt[3]{u^5v^7}}$

e. $\dfrac{2}{\sqrt{2}-1}$ **f.** $\dfrac{\sqrt{2}}{\sqrt{3}-1}$

g. $\dfrac{2x-32}{\sqrt{x}+4}$ **h.** $\dfrac{\sqrt{a}+1}{\sqrt{a}-1}$

21. Rationalize each numerator.

a. $\dfrac{\sqrt{3}}{5}$ **b.** $\dfrac{\sqrt[3]{9}}{3}$

c. $\dfrac{3-\sqrt{x}}{2}$ **d.** $\dfrac{\sqrt{a}-\sqrt{b}}{\sqrt{a}}$

Chapter Test

In Problems 1–4, find each root.

1. $\sqrt{49}$
2. $\sqrt[3]{64}$
3. $\sqrt{4x^2}$
4. $\sqrt[3]{8x^3}$

In Problems 5–6, graph each function and find its domain and range.

5. $f(x) = \sqrt{x-2}$

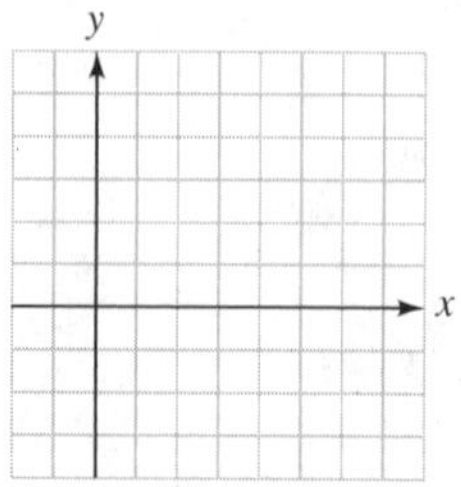

6. $f(x) = \sqrt[3]{x}+3$

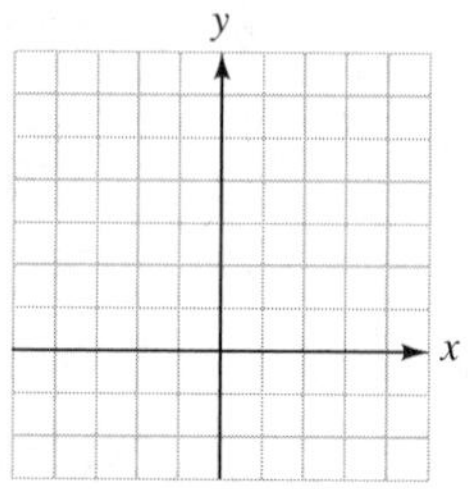

In Problems 7–8, consider the distribution 7, 8, 12, 13.

7. Find the mean of the distribution.

8. Find the standard deviation.

In Problems 9–10, use a calculator.

9. Shipping crates The diagonal brace on the shipping crate in Illustration 1 is 53 inches. Find the height, h, of the crate.

10. Pendulum The 2-meter pendulum in Illustration 2 rises 0.1 meter at the extremes of its swing. Find the width, w, of the swing.

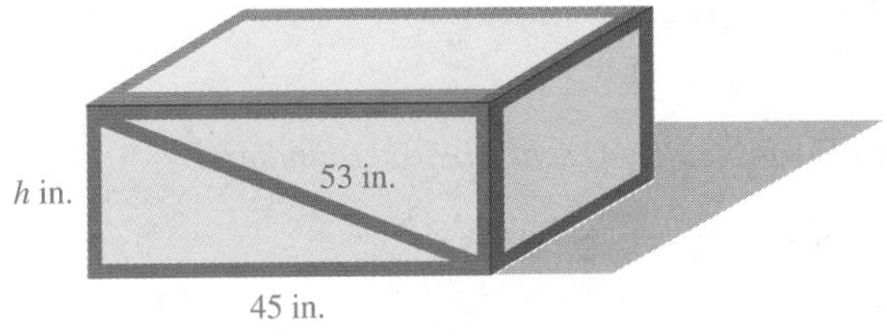

ILLUSTRATION 1

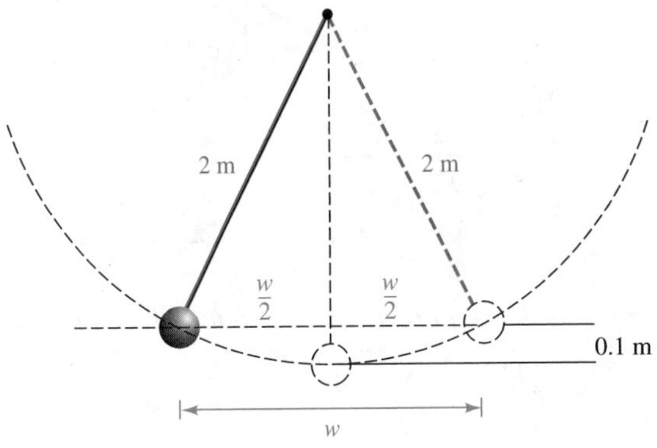

ILLUSTRATION 2

In Problems 11–12, find the distance between points P and Q.

11. $P(6, 8)$ and $Q\ (0, 0)$

12. $P(-2, 5)$ and $Q(22, 12)$

In Problems 13–14, solve and check each equation.

13. $\sqrt[3]{6n + 4} - 4 = 0$

14. $1 - \sqrt{u} = \sqrt{u - 3}$

In Problems 15–20, simplify each expression. Assume that all variables represent positive numbers, and write answers without using negative exponents.

15. $16^{1/4}$

16. $27^{2/3}$

17. $36^{-3/2}$

18. $\left(-\frac{8}{27}\right)^{-2/3}$

19. $\frac{2^{5/3}2^{1/6}}{2^{1/2}}$

20. $\frac{(8x^3y)^{1/2}(8xy^5)^{1/2}}{(x^3y^6)^{1/3}}$

In Problems 21–24, simplify each expression. Assume that all variables represent positive numbers.

21. $\sqrt{48}$

22. $\sqrt{250x^3y^5}$

23. $\frac{\sqrt[3]{24x^{15}y^4}}{\sqrt[3]{y}}$

24. $\sqrt{\frac{3a^5}{48a^7}}$

In Problems 25–28, simplify each expression. Assume that the variables are unrestricted.

25. $\sqrt{12x^2}$ **26.** $\sqrt{8x^6}$ **27.** $\sqrt[3]{81x^3}$ **28.** $\sqrt{18x^4y^9}$

In Problems 29–32, simplify and combine like radicals. Assume that all variables represent positive numbers.

29. $\sqrt{12} - \sqrt{27}$

30. $2\sqrt[3]{40} - \sqrt[3]{5,000} + 4\sqrt[3]{625}$

31. $2\sqrt{48y^5} - 3y\sqrt{12y^3}$

32. $\sqrt[4]{768z^5} + z\sqrt[4]{48z}$

In Problems 33–34, find x to two decimal places.

33.

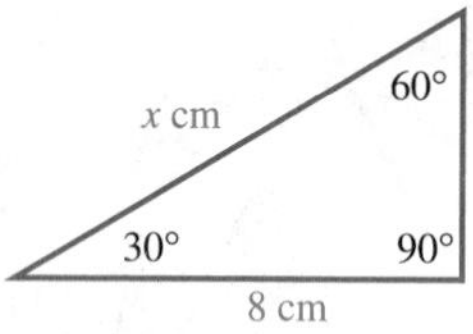

34.

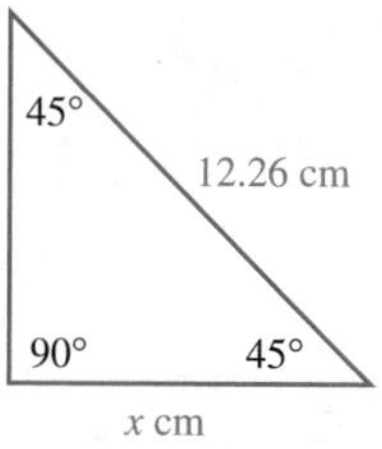

In Problems 35–36, do each operation and simplify, if possible. All variables represent positive numbers.

35. $-2\sqrt{xy}\left(3\sqrt{x} + \sqrt{xy^3}\right)$

36. $\left(3\sqrt{2} + \sqrt{3}\right)\left(2\sqrt{2} - 3\sqrt{3}\right)$

In Problems 37–40, rationalize each denominator.

37. $\dfrac{1}{\sqrt{5}}$ **38.** $\dfrac{6}{\sqrt[3]{9}}$ **39.** $\dfrac{-4\sqrt{2}}{\sqrt{5} + 3}$ **40.** $\dfrac{3t - 1}{\sqrt{3t} - 1}$

In Problems 41–42, rationalize each numerator.

41. $\dfrac{\sqrt{3}}{\sqrt{7}}$

42. $\dfrac{\sqrt{a} + \sqrt{b}}{\sqrt{a} - \sqrt{b}}$

8 Quadratic Functions, Inequalities, and Algebra of Functions

Chemist

MATHEMATICS IN THE WORKPLACE

Chemists search for knowledge about substances and put it to practical use. Most chemists work in research and development. In basic research, chemists investigate the properties, composition, and structure of matter and the laws that govern the combination of elements and the reactions of substances. Their research has resulted in the development of a tremendous variety of synthetic materials, of ingredients that have improved other substances, and of processes that help save energy and reduce pollution. In applied research and development, they create new products or improve existing ones.

SAMPLE APPLICATION ■ A weak acid (0.1 M concentration) breaks down into free cations (the hydrogen ion, H^+) and anions (A^-). When this acid dissociates, the following equilibrium equation is established.

1. $$\frac{[H^+][A^-]}{[HA]} = 4 \times 10^{-4}$$

where $[H^+]$, the hydrogen ion concentration, is equal to $[A^-]$, the anion concentration. [HA] is the concentration of the undissociated acid itself. Find $[H^+]$ at equilibrium.
(See Exercise 85 in Exercise 8.1.)

We have discussed how to solve linear equations and certain quadratic equations in which the quadratic expression is factorable. In this chapter, we will discuss more general methods for solving quadratic equations, and we will consider their graphs.

8.1 Completing the Square and the Quadratic Formula

■ SOLVING QUADRATIC EQUATIONS BY FACTORING ■ THE SQUARE ROOT PROPERTY ■ COMPLETING THE SQUARE ■ SOLVING EQUATIONS BY COMPLETING THE SQUARE ■ THE QUADRATIC FORMULA ■ PROBLEM SOLVING

Getting Ready

Factor each expression.

1. $6x^2 + x - 2$

2. $4x^2 - 4x - 3$

Evaluate $\sqrt{b^2 - 4ac}$ *for the following values.*

3. $a = 6, b = 1, c = -2$

4. $a = 4, b = -4, c = -3$

SOLVING QUADRATIC EQUATIONS BY FACTORING

A *quadratic equation* is an equation of the form $ax^2 + bx + c = 0$ $(a \neq 0)$, where a, b, and c are real numbers. We have discussed how to solve quadratic equations by using factoring. For example, to solve $6x^2 - 7x - 3 = 0$, we proceed as follows:

$$6x^2 - 7x - 3 = 0$$

$$(2x - 3)(3x + 1) = 0 \qquad \text{Factor.}$$

$$2x - 3 = 0 \quad \text{or} \quad 3x + 1 = 0 \qquad \text{Set each factor equal to 0.}$$

$$x = \frac{3}{2} \qquad x = -\frac{1}{3} \qquad \text{Solve each linear equation.}$$

However, many quadratic expressions do not factor easily. For example, it would be difficult to solve $2x^2 + 4x + 1 = 0$ by factoring, because $2x^2 + 4x + 1$ cannot be factored by using only integers.

THE SQUARE ROOT PROPERTY

To develop general methods for solving quadratic equations, we first consider the equation $x^2 = c$. If $c > 0$, we can find the real solutions of $x^2 = c$ as follows:

$$x^2 = c$$

$$x^2 - c = 0 \qquad \text{Subtract } c \text{ from both sides.}$$

$$x^2 - (\sqrt{c})^2 = 0 \qquad c = (\sqrt{c})^2.$$

$$(x + \sqrt{c})(x - \sqrt{c}) = 0 \qquad \text{Factor the difference of two squares.}$$

$$x + \sqrt{c} = 0 \quad \text{or} \quad x - \sqrt{c} = 0 \qquad \text{Set each factor equal to 0.}$$

$$x = -\sqrt{c} \qquad x = \sqrt{c} \qquad \text{Solve each linear equation.}$$

The two solutions of $x^2 = c$ are $x = \sqrt{c}$ and $x = -\sqrt{c}$.

Square Root Property
If $c > 0$, the equation $x^2 = c$ has two real solutions:

$$x = \sqrt{c} \quad \text{and} \quad x = -\sqrt{c}$$

EXAMPLE 1 Solve $x^2 - 12 = 0$.

Solution We can write the equation as $x^2 = 12$ and use the square root property.

$$x^2 - 12 = 0$$

$$x^2 = 12 \qquad \text{Add 12 to both sides.}$$

$$x = \sqrt{12} \quad \text{or} \quad x = -\sqrt{12} \qquad \text{Use the square root property.}$$

$$x = 2\sqrt{3} \qquad x = -2\sqrt{3} \qquad \sqrt{12} = \sqrt{4}\sqrt{3} = 2\sqrt{3}.$$

Verify that each solution satisfies the equation. ■

Self Check Solve $x^2 - 18 = 0$.

Answer $3\sqrt{2}, -3\sqrt{2}$

EXAMPLE 2 Solve $(x - 3)^2 = 16$.

Solution

$$(x - 3)^2 = 16$$

$x - 3 = \sqrt{16}$	or	$x - 3 = -\sqrt{16}$	Use the square root property.
$x - 3 = 4$		$x - 3 = -4$	$\sqrt{16} = 4$.
$x = 3 + 4$		$x = 3 - 4$	Add 3 to both sides.
$x = 7$		$x = -1$	Simplify.

Verify that each solution satisfies the equation. ■

Self Check Solve $(x + 2)^2 = 9$.

Answer 1, −5

COMPLETING THE SQUARE

All quadratic equations can be solved by **completing the square.** This method is based on the special products

$$x^2 + 2ax + a^2 = (x + a)^2 \quad \text{and} \quad x^2 - 2ax + a^2 = (x - a)^2$$

The trinomials $x^2 + 2ax + a^2$ and $x^2 - 2ax + a^2$ are perfect square trinomials, because each one factors as the square of a binomial. In each case, the coefficient of the first term is 1, and if we take one-half of the coefficient of x in the middle term and square it, we obtain the third term.

$$\left[\frac{1}{2}(2a)\right]^2 = a^2$$

$$\left[\frac{1}{2}(-2a)\right]^2 = (-a)^2 = a^2$$

EXAMPLE 3 Add a number to make each binomial a perfect square trinomial: **a.** $x^2 + 10x$, **b.** $x^2 - 6x$, and **c.** $x^2 - 11x$.

Solution **a.** To make $x^2 + 10x$ a perfect square trinomial, we find one-half of 10 to get 5, square 5 to get 25, and add 25 to $x^2 + 10x$.

$$x^2 + 10x + \left[\frac{1}{2}(10)\right]^2 = x^2 + 10x + (5)^2$$

$$= x^2 + 10x + 25 \qquad \text{Note that } x^2 + 10x + 25 = (x + 5)^2.$$

b. To make $x^2 - 6x$ a perfect square trinomial, we find one-half of -6 to get -3, square -3 to get 9, and add 9 to $x^2 - 6x$.

$$x^2 - 6x + \left[\frac{1}{2}(-6)\right]^2 = x^2 - 6x + (-3)^2$$

$$= x^2 - 6x + 9$$

Note that $x^2 - 6x + 9 = (x - 3)^2$.

c. To make $x^2 - 11x$ a perfect square trinomial, we find one-half of -11 to get $-\frac{11}{2}$, square $-\frac{11}{2}$ to get $\frac{121}{4}$, and add $\frac{121}{4}$ to $x^2 - 11x$.

$$x^2 - 11x + \left[\frac{1}{2}(-11)\right]^2 = x^2 - 11x + \left(-\frac{11}{2}\right)^2$$

$$= x^2 - 11x + \frac{121}{4}$$

Note that $x^2 - 11x + \frac{121}{4} = \left(x - \frac{11}{2}\right)^2$. ■

Self Check Change $a^2 - 5a$ into a perfect trinomial square.

Answer $a^2 - 5a + \frac{25}{4}$

SOLVING EQUATIONS BY COMPLETING THE SQUARE

To solve an equation of the form $ax^2 + bx + c = 0$ $(a \neq 0)$ by completing the square, we use the following steps.

Completing the Square

1. Make sure that the coefficient of x^2 is 1. If it is not, make it 1 by dividing both sides of the equation by the coefficient of x^2.
2. If necessary, add a number to both sides of the equation to place the constant term on the right-hand side of the equal sign.
3. Complete the square:
   - **a.** Find one-half of the coefficient of x and square it.
   - **b.** Add the square to both sides of the equation.
4. Factor the trinomial square and combine like terms.
5. Solve the resulting equation by using the square root property.

EXAMPLE 4 Use completing the square to solve $x^2 + 8x + 7 = 0$.

Solution *Step 1:* In this example, the coefficient of x^2 is already 1.

Step 2: We add −7 to both sides to place the constant on the right-hand side of the equal sign:

$$x^2 + 8x + 7 = 0$$
$$x^2 + 8x = -7$$

Step 3: The coefficient of x is 8, one-half of 8 is 4, and $4^2 = 16$. To complete the square, we add 16 to both sides.

$$x^2 + 8x + \mathbf{16} = \mathbf{16} - 7$$

1. $x^2 + 8x + 16 = 9$ $\qquad 16 - 7 = 9.$

Step 4: Since the left-hand side of Equation 1 is a perfect square trinomial, we can factor it to get $(x + 4)^2$.

$$x^2 + 8x + 16 = 9$$

2. $(x + 4)^2 = 9$

Step 5: We then solve Equation 2 by using the square root property.

$$(x + 4)^2 = 9$$
$$x + 4 = \sqrt{9} \quad \text{or} \quad x + 4 = -\sqrt{9}$$
$$x + 4 = 3 \qquad x + 4 = -3$$
$$x = -1 \qquad x = -7$$

Verify that both solutions satisfy the equation. ■

EXAMPLE 5 Solve $6x^2 + 5x - 6 = 0$.

Solution *Step 1:* To make the coefficient of x^2 equal to 1, we divide both sides of the equation by 6.

$$6x^2 + 5x - 6 = 0$$
$$\frac{6x^2}{6} + \frac{5}{6}x - \frac{6}{6} = \frac{0}{6} \qquad \text{Divide both sides by 6.}$$
$$x^2 + \frac{5}{6}x - 1 = 0 \qquad \text{Simplify.}$$

Step 2: We add 1 to both sides to place the constant on the right-hand side of the equal sign:

$$x^2 + \frac{5}{6}x = 1$$

Step 3: The coefficient of x is $\frac{5}{6}$, one-half of $\frac{5}{6}$ is $\frac{5}{12}$, and $\left(\frac{5}{12}\right)^2 = \frac{25}{144}$. To complete the square, we add $\frac{25}{144}$ to both sides.

$$x^2 + \frac{5}{6}x + \frac{\mathbf{25}}{\mathbf{144}} = 1 + \frac{\mathbf{25}}{\mathbf{144}}$$

3. $x^2 + \frac{5}{6}x + \frac{25}{144} = \frac{169}{144}$ $\qquad 1 + \frac{25}{144} = \frac{144}{144} + \frac{25}{144} = \frac{169}{144}.$

Step 4: Since the left-hand side of Equation 3 is a perfect square trinomial, we can factor it to get $\left(x + \frac{5}{12}\right)^2$.

4. $$\left(x + \frac{5}{12}\right)^2 = \frac{169}{144}$$

Step 5: We can solve Equation 4 by using the square root property.

$$x + \frac{5}{12} = \sqrt{\frac{169}{144}} \quad \text{or} \quad x + \frac{5}{12} = -\sqrt{\frac{169}{144}}$$

$$x + \frac{5}{12} = \frac{13}{12} \qquad\qquad x + \frac{5}{12} = -\frac{13}{12}$$

$$x = -\frac{5}{12} + \frac{13}{12} \qquad\qquad x = -\frac{5}{12} - \frac{13}{12}$$

$$x = \frac{8}{12} \qquad\qquad x = -\frac{18}{12}$$

$$x = \frac{2}{3} \qquad\qquad x = -\frac{3}{2}$$

Verify that both solutions satisfy the original equation. ■

EXAMPLE 6 Solve $2x^2 + 4x + 1 = 0$.

Solution

$$2x^2 + 4x + 1 = 0$$

$$x^2 + 2x + \frac{1}{2} = \frac{0}{2}$$ Divide both sides by 2 to make the coefficient of x^2 equal to 1.

$$x^2 + 2x = -\frac{1}{2}$$ Subtract $\frac{1}{2}$ from both sides.

$$x^2 + 2x + \mathbf{1} = \mathbf{1} - \frac{1}{2}$$ Square half the coefficient of x and add it to both sides.

$$(x + 1)^2 = \frac{1}{2}$$ Factor and combine like terms.

$$x + 1 = \sqrt{\frac{1}{2}} \quad \text{or} \quad x + 1 = -\sqrt{\frac{1}{2}}$$

$$x + 1 = \frac{\sqrt{2}}{2} \qquad\qquad x + 1 = -\frac{\sqrt{2}}{2}$$ $\sqrt{\frac{1}{2}} = \frac{1}{\sqrt{2}} = \frac{1 \cdot \sqrt{2}}{\sqrt{2}\sqrt{2}} = \frac{\sqrt{2}}{2}$.

$$x = -1 + \frac{\sqrt{2}}{2} \qquad\qquad x = -1 - \frac{\sqrt{2}}{2}$$

$$x = \frac{-2 + \sqrt{2}}{2} \qquad\qquad x = \frac{-2 - \sqrt{2}}{2}$$

Both values check. ■

Self Check Solve $3x^2 + 6x + 1 = 0$.

Answer $\frac{-3 + \sqrt{6}}{3}, \frac{-3 - \sqrt{6}}{3}$

THE QUADRATIC FORMULA

To develop a formula we can use to solve quadratic equations, we solve the general quadratic equation $ax^2 + bx + c = 0$ $(a \neq 0)$.

$$ax^2 + bx + c = 0$$

$$\frac{ax^2}{a} + \frac{bx}{a} + \frac{c}{a} = \frac{0}{a}$$

Since $a \neq 0$, we can divide both sides by a to make the coefficient of x^2 equal to 1.

$$x^2 + \frac{bx}{a} = -\frac{c}{a}$$

$\frac{0}{a} = 0$; subtract $\frac{c}{a}$ from both sides.

$$x^2 + \frac{b}{a}x + \left(\frac{b}{2a}\right)^2 = \left(\frac{b}{2a}\right)^2 - \frac{c}{a}$$

Complete the square on x by adding $\left(\frac{b}{2a}\right)^2$ to both sides.

$$x^2 + \frac{b}{a}x + \frac{b^2}{4a^2} = \frac{b^2}{4a^2} - \frac{4ac}{4aa}$$

Remove parentheses and get a common denominator on the right-hand side.

5. $$\left(x + \frac{b}{2a}\right)^2 = \frac{b^2 - 4ac}{4a^2}$$

Factor the left-hand side and add the fractions on the right-hand side.

We can solve Equation 5 by using the square root property.

$$x + \frac{b}{2a} = \sqrt{\frac{b^2 - 4ac}{4a^2}} \quad \text{or} \quad x + \frac{b}{2a} = -\sqrt{\frac{b^2 - 4ac}{4a^2}}$$

$$x + \frac{b}{2a} = \frac{\sqrt{b^2 - 4ac}}{2a} \qquad x + \frac{b}{2a} = -\frac{\sqrt{b^2 - 4ac}}{2a}$$

$$x = -\frac{b}{2a} + \frac{\sqrt{b^2 - 4ac}}{2a} \qquad x = -\frac{b}{2a} - \frac{\sqrt{b^2 - 4ac}}{2a}$$

$$= \frac{-b + \sqrt{b^2 - 4ac}}{2a} \qquad = \frac{-b - \sqrt{b^2 - 4ac}}{2a}$$

These two solutions give the **quadratic formula.**

The Quadratic Formula

The solutions of $ax^2 + bx + c = 0$ $(a \neq 0)$ are given by the formula

$$x = \frac{-b \pm \sqrt{b^2 - 4ac}}{2a}$$

Read the symbol $\pm$ as "plus or minus."

WARNING! Be sure to draw the fraction bar under both parts of the numerator, and be sure to draw the radical sign exactly over $b^2 - 4ac$. Do not write the quadratic formula as

$$x = -b \pm \frac{\sqrt{b^2 - 4ac}}{2a} \quad \text{or as} \quad x = -b \pm \sqrt{\frac{b^2 - 4ac}{2a}}$$

EXAMPLE 7 Solve $2x^2 - 3x - 5 = 0$.

Solution In this equation $a = 2$, $b = -3$, and $c = -5$.

$$\begin{aligned} x &= \frac{-b \pm \sqrt{b^2 - 4ac}}{2a} \\ &= \frac{-(-3) \pm \sqrt{(-3)^2 - 4(2)(-5)}}{2(2)} && \text{Substitute 2 for } a, -3 \text{ for } b, \text{ and } -5 \text{ for } c. \\ &= \frac{3 \pm \sqrt{9 + 40}}{4} \\ &= \frac{3 \pm \sqrt{49}}{4} \\ &= \frac{3 \pm 7}{4} \end{aligned}$$

$$\begin{aligned} x &= \frac{3 + 7}{4} \quad \text{or} \quad & x &= \frac{3 - 7}{4} \\ x &= \frac{10}{4} & x &= \frac{-4}{4} \\ x &= \frac{5}{2} & x &= -1 \end{aligned}$$

Verify that both solutions satisfy the original equation. ■

Self Check Solve $3x^2 - 5x - 2 = 0$.

Answer $2, -\frac{1}{3}$

EXAMPLE 8 Solve $2x^2 + 1 = -4x$.

Solution We write the equation in general form before identifying a, b, and c.

$$2x^2 + 4x + 1 = 0$$

In this equation, $a = 2$, $b = 4$, and $c = 1$.

$$x = \frac{-b \pm \sqrt{b^2 - 4ac}}{2a}$$

$$= \frac{-4 \pm \sqrt{4^2 - 4(2)(1)}}{2(2)} \qquad \text{Substitute 2 for } a\text{, 4 for } b\text{, and 1 for } c.$$

$$= \frac{-4 \pm \sqrt{16 - 8}}{4}$$

$$= \frac{-4 \pm \sqrt{8}}{4}$$

$$= \frac{-4 \pm 2\sqrt{2}}{4} \qquad \sqrt{8} = \sqrt{4 \cdot 2} = \sqrt{4}\sqrt{2} = 2\sqrt{2}.$$

$$= \frac{-2 \pm \sqrt{2}}{2} \qquad \frac{-4 \pm 2\sqrt{2}}{4} = \frac{2(-2 \pm \sqrt{2})}{4} = \frac{-2 \pm \sqrt{2}}{2}.$$

Thus, $x = \frac{-2 + \sqrt{2}}{2}$ or $x = \frac{-2 - \sqrt{2}}{2}$. ■

Self Check Solve $3x^2 - 2x - 3 = 0$.

Answer $\frac{1 + \sqrt{10}}{3}, \frac{1 - \sqrt{10}}{3}$

■ PROBLEM SOLVING

EXAMPLE 9 **Dimensions of a rectangle** Find the dimensions of the rectangle shown in Figure 8-1, given that its area is 253 cm^2.

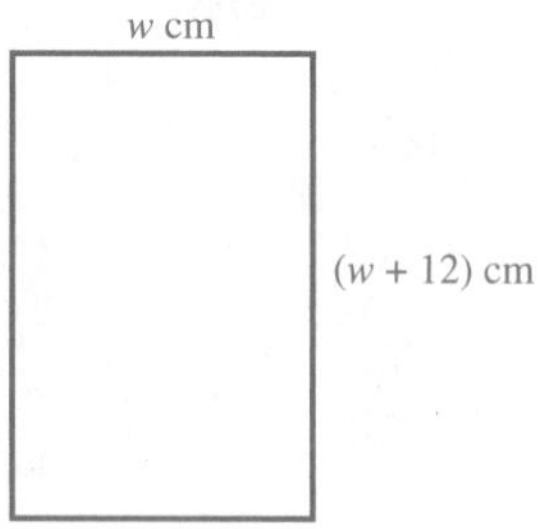

FIGURE 8-1

Solution If we let w represent the width of the rectangle, then $w + 12$ represents its length. Since the area of the rectangle is 253 square centimeters, we can form the equation

$$w(w + 12) = 253 \qquad \text{Area of a rectangle} = \text{width} \times \text{length}.$$

and solve it as follows:

$$w(w + 12) = 253$$

$$w^2 + 12w = 253 \qquad \text{Use the distributive property to remove parentheses.}$$

$$w^2 + 12w - 253 = 0 \qquad \text{Subtract 253 from both sides.}$$

Solution by Factoring

$$(w - 11)(w + 23) = 0$$

$$w - 11 = 0 \quad \text{or} \quad w + 23 = 0$$

$$w = 11 \qquad\qquad w = -23$$

Solution by Formula

$$w = \frac{-12 \pm \sqrt{12^2 - 4(1)(-253)}}{2(1)}$$

$$= \frac{-12 \pm \sqrt{144 + 1{,}012}}{2}$$

$$= \frac{-12 \pm \sqrt{1{,}156}}{2}$$

$$= \frac{-12 \pm 34}{2}$$

$$w = 11 \quad \text{or} \quad w = -23$$

Since the rectangle cannot have a negative width, we discard the solution of -23. Thus, the only solution is $w = 11$. Since the rectangle is 11 centimeters wide and $(11 + 12)$ centimeters long, its dimensions are 11 centimeters by 23 centimeters.

Check: 23 is 12 more than 11, and the area of a rectangle with dimensions of 23 centimeters by 11 centimeters is 253 square centimeters. ■

Orals

Solve each equation.

1. $x^2 = 49$ **2.** $x^2 = 10$

Find the number that must be added to the binomial to make it a perfect square trinomial.

3. $x^2 + 4x$ **4.** $x^2 - 6x$ **5.** $x^2 - 3x$ **6.** $x^2 + 5x$

Identify a, b, and c in each quadratic equation.

7. $3x^2 - 4x + 7 = 0$ **8.** $-2x^2 + x = 5$

EXERCISE 8.1

REVIEW *Solve each equation or inequality.*

1. $\dfrac{t + 9}{2} + \dfrac{t + 2}{5} = \dfrac{8}{5} + 4t$

2. $\dfrac{1 - 5x}{2x} + 4 = \dfrac{x + 3}{x}$

3. $3(t - 3) + 3t - 5 \le 2(t + 1) + t - 4$

4. $-2(y + 4) - 3y + 3 \ge 3(2y - 3) - y - 5$

Solve for the indicated variable.

5. $Ax + By = C$ for B

6. $R = \dfrac{kL}{d^2}$ for L

VOCABULARY AND CONCEPTS *Fill in each blank to make a true statement.*

7. If $c > 0$, the solutions of $x^2 = c$ are ________ and ________.

8. To complete the square on x in $x^2 + 6x$, find one-half of __, square it to get __, and add __ to get ________.

9. The symbol $\pm$ is read as ____________.

10. The solutions of $ax^2 + bx + c = 0$ $(a \neq 0)$ are given by the ________ formula, which is

$x =$ ______________

PRACTICE *In Exercises 11–22, use factoring to solve each equation.*

11. $6x^2 + 12x = 0$
12. $5x^2 + 11x = 0$
13. $2y^2 - 50 = 0$
14. $4y^2 - 64 = 0$
15. $r^2 + 6r + 8 = 0$
16. $x^2 + 9x + 20 = 0$
17. $7x - 6 = x^2$
18. $5t - 6 = t^2$
19. $2z^2 - 5z + 2 = 0$
20. $2x^2 - x - 1 = 0$
21. $6s^2 + 11s - 10 = 0$
22. $3x^2 + 10x - 8 = 0$

In Exercises 23–34, use the square root property to solve each equation.

23. $x^2 = 36$
24. $x^2 = 144$
25. $z^2 = 5$
26. $u^2 = 24$
27. $3x^2 - 16 = 0$
28. $5x^2 - 49 = 0$
29. $(x + 1)^2 = 1$
30. $(x - 1)^2 = 4$
31. $(s - 7)^2 - 9 = 0$
32. $(t + 4)^2 = 16$
33. $(x + 5)^2 - 3 = 0$
34. $(x + 3)^2 - 7 = 0$

In Exercises 35–48, use completing the square to solve each equation.

35. $x^2 + 2x - 8 = 0$
36. $x^2 + 6x + 5 = 0$
37. $x^2 - 6x + 8 = 0$
38. $x^2 + 8x + 15 = 0$
39. $x^2 + 5x + 4 = 0$
40. $x^2 - 11x + 30 = 0$
41. $x + 1 = 2x^2$
42. $-2 = 2x^2 - 5x$
43. $6x^2 + 11x + 3 = 0$
44. $6x^2 + x - 2 = 0$
45. $9 - 6r = 8r^2$
46. $11m - 10 = 3w^2$
47. $\dfrac{7x + 1}{5} = -x^2$
48. $\dfrac{3x^2}{8} = \dfrac{1}{8} - x$

In Exercises 49–60, use the quadratic formula to solve each equation.

49. $x^2 + 3x + 2 = 0$
50. $x^2 - 3x + 2 = 0$
51. $x^2 + 12x = -36$
52. $y^2 - 18y = -81$
53. $5x^2 + 5x + 1 = 0$
54. $4w^2 + 6w + 1 = 0$
55. $8u = -4u^2 - 3$
56. $4t + 3 = 4t^2$
57. $16y^2 + 8y - 3 = 0$
58. $16x^2 + 16x + 3 = 0$
59. $\dfrac{x^2}{2} + \dfrac{5}{2}x = -1$
60. $-3x = \dfrac{x^2}{2} + 2$

In Exercises 61–62, use the quadratic formula and a scientific calculator to solve each equation. Give all answers to the nearest hundredth.

61. $0.7x^2 - 3.5x - 25 = 0$
62. $-4.5x^2 + 0.2x + 3.75 = 0$

63. **Integer problem** The product of two consecutive even positive integers is 288. Find the integers. (*Hint:* If one integer is x, the next consecutive even integer is $x + 2$.)

64. **Integer problem** The product of two consecutive odd negative integers is 143. Find the integers. (*Hint:* If one integer is x, the next consecutive odd integer is $x + 2$.)

65. **Integer problem** The sum of the squares of two consecutive positive integers is 85. Find the integers. (*Hint:* If one integer is x, the next consecutive positive integer is $x + 1$.)

66. **Integer problem** The sum of the squares of three consecutive positive integers is 77. Find the integers. (*Hint:* If one integer is x, the next consecutive positive integer is $x + 1$, and the third is $x + 2$.)

In Exercises 67–70, note that a and b are solutions to the equation $(x - a)(x - b) = 0$.

67. Find a quadratic equation that has a solution set of $\{3, 5\}$.

68. Find a quadratic equation that has a solution set of $\{-4, 6\}$.

69. Find a third-degree equation that has a solution set of $\{2, 3, -4\}$.

70. Find a fourth-degree equation that has a solution set of $\{3, -3, 4, -4\}$.

APPLICATIONS

71. **Dimensions of a rectangle** A rectangle is 4 feet longer than it is wide, and its area is 96 square feet. Find its dimensions.

72. **Dimensions of a rectangle** One side of a rectangle is 3 times as long as another, and its area is 147 square meters. Find its dimensions.

73. **Side of a square** The area of a square is numerically equal to its perimeter. Find the length of each side of the square.

74. **Perimeter of a rectangle** A rectangle is 2 inches longer than it is wide. Numerically, its area exceeds its perimeter by 11. Find the perimeter.

75. **Base of a triangle** The height of a triangle is 5 centimeters longer than three times its base. Find the base of the triangle if its area is 6 square centimeters.

76. **Height of a triangle** The height of a triangle is 4 meters longer than twice its base. Find the height if the area of the triangle is 15 square meters.

77. **Finding rates** A woman drives her snowmobile 150 miles at the rate of r mph. She could have gone the same distance in 2 hours less time if she had increased her speed by 20 mph. Find r.

78. **Finding rates** Jeff bicycles 160 miles at the rate of r mph. The same trip would have taken 2 hours longer if he had decreased his speed by 4 mph. Find r.

79. **Pricing concert tickets** Tickets to a rock concert cost \$4, and the projected attendance is 300 persons. It is further projected that for every 10¢ increase in ticket price, the average attendance will decrease by 5. At what ticket price will the nightly receipts be \$1,248?

80. **Setting bus fares** A bus company has 3,000 passengers daily, paying a 25¢ fare. For each 5¢ increase in fare, the company estimates that it will lose 80 passengers. What increase in fare will produce a \$994 daily revenue?

81. **Computing profit** The *Gazette*'s profit is \$20 per year for each of its 3,000 subscribers. Management estimates that the profit per subscriber will increase by 1¢ for each additional subscriber over the current 3,000. How many subscribers will bring a total profit of \$120,000?

82. **Finding interest rates** A woman invests \$1,000 in a mutual fund for which interest is compounded annually at a rate r. After one year, she deposits an additional \$2,000. After two years, the balance in the account is

$$\$1{,}000(1 + r)^2 + \$2{,}000(1 + r)$$

If this amount is \$3,368.10, find r.

83. Framing a picture The frame around the picture in Illustration 1 has a constant width. How wide is the frame if its area equals the area of the picture?

ILLUSTRATION 1

84. Metal fabrication A box with no top is to be made by cutting a 2-inch square from each corner of the square sheet of metal shown in Illustration 2. After bending up the sides, the volume of the box is to be 200 cubic inches. How large should the piece of metal be?

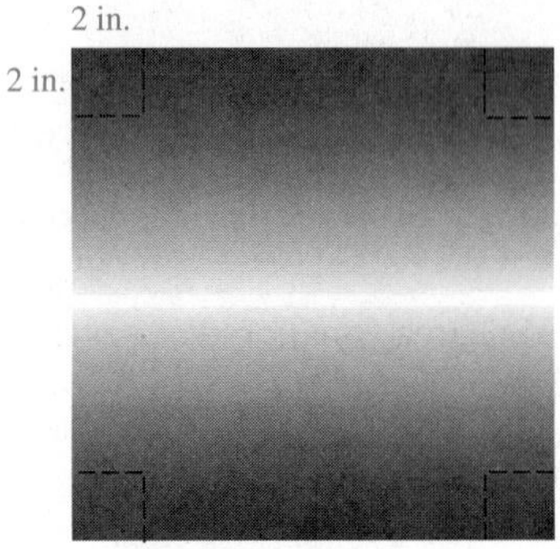

ILLUSTRATION 2

In Exercises 85–86, use a calculator.

85. Chemistry A weak acid (0.1 M concentration) breaks down into free cations (the hydrogen ion, H^+) and anions (A^-). When this acid dissociates, the following equilibrium equation is established:

$$\frac{[H^+][A^-]}{[HA]} = 4 \times 10^{-4}$$

(where $[H^+]$, the hydrogen ion concentration, is equal to $[A^-]$, the anion concentration). $[HA]$ is the concentration of the undissociated acid itself. Find $[H^+]$ at equilibrium. (*Hint:* If $H^+ = x$, then $[HA] = 0.1 - x$.)

86. Chemistry A saturated solution of hydrogen sulfide (0.1 M concentration) dissociates into cation H^+ and anion HS^-, where $H^+ = HS^-$. When this solution dissociates, the following equilibrium equation is established:

$$\frac{[H^+][HS^-]}{[HHS]} = 1.0 \times 10^{-7}$$

Find $[H^+]$. (*Hint:* If $H^+ = x$, then $[HHS] = 0.1 - x$.)

WRITING

87. Explain how to complete the square.

88. Tell why a cannot be 0 in the quadratic equation $ax^2 + bx + c = 0$.

SOMETHING TO THINK ABOUT

89. What number must be added to $x^2 + \sqrt{3}x$ to make it a perfect square trinomial?

90. Solve $x^2 + \sqrt{3}x - \frac{1}{4} = 0$ by completing the square.

8.2 Graphs of Quadratic Functions

■ QUADRATIC FUNCTIONS ■ GRAPHS OF $f(x) = ax^2$ ■ GRAPHS OF $f(x) = ax^2 + c$ ■ GRAPHS OF $f(x) = a(x - h)^2$ ■ GRAPHS OF $f(x) = a(x - h)^2 + k$ ■ GRAPHS OF $f(x) = ax^2 + bx + c$ ■ PROBLEM SOLVING ■ THE VARIANCE

Getting Ready *If* $y = f(x) = 3x^2 + x - 2$, *find each value.*

1. $f(0)$ **2.** $f(1)$ **3.** $f(-1)$ **4.** $f(-2)$

If $x = -\dfrac{b}{2a}$, *find x when a and b have the following values.*

5. $a = 3$ and $b = -6$ **6.** $a = 5$ and $b = -40$

■ QUADRATIC FUNCTIONS

The graph shown in Figure 8-2 shows the height (in relation to time) of a toy rocket launched straight up into the air.

WARNING! Note that the graph describes the height of the rocket, not the path of the rocket. The rocket goes straight up and comes straight down.

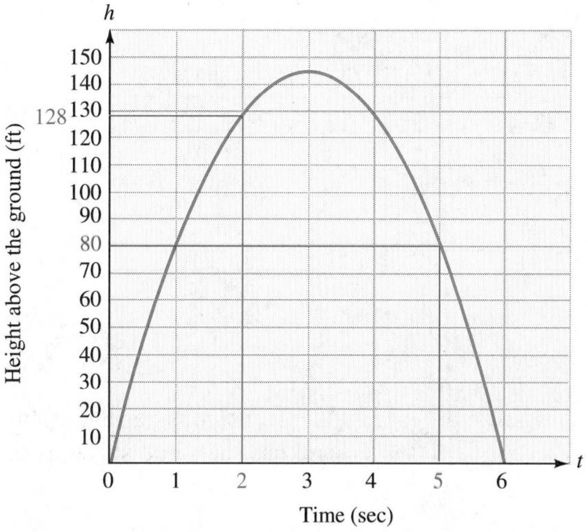

FIGURE 8-2

From the graph, we can see that the height of the rocket 2 seconds after it was launched is about 128 feet and that the height of the rocket 5 seconds after it was launched is 80 feet.

The parabola shown in Figure 8-2 is the graph of a *quadratic function,* a topic of this section.

> **Quadratic Functions**
>
> A **quadratic function** is a second-degree polynomial function of the form
>
> $$y = f(x) = ax^2 + bx + c \quad (a \neq 0)$$
>
> where a, b, and c are real numbers.

We begin the discussion of graphing quadratic functions by considering the graph of $f(x) = ax^2 + bx + c$, where $b = 0$ and $c = 0$.

GRAPHS OF $f(x) = ax^2$

EXAMPLE 1 Graph **a.** $f(x) = x^2$, **b.** $g(x) = 3x^2$, and **c.** $h(x) = \frac{1}{3}x^2$.

Solution We can make a table of ordered pairs that satisfy each equation, plot each point, and join them with a smooth curve, as in Figure 8-3. We note that the graph of $h(x) = \frac{1}{3}x^2$ is wider than the graph of $f(x) = x^2$, and that the graph of $g(x) = 3x^2$ is narrower than the graph of $f(x) = x^2$. In the function $f(x) = ax^2$, the smaller the value of $|a|$, the wider the graph.

$f(x) = x^2$

x	$f(x)$	$(x, f(x))$
-2	4	$(-2, 4)$
-1	1	$(-1, 1)$
0	0	$(0, 0)$
1	1	$(1, 1)$
2	4	$(2, 4)$

$g(x) = 3x^2$

x	$g(x)$	$(x, g(x))$
-2	12	$(-2, 12)$
-1	3	$(-1, 3)$
0	0	$(0, 0)$
1	3	$(1, 3)$
2	12	$(2, 12)$

$h(x) = \frac{1}{3}x^2$

x	$h(x)$	$(x, h(x))$
-2	$\frac{4}{3}$	$\left(-2, \frac{4}{3}\right)$
-1	$\frac{1}{3}$	$\left(-1, \frac{1}{3}\right)$
0	0	$(0, 0)$
1	$\frac{1}{3}$	$\left(1, \frac{1}{3}\right)$
2	$\frac{4}{3}$	$\left(2, \frac{4}{3}\right)$

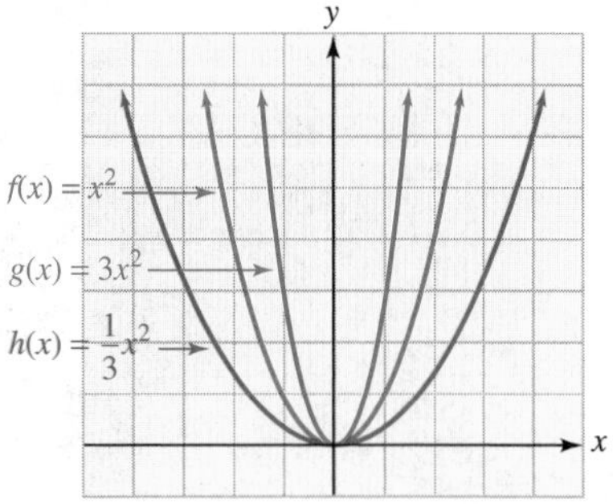

FIGURE 8-3

If we consider the graph of $f(x) = -3x^2$, we will see that it opens downward and has the same shape as the graph of $g(x) = 3x^2$.

EXAMPLE 2 Graph $f(x) = -3x^2$.

Solution We make a table of ordered pairs that satisfy the equation, plot each point, and join them with a smooth curve, as in Figure 8-4.

$f(x) = -3x^2$

x	$f(x)$	$(x, f(x))$
−2	−12	(−2, −12)
−1	−3	(−1, −3)
0	0	(0, 0)
1	−3	(1, −3)
2	−12	(2, −12)

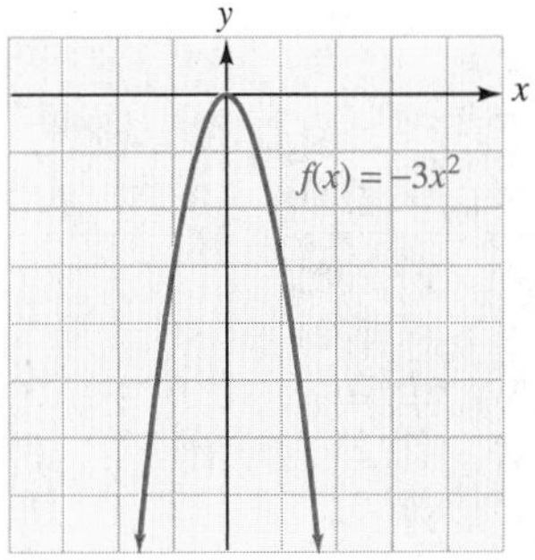

FIGURE 8-4

Self Check Graph $f(x) = -\frac{1}{3}x^2$.

Answer

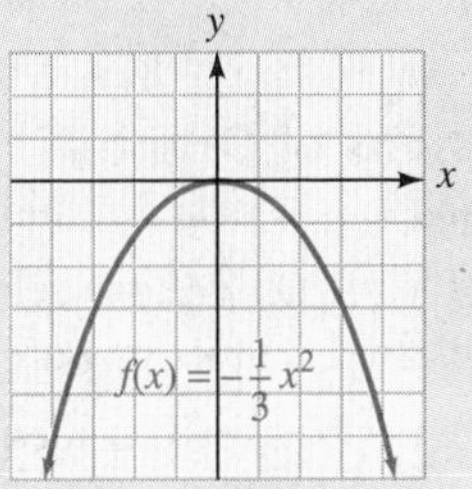

The graphs of quadratic functions are called **parabolas.** They open upward when $a > 0$ and downward when $a < 0$. The lowest point of a parabola that opens upward, or the highest point of a parabola that opens downward, is called the **vertex** of the parabola. The vertex of the parabola shown in Figure 8-4 is the point (0, 0).

The vertical line, called an **axis of symmetry,** that passes through the vertex divides the parabola into two congruent halves. The axis of symmetry of the parabola shown in Figure 8-4 is the y-axis.

GRAPHS OF $f(x) = ax^2 + c$

EXAMPLE 3 Graph **a.** $f(x) = 2x^2$, **b.** $g(x) = 2x^2 + 3$, and **c.** $h(x) = 2x^2 - 3$.

Solution We make a table of ordered pairs that satisfy each equation, plot each point, and join them with a smooth curve, as in Figure 8-5. We note that the graph of $g(x) = 2x^2 + 3$ is identical to the graph of $f(x) = 2x^2$, except that it has been translated 3 units upward. The graph of $h(x) = 2x^2 - 3$ is identical to the graph of $f(x) = 2x^2$, except that it has been translated 3 units downward.

$f(x) = 2x^2$

x	$f(x)$	$(x, f(x))$
−2	8	(−2, 8)
−1	2	(−1, 2)
0	0	(0, 0)
1	2	(1, 2)
2	8	(2, 8)

$g(x) = 2x^2 + 3$

x	$g(x)$	$(x, g(x))$
−2	11	(−2, 11)
−1	5	(−1, 5)
0	3	(0, 3)
1	5	(1, 5)
2	11	(2, 11)

$h(x) = 2x^2 - 3$

x	$h(x)$	$(x, h(x))$
−2	5	(−2, 5)
−1	−1	(−1, −1)
0	−3	(0, −3)
1	−1	(1, −1)
2	5	(2, 5)

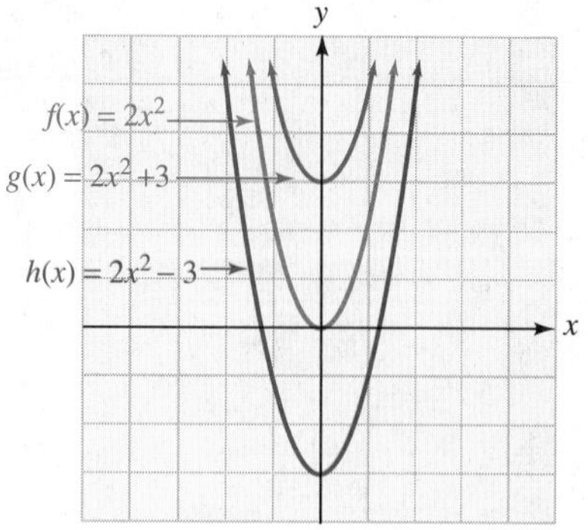

FIGURE 8-5

The results of Example 3 confirm the following facts, which we have previously discussed.

Vertical Translations of Graphs

If f is a function and k is a positive number, then

- The graph of $y = f(x) + k$ is identical to the graph of $y = f(x)$, except that it is translated k units upward.
- The graph of $y = f(x) - k$ is identical to the graph of $y = f(x)$, except that it is translated k units downward.

GRAPHS OF $f(x) = a(x - h)^2$

EXAMPLE 4

Graph **a.** $f(x) = 2x^2$, **b.** $g(x) = 2(x - 3)^2$, and **c.** $h(x) = 2(x + 3)^2$.

Solution We make a table of ordered pairs that satisfy each equation, plot each point, and join them with a smooth curve, as in Figure 8-6. We note that the graph of $g(x) = 2(x - 3)^2$ is identical to the graph of $f(x) = 2x^2$, except that it has been translated 3 units to the right. The graph of $h(x) = 2(x + 3)^2$ is identical to the graph of $f(x) = 2x^2$, except that it has been translated 3 units to the left.

The results of Example 4 confirm the following facts, which we have previously discussed.

Horizontal Translations of Graphs

If f is a function and h is a positive number, then

- The graph of $y = f(x - h)$ is identical to the graph of $y = f(x)$, except that it is translated h units to the right.
- The graph of $y = f(x + h)$ is identical to the graph of $y = f(x)$, except that it is translated h units to the left.

$f(x) = 2x^2$

x	$f(x)$	$(x, f(x))$
−2	8	(−2, 8)
−1	2	(−1, 2)
0	0	(0, 0)
1	2	(1, 2)
2	8	(2, 8)

$g(x) = 2(x - 3)^2$

x	$g(x)$	$(x, g(x))$
1	8	(1, 8)
2	2	(2, 2)
3	0	(3, 0)
4	2	(4, 2)
5	8	(5, 8)

$h(x) = 2(x + 3)^2$

x	$h(x)$	$(x, h(x))$
−5	8	(−5, 8)
−4	2	(−4, 2)
−3	0	(−3, 0)
−2	2	(−2, 2)
−1	8	(−1, 8)

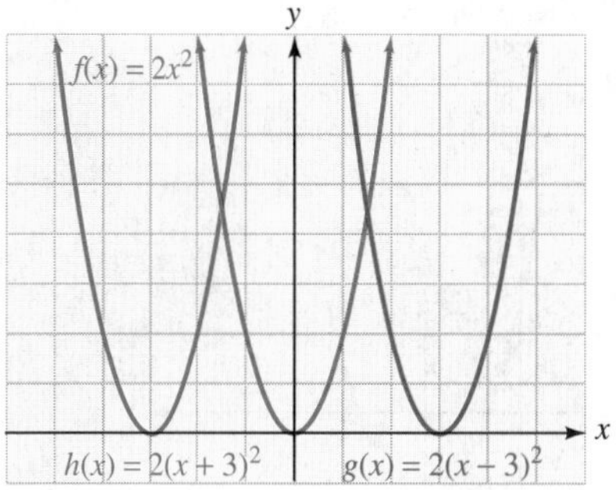

FIGURE 8-6

■

GRAPHS OF $f(x) = a(x - h)^2 + k$

EXAMPLE 5 Graph $f(x) = 2(x - 3)^2 - 4$.

Solution The graph of $f(x) = 2(x - 3)^2 - 4$ is identical to the graph of $g(x) = 2(x - 3)^2$, except that it has been translated 4 units downward. The graph of $g(x) = 2(x - 3)^2$ is identical to the graph of $h(x) = 2x^2$, except that it has been translated 3 units to the right. Thus, to graph $f(x) = 2(x - 3)^2 - 4$, we can graph $h(x) = 2x^2$ and shift it 3 units to the right and then 4 units downward, as shown in Figure 8-7.

The vertex of the graph is the point $(3, -4)$, and the axis of symmetry is the line $x = 3$.

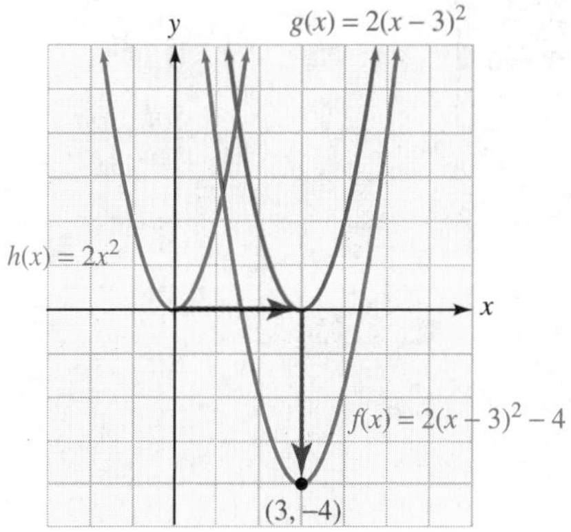

FIGURE 8-7

■

Self Check Graph $f(x) = 2(x + 3)^2 + 1$.

Answer

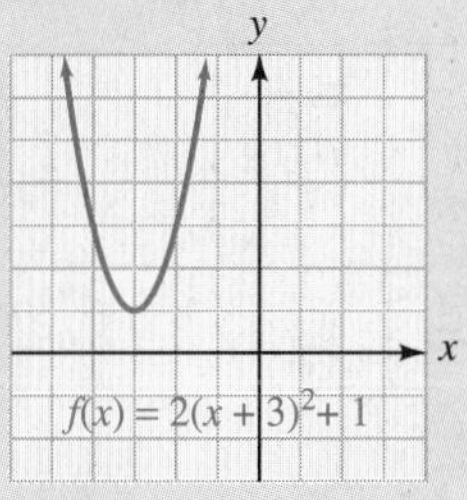

The results of Example 5 confirm the following facts, which we have previously discussed.

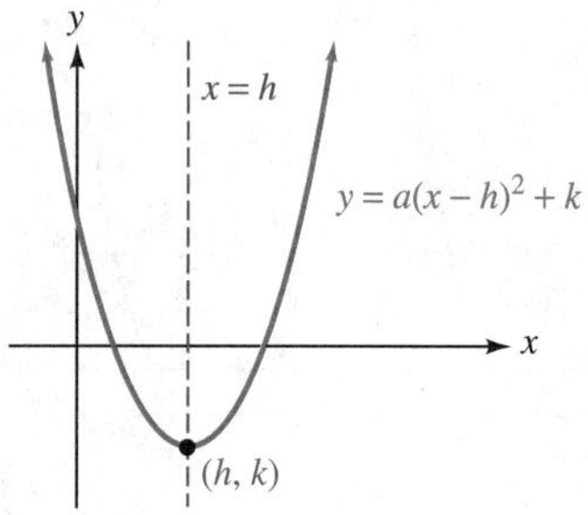

FIGURE 8-8

Vertex and Axis of Symmetry of a Parabola

The graph of the function

$$y = a(x - h)^2 + k \quad (a \neq 0)$$

is a parabola with vertex at (h, k). (See Figure 8-8.)

The parabola opens upward when $a > 0$ and downward when $a < 0$. The axis of symmetry is the line $x = h$.

GRAPHS OF $f(x) = ax^2 + bx + c$

To graph functions of the form $f(x) = ax^2 + bx + c$, we can complete the square to write the function in the form $f(x) = a(x - h)^2 + k$.

EXAMPLE 6 Graph $f(x) = 2x^2 - 4x - 1$.

Solution We complete the square on x to write the function in the form $f(x) = a(x - h)^2 + k$.

$f(x) = 2x^2 - 4x - 1$

$f(x) = 2(x^2 - 2x) - 1$ Factor 2 from $2x^2 - 4x$.

$f(x) = 2(x^2 - 2x + 1) - 1 - 2$ Complete the square on x. Since this adds 2 to the right-hand side, we also subtract 2 from the right-hand side.

1. $f(x) = 2(x - 1)^2 - 3$ Factor $x^2 - 2x + 1$ and combine like terms.

From Equation 1, we can see that the vertex will be at the point $(1, -3)$. We can plot the vertex and a few points on either side of the vertex and draw the graph, which appears in Figure 8-9.

$f(x) = 2x^2 - 4x - 1$

x	$f(x)$	$(x, f(x))$
−1	5	(−1, 5)
0	−1	(0, −1)
1	−3	(1, −3)
2	−1	(2, −1)
3	5	(3, 5)

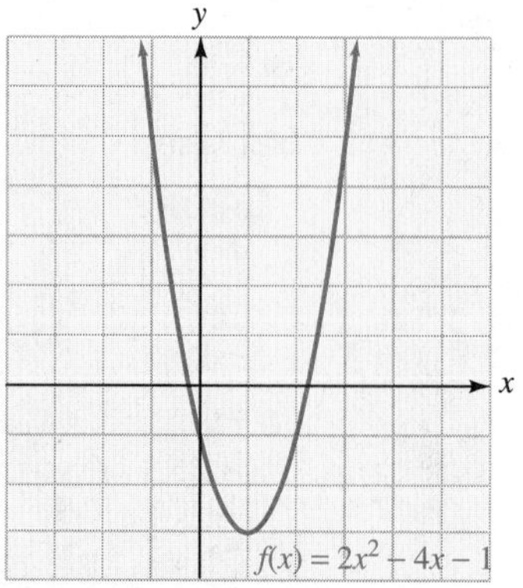

FIGURE 8-9

Self Check Graph $f(x) = 2x^2 - 4x + 1$.

Answer

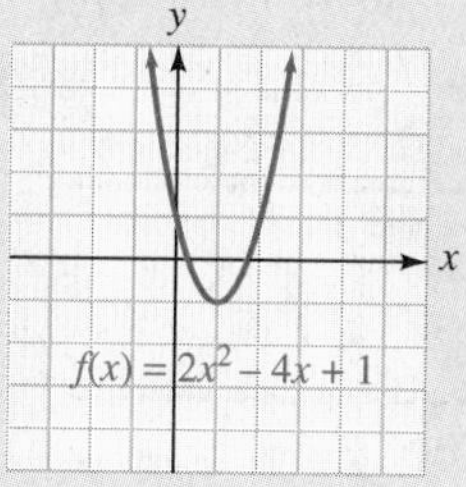

Graphing Quadratic Functions

GRAPHING CALCULATORS

To graph $f(x) = 2x^2 + 6x - 3$ and find the coordinates of the vertex and the axis of symmetry of the parabola, we use a graphing calculator with window settings of $[-10, 10]$ for x and $[-10, 10]$ for y. If we enter the function, we will obtain the graph shown in Figure 8-10(a).

We then trace to move the cursor to the lowest point on the graph, as shown in Figure 8-10(b). By zooming in, we can determine that the vertex is the point $(-1.5, -7.5)$ and that the line $x = -1.5$ is the axis of symmetry.

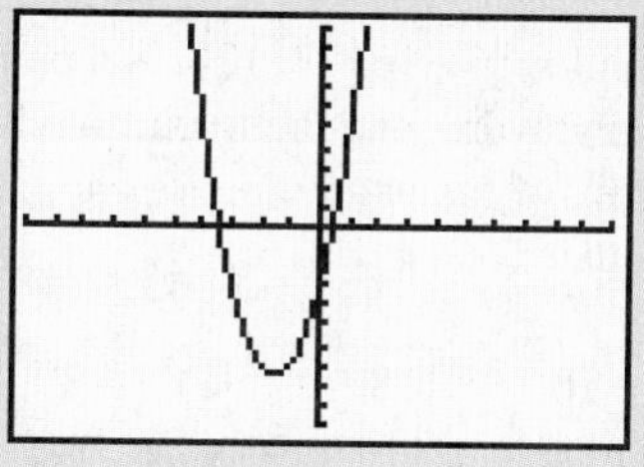

(a)

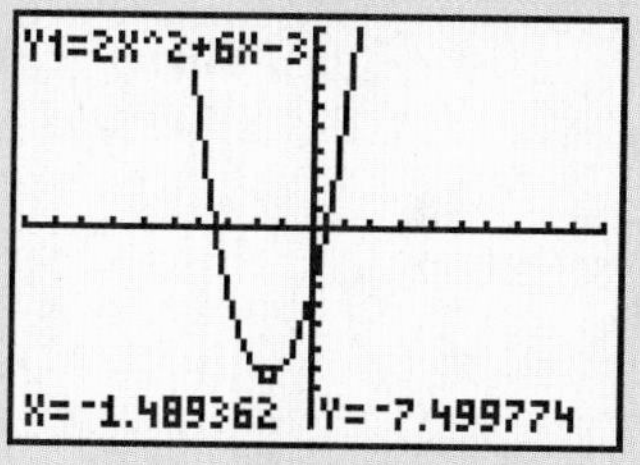

(b)

FIGURE 8-10

Because it is easy to graph quadratic functions with a graphing calculator, we can use graphing to find approximate solutions of quadratic equations. For example, the solutions of $0.7x^2 + 2x - 3.5 = 0$ are the numbers x that will make $y = 0$ in the quadratic function $y = f(x) = 0.7x^2 + 2x - 3.5$. To approximate these numbers, we graph the quadratic function and read the x-intercepts from the graph.

We can use the standard window settings of $[-10, 10]$ for x and $[-10, 10]$ for y and graph the function, as in Figure 8-11(a). We then trace to move the cursor to each x-intercept, as in Figures 8-11(b) and 8-11(c). From the graph, we can read the approximate value of the x-coordinate of each x-intercept. For better results, we can zoom in.

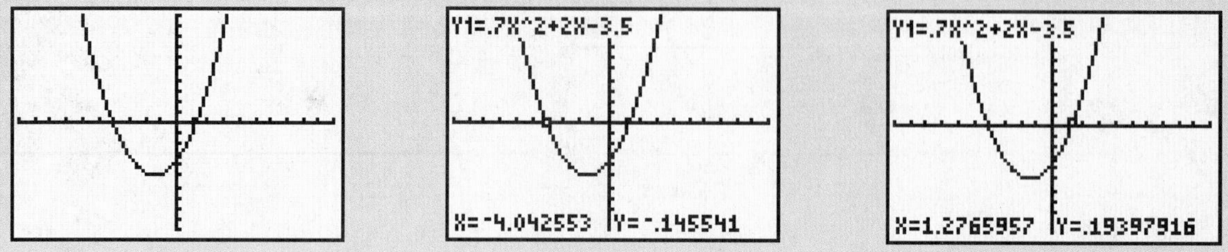

(a) (b) (c)

FIGURE 8-11

PROBLEM SOLVING

EXAMPLE 7

Ballistics The ball shown in Figure 8-12(a) is thrown straight up with a velocity of 128 feet per second. The function $s = h(t) = -16t^2 + 128t$ gives the relation between s (the number of feet the ball is above the ground) and t (the time measured in seconds). How high does the ball travel, and when will it hit the ground?

Solution The graph of $s = -16t^2 + 128t$ is a parabola. Since the coefficient of t^2 is negative, it opens downward, and the maximum height of the ball is given by the s-coordinate of the vertex of the parabola. We can find the coordinates of the vertex by completing the square:

$$\begin{aligned} s &= -16t^2 + 128t \\ &= -16(t^2 - 8t) && \text{Factor out } -16. \\ &= -16(t^2 - 8t + \mathbf{16} - \mathbf{16}) && \text{Add and subtract 16.} \\ &= -16(t^2 - 8t + 16) + 256 && (-16)(-16) = 256. \\ &= -16(t - \mathbf{4})^2 + \mathbf{256} && \text{Factor } t^2 - 8t + 16. \end{aligned}$$

From the result, we can see that the coordinates of the vertex are (**4**, **256**). Since $t = 4$ and $s = 256$ are the coordinates of the vertex, the ball reaches a maximum height of 256 feet in 4 seconds.

From the graph, we can see that the ball will hit the ground in 8 seconds, because the height is 0 when $t = 8$.

To solve this problem with a graphing calculator with window settings of [0, 10] for x and [0, 300] for y, we graph the function $h(t) = -16t^2 + 128t$ to get the graph in Figure 8-12(b). By using trace and zoom, we can determine that the ball reaches a height of 256 feet in 4 seconds and that the ball will hit the ground in 8 seconds.

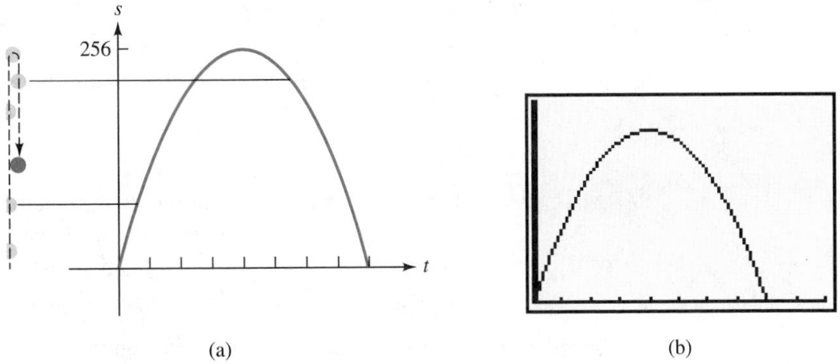

FIGURE 8-12

EXAMPLE 8

Maximizing area A man wants to build the rectangular pen shown in Figure 8-13(a) to house his dog. If he uses one side of his barn, find the maximum area that he can enclose with 80 feet of fencing.

Solution If we let the width of the area be w, the length is represented by $80 - 2w$.

We can find the maximum value of A as follows:

$A = (80 - 2w)w$	$A = lw$.
$= 80w - 2w^2$	Remove parentheses.
$= -2(w^2 - 40w)$	Factor out -2.
$= -2(w^2 - 40w + \mathbf{400} - \mathbf{400})$	Subtract and add 400.
$= -2(w^2 - 40w + 400) + 800$	$-2(-400) = 800$.
$= -2(w - 20)^2 + 800$	Factor $w^2 - 40w + 400$.

Thus, the coordinates of the vertex of the graph of the quadratic function are (20, 800), and the maximum area is 800 square feet.

To solve this problem using a graphing calculator with window settings of [0, 50] for x and [0, 1,000] for y, we graph the function $s(t) = -2w^2 + 80w$ to get the graph in Figure 8-13(b). By using trace and zoom, we can determine that the maximum area is 800 square feet when the width is 20 feet.

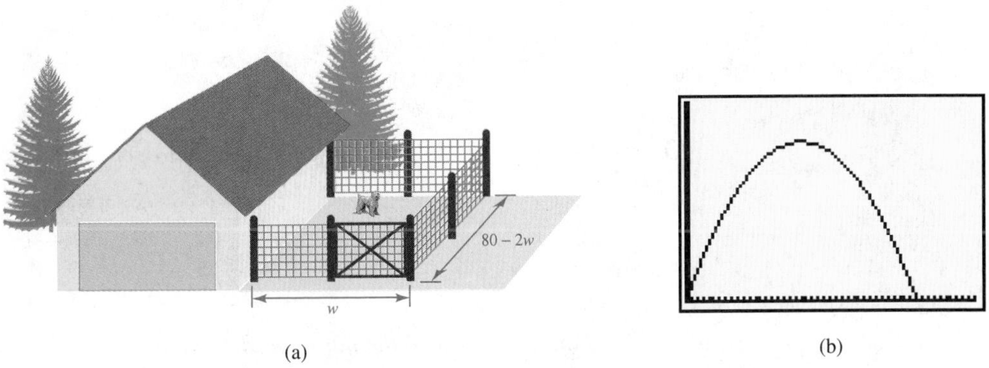

FIGURE 8-13

THE VARIANCE

In statistics, the square of the standard deviation is called the **variance.**

EXAMPLE 9

If p is the probability that a person selected at random has AIDS, then $1 - p$ is the probability that the person does not have AIDS. If 100 people in Minneapolis are randomly sampled, we know from statistics that the variance of this type of sample distribution will be $100p(1 - p)$. What value of p will maximize the variance?

Solution The variance is given by the function

$$v(p) = 100p(1 - p) \qquad \text{or} \qquad v(p) = -100p^2 + 100p$$

Since all probabilities have values between 0 to 1, including 0 and 1, we use window settings of [0, 1] for x when graphing the function $v(p) = -100p^2 + 100p$ on a graphing calculator. If we also use window settings of [0, 30] for y, we will obtain the graph shown in Figure 8-14(a). After using trace and zoom to obtain Figure 8-14(b), we can see that a probability of 0.5 will give the maximum variance.

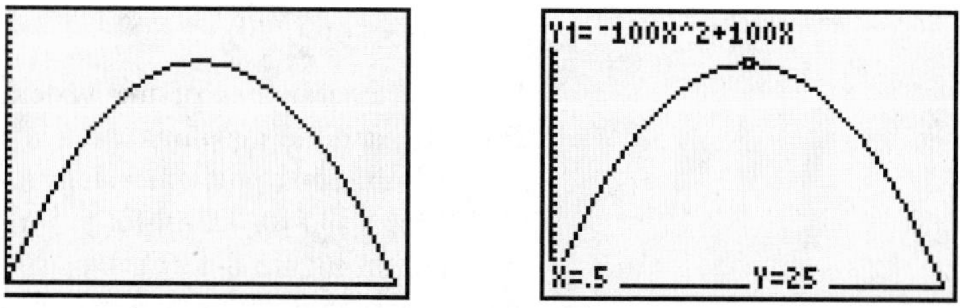

(a) (b)

FIGURE 8-14

Orals *Tell whether the graph of each equation opens up or down.*

1. $y = -3x^2 + x - 5$

2. $y = 4x^2 + 2x - 3$

3. $y = 2(x - 3)^2 - 1$

4. $y = -3(x + 2)^2 + 2$

Find the vertex of the parabola determined by each equation.

5. $y = 2(x - 3)^2 - 1$

6. $y = -3(x + 2)^2 + 2$

EXERCISE 8.2

REVIEW *In Exercises 1–2, find the value of x.*

1.

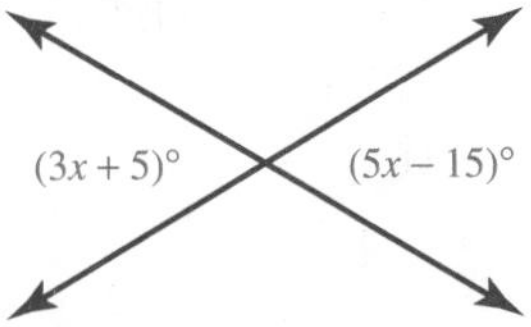

2. Lines r and s are parallel.

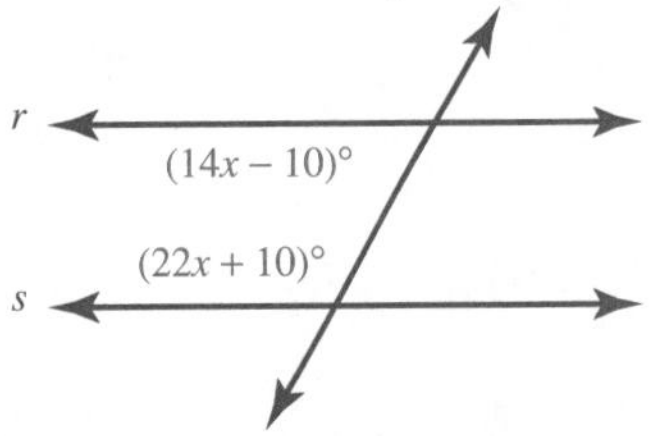

3. Madison and St. Louis are 385 miles apart. One train leaves Madison and heads toward St. Louis at the rate of 30 mph. Three hours later, a second train leaves Madison, bound for St. Louis. If the second train travels at the rate of 55 mph, in how many hours will the faster train overtake the slower train?

4. A woman invests \$25,000, some at 7% annual interest and the rest at 8%. If the annual income from both investments is \$1,900, how much is invested at the higher rate?

VOCABULARY AND CONCEPTS *Fill in each blank to make a true statement.*

5. A quadratic function is a second-degree polynomial function that can be written in the form ________________, where ______.

6. The graphs of quadratic functions are called ________.

7. The highest (or the lowest) point on a parabola is called the ______.

8. A vertical line that divides a parabola into two halves is called an _____ of symmetry.

9. The graph of $y = f(x) + k$ $(k > 0)$ is identical to the graph of $y = f(x)$, except that it is translated k units ______.

10. The graph of $y = f(x) - k$ $(k > 0)$ is identical to the graph of $y = f(x)$, except that it is translated k units ________.

11. The graph of $y = f(x - h)$ $(h > 0)$ is identical to the graph of $y = f(x)$, except that it is translated h units ________.

12. The graph of $y = f(x + h)$ $(h > 0)$ is identical to the graph of $y = f(x)$, except that it is translated h units ________.

13. The graph of $y = f(x) = ax^2 + bx + c$ $(a \neq 0)$ opens ______ when $a > 0$.

14. In statistics, the square of the standard deviation is called the ________.

PRACTICE *In Exercises 15–26, graph each function.*

15. $f(x) = x^2$

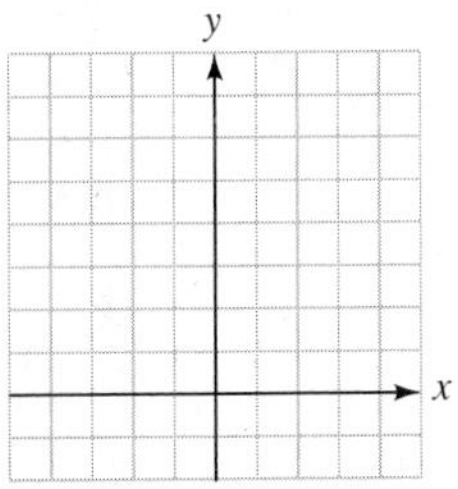

16. $f(x) = -x^2$

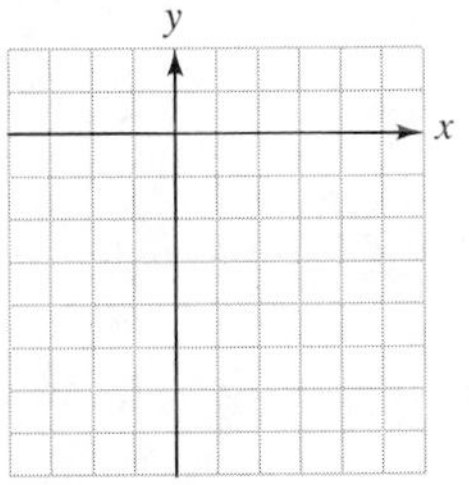

17. $f(x) = x^2 + 2$

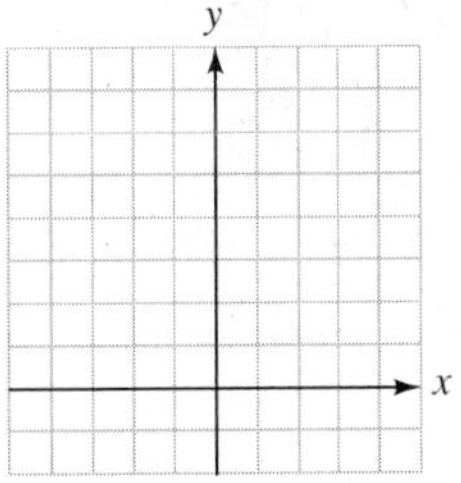

18. $f(x) = x^2 - 3$

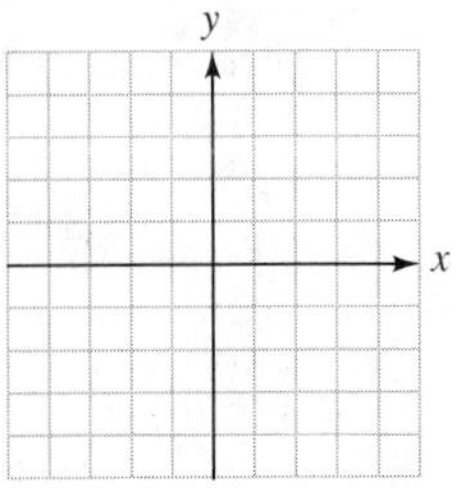

19. $f(x) = -(x - 2)^2$

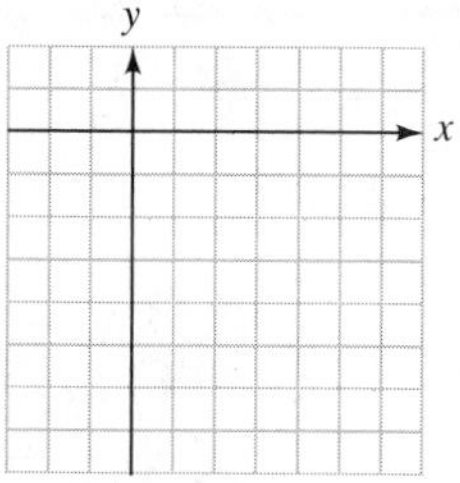

20. $f(x) = (x + 2)^2$

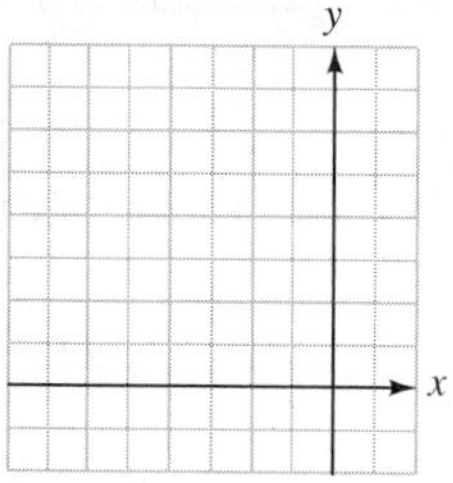

21. $f(x) = (x - 3)^2 + 2$

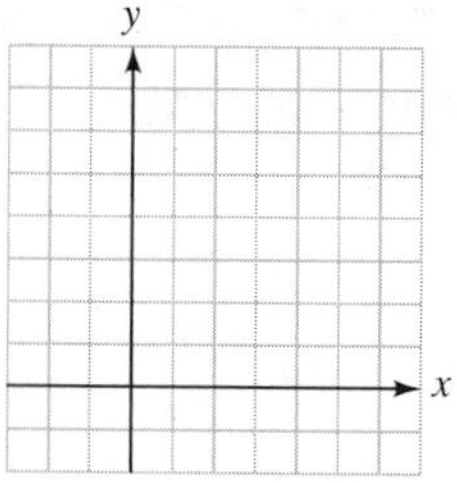

22. $f(x) = (x + 1)^2 - 2$

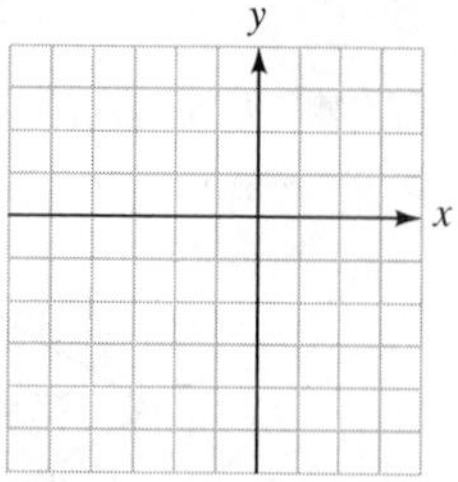

23. $f(x) = x^2 + x - 6$

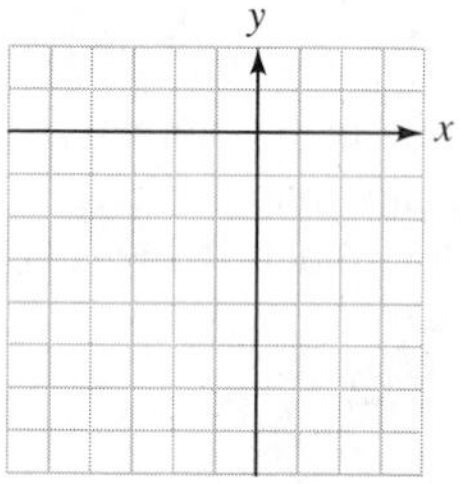

24. $f(x) = x^2 - x - 6$

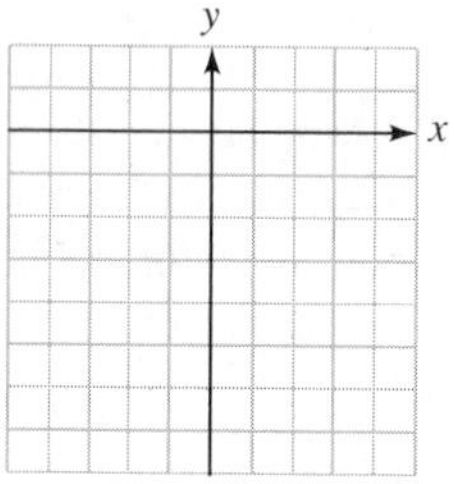

25. $f(x) = 12x^2 + 6x - 6$

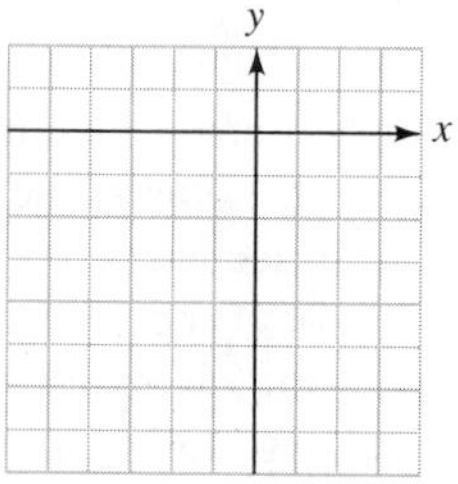

26. $f(x) = -2x^2 + 4x + 3$

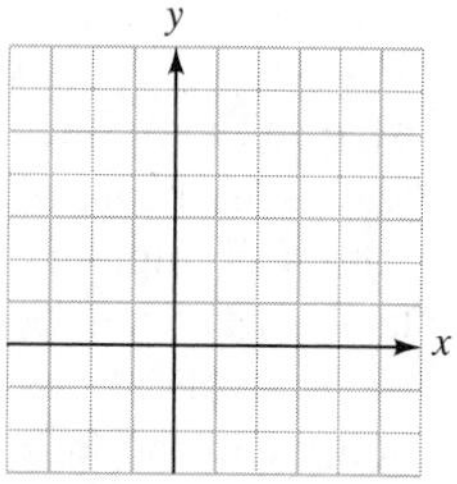

In Exercises 27–38, find the coordinates of the vertex and the axis of symmetry of the graph of each equation. If necessary, complete the square on x to write the equation in the form $y = a(x - h)^2 + k$. ***Do not graph the equation.***

27. $y = (x - 1)^2 + 2$

28. $y = 2(x - 2)^2 - 1$

29. $y = 2(x + 3)^2 - 4$

30. $y = -3(x + 1)^2 + 3$

31. $y = -3x^2$

32. $y = 3x^2 - 3$

33. $y = 2x^2 - 4x$

34. $y = 3x^2 + 6x$

35. $y = -4x^2 + 16x + 5$

36. $y = 5x^2 + 20x + 25$

37. $y - 7 = 6x^2 - 5x$

38. $y - 2 = 3x^2 + 4x$

39. The equation $y - 2 = (x - 5)^2$ represents a quadratic function whose graph is a parabola. Find its vertex.

40. Show that $y = ax^2$, where $a \neq 0$, represents a quadratic function whose vertex is at the origin.

In Exercises 41–44, use a graphing calculator to find the coordinates of the vertex of the graph of each quadratic function. Give results to the nearest hundredth.

41. $y = 2x^2 - x + 1$

42. $y = x^2 + 5x - 6$

43. $y = 7 + x - x^2$

44. $y = 2x^2 - 3x + 2$

In Exercises 45–48, use a graphing calculator to solve each equation. If a result is not exact, give the result to the nearest hundredth.

45. $x^2 + x - 6 = 0$

46. $2x^2 - 5x - 3 = 0$

47. $0.5x^2 - 0.7x - 3 = 0$

48. $2x^2 - 0.5x - 2 = 0$

APPLICATIONS

49. Ballistics If a ball is thrown straight up with an initial velocity of 48 feet per second, its height s after t seconds is given by the equation $s = 48t - 16t^2$. Find the maximum height attained by the ball and the time it takes for the ball to return to earth.

50. Ballistics From the top of the building in Illustration 1, a ball is thrown straight up with an initial velocity of 32 feet per second. The equation $s = -16t^2 + 32t + 48$ gives the height s of the ball t seconds after it is thrown. Find the maximum height reached by the ball and the time it takes for the ball to hit the ground.

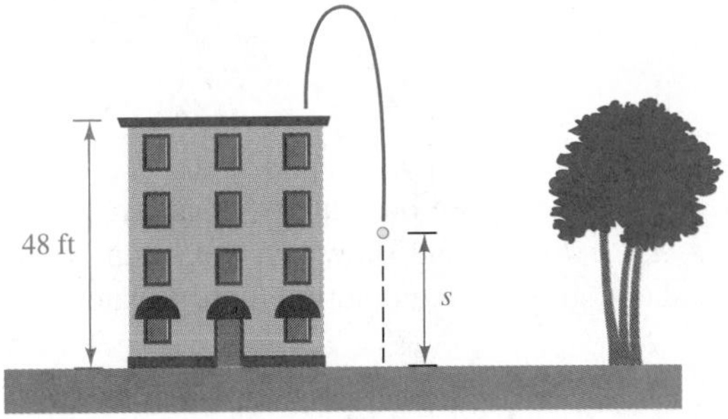

ILLUSTRATION 1

51. Maximizing area Find the dimensions of the rectangle of maximum area that can be constructed with 200 feet of fencing. Find the maximum area.

52. Fencing a field A farmer wants to fence in three sides of a rectangular field (shown in Illustration 2) with 1,000 feet of fencing. The other side of the rectangle will be a river. If the enclosed area is to be maximum, find the dimensions of the field.

ILLUSTRATION 2

In Exercises 53–60, use a graphing calculator to help solve each problem.

53. Maximizing revenue The revenue R received for selling x stereos is given by the equation

$$R = -\frac{x^2}{1,000} + 10x$$

Find the number of stereos that must be sold to obtain the maximum revenue.

54. Maximizing revenue In Exercise 53, find the maximum revenue.

55. Maximizing revenue The revenue received for selling x radios is given by the formula

$$R = -\frac{x^2}{728} + 9x$$

How many radios must be sold to obtain the maximum revenue? Find the maximum revenue.

56. Maximizing revenue The revenue received for selling x stereos is given by the formula

$$R = -\frac{x^2}{5} + 80x - 1,000$$

How many stereos must be sold to obtain the maximum revenue? Find the maximum revenue.

57. Maximizing revenue When priced at \$30 each, a toy has annual sales of 4,000 units. The manufacturer estimates that each \$1 increase in cost will decrease sales by 100 units. Find the unit price that will maximize total revenue. (*Hint:* Total revenue = price · the number of units sold.)

58. Maximizing revenue When priced at \$57, one type of camera has annual sales of 525 units. For each \$1 the camera is reduced in price, management expects to sell an additional 75 cameras. Find the unit price that will maximize total revenue. (*Hint:* Total revenue = price · the number of units sold.)

59. Finding the variance If p is the probability that a person sampled at random has high blood pressure, then $1 - p$ is the probability that the person doesn't. If 50 people are sampled at random, the variance of the sample will be $50p(1 - p)$. What two probabilities p will give a variance of 9.375?

60. Finding the variance If p is the probability that a person sampled at random smokes, then $1 - p$ is the probability that the person doesn't. If 75 people are sampled at random, the variance of the sample will be $75p(1 - p)$. What two probabilities p will give a variance of 12?

WRITING

61. The graph of $y = ax^2 + bx + c$ $(a \neq 0)$ passes the vertical line test. Explain why this shows that the equation defines a function.

62. The graph of $x = y^2 - 2y$ is a parabola. Explain why its graph does not represent a function.

SOMETHING TO THINK ABOUT

63. Can you use a graphing calculator to find solutions of the equation $x^2 + x + 1 = 0$? What is the problem? How do you interpret the result?

64. Complete the square on x in the equation $y = ax^2 + bx + c$ and show that the vertex of the parabolic graph is the point with coordinates of

$$\left(-\frac{b}{2a}, c - \frac{b^2}{4a}\right)$$

8.3 Complex Numbers

■ IMAGINARY NUMBERS ■ POWERS OF i ■ SIMPLIFYING IMAGINARY NUMBERS ■ COMPLEX NUMBERS ■ ARITHMETIC OF COMPLEX NUMBERS ■ RATIONALIZING THE DENOMINATOR ■ ABSOLUTE VALUE OF A COMPLEX NUMBER

Getting Ready *Do the following operations.*

1. $(3x + 5) + (4x - 5)$

2. $(3x + 5) - (4x - 5)$

3. $(3x + 5)(4x - 5)$

4. $(3x + 5)(3x - 5)$

■ IMAGINARY NUMBERS

So far, all of our work with quadratic equations has involved real numbers only. However, the solutions of many quadratic equations are not real numbers.

EXAMPLE 1 Solve $x^2 + x + 1 = 0$.

Solution Because the quadratic trinomial is prime and cannot be factored, we will use the quadratic formula, with $a = 1$, $b = 1$, and $c = 1$:

$$x = \frac{-b \pm \sqrt{b^2 - 4ac}}{2a}$$

$$= \frac{-\mathbf{1} \pm \sqrt{\mathbf{1}^2 - 4(\mathbf{1})(\mathbf{1})}}{2(\mathbf{1})}$$ Substitute 1 for a, 1 for b, and 1 for c.

$$= \frac{-1 \pm \sqrt{1 - 4}}{2}$$

$$= \frac{-1 \pm \sqrt{-3}}{2}$$

$$x = \frac{-1 \pm \sqrt{-3}}{2} \quad \text{or} \quad x = \frac{-1 - \sqrt{-3}}{2}$$ ■

Self Check Solve $a^2 + 3a + 3 = 0$.

Answer $\frac{-3 + \sqrt{-3}}{2}, \frac{-3 - \sqrt{-3}}{2}$

Each solution in Example 1 contains the number $\sqrt{-3}$. Since no real number squared is -3, $\sqrt{-3}$ is not a real number. For years, people believed that numbers such as

$$\sqrt{-1}, \sqrt{-3}, \sqrt{-4}, \text{ and } \sqrt{-9}$$

were nonsense. In the 17th century, René Descartes (1596–1650) called them **imaginary numbers.** Today, imaginary numbers have many important uses, such as describing the behavior of alternating current in electronics.

The imaginary number $\sqrt{-1}$ is often denoted by the letter i:

$$i = \sqrt{-1}$$

Because i represents the square root of -1, it follows that

$$i^2 = -1$$

POWERS OF i

The powers of i produce an interesting pattern:

$$i = \sqrt{-1} = i \qquad i^5 = i^4 i = 1i = i$$
$$i^2 = \left(\sqrt{-1}\right)^2 = -1 \qquad i^6 = i^4 i^2 = 1(-1) = -1$$
$$i^3 = i^2 i = -1i = -i \qquad i^7 = i^4 i^3 = 1(-i) = -i$$
$$i^4 = i^2 i^2 = (-1)(-1) = 1 \qquad i^8 = i^4 i^4 = (1)(1) = 1$$

The pattern continues: $i, -1, -i, 1, \ldots$.

EXAMPLE 2 Simplify i^{29}.

Solution We note that 29 divided by 4 gives a quotient of 7 and a remainder of 1. Thus, $29 = 4 \cdot 7 + 1$, and

$$\begin{aligned} i^{29} &= i^{4\cdot 7+1} && 4\cdot 7+1 = 29. \\ &= (i^4)^7 \cdot i && i^{4\cdot 7+1} = i^{4\cdot 7}\cdot i^1 = (i^4)^7 \cdot i. \\ &= 1^7 \cdot i && i^4 = 1. \\ &= i \end{aligned}$$

■

Self Check Simplify i^{31}.

Answer $-i$

PERSPECTIVE

The Pythagoreans (ca. 500 B.C.) understood the universe as a harmony of whole numbers. They did not classify fractions as numbers, and were upset that $\sqrt{2}$ was not the ratio of whole numbers. For 2,000 years, little progress was made in the understanding of the various kinds of numbers.

The father of algebra, François Vieta (1540–1603), understood the whole numbers, fractions, and certain irrational numbers. But he was unable to accept negative numbers, and certainly not imaginary numbers.

René Descartes (1596–1650) thought these numbers to be nothing more than figments of his imagination, so he called them *imaginary numbers.* Leonhard Euler (1707–1783) used the letter *i* for $\sqrt{-1}$; Augustin Cauchy (1789–1857) used the term *conjugate*; and Karl Gauss (1777–1855) first used the word *complex.*

Today, we accept complex numbers without question, and we use them in science, economics, medicine, and industry. But it took many centuries and the work of many mathematicians to make them respectable.

The results of Example 2 illustrate the following fact.

Powers of *i*

If n is a natural number that has a remainder of r when divided by 4, then

$$i^n = i^r$$

When n is divisible by 4, the remainder r is 0 and $i^0 = 1$.

EXAMPLE 3 Simplify i^{55}.

Solution We divide 55 by 4 and get a remainder of 3. Therefore,

$$i^{55} = i^3 = -i$$

■

Self Check Simplify i^{62}.

Answer -1

SIMPLIFYING IMAGINARY NUMBERS

If we assume that multiplication of imaginary numbers is commutative and associative, then

$$\begin{aligned}(2i)^2 &= 2^2 i^2 \\ &= 4(-1) \qquad i^2 = -1. \\ &= -4\end{aligned}$$

Since $(2i)^2 = -4$, $2i$ is a square root of -4, and we can write

$$\sqrt{-4} = 2i$$

This result can also be obtained by using the multiplication property of radicals:

$$\sqrt{-4} = \sqrt{4(-1)} = \sqrt{4}\sqrt{-1} = 2i$$

We can use the multiplication property of radicals to simplify any imaginary number. For example,

$$\sqrt{-25} = \sqrt{25(-1)} = \sqrt{25}\sqrt{-1} = 5i$$

$$\sqrt{\frac{-100}{49}} = \sqrt{\frac{100}{49}(-1)} = \frac{\sqrt{100}}{\sqrt{49}}\sqrt{-1} = \frac{10}{7}i$$

These examples illustrate the following rule.

Properties of Radicals

If at least one of a and b is a nonnegative real number, then

$$\sqrt{ab} = \sqrt{a}\sqrt{b} \qquad \text{and} \qquad \sqrt{\frac{a}{b}} = \frac{\sqrt{a}}{\sqrt{b}} \quad (b \neq 0)$$

WARNING! If a and b are negative, then $\sqrt{ab} \neq \sqrt{a}\sqrt{b}$. For example, if $a = -16$ and $b = -4$,

$$\sqrt{(-16)(-4)} = \sqrt{64} = 8 \quad \text{but} \quad \sqrt{-16}\sqrt{-4} = (4i)(2i) = 8i^2 = 8(-1) = -8$$

The correct solution is -8.

COMPLEX NUMBERS

The imaginary numbers are a subset of a set of numbers called the **complex numbers.**

Complex Numbers

A **complex number** is any number that can be written in the form $a + bi$, where a and b are real numbers and $i = \sqrt{-1}$.

In the complex number $a + bi$, a is called the **real part,** and b is called the **imaginary part.**

If $b = 0$, the complex number $a + bi$ is a real number. If $b \neq 0$ and $a = 0$, the complex number $0 + bi$ (or just bi) is an imaginary number.

Any imaginary number can be expressed in bi form. For example,

$$\sqrt{-1} = i$$
$$\sqrt{-9} = \sqrt{9(-1)} = \sqrt{9}\sqrt{-1} = 3i$$
$$\sqrt{-3} = \sqrt{3(-1)} = \sqrt{3}\sqrt{-1} = \sqrt{3}i$$

WARNING! The expression $\sqrt{3}i$ is often written as $i\sqrt{3}$ to make it clear that i is not part of the radicand. Do not confuse $\sqrt{3}i$ with $\sqrt{3i}$.

The relationship between the real numbers, the imaginary numbers, and the complex numbers is shown in Figure 8-15.

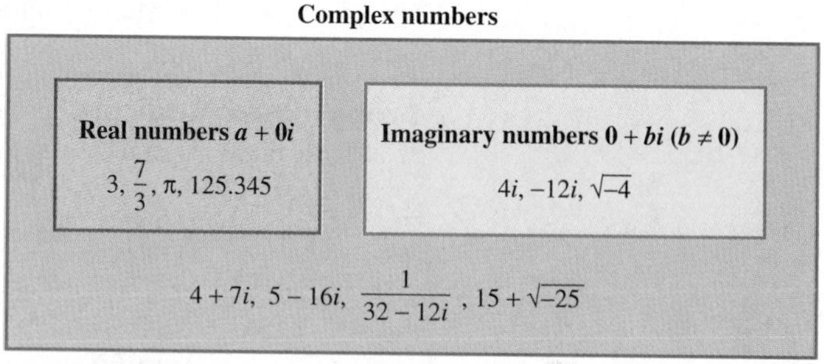

FIGURE 8-15

Equality of Complex Numbers

The complex numbers $a + bi$ and $c + di$ are equal if and only if

$$a = c \quad \text{and} \quad b = d$$

Because of the previous definition, complex numbers are equal when their real parts are equal and their imaginary parts are equal.

EXAMPLE 4

a. $2 + 3i = \sqrt{4} + \frac{6}{2}i$ because $2 = \sqrt{4}$ and $3 = \frac{6}{2}$.

b. $4 - 5i = \frac{12}{3} - \sqrt{25}i$ because $4 = \frac{12}{3}$ and $-5 = -\sqrt{25}$.

c. $x + yi = 4 + 7i$ if and only if $x = 4$ and $y = 7$. ■

ARITHMETIC OF COMPLEX NUMBERS

Addition and Subtraction of Complex Numbers

Complex numbers are added and subtracted as if they were binomials:

$$(a + bi) + (c + di) = (a + c) + (b + d)i$$

$$(a + bi) - (c + di) = (a + bi) + (-c - di) = (a - c) + (b - d)i$$

EXAMPLE 5

Do the operations.

a. $(8 + 4i) + (12 + 8i) = 8 + 4i + 12 + 8i$
$= 20 + 12i$

b. $(7 - 4i) + (9 + 2i) = 7 - 4i + 9 + 2i$
$= 16 - 2i$

c. $(-6 + i) - (3 - 4i) = -6 + i - 3 + 4i$
$= -9 + 5i$

d. $(2 - 4i) - (-4 + 3i) = 2 - 4i + 4 - 3i$
$= 6 - 7i$ ■

Self Check

Do the operations: **a.** $(3 - 5i) + (-2 + 7i)$ and **b.** $(3 - 5i) - (-2 + 7i)$.

Answers

a. $1 + 2i$, **b.** $5 - 12i$

To multiply a complex number by an imaginary number, we use the distributive property to remove parentheses and then simplify. For example,

$$
\begin{aligned}
-5i(4 - 8i) &= -5i(4) - (-5i)8i && \text{Use the distributive property.}\\
&= -20i + 40i^2 && \text{Simplify.}\\
&= -20i + 40(-1) && \text{Remember that } i^2 = -1.\\
&= -40 - 20i
\end{aligned}
$$

To multiply two complex numbers, we use the following definition.

Multiplying Complex Numbers

Complex numbers are multiplied as if they were binomials, with $i^2 = -1$:

$$\begin{aligned}(a + bi)(c + di) &= ac + adi + bci + bdi^2 \\ &= ac + adi + bci + bd(-1) \\ &= (ac - bd) + (ad + bc)i\end{aligned}$$

c

EXAMPLE 6 Multiply the complex numbers.

a. $(2 + 3i)(3 - 2i) = 6 - 4i + 9i - 6i^2$ Use the FOIL method.

$= 6 + 5i + 6$ $i^2 = -1$, and combine $-4i$ and $9i$.

$= 12 + 5i$

b. $(3 + i)(1 + 2i) = 3 + 6i + i + 2i^2$ Use the FOIL method.

$= 3 + 7i - 2$ $i^2 = -1$, and combine $6i$ and i.

$= 1 + 7i$

c. $(-4 + 2i)(2 + i) = -8 - 4i + 4i + 2i^2$ Use the FOIL method.

$= -8 - 2$ $i^2 = -1$, and combine $-4i$ and $4i$.

$= -10$ ■

Self Check Multiply $(-2 + 3i)(3 - 2i)$.

Answer $13i$

The next two examples show how to write complex numbers in $a + bi$ form. It is common to use $a - bi$ as a substitute for $a + (-b)i$.

EXAMPLE 7 Write each number in $a + bi$ form.

a. $7 = 7 + 0i$

b. $3i = 0 + 3i$

c. $4 - \sqrt{-16} = 4 - \sqrt{-1(16)}$

$= 4 - \sqrt{16}\sqrt{-1}$

$= 4 - 4i$

d. $5 + \sqrt{-11} = 5 + \sqrt{-1(11)}$

$= 5 + \sqrt{11}\sqrt{-1}$

$= 5 + \sqrt{11}i$ ■

Self Check Write $3 - \sqrt{-25}$ in $a + bi$ form.

Answer $3 - 5i$

EXAMPLE 8 Simplify each expression.

a. $2i^2 + 4i^3 = 2(-1) + 4(-i)$

$= -2 - 4i$

b. $\frac{3}{2i} = \frac{3}{2i} \cdot \frac{i}{i}$ $\quad \frac{i}{i} = 1.$

$= \frac{3i}{2i^2}$

$= \frac{3i}{2(-1)}$

$= \frac{3i}{-2}$

$= 0 - \frac{3}{2}i$

c. $-\frac{5}{i} = -\frac{5}{i} \cdot \frac{i^3}{i^3}$ $\quad \frac{i^3}{i^3} = 1.$

$= -\frac{5(-i)}{1}$

$= 5i$

$= 0 + 5i$

d. $\frac{6}{i^3} = \frac{6i}{i^3 i}$ $\quad \frac{i}{i} = 1.$

$= \frac{6i}{i^4}$

$= \frac{6i}{1}$

$= 6i$

$= 0 + 6i$ ■

Self Check Simplify **a.** $3i^3 - 2i^2$ and **b.** $\frac{2}{3i}$.

Answers **a.** $2 - 3i$, **b.** $0 - \frac{2}{3}i$

Complex Conjugates
The complex numbers $a + bi$ and $a - bi$ are called **complex conjugates.**

For example,

$3 + 4i$ and $3 - 4i$ are complex conjugates.

$5 - 7i$ and $5 + 7i$ are complex conjugates.

EXAMPLE 9 Find the product of $3 + i$ and its complex conjugate.

Solution The complex conjugate of $3 + i$ is $3 - i$. We can find the product as follows:

$(3 + i)(3 - i) = 9 - 3i + 3i - i^2$ $\quad$ Use the FOIL method.

$= 9 - i^2$ $\quad$ Combine like terms.

$= 9 - (-1)$ $\quad$ $i^2 = -1.$

$= 10$ ■

Self Check Multiply $(2 + 3i)(2 - 3i)$.

Answer 13

The product of the complex number $a + bi$ and its complex conjugate $a - bi$ is the real number $a^2 + b^2$, as the following work shows:

$$\begin{aligned}(a + bi)(a - bi) &= a^2 - abi + abi - b^2i^2 && \text{Use the FOIL method.}\\ &= a^2 - b^2(-1) && i^2 = -1.\\ &= a^2 + b^2\end{aligned}$$

RATIONALIZING THE DENOMINATOR

To divide complex numbers, we often have to rationalize a denominator.

EXAMPLE 10 Divide and write the result in $a + bi$ form: $\dfrac{1}{3 + i}$.

Solution We can rationalize the denominator by multiplying the numerator and the denominator by the complex conjugate of the denominator.

$$\begin{aligned}\frac{1}{3+i} &= \frac{1}{3+i}\cdot\frac{3-i}{3-i} && \tfrac{3-i}{3-i} = 1.\\ &= \frac{3-i}{9 - 3i + 3i - i^2} && \text{Multiply the numerators and multiply the denominators.}\\ &= \frac{3-i}{9-(-1)} && i^2 = -1.\\ &= \frac{3-i}{10}\\ &= \frac{3}{10} - \frac{1}{10}i\end{aligned}$$

■

Self Check Rationalize the denominator: $\frac{1}{5 - i}$.

Answer $\frac{5}{26} + \frac{1}{26}i$

EXAMPLE 11 Write $\dfrac{3 - i}{2 + i}$ in $a + bi$ form.

Solution We multiply both the numerator and the denominator of the fraction by the complex conjugate of the denominator.

$$\frac{3-i}{2+i} = \frac{3-i}{2+i} \cdot \frac{2-i}{2-i} \qquad \frac{2-i}{2-i} = 1.$$

$$= \frac{6-3i-2i+i^2}{4-2i+2i-i^2} \qquad \text{Multiply the numerators and multiply the denominators.}$$

$$= \frac{5-5i}{4-(-1)} \qquad i^2 = -1.$$

$$= \frac{5(1-i)}{5} \qquad \text{Factor out 5 in the numerator.}$$

$$= 1 - i \qquad \text{Simplify.}$$

■

Self Check Rationalize the denominator: $\frac{2+i}{5-i}$.

Answer $\frac{9}{26} + \frac{7}{26}i$

EXAMPLE 12 Write $\dfrac{4+\sqrt{-16}}{2+\sqrt{-4}}$ in $a + bi$ form.

Solution

$$\frac{4+\sqrt{-16}}{2+\sqrt{-4}} = \frac{4+4i}{2+2i} \qquad \text{Write each number in } a + bi \text{ form.}$$

$$= \frac{2\overset{1}{\cancel{(2+2i)}}}{\underset{1}{\cancel{2+2i}}} \qquad \text{Factor out 2 in the numerator and simplify.}$$

$$= 2 + 0i$$

■

Self Check Divide: $\dfrac{3+\sqrt{-25}}{2+\sqrt{-9}}$.

Answer $\frac{21}{13} + \frac{1}{13}i$

WARNING! To avoid mistakes, always put complex numbers in $a + bi$ form before doing any complex number arithmetic.

ABSOLUTE VALUE OF A COMPLEX NUMBER

Absolute Value of a Complex Number

The **absolute value** of the complex number $a + bi$ is $\sqrt{a^2 + b^2}$. In symbols,

$$|a + bi| = \sqrt{a^2 + b^2}$$

EXAMPLE 13 Find each absolute value.

a. $|3 + 4i| = \sqrt{3^2 + 4^2}$
$= \sqrt{9 + 16}$
$= \sqrt{25}$
$= 5$

b. $|3 - 4i| = \sqrt{3^2 + (-4)^2}$
$= \sqrt{9 + 16}$
$= \sqrt{25}$
$= 5$

c. $|-5 - 12i| = \sqrt{(-5)^2 + (-12)^2}$
$= \sqrt{25 + 144}$
$= \sqrt{169}$
$= 13$

d. $|a + 0i| = \sqrt{a^2 + 0^2}$
$= \sqrt{a^2}$
$= |a|$

■

Self Check Evaluate $|5 + 12i|$.

Answer 13

WARNING! Note that $|a + bi| = \sqrt{a^2 + b^2}$, not ~~$|a + bi| = \sqrt{a^2 + (bi)^2}$~~.

Orals *Simplify each power of i.*

1. i^3 **2.** i^2 **3.** i^4 **4.** i^5

Write each imaginary number in bi form.

5. $\sqrt{-49}$ **6.** $\sqrt{-64}$ **7.** $\sqrt{-100}$ **8.** $\sqrt{-81}$

Find each absolute value.

9. $|3 + 4i|$ **10.** $|5 - 12i|$

EXERCISE 8.3

REVIEW *In Exercises 1–2, do each operation.*

1. $\dfrac{x^2 - x - 6}{9 - x^2} \cdot \dfrac{x^2 + x - 6}{x^2 - 4}$

2. $\dfrac{3x + 4}{x - 2} + \dfrac{x - 4}{x + 2}$

3. Wind speed A plane that can fly 200 mph in still air makes a 330-mile flight with a tail wind and returns, flying into the same wind. Find the speed of the wind if the total flying time is $3\frac{1}{3}$ hours.

4. Finding rates A student drove a distance of 135 miles at an average speed of 50 mph. How much faster would he have to drive on the return trip to save 30 minutes of driving time?

VOCABULARY AND CONCEPTS *Fill in each blank to make a true statement.*

5. $\sqrt{-1}$, $\sqrt{-3}$, and $\sqrt{-4}$ are examples of ______ numbers.
6. $\sqrt{-1} =$ __
7. $i^2 =$ ___
8. $i^3 =$ ___
9. $i^4 =$ __
10. $\sqrt{ab} =$ ______, provided a and b are not both negative.
11. $\sqrt{\frac{a}{b}} = \frac{}{}$ __, provided a and b are not both negative.
12. $3 + 5i$, $2 - 7i$, and $5 - \frac{1}{2}i$ are examples of ______ numbers.
13. The real part of $5 + 7i$ is __. The imaginary part is __.
14. $a + bi = c + di$ if and only if $a =$ __ and $b =$ __.
15. $a + bi$ and $a - bi$ are called complex ______.
16. $|a + bi| =$ ______

PRACTICE *In Exercises 17–28, solve each equation. Write all roots in bi or a + bi form.*

17. $x^2 + 9 = 0$
18. $x^2 + 16 = 0$
19. $3x^2 = -16$
20. $2x^2 = -25$
21. $x^2 + 2x + 2 = 0$
22. $x^2 + 3x + 3 = 0$
23. $2x^2 + x + 1 = 0$
24. $3x^2 + 2x + 1 = 0$
25. $3x^2 - 4x = -2$
26. $2x^2 + 3x = -3$
27. $3x^2 - 2x = -3$
28. $5x^2 = 2x - 1$

In Exercises 29–36, simplify each expression.

29. i^{21}
30. i^{19}
31. i^{27}
32. i^{22}
33. i^{100}
34. i^{42}
35. i^{97}
36. i^{200}

In Exercises 37–42, tell whether the complex numbers are equal.

37. $3 + 7i$, $\sqrt{9} + (5 + 2)i$
38. $\sqrt{4} + \sqrt{25}i$, $2 - (-5)i$
39. $8 + 5i$, $2^3 + \sqrt{25}i^3$
40. $4 - 7i$, $-4i^2 + 7i^3$
41. $\sqrt{4} + \sqrt{-4}$, $2 - 2i$
42. $\sqrt{-9} - i$, $4i$

In Exercises 43–76, do the operations. Write all answers in a + bi form.

43. $(3 + 4i) + (5 - 6i)$
44. $(5 + 3i) - (6 - 9i)$
45. $(7 - 3i) - (4 + 2i)$
46. $(8 + 3i) + (-7 - 2i)$
47. $(8 + 5i) + (7 + 2i)$
48. $(-7 + 9i) - (-2 - 8i)$
49. $(1 + i) - 2i + (5 - 7i)$
50. $(-9 + i) - 5i + (2 + 7i)$
51. $(5 + 3i) - (3 - 5i) + \sqrt{-1}$
52. $(8 + 7i) - \left(-7 - \sqrt{-64}\right) + (3 - i)$
53. $\left(-8 - \sqrt{3}i\right) - \left(7 - 3\sqrt{3}i\right)$
54. $\left(2 + 2\sqrt{2}i\right) + \left(-3 - \sqrt{2}i\right)$
55. $3i(2 - i)$
56. $-4i(3 + 4i)$
57. $-5i(5 - 5i)$
58. $2i(7 + 2i)$
59. $(2 + i)(3 - i)$
60. $(4 - i)(2 + i)$

61. $(2 - 4i)(3 + 2i)$

62. $(3 - 2i)(4 - 3i)$

63. $(2 + \sqrt{2}i)(3 - \sqrt{2}i)$

64. $(5 + \sqrt{3}i)(2 - \sqrt{3}i)$

65. $(8 - \sqrt{-1})(-2 - \sqrt{-16})$

66. $(-1 + \sqrt{-4})(2 + \sqrt{-9})$

67. $(2 + i)^2$

68. $(3 - 2i)^2$

69. $(2 + 3i)^2$

70. $(1 - 3i)^2$

71. $i(5 + i)(3 - 2i)$

72. $i(-3 - 2i)(1 - 2i)$

73. $(2 + i)(2 - i)(1 + i)$

74. $(3 + 2i)(3 - 2i)(i + 1)$

75. $(3 + i)[(3 - 2i) + (2 + i)]$

76. $(2 - 3i)[(5 - 2i) - (2i + 1)]$

In Exercises 77–104, write each expression in $a + bi$ form.

77. $\frac{1}{i}$

78. $\frac{1}{i^3}$

79. $\frac{4}{5i^3}$

80. $\frac{3}{2i}$

81. $\frac{3i}{8\sqrt{-9}}$

82. $\frac{5i^3}{2\sqrt{-4}}$

83. $\frac{-3}{5i^5}$

84. $\frac{-4}{6i^7}$

85. $\frac{5}{2 - i}$

86. $\frac{26}{3 - 2i}$

87. $\frac{13i}{5 + i}$

88. $\frac{2i}{5 + 3i}$

89. $\frac{-12}{7 - \sqrt{-1}}$

90. $\frac{4}{3 + \sqrt{-1}}$

91. $\frac{5i}{6 + 2i}$

92. $\frac{-4i}{2 - 6i}$

93. $\frac{3 - 2i}{3 + 2i}$

94. $\frac{2 + 3i}{2 - 3i}$

95. $\frac{3 + 2i}{3 + i}$

96. $\frac{2 - 5i}{2 + 5i}$

97. $\frac{\sqrt{5} - \sqrt{3}i}{\sqrt{5} + \sqrt{3}i}$

98. $\frac{\sqrt{3} + \sqrt{2}i}{\sqrt{3} - \sqrt{2}i}$

99. $\left(\frac{i}{3 + 2i}\right)^2$

100. $\left(\frac{5 + i}{2 + i}\right)^2$

101. $\frac{i(3 - i)}{3 + i}$

102. $\frac{5 + 3i}{i(3 - 5i)}$

103. $\frac{(2 - 5i) - (5 - 2i)}{5 - i}$

104. $\frac{5i}{(5 + 2i) + (2 + i)}$

In Exercises 105–112, find each value.

105. $|6 + 8i|$

106. $|12 + 5i|$

107. $|12 - 5i|$

108. $|3 - 4i|$

109. $|5 + 7i|$

110. $|6 - 5i|$

111. $\left|\frac{3}{5} - \frac{4}{5}i\right|$

112. $\left|\frac{5}{13} + \frac{12}{13}i\right|$

113. Show that $1 - 5i$ is a solution of $x^2 - 2x + 26 = 0$.

114. Show that $3 - 2i$ is a solution of $x^2 - 6x + 13 = 0$.

115. Show that i is a solution of $x^4 - 3x^2 - 4 = 0$.

116. Show that $2 + i$ is *not* a solution of $x^2 + x + 1 = 0$.

WRITING

117. Tell how to decide whether two complex numbers are equal.

118. Define the complex conjugate of a complex number.

SOMETHING TO THINK ABOUT

119. Rationalize the numerator of $\dfrac{3 - i}{2}$.

120. Rationalize the numerator of $\dfrac{2 + 3i}{2 - 3i}$

8.4 The Discriminant and Equations That Can Be Written in Quadratic Form

■ THE DISCRIMINANT ■ EQUATIONS THAT CAN BE WRITTEN IN QUADRATIC FORM ■ SOLUTIONS OF A QUADRATIC EQUATION

Getting Ready *Evaluate $b^2 - 4ac$ for the following values.*

1. $a = 2$, $b = 3$, and $c = -1$

2. $a = -2$, $b = 4$, and $c = -3$

■ THE DISCRIMINANT

We can predict the type of solutions a quadratic equation will have without solving it. To see how, we suppose that the coefficients a, b, and c in the equation $ax^2 + bx + c = 0$ $(a \neq 0)$ are real numbers. Then the solutions of the equation are given by the quadratic formula

$$x = \frac{-b \pm \sqrt{b^2 - 4ac}}{2a} \qquad (a \neq 0)$$

If $b^2 - 4ac \geq 0$, the solutions are real numbers. If $b^2 - 4ac < 0$, the solutions are nonreal complex numbers. Thus, the value of $b^2 - 4ac$, called the **discriminant,** determines the type of solutions for a particular quadratic equation.

The Discriminant

If a, b, and c are real numbers and

If $b^2 - 4ac$ is . . .	***then the solutions are . . .***
positive,	real numbers and unequal.
0,	real numbers and equal.
negative,	nonreal complex numbers and complex conjugates.

If a, b, and c are rational numbers and

If $b^2 - 4ac$ is . . .	***then the solutions are . . .***
a perfect square greater than 0,	rational numbers and unequal.
positive and not a perfect square,	irrational numbers and unequal.

EXAMPLE 1 Determine the type of solutions for the equation $x^2 + x + 1 = 0$.

Solution We calculate the discriminant:

$$\begin{aligned} b^2 - 4ac &= \mathbf{1}^2 - 4(\mathbf{1})(\mathbf{1}) && a = 1,\ b = 1, \text{ and } c = 1. \\ &= -3 \end{aligned}$$

Since $b^2 - 4ac < 0$, the solutions are nonreal complex conjugates. ■

Self Check Determine the type of solutions for $x^2 + x - 1 = 0$.

Answer real numbers that are irrational and unequal

EXAMPLE 2 Determine the type of solutions for the equation $3x^2 + 5x + 2 = 0$.

Solution We calculate the discriminant:

$$\begin{aligned} b^2 - 4ac &= \mathbf{5}^2 - 4(\mathbf{3})(\mathbf{2}) && a = 3,\ b = 5, \text{ and } c = 2. \\ &= 25 - 24 \\ &= 1 \end{aligned}$$

Since $b^2 - 4ac > 0$ and $b^2 - 4ac$ is a perfect square, the solutions are rational and unequal. ■

Self Check Determine the type of solutions for $4x^2 - 10x + 25 = 0$.

Answer nonreal numbers that are complex conjugates

EXAMPLE 3 What value of k will make the solutions of the equation $kx^2 - 12x + 9 = 0$ equal?

Solution We calculate the discriminant:

$$\begin{aligned} b^2 - 4ac &= (\mathbf{-12})^2 - 4(\mathbf{k})(\mathbf{9}) && a = k,\ b = -12, \text{ and } c = 9. \\ &= 144 - 36k \\ &= -36k + 144 \end{aligned}$$

Since the solutions are to be equal, we let $-36k + 144 = 0$ and solve for k.

$$\begin{aligned} -36k + 144 &= 0 \\ -36k &= -144 && \text{Subtract 144 from both sides.} \\ k &= 4 && \text{Divide both sides by } -36. \end{aligned}$$

If $k = 4$, the solutions will be equal. Verify this by solving $4x^2 - 12x + 9 = 0$ and showing that the solutions are equal. ■

Self Check What value of k will make the solutions of $kx^2 - 20x + 25 = 0$ equal?

Answer 4

EQUATIONS THAT CAN BE WRITTEN IN QUADRATIC FORM

Many equations can be written in quadratic form and then solved with the techniques used for solving quadratic equations. For example, we can solve $x^4 - 5x^2 + 4 = 0$ as follows:

$$x^4 - 5x^2 + 4 = 0$$
$$(x^2)^2 - 5(x^2) + 4 = 0$$
$$y^2 - 5y + 4 = 0 \quad \text{Let } y = x^2.$$
$$(y - 4)(y - 1) = 0 \quad \text{Factor } y^2 - 5y + 4.$$
$$y - 4 = 0 \quad \text{or} \quad y - 1 = 0 \quad \text{Set each factor equal to 0.}$$
$$y = 4 \qquad\qquad y = 1$$

Since $x^2 = y$, it follows that $x^2 = 4$ or $x^2 = 1$. Thus,

$$x^2 = 4 \quad \text{or} \quad x^2 = 1$$
$$x = 2 \quad \text{or} \quad x = -2 \qquad x = 1 \quad \text{or} \quad x = -1$$

This equation has four solutions: 1, -1, 2, and -2. Verify that each one satisfies the original equation. Note that this equation can be solved by factoring.

EXAMPLE 4 Solve $x - 7x^{1/2} + 12 = 0$.

Solution If y^2 is substituted for x and y is substituted for $x^{1/2}$, the equation

$$x - 7x^{1/2} + 12 = 0$$

becomes a quadratic equation that can be solved by factoring:

$$y^2 - 7y + 12 = 0 \quad \text{Substitute } y^2 \text{ for } x \text{ and } y \text{ for } x^{1/2}.$$
$$(y - 3)(y - 4) = 0 \quad \text{Factor.}$$
$$y - 3 = 0 \quad \text{or} \quad y - 4 = 0 \quad \text{Set each factor equal to 0.}$$
$$y = 3 \qquad\qquad y = 4$$

Because $x = y^2$, it follows that

$$x = 3^2 \quad \text{or} \quad x = 4^2$$
$$= 9 \qquad\qquad = 16$$

Verify that both solutions satisfy the original equation. ■

Self Check Solve $x + x^{1/2} - 6 = 0$. Be sure to check your solutions.

Answer 4

EXAMPLE 5 Solve $\dfrac{24}{x} + \dfrac{12}{x+1} = 11$.

Solution Since the denominator cannot be 0, x cannot be 0 or -1. If either 0 or -1 appears as a suspected solution, it is extraneous and must be discarded.

$$\frac{24}{x} + \frac{12}{x+1} = 11$$

$$x(x+1)\left(\frac{24}{x} + \frac{12}{x+1}\right) = x(x+1)11$$ Multiply both sides by $x(x+1)$.

$$24(x+1) + 12x = (x^2+x)11$$ Simplify.

$$24x + 24 + 12x = 11x^2 + 11x$$ Use the distributive property to remove parentheses.

$$36x + 24 = 11x^2 + 11x$$ Combine like terms.

$$0 = 11x^2 - 25x - 24$$ Subtract $36x$ and 24 from both sides.

$$0 = (11x+8)(x-3)$$ Factor $11x^2 - 25x - 24$.

$$11x + 8 = 0 \quad \text{or} \quad x - 3 = 0$$ Set each factor equal to 0.

$$x = -\frac{8}{11} \qquad\qquad x = 3$$

Verify that $-\frac{8}{11}$ and 3 satisfy the original equation. ■

Self Check Solve $\frac{12}{x} + \frac{6}{x+3} = 5$.

Answer $3, -\frac{12}{5}$

EXAMPLE 6 Solve the formula $s = 16t^2 - 32$ for t.

Solution We proceed as follows:

$$s = 16t^2 - 32$$

$$s + 32 = 16t^2$$ Add 32 to both sides.

$$\frac{s+32}{16} = t^2$$ Divide both sides by 16.

$$t^2 = \frac{s+32}{16}$$ Write t^2 on the left-hand side.

$$t = \pm\sqrt{\frac{s+32}{16}}$$ Apply the square root property.

$$t = \pm\frac{\sqrt{s+32}}{\sqrt{16}}$$ $\sqrt{\frac{a}{b}} = \frac{\sqrt{a}}{\sqrt{b}}$.

$$t = \pm\frac{\sqrt{s+32}}{4}$$ ■

Self Check Solve $a^2 + b^2 = c^2$ for a.

Answer $a = \pm\sqrt{c^2 - b^2}$

SOLUTIONS OF A QUADRATIC EQUATION

Solutions of a Quadratic Equation

If r_1 and r_2 are the solutions of the quadratic equation $ax^2 + bx + c = 0$, with $a \neq 0$, then

$$r_1 + r_2 = -\frac{b}{a} \quad \text{and} \quad r_1r_2 = \frac{c}{a}$$

Proof We note that the solutions to the equation are given by the quadratic formula

$$r_1 = \frac{-b + \sqrt{b^2 - 4ac}}{2a} \quad \text{and} \quad r_2 = \frac{-b - \sqrt{b^2 - 4ac}}{2a}$$

Thus,

$$r_1 + r_2 = \frac{-b + \sqrt{b^2 - 4ac}}{2a} + \frac{-b - \sqrt{b^2 - 4ac}}{2a}$$

$$= \frac{-b + \sqrt{b^2 - 4ac} - b - \sqrt{b^2 - 4ac}}{2a}$$ Keep the denominator and add the numerators.

$$= -\frac{2b}{2a}$$

$$= -\frac{b}{a}$$

and

$$r_1r_2 = \frac{-b + \sqrt{b^2 - 4ac}}{2a} \cdot \frac{-b - \sqrt{b^2 - 4ac}}{2a}$$

$$= \frac{b^2 - (b^2 - 4ac)}{4a^2}$$ Multiply the numerators and multiply the denominators.

$$= \frac{b^2 - b^2 + 4ac}{4a^2}$$

$$= \frac{4ac}{4a^2}$$ $b^2 - b^2 = 0$.

$$= \frac{c}{a}$$

□

It can also be shown that if

$$r_1 + r_2 = -\frac{b}{a} \quad \text{and} \quad r_1 r_2 = \frac{c}{a}$$

then r_1 and r_2 are solutions of $ax^2 + bx + c = 0$. We can use this fact to check the solutions of quadratic equations.

EXAMPLE 7 Show that $\frac{3}{2}$ and $-\frac{1}{3}$ are solutions of $6x^2 - 7x - 3 = 0$.

Solution Since $a = 6$, $b = -7$, and $c = -3$, we have

$$-\frac{b}{a} = -\frac{-7}{6} = \frac{7}{6} \quad \text{and} \quad \frac{c}{a} = \frac{-3}{6} = -\frac{1}{2}$$

Since $\frac{3}{2} + \left(-\frac{1}{3}\right) = \frac{7}{6}$ and $\left(\frac{3}{2}\right)\left(-\frac{1}{3}\right) = -\frac{1}{2}$, these numbers are solutions. Solve the equation to see that the roots are $\frac{3}{2}$ and $-\frac{1}{3}$. ■

Self Check Are $-\frac{3}{2}$ and $\frac{1}{3}$ solutions of $6x^2 + 7x - 3 = 0$?

Answer yes

Orals *Find $b^2 - 4ac$ when*

1. $a = 1, b = 1, c = 1$
2. $a = 2, b = 1, c = 1$

Determine the type of solutions for

3. $x^2 - 4x + 1 = 0$
4. $8x^2 - x + 2 = 0$

Are the following numbers solutions of $x^2 - 7x + 6 = 0$?

5. 1, 5
6. 1, 6

EXERCISE 8.4

REVIEW *Solve each equation.*

1. $\frac{1}{4} + \frac{1}{t} = \frac{1}{2t}$
2. $\frac{p - 3}{3p} + \frac{1}{2p} = \frac{1}{4}$
3. Find the slope of the line passing through $P(-2, -4)$ and $Q(3, 5)$.
4. Write the equation of the line passing through $P(-2, -4)$ and $Q(3, 5)$ in general form.

VOCABULARY AND CONCEPTS *Consider the equation $ax^2 + bx + c = 0$ $(a \neq 0)$, and fill in each blank to make a true statement.*

5. The discriminant is ________.
6. If $b^2 - 4ac < 0$, the solutions of the equation are nonreal complex ________.

7. If $b^2 - 4ac$ is a perfect square, the solutions are ______ numbers and ______.

8. If r_1 and r_2 are the solutions of the equation, then

$r_1 + r_2 =$ ___ and $r_1 r_2 =$ __

PRACTICE *In Exercises 9–16, use the discriminant to determine what type of solutions exist for each quadratic equation.* **Do not solve the equation.**

9. $4x^2 - 4x + 1 = 0$

10. $6x^2 - 5x - 6 = 0$

11. $5x^2 + x + 2 = 0$

12. $3x^2 + 10x - 2 = 0$

13. $2x^2 = 4x - 1$

14. $9x^2 = 12x - 4$

15. $x(2x - 3) = 20$

16. $x(x - 3) = -10$

In Exercises 17–24, find the value(s) of k that will make the solutions of each given quadratic equation equal.

17. $x^2 + kx + 9 = 0$

18. $kx^2 - 12x + 4 = 0$

19. $9x^2 + 4 = -kx$

20. $9x^2 - kx + 25 = 0$

21. $(k - 1)x^2 + (k - 1)x + 1 = 0$

22. $(k + 3)x^2 + 2kx + 4 = 0$

23. $(k + 4)x^2 + 2kx + 9 = 0$

24. $(k + 15)x^2 + (k - 30)x + 4 = 0$

25. Use the discriminant to determine whether the solutions of $1{,}492x^2 + 1{,}776x - 1{,}984 = 0$ are real numbers.

26. Use the discriminant to determine whether the solutions of $1{,}776x^2 - 1{,}492x + 1{,}984 = 0$ are real numbers.

27. Determine k such that the solutions of $3x^2 + 4x = k$ are nonreal complex numbers.

28. Determine k such that the solutions of $kx^2 - 4x = 7$ are nonreal complex numbers.

In Exercises 29–56, solve each equation.

29. $x^4 - 17x^2 + 16 = 0$

30. $x^4 - 10x^2 + 9 = 0$

31. $x^4 - 3x^2 = -2$

32. $x^4 - 29x^2 = -100$

33. $x^4 = 6x^2 - 5$

34. $x^4 = 8x^2 - 7$

35. $2x^4 - 10x^2 = -8$

36. $2x^4 + 24 = 26x^2$

37. $2x + x^{1/2} - 3 = 0$

38. $2x - x^{1/2} - 1 = 0$

39. $3x + 5x^{1/2} + 2 = 0$

40. $3x - 4x^{1/2} + 1 = 0$

41. $x^{2/3} + 5x^{1/3} + 6 = 0$

42. $x^{2/3} - 7x^{1/3} + 12 = 0$

43. $x^{2/3} - 2x^{1/3} - 3 = 0$

44. $x^{2/3} + 4x^{1/3} - 5 = 0$

45. $x + 5 + \dfrac{4}{x} = 0$

46. $x - 4 + \dfrac{3}{x} = 0$

47. $x + 1 = \dfrac{20}{x}$

48. $x + \dfrac{15}{x} = 8$

49. $\dfrac{1}{x - 1} + \dfrac{3}{x + 1} = 2$

50. $\dfrac{6}{x - 2} - \dfrac{12}{x - 1} = -1$

51. $\dfrac{1}{x + 2} + \dfrac{24}{x + 3} = 13$

52. $\dfrac{3}{x} + \dfrac{4}{x + 1} = 2$

53. $x^{-4} - 2x^{-2} + 1 = 0$

54. $4x^{-4} + 1 = 5x^{-2}$

55. $x + \dfrac{2}{x - 2} = 0$

56. $x + \dfrac{x + 5}{x - 3} = 0$

In Exercises 57–64, solve each equation for the indicated variable.

57. $x^2 + y^2 = r^2$ for x

58. $x^2 + y^2 = r^2$ for y

59. $I = \dfrac{k}{d^2}$ for d

60. $V = \dfrac{1}{3}\pi r^2 h$ for r

61. $xy^2 + 3xy + 7 = 0$ for y

62. $kx = ay - x^2$ for x

63. $\sigma = \sqrt{\dfrac{\Sigma x^2}{N} - \mu^2}$ for μ^2

64. $\sigma = \sqrt{\dfrac{\Sigma x^2}{N} - \mu^2}$ for N

In Exercises 65–72, solve each equation and verify that the sum of the solutions is $-\frac{b}{a}$ and that the product of the solutions is $\frac{c}{a}$.

65. $12x^2 - 5x - 2 = 0$

66. $8x^2 - 2x - 3 = 0$

67. $2x^2 + 5x + 1 = 0$

68. $3x^2 + 9x + 1 = 0$

69. $3x^2 - 2x + 4 = 0$

70. $2x^2 - x + 4 = 0$

71. $x^2 + 2x + 5 = 0$

72. $x^2 - 4x + 13 = 0$

WRITING

73. Describe how to predict what type of solutions the equation $3x^2 - 4x + 5 = 0$ will have.

74. How is the discriminant related to the quadratic formula?

SOMETHING TO THINK ABOUT

75. Can a quadratic equation with integer coefficients have one real and one complex solution? Why?

76. Can a quadratic equation with complex coefficients have one real and one complex solution? Why?

8.5 Quadratic and Other Nonlinear Inequalities

■ SOLVING QUADRATIC INEQUALITIES ■ SOLVING OTHER INEQUALITIES ■ GRAPHS OF NONLINEAR INEQUALITIES IN TWO VARIABLES

Getting Ready *Factor each trinomial.*

1. $x^2 + 2x - 15$

2. $x^2 - 3x + 2$

SOLVING QUADRATIC INEQUALITIES

To solve the inequality $x^2 + x - 6 < 0$, we must find the values of x that make the inequality true. To find these values, we can factor the trinomial to obtain

$$(x + 3)(x - 2) < 0$$

Since the product of $x + 3$ and $x - 2$ is to be less than 0, their values must be opposite in sign. To find the intervals where this is true, we keep track of their signs by constructing the chart in Figure 8-16. The chart shows that

- $x - 2$ is 0 when $x = 2$, is positive when $x > 2$, and is negative when $x < 2$.
- $x + 3$ is 0 when $x = -3$, is positive when $x > -3$, and is negative when $x < -3$.

The only place where the values of the binomial are opposite in sign is in the interval $(-3, 2)$. Therefore, the product $(x + 3)(x - 2)$ will be less than 0 when

$$-3 < x < 2$$

The graph of the solution set is shown on the number line in Figure 8-16.

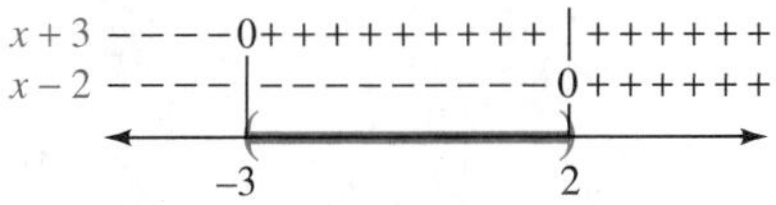

FIGURE 8-16

EXAMPLE 1 Solve $x^2 + 2x - 3 \geq 0$.

Solution We factor the trinomial to get $(x - 1)(x + 3)$ and construct a sign chart, as in Figure 8-17.

- $x - 1$ is 0 when $x = 1$, is positive when $x > 1$, and is negative when $x < 1$.
- $x + 3$ is 0 when $x = -3$, is positive when $x > -3$, and is negative when $x < -3$.

The product of $x - 1$ and $x + 3$ will be greater than 0 when the signs of the binomial factors are the same. This occurs in the intervals $(-\infty, -3)$ and $(1, \infty)$. The numbers -3 and 1 are also included, because they make the product equal to 0. Thus, the solution set is

$$(-\infty, -3] \cup [1, \infty) \qquad \text{or} \qquad x \leq -3 \text{ or } x \geq 1$$

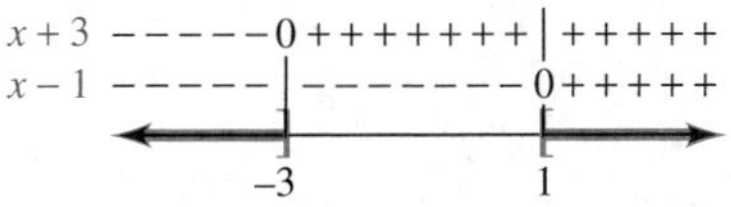

FIGURE 8-17

The graph of the solution set is shown on the number line in Figure 8-17. ■

Self Check Solve $x^2 + 2x - 15 > 0$ and graph the solution set.

Answer $(-\infty, -5) \cup (3, \infty)$

SOLVING OTHER INEQUALITIES

Making a sign chart is useful for solving many inequalities that are neither linear nor quadratic.

EXAMPLE 2 Solve $\frac{1}{x} < 6$.

Solution We subtract 6 from both sides to make the right-hand side equal to 0. We then find a common denominator and add the fractions:

$$\frac{1}{x} < 6$$

$$\frac{1}{x} - 6 < 0 \quad \text{Subtract 6 from both sides.}$$

$$\frac{1}{x} - \frac{6x}{x} < 0 \quad \text{Get a common denominator.}$$

$$\frac{1 - 6x}{x} < 0 \quad \text{Subtract the numerators and keep the common denominator.}$$

We now make a sign chart, as in Figure 8-18.

- The denominator x is 0 when $x = 0$, is positive when $x > 0$, and is negative when $x < 0$.
- The numerator $1 - 6x$ is 0 when $x = \frac{1}{6}$, is positive when $x < \frac{1}{6}$, and is negative when $x > \frac{1}{6}$.

The fraction $\frac{1-6x}{x}$ will be less than 0 when the numerator and denominator are opposite in sign. This occurs in the interval

$$(-\infty, 0) \cup \left(\frac{1}{6}, \infty\right) \quad \text{or} \quad x < 0 \text{ or } x > \frac{1}{6}$$

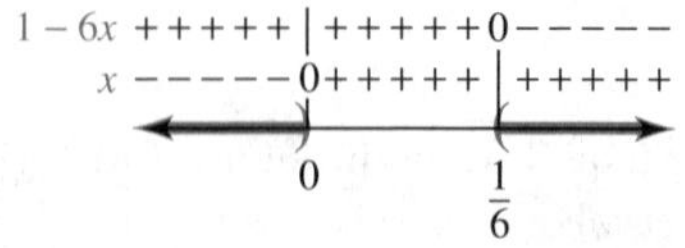

FIGURE 8-18

The graph of this interval is shown in Figure 8-18. ■

Self Check Solve $\frac{3}{x} > 5$.

Answer $\left(0, \frac{3}{5}\right)$

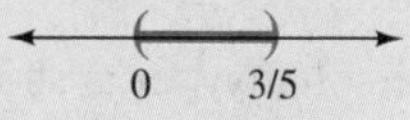

WARNING! Since we don't know whether x is positive, 0, or negative, multiplying both sides of the inequality $\frac{1}{x} < 6$ by x is a three-case situation:

- If $x > 0$, then $1 < 6x$.
- If $x = 0$, then the fraction $\frac{1}{x}$ is undefined.
- If $x < 0$, then $1 > 6x$.

If you multiply both sides by x and solve $1 < 6x$, you are only considering one case and will get only part of the answer.

EXAMPLE 3 Solve $\dfrac{x^2 - 3x + 2}{x - 3} \geq 0$.

Solution We write the fraction with the numerator in factored form.

$$\frac{(x-2)(x-1)}{x-3} \geq 0$$

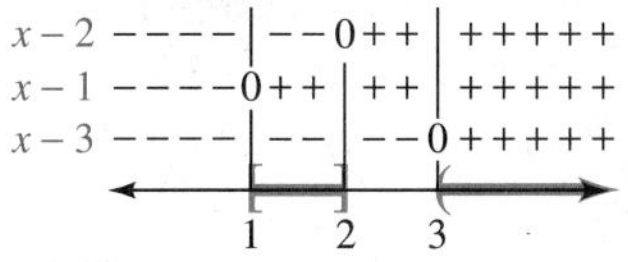

FIGURE 8-19

To keep track of the signs of the binomials, we construct the sign chart shown in Figure 8-19. The fraction will be positive in the intervals where all factors are positive, or where two factors are negative. The numbers 1 and 2 are included, because they make the numerator (and thus the fraction) equal to 0. The number 3 is not included, because it gives a 0 in the denominator.

The solution is the interval $[1, 2] \cup (3, \infty)$. The graph appears in Figure 8-19. ■

Self Check Solve $\dfrac{x + 2}{x^2 - 2x - 3} > 0$ and graph the solution set.

Answer $(-2, -1) \cup (3, \infty)$

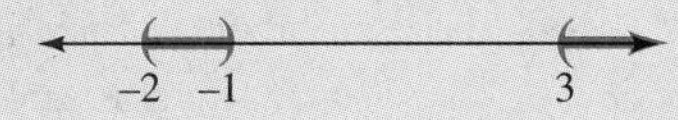

EXAMPLE 4 Solve $\dfrac{3}{x - 1} < \dfrac{2}{x}$.

Solution We subtract $\frac{2}{x}$ from both sides to get 0 on the right-hand side and proceed as follows:

$$\frac{3}{x-1} < \frac{2}{x}$$

$$\frac{3}{x-1} - \frac{2}{x} < 0 \qquad \text{Subtract } \tfrac{2}{x} \text{ from both sides.}$$

$$\frac{3x}{(x-1)x} - \frac{2(x-1)}{x(x-1)} < 0 \qquad \text{Get a common denominator.}$$

$$\frac{3x - 2x + 2}{x(x-1)} < 0 \qquad \text{Keep the denominator and subtract the numerators.}$$

$$\frac{x+2}{x(x-1)} < 0 \qquad \text{Combine like terms.}$$

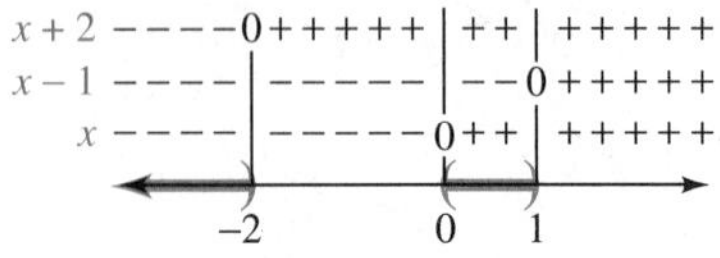

FIGURE 8-20.

We can keep track of the signs of the three factors with the sign chart shown in Figure 8-20. The fraction will be negative in the intervals with either one or three negative factors. The numbers 0 and 1 are not included, because they give a 0 in the denominator, and the number -2 is not included, because it does not satisfy the inequality.

The solution is the interval $(-\infty, -2) \cup (0, 1)$, as shown in Figure 8-20. ■

Self Check Solve $\frac{2}{x+1} > \frac{1}{x}$ and graph the solution set.

Answer $(-1, 0) \cup (1, \infty)$

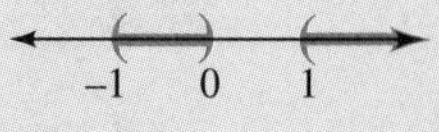

■ ■ ■ ■ ■ ■ ■ ■ ■

Solving Inequalities

GRAPHING CALCULATORS

To approximate the solutions of the inequality $x^2 + 2x - 3 \geq 0$ (Example 1) by graphing, we can use window settings of $[-10, 10]$ for x and $[-10, 10]$ for y and graph the quadratic function $y = x^2 + 2x - 3$, as in Figure 8-21. The solution of the inequality will be those numbers x for which the graph of $y = x^2 + 2x - 3$ lies above or on the x-axis. We can trace to find that this interval is $(-\infty, -3] \cup [1, \infty)$.

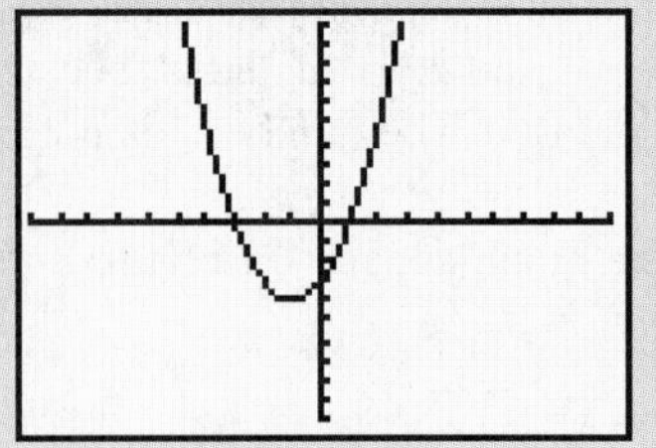

FIGURE 8-21

To approximate the solutions of $\frac{3}{x-1} < \frac{2}{x}$ (Example 4), we first write the inequality in the form

$$\frac{3}{x-1} - \frac{2}{x} < 0$$

Then we use window settings of $[-5, 5]$ for x and $[-3, 3]$ for y and graph the function $y = \frac{3}{x-1} - \frac{2}{x}$, as in Figure 8-22(a). The solution of the inequality will be those numbers x for which the graph lies below the x-axis.

We can trace to see that the graph is below the x-axis when x is less than -2. Since we cannot see the graph in the interval $0 < x < 1$, we redraw the graph using window settings of $[-1, 2]$ for x and $[-25, 10]$ for y. See Figure 8-22(b).

We can now see that the graph is below the x-axis in the interval $(0, 1)$. Thus, the solution of the inequality is the union of two intervals:

$$(-\infty, -2) \cup (0, 1)$$

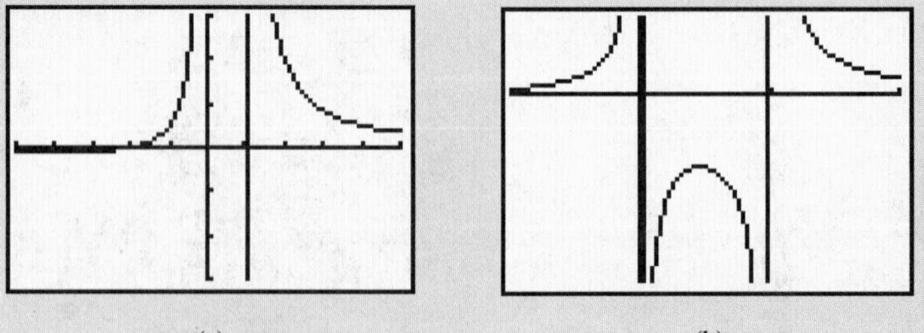

(a) (b)

FIGURE 8-22

GRAPHS OF NONLINEAR INEQUALITIES IN TWO VARIABLES

We now consider the graphs of nonlinear inequalities in two variables.

EXAMPLE 5 Graph $y < -x^2 + 4$.

Solution The graph of $y = -x^2 + 4$ is the parabolic boundary separating the region representing $y < -x^2 + 4$ and the region representing $y > -x^2 + 4$.

We graph $y = -x^2 + 4$ as a broken parabola, because equality is not permitted. Since the coordinates of the origin satisfy the inequality $y < -x^2 + 4$, the point (0, 0) is in the graph. The complete graph is shown in Figure 8-23.

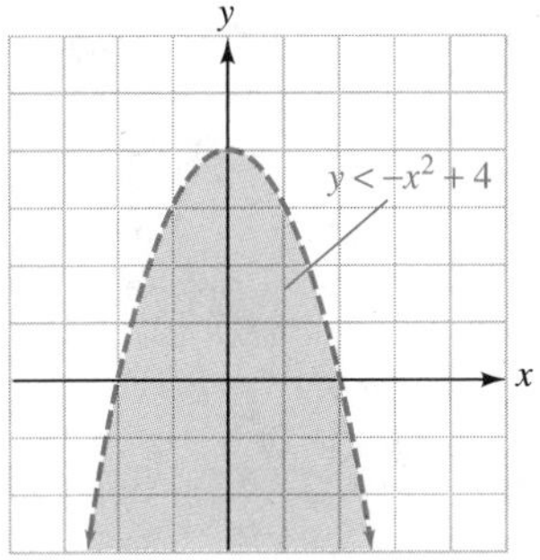

FIGURE 8-23

Self Check Graph $y \geq -x^2 + 4$.

Answer

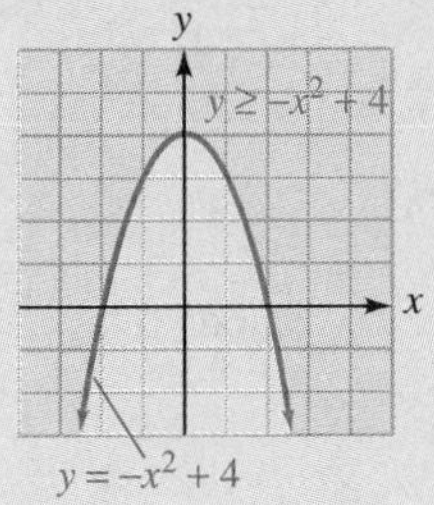

EXAMPLE 6 Graph $x \leq |y|$.

Solution We first graph $x = |y|$ as in Figure 8-24(a), using a solid line because equality is permitted. Since the origin is on the graph, we cannot use the origin as a test point. However, another point, such as (1, 0), will do. We substitute 1 for x and 0 for y into the inequality to get

$$x \leq |y|$$
$$\mathbf{1} \leq |\mathbf{0}|$$
$$1 \leq 0$$

Since $1 \leq 0$ is a false statement, the point (1, 0) does not satisfy the inequality and is not part of the graph. Thus, the graph of $x \leq |y|$ is to the left of the boundary.

The complete graph is shown in Figure 8-24(b).

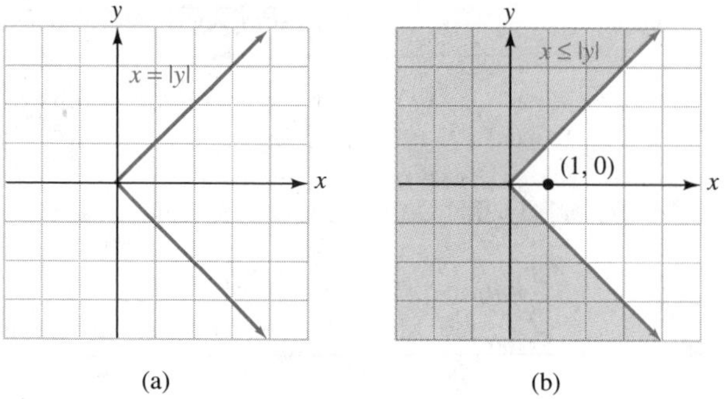

FIGURE 8-24

■

Self Check Graph $x \geq -|y|$.

Answer

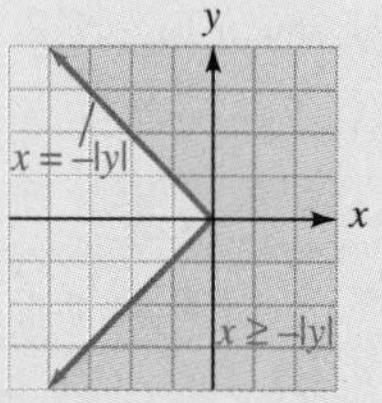

Orals *Tell where $x - 2$ is*

1. 0 **2.** positive **3.** negative

Tell where $x + 3$ is

4. 0 **5.** positive **6.** negative

Multiply both sides of the equation $\frac{1}{x} < 2$ by x when x is

7. positive **8.** negative

EXERCISE 8.5

REVIEW *Write each expression as an equation.*

1. y varies directly with x.

2. y varies inversely with t.

3. t varies jointly with x and y.

4. d varies directly with t but inversely with u^2.

Find the slope of the graph of each equation.

5. $y = 3x - 4$

6. $\dfrac{2x - y}{5} = 8$

VOCABULARY AND CONCEPTS *Fill in each blank to make a true statement.*

7. When $x > 3$, the binomial $x - 3$ is ______ than zero.

8. When $x < 3$, the binomial $x - 3$ is ____ than zero.

9. If $x = 0$, the fraction $\frac{1}{x}$ is ________.

10. To keep track of the signs of factors in a product or quotient, we can use a ____ chart.

PRACTICE *In Exercises 11–50, solve each inequality. Give each result in interval notation and graph the solution set.*

11. $x^2 - 5x + 4 < 0$

12. $x^2 - 3x - 4 > 0$

13. $x^2 - 8x + 15 > 0$

14. $x^2 + 2x - 8 < 0$

15. $x^2 + x - 12 \leq 0$

16. $x^2 + 7x + 12 \geq 0$

17. $x^2 + 2x \geq 15$

18. $x^2 - 8x \leq -15$

19. $x^2 + 8x < -16$

20. $x^2 + 6x \geq -9$

21. $x^2 \geq 9$

22. $x^2 \geq 16$

23. $2x^2 - 50 < 0$

24. $3x^2 - 243 < 0$

25. $\frac{1}{x} < 2$

26. $\frac{1}{x} > 3$

27. $\frac{4}{x} \geq 2$

28. $-\frac{6}{x} < 12$

29. $-\frac{5}{x} < 3$

30. $\frac{4}{x} \geq 8$

31. $\frac{x^2 - x - 12}{x - 1} < 0$

32. $\frac{x^2 + x - 6}{x - 4} \geq 0$

33. $\frac{x^2 + x - 20}{x + 2} \geq 0$

34. $\frac{x^2 - 10x + 25}{x + 5} < 0$

35. $\dfrac{x^2 - 4x + 4}{x + 4} < 0$

36. $\dfrac{2x^2 - 5x + 2}{x + 2} > 0$

37. $\dfrac{6x^2 - 5x + 1}{2x + 1} > 0$

38. $\dfrac{6x^2 + 11x + 3}{3x - 1} < 0$

39. $\dfrac{3}{x - 2} < \dfrac{4}{x}$

40. $\dfrac{-6}{x + 1} \geq \dfrac{1}{x}$

41. $\dfrac{-5}{x + 2} \geq \dfrac{4}{2 - x}$

42. $\dfrac{-6}{x - 3} < \dfrac{5}{3 - x}$

43. $\dfrac{7}{x - 3} \geq \dfrac{2}{x + 4}$

44. $\dfrac{-5}{x - 4} < \dfrac{3}{x + 1}$

45. $\dfrac{x}{x + 4} \leq \dfrac{1}{x + 1}$

46. $\dfrac{x}{x + 9} \geq \dfrac{1}{x + 1}$

47. $\dfrac{x}{x + 16} > \dfrac{1}{x + 1}$

48. $\dfrac{x}{x + 25} < \dfrac{1}{x + 1}$

49. $(x + 2)^2 > 0$

50. $(x - 3)^2 < 0$

In Exercises 51–54, use a graphing calculator to solve each inequality. Give the answer in interval notation.

51. $x^2 - 2x - 3 < 0$

52. $x^2 + x - 6 > 0$

53. $\dfrac{x + 3}{x - 2} > 0$

54. $\dfrac{3}{x} < 2$

In Exercises 55–66, graph each inequality.

55. $y < x^2 + 1$

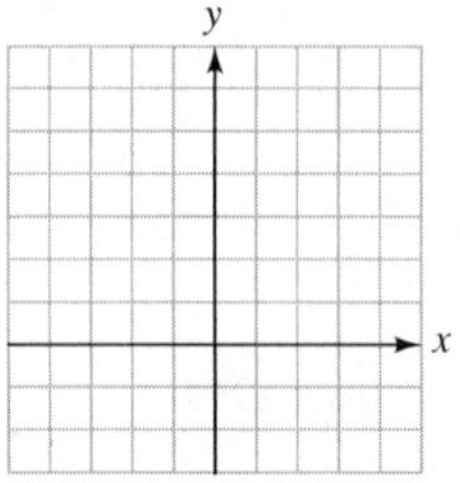

56. $y > x^2 - 3$

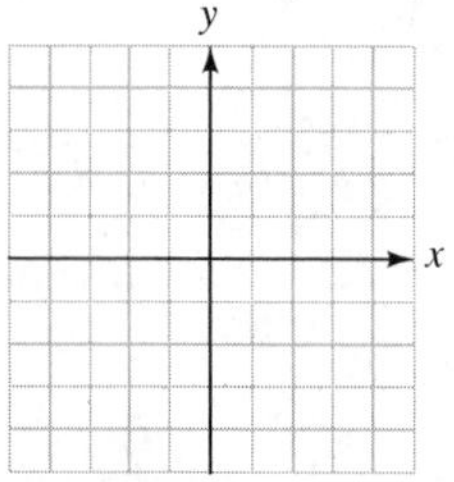

57. $y \leq x^2 + 5x + 6$

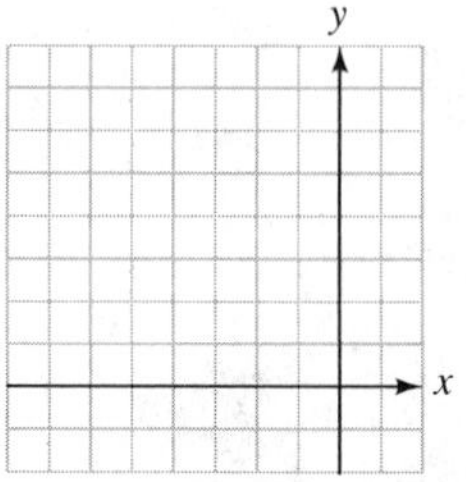

58. $y \geq x^2 + 5x + 4$

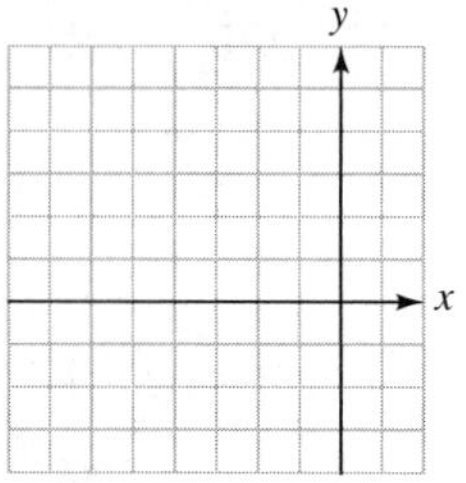

59. $y \geq (x - 1)^2$

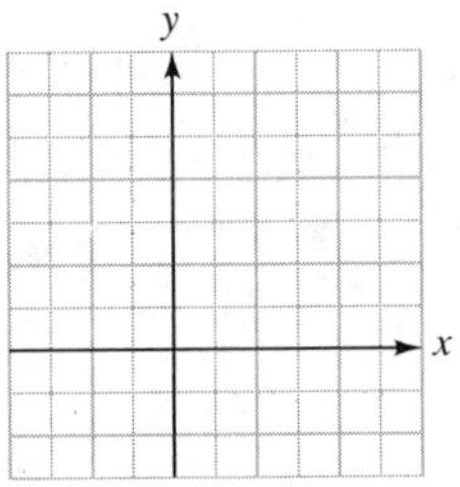

60. $y \leq (x + 2)^2$

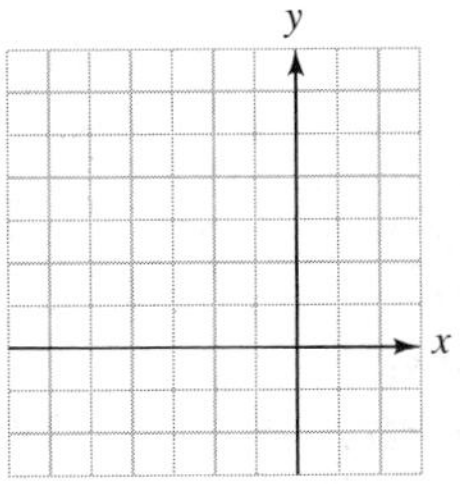

61. $-x^2 - y + 6 > -x$

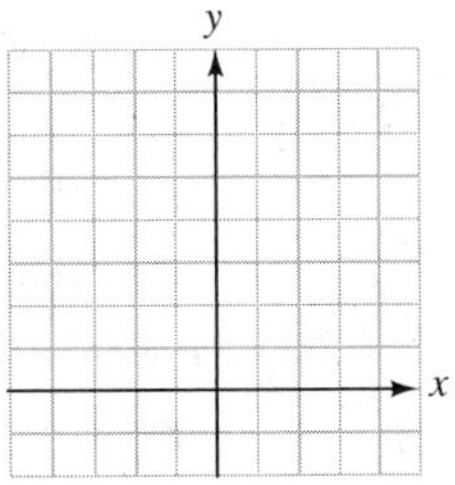

62. $y > (x + 3)(x - 2)$

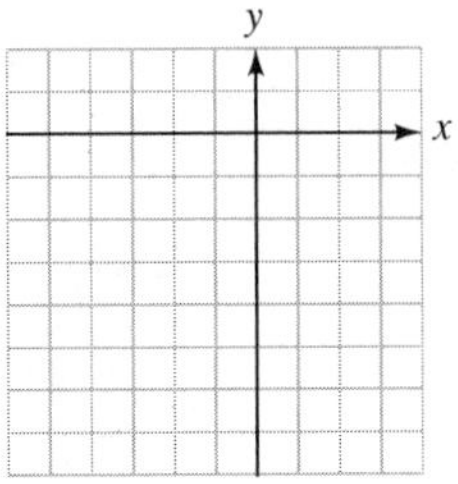

63. $y < |x + 4|$

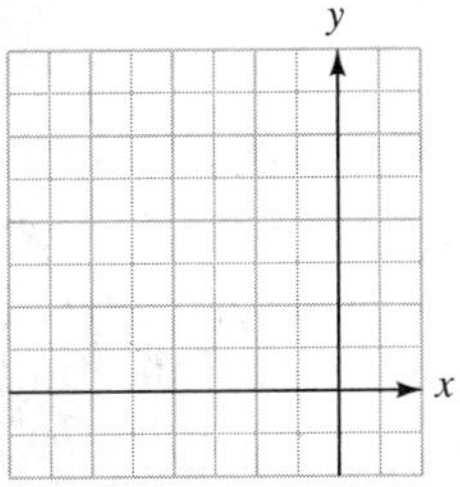

64. $y \geq |x - 3|$

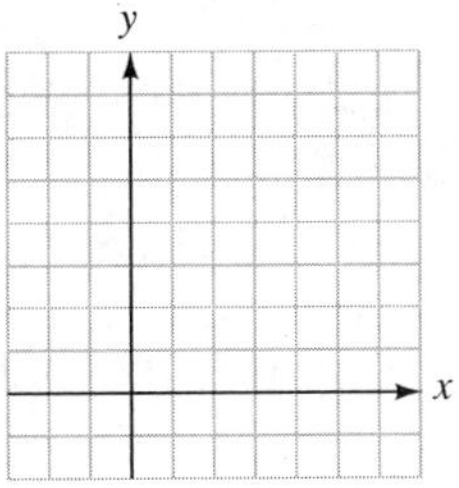

65. $y \leq -|x| + 2$

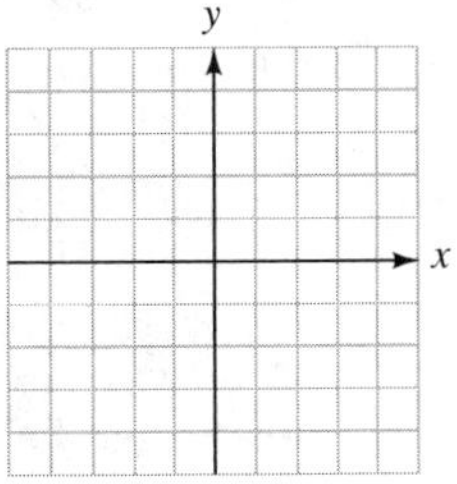

66. $y > |x| - 2$

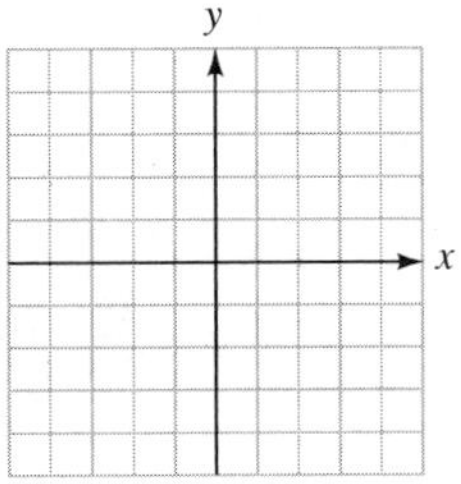

WRITING

67. Explain why $(x - 4)(x + 5)$ will be positive only when the signs of $x - 4$ and $x + 5$ are the same.

68. Tell how to find the graph of $y \geq x^2$.

SOMETHING TO THINK ABOUT

69. Under what conditions will the fraction $\frac{(x - 1)(x + 4)}{(x + 2)(x + 1)}$ be positive?

70. Under what conditions will the fraction $\frac{(x - 1)(x + 4)}{(x + 2)(x + 1)}$ be negative?

8.6 Algebra and Composition of Functions

■ ALGEBRA OF FUNCTIONS ■ COMPOSITION OF FUNCTIONS ■ THE IDENTITY FUNCTION
■ THE DIFFERENCE QUOTIENT ■ PROBLEM SOLVING

Getting Ready *Assume that* $P(x) = 2x + 1$ *and* $Q(x) = x - 2$. *Find each expression.*

1. $P(x) + Q(x)$

2. $P(x) - Q(x)$

3. $P(x) \cdot Q(x)$

4. $\dfrac{P(x)}{Q(x)}$

■ ALGEBRA OF FUNCTIONS

We now consider how functions can be added, subtracted, multiplied, and divided.

Operations on Functions

If the domains and ranges of functions f and g are subsets of the real numbers, then

The **sum** of f and g, denoted as $\mathbf{f + g}$, is defined by

$$(f + g)(x) = f(x) + g(x)$$

The **difference** of f and g, denoted as $\mathbf{f - g}$, is defined by

$$(f - g)(x) = f(x) - g(x)$$

The **product** of f and g, denoted as $\mathbf{f \cdot g}$, is defined by

$$(f \cdot g)(x) = f(x)g(x)$$

The **quotient** of f and g, denoted as $\mathbf{f/g}$, is defined by

$$(f/g)(x) = \frac{f(x)}{g(x)} \quad (g(x) \neq 0)$$

The domain of each of these functions is the set of real numbers x that are in the domain of both f and g. In the case of the quotient, there is the further restriction that $g(x) \neq 0$.

EXAMPLE 1 Let $f(x) = 2x^2 + 1$ and $g(x) = 5x - 3$. Find each function and its domain: **a.** $f + g$ and **b.** $f - g$.

Solution **a.**

$$\begin{aligned}(f + g)(x) &= f(x) + g(x)\\ &= (2x^2 + 1) + (5x - 3)\\ &= 2x^2 + 5x - 2\end{aligned}$$

The domain of $f + g$ is the set of real numbers that are in the domain of both f and g. Since the domain of both f and g is interval $(-\infty, \infty)$, the domain of $f + g$ is also the interval $(-\infty, \infty)$.

b.

$$\begin{aligned}(f - g)(x) &= f(x) - g(x)\\ &= (2x^2 + 1) - (5x - 3)\\ &= 2x^2 + 1 - 5x + 3 && \text{Remove parentheses.}\\ &= 2x^2 - 5x + 4 && \text{Combine like terms.}\end{aligned}$$

Since the domain of both f and g is $(-\infty, \infty)$, the domain of $f - g$ is also the interval $(-\infty, \infty)$. ■

Self Check Let $f(x) = 3x - 2$ and $g(x) = 2x^2 + 3x$. Find **a.** $f + g$ and **b.** $f - g$.

Answers **a.** $2x^2 + 6x - 2$, **b.** $-2x^2 - 2$

EXAMPLE 2 Let $f(x) = 2x^2 + 1$ and $g(x) = 5x - 3$. Find each function and its domain: **a.** $f \cdot g$ and **b.** f/g.

Solution **a.** $(f \cdot g)(x) = f(x)g(x)$

$$= (2x^2 + 1)(5x - 3)$$

$$= 10x^3 - 6x^2 + 5x - 3 \qquad \text{Multiply.}$$

The domain of $f \cdot g$ is the set of real numbers that are in the domain of both f and g. Since the domain of both f and g is the interval $(-\infty, \infty)$, the domain of $f \cdot g$ is also the interval $(-\infty, \infty)$.

b. $(f/g)(x) = \dfrac{f(x)}{g(x)}$

$$= \frac{2x^2 + 1}{5x - 3}$$

Since the denominator of the fraction cannot be 0, $x \neq \frac{3}{5}$. The domain of f/g is the interval $\left(-\infty, \frac{3}{5}\right) \cup \left(\frac{3}{5}, \infty\right)$. ■

Self Check Let $f(x) = 2x^2 - 3$ and $g(x) = x^2 + 1$. Find **a.** $f \cdot g$ and **b.** f/g.

Answers **a.** $2x^4 - x^2 - 3$, **b.** $\dfrac{2x^2 - 3}{x^2 + 1}$

COMPOSITION OF FUNCTIONS

We have seen that a function can be represented by a machine: We put in a number from the domain, and a number from the range comes out. For example, if we put the number 2 into the machine shown in Figure 8-25(a), the number $f(2) = 5(2) - 2 = 8$ comes out. In general, if we put x into the machine shown in Figure 8-25(b), the value $f(x)$ comes out.

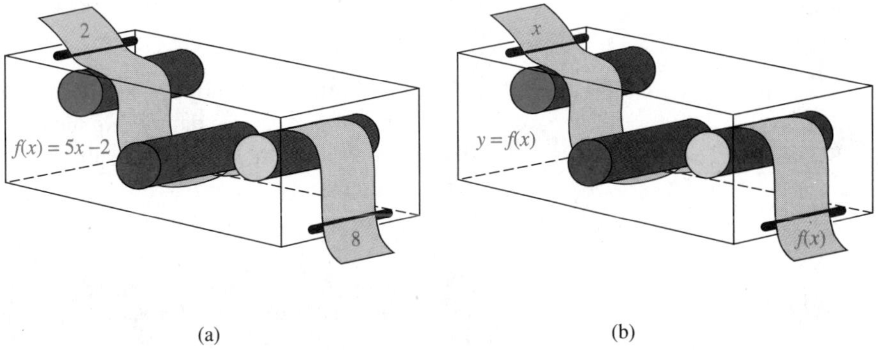

FIGURE 8-25

Often one quantity is a function of a second quantity that depends, in turn, on a third quantity. For example, the cost of a car trip is a function of the gasoline consumed. The amount of gasoline consumed, in turn, is a function of the number of miles driven. Such chains of dependence can be analyzed mathematically as **compositions of functions.**

Suppose that $y = f(x)$ and $y = g(x)$ define two functions. Any number x in the domain of g will produce the corresponding value $g(x)$ in the range of g. If $g(x)$ is in the domain of the function f, then $g(x)$ can be substituted into f, and a corresponding value $f(g(x))$ will be determined. This two-step process defines a new function, called a **composite function,** denoted by $f \circ g$.

The function machines shown in Figure 8-26 illustrate the composition $f \circ g$. When we put a number x into the function g, $g(x)$ comes out. The value $g(x)$ goes into function f, which transforms $g(x)$ into $f(g(x))$. If the function machines for g and f were connected to make a single machine, that machine would be named $f \circ g$.

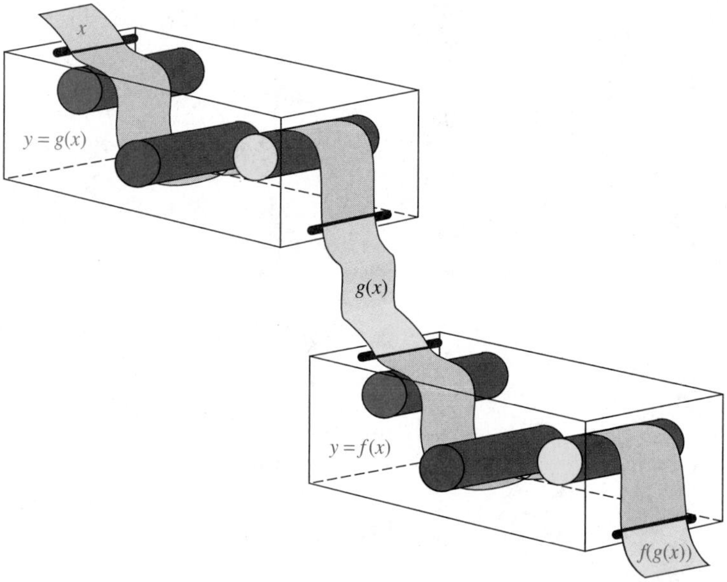

FIGURE 8-26

To be in the domain of the composite function $f \circ g$, a number x has to be in the domain of g. Also, the output of g must be in the domain of f. Thus, the domain of $f \circ g$ consists of those numbers x that are in the domain of g, and for which $g(x)$ is in the domain of f.

Composite Functions

The **composite function $f \circ g$** is defined by

$$(f \circ g)(x) = f(g(x))$$

For example, if $f(x) = 4x - 5$ and $g(x) = 3x + 2$, then

$$\begin{aligned} (f \circ g)(x) &= f(g(x)) \\ &= f(3x + 2) \\ &= 4(3x + 2) - 5 \\ &= 12x + 8 - 5 \\ &= 12x + 3 \end{aligned} \qquad \text{or} \qquad \begin{aligned} (g \circ f)(x) &= g(f(x)) \\ &= g(4x - 5) \\ &= 3(4x - 5) + 2 \\ &= 12x - 15 + 2 \\ &= 12x - 13 \end{aligned}$$

WARNING! Note that in the previous example, $(f \circ g)(x) \neq (g \circ f)(x)$. This shows that the composition of functions is not commutative.

EXAMPLE 3 Let $f(x) = 2x + 1$ and $g(x) = x - 4$. Find **a.** $(f \circ g)(9)$, **b.** $(f \circ g)(x)$, and **c.** $(g \circ f)(-2)$.

Solution **a.** $(f \circ g)(9)$ means $f(g(9))$. In Figure 8-27(a), function g receives the number 9, subtracts 4, and releases the number $g(9) = 5$. The 5 then goes into the f function, which doubles 5 and adds 1. The final result, 11, is the output of the composite function $f \circ g$:

$$(f \circ g)(9) = f(g(9)) = f(5) = 2(5) + 1 = 11$$

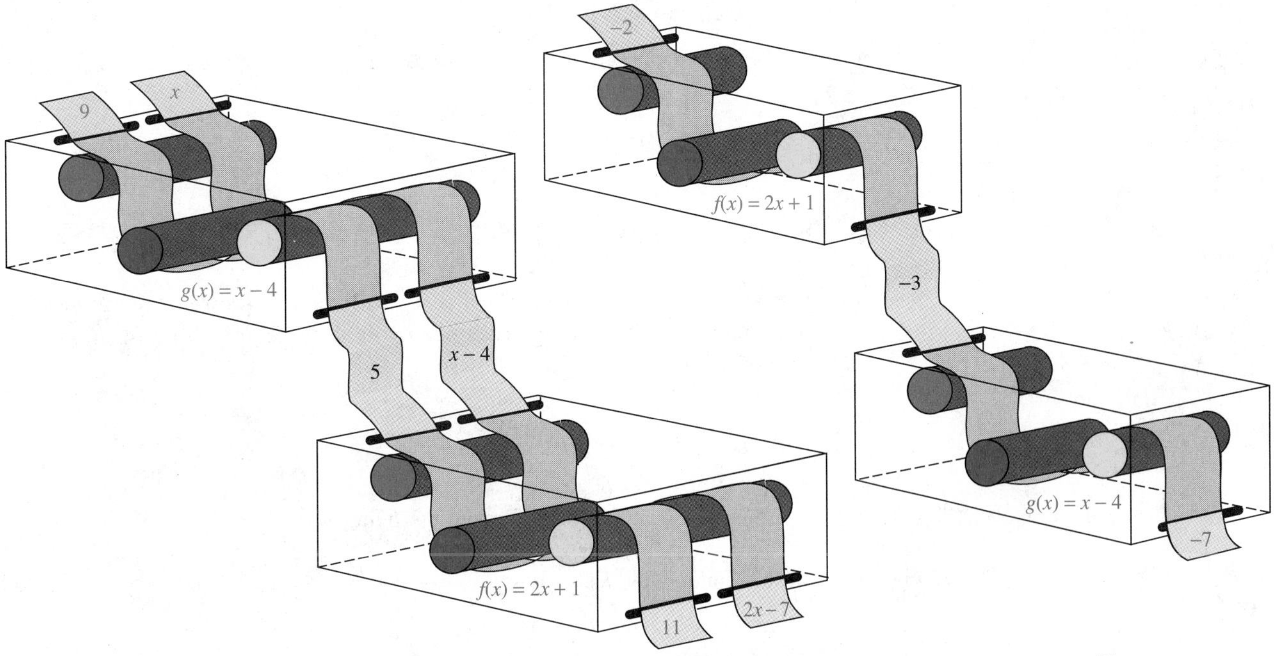

(a) (b)

FIGURE 8-27

b. $(f \circ g)(x)$ means $f(g(x))$. In Figure 8-27(a), function g receives the number x, subtracts 4, and releases the number $x - 4$. The $x - 4$ then goes into the f function, which doubles $x - 4$ and adds 1. The final result, $2x - 7$, is the output of the composite function $f \circ g$.

$$(f \circ g)(x) = f(g(x)) = f(x - 4) = 2(x - 4) + 1 = 2x - 7$$

c. $(g \circ f)(-2)$ means $g(f(-2))$. In Figure 8-27(b), function f receives the number -2, doubles it and adds 1, and releases -3 into the g function. Function g subtracts 4 from -3 and releases a final result of -7. Thus,

$$(g \circ f)(-2) = g(f(-2)) = g(-3) = -3 - 4 = -7$$

■

THE IDENTITY FUNCTION

The **identity function** is defined by the equation $I(x) = x$. Under this function, the value that corresponds to any real number x is x itself. If f is any function, the composition of f with the identity function is the function f:

$$(f \circ I)(x) = (I \circ f)(x) = f(x)$$

EXAMPLE 4 Let f be any function and I be the identity function, $I(x) = x$. Show that **a.** $(f \circ I)(x) = f(x)$ and **b.** $(I \circ f)(x) = f(x)$.

Solution **a.** $(f \circ I)(x)$ means $f(I(x))$. Because $I(x) = x$, we have

$$(f \circ I)(x) = f(I(x)) = f(x)$$

b. $(I \circ f)(x)$ means $I(f(x))$. Because I passes any number through unchanged, we have $I(f(x)) = f(x)$ and

$$(I \circ f)(x) = I(f(x)) = f(x)$$

■

THE DIFFERENCE QUOTIENT

An important function in calculus, called the **difference quotient,** represents the slope of a line that passes through two given points on the graph of a function. The difference quotient is defined as follows:

$$\frac{f(x + h) - f(x)}{h}$$

EXAMPLE 5 If $f(x) = x^2 - 4$, evaluate the difference quotient.

Solution First, we evaluate $f(x + h)$.

$$f(x) = x^2 - 4$$

$$f(x + h) = (x + h)^2 - 4 \qquad \text{Substitute } x + h \text{ for } x.$$

$$= x^2 + 2xh + h^2 - 4 \qquad (x + h)^2 = x^2 + 2hx + h^2.$$

Then we note that $f(x) = x^2 - 4$. We can now substitute the values of $f(x + h)$ and $f(x)$ into the difference quotient and simplify.

$$\frac{f(x+h) - f(x)}{h} = \frac{(x^2 + 2xh + h^2 - 4) - (x^2 - 4)}{h}$$

$$= \frac{x^2 + 2xh + h^2 - 4 - x^2 + 4}{h} \qquad \text{Remove parentheses.}$$

$$= \frac{2xh + h^2}{h} \qquad \text{Combine like terms.}$$

$$= \frac{h(2x + h)}{h} \qquad \text{Factor out } h \text{ in the numerator.}$$

$$= 2x + h \qquad \text{Divide out } h; \tfrac{h}{h} = 1.$$

The difference quotient simplifies as $2x + h$. ■

■ PROBLEM SOLVING

EXAMPLE 6

Temperature change A laboratory sample is removed from a cooler at a temperature of 15° Fahrenheit. Technicians are warming the sample at a controlled rate of 3° F per hour. Express the sample's Celsius temperature as a function of the time, t, since it was removed from refrigeration.

Solution The temperature of the sample is 15° F when $t = 0$. Because it warms at 3° F per hour after that, it warms $3t$°, and the Fahrenheit temperature is given by the function

$$F(t) = 3t + 15$$

The Celsius temperature, C, is a function of this Fahrenheit temperature, F, given by the formula

$$C(F) = \frac{5}{9}(F - 32)$$

To express the sample's Celsius temperature as a function of time, we find the composition function

$$(C \circ F)(t) = C(F(t))$$

$$= \frac{5}{9}(F(t) - 32) \qquad \text{Substitute } F(t) \text{ for } F \text{ in } \tfrac{5}{9}(F - 32).$$

$$= \frac{5}{9}[(3t + 15) - 32] \qquad \text{Substitute } 3t + 15 \text{ for } F(t).$$

$$= \frac{5}{9}(3t - 17) \qquad \text{Simplify.}$$

■

Orals *If $f(x) = 2x$, $g(x) = 3x$, and $h(x) = 4x$, find*

1. $f + g$ **2.** $h - g$ **3.** $f \cdot h$

4. g/f **5.** h/f **6.** $g \cdot h$

7. $(f \circ h)(x)$ **8.** $(f \circ g)(x)$ **9.** $(g \circ h)(x)$

EXERCISE 8.6

REVIEW *Simplify each expression.*

1. $\dfrac{3x^2 + x - 14}{4 - x^2}$

2. $\dfrac{2x^3 + 14x^2}{3 + 2x - x^2} \cdot \dfrac{x^2 - 3x}{x}$

3. $\dfrac{8 + 2x - x^2}{12 + x - 3x^2} \div \dfrac{3x^2 + 5x - 2}{3x - 1}$

4. $\dfrac{x - 1}{1 + \dfrac{x}{x - 2}}$

VOCABULARY AND CONCEPTS *Fill in each blank to make a true statement.*

5. $(f + g)(x) =$ ________

6. $(f - g)(x) =$ ________

7. $(f \cdot g)(x) =$ ________

8. $(f/g)(x) =$ ___ $(g(x) \neq 0)$

9. In Exercises 5–7, the domain of each function is the set of real numbers x that are in the ________ of both f and g.

10. $(f \circ g)(x) =$ ________

11. If I is the identity function, then $(f \circ I)(x) =$ ____.

12. If I is the identity function, then $(I \circ f)(x) =$ ____.

PRACTICE *In Exercises 13–20, $f(x) = 3x$ and $g(x) = 4x$. Find each function and its domain.*

13. $f + g$ **14.** $f - g$ **15.** $f \cdot g$ **16.** f/g

17. $g - f$ **18.** $g + f$ **19.** g/f **20.** $g \cdot f$

In Exercises 21–28, $f(x) = 2x + 1$ and $g(x) = x - 3$. Find each function and its domain.

21. $f + g$ **22.** $f - g$ **23.** $f \cdot g$ **24.** f/g

25. $g - f$ **26.** $g + f$ **27.** g/f **28.** $g \cdot f$

In Exercises 29–32, $f(x) = 3x - 2$ and $g(x) = 2x^2 + 1$. Find each function and its domain.

29. $f - g$ **30.** $f + g$ **31.** f/g **32.** $f \cdot g$

In Exercises 33–36, $f(x) = x^2 - 1$ *and* $g(x) = x^2 - 4$. *Find each function and its domain.*

33. $f - g$
34. $f + g$
35. g/f
36. $g \cdot f$

In Exercises 37–48, $f(x) = 2x + 1$ *and* $g(x) = x^2 - 1$. *Find each value.*

37. $(f \circ g)(2)$
38. $(g \circ f)(2)$
39. $(g \circ f)(-3)$
40. $(f \circ g)(-3)$
41. $(f \circ g)(0)$
42. $(g \circ f)(0)$
43. $(f \circ g)\left(\frac{1}{2}\right)$
44. $(g \circ f)\left(\frac{1}{3}\right)$
45. $(f \circ g)(x)$
46. $(g \circ f)(x)$
47. $(g \circ f)(2x)$
48. $(f \circ g)(2x)$

In Exercises 49–56, $f(x) = 3x - 2$ *and* $g(x) = x^2 + x$. *Find each value.*

49. $(f \circ g)(4)$
50. $(g \circ f)(4)$
51. $(g \circ f)(-3)$
52. $(f \circ g)(-3)$
53. $(g \circ f)(0)$
54. $(f \circ g)(0)$
55. $(g \circ f)(x)$
56. $(f \circ g)(x)$

In Exercises 57–68, find $\dfrac{f(x + h) - f(x)}{h}$.

57. $f(x) = 2x + 3$
58. $f(x) = 3x - 5$
59. $f(x) = x^2$
60. $f(x) = x^2 - 1$
61. $f(x) = 2x^2 - 1$
62. $f(x) = 3x^2$
63. $f(x) = x^2 + x$
64. $f(x) = x^2 - x$
65. $f(x) = x^2 + 3x - 4$
66. $f(x) = x^2 - 4x + 3$
67. $f(x) = 2x^2 + 3x - 7$
68. $f(x) = 3x^2 - 2x + 4$

In Exercises 69–80, find $\dfrac{f(x) - f(a)}{x - a}$.

69. $f(x) = 2x + 3$
70. $f(x) = 3x - 5$
71. $f(x) = x^2$
72. $f(x) = x^2 - 1$
73. $f(x) = 2x^2 - 1$
74. $f(x) = 3x^2$
75. $f(x) = x^2 + x$
76. $f(x) = x^2 - x$
77. $f(x) = x^2 + 3x - 4$
78. $f(x) = x^2 - 4x + 3$
79. $f(x) = 2x^2 + 3x - 7$
80. $f(x) = 3x^2 - 2x + 4$

81. If $f(x) = x + 1$ and $g(x) = 2x - 5$, show that $(f \circ g)(x) \neq (g \circ f)(x)$.

82. If $f(x) = x^2 + 1$ and $g(x) = 3x^2 - 2$, show that $(f \circ g)(x) \neq (g \circ f)(x)$.

83. If $f(x) = x^2 + 2x - 3$, find $f(a)$, $f(h)$, and $f(a + h)$. Then show that $f(a + h) \neq f(a) + f(h)$.

84. If $g(x) = 2x^2 + 10$, find $g(a)$, $g(h)$, and $g(a + h)$. Then show that $g(a + h) \neq g(a) + g(h)$.

85. If $f(x) = x^3 - 1$, find $\dfrac{f(x + h) - f(x)}{h}$.

86. If $f(x) = x^3 + 2$, find $\dfrac{f(x + h) - f(x)}{h}$.

APPLICATIONS

87. **Alloys** A molten alloy must be cooled slowly to control crystallization. When removed from the furnace, its temperature is 2,700° F, and it will be cooled at 200° per hour. Express the Celsius temperature as a function of the number of hours, t, since cooling began.

88. **Weather forecasting** A high pressure area promises increasingly warmer weather for the next 48 hours. The temperature is now 34° Celsius and is expected to rise 1° every 6 hours. Express the Fahrenheit temperature as a function of the number of hours from now. (*Hint:* $F = \frac{9}{5}C + 32$.)

WRITING

89. Explain how to find the domain of f/g.

90. Explain why the difference quotient represents the slope of a line passing through $(x, f(x))$ and $(x + h, f(x + h))$.

SOMETHING TO THINK ABOUT

91. Is composition of functions associative? Choose functions f, g, and h and determine whether $[f \circ (g \circ h)](x) = [(f \circ g) \circ h](x)$.

92. Choose functions f, g, and h and determine whether $f \circ (g + h) = f \circ g + f \circ h$.

8.7 Inverses of Functions

■ INTRODUCTION TO INVERSES OF FUNCTIONS ■ ONE-TO-ONE FUNCTIONS ■ THE HORIZONTAL LINE TEST ■ FINDING INVERSES OF FUNCTIONS

Getting Ready *Solve each equation for x.*

1. $y = 3x + 2$

2. $2x - 3y = 10$

■ INTRODUCTION TO INVERSES OF FUNCTIONS

The function defined by $C = \frac{5}{9}(F - 32)$ is the formula that we use to convert degrees Fahrenheit to degrees Celsius. If we substitute a Fahrenheit reading into the formula, a Celsius reading comes out. For example, if we substitute 41° for F, we obtain a Celsius reading of 5°:

$$\begin{aligned} C &= \frac{5}{9}(F - 32) \\ &= \frac{5}{9}(41 - 32) && \text{Substitute 41 for } F. \\ &= \frac{5}{9}(9) \\ &= 5 \end{aligned}$$

If we want to find a Fahrenheit reading from a Celsius reading, we need a formula into which we can substitute a Celsius reading and have a Fahrenheit reading come out. Such a formula is $F = \frac{9}{5}C + 32$, which takes the Celsius reading of 5° and turns it back into a Fahrenheit reading of 41°.

$$\begin{aligned} F &= \frac{9}{5}C + 32 \\ &= \frac{9}{5}(5) + 32 \qquad \text{Substitute 5 for } C. \\ &= 41 \end{aligned}$$

The functions defined by these two formulas do opposite things. The first turns 41° F into 5° Celsius, and the second turns 5° Celsius back into 41° F. For this reason, we say that the functions are *inverses* of each other.

ONE-TO-ONE FUNCTIONS

Recall that for each input into a function, there is a single output. For some functions, different inputs have the same output, as shown in Figure 8-28(a). For other functions, different inputs have different outputs, as shown in Figure 8-28(b).

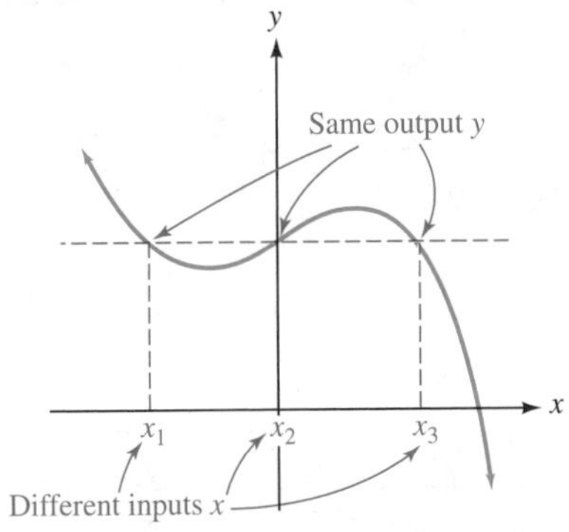

Not a one-to-one function

(a)

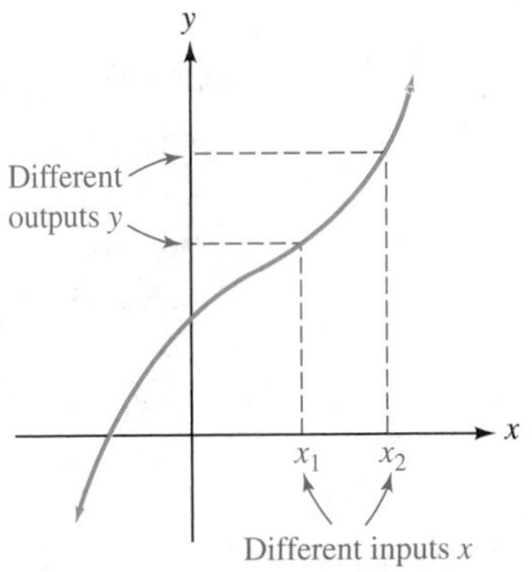

A one-to-one function

(b)

FIGURE 8-28

When every output of a function corresponds to exactly one input, we say that the function is *one-to-one.*

One-to-One Functions

A function is called **one-to-one** if each input value of x in the domain determines a different output value of y in the range.

EXAMPLE 1 Determine whether the functions **a.** $f(x) = x^2$ and **b.** $f(x) = x^3$ are one-to-one.

Solution **a.** The function $f(x) = x^2$ is not one-to-one, because different input values x can determine the same output value y. For example, inputs of 3 and -3 produce the same output value of 9.

$$f(3) = 3^2 = 9 \qquad \text{and} \qquad f(-3) = (-3)^2 = 9$$

b. The function $f(x) = x^3$ is one-to-one, because different input values x determine different output values of y for all x. This is because different numbers have different cubes. ■

Self Check Determine whether $f(x) = 2x + 3$ is one-to-one.

Answer yes

THE HORIZONTAL LINE TEST

The **horizontal line test** can be used to decide whether the graph of a function represents a one-to-one function. If any horizontal line that intersects the graph of a function does so only once, the function is one-to-one. Otherwise, the function is not one-to-one. See Figure 8-29.

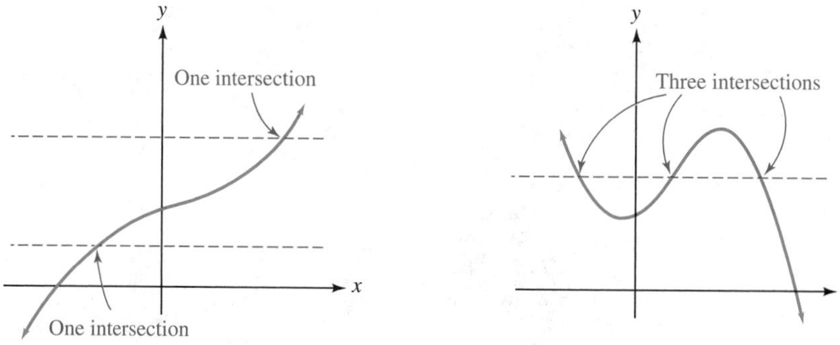

A one-to-one function

Not a one-to-one function

FIGURE 8-29

EXAMPLE 2 Use the horizontal line test to decide whether the graphs in Figure 8-30 represent one-to-one functions.

Solution **a.** Because many horizontal lines intersect the graph shown in Figure 8-30(a) twice, the graph does not represent a one-to-one function.

b. Because each horizontal line that intersects the graph in Figure 8-30(b) does so exactly once, the graph does represent a one-to-one function.

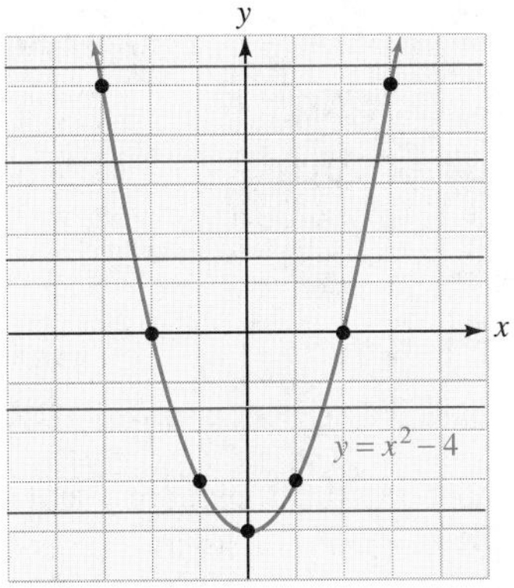

(a)

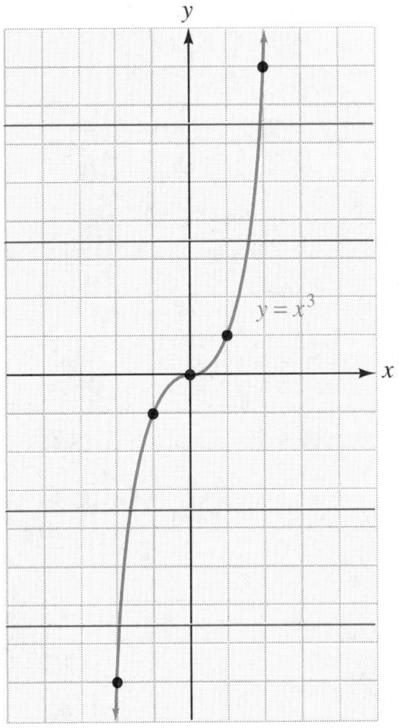

(b)

FIGURE 8-30

Self Check Determine whether the following graph represents a one-to-one function.

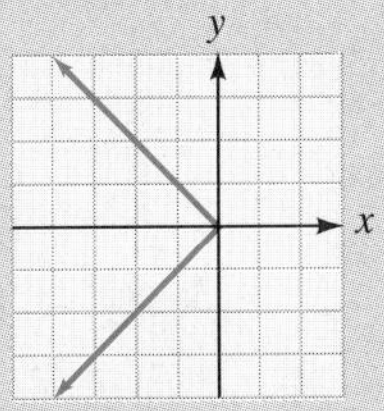

Answer no

WARNING! Make sure to use the vertical line test to determine whether the graph represents a function. If it is, use the horizontal line test to determine whether the function is one-to-one.

FINDING INVERSES OF FUNCTIONS

If f is the function determined by the table shown in Figure 8-31(a), it turns the number 1 into 10, 2 into 20, and 3 into 30. Since the inverse of f must turn 10 back into 1, 20 back into 2, and 30 back into 3, it consists of the ordered pairs shown in Figure 8-31(b).

Function f

x	y
1	10
2	20
3	30
↑ Domain	↑ Range

(a)

Inverse of f

x	y
10	1
20	2
30	3
↑ Domain	↑ Range

(b)

Note that the inverse of f is also a function.

FIGURE 8-31

We note that the domain of f and the range of its inverse is $\{1, 2, 3\}$. The range of f and the domain of its inverse is $\{10, 20, 30\}$.

This example suggests that to form the inverse of a function f, we simply interchange the coordinates of each ordered pair that determines f. When the inverse of a function is also a function, we call it ***f* inverse** and denote it with the symbol f^{-1}.

WARNING! The symbol $f^{-1}(x)$ is read as "the inverse of $f(x)$" or just "f inverse." The -1 in the notation $f^{-1}(x)$ is not an exponent. Remember that $f^{-1}(x) \neq \frac{1}{f(x)}$.

Finding the Inverse of a One-to-One Function

If a function is one-to-one, we find its inverse as follows:

1. Replace $f(x)$ with y, if necessary.
2. Interchange the variables x and y.
3. Solve the resulting equation for y.
4. This equation is $y = f^{-1}(x)$.

EXAMPLE 3 If $f(x) = 4x + 2$, find the inverse of f and tell whether it is a function.

Solution We proceed as follows:

$f(x) = 4x + 2$

$y = 4x + 2$ Replace $f(x)$ with y.

$x = 4y + 2$ Interchange the variables x and y.

To decide whether the inverse $x = 4y + 2$ is a function, we solve the equation for y.

$x = 4y + 2$

$x - 2 = 4y$ Subtract 2 from both sides.

1. $y = \dfrac{x - 2}{4}$ Divide both sides by 4 and write y on the left-hand side.

Because each input x that is substituted into Equation 1 gives one output y, the inverse of f is a function, so we can express it in the form

$$f^{-1}(x) = \frac{x-2}{4}$$

■

Self Check If $y = f(x) = -5x - 3$, find the inverse of f and tell whether it is a function.

Answers $y = -\frac{1}{5}x - \frac{3}{5}$, yes

To emphasize an important relationship between a function and its inverse, we substitute some number x, such as $x = 3$, into the function $f(x) = 4x + 2$ of Example 3. The corresponding value of y produced is

$$y = f(3) = 4(3) + 2 = 14$$

If we substitute 14 into the inverse function, f^{-1}, the corresponding value of y that is produced is

$$y = f^{-1}(14) = \frac{14-2}{4} = 3$$

Thus, the function f turns 3 into 14, and the inverse function f^{-1} turns 14 back into 3. In general, the composition of a function and its inverse is the identity function.

To prove that $f(x) = 4x + 2$ and $f^{-1}(x) = \frac{x-2}{4}$ are inverse functions, we must show that their composition (in both directions) is the identity function:

$$\begin{aligned}(f \circ f^{-1})(x) &= f(f^{-1}(x)) \\ &= f\left(\frac{x-2}{4}\right) \\ &= 4\left(\frac{x-2}{4}\right) + 2 \\ &= x - 2 + 2 \\ &= x\end{aligned}$$

$$\begin{aligned}(f^{-1} \circ f)(x) &= f^{-1}(f(x)) \\ &= f^{-1}(4x + 2) \\ &= \frac{4x + 2 - 2}{4} \\ &= \frac{4x}{4} \\ &= x\end{aligned}$$

Thus, $(f \circ f^{-1})(x) = (f^{-1} \circ f)(x) = x$, which is the identity function $I(x)$.

EXAMPLE 4 The set of all pairs (x, y) determined by $3x + 2y = 6$ is a function. Find its inverse function, and graph the function and its inverse on one coordinate system.

Solution To find the inverse function of $3x + 2y = 6$, we interchange x and y to obtain

$$3y + 2x = 6$$

and then solve the equation for y.

$$3y + 2x = 6$$

$$3y = -2x + 6 \qquad \text{Subtract } 2x \text{ from both sides.}$$

$$y = -\frac{2}{3}x + 2 \qquad \text{Divide both sides by 3.}$$

Thus, $y = f^{-1}(x) = -\frac{2}{3}x + 2$. The graphs of $3x + 2y = 6$ and $y = f^{-1}(x) = -\frac{2}{3}x + 2$ appear in Figure 8-32.

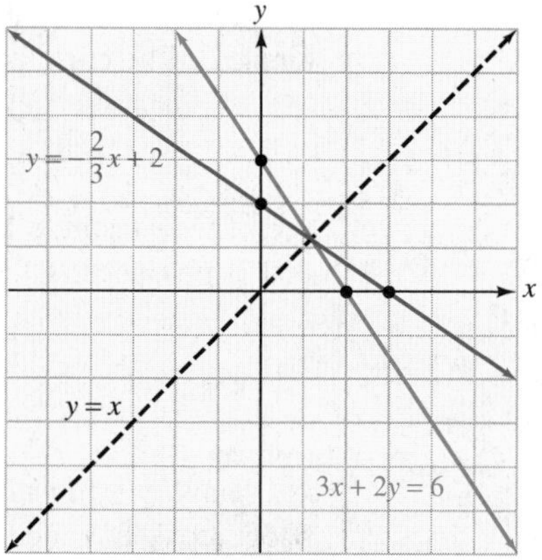

FIGURE 8-32

Self Check Find the inverse of the function defined by $2x - 3y = 6$. Graph the function and its inverse on one coordinate system.

Answer $f^{-1}(x) = \frac{3}{2}x + 3$

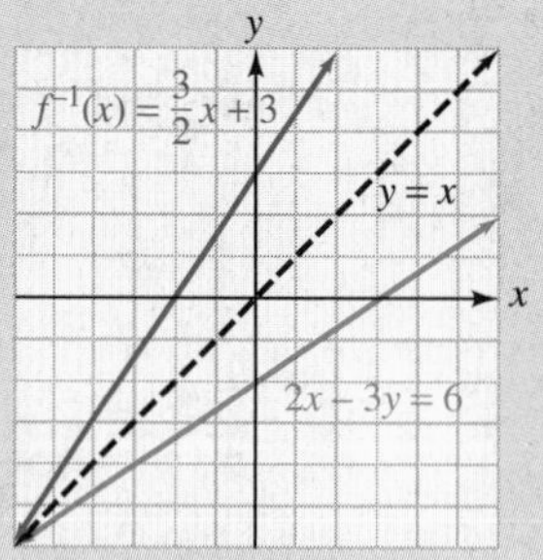

In Example 4, the graph of $3x + 2y = 6$ and $y = f^{-1}(x) = -\frac{2}{3}x + 2$ are symmetric about the line $y = x$. This is always the case, because when the coordinates (a, b) satisfy an equation, the coordinates (b, a) will satisfy its inverse.

In each example so far, the inverse of a function has been another function. This is not always true, as the following example will show.

EXAMPLE 5 Find the inverse of the function determined by $f(x) = x^2$.

Solution

$y = x^2$ Replace $f(x)$ with y.

$x = y^2$ Interchange x and y.

$y = \pm\sqrt{x}$ Use the square root property and write y on the left-hand side.

When the inverse $y = \pm\sqrt{x}$ is graphed as in Figure 8-33, we see that the graph does not pass the vertical line test. Thus, it is not a function.

The graph of $y = x^2$ is also shown in the figure. As expected, the graphs of $y = x^2$ and $y = \pm\sqrt{x}$ are symmetric about the line $y = x$.

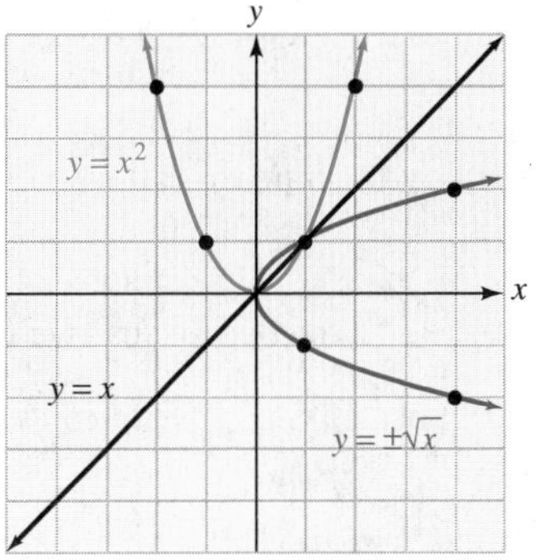

FIGURE 8-33

Self Check Find the inverse of the function determined by $f(x) = 4x^2$.

Answer $y = \pm\dfrac{\sqrt{x}}{2}$

EXAMPLE 6 Find the inverse of $f(x) = x^3$.

Solution To find the inverse, we proceed as follows:

$y = x^3$ Replace $f(x)$ with y.

$x = y^3$ Interchange the variables x and y.

$\sqrt[3]{x} = y$ Take the cube root of both sides.

We note that to each number x there corresponds one real cube root. Thus, $y = \sqrt[3]{x}$ represents a function. In $f^{-1}(x)$ notation, we have

$$f^{-1}(x) = \sqrt[3]{x}$$

Self Check Find the inverse of $f(x) = x^5$.

Answer $f^{-1}(x) = \sqrt[5]{x}$

If a function is not one-to-one, we can often make it one-to-one by restricting its domain.

EXAMPLE 7 Find the inverse of the function defined by $y = x^2$ with $x \geq 0$. Then tell whether the inverse is a function. Graph the function and its inverse on one set of coordinate axes.

Solution The inverse of the function $y = x^2$ with $x \geq 0$ is

$x = y^2$ with $y \geq 0$ Interchange the variables x and y.

Before considering the restriction, this equation can be written in the form

$$y = \pm\sqrt{x} \quad \text{with} \quad y \geq 0$$

Since $y \geq 0$, each number x gives only one value of y: $y = \sqrt{x}$. Thus, the inverse is a function.

The graphs of the two functions appear in Figure 8-34. The line $y = x$ is included so that we can see that the graphs are symmetric about the line $y = x$.

$y = x^2$ and $x \geq 0$

x	y	(x, y)
0	0	(0, 0)
1	1	(1, 1)
2	4	(2, 4)
3	9	(3, 9)

$x = y^2$ and $y \geq 0$

x	y	(x, y)
0	0	(0, 0)
1	1	(1, 1)
4	2	(4, 2)
9	3	(9, 3)

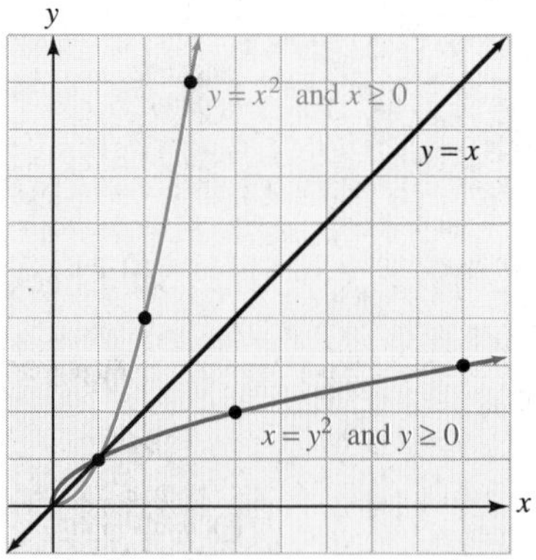

FIGURE 8-34

■

Orals

Find the inverse of each set of ordered pairs.

1. $\{(1, 2), (2, 3), (5, 10)\}$

2. $\{(1, 1), (2, 8), (4, 64)\}$

Find the inverse function of each linear function.

3. $y = \frac{1}{2}x$

4. $y = 2x$

Tell whether each function is one-to-one.

5. $y = x^2 - 2$

6. $y = x^3$

EXERCISE 8.7

REVIEW *Write each complex number in a + bi form or find each value.*

1. $3 - \sqrt{-64}$

2. $(2 - 3i) + (4 + 5i)$

3. $(3 + 4i)(2 - 3i)$

4. $\dfrac{6 + 7i}{3 - 4i}$

5. $|6 - 8i|$

6. $\left|\dfrac{2 + i}{3 - i}\right|$

VOCABULARY AND CONCEPTS *Fill in each blank to make a true statement.*

7. A function is called ________ if each input determines a different output.

8. If every ________ line that intersects a graph of a function does so only once, the function is one-to-one.

9. If a one-to-one function turns an input of 2 into an output of 5, the inverse function will turn 5 into __.

10. The symbol $f^{-1}(x)$ is read as ____________ or ________.

11. $(f \circ f^{-1})(x) =$ __

12. The graphs of a function and its inverse are symmetrical about the line _____.

PRACTICE *In Exercises 13–16, determine whether each function is one-to-one.*

13. $f(x) = 2x$

14. $f(x) = |x|$

15. $f(x) = x^4$

16. $f(x) = x^3 + 1$

In Exercises 17–24, each graph represents a function. Use the horizontal line test to decide whether the function is one-to-one.

17.

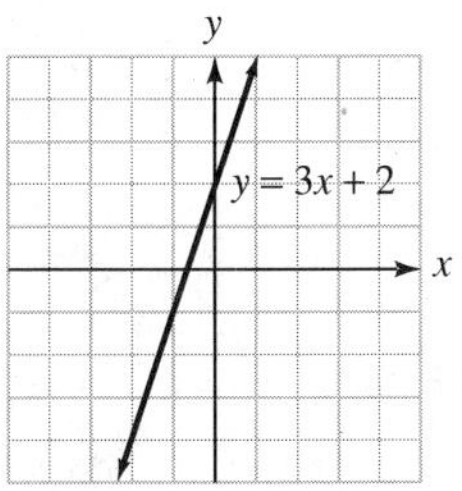

18.

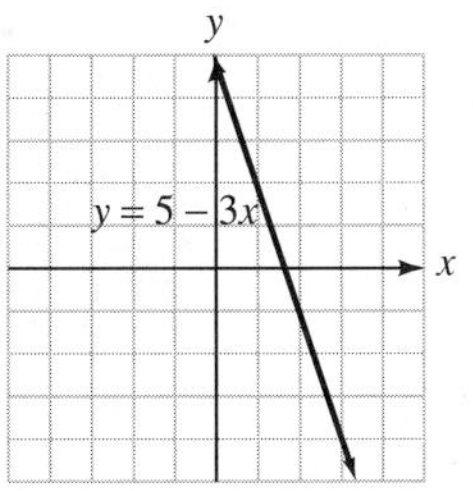

19.

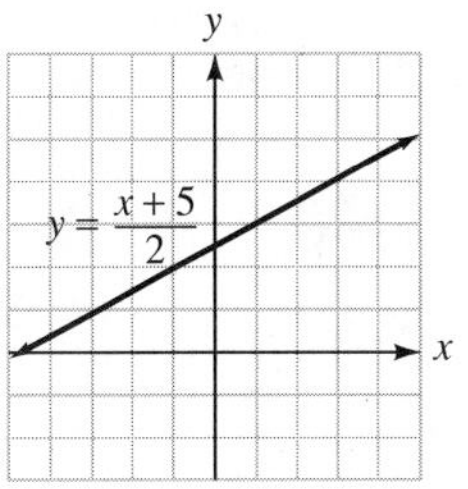

20.

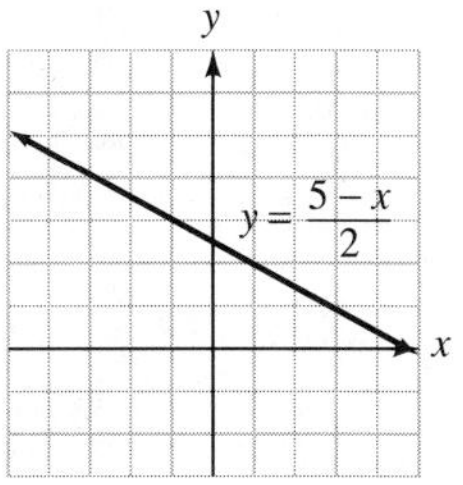

21.

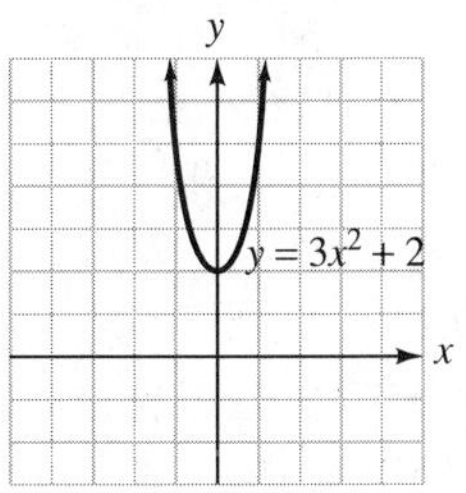

22.

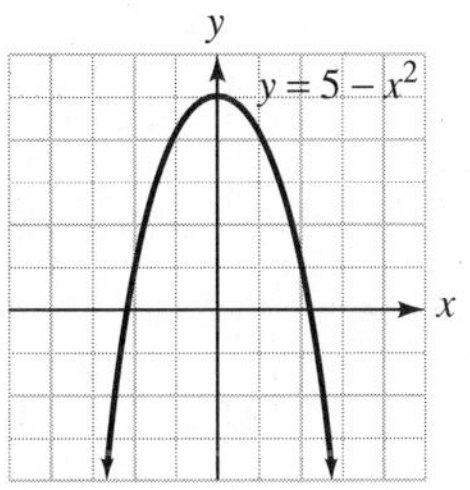

23.

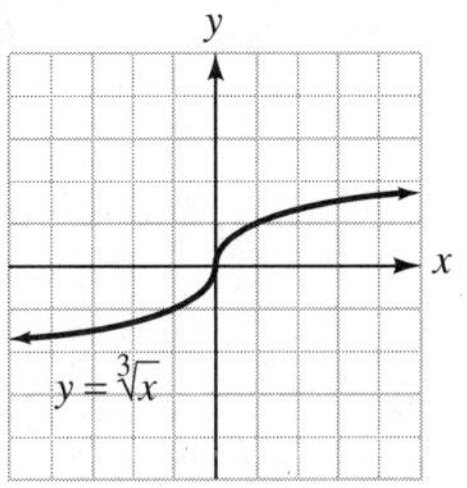

24.

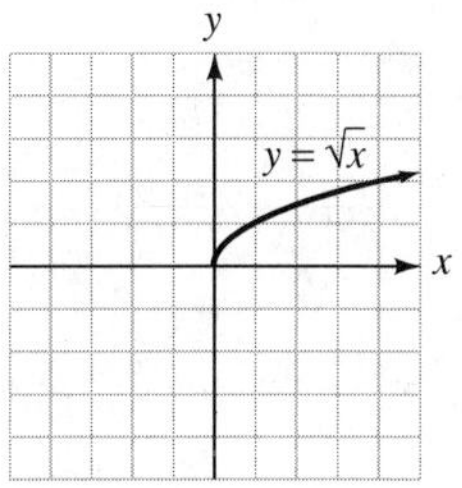

In Exercises 25–30, find the inverse of each set of ordered pairs (x, y) and tell whether the inverse is a function.

25. $\{(3, 2), (2, 1), (1, 0)\}$

26. $\{(4, 1), (5, 1), (6, 1), (7, 1)\}$

27. $\{(1, 2), (2, 3), (1, 3), (1, 5)\}$

28. $\{(-1, -1), (0, 0), (1, 1), (2, 2)\}$

29. $\{(1, 1), (2, 4), (3, 9), (4, 16)\}$

30. $\{(1, 1), (2, 1), (3, 1), (4, 1)\}$

In Exercises 31–38, find the inverse of each function and express it in the form $y = f^{-1}(x)$. Verify each result by showing that $(f \circ f^{-1})(x) = (f^{-1} \circ f)(x) = I(x)$.

31. $f(x) = 3x + 1$

32. $y + 1 = 5x$

33. $x + 4 = 5y$

34. $x = 3y + 1$

35. $f(x) = \dfrac{x - 4}{5}$

36. $f(x) = \dfrac{2x + 6}{3}$

37. $4x - 5y = 20$

38. $3x + 5y = 15$

In Exercises 39–46, find the inverse of each function. Then graph the function and its inverse on one coordinate system. Find the equation of the line of symmetry.

39. $y = 4x + 3$

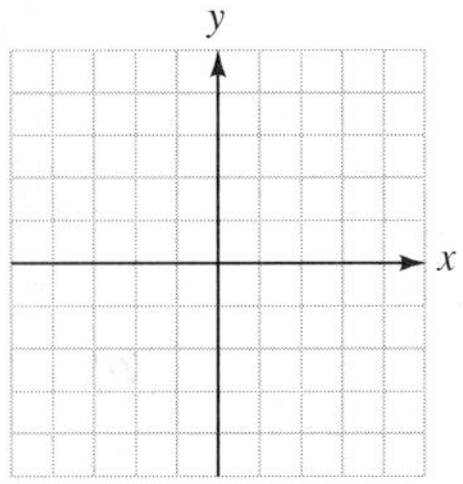

40. $x = 3y - 1$

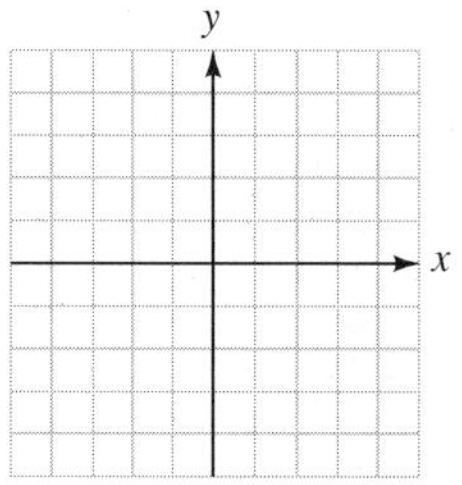

41. $x = \dfrac{y - 2}{3}$

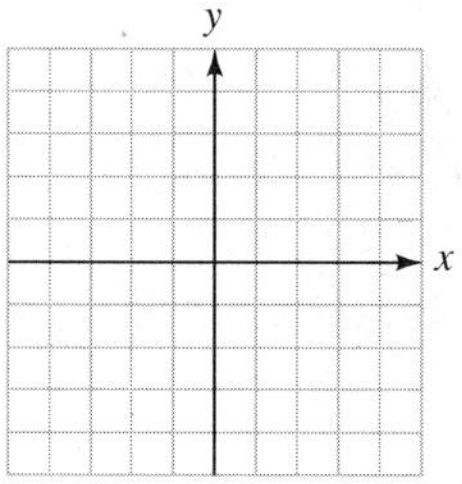

42. $y = \dfrac{x + 3}{4}$

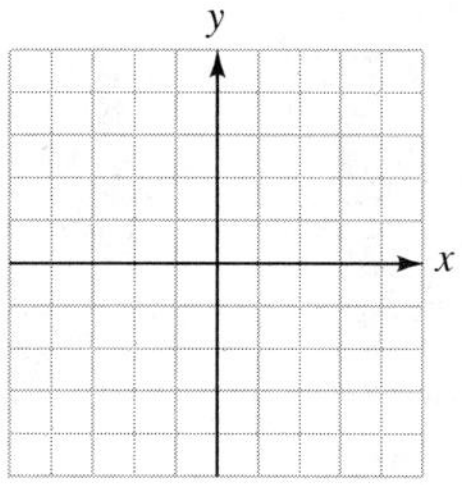

43. $3x - y = 5$

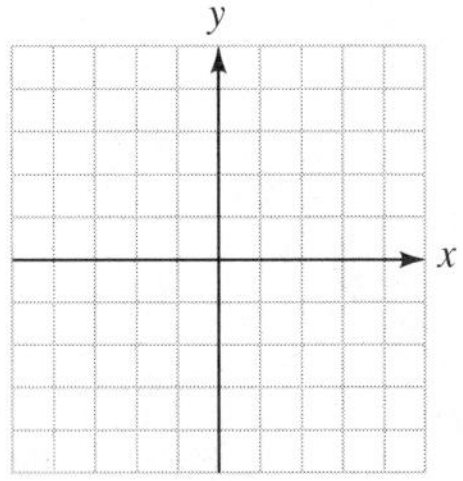

44. $2x + 3y = 9$

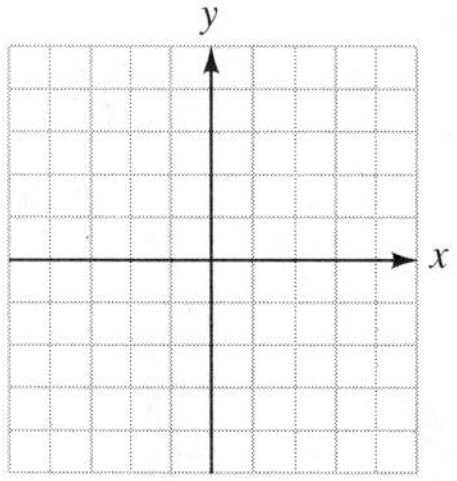

45. $3(x + y) = 2x + 4$

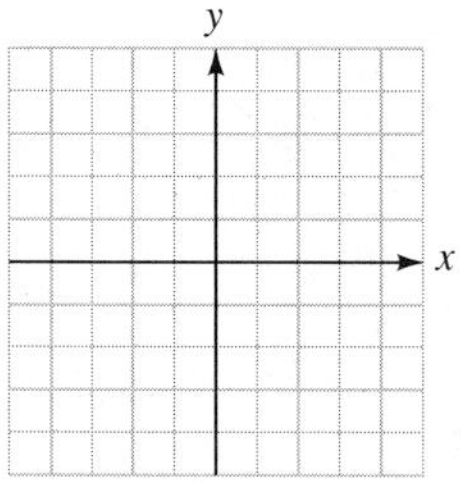

46. $-4(y - 1) + x = 2$

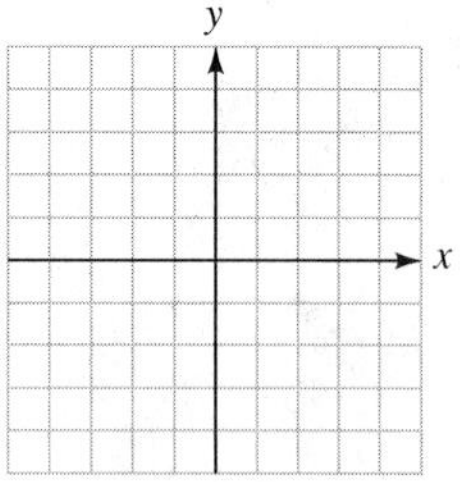

In Exercises 47–52, find the inverse of each function and tell whether it is a function.

47. $y = x^2 + 4$

48. $y = x^2 + 5$

49. $y = x^3$

50. $xy = 4$

51. $y = |x|$

52. $y = \sqrt[3]{x}$

In Exercises 53–54, show that the inverse of the function determined by each equation is also a function. Express it using $f^{-1}(x)$ notation.

53. $f(x) = 2x^3 - 3$

54. $f(x) = \dfrac{3}{x^3} - 1$

In Exercises 55–58, graph each equation and its inverse on one set of coordinate axes. Find the axis of symmetry.

55. $y = x^2 + 1$

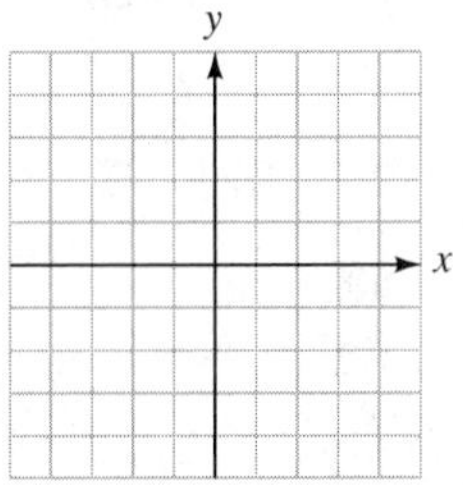

56. $y = \frac{1}{4}x^2 - 3$

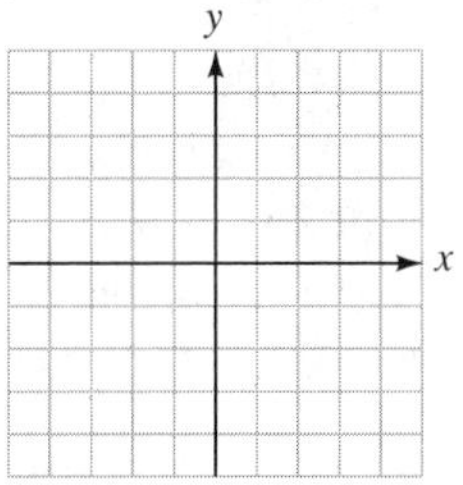

57. $y = \sqrt{x}$

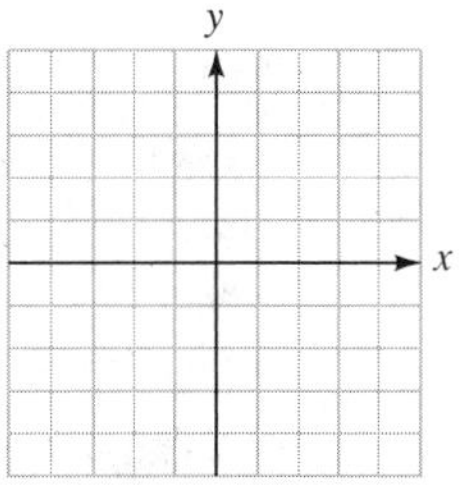

58. $y = |x|$

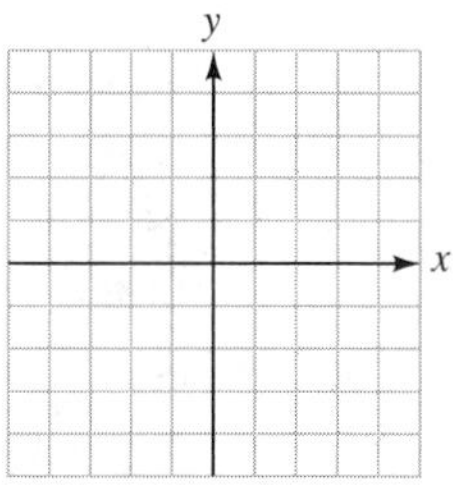

WRITING

59. Explain the purpose of the vertical line test.

60. Explain the purpose of the horizontal line test.

SOMETHING TO THINK ABOUT

61. Find the inverse of $y = \frac{x + 1}{x - 1}$.

62. Using the functions of Exercise 61, show that $(f \circ f^{-1})(x) = x$.

PROJECTS

PROJECT 1

Ballistics is the study of how projectiles fly. The general formula for the height above the ground of an object thrown straight up or straight down is given by the function

$$h(t) = -16t^2 + v_0t + h_0$$

where h is the object's height (in feet) above the ground t seconds after it is thrown. The initial velocity v_0 is the velocity with which the object is thrown, measured in feet per second. The initial height h_0 is the object's height (in feet) above the ground when it is thrown. (If $v_0 > 0$, the object is thrown upward; if $v_0 < 0$, the object is thrown downward.)

This formula takes into account the force of gravity, but it disregards the force of air resistance. It is much more accurate for a smooth, dense ball than for a crumpled piece of paper.

One of the most popular acts of the Bungling Brothers Circus is the Amazing Glendo and his cannonball catching act. A cannon fires a ball vertically into the air; Glendo, standing on a platform above the cannon, uses his catlike reflexes to catch the ball as it passes by on its way toward the roof of the big top. As the balls fly past, they are within Glendo's reach only during a two-foot interval of their upward path.

(continued)

PROJECTS *(continued)*

As an investigator for the company that insures the circus, you have been asked to find answers to the following questions. The answers will determine whether or not Bungling Brothers' insurance policy will be renewed.

a. In the first part of the act, cannonballs are fired from the end of a six-foot cannon with an initial velocity of 80 feet per second. Glendo catches one ball between 40 and 42 feet above the ground. Then he lowers his platform and catches another ball between 25 and 27 feet above the ground.

i. Show that if Glendo missed a cannonball, it would hit the roof of the 56-foot-tall big top. How long would it take for a ball to hit the big top? To prevent this from happening, a special net near the roof catches and holds any missed cannonballs.

ii. Find (to the nearest thousandth of a second) how long the cannonballs are within Glendo's reach for each of his catches. Which catch is easier? Why does your answer make sense? Your company is willing to insure against injuries to Glendo if he has at least 0.025 second to make each catch. Should the insurance be offered?

b. For Glendo's grand finale, the special net at the roof of the big top is removed, making Glendo's catch more significant to the people in the audience, who worry that if Glendo misses, the tent will collapse around them. To make it even more dramatic, Glendo's arms are tied to restrict his reach to a one-foot interval of the ball's flight, and he stands on a platform just under the peak of the big top, so that his catch is made at the very last instant (between 54 and 55 feet above the ground). For this part of the act, however, Glendo has the cannon charged with less gunpowder, so that the muzzle velocity of the cannon is 56 feet per second. Show work to prove that Glendo's big finale is in fact his *easiest* catch, and that even if he misses, the big top is never in any danger of collapsing, so insurance should be offered against injury to the audience.

PROJECT 2

The center of Sterlington is the intersection of Main Street (running east–west) and DueNorth Road (running north–south). The recreation area for the townspeople is Robin Park, a few blocks from there. The park is bounded on the south by Main Street and on every other side by Parabolic Boulevard, named for its distinctive shape. In fact, if Main Street and DueNorth Road were used as the axes of a rectangular coordinate system, Parabolic Boulevard would have the equation $y = -(x - 4)^2 + 5$, where each unit on the axes is 100 yards.

The city council has recently begun to consider whether or not to put two walkways through the park. (See Illustration 1.) The walkways would run from two points on Main Street and converge at the northernmost point of the park, dividing the area of the park exactly into thirds.

The city council is pleased with the esthetics of this arrangement but needs to know two important facts.

a. For planning purposes, they need to know exactly where on Main Street the walkways would begin.

b. In order to budget for the construction, they need to know how long the walkways will be.

Provide answers for the city council, along with explanations and work to show that your answers are correct. You will need to use the formula shown in Illustration 2, due to Archimedes (287–212 B.C.), for the area under a parabola but above a line perpendicular to the axis of symmetry of the parabola.

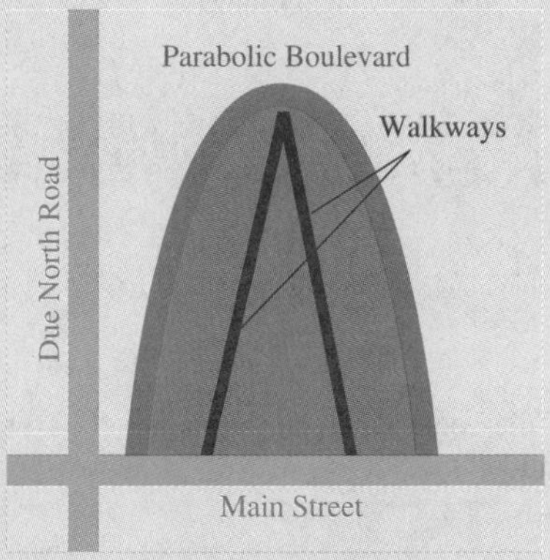

ILLUSTRATION 1

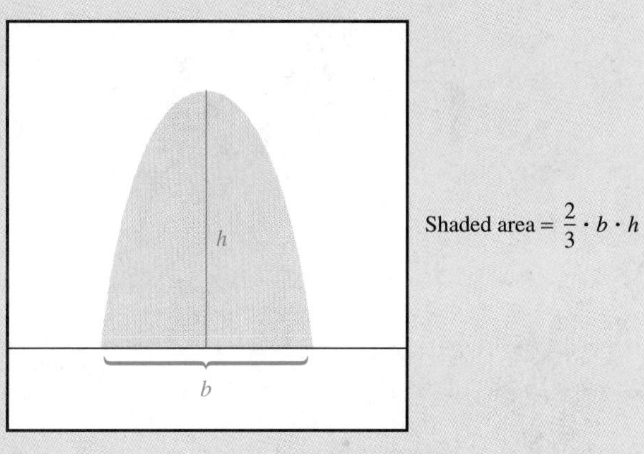

ILLUSTRATION 2

CHAPTER SUMMARY

CONCEPT | REVIEW EXERCISES

SECTION 8.1 Completing the Square and the Quadratic Formula

Square root property:
If $c > 0$, the equation $x^2 = c$ has two real solutions:

$$x = \sqrt{c} \quad \text{and} \quad x = -\sqrt{c}$$

1. Solve each equation by factoring or by using the square root property.

a. $12x^2 + x - 6 = 0$

b. $6x^2 + 17x + 5 = 0$

c. $15x^2 + 2x - 8 = 0$

d. $(x + 2)^2 = 36$

To complete the square, add the square of one-half of the coefficient of x.

2. Solve each equation by completing the square.

a. $x^2 + 6x + 8 = 0$

b. $2x^2 - 9x + 7 = 0$

Quadratic formula:

$$x = \frac{-b \pm \sqrt{b^2 - 4ac}}{2a}$$

$(a \neq 0)$

3. Solve each equation by using the quadratic formula.

a. $x^2 - 8x - 9 = 0$ **b.** $x^2 - 10x = 0$

c. $2x^2 + 13x - 7 = 0$ **d.** $3x^2 - 20x - 7 = 0$

4. Dimensions of a rectangle A rectangle is 2 centimeters longer than it is wide. If both the length and width are doubled, its area is increased by 72 square centimeters. Find the dimensions of the original rectangle.

5. Dimensions of a rectangle A rectangle is 1 foot longer than it is wide. If the length is tripled and the width is doubled, its area is increased by 30 square feet. Find the dimensions of the original rectangle.

6. Ballistics If a rocket is launched straight up into the air with an initial velocity of 112 feet per second, its height after t seconds is given by the formula

$$h = 112t - 16t^2$$

where h represents the height of the rocket in feet. After launch, how long will it be before it hits the ground?

7. Ballistics What is the maximum height of the rocket discussed in Exercise 6?

SECTION 8.2 *Graphs of Quadratic Functions*

If f is a function and k is a positive number, then

The graph of $y = f(x) + k$ is identical to the graph of $y = f(x)$, except that it is translated k units upward.

The graph of $y = f(x) - k$ is identical to the graph of $y = f(x)$, except that it is translated k units downward.

8. Graph each function and give the coordinates of the vertex of the resulting parabola.

a. $y = 2x^2 - 3$

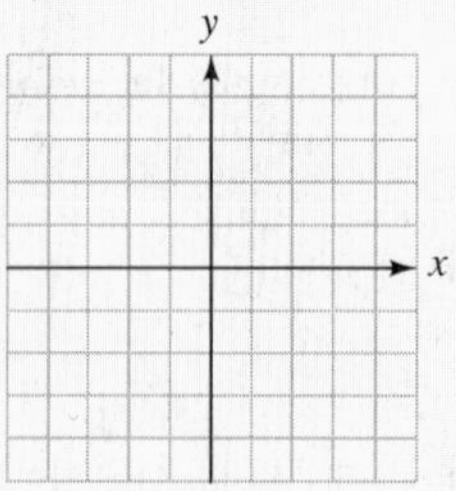

b. $y = -2x^2 - 1$

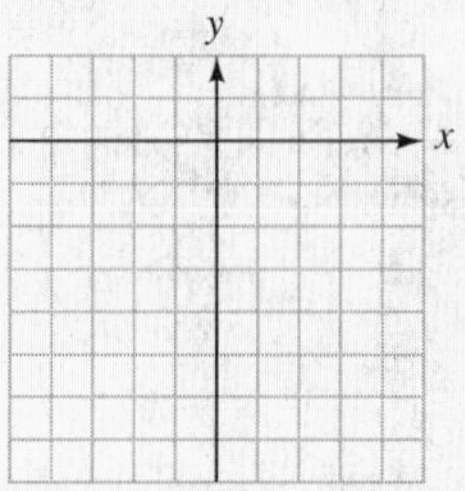

The graph of $y = f(x - h)$ is identical to the graph of $y = f(x)$, except that it is translated h units to the right.

The graph of $y = f(x + h)$ is identical to the graph of $y = f(x)$, except that it is translated h units to the left.

If $a \neq 0$, the graph of $y = a(x - h)^2 + k$ is a parabola with vertex at (h, k). It opens upward when $a > 0$ and downward when $a < 0$.

c. $y = -4(x - 2)^2 + 1$

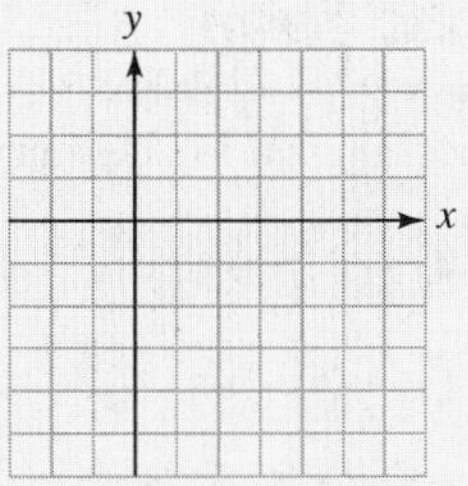

d. $y = 5x^2 + 10x - 1$

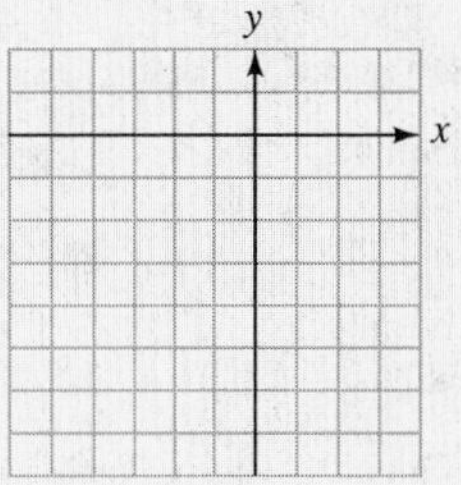

SECTION 8.3 Complex Numbers

Complex numbers: If a, b, c, and d are real numbers and $i^2 = -1$,

$a + bi = c + di$ if and only if $a = c$ and $b = d$

$(a + bi) + (c + di) = (a + c) + (b + d)i$

$(a + bi)(c + di) = (ac - bd) + (ad + bc)i$

$|a + bi| = \sqrt{a^2 + b^2}$

9. Do the operations and give all answers in $a + bi$ form.

a. $(5 + 4i) + (7 - 12i)$

b. $(-6 - 40i) - (-8 + 28i)$

c. $\left(-32 + \sqrt{-144}\right) - \left(64 + \sqrt{-81}\right)$

d. $\left(-8 + \sqrt{-8}\right) + \left(6 - \sqrt{-32}\right)$

e. $(2 - 7i)(-3 + 4i)$

f. $(-5 + 6i)(2 + i)$

g. $\left(5 - \sqrt{-27}\right)\left(-6 + \sqrt{-12}\right)$

h. $\left(2 + \sqrt{-128}\right)\left(3 - \sqrt{-98}\right)$

i. $\dfrac{3}{4i}$ **j.** $\dfrac{-2}{5i^3}$

k. $\dfrac{6}{2 + i}$ **l.** $\dfrac{7}{3 - i}$

m. $\dfrac{4 + i}{4 - i}$ **n.** $\dfrac{3 - i}{3 + i}$

o. $\dfrac{3}{5 + \sqrt{-4}}$ **p.** $\dfrac{2}{3 - \sqrt{-9}}$

q. $|9 + 12i|$ **r.** $|24 - 10i|$

SECTION 8.4 The Discriminant and Equations That Can Be Written in Quadratic Form

The discriminant:
If $b^2 - 4ac > 0$, the solutions of $ax^2 + bx + c = 0$ are unequal real numbers.

If $b^2 - 4ac = 0$, the solutions of $ax^2 + bx + c = 0$ are equal real numbers.

If $b^2 - 4ac < 0$, the solutions of $ax^2 + bx + c = 0$ are complex conjugates.

10. Use the discriminant to determine what type of solutions exist for each equation.

a. $3x^2 + 4x - 3 = 0$

b. $4x^2 - 5x + 7 = 0$

c. Find the values of k that will make the solutions of $(k - 8)x^2 + (k + 16)x = -49$ equal.

d. Find the values of k such that the solutions of $3x^2 + 4x = k + 1$ will be real numbers.

If r_1 and r_2 are solutions of $ax^2 + bx + c = 0$, then

$$r_1 + r_2 = -\frac{b}{a}$$

$$r_1 r_2 = \frac{c}{a}$$

11. Solve each equation.

a. $x - 13x^{1/2} + 12 = 0$

b. $a^{2/3} + a^{1/3} - 6 = 0$

c. $\dfrac{1}{x+1} - \dfrac{1}{x} = -\dfrac{1}{x+1}$

d. $\dfrac{6}{x+2} + \dfrac{6}{x+1} = 5$

12. Find the sum of the solutions of the equation $3x^2 - 14x + 3 = 0$.

13. Find the product of the solutions of the equation $3x^2 - 14x + 3 = 0$.

SECTION 8.5 Quadratic and Other Nonlinear Inequalities

To solve a quadratic inequality in one variable, make a sign chart.

14. Solve each inequality. Give each result in interval notation and graph the solution set.

a. $x^2 + 2x - 35 > 0$

b. $x^2 + 7x - 18 < 0$

To solve inequalities with rational expressions, get 0 on the right-hand side, add the fractions, and then factor the numerator and denominator. Then use a sign chart.

c. $\frac{3}{x} \leq 5$

d. $\frac{2x^2 - x - 28}{x - 1} > 0$

15. Use a graphing calculator to solve each inequality. Compare the results with Review Exercise 14.

a. $x^2 + 2x - 35 > 0$ **b.** $x^2 + 7x - 18 < 0$

c. $\frac{3}{x} \leq 5$ **d.** $\frac{2x^2 - x - 28}{x - 1} > 0$

To graph an inequality such as $y > 4x - 3$, first graph the equation $y = 4x - 3$. Then determine which half-plane represents the graph of $y > 4x - 3$.

16. Graph each inequality.

a. $y < \frac{1}{2}x^2 - 1$

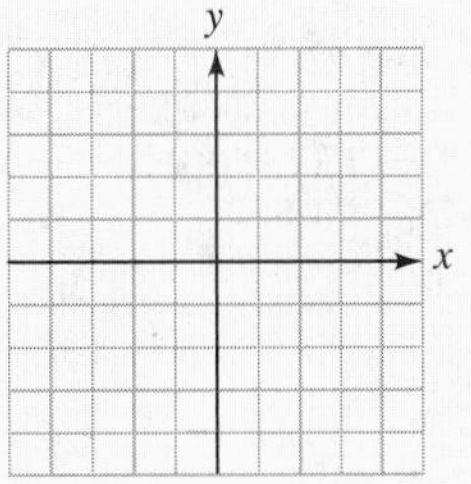

b. $y \geq -|x|$

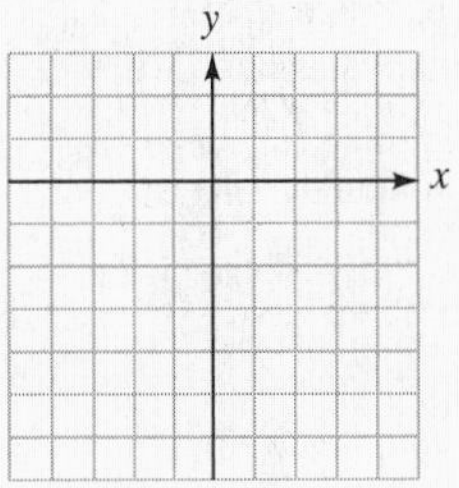

SECTION 8.6 Algebra and Composition of Functions

Operations with functions:

$(f + g)(x) = f(x) + g(x)$

$(f - g)(x) = f(x) - g(x)$

$(f \cdot g)(x) = f(x)g(x)$

$(f/g)(x) = \frac{f(x)}{g(x)} \quad (g(x) \neq 0)$

$(f \circ g)(x) = f(g(x))$

17. *Let* $f(x) = 2x$ *and* $g(x) = x + 1$. *Find each function or value.*

a. $f + g$ **b.** $f - g$

c. $f \cdot g$ **d.** f/g

e. $(f \circ g)(2)$ **f.** $(g \circ f)(-1)$

g. $(f \circ g)(x)$ **h.** $(g \circ f)(x)$

SECTION 8.7 Inverses of Functions

Horizontal line test:
If every horizontal line that intersects the graph of a function does so only once, the function is one-to-one.

18. Graph each function and use the horizontal line test to decide whether the function is one-to-one.

a. $f(x) = 2(x - 3)$

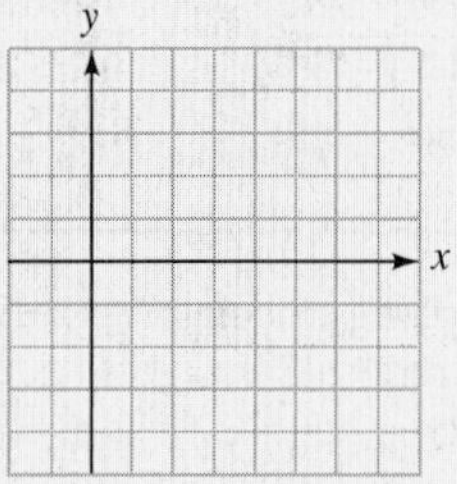

b. $f(x) = x(2x - 3)$

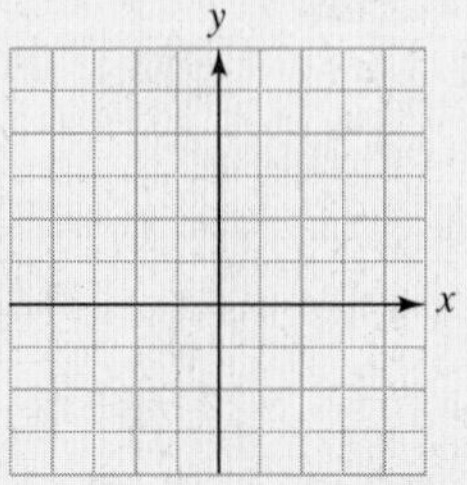

c. $f(x) = -3(x - 2)^2 + 5$

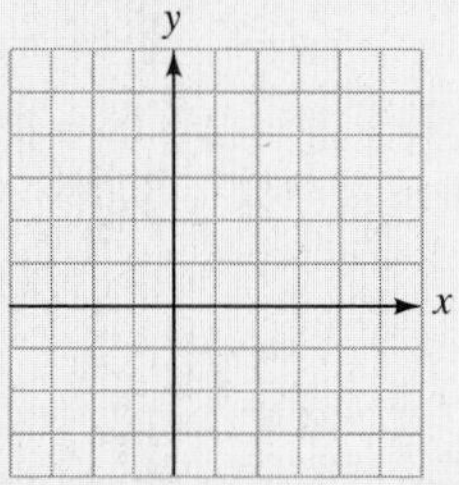

d. $f(x) = |x|$

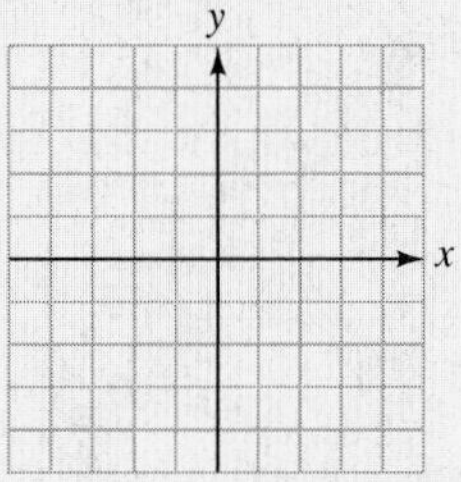

To find the inverse of a function, interchange the positions of variables x and y and solve for y.

19. Find the inverse of each function.

a. $f(x) = 6x - 3$

b. $f(x) = 4x + 5$

c. $y = 2x^2 - 1 \ (x \geq 0)$

d. $y = |x|$

Chapter Test

In Problems 1–2, solve each equation.

1. $x^2 + 3x - 18 = 0$

2. $x(6x + 19) = -15$

In Problems 3–4, determine what number must be added to each binomial to make it a perfect square.

3. $x^2 + 24x$

4. $x^2 - 50x$

In Problems 5–6, solve each equation.

5. $x^2 + 4x + 1 = 0$

6. $x^2 - 5x - 3 = 0$

In Problems 7–12, do the operations. Give all answers in a + bi form.

7. $(2 + 4i) + (-3 + 7i)$

8. $\left(3 - \sqrt{-9}\right) - \left(-1 + \sqrt{-16}\right)$

9. $2i(3 - 4i)$

10. $(3 + 2i)(-4 - i)$

11. $\dfrac{1}{i\sqrt{2}}$

12. $\dfrac{2 + i}{3 - i}$

13. Determine whether the solutions of $3x^2 + 5x + 17 = 0$ are real or nonreal numbers.

14. For what value(s) of k are the solutions of $4x^2 - 2kx + k - 1 = 0$ equal?

15. One leg of a right triangle is 14 inches longer than the other, and the hypotenuse is 26 inches. Find the length of the shorter leg.

16. Solve the equation $2y - 3y^{1/2} + 1 = 0$.

17. Graph the function $f(x) = \frac{1}{2}x^2 - 4$ and give the coordinates of its vertex.

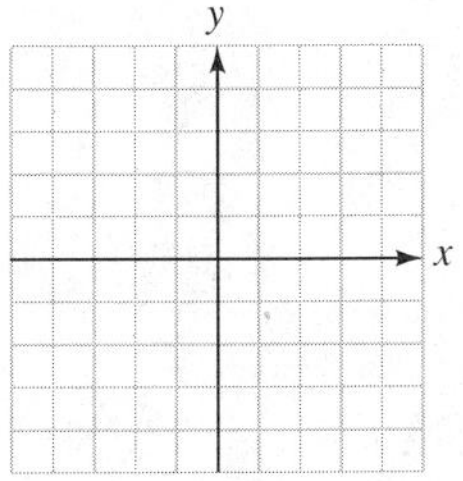

18. Graph the inequality $y \le -x^2 + 3$

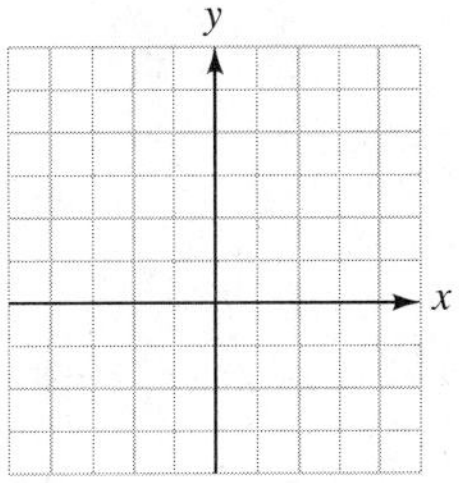

In Problems 19–20, solve each inequality and graph the solution set.

19. $x^2 - 2x - 8 > 0$

20. $\dfrac{x - 2}{x + 3} \le 0$

In Problems 21–24, $f(x) = 4x$ and $g(x) = x - 1$. Find each function.

21. $g + f$

22. $f - g$

23. $g \cdot f$

24. g/f

In Problems 25–28, $f(x) = 4x$ and $g(x) = x - 1$. Find each value.

25. $(g \circ f)(1)$

26. $(f \circ g)(0)$

27. $(f \circ g)(-1)$

28. $(g \circ f)(-2)$

In Problems 29–30, $f(x) = 4x$ and $g(x) = x - 1$. Find each function.

29. $(f \circ g)(x)$

30. $(g \circ f)(x)$

In Problems 31–32, find the inverse of each function.

31. $3x + 2y = 12$

32. $y = 3x^2 + 4 \ (x \le 0)$

Cumulative Review Exercises

In Exercises 1–2, find the domain and range of each function.

1. $y = f(x) = 2x^2 - 3$

2. $y = f(x) = -|x - 4|$

In Exercises 3–4, write the equation of the line with the given properties.

3. $m = 3$, passing through $(-2, -4)$

4. Parallel to the graph of $2x + 3y = 6$ and passing through $(0, -2)$

In Exercises 5–6, do each operation.

5. $(2a^2 + 4a - 7) - 2(3a^2 - 4a)$

6. $(3x + 2)(2x - 3)$

In Exercises 7–8, factor each expression.

7. $x^4 - 16y^4$

8. $15x^2 - 2x - 8$

In Exercises 9–10, solve each equation.

9. $x^2 - 5x - 6 = 0$

10. $6a^3 - 2a = a^2$

In Exercises 11–18, simplify each expression. Assume that all variables represent positive numbers.

11. $\sqrt{25x^4}$

12. $\sqrt{48t^3}$

13. $\sqrt[3]{-27x^3}$

14. $\sqrt[3]{\dfrac{128x^4}{2x}}$

15. $8^{-1/3}$

16. $64^{2/3}$

17. $\dfrac{y^{2/3}y^{5/3}}{y^{1/3}}$

18. $\dfrac{x^{5/3}x^{1/2}}{x^{3/4}}$

In Exercises 19–20, graph each function and give the domain and the range.

19. $f(x) = \sqrt{x - 2}$

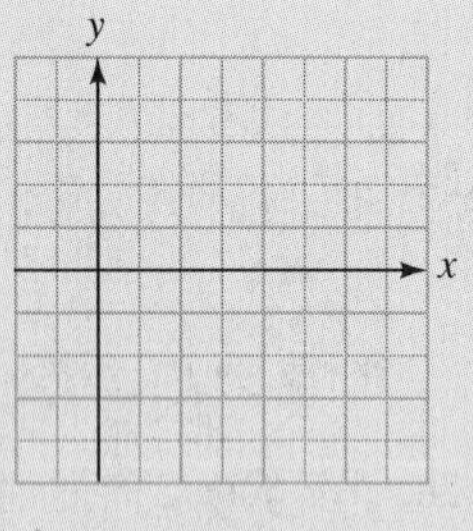

20. $f(x) = -\sqrt{x + 2}$

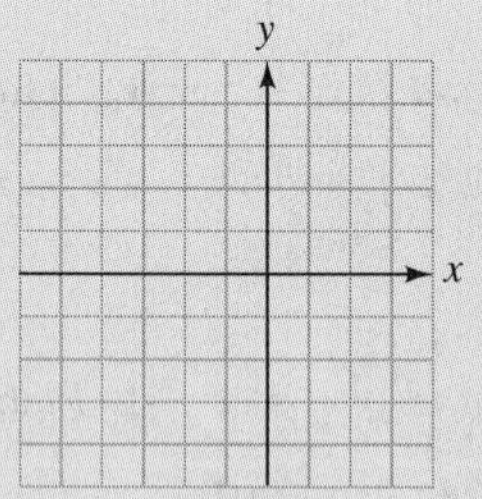

In Exercises 21–22, do the operations.

21. $(x^{2/3} - x^{1/3})(x^{2/3} + x^{1/3})$

22. $(x^{-1/2} + x^{1/2})^2$

In Exercises 23–28, simplify each statement.

23. $\sqrt{50} - \sqrt{8} + \sqrt{32}$

24. $-3\sqrt[4]{32} - 2\sqrt[4]{162} + 5\sqrt[4]{48}$

25. $3\sqrt{2}(2\sqrt{3} - 4\sqrt{12})$

26. $\dfrac{5}{\sqrt[3]{x}}$

27. $\dfrac{\sqrt{x} + 2}{\sqrt{x} - 1}$

28. $\sqrt[6]{x^3y^3}$

In Exercises 29–30, solve each equation.

29. $5\sqrt{x + 2} = x + 8$

30. $\sqrt{x} + \sqrt{x + 2} = 2$

31. Find the length of the hypotenuse of the right triangle shown in Illustration 1.

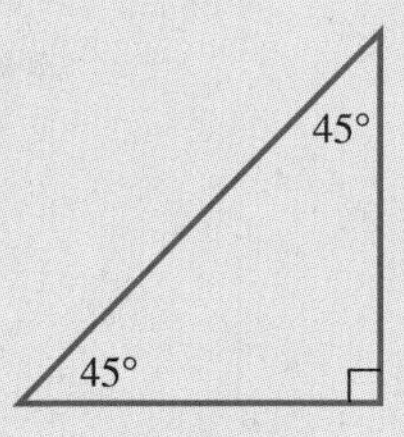

ILLUSTRATION 1

32. Find the length of the hypotenuse of the right triangle shown in Illustration 2.

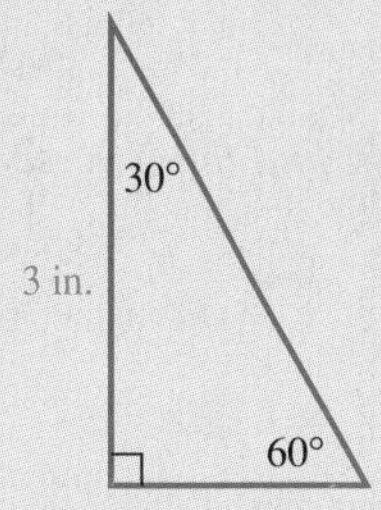

ILLUSTRATION 2

33. Find the distance between $P(-2, 6)$ and $Q(4, 14)$.

34. What number must be added to $x^2 + 6x$ to make a trinomial square?

35. Use the method of completing the square to solve $2x^2 + x - 3 = 0$.

36. Use the quadratic formula to solve $3x^2 + 4x - 1 = 0$.

37. Graph $y = f(x) = \frac{1}{2}x^2 + 5$ and find the coordinates of its vertex.

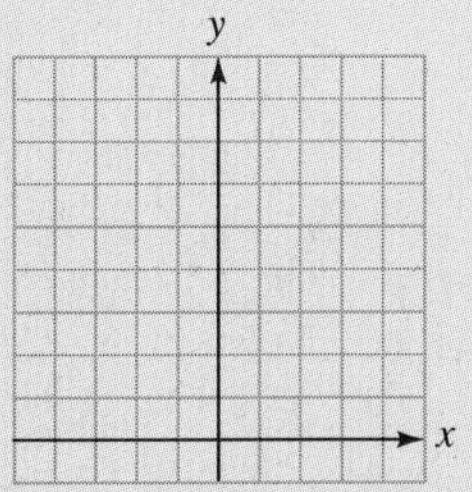

38. Graph $y \leq -x^2 + 3$ and find the coordinates of its vertex.

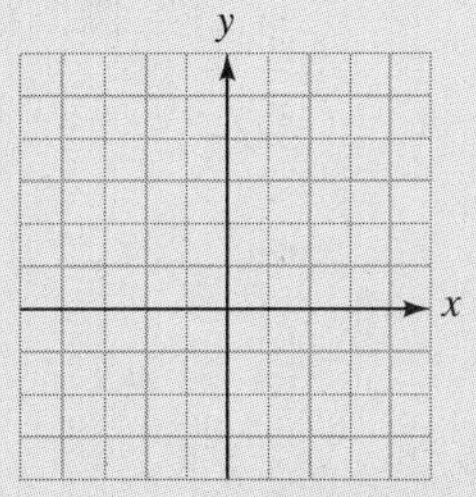

In Exercises 39–46, write each expression as a real number or as a complex number in a + bi form.

39. $(3 + 5i) + (4 - 3i)$

40. $(7 - 4i) - (12 + 3i)$

41. $(2 - 3i)(2 + 3i)$

42. $(3 + i)(3 - 3i)$

43. $(3 - 2i) - (4 + i)^2$

44. $\dfrac{5}{3 - i}$

45. $|3 + 2i|$

46. $|5 - 6i|$

47. For what values of k will the solutions of $2x^2 + 4x = k$ be equal?

48. Solve $a - 7a^{1/2} + 12 = 0$

In Exercises 49–50, solve each inequality and graph the solution set on the number line.

49. $x^2 - x - 6 > 0$

50. $x^2 - x - 6 \leq 0$

In Exercises 51–54, $f(x) = 3x^2 + 2$ and $g(x) = 2x - 1$. Find each value or composite function.

51. $f(-1)$

52. $(g \circ f)(2)$

53. $(f \circ g)(x)$

54. $(g \circ f)(x)$

In Exercises 55–56, find the inverse of each function.

55. $f(x) = 3x + 2$

56. $f(x) = x^3 + 4$

9 Exponential and Logarithmic Functions

MATHEMATICS IN THE WORKPLACE

Medical Laboratory Worker

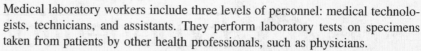

Medical laboratory workers include three levels of personnel: medical technologists, technicians, and assistants. They perform laboratory tests on specimens taken from patients by other health professionals, such as physicians.

Employment of medical laboratory workers is expected to expand faster than the average rate for all occupations through the year 2005, as physicians continue to use laboratory tests in routine physical checkups and in the diagnosis and treatment of disease.

SAMPLE APPLICATION ■ If a medium is inoculated with a bacterial culture that contains 1,000 cells per milliliter, how many generations will pass by the time the culture has grown to a population of 1 million cells per milliliter? (See Example 9 in Section 9.6.)

In this chapter, we will discuss two functions that are important in many applications of mathematics. **Exponential functions** are used to compute compound interest, find radioactive decay, and model population growth. **Logarithmic functions** are used to measure acidity of solutions, drug dosage, gain of an amplifier, intensity of earthquakes, and safe noise levels in factories.

9.1 Exponential Functions

■ IRRATIONAL EXPONENTS ■ EXPONENTIAL FUNCTIONS ■ GRAPHING EXPONENTIAL FUNCTIONS ■ VERTICAL AND HORIZONTAL TRANSLATIONS ■ COMPOUND INTEREST

Getting Ready

Find each value.

1. 2^3 **2.** $25^{1/2}$ **3.** 5^{-2} **4.** $\left(\frac{3}{2}\right)^{-3}$

The graph in Figure 9-1 shows the balance in a bank account in which \$10,000 was invested in 1998 at 9% annual interest, compounded monthly. The graph shows that in the year 2008, the value of the account will be approximately \$25,000, and in the year 2028, the value will be approximately \$147,000. The curve shown in Figure 9-1 is the graph of a function called an *exponential function.*

Before we can discuss exponential functions, we must define irrational exponents.

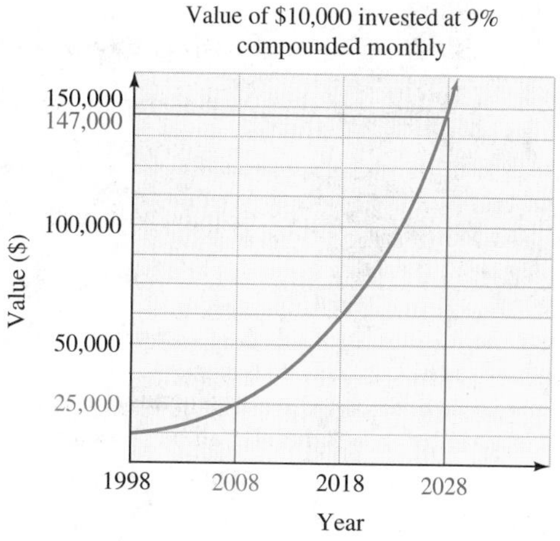

FIGURE 9-1

IRRATIONAL EXPONENTS

We have discussed expressions of the form b^x, where x is a rational number.

$8^{1/2}$ means "the square root of 8."

$5^{1/3}$ means "the cube root of 5."

$3^{-2/5} = \dfrac{1}{3^{2/5}}$ means "the reciprocal of the fifth root of 3^2."

To give meaning to b^x when x is an irrational number, we consider the expression

$5^{\sqrt{2}}$ where $\sqrt{2}$ is the irrational number 1.414213562 . . .

Each number in the following list is defined, because each exponent is a rational number.

$$5^{1.4}, 5^{1.41}, 5^{1.414}, 5^{1.4142}, 5^{1.41421}, \ldots$$

Since the exponents are getting closer to $\sqrt{2}$, the numbers in this list are successively better approximations of $5^{\sqrt{2}}$. We can use a calculator to obtain a very good approximation.

CALCULATORS

Evaluating Exponential Expressions

To find the value of $5^{\sqrt{2}}$ with a scientific calculator, we enter these numbers and press these keys:

5 y^x 2 $\sqrt{\ }$ =

The display will read 9.738517742 .

(continued)

Evaluating Exponential Expressions *(continued)*

With a graphing calculator, we enter these numbers and press these keys:

5 ^ √ 2 Enter

The display will read 5^√(2)
9.738517742

In general, if b is positive and x is a real number, b^x represents a positive number. It can be shown that all of the familiar rules of exponents hold true for irrational exponents.

EXAMPLE 1 Use the rules of exponents to simplify **a.** $(5^{\sqrt{2}})^{\sqrt{2}}$ and **b.** $b^{\sqrt{3}} \cdot b^{\sqrt{12}}$.

Solution **a.**

$$(5^{\sqrt{2}})^{\sqrt{2}} = 5^{\sqrt{2}\sqrt{2}} \quad \text{Keep the base and multiply the exponents.}$$
$$= 5^2 \quad \sqrt{2}\sqrt{2} = \sqrt{4} = 2.$$
$$= 25$$

b.

$$b^{\sqrt{3}} \cdot b^{\sqrt{12}} = b^{\sqrt{3}+\sqrt{12}} \quad \text{Keep the base and add the exponents.}$$
$$= b^{\sqrt{3}+2\sqrt{3}} \quad \sqrt{12} = \sqrt{4}\sqrt{3} = 2\sqrt{3}.$$
$$= b^{3\sqrt{3}} \quad \sqrt{3} + 2\sqrt{3} = 3\sqrt{3}.$$

■

Self Check Simplify **a.** $(3^{\sqrt{2}})^{\sqrt{8}}$ and **b.** $b^{\sqrt{2}} \cdot b^{\sqrt{18}}$.

Answers **a.** 81, **b.** $b^{4\sqrt{2}}$

EXPONENTIAL FUNCTIONS

If $b > 0$ and $b \neq 1$, the function $y = f(x) = b^x$ is an **exponential function.** Since x can be any real number, its domain is the set of real numbers. This is the interval $(-\infty, \infty)$. Since b is positive, the value of $f(x)$ is positive, and the range is the set of positive numbers. This is the interval $(0, \infty)$.

Since $b \neq 1$, an exponential function cannot be the constant function $y = f(x) = 1^x$, in which $f(x) = 1$ for every real number x.

Exponential Functions

An **exponential function with base b** is defined by the equation

$$y = f(x) = b^x \quad (b > 0,\ b \neq 1, \text{ and } x \text{ is a real number})$$

The **domain of any exponential function** is the interval $(-\infty, \infty)$. The **range** is the interval $(0, \infty)$.

GRAPHING EXPONENTIAL FUNCTIONS

Since the domain and range of $y = f(x) = b^x$ are subsets of real numbers, we can graph exponential functions on a rectangular coordinate system.

EXAMPLE 2 Graph $y = f(x) = 2^x$.

Solution To graph $y = f(x) = 2^x$, we find several points (x, y) whose coordinates satisfy the equation, plot the points, and join them with a smooth curve, as shown in Figure 9-2.

$f(x) = 2^x$

x	$f(x)$	$(x, f(x))$
-1	$\frac{1}{2}$	$\left(-1, \frac{1}{2}\right)$
0	1	(0, 1)
1	2	(1, 2)
2	4	(2, 4)
3	8	(3, 8)

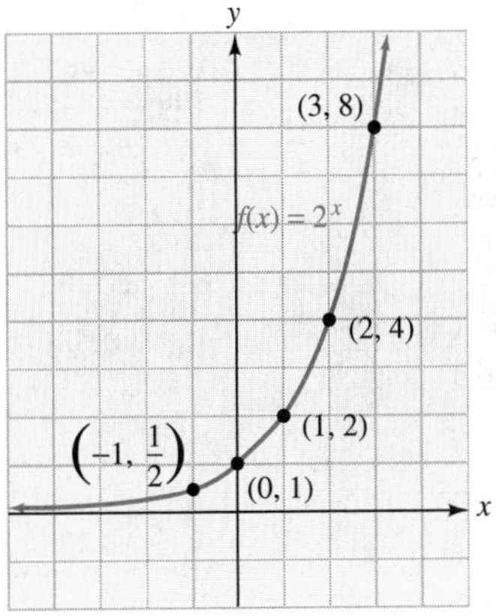

FIGURE 9-2

By looking at the graph, we can verify that the domain is the interval $(-\infty, \infty)$ and that the range is the interval $(0, \infty)$.

Note that as x decreases, the values of $f(x)$ decrease and approach 0. Thus, the x-axis is a horizontal asymptote of the graph.

Also note that the graph of $f(x) = 2^x$ passes through the points (0, 1) and (1, 2). ■

Self Check Graph $y = f(x) = 4^x$.

Answer

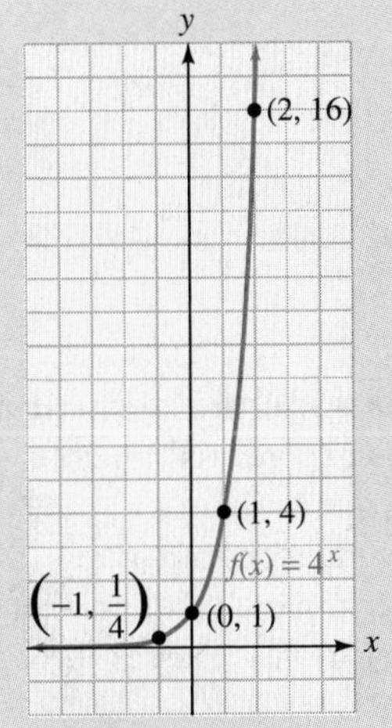

EXAMPLE 3 Graph $y = f(x) = \left(\frac{1}{2}\right)^x$.

Solution We find and plot pairs (x, y) that satisfy the equation. The graph of $y = f(x) = \left(\frac{1}{2}\right)^x$ appears in Figure 9-3.

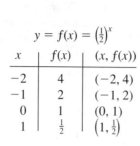

$y = f(x) = \left(\frac{1}{2}\right)^x$

x	$f(x)$	$(x, f(x))$
-2	4	$(-2, 4)$
-1	2	$(-1, 2)$
0	1	$(0, 1)$
1	$\frac{1}{2}$	$\left(1, \frac{1}{2}\right)$

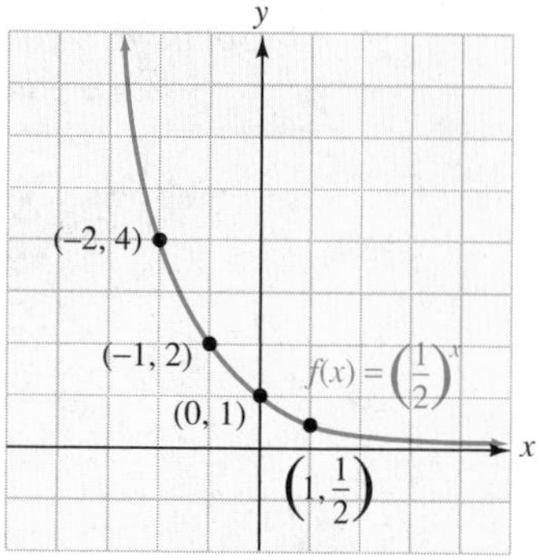

FIGURE 9-3

By looking at the graph, we can verify that the domain is the interval $(-\infty, \infty)$ and that the range is the interval $(0, \infty)$.

In this case, as x increases, the values of $f(x)$ decrease and approach 0. The x-axis is a horizontal asymptote. Note that the graph of $f(x) = \left(\frac{1}{2}\right)^x$ passes through the points $(0, 1)$ and $\left(1, \frac{1}{2}\right)$. ■

Self Check Graph $y = f(x) = \left(\frac{1}{4}\right)^x$.

Answer

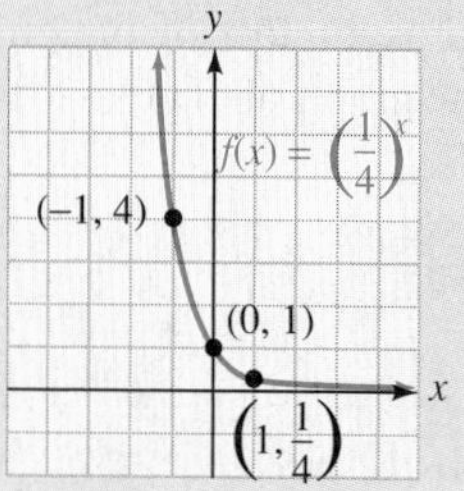

Examples 2 and 3 illustrate the following general properties of exponential functions.

Properties of Exponential Functions

The **domain** of the exponential function $y = f(x) = b^x$ is the interval $(-\infty, \infty)$.
The **range** is the interval $(0, \infty)$.
The graph has a y-intercept of $(0, 1)$.
The x-axis is an asymptote of the graph.
The graph of $y = f(x) = b^x$ passes through the point $(1, b)$.

EXAMPLE 4 From the graph of $y = f(x) = b^x$ shown in Figure 9-4, find the value of b.

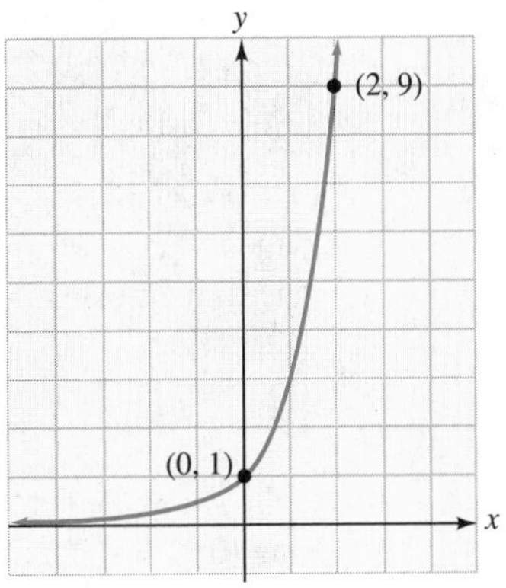

FIGURE 9-4

Solution We first note that the graph passes through (0, 1). Since the point (2, 9) is on the graph, we substitute 9 for y and 2 for x in the equation $y = b^x$ to get

$$y = b^x$$
$$9 = b^2$$
$$3 = b \quad \text{Take the positive square root of both sides.}$$

The base b is 3. ■

Self Check Is this the graph of an exponential function?

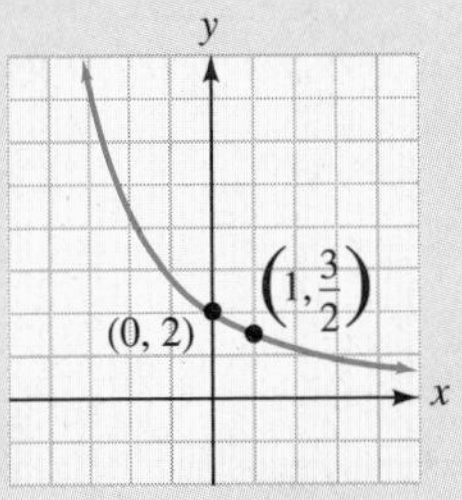

Answer No; it doesn't pass through the point (0, 1).

In Example 2 (where $b = 2$), the values of y increase as the values of x increase. Since the graph rises as we move to the right, we call the function an *increasing function.* When $b > 1$, the larger the value of b, the steeper the curve.

In Example 3 $\left(\text{where } b = \frac{1}{2}\right)$, the values of y decrease as the values of x increase. Since the graph drops as we move to the right, we call the function a *decreasing function.* When $0 < b < 1$, the smaller the value of b, the steeper the curve.

In general, the following is true.

Increasing and Decreasing Functions

If $b > 1$, then $f(x) = b^x$ is an **increasing function.**

If $0 < b < 1$, then $f(x) = b^x$ is a **decreasing function.**

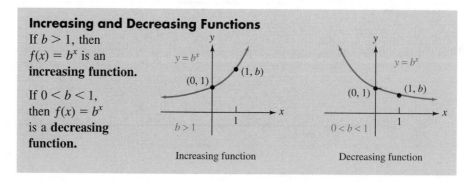

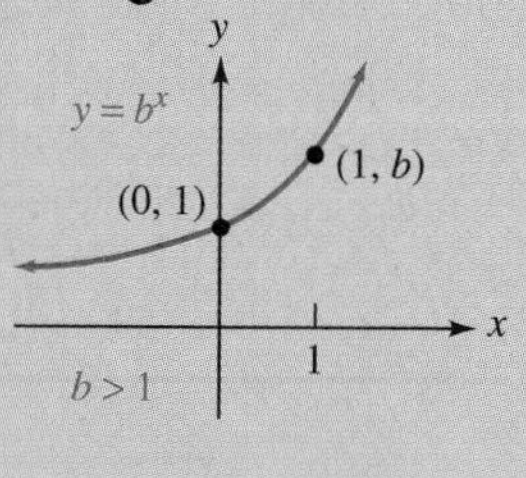

Increasing function

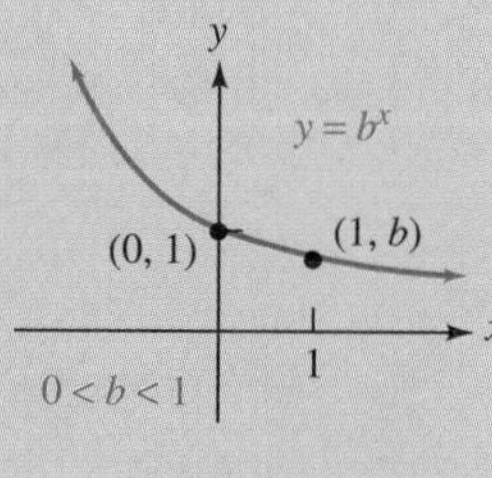

Decreasing function

An exponential function with base b is either increasing (for $b > 1$) or decreasing ($0 < b < 1$). Since different real numbers x determine different values of b^x, exponential functions are one-to-one.

The exponential function defined by

$$y = f(x) = b^x \quad \text{where } b > 0 \text{ and } b \neq 1$$

is one-to-one. Thus,

1. If $b^r = b^s$, then $r = s$.

2. If $r \neq s$, then $b^r \neq b^s$.

Graphing Exponential Functions

GRAPHING CALCULATORS

To use a graphing calculator to graph $y = f(x) = \left(\frac{2}{3}\right)^x$ and $y = f(x) = \left(\frac{3}{2}\right)^x$, we enter the right-hand sides of the equations. The screen will show the following equations.

$\backslash Y_1 = (2/3) \wedge X$

$\backslash Y_2 = (3/2) \wedge X$

If we use window settings of $[-10, 10]$ for x and $[-2, 10]$ for y and press the GRAPH key, we will obtain the graph shown in Figure 9-5.

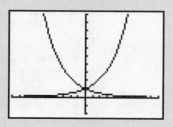

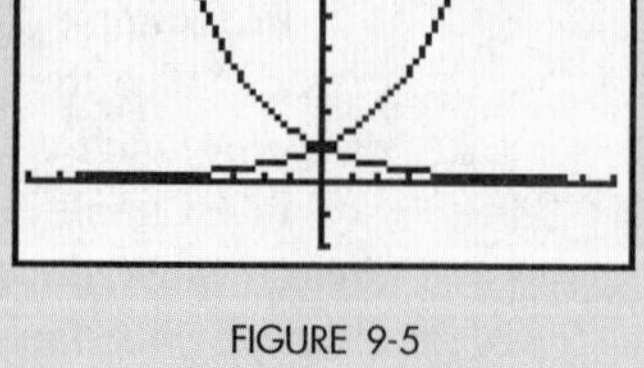

FIGURE 9-5

The graph of $y = f(x) = \left(\frac{2}{3}\right)^x$ passes through the points (0, 1) and $\left(1, \frac{2}{3}\right)$. Since $\frac{2}{3} < 1$, the function is decreasing.

The graph of $y = f(x) = \left(\frac{3}{2}\right)^x$ passes through the points (0, 1) and $\left(1, \frac{3}{2}\right)$. Since $\frac{3}{2} > 1$, the function is increasing.

Since both graphs pass the horizontal line test, each function is one-to-one.

VERTICAL AND HORIZONTAL TRANSLATIONS

We have seen that when $k > 0$ the graph of

$y = f(x) + k$ is the graph of $y = f(x)$ translated k units upward.

$y = f(x) - k$ is the graph of $y = f(x)$ translated k units downward.

$y = f(x - k)$ is the graph of $y = f(x)$ translated k units to the right.

$y = f(x + k)$ is the graph of $y = f(x)$ translated k units to the left.

EXAMPLE 5 On one set of axes, graph $y = f(x) = 2^x$ and $y = f(x) = 2^x + 3$.

Solution The graph of $y = f(x) = 2^x + 3$ is identical to the graph of $y = f(x) = 2^x$, except that it is translated 3 units upward. (See Figure 9-6.)

$y = f(x) = 2^x$

x	y	(x, y)
-4	$\frac{1}{16}$	$\left(-4, \frac{1}{16}\right)$
0	1	(0, 1)
2	4	(2, 4)

$y = f(x) = 2^x + 3$

x	y	(x, y)
-4	$3\frac{1}{16}$	$\left(-4, 3\frac{1}{16}\right)$
0	4	(0, 4)
2	7	(2, 7)

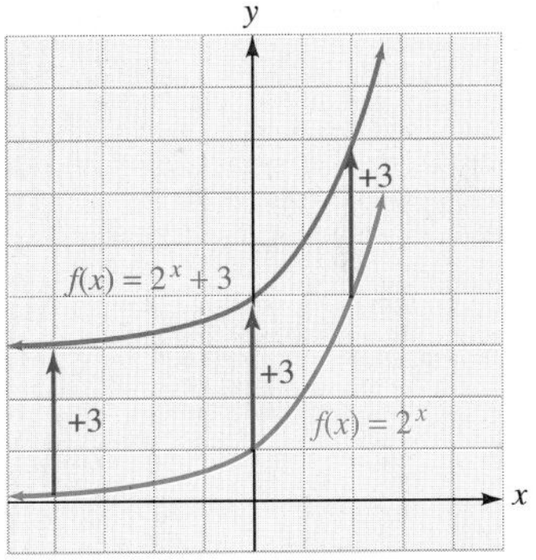

FIGURE 9-6

Self Check Graph $y = f(x) = 4^x$ and $y = f(x) = 4^x - 3$.

Answer

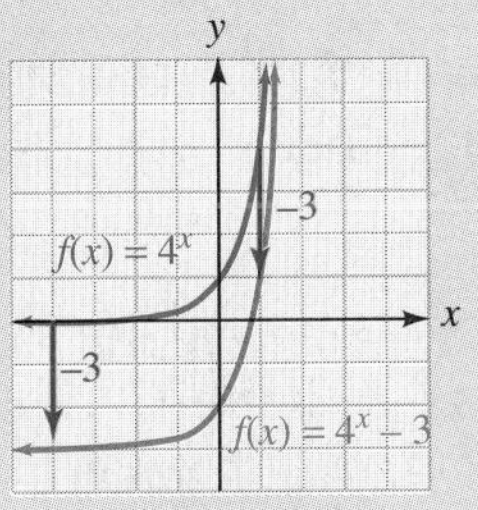

EXAMPLE 6 On one set of axes, graph $y = f(x) = 2^x$ and $y = f(x) = 2^{x+3}$.

Solution The graph of $y = 2^{x+3}$ is identical to the graph of $y = f(x) = 2^x$, except that it is translated 3 units to the left. (See Figure 9-7.)

$y = f(x) = 2^x$

x	y	(x, y)
0	1	(0, 1)
2	4	(2, 4)
4	16	(4, 16)

$y = f(x) = 2^{x+3}$

x	y	(x, y)
0	8	(0, 8)
−1	4	(−1, 4)
−2	2	(−2, 2)

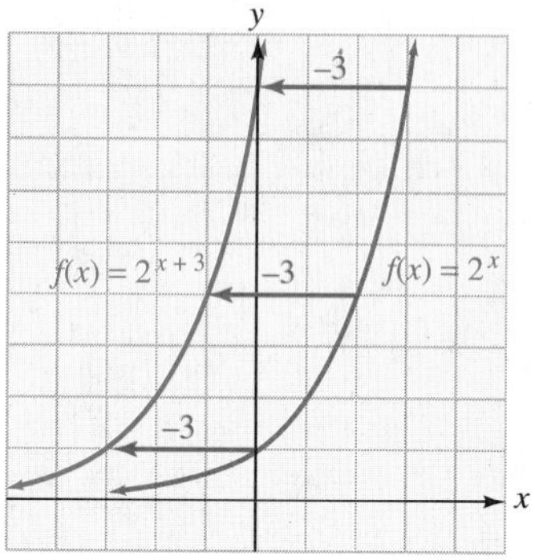

FIGURE 9-7

Self Check On one set of axes, graph $y = f(x) = 4^x$ and $y = f(x) = 4^{x-3}$.

Answer

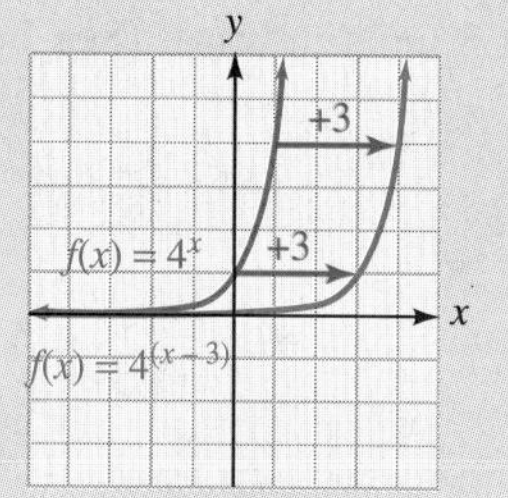

The graphs of $y = f(x) = kb^x$ and $y = f(x) = b^{kx}$ are vertical and horizontal stretchings of the graph of $y = f(x) = b^x$. To graph these functions, we can plot several points and join them with a smooth curve or use a graphing calculator.

Graphing Exponential Functions

GRAPHING CALCULATORS

To use a graphing calculator to graph the exponential function $y = f(x) = 3(2^{x/3})$, we enter the right-hand side of the equation. The display will show the equation

$$Y_1 = 3(2 \wedge (X/3))$$

If we use window settings of [−10, 10] for x and [−2, 18] for y and press the GRAPH key, we will obtain the graph shown in Figure 9-8.

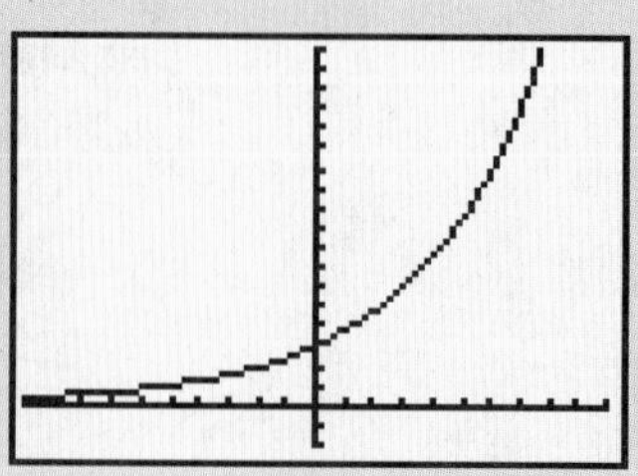

FIGURE 9-8

COMPOUND INTEREST

If we deposit \$$P$ in an account paying an annual simple interest rate r, we can find the amount A in the account at the end of t years by using the formula $A = P + Prt$, or $A = P(1 + rt)$.

Suppose that we deposit \$500 in an account that pays interest every six months. Then $P = 500$, and after six months ($\frac{1}{2}$ year), the amount in the account will be

$$\begin{aligned} A &= 500(1 + rt) \\ &= 500\left(1 + r \cdot \frac{1}{2}\right) && \text{Substitute } \tfrac{1}{2} \text{ for } t. \\ &= 500\left(1 + \frac{r}{2}\right) \end{aligned}$$

The account will begin the second six-month period with a value of \$$500(1 + \frac{r}{2})$. After the second six-month period, the amount will be

$$\begin{aligned} A &= P(1 + rt) \\ A &= \left[500\left(1 + \frac{r}{2}\right)\right]\left(1 + r \cdot \frac{1}{2}\right) && \text{Substitute } 500\left(1 + \tfrac{r}{2}\right) \text{ for } P \text{ and } \tfrac{1}{2} \text{ for } t. \\ &= 500\left(1 + \frac{r}{2}\right)\left(1 + \frac{r}{2}\right) \\ &= 500\left(1 + \frac{r}{2}\right)^2 \end{aligned}$$

At the end of a third six-month period, the amount in the account will be

$$A = 500\left(1 + \frac{r}{2}\right)^3$$

In this discussion, the earned interest is deposited back in the account and also earns interest. When this is the case, we say that the account is earning **compound interest.**

Formula for Compound Interest

If \$$P$ is deposited in an account and interest is paid k times a year at an annual rate r, the amount A in the account after t years is given by

$$A = P\left(1 + \frac{r}{k}\right)^{kt}$$

EXAMPLE 7

Saving for college To save for college, parents invest \$12,000 for their newborn child in a mutual fund that should average a 10% annual return. If the quarterly interest is reinvested, how much will be available in 18 years?

Solution We substitute 12,000 for P, 0.10 for r, and 18 for t into the formula for compound interest and find A. Since interest is paid quarterly, $k = 4$.

$$A = P\left(1 + \frac{r}{k}\right)^{kt}$$

$$A = \mathbf{12{,}000}\left(1 + \frac{\mathbf{0.10}}{\mathbf{4}}\right)^{\mathbf{4(18)}}$$

$$= 12{,}000(1 + 0.025)^{72}$$

$$= 12{,}000(1.025)^{72}$$

$$= 71{,}006.74$$

Use a calculator, and press these keys: 1.025 y^x 72 = × 12,000 = .

In 18 years, the account will be worth $71,006.74. ■

Self Check In Example 7, how much would be available if the parents invest $20,000?

Answer $118,344.56

In business applications, the initial amount of money deposited is often called the **present value** (PV). The amount to which the money will grow is called the **future value** (FV). The interest rate used for each compounding period is the **periodic interest rate** (i), and the number of times interest is compounded is the number of **compounding periods** (n). Using these definitions, we have an alternate formula for compound interest.

Formula for Compound Interest

$$FV = PV(1 + i)^n$$

This alternate formula appears on business calculators. To use this formula to solve Example 7, we proceed as follows:

$$FV = PV(1 + i)^n$$

$$FV = \mathbf{12{,}000}(1 + \mathbf{0.025})^{\mathbf{72}}$$

$$= 71{,}006.74$$

$i = \frac{0.10}{4} = 0.025$ and $n = 4(18) = 72$.

Solving Investment Problems

GRAPHING CALCULATORS

Suppose $1 is deposited in an account earning 6% annual interest, compounded monthly. To use a graphing calculator to estimate how much money will be in the account in 100 years, we can substitute 1 for P, 0.06 for r, and 12 for k into the formula

$$A = P\left(1 + \frac{r}{k}\right)^{kt}$$

$$A = \mathbf{1}\left(1 + \frac{\mathbf{0.06}}{\mathbf{12}}\right)^{\mathbf{12}t}$$

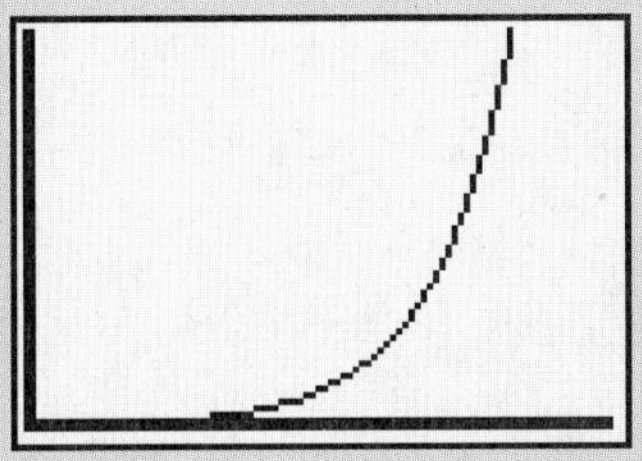

FIGURE 9-9

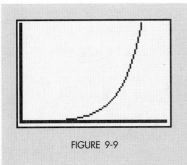

and simplify to get

$$A = (1.005)^{12t}$$

We now graph $A = (1.005)^{12t}$ using window settings of [0, 120] for t and [0, 400] for A to obtain the graph shown in Figure 9-9. We can then trace and zoom to estimate that \$1 grows to be approximately \$397 in 100 years. From the graph, we can see that the money grows slowly in the early years and rapidly in the later years.

Orals *If $x = 2$, evaluate each expression.*

1. 2^x **2.** 5^x **3.** $2(3^x)$ **4.** 3^{x-1}

If $x = -2$, evaluate each expression.

5. 2^x **6.** 5^x **7.** $2(3^x)$ **8.** 3^{x-1}

EXERCISE 9.1

REVIEW *In Illustration 1, lines r and s are parallel.*

1. Find x.

2. Find the measure of $\angle 1$.

3. Find the measure of $\angle 2$.

4. Find the measure of $\angle 3$.

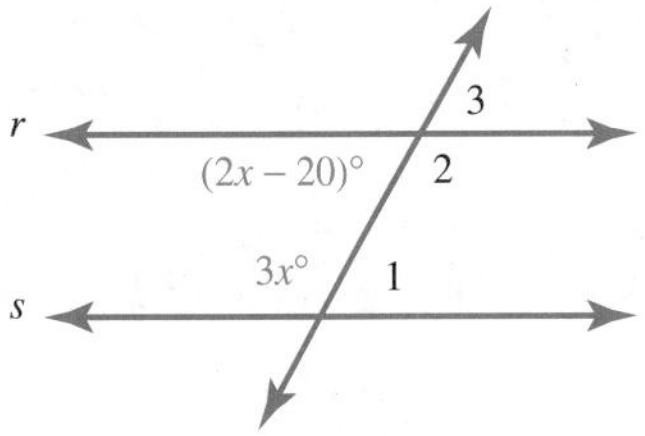

ILLUSTRATION 1

VOCABULARY AND CONCEPTS *Fill in each blank to make a true statement.*

5. If $b > 0$ and $b \neq 1$, $y = f(x) = b^x$ is called an __________ function.

6. The ______ of an exponential function is $(-\infty, \infty)$.

7. The range of an exponential function is the interval ______

8. The graph of $y = f(x) = 3^x$ passes through the points (0, __) and (1, __).

9. If $b > 1$, then $y = f(x)= b^x$ is an ________ function.

10. If $0 < b < 1$, then $y = f(x) = b^x$ is a ________ function.

11. The formula for compound interest is $A =$ ________.

12. An alternate formula for compound interest is $FV =$ ________.

PRACTICE *In Exercises 13–16, find each value to four decimal places.*

13. $2^{\sqrt{2}}$ **14.** $7^{\sqrt{2}}$ **15.** $5^{\sqrt{5}}$ **16.** $6^{\sqrt{3}}$

In Exercises 17–20, simplify each expression.

17. $\left(2^{\sqrt{3}}\right)^{\sqrt{3}}$ **18.** $3^{\sqrt{2}}3^{\sqrt{18}}$ **19.** $7^{\sqrt{3}}7^{\sqrt{12}}$ **20.** $\left(3^{\sqrt{5}}\right)^{\sqrt{5}}$

In Exercises 21–28, graph each exponential function. Check your work with a graphing calculator.

21. $y = f(x) = 3^x$

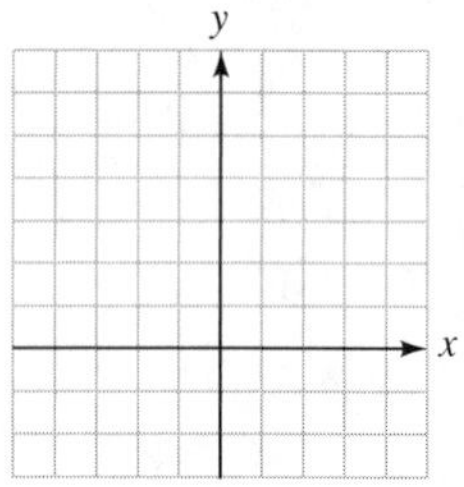

22. $y = f(x) = 5^x$

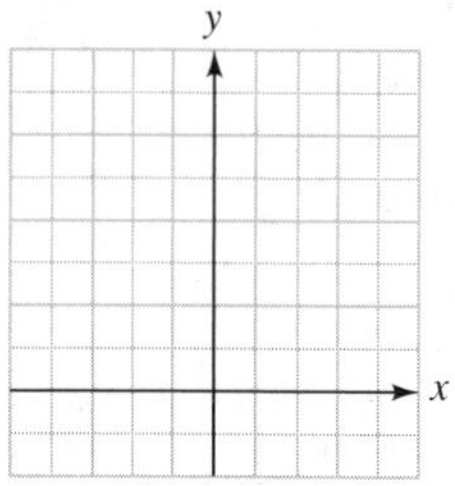

23. $y = f(x) = \left(\frac{1}{3}\right)^x$

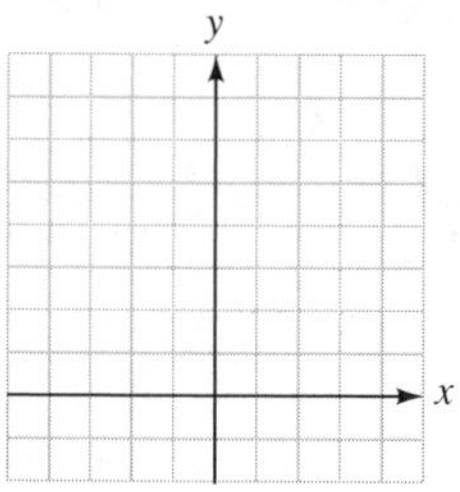

24. $y = f(x) = \left(\frac{1}{5}\right)^x$

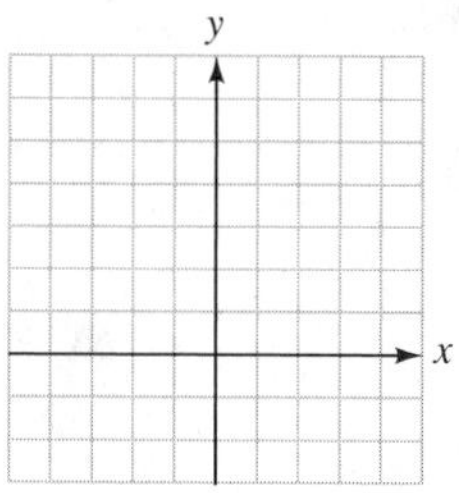

25. $y = f(x) = 3^x - 2$

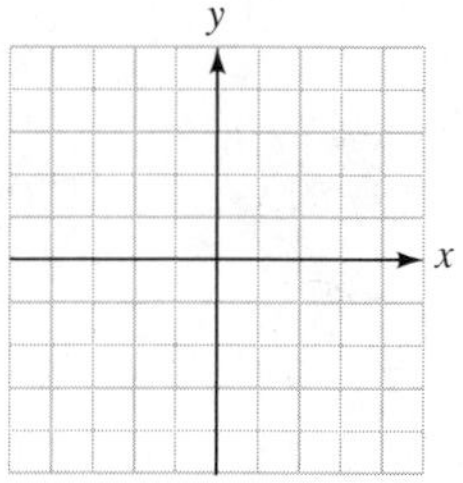

26. $y = f(x) = 2^x + 1$

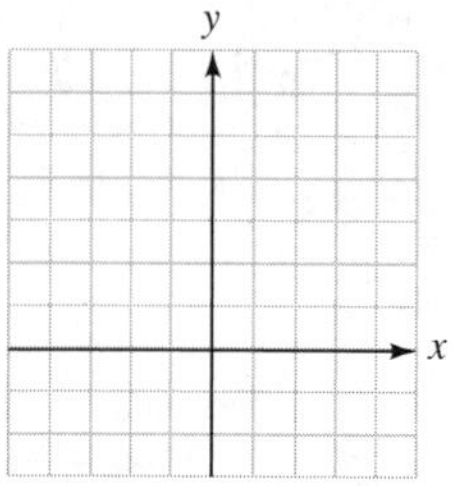

27. $y = f(x) = 3^{x-1}$

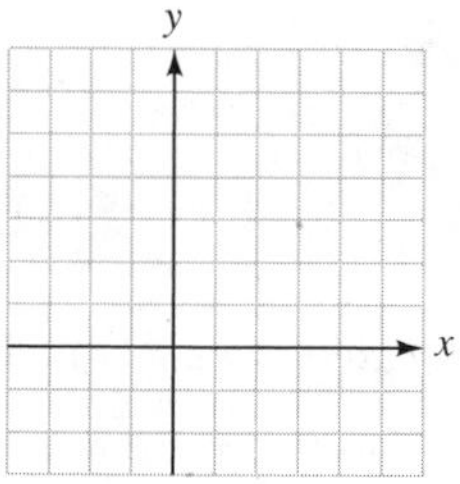

28. $y = f(x) = 2^{x+1}$

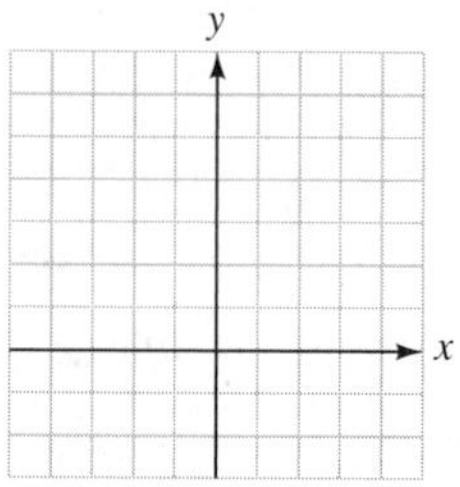

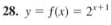

In Exercises 29–36, find the value of b, if any, that would cause the graph of $y = b^x$ to look like the graph indicated.

29.

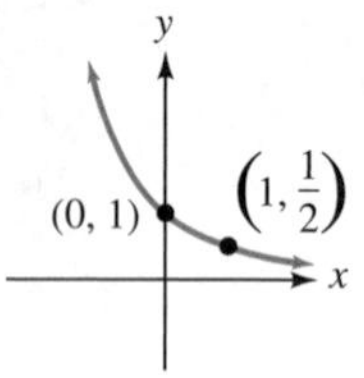

30.

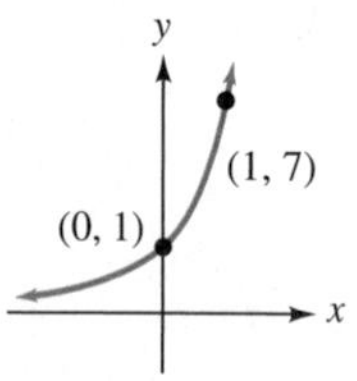

31.

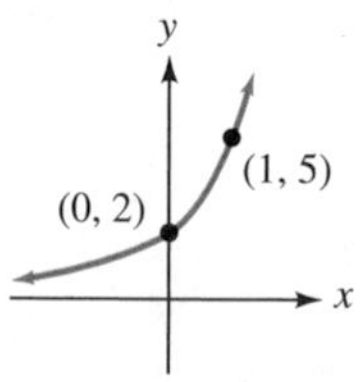

32.

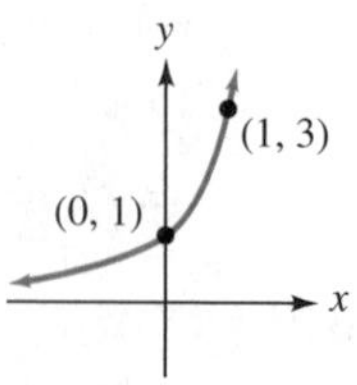

33.

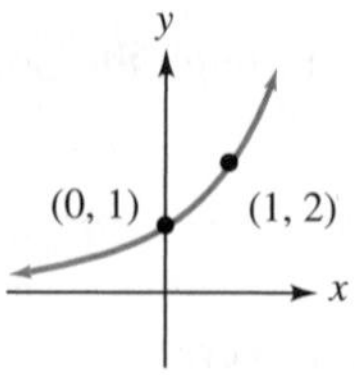

34.

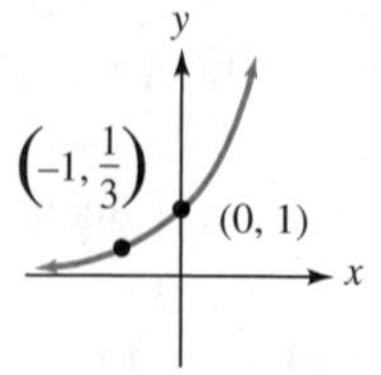

35.

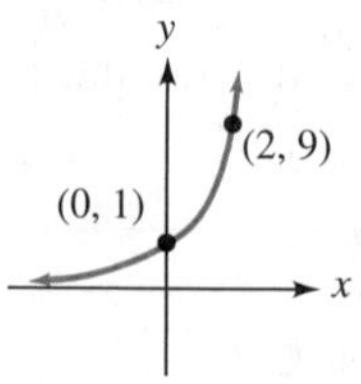

36.

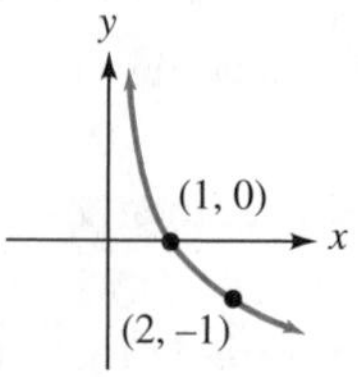

In Exercises 37–40, use a graphing calculator to graph each function. Tell whether the function is an increasing or a decreasing function.

37. $y = f(x) = \frac{1}{2}\left(3^{x/2}\right)$

38. $y = f(x) = -3\left(2^{x/3}\right)$

39. $y = f(x) = 2\left(3^{-x/2}\right)$

40. $y = f(x) = -\frac{1}{4}(2^{-x/2})$

APPLICATIONS

In Exercises 41–46, assume that there are no deposits or withdrawals.

41. Compound interest An initial deposit of \$10,000 earns 8% interest, compounded quarterly. How much will be in the account after 10 years?

42. Compound interest An initial deposit of \$10,000 earns 8% interest, compounded monthly. How much will be in the account after 10 years?

43. Comparing interest rates How much more interest could \$1,000 earn in 5 years, compounded quarterly, if the annual interest rate were $5\frac{1}{2}\%$ instead of 5%?

44. Comparing savings plans Which institution in Illustration 2 provides the better investment?

Fidelity Savings & Loan
Earn 5.25%
compounded monthly

Union Trust
Money Market Account
paying 5.35%,
compounded annually

ILLUSTRATION 2

45. Compound interest If \$1 had been invested on July 4, 1776, at 5% interest, compounded annually, what would it be worth on July 4, 2076?

46. Frequency of compounding \$10,000 is invested in each of two accounts, both paying 6% annual interest. In the first account, interest compounds quarterly, and in the second account, interest compounds daily. Find the difference between the accounts after 20 years.

47. Radioactive decay A radioactive material decays according to the formula $A = A_0\left(\frac{2}{3}\right)^t$, where A_0 is the initial amount present and t is measured in years. Find the amount present in 5 years.

48. Bacteria cultures A colony of 6 million bacteria is growing in a culture medium. (See Illustration 3.) The population P after t hours is given by the formula $P = (6 \times 10^6)(2.3)^t$. Find the population after 4 hours.

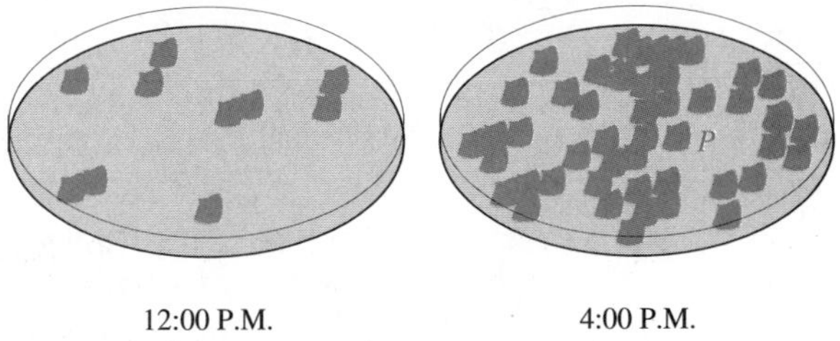

ILLUSTRATION 3

49. Discharging a battery The charge remaining in a battery decreases as the battery discharges. The charge C (in coulombs) after t days is given by the formula $C = (3 \times 10^{-4})(0.7)^t$. Find the charge after 5 days.

50. Town population The population of North Rivers is decreasing exponentially according to the formula $P = 3{,}745(0.93)^t$, where t is measured in years from the present date. Find the population in 6 years, 9 months.

51. Salvage value A small business purchases a computer for \$4,700. It is expected that its value each year will be 75% of its value in the preceding year. If the business disposes of the computer after 5 years, find its salvage value (the value after 5 years).

52. Louisiana Purchase In 1803, the United States acquired territory from France in the Louisiana Purchase. The country doubled its territory by adding 827,000 square miles of land for \$15 million. If the land has appreciated at the rate of 6% each year, what would one square mile of land be worth in 1996?

WRITING

53. If world population is increasing exponentially, why is there cause for concern?

54. How do the graphs of $y = b^x$ differ when $b > 1$ and $0 < b < 1$?

SOMETHING TO THINK ABOUT

55. In the definition of the exponential function, b could not equal 0. Why not?

56. In the definition of the exponential function, b could not be negative. Why not?

9.2 Base-*e* Exponential Functions

■ CONTINUOUS COMPOUND INTEREST ■ GRAPHING THE EXPONENTIAL FUNCTION ■ VERTICAL AND HORIZONTAL TRANSLATIONS ■ MALTHUSIAN POPULATION GROWTH ■ THE MALTHUSIAN THEORY

Getting Ready *Evaluate $\left(1 + \frac{1}{n}\right)^n$ for the following values. Round each answer to the nearest hundredth.*

1. $n = 1$ **2.** $n = 2$ **3.** $n = 4$ **4.** $n = 10$

■ CONTINUOUS COMPOUND INTEREST

If a bank pays interest twice a year, we say that interest is compounded semiannually. If it pays interest four times a year, we say that interest is compounded quarterly. If it pays interest continuously (infinitely many times in a year), we say that interest is compounded continuously.

To develop the formula for continuous compound interest, we start with the formula

$$A = P\left(1 + \frac{r}{k}\right)^{kt}$$ The formula for compound interest.

and substitute rn for k. Since r and k are positive numbers, so is n.

$$A = P\left(1 + \frac{r}{rn}\right)^{rnt}$$

We can then simplify the fraction $\frac{r}{rn}$ and use the commutative property of multiplication to change the order of the exponents.

$$A = P\left(1 + \frac{1}{n}\right)^{nrt}$$

Finally, we can use a property of exponents to write this formula as

1. $$A = P\left[\left(1 + \frac{1}{n}\right)^{n}\right]^{rt}$$ Use the property $a^{mn} = (a^m)^n$.

To find the value of $\left(1 + \frac{1}{n}\right)^n$, we evaluate it for several values of n, as shown in Table 9-1.

n	$\left(1 + \frac{1}{n}\right)^n$
1	2
2	2.25
4	2.44140625. . .
12	2.61303529. . .
365	2.71456748. . .
1,000	2.71692393. . .
100,000	2.71826823. . .
1,000,000	2.71828046. . .

TABLE 9-1

The results suggest that as n gets larger, the value of $\left(1 + \frac{1}{n}\right)^n$ approaches the number 2.71828. . . . This number is called e, which has the following value.

$e =$ **2.718281828459. . .**

In continuous compound interest, k (the number of compoundings) is infinitely large. Since k, r and n are all positive and $k = rn$, as k gets very large (approaches infinity), then so does n. Therefore, we can replace $\left(1 + \frac{1}{n}\right)^n$ in Equation 1 with e to get

$$A = Pe^{rt}$$

Formula for Exponential Growth

If a quantity P increases or decreases at an annual rate r, compounded continuously, then the amount A after t years is given by

$$A = Pe^{rt}$$

If time is measured in years, then r is called the *annual growth rate*. If r is negative, the "growth" represents a decrease.

To compute the amount to which \$12,000 will grow if invested for 18 years at 10% annual interest, compounded continuously, we substitute 12,000 for P, 0.10 for r, and 18 for t in the formula for exponential growth:

$$A = Pe^{rt}$$
$$A = 12{,}000e^{0.10(18)}$$
$$= 12{,}000e^{1.8}$$
$$\approx 72{,}595.76957$$

Use a scientific calculator and press these keys: 1.8 e^x × 12,000 = .

After 18 years, the account will contain \$72,595.77. This is \$1,589.03 more than the result in Example 7 in the previous section, where interest was compounded quarterly.

EXAMPLE 1 If \$25,000 accumulates interest at an annual rate of 8%, compound continuously, find the balance in the account in 50 years.

Solution We substitute 25,000 for P, 0.08 for r, and 50 for t.

$$A = Pe^{rt}$$
$$A = 25{,}000e^{(0.08)(50)}$$
$$= 25{,}000e^{4}$$
$$\approx 1{,}364{,}953.751$$

Use a calculator.

In 50 years, the balance will be \$1,364,953.75—over one million dollars. ■

Self Check In Example 1, find the balance in 60 years.

Answer \$3,037,760.44

The exponential function $y = f(x) = e^x$ is so important that it is often called **the exponential function.**

■ GRAPHING THE EXPONENTIAL FUNCTION

To graph the exponential function, we plot several points and join them with a smooth curve, as shown in Figure 9-10.

$y = f(x) = e^x$

x	$f(x)$	$(x, f(x))$
−2	0.1	(−2, 0.1)
−1	0.4	(−1, 0.4)
0	1	(0, 1)
1	2.7	(1, 2.7)
2	7.4	(2, 7.4)

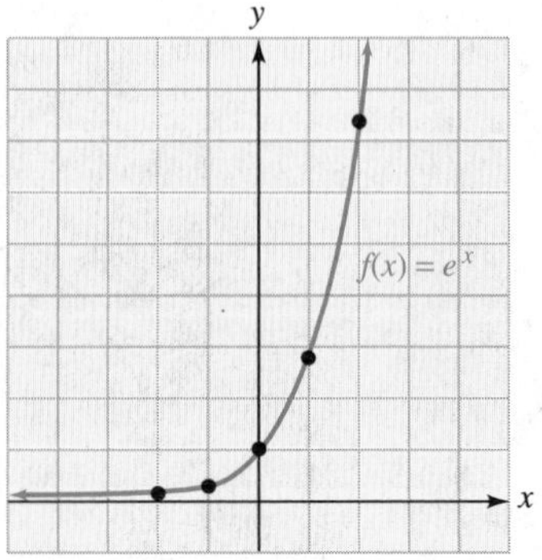

FIGURE 9-10

VERTICAL AND HORIZONTAL TRANSLATIONS

Translations of the Exponential Function

GRAPHING CALCULATORS

Figure 9-11(a) shows the graphs of $y = f(x) = e^x$, $y = f(x) = e^x + 5$, and $y = f(x) = e^x - 3$. To graph these functions with window settings of $[-3, 6]$ for x and $[-5, 15]$ for y, we enter the right-hand sides of the equations after the symbols $Y_1 =$, $Y_2 =$, and $Y_3 =$. The display will show

$$Y_1 = e \wedge (x)$$
$$Y_2 = e \wedge (x) + 5$$
$$Y_3 = e \wedge (x) - 3$$

After graphing these functions, we can see that the graph of $y = f(x) = e^x + 5$ is 5 units above the graph of $y = f(x) = e^x$, and that the graph of $y = f(x) = e^x - 3$ is 3 units below the graph of $y = f(x) = e^x$.

Figure 9-11(b) shows the calculator graphs of $y = f(x) = e^x$, $y = f(x) = e^{x+5}$, and $y = f(x) = e^{x-3}$. To graph these functions with window settings of $[-7, 10]$ for x and $[-5, 15]$ for y, we enter the right-hand sides of the equations after the symbols $Y_1 =$, $Y_2 =$, $Y_3 =$. The display will show

$$Y_1 = e \wedge (x)$$
$$Y_2 = e \wedge (x + 5)$$
$$Y_3 = e \wedge (x - 3)$$

After graphing these functions, we can see that the graph of $y = f(x) = e^{x+5}$ is 5 units to the left of the graph of $y = f(x) = e^x$, and that the graph of $y = f(x) = e^{x-3}$ is 3 units to the right of the graph of $y = f(x) = e^x$.

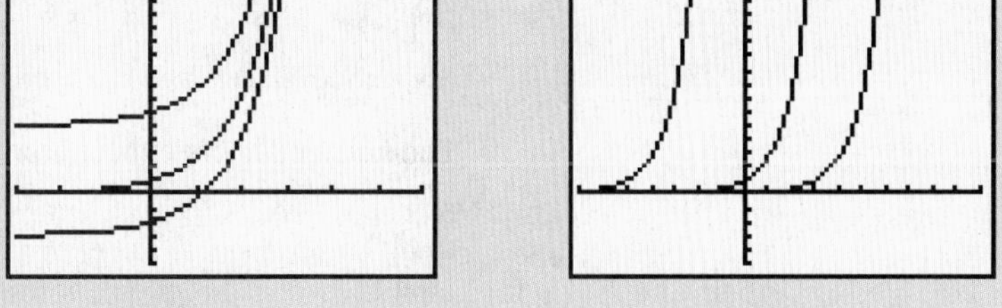

(a) (b)

FIGURE 9-11

Graphing Exponential Functions

GRAPHING CALCULATORS

Figure 9-12 shows the calculator graph $y = f(x) = 3e^{-x/2}$. To graph this function with window settings of $[-7, 10]$ for x and $[-5, 15]$ for y, we enter the right-hand side of the equation after the symbol $Y_1 =$. The display will show the equation

$$Y_1 = 3(e \wedge (-x/2))$$

The calculator graph appears in Figure 9-12. Explain why the graph has a y-intercept of (0, 3).

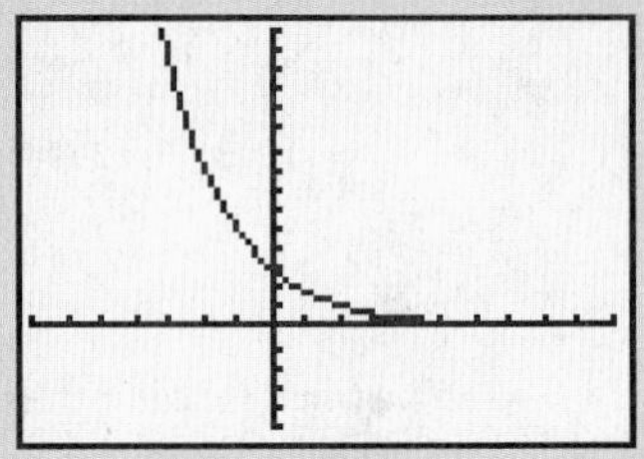

FIGURE 9-12

MALTHUSIAN POPULATION GROWTH

An equation based on the exponential function provides a model for **population growth.** In the **Malthusian model for population growth,** the future population of a colony is related to the present population by the formula $A = Pe^{rt}$.

EXAMPLE 2

City planning The population of a city is currently 15,000, but changing economic conditions are causing the population to decrease by 2% each year. If this trend continues, find the population in 30 years.

Solution Since the population is decreasing by 2% each year, the annual growth rate is -2%, or -0.02. We can substitute -0.02 for r, 30 for t, and 15,000 for P in the formula for exponential growth and find A.

$$A = Pe^{rt}$$
$$A = \mathbf{15{,}000}e^{\mathbf{-0.02(30)}}$$
$$= 15{,}000e^{-0.6}$$
$$= 8{,}232.174541$$

In 30 years, city planners expect a population of approximately 8,232 persons. ■

Self Check In Example 2, find the population in 50 years.

Answer 5,518

THE MALTHUSIAN THEORY

The English economist Thomas Robert Malthus (1766–1834) pioneered in population study. He believed that poverty and starvation were unavoidable, because the human population tends to grow exponentially, but the food supply tends to grow linearly.

EXAMPLE 3 Suppose that a country with a population of 1,000 people is growing exponentially according to the formula

$$P = 1{,}000e^{0.02t}$$

where t is in years. Furthermore, assume that the food supply measured in adequate food per day per person is growing linearly according to the formula

$$y = 30.625x + 2{,}000$$

where x is in years. In how many years will the population outstrip the food supply?

Solution We can use a graphing calculator, with window settings of [0, 100] for x and [0, 10,000] for y. After graphing the functions, we obtain Figure 9-13(a). If we trace, as in Figure 9-13(b), we can find the point where the two graphs intersect. From the graph, we can see that the food supply will be adequate for about 71 years. At that time, the population of approximately 4,200 people will begin to have problems.

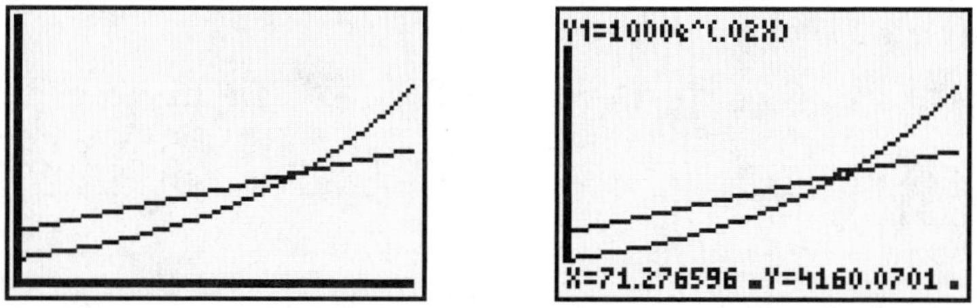

(a) (b)

FIGURE 9-13

Self Check In Example 3, suppose that the population grows at a 3% rate. For how many years will the food supply be adequate?

Answer about 38 years

Orals *Use a calculator to find each value to the nearest hundredth.*

1. e^0 **2.** e^1 **3.** e^2 **4.** e^3

Fill in each blank to make a true statement.

5. The graph of $y = f(x) = e^x + 2$ is __ units above the graph of $y = f(x) = e^x$.

6. The graph of $y = f(x) = e^{(x-2)}$ is __ units to the right of the graph of $y = f(x) = e^x$.

EXERCISE 9.2

REVIEW *Simplify each expression. Assume that all variables represent positive numbers.*

1. $\sqrt{240x^5}$

2. $\sqrt[3]{-125x^5y^4}$

3. $4\sqrt{48y^3} - 3y\sqrt{12y}$

4. $\sqrt[4]{48z^5} + \sqrt[4]{768z^5}$

VOCABULARY AND CONCEPTS *Fill in each blank to make a true statement.*

5. To two decimal places, the value of e is ____.

6. The formula for continuous compound interest is $A =$ ____.

7. Since $e > 1$, the base-e exponential function is a(n) ________ function.

8. The graph of the exponential function $y = e^x$ passes through the points (0, 1) and ____.

9. The Malthusian population growth formula is ________.

10. The Malthusian prediction is pessimistic, because ________ grows exponentially, but food supplies grow ________.

PRACTICE *In Exercises 11–18, graph each function. Check your work with a graphing calculator. Compare each graph to the graph of $y = e^x$.*

11. $y = e^x + 1$

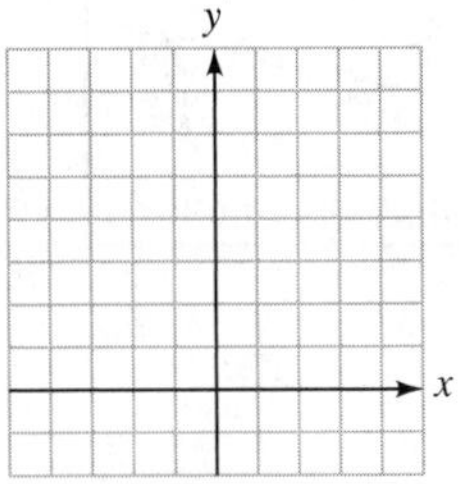

12. $y = e^x - 2$

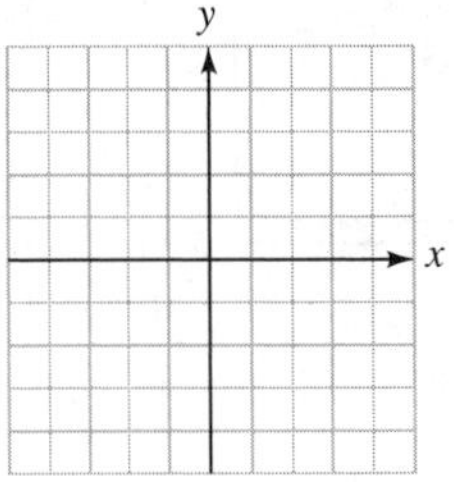

13. $y = e^{(x+3)}$

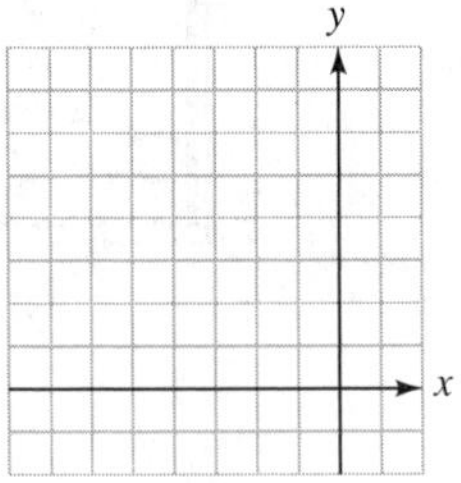

14. $y = e^{(x-5)}$

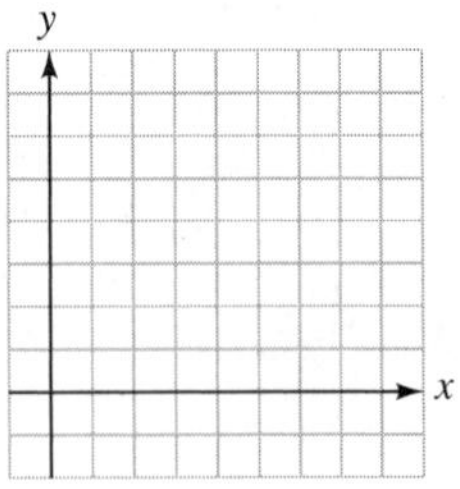

15. $y = -e^x$

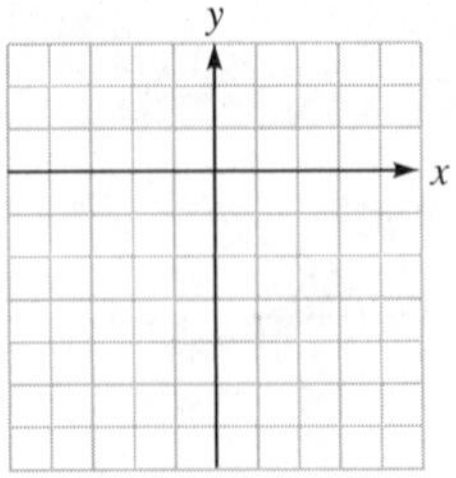

16. $y = -e^x + 1$

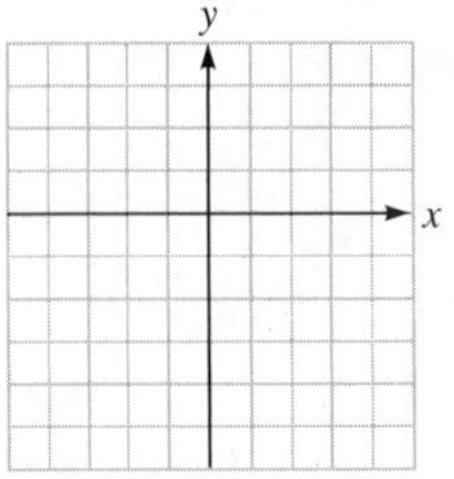

17. $y = 2e^x$

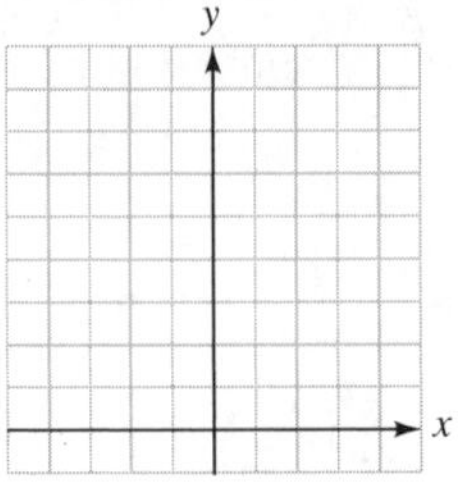

18. $y = \frac{1}{2}e^x$

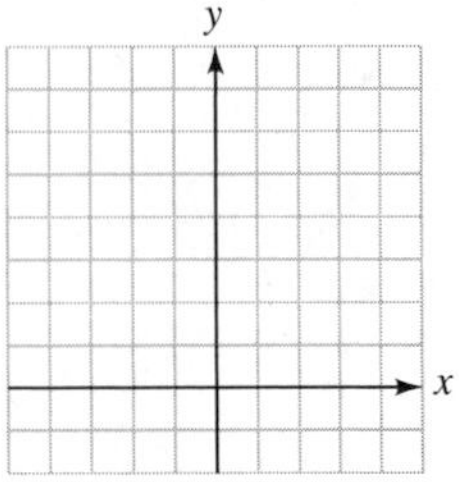

In Exercises 19–22, tell whether the graph of $y = e^x$ could look like the graph shown here.

19.

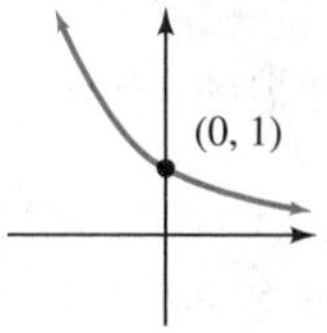

20.

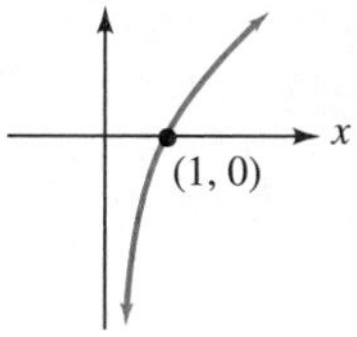

21.

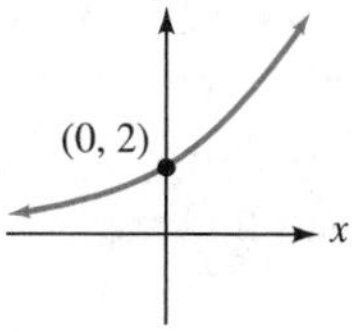

22.

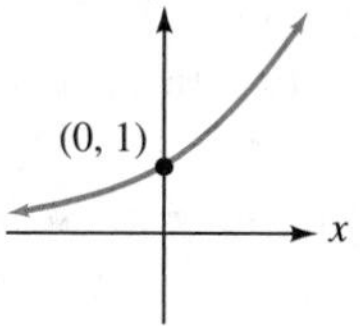

APPLICATIONS *In Exercises 23–28, assume that there are no deposits or withdrawals.*

23. Continuous compound interest An investment of \$5,000 earns 6% interest, compounded continuously. What will the investment be worth in 12 years?

24. Continuous compound interest An investment of \$6,000 earns 7% interest, compounded continuously. What will the investment be worth in 35 years?

25. Determining the initial deposit An account now contains \$12,000 and has been accumulating 7% annual interest, compounded continuously, for 9 years. Find the initial deposit.

26. Determining a previous balance An account now contains \$8,000 and has been accumulating 8% annual interest, compounded continuously. How much was in the account 6 years ago?

27. Comparison of compounding methods An initial deposit of \$5,000 grows at an annual rate of 8.5% for 5 years. Compare the final balances resulting from continuous compounding and annual compounding.

28. Comparison of compounding methods An initial deposit of \$30,000 grows at an annual rate of 8% for 20 years. Compare the final balances resulting from continuous compounding and annual compounding.

Solve each problem.

29. World population growth The population of the earth is approximately 6 billion people and is growing at an annual rate of 1.9%. Assuming a Malthusian growth model, find the world population in 30 years.

30. World population growth The population of the earth is approximately 6 billion people and is growing at an annual rate of 1.9%. Assuming a Malthusian growth model, find the world population in 40 years.

31. World population growth Assuming a Malthusian growth model and an annual growth rate of 1.9%, by what factor will the current population of the earth increase in 50 years? (See Exercise 29.)

32. Population growth The growth of a population is modeled by

$$P = 173e^{0.03t}$$

How large will the population be when $t = 30$?

33. Population decline The decline of a population is modeled by

$$P = 8{,}000e^{-0.008t}$$

How large will the population be when $t = 20$?

34. Epidemics The spread of ungulate fever through a herd of cattle can be modeled by the formula

$$P = P_0e^{0.27t} \quad (t \text{ is in days})$$

If a rancher does not act quickly to treat two cases, how many cattle will have the disease in 10 days?

35. Alcohol absorption In one individual, the blood alcohol level t minutes after drinking two shots of whiskey is given by $P = 0.3(1 - e^{-0.05t})$. Find the blood alcohol level after 15 minutes.

36. Medicine The concentration, x, of a certain drug in an organ after t minutes is given by $x = 0.08(1 - e^{-0.1t})$. Find the concentration of the drug after 30 minutes.

37. Medicine Refer to Exercise 36. Find the initial concentration of the drug.

38. Skydiving Before her parachute opens, a skydiver's velocity v in meters per second is given by $v = 50(1 - e^{-0.2t})$. Find the initial velocity.

39. Skydiving Refer to Exercise 38 and find the velocity after 20 seconds.

40. Free-falling objects After t seconds, a certain falling object has a velocity v given by $v = 50(1 - e^{-0.3t})$. Which is falling faster after 2 seconds, this object or the skydiver in Exercise 38?

41. Depreciation A camping trailer originally purchased for \$4,570 is continuously losing value at the rate of 6% per year. Find its value when it is $6\frac{1}{2}$ years old.

42. Depreciation A boat purchased for \$7,500 has been continuously decreasing in value at the rate of 2% each year. It is now 8 years, 3 months old. Find its value.

In Exercises 43–44, use a graphing calculator to solve each problem.

43. In Example 3, suppose that better farming methods change the formula for food growth to $y = 31x + 2{,}000$. How long will the food supply be adequate?

44. In Example 3, suppose that a birth-control program changed the formula for population growth to $P = 1{,}000e^{0.01t}$. How long will the food supply be adequate?

WRITING

45. Explain why the graph of $y = f(x) = e^x - 5$ is 5 units below the graph of $y = f(x) = e^x$.

46. Explain why the graph of $y = f(x) = e^{(x+5)}$ is 5 units to the left of the graph of $y = f(x) = e^x$.

SOMETHING TO THINK ABOUT

47. The value of e can be calculated to any degree of accuracy by adding the first several terms of the following list:

$$1, 1, \frac{1}{2}, \frac{1}{2 \cdot 3}, \frac{1}{2 \cdot 3 \cdot 4}, \frac{1}{2 \cdot 3 \cdot 4 \cdot 5}, \ldots$$

The more terms that are added, the closer the sum will be to e. Add the first six numbers in the preceding list. To how many decimal places is the sum accurate?

48. Graph the function defined by the equation

$$y = f(x) = \frac{e^x + e^{-x}}{2}$$

from $x = -2$ to $x = 2$. The graph will look like a parabola, but it is not. The graph, called a **catenary,** is important in the design of power distribution networks, because it represents the shape of a uniform flexible cable whose ends are suspended from the same height. The function is called the **hyperbolic cosine function.**

49. If $e^{t+5} = ke^t$, find k.

50. If $e^{5t} = k^t$, find k.

9.3 Logarithmic Functions

■ LOGARITHMS ■ GRAPHS OF LOGARITHMIC FUNCTIONS ■ VERTICAL AND HORIZONTAL TRANSLATIONS ■ BASE-10 LOGARITHMS ■ ELECTRONICS ■ SEISMOLOGY

Getting Ready *Find each value.*

1. 7^0 **2.** 5^2 **3.** 5^{-2} **4.** $16^{1/2}$

LOGARITHMS

Because an exponential function defined by $y = b^x$ is one-to-one, it has an inverse function that is defined by the equation $x = b^y$. To express this inverse function in the form $y = f^{-1}(x)$, we must solve the equation $x = b^y$ for y. For this, we need the following definition.

Logarithmic Functions

If $b > 0$ and $b \neq 1$, the **logarithmic function with base *b*** is defined by

$$y = \log_b x \quad \text{if and only if} \quad x = b^y$$

The **domain of the logarithmic function** is the interval $(0, \infty)$. The **range** is the interval $(-\infty, \infty)$.

Since the function $y = \log_b x$ is the inverse of the one-to-one exponential function $y = b^x$, the logarithmic function is also one-to-one.

WARNING! Since the domain of the logarithmic function is the set of positive numbers, it is impossible to find the logarithm of 0 or the logarithm of a negative number.

The previous definition guarantees that any pair (x, y) that satisfies the equation $y = \log_b x$ also satisfies the equation $x = b^y$.

$\log_4 1 = 0$ because $1 = 4^0$

$\log_5 25 = 2$ because $25 = 5^2$

$\log_5 \frac{1}{25} = -2$ because $\frac{1}{25} = 5^{-2}$

$\log_{16} 4 = \frac{1}{2}$ because $4 = 16^{1/2}$

$\log_2 \frac{1}{8} = -3$ because $\frac{1}{8} = 2^{-3}$

$\log_b x = y$ because $x = b^y$

In each of these examples, the logarithm of a number is an exponent. In fact,

$\log_b x$ is the exponent to which b is raised to get x.

In equation form, we write

$$b^{\log_b x} = x$$

EXAMPLE 1 Find y in each equation: **a.** $\log_6 1 = y$, **b.** $\log_3 27 = y$, and **c.** $\log_5 \frac{1}{5} = y$.

Solution **a.** We can change the equation $\log_6 1 = y$ into the equivalent exponential equation $6^y = 1$. Since $6^0 = 1$, it follows that $y = 0$. Thus,

$$\log_6 1 = 0$$

b. $\log_3 27 = y$ is equivalent to $3^y = 27$. Since $3^3 = 27$, it follows that $3^y = 3^3$, and $y = 3$. Thus,

$$\log_3 27 = 3$$

c. $\log_5 \frac{1}{5} = y$ is equivalent to $5^y = \frac{1}{5}$. Since $5^{-1} = \frac{1}{5}$, it follows that $5^y = 5^{-1}$, and $y = -1$. Thus,

$$\log_5 \frac{1}{5} = -1$$

■

Self Check Find y in each equation: **a.** $\log_3 9 = y$, **b.** $\log_2 64 = y$, and **c.** $\log_5 \frac{1}{125} = y$.

Answers **a.** 2, **b.** 6, **c.** -3

c

EXAMPLE 2

Find the value of x in each equation: **a.** $\log_3 81 = x$, **b.** $\log_x 125 = 3$, and **c.** $\log_4 x = 3$.

Solution

a. $\log_3 81 = x$ is equivalent to $3^x = 81$. Because $3^4 = 81$, it follows that $3^x = 3^4$. Thus, $x = 4$.

b. $\log_x 125 = 3$ is equivalent to $x^3 = 125$. Because $5^3 = 125$, it follows that $x^3 = 5^3$. Thus, $x = 5$.

c. $\log_4 x = 3$ is equivalent to $4^3 = x$. Because $4^3 = 64$, it follows that $x = 64$.

■

Self Check Find x in each equation: **a.** $\log_2 32 = x$, **b.** $\log_x 8 = 3$, and **c.** $\log_5 x = 2$.

Answers **a.** 5, **b.** 2, **c.** 25

EXAMPLE 3

Find the value of x in each equation: **a.** $\log_{1/3} x = 2$, **b.** $\log_{1/3} x = -2$, and **c.** $\log_{1/3} \frac{1}{27} = x$.

Solution

a. $\log_{1/3} x = 2$ is equivalent to $\left(\frac{1}{3}\right)^2 = x$. Thus, $x = \frac{1}{9}$.

b. $\log_{1/3} x = -2$ is equivalent to $\left(\frac{1}{3}\right)^{-2} = x$. Thus,

$$x = \left(\frac{1}{3}\right)^{-2} = 3^2 = 9$$

c. $\log_{1/3} \frac{1}{27} = x$ is equivalent to $\left(\frac{1}{3}\right)^x = \frac{1}{27}$. Because $\left(\frac{1}{3}\right)^3 = \frac{1}{27}$, it follows that $x = 3$.

■

Self Check Find x in each equation: **a.** $\log_{1/4} x = 3$ and **b.** $\log_{1/4} x = -2$.

Answers **a.** $\frac{1}{64}$, **b.** 16

GRAPHS OF LOGARITHMIC FUNCTIONS

To graph the logarithmic function $y = \log_2 x$, we calculate and plot several points with coordinates (x, y) that satisfy the equation $x = 2^y$. After joining these points with a smooth curve, we have the graph shown in Figure 9-14(a).

To graph $y = \log_{1/2} x$, we calculate and plot several points with coordinates (x, y) that satisfy the equation $x = \left(\frac{1}{2}\right)^y$. After joining these points with a smooth curve, we have the graph shown in Figure 9-14(b).

$y = \log_2 x$

x	y	(x, y)
$\frac{1}{4}$	-2	$\left(\frac{1}{4}, -2\right)$
$\frac{1}{2}$	-1	$\left(\frac{1}{2}, -1\right)$
1	0	(1, 0)
2	1	(2, 1)
4	2	(4, 2)
8	3	(8, 3)

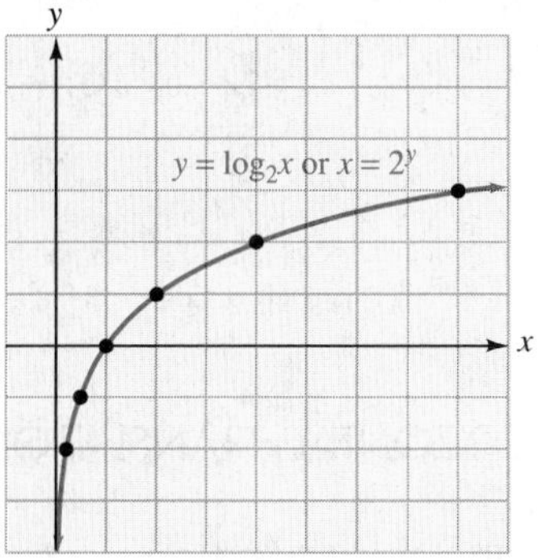

(a)

$y = \log_{1/2} x$

x	y	(x, y)
$\frac{1}{4}$	2	$\left(\frac{1}{4}, 2\right)$
$\frac{1}{2}$	1	$\left(\frac{1}{2}, 1\right)$
1	0	(1, 0)
2	-1	$(2, -1)$
4	-2	$(4, -2)$
8	-3	$(8, -3)$

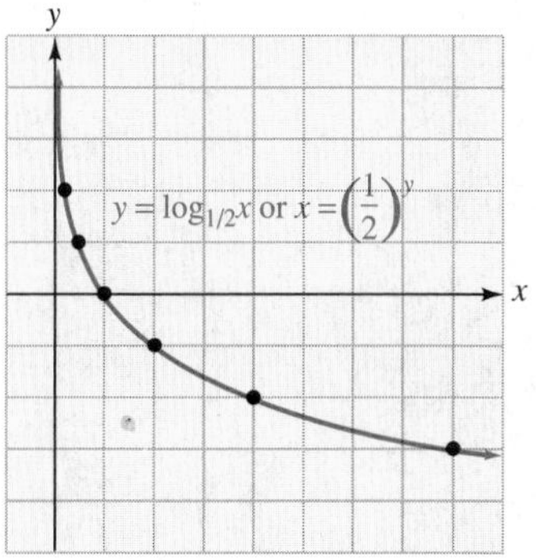

(b)

FIGURE 9-14

The graphs of all logarithmic functions are similar to those in Figure 9-15. If $b > 1$, the logarithmic function is increasing, as in Figure 9-15(a). If $0 < b < 1$, the logarithmic function is decreasing, as in Figure 9-15(b).

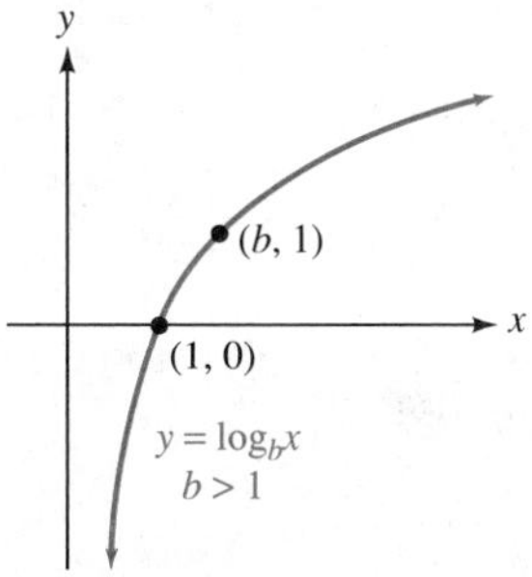

(a)

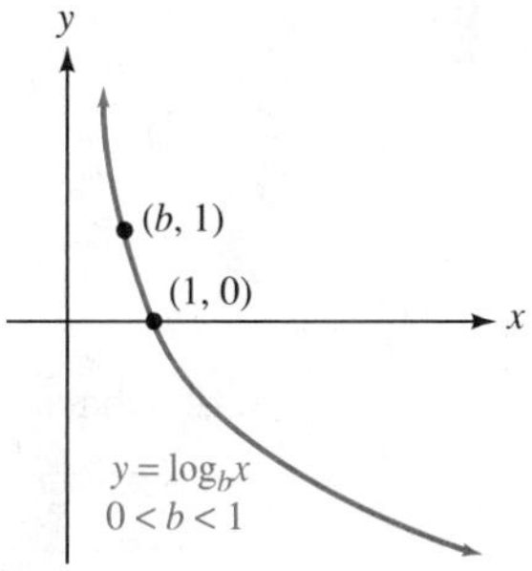

(b)

FIGURE 9-15

The graph of $y = f(x) = \log_b x$ has the following properties.

1. It passes through the point $(1, 0)$.
2. It passes through the point $(b, 1)$
3. The y-axis is an asymptote.
4. The domain is $(0, \infty)$ and the range is $(-\infty, \infty)$.

The exponential and logarithmic functions are inverses of each other and, therefore, have symmetry about the line $y = x$. The graphs of $y = \log_b x$ and $y = b^x$ are shown in Figure 9-16(a) when $b > 1$, and in Figure 9-16(b) when $0 < b < 1$.

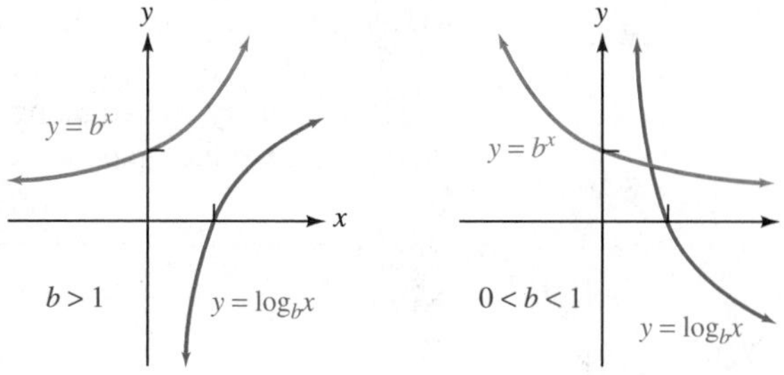

(a) (b)

FIGURE 9-16

VERTICAL AND HORIZONTAL TRANSLATIONS

The graphs of many functions involving logarithms are translations of the basic logarithmic graphs.

EXAMPLE 4 Graph the function defined by $y = 3 + \log_2 x$.

Solution The graph of $y = 3 + \log_2 x$ is identical to the graph of $y = \log_2 x$, except that it is translated 3 units upward. (See Figure 9-17.)

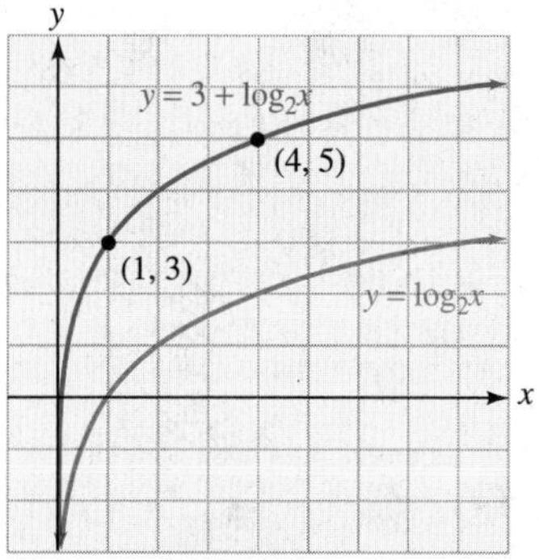

FIGURE 9-17

Self Check Graph $y = \log_3 x - 2$.

Answer

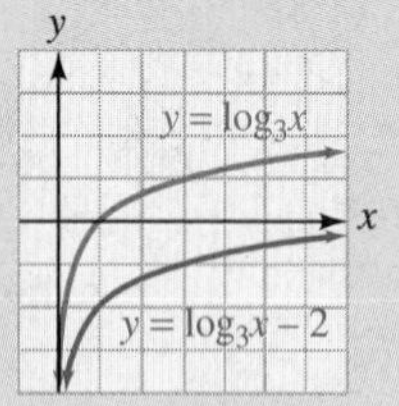

EXAMPLE 5 Graph $y = \log_{1/2} (x - 1)$.

Solution The graph of $y = \log_{1/2} (x - 1)$ is identical to the graph of $y = \log_{1/2} x$, except that it is translated 1 unit to the right. (See Figure 9-18.)

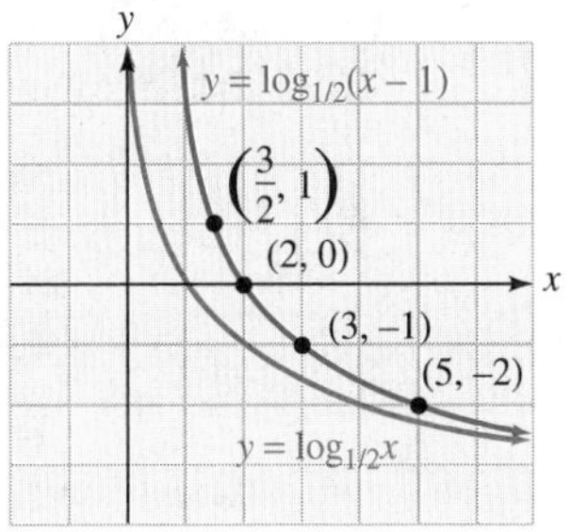

FIGURE 9-18

Self Check Graph $y = \log_{1/3}(x + 2)$.

Answer

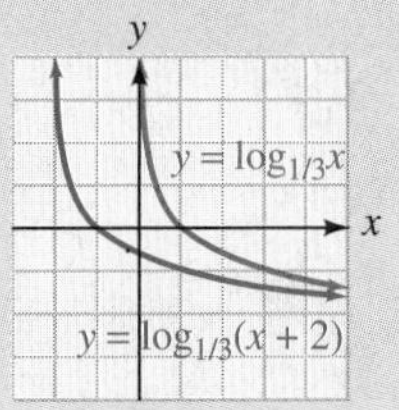

Graphing Logarithmic Functions

GRAPHING CALCULATORS

Graphing calculators can draw graphs of logarithmic functions directly if the base of the logarithmic function is 10 or e. To use a calculator to graph the logarithmic function $y = f(x) = -2 + \log_{10}\left(\frac{1}{2}x\right)$, we enter the right-hand side of the equation after the symbol $Y_1 =$. The display will show the equation

$$Y_1 = -2 + \log(1/2*x)$$

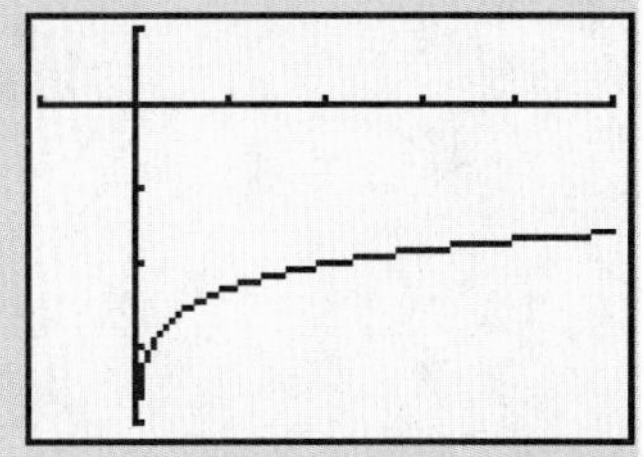

FIGURE 9-19

If we use window settings of $[-1, 5]$ for x and $[-4, 1]$ for y and press the GRAPH key, we will obtain the graph shown in Figure 9-19.

BASE-10 LOGARITHMS

For computational purposes and in many applications, we will use base-10 logarithms (also called **common logarithms**). When the base b is not indicated in the notation $\log x$, we assume that $b = 10$:

$$\textbf{log } x \quad \textbf{means} \quad \textbf{log}_{10}\, x$$

Because base-10 logarithms appear so often, it is a good idea to become familiar with the following base-10 logarithms:

$\log_{10} \frac{1}{100} = -2$ because $10^{-2} = \frac{1}{100}$

$\log_{10} \frac{1}{10} = -1$ because $10^{-1} = \frac{1}{10}$

$\log_{10} 1 = 0$ because $10^0 = 1$

$\log_{10} 10 = 1$ because $10^1 = 10$

$\log_{10} 100 = 2$ because $10^2 = 100$

$\log_{10} 1{,}000 = 3$ because $10^3 = 1{,}000$

In general, we have

$$\mathbf{\log_{10} 10^x = x}$$

CALCULATORS

Finding Logarithms

Before calculators, extensive tables provided logarithms of numbers. Today, logarithms are easy to find with a calculator. For example, to find log 32.58 with a scientific calculator, we enter these numbers and press these keys:

32.58 LOG

The display will read 1.51295108. To four decimal places, log 32.58 = 1.5130.

To use a graphing calculator, we enter these numbers and press these keys:

LOG 32.58 ENTER

The display will read log (32.58
1.51295108

EXAMPLE 6

Find x in the equation $\log x = 0.3568$. Round to four decimal places.

Solution The equation $\log x = 0.3568$ is equivalent to $10^{0.3568} = x$. To find x with a scientific calculator, we enter these numbers and press these keys:

10 y^x .3568 =

The display will read 2.274049951. To four decimal places,

$x = 2.2740$

If your calculator has a 10^x key, enter .3568 and press it to get the same result. ■

Self Check Solve $\log x = 2.7$. Round to four decimal places.

Answer 501.1872

ELECTRONICS

Common logarithms are used in electrical engineering to express the voltage gain (or loss) of an electronic device such as an amplifier. The unit of gain (or loss), called the **decibel,** is defined by a logarithmic relation.

Decibel Voltage Gain

If E_O is the output voltage of a device and E_I is the input voltage, the decibel voltage gain is given by

$$\text{db gain} = 20 \log \frac{E_O}{E_I}$$

EXAMPLE 7 **Finding db gain** If the input to an amplifier is 0.4 volt and the output is 50 volts, find the decibel voltage gain of the amplifier.

Solution We can find the decibel voltage gain by substituting 0.4 for E_I and 50 for E_O into the formula for db gain:

$$\text{db gain} = 20 \log \frac{E_O}{E_I}$$

$$\text{db gain} = 20 \log \frac{50}{0.4}$$

$$= 20 \log 125$$

$$\approx 42 \qquad \text{Use a calculator.}$$

The amplifier provides a 42-decibel voltage gain. ■

SEISMOLOGY

In seismology, common logarithms are used to measure the intensity of earthquakes on the **Richter scale.** The intensity of an earthquake is given by the following logarithmic function.

Richter Scale

If R is the intensity of an earthquake, A is the amplitude (measured in micrometers), and P is the period (the time of one oscillation of the earth's surface, measured in seconds), then

$$R = \log \frac{A}{P}$$

EXAMPLE 8 **Measuring earthquakes** Find the measure on the Richter scale of an earthquake with an amplitude of 10,000 micrometers (1 centimeter) and a period of 0.1 second.

Solution We substitute 10,000 for A and 0.1 for P in the Richter scale formula and simplify:

$$R = \log \frac{A}{P}$$

$$\begin{aligned} R &= \log \frac{\mathbf{10{,}000}}{\mathbf{0.1}} \\ &= \log 100{,}000 \\ &= 5 \end{aligned}$$

The earthquake measures 5 on the Richter scale. ■

Orals *Find the value of x in each equation.*

1. $\log_2 8 = x$ **2.** $\log_3 9 = x$ **3.** $\log_x 125 = 3$

4. $\log_x 8 = 3$ **5.** $\log_4 16 = x$ **6.** $\log_x 32 = 5$

7. $\log_{1/2} x = 2$ **8.** $\log_9 3 = x$ **9.** $\log_x \frac{1}{4} = -2$

EXERCISE 9.3

REVIEW *Solve each equation.*

1. $\sqrt[3]{6x + 4} = 4$

2. $\sqrt{3x - 4} = \sqrt{-7x + 2}$

3. $\sqrt{a + 1} - 1 = 3a$

4. $3 - \sqrt{t - 3} = \sqrt{t}$

VOCABULARY AND CONCEPTS *Fill in each blank to make a true statement.*

5. The equation $y = \log_b x$ is equivalent to ______.

6. The domain of the logarithmic function is the interval ______.

7. The ______ of the logarithmic function is the interval $(-\infty, \infty)$.

8. $b^{\log_b x} =$ __

9. Because an exponential function is one-to-one, it has an ______ function.

10. The inverse of an exponential function is called a ______ function.

11. $\text{Log}_b\, x$ is the ______ to which b is raised to get x.

12. The y-axis is an ______ to the graph of $y = f(x) = \log_b x$.

13. The graph of $y = f(x) = \log_b x$ passes through the points ______ and ______.

14. $\log_{10} 10^x =$ __.

15. db gain = ______

16. The intensity of an earthquake is measured by the formula $R =$ ______.

PRACTICE *In Exercises 17–24, write each equation in exponential form.*

17. $\log_3 27 = 3$ **18.** $\log_8 8 = 1$ **19.** $\log_{1/2} \frac{1}{4} = 2$ **20.** $\log_{1/5} 1 = 0$

21. $\log_4 \frac{1}{64} = -3$

22. $\log_6 \frac{1}{36} = -2$

23. $\log_{1/2} \frac{1}{8} = 3$

24. $\log_{1/5} 1 = 0$

In Exercises 25–32, write each equation in logarithmic form.

25. $6^2 = 36$

26. $10^3 = 1{,}000$

27. $5^{-2} = \frac{1}{25}$

28. $3^{-3} = \frac{1}{27}$

29. $\left(\frac{1}{2}\right)^{-5} = 32$

30. $\left(\frac{1}{3}\right)^{-3} = 27$

31. $x^y = z$

32. $m^n = p$

In Exercises 33–72, find each value of x.

33. $\log_2 16 = x$

34. $\log_3 9 = x$

35. $\log_4 16 = x$

36. $\log_6 216 = x$

37. $\log_{1/2} \frac{1}{8} = x$

38. $\log_{1/3} \frac{1}{81} = x$

39. $\log_9 3 = x$

40. $\log_{125} 5 = x$

41. $\log_{1/2} 8 = x$

42. $\log_{1/2} 16 = x$

43. $\log_7 x = 2$

44. $\log_5 x = 0$

45. $\log_6 x = 1$

46. $\log_2 x = 4$

47. $\log_{25} x = \frac{1}{2}$

48. $\log_4 x = \frac{1}{2}$

49. $\log_5 x = -2$

50. $\log_3 x = -2$

51. $\log_{36} x = -\frac{1}{2}$

52. $\log_{27} x = -\frac{1}{3}$

53. $\log_{100} \frac{1}{1{,}000} = x$

54. $\log_{5/2} \frac{4}{25} = x$

55. $\log_{27} 9 = x$

56. $\log_{12} x = 0$

57. $\log_x 5^3 = 3$

58. $\log_x 5 = 1$

59. $\log_x \frac{9}{4} = 2$

60. $\log_x \frac{\sqrt{3}}{3} = \frac{1}{2}$

61. $\log_x \frac{1}{64} = -3$

62. $\log_x \frac{1}{100} = -2$

63. $\log_{2\sqrt{2}} x = 2$

64. $\log_4 8 = x$

65. $2^{\log_2 4} = x$

66. $3^{\log_3 5} = x$

67. $x^{\log_4 6} = 6$

68. $x^{\log_3 8} = 8$

69. $\log 10^3 = x$

70. $\log 10^{-2} = x$

71. $10^{\log x} = 100$

72. $10^{\log x} = \frac{1}{10}$

In Exercises 73–76, use a calculator to find each value, if possible. Give answers to four decimal places.

73. $\log 8.25$

74. $\log 0.77$

75. $\log 0.00867$

76. $\log 375.876$

In Exercises 77–84, use a calculator to find each value of y, if possible. If an answer is not exact, give the answer to two decimal places.

77. $\log y = 1.4023$

78. $\log y = 2.6490$

79. $\log y = 4.24$

80. $\log y = 0.926$

81. $\log y = -3.71$

82. $\log y = -0.28$

83. $\log y = \log 8$

84. $\log y = \log 7$

In Exercises 85–88, graph each function. Tell whether each function is an increasing or decreasing function.

85. $y = f(x) = \log_3 x$

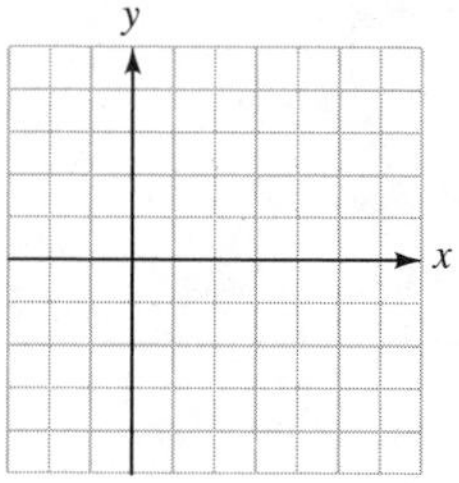

86. $y = f(x) = \log_{1/3} x$

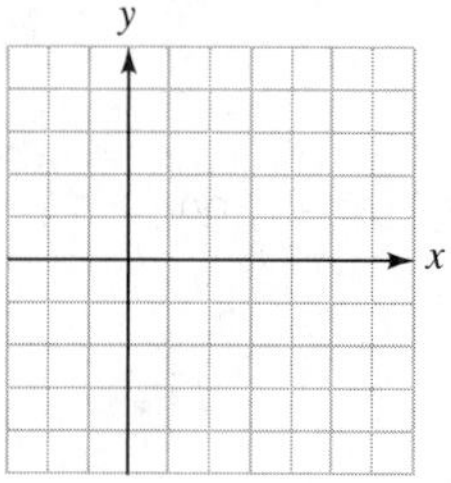

87. $y = f(x) = \log_{1/2} x$

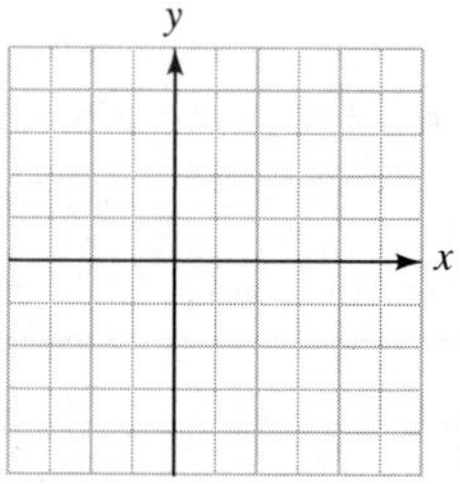

88. $y = f(x) = \log_4 x$

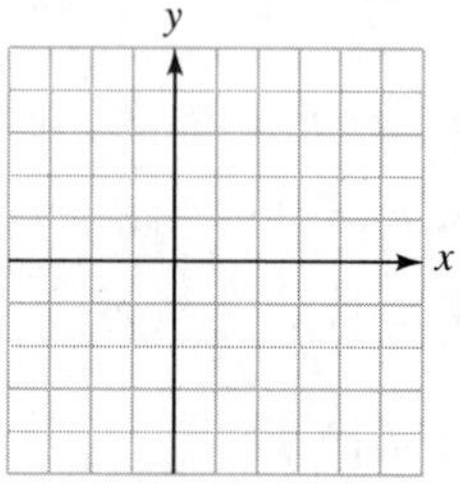

In Exercises 89–92, graph each function.

89. $y = f(x) = 3 + \log_3 x$

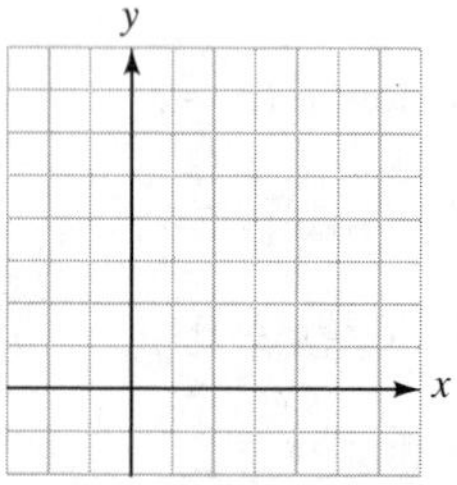

90. $y = f(x) = \log_{1/3} x - 1$

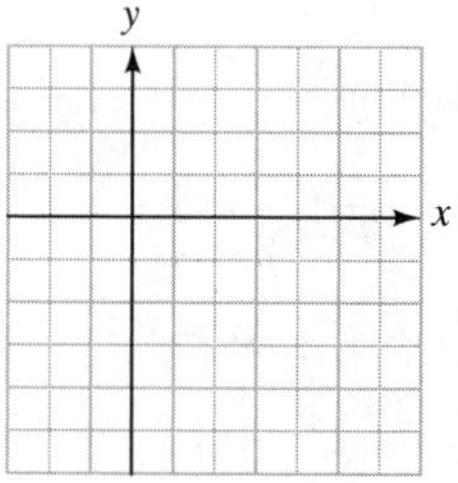

91. $y = f(x) = \log_{1/2} (x - 2)$

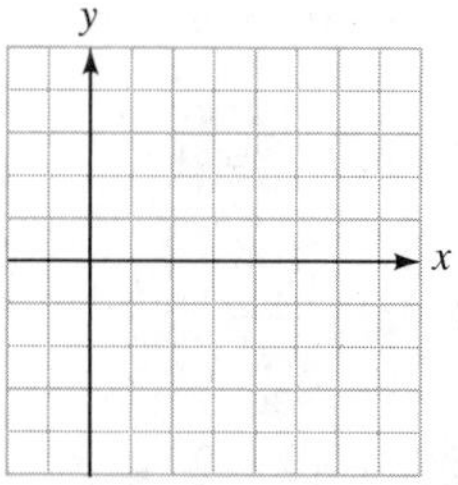

92. $y = f(x) = \log_4 (x + 2)$

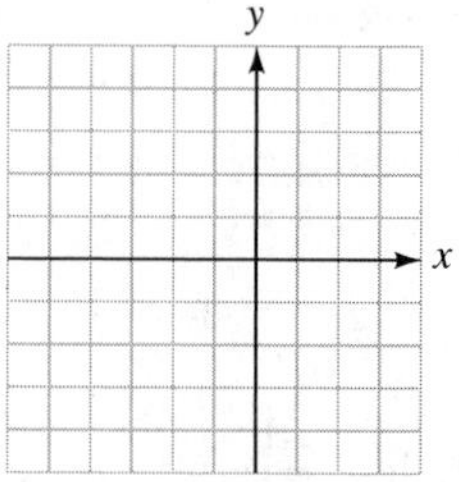

In Exercises 93–96, graph each pair of inverse functions on a single coordinate system.

93. $y = f(x) = 2^x$
$y = g(x) = \log_2 x$

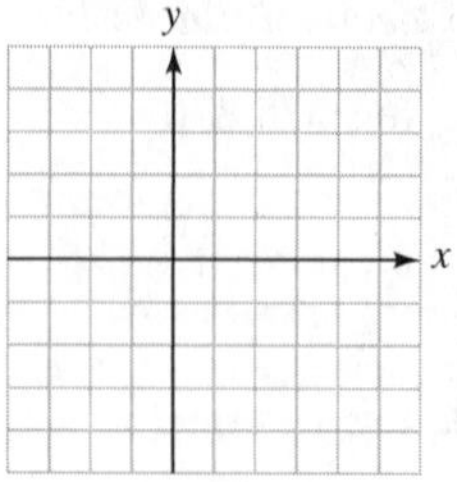

94. $y = f(x) = \left(\frac{1}{2}\right)^x$
$y = g(x) = \log_{1/2} x$

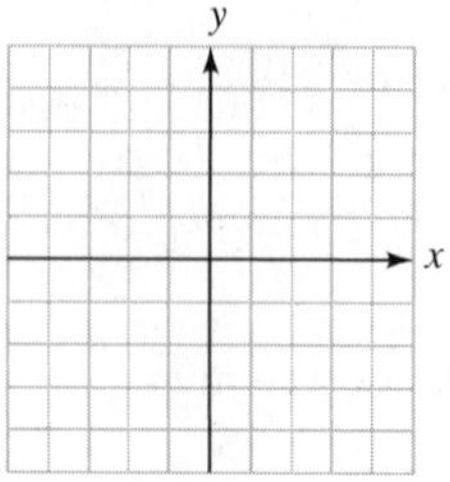

95. $y = f(x) = \left(\frac{1}{4}\right)^x$
$y = g(x) = \log_{1/4} x$

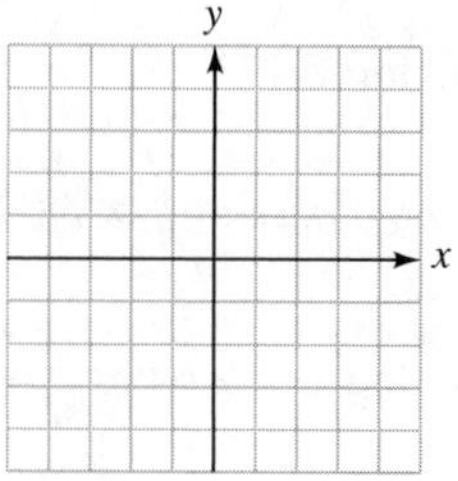

96. $y = f(x) = 4^x$
$y = g(x) = \log_4 x$

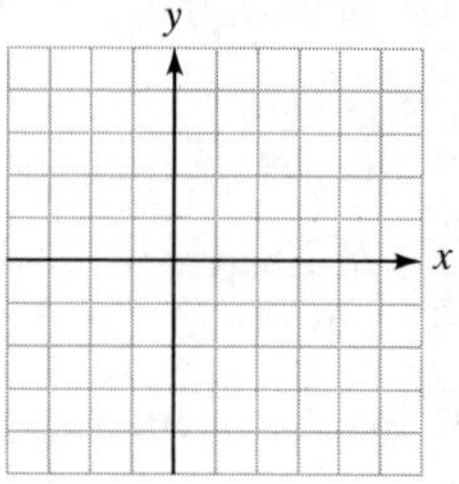

In Exercises 97–100, find the value of b, if any, that would cause the graph of $f(x) = \log_b x$ to look like the graph indicated.

97.

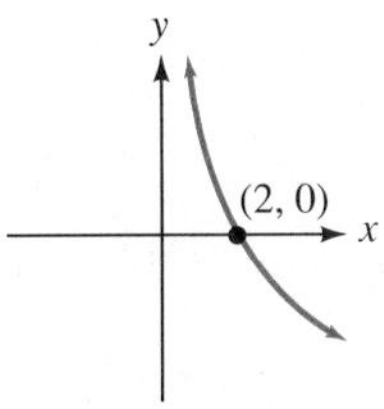

98.

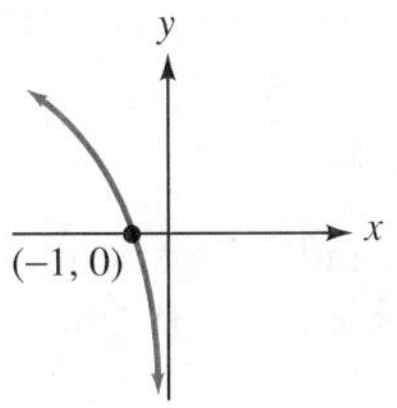

99.

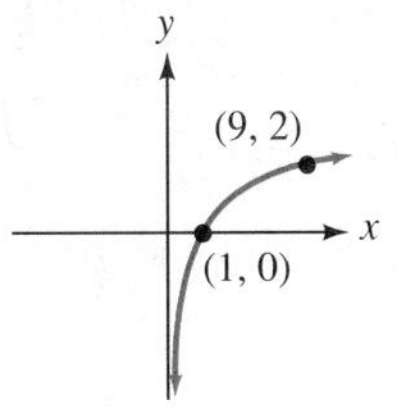

100.

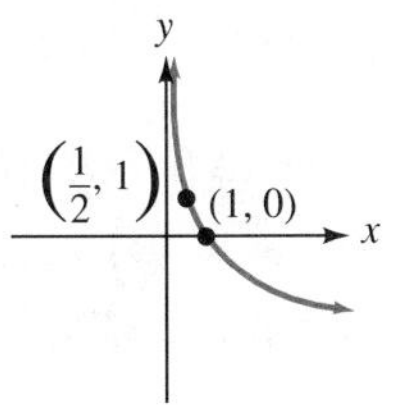

APPLICATIONS *In Exercises 101—112, if an answer is not exact, round to the nearest tenth.*

101. Finding the gain of an amplifier Find the db gain of an amplifier if input voltage is 0.71 volt when the output voltage is 20 volts.

102. Finding the gain of an amplifier Find the db gain of an amplifier if the output voltage is 2.8 volts when the input voltage is 0.05 volt.

103. db gain of an amplifier Find the db gain of the amplifier shown in Illustration 1.

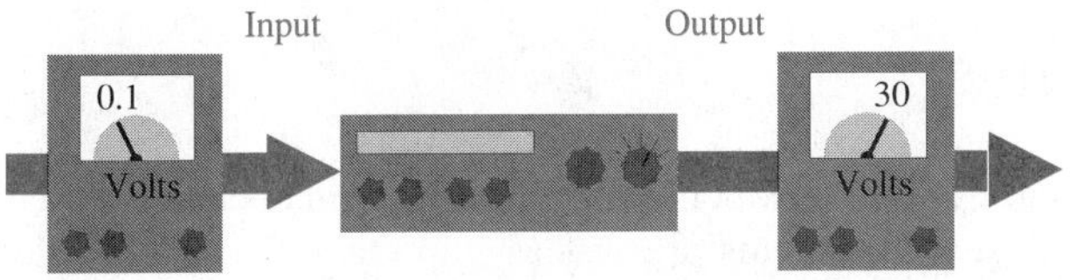

ILLUSTRATION 1

104. db gain of an amplifier An amplifier produces an output of 80 volts when driven by an input of 0.12 volts. Find the amplifier's db gain.

105. Earthquakes An earthquake has an amplitude of 5,000 micrometers and a period of 0.2 second. Find its measure on the Richter scale.

106. Earthquakes The period of an earthquake with amplitude of 80,000 micrometers is 0.08 second. Find its measure on the Richter scale.

107. Earthquakes An earthquake has a period of $\frac{1}{4}$ second and an amplitude of 2,500 micrometers. Find its measure on the Richter scale.

108. Earthquakes By what factor must the amplitude of an earthquake change to increase its severity by 1 point on the Richter scale? Assume that the period remains constant.

109. Depreciation Business equipment is often depreciated using the double declining-balance method. In this method, a piece of equipment with a life expectancy of N years, costing \$$C$, will depreciate to a value of \$$V$ in n years, where n is given by the formula

$$n = \frac{\log V - \log C}{\log\left(1 - \frac{2}{N}\right)}$$

A computer that cost \$17,000 has a life expectancy of 5 years. If it has depreciated to a value of \$2,000, how old is it?

110. Depreciation See Exercise 109. A printer worth \$470 when new had a life expectancy of 12 years. If it is now worth \$189, how old is it?

111. Time for money to grow If \$$P$ is invested at the end of each year in an annuity earning annual interest at a rate r, the amount in the account will be \$$A$ after n years, where

$$n = \frac{\log\left[\frac{Ar}{P} + 1\right]}{\log(1 + r)}$$

If \$1,000 is invested each year in an annuity earning 12% annual interest, how long will it take for the account to be worth \$20,000?

112. Time for money to grow If \$5,000 is invested each year in an annuity earning 8% annual interest, how long will it take for the account to be worth \$50,000? (See Exercise 111.)

WRITING

113. Describe the appearance of the graph of $y = f(x) = \log_b x$ when $0 < b < 1$ and when $b > 1$.

114. Explain why it is impossible to find the logarithm of a negative number.

SOMETHING TO THINK ABOUT

115. Graph $y = f(x) = -\log_3 x$. How does the graph compare to the graph of $y = f(x) = \log_3 x$?

116. Find a logarithmic function that passes through the points (1, 0) and (5, 1).

9.4 Base-*e* Logarithms

■ BASE-*e* LOGARITHMS ■ GRAPHING BASE-*e* LOGARITHMS ■ DOUBLING TIME

Getting Ready *Evaluate each expression.*

1. $\log_4 16$

2. $\log_2 \frac{1}{8}$

3. $\log_5 5$

4. $\log_7 1$

■ BASE-*e* LOGARITHMS

We have seen the importance of base-*e* exponential functions in mathematical models of events in nature. Base-*e* logarithms are just as important. They are called **natural logarithms** or **Napierian logarithms,** after John Napier (1550–1617), and are usually written as $\ln x$, rather than $\log_e x$:

$$\textbf{ln}\ x \quad \textbf{means} \quad \textbf{log}_e\ x$$

As with all logarithmic functions, the domain of $y = f(x) = \ln x$ is the interval $(0, \infty)$, and the range is the interval $(-\infty, \infty)$.

To find the base-*e* logarithms of numbers, we can use a calculator.

Evaluating Logarithms

CALCULATORS

To use a scientific calculator to find the value of ln 9.87, we enter these numbers and press these keys:

9.87 LN

The display will read 2.289499853. To four decimal places, $\ln 9.87 = 2.2895$.

To use a graphing calculator, we enter these numbers and press these keys:

LN 9.87 ENTER

The display will read `ln 9.87`
`2.289499853`

EXAMPLE 1 Use a calculator to find each value: **a.** ln 17.32 and **b.** ln (log 0.05).

Solution **a.** Enter these numbers and press these keys:

Scientific Calculator	***Graphing Calculator***
17.32 LN	LN 17.32 ENTER

Either way, the result is 2.851861903.

b. Enter these numbers and press these keys:

Scientific Calculator	***Graphing Calculator***
0.05 LOG LN	LN (LOG 0.05) ENTER

Either way, we obtain an error, because log 0.05 is a negative number, and we cannot take the logarithm of a negative number. ■

Self Check Find each value to four decimal places: **a.** $\ln \pi$ and **b.** $\ln\left(\log \frac{1}{2}\right)$.

Answers **a.** 1.1447, **b.** no value

EXAMPLE 2 Solve each equation: **a.** $\ln x = 1.335$ and **b.** $\ln x = \log 5.5$. Give each result to four decimal places.

Solution **a.** The equation $\ln x = 1.335$ is equivalent to $e^{1.335} = x$. To use a scientific calculator to find x, enter these numbers and press these keys:

1.335 e^x

The display will read 3.799995946. To four decimal places,

$$x = 3.8000$$

b. The equation $\ln x = \log 5.5$ is equivalent to $e^{\log 5.5} = x$. To use a scientific calculator to find x, press these keys:

5.5 LOG e^x

The display will read 2.096695826. To four decimal places,

$$x = 2.0967$$ ■

Self Check Solve **a.** $\ln x = 2.5437$ and **b.** $\log x = \ln 5$.

Answers **a.** 12.7267, **b.** 40.6853

GRAPHING BASE-*e* LOGARITHMS

The equation $y = \ln x$ is equivalent to the equation $x = e^y$. To get the graph of $\ln x$, we can plot points that satisfy the equation $x = e^y$ and join them with a smooth curve, as shown in Figure 9-20(a). Figure 9-20(b) shows the calculator graph.

$y = \ln x$

x	y	(x, y)
$\frac{1}{e} \approx 0.4$	-1	$(0.4, -1)$
1	0	$(1, 0)$
$e \approx 2.7$	1	$(2.7, 1)$
$e^2 \approx 7.4$	2	$(7.4, 2)$

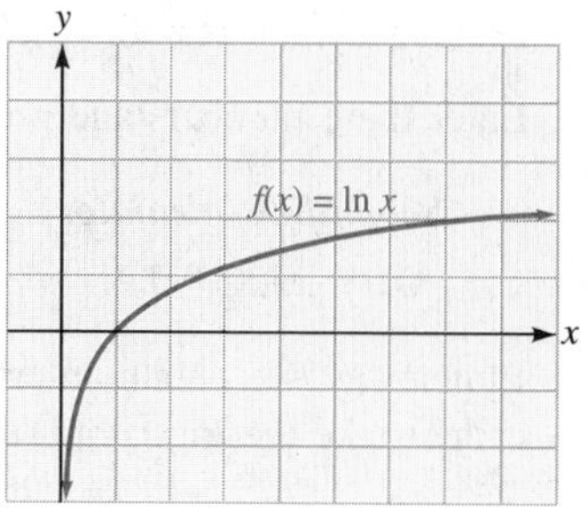

(a)

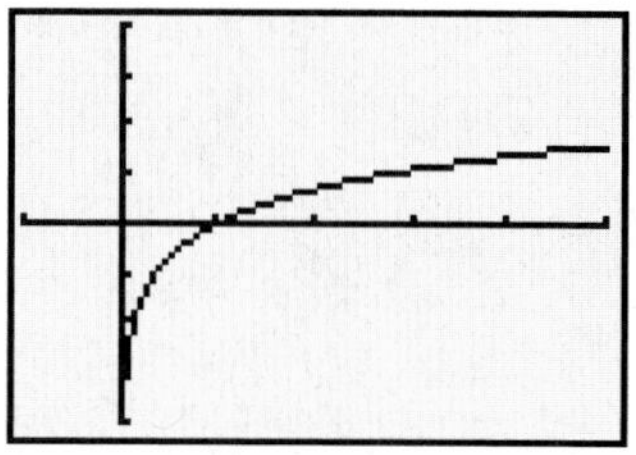

(b)

FIGURE 9-20

Graphing Logarithmic Functions

GRAPHING CALCULATORS

Many graphs of logarithmic functions involve translations of the graph of $y = f(x) = \ln x$. For example, Figure 9-21 shows calculator graphs of the functions $y = \ln x$, $y = \ln x + 2$, and $y = \ln x - 3$.

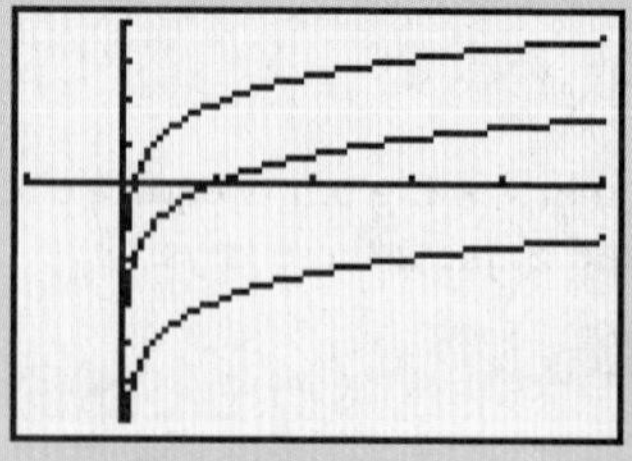

The graph of $y = \ln x + 2$ is 2 units above the graph of $y = \ln x$.

The graph of $y = \ln x - 3$ is 3 units below the graph of $y = \ln x$.

FIGURE 9-21

Figure 9-22 shows the calculator graphs of the functions $y = \ln x$, $y = \ln(x - 2)$, and $y = \ln(x + 3)$.

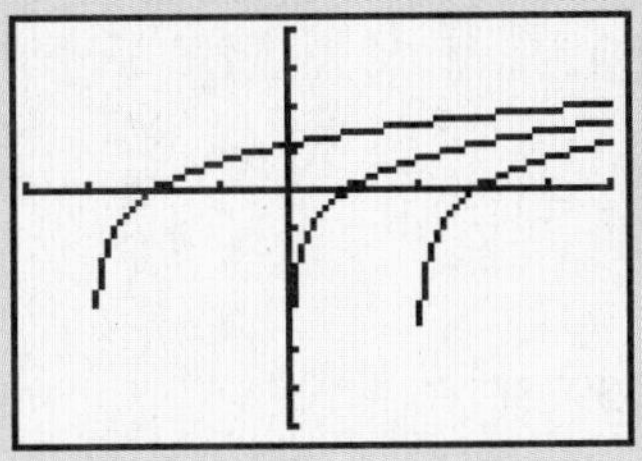

The graph of $y = \ln(x - 2)$ is 2 units to the right of the graph of $y = \ln x$.

The graph of $y = \ln(x + 3)$ is 3 units to the left of the graph of $y = \ln x$.

FIGURE 9-22

Base-e logarithms have many applications.

DOUBLING TIME

If a population grows exponentially at a certain annual rate, the time required for the population to double is called the **doubling time.** It is given by the following formula.

Formula for Doubling Time

If r is the annual rate (compounded continuously) and t is the time required for a population to double, then

$$t = \frac{\ln 2}{r}$$

EXAMPLE 3 The population of the earth is growing at the approximate rate of 2% per year. If this rate continues, in how many years will the population double?

Solution Because the population is growing at the rate of 2% per year, we substitute 0.02 for r into the formula for doubling time and simplify.

$$t = \frac{\ln 2}{r}$$

$$t = \frac{\ln 2}{\mathbf{0.02}}$$

$$\approx 34.65735903$$

The population will double in about 35 years.

Self Check In Example 3, if the world population's annual growth rate could be reduced to 1.5% per year, what would be the doubling time?

Answer 46 years

EXAMPLE 4 **Doubling time** How long will it take \$1,000 to double at an annual rate of 8%, compounded continuously?

Solution We substitute 0.08 for r and simplify:

$$t = \frac{\ln 2}{r}$$

$$t = \frac{\ln 2}{\mathbf{0.08}}$$

$$\approx 8.664339757$$

It will take about $8\frac{2}{3}$ years for the money to double. ■

Self Check In Example 4, how long will it take at 9%, compounded continuously?

Answer about 7.7 years

Orals

1. Write $y = \ln x$ as an exponential equation.
2. Write $e^a = b$ as a logarithmic equation.
3. Write the formula for doubling time.

EXERCISE 9.4

REVIEW *Write the equation of the required line.*

1. Parallel to $y = 5x + 8$ and passing through the origin
2. Having a slope of 9 and a y-intercept of (0, 5)
3. Passing through the point (3, 2) and perpendicular to the line $y = \frac{2}{3}x - 12$
4. Parallel to the line $3x + 2y = 9$ and passing through the point $(-3, 5)$
5. Vertical line through the point (5, 3)
6. Horizontal line through the point (2, 5)

Simplify each expression.

7. $\dfrac{2x + 3}{4x^2 - 9}$
8. $\dfrac{x + 1}{x} + \dfrac{x - 1}{x + 1}$
9. $\dfrac{x^2 + 3x + 2}{3x + 12} \cdot \dfrac{x + 4}{x^2 - 4}$
10. $\dfrac{1 + \frac{y}{x}}{\frac{y}{x} - 1}$

VOCABULARY AND CONCEPTS *Fill in each blank to make a true statement.*

11. $\ln x$ means ______.
12. The domain of the function $y = f(x) = \ln x$ is the interval ______.

13. The range of the function $y = f(x) = \ln x$ is the interval ________.

14. The graph of $y = f(x) = \ln x$ has the ________ as an asymptote.

15. In the expression $\log x$, the base is understood to be ___.

16. In the expression $\ln x$, the base is understood to be ___.

17. If a population grows exponentially at a rate r, the time it will take the population to double is given by the formula $t =$ ___.

18. The logarithm of a negative number is ________.

PRACTICE *In Exercises 19–26, use a calculator to find each value, if possible. Express all answers to four decimal places.*

19. $\ln 25.25$

20. $\ln 0.523$

21. $\ln 9.89$

22. $\ln 0.00725$

23. $\log (\ln 2)$

24. $\ln (\log 28.8)$

25. $\ln (\log 0.5)$

26. $\log (\ln 0.2)$

In Exercises 27–34, use a calculator to find y, if possible. Express all answers to four decimal places.

27. $\ln y = 2.3015$

28. $\ln y = 1.548$

29. $\ln y = 3.17$

30. $\ln y = 0.837$

31. $\ln y = -4.72$

32. $\ln y = -0.48$

33. $\log y = \ln 6$

34. $\ln y = \log 5$

In Exercises 35–38, tell whether the graph could represent the graph of $y = \ln x$.

35.

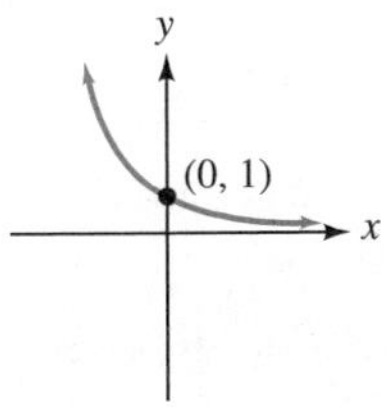

36.

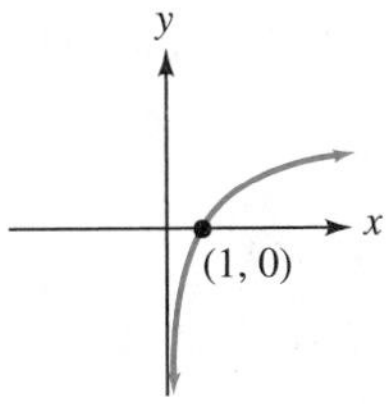

37.

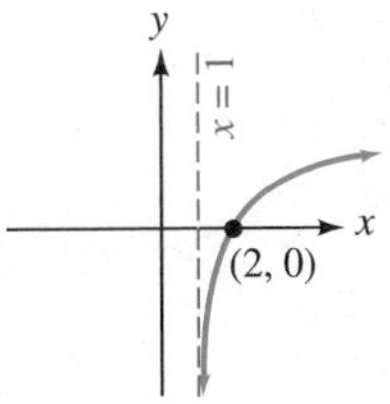

38.

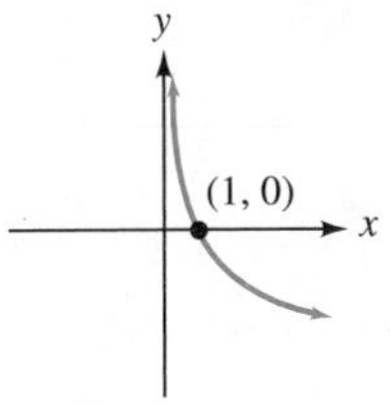

In Exercises 39–42, use a graphing calculator to graph each function.

39. $y = -\ln x$

40. $y = \ln x^2$

41. $y = \ln (-x)$

42. $y = \ln \left(\frac{1}{2}x\right)$

APPLICATIONS *Use a calculator to solve each problem. Round each answer to the nearest tenth.*

43. Population growth See Illustration 1. How long will it take the population of River City to double?

44. Doubling money How long will it take \$1,000 to double if it is invested at an annual rate of 5%, compounded continuously?

River City

A growing community

• 6 parks • 12% annual growth

• 10 churches • Low crime rate

ILLUSTRATION 1

45. Population growth A population growing at an annual rate r will triple in a time t given by the formula

$$t = \frac{\ln 3}{r}$$

How long will it take the population of a town growing at the rate of 12% per year to triple?

46. Tripling money Find the length of time for \$25,000 to triple if invested at 6% annual interest, compounded continuously. (See Exercise 45.)

WRITING

47. Explain why an earthquake measuring 7 on the Richter scale is much worse than an earthquake measuring 6.

48. The time it takes money to double at an annual rate r, compounded continuously, is given by the formula $t = (\ln 2)/r$. Explain why money doubles more quickly as the rate increases.

SOMETHING TO THINK ABOUT

49. Use the formula $P = P_0e^{rt}$ to verify that P will be twice P_0 when $t = \frac{\ln 2}{r}$.

50. Use the formula $P = P_0e^{rt}$ to verify that P will be three times as large as P_0 when $t = \frac{\ln 3}{r}$.

51. Find a formula to find how long it will take money to quadruple.

52. Use a graphing calculator to graph

$$y = f(x) = \frac{1}{1 + e^{-2x}}$$

and discuss the graph.

9.5 Properties of Logarithms

■ PROPERTIES OF LOGARITHMS ■ THE CHANGE-OF-BASE FORMULA ■ CHEMISTRY ■ PHYSIOLOGY

Getting Ready *Simplify each expression.*

1. x^mx^n

2. x^0

3. $(x^m)^n$

4. $\frac{x^m}{x^n}$

■ PROPERTIES OF LOGARITHMS

Since logarithms are exponents, the properties of exponents have counterparts in the theory of logarithms. We begin with four basic properties.

Properties of Logarithms

If b is a positive number and $b \neq 1$, then

1. $\log_b 1 = 0$

2. $\log_b b = 1$

3. $\log_b b^x = x$

4. $b^{\log_b x} = x \quad (x > 0)$

Properties 1 through 4 follow directly from the definition of a logarithm.

1. $\log_b 1 = 0$, because $b^0 = 1$.
2. $\log_b b = 1$, because $b^1 = b$.
3. $\log_b b^x = x$, because $b^x = b^x$.
4. $b^{\log_b x} = x$, because $\log_b x$ is the exponent to which b is raised to get x.

Properties 3 and 4 also indicate that the composition of the exponential and logarithmic functions with the same base (in both directions) is the identity function. This is expected, because the exponential and logarithmic functions are inverse functions.

EXAMPLE 1 Simplify each expression: **a.** $\log_5 1$, **b.** $\log_3 3$, **c.** $\log_7 7^3$, and **d.** $b^{\log_b 7}$.

Solution **a.** By Property 1, $\log_5 1 = 0$, because $5^0 = 1$.

b. By Property 2, $\log_3 3 = 1$, because $3^1 = 3$.

c. By Property 3, $\log_7 7^3 = 3$, because $7^3 = 7^3$.

d. By Property 4, $b^{\log_b 7} = 7$, because $\log_b 7$ is the power to which b is raised to get 7. ■

Self Check Simplify **a.** $\log_4 1$, **b.** $\log_5 5$, **c.** $\log_2 2^4$, and **d.** $5^{\log_5 2}$.

Answers **a.** 0, **b.** 1, **c.** 4, **d.** 2

The next two properties state that

The logarithm of a product is the sum of the logarithms.

The logarithm of a quotient is the difference of the logarithms.

Properties of Logarithms

If M, N, and b are positive numbers and $b \neq 1$, then

5. $\log_b MN = \log_b M + \log_b N$ **6.** $\log_b \frac{M}{N} = \log_b M - \log_b N$

Proof To prove Property 5, we let $x = \log_b M$ and $y = \log_b N$. We use the definition of logarithm to write each equation in exponential form.

$$M = b^x \qquad \text{and} \qquad N = b^y$$

Then $MN = b^x b^y$, and a property of exponents gives

$$MN = b^{x+y} \qquad b^x b^y = b^{x+y}\text{: Keep the base and add the exponents.}$$

We write this exponential equation in logarithmic form as

$$\log_b MN = x + y$$

Substituting the values of x and y completes the proof.

$$\log_b MN = \log_b M + \log_b N \quad \square$$

The proof of Property 6 is similar.

WARNING! By Property 5 of logarithms, the logarithm of a *product* is equal to the *sum* of the logarithms. The logarithm of a sum or a difference usually does not simplify. In general,

$$\log_b (M + N) \neq \log_b M + \log_b N$$
$$\log_b (M - N) \neq \log_b M - \log_b N$$

By Property 6, the logarithm of a *quotient* is equal to the *difference* of the logarithms. The logarithm of a quotient is not the quotient of the logarithms:

$$\log_b \frac{M}{N} \neq \frac{\log_b M}{\log_b N}$$

Verifying Properties

CALCULATORS

We can use a calculator to illustrate Property 5 of logarithms by showing that

$$\ln [(3.7)(15.9)] = \ln 3.7 + \ln 15.9$$

We calculate the left- and right-hand sides of the equation separately and compare the results. To use a scientific calculator to find ln [(3.7)(15.9)], we enter these numbers and press these keys:

3.7 × 15.9 = LN

The display will read 4.074651929 .

To find ln 3.7 + ln 15.9, we enter these numbers and press these keys:

3.7 LN + 15.9 LN =

The display will read 4.074651929 . Since the left- and right-hand sides are equal, the equation is true.

Two more properties state that

The logarithm of an expression to a power is the power times the logarithm of the expression.

If the logarithms of two numbers are equal, the numbers are equal.

Properties of Logarithms

If M, p, and b are positive numbers and $b \neq 1$, then

7. $\log_b M^p = p \log_b M$

8. If $\log_b x = \log_b y$, then $x = y$.

Proof To prove Property 7, we let $x = \log_b M$, write the expression in exponential form, and raise both sides to the pth power:

$$M = b^x$$

$$(M)^p = (b^x)^p \quad \text{Raise both sides to the } p\text{th power.}$$

$$M^p = b^{px} \quad \text{Keep the base and multiply the exponents.}$$

Using the definition of logarithms gives

$$\log_b M^p = px$$

Substituting the value for x completes the proof.

$$\log_b M^p = p \log_b M$$

□

Property 8 follows from the fact that the logarithmic function is a one-to-one function. Property 8 will be important in the next section when we solve logarithmic equations.

We can use the properties of logarithms to write a logarithm as the sum or difference of several logarithms.

EXAMPLE 2 Assume that b $(b \neq 1)$, x, y, and z are positive numbers. Write each expression in terms of the logarithms of x, y, and z: **a.** $\log_b xyz$ and **b.** $\log_b \dfrac{xy}{z}$.

Solution **a.** $\log_b xyz$

$= \log_b (xy)z$

$= \log_b (xy) + \log_b z$ — The log of a product is the sum of the logs.

$= \log_b x + \log_b y + \log_b z$ — The log of a product is the sum of the logs.

b. $\log_b \dfrac{xy}{z}$

$= \log_b (xy) - \log_b z$ — The log of a quotient is the difference of the logs.

$= (\log_b x + \log_b y) - \log_b z$ — The log of a product is the sum of the logs.

$= \log_b x + \log_b y - \log_b z$ — Remove parentheses. ■

Self Check Find $\log_b \frac{x}{yz}$.

Answer $\log_b x - \log_b y - \log_b z$

EXAMPLE 3 Assume that b $(b \neq 1)$, x, y, and z are positive numbers. Write each expression in terms of the logarithms of x, y, and z.

a. $\log_b (x^2y^3z)$ and **b.** $\log_b \dfrac{\sqrt{x}}{y^3z}$

Solution **a.** $\log_b (x^2y^3z) = \log_b x^2 + \log_b y^3 + \log_b z$ — The log of a product is the sum of the logs.

$= 2 \log_b x + 3 \log_b y + \log_b z$ — The log of an expression to a power is the power times the log of the expression.

b. $\log_b \dfrac{\sqrt{x}}{y^3z} = \log_b \sqrt{x} - \log (y^3z)$ — The log of a quotient is the difference of the logs.

$= \log x^{1/2} - (\log y^3 + \log z)$ — $\sqrt{x} = x^{1/2}$; The log of a product is the sum of the logs.

$= \dfrac{1}{2} \log x - (3 \log y + \log z)$ — The log of a power is the power times the log.

$= \dfrac{1}{2} \log x - 3 \log y - \log z$ — Use the distributive property to remove parentheses. ■

Self Check Find $\log_b \sqrt[4]{\dfrac{x^3y}{z}}$.

Answer $\frac{1}{4}(3 \log_b x + \log_b y - \log_b z)$

We can use the properties of logarithms to combine several logarithms into one logarithm.

b

EXAMPLE 4 Assume that b ($b \neq 1$), x, y, and z are positive numbers. Write each expression as one logarithm: **a.** $3 \log_b x + \frac{1}{2} \log_b y$ and **b.** $\frac{1}{2} \log_b (x - 2) - \log_b y + 3 \log_b z$.

Solution **a.** $3 \log_b x + \dfrac{1}{2} \log_b y = \log_b x^3 + \log_b y^{1/2}$ — A power times a log is the log of the power.

$= \log_b (x^3y^{1/2})$ — The sum of two logs is the log of a product.

b. $\dfrac{1}{2} \log_b (x - 2) - \log_b y + 3 \log_b z$

$= \log_b (x - 2)^{1/2} - \log_b y + \log_b z^3$ — A power times a log is the log of the power.

$= \log_b \dfrac{(x - 2)^{1/2}}{y} + \log_b z^3$ — The difference of two logs is the log of the quotient.

$= \log_b \dfrac{z^3\sqrt{x - 2}}{y}$ — The sum of two logs is the log of a product. ■

Self Check Write the expression as one logarithm: $2\log_b x + \frac{1}{2}\log_b y - 2\log_b (x - y)$.

Answer $\log_b \dfrac{x^2\sqrt{y}}{(x - y)^2}$

We summarize the properties of logarithms as follows.

Properties of Logarithms

If b, M, and N are positive numbers and $b \neq 1$, then

1. $\log_b 1 = 0$ **2.** $\log_b b = 1$

3. $\log_b b^x = x$ **4.** $b^{\log_b x} = x$

5. $\log_b MN = \log_b M + \log_b N$ **6.** $\log_b \dfrac{M}{N} = \log_b M - \log_b N$

7. $\log_b M^p = p \log_b M$ **8.** If $\log_b x = \log_b y$, then $x = y$.

EXAMPLE 5 Given that $\log 2 \approx 0.3010$ and $\log 3 \approx 0.4771$, find approximations for **a.** $\log 6$, **b.** $\log 9$, **c.** $\log 18$, and **d.** $\log 2.5$.

Solution **a.**
$$\log 6 = \log (2 \cdot 3)$$
$$= \log 2 + \log 3$$ The log of a product is the sum of the logs.
$$\approx 0.3010 + 0.4771$$ Substitute the value of each logarithm.
$$\approx 0.7781$$

b.
$$\log 9 = \log (3^2)$$
$$= 2 \log 3$$ The log of a power is the power times the log.
$$\approx 2(0.4771)$$ Substitute the value of log 3.
$$\approx 0.9542$$

c.
$$\log 18 = \log (2 \cdot 3^2)$$
$$= \log 2 + \log 3^2$$ The log of a product is the sum of the logs.
$$= \log 2 + 2 \log 3$$ The log of a power is the power times the log.
$$\approx 0.3010 + 2(0.4771)$$
$$\approx 1.2552$$

d.
$$\log 2.5 = \log \left(\frac{5}{2}\right)$$
$$= \log 5 - \log 2$$ The log of a quotient is the difference of the logs.
$$= \log \frac{10}{2} - \log 2$$ Write 5 as $\frac{10}{2}$.
$$= \log 10 - \log 2 - \log 2$$ The log of a quotient is the difference of the logs.
$$= 1 - 2 \log 2$$ $\log_{10} 10 = 1$.
$$\approx 1 - 2(0.3010)$$
$$\approx 0.3980$$

■

Self Check Use the values given in Example 5 and find **a.** log 1.5 and **b.** log 0.2.

Answers **a.** 0.1761, **b.** −0.6990

THE CHANGE-OF-BASE FORMULA

If we know the base-a logarithm of a number, we can find its logarithm to some other base b with a formula called the **change-of-base formula.**

Change-of-Base Formula

If a, b, and x are real numbers, then

$$\log_b x = \frac{\log_a x}{\log_a b}$$

Proof To prove this formula, we begin with the equation $\log_b x = y$.

$y = \log_b x$	
$x = b^y$	Change the equation from logarithmic to exponential form.
$\log_a x = \log_a b^y$	Take the base-a logarithm of both sides.
$\log_a x = y \log_a b$	The log of a power is the power times the log.
$y = \dfrac{\log_a x}{\log_a b}$	Divide both sides by $\log_a b$.
$\log_b x = \dfrac{\log_a x}{\log_a b}$	Refer to the first equation and substitute $\log_b x$ for y. □

If we know logarithms to base a (for example, $a = 10$), we can find the logarithm of x to a new base b. We simply divide the base-a logarithm of x by the base-a logarithm of b.

WARNING! $\dfrac{\log_a x}{\log_a b}$ means that one logarithm is to be divided by the other. They are not to be subtracted.

EXAMPLE 6 Find $\log_4 9$ using base-10 logarithms.

Solution We can substitute 4 for b, 10 for a, and 9 for x into the change-of-base formula:

$$\log_b x = \frac{\log_a x}{\log_a b}$$

$$\log_4 9 = \frac{\log_{10} 9}{\log_{10} 4}$$

$$\approx 1.584962501$$

To four decimal places, $\log_4 9 = 1.5850$. ■

Self Check Find $\log_5 3$ using base-10 logarithms.

Answer 0.6826

CHEMISTRY

Common logarithms are used to express the acidity of solutions. The more acidic a solution, the greater the concentration of hydrogen ions. This concentration is indicated indirectly by the *pH scale,* or *hydrogen ion index.* The pH of a solution is defined by the following equation.

pH of a Solution

If $[H^+]$ is the hydrogen ion concentration in gram-ions per liter, then

$$\text{pH} = -\log [H^+]$$

EXAMPLE 7 **Finding pH of a solution** Find the pH of pure water, which has a hydrogen ion concentration of 10^{-7} gram-ions per liter.

Solution Since pure water has approximately 10^{-7} gram-ions per liter, its pH is

$$\text{pH} = -\log [H^+]$$

$$\text{pH} = -\log 10^{-7}$$

$$= -(-7) \log 10 \qquad \text{The log of a power is the power times the log.}$$

$$= -(-7) \cdot 1 \qquad \log 10 = 1.$$

$$= 7$$

■

EXAMPLE 8 **Finding hydrogen ion concentration** Find the hydrogen ion concentration of seawater if its pH is 8.5.

Solution To find its hydrogen ion concentration, we substitute 8.5 for the pH and find $[H^+]$.

$$8.5 = -\log [H^+]$$

$$-8.5 = \log [H^+] \quad \text{Multiply both sides by } -1.$$

$$[H^+] = 10^{-8.5} \quad \text{Change the equation to exponential form.}$$

We can use a calculator to find that

$$[H^+] \approx 3.2 \times 10^{-9} \text{ gram-ions per liter.}$$

■

PHYSIOLOGY

In physiology, experiments suggest that the relationship between the loudness and the intensity of sound is a logarithmic one known as the **Weber–Fechner law.**

Weber-Fechner Law

If L is the apparent loudness of a sound, I is the actual intensity, and k is a constant, then

$$L = k \ln I$$

EXAMPLE 9 **Weber-Fechner law** Find the increase in intensity that will cause the apparent loudness of a sound to double.

Solution If the original loudness L_O is caused by an actual intensity I_O, then

1. $L_0 = k \ln I_0$

To double the apparent loudness, we multiply both sides of Equation 1 by 2 and use the power rule of logarithms:

$$2L_0 = 2k \ln I_0$$

$$= k \ln (I_0)^2$$

To double the loudness of a sound, the intensity must be squared. ■

Self Check In Example 9, what decrease in intensity will cause a sound to be half as loud?

Answer Find the square root of the intensity.

Orals *Find the value of x in each equation.*

1. $\log_3 9 = x$ **2.** $\log_x 5 = 1$ **3.** $\log_7 x = 3$

4. $\log_2 x = -2$ **5.** $\log_4 x = \frac{1}{2}$ **6.** $\log_x 4 = 2$

7. $\log_{1/2} x = 2$ **8.** $\log_9 3 = x$ **9.** $\log_x \frac{1}{4} = -2$

EXERCISE 9.5

REVIEW *Consider the line that passes through P(−2, 3) and Q(4, −4).*

1. Find the slope of line *PQ*.

2. Find the distance *PQ*.

3. Find the midpoint of segment *PQ*.

4. Write the equation of line *PQ*.

VOCABULARY AND CONCEPTS *Fill in each blank to make a true statement.*

5. $\log_b 1 =$ ___

6. $\log_b b =$ ___

7. $\log_b MN = \log_b$ ___ $+ \log_b$ ___

8. $b^{\log_b x} =$ ___

9. If $\log_b x = \log_b y$, then ___ = ___.

10. $\log_b \frac{M}{N} = \log_b M$ ___ $\log_b N$

11. $\log_b x^p = p \cdot \log_b$ ___

12. $\log_b b^x =$ ___

13. $\log_b (A + B)$ ___ $\log_b A + \log_b B$

14. $\log_b A + \log_b B$ ___ $\log_b AB$

15. $\log_4 1 =$ ___ **16.** $\log_4 4 =$ ___ **17.** $\log_4 4^7 =$ ___ **18.** $4^{\log_4 8} =$ ___

19. $5^{\log_5 10} =$ ___ **20.** $\log_5 5^2 =$ ___ **21.** $\log_5 5 =$ ___ **22.** $\log_5 1 =$ ___

23. $\log_7 1 =$ ___ **24.** $\log_9 9 =$ ___ **25.** $\log_3 3^7 =$ ___ **26.** $5^{\log_5 8} =$ ___

27. $8^{\log_8 10} =$ ___ **28.** $\log_4 4^2 =$ ___ **29.** $\log_9 9 =$ ___ **30.** $\log_3 1 =$ ___

PRACTICE *In Exercises 31–36, use a calculator to verify each equation.*

31. $\log [(2.5)(3.7)] = \log 2.5 + \log 3.7$

32. $\ln \frac{11.3}{6.1} = \ln 11.3 - \ln 6.1$

33. $\ln (2.25)^4 = 4 \ln 2.25$

34. $\log 45.37 = \frac{\ln 45.37}{\ln 10}$

35. $\log \sqrt{24.3} = \frac{1}{2} \log 24.3$

36. $\ln 8.75 = \frac{\log 8.75}{\log e}$

In Exercises 37–48, assume that x, y, z, and b (b ≠ 1) are positive numbers. Use the properties of logarithms to write each expression in terms of the logarithms of x, y, and z.

37. $\log_b xyz$

38. $\log_b 4xz$

39. $\log_b \frac{2x}{y}$

40. $\log_b \frac{x}{yz}$

41. $\log_b x^3y^2$

42. $\log_b xy^2z^3$

43. $\log_b (xy)^{1/2}$

44. $\log_b x^3y^{1/2}$

45. $\log_b x\sqrt{z}$

46. $\log_b \sqrt{xy}$

47. $\log_b \dfrac{\sqrt[3]{x}}{\sqrt[4]{yz}}$

48. $\log_b \sqrt[4]{\dfrac{x^3y^2}{z^4}}$

In Exercises 49–56, assume that x, y, z, and b (b ≠ 1) are positive numbers. Use the properties of logarithms to write each expression as the logarithm of a single quantity.

49. $\log_b (x + 1) - \log_b x$

50. $\log_b x + \log_b (x + 2) - \log_b 8$

51. $2 \log_b x + \dfrac{1}{2}\log_b y$

52. $-2 \log_b x - 3 \log_b y + \log_b z$

53. $-3 \log_b x - 2 \log_b y + \dfrac{1}{2} \log_b z$

54. $3 \log_b (x + 1) - 2 \log_b (x + 2) + \log_b x$

55. $\log_b \left(\dfrac{x}{z} + x\right) - \log_b \left(\dfrac{y}{z} + y\right)$

56. $\log_b (xy + y^2) - \log_b (xz + yz) + \log_b z$

In Exercises 57–68, tell whether the given statement is true. If a statement is false, explain why.

57. $\log_b 0 = 1$

58. $\log_b (x + y) \neq \log_b x + \log_b y$

59. $\log_b xy = (\log_b x)(\log_b y)$

60. $\log_b ab = \log_b a + 1$

61. $\log_7 7^7 = 7$

62. $7^{\log_7 7} = 7$

63. $\dfrac{\log_b A}{\log_b B} = \log_b A - \log_b B$

64. $\log_b (A - B) = \dfrac{\log_b A}{\log_b B}$

65. $3 \log_b \sqrt[3]{a} = \log_b a$

66. $\dfrac{1}{3} \log_b a^3 = \log_b a$

67. $\log_b \dfrac{1}{a} = -\log_b a$

68. $\log_b 2 = \log_2 b$

In Exercises 69–80, assume that log 4 = 0.6021, log 7 = 0.8451, and log 9 = 0.9542. Use these values and the properties of logarithms to find each value. ***Do not use a calculator.***

69. $\log 28$

70. $\log \dfrac{7}{4}$

71. $\log 2.25$

72. $\log 36$

73. $\log \dfrac{63}{4}$

74. $\log \dfrac{4}{63}$

75. $\log 252$

76. $\log 49$

77. $\log 112$

78. $\log 324$

79. $\log \dfrac{144}{49}$

80. $\log \dfrac{324}{63}$

In Exercises 81–88, use a calculator and the change-of-base formula to find each logarithm.

81. $\log_3 7$

82. $\log_7 3$

83. $\log_{1/3} 3$

84. $\log_{1/2} 6$

85. $\log_3 8$

86. $\log_5 10$

87. $\log_{\sqrt{2}} \sqrt{5}$

88. $\log_\pi e$

APPLICATIONS

89. pH of a solution Find the pH of a solution with a hydrogen ion concentration of 1.7×10^{-5} gram-ions per liter.

90. Hydrogen ion concentration Find the hydrogen ion concentration of a saturated solution of calcium hydroxide whose pH is 13.2.

91. Aquariums To test for safe pH levels in a freshwater aquarium, a test strip is compared with the scale shown in Illustration 1. Find the corresponding range in the hydrogen ion concentration.

92. pH of pickles The hydrogen ion concentration of sour pickles is 6.31×10^{-4}. Find the pH.

93. Change in loudness If the intensity of a sound is doubled, find the apparent change in loudness.

94. Change in loudness If the intensity of a sound is tripled, find the apparent change in loudness.

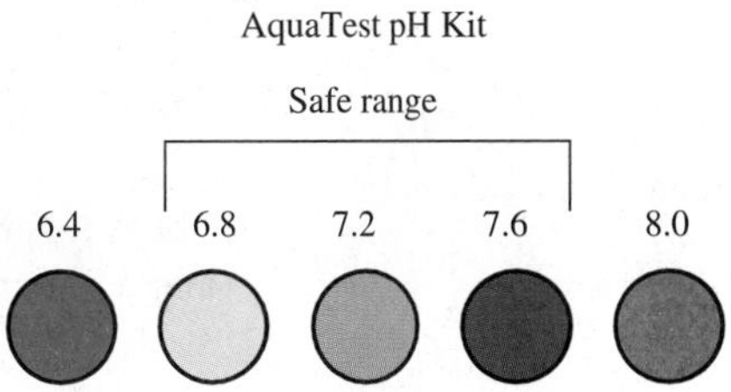

ILLUSTRATION 1

95. Change in intensity What change in intensity of sound will cause an apparent tripling of the loudness?

96. Change in intensity What increase in the intensity of a sound will cause the apparent loudness to be multiplied by 4?

WRITING

97. Explain why ln (log 0.9) is undefined.

98. Explain why $\log_b$ (ln 1) is undefined.

SOMETHING TO THINK ABOUT

99. Show that $\ln (e^x) = x$.

100. If $\log_b 3x = 1 + \log_b x$, find b.

101. Show that $\log_{b^2} x = \frac{1}{2} \log_b x$.

102. Show that $e^{x \ln a} = a^x$.

9.6 Exponential and Logarithmic Equations

■ SOLVING EXPONENTIAL EQUATIONS ■ SOLVING LOGARITHMIC EQUATIONS ■ RADIOACTIVE DECAY ■ POPULATION GROWTH

Getting Ready *Write each expression without using exponents.*

1. $\log x^2$

2. $\log x^{1/2}$

3. $\log x^0$

4. $\log a^b + b \log a$

An **exponential equation** is an equation that contains a variable in one of its exponents. Some examples of exponential equations are

$$3^x = 5, \qquad 6^{x-3} = 2^x, \qquad \text{and} \qquad 3^{2x+1} - 10(3^x) + 3 = 0$$

A **logarithmic equation** is an equation with logarithmic expressions that contain a variable. Some examples of logarithmic equations are

$$\log 2x = 25, \qquad \ln x - \ln (x - 12) = 24, \qquad \text{and} \qquad \log x = \log \frac{1}{x} + 4$$

In this section, we will learn how to solve many of these equations.

SOLVING EXPONENTIAL EQUATIONS

EXAMPLE 1 Solve $4^x = 7$.

Solution Since logarithms of equal numbers are equal, we can take the common logarithm of each side of the equation. The power rule of logarithms then provides a way of moving the variable x from its position as an exponent to a position as a coefficient.

$$4^x = 7$$

$$\log 4^x = \log 7 \qquad \text{Take the common logarithm of each side.}$$

$$x \log 4 = \log 7 \qquad \text{The log of a power is the power times the log.}$$

1. $$x = \frac{\log 7}{\log 4} \qquad \text{Divide both sides by log 4.}$$

$$\approx 1.403677461 \qquad \text{Use a calculator.}$$

To four decimal places, $x = 1.4037$. ■

Self Check Solve $5^x = 4$.

Answer 0.8614

WARNING! A careless reading of Equation 1 leads to a common error. The right-hand side of Equation 1 calls for a division, not a subtraction.

$$\frac{\log 7}{\log 4} \quad \text{means} \quad (\log 7) \div (\log 4)$$

It is the expression $\log \frac{7}{4}$ that means $\log 7 - \log 4$.

EXAMPLE 2 Solve $6^{x-3} = 2^x$.

Solution

$$6^{x-3} = 2^x$$

$$\log 6^{x-3} = \log 2^x \quad \text{Take the common logarithm of each side.}$$

$$(x - 3)\log 6 = x \log 2 \quad \text{The log of a power is the power times the log.}$$

$$x \log 6 - 3 \log 6 = x \log 2 \quad \text{Use the distributive property.}$$

$$x \log 6 - x \log 2 = 3 \log 6 \quad \text{Add 3 log 6 and subtract } x \log 2 \text{ from both sides.}$$

$$x(\log 6 - \log 2) = 3 \log 6 \quad \text{Factor out } x \text{ on the left-hand side.}$$

$$x = \frac{3 \log 6}{\log 6 - \log 2} \quad \text{Divide both sides by } \log 6 - \log 2.$$

$$x \approx 4.892789261 \quad \text{Use a calculator.}$$

■

Self Check Solve $5^{x-2} = 3^x$.

Answer $\frac{2 \log 5}{\log 5 - \log 3} \approx 6.301320206$

EXAMPLE 3 Solve $2^{x^2 + 2x} = \frac{1}{2}$.

Solution Since $\frac{1}{2} = 2^{-1}$, we can write the equation in the form

$$2^{x^2+2x} = 2^{-1}$$

Since equal quantities with equal bases have equal exponents, we have

$$x^2 + 2x = -1$$

$$x^2 + 2x + 1 = 0 \quad \text{Add 1 to both sides.}$$

$$(x + 1)(x + 1) = 0 \quad \text{Factor the trinomial.}$$

$$x + 1 = 0 \quad \text{or} \quad x + 1 = 0 \quad \text{Set each factor equal to 0.}$$

$$x = -1 \qquad\qquad x = -1$$

Verify that -1 satisfies the equation. ■

Self Check Solve $3^{x^2-2x} = \frac{1}{3}$.

Answer 1, 1

Solving Exponential Equations

GRAPHING CALCULATORS

To use a graphing calculator to approximate the solutions of $2^{x^2+2x} = \frac{1}{2}$ (see Example 3), we can subtract $\frac{1}{2}$ from both sides of the equation to get

$$2^{x^2+2x} - \frac{1}{2} = 0$$

and graph the corresponding function

$$y = f(x) = 2^{x^2+2x} - \frac{1}{2}$$

If we use window settings of $[-4, 4]$ for x and $[-2, 6]$ for y, we obtain the graph shown in Figure 9-23(a).

Since the solutions of the equation are its x-intercepts, we can approximate the solutions by zooming in on the values of the x-intercepts, as in Figure 9-23(b). Since $x = -1$ is the only x-intercept, -1 is the only solution. In this case, we have found an exact solution.

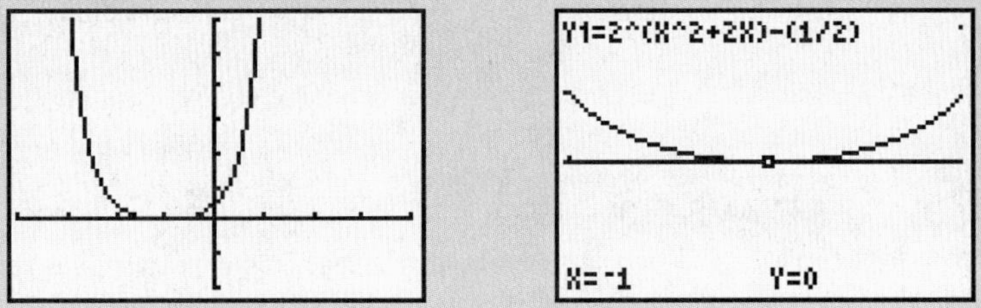

(a) (b)

FIGURE 9-23

SOLVING LOGARITHMIC EQUATIONS

In each of the following examples, we use the properties of logarithms to change a logarithmic equation into an algebraic equation.

EXAMPLE 4 Solve $\log_b (3x + 2) - \log_b (2x - 3) = 0$.

Solution

$$\log_b (3x + 2) - \log_b (2x - 3) = 0$$

$$\log_b (3x + 2) = \log_b (2x - 3)$$ Add $\log_b (2x - 3)$ to both sides.

$$3x + 2 = 2x - 3$$ If $\log_b r = \log_b s$, then $r = s$.

$$x = -5$$ Subtract $2x$ and 2 from both sides.

Check:

$$\log_b (3x + 2) - \log_b (2x - 3) = 0$$
$$\log_b [3(-5) + 2] - \log_b [2(-5) - 3] \stackrel{?}{=} 0$$
$$\log_b (-13) - \log_b (-13) \stackrel{?}{=} 0$$

Since the logarithm of a negative number does not exist, the apparent solution of -5 must be discarded. This equation has no solutions. ■

Self Check Solve $\log_b (5x + 2) - \log_b (7x - 2) = 0$.

Answer 2

WARNING! Example 4 illustrates that you must check the solutions of a logarithmic equation.

EXAMPLE 5 Solve $\log x + \log (x - 3) = 1$.

Solution

$$\log x + \log (x - 3) = 1$$

$$\log x(x - 3) = 1$$ The sum of two logs is the log of a product.

$$x(x - 3) = 10^1$$ Use the definition of logarithms to change the equation to exponential form.

$$x^2 - 3x - 10 = 0$$ Remove parentheses and subtract 10 from both sides.

$$(x + 2)(x - 5) = 0$$ Factor the trinomial.

$$x + 2 = 0 \quad \text{or} \quad x - 5 = 0$$ Set each factor equal to 0.

$$x = -2 \qquad\qquad x = 5$$

Check: The number -2 is not a solution, because it does not satisfy the equation (a negative number does not have a logarithm). We will check the remaining number, 5.

$$\log x + \log (x - 3) = 1$$

$$\log 5 + \log (5 - 3) \stackrel{?}{=} 1$$ Substitute 5 for x.

$$\log 5 + \log 2 \stackrel{?}{=} 1$$

$$\log 10 \stackrel{?}{=} 1$$ The sum of two logs is the log of a product.

$$1 = 1$$ $\log 10 = 1$.

Since 5 satisfies the equation, it is a solution. ■

Self Check Solve $\log x + \log (x + 3) = 1$.

Answer 2

EXAMPLE 6 Solve $\dfrac{\log(5x-6)}{\log x} = 2.$

Solution We can multiply both sides of the equation by $\log x$ to get

$$\log(5x-6) = 2\log x$$

and apply the power rule of logarithms to get

$$\log(5x-6) = \log x^2$$

By Property 8 of logarithms, $5x - 6 = x^2$, because they have equal logarithms. Thus,

$$\begin{aligned} 5x - 6 &= x^2 \\ 0 &= x^2 - 5x + 6 \\ 0 &= (x-3)(x-2) \end{aligned}$$

$$x - 3 = 0 \quad \text{or} \quad x - 2 = 0$$
$$x = 3 \qquad\qquad x = 2$$

Verify that both 2 and 3 satisfy the equation. ■

Self Check Solve $\dfrac{\log(5x+6)}{\log x} = 2.$

Answer 6

Solving Logarithmic Equations

GRAPHING CALCULATORS

To use a graphing calculator to approximate the solutions of $\log x + \log(x-3) = 1$ (see Example 5), we can subtract 1 from both sides of the equation to get

$$\log x + \log(x-3) - 1 = 0$$

and graph the corresponding function

$$y = f(x) = \log x + \log(x-3) - 1$$

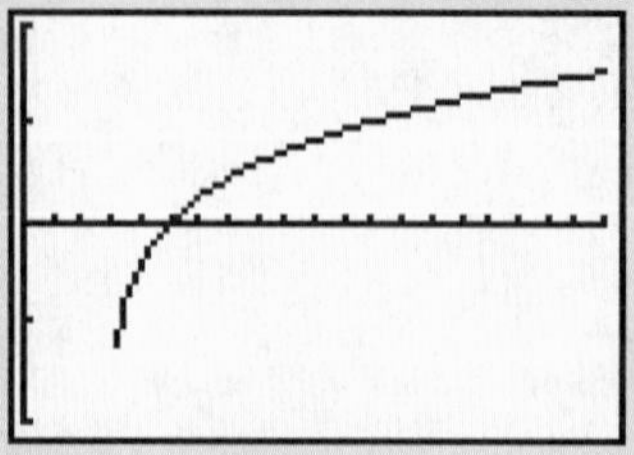

FIGURE 9-24

If we use window settings of [0, 20] for x and [−2, 2] for y, we obtain the graph shown in Figure 9-24. Since the solution of the equation is the x-intercept, we can find the solution by zooming in on the value of the x-intercept. The solution is $x = 5$.

RADIOACTIVE DECAY

Experiments have determined the time that it takes for half of a sample of a given radioactive material to decompose. This time is a constant, called the material's **half-life.**

When living organisms die, the oxygen/carbon dioxide cycle common to all living things stops, and carbon-14, a radioactive isotope with a half-life of 5,700 years, is no longer absorbed. By measuring the amount of carbon-14 present in an ancient object, archaeologists can estimate the object's age by using the radioactive decay formula.

Radioactive Decay Formula

If A is the amount of radioactive material present at time t, A_0 was the amount present at $t = 0$, and h is the material's half-life, then

$$A = A_0 2^{-t/h}$$

EXAMPLE 7 **Carbon-14 dating** How old is a wooden statue that retains only one-third of its original carbon-14 content?

Solution To find the time t when $A = \frac{1}{3}A_0$, we substitute $\frac{A^0}{3}$ for A and 5,700 for h into the radioactive decay formula and solve for t:

$$A = A_0 2^{-t/h}$$

$$\frac{A_0}{3} = A_0 2^{-t/5,700}$$

$$1 = 3(2^{-t/5,700})$$ Divide both sides by A_0 and multiply both sides by 3.

$$\log 1 = \log 3(2^{-t/5,700})$$ Take the common logarithm of each side.

$$0 = \log 3 + \log 2^{-t/5,700}$$ Log 1 = 0, and the log of a product is the sum of the logs.

$$-\log 3 = -\frac{t}{5,700} \log 2$$ Subtract log 3 from both sides and use the power rule of logarithms.

$$5,700\left(\frac{\log 3}{\log 2}\right) = t$$ Multiply both sides by $-\frac{5,700}{\log 2}$.

$$t \approx 9,034.286254$$ Use a calculator.

The statue is approximately 9,000 years old. ■

Self Check How old is a statue that retains 25% of its original carbon-14 content?

Answer about 11,400 years

POPULATION GROWTH

Recall that when there is sufficient food and space, populations of living organisms tend to increase exponentially according to the Malthusian growth model.

Malthusian Growth Model

If P is the population at some time t, P_0 is the initial population at $t = 0$, and k depends on the rate of growth, then

$$P = P_0 e^{kt}$$

EXAMPLE 8

Population growth The bacteria in a laboratory culture increased from an initial population of 500 to 1,500 in 3 hours. How long will it take for the population to reach 10,000?

Solution We substitute 500 for P_0, 1,500 for P, and 3 for t and simplify to find k:

$$P = P_0 e^{kt}$$

$$\mathbf{1{,}500} = \mathbf{500}(e^{k3}) \quad \text{Substitute 1,500 for } P\text{, 500 for } P_0\text{, and 3 for } t.$$

$$3 = e^{3k} \quad \text{Divide both sides by 500.}$$

$$3k = \ln 3 \quad \text{Change the equation from exponential to logarithmic form.}$$

$$k = \frac{\ln 3}{3} \quad \text{Divide both sides by 3.}$$

To find when the population will reach 10,000, we substitute 10,000 for P, 500 for P_0, and $\frac{\ln 3}{3}$ for k in the equation $P = P_0 e^{kt}$ and solve for t:

$$P = P_0 e^{kt}$$

$$\mathbf{10{,}000} = \mathbf{500} e^{[(\ln 3)/3]t}$$

$$20 = e^{[(\ln 3)/3]t} \quad \text{Divide both sides by 500.}$$

$$\left(\frac{\ln 3}{3}\right)t = \ln 20 \quad \text{Change the equation to logarithmic form.}$$

$$t = \frac{3 \ln 20}{\ln 3} \quad \text{Multiply both sides by } \frac{3}{\ln 3}.$$

$$\approx 8.180499084 \quad \text{Use a calculator.}$$

The culture will reach 10,000 bacteria in about 8 hours. ■

Self Check In Example 8, how long will it take to reach 20,000?

Answer about 10 hours

EXAMPLE 9

Generation time If a medium is inoculated with a bacterial culture that contains 1,000 cells per milliliter, how many generations will pass by the time the culture has grown to a population of 1 million cells per milliliter?

Solution During bacterial reproduction, the time required for a population to double is called the *generation time.* If b bacteria are introduced into a medium, then after the generation time of the organism has elapsed, there are $2b$ cells. After another generation, there are $2(2b)$, or $4b$ cells, and so on. After n generations, the number of cells present will be

1. $B = b \cdot 2^n$

To find the number of generations that have passed while the population grows from b bacteria to B bacteria, we solve Equation 1 for n.

$$\log B = \log (b \cdot 2^n)$$ Take the common logarithm of both sides.

$$\log B = \log b + n \log 2$$ Apply the product and power rules of logarithms.

$$\log B - \log b = n \log 2$$ Subtract $\log b$ from both sides.

$$n = \frac{1}{\log 2}(\log B - \log b)$$ Multiply both sides by $\frac{1}{\log 2}$.

2. $$n = \frac{1}{\log 2}\left(\log \frac{B}{b}\right)$$ Use the quotient rule of logarithms.

Equation 2 is a formula that gives the number of generations that will pass as the population grows from b bacteria to B bacteria.

To find the number of generations that have passed while a population of 1,000 cells per milliliter has grown to a population of 1 million cells per milliliter, we substitute 1,000 for b and 1,000,000 for B in Equation 2 and solve for n.

$$n = \frac{1}{\log 2} \log \frac{1{,}000{,}000}{1{,}000}$$

$$= \frac{1}{\log 2} \log 1{,}000$$ Simplify.

$$= 3.321928095(3)$$ $\frac{1}{\log 2} \approx 3.321928095$ and $\log 1{,}000 = 3$.

$$= 9.965784285$$

Approximately 10 generations will have passed. ■

Orals *Solve each equation for x.* **Do not simplify answers.**

1. $3^x = 5$

2. $5^x = 3$

3. $2^{-x} = 7$

4. $6^{-x} = 1$

5. $\log 2x = \log (x + 2)$

6. $\log 2x = 0$

7. $\log x^4 = 4$

8. $\log \sqrt{x} = \frac{1}{2}$

EXERCISE 9.6

REVIEW *Solve each equation.*

1. $5x^2 - 25x = 0$

2. $4y^2 - 25 = 0$

3. $3p^2 + 10p = 8$

4. $4t^2 + 1 = -6t$

VOCABULARY AND CONCEPTS *Fill in each blank to make a true statement.*

5. An equation with a variable in its exponent is called a(n) ________ equation.

6. An equation with a logarithmic expression that contains a variable is a(n) ________ equation.

7. The formula for carbon dating is $A =$ ______.

8. The formula for population growth is $P =$ _____.

PRACTICE *In Exercises 9–28, solve each exponential equation. If an answer is not exact, give the answer to four decimal places.*

9. $4^x = 5$

10. $7^x = 12$

11. $13^{x-1} = 2$

12. $5^{x+1} = 3$

13. $2^{x+1} = 3^x$

14. $5^{x-3} = 3^{2x}$

15. $2^x = 3^x$

16. $3^{2x} = 4^x$

17. $7^{x^2} = 10$

18. $8^{x^2} = 11$

19. $8^{x^2} = 9^x$

20. $5^{x^2} = 2^{5x}$

21. $2^{x^2-2x} = 8$

22. $3^{x^2-3x} = 81$

23. $3^{x^2+4x} = \frac{1}{81}$

24. $7^{x^2+3x} = \frac{1}{49}$

25. $4^{x+2} - 4^x = 15$ (*Hint:* $4^{x+2} = 4^x4^2$.)

26. $3^{x+3} + 3^x = 84$ (*Hint:* $3^{x+3} = 3^x3^3$.)

27. $2(3^x) = 6^{2x}$

28. $2(3^{x+1}) = 3(2^{x-1})$

In Exercises 29–32, use a calculator to solve each equation, if possible. Give all answers to the nearest tenth.

29. $2^{x+1} = 7$

30. $3^{x-1} = 2^x$

31. $2^{x^2-2x} - 8 = 0$

32. $3^x - 10 = 3^{-x}$

In Exercises 33–62, solve each logarithmic equation. Check all solutions.

33. $\log 2x = \log 4$

34. $\log 3x = \log 9$

35. $\log (3x + 1) = \log (x + 7)$

36. $\log (x^2 + 4x) = \log (x^2 + 16)$

37. $\log (3 - 2x) - \log (x + 24) = 0$

38. $\log (3x + 5) - \log (2x + 6) = 0$

39. $\log \dfrac{4x + 1}{2x + 9} = 0$

40. $\log \dfrac{2 - 5x}{2(x + 8)} = 0$

41. $\log x^2 = 2$

42. $\log x^3 = 3$

43. $\log x + \log (x - 48) = 2$

44. $\log x + \log (x + 9) = 1$

45. $\log x + \log (x - 15) = 2$

46. $\log x + \log (x + 21) = 2$

47. $\log (x + 90) = 3 - \log x$

48. $\log (x - 90) = 3 - \log x$

49. $\log (x - 6) - \log (x - 2) = \log \dfrac{5}{x}$

50. $\log (3 - 2x) - \log (x + 9) = 0$

51. $\log x^2 = (\log x)^2$

52. $\log (\log x) = 1$

53. $\dfrac{\log (3x - 4)}{\log x} = 2$

54. $\dfrac{\log (8x - 7)}{\log x} = 2$

55. $\dfrac{\log (5x + 6)}{2} = \log x$

56. $\dfrac{1}{2} \log (4x + 5) = \log x$

57. $\log_3 x = \log_3 \left(\dfrac{1}{x}\right) + 4$

58. $\log_5 (7 + x) + \log_5 (8 - x) - \log_5 2 = 2$

59. $2 \log_2 x = 3 + \log_2 (x - 2)$

60. $2 \log_3 x - \log_3 (x - 4) = 2 + \log_3 2$

61. $\log (7y + 1) = 2 \log (y + 3) - \log 2$

62. $2 \log (y + 2) = \log (y + 2) - \log 12$

In Exercises 63–66, use a graphing calculator to solve each equation. If an answer is not exact, give all answers to the nearest tenth.

63. $\log x + \log (x - 15) = 2$

64. $\log x + \log (x + 3) = 1$

65. $\ln (2x + 5) - \ln 3 = \ln (x - 1)$

66. $2 \log(x^2 + 4x) = 1$

APPLICATIONS

67. **Tritium decay** The half-life of tritium is 12.4 years. How long will it take for 25% of a sample of tritium to decompose?

68. **Radioactive decay** In two years, 20% of a radioactive element decays. Find its half-life.

69. **Thorium decay** An isotope of thorium, ^{227}Th, has a half-life of 18.4 days. How long will it take for 80% of the sample to decompose?

70. **Lead decay** An isotope of lead, ^{201}Pb, has a half-life of 8.4 hours. How many hours ago was there 30% more of the substance?

71. **Carbon-14 dating** The bone fragment shown in Illustration 1 contains 60% of the carbon-14 that it is assumed to have had initially. How old is it?

ILLUSTRATION 1

72. **Carbon-14 dating** Only 10% of the carbon-14 in a small wooden bowl remains. How old is the bowl?

73. Compound interest If \$500 is deposited in an account paying 8.5% annual interest, compounded semiannually, how long will it take for the account to increase to \$800?

74. Continuous compound interest In Exercise 73, how long will it take if the interest is compounded continuously?

75. Compound interest If \$1,300 is deposited in a savings account paying 9% interest, compounded quarterly, how long will it take the account to increase to \$2,100?

76. Compound interest A sum of \$5,000 deposited in an account grows to \$7,000 in 5 years. Assuming annual compounding, what interest rate is being paid?

77. Rule of seventy A rule of thumb for finding how long it takes an investment to double is called the **rule of seventy.** To apply the rule, divide 70 by the interest rate written as a percent. At 5%, it takes $\frac{70}{5} = 14$ years to double an investment. At 7%, it takes $\frac{70}{7} = 10$ years. Explain why this formula works.

78. Bacterial growth A bacterial culture grows according to the formula

$$P = P_0 a^t$$

If it takes 5 days for the culture to triple in size, how long will it take to double in size?

79. Rodent control The rodent population in a city is currently estimated at 30,000. If it is expected to double every 5 years, when will the population reach 1 million?

80. Population growth The population of a city is expected to triple every 15 years. When can the city planners expect the present population of 140 persons to double?

81. Bacterial culture A bacterial culture doubles in size every 24 hours. By how much will it have increased in 36 hours?

82. Oceanography The intensity I of a light a distance x meters beneath the surface of a lake decreases exponentially. From Illustration 2, find the depth at which the intensity will be 20%.

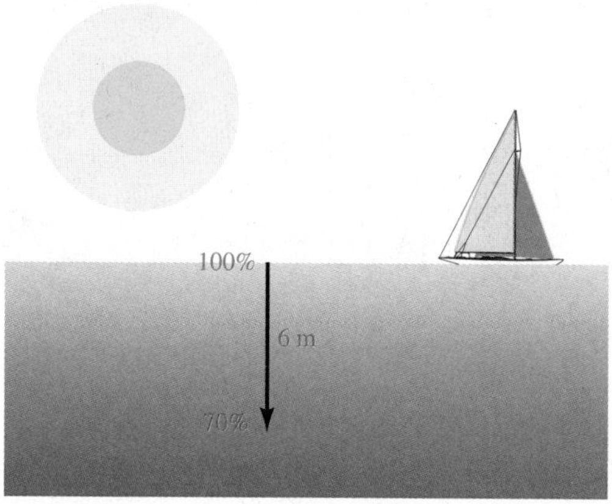

ILLUSTRATION 2

83. Medicine If a medium is inoculated with a bacterial culture containing 500 cells per milliliter, how many generations will have passed by the time the culture contains 5×10^6 cells per milliliter?

84. Medicine If a medium is inoculated with a bacterial culture containing 800 cells per milliliter, how many generations will have passed by the time the culture contains 6×10^7 cells per milliliter?

WRITING

85. Explain how to solve the equation $2^x = 7$.

86. Explain how to solve the equation $x^2 = 7$.

SOMETHING TO THINK ABOUT

87. Without solving the following equation, find the values of x that cannot be a solution:

$$\log (x - 3) - \log (x^2 + 2) = 0$$

88. Solve the equation $x^{\log x} = 10{,}000$.

PROJECTS

PROJECT 1

When an object moves through air, it encounters air resistance. So far, all ballistics problems in this text have ignored air resistance. We now consider the case where an object's fall is affected by air resistance.

At relatively low velocities ($v < 200$ feet per second), the force resisting an object's motion is a constant multiple of the object's velocity:

$$\text{Resisting force} = f_r = bv$$

where b is a constant that depends on the size, shape, and texture of the object, and has units of kilograms per second. This is known as *Stokes' law of resistance.*

In a vacuum, the downward velocity of an object dropped with an initial velocity of 0 feet per second is

$$v(t) = 32t \quad \text{(no air resistance)}$$

t seconds after it is released. However, with air resistance, the velocity is given by the formula

$$v(t) = \frac{32m}{b}\left(1 - e^{-(b/m)t}\right)$$

where m is the object's mass (in kilograms). There is also a formula for the distance an object falls (in feet) during the first t seconds after release, taking into account air resistance:

$$d(t) = \frac{32m}{b}t - \frac{32m^2}{b^2}\left(1 - e^{-(b/m)t}\right)$$

Without air resistance, the formula would be

$$d(t) = 16t^2$$

a. Fearless Freda, a renowned skydiving daredevil, performs a practice dive from a hot-air balloon with an altitude of 5,000 feet. With her parachute on, Freda has a mass of 75 kg, so that $b = 15$ kg/sec. How far (to the nearest foot) will Freda fall in 5 seconds? Compare this with the answer you get by disregarding air resistance.

b. What downward velocity (to the nearest ft/sec) does Freda have after she has fallen for 2 seconds? For 5 seconds? Compare these answers with the answers you get by disregarding air resistance.

(continued)

PROJECTS *(continued)*

c. Find Freda's downward velocity after falling for 20, 22, and 25 seconds (without air resistance, Freda would hit the ground in less than 18 seconds). Note that Freda's velocity increases only slightly. This is because for a large enough velocity, the force of air resistance almost counteracts the force of gravity; after Freda has been falling for a few seconds, her velocity becomes nearly constant. The constant velocity that a falling object approaches is called the *terminal velocity.*

$$\text{Terminal velocity} = \frac{32m}{b}$$

Find Freda's terminal velocity for her practice dive.

d. In Freda's show, she dives from a hot-air balloon with an altitude of only 550 feet, and pulls her ripcord when her velocity is 100 feet per second. (She can't tell her speed, but she knows how long it takes to reach that speed.) It takes a fall of 80 more feet for the chute to open fully, but then the chute increases the force of air resistance, making $b = 80$. After that, Freda's velocity approaches the terminal velocity of an object with this new b value.

To the nearest hundredth of a second, how long should Freda fall before she pulls the ripcord? To the nearest foot, how close is she to the ground when she pulls the ripcord? How close to the ground is she when the chute takes full effect? At what velocity will Freda hit the ground?

PROJECT 2

If an object at temperature T_0 is surrounded by a constant temperature T_s (for instance, an oven or a large amount of fluid that has a constant temperature), the temperature of the object will change with time t according to the formula

$$T(t) = T_s + (T_0 - T_s)e^{-kt}$$

This is *Newton's law of cooling and warming.* The number k is a constant that depends on how well the object absorbs and dispels heat.

In the course of brewing "yo ho! grog," the dread pirates of Hancock Isle have learned that it is important that their rather disgusting, soupy mash be heated slowly to allow all of the ingredients a chance to add their particular offensiveness to the mixture. However, after the mixture has simmered for several hours, it is equally important that the grog be cooled very quickly, so that it retains its potency. The kegs of grog are then stored in a cool spring.

By trial and error, the pirates have learned that by placing the mash pot into a tub of boiling water (100° C), they can heat the mash in the correct amount of time. They have also learned that they can cool the grog to the temperature of the spring by placing it in ice caves for 1 hour.

With a thermometer, you find that the pirates heat the mash from 20° C to 95° C and then cool the grog from 95° C to 7° C. Calculate how long the pirates cook the mash, and how cold the ice caves are. Assume that $k = 0.5$, and t is measured in hours.

CHAPTER SUMMARY

CONCEPT | **REVIEW EXERCISES**

SECTION 9.1 Exponential Functions

An exponential function with base b is defined by the equation

$y = f(x) = b^x$
$(b > 0, b \neq 1)$

1. Use properties of exponents to simplify.

a. $5^{\sqrt{2}} \cdot 5^{\sqrt{2}}$ **b.** $\left(2^{\sqrt{5}}\right)^{\sqrt{2}}$

2. Graph the function defined by each equation.

a. $y = 3^x$ **b.** $y = \left(\frac{1}{3}\right)^x$

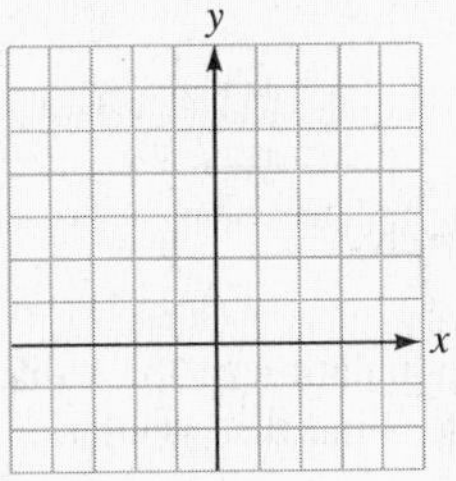

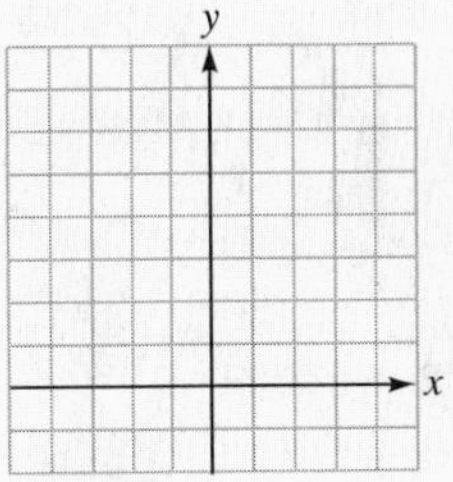

3. The graph of $f(x) = 6^x$ will pass through the points $(0, x)$ and $(1, y)$. Find x and y.

4. Give the domain and range of the function $f(x) = b^x$.

5. Graph each function by using a translation.

a. $y = f(x) = \left(\frac{1}{2}\right)^x - 2$ **b.** $y = f(x) = \left(\frac{1}{2}\right)^{x+2}$

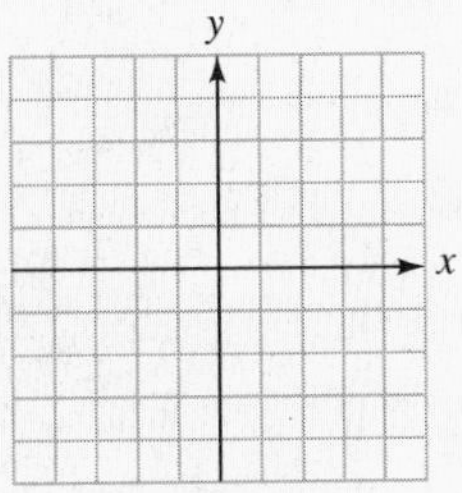

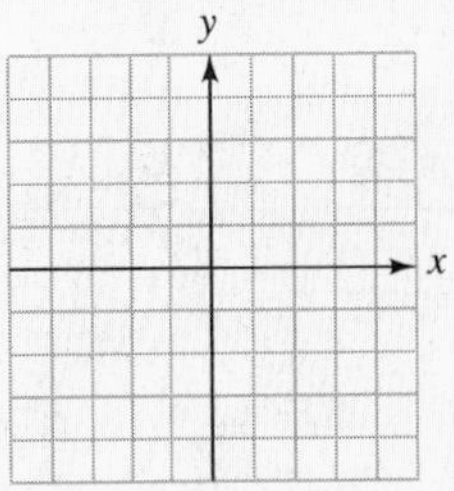

Compound interest

$A = P\left(1 + \frac{r}{k}\right)^{kt}$

6. How much will \$10,500 become if it earns 9% per year for 60 years, compounded quarterly?

SECTION 9.2 Base-*e* Exponential Functions

$e = 2.71828182845904...$

Continuous compound interest

$A = Pe^{rt}$

Malthusian population growth

$P = P_0e^{kt}$

7. If \$10,500 accumulates interest at an annual rate of 9%, compounded continuously, how much will be in the account in 60 years?

8. Graph each function.

a. $y = f(x) = e^x + 1$

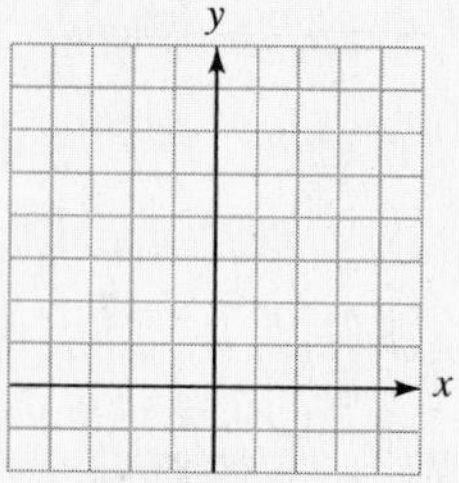

b. $y = f(x) = e^{x-3}$

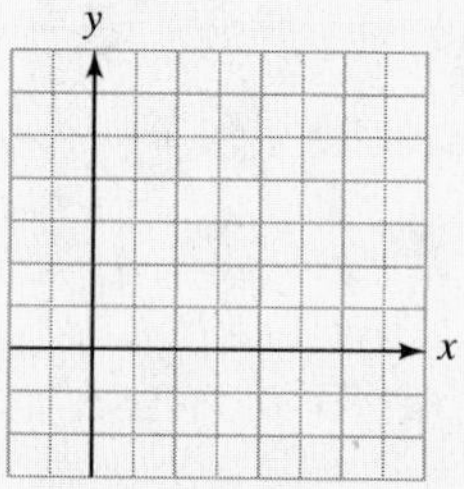

9. The population of the United States is approximately 275,000,000 people. Find the population in 50 years if $k = 0.015$.

SECTION 9.3 Logarithmic Functions

If $b > 0$ and $b \neq 1$, then

$y = \log_b x$ means $x = b^y$

10. Give the domain and range of the logarithmic function.

11. Find each value.

a. $\log_3 9$

b. $\log_9 \frac{1}{3}$

c. $\log_\pi 1$

d. $\log_5 0.04$

e. $\log_a \sqrt{a}$

f. $\log_a \sqrt[3]{a}$

12. Find x.

a. $\log_2 x = 5$

b. $\log_{\sqrt{3}} x = 4$

c. $\log_{\sqrt{3}} x = 6$

d. $\log_{0.1} 10 = x$

e. $\log_x 2 = -\frac{1}{3}$

f. $\log_x 32 = 5$

g. $\log_{0.25} x = -1$

h. $\log_{0.125} x = -\frac{1}{3}$

i. $\log_{\sqrt{2}} 32 = x$

j. $\log_{\sqrt{5}} x = -4$

k. $\log_{\sqrt{3}} 9\sqrt{3} = x$

l. $\log_{\sqrt{5}} 5\sqrt{5} = x$

13. Graph each function.

a. $y = f(x) = \log(x - 2)$

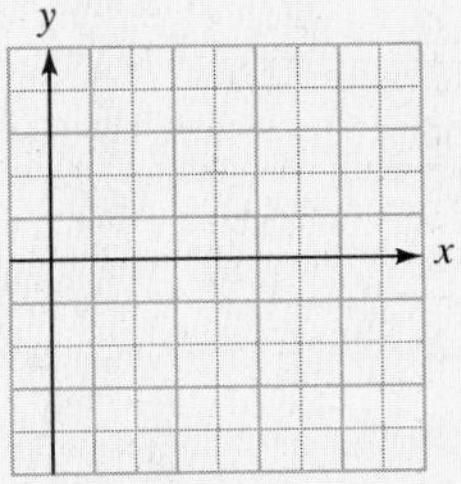

b. $y = f(x) = 3 + \log x$

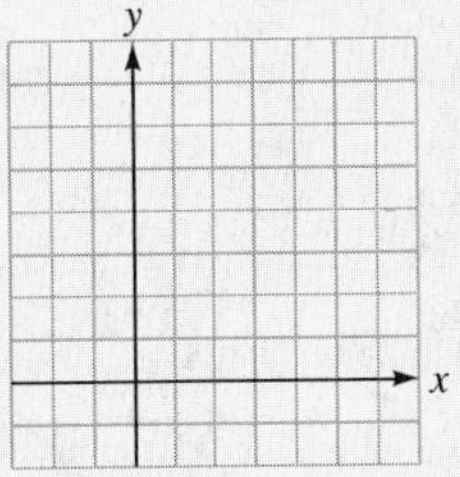

14. Graph each pair of equations on one set of coordinate axes.

a. $y = 4^x$ and $y = \log_4 x$

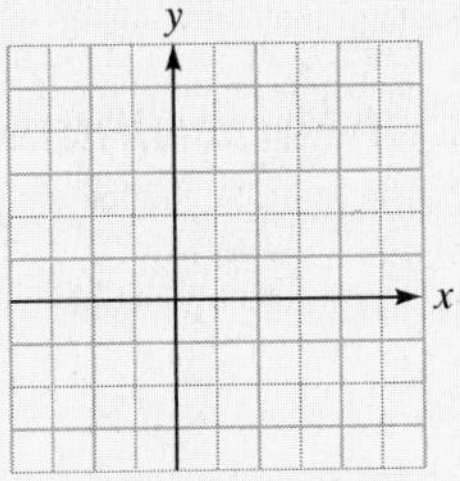

b. $y = \left(\frac{1}{3}\right)^x$ and $y = \log_{1/3} x$

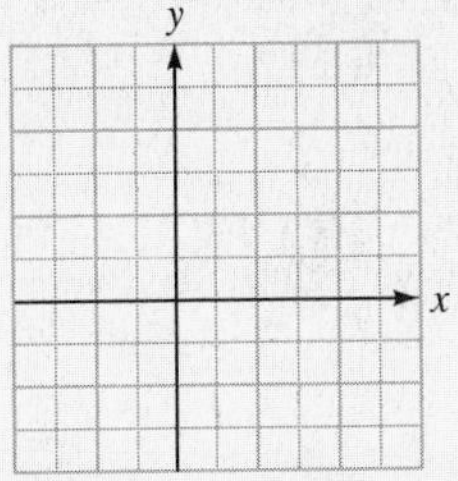

Decibel voltage gain

$$\text{db gain} = 20 \log \frac{E_O}{E_I}$$

15. An amplifier has an output of 18 volts when the input is 0.04 volt. Find the db gain.

Richter scale

$$R = \log \frac{A}{P}$$

16. An earthquake had a period of 0.3 second and an amplitude of 7,500 micrometers. Find its measure on the Richter scale.

SECTION 9.4 Base-e Logarithms

$\ln x$ means $\log_e x$.

17. Use a calculator to find each value to four decimal places.

a. $\ln 452$

b. $\ln(\log 7.85)$

18. Solve each equation.

a. $\ln x = 2.336$

b. $\ln x = \log 8.8$

19. Graph each function.

a. $y = f(x) = 1 + \ln x$

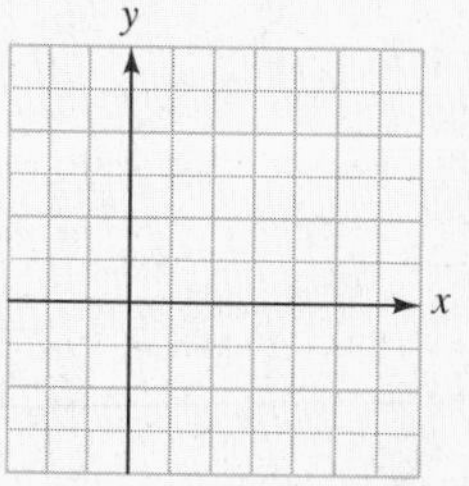

b. $y = f(x) = \ln(x + 1)$

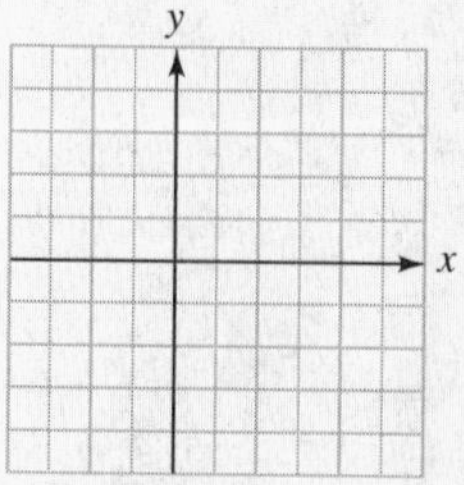

Population doubling time

$$t = \frac{\ln 2}{r}$$

20. How long will it take the population of the United States to double if the growth rate is 3% per year?

SECTION 9.5 Properties of Logarithms

Properties of logarithms
If b is a positive number and $b \neq 1$,

1. $\log_b 1 = 0$
2. $\log_b b = 1$
3. $\log_b b^x = x$
4. $b^{\log_b x} = x$
5. $\log_b MN = \log_b M + \log_b N$
6. $\log_b \frac{M}{N} = \log_b M - \log_b N$
7. $\log_b M^p = p \log_b M$
8. If $\log_b x = \log_b y$, then $x = y$.

21. Simplify each expression.

a. $\log_7 1$ **b.** $\log_7 7$
c. $\log_7 7^3$ **d.** $7^{\log_7 4}$

22. Simplify each expression.

a. $\ln e^4$ **b.** $\ln 1$
c. $10^{\log_{10} 7}$ **d.** $e^{\ln 3}$
e. $\log_b b^4$ **f.** $\ln e^9$

23. Write each expression in terms of the logarithms of x, y, and z.

a. $\log_b \frac{x^2y^3}{z^4}$ **b.** $\log_b \sqrt{\frac{x}{yz^2}}$

24. Write each expression as the logarithm of one quantity.

a. $3 \log_b x - 5 \log_b y + 7 \log_b z$

b. $\frac{1}{2} \log_b x + 3 \log_b y - 7 \log_b z$

25. Assume that $\log a = 0.6$, $\log b = 0.36$, and $\log c = 2.4$. Find the value of each expression.

a. $\log abc$ **b.** $\log a^2b$
c. $\log \frac{ac}{b}$ **d.** $\log \frac{a^2}{c^3b^2}$

Change-of-base formula

$$\log_b y = \frac{\log_a y}{\log_a b}$$

26. To four decimal places, find $\log_5 17$.

pH scale

$\text{pH} = -\log[\text{H}^+]$

27. pH of grapefruit The pH of grapefruit juice is about 3.1. Find its hydrogen ion concentration.

Weber-Fechner law

$L = k \ln I$

28. Find the decrease in loudness if the intensity is cut in half.

SECTION 9.6 *Exponential and Logarithmic Equations*

29. Solve each equation for x, if possible.

a. $3^x = 7$ **b.** $5^{x+2} = 625$

c. $2^x = 3^{x-1}$ **d.** $2^{x^2+4x} = \frac{1}{8}$

30. Solve each equation for x.

a. $\log x + \log (29 - x) = 2$

b. $\log_2 x + \log_2 (x - 2) = 3$

c. $\log_2 (x + 2) + \log_2 (x - 1) = 2$

d. $\dfrac{\log (7x - 12)}{\log x} = 2$

e. $\log x + \log (x - 5) = \log 6$

f. $\log 3 - \log (x - 1) = -1$

g. $e^{x \ln 2} = 9$

h. $\ln x = \ln (x - 1)$

i. $\ln x = \ln (x - 1) + 1$

j. $\ln x = \log_{10} x$ (*Hint:* Use the change-of-base formula.)

Carbon dating.

$A = A_0 2^{-t/h}$

31. Carbon-14 dating A wooden statue found in Egypt has a carbon-14 content that is two-thirds of that found in living wood. If the half-life of carbon-14 is 5,700 years, how old is the statue?

Chapter Test

In Problems 1–2, graph each function.

1. $f(x) = 2^x + 1$

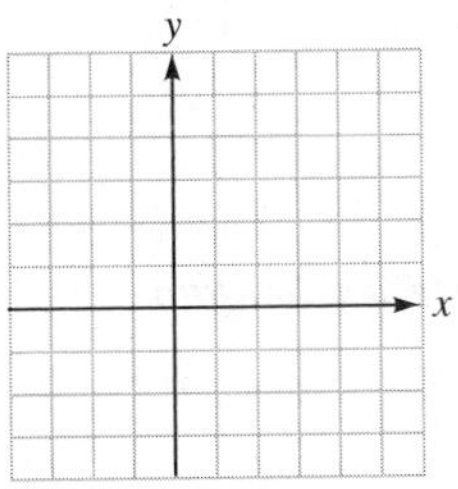

2. $f(x) = 2^{-x}$

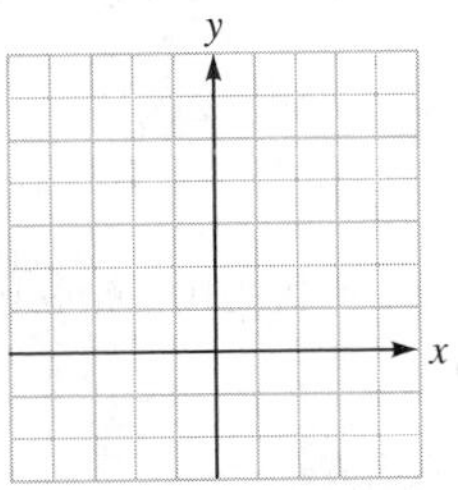

In Problems 3–4, solve each equation.

3. A radioactive material decays according to the formula $A = A_0(2)^{-t}$. How much of a 3-gram sample will be left in 6 years?

4. An initial deposit of \$1,000 earns 6% interest, compounded twice a year. How much will be in the account in one year?

5. Graph the function $f(x) = e^x$.

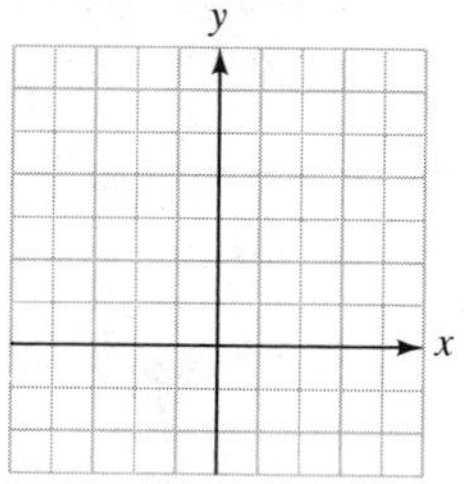

6. An account contains \$2,000 and has been earning 8% interest, compounded continuously. How much will be in the account in 10 years?

In Problems 7–12, find x.

7. $\log_4 16 = x$

8. $\log_x 81 = 4$

9. $\log_3 x = -3$

10. $\log_x 100 = 2$

11. $\log_{3/2} \frac{9}{4} = x$

12. $\log_{2/3} x = -3$

In Problems 13–14, graph each function.

13. $f(x) = -\log_3 x$

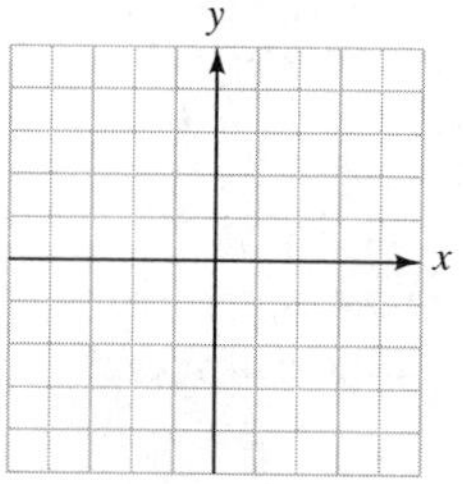

14. $f(x) = \ln x$

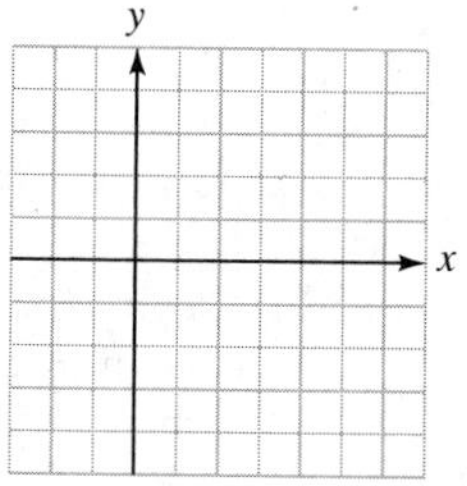

In Problems 15–16, write each expression in terms of the logarithms of a, b, and c.

15. $\log a^2bc^3$

16. $\ln \sqrt{\frac{a}{b^2c}}$

In Problems 17–18, write each expression as a logarithm of a single quantity.

17. $\frac{1}{2} \log (a + 2) + \log b - 3 \log c$

18. $\frac{1}{3} (\log a - 2 \log b) - \log c$

In Problems 19–20, assume that log 2 = 0.3010 and log 3 = 0.4771. Find each value. **Do not use a calculator.**

19. $\log 24$

20. $\log \frac{8}{3}$

In Problems 21–22, use the change-of-base formula to find each logarithm. **Do not attempt to simplify the answer.**

21. $\log_7 3$

22. $\log_\pi e$

In Problems 23–26, tell whether each statement is true. If a statement is not true, explain why.

23. $\log_a ab = 1 + \log_a b$

24. $\dfrac{\log a}{\log b} = \log a - \log b$

25. $\log a^{-3} = \dfrac{1}{3 \log a}$

26. $\ln(-x) = -\ln x$

27. Find the pH of a solution with a hydrogen ion concentration of 3.7×10^{-7}. (*Hint:* $\text{pH} = -\log [H^+]$.)

28. Find the db gain of an amplifier when $E_O = 60$ volts and $E_I = 0.3$ volt. (*Hint:* db gain $= 20 \log (E_O/E_I)$.)

In Problems 29–30, solve each equation. **Do not simplify the logarithms.**

29. $5^x = 3$

30. $3^{x-1} = 100^x$

In Problems 31–32, solve each equation.

31. $\log (5x + 2) = \log (2x + 5)$

32. $\log x + \log (x - 9) = 1$

10 More Graphing and Conic Sections

TEACHER

MATHEMATICS IN THE WORKPLACE

Secondary school teachers help students delve more deeply into subjects introduced in elementary school and learn more about the world and about themselves. Teachers plan and evaluate lessons, prepare tests, grade papers, prepare report cards, oversee study halls and homerooms, supervise extracurricular activities, and meet with parents and school staff. In recent years, teachers have become more involved in curriculum design, such as choosing textbooks and evaluating teaching methods.

EXAMPLE APPLICATION ■ An art teacher wants to paint a mural on an elliptical background that is 10 feet wide and 6 feet high. To see how to do this construction, see page 747.

We have seen that the graphs of linear functions are straight lines, and that the graphs of quadratic functions are parabolas. In this chapter, we will discuss some special curves, called **conic sections**. We will then discuss **piecewise-defined functions** and **step functions**.

10.1 THE CIRCLE AND THE PARABOLA

■ INTRODUCTION TO THE CONIC SECTIONS ■ THE CIRCLE ■ PROBLEM SOLVING ■ THE PARABOLA ■ PROBLEM SOLVING

Getting Ready

Square each binomial.

1. $(x - 2)^2$

2. $(x + 4)^2$

What number must be added to each binomial to make it a perfect square trinomial?

3. $x^2 + 9x$

4. $x^2 - 12x$

■ INTRODUCTION TO THE CONIC SECTIONS

The graphs of second-degree equations in x and y represent figures that were fully investigated in the 17th century by René Descartes (1596–1650) and Blaise Pascal (1623–1662). Descartes discovered that graphs of second-degree equations fall into one of several categories: a pair of lines, a point, a circle, a parabola, an ellipse, a hyperbola, or no graph at all. Because all of these graphs can be formed by the intersection of a plane and a right-circular cone, they are called **conic sections**. See Figure 10-1.

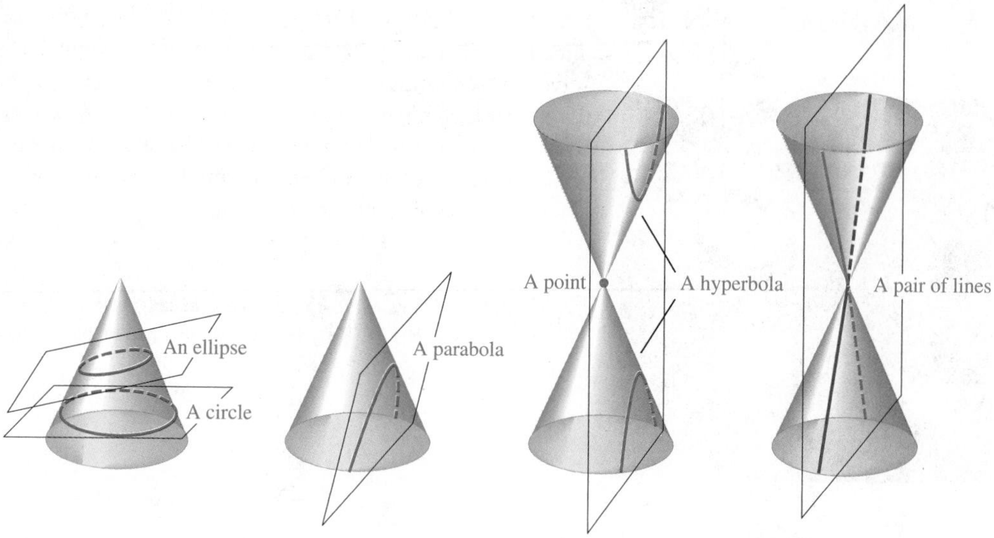

FIGURE 10-1

Conic sections have many applications. For example, everyone knows the importance of circular wheels and gears, pizza cutters, and ferris wheels.

Parabolas can be rotated to generate dish-shaped surfaces called *paraboloids*. Any light or sound placed at the *focus* of a paraboloid is reflected outward in parallel paths, as shown in Figure 10-2(a). This property makes parabolic surfaces ideal for flashlight and headlight reflectors. It also makes parabolic surfaces good antennas, because signals captured by such antennas are concentrated at the focus. Parabolic mirrors are capable of concentrating the rays of the sun at a single point and thereby generating tremendous heat. This property is used in the design of certain solar furnaces.

Any object thrown upward and outward travels in a parabolic path, as shown in Figure 10-2(b). In architecture, many arches are parabolic in shape, because this gives strength. Cables that support suspension bridges hang in the form of a parabola. (See Figure 10-2(c).)

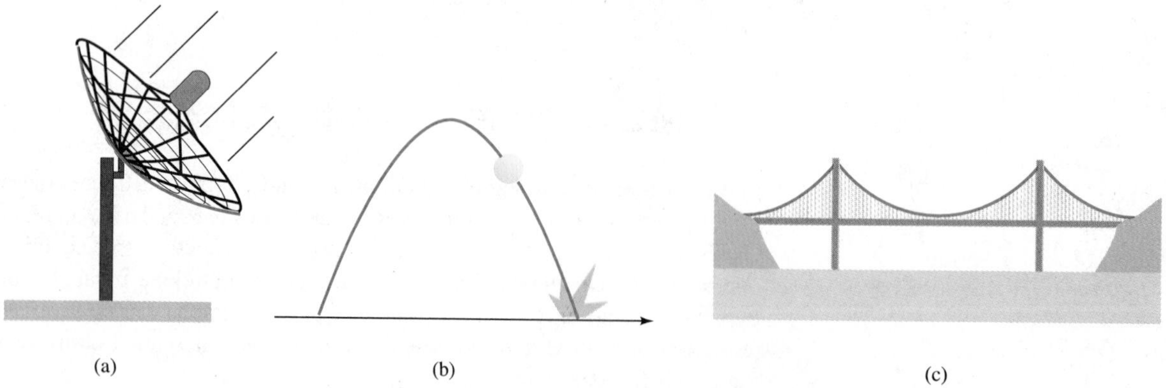

FIGURE 10-2

Ellipses have optical and acoustical properties that are useful in architecture and engineering. For example, many arches are portions of an ellipse, because the shape is pleasing to the eye. (See Figure 10-3(a).) The planets and many comets have elliptical orbits. (See Figure 10-3(b).) Gears are often cut into elliptical shapes to provide nonuniform motion. (See Figure 10-3(c).)

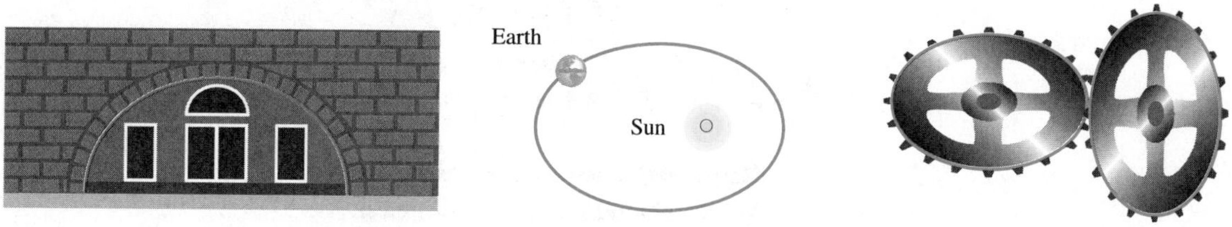

(a) (b) (c)

FIGURE 10-3

Hyperbolas serve as the basis of a navigational system known as LORAN (LOng RAnge Navigation). (See Figure 10-4.) They are also used to find the source of a distress signal, are the basis for the design of hypoid gears, and describe the orbits of some comets.

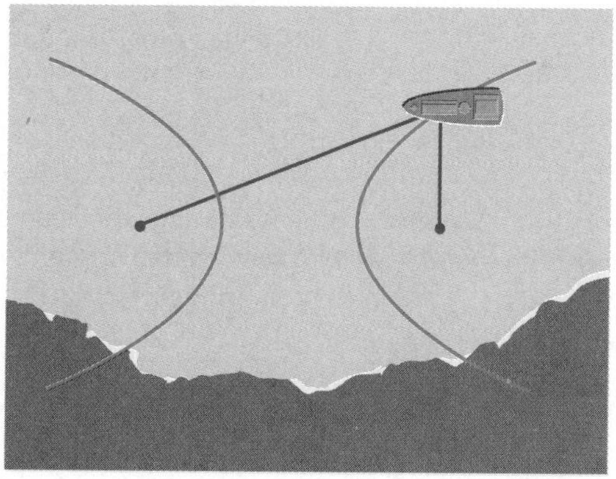

FIGURE 10-4

THE CIRCLE

The Circle

A **circle** is the set of all points in a plane that are a fixed distance from a point called its **center**.

The fixed distance is the **radius** of the circle.

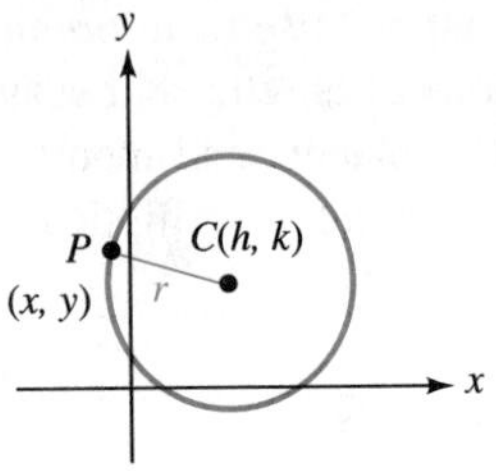

FIGURE 10-5

To develop the general equation of a circle, we must write the equation of a circle with a radius of r and with center at some point $C(h, k)$, as in Figure 10-5. This task is equivalent to finding all points $P(x, y)$ such that the length of line segment CP is r. We can use the distance formula to find r.

$$r = \sqrt{(x - h)^2 + (y - k)^2}$$

We then square both sides to obtain

1. $r^2 = (x - h)^2 + (y - k)^2$

Equation 1 is called the **standard form of the equation of a circle** with a radius of r and center at the point with coordinates (h, k).

Standard Equation of a Circle with Center at (*h*, *k*)

Any equation that can be written in the form

$$(x - h)^2 + (y - k)^2 = r^2$$

has a graph that is a circle with radius r and center at point (h, k).

If $r = 0$, the graph reduces to a single point called a **point circle**. If $r^2 < 0$, a circle does not exist. If both coordinates of the center are 0, the center of the circle is the origin.

Standard Equation of a Circle with Center at (0, 0)

Any equation that can be written in the form

$$x^2 + y^2 = r^2$$

has a graph that is a circle with radius r and center at the origin.

EXAMPLE 1 Graph $x^2 + y^2 = 25$.

Solution Because this equation can be written in the form $x^2 + y^2 = r^2$, its graph is a circle with center at the origin. Since $r^2 = 25 = 5^2$, the circle has a radius of 5. The graph appears in Figure 10-6.

$x^2 + y^2 = 25$

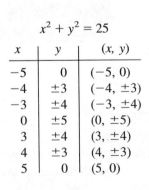

x	y	(x, y)
-5	0	$(-5, 0)$
-4	± 3	$(-4, \pm 3)$
-3	± 4	$(-3, \pm 4)$
0	± 5	$(0, \pm 5)$
3	± 4	$(3, \pm 4)$
4	± 3	$(4, \pm 3)$
5	0	$(5, 0)$

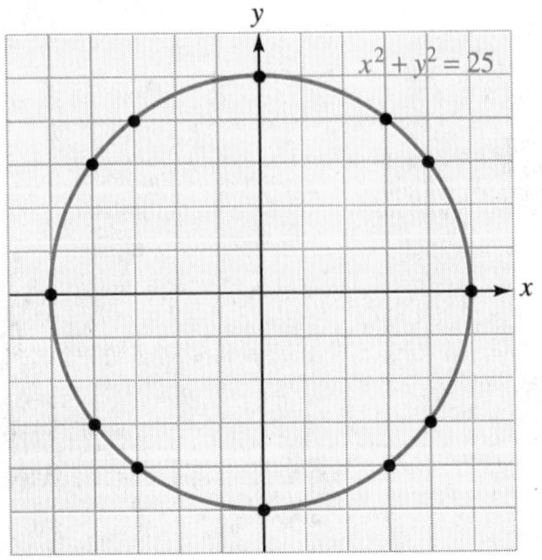

FIGURE 10-6

■

Self Check Graph $x^2 + y^2 = 4$.

Answer

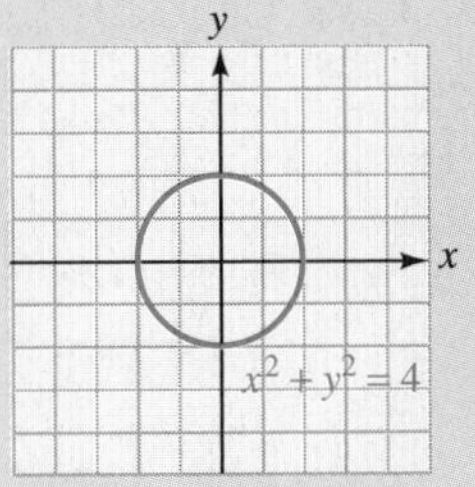

EXAMPLE 2 Find the equation of the circle with radius 5 and center at $C(3, 2)$.

Solution We substitute 5 for r, 3 for h, and 2 for k in the standard form of a circle and simplify.

$$(x - h)^2 + (y - k)^2 = r^2$$
$$(x - 3)^2 + (y - 2)^2 = 5^2$$
$$x^2 - 6x + 9 + y^2 - 4y + 4 = 25$$
$$x^2 + y^2 - 6x - 4y - 12 = 0$$

The equation is $x^2 + y^2 - 6x - 4y - 12 = 0$. ■

Self Check Find the equation of the circle with radius 6 and center at (2, 3).

Answer $x^2 + y^2 - 4x - 6y - 23 = 0$

EXAMPLE 3 Graph the equation $x^2 + y^2 - 4x + 2y = 20$.

Solution To identify the curve, we complete the square on x and y and write the equation in standard form.

$$x^2 + y^2 - 4x + 2y = 20$$
$$x^2 - 4x + y^2 + 2y = 20$$

To complete the square on x and y, add 4 and 1 to both sides.

$$x^2 - 4x + 4 + y^2 + 2y + 1 = 20 + 4 + 1$$
$$(x - 2)^2 + (y + 1)^2 = 25 \qquad \text{Factor } x^2 - 4x + 4 \text{ and } y^2 + 2y + 1.$$
$$(x - 2)^2 + [y - (-1)]^2 = 5^2$$

We can now see that this result is the equation of a circle with a radius of 5 and center at $h = 2$ and $k = -1$. If we plot the center and draw a circle with a radius of 5 units, we will obtain the circle shown in Figure 10-7. ■

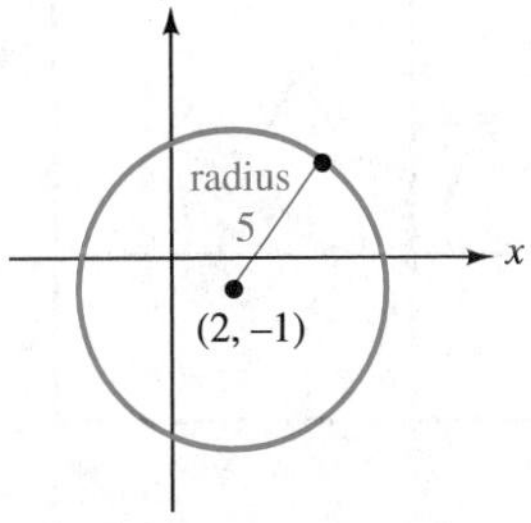

$x^2 + y^2 - 4x + 2y = 20$

FIGURE 10-7

Self Check Write the equation $x^2 + y^2 + 2x - 4y - 11 = 0$ in standard form and graph it.

Answer $(x + 1)^2 + (y - 2)^2 = 16$

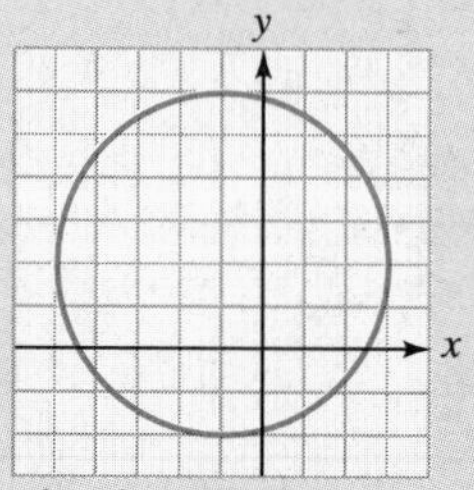

$x^2 + y^2 + 2x - 4y - 11 = 0$

GRAPHING CALCULATORS

Graphing Circles

Since the graphs of circles fail the vertical line test, their equations do not represent functions. It is somewhat more difficult to use a graphing calculator to graph equations that are not functions. For example, to graph the circle described by $(x - 1)^2 + (y - 2)^2 = 4$, we must split the equation into two functions and graph each one separately. We begin by solving the equation for y.

$$(x - 1)^2 + (y - 2)^2 = 4$$

$$(y - 2)^2 = 4 - (x - 1)^2 \quad \text{Subtract } (x-1)^2 \text{ from both sides.}$$

$$y - 2 = \pm\sqrt{4 - (x - 1)^2} \quad \text{Take the square root of both sides.}$$

$$y = 2 \pm \sqrt{4 - (x - 1)^2} \quad \text{Add 2 to both sides.}$$

This equation defines two functions. If we use window settings of $[-3, 5]$ for x and $[-3, 5]$ for y and graph the functions

$$y = 2 + \sqrt{4 - (x - 1)^2} \quad \text{and} \quad y = 2 - \sqrt{4 - (x - 1)^2}$$

we get the distorted circle shown in Figure 10-8(a). To get a better circle, graphing calculators have a squaring feature that gives an equal unit distance on both the x- and y-axes. After using this feature, we get the circle shown in Figure 10-8(b).

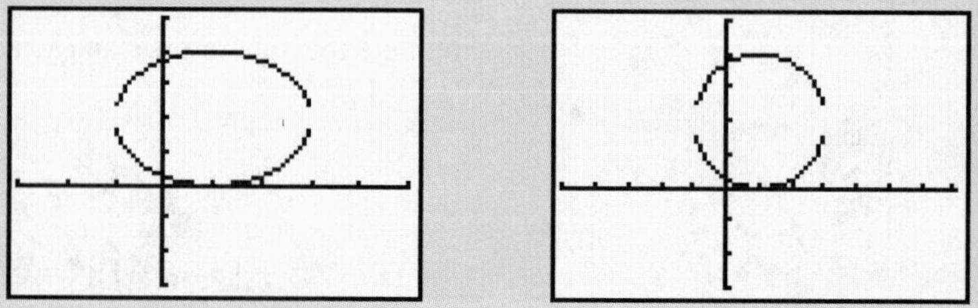

(a) (b)

FIGURE 10-8

PROBLEM SOLVING

EXAMPLE 4 **Radio translators** The broadcast area of a television station is bounded by the circle $x^2 + y^2 = 3{,}600$, where x and y are measured in miles. A translator station picks up the signal and retransmits it from the center of a circular area bounded by $(x + 30)^2 + (y - 40)^2 = 1{,}600$. Find the location of the translator and the greatest distance from the main transmitter that the signal can be received.

Solution The coverage of the television station is bounded by $x^2 + y^2 = 60^2$, a circle centered at the origin with a radius of 60 miles, as shown in Figure 10-9. Because the translator is at the center of the circle $(x + 30)^2 + (y - 40)^2 = 1{,}600$, it is located at $(-30, 40)$, a point 30 miles west and 40 miles north of the television station. The radius of the translator's coverage is $\sqrt{1{,}600}$, or 40 miles.

As shown in Figure 10-9, the greatest distance of reception is the sum of A, the distance from the translator to the television station, and 40 miles, the radius of the translator's coverage.

To find A, we use the distance formula to find the distance between $(x_1, y_1) = (-30, 40)$ and the origin, $(x_2, y_2) = (0, 0)$.

$$A = \sqrt{(x_1 - x_2)^2 + (y_1 - y_2)^2}$$

$$\begin{aligned} A &= \sqrt{(-30 - 0)^2 + (40 - 0)^2} \\ &= \sqrt{(-30)^2 + 40^2} \\ &= \sqrt{2{,}500} \\ &= 50 \end{aligned}$$

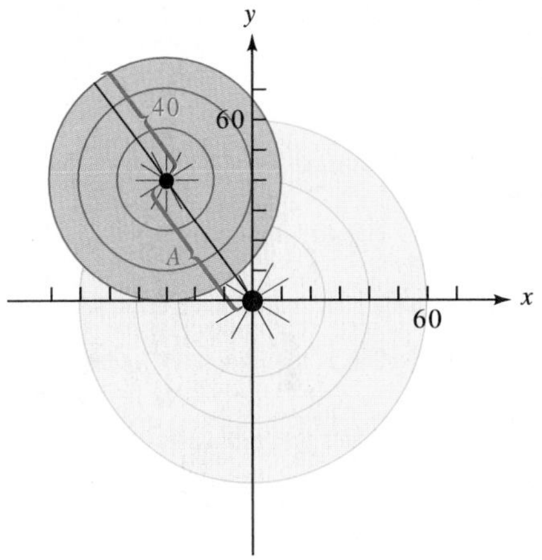

FIGURE 10-9

The translator is located 50 miles from the television station, and it broadcasts the signal an additional 40 miles. The greatest reception distance is $50 + 40$, or 90 miles. ■

THE PARABOLA

We have seen that equations of the form $y = a(x - h)^2 + k$, with $a \neq 0$, represent parabolas with the vertex at the point (h, k). They open upward when $a > 0$ and downward when $a < 0$.

Equations of the form $x = a(y - k)^2 + h$ also represent parabolas with vertex at point (h, k). However, they open to the right when $a > 0$ and to the left when

$a < 0$. Parabolas that open to the right or left do not represent functions, because their graphs fail the vertical line test.

Several types of parabolas are summarized in the following chart.

Equations of Parabolas ($a > 0$)

Parabola opening	*Vertex at origin*	*Vertex at (h, k)*
Up	$y = ax^2$	$y = a(x - h)^2 + k$
Down	$y = -ax^2$	$y = -a(x - h)^2 + k$
Right	$x = ay^2$	$x = a(y - k)^2 + h$
Left	$x = -ay^2$	$x = -a(y - k)^2 + h$

EXAMPLE 5 Graph **a.** $x = \frac{1}{2}y^2$ and **b.** $x = -2(y - 2)^2 + 3$.

Solution **a.** We make a table of ordered pairs that satisfy the equation, plot each pair, and draw the parabola, as in Figure 10-10(a). Because the equation is of the form $x = ay^2$ with $a > 0$, the parabola opens to the right and has its vertex at the origin.

b. We make a table of ordered pairs that satisfy the equation, plot each pair, and draw the parabola, as in Figure 10-10(b). Because the equation is of the form $x = -a(y - k)^2 + h$, the parabola opens to the left and has its vertex at the point with coordinates (3, 2).

$x = \frac{1}{2}y^2$

x	y	(x, y)
0	0	(0, 0)
2	2	(2, 2)
2	−2	(2, −2)
8	4	(8, 4)
8	−4	(8, −4)

$x = -2(y - 2)^2 + 3$

x	y	(x, y)
−5	0	(−5, 0)
1	1	(1, 1)
3	2	(3, 2)
1	3	(1, 3)
−5	4	(−5, 4)

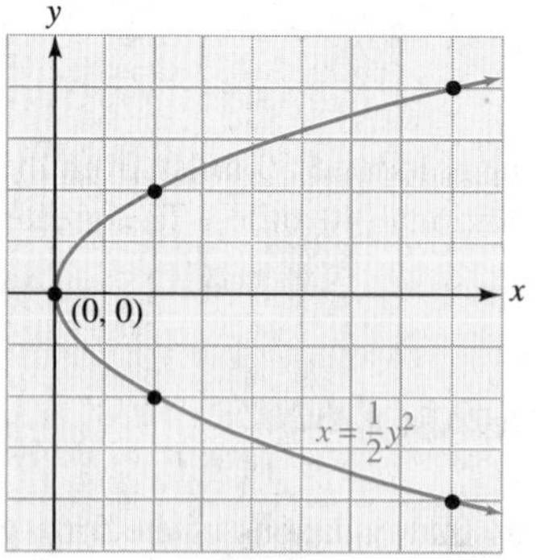

(a)

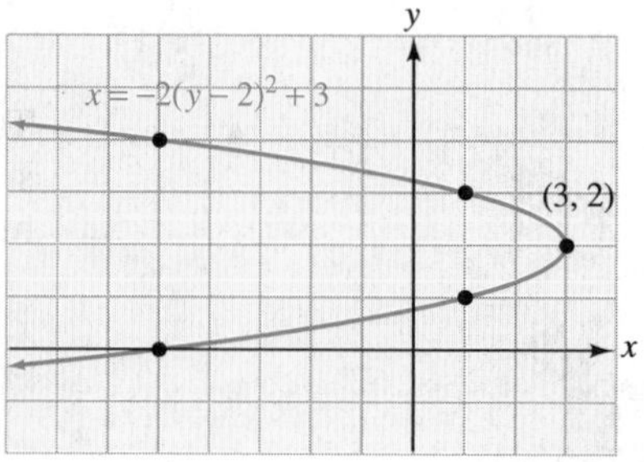

(b)

FIGURE 10-10

■

Self Check Graph $x = \frac{1}{2}(y-1)^2 - 2$.

Answer

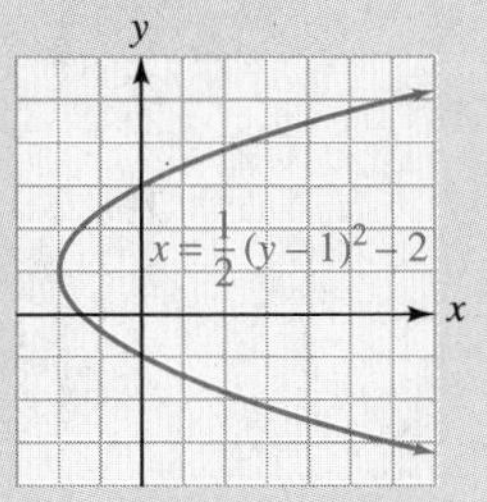

EXAMPLE 6 Graph $y = -2x^2 + 12x - 15$.

Solution Because the equation is not in standard form, the coordinates of the vertex of its graph are not obvious. To write the equation in standard form, we complete the square on x.

$$y = -2x^2 + 12x - 15$$

$$y = -2(x^2 - 6x) - 15 \qquad \text{Factor out } -2 \text{ from } -2x^2 + 12x.$$

$$y = \mathbf{-2}(x^2 - 6x + \mathbf{9}) - 15 + \mathbf{18} \qquad \text{Subtract and add 18; } -2(9) = -18.$$

$$y = -2(x-3)^2 + 3$$

Because the equation is written in the form $y = -a(x-h)^2 + k$, we can see that the parabola opens downward and has its vertex at (3, 3). The graph of the function is shown in Figure 10-11.

$y = -2x^2 + 12x - 15$

x	y	(x, y)
1	−5	(1, −5)
2	1	(2, 1)
3	3	(3, 3)
4	1	(4, 1)
5	−5	(5, −5)

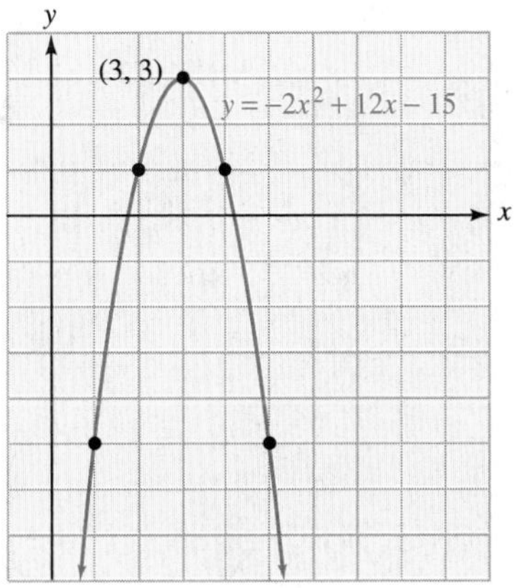

FIGURE 10-11

■

Self Check Graph $y = 0.5x^2 - x - 1$.

Answer

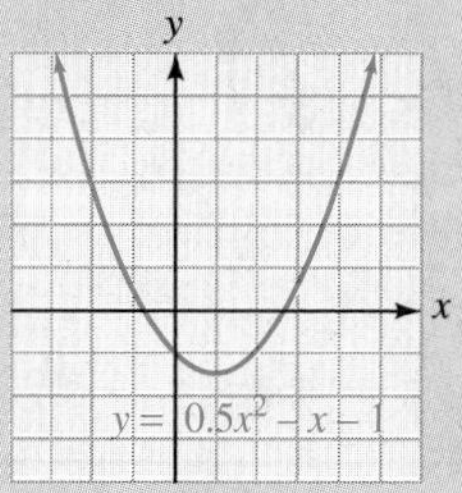

PROBLEM SOLVING

EXAMPLE 7 **Gateway Arch** The shape of the Gateway Arch in St. Louis is approximately a parabola, as shown in Figure 10-12(a). How high is the arch 100 feet from its foundation?

Solution We place the parabola in a coordinate system as in Figure 10-12(b), with ground level on the x-axis and the vertex of the parabola at the point $(h, k) = (0, 630)$. The equation of this downward-opening parabola has the form

$$y = -a(x - h)^2 + k$$

$$y = -a(x - \mathbf{0})^2 + \mathbf{630} \quad \text{Substitute } h = 0 \text{ and } y = 630.$$

$$y = -ax^2 + 630 \quad \text{Simplify.}$$

Because the Gateway Arch is 630 feet wide at its base, the parabola passes through the point $\left(\frac{630}{2}, 0\right)$, or $(315, 0)$. To find a in the equation of the parabola, we proceed as follows:

$$y = -ax^2 + 630$$

$$\mathbf{0} = -a(\mathbf{315})^2 + 630 \quad \text{Substitute 315 for } x \text{ and 0 for } y.$$

$$\frac{-630}{315^2} = -a \quad \text{Subtract 630 from both sides and divide both sides by } 315^2.$$

$$\frac{2}{315} = a \quad \text{Multiply both sides by } -1;\ \frac{630}{315^2} = \frac{2}{315}.$$

The equation of the parabola that approximates the shape of the Gateway Arch is

$$y = -\frac{2}{315}x^2 + 630$$

To find the height of the arch at a point 100 feet from its foundation, we substitute $315 - 100$, or 215, for x in the equation of the parabola and solve for y.

$$y = -\frac{2}{315}x^2 + 630$$

$$y = -\frac{2}{315}(215)^2 + 630$$

$$= 336.5079365$$

At a point 100 feet from the foundation, the height of the arch is about 337 feet.

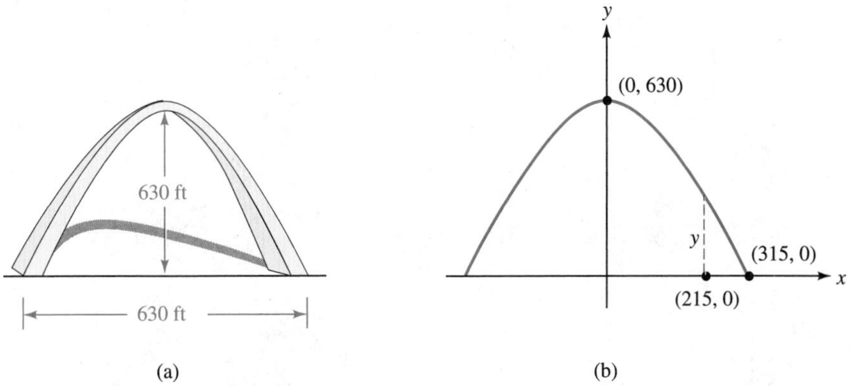

FIGURE 10-12

Orals

Find the center and the radius of each circle.

1. $x^2 + y^2 = 144$
2. $x^2 + y^2 = 121$
3. $(x - 2)^2 + y^2 = 16$
4. $x^2 + (y + 1)^2 = 9$

Tell whether each parabola opens up or down or left or right.

5. $y = -3x^2 - 2$
6. $y = 7x^2 - 5$
7. $x = -3y^2$
8. $x = (y - 3)^2$

EXERCISE 10.1

REVIEW *Solve each equation.*

1. $|3x - 4| = 11$
2. $\left|\frac{4 - 3x}{5}\right| = 12$
3. $|3x + 4| = |5x - 2|$
4. $|6 - 4x| = |x + 2|$

VOCABULARY AND CONCEPTS *Fill in each blank to make a true statement.*

5. A ______ is the set of all points in a ______ that are a fixed distance from a given point.
6. The fixed distance in Exercise 5 is called the ______ of the circle, and the point is called its ______.

7. If ______ in the equation $x^2 + y^2 = r^2$, no circle exists.

8. The graph of $y = ax^2$ $(a > 0)$ is a ______ with vertex at the ______ that opens ______.

9. The graph of $x = a(y - 2)^2 + 3$ $(a > 0)$ is a ______ with vertex at ______ that opens to the ______.

10. The graph of $x = -a(y - 1)^2 - 3$ $(a > 0)$ is a ______ with vertex at ______ that opens to the ______.

PRACTICE *In Exercises 11–20, graph each equation.*

11. $x^2 + y^2 = 9$

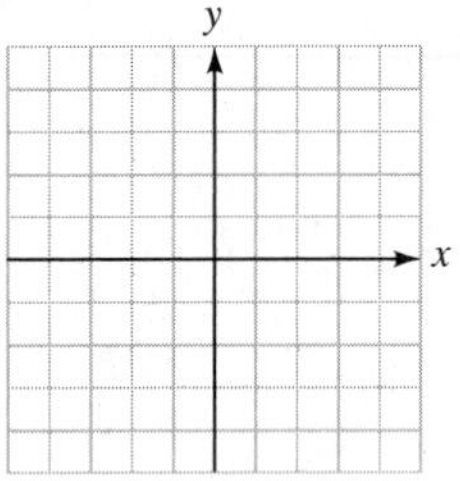

12. $x^2 + y^2 = 16$

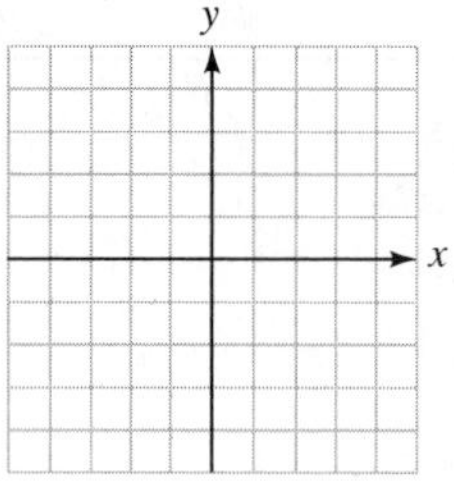

13. $(x - 2)^2 + y^2 = 9$

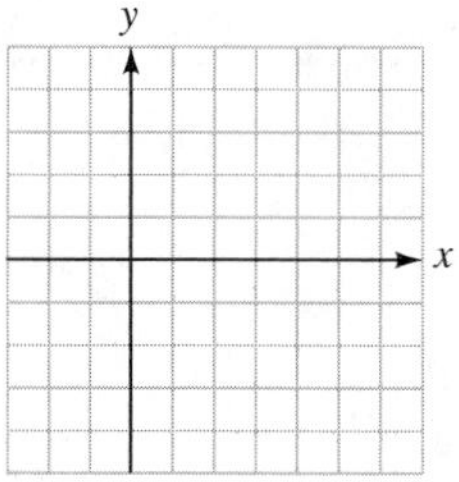

14. $x^2 + (y - 3)^2 = 4$

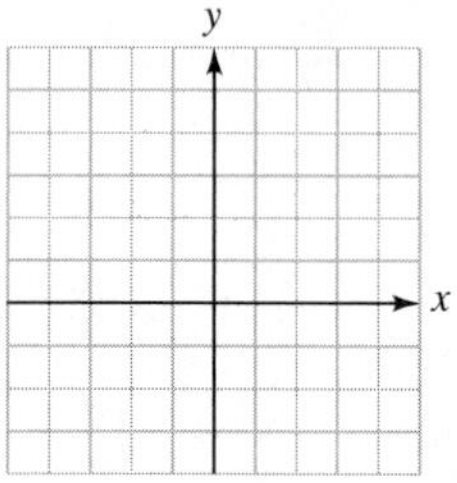

15. $(x - 2)^2 + (y - 4)^2 = 4$

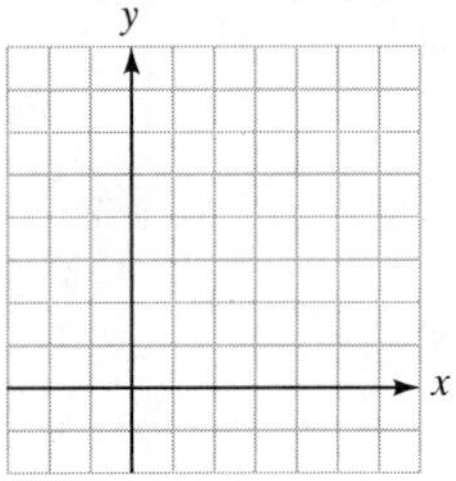

16. $(x - 3)^2 + (y - 2)^2 = 4$

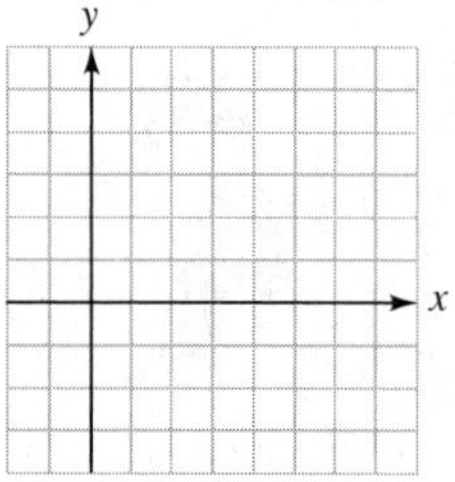

17. $(x + 3)^2 + (y - 1)^2 = 16$

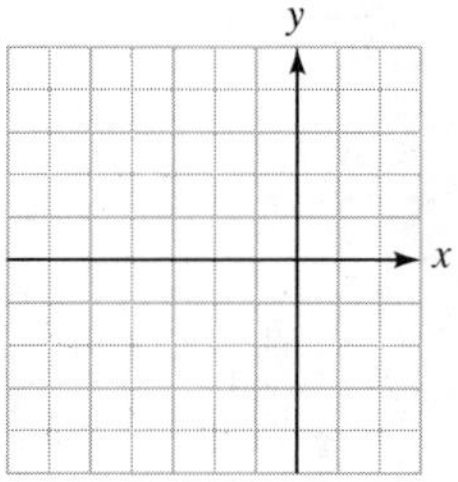

18. $(x - 1)^2 + (y + 4)^2 = 9$

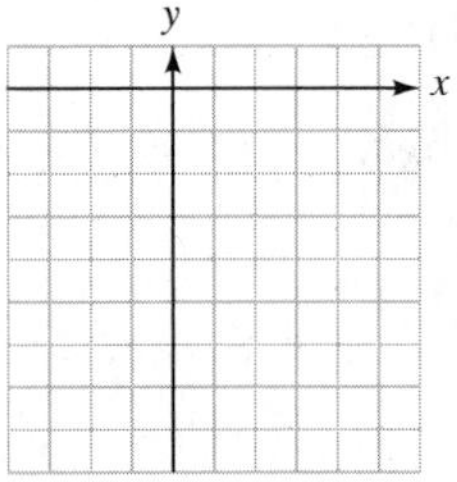

19. $x^2 + (y + 3)^2 = 1$

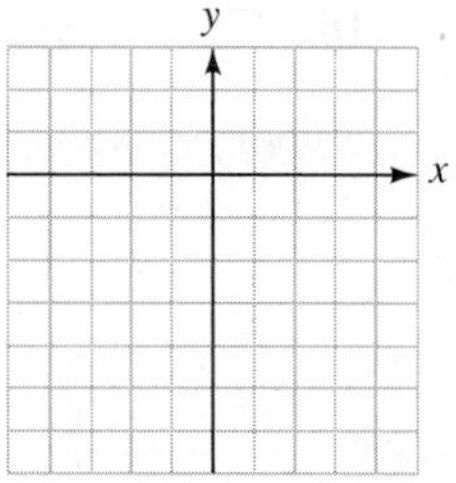

20. $(x + 4)^2 + y^2 = 1$

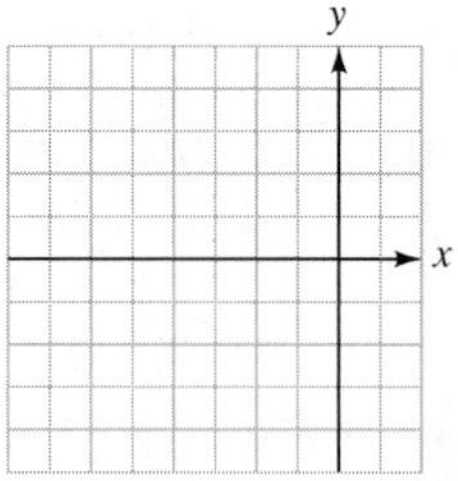

In Exercises 21–24, use a graphing calculator to graph each equation.

21. $3x^2 + 3y^2 = 16$

22. $2x^2 + 2y^2 = 9$

23. $(x + 1)^2 + y^2 = 16$

24. $x^2 + (y - 2)^2 = 4$

In Exercises 25–32, write the equation of the circle with the following properties.

25. Center at origin; radius 1

26. Center at origin; radius 4

27. Center at (6, 8); radius 5

28. Center at (5, 3); radius 2

29. Center at $(-2, 6)$; radius 12

30. Center at $(5, -4)$; radius 6

31. Center at the origin; diameter of $2\sqrt{2}$

32. Center at the origin; diameter of $8\sqrt{3}$

In Exercises 33–40, graph each circle. Give the coordinates of the center.

33. $x^2 + y^2 + 2x - 8 = 0$

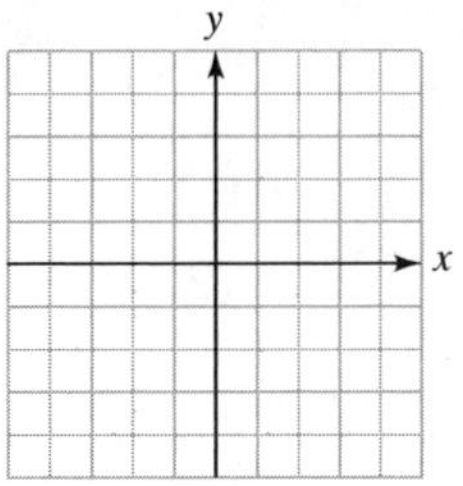

34. $x^2 + y^2 - 4y = 12$

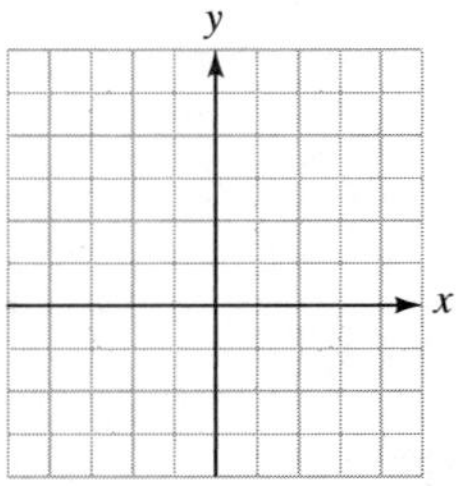

35. $9x^2 + 9y^2 - 12y = 5$

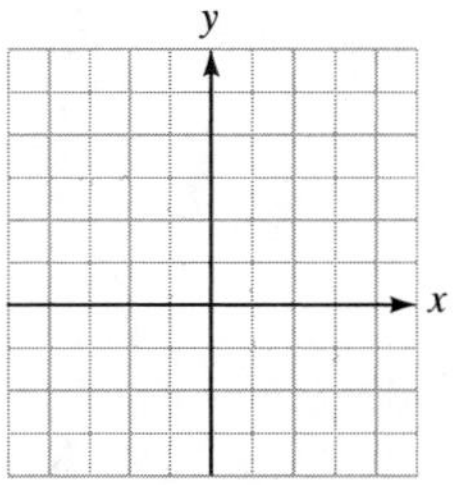

36. $4x^2 + 4y^2 + 4y = 15$

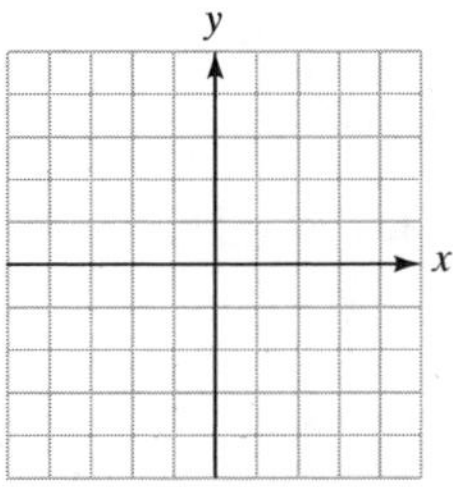

37. $x^2 + y^2 - 2x + 4y = -1$

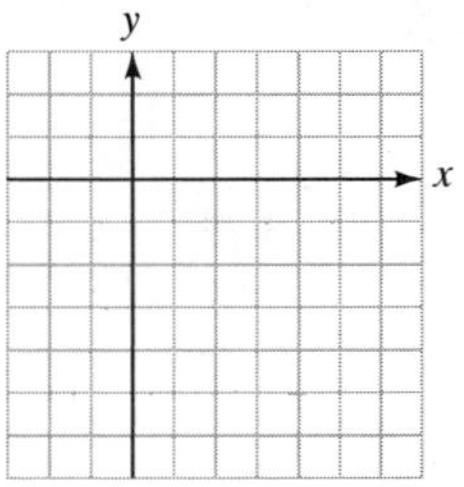

38. $x^2 + y^2 + 4x + 2y = 4$

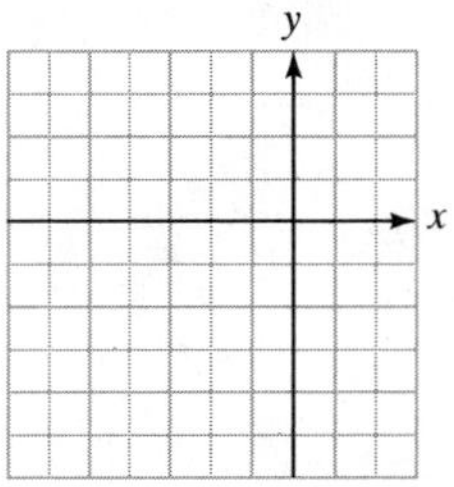

39. $x^2 + y^2 + 6x - 4y = -12$

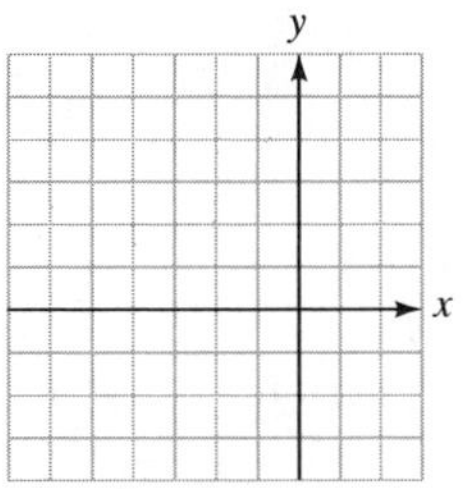

40. $x^2 + y^2 + 8x + 2y = -13$

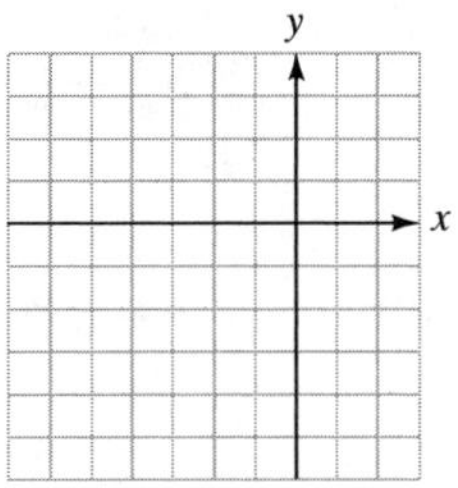

In Exercises 41–52, find the vertex of each parabola and graph it.

41. $x = y^2$

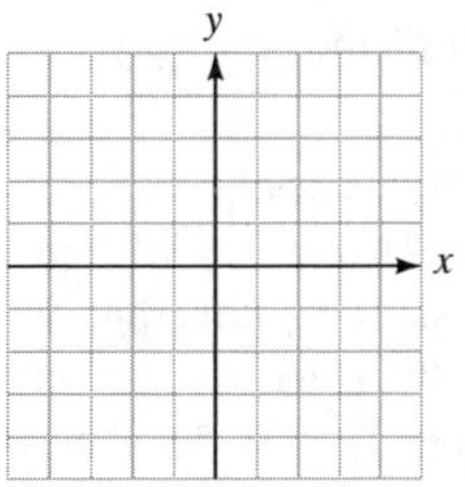

42. $x = -y^2 + 1$

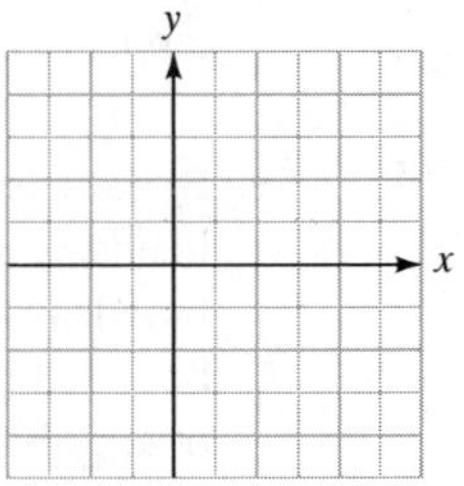

43. $x = -\frac{1}{4}y^2$

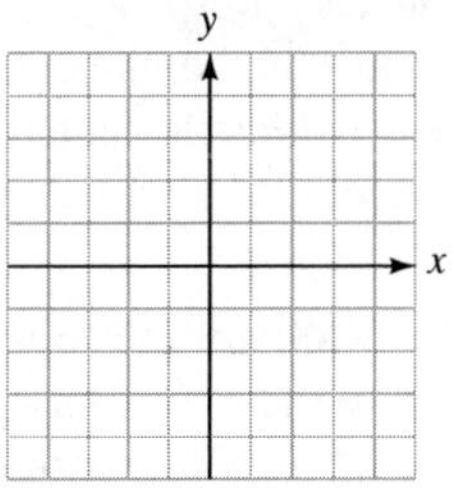

44. $x = 4y^2$

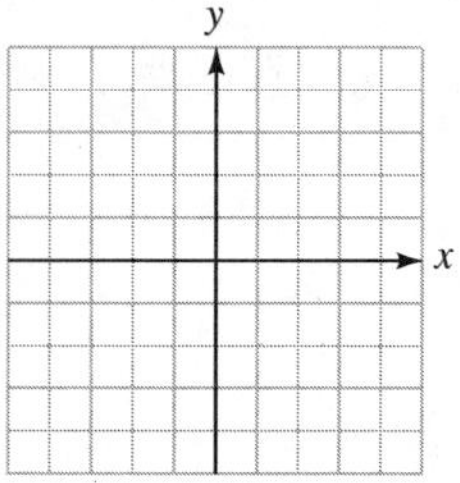

45. $y = x^2 + 4x + 5$

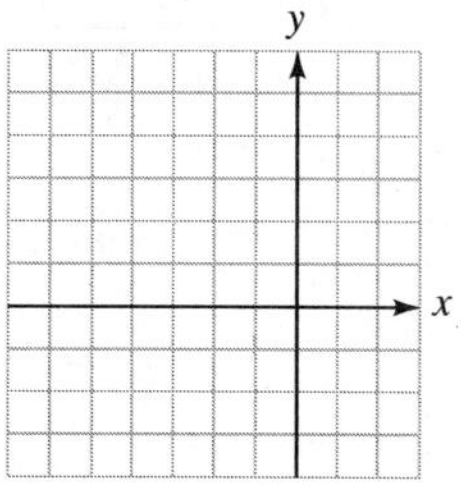

46. $y = -x^2 - 2x + 3$

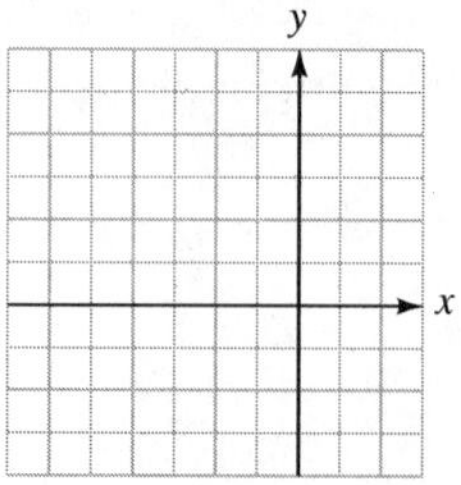

47. $y = -x^2 - x + 1$

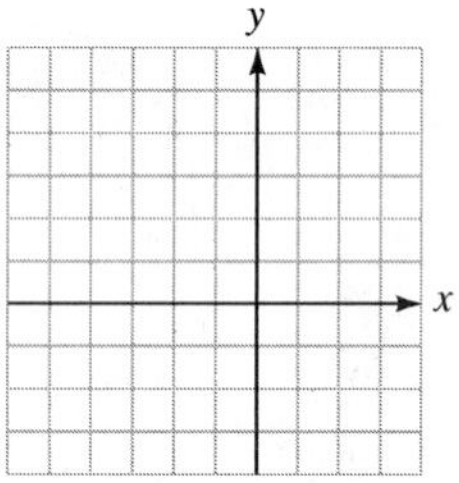

48. $x = \frac{1}{2}y^2 + 2y$

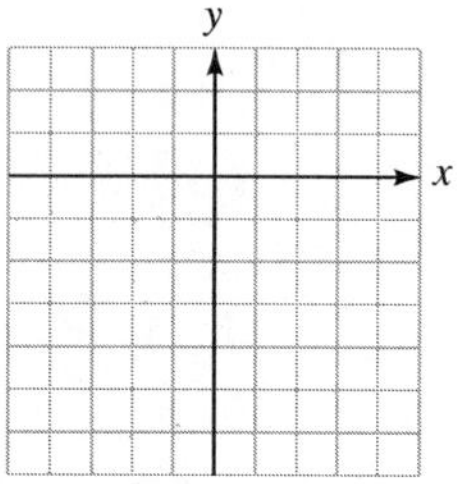

49. $y^2 + 4x - 6y = -1$

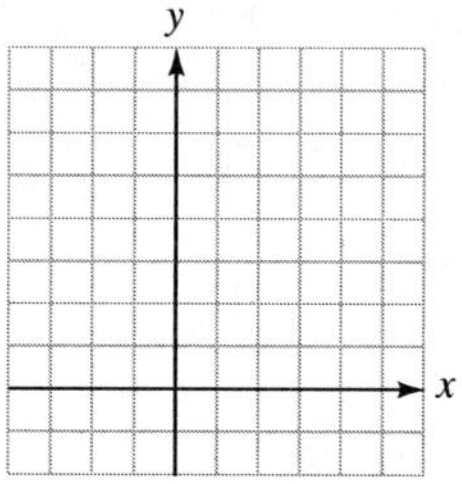

50. $x^2 - 2y - 2x = -7$

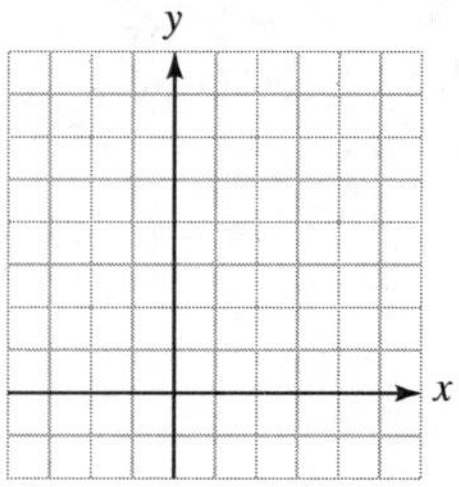

51. $y = 2(x - 1)^2 + 3$

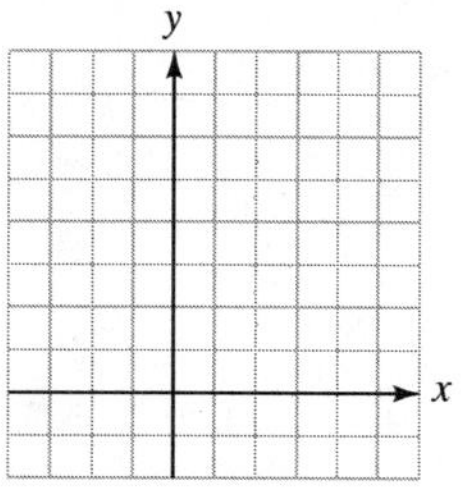

52. $y = -2(x + 1)^2 + 2$

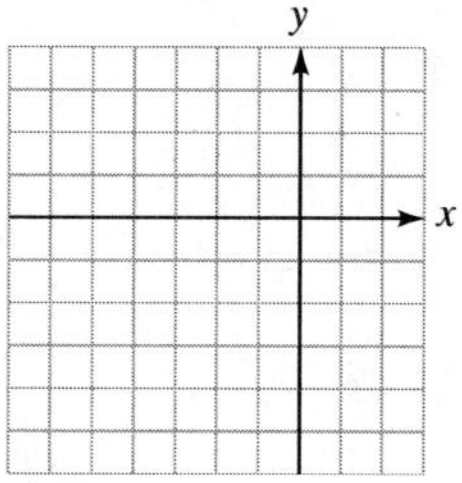

In Exercises 53–56, use a graphing calculator to graph each equation.

53. $x = 2y^2$

54. $x = y^2 - 4$

55. $x^2 - 2x + y = 6$

56. $x = -2(y - 1)^2 + 2$

APPLICATIONS

57. Meshing gears For design purposes, the large gear in Illustration 1 is the circle $x^2 + y^2 = 16$. The smaller gear is a circle centered at $(7, 0)$ and tangent to the larger circle. Find the equation of the smaller gear.

58. Width of a walkway The walkway in Illustration 2 is bounded by the two circles $x^2 + y^2 = 2{,}500$ and $(x - 10)^2 + y^2 = 900$, measured in feet. Find the largest and the smallest width of the walkway.

59. Broadcast ranges Radio stations applying for licensing may not use the same frequency if their broadcast areas overlap. One station's coverage is bounded by $x^2 + y^2 - 8x - 20y + 16 = 0$, and the other's by $x^2 + y^2 + 2x + 4y - 11 = 0$. May they be licensed for the same frequency?

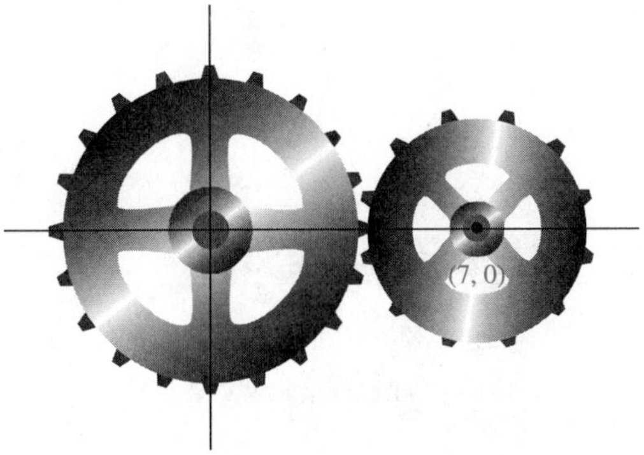

ILLUSTRATION 1

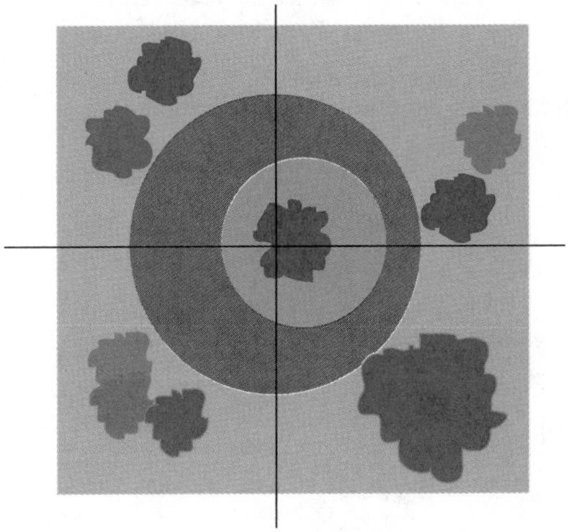

ILLUSTRATION 2

60. Highway design Engineers want to join two sections of highway with a curve that is one-quarter of a circle, as in Illustration 3. The equation of the circle is $x^2 + y^2 - 16x - 20y + 155 = 0$, where distances are measured in kilometers. Find the locations (relative to the center of town) of the intersections of the highway with State and with Main.

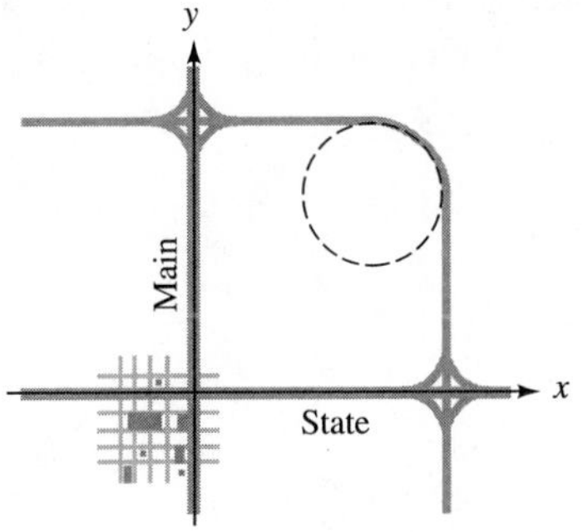

ILLUSTRATION 3

61. Flight of a projectile The cannonball in Illustration 4 follows the parabolic trajectory $y = 30x - x^2$. Where does it land?

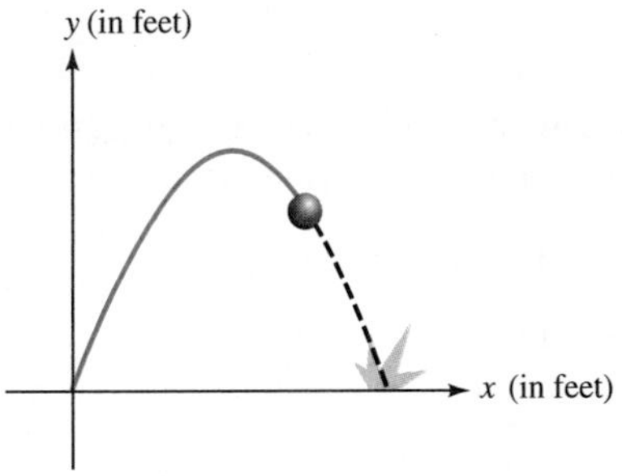

ILLUSTRATION 4

62. Flight of a projectile In Exercise 61, how high does the cannonball get?

63. Orbit of a comet If the orbit of the comet shown in Illustration 5 is given by the equation $2y^2 - 9x = 18$, how far is it from the sun at the vertex of the orbit? Distances are measured in astronomical units (AU).

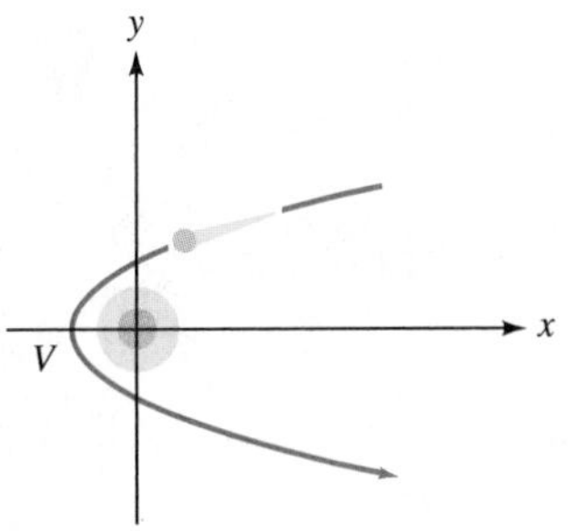

ILLUSTRATION 5

64. Satellite antenna The cross section of the satellite antenna in Illustration 6 is a parabola given by the equation $y = \frac{1}{16}x^2$, with distances measured in feet. If the dish is 8 feet wide, how deep is it?

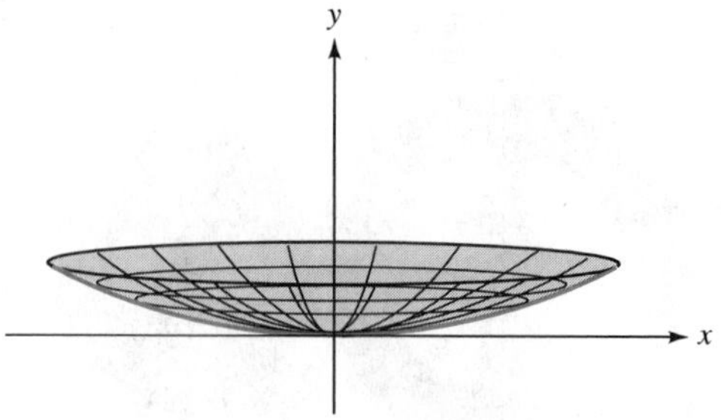

ILLUSTRATION 6

WRITING

65. Explain how to decide from its equation whether the graph of a parabola opens up, down, right, or left.

66. From the equation of a circle, explain how to determine the radius and the coordinates of the center.

SOMETHING TO THINK ABOUT

67. From the values of *a, h,* and *k*, explain how to determine the number of *x*-intercepts of the graph of $y = a(x - h)^2 + k$.

68. Under what conditions will the graph of $x = a(y - k)^2 + h$ have no *y*-intercepts?

10.2 The Ellipse

■ THE ELLIPSE ■ CONSTRUCTING AN ELLIPSE ■ GRAPHING ELLIPSES ■ PROBLEM SOLVING

Getting Ready *Solve each equation for the indicated variable.*

1. $\frac{y^2}{b^2} = 1$ for y

2. $\frac{x^2}{a^2} = 1$ for x

■ THE ELLIPSE

The Ellipse

An **ellipse** is the set of all points *P* in the plane the sum of whose distances from two fixed points is a constant. See Figure 10-13, in which $d_1 + d_2$ is a constant.

Each of the two points is called a **focus**. Midway between the foci is the *center* of the ellipse.

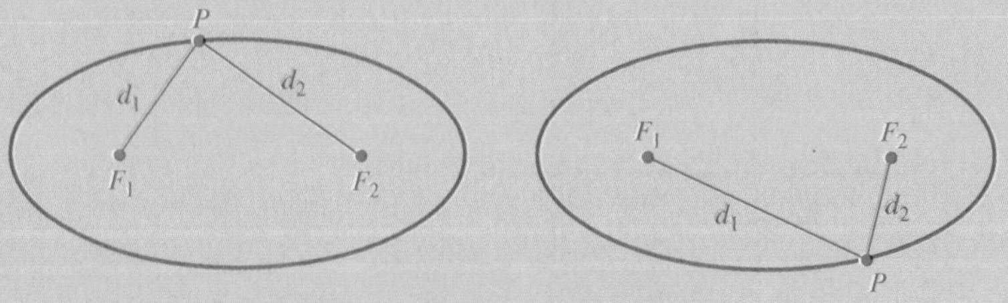

FIGURE 10-13

CONSTRUCTING AN ELLIPSE

We can construct an ellipse by placing two thumbtacks fairly close together, as in Figure 10-14. We then tie each end of a piece of string to a thumbtack, catch the loop with the point of a pencil, and, while keeping the string taut, draw the ellipse.

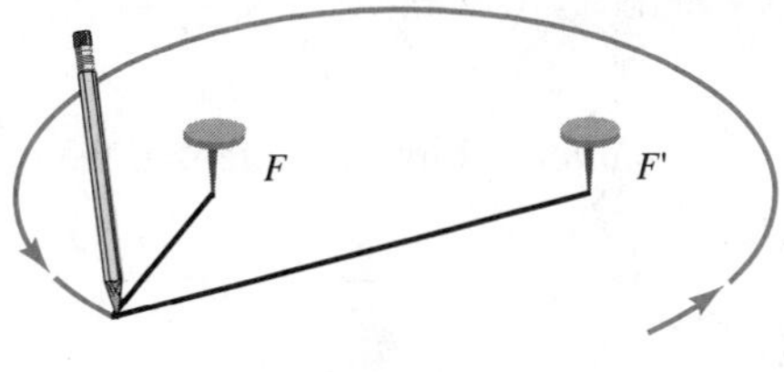

FIGURE 10-14

Using this method, we can construct an ellipse of any specific size. For example, to construct an ellipse that is 10 feet wide and 6 feet high, we must find the length of string to use and the distance between thumbtacks.

To do this, we will let a represent the distance between the center and vertex V, as shown in Figure 10-15(a). We will also let c represent the distance between the center of the ellipse and either focus. When the pencil is at vertex V, the length of the string is $c + a + (a - c)$, or just $2a$. Because $2a$ is the 10-foot width of the ellipse, the string needs to be 10 feet long. The distance $2a$ is constant for any point on the ellipse, including point B shown in Figure 10-15(b).

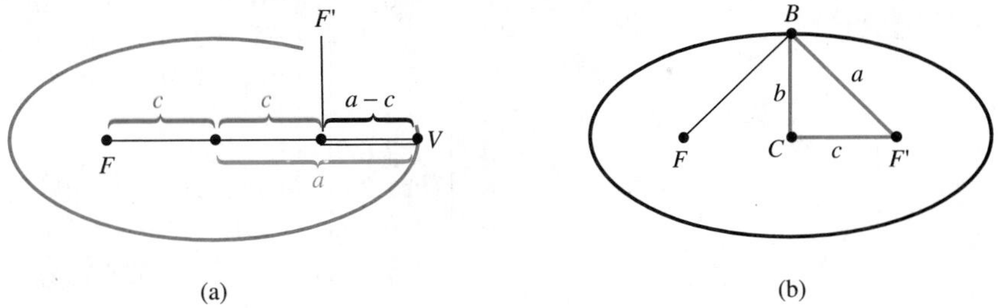

FIGURE 10-15

From right triangle CBF' and the Pythagorean theorem, we can find c as follows:

$$a^2 = b^2 + c^2 \qquad \text{or} \qquad c = \sqrt{a^2 - b^2}$$

Since distance b is one-half of the height of the ellipse, $b = 3$. Since $2a = 10$, $a = 5$. We can now substitute $a = 5$ and $b = 3$ into the formula to find c:

$$\begin{aligned} c &= \sqrt{5^2 - 3^2} \\ &= \sqrt{25 - 9} \\ &= \sqrt{16} \\ &= 4 \end{aligned}$$

Since $c = 4$, the distance between the thumbtacks must be 8 feet. We can construct the ellipse by tying a 10-foot string to thumbtacks that are 8 feet apart.

GRAPHING ELLIPSES

To graph ellipses, we can make a table of ordered pairs that satisfy the equation, plot them, and join the points with a smooth curve.

EXAMPLE 1 Graph $\frac{x^2}{36} + \frac{y^2}{9} = 1$.

Solution First we note that the equation can be written in the form

$$\frac{x^2}{6^2} + \frac{y^2}{3^2} = 1 \qquad 36 = 6^2 \text{ and } 9 = 3^2.$$

After making a table of ordered pairs that satisfy the equation, plotting each of them, and joining the points with a curve, we obtain the ellipse shown in Figure 10-16.

We note that the center of the ellipse is the origin, the ellipse intersects the x-axis at points (6, 0) and (−6, 0), and the ellipse intersects the y-axis at points (0, 3) and (0, −3).

$$\frac{x^2}{36} + \frac{y^2}{9} = 1$$

x	y	(x, y)
−6	0	(−6, 0)
−4	±2.2	(−4, ±2.2)
−2	±2.8	(−2, ±2.8)
0	±3	(0, ±3)
2	±2.8	(2, ±2.8)
4	±2.2	(4, ±2.2)
6	0	(6, 0)

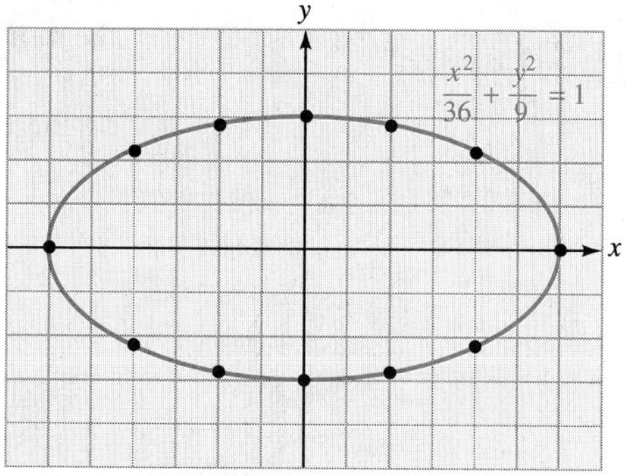

FIGURE 10-16

Self Check Graph $\frac{x^2}{4} + \frac{y^2}{16} = 1$.

Answer

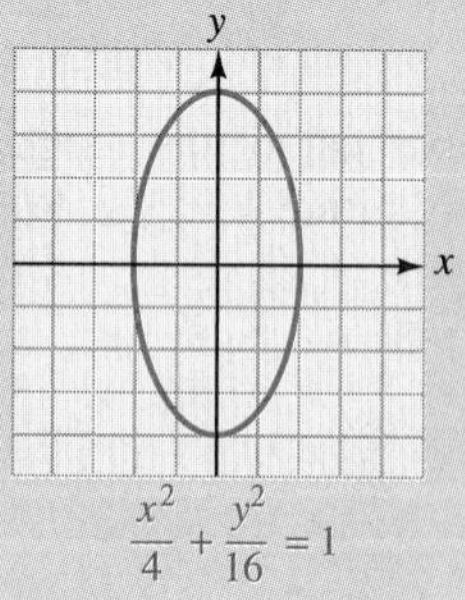

Example 1 illustrates that the graph of

$$\frac{x^2}{a^2} + \frac{y^2}{b^2} = 1$$

is an ellipse centered at the origin. To find the x-intercepts of the graph, we can let $y = 0$ and solve for x.

$$\frac{x^2}{a^2} + \frac{0^2}{b^2} = 1$$

$$\frac{x^2}{a^2} + 0 = 1$$

$$x^2 = a^2$$

$$x = a \quad \text{or} \quad x = -a$$

The x-intercepts are $(a, 0)$ and $(-a, 0)$.

To find the y-intercepts, we let $x = 0$ and solve for y.

$$\frac{0^2}{a^2} + \frac{y^2}{b^2} = 1$$

$$0 + \frac{y^2}{b^2} = 1$$

$$y^2 = b^2$$

$$y = b \quad \text{or} \quad y = -b$$

The y-intercepts are $(0, b)$ and $(0, -b)$.

In general, we have the following results.

Equations of an Ellipse Centered at the Origin

The equation of an ellipse centered at the origin, with x-intercepts at $V_1(a, 0)$ and $V_2(-a, 0)$ and with y-intercepts of $(0, b)$ and $(0, -b)$, is

$$\frac{x^2}{a^2} + \frac{y^2}{b^2} = 1 \quad (a > b > 0)$$ See Figure 10-17(a).

The equation of an ellipse centered at the origin, with y-intercepts at $V_1(0, a)$ and $V_2(0, -a)$ and x-intercepts at $(b, 0)$ and $(-b, 0)$, is

$$\frac{x^2}{b^2} + \frac{y^2}{a^2} = 1 \quad (a > b > 0)$$ See Figure 10-17(b).

In Figure 10-17, the points V_1 and V_2 are the **vertices** of the ellipse, the midpoint of V_1V_2 is the **center** of the ellipse, and the distance between the center of the ellipse and either vertex is a. The segment V_1V_2 is called the **major axis**, and the segment joining either $(0, b)$ and $(0, -b)$ or $(b, 0)$ and $(-b, 0)$ is called the **minor axis**.

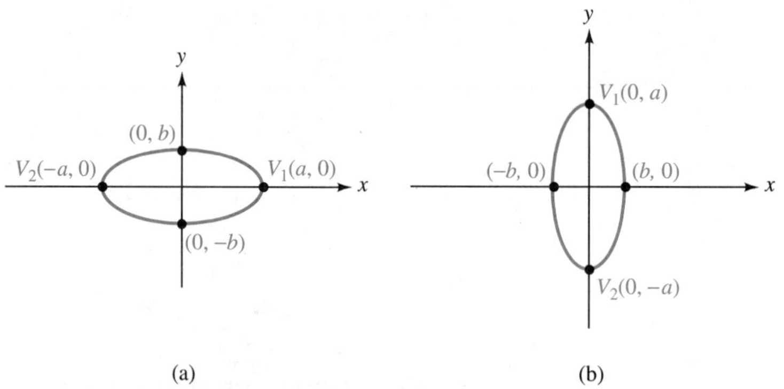

FIGURE 10-17

The equations for ellipses centered at (h, k) are as follows.

Equations of an Ellipse Centered at (*h, k*)

The equation of an ellipse centered at (h, k), with major axis parallel to the x-axis, is

1. $$\frac{(x-h)^2}{a^2} + \frac{(y-k)^2}{b_2} = 1 \quad (a > b > 0)$$

The equation of an ellipse centered at (h, k), with major axis parallel to the y-axis, is

2. $$\frac{(x-h)^2}{b^2} + \frac{(y-k)^2}{a^2} = 1 \quad (a > b > 0)$$

EXAMPLE 2 Graph the ellipse $\dfrac{(x-2)^2}{16} + \dfrac{(y+3)^2}{25} = 1$.

Solution We first write the equation in the form

$$\frac{(x-2)^2}{4^2} + \frac{[y-(-3)]^2}{5^2} = 1 \quad (5 > 4)$$

This is the equation of an ellipse centered at $(h, k) = (2, -3)$ with major axis parallel to the y-axis and with $b = 4$ and $a = 5$. We first plot the center, as shown in Figure 10-18. Since a is the distance from the center to the vertex, we can locate the vertices by counting 5 units above and 5 units below the center. The vertices are at points $(2, 2)$ and $(2, -8)$.

Since $b = 4$, we can locate two more points on the ellipse by counting 4 units to the left and 4 units to the right of the center. The points $(-2, -3)$ and $(6, -3)$ are also on the graph.

Using these four points as guides, we can draw the ellipse.

$$\frac{(x-2)^2}{16} + \frac{(y+3)^2}{25} = 1$$

x	y	(x, y)
2	2	(2, 2)
2	−8	(2, −8)
6	−3	(6, −3)
−2	−3	(−2, −3)

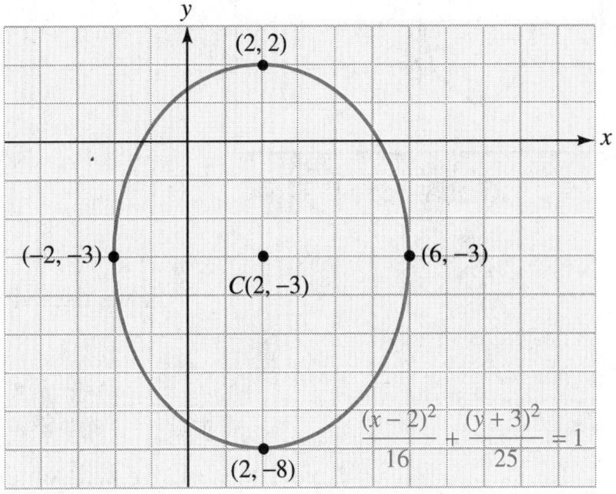

FIGURE 10-18

Self Check Graph $\frac{(x-1)^2}{9} + \frac{(y+2)^2}{16} = 1$.

Answer

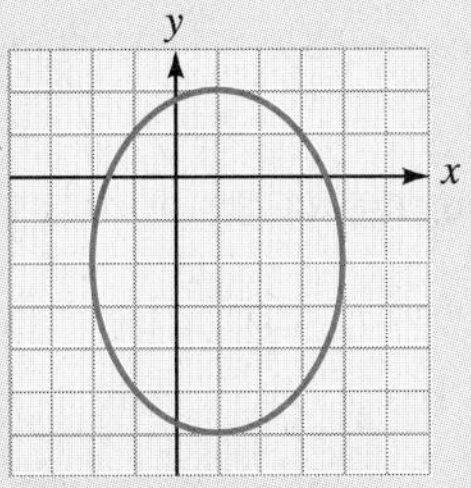

Graphing Ellipses

GRAPHING CALCULATORS

To use a graphing calculator to graph

$$\frac{(x+2)^2}{4} + \frac{(y-1)^2}{25} = 1$$

(continued)

Graphing Ellipses *(continued)*

we first clear the fractions by multiplying both sides by 100 and solving for y.

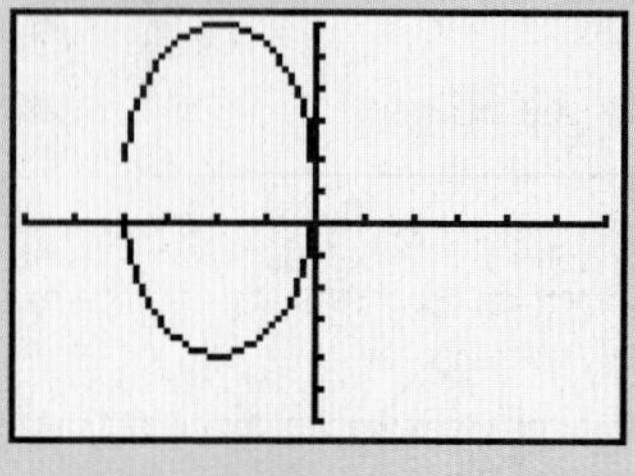

FIGURE 10-19

$$25(x+2)^2 + 4(y-1)^2 = 100 \quad \text{Multiply both sides by 100.}$$

$$4(y-1)^2 = 100 - 25(x+2)^2 \quad \text{Subtract } 25(x+2)^2 \text{ from both sides.}$$

$$(y-1)^2 = \frac{100 - 25(x+2)^2}{4} \quad \text{Divide both sides by 4.}$$

$$y - 1 = \pm\frac{\sqrt{100 - 25(x+2)^2}}{2} \quad \text{Take the square root of both sides.}$$

$$y = 1 \pm \frac{\sqrt{100 - 25(x+2)^2}}{2} \quad \text{Add 1 to both sides.}$$

If we use window settings $[-6, 6]$ for x and $[-6, 6]$ for y and graph the functions

$$y = 1 + \frac{\sqrt{100 - 25(x+2)^2}}{2} \quad \text{and} \quad y = 1 - \frac{\sqrt{100 - 25(x+2)^2}}{2}$$

we will obtain the ellipse shown in Figure 10-19.

We can use completing the square to write the equations of many ellipses in standard form.

EXAMPLE 3 Write $4x^2 + 9y^2 - 16x - 18y = 11$ in standard form to show that the equation represents an ellipse.

Solution We write the equation in standard form by completing the square on x and y:

$$4x^2 + 9y^2 - 16x - 18y = 11$$

$$4x^2 - 16x + 9y^2 - 18y = 11 \quad \text{Use the commutative property to rearrange terms.}$$

$$4(x^2 - 4x) + 9(y^2 - 2y) = 11 \quad \text{Factor 4 from } 4x^2 - 16x \text{ and factor 9 from } 9y^2 - 18y \text{ to get coefficients of 1 for the squared terms.}$$

$$4(x^2 - 4x + 4) + 9(y^2 - 2y + 1) = 11 + 16 + 9 \quad \text{Complete the square to make } x^2 - 4x \text{ and } y^2 - 2y \text{ perfect trinomial squares. Since } 16 + 9 \text{ is added to the left-hand side, add } 16 + 9 \text{ to the right-hand side.}$$

$$4(x-2)^2 + 9(y-1)^2 = 36 \qquad \text{Factor } x^2 - 4x + 4 \text{ and } y^2 - 2y + 1.$$

$$\frac{(x-2)^2}{9} + \frac{(y-1)^2}{4} = 1 \qquad \text{Divide both sides by 36.}$$

Since this equation matches Equation 1 on page 750, it represents an ellipse. Its graph is shown in Figure 10-20. ■

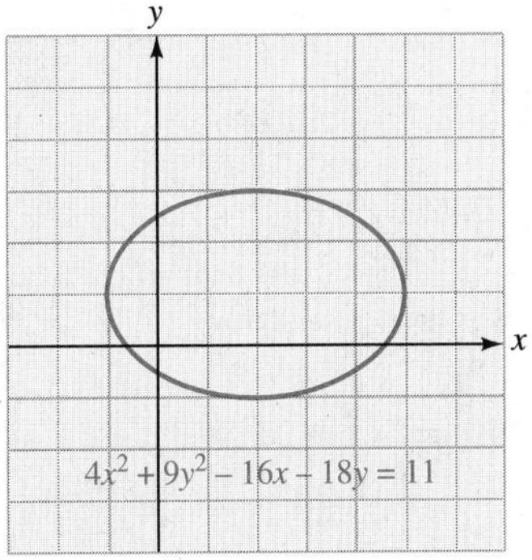

FIGURE 10-20

Self Check Graph $4x^2 - 8x + 9y^2 - 36y = -4$.

Answer

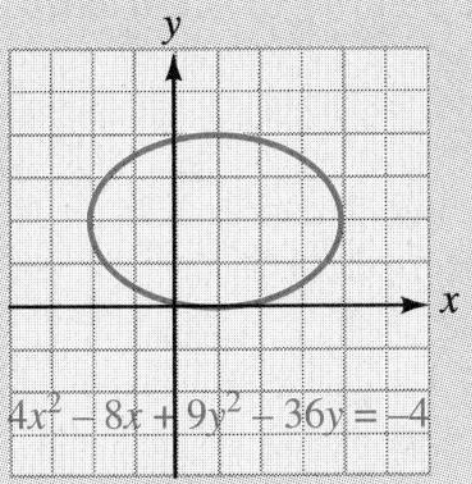

■ PROBLEM SOLVING

EXAMPLE 4 **Landscape design** A landscape architect is designing an elliptical pool that will fit in the center of a 20-by-30-foot rectangular garden, leaving at least 5 feet of space on all sides. Find the equation of the ellipse.

Solution We place the rectangular garden in a coordinate system, as in Figure 10-21. To maintain 5 feet of clearance at the ends of the ellipse, the vertices must be the points $V_1(10, 0)$ and $V_2(-10, 0)$. Similarly, the y-intercepts are the points $(0, 5)$ and $(0, -5)$.

The equation of the ellipse has the form

$$\frac{x^2}{a^2} + \frac{y^2}{b^2} = 1$$

with $a = 10$ and $b = 5$. Thus, the equation of the boundary of the pool is

$$\frac{x^2}{100} + \frac{y^2}{25} = 1$$

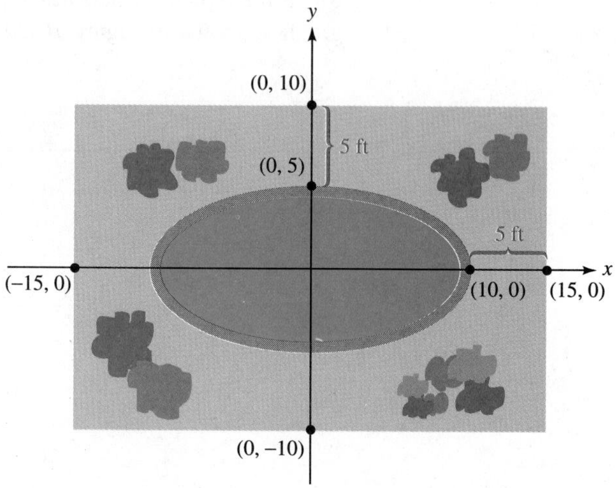

FIGURE 10-21

■

Orals *Find the x- and y-intercepts of each ellipse.*

1. $\frac{x^2}{9} + \frac{y^2}{16} = 1$

2. $\frac{x^2}{25} + \frac{y^2}{36} = 1$

Find the center of each ellipse.

3. $\frac{(x-2)^2}{9} + \frac{y^2}{16} = 1$

4. $\frac{x^2}{25} + \frac{(y+1)^2}{36} = 1$

EXERCISE 10.2

REVIEW *Find each product.*

1. $3x^{-2}y^2(4x^2 + 3y^{-2})$

2. $(2a^{-2} - b^{-2})(2a^{-2} + b^{-2})$

Write each expression without using negative exponents.

3. $\frac{x^{-2} + y^{-2}}{x^{-2} - y^{-2}}$

4. $\frac{2x^{-3} - 2y^{-3}}{4x^{-3} + 4y^{-3}}$

VOCABULARY AND CONCEPTS *Fill in each blank to make a true statement.*

5. An ______ is the set of all points in a plane the ______ of whose distances from two fixed points is a constant.

6. The fixed points in Exercise 5 are the ______ of the ellipse.

7. The midpoint of the line segment joining the foci of an ellipse is called the ______ of the ellipse.

8. The elliptical graph of

$$\frac{x^2}{a^2} + \frac{y^2}{b^2} = 1$$

has x-intercepts of ______ and y-intercepts of ______.

9. The center of the ellipse with an equation of

$$\frac{x^2}{a^2} + \frac{y^2}{b^2} = 1$$

is the point ______.

10. The center of the ellipse with an equation of

$$\frac{(x-h)^2}{a^2} + \frac{(y-k)^2}{b^2} = 1$$

is the point ______.

PRACTICE *In Exercises 11–20, graph each equation.*

11. $\frac{x^2}{4} + \frac{y^2}{9} = 1$

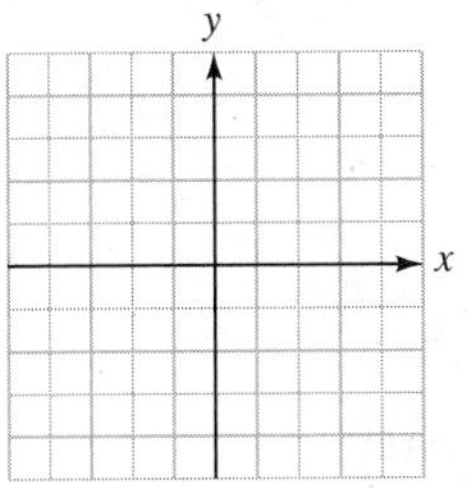

12. $x^2 + \frac{y^2}{9} = 1$

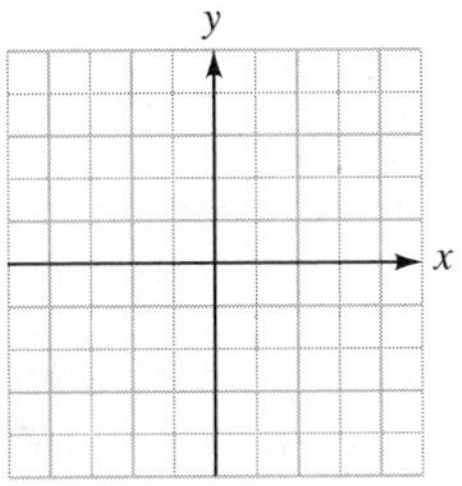

13. $x^2 + 9y^2 = 9$

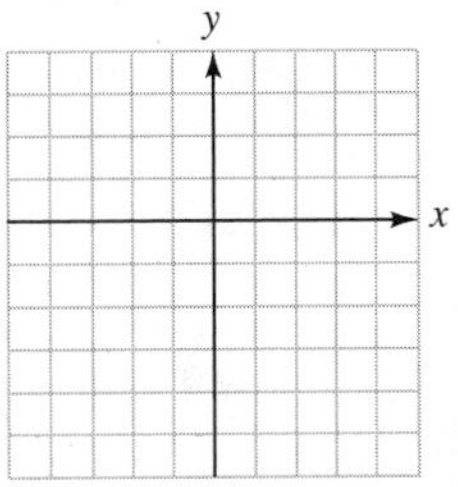

14. $25x^2 + 9y^2 = 225$

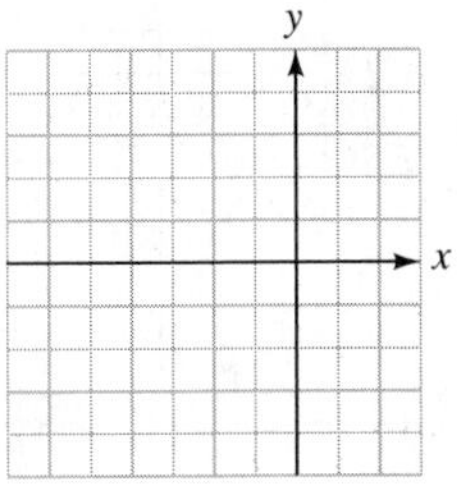

15. $16x^2 + 4y^2 = 64$

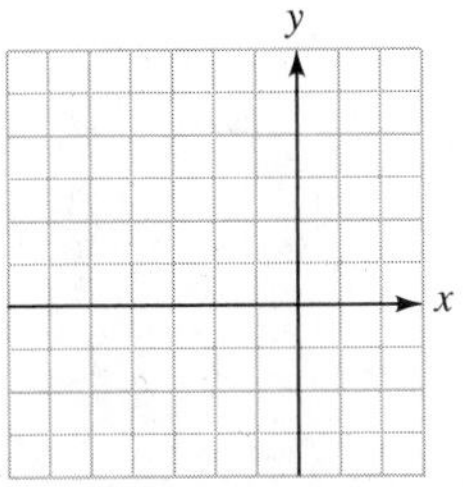

16. $4x^2 + 9y^2 = 36$

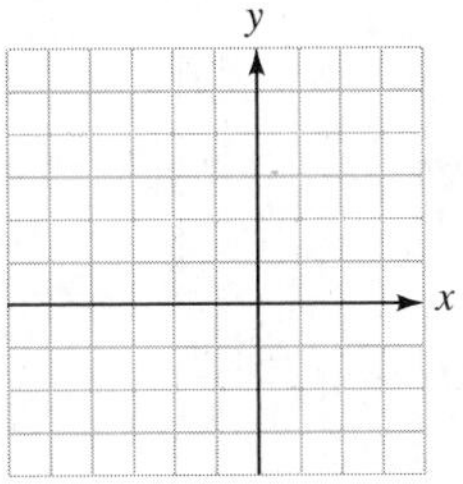

17. $\frac{(x-2)^2}{9} + \frac{(y-1)^2}{4} = 1$

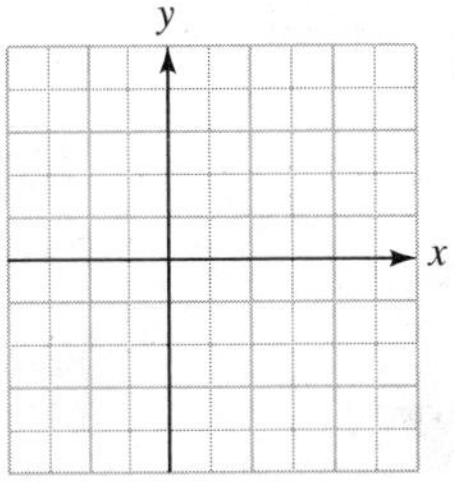

18. $\frac{(x-1)^2}{9} + \frac{(y-3)^2}{4} = 1$

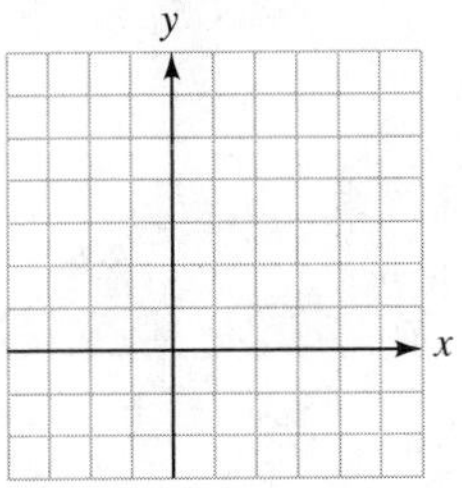

19. $(x+1)^2 + 4(y+2)^2 = 4$

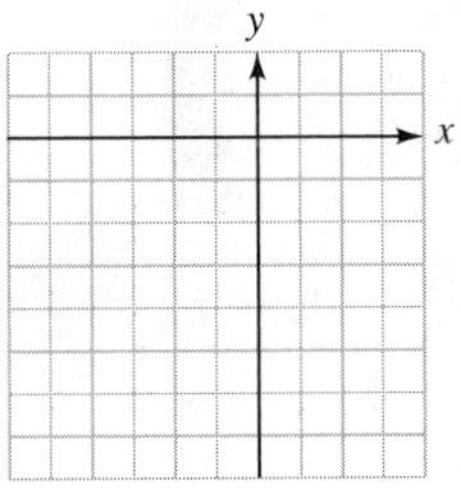

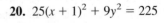

20. $25(x+1)^2 + 9y^2 = 225$

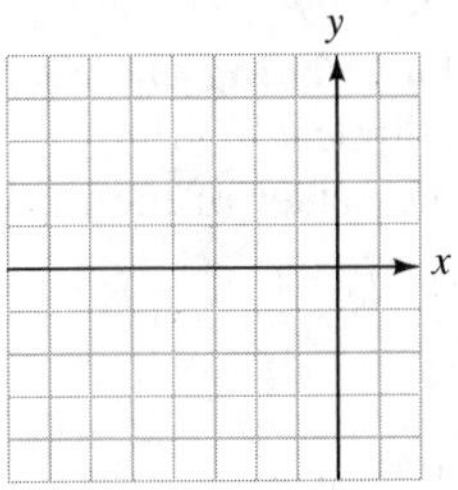

In Exercises 21–24, use a graphing calculator to graph each equation.

21. $\frac{x^2}{9} + \frac{y^2}{4} = 1$

22. $x^2 + 16y^2 = 16$

23. $\frac{x^2}{4} + \frac{(y-1)^2}{9} = 1$

24. $\frac{(x+1)^2}{9} + \frac{(y-2)^2}{4} = 1$

Write each equation in standard form and graph it.

25. $x^2 + 4y^2 - 4x + 8y + 4 = 0$

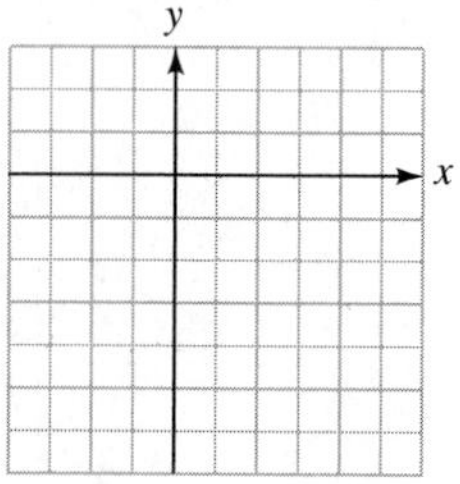

26. $x^2 + 4y^2 - 2x - 16y = -13$

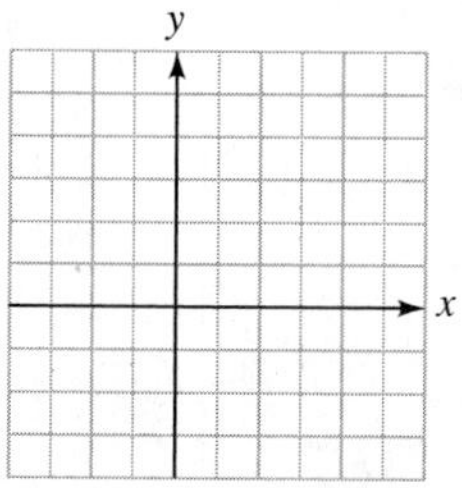

27. $9x^2 + 4y^2 - 18x + 16y = 11$

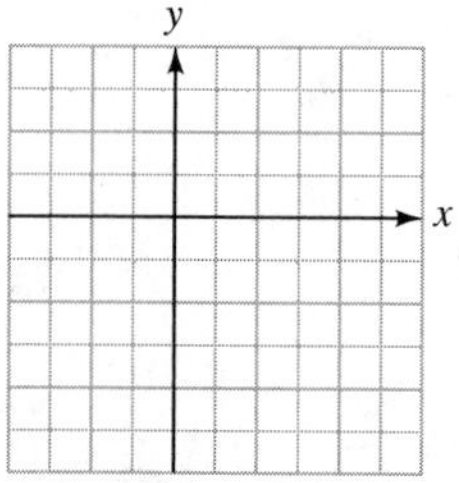

28. $16x^2 + 25y^2 - 160x - 200y + 400 = 0$

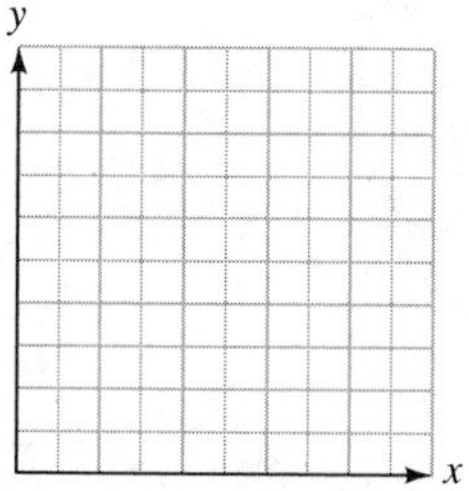

APPLICATIONS

29. Designing an underpass The arch of the underpass in Illustration 1 is a part of an ellipse. Find the equation of the arch.

30. Calculating clearance Find the height of the elliptical arch in Exercise 29 at a point 10 feet from the center of the roadway.

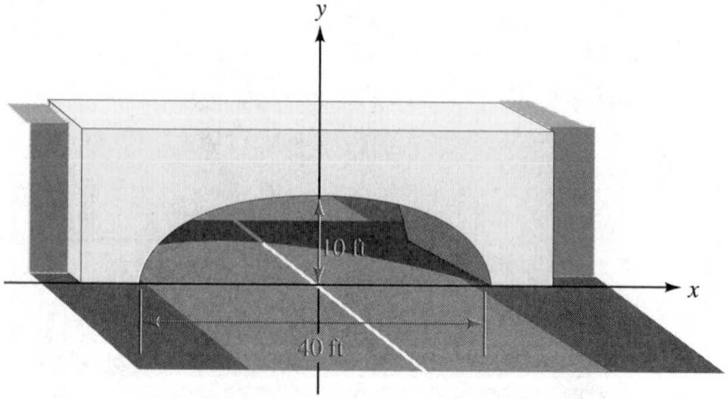

ILLUSTRATION 1

31. Area of an ellipse The area A of the ellipse

$$\frac{x^2}{a^2} + \frac{y^2}{b^2} = 1$$

is given by $A = \pi ab$. Find the area of the ellipse $9x^2 + 16y^2 = 144$.

32. Area of a track The elliptical track in Illustration 2 is bounded by the ellipses $4x^2 + 9y^2 = 576$ and $9x^2 + 25y^2 = 900$. Find the area of the track. (See Exercise 31.)

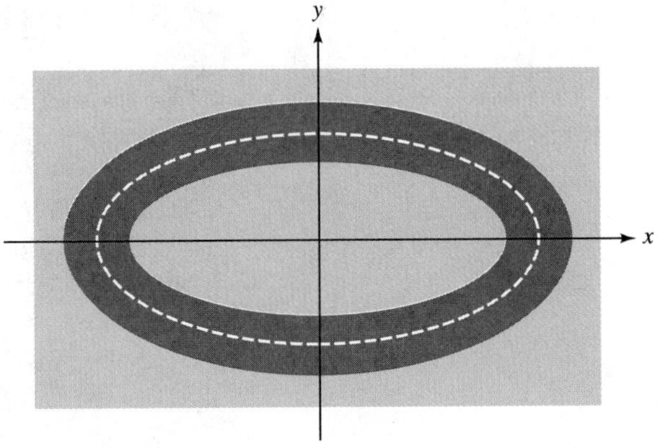

ILLUSTRATION 2

WRITING

33. Explain how to find the x- and the y-intercepts of the graph of the ellipse

$$\frac{x^2}{a^2} + \frac{y^2}{b^2} = 1$$

34. Explain the relationship between the center, focus, and vertex of an ellipse.

SOMETHING TO THINK ABOUT

35. What happens to the graph of

$$\frac{x^2}{a^2} + \frac{y^2}{b^2} = 1$$

when $a = b$?

36. Explain why the graph of $x^2 - 2x + y^2 + 4y + 20 = 0$ does not exist.

10.3 THE HYPERBOLA

■ THE HYPERBOLA ■ PROBLEM SOLVING

Getting Ready *Find the value of y when* $\frac{x^2}{25} - \frac{y^2}{9} = 1$ *and x is the given value. Give each result to the nearest tenth.*

1. $x = 6$

2. $x = -7$

THE HYPERBOLA

The Hyperbola

A **hyperbola** is the set of all points P in the plane for which the difference of the distances of each point from two fixed points is a constant. See Figure 10-22, in which $d_1 - d_2$ is a constant.

Each of the two points is called a **focus**. Midway between the foci is the **center** of the hyperbola.

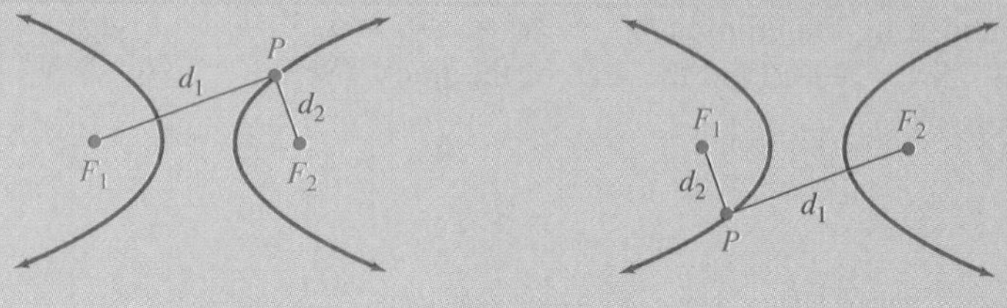

FIGURE 10-22

The graph of the equation

$$\frac{x^2}{25} - \frac{y^2}{9} = 1$$

is a hyperbola. To graph the equation, we make a table of ordered pairs that satisfy the equation, plot each pair, and join the points with a smooth curve as in Figure 10-23.

$$\frac{x^2}{25} - \frac{y^2}{9} = 1$$

x	y	(x, y)
-7	± 2.9	$(-7, \pm 2.9)$
-6	± 2.0	$(-6, \pm 2.0)$
-5	0	$(-5, 0)$
5	0	$(5, 0)$
6	± 2.0	$(6, \pm 2.0)$
7	± 2.9	$(7, \pm 2.9)$

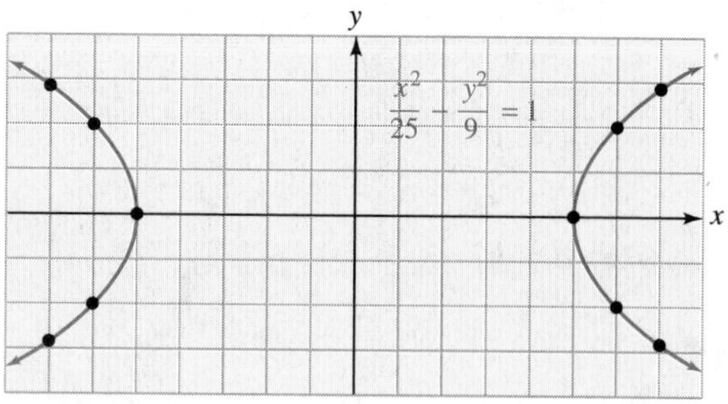

FIGURE 10-23

This graph is centered at the origin and intersects the x-axis at $(\sqrt{25}, 0)$ and $(-\sqrt{25}, 0)$. After simplifying, these points are $(5, 0)$ and $(-5, 0)$. We also note that the graph does not intersect the y-axis.

It is possible to draw a hyperbola without plotting points. For example, if we want to graph the hyperbola with an equation of

$$\frac{x^2}{a^2} - \frac{y^2}{b^2} = 1$$

we first look at the x- and y-intercepts. To find the x-intercepts, we let $y = 0$ and solve for x:

$$\frac{x^2}{a^2} - \frac{0^2}{b^2} = 1$$
$$x^2 = a^2$$
$$x = \pm a$$

Thus, the hyperbola crosses the x-axis at the points $V_1(a, 0)$ and $V_2(-a, 0)$, called the **vertices** of the hyperbola. See Figure 10-24.

To attempt to find the y-intercepts, we let $x = 0$ and solve for y:

$$\frac{0^2}{a^2} - \frac{y^2}{b^2} = 1$$
$$y^2 = -b^2$$
$$y = \pm\sqrt{-b^2}$$

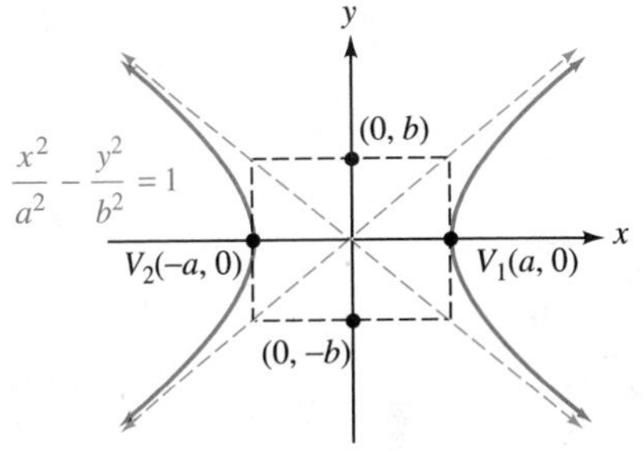

FIGURE 10-24

Since b^2 is always positive, $\sqrt{-b^2}$ is an imaginary number. This means that the hyperbola does not cross the y-axis.

If we construct a rectangle, called the **fundamental rectangle**, whose sides pass horizontally through $\pm b$ on the y-axis and vertically through $\pm a$ on the x-axis, the extended diagonals of the rectangle will be asymptotes of the hyperbola.

Equation of a Hyperbola Centered at the Origin

Any equation that can be written in the form

$$\frac{x^2}{a^2} - \frac{y^2}{b^2} = 1$$

has a graph that is a hyperbola centered at the origin, as in Figure 10-25. The x-intercepts are the vertices $V_1(a, 0)$ and $V_2(-a, 0)$. There are no y-intercepts.

The asymptotes of the hyperbola are the extended diagonals of the rectangle in the figure.

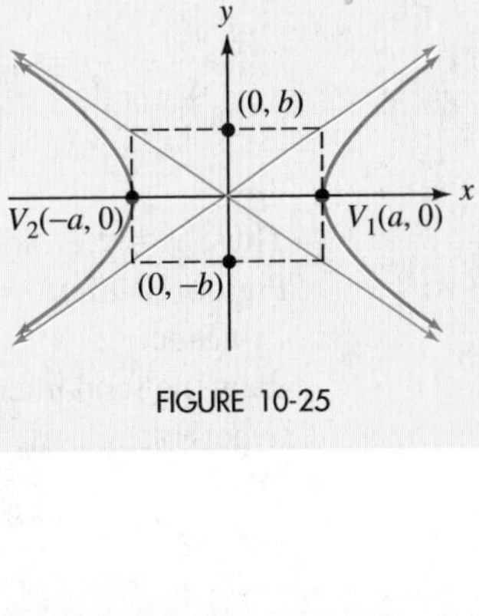

FIGURE 10-25

The branches of the hyperbola in previous discussions open to the left and to the right. It is possible for hyperbolas to have different orientations with respect to the x- and y-axes. For example, the branches of a hyperbola can open upward and downward. In that case, the following equation applies.

Equation of a Hyperbola Centered at the Origin

Any equation that can be written in the form

$$\frac{y^2}{a^2} - \frac{x^2}{b^2} = 1$$

has a graph that is a hyperbola centered at the origin, as in Figure 10-26. The y-intercepts are the vertices $V_1(0, a)$ and $V_2(0, -a)$. There are no x-intercepts.

The asymptotes of the hyperbola are the extended diagonals of the rectangle shown in the figure.

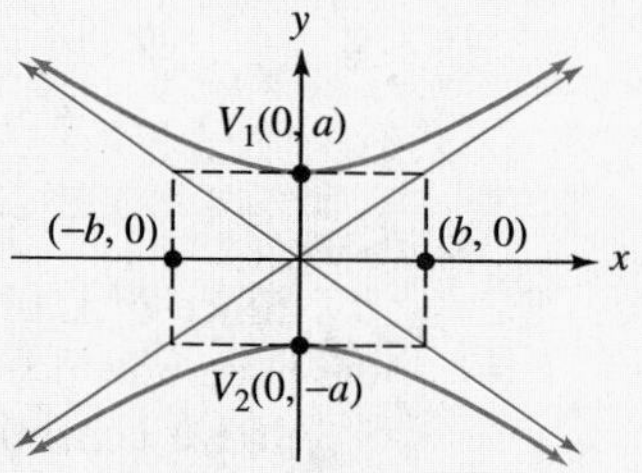

FIGURE 10-26

EXAMPLE 1 Graph $9y^2 - 4x^2 = 36$.

Solution To write the equation in standard form, we divide both sides by 36 to obtain

$$\frac{9y^2}{36} - \frac{4x^2}{36} = 1$$

$$\frac{y^2}{4} - \frac{x^2}{9} = 1 \quad \text{Simplify each fraction.}$$

We then find the y-intercepts by letting $x = 0$ and solving for y:

$$\frac{y^2}{4} - \frac{0^2}{9} = 1$$

$$y^2 = 4$$

Thus, $y = \pm 2$, and the vertices of the hyperbola are $V_1(0, 2)$ and $V_2(0, -2)$. (See Figure 10-27.)

Since $\pm\sqrt{9} = \pm 3$, we use the points $(3, 0)$ and $(-3, 0)$ on the x-axis to help draw the fundamental rectangle. We then draw its extended diagonals and sketch the hyperbola.

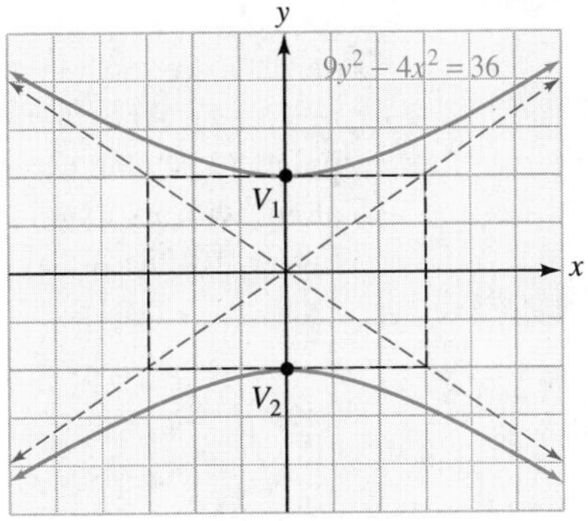

FIGURE 10-27

Self Check Graph $9x^2 - 4y^2 = 36$.

Answer

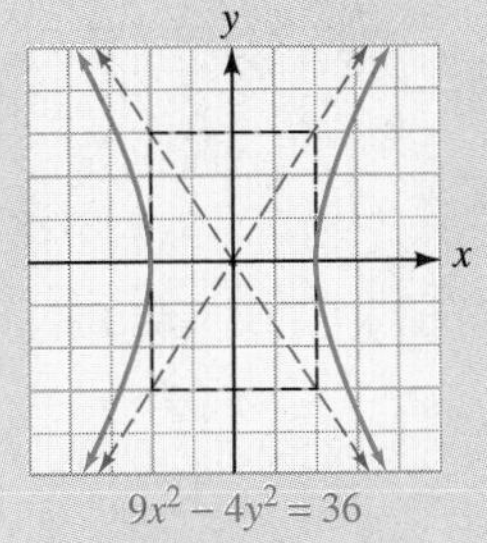

If a hyperbola is centered at a point with coordinates (h, k), the following equations apply.

Equations of Hyperbolas Centered at (h, k)

Any equation that can be written in the form

$$\frac{(x-h)^2}{a^2} - \frac{(y-k)^2}{b^2} = 1$$

is a hyperbola with center at (h, k) that opens left and right.

Any equation of the form

$$\frac{(y-k)^2}{a^2} - \frac{(x-h)^2}{b^2} = 1$$

is a hyperbola with center at (h, k) that opens up and down.

EXAMPLE 2 Graph $\frac{(x-3)^2}{16} - \frac{(y+1)^2}{4} = 1$.

Solution We write the equation in the form

$$\frac{(x-3)^2}{4^2} - \frac{[y-(-1)]^2}{2^2} = 1$$

to see that its graph will be a hyperbola centered at the point $(h, k) = (3, -1)$. Its vertices are located at $a = 4$ units to the right and left of the center, at $(7, -1)$ and $(-1, -1)$. Since $b = 2$, we can count 2 units above and below the center to locate points $(3, 1)$ and $(3, -3)$. With these points, we can draw the fundamental rectangle along with its extended diagonals. We can then sketch the hyperbola, as shown in Figure 10-28.

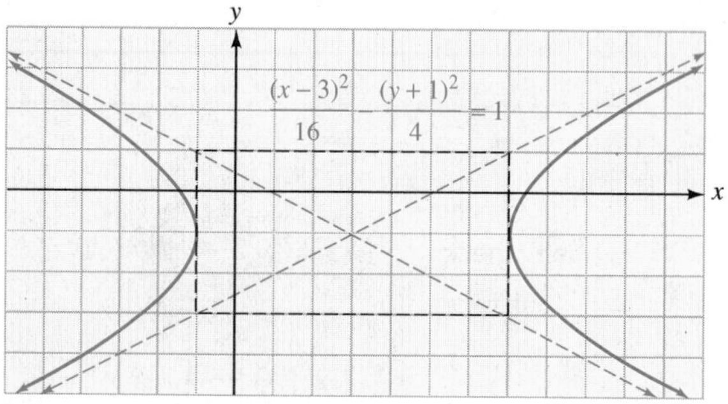

FIGURE 10-28

Self Check Graph $\frac{(x+2)^2}{9} - \frac{(y-1)^2}{4} = 1$.

Answer

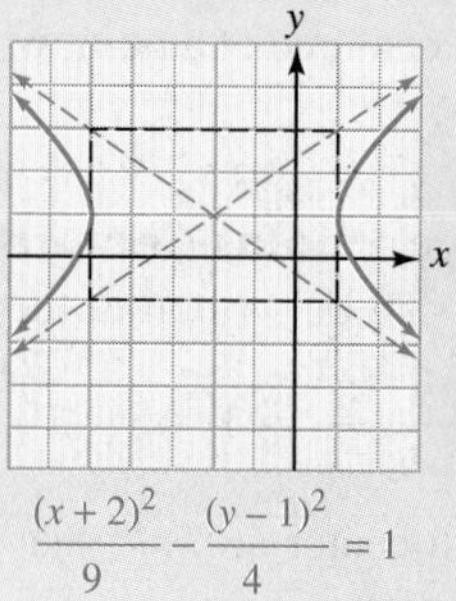

EXAMPLE 3 Write the equation $x^2 - y^2 - 2x + 4y = 12$ in standard form to show that the equation represents a hyperbola. Then graph it.

Solution We proceed as follows.

$$x^2 - y^2 - 2x + 4y = 12$$

$$x^2 - 2x - y^2 + 4y = 12 \quad \text{Use the commutative property to rearrange terms.}$$

$$x^2 - 2x - (y^2 - 4y) = 12 \quad \text{Factor } -1 \text{ from } -y^2 + 4y.$$

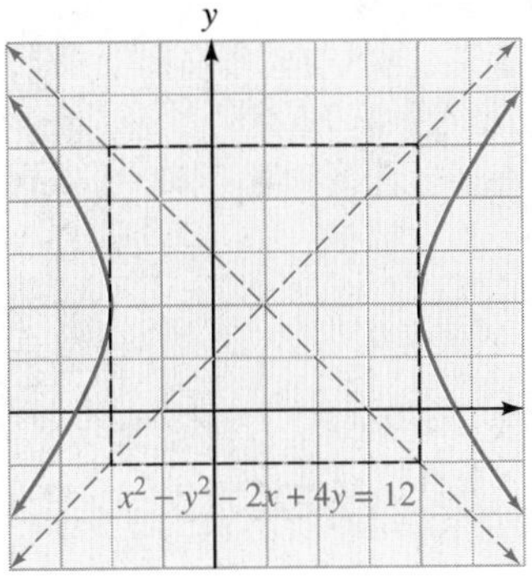

FIGURE 10-29

We then complete the square on x and y to make $x^2 - 2x$ and $y^2 - 4y$ perfect trinomial squares.

$$x^2 - 2x + \mathbf{1} - (y^2 - 4y + \mathbf{4}) = 12 + \mathbf{1} - \mathbf{4}$$

We then factor $x^2 - 2x + 1$ and $y^2 - 4y + 4$ to get

$$(x - 1)^2 - (y - 2)^2 = 9$$

$$\frac{(x - 1)^2}{9} - \frac{(y - 2)^2}{9} = 1 \qquad \text{Divide both sides by 9.}$$

This is the equation of a hyperbola with center at (1, 2). Its graph is shown in Figure 10-29. ■

Self Check Graph $x^2 - 4y^2 + 2x - 8y = 7$.

Answer

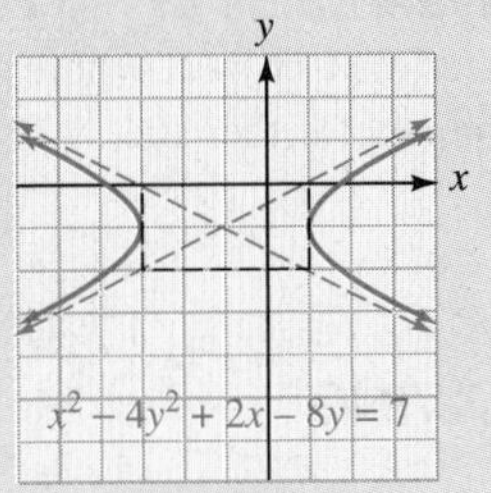

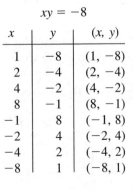

There is a special type of hyperbola (also centered at the origin) that does not intersect either the x- or the y-axis. These hyperbolas have equations of the form $xy = k$, where $k \neq 0$.

EXAMPLE 4 Graph $xy = -8$.

Solution We make a table of ordered pairs, plot each pair, and join the points with a smooth curve to obtain the hyperbola in Figure 10-30.

$xy = -8$

x	y	(x, y)
1	−8	(1, −8)
2	−4	(2, −4)
4	−2	(4, −2)
8	−1	(8, −1)
−1	8	(−1, 8)
−2	4	(−2, 4)
−4	2	(−4, 2)
−8	1	(−8, 1)

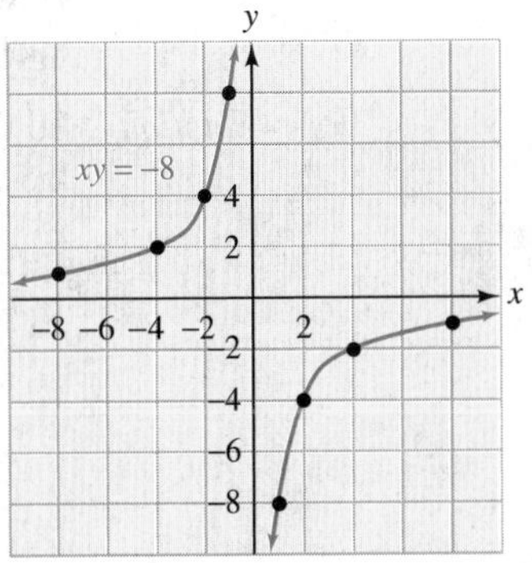

FIGURE 10-30

■

Self Check Graph $xy = 6$.

Answer

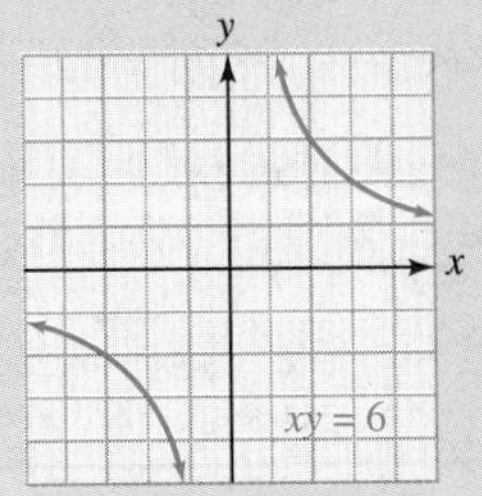

The result in Example 4 illustrates the following general equation.

Equations of Hyperbolas of the Form $xy = k$

Any equation of the form $xy = k$, where $k \neq 0$, has a graph that is a **hyperbola**, which does not intersect either the x- or the y-axis.

PROBLEM SOLVING

EXAMPLE 5 **Atomic structure** In an experiment that led to the discovery of the atomic structure of matter, Lord Rutherford (1871–1937) shot high-energy alpha particles toward a thin sheet of gold. Because many were reflected, Rutherford showed the existence of the nucleus of a gold atom. The alpha particle in Figure 10-31 is repelled by the nucleus at the origin; it travels along the hyperbolic path given by $4x^2 - y^2 = 16$. How close does the particle come to the nucleus?

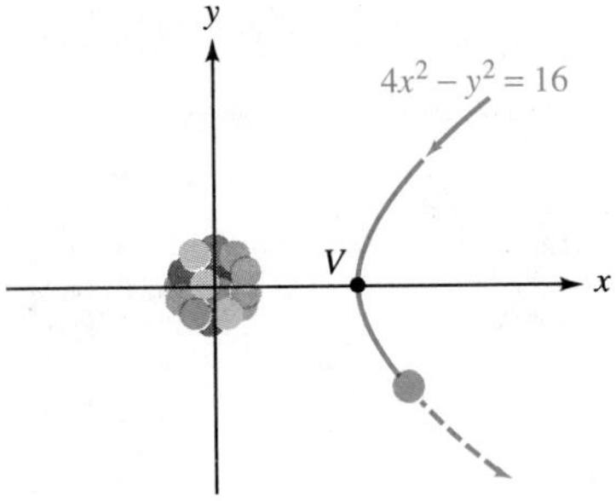

FIGURE 10-31

Solution To find the distance from the nucleus at the origin, we must find the coordinates of the vertex V. To do so, we write the equation of the particle's path in standard form:

$$4x^2 - y^2 = 16$$

$$\frac{4x^2}{16} - \frac{y^2}{16} = \frac{16}{16} \quad \text{Divide both sides by 16.}$$

$$\frac{x^2}{4} - \frac{y^2}{16} = 1 \quad \text{Simplify.}$$

$$\frac{x^2}{2^2} - \frac{y^2}{4^2} = 1 \quad \text{Write 4 as } 2^2 \text{ and 16 as } 4^2.$$

This equation is in the form

$$\frac{x^2}{a^2} - \frac{y^2}{b^2} = 1$$

with $a = 2$. Thus, the vertex of the path is (2, 0). The particle is never closer than 2 units from the nucleus. ■

Orals *Find the x- or y-intercepts of each hyperbola.*

1. $\frac{x^2}{9} - \frac{y^2}{16} = 1$

2. $\frac{y^2}{25} - \frac{x^2}{36} = 1$

EXERCISE 10.3

REVIEW *Factor each expression.*

1. $-6x^4 + 9x^3 - 6x^2$

2. $4a^2 - b^2$

3. $15a^2 - 4ab - 4b^2$

4. $8p^3 - 27q^3$

VOCABULARY AND CONCEPTS *Fill in each blank to make a true statement.*

5. A ________ is the set of all points in a plane for which the ________ of the distances from two fixed points is a constant.

6. The fixed points in Exercise 5 are the _____ of the hyperbola.

7. The midpoint of the line segment joining the foci of a hyperbola is called the _____ of the hyperbola.

8. The hyperbolic graph of

$$\frac{x^2}{a^2} - \frac{y^2}{b^2} = 1$$

has x-intercepts of _______. There are no ________.

9. The center of the hyperbola with an equation of

$$\frac{x^2}{a^2} - \frac{y^2}{b^2} = 1$$

is the point _____.

10. The center of the hyperbola with an equation of

$$\frac{(x-h)^2}{a^2} - \frac{(y-k)^2}{b^2} = 1$$

is the point _____.

PRACTICE *In Exercises 11–22, graph each hyperbola.*

11. $\frac{x^2}{9} - \frac{y^2}{4} = 1$

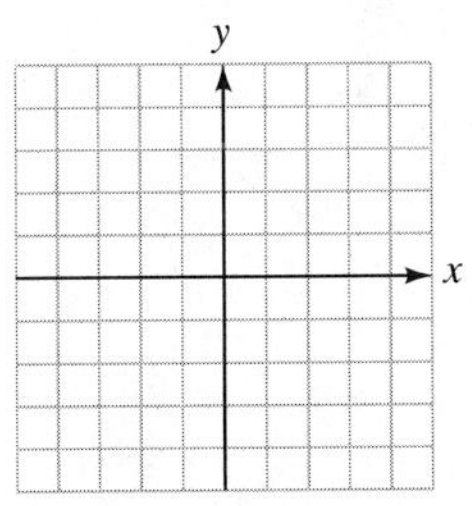

12. $\frac{x^2}{4} - \frac{y^2}{4} = 1$

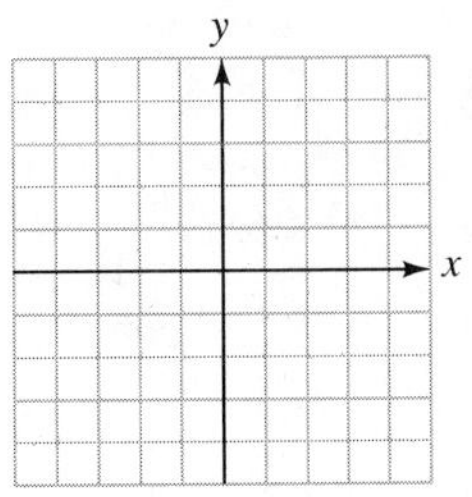

13. $\frac{y^2}{4} - \frac{x^2}{9} = 1$

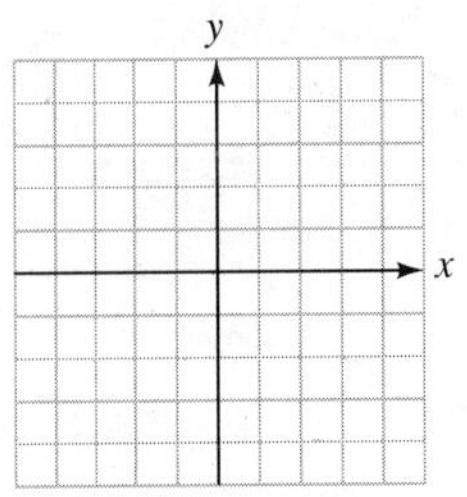

14. $\frac{y^2}{4} - \frac{x^2}{64} = 1$

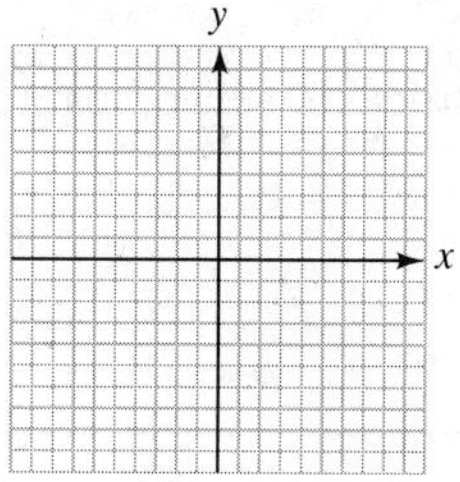

15. $25x^2 - y^2 = 25$

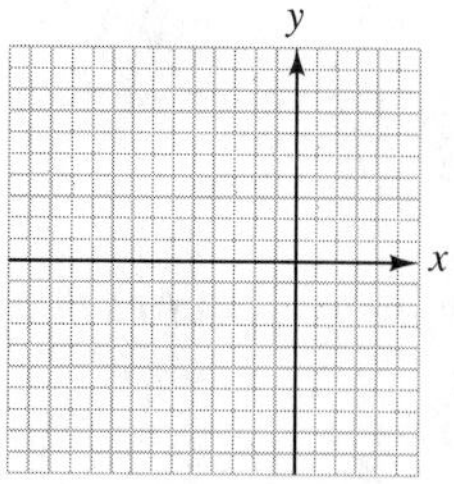

16. $9x^2 - 4y^2 = 36$

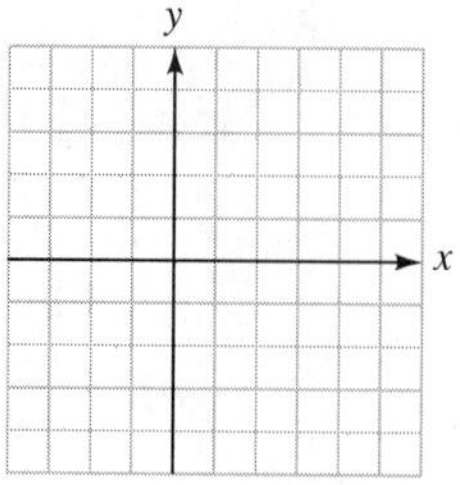

17. $\frac{(x-2)^2}{9} - \frac{y^2}{16} = 1$

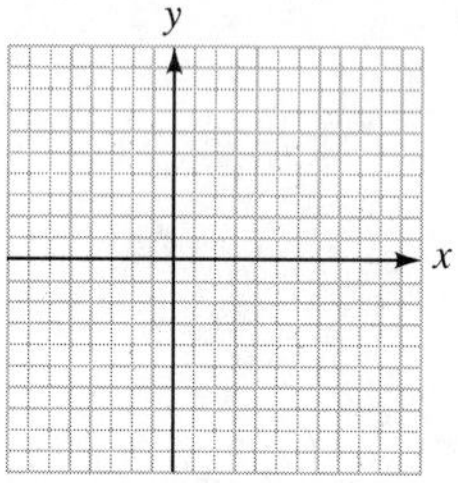

18. $\frac{(x+2)^2}{16} - \frac{(y-3)^2}{25} = 1$

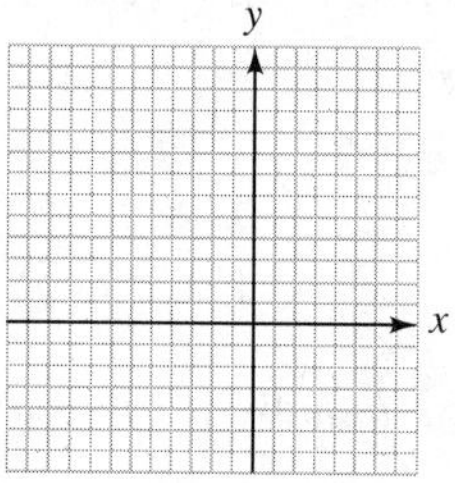

19. $4(x+3)^2 - (y-1)^2 = 4$

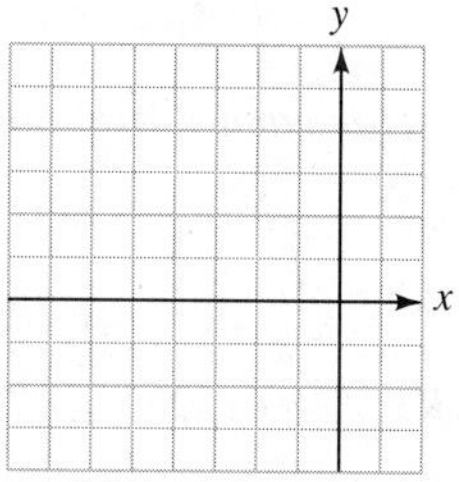

20. $(x+5)^2 - 16y^2 = 16$

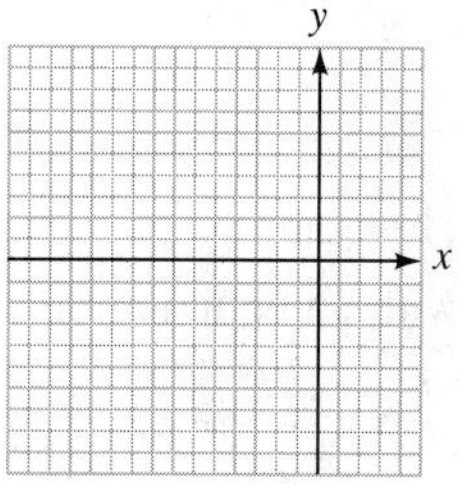

21. $xy = 8$

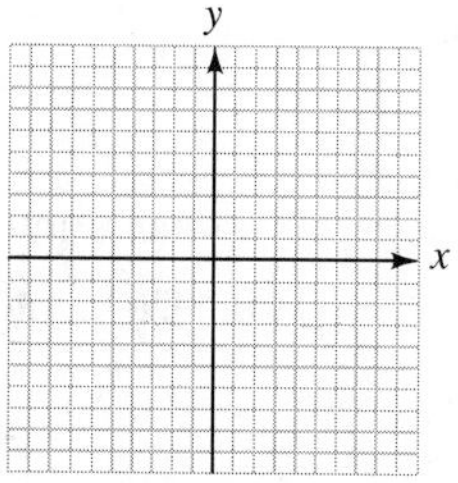

22. $xy = -10$

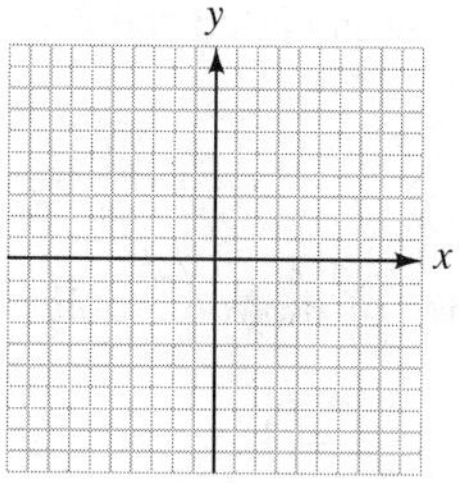

In Exercises 23–26, use a graphing calculator to graph each equation.

23. $\frac{x^2}{9} - \frac{y^2}{4} = 1$

24. $y^2 - 16x^2 = 16$

25. $\frac{x^2}{4} - \frac{(y-1)^2}{9} = 1$

26. $\frac{(y+1)^2}{9} - \frac{(x-2)^2}{4} = 1$

In Exercises 27–30, write each equation in standard form and graph it.

27. $4x^2 - y^2 + 8x - 4y = 4$

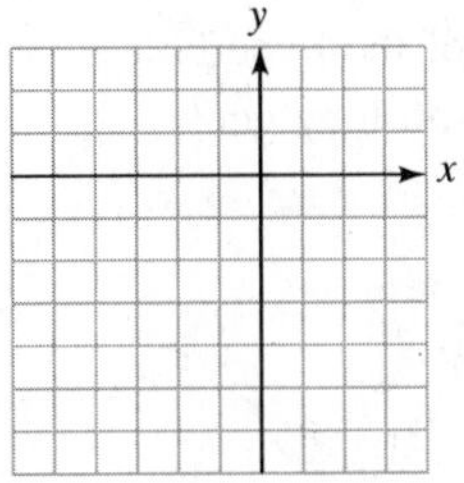

28. $x^2 - 9y^2 - 4x - 54y = 86$

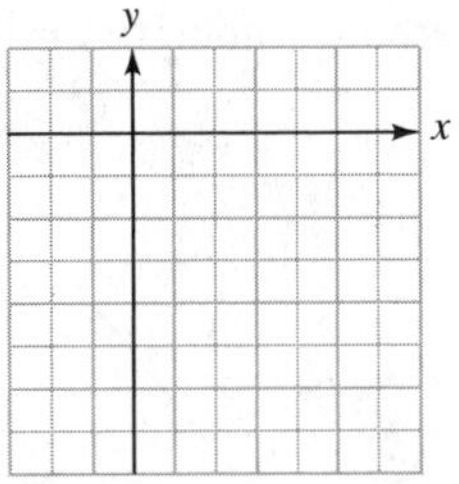

29. $4y^2 - x^2 + 8y + 4x = 4$

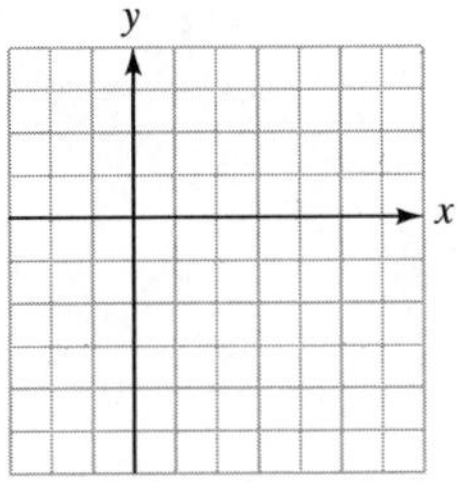

30. $y^2 - 4x^2 - 4y - 8x = 4$

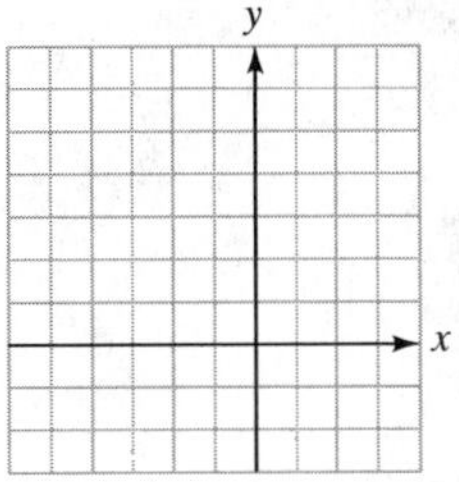

APPLICATIONS

31. Alpha particles The particle in Illustration 1 approaches the nucleus at the origin along the path $9y^2 - x^2 = 81$. How close does the particle come to the nucleus?

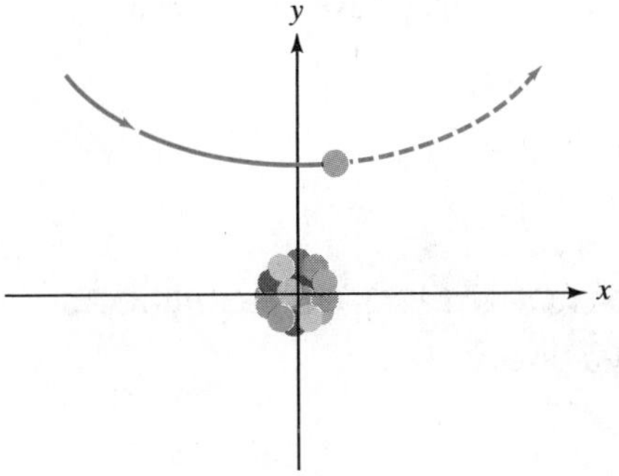

ILLUSTRATION 1

32. LORAN By determining the difference of the distances between the ship in Illustration 2 and two radio transmitters, the LORAN system places the ship on the hyperbola $x^2 - 4y^2 = 576$. If the ship is also 5 miles out to sea, find its coordinates.

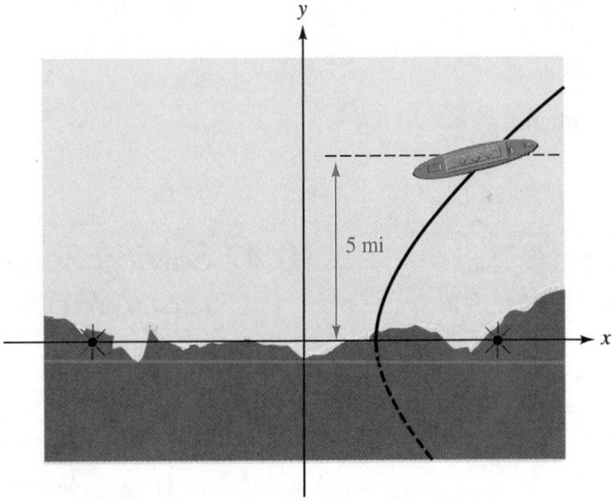

ILLUSTRATION 2

33. Sonic boom The position of the sonic boom caused by the faster-than-sound aircraft in Illustration 3 is the hyperbola $y^2 - x^2 = 25$ in the coordinate system shown. How wide is the hyperbola 5 units from its vertex?

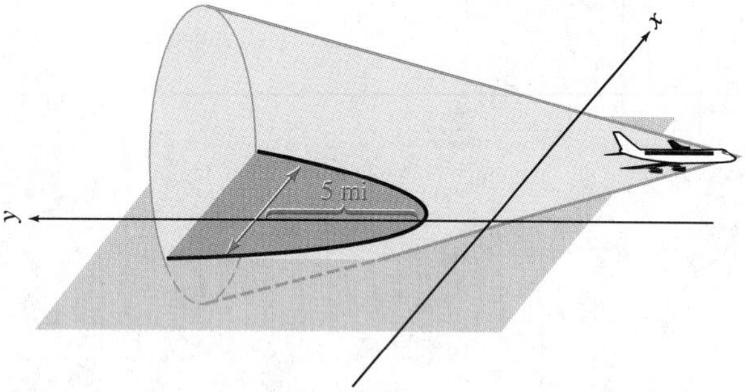

ILLUSTRATION 3

34. Electrostatic repulsion Two similarly charged particles are shot together for an almost head-on collision, as in Illustration 4. They repel each other and travel the two branches of the hyperbola given by $x^2 - 4y^2 = 4$. How close do they get?

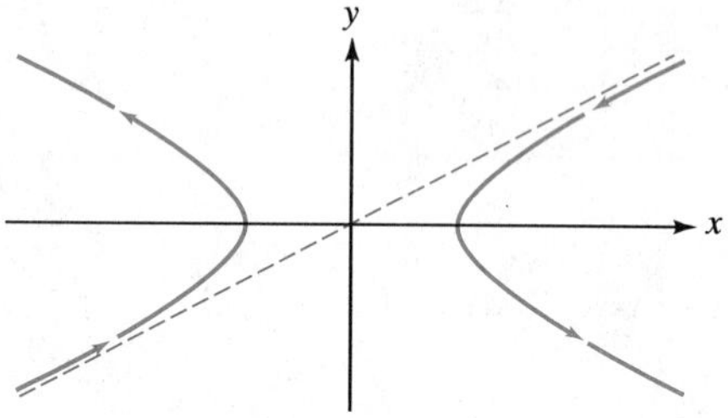

ILLUSTRATION 4

WRITING

35. Explain how to find the x- and the y-intercepts of the graph of the hyperbola

$$\frac{x^2}{a^2} - \frac{y^2}{b^2} = 1$$

36. Explain why the graph of the hyperbola

$$\frac{x^2}{a^2} - \frac{y^2}{b^2} = 1$$

has no y-intercept.

SOMETHING TO THINK ABOUT

37. Describe the fundamental rectangle of

$$\frac{x^2}{a^2} - \frac{y^2}{b^2} = 1$$

when $a = b$.

38. The hyperbolas $x^2 - y^2 = 1$ and $y^2 - x^2 = 1$ are called **conjugate** hyperbolas. Graph both on the same axes. What do they have in common?

10.4 Solving Simultaneous Second-Degree Equations

■ SOLUTION BY GRAPHING ■ SOLUTION BY ELIMINATION

Getting Ready *Add the left-hand sides and the right-hand sides of the following equations.*

1. $3x^2 + 2y^2 = 12$
$4x^2 - 3y^2 = 32$

2. $-7x^2 - 5y^2 = -17$
$12x^2 + 2y^2 = 25$

SOLUTION BY GRAPHING

We now discuss ways to solve systems of two equations in two variables where at least one of the equations is of second degree.

EXAMPLE 1 Solve $\begin{cases} x^2 + y^2 = 25 \\ 2x + y = 10 \end{cases}$ by graphing.

Solution The graph of $x^2 + y^2 = 25$ is a circle with center at the origin and radius of 5. The graph of $2x + y = 10$ is a line. Depending on whether the line is a secant (intersecting the circle at two points) or a tangent (intersecting the circle at one point) or does not intersect the circle at all, there are two, one, or no solutions to the system, respectively.

After graphing the circle and the line, as shown in Figure 10-32, we see that there are two intersection points, $P(3, 4)$ and $P'(5, 0)$. These are the solutions of the system:

$$\begin{cases} x = 3 \\ y = 4 \end{cases} \quad \text{and} \quad \begin{cases} x = 5 \\ y = 0 \end{cases}$$

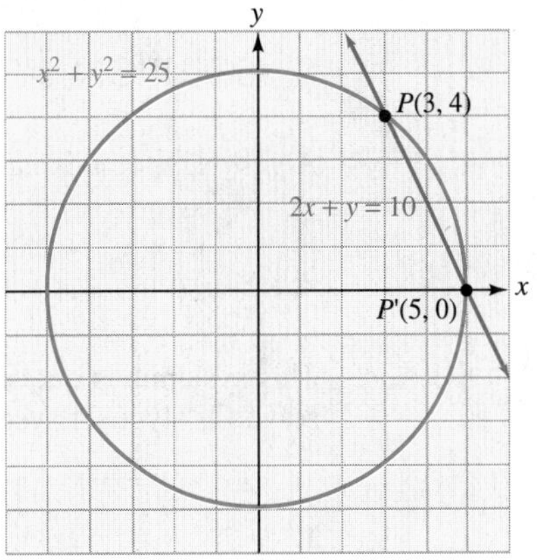

FIGURE 10-32

■

Self Check Solve $\begin{cases} x^2 + y^2 = 13 \\ y = -\frac{1}{5}x + \frac{13}{5}. \end{cases}$

Answer (3, 2), (−2, 3)

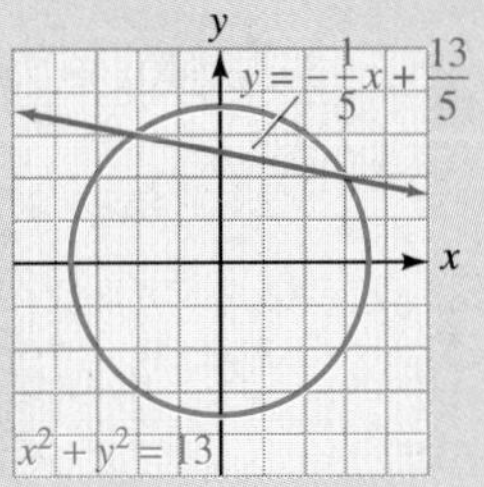

Solving Systems of Equations

GRAPHING CALCULATORS

To solve Example 1 with a graphing calculator, we graph the circle and the line on one set of coordinate axes (see Figure 10-33(a)). We then trace to find the coordinates of the intersection points of the graphs (see Figure 10-33(b) and Figure 10-33(c)).

We can zoom for better results.

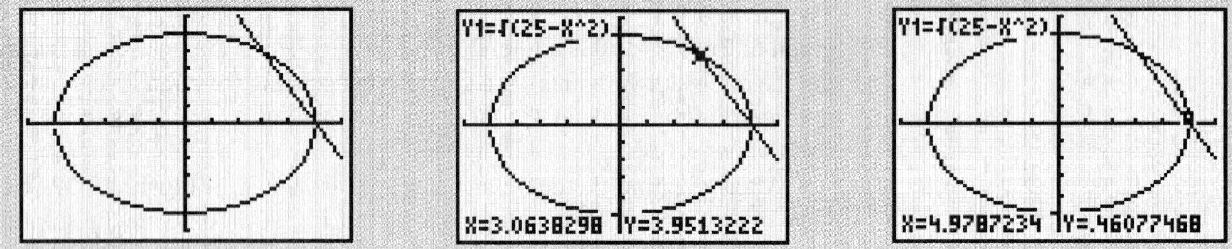

(a) (b) (c)

FIGURE 10-33

SOLUTION BY ELIMINATION

Algebraic methods can also be used to solve systems of equations.

EXAMPLE 2 Solve $\begin{cases} x^2 + y^2 = 25 \\ 2x + y = 10 \end{cases}$.

Solution This system has one second-degree equation and one first-degree equation. We can solve this type of system by substitution. Solving the linear equation for y gives

$$2x + y = 10$$

1. $$y = -2x + 10$$

We can substitute $-2x + 10$ for y in the second-degree equation and solve the resulting quadratic equation for x:

$$x^2 + y^2 = 25$$

$$x^2 + (-2x + 10)^2 = 25$$

$$x^2 + 4x^2 - 40x + 100 = 25 \quad (-2x + 10)(-2x + 10) = 4x^2 - 40x + 100.$$

$$5x^2 - 40x + 75 = 0 \quad \text{Combine like terms and subtract 25 from both sides.}$$

$$x^2 - 8x + 15 = 0 \quad \text{Divide both sides by 5.}$$

$$(x - 5)(x - 3) = 0 \quad \text{Factor } x^2 - 8x + 15.$$

$$x - 5 = 0 \quad \text{or} \quad x - 3 = 0 \quad \text{Set each factor equal to 0.}$$

$$x = 5 \qquad\qquad x = 3$$

If we substitute 5 for x in Equation 1, we get $y = 0$. If we substitute 3 for x in Equation 1, we get $y = 4$. The two solutions are

$$\begin{cases} x = 5 \\ y = 0 \end{cases} \quad \text{or} \quad \begin{cases} x = 3 \\ y = 4 \end{cases}$$

■

Self Check Solve by substitution: $\begin{cases} x^2 + y^2 = 13 \\ y = -\frac{1}{5}x + \frac{13}{5} \end{cases}$.

Answer (3, 2), (−2, 3)

EXAMPLE 3 Solve $\begin{cases} 4x^2 + 9y^2 = 5 \\ y = x^2 \end{cases}$.

Solution We can solve this system by substitution.

$$\begin{aligned} 4x^2 + 9y^2 &= 5 & & \\ 4y + 9y^2 &= 5 & & \text{Substitute } y \text{ for } x^2. \\ 9y^2 + 4y - 5 &= 0 & & \text{Subtract 5 from both sides.} \\ (9y - 5)(y + 1) &= 0 & & \text{Factor } 9y^2 + 4y - 5. \\ 9y - 5 = 0 \quad &\text{or} \quad y + 1 = 0 & & \text{Set each factor equal to 0.} \\ y = \frac{5}{9} \qquad & \qquad\quad y = -1 & & \end{aligned}$$

Since $y = x^2$, the values of x are found by solving the equations

$$x^2 = \frac{5}{9} \quad \text{and} \quad x^2 = -1$$

Because $x^2 = -1$ has no real solutions, this possibility is discarded. The solutions of $x^2 = \frac{5}{9}$ are

$$x = \frac{\sqrt{5}}{3} \quad \text{or} \quad x = -\frac{\sqrt{5}}{3}$$

The solutions of the system are

$$\left(\frac{\sqrt{5}}{3}, \frac{5}{9}\right) \quad \text{and} \quad \left(-\frac{\sqrt{5}}{3}, \frac{5}{9}\right)$$

■

Self Check Solve $\begin{cases} x^2 + y^2 = 20 \\ y = x^2 \end{cases}$.

Answer (2, 4), (−2, 4)

EXAMPLE 4 Solve $\begin{cases} 3x^2 + 2y^2 = 36 \\ 4x^2 - y^2 = 4 \end{cases}$.

Solution Since both equations are in the form $ax^2 + by^2 = c$, we can solve the system by addition.

We can copy the first equation and multiply the second equation by 2 to obtain the equivalent system

$$\begin{cases} 3x^2 + 2y^2 = 36 \\ 8x^2 - 2y^2 = 8 \end{cases}$$

We add the equations to eliminate y and solve the resulting equation for x:

$$11x^2 = 44$$
$$x^2 = 4$$
$$x = 2 \quad \text{or} \quad x = -2$$

To find y, we substitute 2 for x and then -2 for x in the first equation and proceed as follows:

For $x = 2$

$$3x^2 + 2y^2 = 36$$
$$3(2)^2 + 2y^2 = 36$$
$$12 + 2y^2 = 36$$
$$2y^2 = 24$$
$$y^2 = 12$$
$$y = +\sqrt{12} \quad \text{or} \quad y = -\sqrt{12}$$
$$y = 2\sqrt{3} \qquad\qquad y = -2\sqrt{3}$$

For $x = -2$

$$3x^2 + 2y^2 = 36$$
$$3(-2)^2 + 2y^2 = 36$$
$$12 + 2y^2 = 36$$
$$2y^2 = 24$$
$$y^2 = 12$$
$$y = +\sqrt{12} \quad \text{or} \quad y = -\sqrt{12}$$
$$y = 2\sqrt{3} \qquad\qquad y = -2\sqrt{3}$$

The four solutions of this system are

$$\left(2, 2\sqrt{3}\right), \quad \left(2, -2\sqrt{3}\right), \quad \left(-2, 2\sqrt{3}\right), \quad \text{and} \quad \left(-2, -2\sqrt{3}\right)$$

■

Self Check Solve $\begin{cases} x^2 + 4y^2 = 16 \\ x^2 - y^2 = 1 \end{cases}$.

Answer $(2, \sqrt{3}), (2, -\sqrt{3}), (-2, \sqrt{3}), (-2, -\sqrt{3})$

Orals *Give the possible number of solutions of a system when the graphs of the equations are*

1. A line and a parabola

2. A line and a hyperbola

3. A circle and a parabola

4. A circle and a hyperbola

EXERCISE 10.4

REVIEW *Simplify each radical expression. Assume that all variables represent positive numbers.*

1. $\sqrt{200x^2} - 3\sqrt{98x^2}$

2. $a\sqrt{112a} - 5\sqrt{175a^3}$

3. $\dfrac{3t\sqrt{2t} - 2\sqrt{2t^3}}{\sqrt{18t} - \sqrt{2t}}$

4. $\sqrt[3]{\dfrac{x}{4}} + \sqrt[3]{\dfrac{x}{32}} - \sqrt[3]{\dfrac{x}{500}}$

VOCABULARY AND CONCEPTS *Fill in each blank to make a true statement.*

5. We can solve systems of equations by ________, addition, or ________.

6. A line can intersect an ellipse in at most ____ points.

PRACTICE *In Exercises 7–14, solve each system of equations by graphing.*

7. $\begin{cases} 8x^2 + 32y^2 = 256 \\ x = 2y \end{cases}$

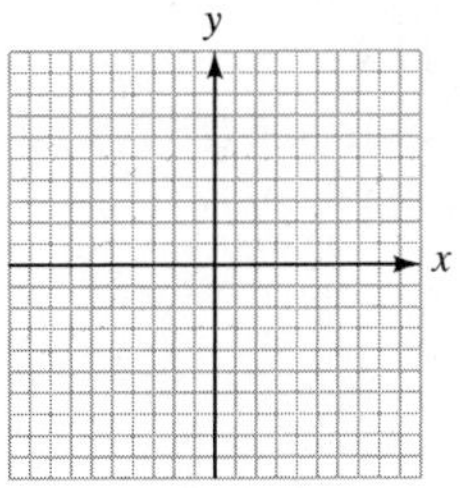

8. $\begin{cases} x^2 + y^2 = 2 \\ x + y = 2 \end{cases}$

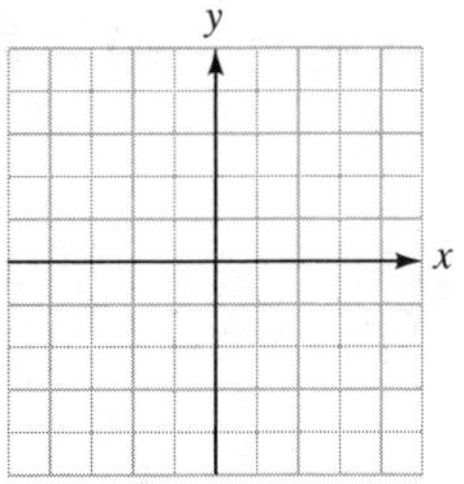

9. $\begin{cases} x^2 + y^2 = 10 \\ y = 3x^2 \end{cases}$

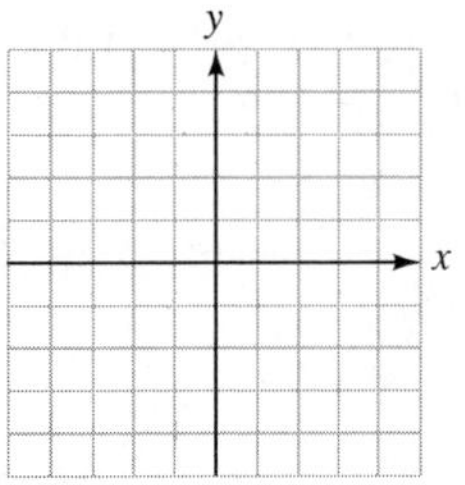

10. $\begin{cases} x^2 + y^2 = 5 \\ x + y = 3 \end{cases}$

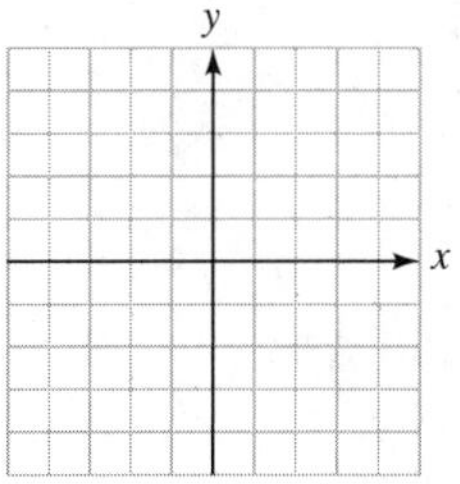

11. $\begin{cases} x^2 + y^2 = 25 \\ 12x^2 + 64y^2 = 768 \end{cases}$

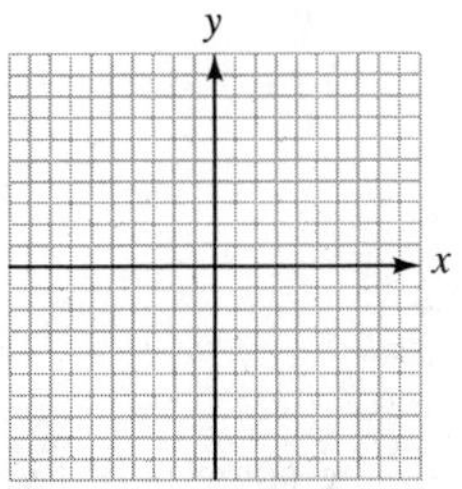

12. $\begin{cases} x^2 + y^2 = 13 \\ y = x^2 - 1 \end{cases}$

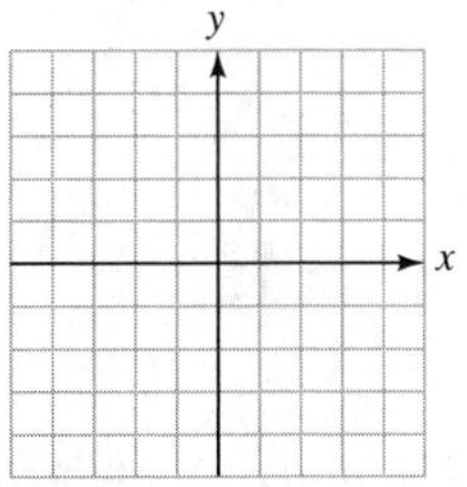

13. $\begin{cases} x^2 - 13 = -y^2 \\ y = 2x - 4 \end{cases}$

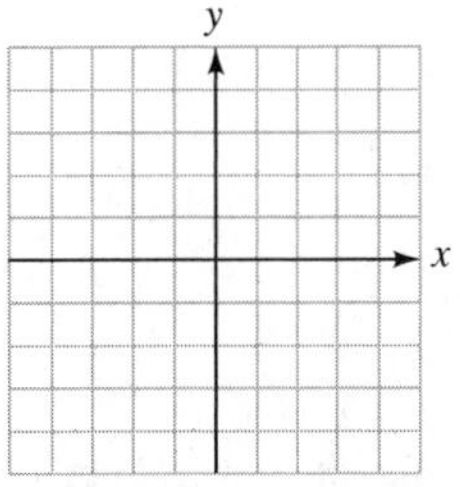

14. $\begin{cases} x^2 + y^2 = 20 \\ y = x^2 \end{cases}$

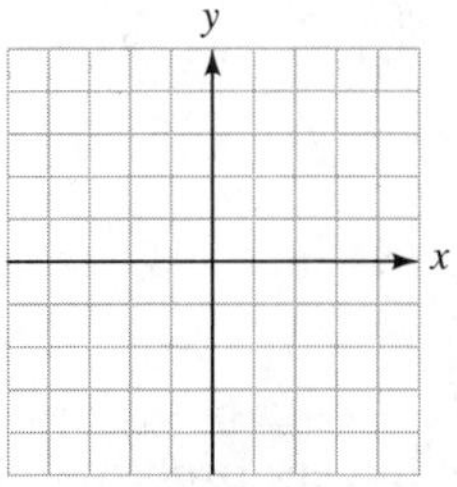

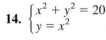

In Exercises 15–16, use a graphing calculator to solve each system.

15. $\begin{cases} x^2 - 6x - y = -5 \\ x^2 - 6x + y = -5 \end{cases}$

16. $\begin{cases} x^2 - y^2 = -5 \\ 3x^2 + 2y^2 = 30 \end{cases}$

In Exercises 17–42, solve each system of equations algebraically for real values of x and y.

17. $\begin{cases} 25x^2 + 9y^2 = 225 \\ 5x + 3y = 15 \end{cases}$

18. $\begin{cases} x^2 + y^2 = 20 \\ y = x^2 \end{cases}$

19. $\begin{cases} x^2 + y^2 = 2 \\ x + y = 2 \end{cases}$

20. $\begin{cases} x^2 + y^2 = 36 \\ 49x^2 + 36y^2 = 1{,}764 \end{cases}$

21. $\begin{cases} x^2 + y^2 = 5 \\ x + y = 3 \end{cases}$

22. $\begin{cases} x^2 - x - y = 2 \\ 4x - 3y = 0 \end{cases}$

23. $\begin{cases} x^2 + y^2 = 13 \\ y = x^2 - 1 \end{cases}$

24. $\begin{cases} x^2 + y^2 = 25 \\ 2x^2 - 3y^2 = 5 \end{cases}$

25. $\begin{cases} x^2 + y^2 = 30 \\ y = x^2 \end{cases}$

26. $\begin{cases} 9x^2 - 7y^2 = 81 \\ x^2 + y^2 = 9 \end{cases}$

27. $\begin{cases} x^2 + y^2 = 13 \\ x^2 - y^2 = 5 \end{cases}$

28. $\begin{cases} 2x^2 + y^2 = 6 \\ x^2 - y^2 = 3 \end{cases}$

29. $\begin{cases} x^2 + y^2 = 20 \\ x^2 - y^2 = -12 \end{cases}$

30. $\begin{cases} xy = -\dfrac{9}{2} \\ 3x + 2y = 6 \end{cases}$

31. $\begin{cases} y^2 = 40 - x^2 \\ y = x^2 - 10 \end{cases}$

32. $\begin{cases} x^2 - 6x - y = -5 \\ x^2 - 6x + y = -5 \end{cases}$

33. $\begin{cases} y = x^2 - 4 \\ x^2 - y^2 = -16 \end{cases}$

34. $\begin{cases} 6x^2 + 8y^2 = 182 \\ 8x^2 - 3y^2 = 24 \end{cases}$

35. $\begin{cases} x^2 - y^2 = -5 \\ 3x^2 + 2y^2 = 30 \end{cases}$

36. $\begin{cases} \dfrac{1}{x} + \dfrac{1}{y} = 5 \\ \dfrac{1}{x} - \dfrac{1}{y} = -3 \end{cases}$

37. $\begin{cases} \dfrac{1}{x} + \dfrac{2}{y} = 1 \\ \dfrac{2}{x} - \dfrac{1}{y} = \dfrac{1}{3} \end{cases}$

38. $\begin{cases} \dfrac{1}{x} + \dfrac{3}{y} = 4 \\ \dfrac{2}{x} - \dfrac{1}{y} = 7 \end{cases}$

39. $\begin{cases} 3y^2 = xy \\ 2x^2 + xy - 84 = 0 \end{cases}$

40. $\begin{cases} x^2 + y^2 = 10 \\ 2x^2 - 3y^2 = 5 \end{cases}$

41. $\begin{cases} xy = \dfrac{1}{6} \\ y + x = 5xy \end{cases}$

42. $\begin{cases} xy = \dfrac{1}{12} \\ y + x = 7xy \end{cases}$

43. Integer problem The product of two integers is 32, and their sum is 12. Find the integers.

44. Number problem The sum of the squares of two numbers is 221, and the sum of the numbers is 9. Find the numbers.

APPLICATIONS

45. Geometry problem The area of a rectangle is 63 square centimeters, and its perimeter is 32 centimeters. Find the dimensions of the rectangle.

46. Investing money Grant receives $225 annual income from one investment. Jeff invested $500 more than Grant, but at an annual rate of 1% less. Jeff's annual income is $240. What is the amount and rate of Grant's investment?

PERSPECTIVE

Focus on Conics

Satellite dishes and flashlight reflectors are familiar examples of a conic's ability to reflect a beam of light or to concentrate incoming satellite signals at one point. That property is shown in Illustration 1.

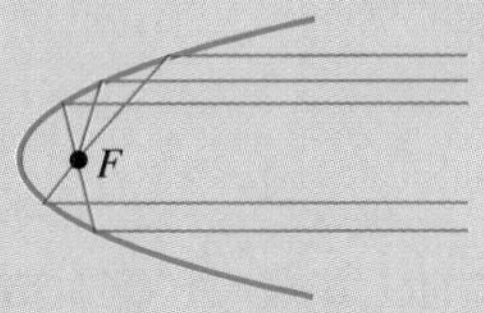

ILLUSTRATION 1

An ellipse has two foci, the points labeled F in Illustration 2. Any light or signal that starts at one focus will be reflected to the other. This property is the basis of whispering galleries, where a person standing at one focus can clearly hear another person speaking at the other focus.

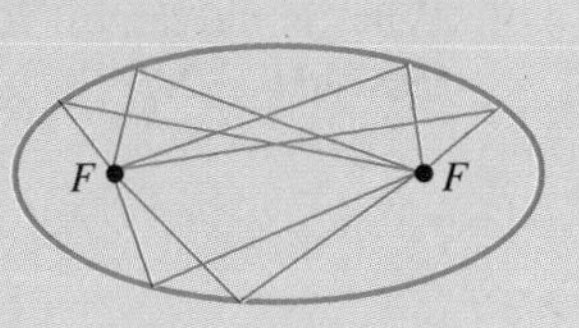

ILLUSTRATION 2

The focal property of the ellipse is also used in **lithotripsy**, a medical procedure for treating kidney stones. The patient is placed in an elliptical tank of water with the kidney stone at one focus. Shock waves from a small controlled explosion at the other focus are concentrated on the stone, pulverizing it.

The hyperbola also has two foci, the two points labeled F in Illustration 3. As in the ellipse, light aimed at one focus is reflected toward the other. Hyperbolic mirrors are used in some reflecting telescopes.

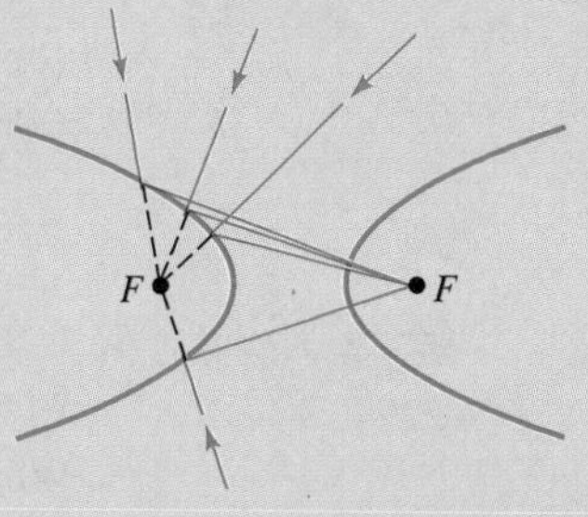

ILLUSTRATION 3

47. Investing money Carol receives \$67.50 annual income from one investment. John invested \$150 more than Carol at an annual rate of $1\frac{1}{2}\%$ more. John's annual income is \$94.50. What is the amount and rate of Carol's investment? (*Hint:* There are two answers.)

48. Artillery The shell fired from the base of the hill in Illustration 1 follows the parabolic path $y = -\frac{1}{6}x^2 + 2x$, with distances measured in miles. The hill has a slope of $\frac{1}{3}$. How far from the gun is the point of impact? (*Hint*: Find the coordinates of the point and then the distance.)

49. Driving rates Jim drove 306 miles. Jim's brother made the same trip at a speed 17 mph slower than Jim did and required an extra $1\frac{1}{2}$ hours. What was Jim's rate and time?

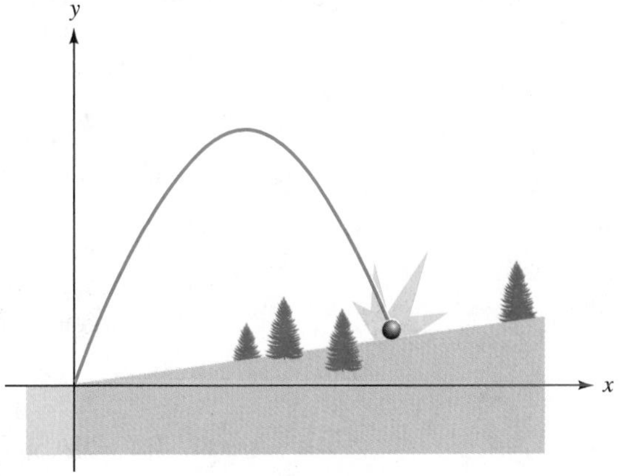

ILLUSTRATION 1

WRITING

50. Describe the benefits of the graphical method for solving a system of equations.

51. Describe the drawbacks of the graphical method.

SOMETHING TO THINK ABOUT

52. The graphs of the two independent equations of a system are parabolas. How many solutions might the system have?

53. The graphs of the two independent equations of a system are hyperbolas. How many solutions might the system have?

10.5 Piecewise-Defined Functions and the Greatest Integer Function

■ PIECEWISE-DEFINED FUNCTIONS ■ INCREASING AND DECREASING FUNCTIONS ■ THE GREATEST INTEGER FUNCTION

Getting Ready

1. Is $f(x) = x^2$ positive or negative when $x > 0$?

2. Is $f(x) = -x^2$ positive or negative when $x > 0$?

3. What is the largest integer that is less than 98.6?

4. What is the largest integer that is less than -2.7?

■ PIECEWISE-DEFINED FUNCTIONS

In this section, we will discuss functions that are defined by using different equations for different parts of their domains. Such functions are called **piecewise-defined functions**.

A simple piecewise-defined function is the absolute value function, $f(x) = |x|$, which can be written in the form

$$f(x) = \begin{cases} x \text{ when } x \geq 0 \\ -x \text{ when } x < 0 \end{cases}$$

When x is in the interval $[0, \infty)$, we use the function $f(x) = x$ to evaluate $|x|$. However, when x is in the interval $(-\infty, 0)$, we use the function $f(x) = -x$ to evaluate $|x|$. The graph of the absolute value function is shown in Figure 10-34.

■ INCREASING AND DECREASING FUNCTIONS

We have seen that if the values of $f(x)$ increase as x increases on an interval, we say that the function is *increasing on the interval* (see Figure 10-35(a)). If the values of

For $x \geq 0$

x	y	(x, y)
0	0	(0, 0)
1	1	(1, 1)
2	2	(2, 2)
3	3	(3, 3)

For $x \leq 0$

x	y	(x, y)
−4	4	(−4, 4)
−3	3	(−3, 3)
−2	2	(−2, 2)
−1	1	(−1, 1)

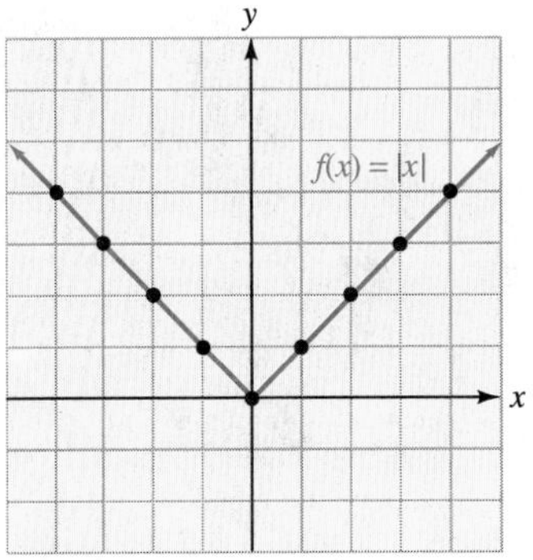

FIGURE 10-34

$f(x)$ decrease as x increases on an interval, we say that the function is *decreasing on the interval* (see Figure 10-35(b)). If the values of $f(x)$ remain constant as x increases on an interval, we say that the function is *constant on the interval* (see Figure 10-35(c)).

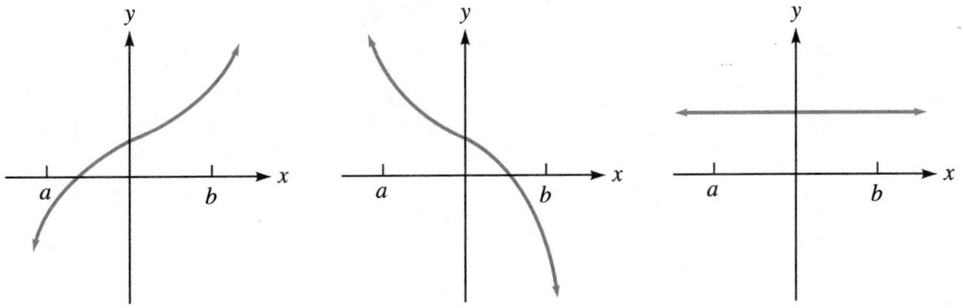

Increasing on (a, b)
(a)

Decreasing on (a, b)
(b)

Constant on (a, b)
(c)

FIGURE 10-35

The absolute value function, pictured in Figure 10-34, is decreasing on the interval $(-\infty, 0)$ and is increasing on the interval $(0, \infty)$.

EXAMPLE 1 Graph the piecewise-defined function given by

$$f(x) = \begin{cases} x^2 \text{ when } x \leq 0 \\ x \text{ when } 0 < x < 2 \\ -1 \text{ when } x \geq 2 \end{cases}$$

and tell where the function is increasing, decreasing, or constant.

Solution For each number x, we decide which of the three equations will be used to find the corresponding value of y:

- For numbers $x \leq 0$, $f(x)$ is determined by $f(x) = x^2$, and the graph is the left half of a parabola. See Figure 10-36. Since the values of $f(x)$ decrease on this graph as x increases, the function is decreasing on the interval $(-\infty, 0)$.
- For numbers $0 < x < 2$, $f(x)$ is determined by $f(x) = x$, and the graph is part of a line. Since the values of $f(x)$ increase on this graph as x increases, the function is increasing on the interval $(0, 2)$.
- For numbers $x \geq 2$, $f(x)$ is the constant -1, and the graph is part of a horizontal line. Since the values of $f(x)$ remain constant on this line, the function is constant on the interval $(2, \infty)$.

The use of solid and open circles on the graph indicates that $f(x) = -1$ when $x = 2$.

Since every number x determines one value y, the domain of this function is the interval $(-\infty, \infty)$. The range is the interval $[0, \infty) \cup \{-1\}$.

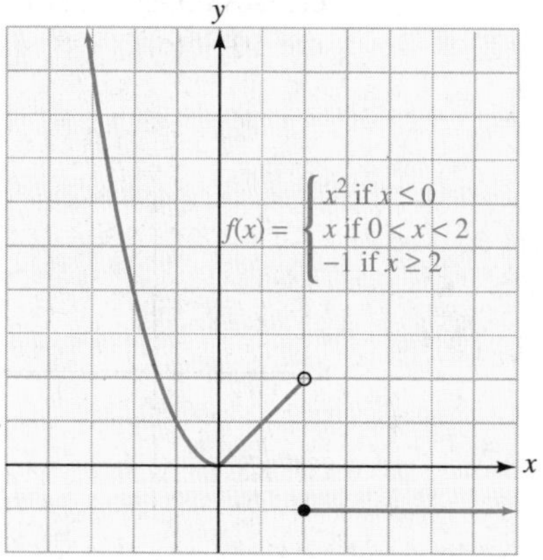

FIGURE 10-36

■

Self Check Graph the function

$$f(x) = \begin{cases} 2x \text{ when } x \leq 0 \\ \frac{1}{2}x \text{ when } x > 0 \end{cases}.$$

Answer

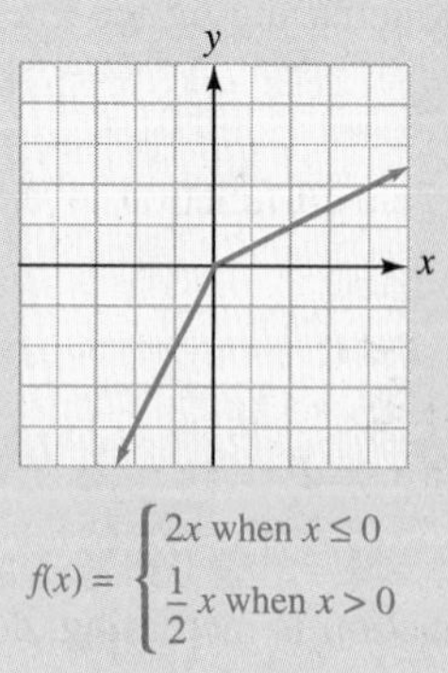

THE GREATEST INTEGER FUNCTION

The **greatest integer function** is important in computer applications. It is a function determined by the equation

$y = f(x) = [\![x]\!]$ Read as "y equals the greatest integer in x."

where the value of y that corresponds to x is the greatest integer that is less than or equal to x. For example,

$$[\![4.7]\!] = 4, \quad \left[\!\!\left[2\tfrac{1}{2}\right]\!\!\right] = 2, \quad [\![\pi]\!] = 3, \quad [\![-3.7]\!] = -4, \quad [\![-5.7]\!] = -6$$

EXAMPLE 2 Graph $y = [\![x]\!]$.

Solution We list several intervals and the corresponding values of the greatest integer function:

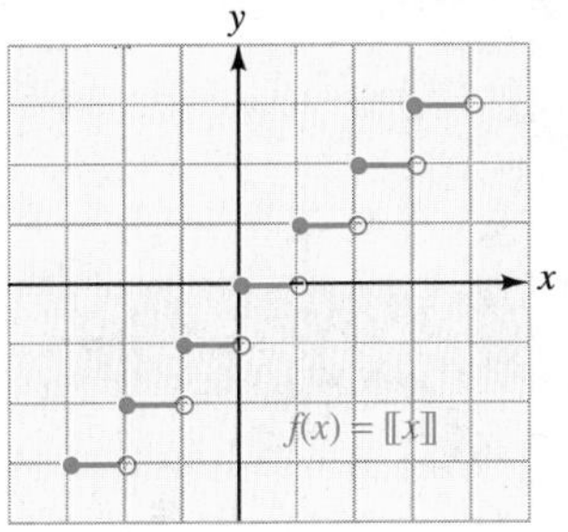

FIGURE 10-37

$[0, 1)$	$y = [\![x]\!] = 0$	For numbers from 0 to 1, not including 1, the greatest integer in the interval is 0.
$[1, 2)$	$y = [\![x]\!] = 1$	For numbers from 1 to 2, not including 2, the greatest integer in the interval is 1.
$[2, 3)$	$y = [\![x]\!] = 2$	For numbers from 2 to 3, not including 3, the greatest integer in the interval is 2.

In each interval, the values of y are constant, but they jump by 1 at integer values of x. The graph is shown in Figure 10-37. From the graph, we see that the domain is $(-\infty, \infty)$, and the range is the set of integers $\{. . ., -3, -2, -1, 0, 1, 2, 3,. . .\}$.

Since the greatest integer function is made up of a series of horizontal line segments, it is an example of a group of functions called **step functions**. ■

EXAMPLE 3 To print stationery, a printer charges \$10 for setup charges, plus \$20 for each box. The printer counts any portion of a box as a full box. Graph this step function.

Solution If we order stationery and cancel the order before it is printed, the cost will be \$10. Thus, the ordered pair (0, 10) will be on the graph.

If we purchase 1 box, the cost will be \$10 for setup plus \$20 for printing, for a total cost of \$30. Thus, the ordered pair (1, 30) will be on the graph.

The cost of $1\frac{1}{2}$ boxes will be the same as the cost of 2 boxes, or \$50. Thus, the ordered pairs (1.5, 50) and (2, 50) will be on the graph.

The complete graph is shown in Figure 10-38.

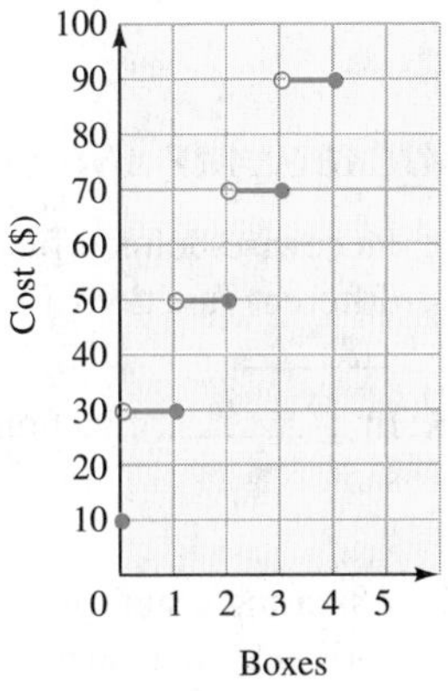

FIGURE 10-38 ■

Self Check In Example 3, how much will $3\frac{1}{2}$ boxes cost?

Answer $90

Orals *Tell whether each function is increasing, decreasing, or constant on the interval* $(-2, 3)$.

1.

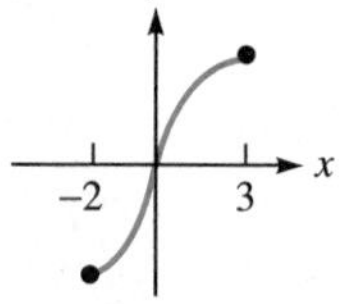

2.

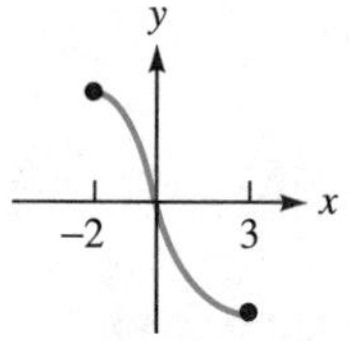

3.

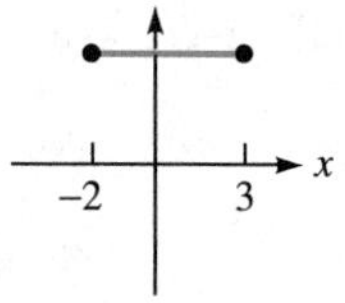

4.

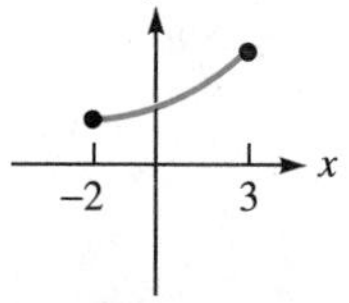

EXERCISE 10.5

REVIEW *Find the value of x. Assume that lines r and s are parallel.*

1.

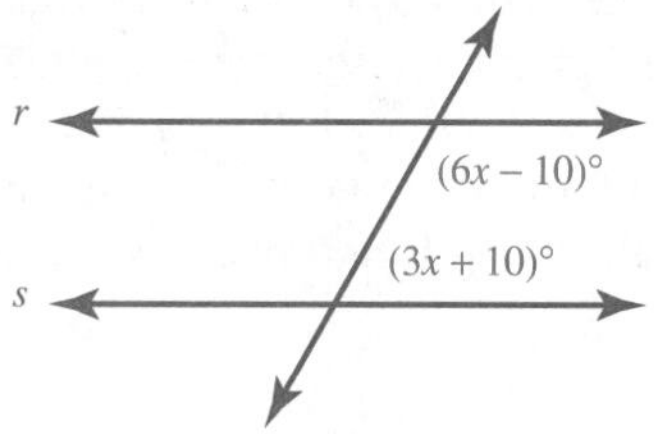

2.

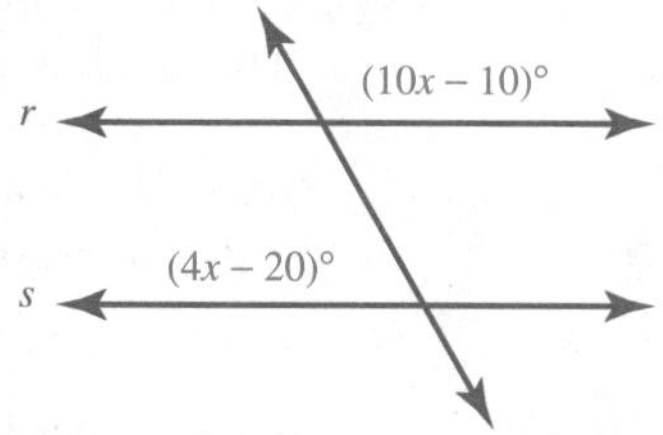

VOCABULARY AND CONCEPTS *Fill in each blank to make a true statement.*

3. Piecewise-defined functions are defined by using different functions for different parts of their ________.

4. When the values of $f(x)$ increase as the values of x increase over an interval, we say that the function is an ________ function over that interval.

5. In a ________ function, the values of _____ are the same.

6. When the values of $f(x)$ decrease as the values of x ________ over an interval, we say that the function is a decreasing function over that interval.

7. When the graph of a function contains a series of horizontal line segments, the function is called a _____ function.

8. The function that gives the largest integer that is less than or equal to a number x is called the ____________ function.

PRACTICE *In Exercises 9–12, give the intervals on which each function is increasing, decreasing, or constant.*

9.

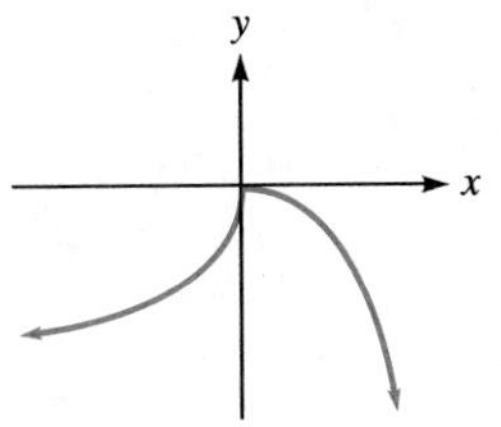

10.

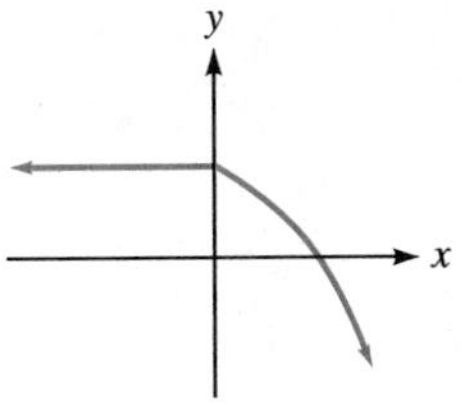

11.

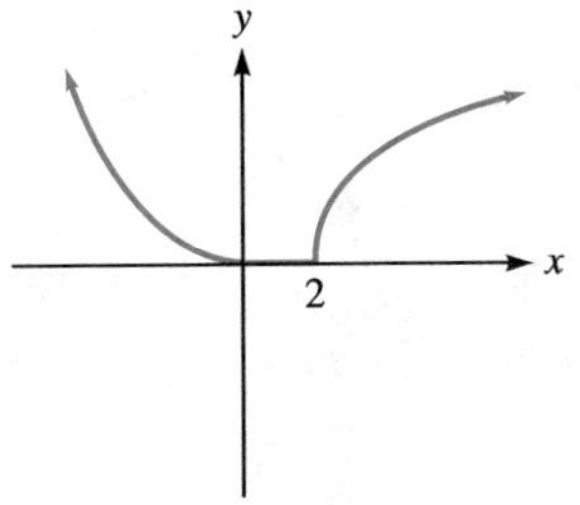

12.

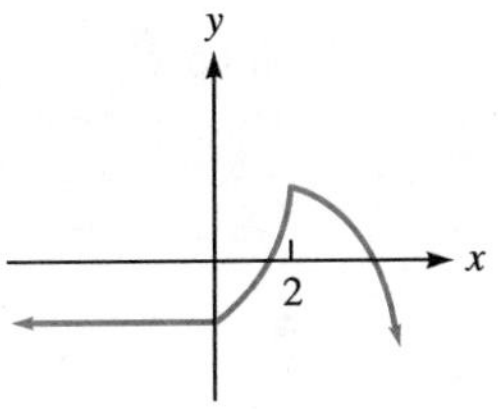

In Exercises 13–16, graph each function and give the intervals on which f is increasing, decreasing, or constant.

13. $f(x) = \begin{cases} -1 \text{ if } x \leq 0 \\ x \text{ if } x > 0 \end{cases}$

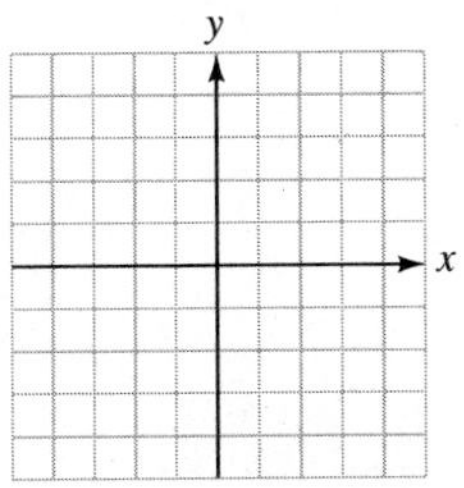

14. $f(x) = \begin{cases} -2 \text{ if } x \leq 0 \\ x^2 \text{ if } x > 0 \end{cases}$

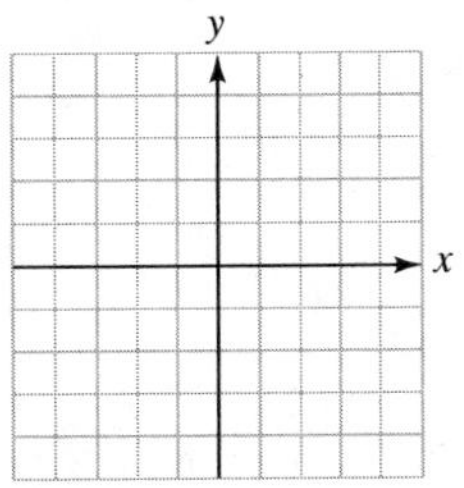

15. $f(x) = \begin{cases} -x \text{ if } x \leq 0 \\ x \text{ if } 0 < x < 2 \\ -x \text{ if } x \geq 2 \end{cases}$

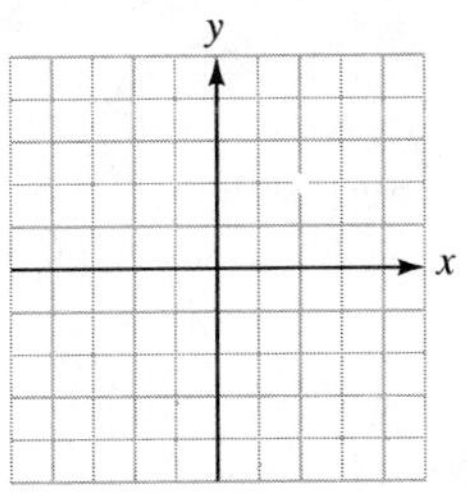

16. $f(x) = \begin{cases} -x \text{ if } x < 0 \\ x^2 \text{ if } 0 \leq x \leq 1 \\ 1 \text{ if } x > 1 \end{cases}$

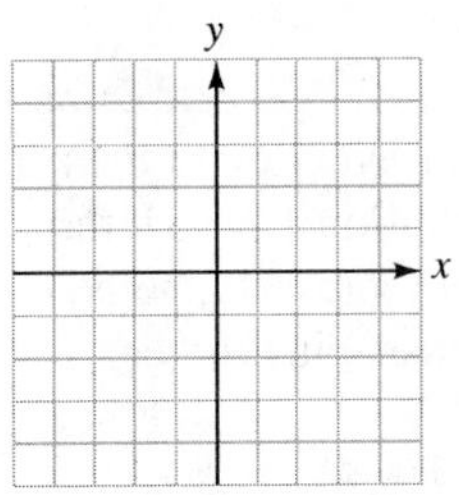

In Exercises 17–20, graph each function.

17. $f(x) = -[\![x]\!]$

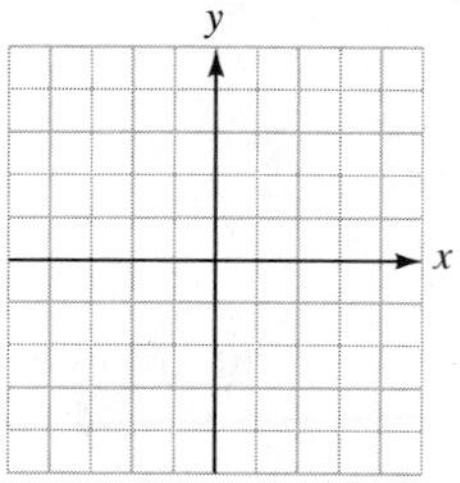

18. $f(x) = [\![x]\!] + 2$

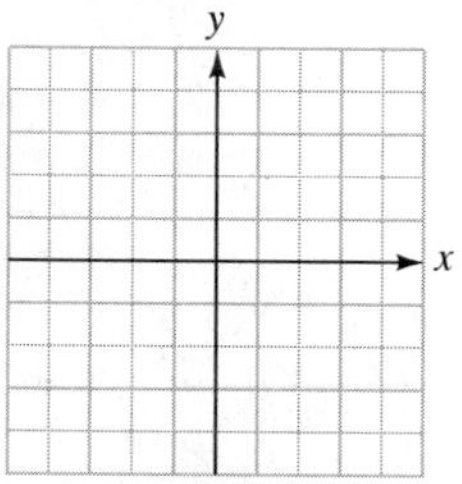

19. $f(x) = 2[\![x]\!]$

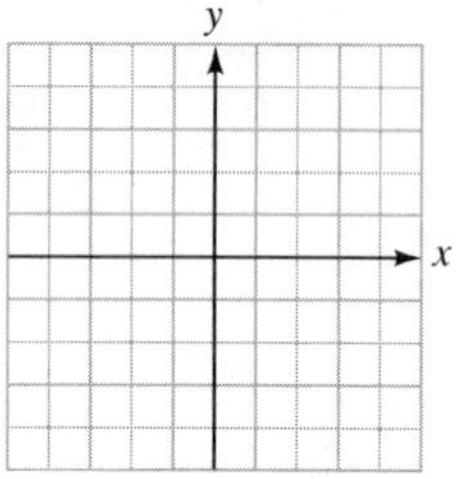

20. $f(x) = \left[\!\left[\frac{1}{2}x\right]\!\right]$

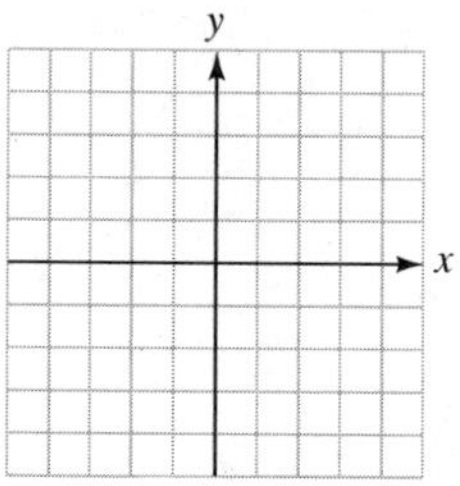

21. Signum function Computer programmers use a function denoted by $f(x) = \text{sgn } x$, that is defined in the following way:

$$f(x) = \begin{cases} -1 \text{ if } x < 0 \\ 0 \text{ if } x = 0 \\ 1 \text{ if } x > 0 \end{cases}$$

Graph this function.

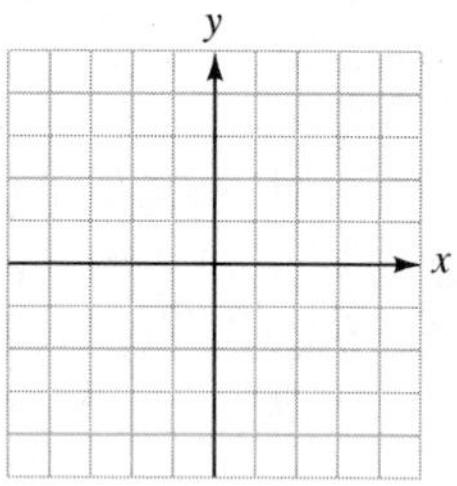

22. Heaviside unit step function This function, used in calculus, is defined by

$$f(x) = \begin{cases} 1 \text{ if } x > 0 \\ 0 \text{ if } x < 0 \end{cases}$$

Graph this function.

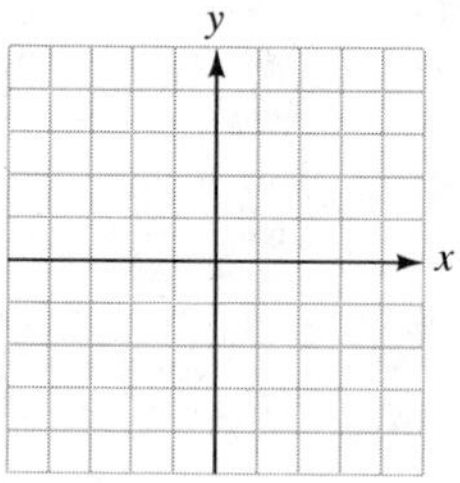

APPLICATIONS

23. Renting a jet ski A marina charges \$20 to rent a jet ski for 1 hour, plus \$5 for every extra hour (or portion of an hour). In Illustration 1, graph the ordered pairs (h, c), where h represents the number of hours and c represents the cost. Find the cost if the ski is used for 2.5 hours.

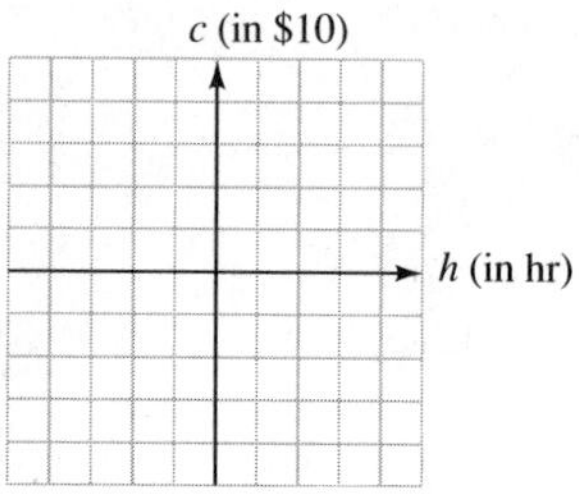

ILLUSTRATION 1

24. Riding in a taxi A cab company charges \$3 for a trip up to 1 mile, and \$2 for every extra mile (or portion of a mile). In Illustration 2, graph the ordered pairs (m, c), where m represents the number of miles traveled and c represents the cost. Find the cost to ride $10\frac{1}{4}$ miles.

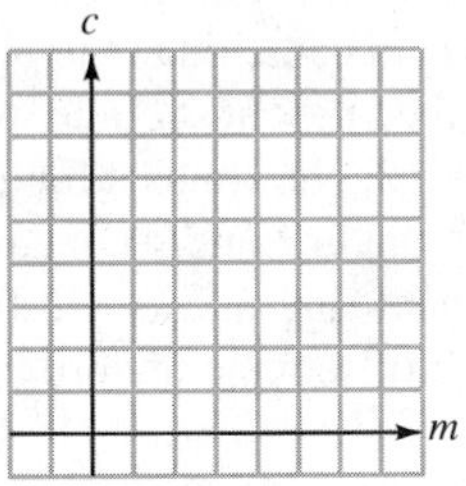

ILLUSTRATION 2

25. Information access Computer access to international data network A costs \$10 per day plus \$8 per hour or fraction of an hour. Network B charges \$15 per day, but only \$6 per hour or fraction of an hour. For each network, use Illustration 3 to graph the ordered pairs (t, C), where t represents the connect time and C represents the total cost. Find the minimal daily usage at which it would be more economical to use network B.

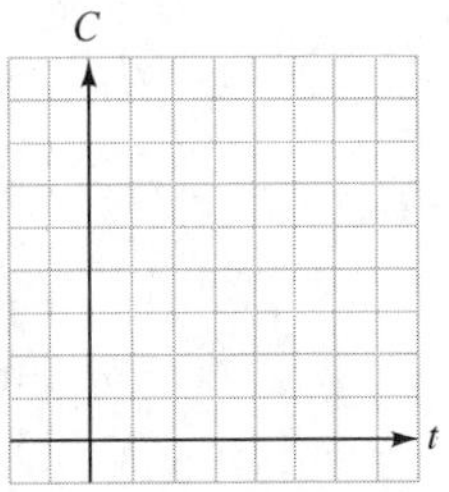

ILLUSTRATION 3

26. Royalties A publisher has agreed to pay the author of a novel 7% royalties on sales of the first 50,000 copies and 10% on sales thereafter. If the book sells for \$10, express the royalty income, I, as a function of s, the number of copies sold, and use Illustration 4 to graph the function. (*Hint*: When sales are into the second 50,000 copies, how much was earned on the first 50,000?)

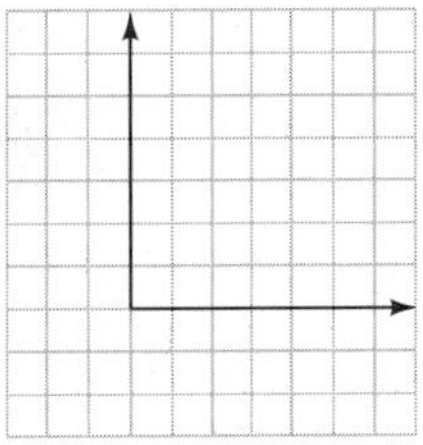

ILLUSTRATION 4

WRITING

27. Tell how to decide whether a function is increasing on the interval (a, b).

28. Describe the greatest integer function.

SOMETHING TO THINK ABOUT

29. Find a piecewise-defined function that is increasing on the interval $(-\infty, -2)$ and decreasing on the interval $(-2, \infty)$.

30. Find a piecewise-defined function that is constant on the interval $(-\infty, 0)$, increasing on the interval $(0, 5)$, and decreasing on the interval $(5, \infty)$.

PROJECT

The zillionaire G.I. Luvmoney is known for his love of flowers. On his estate, he recently set aside a circular plot of land with a radius of 100 yards to be made into a flower garden. He has hired your landscape design firm to do the job. If Luvmoney is satisfied, he will hire your firm to do more lucrative jobs. Here is Luvmoney's plan.

The center of the circular plot of land is to be the origin of a rectangular coordinate system. You are to make 100 circles, all centered at the origin, with radii of 1 yard, 2 yards, 3 yards, and so on, up to the outermost circle, which will have a radius of 100 yards. Inside the innermost circle, he wants a fountain with a circular walkway around it. In the ring between the first and second circle, he wants to plant his favorite kind of flower; in the next ring his second favorite, and so on, until you reach the edge of the circular plot. Luvmoney provides you with a list ranking his 99 favorite flowers.

The first thing he wants to know is the area of each ring, so that he will know how many of each plant to order. Then he wants a simple formula that will give the area of any ring just by substituting in the number of the ring.

He also wants a walkway to go through the garden in the form of a hyperbolic path, following the equation

$$x^2 - \frac{y^2}{9} = 1$$

Luvmoney wants to know the x- and y-coordinates of the points where the path will intersect the circles, so that those points can be marked with stakes to keep gardeners from planting flowers where the walkway will later be built. He wants a formula (or two) that will enable him to put in the number of a circle and get out the intersection points.

Finally, although cost has no importance for Luvmoney, his accountants will want an estimate of the total cost of all of the flowers.

You go back to your office with Luvmoney's list. You find that because the areas of the rings grow from the inside of the garden to the outside, and because of Luvmoney's ranking of flowers, a strange thing happens. The first ring of flowers will cost \$360, and the flowers in every ring after that will cost 110% as much as the flowers in the previous ring. That is, the second ring of flowers will cost \$360(1.1) = \$396, the third will cost \$435.60, and so on.

Find answers to all of Luvmoney's questions, and show work that will convince him that you are right.

CHAPTER SUMMARY

CONCEPTS

REVIEW EXERCISES

SECTION 10.1 The Circle and the Parabola

Equations of a circle:

$(x - h)^2 + (y - k)^2 = r^2$
center (h, k), radius r

$x^2 + y^2 = r^2$
center $(0, 0)$, radius r

Equations of parabolas ($a > 0$):

Parabola opening	*Vertex at origin*
Up	$y = ax^2$
Down	$y = -ax^2$
Right	$x = ay^2$
Left	$x = -ay^2$

Parabola opening	*Vertex at (h, k)*
Up	$y = a(x - h)^2 + k$
Down	$y = -a(x - h)^2 + k$
Right	$x = a(y - k)^2 + h$
Left	$x = -a(y - k)^2 + h$

1. Graph each equation.

a. $(x - 1)^2 + (y + 2)^2 = 9$

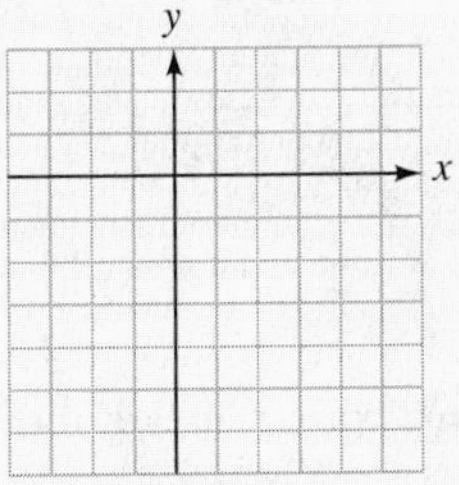

b. $x^2 + y^2 = 16$

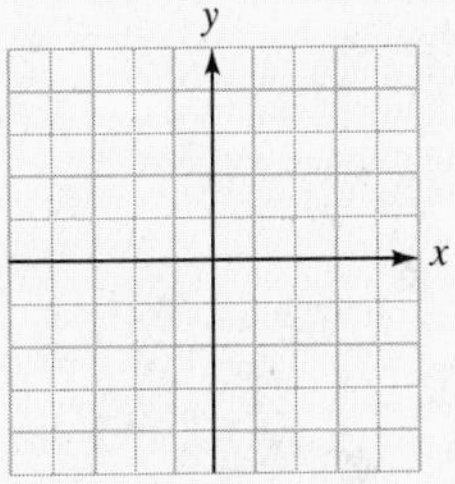

2. Write the equation in standard form and graph it.
$x^2 + y^2 + 4x - 2y = 4$

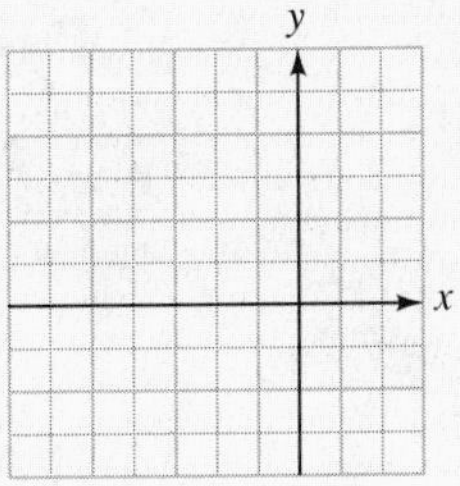

3. Graph each equation.

a. $x = -3(y - 2)^2 + 5$

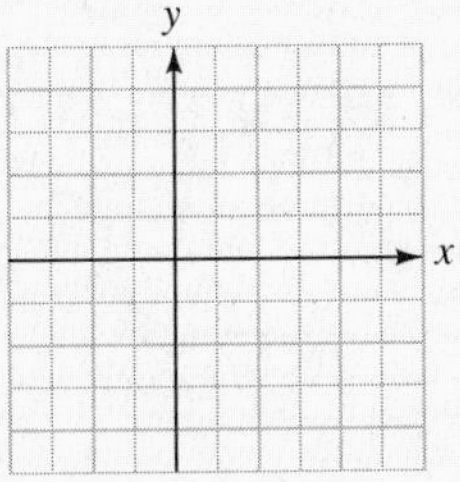

b. $x = 2(y + 1)^2 - 2$

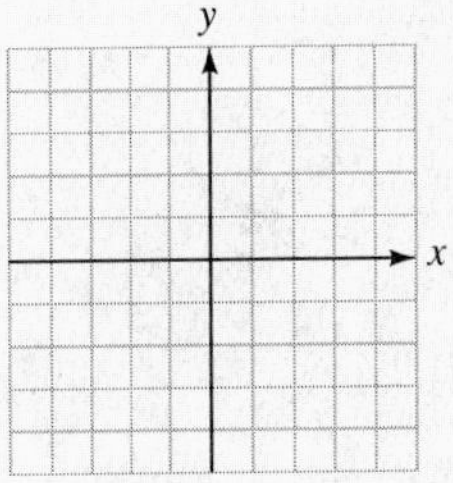

SECTION 10.2 The Ellipse

Equations of an ellipse:

Center at (0, 0)

$$\frac{x^2}{a^2} + \frac{y^2}{b^2} = 1 \quad a > b > 0$$

$$\frac{x^2}{b^2} + \frac{y^2}{a^2} = 1 \quad a > b > 0$$

Center at (*h*, *k*)

$$\frac{(x-h)^2}{a^2} + \frac{(y-k)^2}{b^2} = 1$$

$$\frac{(x-h)^2}{b^2} + \frac{(y-k)^2}{a^2} = 1$$

4. Graph each ellipse.

a. $9x^2 + 16y^2 = 144$

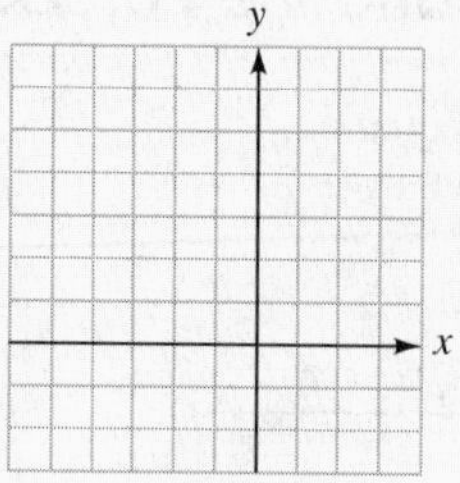

b. $\frac{(x-2)^2}{4} + \frac{(y-1)^2}{9} = 1$

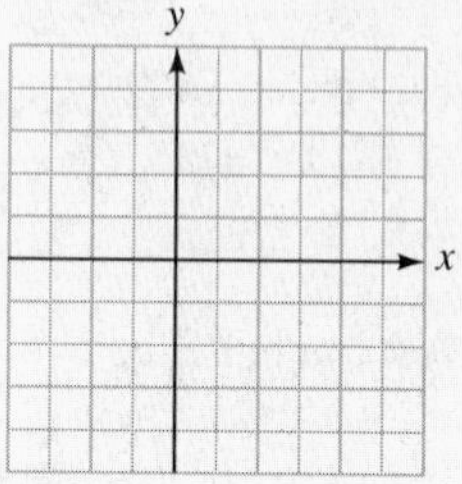

5. Write the equation in standard form and graph it.
$4x^2 + 9y^2 + 8x - 18y = 23$

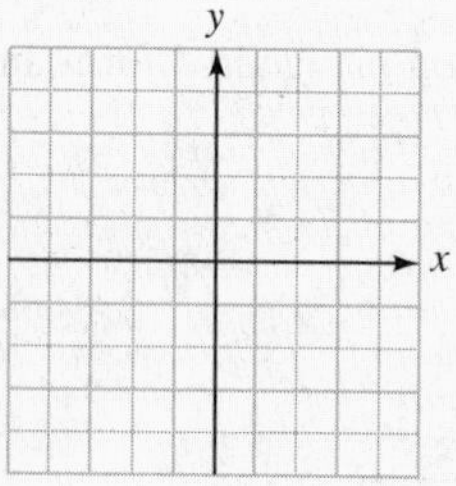

SECTION 10.3 The Hyperbola

Equations of a hyperbola:

Center at (0, 0)

$$\frac{x^2}{a^2} - \frac{y^2}{b^2} = 1$$

$$\frac{y^2}{a^2} - \frac{x^2}{b^2} = 1$$

Center at (*h*, *k*)

$$\frac{(x-h)^2}{a^2} - \frac{(y-k)^2}{b^2} = 1$$

$$\frac{(y-k)^2}{a^2} - \frac{(x-h)^2}{b^2} = 1$$

6. Graph each hyperbola.

a. $9x^2 - y^2 = -9$

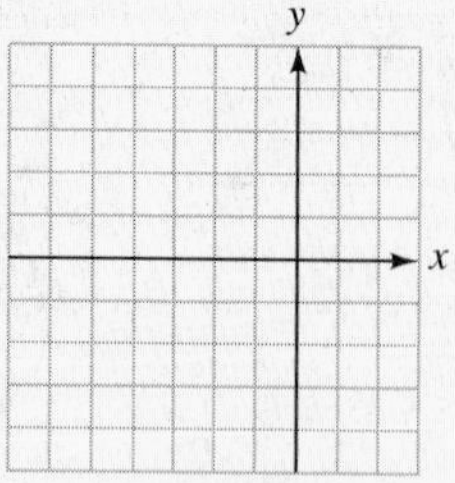

b. $xy = 9$

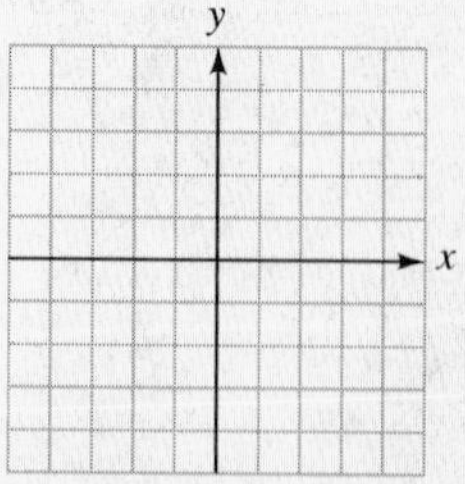

7. Write the equation $4x^2 - 2y^2 + 8x - 8y = 8$ in standard form and tell whether its graph will be an ellipse or a hyperbola.

8. Write the equation in standard form and graph it.
$9x^2 - 4y^2 - 18x - 8y = 31$

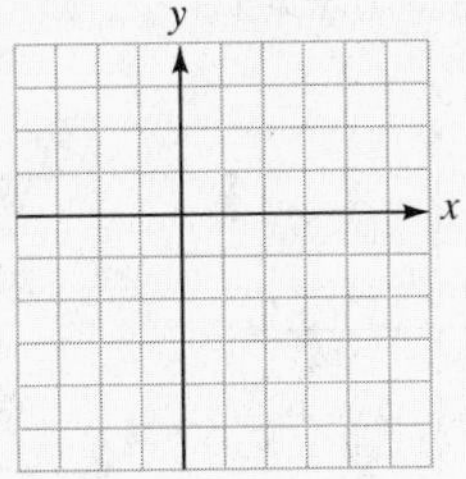

SECTION 10.4 Solving Simultaneous Second-Degree Equations

9. Solve each system.

a. $\begin{cases} 3x^2 + y^2 = 52 \\ x^2 - y^2 = 12 \end{cases}$

b. $\begin{cases} \dfrac{x^2}{16} + \dfrac{y^2}{12} = 1 \\ x^2 - \dfrac{y^2}{3} = 1 \end{cases}$

SECTION 10.5 Piecewise-Defined Functions and the Greatest Integer Function

A function is increasing on the interval (a, b) if the values of $f(x)$ increase as x increases from a to b.

A function is decreasing on the interval (a, b) if the values of $f(x)$ decrease as x increases from a to b.

A function is constant on the interval (a, b) if the value of $f(x)$ is constant as x increases from a to b.

10. Tell when the function is increasing, decreasing, or constant.

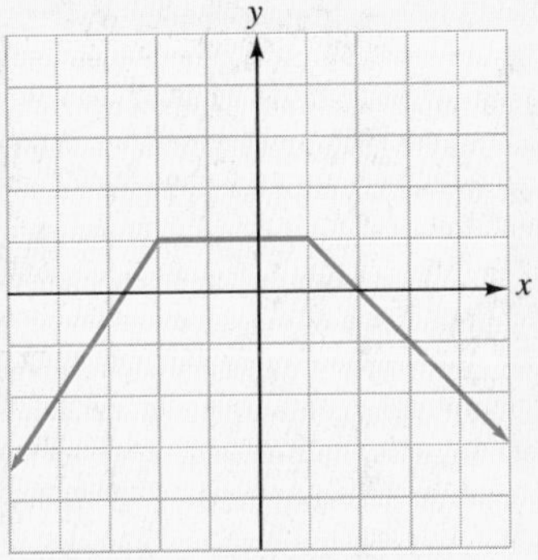

11. Graph each function.

a. $f(x) = \begin{cases} x \text{ if } x \leq 1 \\ -x^2 \text{ if } x > 1 \end{cases}$

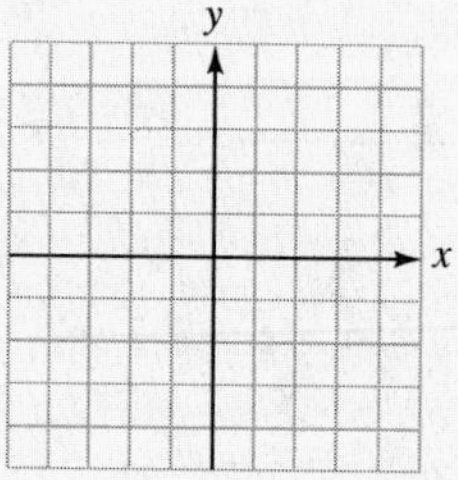

b. $f(x) = 3[\![x]\!]$

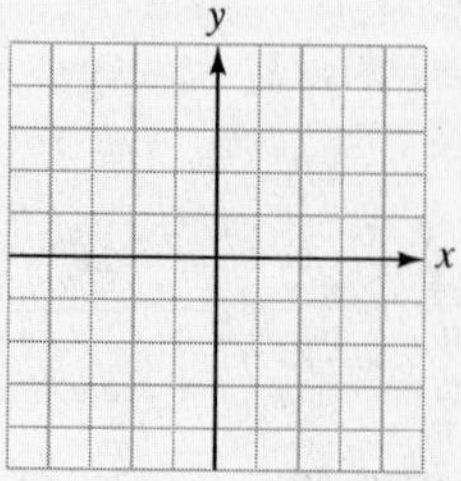

Chapter Test

1. Find the center and the radius of the circle $(x - 2)^2 + (y + 3)^2 = 4$.

2. Find the center and the radius of the circle $x^2 + y^2 + 4x - 6y = 3$.

In Problems 3–6, graph each equation.

3. $(x + 1)^2 + (y - 2)^2 = 9$

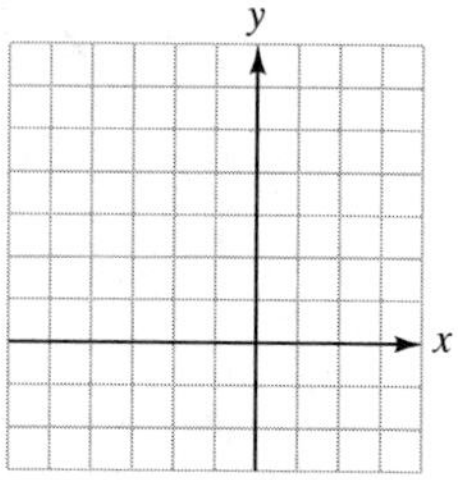

4. $x = (y - 2)^2 - 1$

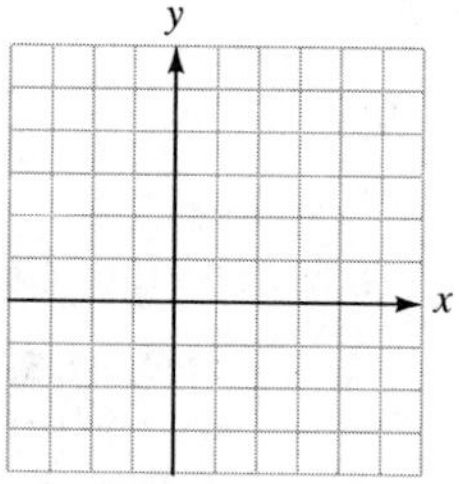

5. $9x^2 + 4y^2 = 36$

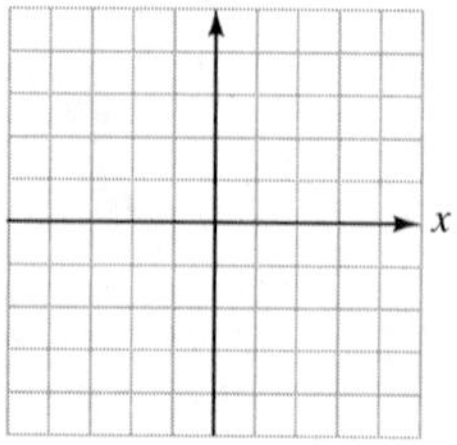

6. $\dfrac{(x - 2)^2}{9} - y^2 = 1$

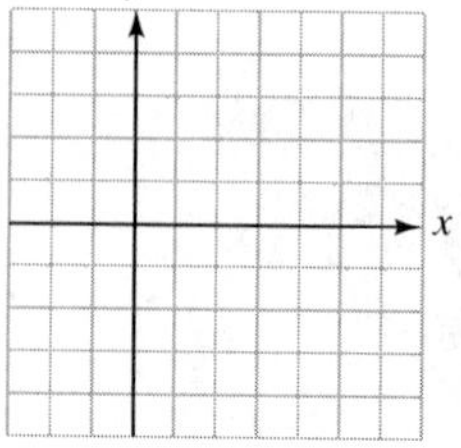

In Problems 7–8, write each equation in standard form and graph the equation.

7. $4x^2 + y^2 - 24x + 2y = -33$

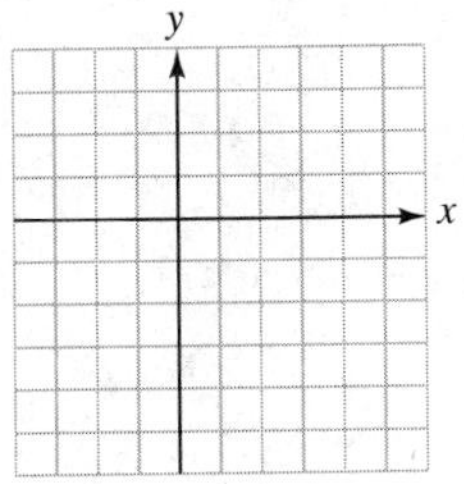

8. $x^2 - 9y^2 + 2x + 36y = 44$

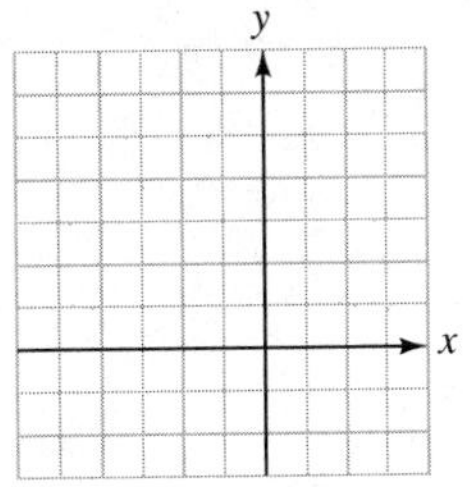

In Problems 9–10, solve each system.

9. $\begin{cases} 2x - y = -2 \\ x^2 + y^2 = 16 + 4y \end{cases}$

10. $\begin{cases} x^2 + y^2 = 25 \\ 4x^2 - 9y = 0 \end{cases}$

11. Tell where the function is increasing, decreasing, or constant.

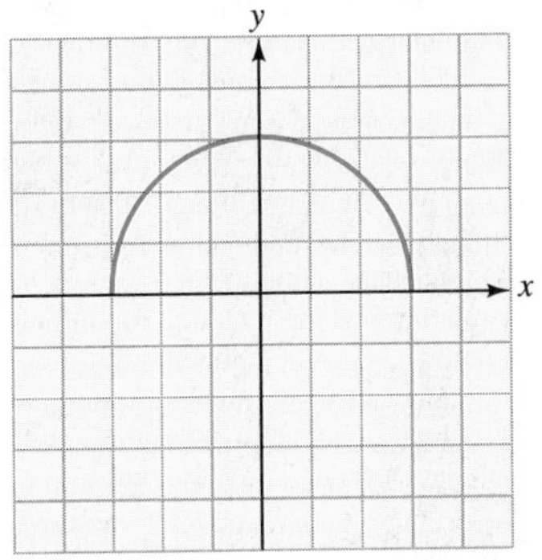

12. Graph $f(x) = \begin{cases} -x^2, \text{ when } x < 0 \\ -x, \text{ when } x \geq 0 \end{cases}$.

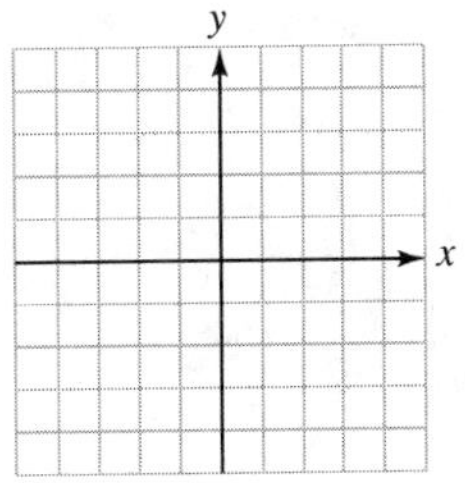

Cumulative Review Exercises

In Exercises 1–2, do the operations.

1. $(4x - 3y)(3x + y)$

2. $(a^n + 1)(a^n - 3)$

In Exercises 3–4, simplify each fraction.

3. $\dfrac{5a - 10}{a^2 - 4a + 4}$

4. $\dfrac{a^4 - 5a^2 + 4}{a^2 + 3a + 2}$

In Exercises 5–6, do the operations and simplify the result, if possible.

5. $\dfrac{a^2 - a - 6}{a^2 - 4} \div \dfrac{a^2 - 9}{a^2 + a - 6}$

6. $\dfrac{2}{a - 2} + \dfrac{3}{a + 2} - \dfrac{a - 1}{a^2 - 4}$

In Exercises 7–8, tell whether the graphs of the linear equations are parallel, perpendicular, or neither.

7. $3x - 4y = 12$, $y = \dfrac{3}{4}x - 5$

8. $y = 3x + 4$, $x = -3y + 4$

In Exercises 9–10, write the equation of each line with the following properties.

9. $m = -2$, passing through $(0, 5)$

10. Passing through $P(8, -5)$ and $Q(-5, 4)$

In Exercises 11–12, graph each inequality.

11. $2x - 3y < 6$

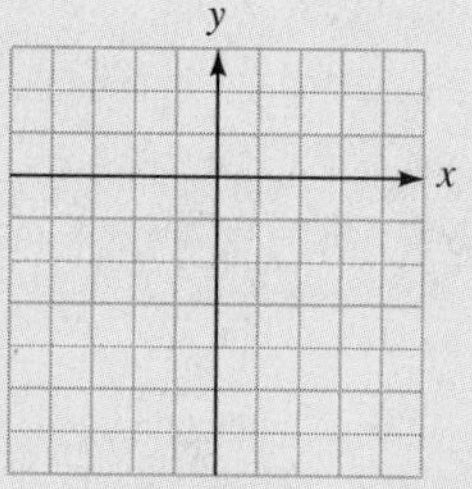

12. $y \geq x^2 - 4$

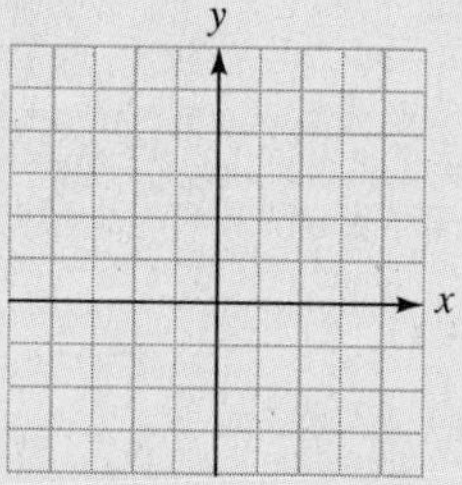

In Exercises 13–14, simplify each expression.

13. $\sqrt{98} + \sqrt{8} - \sqrt{32}$

14. $12\sqrt[3]{648x^4} + 3\sqrt[3]{81x^4}$

In Exercises 15–18, solve each equation.

15. $\sqrt{3a + 1} = a - 1$

16. $\sqrt{x + 3} - \sqrt{3} = \sqrt{x}$

17. $6a^2 + 5a - 6 = 0$

18. $3x^2 + 8x - 1 = 0$

19. If $f(x) = x^2 - 2$ and $g(x) = 2x + 1$, find $(f \circ g)(x)$.

20. Find the inverse function of $y = 2x^3 - 1$.

21. Graph $y = \left(\frac{1}{2}\right)^x$.

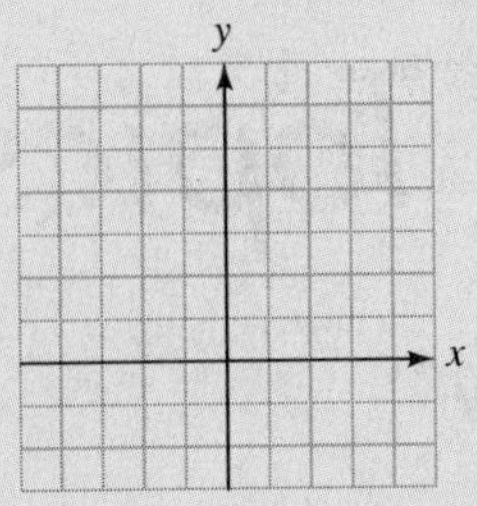

22. Write $y = \log_2 x$ as an exponential equation.

In Exercises 23–24, solve each equation.

23. $2^{x+2} = 3^x$

24. $2 \log 5 + \log x - \log 4 = 2$

In Exercises 25–26, graph each equation.

25. $x^2 + (y + 1)^2 = 9$

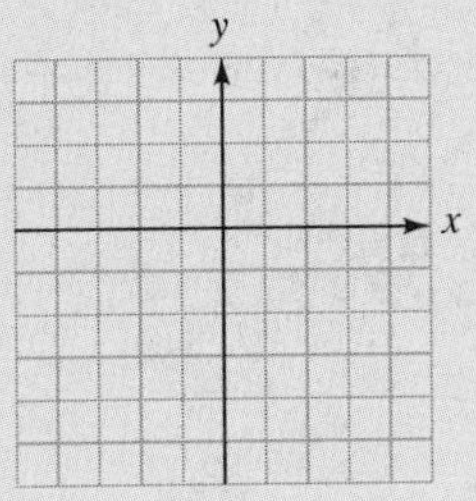

26. $x^2 - 9(y + 1)^2 = 9$

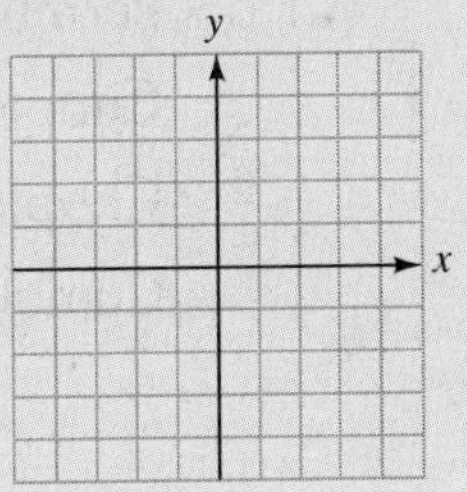

11 Miscellaneous Topics

Actuary

MATHEMATICS IN THE WORKPLACE

Why do young people pay more for automobile insurance than older people? How much should an insurance policy cost? How much should an organization contribute each year to its pension fund? Answers to these questions are provided by actuaries, who design insurance and pension plans and follow their performance to make sure that the plans are maintained on a sound financial basis.

SAMPLE APPLICATION ■ An *annuity* is a sequence of equal payments made periodically over a length of time. The sum of the payments and the interest earned during the *term* of the annuity is called the *amount* of the annuity.

After a sales clerk works six months, her employer will begin an annuity for her and will contribute \$500 semiannually to a fund that pays 8% annual interest. After she has been employed for two years, what will be the amount of her annuity?
(See Example 7 in Section 11.4.)

In this chapter, we introduce several topics that have applications in advanced mathematics and in many occupational areas. The binomial theorem, permutations, and combinations are used in statistics. Arithmetic sequences and geometric sequences are used in the mathematics of finance.

11.1 The Binomial Theorem

■ RAISING BINOMIALS TO POWERS ■ PASCAL'S TRIANGLE ■ FACTORIAL NOTATION ■ THE BINOMIAL THEOREM

Getting Ready *Raise each binomial to the indicated power.*

1. $(x + 2)^2$

2. $(x - 3)^2$

3. $(x + 1)^3$

4. $(x - 2)^3$

■ RAISING BINOMIALS TO POWERS

We have discussed how to raise binomials to positive integral powers. For example, we know that

$$(a + b)^2 = a^2 + 2ab + b^2$$

and that

$$\begin{aligned}(a+b)^3 &= (a+b)(a+b)^2 \\ &= (a+b)(a^2+2ab+b^2) \\ &= a^3+2a^2b+ab^2+a^2b+2ab^2+b^3 \\ &= a^3+3a^2b+3ab^2+b^3\end{aligned}$$

To show how to raise binomials to positive-integer powers without doing the actual multiplications, we consider the following binomial expansions:

$$\begin{aligned}(a+b)^0 &= 1 \\ (a+b)^1 &= a+b \\ (a+b)^2 &= a^2+2ab+b^2 \\ (a+b)^3 &= a^3+3a^2b+3ab^2+b^3 \\ (a+b)^4 &= a^4+4a^3b+6a^2b^2+4ab^3+b^4 \\ (a+b)^5 &= a^5+5a^4b+10a^3b^2+10a^2b^3+5ab^4+b^5 \\ (a+b)^6 &= a^6+6a^5b+15a^4b^2+20a^3b^3+15a^2b^4+6ab^5+b^6\end{aligned}$$

Several patterns appear in these expansions:

1. Each expansion has one more term than the power of the binomial.
2. The degree of each term in each expansion is equal to the exponent of the binomial that is being expanded.
3. The first term in each expansion is a, raised to the power of the binomial.
4. The exponents of a decrease by 1 in each successive term. The exponents of b, beginning with $b^0 = 1$ in the first term, increase by 1 in each successive term. Thus, the variables have the pattern

$$a^n, a^{n-1}b, a^{n-2}b^2, \ldots, ab^{n-1}, b^n$$

PASCAL'S TRIANGLE

To see another pattern, we write the coefficients of each expansion in the following triangular array:

$$\begin{array}{ccccccccccccc} & & & & & & 1 & & & & & & \\ & & & & & 1 & & 1 & & & & & \\ & & & & 1 & & 2 & & 1 & & & & \\ & & & 1 & & 3 & & 3 & & 1 & & & \\ & & 1 & & 4 & & 6 & & 4 & & 1 & & \\ & 1 & & 5 & & 10 & & 10 & & 5 & & 1 & \\ 1 & & 6 & & 15 & & 20 & & 15 & & 6 & & 1\end{array}$$

In this array, called **Pascal's triangle**, each entry between the 1's is the sum of the closest pair of numbers in the line immediately above it. For example, the first 15 in the bottom row is the sum of the 5 and 10 immediately above it. Pascal's triangle continues with the same pattern forever. The next two lines are

$$\begin{array}{ccccccccccccccc} & 1 & & 7 & & 21 & & 35 & & 35 & & 21 & & 7 & & 1 & \\ 1 & & 8 & & 28 & & 56 & & 70 & & 56 & & 28 & & 8 & & 1 \end{array}$$

EXAMPLE 1 Expand $(x + y)^5$.

Solution The first term in the expansion is x^5, and the exponents of x decrease by 1 in each successive term. A y first appears in the second term, and the exponents on y increase by 1 in each successive term, concluding when the term y^5 is reached. Thus, the variables in the expansion are

$$x^5,\ x^4y,\ x^3y^2,\ x^2y^3,\ xy^4,\ y^5$$

The coefficients of these variables are given in Pascal's triangle in the row whose second entry is 5, the same as the exponent of the binomial being expanded:

1 5 10 10 5 1

Combining this information gives the following expansion:

$$(x + y)^5 = x^5 + 5x^4y + 10x^3y^2 + 10x^2y^3 + 5xy^4 + y^5$$ ■

Self Check Expand $(x + y)^4$.

Answer $x^4 + 4x^3y + 6x^2y^2 + 4xy^3 + y^4$

EXAMPLE 2 Expand $(u - v)^4$.

Solution We note that $(u - v)^4$ can be written in the form $[u + (-v)]^4$. The variables in this expansion are

$$u^4,\ u^3(-v),\ u^2(-v)^2,\ u(-v)^3,\ (-v)^4$$

and the coefficients are given in Pascal's triangle in the row whose second entry is 4:

1 4 6 4 1

Thus, the required expansion is

$$\begin{aligned}(u - v)^4 &= u^4 + 4u^3(-v) + 6u^2(-v)^2 + 4u(-v)^3 + (-v)^4 \\ &= u^4 - 4u^3v + 6u^2v^2 - 4uv^3 + v^4\end{aligned}$$ ■

Self Check Expand $(x - y)^5$.

Answer $x^5 - 5x^4y + 10x^3y^2 - 10x^2y^3 + 5xy^4 - y^5$

FACTORIAL NOTATION

Although Pascal's triangle gives the coefficients of the terms in a binomial expansion, it is not the best way to expand a binomial. To develop a better way, we introduce **factorial notation.**

Factorial Notation

If n is a natural number, the symbol $n!$ (read as "***n* factorial**" or as "**factorial *n***") is defined as

$$n! = n(n-1)(n-2)(n-3) \cdots (3)(2)(1)$$

EXAMPLE 3 Find **a.** $2!$, **b.** $5!$ **c.** $-9!$ and **d.** $(n-2)!$.

Solution

a. $2! = 2 \cdot 1 = 2$

b. $5! = 5 \cdot 4 \cdot 3 \cdot 2 \cdot 1 = 120$

c. $-9! = -9 \cdot 8 \cdot 7 \cdot 6 \cdot 5 \cdot 4 \cdot 3 \cdot 2 \cdot 1 = -362{,}880$

d. $(n-2)! = (n-2)(n-3)(n-4) \cdot \cdots \cdot 3 \cdot 2 \cdot 1$

WARNING! According to the previous definition, part **d** is meaningful only if $n-2$ is a natural number.

■

Self Check Find **a.** $4!$ and **b.** $x!$.

Answers **a.** 24, **b.** $x(x-1)(x-2) \cdot \cdots \cdot 3 \cdot 2 \cdot 1$

We define zero factorial as follows.

Zero Factorial

$$0! = 1$$

We note that

$$5 \cdot 4! = 5 \cdot 4 \cdot 3 \cdot 2 \cdot 1 = 5!$$
$$7 \cdot 6! = 7 \cdot 6 \cdot 5 \cdot 4 \cdot 3 \cdot 2 \cdot 1 = 7!$$
$$10 \cdot 9! = 10 \cdot 9 \cdot 8 \cdot 7 \cdot 6 \cdot 5 \cdot 4 \cdot 3 \cdot 2 \cdot 1 = 10!$$

These examples suggest the following property of factorials.

$n(n-1)!$
If n is a positive integer, then $n(n-1)! = n!$.

THE BINOMIAL THEOREM

We now state the binomial theorem.

The Binomial Theorem
If n is any positive integer, then

$$(a+b)^n = a^n + \frac{n!}{1!(n-1)!}a^{n-1}b + \frac{n!}{2!(n-2)!}a^{n-2}b^2 + \frac{n!}{3!(n-3)!}a^{n-3}b^3 + \cdots + \frac{n!}{r!(n-r)!}a^{n-r}b^r + \cdots + b^n$$

In the binomial theorem, the exponents of the variables follow the familiar pattern:

- The sum of the exponents on a and b in each term is n,
- the exponents on a decrease, and
- the exponents on b increase.

Only the method of finding the coefficients is different. Except for the first and last terms, the numerator of each coefficient is $n!$. If the exponent of b in a particular term is r, the denominator of the coefficient of that term is $r!(n-r)!$.

EXAMPLE 4 Use the binomial theorem to expand $(a+b)^3$.

Solution We can substitute directly into the binomial theorem and simplify:

$$\begin{aligned}(a+b)^3 &= a^3 + \frac{3!}{1!(3-1)!}a^2b + \frac{3!}{2!(3-2)!}ab^2 + b^3 \\ &= a^3 + \frac{3!}{1!\cdot 2!}a^2b + \frac{3!}{2!\cdot 1!}ab^2 + b^3 \\ &= a^3 + \frac{3\cdot 2\cdot 1}{1\cdot 2\cdot 1}a^2b + \frac{3\cdot 2\cdot 1}{2\cdot 1\cdot 1}ab^2 + b^3 \\ &= a^3 + 3a^2b + 3ab^2 + b^3\end{aligned}$$

■

Self Check Use the binomial theorem to expand $(a+b)^4$.

Answer $a^4 + 4a^3b + 6a^2b^2 + 4ab^3 + b^4$

EXAMPLE 5 Use the binomial theorem to expand $(x - y)^4$.

Solution We can write $(x - y)^4$ in the form $[x + (-y)]^4$, substitute directly into the binomial theorem, and simplify:

$$\begin{aligned}(x - y)^4 &= [x + (-y)]^4 \\ &= x^4 + \frac{4!}{1!(4-1)!}x^3(-y) + \frac{4!}{2!(4-2)!}x^2(-y)^2 + \frac{4!}{3!(4-3)!}x(-y)^3 + (-y)^4 \\ &= x^4 - \frac{4 \cdot 3!}{1! \cdot 3!}x^3y + \frac{4 \cdot 3 \cdot 2!}{2! \cdot 2!}x^2y^2 - \frac{4 \cdot 3!}{3! \cdot 1!}xy^3 + y^4 \\ &= x^4 - 4x^3y + 6x^2y^2 - 4xy^3 + y^4\end{aligned}$$

■

Self Check Use the binomial theorem to expand $(x - y)^3$.

Answer $x^3 - 3x^2y + 3xy^2 - y^3$

EXAMPLE 6 Use the binomial theorem to expand $(3u - 2v)^4$.

Solution We write $(3u - 2v)^4$ in the form $[3u + (-2v)]^4$ and let $a = 3u$ and $b = -2v$. Then, we can use the binomial theorem to expand $(a + b)^4$.

$$\begin{aligned}(a + b)^4 &= a^4 + \frac{4!}{1!(4-1)!}a^3b + \frac{4!}{2!(4-2)!}a^2b^2 + \frac{4!}{3!(4-3)!}ab^3 + b^4 \\ &= a^4 + 4a^3b + 6a^2b^2 + 4ab^3 + b^4\end{aligned}$$

Now we can substitute $3u$ for a and $-2v$ for b and simplify:

$$\begin{aligned}(3u - 2v)^4 &= (3u)^4 + 4(3u)^3(-2v) + 6(3u)^2(-2v)^2 + 4(3u)(-2v)^3 + (-2v)^4 \\ &= 81u^4 - 216u^3v + 216u^2v^2 - 96uv^3 + 16v^4\end{aligned}$$

■

Self Check Use the binomial theorem to expand $(2a - 3b)^3$.

Answer $8a^3 - 36a^2b + 54ab^2 - 27b^3$

Orals *Find each value.*

1. $1!$ **2.** $4!$ **3.** $0!$ **4.** $5!$

Expand each binomial.

5. $(m + n)^2$ **6.** $(m - n)^2$

7. $(p + 2q)^2$ **8.** $(2p - q)^2$

EXERCISE 11.1

REVIEW *Find each value of x.*

1. $\log_4 16 = x$

2. $\log_x 49 = 2$

3. $\log_{25} x = \frac{1}{2}$

4. $\log_{1/2} \frac{1}{8} = x$

VOCABULARY AND CONCEPTS *Fill in each blank to make a true statement.*

5. Every binomial expansion has ____ more term than the power of the binomial.

6. The first term in the expansion of $(a + b)^{20}$ is ____.

7. The triangular array that can be used to find the coefficients of a binomial expansion is called ________ triangle.

8. The symbol 5! is read as "____________."

9. $6 \cdot 5 \cdot 4 \cdot 3 \cdot 2 \cdot 1 =$ ____ (Write your answer in factorial notation.)

10. $8! = 8 \cdot$ ____

11. $0! =$ ____

12. According to the binomial theorem, the third term of the expansion of $(a + b)^n$ is ______________.

PRACTICE *In Exercises 13–32, evaluate each expression.*

13. $3!$

14. $7!$

15. $-5!$

16. $-6!$

17. $3! + 4!$

18. $2!(3!)$

19. $3!(4!)$

20. $4! + 4!$

21. $8(7!)$

22. $4!(5)$

23. $\frac{9!}{11!}$

24. $\frac{13!}{10!}$

25. $\frac{49!}{47!}$

26. $\frac{101!}{100!}$

27. $\frac{5!}{3!(5-3)!}$

28. $\frac{6!}{4!(6-4)!}$

29. $\frac{7!}{5!(7-5)!}$

30. $\frac{8!}{6!(8-6)!}$

31. $\frac{5!(8-5)!}{4! \cdot 7!}$

32. $\frac{6! \cdot 7!}{(8-3)!(7-4)!}$

In Exercises 33–48, use the binomial theorem to expand each expression.

33. $(x + y)^3$

34. $(x + y)^4$

35. $(x - y)^4$

36. $(x - y)^3$

37. $(2x + y)^3$

38. $(x + 2y)^3$

39. $(x - 2y)^3$

40. $(2x - y)^3$

41. $(2x + 3y)^3$

42. $(3x - 2y)^3$

43. $\left(\frac{x}{2} - \frac{y}{3}\right)^3$

44. $\left(\frac{x}{3} + \frac{y}{2}\right)^3$

45. $(3 + 2y)^4$

46. $(2x + 3)^4$

47. $\left(\frac{x}{3} - \frac{y}{2}\right)^4$

48. $\left(\frac{x}{2} + \frac{y}{3}\right)^4$

49. Without referring to the text, write the first ten rows of Pascal's triangle.

50. Find the sum of the numbers in each row of the first ten rows of Pascal's triangle. What is the pattern?

51. Find the sum of the numbers in the designated diagonal rows of Pascal's triangle shown in Illustration 1. What is the pattern?

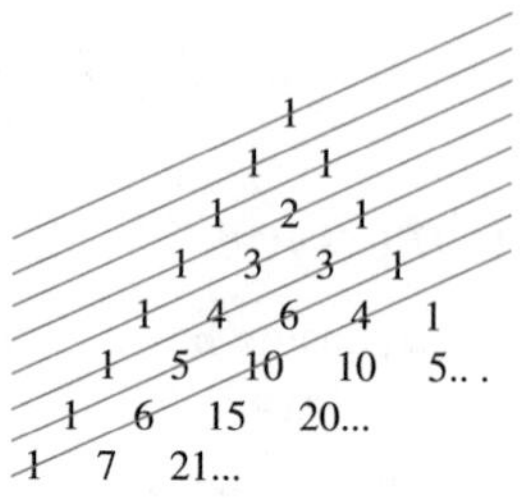

ILLUSTRATION 1

WRITING

52. Tell how to construct Pascal's triangle.

53. Tell how to find the variables of the terms in the expansion of $(r + s)^4$.

SOMETHING TO THINK ABOUT

54. If we apply the pattern of the coefficients to the coefficient of the first term in a binomial expansion, the coefficient would be $\frac{n!}{0!(n-0)!}$. Show that this expression is 1.

55. If we apply the pattern of the coefficients to the coefficient of the last term in a binomial expansion, the coefficient would be $\frac{n!}{n!(n-n)!}$. Show that this expression is 1.

11.2 The *n*th Term of a Binomial Expansion

■ FINDING A PARTICULAR TERM OF A BINOMIAL EXPANSION

Getting Ready *Expand the binomial and find the coefficient of the third term in each expansion.*

1. $(r + s)^4$

2. $(r + s)^5$

Expand the binomial and find the coefficient of the fourth term in each expansion.

3. $(p - q)^6$

4. $(p - q)^4$

■ FINDING A PARTICULAR TERM OF A BINOMIAL EXPANSION

To find the fourth term of the expansion of $(a + b)^9$, we could raise the binomial $a + b$ to the 9th power and look at the fourth term. However, this task would be tedious. By using the binomial theorem, we can construct the fourth term without finding the complete expansion of $(a + b)^9$.

EXAMPLE 1 Find the fourth term in the expansion of $(a + b)^9$.

Solution Since b^1 appears in the second term, b^2 appears in the third term, and so on, the exponent on b in the fourth term is 3. Since the exponent on b added to the exponent on a must equal 9, the exponent on a must be 6. Thus, the variables of the fourth term are

$$a^6b^3 \qquad \text{The sum of the exponents must be 9.}$$

The coefficient of these variables is

$$\frac{n!}{r!(n-r)!} = \frac{9!}{3!(9-3)!}$$

The complete fourth term is

$$\frac{9!}{3!(9-3)!}a^6b^3 = \frac{9 \cdot 8 \cdot 7 \cdot 6!}{3 \cdot 2 \cdot 1 \cdot 6!}a^6b^3$$
$$= 84a^6b^3$$

■

Self Check Find the third term of the expansion of $(a + b)^9$.

Answer $36a^7b^2$

EXAMPLE 2 Find the sixth term in the expansion of $(x - y)^7$.

Solution We first find the sixth term of $[x + (-y)]^7$. In the sixth term, the exponent on $(-y)$ is 5. Thus, the variables in the sixth term are

$$x^2(-y)^5 \qquad \text{The sum of the exponents must be 7.}$$

The coefficient of these variables is

$$\frac{n!}{r!(n-r)!} = \frac{7!}{5!(7-5)!}$$

The complete sixth term is

$$\frac{7!}{5!(7-5)!}x^2(-y)^5 = -\frac{7 \cdot 6 \cdot 5!}{5! \cdot 2 \cdot 1}x^2y^5$$
$$= -21x^2y^5$$

■

Self Check Find the fifth term of the expansion of $(a - b)^7$.

Answer $35a^3b^4$

EXAMPLE 3 Find the fourth term of the expansion of $(2x - 3y)^6$.

Solution We can let $a = 2x$ and $b = -3y$ and find the fourth term of the expansion of $(a + b)^6$:

$$\begin{aligned}\frac{6!}{3!(6-3)!}a^3b^3 &= \frac{6 \cdot 5 \cdot 4 \cdot 3!}{3! \cdot 3 \cdot 2 \cdot 1}a^3b^3 \\ &= 20a^3b^3\end{aligned}$$

We can now substitute $2x$ for a and $-3y$ for b and simplify:

$$\begin{aligned}20a^3b^3 &= 20(2x)^3(-3y)^3 \\ &= -4{,}320x^3y^3\end{aligned}$$

The fourth term is $-4{,}320x^3y^3$. ■

Self Check Find the third term of the expansion of $(2a - 3b)^6$.

Answer $2{,}160a^4b^2$

Orals *In the expansion of $(x + y)^8$, find the exponent on y in the*

1. 3rd term **2.** 4th term **3.** 7th term

In the expansion of $(x + y)^8$, find the exponent on x in the

4. 3rd term **5.** 4th term **6.** 7th term

In the expansion of $(x + y)^8$, find the coefficient of the

7. 1st term **8.** 2nd term

EXERCISE 11.2

REVIEW *Solve each system of equations.*

1. $\begin{cases} 3x + 2y = 12 \\ 2x - y = 1 \end{cases}$

2. $\begin{cases} a + b + c = 6 \\ 2a + b + 3c = 11 \\ 3a - b - c = 6 \end{cases}$

Evaluate each determinant.

3. $\begin{vmatrix} 2 & -3 \\ 4 & -2 \end{vmatrix}$

4. $\begin{vmatrix} 1 & 2 & 3 \\ 4 & 5 & 0 \\ -1 & -2 & 1 \end{vmatrix}$

VOCABULARY AND CONCEPTS *Fill in each blank to make a true statement.*

5. The exponent on b in the fourth term of the expansion of $(a + b)^6$ is ___.
6. The exponent on b in the fifth term of the expansion of $(a + b)^6$ is ___.
7. In the expansion of $(a + b)^7$, the sum of the exponents on a and b is ___.
8. The coefficient of the fourth term of the expansion of $(a + b)^9$ is 9! divided by ________.

PRACTICE *In Exercises 9–36, use the binomial theorem to find the required term of each expansion.*

9. $(a + b)^3$; second term
10. $(a + b)^3$; third term
11. $(x - y)^4$; fourth term
12. $(x - y)^5$; second term
13. $(x + y)^6$; fifth term
14. $(x + y)^7$; fifth term
15. $(x - y)^8$; third term
16. $(x - y)^9$; seventh term
17. $(x + 3)^5$; third term
18. $(x - 2)^4$; second term
19. $(4x + y)^5$; third term
20. $(x + 4y)^5$; fourth term
21. $(x - 3y)^4$; second term
22. $(3x - y)^5$; third term
23. $(2x - 5)^7$; fourth term
24. $(2x + 3)^6$; sixth term
25. $(2x - 3y)^5$; fifth term
26. $(3x - 2y)^4$; second term
27. $(\sqrt{2}x + \sqrt{3}y)^6$; third term
28. $(\sqrt{3}x + \sqrt{2}y)^5$; second term
29. $\left(\frac{x}{2} - \frac{y}{3}\right)^4$; second term
30. $\left(\frac{x}{3} + \frac{y}{2}\right)^5$; fourth term
31. $(a + b)^n$; fourth term
32. $(a + b)^n$; third term
33. $(a - b)^n$; fifth term
34. $(a - b)^n$; sixth term
35. $(a + b)^n$; rth term
36. $(a + b)^n$; $(r + 1)$th term

WRITING

37. Tell how to find the coefficients in the expansion of $(x + y)^5$.
38. Explain why the signs alternate in the expansion of $(x - y)^9$.

SOMETHING TO THINK ABOUT

39. Find the constant term in the expansion of $\left(x + \frac{1}{x}\right)^{10}$.
40. Find the coefficient of a^5 in the expansion of $\left(a - \frac{1}{a}\right)^9$.

11.3 Arithmetic Sequences

■ SEQUENCES ■ ARITHMETIC SEQUENCES ■ ARITHMETIC MEANS ■ THE SUM OF THE FIRST n TERMS OF AN ARITHMETIC SEQUENCE ■ SUMMATION NOTATION

Getting Ready *Complete each table.*

1.

n	$2n + 1$
1	
2	
3	
4	

2.

n	$3n - 5$
3	
4	
5	
6	

■ SEQUENCES

A **sequence** is a function whose domain is the set of natural numbers. For example, the function $f(n) = 3n + 2$, where n is a natural number, is a sequence. Because a sequence is a function whose domain is the set of natural numbers, we can write its values as a list. If the natural numbers are substituted for n, the function $f(n) = 3n + 2$ generates the list

$$5, 8, 11, 14, 17, \ldots$$

It is common to call the list, as well as the function, a sequence. Each number in the list is called a **term** of the sequence. Other examples of sequences are

$1^3, 2^3, 3^3, 4^3, \ldots$ The ordered list of the cubes of the natural numbers.

$4, 8, 12, 16, \ldots$ The ordered list of the positive multiples of 4.

$2, 3, 5, 7, 11, \ldots$ The ordered list of prime numbers.

$1, 1, 2, 3, 5, 8, 13, 21, \ldots$ The Fibonacci sequence.

The **Fibonacci sequence** is named after the 12th-century mathematician Leonardo of Pisa—also known as Fibonacci. Beginning with the 2, each term of the sequence is the sum of the two preceding terms.

■ ARITHMETIC SEQUENCES

One common type of sequence is the **arithmetic sequence.**

Arithmetic Sequence

An **arithmetic sequence** is a sequence of the form

$$a, a + d, a + 2d, a + 3d, \ldots, a + (n - 1)d, \ldots$$

where a is the **first term,** $a + (n - 1)d$ is the ***n*th term,** and d is the **common difference.**

We note that the second term of an arithmetic sequence has an addend of $1d$, the third term has an addend of $2d$, the fourth term has an addend of $3d$, and the nth term has an addend of $(n - 1)d$. We also note that the difference between any two consecutive terms in an arithmetic sequence is d.

EXAMPLE 1 An arithmetic sequence has a first term of 5 and a common difference of 4.

a. Write the first six terms of the sequence.

b. Write the 25th term of the sequence.

Solution **a.** Since the first term is $a = 5$ and the common difference is $d = 4$, the first six terms are

$$5, \quad 5 + 4, \quad 5 + 2(4), \quad 5 + 3(4), \quad 5 + 4(4), \quad 5 + 5(4)$$

or

$$5, 9, 13, 17, 21, 25$$

b. The nth term is $a + (n - 1)d$. Since we want the 25th term, we let $n = 25$:

$$\begin{aligned} n\text{th term} &= a + (n - 1)d \\ 25\text{th term} &= 5 + (25 - 1)4 && \text{Remember that } a = 5 \text{ and } d = 4. \\ &= 5 + 24(4) \\ &= 5 + 96 \\ &= 101 \end{aligned}$$

■

Self Check See Example 1. **a.** Write the seventh term of the sequence. **b.** Write the 30th term of the sequence.

Answers **a.** 29, **b.** 121

EXAMPLE 2 The first three terms of an arithmetic sequence are 3, 8, and 13. Find **a.** the 67th term and **b.** the 100th term.

Solution We first find d, the common difference. It is the difference between two successive terms:

$$d = 8 - 3 = 13 - 8 = 5$$

a. We substitute 3 for a, 67 for n, and 5 for d in the formula for the nth term and simplify:

$$\begin{aligned} n\text{th term} &= a + (n - 1)d \\ 67\text{th term} &= 3 + (67 - 1)5 \\ &= 3 + 66(5) \\ &= 333 \end{aligned}$$

b. We substitute 3 for a, 100 for n, and 5 for d in the formula for the nth term, and simplify:

$$\begin{aligned} n\text{th term} &= a + (n - 1)d \\ 100\text{th term} &= 3 + (100 - 1)5 \\ &= 3 + 99(5) \\ &= 498 \end{aligned}$$

■

Self Check: See Example 2. Find the 50th term.

Answer: 248

EXAMPLE 3 The first term of an arithmetic sequence is 12, and the 50th term is 3,099. Write the first six terms of the sequence.

Solution The key is to find the common difference. Because the 50th term of this sequence is 3,099, we can let $n = 50$ and solve the following equation for d:

$$\begin{aligned} 50\text{th term} &= a + (n - 1)d \\ 3{,}099 &= 12 + (50 - 1)d \\ 3{,}099 &= 12 + 49d && \text{Simplify.} \\ 3{,}087 &= 49d && \text{Subtract 12 from both sides.} \\ 63 &= d && \text{Divide both sides by 49.} \end{aligned}$$

The first term of the sequence is 12, and the common difference is 63. Its first six terms are

12, 75, 138, 201, 264, 327 ■

Self Check: The first term of an arithmetic sequence is 15, and the 12th term is 92. Write the first four terms of the sequence.

Answer: 15, 22, 29, 36

ARITHMETIC MEANS

If numbers are inserted between two numbers a and b to form an arithmetic sequence, the inserted numbers are called **arithmetic means** between a and b.

If a single number is inserted between the numbers a and b to form an arithmetic sequence, that number is called **the arithmetic mean** between a and b.

EXAMPLE 4 Insert two arithmetic means between 6 and 27.

Solution Here, the first term is $a = 6$, and the fourth term (or the last term) is $l = 27$. We must find the common difference so that the terms

$$6, 6 + d, 6 + 2d, 27$$

form an arithmetic sequence. To find d, we substitute 6 for a and 4 for n into the formula for the nth term:

$$\begin{aligned} n\text{th term} &= a + (n - 1)d \\ 4\text{th term} &= 6 + (4 - 1)d \\ 27 &= 6 + 3d && \text{Simplify.} \\ 21 &= 3d && \text{Subtract 6 from both sides.} \\ 7 &= d && \text{Divide both sides by 3.} \end{aligned}$$

The two arithmetic means between 6 and 27 are

$$\begin{aligned} 6 + d &= 6 + 7 \\ &= 13 \end{aligned} \qquad \text{or} \qquad \begin{aligned} 6 + 2d &= 6 + 2(7) \\ &= 6 + 14 \\ &= 20 \end{aligned}$$

The numbers 6, 13, 20, and 27 are the first four terms of an arithmetic sequence. ■

Self Check Insert two arithmetic means between 8 and 44.

Answer 20, 32

THE SUM OF THE FIRST n TERMS OF AN ARITHMETIC SEQUENCE

There is a formula that gives the sum of the first n terms of an arithmetic sequence. To develop this formula, we let S_n represent the sum of the first n terms of an arithmetic sequence:

$$S_n = a + [a + d] + [a + 2d] + \cdots + [a + (n - 1)d]$$

We write the same sum again, but in reverse order:

$$S_n = [a + (n - 1)d] + [a + (n - 2)d] + [a + (n - 3)d] + \cdots + a$$

We add these two equations together, term by term, to get

$$2S_n = [2a + (n - 1)d] + [2a + (n - 1)d] + [2a + (n - 1)d] + \cdots + [2a + (n - 1)d]$$

Because there are n equal terms on the right-hand side of the preceding equation, we can write

$$2S_n = n[2a + (n - 1)d]$$

$$2S_n = n[a + a + (n - 1)d]$$

$$2S_n = n[a + l]$$

Substitute l for $a + (n - 1)d$, because $a + (n - 1)d$ is the last term of the sequence.

$$S_n = \frac{n(a + l)}{2}$$

This reasoning establishes the following theorem.

Sum of the First *n* Terms of an Arithmetic Sequence

The sum of the first n terms of an arithmetic sequence is given by the formula

$$S_n = \frac{n(a + l)}{2} \quad \text{with } l = a + (n - 1)d$$

where a is the first term, l is the last (or nth) term, and n is the number of the terms in the sequence.

EXAMPLE 5 Find the sum of the first 40 terms of the arithmetic sequence 4, 10, 16, ...

Solution In this example, we let $a = 4$, $n = 40$, $d = 6$, and $l = 4 + (40 - 1)6 = 238$ and substitute these values into the formula for S_n:

$$S_n = \frac{n(a + l)}{2}$$

$$S_{40} = \frac{40(4 + 238)}{2}$$

$$= 20(242)$$

$$= 4{,}840$$

The sum of the first 40 terms is 4,840. ■

Self Check Find the sum of the first 50 terms of the arithmetic sequence 3, 8, 13, ...

Answer 6,275

SUMMATION NOTATION

There is a shorthand notation for indicating the sum of a finite (ending) number of consecutive terms in a sequence. This notation, called **summation notation**, involves the Greek letter Σ (sigma). The expression

$$\sum_{k=2}^{5} 3k$$

Read as "the summation of $3k$ as k runs from 2 to 5."

designates the sum of all terms obtained if we successively substitute the numbers 2, 3, 4, and 5 for k, called the **index of the summation**. Thus, we have

$$\begin{aligned}\sum_{k=2}^{5} 3k &= 3(2) + 3(3) + 3(4) + 3(5)\\ &= 6 + 9 + 12 + 15\\ &= 42\end{aligned}$$

EXAMPLE 6 Find each sum: **a.** $\sum_{k=3}^{5}(2k + 1)$, **b.** $\sum_{k=2}^{5} k^2$, and **c.** $\sum_{k=1}^{3}(3k^2 + 3)$.

Solution **a.** $$\begin{aligned}\sum_{k=3}^{5}(2k + 1) &= [2(3) + 1] + [2(4) + 1] + [2(5) + 1]\\ &= 7 + 9 + 11\\ &= 27\end{aligned}$$

b. $$\begin{aligned}\sum_{k=2}^{5} k^2 &= 2^2 + 3^2 + 4^2 + 5^2\\ &= 4 + 9 + 16 + 25\\ &= 54\end{aligned}$$

c. $$\begin{aligned}\sum_{k=1}^{3}(3k^2 + 3) &= [3(1^2) + 3] + [3(2^2) + 3] + [3(3^2) + 3]\\ &= 6 + 15 + 30\\ &= 51\end{aligned}$$

■

Self Check Evaluate $\sum_{k=1}^{4}(2k^2 - 2)$.

Answer 52

Orals *Find the next term in each arithmetic sequence.*

1. 2, 6, 10, . . .

2. 10, 7, 4, . . .

Find the common difference in each arithmetic sequence.

3. -2, 3, 8, . . .

4. 5, -1, -7, . . .

Find each sum.

5. $\sum_{k=1}^{2} k$

6. $\sum_{k=2}^{3} k$

EXERCISE 11.3

REVIEW *Do the operations and simplify, if possible.*

1. $3(2x^2 - 4x + 7) + 4(3x^2 + 5x - 6)$

2. $(2p + q)(3p^2 + 4pq - 3q^2)$

3. $\dfrac{3a + 4}{a - 2} + \dfrac{3a - 4}{a + 2}$

4. $2t - 3\overline{)8t^4 - 12t^3 + 8t^2 - 16t + 6}$

VOCABULARY AND CONCEPTS *Fill in each blank to make a true statement.*

5. A ________ is a function whose domain is the set of natural numbers.

6. The sequence 1, 1, 2, 3, 5, 8, 13, 21, . . . is called the ________ sequence.

7. The sequence 3, 9, 15, 21, . . . is an example of an ________ sequence with a common ________ of 6.

8. The last term of an arithmetic sequence is given by the formula ____________.

9. If a number c is inserted between two numbers a and b to form an arithmetic sequence, then c is called the ____________ between a and b.

10. The sum of the first n terms of an arithmetic sequence is given by the formula $S_n = \dfrac{}{\rule{2em}{0.4pt}}$.

11. $\sum_{k=1}^{5} k$ means ____________.

12. In the symbol $\sum_{k=1}^{5} (2k - 5)$, k is called the _____ of the summation.

PRACTICE *In Exercises 13–26, write the first five terms of each arithmetic sequence with the given properties.*

13. $a = 3, d = 2$

14. $a = -2, d = 3$

15. $a = -5, d = -3$

16. $a = 8, d = -5$

17. $a = 5$, fifth term is 29

18. $a = 4$, sixth term is 39

19. $a = -4$, sixth term is -39

20. $a = -5$, fifth term is -37

21. $d = 7$, sixth term is -83

22. $d = 3$, seventh term is 12

23. $d = -3$, seventh term is 16

24. $d = -5$, seventh term is -12

25. The 19th term is 131 and the 20th term is 138.

26. The 16th term is 70 and the 18th term is 78.

27. Find the 30th term of the arithmetic sequence with $a = 7$ and $d = 12$.

28. Find the 55th term of the arithmetic sequence with $a = -5$ and $d = 4$.

29. Find the 37th term of the arithmetic sequence with a second term of -4 and a third term of -9.

30. Find the 40th term of the arithmetic sequence with a second term of 6 and a fourth term of 16.

31. Find the first term of the arithmetic sequence with a common difference of 11 and whose 27th term is 263.

32. Find the common difference of the arithmetic sequence with a first term of -164 if its 36th term is -24.

33. Find the common difference of the arithmetic sequence with a first term of 40 if its 44th term is 556.

34. Find the first term of the arithmetic sequence with a common difference of -5 and whose 23rd term is -625.

35. Insert three arithmetic means between 2 and 11.

36. Insert four arithmetic means between 5 and 25.

37. Insert four arithmetic means between 10 and 20.

38. Insert three arithmetic means between 20 and 30.

39. Find the arithmetic mean between 10 and 19.

40. Find the arithmetic mean between 5 and 23.

41. Find the arithmetic mean between -4.5 and 7.

42. Find the arithmetic mean between -6.3 and -5.2.

In Exercises 43–50, find the sum of the first n terms of each arithmetic sequence.

43. $1, 4, 7, \ldots; n = 30$

44. $2, 6, 10, \ldots; n = 28$

45. $-5, -1, 3, \ldots; n = 17$

46. $-7, -1, 5, \ldots; n = 15$

47. Second term is 7, third term is 12; $n = 12$

48. Second term is 5, fourth term is 9; $n = 16$

49. $f(n) = 2n + 1$, nth term is 31; n is a natural number

50. $f(n) = 4n + 3$, nth term is 23; n is a natural number

51. Find the sum of the first 50 natural numbers.

52. Find the sum of the first 100 natural numbers.

53. Find the sum of the first 50 odd natural numbers.

54. Find the sum of the first 50 even natural numbers.

In Exercises 55–60, find each sum.

55. $\sum_{k=1}^{4} 6k$

56. $\sum_{k=2}^{5} 3k$

57. $\sum_{k=3}^{4} (k^2 + 3)$

58. $\sum_{k=2}^{6} (k^2 + 1)$

59. $\sum_{k=4}^{4} (2k + 4)$

60. $\sum_{k=3}^{5} (3k^2 - 7)$

APPLICATIONS

61. Saving money Yasmeen puts $60 into a safety deposit box. Each month, she puts $50 more in the box. Write the first six terms of an arithmetic sequence that gives the monthly amounts in her savings, and find her savings after 10 years.

62. Installment loan Maria borrowed $10,000, interest-free, from her mother. She agreed to pay back the loan in monthly installments of $275. Write the first six terms of an arithmetic sequence that shows the balance due after each month, and find the balance due after 17 months.

63. Designing a patio Each row of bricks in the triangular patio in Illustration 1 (on page 812) is to have one more brick than the previous row, ending with the longest row of 150 bricks. How many bricks will be needed?

ILLUSTRATION 1

64. Falling object The equation $s = 16t^2$ represents the distance s in feet that an object will fall in t seconds. After 1 second, the object has fallen 16 feet. After 2 seconds, the object has fallen 64 feet, and so on. Find the distance that the object will fall during the second and third seconds.

65. Falling object Refer to Exercise 64. How far will the object fall during the 12th second?

66. Interior angles The sum of the angles of several polygons are given in the table shown in Illustration 2. Assuming that the pattern continues, complete the table.

Figure	Number of sides	Sum of angles
Triangle	3	180°
Quadrilateral	4	360°
Pentagon	5	540°
Hexagon	6	720°
Octagon	8	
Dodecagon	12	

ILLUSTRATION 2

WRITING

67. Define an arithmetic sequence.

68. Develop the formula for finding the sum of the first n terms of an arithmetic sequence.

SOMETHING TO THINK ABOUT

69. Write the first 6 terms of the arithmetic sequence given by

$$\sum_{n=1}^{k}\left(\frac{1}{2}n + 1\right)$$

70. Find the sum of the first 6 terms of the sequence given in Exercise 69.

71. Show that the arithmetic mean between a and b is the average of a and b: $\dfrac{a+b}{2}$

72. Show that the sum of the two arithmetic means between a and b is $a + b$.

73. Show that $\sum_{k=1}^{5} 5k = 5\sum_{k=1}^{5} k$.

74. Show that $\sum_{k=3}^{6}(k^2 + 3k) = \sum_{k=3}^{6} k^2 + \sum_{k=3}^{6} 3k$.

75. Show that $\sum_{k=1}^{n} 3 = 3n$. (*Hint:* Consider 3 to be $3k^0$.)

76. Show that $\sum_{k=1}^{3}\frac{k^2}{k} \neq \dfrac{\sum_{k=1}^{3} k^2}{\sum_{k=1}^{3} k}$.

11.4 Geometric Sequences

■ GEOMETRIC SEQUENCES ■ GEOMETRIC MEANS ■ THE SUM OF THE FIRST *n* TERMS OF A GEOMETRIC SEQUENCE ■ POPULATION GROWTH

Getting Ready *Complete each table.*

1.

n	$5(2^n)$
1	
2	
3	

2.

n	$6(3^n)$
1	
2	
3	

■ GEOMETRIC SEQUENCES

Another common type of sequence is called a **geometric sequence.**

Geometric Sequence

A **geometric sequence** is a sequence of the form

$$a, ar, ar^2, ar^3, \ldots, ar^{n-1}, \ldots$$

where a is the **first term,** ar^{n-1} is the ***n*th term,** and r is the **common ratio.**

We note that the second term of a geometric sequence has a factor of r^1, the third term has a factor of r^2, the fourth term has a factor of r^3, and the nth term has a factor of r^{n-1}. We also note that the quotient obtained when any term is divided by the previous term is r.

EXAMPLE 1 A geometric sequence has a first term of 5 and a common ratio of 3.

a. Write the first five terms of the sequence.

b. Find the ninth term.

Solution **a.** Because the first term is $a = 5$ and the common ratio is $r = 3$, the first five terms are

$$5, \quad 5(3), \quad 5(3^2), \quad 5(3^3), \quad 5(3^4)$$

or

$$5, 15, 45, 135, 405$$

b. The nth term is ar^{n-1} where $a = 5$ and $r = 3$. Because we want the ninth term, we let $n = 9$:

$$\begin{aligned} n\text{th term} &= ar^{n-1} \\ 9\text{th term} &= 5(3)^{9-1} \\ &= 5(3)^8 \\ &= 5(6{,}561) \\ &= 32{,}805 \end{aligned}$$

■

Self Check A geometric sequence has a first term of 3 and a common ratio of 4. **a.** Write the first four terms. **b.** Find the eighth term.

Answers **a.** 3, 12, 48, 192, **b.** 49,152

EXAMPLE 2 The first three terms of a geometric sequence are 16, 4, and 1. Find the seventh term.

Solution We substitute 16 for a, $\frac{1}{4}$ for r, and 7 for n in the formula for the nth term and simplify:

$$\begin{aligned} n\text{th term} &= ar^{n-1} \\ 7\text{th term} &= 16\left(\frac{1}{4}\right)^{7-1} \\ &= 16\left(\frac{1}{4}\right)^{6} \\ &= 16\left(\frac{1}{4{,}096}\right) \\ &= \frac{1}{256} \end{aligned}$$

■

Self Check See Example 2. Find the tenth term.

Answer $\frac{1}{16{,}384}$

GEOMETRIC MEANS

If numbers are inserted between two numbers a and b to form a geometric sequence, the inserted numbers are called **geometric means** between a and b.

If a single number is inserted between the numbers a and b to form a geometric sequence, that number is called a **geometric mean** between a and b.

EXAMPLE 3 Insert two geometric means between 7 and 1,512.

Solution Here the first term is $a = 7$, and the fourth term (or last term) is $l = 1{,}512$. To find the common ratio r so that the terms

$$7,\ 7r,\ 7r^2,\ 1{,}512$$

form a geometric sequence, we substitute 4 for n and 7 for a into the formula for the nth term of a geometric sequence and solve for r.

$$\begin{aligned} n\text{th term} &= ar^{n-1} \\ 4\text{th term} &= 7r^{4-1} \\ 1{,}512 &= 7r^3 \\ 216 &= r^3 && \text{Divide both sides by 7.} \\ 6 &= r && \text{Take the cube root of both sides.} \end{aligned}$$

The two geometric means between 7 and 1,512 are

$$7r = 7(6) = 42$$

and

$$7r^2 = 7(6)^2 = 7(36) = 252$$

The numbers 7, 42, 252, and 1,512 are the first four terms of a geometric sequence. ■

Self Check Insert three positive geometric means between 1 and 16.

Answer 2, 4, 8

EXAMPLE 4 Find a geometric mean between 2 and 20.

Solution We want to find the middle term of the three-term geometric sequence

$$2,\ 2r,\ 20$$

with $a = 2$, $l = 20$, and $n = 3$. To find r, we substitute these values into the formula for the nth term of a geometric sequence:

$$\begin{aligned} n\text{th term} &= ar^{n-1} \\ 3\text{rd term} &= 2r^{3-1} \\ 20 &= 2r^2 \\ 10 &= r^2 && \text{Divide both sides by 2.} \\ \pm\sqrt{10} &= r && \text{Take the square root of both sides.} \end{aligned}$$

Because r can be either $\sqrt{10}$ or $-\sqrt{10}$, there are two values for the geometric mean. They are

$$2r = 2\sqrt{10}$$

and

$$2r = -2\sqrt{10}$$

The numbers 2, $2\sqrt{10}$, 20 and 2, $-2\sqrt{10}$, 20 both form geometric sequences. The common ratio of the first sequence is $\sqrt{10}$, and the common ratio of the second sequence is $-\sqrt{10}$. ■

Self Check Find the positive geometric mean between 2 and 200.

Answer 20

THE SUM OF THE FIRST *n* TERMS OF A GEOMETRIC SEQUENCE

There is a formula that gives the sum of the first n terms of a geometric sequence. To develop this formula, we let S_n represent the sum of the first n terms of a geometric sequence.

1. $S_n = a + ar + ar^2 + ar^3 + \ldots + ar^{n-1}$

We multiply both sides of Equation 1 by r to get

2. $S_n r = \quad ar + ar^2 + ar^3 + \ldots + ar^{n-1} + ar^n$

We now subtract Equation 2 from Equation 1 and solve for S_n:

$$S_n - S_n r = a - ar^n$$

$$S_n(1 - r) = a - ar^n \qquad \text{Factor out } S_n \text{ from the left-hand side.}$$

$$S_n = \frac{a - ar^n}{1 - r} \qquad \text{Divide both sides by } 1 - r.$$

This reasoning establishes the following formula.

Sum of the First *n* Terms of a Geometric Sequence
The sum of the first n terms of a geometric sequence is given by the formula

$$S_n = \frac{a - ar^n}{1 - r} \quad (r \neq 1)$$

where S_n is the sum, a is the first term, r is the common ratio, and n is the number of terms.

EXAMPLE 5 Find the sum of the first six terms of the geometric sequence 250, 50, 10, ...

Solution Here $a = 250$, $r = \frac{1}{5}$, and $n = 6$. We substitute these values into the formula for the sum of the first n terms of a geometric sequence and simplify:

$$S_n = \frac{a - ar^n}{1 - r}$$

$$S_6 = \frac{250 - 250\left(\frac{1}{5}\right)^6}{1 - \frac{1}{5}}$$

$$= \frac{250 - 250\left(\frac{1}{15,625}\right)}{\frac{4}{5}}$$

$$= \frac{5}{4}\left(250 - \frac{250}{15,625}\right)$$

$$= \frac{5}{4}\left(\frac{3,906,000}{15,625}\right)$$

$$= 312.48$$

The sum of the first six terms is 312.48. ■

Self Check Find the sum of the first five terms of the geometric sequence 100, 20, 4, ...

Answer 124.96

POPULATION GROWTH

EXAMPLE 6 **Growth of a town** The mayor of Eagle River (population 1,500) predicts a growth rate of 4% each year for the next ten years. Find the expected population of Eagle River ten years from now.

Solution Let P_0 be the initial population of Eagle River. After 1 year, there will be a different population, P_1. The initial population (P_0) plus the growth (the product of P_0 and the rate of growth, r) will equal this new population, P_1:

$$P_1 = P_0 + P_0 r = P_0(1 + r)$$

The population after 2 years will be P_2, and

$$P_2 = P_1 + P_1 r$$

$$= P_1(1 + r) \quad \text{Factor out } P_1.$$

$$= P_0(1 + r)(1 + r) \quad \text{Remember that } P_1 = P_0(1 + r).$$

$$= P_0(1 + r)^2$$

The population after 3 years will be P_3, and

$$\begin{aligned} P_3 &= P_2 + P_2 r \\ &= P_2(1 + r) && \text{Factor out } P_2. \\ &= P_0(1 + r)^2(1 + r) && \text{Remember that } P_2 = P_0(1 + r)^2. \\ &= P_0(1 + r)^3 \end{aligned}$$

The yearly population figures

$$P_0, \quad P_1, \quad P_2, \quad P_3, \ldots$$

or

$$P_0, \quad P_0(1 + r), \quad P_0(1 + r)^2, \quad P_0(1 + r)^3, \ldots$$

form a geometric sequence with a first term of P_0 and a common ratio of $1 + r$. The population of Eagle River after 10 years is P_{10}, which is the 11th term of this sequence:

$$\begin{aligned} n\text{th term} &= ar^{n-1} \\ P_{10} = 11\text{th term} &= P_0(1 + r)^{10} \\ &= 1{,}500(1 + 0.04)^{10} \\ &= 1{,}500(1.04)^{10} \\ &\approx 1{,}500(1.480244285) \\ &\approx 2{,}220 \end{aligned}$$

The expected population ten years from now is 2,220 people. ■

EXAMPLE 7 **Amount of an annuity** An **annuity** is a sequence of equal payments made periodically over a length of time. The sum of the payments and the interest earned during the *term* of the annuity is called the *amount* of the annuity.

After a sales clerk works six months, her employer will begin an annuity for her and will contribute \$500 every six months to a fund that pays 8% annual interest. After she has been employed for two years, what will be the amount of the annuity?

Solution Because the payments are to be made semiannually, there will be four payments of \$500, each earning a rate of 4% per six-month period. These payments will occur at the end of 6 months, 12 months, 18 months, and 24 months. The first payment, to be made after 6 months, will earn interest for three interest periods. Thus, the amount of the first payment is $\$500(1.04)^3$. The amounts of each of the four payments after two years are shown in Figure 11-1.

The amount of the annuity is the sum of the amounts of the individual payments, a sum of \$2,123.23.

Payment (at the end of period)	Amount of payment at the end of 2 years
1	$\$500(1.04)^3 = \$\ 562.43$
2	$\$500(1.04)^2 = \$\ 540.80$
3	$\$500(1.04)^1 = \$\ 520.00$
4	$\$500 = \$\ 500.00$
	$A_n = \$2{,}123.23$

FIGURE 11-1

■

Orals *Find the next term in each geometric sequence.*

1. 1, 3, 9, . . .

2. $1, \frac{1}{3}, \frac{1}{9}, \ldots$

Find the common ratio in each geometric sequence.

3. 0.2, 0.5, 1.25, . . .

4. $\sqrt{3}, 3, 3\sqrt{3}, \ldots$

Find x in each geometric sequence.

5. $2, x, 18, 54, \ldots$

6. $3, x, \frac{1}{3}, \frac{1}{9}, \ldots$

EXERCISE 11.4

REVIEW *Solve each inequality.*

1. $x^2 - 5x - 6 \leq 0$

2. $a^2 - 7a + 12 \geq 0$

3. $\frac{x-4}{x+3} > 0$

4. $\frac{t^2 + t - 20}{t + 2} < 0$

VOCABULARY AND CONCEPTS *Fill in each blank to make a true statement.*

5. A sequence of the form $a, ar, ar^2, \ldots$ is called a ________ sequence.

6. The formula for the nth term of a geometric sequence is ____________.

7. In a geometric sequence, r is called the ____________.

8. A number inserted between two numbers a and b to form a geometric sequence is called a geometric ______ between a and b.

9. The sum of the first n terms of a geometric sequence is given by the formula ____________

10. In the formula for Exercise 9, a is the _____ term of the sequence.

PRACTICE *In Exercises 11–24, write the first five terms of each geometric sequence with the given properties.*

11. $a = 3, r = 2$

12. $a = -2, r = 2$

13. $a = -5, r = \dfrac{1}{5}$

14. $a = 8, r = \dfrac{1}{2}$

15. $a = 2, r > 0$, third term is 32

16. $a = 3$, fourth term is 24

17. $a = -3$, fourth term is -192

18. $a = 2, r < 0$, third term is 50

19. $a = -64, r < 0$, fifth term is -4

20. $a = -64, r > 0$, fifth term is -4

21. $a = -64$, sixth term is -2

22. $a = -81$, sixth term is $\dfrac{1}{3}$

23. The second term is 10, and the third term is 50.

24. The third term is -27, and the fourth term is 81.

25. Find the tenth term of the geometric sequence with $a = 7$ and $r = 2$.

26. Find the 12th term of the geometric sequence with $a = 64$ and $r = \frac{1}{2}$.

27. Find the first term of the geometric sequence with a common ratio of -3 and an eighth term of -81.

28. Find the first term of the geometric sequence with a common ratio of 2 and a tenth term of 384.

29. Find the common ratio of the geometric sequence with a first term of -8 and a sixth term of $-1{,}944$.

30. Find the common ratio of the geometric sequence with a first term of 12 and a sixth term of $\frac{3}{8}$.

31. Insert three positive geometric means between 2 and 162.

32. Insert four geometric means between 3 and 96.

33. Insert four geometric means between -4 and $-12{,}500$.

34. Insert three geometric means (two positive and one negative) between -64 and $-1{,}024$.

35. Find the negative geometric mean between 2 and 128.

36. Find the positive geometric mean between 3 and 243.

37. Find the positive geometric mean between 10 and 20.

38. Find the negative geometric mean between 5 and 15.

39. Find a geometric mean, if possible, between -50 and 10.

40. Find a negative geometric mean, if possible, between -25 and -5.

In Exercises 41–52, find the sum of the first n terms of each geometric sequence.

41. $2, 6, 18, \ldots; n = 6$

42. $2, -6, 18, \ldots; n = 6$

43. $2, -6, 18, \ldots; n = 5$

44. $2, 6, 18, \ldots; n = 5$

45. $3, -6, 12, \ldots; n = 8$

46. $3, 6, 12, \ldots; n = 8$

47. $3, 6, 12, \ldots; n = 7$

48. $3, -6, 12, \ldots; n = 7$

49. The second term is 1, and the third term is $\frac{1}{5}$; $n = 4$.

50. The second term is 1, and the third term is 4; $n = 5$.

51. The third term is -2, and the fourth term is 1; $n = 6$.

52. The third term is -3, and the fourth term is 1; $n = 5$.

APPLICATIONS *In Exercises 53–62, use a calculator to help solve each problem.*

53. Population growth The population of Union is predicted to increase by 6% each year. What will be the population of Union 5 years from now if its current population is 500?

54. Population decline The population of Bensville is decreasing by 10% each year. If its current population is 98, what will be the population 8 years from now?

55. Declining savings John has \$10,000 in a safety deposit box. Each year he spends 12% of what is left in the box. How much will be in the box after 15 years?

56. Savings growth Sally has \$5,000 in a savings account earning 12% annual interest. How much will be in her account 10 years from now? (Assume that Sally makes no deposits or withdrawals.)

57. House appreciation A house appreciates by 6% each year. If the house is worth \$70,000 today, how much will it be worth 12 years from now?

58. Motorboat depreciation A motorboat that cost \$5,000 when new depreciates at a rate of 9% per year. How much will the boat be worth in 5 years?

59. Inscribed squares Each inscribed square in Illustration 1 joins the midpoints of the next larger square. The area of the first square, the largest, is 1. Find the area of the 12th square.

60. Genealogy The family tree in Illustration 2 spans 3 generations and lists 7 people. How many names would be listed in a family tree that spans 10 generations?

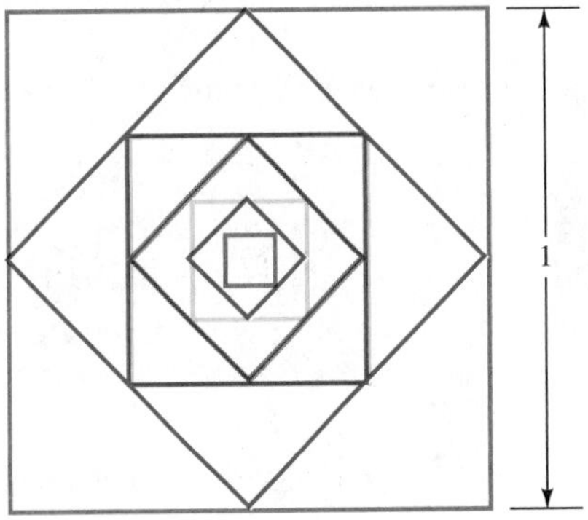

ILLUSTRATION 1

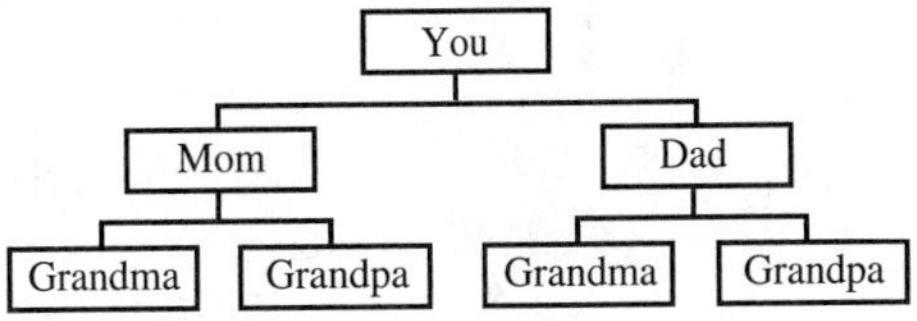

ILLUSTRATION 2

61. Annuities Find the amount of an annuity if \$1,000 is paid semiannually for two years at 6% annual interest. Assume that the first of the four payments is made immediately.

62. Annuities Note that the amounts shown in Figure 11-1 form a geometric sequence. Verify the answer for Example 7 by using the formula for the sum of a geometric sequence.

WRITING

63. Define a geometric sequence.

64. Develop the formula for finding the sum of the first n terms of a geometric sequence.

SOMETHING TO THINK ABOUT

65. Show that the formula for the sum of the first n terms of a geometric sequence can be found by using the formula $S = \frac{a - lr}{1 - r}$.

66. Show that the geometric mean between a and b is $\sqrt{ab}$.

67. If $a > b > 0$, which is larger: the arithmetic mean between a and b or the geometric mean between a and b?

68. Is there a geometric mean between -5 and 5?

69. Show that the formula for the sum of the first n terms of a geometric sequence can be written in the form

$$S_n = \frac{a - lr}{1 - r} \quad \text{where } l = ar^{n-1}$$

70. Show that the formula for the sum of the first n terms of a geometric sequence can be written in the form

$$S_n = \frac{a(1 - r^n)}{1 - r}$$

11.5 Infinite Geometric Sequences

■ INFINITE GEOMETRIC SEQUENCES ■ THE SUM OF AN INFINITE GEOMETRIC SEQUENCE

Getting Ready *Evaluate each expression.*

1. $\dfrac{2}{1 - \frac{1}{2}}$ **2.** $\dfrac{3}{1 - \frac{1}{3}}$ **3.** $\dfrac{\frac{3}{2}}{1 - \frac{1}{2}}$ **4.** $\dfrac{\frac{5}{3}}{1 - \frac{1}{3}}$

■ INFINITE GEOMETRIC SEQUENCES

An **infinite geometric sequence** is a geometric sequence with infinitely many terms. Two examples of infinite geometric sequences are

$$2, 6, 18, 54, 162, \ldots \quad (r = 3)$$

$$\frac{3}{2}, \frac{3}{4}, \frac{3}{8}, \frac{3}{16}, \frac{3}{32}, \ldots \quad \left(r = \frac{1}{2}\right)$$

■ THE SUM OF AN INFINITE GEOMETRIC SEQUENCE

Under certain conditions, we can find the sum of all the terms of an infinite geometric sequence. To define this sum, we consider the geometric sequence

$$a, ar, ar^2, ar^3, \ldots, ar^{n-1}, \ldots$$

- The first **partial sum,** S_1, of the sequence is $S_1 = a$.
- The second partial sum, S_2, of the sequence is $S_2 = a + ar$.
- The third partial sum, S_3, of the sequence is $S_3 = a + ar + ar^2$.
- The nth partial sum, S_n, of the sequence is $S_n = a + ar + ar^2 + \cdots + ar^{n-1}$.

If the nth partial sum, S_n, approaches some number S as n approaches infinity, then S is called the **sum of the infinite geometric sequence.**

To develop a formula for finding the sum (if it exists) of all the terms in an infinite geometric sequence, we consider the formula that gives the sum of the first n terms of a geometric sequence.

$$S_n = \frac{a - ar^n}{1 - r} \quad (r \neq 1)$$

If $|r| < 1$ and a is constant, then the term ar^n in the above formula approaches 0 as n becomes very large. For example,

$$a\left(\frac{1}{4}\right)^1 = \frac{1}{4}a, \quad a\left(\frac{1}{4}\right)^2 = \frac{1}{16}a, \quad a\left(\frac{1}{4}\right)^3 = \frac{1}{64}a$$

and so on. When n is very large, the value of ar^n is negligible, and the term ar^n in the above formula can be ignored. This reasoning justifies the following theorem.

Sum of an Infinite Geometric Sequence

If a is the first term and r is the common ratio of an infinite geometric sequence, and if $|r| < 1$, the sum of the terms of the sequence is given by the formula

$$S = \frac{a}{1 - r}$$

EXAMPLE 1 Find the sum of the terms of the infinite geometric sequence 125, 25, 5, ...

Solution Here $a = 125$ and $r = \frac{1}{5}$. Since $|r| = |\frac{1}{5}| = \frac{1}{5} < 1$, we can find the sum of all the terms of the sequence. We do this by substituting 125 for a and $\frac{1}{5}$ for r in the formula $S = \frac{a}{1-r}$ and simplifying:

$$S = \frac{a}{1 - r} = \frac{125}{1 - \frac{1}{5}} = \frac{125}{\frac{4}{5}} = \frac{5}{4}(125) = \frac{625}{4}$$

The sum of the terms of the sequence 125, 25, 5, . . . is 156.25. ■

Self Check Find the sum of the terms of the infinite geometric sequence 100, 20, 4, ...

Answer 125

EXAMPLE 2 Find the sum of the infinite geometric sequence 64, -4, $\frac{1}{4}$, ...

Solution Here $a = 64$ and $r = -\frac{1}{16}$. Since $|r| = |-\frac{1}{16}| = \frac{1}{16} < 1$, we can find the sum of all the terms of the sequence. We substitute 64 for a and $-\frac{1}{16}$ for r in the formula $S = \frac{a}{1-r}$ and simplify:

$$S = \frac{a}{1 - r} = \frac{64}{1 - \left(-\frac{1}{16}\right)} = \frac{64}{\frac{17}{16}} = \frac{16}{17}(64) = \frac{1,024}{17}$$

The sum of the terms of the geometric sequence 64, -4, $\frac{1}{4}$, . . . is $\frac{1,024}{17}$. ■

Self Check Find the sum of the infinite geometric sequence 81, 27, 9, ...

Answer $\frac{243}{2}$

EXAMPLE 3 Change $0.\overline{8}$ to a common fraction.

Solution The decimal $0.\overline{8}$ can be written as the sum of an infinite geometric sequence:

$$0.\overline{8} = 0.888\ldots = \frac{8}{10} + \frac{8}{100} + \frac{8}{1{,}000} + \cdots$$

where $a = \frac{8}{10}$ and $r = \frac{1}{10}$. Because $|r| = \left|\frac{1}{10}\right| = \frac{1}{10} < 1$, we can find the sum as follows:

$$S = \frac{a}{1-r} = \frac{\frac{8}{10}}{1 - \frac{1}{10}} = \frac{\frac{8}{10}}{\frac{9}{10}} = \frac{8}{9}$$

Thus, $0.\overline{8} = \frac{8}{9}$. Long division will verify that $\frac{8}{9} = 0.888\ldots$. ■

Self Check Change $0.\overline{6}$ to a common fraction.

Answer $\frac{2}{3}$

EXAMPLE 4 Change $0.\overline{25}$ to a common fraction.

Solution The decimal $0.\overline{25}$ can be written as the sum of an infinite geometric sequence:

$$0.\overline{25} = 0.252525\ldots = \frac{25}{100} + \frac{25}{10{,}000} + \frac{25}{1{,}000{,}000} + \cdots$$

where $a = \frac{25}{100}$ and $r = \frac{1}{100}$. Since $|r| = \left|\frac{1}{100}\right| = \frac{1}{100} < 1$, we can find the sum as follows:

$$S = \frac{a}{1-r} = \frac{\frac{25}{100}}{1 - \frac{1}{100}} = \frac{\frac{25}{100}}{\frac{99}{100}} = \frac{25}{99}$$

Thus, $0.\overline{25} = \frac{25}{99}$. Long division will verify that this is true. ■

Self Check Change $0.\overline{15}$ to a common fraction.

Answer $\frac{5}{33}$

Orals *Find the common ratio in each infinite geometric sequence.*

1. $\frac{1}{64}, \frac{1}{8}, 1, \ldots$
2. $1, \frac{1}{8}, \frac{1}{64}, \ldots$
3. $\frac{2}{3}, \frac{1}{3}, \frac{1}{6}, \ldots$
4. $64, 8, 1, \ldots$

Find the sum of the terms in each infinite geometric sequence.

5. $18, 6, 2, \ldots$
6. $12, 3, \frac{3}{4}, \ldots$

EXERCISE 11.5

REVIEW *Determine whether each equation determines y to be a function of x.*

1. $y = 3x^3 - 4$
2. $xy = 12$
3. $3x = y^2 + 4$
4. $x = |y|$

VOCABULARY AND CONCEPTS *Fill in each blank to make a true statement.*

5. If a geometric sequence has infinitely many terms, it is called an ______ geometric sequence.
6. The third partial sum of the sequence 2, 6, 18, 54, . . . is __________ = 26.
7. The formula for the sum of an infinite geometric sequence with $|r| < 1$ is ______.
8. Write $0.\overline{75}$ as the sum of an infinite geometric sequence. ________________.

PRACTICE *In Exercises 9–20, find the sum of each infinite geometric sequence, if possible.*

9. $8, 4, 2, \ldots$
10. $12, 6, 3, \ldots$
11. $54, 18, 6, \ldots$
12. $45, 15, 5, \ldots$
13. $12, -6, 3, \ldots$
14. $8, -4, 2, \ldots$
15. $-45, 15, -5, \ldots$
16. $-54, 18, -6, \ldots$
17. $\frac{9}{2}, 6, 8, \ldots$
18. $-112, -28, -7, \ldots$
19. $-\frac{27}{2}, -9, -6, \ldots$
20. $\frac{18}{25}, \frac{6}{5}, 2, \ldots$

In Exercises 21–28, change each decimal to a common fraction. Then check the answer by doing a long division.

21. $0.\overline{1}$
22. $0.\overline{2}$
23. $-0.\overline{3}$
24. $-0.\overline{4}$
25. $0.\overline{12}$
26. $0.\overline{21}$
27. $0.\overline{75}$
28. $0.\overline{57}$

APPLICATIONS

29. **Bouncing ball** On each bounce, the rubber ball in Illustration 1 (page 826) rebounds to a height one-half of that from which it fell. Find the total distance the ball travels.
30. **Bouncing ball** A golf ball is dropped from a height of 12 feet. On each bounce, it returns to a height two-thirds of that from which it fell. Find the total distance the ball travels.

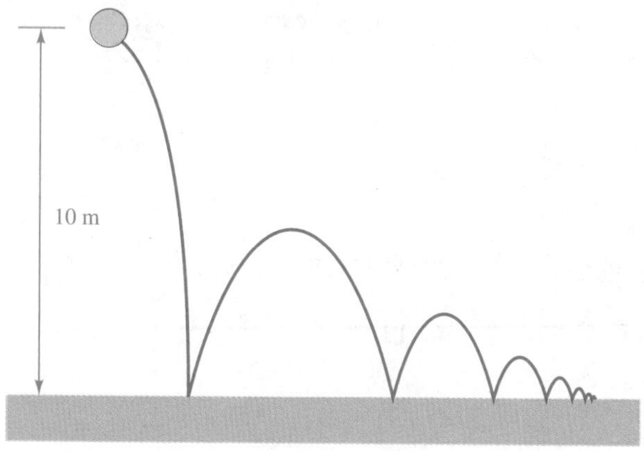

ILLUSTRATION 1

31. **Controlling moths** To reduce the population of a destructive moth, biologists release 1,000 sterilized male moths each day into the environment. If 80% of these moths alive one day survive until the next, then after a long time, the population of sterile males is the sum of the infinite geometric sequence

$$1{,}000 + 1{,}000(0.8) + 1{,}000(0.8)^2 + 1{,}000(0.8)^3 + \cdots$$

Find the long-term population.

32. **Controlling moths** If mild weather increases the day-to-day survival rate of the sterile male moths in Exercise 31 to 90%, find the long-term population.

WRITING

33. Why must the absolute value of the common ratio be less than 1 before an infinite geometric sequence can have a sum?

34. Can an infinite arithmetic sequence have a sum?

SOMETHING TO THINK ABOUT

35. An infinite geometric sequence has a sum of 5 and a first term of 1. Find the common ratio.

36. An infinite geometric sequence has a common ratio of $-\frac{2}{3}$ and a sum of 9. Find the first term.

37. Show that $0.\overline{9} = 1$.

38. Show that $1.\overline{9} = 2$.

39. Does $0.999999 = 1$? Explain.

40. If $f(x) = 1 + x + x^2 + x^3 + x^4 + \cdots$, find $f\left(\frac{1}{2}\right)$ and $f\left(-\frac{1}{2}\right)$.

11.6 Permutations and Combinations

■ THE MULTIPLICATION PRINCIPLE FOR EVENTS ■ PERMUTATIONS ■ COMBINATIONS ■ ALTERNATIVE FORM OF THE BINOMIAL THEOREM

Getting Ready *Evaluate each expression.*

1. $4 \cdot 3 \cdot 2 \cdot 1$

2. $5 \cdot 4 \cdot 3 \cdot 2 \cdot 1$

3. $\dfrac{6 \cdot 5 \cdot 4 \cdot 3 \cdot 2 \cdot 1}{4 \cdot 3 \cdot 2 \cdot 1}$

4. $\dfrac{8 \cdot 7 \cdot 6 \cdot 5 \cdot 4 \cdot 3 \cdot 2 \cdot 1}{2(5 \cdot 4 \cdot 3 \cdot 2 \cdot 1)}$

■ THE MULTIPLICATION PRINCIPLE FOR EVENTS

Steven goes to the cafeteria for lunch. He has a choice of three different sandwiches (hamburger, hot dog, or ham and cheese) and four different beverages (cola, root beer, orange, or milk). How many different lunches can he choose?

He has three choices of sandwich, and for any one of these choices, he has four choices of drink. The different options are shown in the *tree diagram* in Figure 11-2.

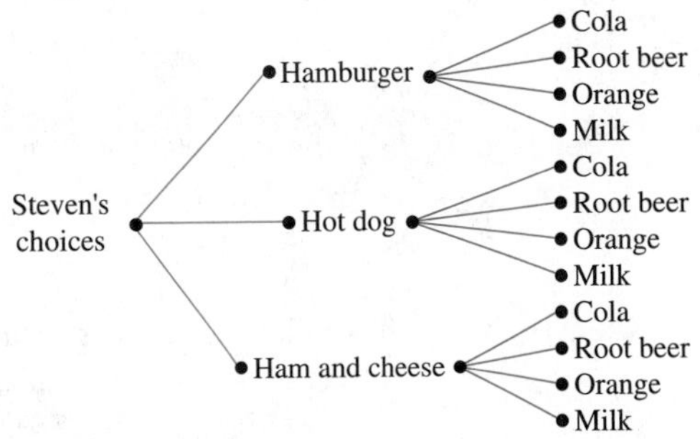

FIGURE 11-2

The tree diagram shows that there is a total of 12 different lunches to choose from. One of the possibilities is a hamburger with a cola, and another is a hot dog with milk.

A situation that can have several different outcomes—such as choosing a sandwich—is called an **event**. Choosing a sandwich and choosing a beverage can be thought of as two events. The preceding example illustrates the **multiplication principle for events.**

Multiplication Principle for Events
Let E_1 and E_2 be two events. If E_1 can be done in a_1 ways, and if—after E_1 has occurred—E_2 can be done in a_2 ways, the event "E_1 followed by E_2" can be done in $a_1 \cdot a_2$ ways.

EXAMPLE 1 Before studying for an exam, Heidi plans to watch the evening news and then a situation comedy on television. If she has a choice of four news broadcasts and two comedies, in how many ways can she choose to watch television?

Solution Let E_1 be the event "watching the news" and E_2 be the event "watching a comedy." Because there are four ways to accomplish E_1 and two ways to accomplish E_2, the number of choices that she has is $4 \cdot 2 = 8$. ■

Self Check If Alex has 7 shirts and 5 pairs of pants, how many outfits could he wear?

Answer 35

The multiplication principle can be extended to any number of events. In Example 2, we use it to complete the number of ways that we can arrange objects in a row.

EXAMPLE 2 In how many ways can we arrange five books on a shelf?

Solution We can fill the first space with any of the 5 books, the second space with any of the remaining 4 books, the third space with any of the remaining 3 books, the fourth space with any of the remaining 2 books, and the fifth space with the remaining 1 (or last) book. By the multiplication principle for events, the number of ways in which the books can be arranged is

$$5 \cdot 4 \cdot 3 \cdot 2 \cdot 1 = 120$$

■

Self Check In how many ways can 4 boys line up in a row?

Answer 24

EXAMPLE 3 If a sailor has six flags, each of a different color, to hang on a flagpole, how many different signals can the sailor send by using four flags?

Solution The sailor must find the number of arrangements of 4 flags when there are 6 flags to choose from. The sailor can hang any one of the 6 flags in the top position, any one of the remaining 5 flags in the second position, any one of the remaining 4 flags in the third position, and any one of the remaining 3 flags in the lowest position. By the multiplication principle for events, the total number of signals that can be sent is

$$6 \cdot 5 \cdot 4 \cdot 3 = 360$$

■

Self Check In Example 3, how many different signals can the sailor send if each signal uses three flags?

Answer 120

PERMUTATIONS

When computing the number of possible arrangements of objects such as books on a shelf or flags on a pole, we are finding the number of **permutations** of those objects. In Example 2, we found that the number of permutations of five books, using all five of them, is 120. In Example 3, we found that the number of permutations of six flags, using four of them, is 360.

The symbol $P(n, r)$, read as "the number of permutations of n things r at a time," is often used to express permutation problems. In Example 2, we found that $P(5, 5) = 120$. In Example 3, we found that $P(6, 4) = 360$.

EXAMPLE 4 If Sarah has seven flags, each of a different color, to hang on a flagpole, how many different signals can she send by using three flags?

Solution She must find $P(7, 3)$ (the number of permutations of 7 things 3 at a time). In the top position Sarah can hang any of the 7 flags, in the middle position any one of the remaining 6 flags, and in the bottom position any one of the remaining 5 flags. According to the multiplication principle for events,

$$P(7, 3) = 7 \cdot 6 \cdot 5 = 210$$

She can send 210 signals using only three of the seven flags. ■

Self Check In Example 4, how many different signals can Sarah send using four flags?

Answer 840

Although it is correct to write $P(7, 3) = 7 \cdot 6 \cdot 5$, there is an advantage in changing the form of this answer to obtain a formula for computing $P(7, 3)$:

$$\begin{aligned} P(7, 3) &= 7 \cdot 6 \cdot 5 \\ &= \frac{7 \cdot 6 \cdot 5 \cdot \mathbf{4 \cdot 3 \cdot 2 \cdot 1}}{\mathbf{4 \cdot 3 \cdot 2 \cdot 1}} \qquad \text{Multiply both the numerator and denominator by } 4 \cdot 3 \cdot 2 \cdot 1. \\ &= \frac{7!}{4!} \\ &= \frac{7!}{(7-3)!} \end{aligned}$$

The generalization of this idea gives the following formula.

Finding *P*(*n*, *r*)

The number of permutations of n things r at a time is given by the formula

$$P(n, r) = \frac{n!}{(n-r)!}$$

EXAMPLE 5 Compute **a.** $P(8, 2)$, **b.** $P(7, 5)$, **c.** $P(n, n)$, and **d.** $P(n, 0)$.

Solution

a. $$P(8, 2) = \frac{8!}{(8-2)!} = \frac{8 \cdot 7 \cdot 6!}{6!} = 8 \cdot 7 = 56$$

b. $$P(7, 5) = \frac{7!}{(7-5)!} = \frac{7 \cdot 6 \cdot 5 \cdot 4 \cdot 3 \cdot 2!}{2!} = 7 \cdot 6 \cdot 5 \cdot 4 \cdot 3 = 2{,}520$$

c. $$P(n, n) = \frac{n!}{(n-n)!} = \frac{n!}{0!} = \frac{n!}{1} = n!$$

d. $$P(n, 0) = \frac{n!}{(n-0)!} = \frac{n!}{n!} = 1$$

■

Self Check Compute **a.** $P(10, 6)$ and **b.** $P(10, 0)$.

Answers **a.** 151,200 **b.** 1

Parts **c** and **d** of Example 5 establish the following formulas.

Finding $P(n, n)$ and $P(n, 0)$

The number of permutations of n things n at a time and n things 0 at a time are given by the formulas

$$P(n, n) = n! \qquad \text{and} \qquad P(n, 0) = 1$$

EXAMPLE 6

a. In how many ways can a television executive arrange the Saturday night lineup of 6 programs if there are 15 programs to choose from?

b. If there are only 6 programs to choose from?

Solution

a. To find the number of permutations of 15 programs 6 at a time, we use the formula $P(n, r) = \frac{n!}{(n-r)!}$ with $n = 15$ and $r = 6$.

$$\begin{aligned} P(15, 6) &= \frac{15!}{(15-6)!} \\ &= \frac{15 \cdot 14 \cdot 13 \cdot 12 \cdot 11 \cdot 10 \cdot 9!}{9!} \\ &= 15 \cdot 14 \cdot 13 \cdot 12 \cdot 11 \cdot 10 \\ &= 3{,}603{,}600 \end{aligned}$$

b. To find the number of permutations of 6 programs 6 at a time, we use the formula $P(n, n) = n!$ with $n = 6$.

$$P(6, 6) = 6! = 720$$

■

Self Check

In Example 6, how many ways are there if the executive has 20 programs to choose from?

Answer

27,907,200

COMBINATIONS

Suppose that Raul must read 4 books from a reading list of 10 books. The order in which he reads them is not important. For the moment, however, let's assume that order is important and find the number of permutations of 10 things 4 at a time:

$$\begin{aligned} P(10, 4) &= \frac{10!}{(10-4)!} \\ &= \frac{10 \cdot 9 \cdot 8 \cdot 7 \cdot 6!}{6!} \\ &= 10 \cdot 9 \cdot 8 \cdot 7 \\ &= 5{,}040 \end{aligned}$$

If order is important, there are 5,040 ways of choosing 4 books when there are 10 books to choose from. However, because the order in which Raul reads the books does not matter, the previous result of 5,040 is too big. Since there are 24 (or 4!) ways of ordering the 4 books that are chosen, the result of 5,040 is exactly 24 (or 4!) times too big. Therefore, the number of choices that Raul has is the number of permutations of 10 things 4 at a time, divided by 24:

$$\frac{P(10, 4)}{24} = \frac{5{,}040}{24} = 210$$

Raul has 210 ways of choosing four books to read from the list of ten books.

In situations where order is *not* important, we are interested in **combinations**, not permutations. The symbols $C(n, r)$ and $\binom{n}{r}$ both mean the number of combinations of n things r at a time.

If a selection of r books is chosen from a total of n books, the number of possible selections is $C(n, r)$ and there are $r!$ arrangements of the r books in each selection. If we consider the selected books as an ordered grouping, the number of orderings is $P(n, r)$. Therefore, we have

1. $r! \cdot C(n, r) = P(n, r)$

We can divide both sides of Equation 1 by $r!$ to get the formula for finding $C(n, r)$:

$$C(n, r) = \binom{n}{r} = \frac{P(n, r)}{r!} = \frac{n!}{r!(n - r)!}$$

Finding $C(n, r)$

The number of combinations of n things r at a time is given by

$$C(n, r) = \frac{n!}{r!(n - r)!}$$

EXAMPLE 7 Compute **a.** $C(8, 5)$, **b.** $\binom{7}{2}$, **c.** $C(n, n)$, and **d.** $C(n, 0)$.

Solution **a.**
$$C(8, 5) = \frac{8!}{5!(8 - 5)!} = \frac{8 \cdot 7 \cdot 6 \cdot 5!}{5! \cdot 3!} = 8 \cdot 7 = 56$$

b.
$$\binom{7}{2} = \frac{7!}{2!(7 - 2)!} = \frac{7 \cdot 6 \cdot 5!}{2 \cdot 1 \cdot 5!} = 21$$

c. $C(n, n) = \dfrac{n!}{n!(n-n)!}$

$= \dfrac{n!}{n!(0!)}$

$= \dfrac{n!}{n!(1)}$

$= 1$

d. $C(n, 0) = \dfrac{n!}{0!(n-0)!}$

$= \dfrac{n!}{0! \cdot n!}$

$= \dfrac{1}{0!}$

$= \dfrac{1}{1}$

$= 1$

The symbol $C(n, 0)$ indicates that we choose 0 things from the available n things. ■

Self Check Compute **a.** $C(9, 6)$ and **b.** $C(10, 10)$.

Answers **a.** 84, **b.** 1

Parts **c** and **d** of Example 7 establish the following formulas.

Finding $C(n, n)$ and $C(n, 0)$

The number of combinations of n things n at a time is 1. The number of combinations of n things 0 at a time is 1.

$C(n, n) = 1$ and $C(n, 0) = 1$

EXAMPLE 8 If 15 students want to pick a committee of 4 students to plan a party, how many different committees are possible?

Solution Since the ordering of people on each possible committee is unimportant, we find the number of combinations of 15 people 4 at a time:

$$C(15, 4) = \frac{15!}{4!(15-4)!}$$

$$= \frac{15 \cdot 14 \cdot 13 \cdot 12 \cdot 11!}{4 \cdot 3 \cdot 2 \cdot 1 \cdot 11!}$$

$$= \frac{15 \cdot 14 \cdot 13 \cdot 12}{4 \cdot 3 \cdot 2 \cdot 1}$$

$$= 1{,}365$$

There are 1,365 possible committees. ■

Self Check In how many ways can 20 students pick a committee of 5 students to plan a party?

Answer 15,504

PERSPECTIVE

Gambling is an occasional diversion for some and an obsession for others. Whether it is horse racing, slot machines, or state lotteries, the lure of instant riches is hard to resist.

Many states conduct lotteries, and the systems vary. One scheme is typical: For \$1, you have two chances to match 6 numbers chosen from 55 numbers and win a grand prize of about \$5 million. How likely are you to win? Is it worth \$1 to play the game?

To match 6 numbers chosen from 55, you must choose the one winning combination out of $C(55, 6)$ possibilities:

$$C(n, r) = \frac{n!}{r!(n - r)!}$$

$$\begin{aligned} C(55, 6) &= \frac{55!}{6!(55 - 6)!} \\ &= \frac{55 \cdot 54 \cdot 53 \cdot 52 \cdot 51 \cdot 50}{6 \cdot 5 \cdot 4 \cdot 3 \cdot 2} \\ &= 28{,}989{,}675 \end{aligned}$$

In this game, you have two chances in about 29 million of winning \$5 million. Over the long haul, you will win $\frac{2}{29,000,000}$ of the time, so each ticket is worth $\frac{2}{29,000,000}$ of \$5,000,000, or about 35¢, if you don't have to share the prize with another winner. For every dollar spent to play the game, you can expect to throw away 65¢. This state lottery is a poor bet. Casinos pay better than 50¢ on the dollar, with some slot machines returning 90¢. "You can't win if you're not in!" is the claim of the lottery promoters. A better claim would be "You won't regret if you don't bet!"

EXAMPLE 9 A committee in Congress consists of ten Democrats and eight Republicans. In how many ways can a subcommittee be chosen if it is to contain five Democrats and four Republicans?

Solution There are $C(10, 5)$ ways of choosing the 5 Democrats and $C(8, 4)$ ways of choosing the 4 Republicans. By the multiplication principle for events, there are $C(10, 5) \cdot C(8, 4)$ ways of choosing the subcommittee:

$$\begin{aligned} C(10, 5) \cdot C(8, 4) &= \frac{10!}{5!(10 - 5)!} \cdot \frac{8!}{4!(8 - 4)!} \\ &= \frac{10 \cdot 9 \cdot 8 \cdot 7 \cdot 6 \cdot 5!}{120 \cdot 5!} \cdot \frac{8 \cdot 7 \cdot 6 \cdot 5 \cdot 4!}{24 \cdot 4!} \\ &= \frac{10 \cdot 9 \cdot 8 \cdot 7 \cdot 6}{120} \cdot \frac{8 \cdot 7 \cdot 6 \cdot 5}{24} \\ &= 17{,}640 \end{aligned}$$

There are 17,640 possible subcommittees. ■

Self Check See Example 9. In how many ways can a subcommittee be chosen if it is to contain four members from each party?

Answer 14,700

ALTERNATIVE FORM OF THE BINOMIAL THEOREM

We have seen that the expansion of $(x + y)^3$ is

$$(x + y)^3 = 1x^3 + 3x^2y + 3xy^2 + 1y^3$$

and that

$$\binom{3}{0} = 1, \binom{3}{1} = 3, \binom{3}{2} = 3, \text{ and } \binom{3}{3} = 1$$

Putting these facts together gives the following way of writing the expansion of $(x + y)^3$:

$$(x + y)^3 = \binom{3}{0}x^3 + \binom{3}{1}x^2y + \binom{3}{2}xy^2 + \binom{3}{3}y^3$$

Likewise, we have

$$(x + y)^4 = \binom{4}{0}x^4 + \binom{4}{1}x^3y + \binom{4}{2}x^2y^2 + \binom{4}{3}xy^3 + \binom{4}{4}y^4$$

The generalization of this idea allows us to state the binomial theorem in an alternative form using combinatorial notation.

The Binomial Theorem

If n is any positive integer, then

$$(a + b)^n = \binom{n}{0}a^n + \binom{n}{1}a^{n-1}b + \binom{n}{2}a^{n-2}b^2 + \cdots + \binom{n}{r}a^{n-r}b^r + \cdots + \binom{n}{n}b^n$$

EXAMPLE 10 Use the alternative form of the binomial theorem to expand $(x + y)^6$.

Solution

$$\begin{aligned}(x + y)^6 &= \binom{6}{0}x^6 + \binom{6}{1}x^5y + \binom{6}{2}x^4y^2 + \binom{6}{3}x^3y^3 + \binom{6}{4}x^2y^4 \\ &\quad + \binom{6}{5}xy^5 + \binom{6}{6}y^6 \\ &= x^6 + 6x^5y + 15x^4y^2 + 20x^3y^3 + 15x^2y^4 + 6xy^5 + y^6\end{aligned}$$

■

Self Check Use the alternative form of the binomial theorem to expand $(a + b)^2$.

Answer $a^2 + 2ab + b^2$

EXAMPLE 11 Use the alternative form of the binomial theorem to expand $(2x - y)^3$.

Solution

$$\begin{aligned}(2x - y)^3 &= [2x + (-y)]^3 \\ &= \binom{3}{0}(2x)^3 + \binom{3}{1}(2x)^2(-y) + \binom{3}{2}(2x)(-y)^2 + \binom{3}{3}(-y)^3 \\ &= 1(2x)^3 + 3(4x^2)(-y) + 3(2x)(y^2) + (-y)^3 \\ &= 8x^3 - 12x^2y + 6xy^2 - y^3\end{aligned}$$

■

Self Check Use the alternative form of the binomial theorem to expand $(3a + b)^3$.

Answer $27a^3 + 27a^2b + 9ab^2 + b^3$

Orals

1. If there are 3 books and 5 records, in how many ways can you pick 1 book and 1 record?
2. In how many ways can 5 soldiers stand in line?
3. Find $P(3, 1)$
4. Find $P(3, 3)$
5. Find $C(3, 0)$
6. $C(3, 3)$

EXERCISE 11.6

REVIEW *Find each value of x.*

1. $|2x - 3| = 9$

2. $2x^2 - x = 15$

3. $\frac{3}{x - 5} = \frac{8}{x}$

4. $\frac{3}{x} = \frac{x - 2}{8}$

VOCABULARY AND CONCEPTS *Fill in each blank to make a true statement.*

5. If an event E_1 can be done in p ways and, after it occurs, a second event E_2 can be done in q ways, the event E_1 followed by E_2 can be done in ____ ways.

6. A ________ is an arrangement of objects.

7. The symbol _____ means the number of permutations of n things taken r at a time.

8. The formula for the number of permutations of n things taken r at a time is __________.

9. $P(n, n) =$ __

10. $P(n, 0) =$ __

11. The symbol $C(n, r)$ or ____ means the number of ________ of n things taken r at a time.

12. The formula for the number of combinations of n things taken r at a time is __________.

13. $C(n, n) =$ __

14. $C(n, 0) =$ __

PRACTICE *In Exercises 15–36, evaluate each permutation or combination.*

15. $P(3, 3)$

16. $P(4, 4)$

17. $P(5, 3)$

18. $P(3, 2)$

19. $P(2, 2) \cdot P(3, 3)$

20. $P(3, 2) \cdot P(3, 3)$

21. $\frac{P(5, 3)}{P(4, 2)}$

22. $\frac{P(6, 2)}{P(5, 4)}$

23. $\frac{P(6, 2) \cdot P(7, 3)}{P(5, 1)}$

24. $\frac{P(8, 3)}{P(5, 3) \cdot P(4, 3)}$

25. $C(5, 3)$

26. $C(5, 4)$

27. $\binom{6}{3}$

28. $\binom{6}{4}$

29. $\binom{5}{4}\binom{5}{3}$

30. $\binom{6}{5}\binom{6}{4}$

31. $\frac{C(38, 37)}{C(19, 18)}$

32. $\frac{C(25, 23)}{C(40, 39)}$

33. $C(12, 0)C(12, 12)$

34. $\frac{C(8, 0)}{C(8, 1)}$

35. $C(n, 2)$

36. $C(n, 3)$

In Exercises 37–42, use the alternative form of the binomial theorem to expand each expression.

37. $(x + y)^4$

38. $(x - y)^2$

39. $(2x + y)^3$

40. $(2x + 1)^4$

41. $(3x - 2)^4$

42. $(3 - x^2)^3$

In Exercises 43–46, find the indicated term of the binomial expansion.

43. $(x - 5y)^5$; fourth term

44. $(2x - y)^5$; third term

45. $(x^2 - y^3)^4$; second term

46. $(x^3 - y^2)^4$; fourth term

APPLICATIONS

47. Arranging an evening Kristy plans to go to dinner and see a movie. In how many ways can she arrange her evening if she has a choice of five movies and seven restaurants?

48. Travel choices Paula has five ways to travel from New York to Chicago, three ways to travel from Chicago to Denver, and four ways to travel from Denver to Los Angeles. How many choices are available to Paula if she travels from New York to Los Angeles?

49. Making license plates How many six-digit license plates can be manufactured? Note that there are ten choices—0, 1, 2, 3, 4, 5, 6, 7, 8, 9—for each digit.

50. Making license plates How many six-digit license plates can be manufactured if no digit can be repeated?

51. Making license plates How many six-digit license plates can be manufactured if no license can begin with 0 and if no digit can be repeated?

52. Making license plates How many license plates can be manufactured with two letters followed by four digits?

53. Phone numbers How many seven-digit phone numbers are available in area code 815 if no phone number can begin with 0 or 1?

54. Phone numbers How many ten-digit phone numbers are available if area codes of 000 and 911 cannot be used and if no local number can begin with 0 or 1?

55. Lining up In how many ways can six people be placed in a line?

56. Arranging books In how many ways can seven books be placed on a shelf?

57. Arranging books In how many ways can four novels and five biographies be arranged on a shelf if the novels are placed on the left?

58. Making a ballot In how many ways can six candidates for mayor and four candidates for the county board be arranged on a ballot if all of the candidates for mayor must be placed on top?

59. Combination locks How many permutations does a combination lock have if each combination has three numbers, no two numbers of any combination are equal, and the lock has 25 numbers?

60. Combination locks How many permutations does a combination lock have if each combination has three numbers, no two numbers of any combination are equal, and the lock has 50 numbers?

61. Arranging appointments The receptionist at a dental office has only three appointment times available before Tuesday, and ten patients have toothaches. In how many ways can the receptionist fill those appointments?

62. **Computers** In many computers, a *word* consists of 32 *bits*—a string of thirty-two 1's and 0's. How many different words are possible?

63. **Palindromes** A palindrome is any word, such as *madam* or *radar*, that reads the same backward and forward. How many five-digit numerical palindromes (such as 13531) are there? (*Hint:* A leading 0 would be dropped.)

64. **Call letters** The call letters of U.S. commercial radio stations have 3 or 4 letters, and the first is always a W or a K. How many radio stations could this system support?

65. **Planning a picnic** A class of 14 students wants to pick a committee of 3 students to plan a picnic. How many committees are possible?

66. **Choosing books** Jeff must read 3 books from a reading list of 15 books. How many choices does he have?

67. **Forming committees** The number of three-person committees that can be formed from a group of persons is ten. How many persons are in the group?

68. **Forming committees** The number of three-person committees that can be formed from a group of persons is 20. How many persons are in the group?

69. **Winning a lottery** In one state lottery, anyone who picks the correct 6 numbers (in any order) wins. With the numbers 0 through 99 available, how many choices are possible?

70. **Taking a test** The instructions on a test read: "Answer any ten of the following fifteen questions. Then choose one of the remaining questions for homework, and turn in its solution tomorrow." In how many ways can the questions be chosen?

71. **Forming a committee** In how many ways can we select a committee of two men and two women from a group containing three men and four women?

72. **Forming a committee** In how many ways can we select a committee of three men and two women from a group containing five men and three women?

73. **Choosing clothes** In how many ways can we select 2 shirts and 3 neckties from a group of 12 shirts and 10 neckties?

74. **Choosing clothes** In how many ways can we select five dresses and two coats from a wardrobe containing nine dresses and three coats?

WRITING

75. State the multiplication principle for events.

76. Explain why *permutation lock* would be a better name for a combination lock.

SOMETHING TO THINK ABOUT

77. How many ways could five people stand in line if two people insist on standing together?

78. How many ways could five people stand in line if two people refuse to stand next to each other?

PROJECTS

PROJECT 1 Baytown is building an auditorium. The city council has already decided on the layout shown in Illustration 1. Each of the sections A, B, C, D, E is to be 60 feet in length from front to back. The aisle widths cannot be changed due to fire regulations. The one thing left to decide is how many rows of seats to put in each section. Based on the following information regarding each section of the auditorium, help the council decide on a final plan.

(continued)

PROJECTS *(continued)*

Sections A and C each have four seats in the front row, five seats in the second row, six seats in the third row and so on, adding one seat per row as we count from front to back.

Section B has eight seats in the front row and adds one seat per row as we count from front to back.

Sections D and E each have 28 seats in the front row and add two seats per row as we count from front to back.

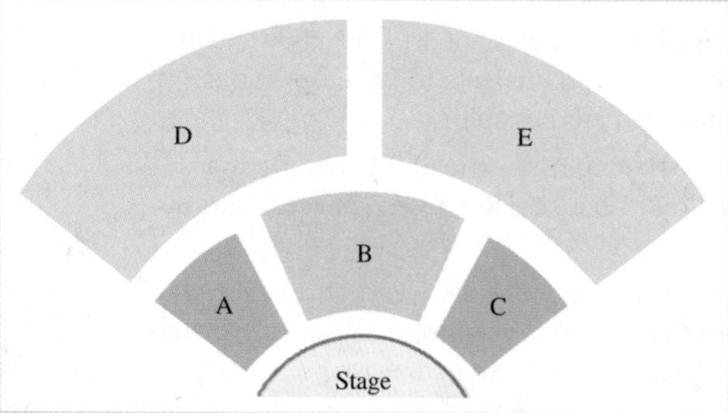

ILLUSTRATION 1

a. One plan calls for a distance of 36 inches, front to back, for each row of seats. Another plan allows for 40 inches (an extra four inches of legroom) for each row. How many seats will the auditorium have under each of these plans?

b. Another plan calls for the higher-priced seats (Sections A, B, and C) to have the extra room afforded by 40-inch rows, but for Sections D and E to have enough rows to make sure that the auditorium holds at least 2,700 seats. Determine how many rows Sections D and E would have to contain for this to work. (This answer should be an integer.) How much space (to the nearest tenth of an inch) would be allotted for each row in Sections D and E?

PROJECT 2

Pascal's triangle contains a wealth of interesting patterns. You have seen two in Exercises 50 and 51 of Section 11.1. Here are a few more.

a. Find the hockey-stick pattern in the numbers in Illustration 2. What would be the missing number in the rightmost hockey stick? Does this pattern work for larger hockey sticks? Experiment.

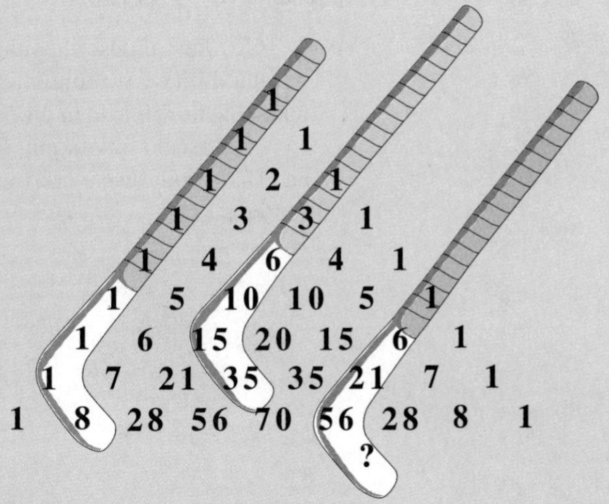

ILLUSTRATION 2

b. In Illustration 3, find the pattern in the sums of increasingly larger portions of Pascal's triangle. Find the sum of all of the numbers up to and including the row that begins 1 10 45 . . . .

1	1 1 1	1 1 1 1 2 1	1 1 1 1 2 1 1 3 3 1	1 1 1 1 2 1 1 3 3 1 1 4 6 4 1
=1	=3	=7	=?	=?

ILLUSTRATION 3

c. In Illustration 4, find the pattern in the sums of the squares of the numbers in each row of the triangle. What is the sum of the squares of the numbers in the row that begins 1 10 45 . . . ? (*Hint:* Calculate $P(2, 1)$, $P(4, 2)$, $P(6, 3)$, ... Do these numbers appear elsewhere in the triangle?)

$$1^2 = 1$$
$$1^2 + 1^2 = 2$$
$$1^2 + 2^2 + 1^2 = 6$$
$$1^2 + 3^2 + 3^2 + 1^2 = 20$$
$$1^2 + 4^2 + 6^2 + 4^2 + 1^2 = 70$$
$$1^2 + 5^2 + 10^2 + 10^2 + 5^2 + 1^2 = ?$$
$$1^2 + 6^2 + 15^2 + 20^2 + 15^2 + 6^2 + 1^2 = ?$$

ILLUSTRATION 4

(continued)

PROJECTS *(continued)*

d. In 1653, Pascal described the triangle in *Treatise on the Arithmetic Triangle*, writing, "I have left out many more properties than I have included. It is remarkable how fertile in properties this triangle is. *Everyone can try his hand.*" Accept Pascal's invitation. Find some of the triangle's patterns for yourself and share your discoveries with your class. Illustration 5 is an idea to get you started.

1
1 1
1 2 1
1 3 3 1
1 4 6 4 1
1 5 10 10 5 1
1 6 15 20 15 6 1
1 7 21 35 35 21 7 1
1 8 28 56 70 56 28 8 1

ILLUSTRATION 5

CHAPTER SUMMARY

CONCEPTS

REVIEW EXERCISES

SECTION 11.1 The Binomial Theorem

The symbol $n!$ (**n factorial**) is defined as

$$n! = n(n-1)(n-2)\cdots 2\cdot 1$$

where n is a natural number.

$0! = 1$

$n(n-1)! = n!$ (n is a natural number)

The binomial theorem:

$$(a+b)^n = a^n + \frac{n!}{1!(n-1)!}a^{n-1}b + \frac{n!}{2!(n-2)!}a^{n-2}b^2 + \cdots + b^n$$

1. Evaluate each expression.

a. $(4!)(3!)$

b. $\frac{5!}{3!}$

c. $\frac{6!}{2!(6-2)!}$

d. $\frac{12!}{3!(12-3)!}$

e. $(n-n)!$

f. $\frac{8!}{7!}$

2. Use the binomial theorem to find each expansion.

a. $(x+y)^5$

b. $(x-y)^4$

c. $(4x-y)^3$

d. $(x+4y)^3$

SECTION 11.2 The nth Term of a Binomial Expansion

The binomial theorem can be used to find the nth term of a binomial expansion.

3. Find the specified term in each expansion.

a. $(x + y)^4$; third term

b. $(x - y)^5$; fourth term

c. $(3x - 4y)^3$; second term

d. $(4x + 3y)^4$; third term

SECTION 11.3 Arithmetic Sequences

An **arithmetic sequence** is a sequence of the form

$$a, a + d, a + 2d, \ldots, a + (n - 1)d$$

where a is the first term, $a + (n - 1)d$ is the nth term, and d is the common difference.

4. Find the eighth term of an arithmetic sequence whose first term is 7 and whose common difference is 5.

5. Write the first five terms of the arithmetic sequence whose ninth term is 242 and whose seventh term is 212.

If numbers are inserted between two given numbers a and b to form an arithmetic sequence, the inserted numbers are **arithmetic means** between a and b.

6. Find two arithmetic means between 8 and 25.

The sum of the first n terms of an arithmetic sequence is given by

$$S_n = \frac{n(a + l)}{2} \quad \text{with} \quad l = a + (n - 1)d$$

where a is the first term, l is the last (or nth) term, and n is the number of terms in the sequence.

7. Find the sum of the first 20 terms of the sequence 11, 18, 25, ...

8. Find the sum of the first ten terms of the sequence 9, $6\frac{1}{2}$, 4, ...

$$\sum_{k=1}^{n} f(k) = f(1) + f(2) + \cdots + f(n)$$

9. Find each sum.

a. $\sum_{k=4}^{6} \frac{1}{2}k$

b. $\sum_{k=2}^{5} 7k^2$

c. $\sum_{k=1}^{4} (3k - 4)$

d. $\sum_{k=10}^{10} 36k$

SECTION 11.4 Geometric Sequences

A **geometric sequence** is a sequence of the form

$$a, ar, ar^2, ar^3, \ldots, ar^{n-1}$$

where a is the first term, ar^{n-1} is the nth term, and r is the common ratio.

If numbers are inserted between a and b to form a geometric sequence, the inserted numbers are **geometric means** between a and b.

The sum of the first n terms of a geometric sequence is given by

$$S_n = \frac{a - ar^n}{1 - r} \quad (r \neq 1)$$

where S_n is the sum, a is the first term, r is the common ratio, and n is the number of terms in the sequence.

10. Write the first five terms of the geometric sequence whose fourth term is 3 and whose fifth term is $\frac{3}{2}$.

11. Find the sixth term of a geometric sequence with a first term of $\frac{1}{8}$ and a common ratio of 2.

12. Find two geometric means between -6 and 384.

13. Find the sum of the first seven terms of the sequence 162, 54, 18, ...

14. Find the sum of the first eight terms of the sequence $\frac{1}{8}, -\frac{1}{4}, \frac{1}{2}, \ldots$.

SECTION 11.5 Infinite Geometric Sequences

If r is the common ratio of an infinite geometric sequence, and if $|r| < 1$, the sum of the terms of the infinite geometric sequence is given by

$$S = \frac{a}{1 - r}$$

where a is the first term and r is the common ratio.

15. Find the sum of the infinite geometric sequence 25, 20, 16, ...

16. Change the decimal $0.\overline{05}$ to a common fraction.

SECTION 11.6 Permutations and Combinations

The multiplication principle for events:
If E_1 and E_2 are two events, and if E_1 can be done in a_1 ways and E_2 can be done in a_2 ways, then the event "E_1 followed by E_2" can be done in $a_1 \cdot a_2$ ways.

17. If there are 17 flights from New York to Chicago, and 8 flights from Chicago to San Francisco, in how many different ways could a passenger plan a trip from New York to San Francisco?

Formulas for permutations:

$$P(n, r) = \frac{n!}{(n-r)!}$$

$$P(n, n) = n! \quad \text{and}$$

$$P(n, 0) = 1$$

18. Evaluate each expression.

a. $P(7, 7)$ **b.** $P(7, 0)$

c. $P(8, 6)$ **d.** $\dfrac{P(9, 6)}{P(10, 7)}$

Formulas for combinations:

$$C(n, r) = \binom{n}{r} = \frac{n!}{r!(n-r)!}$$

$$C(n, n) = \binom{n}{n} = 1$$

$$C(n, 0) = \binom{n}{0} = 1$$

19. Evaluate each expression.

a. $C(7, 7)$ **b.** $C(7, 0)$

c. $\dbinom{8}{6}$ **d.** $\dbinom{9}{6}$

e. $C(6, 3) \cdot C(7, 3)$ **f.** $\dfrac{C(7, 3)}{C(6, 3)}$

20. Car depreciation A $5,000 car depreciates at the rate of 20% of the previous year's value. How much is the car worth after 5 years?

21. Stock appreciation The value of Mia's stock portfolio is expected to appreciate at the rate of 18% per year. How much will the portfolio be worth in 10 years if its current value is $25,700?

22. Planting corn A farmer planted 300 acres in corn this year. He intends to plant an additional 75 acres in corn in each successive year until he has 1,200 acres in corn. In how many years will that be?

23. Falling object If an object is in free fall, the sequence 16, 48, 80, . . . represents the distance in feet that object falls during the first second, during the second second, during the third second, and so on. How far will the object fall during the first ten seconds?

24. Lining up In how many ways can five people be arranged in a line?

25. Lining up In how many ways can three men and five women be arranged in a line if the women are placed ahead of the men?

26. Choosing people In how many ways can we pick three people from a group of ten?

27. Forming committees In how many ways can we pick a committee of two Democrats and two Republicans from a group containing five Democrats and six Republicans?

Chapter Test

1. Evaluate $\frac{7!}{4!}$

2. Evaluate 0!

3. Find the second term in the expansion of $(x - y)^5$.

4. Find the third term in the expansion of $(x + 2y)^4$.

5. Find the tenth term of an arithmetic sequence with the first three terms of 3, 10, and 17.

6. Find the sum of the first 12 terms of the sequence $-2, 3, 8, \ldots$.

7. Find two arithmetic means between 2 and 98.

8. Evaluate $\sum_{k=1}^{3} (2k - 3)$.

9. Find the seventh term of the geometric sequence whose first three terms are $-\frac{1}{9}$, $-\frac{1}{3}$, and -1.

10. Find the sum of the first six terms of the sequence $\frac{1}{27}, \frac{1}{9}, \frac{1}{3}, \ldots$.

11. Find two geometric means between 3 and 648.

12. Find the sum of all of the terms of the infinite geometric sequence 9, 3, 1, ...

In Problems 13–20, find the value of each expression.

13. $P(5, 4)$

14. $P(8, 8)$

15. $C(6, 4)$

16. $C(8, 3)$

17. $C(6, 0) \cdot P(6, 5)$

18. $P(8, 7) \cdot C(8, 7)$

19. $\frac{P(6, 4)}{C(6, 4)}$

20. $\frac{C(9, 6)}{P(6, 4)}$

21. Choosing people In how many ways can we pick three people from a group of seven?

22. Choosing committees From a group of five men and four women, how many three-person committees can be chosen that will include two women?

Cumulative Review Exercises

1. Use graphing to solve $\begin{cases} 2x + y = 5 \\ x - 2y = 0 \end{cases}$.

2. Use substitution to solve $\begin{cases} 3x + y = 4 \\ 2x - 3y = -1 \end{cases}$.

3. Use addition to solve $\begin{cases} x + 2y = -2 \\ 2x - y = 6 \end{cases}$.

4. Use any method to solve $\begin{cases} \frac{x}{10} + \frac{y}{5} = \frac{1}{2} \\ \frac{x}{2} - \frac{y}{5} = \frac{13}{10} \end{cases}$.

5. Evaluate $\begin{vmatrix} 3 & -2 \\ 1 & -1 \end{vmatrix}$.

6. Use Cramer's rule and solve for y only: $\begin{cases} 4x - 3y = -1 \\ 3x + 4y = -7 \end{cases}$

7. Solve $\begin{cases} x + y + z = 1 \\ 2x - y - z = -4. \\ x - 2y + z = 4 \end{cases}$

8. Solve for z only: $\begin{cases} x + 2y + 3z = 6 \\ 3x + 2y + z = 6. \\ 2x + 3y + z = 6 \end{cases}$

9. Solve $\begin{cases} 3x - 2y < 6 \\ y < -x + 2 \end{cases}$.

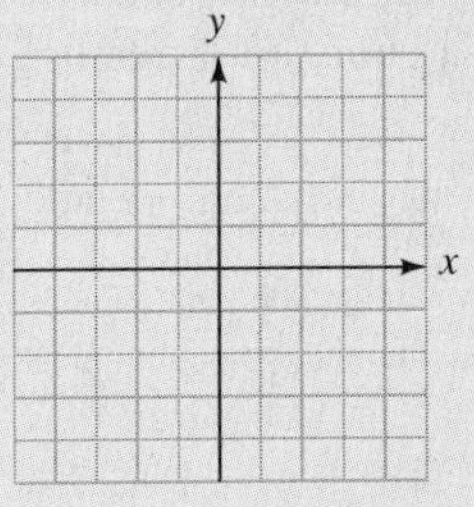

10. Solve $\begin{cases} y < x + 2 \\ 3x + y \le 6 \end{cases}$.

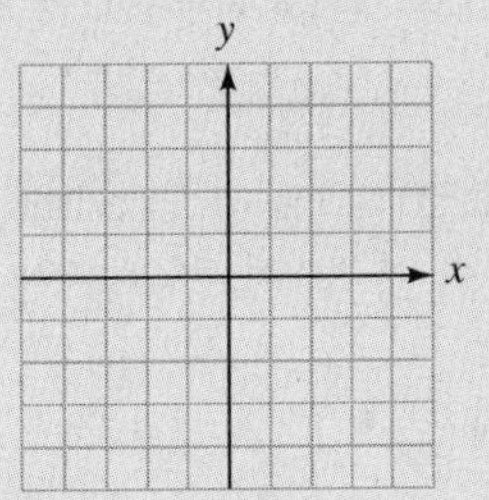

11. Graph $y = \left(\frac{1}{2}\right)^x$.

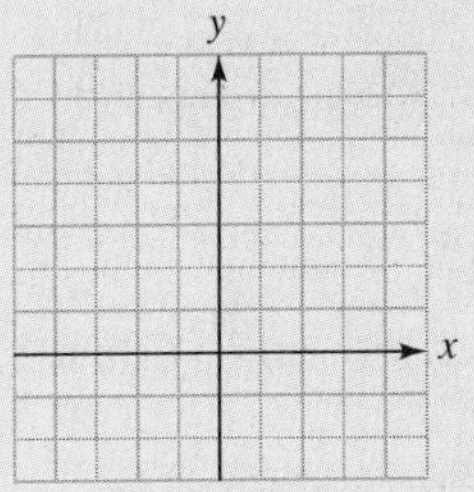

12. Write $y = \log_2 x$ as an exponential equation.

In Exercises 13–16, find x.

13. $\log_x 25 = 2$

14. $\log_5 125 = x$

15. $\log_3 x = -3$

16. $\log_5 x = 0$

17. Find the inverse of $y = \log_2 x$.

18. If $\log_{10} 10^x = y$, then y equals what quantity?

In Exercises 19–22, $\log 7 = 0.8451$ *and* $\log 14 = 1.1461$. *Evaluate each expression without using a calculator or tables.*

19. $\log 98$

20. $\log 2$

21. $\log 49$

22. $\log \frac{7}{5}$ (*Hint:* $\log 10 = 1$.)

23. Solve $2^{x+5} = 3^x$.

24. Solve $\log 5 + \log x - \log 4 = 1$.

In Exercises 25–26, use a calculator.

25. Boat depreciation How much will a $9,000 boat be worth after 9 years if it depreciates 12% per year?

26. Find $\log_6 8$.

27. Evaluate $\frac{6!7!}{5!}$.

28. Use the binomial theorem to expand $(3a - b)^4$.

29. Find the seventh term in the expansion of $(2x - y)^8$.

30. Find the 20th term of an arithmetic sequence with a first term of -11 and a common difference of 6.

31. Find the sum of the first 20 terms of an arithmetic sequence with a first term of 6 and a common difference of 3.

32. Insert two arithmetic means between -3 and 30.

33. Evaluate $\sum_{k=1}^{3} 3k^2$.

34. Evaluate $\sum_{k=3}^{5} (2k + 1)$.

35. Find the seventh term of a geometric sequence with a first term of $\frac{1}{27}$ and a common ratio of 3.

36. Find the sum of the first ten terms of the sequence $\frac{1}{64}, \frac{1}{32}, \frac{1}{16}, \ldots$

37. Insert two geometric means between -3 and 192.

38. Find the sum of all the terms of the sequence 9, 3, 1, ...

39. Evaluate $P(9, 3)$.

40. Evaluate $C(7, 4)$.

41. Evaluate $\frac{C(8, 4)C(8, 0)}{P(6, 2)}$.

42. If $n > 1$, which is smaller: $P(n, n)$ or $C(n, n)$?

43. Lining up In how many ways can seven people stand in a line?

44. Forming a committee In how many ways can a committee of three people be chosen from a group containing nine people?

APPENDIX I
SYMMETRIES OF GRAPHS

There are several ways that a graph can exhibit symmetry about the coordinate axes and the origin. It is often easier to draw graphs of equations if we first find the x- and y-intercepts and find any of the following symmetries of the graph:

1. **y-axis symmetry**: If the point $(-x, y)$ lies on a graph whenever the point (x, y) does, as in Figure I-1(a), we say that the graph is **symmetric about the y-axis**.
2. **Symmetry about the origin**: If the point $(-x, -y)$ lies on the graph whenever the point (x, y) does, as in Figure I-1(b), we say that the graph is **symmetric about the origin**.
3. **x-axis symmetry**: If the point $(x, -y)$ lies on the graph whenever the point (x, y) does, as in Figure I-1(c), we say that the graph is **symmetric about the x-axis**.

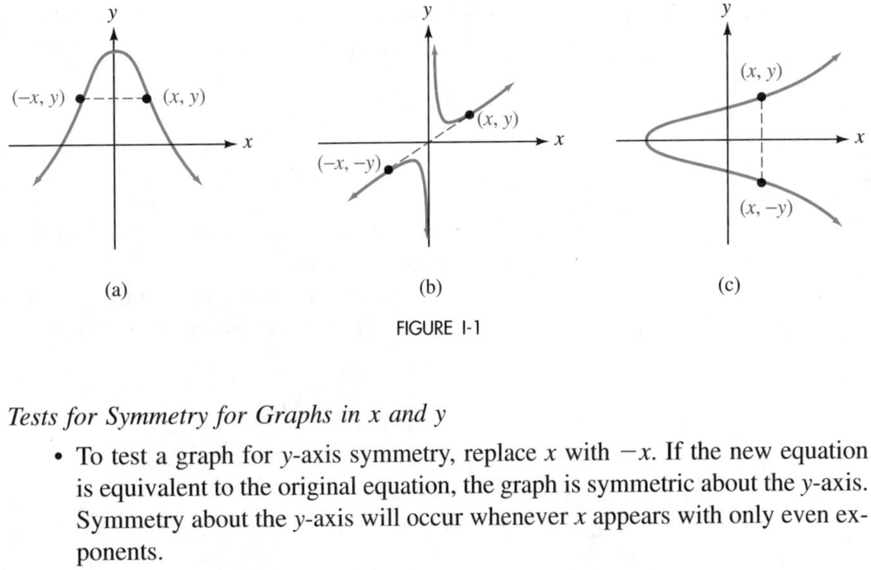

FIGURE I-1

Tests for Symmetry for Graphs in x and y

- To test a graph for y-axis symmetry, replace x with $-x$. If the new equation is equivalent to the original equation, the graph is symmetric about the y-axis. Symmetry about the y-axis will occur whenever x appears with only even exponents.

- To test a graph for symmetry about the origin, replace x with $-x$ and y with $-y$. If the resulting equation is equivalent to the original equation, the graph is symmetric about the origin.
- To test a graph for x-axis symmetry, replace y with $-y$. If the resulting equation is equivalent to the original equation, the graph is symmetric about the x-axis. The only function that is symmetric about the x-axis is $f(x) = 0$.

EXAMPLE 1 Find the intercepts and the symmetries of the graph of $y = f(x) = x^3 - 9x$. Then graph the function.

Solution **x-intercepts**: To find the x-intercepts, we let $y = 0$ and solve for x:

$y = x^3 - 9x$	
$0 = x^3 - 9x$	Substitute 0 for y.
$0 = x(x^2 - 9)$	Factor out x.
$0 = x(x + 3)(x - 3)$	Factor $x^2 - 9$.
$x = 0$ or $x + 3 = 0$ or $x - 3 = 0$	Set each factor equal to 0.
$x = -3$ $\quad$ $x = 3$	

Since the x-coordinates of the x-intercepts are 0, -3, and 3, the graph intersects the x-axis at (0, 0), $(-3, 0)$, and (3, 0).

y-intercepts: To find the y-intercepts, we let $x = 0$ and solve for y.

$y = x^3 - 9x$	
$y = 0^3 - 9(0)$	Substitute 0 for x.
$y = 0$	

Since the y-coordinate of the y-intercept is 0, the graph intersects the y-axis at (0, 0).

Symmetry: We test for symmetry about the y-axis by replacing x with $-x$, simplifying, and comparing the result to the original equation.

1.	$y = x^3 - 9x$	The original equation.
	$y = (-x)^3 - 9(-x)$	Replace x with $-x$.
2.	$y = -x^3 + 9x$	Simplify.

Because Equation 2 is not equivalent to Equation 1, the graph is not symmetric about the y-axis.

We test for symmetry about the origin by replacing x and y with $-x$ and $-y$, respectively, and comparing the result to the original equation.

1.	$y = x^3 - 9x$	The original equation.
	$-y = (-x)^3 - 9(-x)$	Replace x with $-x$, and y with $-y$.
	$-y = -x^3 + 9x$	Simplify.
3.	$y = x^3 - 9x$	Multiply both sides by -1 to solve for y.

Because Equation 3 is equivalent to Equation 1, the graph is symmetric about the origin. Because the equation is the equation of a nonzero function, there is no symmetry about the x-axis.

To graph the equation, we plot the x-intercepts of $(-3, 0)$, $(0, 0)$, and $(3, 0)$ and the y-intercept of $(0, 0)$. We also plot other points for positive values of x and use the symmetry about the origin to draw the rest of the graph, as in Figure I-2(a). (Note that the scale on the x-axis is different from the scale on the y-axis.)

If we graph the equation with a graphing calculator, with window settings of $[-10, 10]$ for x and $[-10, 10]$ for y, we will obtain the graph shown in Figure I-2(b).

From the graph, we can see that the domain is the interval $(-\infty, \infty)$, and the range is the interval $(-\infty, \infty)$.

$y = x^3 - 9x$

x	y
0	0
1	-8
2	-10
3	0
4	28

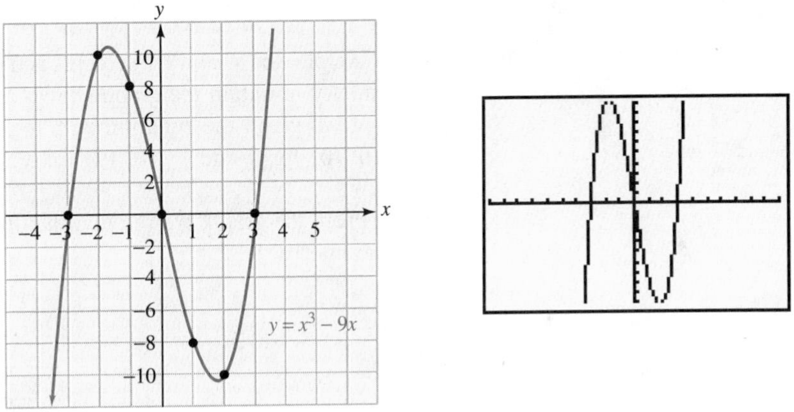

(a) (b)

FIGURE I-2

EXAMPLE 2 Graph the function $y = f(x) = |x| - 2$.

Solution **x-intercepts**: To find the x-intercepts, we let $y = 0$ and solve for x:

$$\begin{aligned} y &= |x| - 2 \\ \mathbf{0} &= |x| - 2 \\ 2 &= |x| \\ x = -2 \quad &\text{or} \quad x = 2 \end{aligned}$$

Since -2 and 2 are solutions, the points $(-2, 0)$ and $(2, 0)$ are the x-intercepts, and the graph passes through $(-2, 0)$ and $(2, 0)$.

y-intercepts: To find the y-intercepts, we let $x = 0$ and solve for y:

$$\begin{aligned} y &= |x| - 2 \\ y &= |\mathbf{0}| - 2 \end{aligned}$$

Since $y = -2$, $(0, -2)$ is the y-intercept, and the graph passes through the point $(0, -2)$.

Symmetry: To test for y-axis symmetry, we replace x with $-x$.

4.	$y = \lvert x \rvert - 2$	The original equation.
	$y = \lvert -x \rvert - 2$	Replace x with $-x$.
5.	$y = \lvert x \rvert - 2$	$\lvert -x \rvert = \lvert x \rvert$.

Since Equation 5 is equivalent to Equation 4, the graph is symmetric about the y-axis. The graph has no other symmetries.

We plot the x- and y-intercepts and several other points (x, y), and use the y-axis symmetry to obtain the graph shown in Figure I-3(a).

If we graph the equation with a graphing calculator, with window settings of $[-10, 10]$ for x and $[-10, 10]$ for y, we will obtain the graph shown in Figure I-3(b).

From the graph, we see that the domain is the interval $(-\infty, \infty)$, and the range is the interval $[-2, \infty)$.

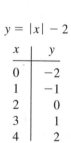

$y = \lvert x \rvert - 2$

x	y
0	−2
1	−1
2	0
3	1
4	2

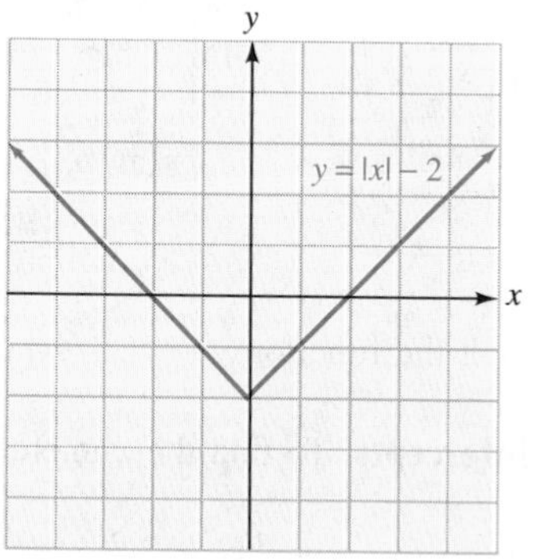

(a)

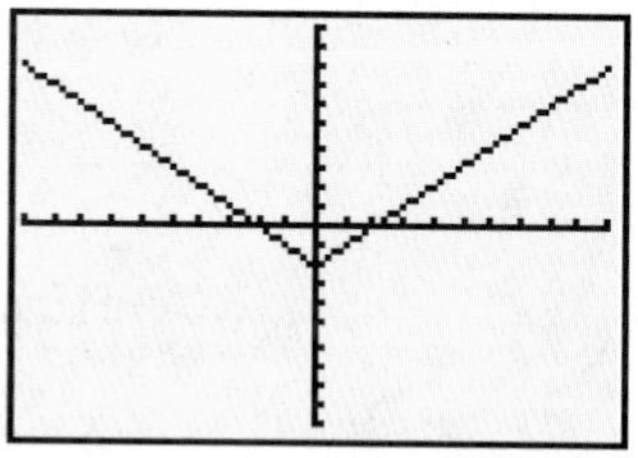

(b)

FIGURE I-3

EXERCISE I.1

In Exercises 1–12, find the symmetries of the graph of each relation. ***Do not draw the graph.***

1. $y = x^2 - 1$
2. $y = x^3$
3. $y = x^5$
4. $y = x^4$
5. $y = -x^2 + 2$
6. $y = x^3 + 1$
7. $y = x^2 - x$
8. $y^2 = x + 7$
9. $y = -\lvert x + 2 \rvert$
10. $y = \lvert x \rvert - 3$
11. $\lvert y \rvert = x$
12. $y = 2\sqrt{x}$

In Exercises 13–24, graph each function and give its domain and range. Check each graph with a graphing calculator.

13. $y = x^4 - 4$

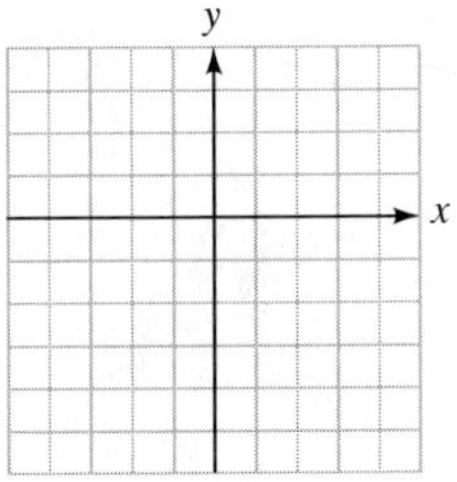

14. $y = \frac{1}{2}x^4 - 1$

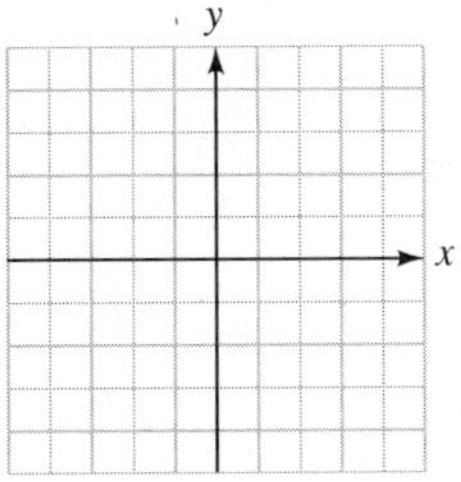

15. $y = -x^3$

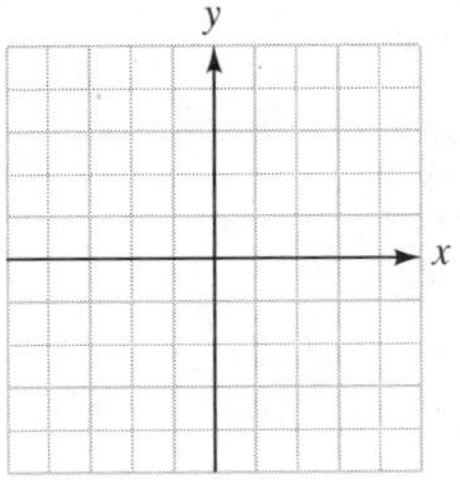

16. $y = x^3 + 2$

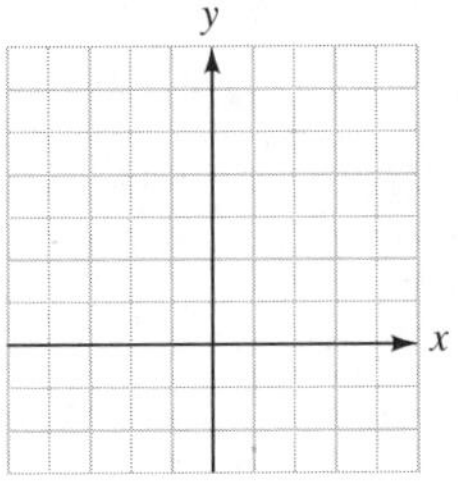

17. $y = x^4 + x^2$

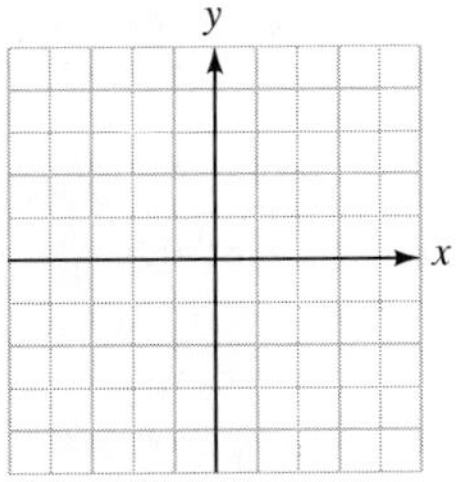

18. $y = 3 - x^4$

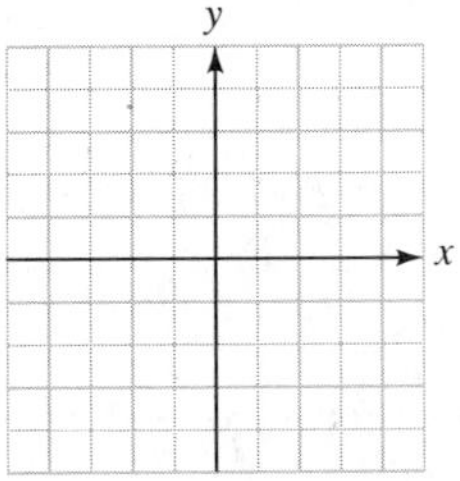

19. $y = x^3 - x$

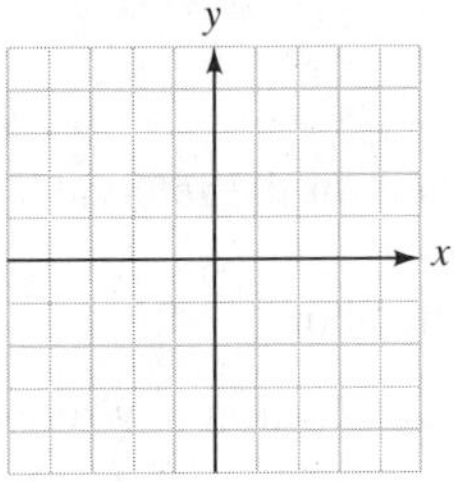

20. $y = x^3 + x$

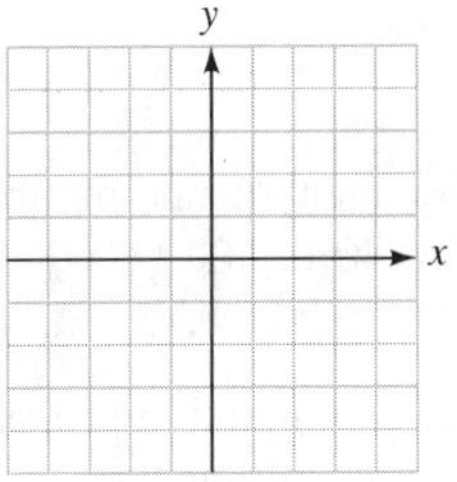

21. $y = \frac{1}{2}|x| - 1$

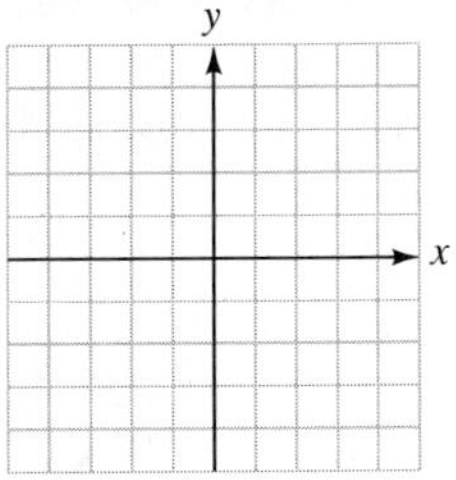

22. $y = -|x| + 1$

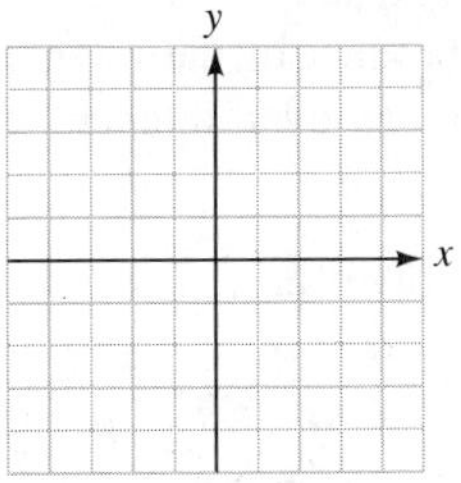

23. $y = -|x + 2|$

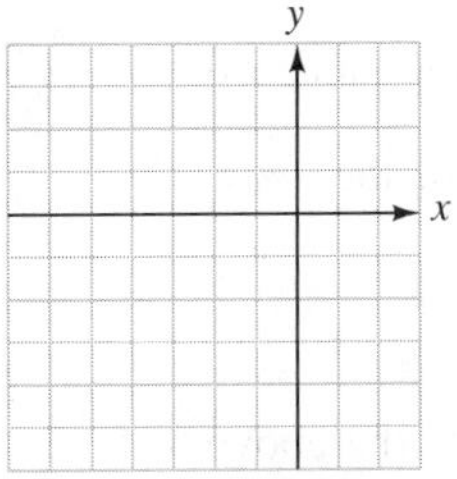

24. $y = |x - 2|$

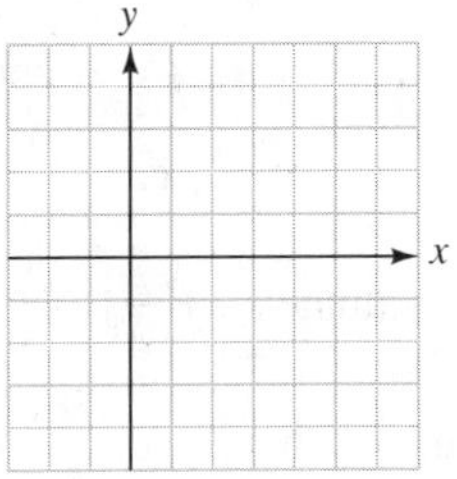

APPENDIX II
SAMPLE FINAL EXAMINATION

1. How many prime numbers are there in the interval from 40 to 50?
a. 3 **b.** 4 **c.** 5 **d.** none of the above

2. The commutative property of multiplication is written symbolically as
a. $ab = ba$ **b.** $(ab)c = a(bc)$ **c.** If $a = b$, then $b = a$ **d.** none of the above

3. If $a = 3$, $b = -2$, and $c = 6$, the value of $\frac{c - ab}{bc}$ is
a. 1 **b.** 0 **c.** -1 **d.** none of the above

4. Evaluate $\frac{1}{2} + \frac{3}{4} \div \frac{5}{6}$.
a. $\frac{3}{2}$ **b.** $\frac{7}{2}$ **c.** $\frac{9}{8}$ **d.** none of the above

5. The expression $\left(\frac{a^2}{a^5}\right)^{-5}$ can be written as
a. 15 **b.** a^{-15} **c.** a^{15} **d.** none of the above

6. Write 0.0000234 in scientific notation.
a. 2.34×10^{-5} **b.** 2.34×10^{5} **c.** $234. \times 10^{-7}$
d. none of the above

7. If $P(x) = 2x^2 - x - 1$, find $P(-1)$.
a. 4 **b.** 0 **c.** 2 **d.** none of the above

8. Simplify: $(3x + 2) - (2x - 1) + (x - 3)$.
a. $2x - 2$ **b.** $2x$ **c.** 2 **d.** none of the above

9. Multiply: $(3x - 2)(2x + 3)$.
a. $6x^2 + 5x - 6$ **b.** $6x^2 - 5x - 6$ **c.** $6x^2 + 5x + 6$
d. none of the above

10. Divide: $2x + 1\overline{)2x^2 - 3x - 2}$.
a. $x + 2$ **b.** $x - 1$ **c.** $x - 2$ **d.** none of the above

11. Solve for x: $5x - 3 = -2x + 10$.
a. $\frac{11}{7}$ **b.** $\frac{13}{3}$ **c.** 1 **d.** none of the above

12. The sum of two consecutive odd integers is 44. The product of the integers is
a. an even integer **b.** 463 **c.** 483 **d.** none of the above

13. Solve for x: $2ax - a = b + x$.
a. $\frac{a + b}{2a}$ **b.** $\frac{a + b}{2a - 1}$ **c.** $a + b - 2a$
d. $-a + b + 1$

14. The sum of the solutions of $|2x + 5| = 13$ is
a. 4 **b.** 8 **c.** 12 **d.** none of the above

15. Solve for x: $-2x + 5 > 9$.
a. $x > 7$ **b.** $x < 7$ **c.** $x < -2$ **d.** $x > -2$

16. Solve for x: $|2x - 5| \leq 9$.
a. $2 \leq x \leq 7$ **b.** $-2 \geq x$ and $x \leq 7$ **c.** $x \leq 7$
d. $-2 \leq x \leq 7$

17. Factor completely: $3ax^2 + 6a^2x$.
a. $3(ax^2 + 2a^2x)$ **b.** $3a(x^2 + 2ax)$ **c.** $3x(ax + 2a^2)$
d. none of the above

18. The sum of the prime factors of $x^4 - 16$ is
a. $2x^2$ **b.** $x^2 + 2x + 4$ **c.** $4x + 4$ **d.** none of the above

19. The sum of the factors of $8x^2 - 2x - 3$ is
a. $6x - 2$ **b.** $8x - 3$ **c.** $6x + 2$ **d.** none of the above

20. One of the factors of $27a^3 + 8$ is
a. $3a - 2$ **b.** $9a^2 + 12a + 4$ **c.** $9a^2 + 6a + 4$
d. none of the above

21. The smallest solution of the equation $6x^2 - 5x - 6 = 0$ is
a. $-\frac{3}{2}$ **b.** $\frac{2}{3}$ **c.** $-\frac{2}{3}$ **d.** none of the above

22. Simplify $\frac{x^2 + 5x + 6}{x^2 - 9}$.
a. $-\frac{2}{3}$ **b.** $\frac{x + 2}{x - 3}$ **c.** $\frac{-5x + 6}{9}$ **d.** none of the above

23. Simplify $\frac{3x + 6}{x + 3} - \frac{x^2 - 4}{x^2 + x - 6}$.
a. $\frac{3(x + 2)(x + 2)}{(x + 3)(x + 3)}$ **b.** 3 **c.** $\frac{1}{3}$ **d.** none of these

24. The numerator of the sum $\frac{y}{x + y} + \frac{x}{x - y}$ is
a. $x^2 + 2xy - y^2$ **b.** $y + x$ **c.** $y - x$ **d.** none of the above

25. Simplify $\frac{\frac{1}{x} + \frac{1}{y}}{\frac{1}{y}}$.
a. $\frac{1}{xy}$ **b.** 1 **c.** $\frac{y + x}{x}$ **d.** none of the above

26. Solve for y: $\frac{2}{y + 1} = \frac{1}{y + 1} - \frac{1}{3}$.
a. 4 **b.** 3 **c.** -4 **d.** none of the above

27. The sum of the x- and y-intercepts of the graph of $2x + 3y = 6$ is
a. $-\frac{2}{3}$ **b.** 0 **c.** 5 **d.** none of the above

28. The slope of the line passing through $(3, -2)$ and $(5, -1)$ is
a. $-\frac{1}{2}$ **b.** 2 **c.** -2 **d.** $\frac{1}{2}$

29. The graphs of the equations $\begin{cases} 2x - 3y = 4 \\ 3x + 2y = 1 \end{cases}$
a. are parallel **b.** are perpendicular **c.** do not intersect **d.** are the same line

30. The equation of the line passing through $(-2, 5)$ and $(6, 7)$ is
a. $y = -\frac{1}{4}x - \frac{11}{2}$ **b.** $y = \frac{1}{4}x + \frac{11}{2}$
c. $y = \frac{1}{4}x - \frac{11}{2}$ **d.** $y = -\frac{1}{4}x + \frac{11}{2}$

31. If $g(x) = x^2 - 3$, then $g(t + 1)$ is
a. $t^2 - 2$ **b.** -2 **c.** $t - 2$ **d.** $t^2 + 2t - 2$

32. Assume that d varies directly with t. Find the constant of variation if $d = 12$ when $t = 3$.
a. 36 **b.** 4 **c.** -36 **d.** none of the above

33. The expression $x^{a/2}x^{a/5}$ can be expressed as
a. $x^{a/7}$ **b.** $x^{7a/10}$ **c.** $x^{a/10}$ **d.** $x^{a/5}$

34. The product of $(x^{1/2} + 2)$ and $(x^{-1/2} - 2)$ is
a. -3 **b.** $-3 + 2x$ **c.** $-3 - 2x^{1/2} + 2x^{-1/2}$
d. none of the above

35. Completely simplify $\sqrt{112a^3}$ $(a \geq 0)$.
a. $a\sqrt{112a}$ **b.** $4a^2\sqrt{7a}$ **c.** $4\sqrt{7a^2}$ **d.** none of the above

36. Simplify and combine terms: $\sqrt{50} - \sqrt{98} + \sqrt{128}$.
a. $-2\sqrt{2} + \sqrt{128}$ **b.** $6\sqrt{2}$ **c.** $20\sqrt{2}$
d. $-4\sqrt{2}$

37. Rationalize the denominator and simplify: $\frac{3}{2 - \sqrt{3}}$.
a. $-6 - \sqrt{3}$ **b.** $6 + 3\sqrt{3}$ **c.** $2(2 + \sqrt{3})$
d. $6(2 + \sqrt{3})$

38. The distance between the points $(-2, 3)$ and $(6, -8)$ is
a. $\sqrt{185}$ **b.** $\sqrt{41}$ **c.** $\sqrt{57}$ **d.** none of the above

39. Solve for x: $\sqrt{x + 7} - 2x = -1$.
a. 2 **b.** $-\frac{3}{4}$ **c.** solutions are extraneous
d. $2, -\frac{3}{4}$

40. The graph of $y > 3x + 2$ contains no points in
a. quadrant I **b.** quadrant II **c.** quadrant III
d. quadrant IV

41. The quadratic formula is
a. $x = \frac{b \pm \sqrt{b^2 - 4ac}}{2a}$ **b.** $x = \frac{-b \pm \sqrt{b^2 - 4ac}}{a}$
c. $x = \frac{-b \pm \sqrt{b - 4ac}}{a}$ **d.** $x = \frac{-b \pm \sqrt{b^2 - 4ac}}{2a}$

42. Write the complex number $(2 + 3i)^2$ in $a + bi$ form.
a. $-5 + 12i$ **b.** -5 **c.** $13 + 12i$ **d.** 13

43. Write the complex number $\frac{i}{3 + i}$ in $a + bi$ form.

a. $\frac{1}{3} + 0i$ **b.** $\frac{1}{10} + \frac{3}{10}i$ **c.** $\frac{1}{8} + \frac{3}{8}i$ **d.** none of the above

44. The vertex of the parabola determined by $y = 2x^2 + 4x - 3$ is at point
a. $(0, -3)$ **b.** $(2, 13)$ **c.** $(-1, -5)$ **d.** $(1, 3)$

45. Solve for x: $\frac{2}{x} < 3$.

a. $0 < x < \frac{2}{3}$ **b.** $x > \frac{2}{3}$ **c.** $x < 0$ or $x > \frac{2}{3}$

d. $x < 0$ and $x > \frac{2}{3}$

46. If $f(x) = 2x^2 + 1$, then $f(3) =$
a. 7 **b.** 19 **c.** 17 **d.** 37

47. The inverse function of $y = 3x + 2$ is

a. $y = \frac{x + 2}{3}$ **b.** $x = 3y - 2$ **c.** $y = \frac{x - 2}{3}$

d. $x = \frac{y - 2}{3}$

48. The equation of a circle with center at $(-2, 4)$ and radius of 4 units is
a. $(x + 2)^2 + (y - 4)^2 = 16$
b. $(x + 2)^2 + (y - 4)^2 = 4$
c. $(x + 2)^2 + (y - 4)^2 = 2$
d. $(x - 2)^2 + (y + 4)^2 = 16$

49. The sum of the solutions of the system
$$\begin{cases} \frac{4}{x} + \frac{2}{y} = 2 \\ \frac{2}{x} - \frac{3}{y} = -1 \end{cases}$$
is

a. 6 **b.** -6 **c.** 2 **d.** -4

50. The y value of the solution of the system
$$\begin{cases} 4x + 6y = 5 \\ 8x - 9y = 3 \end{cases}$$
is

a. $\frac{3}{4}$ **b.** $\frac{1}{3}$ **c.** $\frac{1}{2}$ **d.** $\frac{2}{3}$

51. The value of the determinant $\begin{vmatrix} 2 & -3 \\ 4 & 4 \end{vmatrix}$ is

a. 0 **b.** 20 **c.** -4 **d.** -20

52. The value of z in the system
$$\begin{cases} x + y + z = 4 \\ 2x + y + z = 6 \\ 3x + y + 2z = 8 \end{cases}$$
is

a. 0 **b.** 1 **c.** 2 **d.** 3

53. $\log_a N = x$ means
a. $a^x = N$ **b.** $a^N = x$ **c.** $x^N = a$ **d.** $N^a = x$

54. $\log_2 \frac{1}{32} =$

a. 5 **b.** $\frac{1}{5}$ **c.** -5 **d.** $-\frac{1}{5}$

55. $\log 7 + \log 5 =$

a. $\log 12$ **b.** $\log \frac{7}{12}$ **c.** $\log 2$ **d.** $\log 35$

56. $b^{\log_b x} =$
a. b **b.** x **c.** 10 **d.** 0

57. Solve for y: $\log y + \log (y + 3) = 1$.
a. $-5, 2$ **b.** 2 **c.** 5 **d.** none of the above

58. The coefficient of the third term in the expansion of $(a + b)^6$ is
a. 30 **b.** 2! **c.** 15 **d.** 120

59. Compute $P(7, 3)$.
a. 35 **b.** 210 **c.** 21 **d.** none of the above

60. Compute $C(7, 3)$.
a. 35 **b.** 210 **c.** $7! \cdot 3!$ **d.** none of the above

61. Find the 100th term of 2, 5, 8, 11, ...
a. 301 **b.** 299 **c.** 297 **d.** 295

62. Find the sum of all the terms of $1, \frac{1}{3}, \frac{1}{9}, \frac{1}{27}, \ldots$.

a. 2 **b.** $\frac{2}{3}$ **c.** $\frac{3}{2}$ **d.** $\frac{5}{3}$

APPENDIX III
TABLES

Table A Powers and Roots

n	n^2	$\sqrt{n}$	n^3	$\sqrt[3]{n}$	n	n^2	$\sqrt{n}$	n^3	$\sqrt[3]{n}$
1	1	1.000	1	1.000	51	2,601	7.141	132,651	3.708
2	4	1.414	8	1.260	52	2,704	7.211	140,608	3.733
3	9	1.732	27	1.442	53	2,809	7.280	148,877	3.756
4	16	2.000	64	1.587	54	2,916	7.348	157,464	3.780
5	25	2.236	125	1.710	55	3,025	7.416	166,375	3.803
6	36	2.449	216	1.817	56	3,136	7.483	175,616	3.826
7	49	2.646	343	1.913	57	3,249	7.550	185,193	3.849
8	64	2.828	512	2.000	58	3,364	7.616	195,112	3.871
9	81	3.000	729	2.080	59	3,481	7.681	205,379	3.893
10	100	3.162	1,000	2.154	60	3,600	7.746	216,000	3.915
11	121	3.317	1,331	2.224	61	3,721	7.810	226,981	3.936
12	144	3.464	1,728	2.289	62	3,844	7.874	238,328	3.958
13	169	3.606	2,197	2.351	63	3,969	7.937	250,047	3.979
14	196	3.742	2,744	2.410	64	4,096	8.000	262,144	4.000
15	225	3.873	3,375	2.466	65	4,225	8.062	274,625	4.021
16	256	4.000	4,096	2.520	66	4,356	8.124	287,496	4.041
17	289	4.123	4,913	2.571	67	4,489	8.185	300,763	4.062
18	324	4.243	5,832	2.621	68	4,624	8.246	314,432	4.082
19	361	4.359	6,859	2.668	69	4,761	8.307	328,509	4.102
20	400	4.472	8,000	2.714	70	4,900	8.367	343,000	4.121
21	441	4.583	9,261	2.759	71	5,041	8.426	357,911	4.141
22	484	4.690	10,648	2.802	72	5,184	8.485	373,248	4.160
23	529	4.796	12,167	2.844	73	5,329	8.544	389,017	4.179
24	576	4.899	13,824	2.884	74	5,476	8.602	405,224	4.198
25	625	5.000	15,625	2.924	75	5,625	8.660	421,875	4.217
26	676	5.099	17,576	2.962	76	5,776	8.718	438,976	4.236
27	729	5.196	19,683	3.000	77	5,929	8.775	456,533	4.254
28	784	5.292	21,952	3.037	78	6,084	8.832	474,552	4.273
29	841	5.385	24,389	3.072	79	6,241	8.888	493,039	4.291
30	900	5.477	27,000	3.107	80	6,400	8.944	512,000	4.309
31	961	5.568	29,791	3.141	81	6,561	9.000	531,441	4.327
32	1,024	5.657	32,768	3.175	82	6,724	9.055	551,368	4.344
33	1,089	5.745	35,937	3.208	83	6,889	9.110	571,787	4.362
34	1,156	5.831	39,304	3.240	84	7,056	9.165	592,704	4.380
35	1,225	5.916	42,875	3.271	85	7,225	9.220	614,125	4.397
36	1,296	6.000	46,656	3.302	86	7,396	9.274	636,056	4.414
37	1,369	6.083	50,653	3.332	87	7,569	9.327	658,503	4.431
38	1,444	6.164	54,872	3.362	88	7,744	9.381	681,472	4.448
39	1,521	6.245	59,319	3.391	89	7,921	9.434	704,969	4.465
40	1,600	6.325	64,000	3.420	90	8,100	9.487	729,000	4.481
41	1,681	6.403	68,921	3.448	91	8,281	9.539	753,571	4.498
42	1,764	6.481	74,088	3.476	92	8,464	9.592	778,688	4.514
43	1,849	6.557	79,507	3.503	93	8,649	9.644	804,357	4.531
44	1,936	6.633	85,184	3.530	94	8,836	9.695	830,584	4.547
45	2,025	6.708	91,125	3.557	95	9,025	9.747	857,375	4.563
46	2,116	6.782	97,336	3.583	96	9,216	9.798	884,736	4.579
47	2,209	6.856	103,823	3.609	97	9,409	9.849	912,673	4.595
48	2,304	6.928	110,592	3.634	98	9,604	9.899	941,192	4.610
49	2,401	7.000	117,649	3.659	99	9,801	9.950	970,299	4.626
50	2,500	7.071	125,000	3.684	100	10,000	10.000	1,000,000	4.642

Table B Base-10 Logarithms

N	0	1	2	3	4	5	6	7	8	9
1.0	.0000	.0043	.0086	.0128	.0170	.0212	.0253	.0294	.0334	.0374
1.1	.0414	.0453	.0492	.0531	.0569	.0607	.0645	.0682	.0719	.0755
1.2	.0792	.0828	.0864	.0899	.0934	.0969	.1004	.1038	.1072	.1106
1.3	.1139	.1173	.1206	.1239	.1271	.1303	.1335	.1367	.1399	.1430
1.4	.1461	.1492	.1523	.1553	.1584	.1614	.1644	.1673	.1703	.1732
1.5	.1761	.1790	.1818	.1847	.1875	.1903	.1931	.1959	.1987	.2014
1.6	.2041	.2068	.2095	.2122	.2148	.2175	.2201	.2227	.2253	.2279
1.7	.2304	.2330	.2355	.2380	.2405	.2430	.2455	.2480	.2504	.2529
1.8	.2553	.2577	.2601	.2625	.2648	.2672	.2695	.2718	.2742	.2765
1.9	.2788	.2810	.2833	.2856	.2878	.2900	.2923	.2945	.2967	.2989
2.0	.3010	.3032	.3054	.3075	.3096	.3118	.3139	.3160	.3181	.3201
2.1	.3222	.3243	.3263	.3284	.3304	.3324	.3345	.3365	.3385	.3404
2.2	.3424	.3444	.3464	.3483	.3502	.3522	.3541	.3560	.3579	.3598
2.3	.3617	.3636	.3655	.3674	.3692	.3711	.3729	.3747	.3766	.3784
2.4	.3802	.3820	.3838	.3856	.3874	.3892	.3909	.3927	.3945	.3962
2.5	.3979	.3997	.4014	.4031	.4048	.4065	.4082	.4099	.4116	.4133
2.6	.4150	.4166	.4183	.4200	.4216	.4232	.4249	.4265	.4281	.4298
2.7	.4314	.4330	.4346	.4362	.4378	.4393	.4409	.4425	.4440	.4456
2.8	.4472	.4487	.4502	.4518	.4533	.4548	.4564	.4579	.4594	.4609
2.9	.4624	.4639	.4654	.4669	.4683	.4698	.4713	.4728	.4742	.4757
3.0	.4771	.4786	.4800	.4814	.4829	.4843	.4857	.4871	.4886	.4900
3.1	.4914	.4928	.4942	.4955	.4969	.4983	.4997	.5011	.5024	.5038
3.2	.5051	.5065	.5079	.5092	.5105	.5119	.5132	.5145	.5159	.5172
3.3	.5185	.5198	.5211	.5224	.5237	.5250	.5263	.5276	.5289	.5302
3.4	.5315	.5328	.5340	.5353	.5366	.5378	.5391	.5403	.5416	.5428
3.5	.5441	.5453	.5465	.5478	.5490	.5502	.5514	.5527	.5539	.5551
3.6	.5563	.5575	.5587	.5599	.5611	.5623	.5635	.5647	.5658	.5670
3.7	.5682	.5694	.5705	.5717	.5729	.5740	.5752	.5763	.5775	.5786
3.8	.5798	.5809	.5821	.5832	.5843	.5855	.5866	.5877	.5888	.5899
3.9	.5911	.5922	.5933	.5944	.5955	.5966	.5977	.5988	.5999	.6010
4.0	.6021	.6031	.6042	.6053	.6064	.6075	.6085	.6096	.6107	.6117
4.1	.6128	.6138	.6149	.6160	.6170	.6180	.6191	.6201	.6212	.6222
4.2	.6232	.6243	.6253	.6263	.6274	.6284	.6294	.6304	.6314	.6325
4.3	.6335	.6345	.6355	.6365	.6375	.6385	.6395	.6405	.6415	.6425
4.4	.6435	.6444	.6454	.6464	.6474	.6484	.6493	.6503	.6513	.6522
4.5	.6532	.6542	.6551	.6561	.6571	.6580	.6590	.6599	.6609	.6618
4.6	.6628	.6637	.6646	.6656	.6665	.6675	.6684	.6693	.6702	.6712
4.7	.6721	.6730	.6739	.6749	.6758	.6767	.6776	.6785	.6794	.6803
4.8	.6812	.6821	.6830	.6839	.6848	.6857	.6866	.6875	.6884	.6893
4.9	.6902	.6911	.6920	.6928	.6937	.6946	.6955	.6964	.6972	.6981
5.0	.6990	.6998	.7007	.7016	.7024	.7033	.7042	.7050	.7059	.7067
5.1	.7076	.7084	.7093	.7101	.7110	.7118	.7126	.7135	.7143	.7152
5.2	.7160	.7168	.7177	.7185	.7193	.7202	.7210	.7218	.7226	.7235
5.3	.7243	.7251	.7259	.7267	.7275	.7284	.7292	.7300	.7308	.7316
5.4	.7324	.7332	.7340	.7348	.7356	.7364	.7372	.7380	.7388	.7396

Table B *(continued)*

N	0	1	2	3	4	5	6	7	8	9
5.5	.7404	.7412	.7419	.7427	.7435	.7443	.7451	.7459	.7466	.7474
5.6	.7482	.7490	.7497	.7505	.7513	.7520	.7528	.7536	.7543	.7551
5.7	.7559	.7566	.7574	.7582	.7589	.7597	.7604	.7612	.7619	.7627
5.8	.7634	.7642	.7649	.7657	.7664	.7672	.7679	.7686	.7694	.7701
5.9	.7709	.7716	.7723	.7731	.7738	.7745	.7752	.7760	.7767	.7774
6.0	.7782	.7789	.7796	.7803	.7810	.7818	.7825	.7832	.7839	.7846
6.1	.7853	.7860	.7868	.7875	.7882	.7889	.7896	.7903	.7910	.7917
6.2	.7924	.7931	.7938	.7945	.7952	.7959	.7966	.7973	.7980	.7987
6.3	.7993	.8000	.8007	.8014	.8021	.8028	.8035	.8041	.8048	.8055
6.4	.8062	.8069	.8075	.8082	.8089	.8096	.8102	.8109	.8116	.8122
6.5	.8129	.8136	.8142	.8149	.8156	.8162	.8169	.8176	.8182	.8189
6.6	.8195	.8202	.8209	.8215	.8222	.8228	.8235	.8241	.8248	.8254
6.7	.8261	.8267	.8274	.8280	.8287	.8293	.8299	.8306	.8312	.8319
6.8	.8325	.8331	.8338	.8344	.8351	.8357	.8363	.8370	.8376	.8382
6.9	.8388	.8395	.8401	.8407	.8414	.8420	.8426	.8432	.8439	.8445
7.0	.8451	.8457	.8463	.8470	.8476	.8482	.8488	.8494	.8500	.8506
7.1	.8513	.8519	.8525	.8531	.8537	.8543	.8549	.8555	.8561	.8567
7.2	.8573	.8579	.8585	.8591	.8597	.8603	.8609	.8615	.8621	.8627
7.3	.8633	.8639	.8645	.8651	.8657	.8663	.8669	.8675	.8681	.8686
7.4	.8692	.8698	.8704	.8710	.8716	.8722	.8727	.8733	.8739	.8745
7.5	.8751	.8756	.8762	.8768	.8774	.8779	.8785	.8791	.8797	.8802
7.6	.8808	.8814	.8820	.8825	.8831	.8837	.8842	.8848	.8854	.8859
7.7	.8865	.8871	.8876	.8882	.8887	.8893	.8899	.8904	.8910	.8915
7.8	.8921	.8927	.8932	.8938	.8943	.8949	.8954	.8960	.8965	.8971
7.9	.8976	.8982	.8987	.8993	.8998	.9004	.9009	.9015	.9020	.9025
8.0	.9031	.9036	.9042	.9047	.9053	.9058	.9063	.9069	.9074	.9079
8.1	.9085	.9090	.9096	.9101	.9106	.9112	.9117	.9122	.9128	.9133
8.2	.9138	.9143	.9149	.9154	.9159	.9165	.9170	.9175	.9180	.9186
8.3	.9191	.9196	.9201	.9206	.9212	.9217	.9222	.9227	.9232	.9238
8.4	.9243	.9248	.9253	.9258	.9263	.9269	.9274	.9279	.9284	.9289
8.5	.9294	.9299	.9304	.9309	.9315	.9320	.9325	.9330	.9335	.9340
8.6	.9345	.9350	.9355	.9360	.9365	.9370	.9375	.9380	.9385	.9390
8.7	.9395	.9400	.9405	.9410	.9415	.9420	.9425	.9430	.9435	.9440
8.8	.9445	.9450	.9455	.9460	.9465	.9469	.9474	.9479	.9484	.9489
8.9	.9494	.9499	.9504	.9509	.9513	.9518	.9523	.9528	.9533	.9538
9.0	.9542	.9547	.9552	.9557	.9562	.9566	.9571	.9576	.9581	.9586
9.1	.9590	.9595	.9600	.9605	.9609	.9614	.9619	.9624	.9628	.9633
9.2	.9638	.9643	.9647	.9652	.9657	.9661	.9666	.9671	.9675	.9680
9.3	.9685	.9689	.9694	.9699	.9703	.9708	.9713	.9717	.9722	.9727
9.4	.9731	.9736	.9741	.9745	.9750	.9754	.9759	.9763	.9768	.9773
9.5	.9777	.9782	.9786	.9791	.9795	.9800	.9805	.9809	.9814	.9818
9.6	.9823	.9827	.9832	.9836	.9841	.9845	.9850	.9854	.9859	.9863
9.7	.9868	.9872	.9877	.9881	.9886	.9890	.9894	.9899	.9903	.9908
9.8	.9912	.9917	.9921	.9926	.9930	.9934	.9939	.9943	.9948	.9952
9.9	.9956	.9961	.9965	.9969	.9974	.9978	.9983	.9987	.9991	.9996

Table C Base-*e* Logarithms

N	0	1	2	3	4	5	6	7	8	9
1.0	.0000	.0100	.0198	.0296	.0392	.0488	.0583	.0677	.0770	.0862
1.1	.0953	.1044	.1133	.1222	.1310	.1398	.1484	.1570	.1655	.1740
1.2	.1823	.1906	.1989	.2070	.2151	.2231	.2311	.2390	.2469	.2546
1.3	.2624	.2700	.2776	.2852	.2927	.3001	.3075	.3148	.3221	.3293
1.4	.3365	.3436	.3507	.3577	.3646	.3716	.3784	.3853	.3920	.3988
1.5	.4055	.4121	.4187	.4253	.4318	.4383	.4447	.4511	.4574	.4637
1.6	.4700	.4762	.4824	.4886	.4947	.5008	.5068	.5128	.5188	.5247
1.7	.5306	.5365	.5423	.5481	.5539	.5596	.5653	.5710	.5766	.5822
1.8	.5878	.5933	.5988	.6043	.6098	.6152	.6206	.6259	.6313	.6366
1.9	.6419	.6471	.6523	.6575	.6627	.6678	.6729	.6780	.6831	.6881
2.0	.6931	.6981	.7031	.7080	.7129	.7178	.7227	.7275	.7324	.7372
2.1	.7419	.7467	.7514	.7561	.7608	.7655	.7701	.7747	.7793	.7839
2.2	.7885	.7930	.7975	.8020	.8065	.8109	.8154	.8198	.8242	.8286
2.3	.8329	.8372	.8416	.8459	.8502	.8544	.8587	.8629	.8671	.8713
2.4	.8755	.8796	.8838	.8879	.8920	.8961	.9002	.9042	.9083	.9123
2.5	.9163	.9203	.9243	.9282	.9322	.9361	.9400	.9439	.9478	.9517
2.6	.9555	.9594	.9632	.9670	.9708	.9746	.9783	.9821	.9858	.9895
2.7	.9933	.9969	1.0006	.0043	.0080	.0116	.0152	.0188	.0225	.0260
2.8	1.0296	.0332	.0367	.0403	.0438	.0473	.0508	.0543	.0578	.0613
2.9	.0647	.0682	.0716	.0750	.0784	.0818	.0852	.0886	.0919	.0953
3.0	1.0986	.1019	.1053	.1086	.1119	.1151	.1184	.1217	.1249	.1282
3.1	.1314	.1346	.1378	.1410	.1442	.1474	.1506	.1537	.1569	.1600
3.2	.1632	.1663	.1694	.1725	.1756	.1787	.1817	.1848	.1878	.1909
3.3	.1939	.1969	.2000	.2030	.2060	.2090	.2119	.2149	.2179	.2208
3.4	.2238	.2267	.2296	.2326	.2355	.2384	.2413	.2442	.2470	.2499
3.5	1.2528	.2556	.2585	.2613	.2641	.2669	.2698	.2726	.2754	.2782
3.6	.2809	.2837	.2865	.2892	.2920	.2947	.2975	.3002	.3029	.3056
3.7	.3083	.3110	.3137	.3164	.3191	.3218	.3244	.3271	.3297	.3324
3.8	.3350	.3376	.3403	.3429	.3455	.3481	.3507	.3533	.3558	.3584
3.9	.3610	.3635	.3661	.3686	.3712	.3737	.3762	.3788	.3813	.3838
4.0	1.3863	.3888	.3913	.3938	.3962	.3987	.4012	.4036	.4061	.4085
4.1	.4110	.4134	.4159	.4183	.4207	.4231	.4255	.4279	.4303	.4327
4.2	.4351	.4375	.4398	.4422	.4446	.4469	.4493	.4516	.4540	.4563
4.3	.4586	.4609	.4633	.4656	.4679	.4702	.4725	.4748	.4770	.4793
4.4	.4816	.4839	.4861	.4884	.4907	.4929	.4951	.4974	.4996	.5019
4.5	1.5041	.5063	.5085	.5107	.5129	.5151	.5173	.5195	.5217	.5239
4.6	.5261	.5282	.5304	.5326	.5347	.5369	.5390	.5412	.5433	.5454
4.7	.5476	.5497	.5518	.5539	.5560	.5581	.5602	.5623	.5644	.5665
4.8	.5686	.5707	.5728	.5748	.5769	.5790	.5810	.5831	.5851	.5872
4.9	.5892	.5913	.5933	.5953	.5974	.5994	.6014	.6034	.6054	.6074
5.0	1.6094	.6114	.6134	.6154	.6174	.6194	.6214	.6233	.6253	.6273
5.1	.6292	.6312	.6332	.6351	.6371	.6390	.6409	.6429	.6448	.6467
5.2	.6487	.6506	.6525	.6544	.6563	.6582	.6601	.6620	.6639	.6658
5.3	.6677	.6696	.6715	.6734	.6752	.6771	.6790	.6808	.6827	.6845
5.4	.6864	.6882	.6901	.6919	.6938	.6956	.6974	.6993	.7011	.7029

Use the properties of logarithms and ln 10 ≈ 2.3026 to find logarithms of numbers less than 1 or greater than 10.

Table C (*continued*)

N	0	1	2	3	4	5	6	7	8	9
5.5	1.7047	.7066	.7084	.7102	.7120	.7138	.7156	.7174	.7192	.7210
5.6	.7228	.7246	.7263	.7281	.7299	.7317	.7334	.7352	.7370	.7387
5.7	.7405	.7422	.7440	.7457	.7475	.7492	.7509	.7527	.7544	.7561
5.8	.7579	.7596	.7613	.7630	.7647	.7664	.7681	.7699	.7716	.7733
5.9	.7750	.7766	.7783	.7800	.7817	.7834	.7851	.7867	.7884	.7901
6.0	1.7918	.7934	.7951	.7967	.7984	.8001	.8017	.8034	.8050	.8066
6.1	.8083	.8099	.8116	.8132	.8148	.8165	.8181	.8197	.8213	.8229
6.2	.8245	.8262	.8278	.8294	.8310	.8326	.8342	.8358	.8374	.8390
6.3	.8405	.8421	.8437	.8453	.8469	.8485	.8500	.8516	.8532	.8547
6.4	.8563	.8579	.8594	.8610	.8625	.8641	.8656	.8672	.8687	.8703
6.5	1.8718	.8733	.8749	.8764	.8779	.8795	.8810	.8825	.8840	.8856
6.6	.8871	.8886	.8901	.8916	.8931	.8946	.8961	.8976	.8991	.9006
6.7	.9021	.9036	.9051	.9066	.9081	.9095	.9110	.9125	.9140	.9155
6.8	.9169	.9184	.9199	.9213	.9228	.9242	.9257	.9272	.9286	.9301
6.9	.9315	.9330	.9344	.9359	.9373	.9387	.9402	.9416	.9430	.9445
7.0	1.9459	.9473	.9488	.9502	.9516	.9530	.9544	.9559	.9573	.9587
7.1	.9601	.9615	.9629	.9643	.9657	.9671	.9685	.9699	.9713	.9727
7.2	.9741	.9755	.9769	.9782	.9796	.9810	.9824	.9838	.9851	.9865
7.3	.9879	.9892	.9906	.9920	.9933	.9947	.9961	.9974	.9988	2.0001
7.4	2.0015	.0028	.0042	.0055	.0069	.0082	.0096	.0109	.0122	.0136
7.5	2.0149	.0162	.0176	.0189	.0202	.0215	.0229	.0242	.0255	.0268
7.6	.0281	.0295	.0308	.0321	.0334	.0347	.0360	.0373	.0386	.0399
7.7	.0412	.0425	.0438	.0451	.0464	.0477	.0490	.0503	.0516	.0528
7.8	.0541	.0554	.0567	.0580	.0592	.0605	.0618	.0631	.0643	.0656
7.9	.0669	.0681	.0694	.0707	.0719	.0732	.0744	.0757	.0769	.0782
8.0	2.0794	.0807	.0819	.0832	.0844	.0857	.0869	.0882	.0894	.0906
8.1	.0919	.0931	.0943	.0956	.0968	.0980	.0992	.1005	.1017	.1029
8.2	.1041	.1054	.1066	.1078	.1090	.1102	.1114	.1126	.1138	.1150
8.3	.1163	.1175	.1187	.1199	.1211	.1223	.1235	.1247	.1258	.1270
8.4	.1282	.1294	.1306	.1318	.1330	.1342	.1353	.1365	.1377	.1389
8.5	2.1401	.1412	.1424	.1436	.1448	.1459	.1471	.1483	.1494	.1506
8.6	.1518	.1529	.1541	.1552	.1564	.1576	.1587	.1599	.1610	.1622
8.7	.1633	.1645	.1656	.1668	.1679	.1691	.1702	.1713	.1725	.1736
8.8	.1748	.1759	.1770	.1782	.1793	.1804	.1815	.1827	.1838	.1849
8.9	.1861	.1872	.1883	.1894	.1905	.1917	.1928	.1939	.1950	.1961
9.0	2.1972	.1983	.1994	.2006	.2017	.2028	.2039	.2050	.2061	.2072
9.1	.2083	.2094	.2105	.2116	.2127	.2138	.2148	.2159	.2170	.2181
9.2	.2192	.2203	.2214	.2225	.2235	.2246	.2257	.2268	.2279	.2289
9.3	.2300	.2311	.2322	.2332	.2343	.2354	.2364	.2375	.2386	.2396
9.4	.2407	.2418	.2428	.2439	.2450	.2460	.2471	.2481	.2492	.2502
9.5	2.2513	.2523	.2534	.2544	.2555	.2565	.2576	.2586	.2597	.2607
9.6	.2618	.2628	.2638	.2649	.2659	.2670	.2680	.2690	.2701	.2711
9.7	.2721	.2732	.2742	.2752	.2762	.2773	.2783	.2793	.2803	.2814
9.8	.2824	.2834	.2844	.2854	.2865	.2875	.2885	.2895	.2905	.2915
9.9	.2925	.2935	.2946	.2956	.2966	.2976	.2986	.2996	.3006	.3016

APPENDIX IV
ANSWERS TO SELECTED EXERCISES

Orals (page 11)

5. 2, 3, 5, 7 **6.** 2, 4, 6, 8, 10 **7.** 6 **8.** 10

Exercise 1.1 (page 12)

1. $\frac{3}{4}$ **3.** $\frac{4}{5}$ **5.** $\frac{3}{20}$ **7.** $\frac{14}{9}$ **9.** 1 **11.** $\frac{22}{15}$ **13.** set **15.** even **17.** natural, 1, itself **19.** 0 **21.** rational **23.** $<$ **25.** 1, 2, 9 **27.** $-3, 0, 1, 2, 9$ **29.** $\sqrt{3}$ **31.** 2 **33.** 2 **35.** 9 **37.** 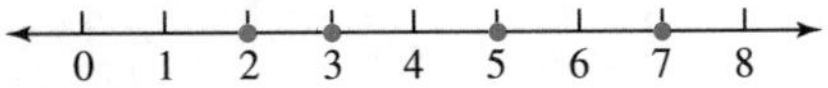

39. 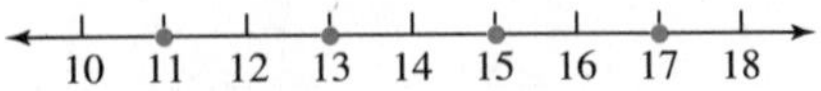

41. 0.875, terminating **43.** $-0.7\overline{3}$, repeating **45.** $<$ **47.** $>$ **49.** $<$ **51.** $>$ **53.** $12 < 19$ **55.** $-5 \geq -6$ **57.** $-3 \leq 5$ **59.** $0 > -10$ **61.** 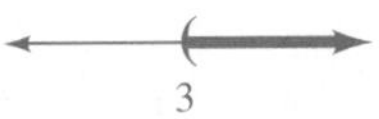

63. 

65. 

67. 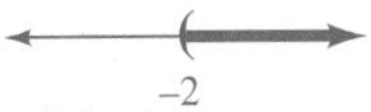

69. 

71. 

73. 20 **75.** -6 **77.** 7 **79.** 20 **81.** 3 or -3 **83.** $x \geq 0$ **89.** 99

Getting Ready (page 14)

1. 9 **2.** 9 **3.** 12 **4.** 12 **5.** 5 **6.** 5 **7.** 3 **8.** 6

Orals (page 26)

1. -2 **2.** -7 **3.** -28 **4.** 28 **5.** -3 **6.** -3

Exercise 1.2 (page 26)

1. 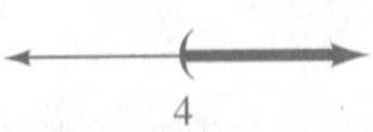

3. 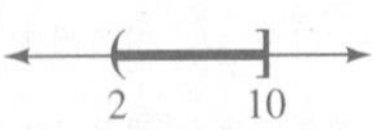

5. \$45.53 **7.** absolute, common **9.** change, add **11.** negative **13.** mean, median, mode **15.** $(a \cdot b) \cdot c = a \cdot (b \cdot c)$ **17.** $a(b + c) = ab + ac$ **19.** 1 **21.** -8 **23.** -5 **25.** -7 **27.** 0 **29.** -12 **31.** 21 **33.** -2 **35.** 4 **37.** $\frac{1}{6}$ **39.** $\frac{11}{10}$ **41.** $-\frac{1}{6}$ **43.** $-\frac{6}{7}$ **45.** -2 **47.** $\frac{24}{25}$ **49.** 23 **51.** 0 **53.** 2 **55.** -13 **57.** 1 **59.** 4 **61.** -1 **63.** -4 **65.** -12 **67.** -20 **69.** 8 **71.** 9 **73.** 12 **75.** -8 **77.** -9 **79.** $-\frac{5}{4}$ **81.** $-\frac{1}{8}$ **83.** 19,900 **85.** 100.4 ft **87.** comm. prop. of add. **89.** distrib. prop. **91.** additive identity prop. **93.** mult. inverse prop. **95.** assoc. prop. of add. **97.** comm. prop. of mult. **99.** assoc. prop. of add. **101.** distrib. prop. **103.** \$62 **105.** $+4°$ **107.** 12° **109.** 6,900 gal **111.** +1,325 m **113.** \$421.88 **115.** \$1,211 **117.** 80 **119.** not really **121.** 30 cm

Getting Ready (page 30)

1. 4 **2.** 27 **3.** −64 **4.** 81 **5.** $\frac{1}{27}$ **6.** $-\frac{16}{625}$

Orals (page 41)

1. 16 **2.** 27 **3.** x^5 **4.** y^7 **5.** 1 **6.** x^6 **7.** a^6b^3 **8.** $\frac{b^2}{a^4}$ **9.** $\frac{1}{25}$ **10.** x^2 **11.** x^3 **12.** $\frac{1}{x^3}$

Exercise 1.3 (page 41)

1. 7 **3.** 1 **5.** base, exponent **7.** x^{m+n} **9.** x^ny^n **11.** 1 **13.** x^{m-n} **15.** $A = s^2$ **17.** $A = \frac{1}{2}bh$ **19.** $A = \pi r^2$ **21.** $V = lwh$ **23.** $V = Bh$ **25.** $V = \frac{1}{3}Bh$ **27.** base is 5, exponent is 3 **29.** base is x, exponent is 5 **31.** base is b, exponent is 6 **33.** base is $-mn^2$, exponent is 3 **35.** 9 **37.** −9 **39.** 9 **41.** $\frac{1}{25}$ **43.** $-\frac{1}{25}$ **45.** $\frac{1}{25}$ **47.** 1 **49.** 1 **51.** $-32x^5$ **53.** $64x^6$ **55.** x^5 **57.** k^7 **59.** x^{10} **61.** p^{10} **63.** a^4b^5 **65.** x^5y^4 **67.** x^{28} **69.** $\frac{1}{b^{72}}$ **71.** $x^{12}y^8$ **73.** $\frac{s^3}{r^9}$ **75.** a^{20} **77.** $-\frac{1}{d^3}$ **79.** $27x^9y^{12}$ **81.** $\frac{1}{729}m^6n^{12}$ **83.** $\frac{a^{15}}{b^{10}}$ **85.** $\frac{a^6}{b^4}$ **87.** a^5 **89.** c^7 **91.** m **93.** a^4 **95.** $3m$ **97.** $\frac{64b^{12}}{27a^9}$ **99.** 1 **101.** $\frac{-b^3}{8a^{21}}$ **103.** $\frac{27}{8a^{18}}$ **105.** $\frac{1}{9x^3}$ **107.** a^{n-1} **109.** b^{3n-9} **111.** $\frac{1}{a^{n+1}}$ **113.** a^{2-n} **115.** 3.462825992 **117.** −244.140625 **127.** 108 **129.** $-\frac{1}{216}$ **131.** $\frac{1}{324}$ **133.** $\frac{27}{8}$ **135.** 15 m^2 **137.** 113 cm^2 **139.** 45 cm^2 **141.** 300 cm^2 **143.** 343 m^3 **145.** 360 ft^3 **147.** 168 ft^3 **149.** 2,714 m^3 **155.** $\frac{7}{12}$

Getting Ready (page 44)

1. 10 **2.** 100 **3.** 1,000 **4.** 10,000 **5.** $\frac{1}{100}$ **6.** $\frac{1}{10,000}$ **7.** 4,000 **8.** $\frac{7}{10,000}$

Orals (page 50)

1. 3.52×10^2 **2.** 5.13×10^3 **3.** 2×10^{-3} **4.** 2.5×10^{-4} **5.** 350 **6.** 4,300 **7.** 0.27 **8.** 0.085

Exercise 1.4 (page 50)

1. 0.75 **3.** $1.\overline{4}$ **5.** 89 **7.** 10^n **9.** left **11.** 3.9×10^3 **13.** 7.8×10^{-3} **15.** -4.5×10^4 **17.** -2.1×10^{-4} **19.** 1.76×10^7 **21.** 9.6×10^{-6} **23.** 3.23×10^7 **25.** 6.0×10^{-4} **27.** 5.27×10^3 **29.** 3.17×10^{-4} **31.** 270 **33.** 0.00323 **35.** 796,000 **37.** 0.00037 **39.** 5.23 **41.** 23,650,000 **43.** 2×10^{11} **45.** 1.44×10^7 **47.** 0.04 **49.** 6,000 **51.** 0.64 **53.** 1.2874×10^{13} **55.** 5.671×10^{10} **57.** 3.6×10^{25} **59.** 1.19×10^8 cm/hr **61.** 1.67248×10^{-18} g **63.** 1.49×10^{10} in. **65.** 5.9×10^4 mi **67.** 3×10^8 **69.** almost 23 years **71.** 2.5×10^{13} mi **75.** 332

Getting Ready (page 53)

1. 2 **2.** 4 **3.** 3 **4.** 6

Orals (page 61)

1. $9x$ **2.** $2s^2$ **3.** no **4.** yes **5.** no **6.** no **7.** 3 **8.** 12 **9.** 5 **10.** 3

Exercise 1.5 (page 61)

1. −64 **3.** 1 **5.** x^8 **7.** $\frac{1}{8x^3}$ **9.** equation **11.** equivalent **13.** $c, \frac{b}{c}$ **15.** like **17.** identity **19.** yes **21.** yes **23.** 2 **25.** 25 **27.** 3 **29.** 28 **31.** $\frac{2}{3}$ **33.** 6 **35.** −8 **37.** $\frac{8}{3}$ **39.** yes, $8x$ **41.** no **43.** yes, $-2x^2$ **45.** no **47.** 3 **49.** 13 **51.** 9 **53.** −4 **55.** −6 **57.** −11 **59.** 13 **61.** −8 **63.** −2 **65.** 24 **67.** 6 **69.** 4 **71.** 3 **73.** 0 **75.** 6 **77.** identity **79.** −6 **81.** impossible **83.** identity **85.** $w = \frac{A}{l}$ **87.** $B = \frac{3V}{h}$ **89.** $t = \frac{I}{pr}$ **91.** $w = \frac{p - 2l}{2}$ **93.** $B = \frac{2A}{h} - b$ **95.** $x = \frac{y - b}{m}$ **97.** $n = \frac{l - a + d}{d}$

99. $l = \frac{a - S + Sr}{r}$ **101.** $l = \frac{2S - na}{n}$ **103.** $m = Fd^2/(GM)$ **105.** 0°, 21.1°, 100° **107.** $n = (C - 6.50)/0.07$; 621, 1,000, 1,692.9 kwh **109.** $R = E/I$; $R = 8$ ohms **111.** $n = 360°/(180° - a)$; 8

Getting Ready (page 65)

1. 20 **2.** 72 **3.** 41 **4.** 7

Orals (page 72)

1. 100 **2.** 200 **3.** \$270 **4.** \$54x **5.** 24 m^2 **6.** $l(l - 5)$ m^2

Exercise 1.6 (page 72)

1. $\frac{256x^{20}}{81}$ **3.** a^{m+1} **5.** $5x + 4$ **7.** $40x$ **9.** 180° **11.** complementary **13.** right **15.** vertex **17.** 32 ft **19.** 7 ft, 15 ft **21.** \$355 **23.** 20% **25.** 233% **27.** 300 shares of BB, 200 shares of SS **29.** 35 \$15 calculators, 50 \$67 calculators **31.** 13 **33.** 60 mi, 120 mi **35.** 12 m by 24 m **37.** 156 ft by 312 ft **39.** 10 ft **41.** 72.5° **43.** 56° **45.** 50 **47.** 60° **49.** 40° **51.** 12 in. **53.** 5 ft **55.** 90 lb **57.** 4 ft **59.** −40°

Getting Ready (page 77)

1. \$40 **2.** \$0.08x **3.** \$6,000 **4.** \$(15,000 − x) **5.** 200 mi **6.** \$20

Orals (page 81)

1. \$90 **2.** \$0.05x **3.** \$(30,000 − x) **4.** \$20x **5.** 0.08 gal **6.** 0.04x gal

Exercise 1.7 (page 82)

1. 1 **3.** 8 **5.** principal, rate, time **7.** value, price, number **9.** \$2,000 at 8%, \$10,000 at 9% **11.** 10.8% **13.** \$8,250 **15.** 25 **17.** $3\frac{1}{3}$ hr **19.** $1\frac{1}{2}$ hr **21.** $1\frac{1}{2}$ hr **23.** 4 hr at each rate **25.** 20 lb of 95¢ candy; 10 lb of \$1.10 candy **27.** 10 oz **29.** 2 gal **31.** 15 **33.** 30 **35.** \$50 **37.** 6 in.

Chapter Summary (page 87)

1. a. 0, 1, 2, 4 **b.** 1, 2, 4 **c.** −4, $-\frac{2}{3}$, 0, 1, 2, 4 **d.** −4, 0, 1, 2, 4 **e.** π **f.** −4, $-\frac{2}{3}$, 0, 1, 2, π, 4 **g.** −4, $-\frac{2}{3}$ **h.** 1, 2, π, 4 **i.** 2 **j.** 4 **k.** −4, 0, 2, 4 **l.** 1 **2.** 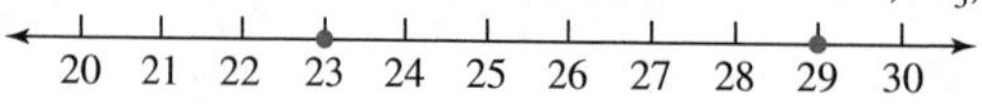

3. 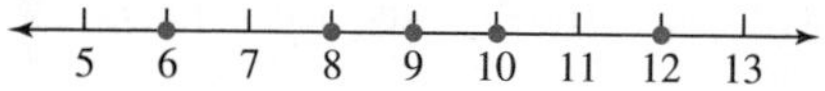

4. a. 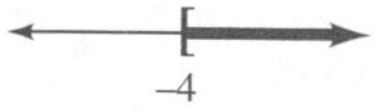

b. 

c. 

d. 

e. 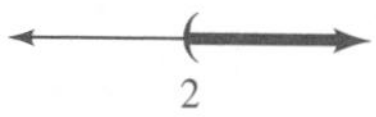

f. 

g. 

5. a. 0 **b.** 1 **c.** 8 **d.** −8 **6. a.** 8 **b.** −9 **c.** −28 **d.** 57 **e.** 3 **f.** −9 **g.** 5 **h.** −2 **i.** −13 **j.** 5 **k.** −13 **l.** −5 **m.** 39 **n.** 53 **o.** 35 **p.** 42 **q.** 4 **r.** 5 **s.** −12 **t.** −24 **u.** −4 **v.** −2 **7. a.** 12 **b.** 27 **c.** 0 **d.** −60 **e.** −1 **f.** 4 **8. a.** 15.4 **b.** 15 **c.** 15 **d.** yes **9. a.** $-\frac{19}{6}$ **b.** $\frac{2}{15}$ **c.** $-\frac{4}{3}$ **d.** $\frac{11}{5}$ **10. a.** distrib. prop. **b.** comm. prop. of add. **c.** assoc. prop. of add. **d.** add. identity prop. **e.** add. inverse prop. **f.** comm. prop. of mult. **g.** assoc. prop. of mult. **h.** mult. identity prop. **i.** mult. inverse prop. **j.** double negative rule **11. a.** 729 **b.** −64 **c.** −64 **d.** $-\frac{1}{625}$ **e.** $-6x^6$ **f.** $-3x^8$ **g.** $\frac{1}{x}$ **h.** x^2 **i.** $27x^6$ **j.** $256x^{16}$ **k.** $-32x^{10}$ **l.** $243x^{15}$ **m.** $\frac{1}{x^{10}}$ **n.** x^{20} **o.** $\frac{x^6}{9}$ **p.** $\frac{16}{x^{16}}$ **q.** x^2 **r.** x^5 **s.** $\frac{1}{a^5}$ **t.** $\frac{1}{a^3}$ **u.** $\frac{1}{y^7}$ **v.** y^9 **w.** $\frac{1}{x}$ **x.** x^3 **12. a.** $9x^4y^6$ **b.** $\frac{1}{81a^{12}b^8}$ **c.** $\frac{64y^9}{27x^6}$ **d.** $\frac{64y^3}{125}$ **13. a.** 1.93×10^{10} **b.** 2.73×10^{-8} **14. a.** 72,000,000 **b.** 0.0000000083 **15. a.** 5 **b.** −9 **c.** 8 **d.** 7 **e.** 19 **f.** 8 **g.** 12 **h.** 5 **16. a.** $r^3 = \frac{3V}{4\pi}$ **b.** $h = \frac{3V}{\pi r^2}$

c. $x = \frac{6v}{ab} - y$ or $x = \frac{6v - aby}{ab}$ **d.** $r = \frac{V}{\pi h^2} + \frac{h}{3}$ **17.** 5 ft from one end **18.** 45 m^2 **19.** $2\frac{2}{3}$ ft
20. \$18,000 at 10%, \$7,000 at 9% **21.** 10 liters **22.** $\frac{1}{3}$ hr

Chapter 1 Test (page 91)

1. 1, 2, 5 **2.** $\sqrt{7}$ **3.** 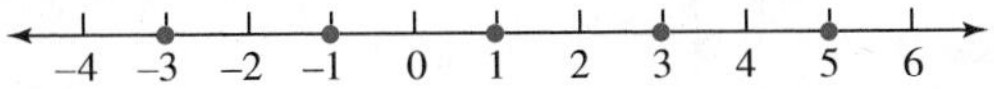

4. 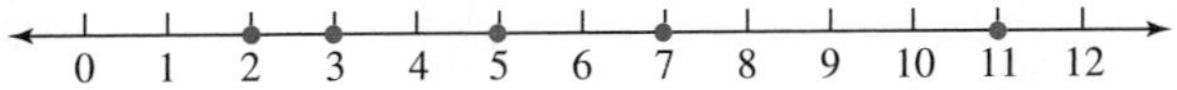

5. 

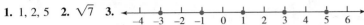

6. 

7. −8

8. 5 **9.** 2 **10.** 20 **11.** −4 **12.** 1 **13.** 0.3 **14.** 0.5 **15.** −6 **16.** −10 **17.** 6 **18.** 1 **19.** comm. prop. of add.
20. distrib. prop. **21.** x^8 **22.** x^6y^9 **23.** $\frac{1}{m^8}$ **24.** $\frac{m^4}{n^{10}}$ **25.** 4.7×10^6 **26.** 2.3×10^{-7} **27.** 653,000 **28.** 0.0245 **29.** −12
30. 6 **31.** $i = \frac{f(P - L)}{s}$ **32.** 36 cm^2 **33.** \$4,000 **34.** 80 liters

Getting Ready (page 95)

1. 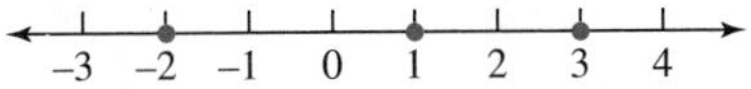

2. 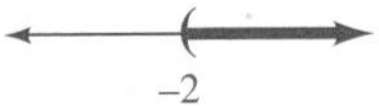

3. 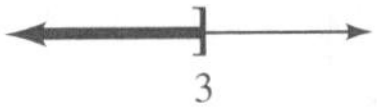

4. 

Orals (page 104)

2. the origin **3.** IV **4.** y-axis

Exercise 2.1 (page 104)

1. 12 **3.** $\frac{1}{2}$ **5.** 7 **7.** ordered pair **9.** origin **11.** rectangular coordinate **13.** no **15.** origin, right, down **17.** II
19. −3, −1, 1, 3 **21–28.** 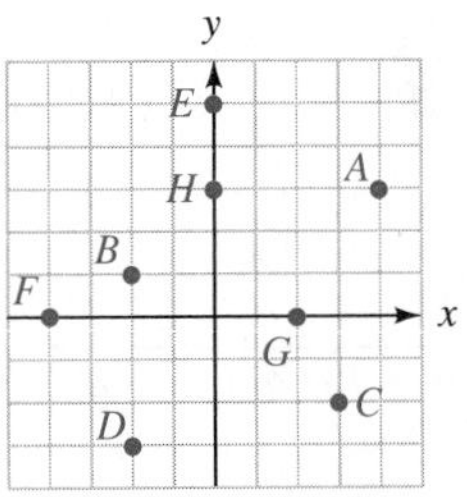

29. (2, 4) **31.** (−2, −1) **33.** (4, 0) **35.** (0, 0)

37. 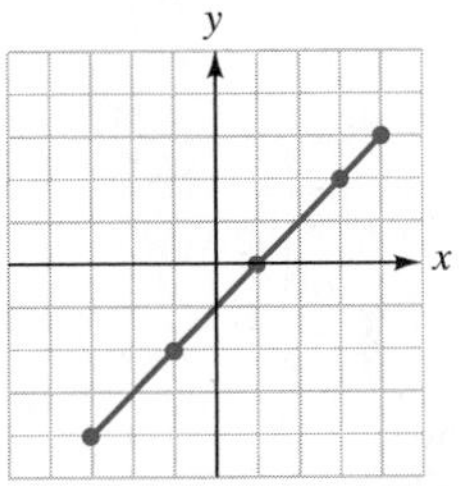

39. a. on the surface **b.** diving **c.** 1,000 ft **41.** 1 **43.** 1 **45.** 1 is running, 2 is stopped
47. 2, 4, 3, 1 **49.** Carbondale (3, J), Chicago (5, B), Rockford (3, A)
51. A(2, 3), B(6, 3), C(4, 0), D(9, 2) **53. a.** \$3 **b.** \$5 **c.** \$8 **d.** \$10

Getting Ready (page 109)

1. −3 **2.** 3 **3.** −7 **4.** −8

Orals (page 118)

1. (3, 0), (0, 3) **2.** (2, 0), (0, 6) **3.** (8, 0), (0, 2) **4.** (4, 0), (0, −3) **5.** vertical **6.** horizontal **7.** (4, 6) **8.** (0, −1)

Exercise 2.2 (page 119)

1. 22 **3.** $\frac{5}{2}$ **5.** 11, 13, 17, 19, 23, 29 **7.** satisfy **9.** the y-intercept **11.** vertical **13.** sub 1 **15.** 5, 4, 2 **17.** −5, −3, 3

19.

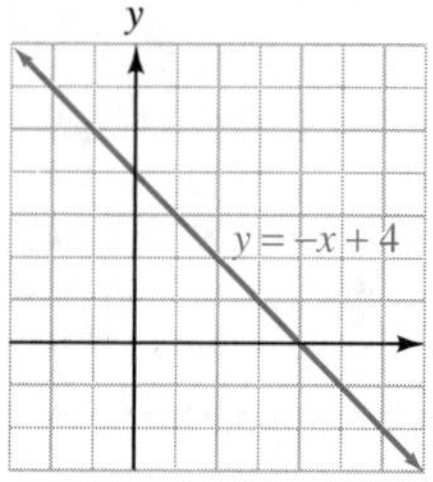

21.

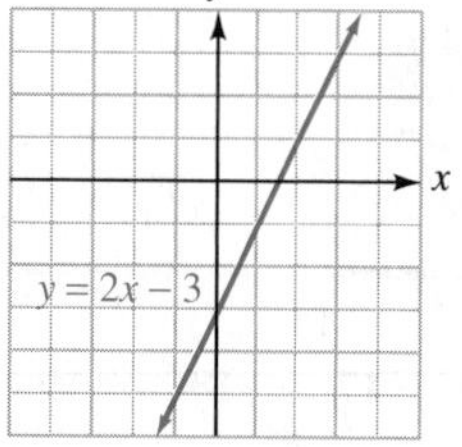

23.

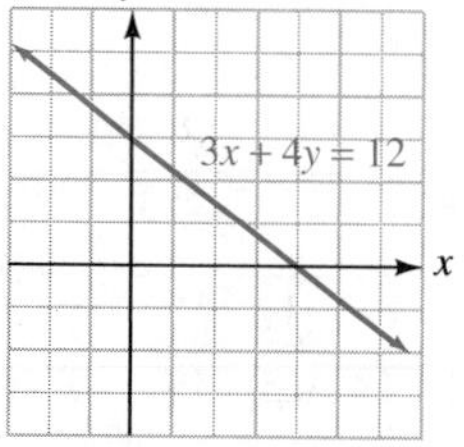

25.

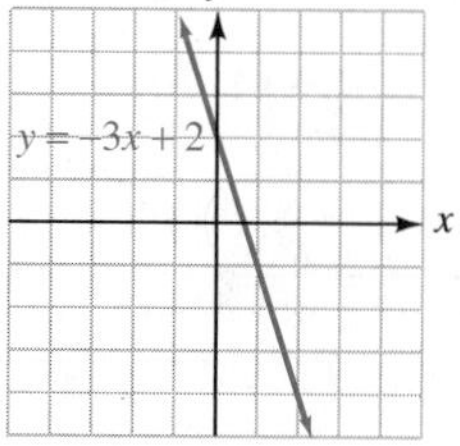

27.

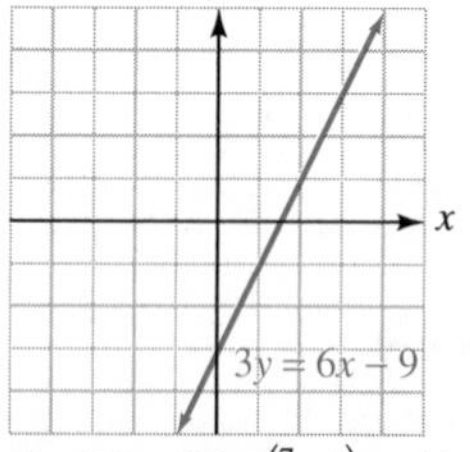

29.

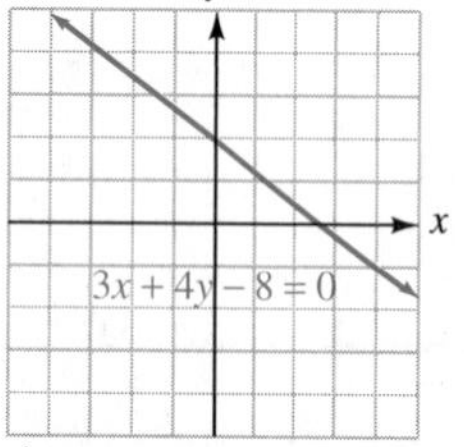

31.

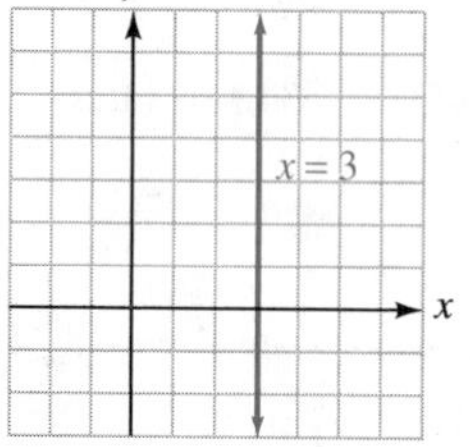

33.

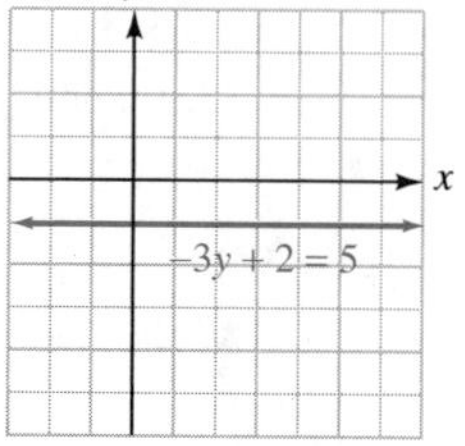

35. (3, 4)

37. (9, 12) **39.** $\left(\frac{7}{2}, 6\right)$ **41.** $\left(\frac{1}{2}, -2\right)$ **43.** (−4, 0) **45.** (4, 1) **47.** 1.22 **49.** 4.67 **51.** \$48 **53.** \$3,000 **55.** \$162,500 **57.** 200 **59.** 12 mi **61. a.** $y = 0.25x + 5$ **b.** 6, 7, 8, 9 **c.** \$10

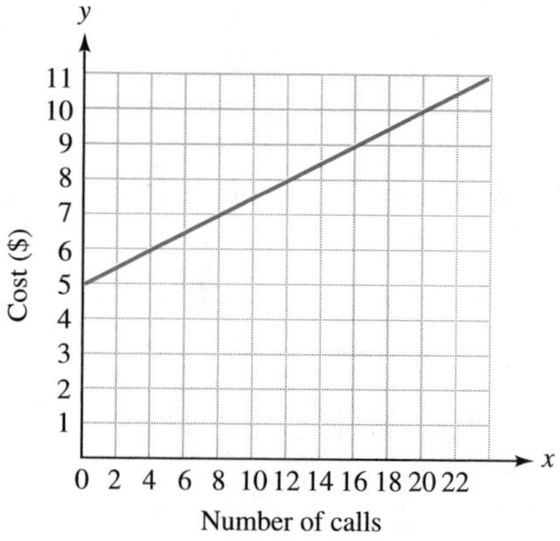

65. $a = 0, b > 0$

Getting Ready (page 123)

1. 1 **2.** −1 **3.** $\frac{13}{14}$ **4.** $\frac{3}{14}$

Orals (page 131)

1. 3 **2.** 2 **3.** yes **4.** 5 **5.** yes

Exercise 2.3 (page 131)

1. x^9y^6 **3.** $\frac{x^{12}}{y^8}$ **5.** 1 **7.** y, x **9.** $\frac{y_2 - y_1}{x_2 - x_1}$ **11.** run **13.** vertical **15.** parallel **17.** 3 **19.** −1 **21.** $-\frac{1}{3}$ **23.** 0 **25.** undefined **27.** −1 **29.** $-\frac{3}{2}$ **31.** $\frac{3}{4}$ **33.** $\frac{1}{2}$ **35.** 0 **37.** negative **39.** positive **41.** undefined **43.** perpendicular **45.** neither **47.** parallel **49.** parallel **51.** perpendicular **53.** neither **55.** not the same line **57.** not the same line **59.** same line **61.** $y = 0, m = 0$ **69.** $\frac{1}{165}$ **71.** $\frac{18}{5}$ **73.** \$20,000 per yr **79.** 4

Getting Ready (page 134)

1. 14 **2.** $-\frac{5}{3}$ **3.** $y = 3x - 4$ **4.** $x = \frac{-By - 3}{A}$

Orals (page 144)

1. $y - 3 = 2(x - 2)$ **2.** $y - 8 = 2(x + 3)$ **3.** $y = -3x + 5$ **4.** $y = -3x - 7$ **5.** parallel **6.** perpendicular

Exercise 2.4 (page 145)

1. 6 **3.** -1 **5.** 20 oz **7.** $y - y_1 = m(x - x_1)$. **9.** $Ax + By = C$ **11.** perpendicular **13.** $5x - y = -7$ **15.** $3x + y = 6$ **17.** $2x - 3y = -11$ **19.** $y = x$ **21.** $y = \frac{7}{3}x - 3$ **23.** $y = -\frac{9}{5}x + \frac{2}{5}$ **25.** $y = 3x + 17$ **27.** $y = -7x + 54$ **29.** $y = -4$ **31.** $y = -\frac{1}{2}x + 11$ **33.** 1, $(0, -1)$

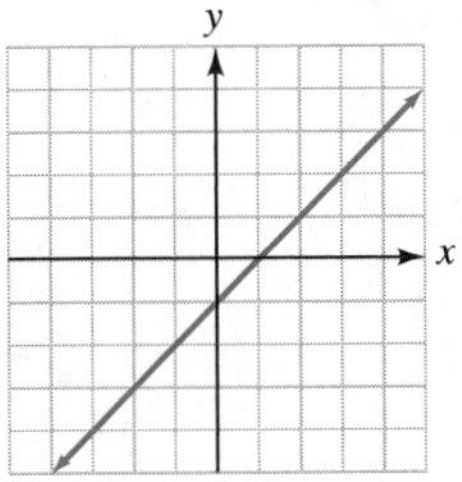

35. $\frac{2}{3}$, $(0, 2)$

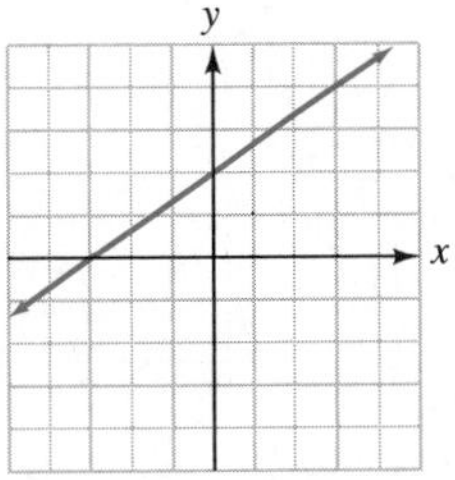

37. $-\frac{2}{3}$, $(0, 6)$

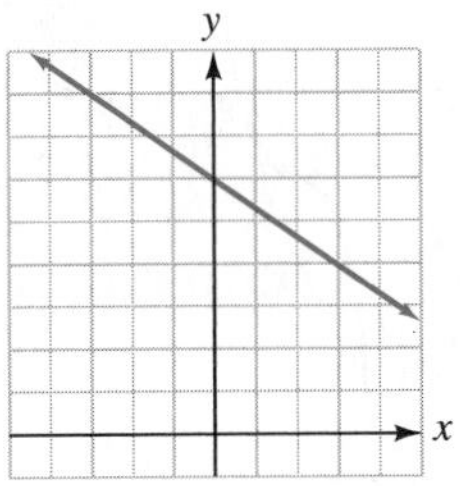

39. $\frac{3}{2}$, $(0, -4)$ **41.** $-\frac{1}{3}$, $\left(0, -\frac{5}{6}\right)$ **43.** $\frac{7}{2}$, $(0, 2)$ **45.** parallel **47.** perpendicular **49.** parallel **51.** perpendicular **53.** perpendicular **55.** perpendicular **57.** $y = 4x$ **59.** $y = 4x - 3$ **61.** $y = \frac{4}{5}x - \frac{26}{5}$ **63.** $y = -\frac{1}{4}x$ **65.** $y = -\frac{1}{4}x + \frac{11}{2}$ **67.** $y = -\frac{5}{4}x + 3$ **69.** perpendicular **71.** parallel **73.** $x = -2$ **75.** $x = 5$ **77.** $y = -\frac{A}{B}x + \frac{C}{B}$ **79.** $y = -2{,}298x + 19{,}984$ **81.** $y = 50{,}000x + 250{,}000$ **83.** $y = -\frac{950}{3}x + 1{,}750$ **85.** \$490 **87.** \$154,000 **89.** \$180 **99.** $a < 0$, $b > 0$

Getting Ready (page 150)

1. 1 **2.** 7 **3.** -20 **4.** $-\frac{11}{4}$

Orals (page 157)

1. yes **2.** no **3.** no **4.** 1 **5.** 3 **6.** -3

Exercise 2.5 (page 157)

1. -2 **3.** 2 **5.** input **7.** x **9.** function, input, output **11.** range **13.** 0 **15.** $mx + b$ **17.** yes **19.** yes **21.** yes **23.** no **25.** 9, -3 **27.** 3, -5 **29.** 22, 2 **31.** 3, 11 **33.** 4, 9 **35.** 7, 26 **37.** 9, 16 **39.** 6, 15 **41.** 4, 4 **43.** 2, 2 **45.** $\frac{1}{5}$, 1 **47.** -2, $\frac{2}{5}$ **49.** $2w$, $2w + 2$ **51.** $3w - 5$, $3w - 2$ **53.** 12 **55.** $2b - 2a$ **57.** $2b$ **59.** 1 **61.** D = $\{-2, 4, 6\}$; R = $\{3, 5, 7\}$ **63.** D = $(-\infty, 4) \cup (4, \infty)$; R = $(-\infty, 0) \cup (0, \infty)$ **65.** not a function **67.** a function; D = $(-\infty, \infty)$; R = $(-\infty, \infty)$

69. D = $(-\infty, \infty)$; R = $(-\infty, \infty)$

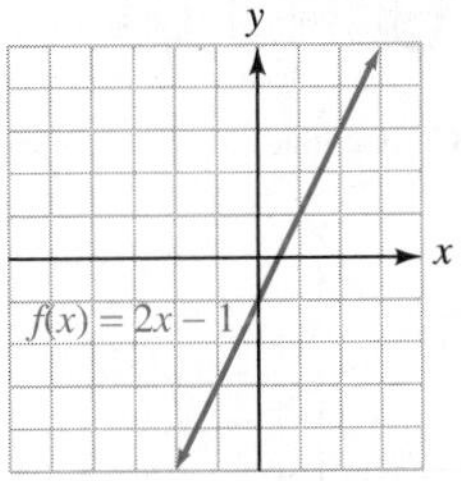

71. D = $(-\infty, \infty)$; R = $(-\infty, \infty)$

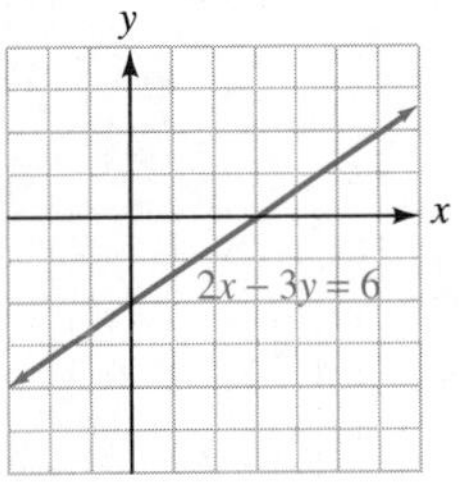

73. no

75. yes **77.** 624 ft **79.** 77° F **81.** 192 **85.** yes

Getting Ready (page 160)

1. 2, $(0, -3)$ **2.** -3, $(0, 4)$ **3.** 6, -9 **4.** 4, $\frac{5}{2}$

Exercise 2.6 (page 171)

1. 41, 43, 47 **3.** $a \cdot b = b \cdot a$ **5.** 1 **7.** squaring **9.** absolute value **11.** horizontal **13.** 2, down **15.** 4, to the left

17.
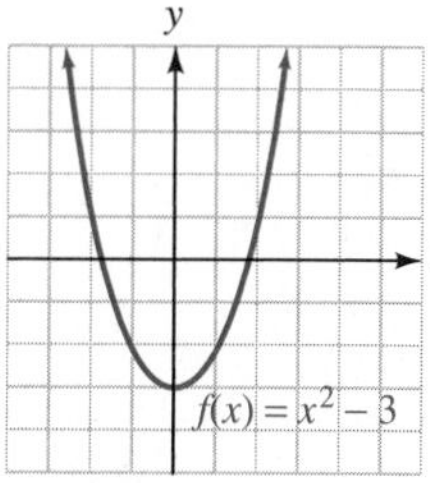

19.
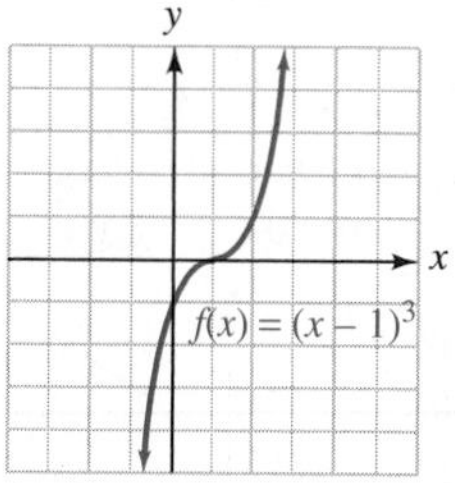

21.
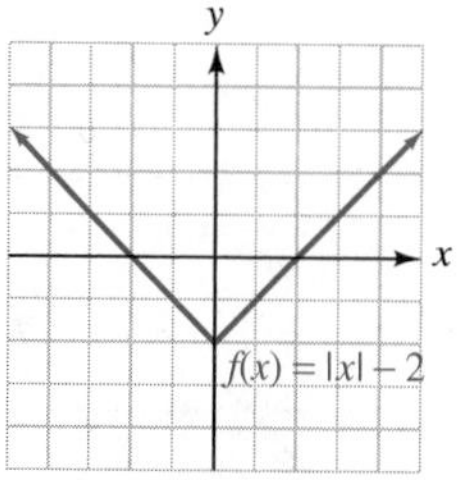

23.
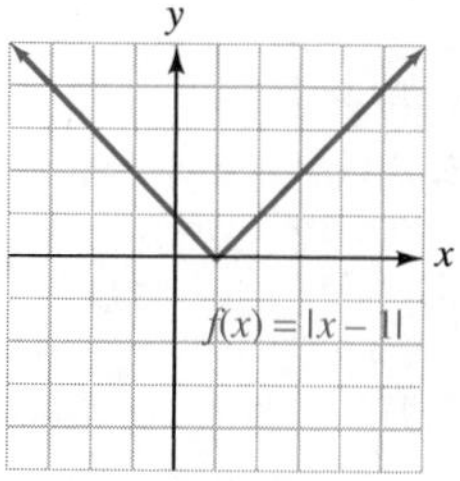

25.
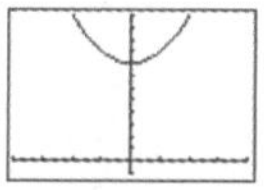

27.
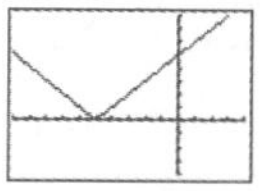

29.
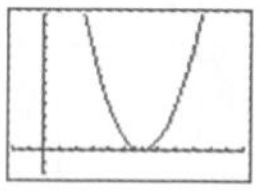

31.
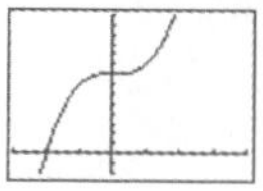

33.
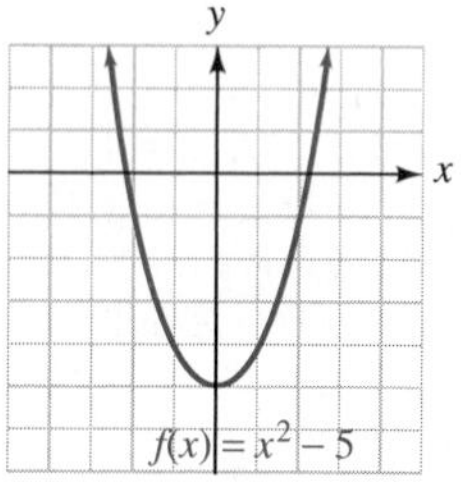

35.
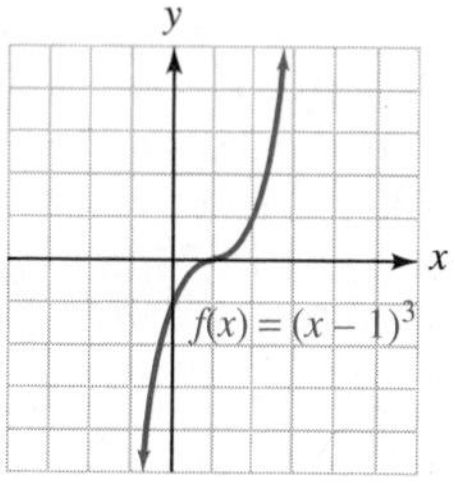

37.
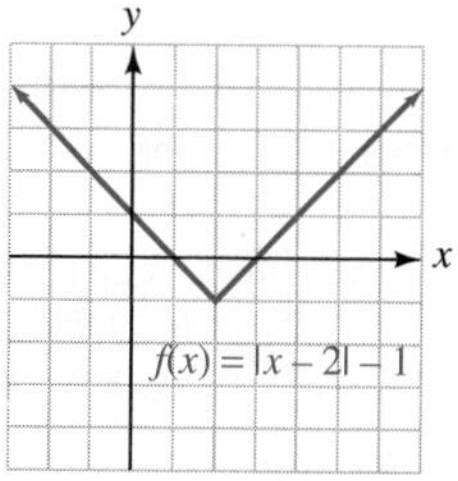

39.
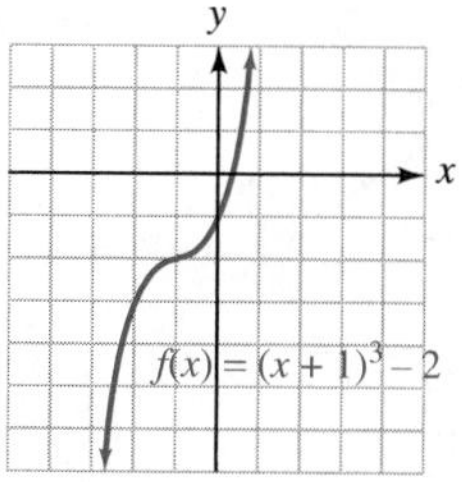

41.
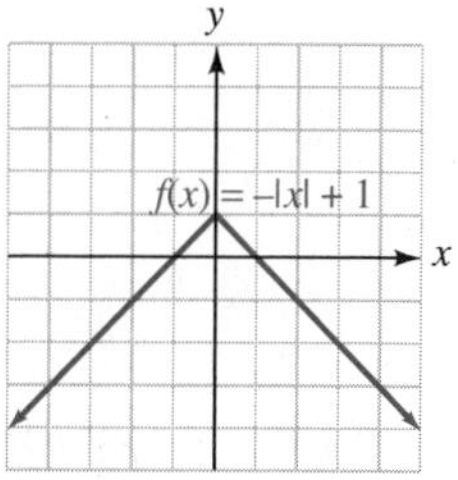

43.
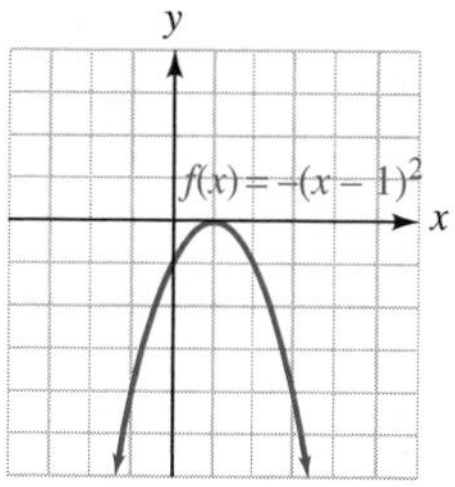

45. -2 **47.** 4 **49.** -3

Chapter Summary (page 176)

1. $-5, 0, -1, 4, 3, 2, 3$ **2. a.** 20 **b.** May **c.** February and March

3. a. 2, 0, 2, 12, 8, 5, 1 **b.**
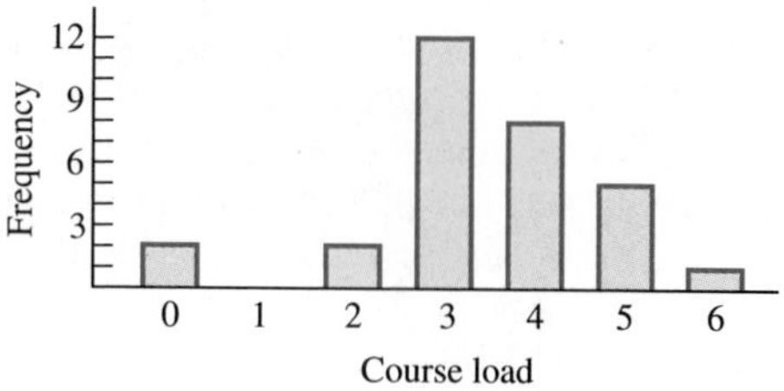

4. a. \$40 **b.** \$70

5. a.
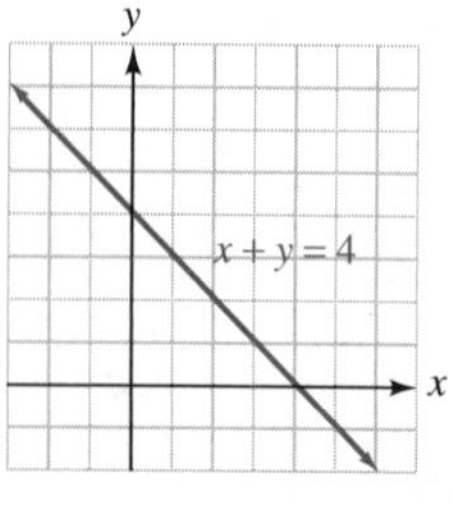

b.
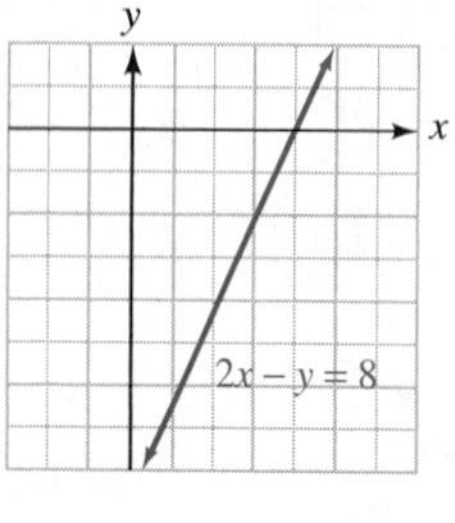

c.
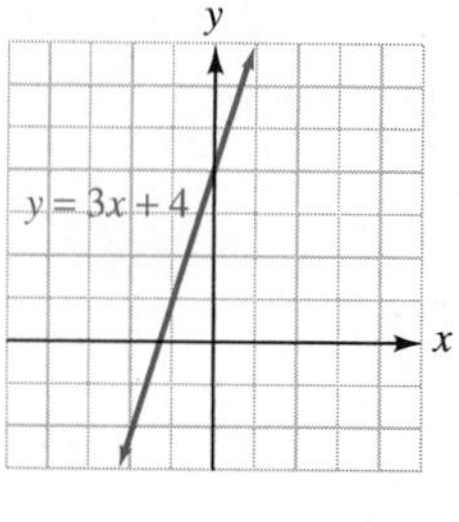

d.
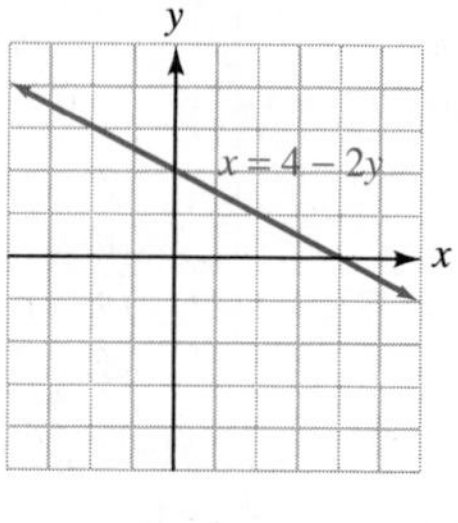

e.
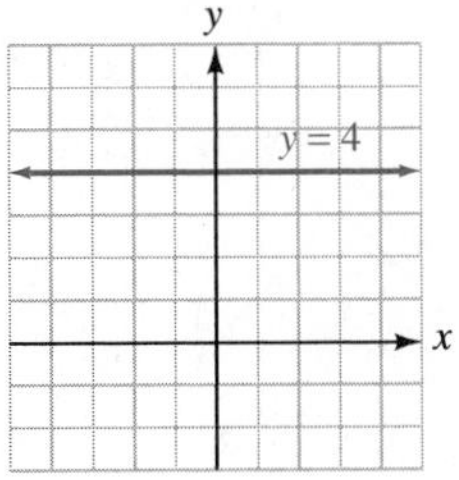

f.
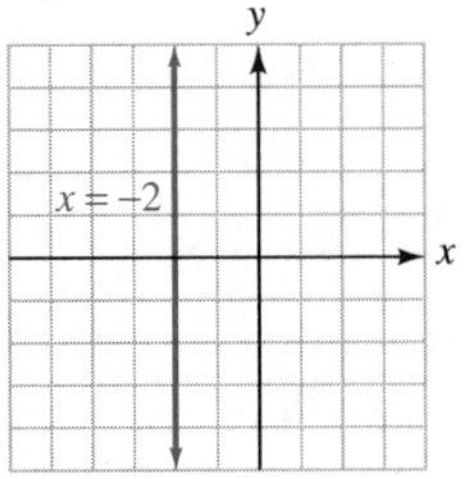

g.
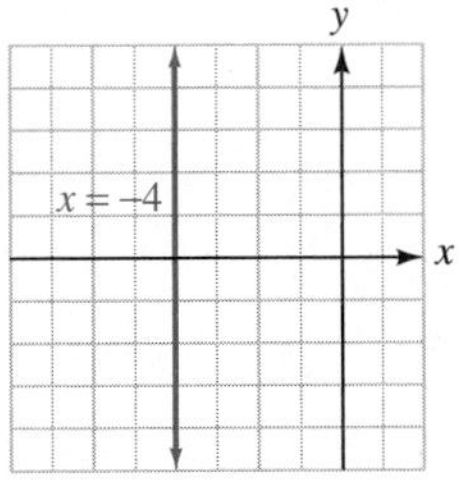

h.
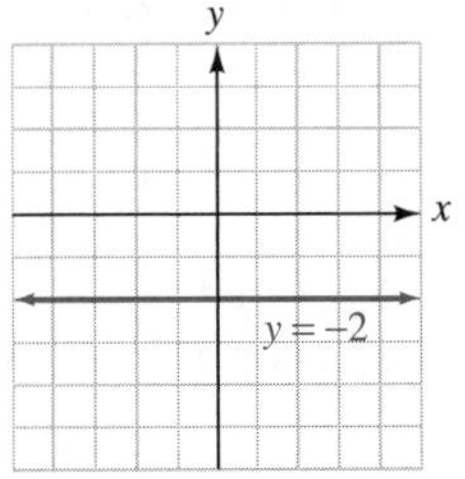

6. $\left(\frac{3}{2}, 8\right)$ **7. a.** 1 **b.** $\frac{14}{9}$ **c.** 5 **d.** $\frac{5}{11}$ **e.** 0 **f.** no defined slope **8. a.** $\frac{2}{3}$ **b.** -2 **c.** undefined **d.** 0 **9. a.** perpendicular **b.** parallel **c.** neither **d.** perpendicular **10.** \$21,666.67 **11. a.** $3x - y = -29$ **b.** $13x + 8y = 6$ **c.** $3x - 2y = 1$ **d.** $2x + 3y = -21$ **12.** $y = -1{,}720x + 8{,}700$ **13. a.** yes **b.** yes **c.** no **d.** no **14. a.** -7 **b.** 60 **c.** 0 **d.** 17 **15. a.** $D = (-\infty, \infty)$; $R = (-\infty, \infty)$ **b.** $D = (-\infty, \infty)$; $R = (-\infty, \infty)$ **c.** $D = (-\infty, \infty)$; $R = [1, \infty)$ **d.** $D = (-\infty, 2) \cup (2, \infty)$; $R = (-\infty, 0) \cup (0, \infty)$ **e.** $D = (-\infty, 3) \cup (3, \infty)$; $R = (-\infty, 0) \cup (0, \infty)$ **f.** $D = (-\infty, \infty)$; $R = \{7\}$ **16. a.** a function **b.** not a function **c.** not a function **d.** a function

17. a.
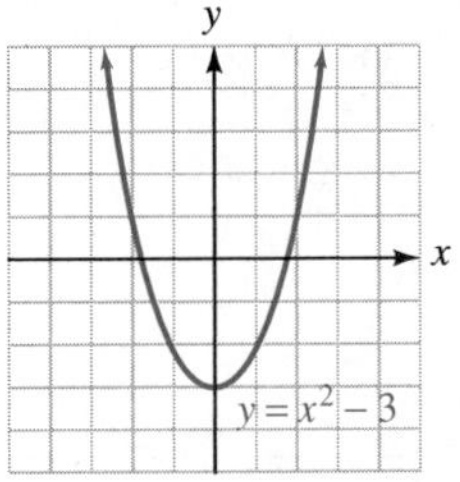

b.
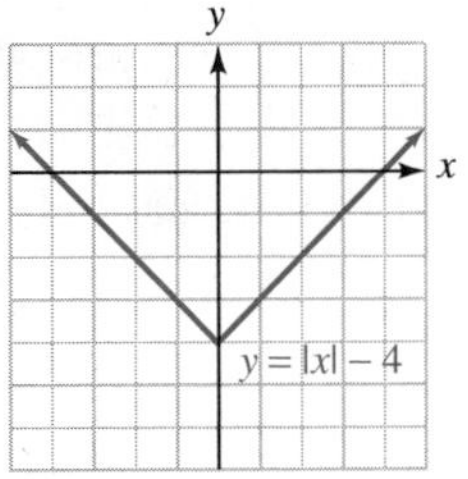

c.
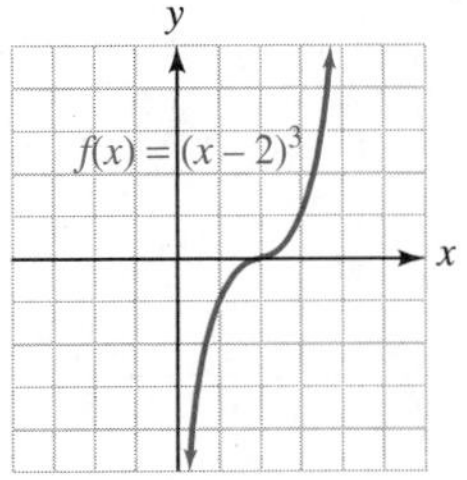

d.
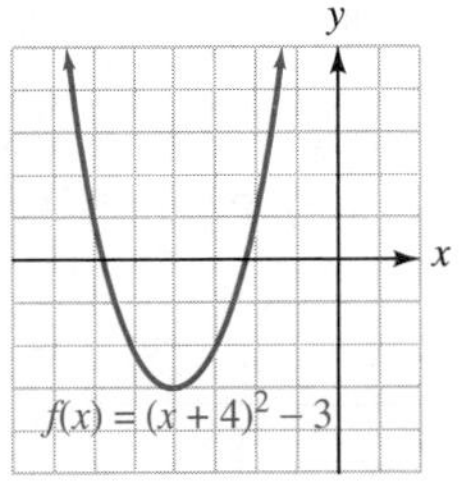

18. a.
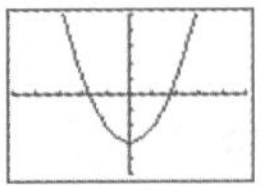

b.
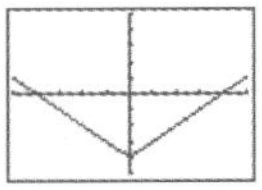

c.
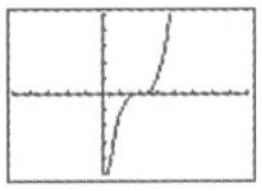

d.
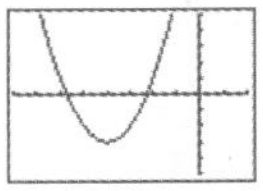

19. a. yes **b.** yes **c.** yes **d.** no

20.
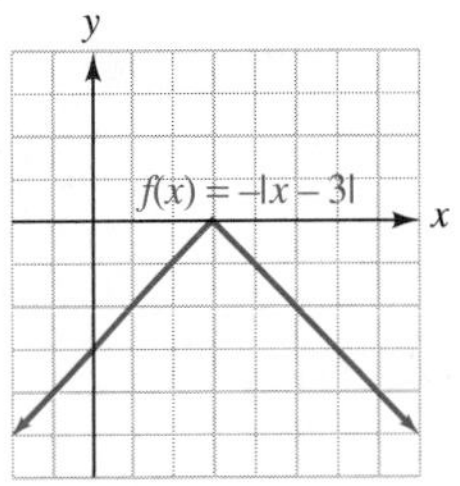

Chapter 2 Test (page 182)

1. a. 220 ft **b.** 1 sec and 7 sec **c.** about 238 ft **d.** 8 sec **2.**
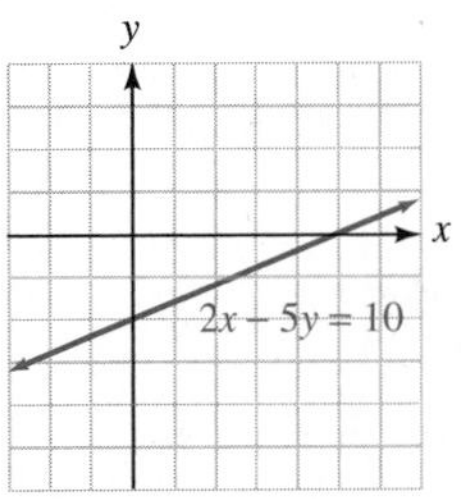

3. x-intercept $(3, 0)$, y-intercept $\left(0, -\frac{3}{5}\right)$ **4.** $\left(\frac{1}{2}, \frac{1}{2}\right)$ **5.** $\frac{1}{2}$ **6.** $\frac{2}{3}$ **7.** undefined **8.** 0 **9.** $y = \frac{2}{3}x - \frac{23}{3}$ **10.** $8x - y = -22$ **11.** $m = -\frac{1}{3}$, $\left(0, -\frac{3}{2}\right)$ **12.** neither **13.** perpendicular **14.** $y = \frac{3}{2}x$ **15.** $y = \frac{3}{2}x + \frac{21}{2}$ **16.** no **17.** $D = (-\infty, \infty)$; $R = [0, \infty)$ **18.** $D = (-\infty, \infty)$; $R = (-\infty, \infty)$

19. 10 **20.** -2 **21.** $3a + 1$ **22.** $x^2 - 2$ **23.** yes **24.** no **25.**

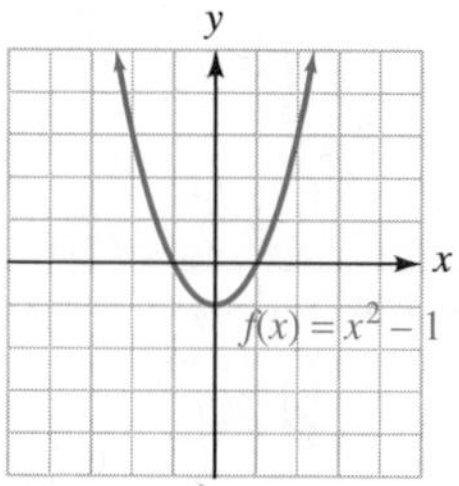

26.

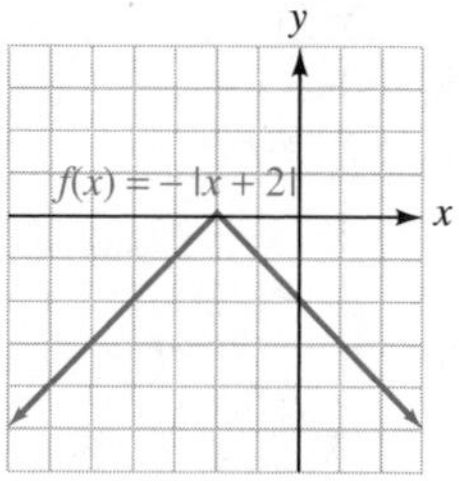

Cumulative Review Exercises (page 183)

1. 1, 2, 6, 7 **2.** 0, 1, 2, 6, 7 **3.** $-2, 0, 1, 2, \frac{13}{12}, 6, 7$ **4.** $\sqrt{5}, \pi$ **5.** -2 **6.** $-2, 0, 1, 2, \frac{13}{12}, 6, 7, \sqrt{5}, \pi$ **7.** 2, 7 **8.** 6 **9.** $-2, 0, 2, 6$ **10.** 1, 7 **11.**

12.

13. -2 **14.** -2 **15.** 22 **16.** -2 **17.** -4 **18.** -3 **19.** 4 **20.** -5 **21.** assoc. prop. of add. **22.** distrib. prop. **23.** comm. prop of add. **24.** assoc. prop. of mult. **25.** x^8y^{12} **26.** c^2 **27.** $-\frac{b^3}{a^2}$ **28.** 1 **29.** 4.97×10^{-6} **30.** 932,000,000 **31.** 8 **32.** -27 **33.** -1 **34.** 6 **35.** $a = \frac{2S}{n} - l$ **36.** $h = \frac{2A}{b_1 + b_2}$ **37.** 28, 30, 32 **38.** 14 cm by 42 cm **39.** yes **40.** $-\frac{7}{5}$ **41.** $y = -\frac{7}{5}x + \frac{11}{5}$ **42.** $y = -3x - 3$ **43.** 5 **44.** -1 **45.** $2t - 1$ **46.** $3r^2 + 2$

47. yes; D = $(-\infty, \infty)$; R = $(-\infty, 1]$

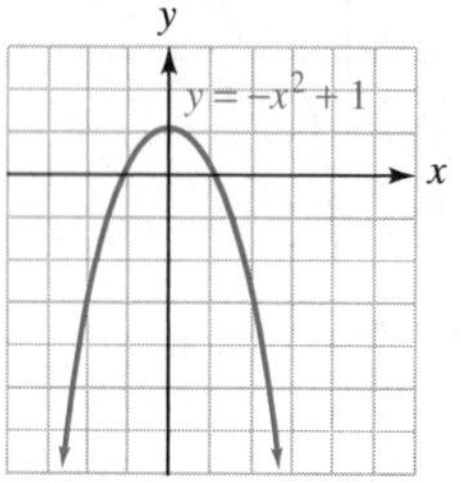

48. yes; D = $(-\infty, \infty)$; R = $[0, \infty)$

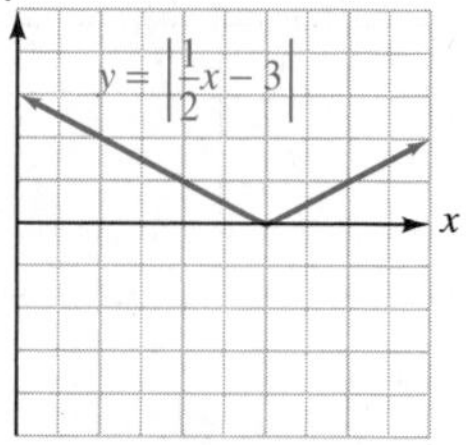

Getting Ready (page 187)

1. 2 **2.** -7 **3.** 11 **4.** 3

Orals (page 193)

1. no solution **2.** infinitely many **3.** one solution **4.** no solution

Exercise 3.1 (page 193)

1. 9.3×10^7 **3.** 3.45×10^4 **5.** system **7.** inconsistent **9.** dependent **11.** yes **13.** no

15.

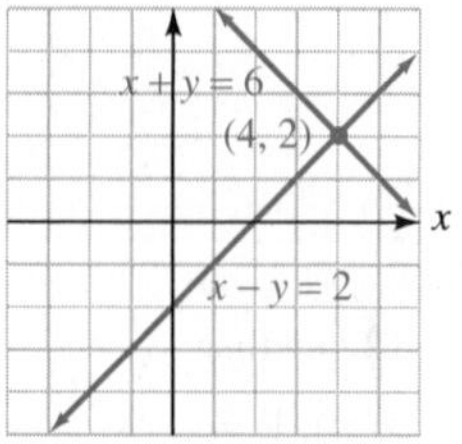

17.

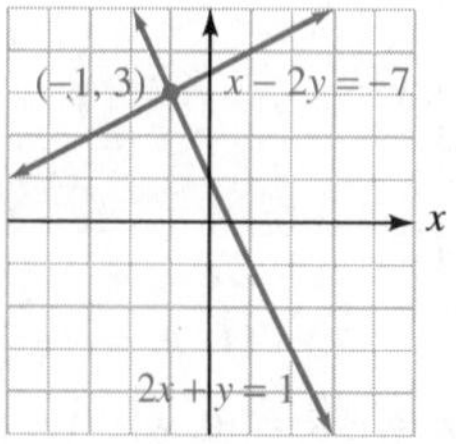

19.

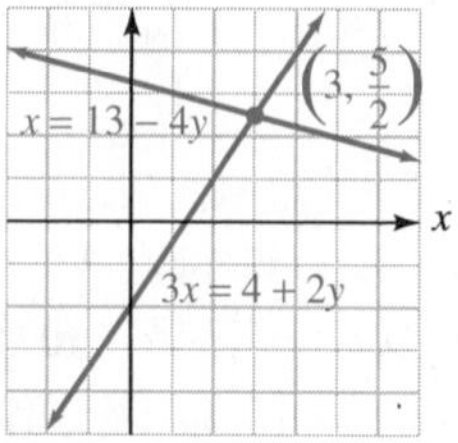

21.

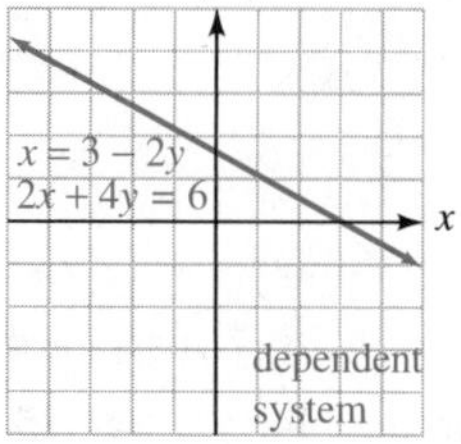

23.
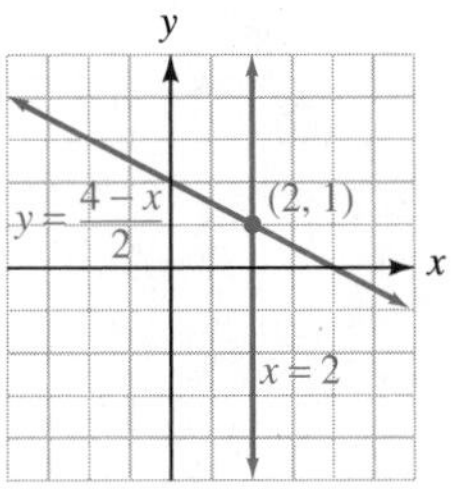

25.
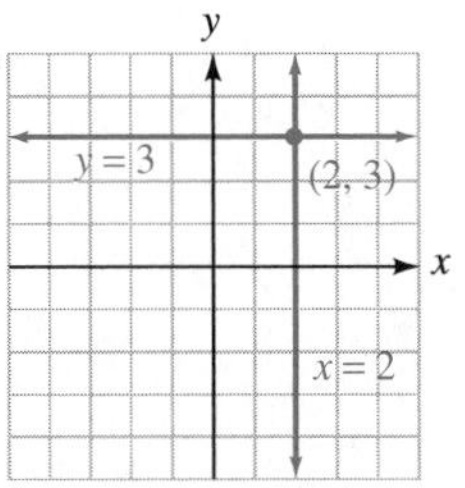

27.
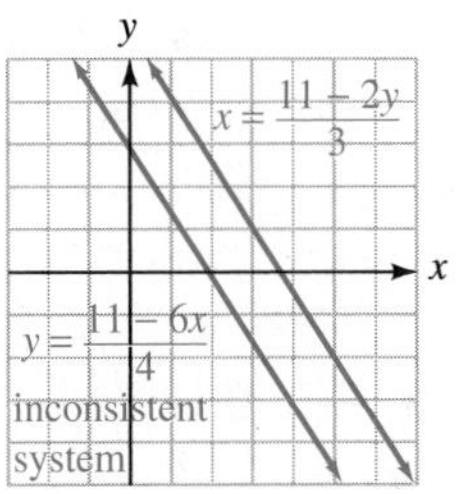

29.
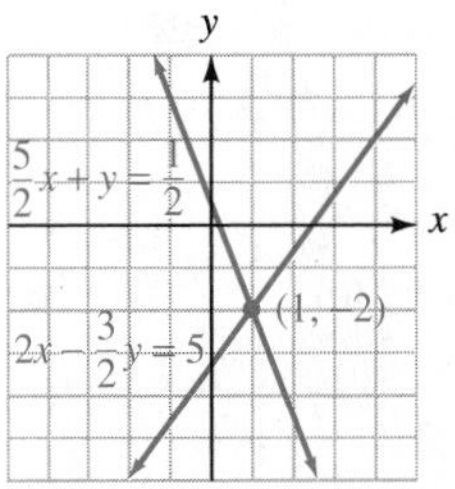

31.
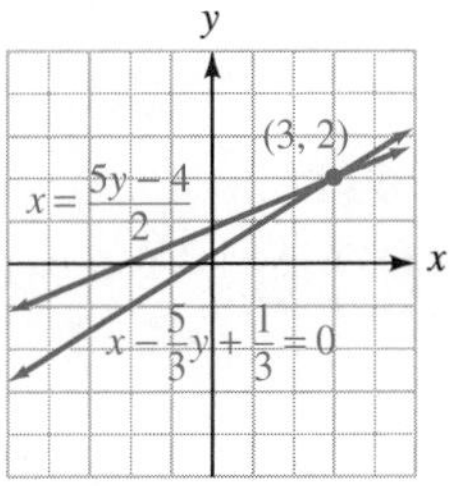

33.
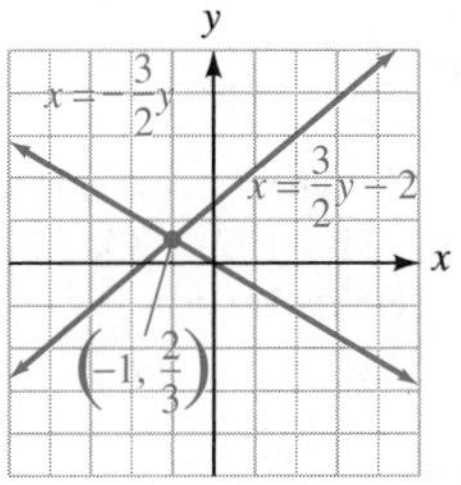

35. (−0.37, −2.69) **37.** (−7.64, 7.04) **39.** **a.** yes **b.** (3.75, −0.5) **c.** no **43.** One possible answer is $\begin{cases} x + y = -3 \\ x - y = -7 \end{cases}$.

Getting Ready (page 195)

1. $6x - 21$ **2.** $-12x - 20$ **3.** $3x - 9$ **4.** $-2x + 2$ **5.** $7x = 15$ **6.** $-9b = 27$

Orals (page 207)

1. 2 **2.** 4 **3.** 4 **4.** 3

Exercise 3.2 (page 208)

1. a^{22} **3.** $\dfrac{1}{81x^{32}y^4}$ **5.** setup, unit **7.** parallelogram **9.** opposite **11.** (2, 2) **13.** (5, 3) **15.** (−2, 4) **17.** no solution **19.** $\left(5, \frac{3}{2}\right)$ **21.** $\left(-2, \frac{3}{2}\right)$ **23.** (5, 2) **25.** (−4, −2) **27.** (1, 2) **29.** $\left(\frac{1}{2}, \frac{2}{3}\right)$ **31.** dependent equations **33.** no solution **35.** (4, 8) **37.** (20, −12) **39.** $\left(\frac{2}{3}, \frac{3}{2}\right)$ **41.** $\left(\frac{1}{2}, -3\right)$ **43.** $\frac{1}{3}$ **45.** $-\frac{691}{1{,}980}$ **47.** (2, 3) **49.** $\left(-\frac{1}{3}, 1\right)$ **51.** \$57 **53.** 625Ω, 750Ω **55.** 16 m by 20 m **57.** \$3,000 at 10%, \$5,000 at 12% **59.** 40 oz of 8% solution, 60 oz of 15% solution **61.** 55 mph **63.** 85 racing bikes, 120 mountain bikes **65.** 200 plates **67.** 21 **69.** 750 **71.** 6,500 gal per month **73.** A (a smaller loss) **75.** 590 units per month **77.** A (smaller loss) **79.** A **81.** 35°, 145° **83.** $x = 22.5$, $y = 67.5$ **85.** $f^2 = \dfrac{1}{4\pi^2LC}$

Getting Ready (page 212)

1. yes **2.** yes **3.** no **4.** yes

Orals (page 220)

1. yes **2.** no

Exercise 3.3 (page 220)

1. $\frac{9}{5}$ **3.** 1 **5.** $2s^2 + 1$ **7.** plane **9.** infinitely **11.** yes **13.** (1, 1, 2) **15.** (0, 2, 2) **17.** (3, 2, 1) **19.** inconsistent system **21.** dependent equations **23.** (2, 6, 9) **25.** −2, 4, 16 **27.** $A = 40°$, $B = 60°$, $C = 80°$ **29.** 1, 2, 3 **31.** 30 expensive, 50 middle-priced, 100 inexpensive **33.** 250 \$5 tickets, 375 \$3 tickets, 125 \$2 tickets **35.** 3 poles, 2 bears, 4 deer **37.** $y = x^2 - 4x$ **39.** $x^2 + y^2 - 2x - 2y - 2 = 0$ **43.** (1, 1, 0, 1)

Getting Ready (page 224)

1. 5 8 13 **2.** 0 3 7 **3.** −1 −1 −2 **4.** 3 3 −5

Orals (page 230)

1. $\begin{bmatrix} 3 & 2 \\ 4 & -3 \end{bmatrix}$ **2.** $\begin{bmatrix} 3 & 2 & 8 \\ 4 & -3 & 6 \end{bmatrix}$ **3.** yes **4.** no

Exercise 3.4 (page 231)

1. 9.3×10^7 **3.** 6.3×10^4 **5.** matrix **7.** 3, columns **9.** augmented **11.** type 1 **13.** nonzero **15.** 0 **17.** 8 **19.** (1, 1) **21.** (2, −3) **23.** (0, −3) **25.** (1, 2, 3) **27.** (−1, −1, 2) **29.** (2, 1, 0) **31.** (1, 2) **33.** (2, 0) **35.** no solution **37.** (1, 2) **39.** $(-6 - z, 2 - z, z)$ **41.** $(2 - z, 1 - z, z)$ **43.** 22°, 68° **45.** 40°, 65°, 75° **47.** $y = 2x^2 - x + 1$ **51.** $k \neq 0$

Getting Ready (page 233)

1. −22 **2.** 22 **3.** −13 **4.** −13

Orals (page 241)

1. 1 **2.** −2 **3.** 0 **4.** $\begin{vmatrix} 1 & 2 \\ 2 & -1 \end{vmatrix}$ **5.** $\begin{vmatrix} 5 & 2 \\ 4 & -1 \end{vmatrix}$ **6.** $\begin{vmatrix} 1 & 5 \\ 2 & 4 \end{vmatrix}$

Exercise 3.5 (page 242)

1. −3 **3.** 0 **5.** number **7.** $\begin{vmatrix} a_2 & c_2 \\ a_3 & c_3 \end{vmatrix}$ **9.** $\begin{vmatrix} 3 & 4 \\ 2 & -3 \end{vmatrix}$ **11.** 8 **13.** −2 **15.** $x^2 - y^2$ **17.** 0 **19.** −13 **21.** 26 **23.** 0 **25.** $10a$ **27.** 0 **29.** (4, 2) **31.** (−1, 3) **33.** $\left(-\frac{1}{2}, \frac{1}{3}\right)$ **35.** (2, −1) **37.** no solution **39.** $\left(5, \frac{14}{5}\right)$ **41.** (1, 1, 2) **43.** (3, 2, 1) **45.** no solution **47.** (3, −2, 1) **49.** dependent equations **51.** (−2, 3, 1) **53.** no solution **55.** 2 **57.** 2 **59.** \$5,000 in HiTech, \$8,000 in SaveTel, \$7,000 in HiGas **61.** −23 **63.** 26 **69.** −4

Chapter Summary (page 247)

1. a.

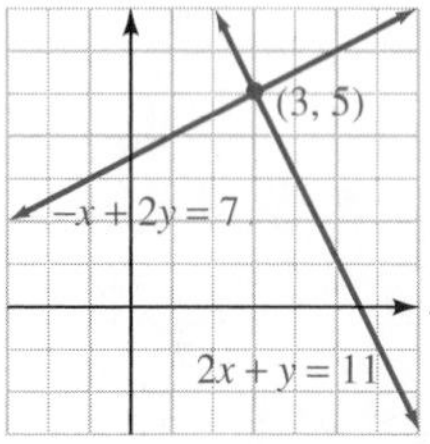

b.

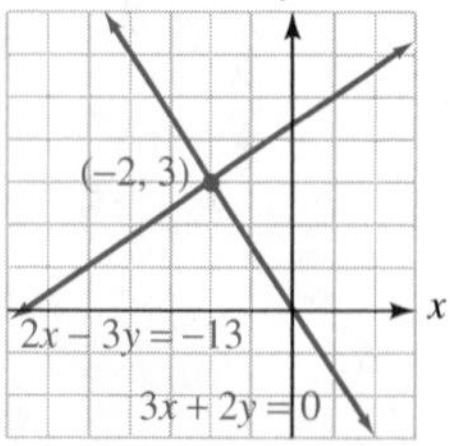

c.

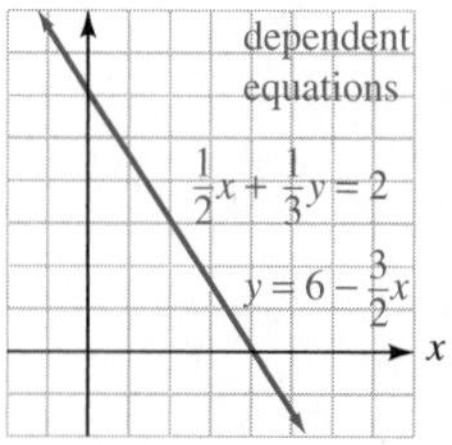

d.

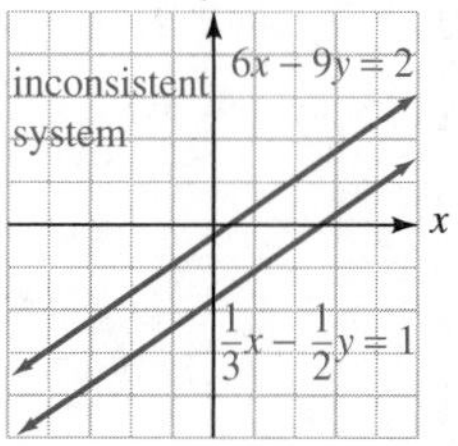

2. a. (−1, 3) **b.** (−3, −1) **c.** (3, 4) **d.** (−4, 2) **3. a.** (−3, 1) **b.** (1, −1) **c.** (9, −4) **d.** $\left(4, \frac{1}{2}\right)$ **4. a.** (1, 2, 3) **b.** inconsistent system **5. a.** (2, 1) **b.** (1, 3, 2) **c.** (1, 2) **d.** $(3z, 1 - 2z, z)$ **6. a.** 18 **b.** 38 **c.** −3 **d.** 28 **7. a.** (2, 1) **b.** (−1, 3) **c.** (1, −2, 3) **d.** (−3, 2, 2)

Chapter 3 Test (page 249)

1. 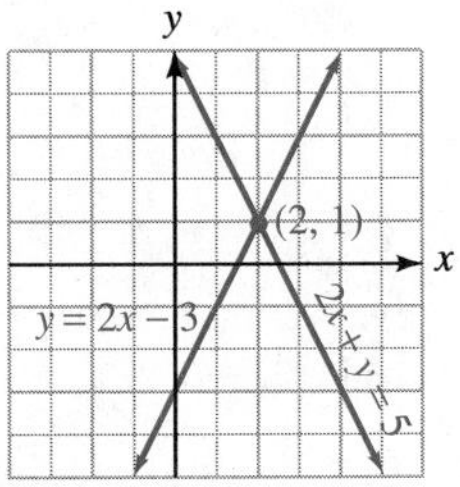

2. (7, 0) **3.** (2, −3) **4.** (−6, 4) **5.** dependent **6.** consistent **7.** 6 **8.** −8

9. $\begin{bmatrix} 1 & 1 & 1 & 4 \\ 1 & 1 & -1 & 6 \\ 2 & -3 & 1 & -1 \end{bmatrix}$ **10.** $\begin{bmatrix} 1 & 1 & 1 \\ 1 & 1 & -1 \\ 2 & -3 & 1 \end{bmatrix}$ **11.** (2, 2) **12.** (−1, 3) **13.** 22 **14.** −17 **15.** 4 **16.** 13

17. $\begin{vmatrix} -6 & -1 \\ -6 & 1 \end{vmatrix}$ **18.** $\begin{vmatrix} 1 & -1 \\ 3 & 1 \end{vmatrix}$ **19.** −3 **20.** 3 **21.** 3 **22.** −1

Getting Ready (page 252)

1. 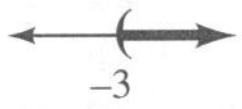

2. 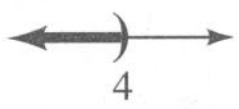

3. 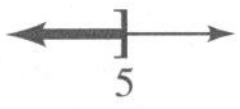

4. 

Orals (page 261)

1. $x < 2$ **2.** $x \geq 3$ **3.** $x < -4$ **4.** $x \geq -8$ **5.** $-1 < x < 4$ **6.** $9 \leq x \leq 12$

Exercise 4.1 (page 261)

1. $\frac{1}{t^{12}}$ **3.** 471 or more **5.** ≠ **7.** < **9.** ≥ **11.** $a < c$ **13.** reversed **15.** $c < x, x < d$ **17.** (−∞, 1) 

19. [−2, ∞) 

21. (2, ∞) 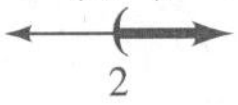

23. (−∞, 20] 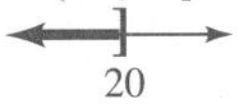

25. (−∞, 10) 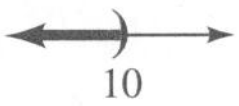

27. [−36, ∞) 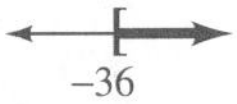

29. (−∞, 45/7] 

31. (−2, 5) 

33. (8, 11) 

35. [−4, 6) 

37. no solution **39.** [−2, 4] 

41. (2, 3) 

43. [1, 9/4] 

45. (−∞, −15) 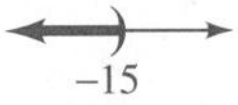

47. (−∞, 2) ∪ (7, ∞) 

49. (−∞, 1) 

51. no solution **53.** no

55. 5 hr **57.** more than \$5,000 **59.** 18 **61.** 88 or higher **63.** 13 **65.** anything over \$900 **67.** $x < 1$ **69.** $x \geq -4$ **71.** 139 **75.** a, b, c

Getting Ready (page 264)

1. 5 **2.** 0 **3.** 10 **4.** −5 **5.** (−5, 4) **6.** (−∞, −3) ∪ (3, ∞)

Orals (page 274)

1. 5 **2.** −5 **3.** −6 **4.** −4 **5.** 8 or −8 **6.** no solution **7.** $-8 < x < 8$ **8.** $x < -8$ or $x > 8$ **9.** $x \le -4$ or $x \ge 4$ **10.** $-7 \le x \le 7$

Exercise 4.2 (page 275)

1. $\frac{3}{4}$ **3.** 6 **5.** $t = \frac{A - p}{pr}$ **7.** x **9.** 0 **11.** reflected **13.** $a = b$ or $a = -b$ **15.** $x \le -k$ or $x \ge k$ **17.** 8 **19.** −2 **21.** −30 **23.** $4 - \pi$ **25.**

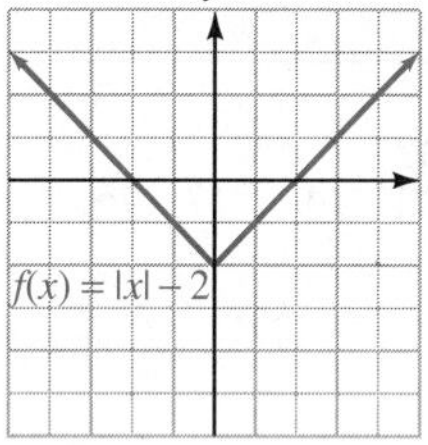

27.

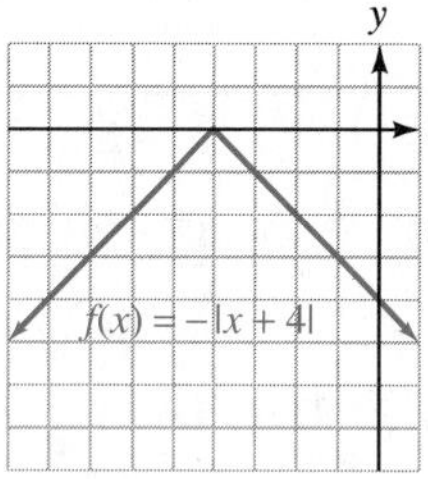

29. 4, −4 **31.** 9, −3 **33.** 4, −1 **35.** $\frac{14}{3}$, −6

37. no solution **39.** 8, −4 **41.** 2, $-\frac{1}{2}$ **43.** −8 **45.** −4, −28 **47.** 0, −6 **49.** $\frac{20}{3}$ **51.** −2, $-\frac{4}{5}$ **53.** 3, −1 **55.** 0, −2 **57.** 0 **59.** $\frac{4}{3}$ **61.** no solution **63.** (−4, 4)

65. [−21, 3]

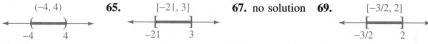

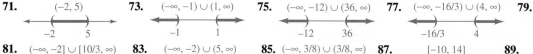

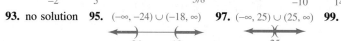

67. no solution **69.** [−3/2, 2]

71. (−2, 5)

73. (−∞, −1) ∪ (1, ∞)

75. (−∞, −12) ∪ (36, ∞)

77. (−∞, −16/3) ∪ (4, ∞)

79. (−∞, ∞)

81. (−∞, −2] ∪ [10/3, ∞)

83. (−∞, −2) ∪ (5, ∞)

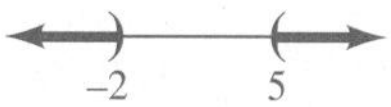

85. (−∞, 3/8) ∪ (3/8, ∞)

87. [−10, 14]

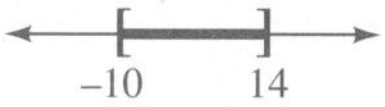

89. (−5/3, 1)

91. (−∞, −4] ∪ [−1, ∞)

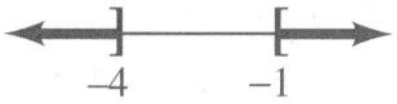

93. no solution **95.** (−∞, −24) ∪ (−18, ∞)

97. (−∞, 25) ∪ (25, ∞)

99. [−7, −7]

101. [5, 5]

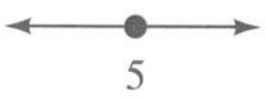

103. $|x| < 4$ **105.** $|x + 3| > 6$ **107.** $70° \le t \le 86°$ **109.** $|c - 0.6°| \le 0.5°$ **115.** $k < 0$ **117.** x and y must have different signs.

Getting Ready (page 278)

1. yes **2.** yes **3.** no **4.** yes **5.** yes **6.** no **7.** no **8.** yes

Orals (page 284)

1. yes **2.** no **3.** no **4.** yes **5.** no **6.** no **7.** yes **8.** yes

Exercise 4.3 (page 284)

1. (3, 1) **3.** (2, −3) **5.** linear **7.** edge

9.

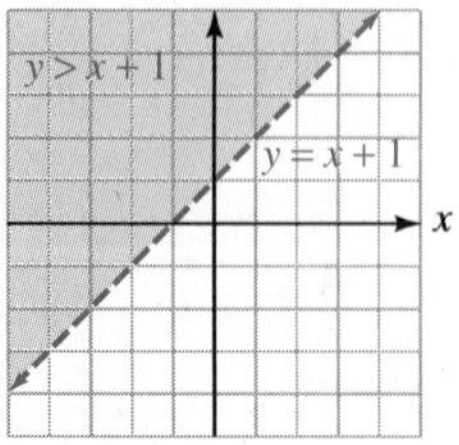

11.

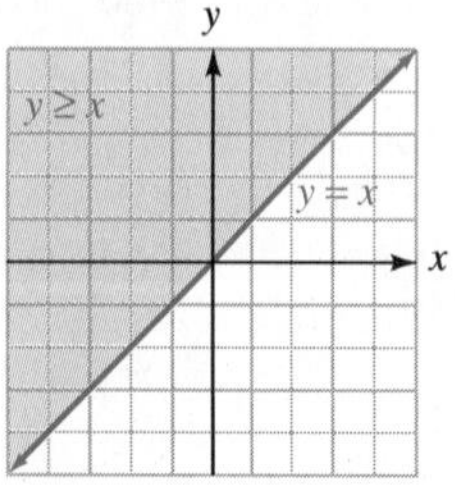

13.

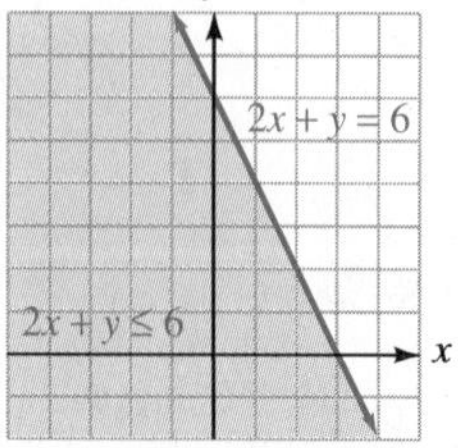

15.

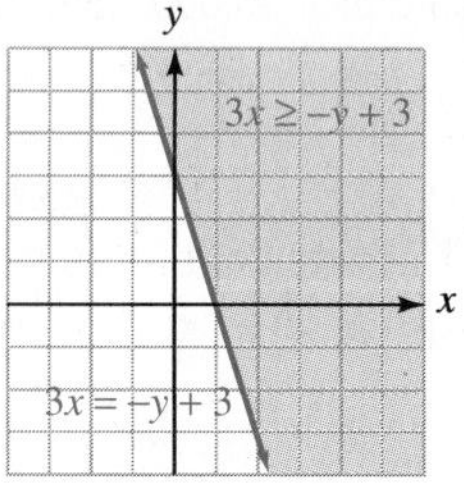

17. 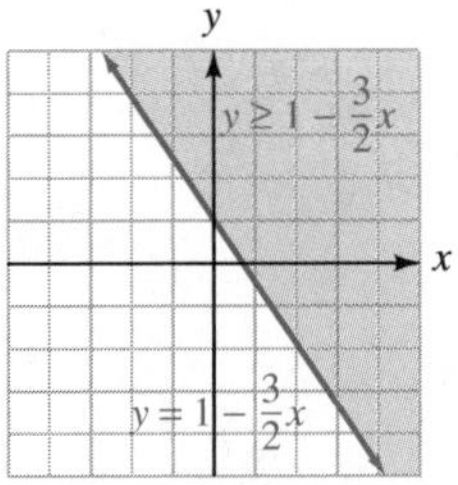

19. 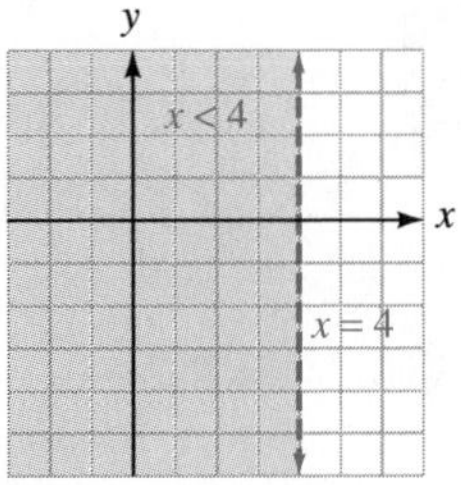

21. 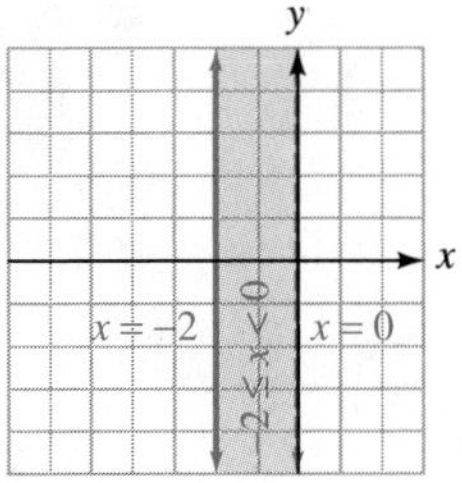

23. 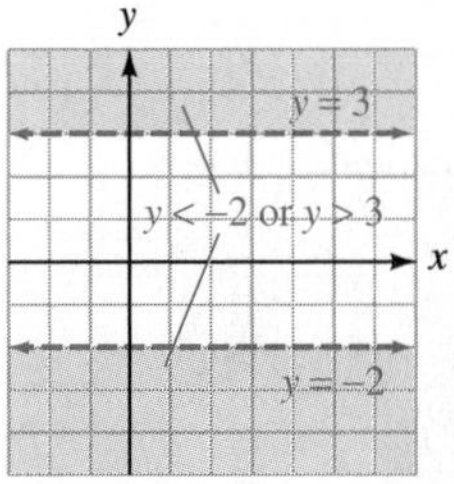

25. $3x + 2y > 6$ **27.** $x \leq 3$ **29.** $y \leq x$ **31.** $-2 \leq x \leq 3$ **33.** $y > -1$ or $y \leq -3$ **35.** 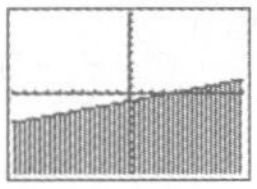 **37.** 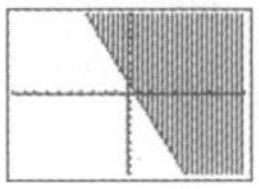

39. (1, 1), (2, 1), (2, 2)

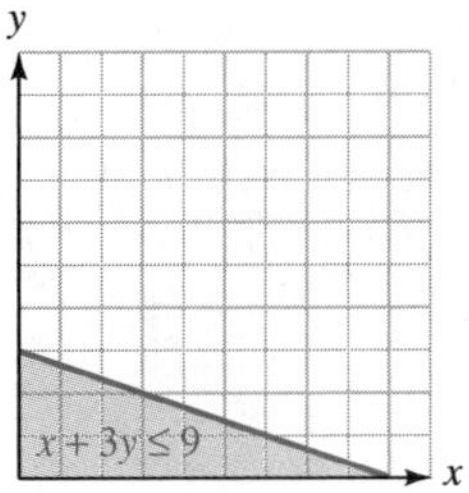

41. (2, 2), (3, 3), (5, 1)

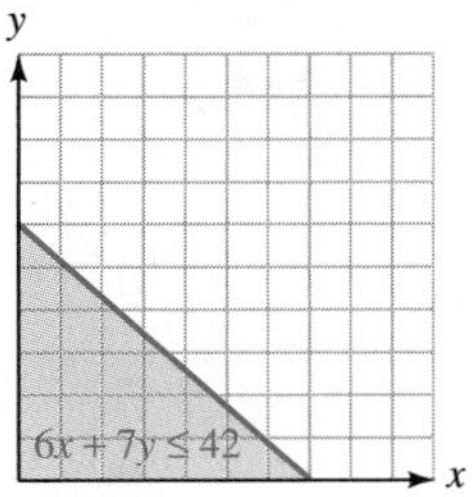

43. (40, 20), (60, 40), (80, 20)

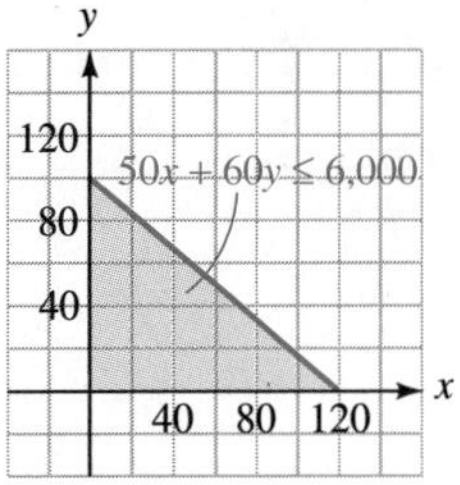

Getting Ready (page 288)

1. yes **2.** yes **3.** no **4.** no **5.** no **6.** no **7.** yes **8.** no

Orals (page 293)

1. yes **2.** yes

Exercise 4.4 (page 293)

1. $r = \frac{A - p}{pt}$ **3.** $x = z\sigma + \mu$ **5.** $d = \frac{l - a}{n - 1}$ **7.** intersect **9.**

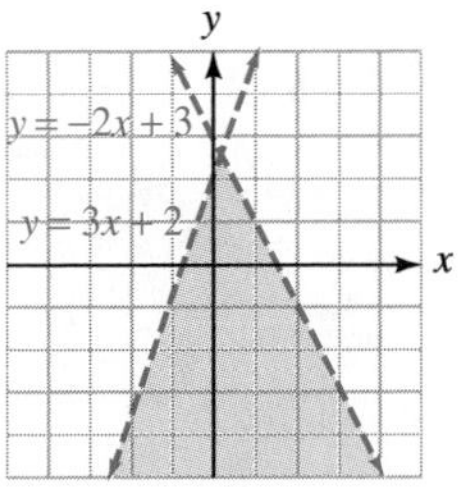

11. 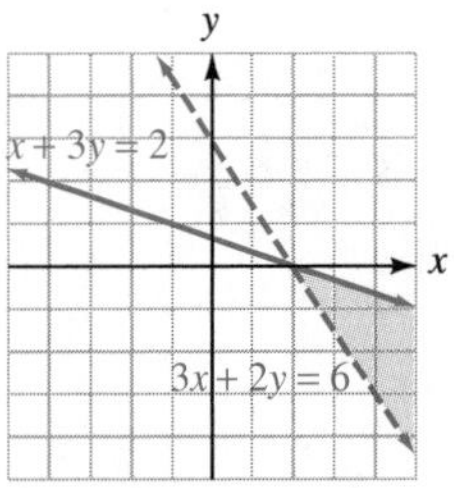

13. 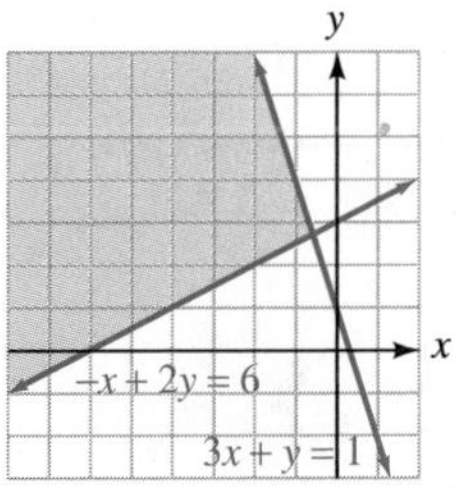

15. 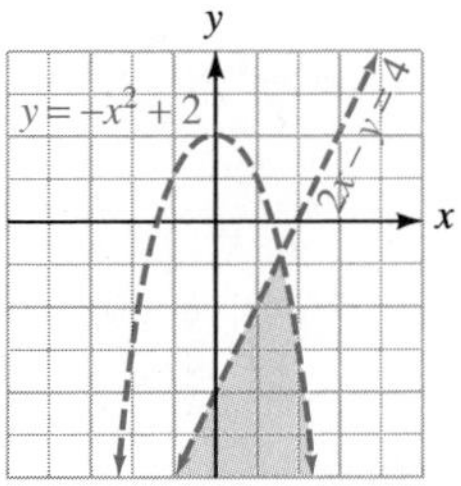

17. 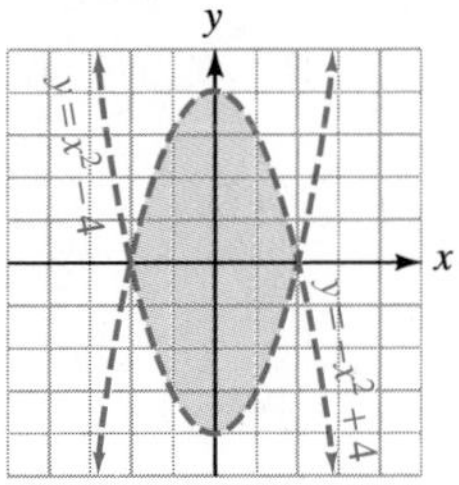

19. 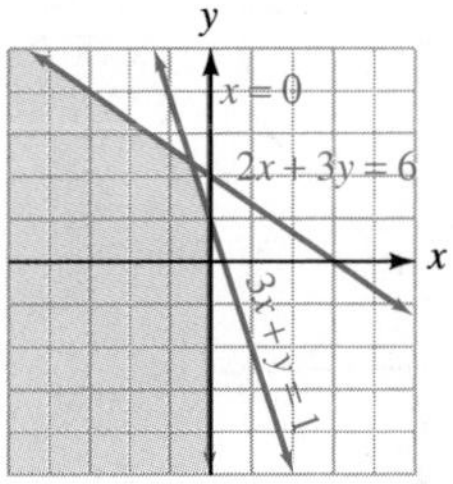

21.
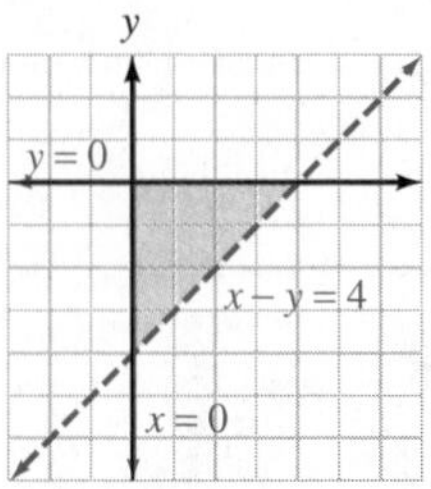

23.
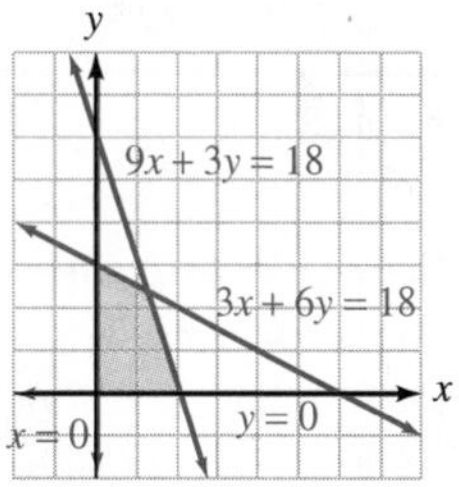

25.
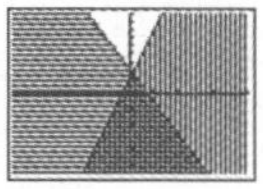

27. 1 \$10 CD and 2 \$15 CDs, 4 \$10 CDs and 1 \$15 CD
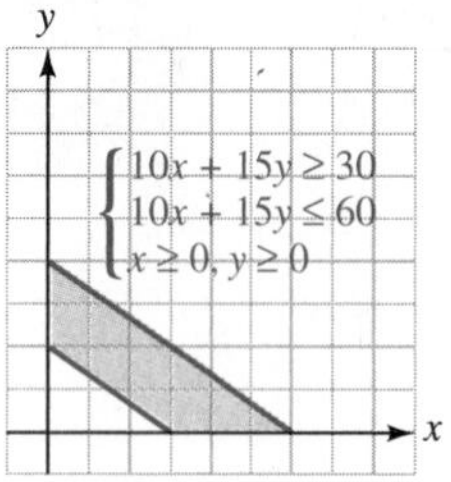

29. 2 desk chairs and 4 side chairs, 1 desk chair and 5 side chairs
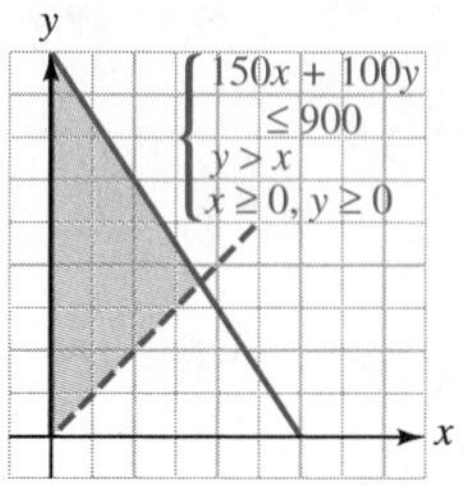

33. no

Getting Ready (page 296)

1. 0 **2.** 6 **3.** 10 **4.** 12

Orals (page 303)

1. 25 **2.** 9 **3.** (0, 0), (0, 3), (3, 0) **4.** (0, 0), (0, 2), (4, 0)

Exercise 4.5 (page 303)

1. $\frac{3}{7}$ **3.** $y = \frac{3}{7}x + \frac{34}{7}$ **5.** constraints **7.** objective **9.** $P = 12$ at (0, 4) **11.** $P = \frac{13}{6}$ at $\left(\frac{5}{3}, \frac{4}{3}\right)$ **13.** $P = \frac{18}{7}$ at $\left(\frac{3}{7}, \frac{12}{7}\right)$ **15.** $P = 3$ at (1, 0) **17.** $P = 0$ at (0, 0) **19.** $P = 0$ at (0, 0) **21.** $P = -12$ at $(-2, 0)$ **23.** $P = -2$ at (1, 2) and $(-1, 0)$ **25.** 3 tables, 12 chairs, \$1,260 **27.** 30 IBMs, 30 Macs, \$2,700 **29.** 15 VCRs, 30 TVs, \$1,560 **31.** \$150,000 in stocks, \$50,000 in bonds; \$17,000

Chapter Summary (page 307)

1. a.
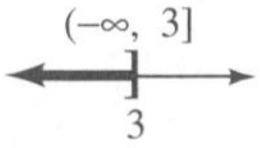

b.
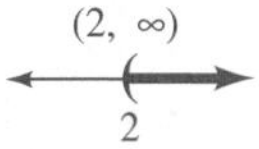

c.
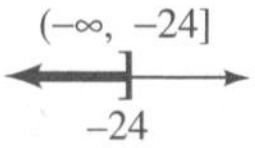

d.
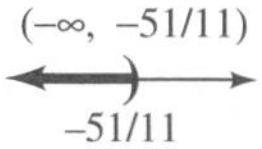

e.
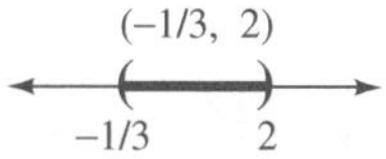

f.
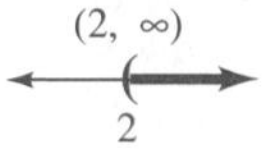

g.
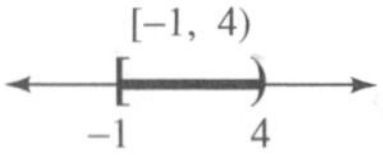

2. \$20,000 or more **3. a.** 7 **b.** 8 **c.** −7 **d.** −12 **4. a.**
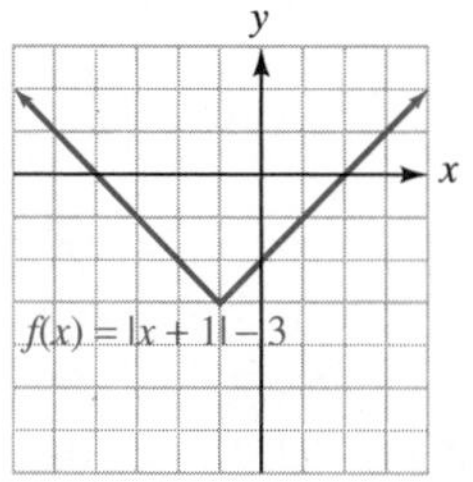

b.
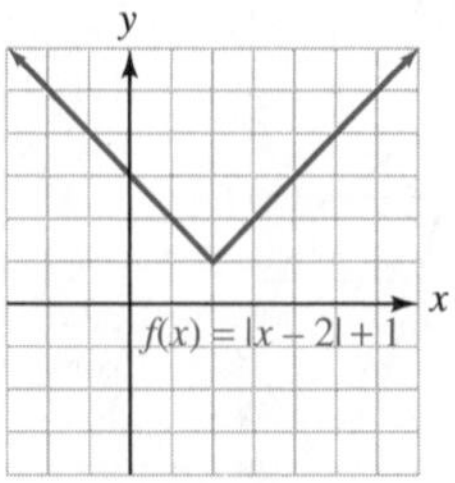

5. a. $3, -\frac{11}{3}$ **b.** $\frac{26}{3}, -\frac{10}{3}$ **c.** $14, -10$ **d.** $\frac{1}{5}, -5$ **e.** $-1, 1$ **f.** $\frac{13}{12}$ **6. a.**
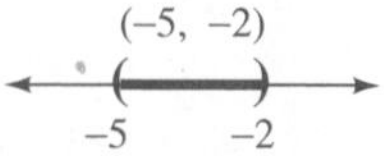

b. [–3, 19/3]
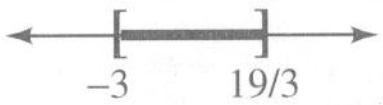

c. no solutions **d.** (–∞, –4) ∪ (22/5, ∞)

e. (–∞, 4/3] ∪ [4, ∞)

f. (–∞, ∞)

7. a.
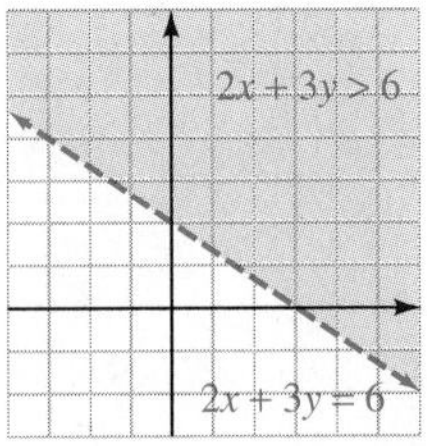

b.
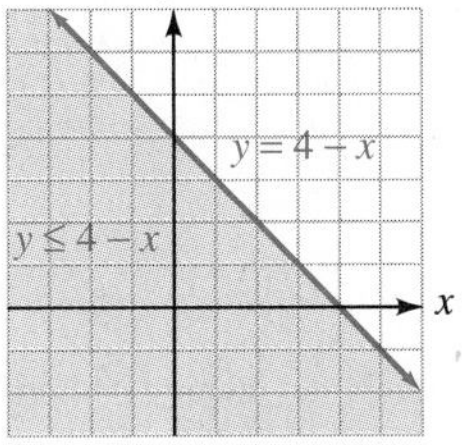

c.
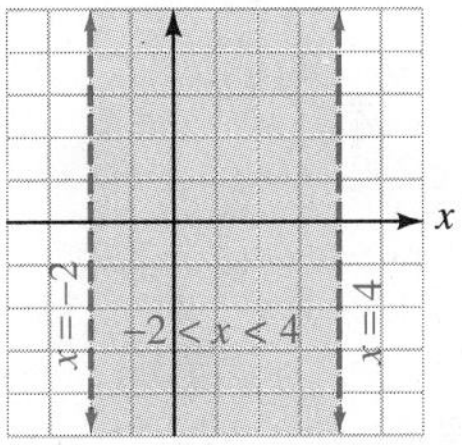

d.
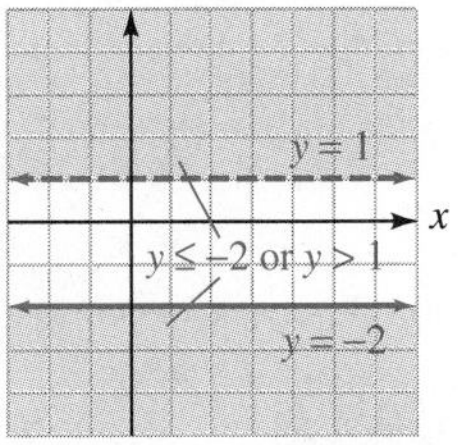

8. a.
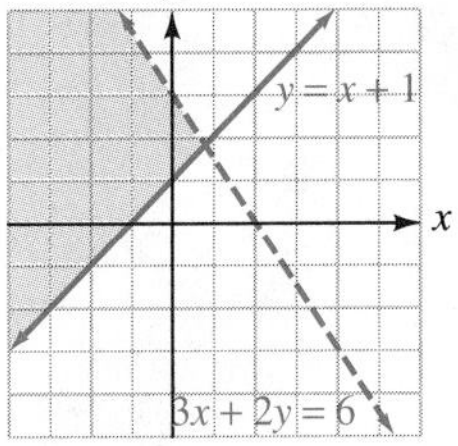

b.
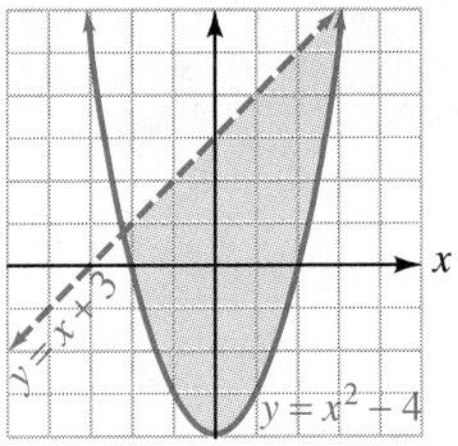

9. max. of 6 at (3, 0) **10.** 1,000 bags of X, 1,400 bags of Y

Chapter 4 Test (page 310)

1. (–∞, –5]
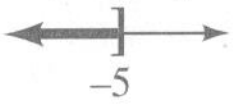

2. (–2, 16)

3. 3 **4.** $4\pi - 4$ **5.**
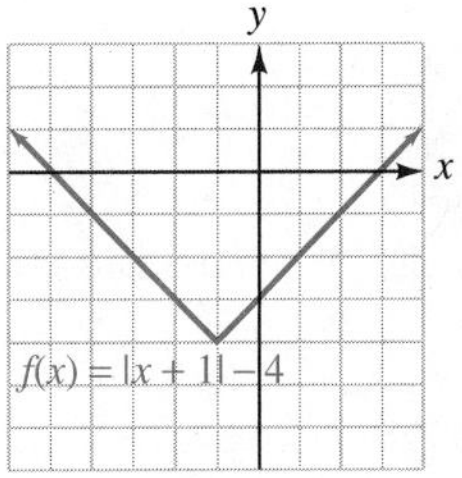

6.
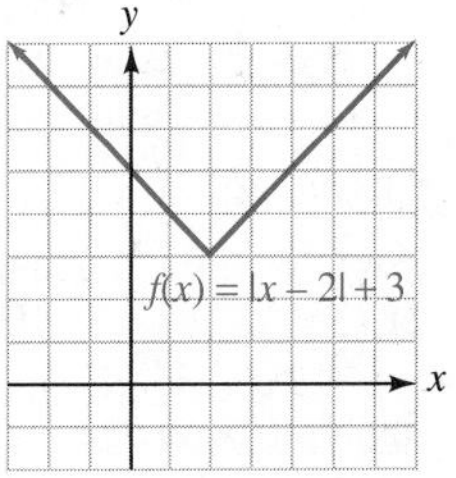

7. 4, –7

8. $-5, \frac{23}{3}$
9. 4, –4
10. 0
11. [–7, 1]

12. (–∞, –9) ∪ (13, ∞)
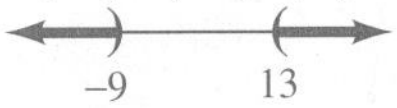

13. (–∞, 1) ∪ (3, ∞)

14. [1, 3]

15.
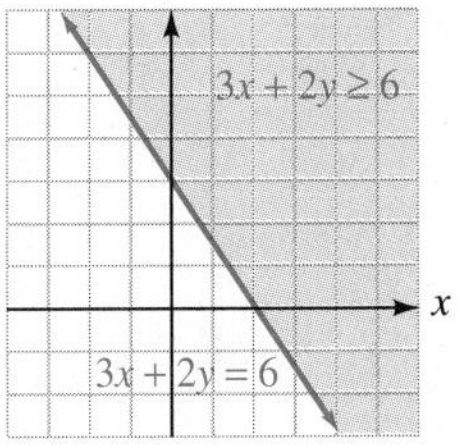

16.
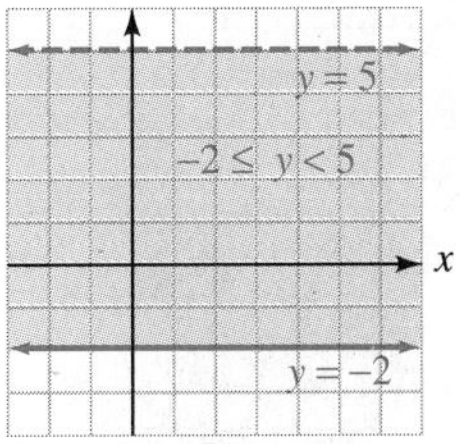

17.
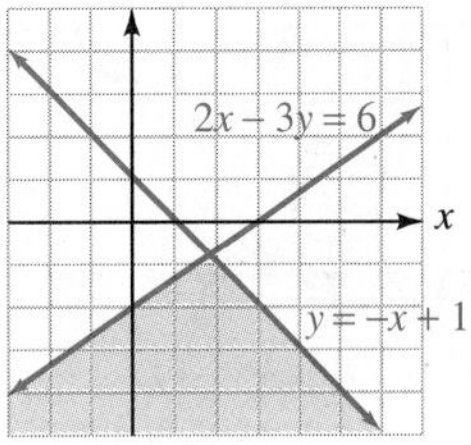

18.
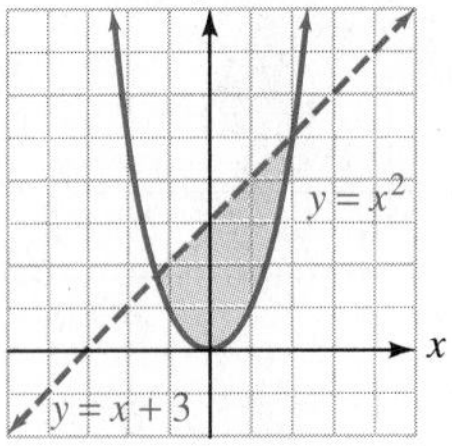

19. $P = 2$ at (1, 1)

Cumulative Review Exercises (page 312)

1.
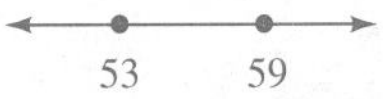

2. 5 **3.** 10 **4.** –6 **5.** x^{10} **6.** x^{14} **7.** x^4 **8.** $a^{2-n}b^{n-2}$ **9.** 3.26×10^7 **10.** 1.2×10^{-5} **11.** $\frac{26}{3}$

12. 3 **13.** 6 **14.** impossible **15.** perpendicular **16.** parallel **17.** $y = \frac{1}{3}x + \frac{11}{3}$ **18.** $h = \dfrac{2A}{b_1 + b_2}$ **19.** 10 **20.** 14

21. (2, 1)

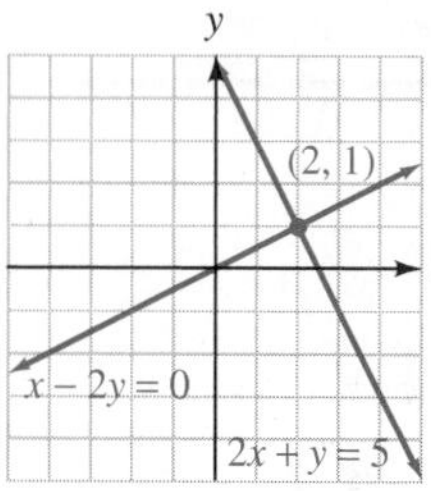

22. (1, 1) **23.** (2, −2) **24.** (3, 1) **25.** (−1, −1, 3) **26.** (0, −1, 1) **27.** −1 **28.** 16

29. (−1, −1) **30.** (1, 2, −1) **31.** $x \le 11$ **32.** $-3 < x < 3$ **33.** 3, $-\frac{3}{2}$ **34.** −5, $-\frac{3}{5}$ **35.** $-\frac{2}{3} \le x \le 2$ **36.** $x < -4$ or $x > 1$

37.

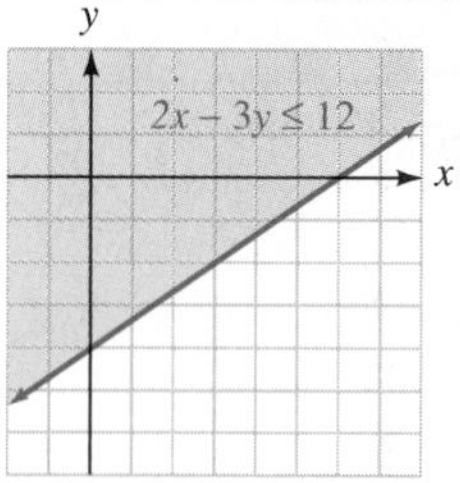

38.

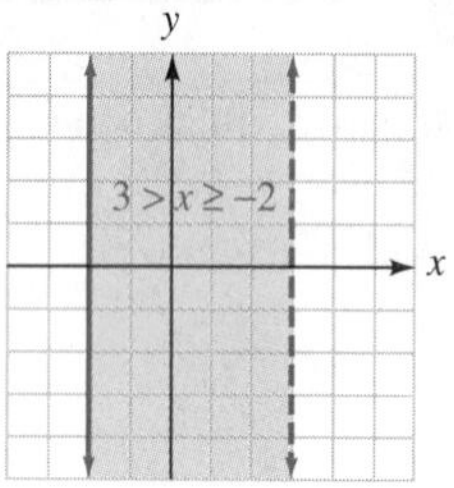

39.

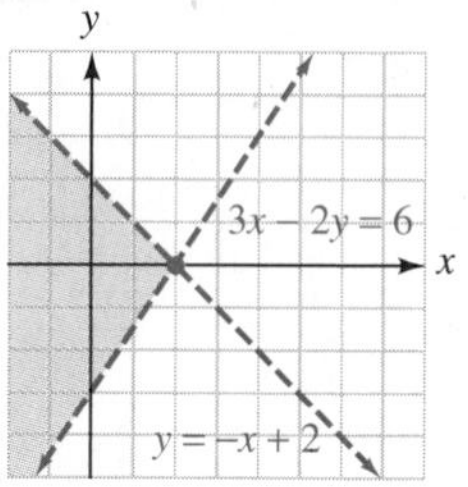

40.

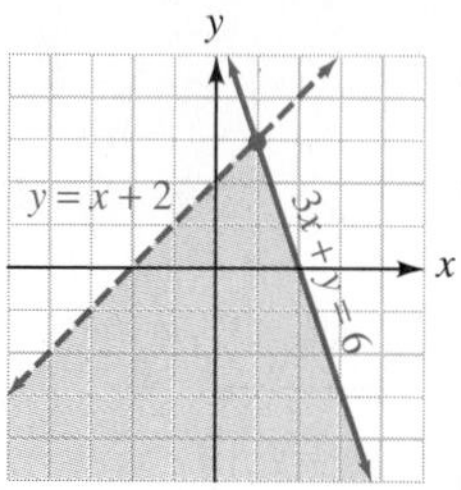

41. 24

Getting Ready (page 316)

1. $3a^2b^2$ **2.** $-5x^3y$ **3.** $4p^2 + 7q^2$ **4.** $a^3 - b^2$

Orals (page 323)

1. 3 **2.** 4 **3.** 3 **4.** 3 **5.** 1 **6.** 5 **7.** −1 **8.** −3

Exercise 5.1 (page 323)

1. a^5 **3.** $3y^{23}$ **5.** 1.14×10^8 **7.** sum, whole **9.** binomial **11.** one **13.** monomial **15.** trinomial **17.** binomial **19.** monomial **21.** 2 **23.** 8 **25.** 10 **27.** 0 **29.** $-2x^4 - 5x^2 + 3x + 7$ **31.** $7a^3x^5 - ax^3 - 5a^3x^2 + a^2x$ **33.** $7y + 4y^2 - 5y^3 - 2y^5$ **35.** $2x^4y - 5x^3y^3 - 2y^4 + 5x^3y^6 + x^5y^7$ **37.** 2 **39.** 8 **41.** 0 ft **43.** 64 ft **45.** 13 **47.** 35 **49.** −90 **51.** −48 **53.** −34.225 **55.** 0.17283171 **57.** −0.12268448

59.

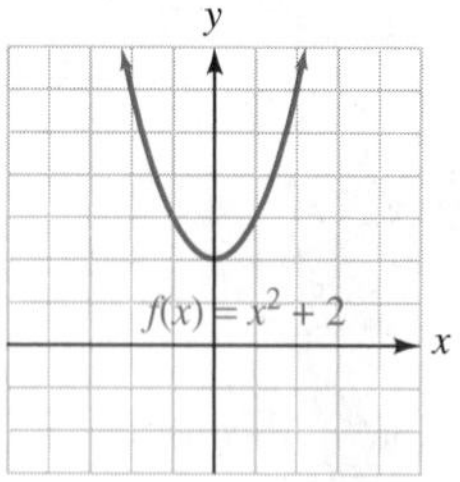

61.

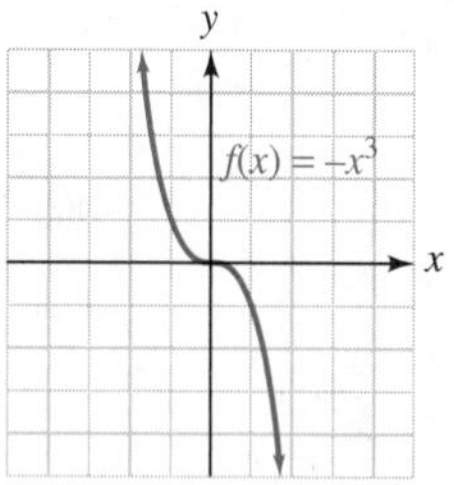

63.

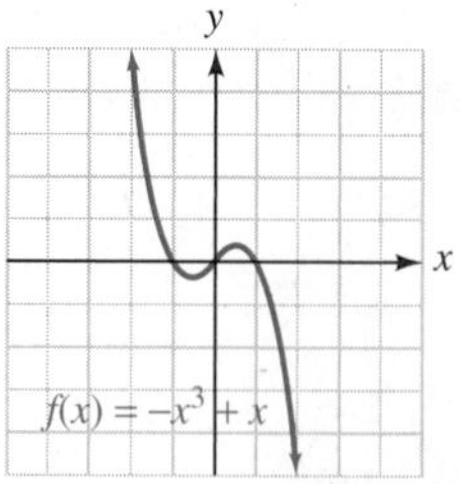

65.

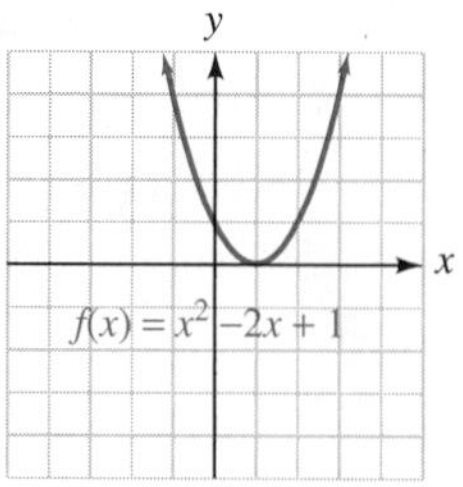

67.

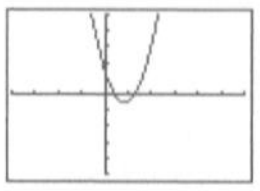

69.

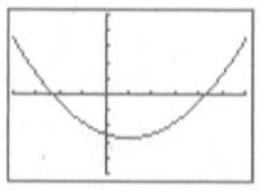

71. 63 ft **73.** 198 ft **75.** 10 in.^{2} **77.** 18 in.^{2} **81.** no **83.** 12

Getting Ready (page 327)

1. $2x + 6$ **2.** $-8x + 20$ **3.** $5x + 15$ **4.** $-2x + 6$

Orals (page 330)

1. $9x^2$ **2.** $-2y^2$ **3.** $3x^2 + 2$ **4.** $-x^2 + 4x$ **5.** $3x^2 - 4x - 4$ **6.** $x^2 + 2x - 2$

Exercise 5.2 (page 330)

1. $(-\infty, 4]$ **3.** $(-1, 9)$ **5.** exponents **7.** coefficients **9.** like terms, $10x$ **11.** unlike terms **13.** like terms, $-5r^2t^3$ **15.** unlike terms **17.** $12x$ **19.** $2x^3y^2z$ **21.** $-7x^2y^3 + 3xy^4$ **23.** $10x^4y^2$ **25.** $x^2 - 5x + 6$ **27.** $-5a^2 + 4a + 4$ **29.** $-y^3 + 4y^2 + 6$ **31.** $4x^2 - 11$ **33.** $3x^3 - x + 13$ **35.** $4y^2 - 9y + 3$ **37.** $6x^3 - 6x^2 + 14x - 17$ **39.** $-9y^4 + 3y^3 - 15y^2 + 20y + 12$ **41.** $x^2 - 8x + 22$ **43.** $-3y^3 + 18y^2 - 28y + 35$ **45.** $5x - 4$ **47.** $-11t + 29$ **49.** $8x^3 - x^2$ **51.** $-4m - 3n$ **53.** $-16x^3 - 27x^2 - 12x$ **55.** $-8z^2 - 40z + 54$ **57.** $14a^2 + 16a - 24$ **59.** $3x + 3$ **61.** \$136,000 **63.** $y = 2{,}500x + 275{,}000$ **65.** $y = -2{,}100x + 16{,}600$ **67.** $y = -4{,}800x + 35{,}800$ **71.** $-2x + 9$ **73.** $-8x^2 - 2x + 2$

Getting Ready (page 333)

1. $12a$ **2.** $4a^4$ **3.** $-7a^4$ **4.** $12a^5$ **5.** $2a + 8$ **6.** $a^2 - 3a$ **7.** $-4a + 12$ **8.** $-2b^2 - 4b$

Orals (page 341)

1. $-6a^3b^3$ **2.** $-8x^2y^3$ **3.** $6a^3 - 3a^2$ **4.** $-16mn^2 + 4n^3$ **5.** $2x^2 + 3x + 1$ **6.** $6y^2 - y - 2$

Exercise 5.3 (page 341)

1. 10 **3.** -15 **5.** \$51,025 **7.** variable **9.** term **11.** $x^2 + 2xy + y^2$ **13.** $x^2 - y^2$ **15.** $-6a^3b$ **17.** $-15a^2b^2c^3$ **19.** $-120a^9b^3$ **21.** $25x^8y^8$ **23.** $-405x^7y^4$ **25.** $3x + 6$ **27.** $-a^2 + ab$ **29.** $3x^3 + 9x^2$ **31.** $-6x^3 + 6x^2 - 4x$ **33.** $10a^6b^4 - 25a^2b^6$ **35.** $7r^3st + 7rs^3t - 7rst^3$ **37.** $-12m^4n^2 - 12m^3n^3$ **39.** $x^2 + 5x + 6$ **41.** $z^2 - 9z + 14$ **43.** $2a^2 - 3a - 2$ **45.** $6t^2 + 5t - 6$ **47.** $6y^2 - 5yz + z^2$ **49.** $2x^2 + xy - 6y^2$ **51.** $9x^2 - 6xy - 3y^2$ **53.** $8a^2 + 14ab - 15b^2$ **55.** $x^2 + 4x + 4$ **57.** $a^2 - 8a + 16$ **59.** $4a^2 + 4ab + b^2$ **61.** $4x^2 - 4xy + y^2$ **63.** $x^2 - 4$ **65.** $a^2 - b^2$ **67.** $4x^2 - 9y^2$ **69.** $x^3 - y^3$ **71.** $6y^3 + 11y^2 + 9y + 2$ **73.** $8a^3 - b^3$ **75.** $4x^3 - 8x^2 - 9x + 6$ **77.** $a^3 - 3a^2b - ab^2 + 3b^3$ **79.** $2x^5 + x$ **81.** $\frac{3x}{y^4z} - \frac{x^6}{y^{10}z^2}$ **83.** $\frac{1}{x^2} - y^2$ **85.** $2x^3y^3 - \frac{2}{x^3y^3} + 3$ **87.** $x^{3n} - x^{2n}$ **89.** $x^{2n} - 1$ **91.** $x^{2n} - \frac{x^n}{y^n} - x^ny^n + 1$ **93.** $x^{4n} - y^{4n}$ **95.** $x^{2n} + x^n + x^ny^n + y^n$ **97.** $3x^2 + 12x$ **99.** $-p^2 + 4pq$ **101.** $m^2 - mn + 2n^2$ **103.** $3x^2 + 3x - 11$ **105.** $5x^2 - 36x + 7$ **107.** $21x^2 - 6xy + 29y^2$ **109.** $24y^2 - 4yz + 21z^2$ **111.** $9.2127x^2 - 7.7956x - 36.0315$ **113.** $299.29y^2 - 150.51y + 18.9225$ **117.** 15 **119.** $r = -\frac{1}{5}p^2 + 90p$ **121.** $x^2 + 2x - 8$ **125.** 2.31×10^7

Getting Ready (page 345)

1. $4a + 8$ **2.** $-5b + 25$ **3.** $a^2 + 5a$ **4.** $-b^2 + 3b$ **5.** $6a^2 + 12ab$ **6.** $-6p^3 - 10p^2$

Orals (page 352)

1. $x(3x - 1)$ **2.** $7t^2(t + 2)$ **3.** $-3a(a + 2)$ **4.** $-4x(x - 3)$ **5.** $(a + b)(3 + x)$ **6.** $(m - n)(a - b)$

Exercise 5.4 (page 352)

1. $a^2 - 16$ **3.** $16r^4 - 9s^2$ **5.** $m^3 + 64$ **7.** factoring **9.** greatest common factor **11.** $2 \cdot 3$ **13.** $3^3 \cdot 5$ **15.** 2^7 **17.** $5^2 \cdot 13$ **19.** 12 **21.** 2 **23.** $4a^2$ **25.** $6xy^2z^2$ **27.** 4 **29.** 1 **31.** $2(x + 4)$ **33.** $2x(x - 3)$ **35.** prime **37.** $5x^2y(3 - 2y)$ **39.** $9x^2y^2(7x + 9y^2)$ **41.** prime **43.** $3z(9z^2 + 4z + 1)$ **45.** $6s(4s^2 - 2st + t^2)$ **47.** $9x^7y^3(5x^3 - 7y^4 + 9x^3y^7)$ **49.** prime **51.** $-3(a + 2)$ **53.** $-x(3x + 1)$ **55.** $-3x(2x + y)$ **57.** $-6ab(3a + 2b)$ **59.** $-7u^2v^3z^2(9uv^3z^7 - 4v^4 + 3uz^2)$ **61.** $x^2(x^n + x^{n+1})$ **63.** $y^n(2y^2 - 3y^3)$ **65.** $x^{-2}(x^6 - 5x^8)$ **67.** $t^{-3}(t^8 + 4t^{-3})$ **69.** $4y^{-2n}(2y^{4n} + 3y^{2n} + 4)$ **71.** $(x + y)(4 + t)$ **73.** $(a - b)(r - s)$ **75.** $(m + n + p)(3 + x)$ **77.** $(x + y)(x + y + z)$ **79.** $(u + v)(u + v - 1)$ **81.** $-(x + y)(a - b)$ **83.** $(x + y)(a + b)$ **85.** $(x + 2)(x + y)$ **87.** $(3 - c)(c + d)$ **89.** $(a + b)(a - 4)$ **91.** $(a + b)(x - 1)$ **93.** $(x + y)(x + y + z)$ **95.** $x(m + n)(p + q)$ **97.** $y(x + y)(x + y + 2z)$ **99.** $n(2n - p + 2m)(n^2p - 1)$ **101.** $r_1 = \frac{rr_2}{r_2 - r}$ **103.** $f = \frac{d_1d_2}{d_2 + d_1}$ **105.** $a^2 = \frac{b^2x^2}{b^2 - y^2}$ **107.** $r = \frac{S - a}{S - l}$ **109.** $a = \frac{Hb}{2b - H}$ **111.** $y = \frac{x + 3}{3x - 2}$ **113.** $x[x(2x + 5) - 2] + 8$ **119.** yes **121.** no **123.** yes

Getting Ready (page 355)

1. $a^2 - b^2$ **2.** $25p^2 - q^2$ **3.** $9m^2 - 4n^2$ **4.** $4a^4 - b^4$ **5.** $a^3 - 27$ **6.** $p^3 + 8$

Orals (page 361)

1. $(x + 1)(x - 1)$ **2.** $(a^2 + 4)(a + 2)(a - 2)$ **3.** $(x + 1)(x^2 - x + 1)$ **4.** $(a - 2)(a^2 + 2a + 4)$ **5.** $2(x + 2)(x - 2)$ **6.** prime

Exercise 5.5 (page 362)

1. $x^2 + 2x + 1$ **3.** $4m^2 + 4mn + n^2$ **5.** $a^2 + 7a + 12$ **7.** $8r^2 - 10rs + 3s^2$ **9.** 1, 4, 9, 16, 25, 36, 49, 64, 81, 100 **11.** cannot **13.** $(p^2 - pq + q^2)$ **15.** $(x + 2)(x - 2)$ **17.** $(3y + 8)(3y - 8)$ **19.** prime **21.** $(25a + 13b^2)(25a - 13b^2)$ **23.** $(9a^2 + 7b)(9a^2 - 7b)$ **25.** $(6x^2y + 7z^2)(6x^2y - 7z^2)$ **27.** $(x + y + z)(x + y - z)$ **29.** $(a - b + c)(a - b - c)$ **31.** $(x^2 + y^2)(x + y)(x - y)$ **33.** $(16x^2y^2 + z^4)(4xy + z^2)(4xy - z^2)$ **35.** $2(x + 12)(x - 12)$ **37.** $2x(x + 4)(x - 4)$ **39.** $5x(x + 5)(x - 5)$ **41.** $t^2(rs + x^2y)(rs - x^2y)$ **43.** $(r + s)(r^2 - rs + s^2)$ **45.** $(x - 2y)(x^2 + 2xy + 4y^2)$ **47.** $(4a - 5b^2)(16a^2 + 20ab^2 + 25b^4)$ **49.** $(5xy^2 + 6z^3)(25x^2y^4 - 30xy^2z^3 + 36z^6)$ **51.** $(x^2 + y^2)(x^4 - x^2y^2 + y^4)$ **53.** $5(x + 5)(x^2 - 5x + 25)$ **55.** $4x^2(x - 4)(x^2 + 4x + 16)$ **57.** $2u^2(4v - t)(16v^2 + 4vt + t^2)$ **59.** $(a + b)(x + 3)(x^2 - 3x + 9)$ **61.** $(x^m + y^{2n})(x^m - y^{2n})$ **63.** $(10a^{2m} + 9b^n)(10a^{2m} - 9b^n)$ **65.** $(x^n - 2)(x^{2n} + 2x^n + 4)$ **67.** $(a^m + b^n)(a^{2m} - a^mb^n + b^{2n})$ **69.** $2(x^{2m} + 2y^m)(x^{4m} - 2x^{2m}y^m + 4y^{2m})$ **71.** $(a + b)(a - b + 1)$ **73.** $(a - b)(a + b + 2)$ **75.** $(2x + y)(1 + 2x - y)$ **79.** $(x^{16} + y^{16})(x^8 + y^8)(x^4 + y^4)(x^2 + y^2)(x + y)(x - y)$

Getting Ready (page 364)

1. $a^2 - a - 12$ **2.** $6a^2 + 7a - 5$ **3.** $a^2 + 3ab + 2b^2$ **4.** $2a^2 - ab - b^2$ **5.** $6a^2 - 13ab + 6b^2$ **6.** $16a^2 - 24ab + 9b^2$

Orals (page 373)

1. $(x + 2)(x + 1)$ **2.** $(x + 4)(x + 1)$ **3.** $(x - 3)(x - 2)$ **4.** $(x + 1)(x - 4)$ **5.** $(2x + 1)(x + 1)$ **6.** $(3x + 1)(x + 1)$

Exercise 5.6 (page 373)

1. 31 **3.** 12 **5.** -3 **7.** $2xy + y^2$ **9.** $x^2 - y^2$ **11.** $x + 2$ **13.** $x - 3$ **15.** $2a + 1$ **17.** $2m + 3n$ **19.** $(x + 1)^2$ **21.** $(a - 9)^2$ **23.** $(2y + 1)^2$ **25.** $(3b - 2)^2$ **27.** $(3z + 4)^2$ **29.** $(x + 8)(x + 1)$ **31.** $(x - 2)(x - 5)$ **33.** prime **35.** $(x + 5)(x - 6)$ **37.** $(a + 10)(a - 5)$ **39.** $(y - 7)(y + 3)$ **41.** $3(x + 7)(x - 3)$ **43.** $b^2(a - 11)(a - 2)$ **45.** $x^2(b - 7)(b - 5)$ **47.** $-(a - 8)(a + 4)$ **49.** $-3(x - 3)(x - 2)$ **51.** $-4(x - 5)(x + 4)$ **53.** $(3y + 2)(2y + 1)$ **55.** $(4a - 3)(2a + 3)$ **57.** $(3x - 4)(2x + 1)$ **59.** prime **61.** $(4x - 3)(2x - 1)$ **63.** $(a + b)(a - 4b)$ **65.** $(2y - 3t)(y + 2t)$ **67.** $x(3x - 1)(x - 3)$ **69.** $-(3a + 2b)(a - b)$ **71.** $-(2x - 3)^2$ **73.** $5(a - 3b)^2$ **75.** $z(8x^2 + 6xy + 9y^2)$ **77.** $x^2(7x - 8)(3x + 2)$ **79.** $(x^2 + 5)(x^2 + 3)$ **81.** $(y^2 - 10)(y^2 - 3)$ **83.** $(a + 3)(a - 3)(a + 2)(a - 2)$ **85.** $(z^2 + 3)(z + 2)(z - 2)$ **87.** $(x^3 + 3)(x + 1)(x^2 - x + 1)$ **89.** $(x^n + 1)^2$ **91.** $(2a^{3n} + 1)(a^{3n} - 2)$ **93.** $(x^{2n} + y^{2n})^2$ **95.** $(3x^n - 1)(2x^n + 3)$ **97.** $(x + 2)^2$ **99.** $(a + b + 4)(a + b - 6)$ **101.** $(3x + 3y + 4)(2x + 2y - 5)$ **103.** $(x + 2 + y)(x + 2 - y)$ **105.** $(x + 1 + 3z)(x + 1 - 3z)$ **107.** $(c + 2a - b)(c - 2a + b)$ **109.** $(a + 4 + b)(a + 4 - b)$ **111.** $(2x + y + z)(2x + y - z)$ **113.** $(a - 16)(a - 1)$ **115.** $(2u + 3)(u + 1)$ **117.** $(5r + 2s)(4r - 3s)$ **119.** $(5u + v)(4u + 3v)$ **123.** yes

Getting Ready (page 376)

1. $3pq(p - q)$ **2.** $(2p + 3q)(2p - 3q)$ **3.** $(p + 6)(p - 1)$ **4.** $(2p - 3q)(3p - 2q)$

Orals (page 379)

1. $(x + y)(x - y)$ **2.** $2x^3(1 - 2x)$ **3.** $(x + 2)^2$ **4.** $(x - 3)(x - 2)$ **5.** $(x - 2)(x^2 + 2x + 4)$ **6.** $(x + 2)(x^2 - 2x + 4)$

Exercise 5.7 (page 379)

1. $7a^2 + a - 7$ **3.** $4y^2 - 11y + 3$ **5.** $m^2 + 2m - 8$ **7.** common factors **9.** trinomial **11.** $(x + 4)^2$ **13.** $(2xy - 3)(4x^2y^2 + 6xy + 9)$ **15.** $(x - t)(y + s)$ **17.** $(5x + 4y)(5x - 4y)$ **19.** $(6x + 5)(2x + 7)$ **21.** $2(3x - 4)(x - 1)$

23. $(8x - 1)(7x - 1)$ **25.** $y^2(2x + 1)(2x + 1)$ **27.** $(x + a^2y)(x^2 - a^2xy + a^4y^2)$ **29.** $2(x - 3)(x^2 + 3x + 9)$ **31.** $(a + b)(f + e)$ **33.** $(2x + 2y + 3)(x + y - 1)$ **35.** $(25x^2 + 16y^2)(5x + 4y)(5x - 4y)$ **37.** $36(x^2 + 1)(x + 1)(x - 1)$ **39.** $2(x^2 + y^2)(x^4 - x^2y^2 + y^4)$ **41.** $(a + 3)(a - 3)(a + 2)(a - 2)$ **43.** $(x + 3 + y)(x + 3 - y)$ **45.** $(2x + 1 + 2y)(2x + 1 - 2y)$ **47.** $(x + y + 1)(x - y - 1)$ **49.** $(x + 1)(x^2 - x + 1)(x + 1)(x - 1)$ **51.** $(x + 3)(x - 3)(x + 2)(x^2 - 2x + 4)$ **53.** $2z(x + y)(x - y)^2(x^2 + xy + y^2)$ **55.** $(x^m - 3)(x^m + 2)$ **57.** $(a^n - b^n)(a^{2n} + a^nb^n + b^{2n})$ **59.** $\left(\frac{1}{x} + 1\right)^2$ **61.** $\left(\frac{3}{x} + 2\right)\left(\frac{2}{x} - 3\right)$ **67.** $(x^2 + x + 1)(x^2 - x + 1)$

Getting Ready (page 381)

1. $2a(a - 2)$ **2.** $(a + 5)(a - 5)$ **3.** $(3a - 2)(2a + 3)$ **4.** $a(3a - 2)(2a + 1)$

Orals (page 387)

1. 2, 3 **2.** −4, 2 **3.** 2, 3, −1 **4.** −3, −2, 5, 6

Exercise 5.8 (page 387)

1. 2, 3, 5, 7 **3.** 40,081.00 cm^3 **5.** $ax^2 + bx + c = 0$ **7.** 0, −2 **9.** 4, −4 **11.** 0, −1 **13.** 0, 5 **15.** −3, −5 **17.** 1, 6 **19.** 3, 4 **21.** −2, −4 **23.** $-\frac{1}{3}$, −3 **25.** $\frac{1}{2}$, 2 **27.** 1, $-\frac{1}{2}$ **29.** 2, $-\frac{1}{3}$ **31.** 3, 3 **33.** $\frac{1}{4}$, $-\frac{3}{2}$ **35.** 2, $-\frac{5}{6}$ **37.** $\frac{1}{3}$, −1 **39.** 2, $\frac{1}{2}$ **41.** $\frac{1}{5}$, $-\frac{5}{3}$ **43.** 0, 0, −1 **45.** 0, 7, −7 **47.** 0, 7, −3 **49.** 3, −3, 2, −2 **51.** 0, −2, −3 **53.** 0, $\frac{5}{6}$, −7 **55.** 1, −1, −3 **57.** 3, −3, $-\frac{3}{2}$ **59.** 2, −2, $-\frac{1}{3}$ **61.** 16, 18, or −18, −16 **63.** 6, 7 **65.** 40 m **67.** 15 ft by 25 ft **69.** 5 ft by 12 ft **71.** 20 ft by 40 ft **73.** 10 sec **75.** 11 sec and 19 sec **77.** 50 m **79.** 3 ft **81.** no solution **83.** 1 **87.** $x^2 - 8x + 15 = 0$ **89.** $x^2 + 5x = 0$

Chapter Summary (page 393)

1. 5 **2.** 8 **3. a.** 6 **b.** 9 **c.** $-t^2 - 4t + 6$ **d.** $-z^2 + 4z + 6$ **4. a.**

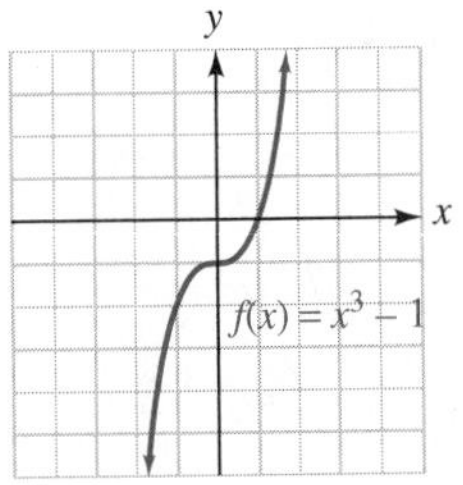

b.

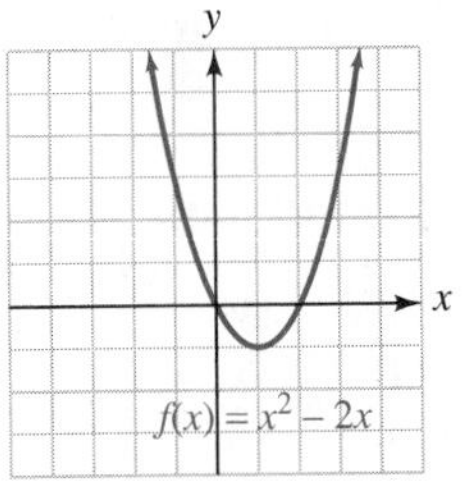

5. a. $5x^2 + 2x + 16$ **b.** $6x^3 + 4x^2 + x + 9$ **c.** $4x^2 - 9x + 19$ **d.** $-7x^3 - 30x^2 - 4x + 3$ **6. a.** $-16a^3b^3c$ **b.** $-6x^2y^2z^4$ **7. a.** $2x^4y^3 - 8x^2y^7$ **b.** $a^4b + 2a^3b^2 + a^2b^3$ **8. a.** $16x^2 + 14x - 15$ **b.** $6x^2 - 8x - 8$ **c.** $15x^2 - 22x + 8$ **d.** $3x^4 - 3x^3 + 4x^2 + 2x - 4$ **9. a.** $4(x + 2)$ **b.** $3x(x - 2)$ **c.** $5xy^2(xy - 2)$ **d.** $7a^3b(ab + 7)$ **e.** $-4x^2y^3z^2(2z^2 + 3x^2)$ **f.** $3a^2b^4c^2(4a^4 + 5c^4)$ **g.** $9x^2y^3z^2(3xz + 9x^2y^2 - 10z^5)$ **h.** $-12a^2b^3c^2(3a^3b - 5a^5b^2c + 2c^5)$ **i.** $x^n(x^n + 1)$ **j.** $y^{2n}(1 - y^{2n})$ **k.** $x^{-2}(x^{-2} - 1)$ **l.** $a^{-3}(a^9 + a^3)$ **m.** $5x^2(x + y)^3(1 - 3x^2 - 3xy)$ **n.** $-7a^2b^2(a - b)^3(7a^2 - 7ab - 9b^2)$ **10. a.** $(x + 2)(y + 4)$ **b.** $(a + b)(c + 3)$ **c.** $(x^2 + 4)(x^2 + y)$ **d.** $(a^3 + c)(a^2 + b^2)$ **11. a.** $h = \frac{S - 2wl}{2w + 2l}$ **b.** $l = \frac{S - 2wh}{2w + 2h}$ **12. a.** $(z + 4)(z - 4)$ **b.** $(y + 11)(y - 11)$ **c.** $(xy^2 + 8z^3)(xy^2 - 8z^3)$ **d.** prime **e.** $(x + z + t)(x + z - t)$ **f.** $(c + a + b)(c - a - b)$ **g.** $2(x^2 + 7)(x^2 - 7)$ **h.** $3x^2(x^2 + 10)(x^2 - 10)$ **13. a.** $(x + 7)(x^2 - 7x + 49)$ **b.** $(a - 5)(a^2 + 5a + 25)$ **c.** $8(y - 4)(y^2 + 4y + 16)$ **d.** $4y(x + 3z)(x^2 - 3xz + 9z^2)$ **14. a.** $(x + 5)(x + 5)$ **b.** $(a - 7)(a - 7)$ **15. a.** $(y + 20)(y + 1)$ **b.** $(z - 5)(z - 6)$ **c.** $-(x + 7)(x - 4)$ **d.** $(y - 8)(y + 3)$ **e.** $(4a - 1)(a - 1)$ **f.** prime **g.** prime **h.** $-(5x + 2)(3x - 4)$ **i.** $y(y + 2)(y - 1)$ **j.** $2a^2(a + 3)(a - 1)$ **k.** $-3(x + 2)(x + 1)$ **l.** $4(2x + 3)(x - 2)$ **m.** $3(5x + y)(x - 4y)$ **n.** $5(6x + y)(x + 2y)$ **o.** $(8x + 3y)(3x - 4y)$ **p.** $(2x + 3y)(7x - 4y)$ **16. a.** $x(x - 1)(x + 6)$ **b.** $3y(x + 3)(x - 7)$ **c.** $(z - 2)(z + x + 2)$ **d.** $(x + 1 + p)(x + 1 - p)$ **e.** $(x + 2 + 2p^2)(x + 2 - 2p^2)$ **f.** $(y + 2)(y + 1 + x)$ **g.** $(x^m + 3)(x^m - 1)$ **h.** $\left(\frac{1}{x} - 2\right)\left(\frac{1}{x} + 1\right)$ **17. a.** 0, $\frac{3}{4}$ **b.** 6, −6 **c.** $\frac{1}{2}$, $-\frac{5}{6}$ **d.** $\frac{2}{7}$, 5 **e.** 0, $-\frac{2}{3}$, $\frac{4}{5}$ **f.** $-\frac{2}{3}$, 7, 0 **18.** 7 cm **19.** 17 m by 20 m

Chapter 5 Test (page 397)

1. 5 **2.** 13 **3.** -9 **4.** -6 **5.**

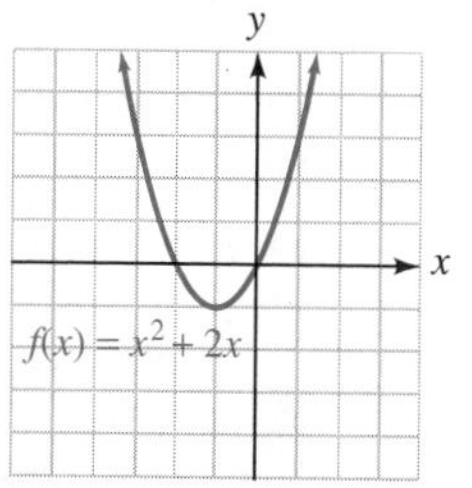

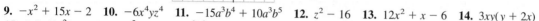

6. $5y^2 + y - 1$ **7.** $-4u^2 + 2u - 14$ **8.** $10a^2 + 22$
9. $-x^2 + 15x - 2$ **10.** $-6x^4yz^4$ **11.** $-15a^3b^4 + 10a^3b^5$ **12.** $z^2 - 16$ **13.** $12x^2 + x - 6$ **14.** $3xy(y + 2x)$
15. $3abc(4a^2b - abc + 2c^2)$ **16.** $y^n(x^2y^2 + 1)$ **17.** $b^n(a^n - ab^{-2n})$ **18.** $(u - v)(r + s)$ **19.** $(a - y)(x + y)$ **20.** $(x + 7)(x - 7)$
21. $2(x + 4)(x - 4)$ **22.** $4(y^2 + 4)(y + 2)(y - 2)$ **23.** $(b + 5)(b^2 - 5b + 25)$ **24.** $(b - 3)(b^2 + 3b + 9)$
25. $3(u - 2)(u^2 + 2u + 4)$ **26.** $(a - 6)(a + 1)$ **27.** $(3b + 2)(2b - 1)$ **28.** $3(u + 2)(2u - 1)$ **29.** $5(4r + 1)(r - 1)$
30. $(x^n + 1)^2$ **31.** $(x + 3 + y)(x + 3 - y)$ **32.** $r = \dfrac{r_1r_2}{r_2 + r_1}$ **33.** 6, -1 **34.** 25 **35.** 11 ft by 22 ft

Getting Ready (page 400)

1. $\frac{3}{4}$ **2.** $\frac{4}{5}$ **3.** $-\frac{5}{13}$ **4.** $\frac{7}{9}$

Orals (page 410)

1. -1 **2.** 1 **3.** $(-\infty, 2) \cup (2, \infty)$ **4.** $(-\infty, -3) \cup (-3, 3) \cup (3, \infty)$ **5.** $\frac{5}{6}$ **6.** $\frac{x}{y}$ **7.** 2 **8.** -1

Exercise 6.1 (page 410)

1. $3x(x - 3)$ **3.** $(3x^2 + 4y)(9x^4 - 12x^2y + 16y^2)$ **5.** rational **7.** asymptote **9.** a **11.** $\frac{a}{b}$, 0 **13.** 20 hr **15.** 12 hr
17. \$5,555.56 **19.** \$50,000 **21.** $c = f(x) = 1.25x + 700$ **23.** \$1,325 **25.** \$1.95 **27.** $c = f(n) = 0.09n + 7.50$ **29.** \$77.25
31. 9.75¢ **33.** almost 8 days **35.** about 2.55 hr
37. $(-\infty, 2) \cup (2, \infty)$

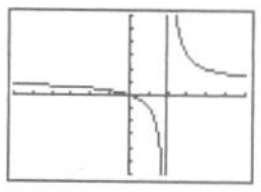

39. $(-\infty, -2) \cup (-2, 2) \cup (2, \infty)$

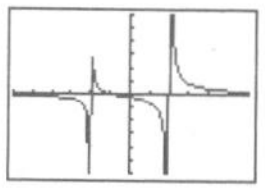

41. $\frac{2}{3}$ **43.** $-\frac{28}{9}$ **45.** $\frac{12}{13}$ **47.** $-\frac{122}{37}$
49. $4x^2$ **51.** $-\dfrac{4y}{3x}$ **53.** x^2 **55.** $-\dfrac{x}{2}$ **57.** $\dfrac{3y}{7(y - z)}$ **59.** 1 **61.** $\dfrac{1}{x - y}$ **63.** $\dfrac{5}{x - 2}$ **65.** $\dfrac{-3(x + 2)}{x + 1}$ **67.** 3 **69.** $x + 2$
71. $\dfrac{x + 1}{x + 3}$ **73.** $\dfrac{m - 2n}{n - 2m}$ **75.** $\dfrac{x + 4}{2(2x - 3)}$ **77.** $\dfrac{3(x - y)}{x + 2}$ **79.** $\dfrac{2x + 1}{2 - x}$ **81.** $\dfrac{a^2 - 3a + 9}{4(a - 3)}$ **83.** in lowest terms **85.** $m + n$
87. $-\dfrac{m + n}{2m + n}$ **89.** $\dfrac{x - y}{x + y}$ **91.** $\dfrac{2a - 3b}{a - 2b}$ **93.** $\dfrac{1}{x^2 + xy + y^2 - 1}$ **95.** $x^2 - y^2$ **97.** $\dfrac{(x + 1)^2}{(x - 1)^3}$ **99.** $\dfrac{3a + b}{y + b}$ **101.** $\frac{1}{6}$ **103.** 0
105. $\frac{1}{2}$ **107.** $\frac{1}{13}$ **109.** $\frac{1}{6{,}000{,}000}$ **111.** $\frac{1}{8}$ **115.** yes **117.** a, d

Getting Ready (page 414)

1. $\frac{7}{3}$ **2.** 1.44×10^9 **3.** linear function

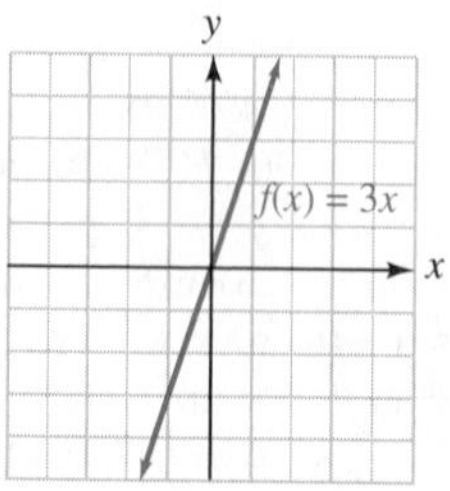

4. rational function

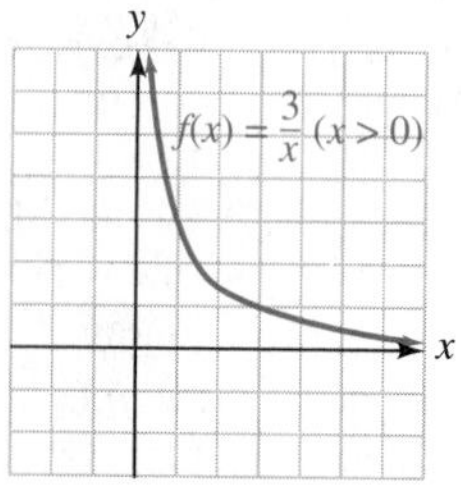

Orals (page 424)

1. 1 **2.** 9 **3.** $\frac{14}{5}$ **4.** $a = kb$ **5.** $a = \frac{k}{b}$ **6.** $a = kbc$ **7.** $a = \frac{kb}{c}$

Exercise 6.2 (page 424)

1. x^{10} **3.** -1 **5.** 3.5×10^4 **7.** 0.0025 **9.** unit costs, rates **11.** extremes, means **13.** direct **15.** rational **17.** joint **19.** direct **21.** neither **23.** 3 **25.** 5 **27.** -3 **29.** 5 **31.** 4, -1 **33.** 2, -2 **35.** 39 **37.** $-\frac{5}{2}$, -1 **39.** $A = kp^2$ **41.** $v = k/r^3$ **43.** $B = kmn$ **45.** $P = ka^2/j^3$ **47.** L varies jointly with m and n **49.** E varies jointly with a and the square of b **51.** X varies directly with x^2 and inversely with y^2 **53.** R varies directly with L and inversely with d^2 **55.** \$62.50 **57.** $7\frac{1}{2}$ gal **59.** 32 ft **61.** 80 ft **63.** 0.18 g **65.** 42 ft **67.** $46\frac{7}{8}$ ft **69.** 6,750 ft **71.** 36π in.^{2} **73.** 432 mi **75.** 25 days **77.** 12 in.^{3} **79.** 85.3 **81.** 12 **83.** 26,437.5 gal **85.** 3 ohms **87.** 0.275 in. **89.** 546 Kelvin

Getting Ready (page 429)

1. $\frac{3}{2}$ **2.** $\frac{2}{5}$ **3.** $\frac{4}{3}$ **4.** $\frac{14}{3}$

Orals (page 435)

1. $\frac{9}{8}$ **2.** $\frac{1}{2}$ **3.** $\frac{x-2}{x+2}$ **4.** $\frac{9}{16}$ **5.** 5 **6.** $\frac{y}{2}$

Exercise 6.3 (page 435)

1. $-6a^5 + 2a^4$ **3.** $m^{2n} - 4$ **5.** $\frac{ac}{bd}$ **7.** 0 **9.** $\frac{10}{7}$ **11.** $-\frac{5}{6}$ **13.** $\frac{xy^2d}{c^3}$ **15.** $-\frac{x^{10}}{y^2}$ **17.** $x + 1$ **19.** 1 **21.** $\frac{x-4}{x+5}$ **23.** $\frac{(a+7)^2(a-5)}{12x^2}$ **25.** $\frac{t-1}{t+1}$ **27.** $\frac{n+2}{n+1}$ **29.** $\frac{1}{x+1}$ **31.** $(x+1)^2$ **33.** $x - 5$ **35.** $\frac{x+y}{x-y}$ **37.** $-\frac{x+3}{x+2}$ **39.** $\frac{a+b}{(x-3)(c+d)}$ **41.** $-\frac{x+1}{x+3}$ **43.** $-\frac{2x+y}{3x+y}$ **45.** $-\frac{x^7}{18y^4}$ **47.** $x^2(x+3)$ **49.** $\frac{3x}{2}$ **51.** $\frac{t+1}{t}$ **53.** $\frac{x-2}{x-1}$ **55.** $\frac{x+2}{x-2}$ **57.** $\frac{x-1}{3x+2}$ **59.** $\frac{x-7}{x+7}$ **61.** 1 **65.** $\div, \cdot$

Getting Ready (page 437)

1. yes **2.** no **3.** yes **4.** yes **5.** no **6.** yes

Orals (page 445)

1. x **2.** $\frac{a}{2}$ **3.** 1 **4.** 1 **5.** $\frac{7x}{6}$ **6.** $\frac{5y-3x}{xy}$

Exercise 6.4 (page 445)

1.

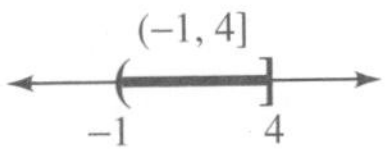

3. $w = \frac{P-2l}{2}$ **5.** $\frac{a+c}{b}$ **7.** subtract, keep **9.** LCD **11.** $\frac{5}{2}$ **13.** $-\frac{1}{3}$ **15.** $\frac{11}{4y}$ **17.** $\frac{3-a}{a+b}$ **19.** 2 **21.** 3 **23.** 3 **25.** $\frac{6x}{(x-3)(x-2)}$ **27.** 72 **29.** $x(x+3)(x-3)$ **31.** $(x+3)^2(x^2-3x+9)$ **33.** $(2x+3)^2(x+1)^2$ **35.** $\frac{5}{6}$ **37.** $-\frac{16}{75}$ **39.** $\frac{9a}{10}$ **41.** $\frac{21a-8b}{14}$ **43.** $\frac{17}{12x}$ **45.** $\frac{9a^2-4b^2}{6ab}$ **47.** $\frac{10a+4b}{21}$ **49.** $\frac{8x-2}{(x+2)(x-4)}$ **51.** $\frac{7x+29}{(x+5)(x+7)}$ **53.** $\frac{x^2+1}{x}$ **55.** 2 **57.** $\frac{9a^2-11a-10}{(3a+2)(3a-2)}$ **59.** $\frac{2x^2+x}{(x+3)(x+2)(x-2)}$ **61.** $\frac{-x^2+11x+8}{(3x+2)(x+1)(x-3)}$ **63.** $\frac{-4x^2+14x+54}{x(x+3)(x-3)}$ **65.** $\frac{x^3+x^2-1}{x(x+1)(x-1)}$ **67.** $\frac{2x^2+5x+4}{x+1}$ **69.** $\frac{x^2-5x-5}{x-5}$ **71.** $\frac{-x^3-x^2+5x}{x-1}$ **73.** $\frac{-y^2+48y+161}{(y+4)(y+3)}$ **75.** $\frac{2}{x+1}$ **77.** $\frac{3x+1}{x(x+3)}$

79. $\frac{2x^3 + x^2 - 43x - 35}{(x+5)(x-5)(2x+1)}$ **81.** $\frac{3x^2 - 2x - 17}{(x-3)(x-2)}$ **83.** $\frac{2a}{a-1}$ **85.** $\frac{x^2 - 6x - 1}{2(x+1)(x-1)}$ **87.** $\frac{2b}{a+b}$ **89.** $\frac{7mn^2 - 7n^3 - 6m^2 + 3mn + n}{(m-n)^2}$ **91.** $\frac{2}{m-1}$ **93.** 0 **95.** 1

Getting Ready (page 449)

1. 3 **2.** -34 **3.** $3 + 2a$ **4.** $5 - 2b$

Orals (page 456)

1. $\frac{3}{5}$ **2.** $\frac{a}{d}$ **3.** 2 **4.** $\frac{x+y}{x-y}$ **5.** $\frac{x-y}{x}$ **6.** $\frac{b+a}{a}$

Exercise 6.5 (page 456)

1. 8 **3.** 2, -2, 3, -3 **5.** complex **7.** $\frac{2}{3}$ **9.** -1 **11.** $\frac{10}{3}$ **13.** $-\frac{1}{7}$ **15.** $\frac{2y}{3z}$ **17.** $125b$ **19.** $-\frac{1}{y}$ **21.** $\frac{y-x}{x^2y^2}$ **23.** $\frac{b+a}{b}$ **25.** $\frac{y+x}{y-x}$ **27.** $y - x$ **29.** $\frac{-1}{a+b}$ **31.** $x^2 + x - 6$ **33.** $\frac{5x^2y^2}{xy+1}$ **35.** $\frac{x+2}{x-3}$ **37.** $\frac{a-1}{a+1}$ **39.** $\frac{y+x}{x^2y}$ **41.** $\frac{xy^2}{y-x}$ **43.** $\frac{y+x}{y-x}$ **45.** xy **47.** $\frac{x^2(xy^2-1)}{y^2(x^2y-1)}$ **49.** $\frac{(b+a)(b-a)}{b(b-a-ab)}$ **51.** $\frac{x-1}{x}$ **53.** $\frac{5b}{5b+4}$ **55.** $\frac{3a^2+2a}{2a+1}$ **57.** $\frac{(-x^2+2x-2)(3x+2)}{(2-x)(-3x^2-2x+9)}$ **59.** $\frac{2(x^2-4x-1)}{-x^2+4x+8}$ **61.** $\frac{-2}{x^2-3x-7}$ **63.** $\frac{k_1k_2}{k_2+k_1}$ **67.** $\frac{1}{y+x}$

Getting Ready (page 459)

1. $\frac{9}{2}$ **2.** 3, -4

Orals (page 467)

1. 2 **2.** 3 **3.** 1 **4.** 3 **5.** 1 **6.** 2

Exercise 6.6 (page 467)

1. $\frac{n^6}{m^4}$ **3.** 0 **5.** rational **7.** 12 **9.** 40 **11.** $\frac{1}{2}$ **13.** no solution **15.** $\frac{17}{25}$ **17.** 0 **19.** 1 **21.** 2 **23.** 2 **25.** $\frac{1}{3}$ **27.** 0 **29.** 2, -5 **31.** -4, 3 **33.** 6, $\frac{17}{3}$ **35.** 1, -11 **37.** $f = \frac{pq}{q+p}$ **39.** $r = \frac{S-a}{S-l}$ **41.** $R = \frac{r_1r_2r_3}{r_1r_3 + r_1r_2 + r_2r_3}$ **43.** $4\frac{8}{13}$ in. **45.** $1\frac{7}{8}$ days **47.** $5\frac{5}{6}$ min **49.** $1\frac{1}{5}$ days **51.** $2\frac{4}{13}$ weeks **53.** 3 mph **55.** 60 mph and 40 mph **57.** 3 mph **59.** 60 mph **61.** 7 **63.** 12 days

Getting Ready (page 470)

1. $\frac{2}{3}$ **2.** $-\frac{5}{9}$ **3.** 5 **4.** 16

Orals (page 476)

1. $3xy$ **2.** $2b + 4a$ **3.** $x + 1$ **4.** $x + 2$

Exercise 6.7 (page 476)

1. $8x^2 + 2x + 4$ **3.** $-2y^3 - 3y^2 + 6y - 6$ **5.** $\frac{1}{b}$ **7.** quotient **9.** $\frac{y}{2x^3}$ **11.** $\frac{3b^4}{4a^4}$ **13.** $\frac{-5}{7xy^7t^2}$ **15.** $\frac{13a^nb^{2n}c^{3n-1}}{3}$
17. $\frac{2x}{3} - \frac{x^2}{6}$ **19.** $\frac{2xy^2}{3} + \frac{x^2y}{6}$ **21.** $\frac{x^4y^4}{2} - \frac{x^3y^9}{4} + \frac{3}{4xy^2}$ **23.** $\frac{b^2}{4a} - \frac{a^3}{2b^4} + \frac{3}{4ab}$ **25.** $1 - 3x^ny^n + 6x^{2n}y^{2n}$ **27.** $x + 2$
29. $x + 7$ **31.** $3x - 5 + \frac{3}{2x+3}$ **33.** $3x^2 + x + 2 + \frac{8}{x-1}$ **35.** $2x^2 + 5x + 3 + \frac{4}{3x-2}$ **37.** $3x^2 + 4x + 3$ **39.** $a + 1$ **41.** $2y + 2$
43. $6x - 12$ **45.** $3x^2 - x + 2$ **47.** $4x^3 - 3x^2 + 3x + 1$ **49.** $a^2 + a + 1 + \frac{2}{a-1}$ **51.** $5a^2 - 3a - 4$ **53.** $6y - 12$
55. $16x^4 - 8x^3y + 4x^2y^2 - 2xy^3 + y^4$ **57.** $x^4 + x^2 + 4$ **59.** $x^2 + x + 1$ **61.** $x^2 + x + 2$ **63.** $9.8x + 16.4 + \frac{-36.5}{x-2}$ **65.** $3x + 5$
69. It is.

Getting Ready (page 478)

1. $x + 1$ with a remainder of 1, 1 **2.** $x + 3$ with a remainder of 9, 9

Orals (page 484)

1. 9 **2.** -3 **3.** yes **4.** no

Exercise 6.8 (page 484)

1. 4 **3.** $12a^2 + 4a - 1$ **5.** $8x^2 + 2x + 4$ **7.** $P(r)$ **9.** $x + 2$ **11.** $x - 3$ **13.** $x + 2$ **15.** $x - 7 + \frac{28}{x+2}$ **17.** $3x^2 - x + 2$
19. $2x^2 + 4x + 3$ **21.** $6x^2 - x + 1 + \frac{3}{x+1}$ **23.** $7.2x - 0.66 + \frac{0.368}{x-0.2}$ **25.** $2.7x - 3.59 + \frac{0.903}{x+1.7}$
27. $9x^2 - 513x + 29{,}241 + \frac{-1{,}666{,}762}{x+57}$ **29.** -1 **31.** -37 **33.** 23 **35.** -1 **37.** 2 **39.** -1 **41.** 18 **43.** 174 **45.** -8
47. 59 **49.** 44 **51.** $\frac{29}{32}$ **53.** yes **55.** no **57.** 64 **61.** 1

Chapter Summary (page 488)

1.

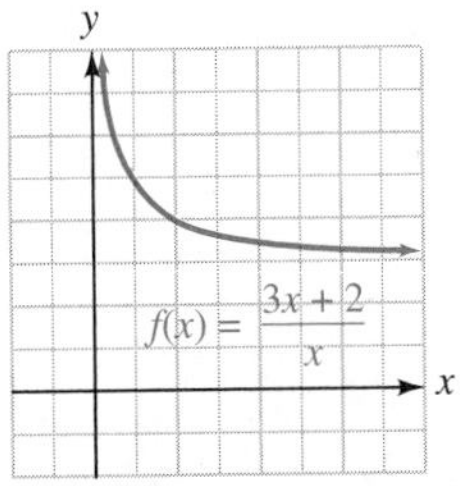

Horizontal asymptote: $y = 3$
Vertical asymptote: $x = 0$

2. a. $\frac{31x}{72y}$ **b.** $\frac{53m}{147n^2}$ **c.** $\frac{x-7}{x+7}$ **d.** $\frac{1}{x-6}$ **e.** $\frac{1}{2x+4}$ **f.** -1
g. -2 **h.** $\frac{-a-b}{c+d}$ **3. a.** $\frac{1}{18}$ **b.** $\frac{1}{8}$ **4. a.** 5 **b.** $-4, -12$ **5.** 70.4 ft **6.** 72 **7.** 6 **8.** 2 **9.** 16 **10. a.** 1 **b.** 1
c. $\frac{3x(x-1)}{(x-3)(x+1)}$ **d.** 1 **11. a.** $\frac{5y-3}{x-y}$ **b.** $\frac{6x-7}{x^2+2}$ **c.** $\frac{5x+13}{(x+2)(x+3)}$ **d.** $\frac{4x^2+9x+12}{(x-4)(x+3)}$ **e.** $\frac{5x^2+11x}{(x+1)(x+2)}$ **f.** $\frac{2(3x+1)}{x-3}$
g. $\frac{5x^2+23x+4}{(x+1)(x-1)(x-1)}$ **h.** $\frac{x^2+26x+3}{(x+3)(x-3)^2}$ **12. a.** $\frac{3y-2x}{x^2y^2}$ **b.** $\frac{y+2x}{2y-x}$ **c.** $\frac{2x+1}{x+1}$ **d.** $\frac{3x+2}{3x-2}$ **e.** $\frac{x-2}{x+3}$ **f.** $\frac{1}{x}$ **g.** $\frac{y-x}{y+x}$
h. $\frac{x^2y^2}{(x-y)^2(y^2-x^2)}$ **13. a.** 5 **b.** $-1, -2$ **c.** 2 and -2 are extraneous. **d.** $-1, -12$
14. a. $y^2 = \frac{x^2b^2 - a^2b^2}{a^2}$ **b.** $b = \frac{Ha}{2a-H}$ **15.** 50 mph **16.** 200 mph **17.** $14\frac{2}{5}$ hr **18.** $18\frac{2}{3}$ days **19. a.** $-\frac{x^3}{2y^3}$
b. $-3x^2y + \frac{3x}{2} + y$ **c.** $x + 5y$ **d.** $x^2 + 2x - 1 + \frac{6}{2x+3}$ **20. a.** yes **b.** no

Chapter 6 Test (page 491)

1. $\frac{-2}{3xy}$ **2.** $\frac{2}{x-2}$ **3.** -3 **4.** $\frac{2x+1}{4}$ **5.** $\frac{1}{8}$ **6.** 18 ft **7.** 6, -1 **8.** $\frac{44}{3}$ **9.** $\frac{xz}{y^4}$ **10.** $\frac{x+1}{2}$ **11.** 1 **12.** $\frac{(x+y)^2}{2}$ **13.** $\frac{2}{x+1}$ **14.** -1 **15.** 3 **16.** 2 **17.** $\frac{2s+r^2}{rs}$ **18.** $\frac{2x+3}{(x+1)(x+2)}$ **19.** $\frac{u^2}{2vw}$ **20.** $\frac{2x+y}{xy-2}$ **21.** $\frac{5}{2}$ **22.** 5; 3 is extraneous **23.** $a^2 = \frac{x^2b^2}{b^2-y^2}$ **24.** $r_2 = \frac{rr_1}{r_1-r}$ **25.** 10 days **26.** \$5,000 at 6% and \$3,000 at 10% **27.** $\frac{-6x}{y} + \frac{4x^2}{y^2} - \frac{3}{y^3}$ **28.** $3x^2+4x+2$ **29.** -7 **30.** 47

Cumulative Review Exercises (page 493)

1. a^8b^4 **2.** $\frac{b^4}{a^4}$ **3.** $\frac{81b^{16}}{16a^8}$ **4.** $x^{21}y^3$ **5.** 42,500 **6.** 0.000712 **7.** 1 **8.** -2 **9.** $\frac{5}{6}$ **10.** $-\frac{3}{4}$ **11.** 3 **12.** $-\frac{1}{3}$ **13.** 0 **14.** 8 **15.** $-\frac{16}{25}$ **16.** t^2-4t+3 **17.** $y = \frac{kxy}{r}$ **18.** no **19.** $[\frac{5}{2}, \infty)$ 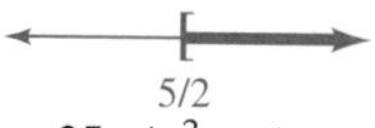

20. $(-\infty, 3] \cup [\frac{11}{3}, \infty)$ 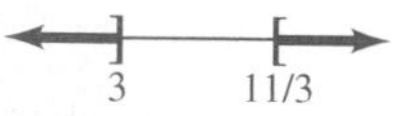

21. trinomial **22.** 7 **23.** 18 **24.**

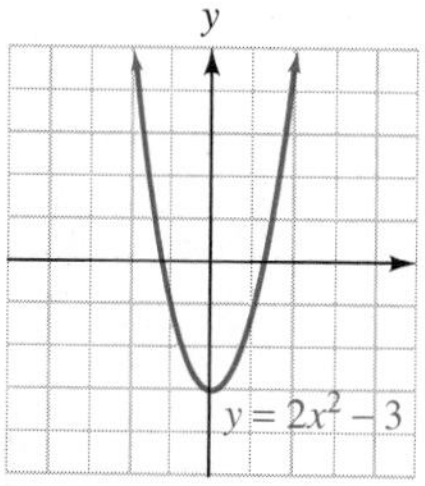

25. $4x^2-4x+14$ **26.** $-3x^2-3$ **27.** $6x^2-7x-20$ **28.** $2x^{2n}+3x^n-2$ **29.** $3rs^3(r-2s)$ **30.** $(x-y)(5-a)$ **31.** $(x+y)(u+v)$ **32.** $(9x^2+4y^2)(3x+2y)(3x-2y)$ **33.** $(2x-3y^2)(4x^2+6xy^2+9y^4)$ **34.** $(2x+3)(3x-2)$ **35.** $(3x-5)^2$ **36.** $(5x+3)(3x-2)$ **37.** $(3a+2b)(9a^2-6ab+4b^2)$ **38.** $(2x+5)(3x-7)$ **39.** $(x+5+y^2)(x+5-y^2)$ **40.** $(y+x-2)(y-x+2)$ **41.** 0, 2, -2 **42.** $-\frac{1}{3}, -\frac{7}{2}$ **43.** $\frac{2x-3}{3x-1}$ **44.** $\frac{x-2}{x+3}$ **45.** $\frac{4}{x-y}$ **46.** $\frac{a^2+ab^2}{a^2b-b^2}$ **47.** 0 **48.** -17 **49.** $x+4$ **50.** $-x^2+x+5+\frac{8}{x-1}$

Getting Ready (page 497)

1. 0 **2.** 16 **3.** 16 **4.** -16 **5.** $\frac{8}{125}$ **6.** $\frac{81}{256}$ **7.** $49x^2y^2$ **8.** $343x^3y^3$

Orals (page 508)

1. 3 **2.** -4 **3.** -2 **4.** 2 **5.** $8|x|$ **6.** $-3x$ **7.** not a real number **8.** $(x+1)^2$

Exercise 7.1 (page 508)

1. $\frac{x+3}{x-4}$ **3.** 1 **5.** $\frac{3(m^2+2m-1)}{(m+1)(m-1)}$ **7.** $(5x^2)^2$ **9.** positive **11.** 3, up **13.** $y^3 = x$ **15.** odd **17.** 0 **19.** $3x^2$ **21.** a^2+b^3 **23.** 11 **25.** -8 **27.** $\frac{1}{3}$ **29.** $-\frac{5}{7}$ **31.** not real **33.** 0.4 **35.** 4 **37.** not real **39.** 3.4641 **41.** 26.0624 **43.** $2|x|$ **45.** $|t+5|$ **47.** $5|b|$ **49.** $|a+3|$ **51.** 0 **53.** 4 **55.** 4.1231 **57.** 2.5539

59. D: $[-4, \infty)$, R: $[0, \infty)$

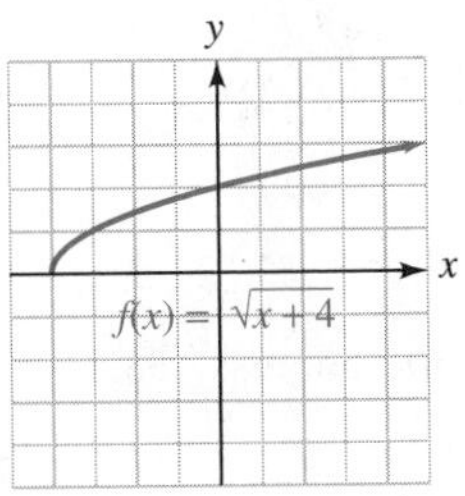

61. D: $[0, \infty)$, R: $(-\infty, -3]$

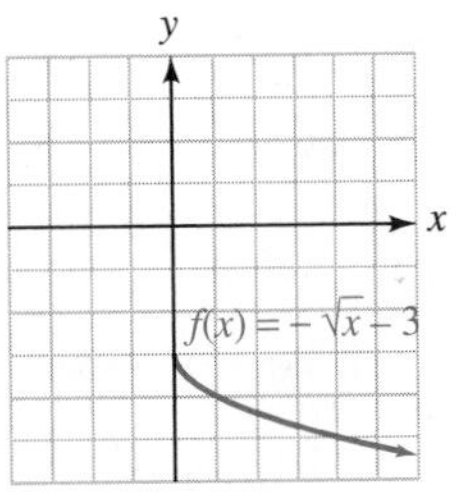

63. 1 **65.** −5

67. $-\frac{2}{3}$ **69.** 0.4 **71.** $2a$ **73.** $-10pq$ **75.** $-\frac{1}{2}m^2n$ **77.** $0.2z^3$ **79.** 3 **81.** −3 **83.** −2 **85.** $\frac{2}{5}$ **87.** $\frac{1}{2}$ **89.** not real **91.** $2|x|$ **93.** $2a$ **95.** $\frac{1}{2}|x|$ **97.** $|x^3|$ **99.** $-x$ **101.** $-3a^2$ **103.** $x + 2$ **105.** $0.1x^2|y|$ **107.** 1.67 **109.** 11.8673 **111.** 3 units **113.** 4 sec **115.** about 7.4 amperes

Getting Ready (page 511)

1. 25 **2.** 169 **3.** 18 **4.** 11,236

Orals (page 516)

1. 5 **2.** 10 **3.** 13 **4.** 5 **5.** 10 **6.** 13 **7.** 4 **8.** 3 **9.** 5

Exercise 7.2 (page 516)

1. $12x^2 - 14x - 10$ **3.** $15t^2 + 2ts - 8s^2$ **5.** hypotenuse **7.** $a^2 + b^2 = c^2$ **9.** distance **11.** 10 ft **13.** 80 m **15.** 9.9 cm **17.** 5 **19.** 5 **21.** 13 **23.** 10 **25.** 10.2 **29.** $x = 7$ **31.** (7, 0) and (3, 0) **33.** 13 ft **35.** about 127 ft **37.** about 135 ft **39.** not quite **41.** yes **43.** 173 yd **45.** 0.05 ft **47.** 8 cm^3 **51.** 54.1 in.

Getting Ready (page 520)

1. a **2.** $5x$ **3.** $x + 4$ **4.** $y - 3$

Orals (page 526)

1. 7 **2.** 3 **3.** 0 **4.** 9 **5.** 17 **6.** 31

Exercise 7.3 (page 526)

1. 2 **3.** 6 **5.** $x^n = y^n$ **7.** square **9.** extraneous **11.** 2 **13.** 4 **15.** 0 **17.** 4 **19.** 8 **21.** $\frac{5}{2}, \frac{1}{2}$ **23.** 1 **25.** 16 **27.** 14, ~~6~~ **29.** 4, 3 **31.** 2, ~~7~~ **33.** 9, ~~−25~~ **35.** 2, −1 **37.** −1, ~~1~~ **39.** ~~1~~, no solutions **41.** 0, ~~4~~ **43.** ~~−3~~, no solutions **45.** 0 **47.** 1, 9 **49.** 4, ~~0~~ **51.** 2, ~~142~~ **53.** 2 **55.** ~~6~~, no solutions **57.** 0, ~~$-\frac{12}{11}$~~ **59.** 1 **61.** 4, ~~−9~~ **63.** 2, ~~$\frac{21}{2}$~~ **65.** 2,010 ft **67.** about 29 mph **69.** \$5 **73.** 0, 4

Getting Ready (page 529)

1. x^7 **2.** a^{12} **3.** a^4 **4.** 1 **5.** $\frac{1}{x^4}$ **6.** a^3b^6 **7.** $\frac{b^6}{c^9}$ **8.** a^{10}

Orals (page 536)

1. 2 **2.** 3 **3.** 3 **4.** 1 **5.** 8 **6.** 4 **7.** $\frac{1}{2}$ **8.** 2 **9.** $2x$ **10.** $2x^2$

Exercise 7.4 (page 536)

1. $x < 3$ **3.** $r > 28$ **5.** $1\frac{2}{3}$ pints **7.** $a \cdot a \cdot a \cdot a$ **9.** a^{mn} **11.** $\frac{a^n}{b^n}$ **13.** $\frac{1}{a^n}$, 0 **15.** $\left(\frac{b}{a}\right)^n$ **17.** $|x|$ **19.** $\sqrt[3]{7}$ **21.** $\sqrt[4]{3x}$ **23.** $\sqrt[4]{\frac{1}{2}x^3y}$ **25.** $\sqrt{x^2+y^2}$ **27.** $11^{1/2}$ **29.** $(3a)^{1/4}$ **31.** $\left(\frac{1}{7}abc\right)^{1/6}$ **33.** $(a^2-b^2)^{1/3}$ **35.** 2 **37.** 2 **39.** 2 **41.** 2 **43.** $\frac{1}{2}$ **45.** $\frac{1}{2}$ **47.** -2 **49.** -3 **51.** not real **53.** 0 **55.** $5|y|$ **57.** $2|x|$ **59.** $3x$ **61.** not real **63.** 216 **65.** 27 **67.** 1,728 **69.** $\frac{1}{4}$ **71.** $125x^6$ **73.** $\frac{4x^2}{9}$ **75.** $\frac{1}{2}$ **77.** $\frac{1}{8}$ **79.** $\frac{1}{64x^3}$ **81.** $\frac{1}{9y^2}$ **83.** $\frac{1}{4p^2}$ **85.** 8 **87.** $\frac{16}{81}$ **89.** $-\frac{3}{2x}$ **91.** $5^{8/9}$ **93.** $4^{3/5}$ **95.** $9^{1/5}$ **97.** $7^{1/2}$ **99.** $\frac{1}{36}$ **101.** $2^{2/3}$ **103.** a **105.** $a^{2/9}$ **107.** $a^{3/4}b^{1/2}$ **109.** $\frac{n^{2/5}}{m^{3/5}}$ **111.** $\frac{2x}{3}$ **113.** $\frac{1}{3}x$ **115.** $y+y^2$ **117.** $x^2-x+x^{3/5}$ **119.** $x-4$ **121.** $x^{4/3}-x^2$ **123.** $x^{4/3}+2x^{2/3}y^{2/3}+y^{4/3}$ **125.** $a^3-2a^{3/2}b^{3/2}+b^3$ **127.** $\sqrt{p}$ **129.** $\sqrt{5b}$ **133.** yes

Getting Ready (page 538)

1. 15 **2.** 24 **3.** 5 **4.** 7 **5.** $4x^2$ **6.** $\frac{8}{11}x^3$ **7.** $3ab^3$ **8.** $-2a^4$

Orals (page 546)

1. 7 **2.** 4 **3.** 3 **4.** $3\sqrt{2}$ **5.** $2\sqrt[3]{2}$ **6.** $\frac{\sqrt[3]{3x^2}}{4b^2}$ **7.** $7\sqrt{3}$ **8.** $3\sqrt{7}$ **9.** $5\sqrt[3]{9}$ **10.** $8\sqrt[5]{4}$

Exercise 7.5 (page 547)

1. $\frac{-15x^5}{y}$ **3.** $9t^2+12t+4$ **5.** $3p+4+\frac{-5}{2p-5}$ **7.** $\sqrt[n]{a}\sqrt[n]{b}$ **9.** 6 **11.** t **13.** $5x$ **15.** 10 **17.** $7x$ **19.** $6b$ **21.** 2 **23.** $3a$ **25.** $2\sqrt{5}$ **27.** $-10\sqrt{2}$ **29.** $2\sqrt[3]{10}$ **31.** $-3\sqrt[3]{3}$ **33.** $2\sqrt[4]{2}$ **35.** $2\sqrt[5]{3}$ **37.** $\frac{\sqrt{7}}{3}$ **39.** $\frac{\sqrt[3]{7}}{4}$ **41.** $\frac{\sqrt[4]{3}}{10}$ **43.** $\frac{\sqrt[5]{3}}{2}$ **45.** $5x\sqrt{2}$ **47.** $4\sqrt{2b}$ **49.** $-4a\sqrt{7a}$ **51.** $5ab\sqrt{7b}$ **53.** $-10\sqrt{3xy}$ **55.** $-3x^2\sqrt[3]{2}$ **57.** $2x^4y\sqrt[3]{2}$ **59.** $2x^3y\sqrt[4]{2}$ **61.** $\frac{z}{4x}$ **63.** $\frac{\sqrt[4]{5x}}{2z}$ **65.** $10\sqrt{2x}$ **67.** $\sqrt[5]{7a^2}$ **69.** $4\sqrt{3}$ **71.** $-\sqrt{2}$ **73.** $2\sqrt{2}$ **75.** $9\sqrt{6}$ **77.** $3\sqrt[3]{3}$ **79.** $-\sqrt[3]{4}$ **81.** -10 **83.** $-17\sqrt[4]{2}$ **85.** $16\sqrt[4]{2}$ **87.** $-4\sqrt{2}$ **89.** $3\sqrt{2}+\sqrt{3}$ **91.** $-11\sqrt[3]{2}$ **93.** $y\sqrt{z}$ **95.** $13y\sqrt{x}$ **97.** $12\sqrt[3]{a}$ **99.** $-7y^2\sqrt{y}$ **101.** $4x\sqrt[5]{xy^2}$ **103.** $2x+2$ **105.** $h = 2.83$, $x = 2.00$ **107.** $x = 8.66$, $h = 10.00$ **109.** $x = 4.69$, $y = 8.11$ **111.** $x = 12.11$, $y = 12.11$ **115.** If $a = 0$, then b can be any real number. If $b = 0$, then a can be any real number.

Getting Ready (page 549)

1. a^7 **2.** b^3 **3.** a^2-2a **4.** $6b^3+9b^2$ **5.** $a^2-3a-10$ **6.** $4a^2-9b^2$

Orals (page 556)

1. 3 **2.** 2 **3.** $3\sqrt{3}$ **4.** $a^2|b|$ **5.** $6+3\sqrt{2}$ **6.** 1 **7.** $\frac{\sqrt{2}}{2}$ **8.** $\frac{\sqrt{3}+1}{2}$

Exercise 7.6 (page 556)

1. 1 **3.** $\frac{1}{3}$ **5.** 2, $\sqrt{7}$, $\sqrt{5}$ **7.** FOIL **9.** conjugate **11.** 4 **13.** $5\sqrt{2}$ **15.** $6\sqrt{2}$ **17.** 5 **19.** 18 **21.** $2\sqrt[3]{3}$ **23.** ab^2 **25.** $5a\sqrt{b}$ **27.** $r\sqrt[3]{10s}$ **29.** $2a^2b^2\sqrt[3]{2}$ **31.** $x^2(x+3)$ **33.** $3x(y+z)\sqrt[3]{4}$ **35.** $12\sqrt{5}-15$ **37.** $12\sqrt{6}+6\sqrt{14}$ **39.** $-8x\sqrt{10}+6\sqrt{15x}$ **41.** $-1-2\sqrt{2}$ **43.** $8x-14\sqrt{x}-15$ **45.** $5z+2\sqrt{15z}+3$ **47.** $3x-2y$ **49.** $6a+5\sqrt{3ab}-3b$

51. $18r - 12\sqrt{2r} + 4$ **53.** $-6x - 12\sqrt{x} - 6$ **55.** $\frac{\sqrt{7}}{7}$ **57.** $\frac{\sqrt{6}}{3}$ **59.** $\frac{\sqrt{10}}{4}$ **61.** 2 **63.** $\frac{\sqrt[3]{4}}{2}$ **65.** $\sqrt[3]{3}$ **67.** $\frac{\sqrt[3]{6}}{3}$ **69.** $2\sqrt{2x}$ **71.** $\frac{\sqrt{5y}}{y}$ **73.** $\frac{\sqrt[3]{2ab^2}}{b}$ **75.** $\frac{\sqrt[4]{4}}{2}$ **77.** $\frac{\sqrt[5]{2}}{2}$ **79.** $\sqrt{2} + 1$ **81.** $\frac{3\sqrt{2} - \sqrt{10}}{4}$ **83.** $2 + \sqrt{3}$ **85.** $\frac{9 - 2\sqrt{14}}{5}$ **87.** $\frac{2(\sqrt{x} - 1)}{x - 1}$ **89.** $\frac{x(\sqrt{x} + 4)}{x - 16}$ **91.** $\sqrt{2z} + 1$ **93.** $\frac{x - 2\sqrt{xy} + y}{x - y}$ **95.** $\frac{1}{\sqrt{3} - 1}$ **97.** $\frac{x - 9}{x(\sqrt{x} - 3)}$ **99.** $\frac{x - y}{\sqrt{x}(\sqrt{x} - \sqrt{y})}$ **101.** $f/4$ **105.** $\frac{x - 9}{4(\sqrt{x} + 3)}$

Chapter Summary (page 561)

1. a. 7 **b.** −11 **c.** −6 **d.** 15 **e.** −3 **f.** −6 **g.** 5 **h.** −2 **i.** $5|x|$ **j.** $|x + 2|$ **k.** $3a^2b$ **l.** $4x^2|y|$

2. a.

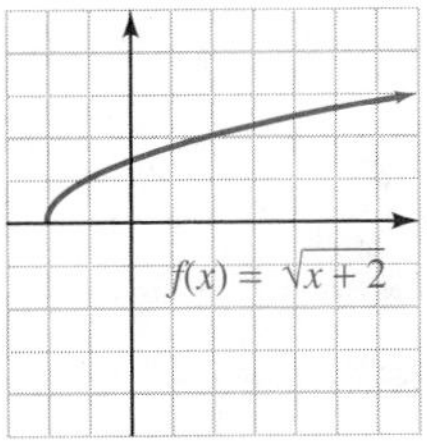

b.

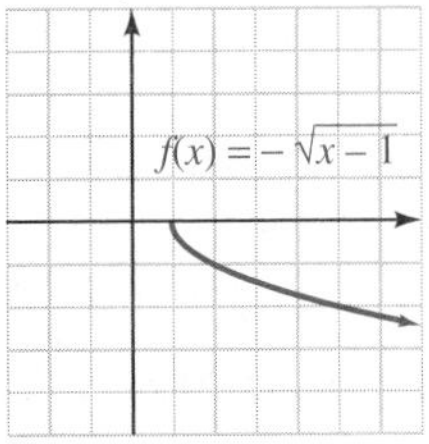

c.

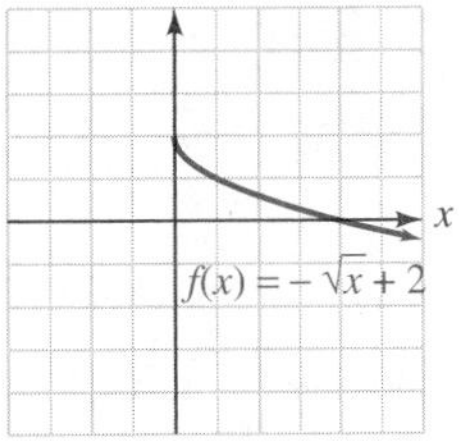

d.

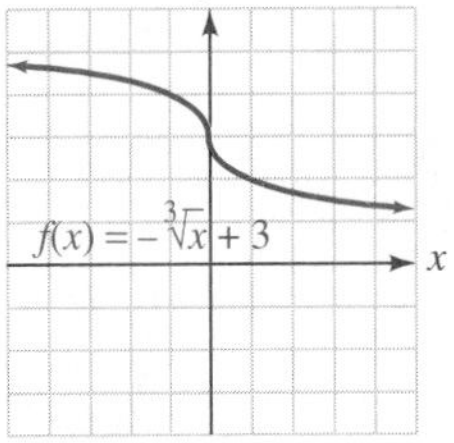

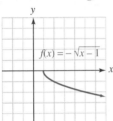

3. a. 12 **b.** about 5.7 **4.** 3 mi **5.** 8.2 ft **6.** 88 yd **7.** 16,000 yd, or about 9 mi **8.** 13 **9.** 2.83 units **10. a.** 22 **b.** 16, 9 **c.** 3, 9 **d.** $\frac{9}{16}$ **e.** 2 **f.** 0, −2 **11. a.** 5 **b.** −6 **c.** 27 **d.** 64 **e.** −2 **f.** −4 **g.** $\frac{1}{4}$ **h.** $\frac{1}{2}$ **i.** −16,807 **j.** $\frac{1}{3{,}125}$ **k.** 8 **l.** $\frac{27}{8}$ **m.** $3xy^{1/3}$ **n.** $3xy^{1/2}$ **o.** $125x^{9/2}y^6$ **p.** $\frac{1}{4u^{4/3}v^2}$ **12. a.** $5^{3/4}$ **b.** $a^{5/7}$ **c.** $u - 1$ **d.** $v + v^2$ **e.** $x + 2x^{1/2}y^{1/2} + y$ **f.** $a^{4/3} - b^{4/3}$ **13. a.** $\sqrt[3]{5}$ **b.** $\sqrt{x}$ **c.** $\sqrt[3]{3ab^2}$ **d.** $\sqrt{5ab}$ **14. a.** $4\sqrt{15}$ **b.** $3\sqrt[3]{2}$ **c.** $2\sqrt[4]{2}$ **d.** $2\sqrt[5]{3}$ **e.** $2|x|\sqrt{2x}$ **f.** $3x^2|y|\sqrt{2y}$ **g.** $2xy\sqrt[3]{2x^2y}$ **h.** $3x^2y\sqrt[3]{2x}$ **i.** $4|x|$ **j.** $2x$ **k.** $\frac{\sqrt[3]{2a^2b}}{3x}$ **l.** $\frac{\sqrt{17xy}}{8a^2}$ **15. a.** $3\sqrt{2}$ **b.** $\sqrt{5}$ **c.** 0 **d.** $8\sqrt[4]{2}$ **e.** $29x\sqrt{2}$ **f.** $32a\sqrt{3a}$ **g.** $13\sqrt[3]{2}$ **h.** $-4x\sqrt[4]{2x}$ **16.** $7\sqrt{2}$ m **17.** $6\sqrt{3}$ cm, 18 cm **18. a.** 7.07 in. **b.** 8.66 cm **19. a.** $6\sqrt{10}$ **b.** 72 **c.** $3x$ **d.** 3 **e.** $-2x$ **f.** $-20x^3y^3\sqrt{xy}$ **g.** $4 - 3\sqrt{2}$ **h.** $2 + 3\sqrt{2}$ **i.** $\sqrt{10} - \sqrt{5}$ **j.** $3 + \sqrt{6}$ **k.** 1 **l.** $5 + 2\sqrt{6}$ **m.** $x - y$ **n.** $6u - 12 + \sqrt{u}$ **20. a.** $\frac{\sqrt{3}}{3}$ **b.** $\frac{\sqrt{15}}{5}$ **c.** $\frac{\sqrt{xy}}{y}$ **d.** $\frac{\sqrt[3]{u^2}}{u^2v^2}$ **e.** $2(\sqrt{2} + 1)$ **f.** $\frac{\sqrt{6} + \sqrt{2}}{2}$ **g.** $2(\sqrt{x} - 4)$ **h.** $\frac{a + 2\sqrt{a} + 1}{a - 1}$ **21. a.** $\frac{3}{5\sqrt{3}}$ **b.** $\frac{1}{\sqrt[3]{3}}$ **c.** $\frac{9 - x}{2(3 + \sqrt{x})}$ **d.** $\frac{a - b}{a + \sqrt{ab}}$

Chapter 7 Test (page 566)

1. 7 **2.** 4 **3.** $2|x|$ **4.** $2x$ **5.**

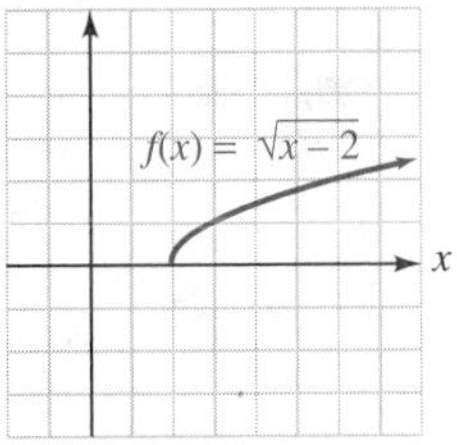

D = [2, ∞), R = [0, ∞) **6.**

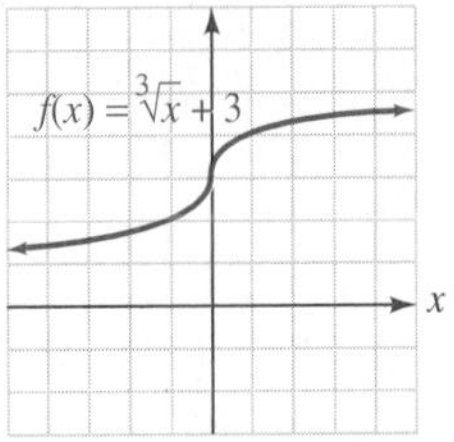

D = (−∞, ∞), R = (−∞, ∞)

7. 10 **8.** about 2.5 **9.** 28 in. **10.** 1.25 m **11.** 10 **12.** 25 **13.** 10 **14.** ~~4~~, no solutions **15.** 2 **16.** 9 **17.** $\frac{1}{216}$ **18.** $\frac{9}{4}$ **19.** $2^{4/3}$ **20.** $8xy$ **21.** $4\sqrt{3}$ **22.** $5xy^2\sqrt{10xy}$ **23.** $2x^5y\sqrt[3]{3}$ **24.** $\frac{1}{4a}$ **25.** $2|x|\sqrt{3}$ **26.** $2|x^3|\sqrt{2}$ **27.** $3x\sqrt[3]{3}$

28. $3x^2y^4\sqrt{2y}$ **29.** $-\sqrt{3}$ **30.** $14\sqrt[3]{5}$ **31.** $2y^2\sqrt{3y}$ **32.** $6z\sqrt[4]{3z}$ **33.** 9.24 cm **34.** 8.67 cm **35.** $-6x\sqrt{y} - 2xy^2$ **36.** $3 - 7\sqrt{6}$ **37.** $\frac{\sqrt{5}}{5}$ **38.** $2\sqrt[3]{3}$ **39.** $\sqrt{2}(\sqrt{5} - 3)$ **40.** $\sqrt{3t} + 1$ **41.** $\frac{3}{\sqrt{21}}$ **42.** $\frac{a - b}{a - 2\sqrt{ab} + b}$

Getting Ready (page 570)

1. $(3x + 2)(2x - 1)$ **2.** $(2x - 3)(2x + 1)$ **3.** 7 **4.** 8

Orals (page 579)

1. ± 7 **2.** $\pm\sqrt{10}$ **3.** 4 **4.** 9 **5.** $\frac{9}{4}$ **6.** $\frac{25}{4}$ **7.** 3, −4, 7 **8.** −2, 1, −5

Exercise 8.1 (page 579)

1. 1 **3.** $t \leq 4$ **5.** $B = \frac{-Ax + C}{y}$ **7.** $x = \sqrt{c}, x = -\sqrt{c}$ **9.** plus or minus **11.** 0, −2 **13.** 5, −5 **15.** −2, −4 **17.** 6, 1 **19.** 2, $\frac{1}{2}$ **21.** $\frac{2}{3}, -\frac{5}{2}$ **23.** ± 6 **25.** $\pm\sqrt{5}$ **27.** $\pm\frac{4\sqrt{3}}{3}$ **29.** 0, −2 **31.** 4, 10 **33.** $-5 \pm \sqrt{3}$ **35.** 2, −4 **37.** 2, 4 **39.** −1, −4 **41.** 1, $-\frac{1}{2}$ **43.** $-\frac{1}{3}, -\frac{3}{2}$ **45.** $\frac{3}{4}, -\frac{3}{2}$ **47.** $\frac{-7 \pm \sqrt{29}}{10}$ **49.** −1, −2 **51.** −6, −6 **53.** $\frac{-5 \pm \sqrt{5}}{10}$ **55.** $-\frac{3}{2}, -\frac{1}{2}$ **57.** $\frac{1}{4}, -\frac{3}{4}$ **59.** $\frac{-5 \pm \sqrt{17}}{2}$ **61.** 8.98, −3.98 **63.** 16, 18 **65.** 6, 7 **67.** $x^2 - 8x + 15 = 0$ **69.** $x^3 - x^2 - 14x + 24 = 0$ **71.** 8 ft by 12 ft **73.** 4 units **75.** $\frac{4}{3}$ cm **77.** 30 mph **79.** \$4.80 or \$5.20 **81.** 4,000 **83.** 2.26 in. **85.** about 6.13×10^{-3} M **89.** $\frac{3}{4}$

Getting Ready (page 583)

1. −2 **2.** 2 **3.** 0 **4.** 8 **5.** 1 **6.** 4

Orals (page 592)

1. down **2.** up **3.** up **4.** down **5.** (3, −1) **6.** (−2, 2)

Exercise 8.2 (page 593)

1. 10 **3.** $3\frac{3}{5}$ hr **5.** $f(x) = ax^2 + bx + c, a \neq 0$ **7.** vertex **9.** upward **11.** to the right **13.** upward

15.
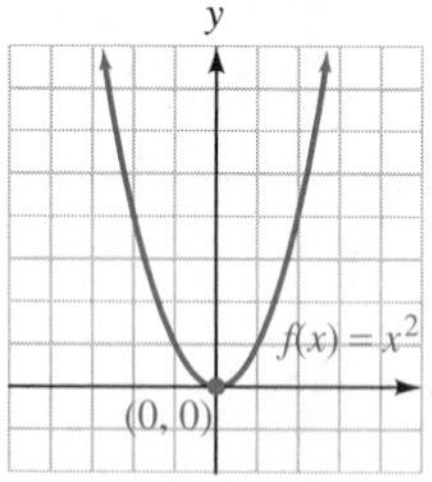

17.
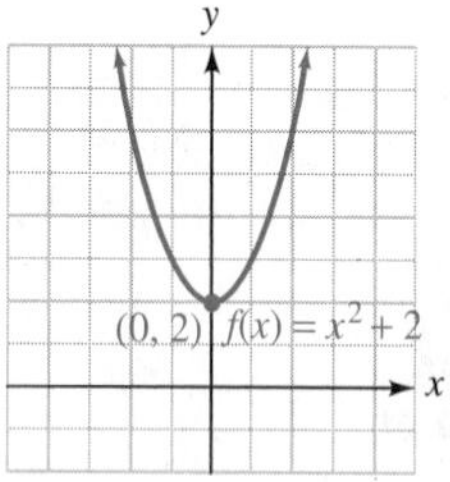

19.
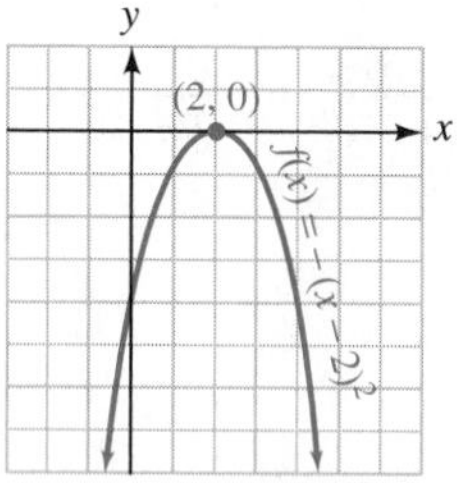

21.
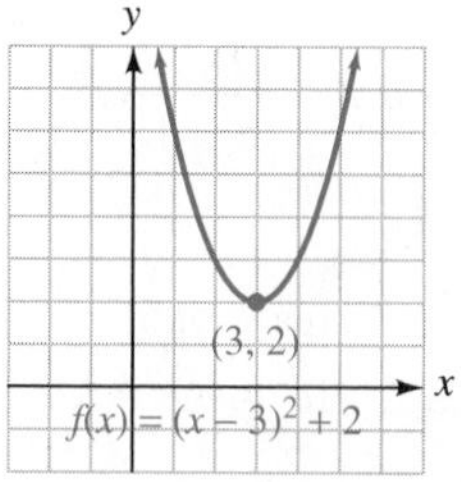

23.
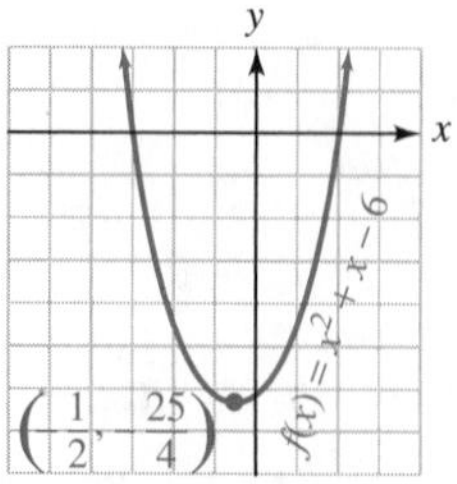

25.
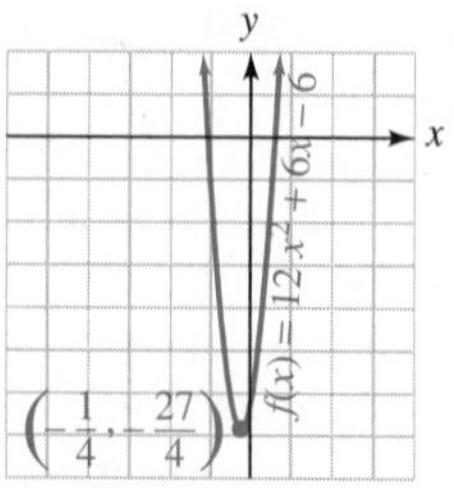

27. (1, 2), $x = 1$ **29.** (−3, −4), $x = -3$ **31.** (0, 0), $x = 0$ **33.** (1, −2), $x = 1$ **35.** (2, 21), $x = 2$ **37.** $\left(\frac{5}{12}, \frac{143}{24}\right)$, $x = \frac{5}{12}$ **39.** (5, 2) **41.** (0.25, 0.88) **43.** (0.5, 7.25) **45.** 2, −3 **47.** −1.85, 3.25 **49.** 36 ft, 3 sec **51.** 50 ft by 50 ft, 2,500 ft^2 **53.** 5,000 **55.** 3,276, \$14,742 **57.** \$35 **59.** 0.25 and 0.75

Getting Ready (page 596)

1. $7x$ **2.** $-x + 10$ **3.** $12x^2 + 5x - 25$ **4.** $9x^2 - 25$

Orals (page 606)

1. $-i$ **2.** -1 **3.** 1 **4.** i **5.** $7i$ **6.** $8i$ **7.** $10i$ **8.** $9i$ **9.** 5 **10.** 13

Exercise 8.3 (page 606)

1. -1 **3.** 20 mph **5.** imaginary **7.** -1 **9.** 1 **11.** $\dfrac{\sqrt{a}}{\sqrt{b}}$ **13.** 5, 7 **15.** conjugates **17.** $\pm 3i$ **19.** $\pm\dfrac{4\sqrt{3}}{3}i$ **21.** $-1 \pm i$ **23.** $-\dfrac{1}{4} \pm \dfrac{\sqrt{7}}{4}i$ **25.** $\dfrac{2}{3} \pm \dfrac{\sqrt{2}}{3}i$ **27.** $\dfrac{1}{3} \pm \dfrac{2\sqrt{2}}{3}i$ **29.** i **31.** $-i$ **33.** 1 **35.** i **37.** yes **39.** no **41.** no **43.** $8 - 2i$ **45.** $3 - 5i$ **47.** $15 + 7i$ **49.** $6 - 8i$ **51.** $2 + 9i$ **53.** $-15 + 2\sqrt{3}i$ **55.** $3 + 6i$ **57.** $-25 - 25i$ **59.** $7 + i$ **61.** $14 - 8i$ **63.** $8 + \sqrt{2}i$ **65.** $-20 - 30i$ **67.** $3 + 4i$ **69.** $-5 + 12i$ **71.** $7 + 17i$ **73.** $5 + 5i$ **75.** $16 + 2i$ **77.** $0 - i$ **79.** $0 + \dfrac{4}{5}i$ **81.** $\dfrac{1}{8} - 0i$ **83.** $0 + \dfrac{3}{5}i$ **85.** $2 + i$ **87.** $\dfrac{1}{2} + \dfrac{5}{2}i$ **89.** $-\dfrac{42}{25} - \dfrac{6}{25}i$ **91.** $\dfrac{1}{4} + \dfrac{3}{4}i$ **93.** $\dfrac{5}{13} - \dfrac{12}{13}i$ **95.** $\dfrac{11}{10} + \dfrac{3}{10}i$ **97.** $\dfrac{1}{4} - \dfrac{\sqrt{15}}{4}i$ **99.** $-\dfrac{5}{169} + \dfrac{12}{169}i$ **101.** $\dfrac{3}{5} + \dfrac{4}{5}i$ **103.** $-\dfrac{6}{13} - \dfrac{9}{13}i$ **105.** 10 **107.** 13 **109.** $\sqrt{74}$ **111.** 1 **119.** $\dfrac{5}{3 + i}$

Getting Ready (page 609)

1. 17 **2.** -8

Orals (page 614)

1. -3 **2.** -7 **3.** irrational and unequal **4.** complex conjugates **5.** no **6.** yes

Exercise 8.4 (page 614)

1. -2 **3.** $\frac{9}{5}$ **5.** $b^2 - 4ac$ **7.** rational, unequal **9.** rational, equal **11.** complex conjugates **13.** irrational, unequal **15.** rational, unequal **17.** 6, -6 **19.** 12, -12 **21.** 5 **23.** 12, -3 **25.** yes **27.** $k < -\frac{4}{3}$ **29.** 1, -1, 4, -4 **31.** 1, -1, $\sqrt{2}$, $-\sqrt{2}$ **33.** 1, -1, $\sqrt{5}$, $-\sqrt{5}$ **35.** 1, -1, 2, -2 **37.** 1 **39.** no solution **41.** -8, -27 **43.** -1, 27 **45.** -1, -4 **47.** 4, -5 **49.** 0, 2 **51.** -1, $-\frac{27}{13}$ **53.** 1, 1, -1, -1 **55.** $1 \pm i$ **57.** $x = \pm\sqrt{r^2 - y^2}$ **59.** $d = \pm\sqrt{\dfrac{k}{I}} = \pm\dfrac{\sqrt{kI}}{I}$ **61.** $y = \dfrac{-3x \pm \sqrt{9x^2 - 28x}}{2x}$ **63.** $\mu^2 = \dfrac{\Sigma x^2}{N} - \sigma^2$ **65.** $\frac{2}{3}$, $-\frac{1}{4}$ **67.** $\dfrac{-5 \pm \sqrt{17}}{4}$ **69.** $\dfrac{1 \pm i\sqrt{11}}{3}$ **71.** $-1 \pm 2i$ **75.** no

Getting Ready (page 616)

1. $(x + 5)(x - 3)$ **2.** $(x - 2)(x - 1)$

Orals (page 622)

1. $x = 2$ **2.** $x > 2$ **3.** $x < 2$ **4.** $x = -3$ **5.** $x > -3$ **6.** $x < -3$ **7.** $1 < 2x$ **8.** $1 > 2x$

Exercise 8.5 (page 622)

1. $y = kx$ **3.** $t = kxy$ **5.** 3 **7.** greater **9.** undefined **11.** $(1, 4)$

13. $(-\infty, 3) \cup (5, \infty)$

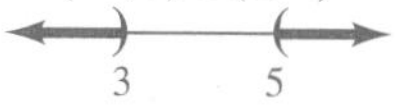

15. $[-4, 3]$

17. $(-\infty, -5] \cup [3, \infty)$

19. no solutions **21.** $(-\infty, -3] \cup [3, \infty)$

23. $(-5, 5)$

25. $(-\infty, 0) \cup (1/2, \infty)$

27. $(0, 2]$

29. $(-\infty, -5/3) \cup (0, \infty)$

31. $(-\infty, -3) \cup (1, 4)$

33. $[-5, -2) \cup [4, \infty)$

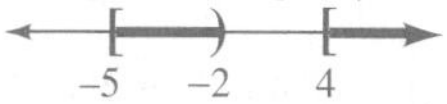

35. $(-\infty, -4)$

37. $(-1/2, 1/3) \cup (1/2, \infty)$

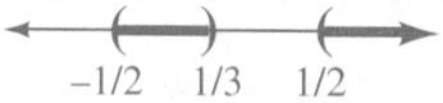

39. $(0, 2) \cup (8, \infty)$

41. $(-\infty, -2) \cup (2, 18]$

43. $[-34/5, -4) \cup (3, \infty)$

45. $(-4, -2] \cup (-1, 2]$

47. $(-\infty, -16) \cup (-4, -1) \cup (4, \infty)$

49. $(-\infty, -2) \cup (-2, \infty)$

51. $(-1, 3)$ **53.** $(-\infty, -3) \cup (2, \infty)$ **55.**

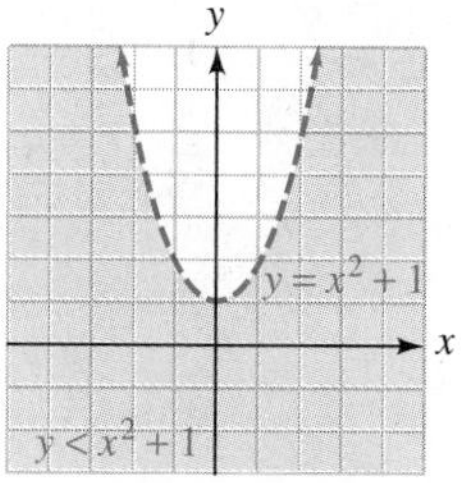

57.

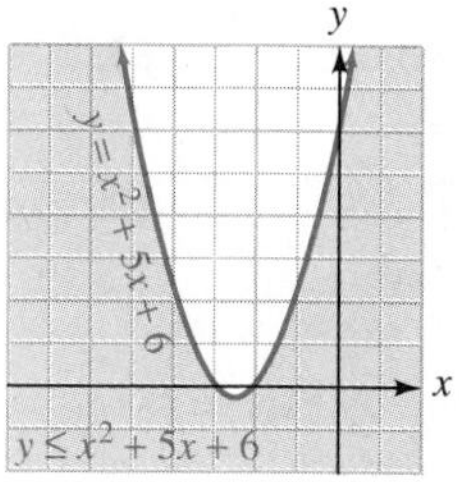

59.

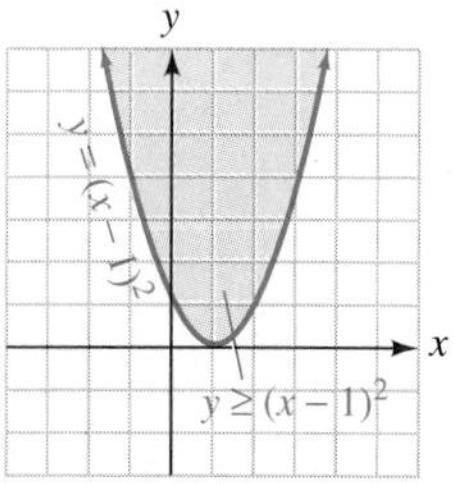

61.

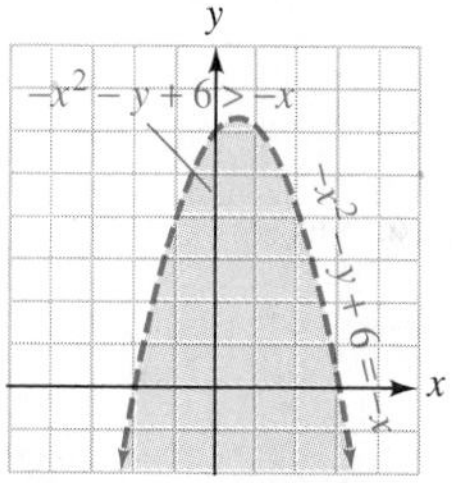

63.

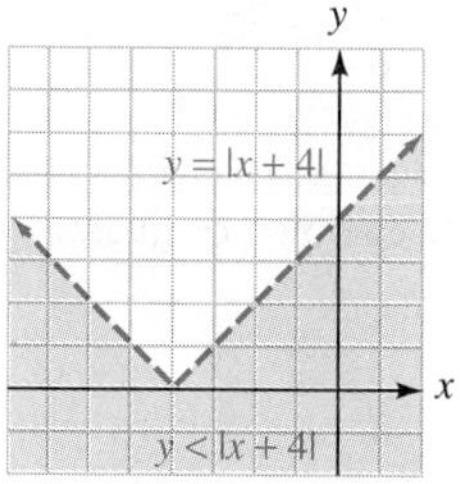

65.

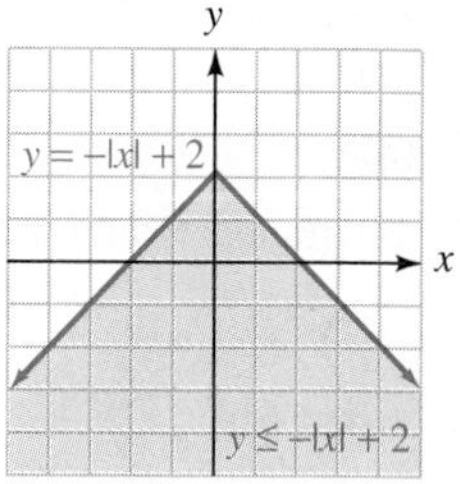

69. when 4 factors are negative, 2 factors are negative, or no factors are negative

Getting Ready (page 625)

1. $3x - 1$ **2.** $x + 3$ **3.** $2x^2 - 3x - 2$ **4.** $\frac{2x+1}{x-2}$

Orals (page 632)

1. $5x$ **2.** x **3.** $8x^2$ **4.** $\frac{3}{2}$ **5.** 2 **6.** $12x^2$ **7.** $8x$ **8.** $6x$ **9.** $12x$

Exercise 8.6 (page 632)

1. $-\dfrac{3x+7}{x+2}$ **3.** $\dfrac{x-4}{3x^2-x-12}$ **5.** $f(x) + g(x)$ **7.** $f(x)g(x)$ **9.** domain **11.** $f(x)$ **13.** $7x, (-\infty, \infty)$ **15.** $12x^2, (-\infty, \infty)$
17. $x, (-\infty, \infty)$ **19.** $\frac{4}{3}, (-\infty, 0) \cup (0, \infty)$ **21.** $3x - 2, (-\infty, \infty)$ **23.** $2x^2 - 5x - 3, (-\infty, \infty)$ **25.** $-x - 4, (-\infty, \infty)$
27. $\frac{x-3}{2x+1}, \left(-\infty, -\frac{1}{2}\right) \cup \left(-\frac{1}{2}, \infty\right)$ **29.** $-2x^2 + 3x - 3, (-\infty, \infty)$ **31.** $(3x - 2)/(2x^2 + 1), (-\infty, \infty)$ **33.** $3, (-\infty, \infty)$
35. $(x^2 - 4)/(x^2 - 1), (-\infty, -1) \cup (-1, 1) \cup (1, \infty)$ **37.** 7 **39.** 24 **41.** -1 **43.** $-\frac{1}{2}$ **45.** $2x^2 - 1$ **47.** $16x^2 + 8x$ **49.** 58

51. 110 **53.** 2 **55.** $9x^2 - 9x + 2$ **57.** 2 **59.** $2x + h$ **61.** $4x + 2h$ **63.** $2x + h + 1$ **65.** $2x + h + 3$ **67.** $4x + 2h + 3$ **69.** 2 **71.** $x + a$ **73.** $2x + 2a$ **75.** $x + a + 1$ **77.** $x + a + 3$ **79.** $2x + 2a + 3$ **85.** $3x^2 + 3xh + h^2$ **87.** $C(t) = \frac{5}{9}(2{,}668 - 200t)$

Getting Ready (page 634)

1. $x = \frac{y - 2}{3}$ **2.** $x = \frac{3y + 10}{2}$

Orals (page 642)

1. {(2, 1), (3, 2), (10, 5)} **2.** {(1, 1), (8, 2), (64, 4)} **3.** $f^{-1}(x) = 2x$ **4.** $f^{-1}(x) = \frac{1}{2}x$ **5.** no **6.** yes

Exercise 8.7 (page 643)

1. $3 - 8i$ **3.** $18 - i$ **5.** 10 **7.** one-to-one **9.** 2 **11.** x **13.** yes **15.** no

17.

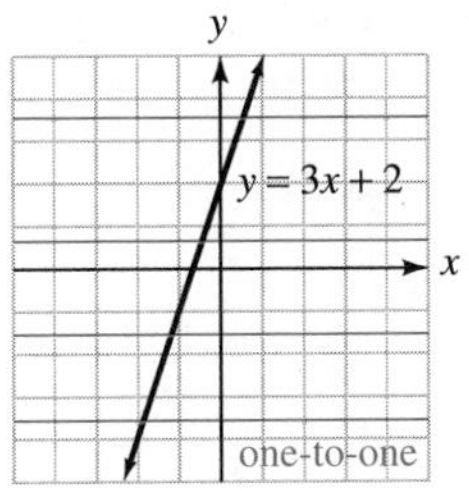

19.

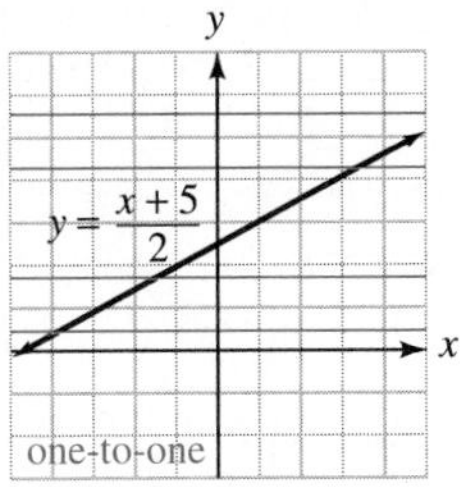

21.

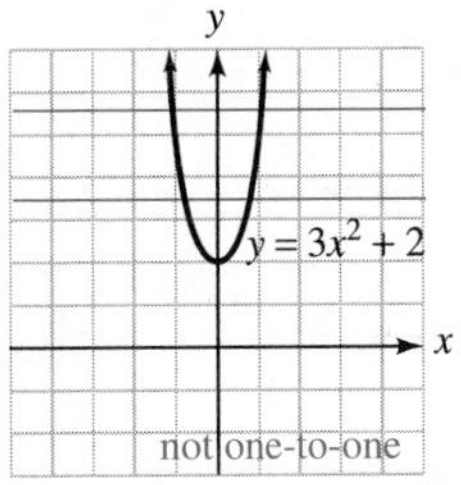

23.

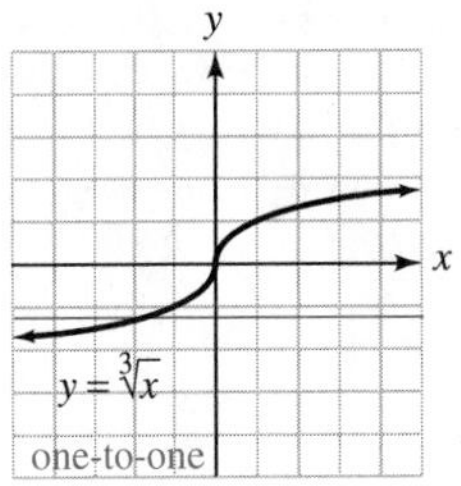

25. {(2, 3), (1, 2), (0, 1)}; yes **27.** {(2, 1), (3, 2), (3, 1), (5, 1)}; no **29.** {(1, 1), (4, 2), (9, 3), (16, 4)}; yes **31.** $f^{-1}(x) = \frac{1}{3}x - \frac{1}{3}$ **33.** $f^{-1}(x) = 5x - 4$ **35.** $f^{-1}(x) = 5x + 4$ **37.** $f^{-1}(x) = \frac{5}{4}x + 5$

39.

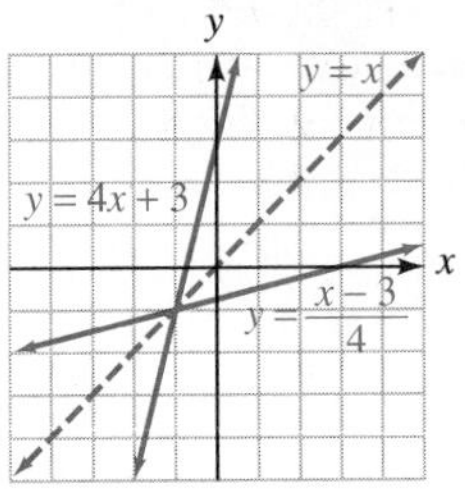

41.

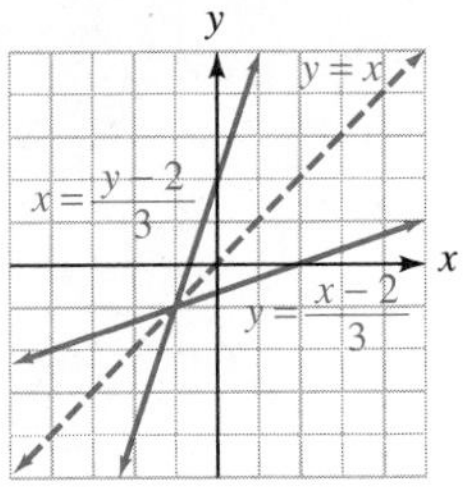

43.

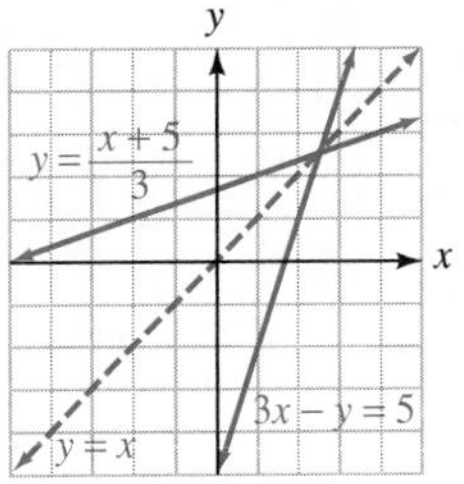

45.

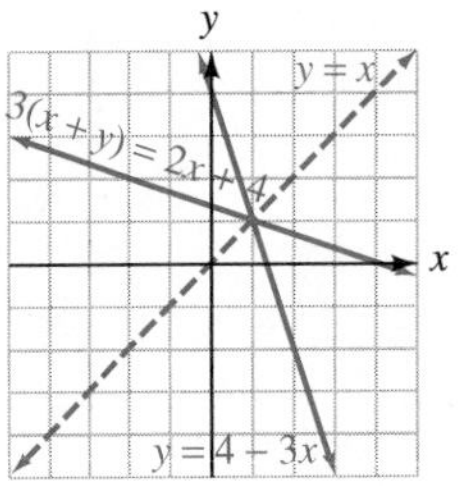

47. $y = \pm\sqrt{x - 4}$, no **49.** $y = \sqrt[3]{x}$, yes **51.** $x = |y|$, no **53.** $f^{-1}(x) = \sqrt[3]{\frac{x + 3}{2}}$ **55.**

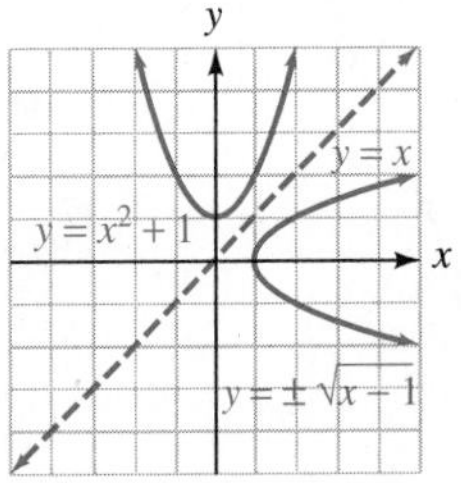

57.

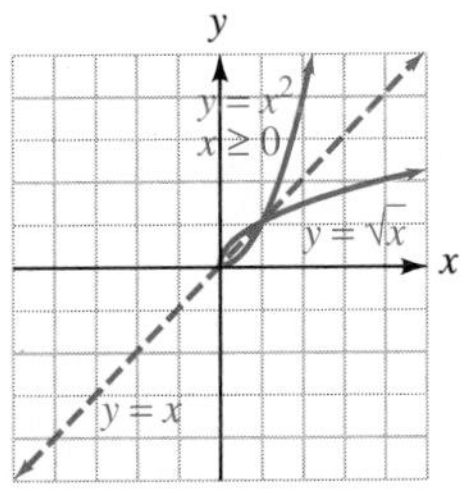

61. $f^{-1}(x) = \frac{x + 1}{x - 1}$

Chapter Summary (page 647)

1. a. $\frac{2}{3}, -\frac{3}{4}$ **b.** $-\frac{1}{3}, -\frac{5}{2}$ **c.** $\frac{2}{3}, -\frac{4}{5}$ **d.** $4, -8$ **2. a.** $-4, -2$ **b.** $\frac{7}{2}, 1$ **3. a.** $9, -1$ **b.** $0, 10$ **c.** $\frac{1}{2}, -7$ **d.** $7, -\frac{1}{3}$
4. 4 cm by 6 cm **5.** 2 ft by 3 ft **6.** 7 sec **7.** 196 ft
8. a.

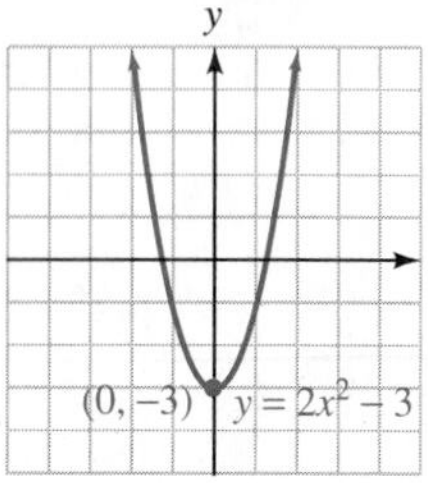

b.

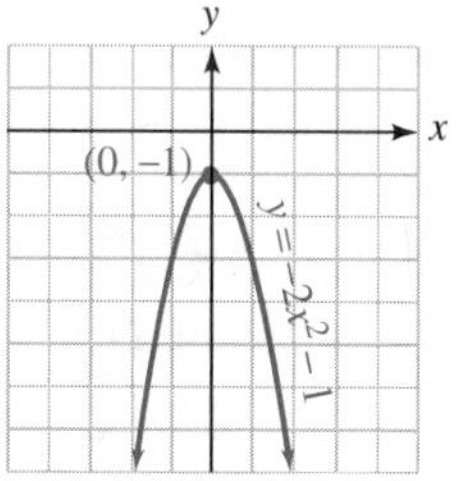

c.

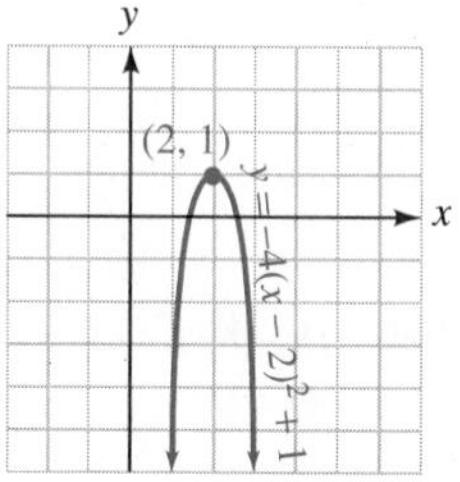

d.

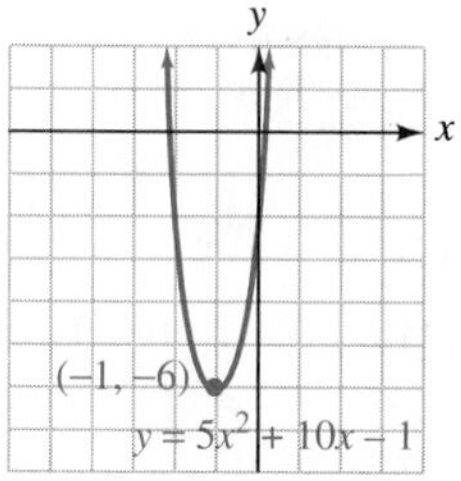

9. a. $12 - 8i$ **b.** $2 - 68i$ **c.** $-96 + 3i$ **d.** $-2 - 2\sqrt{2}i$ **e.** $22 + 29i$ **f.** $-16 + 7i$ **g.** $-12 + 28\sqrt{3}i$ **h.** $118 + 10\sqrt{2}i$
i. $0 - \frac{3}{4}i$ **j.** $0 - \frac{2}{5}i$ **k.** $\frac{12}{5} - \frac{6}{5}i$ **l.** $\frac{21}{10} + \frac{7}{10}i$ **m.** $\frac{15}{17} + \frac{8}{17}i$ **n.** $\frac{4}{5} - \frac{3}{5}i$ **o.** $\frac{15}{29} - \frac{6}{29}i$ **p.** $\frac{1}{3} + \frac{1}{3}i$ **q.** $15 + 0i$ **r.** $26 + 0i$
10. a. irrational, unequal **b.** complex conjugates **c.** 12, 152 **d.** $k \geq -\frac{7}{3}$ **11. a.** 1, 144 **b.** 8, -27 **c.** 1 **d.** $1, -\frac{8}{5}$ **12.** $\frac{14}{3}$
13. 1 **14. a.**

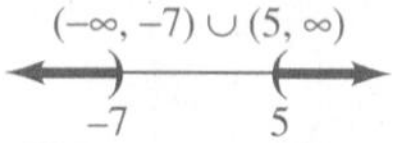

b.

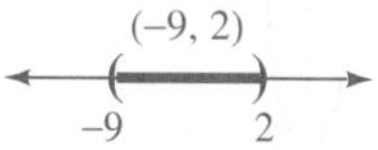

c.

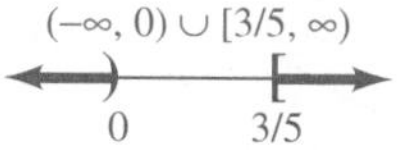

d.

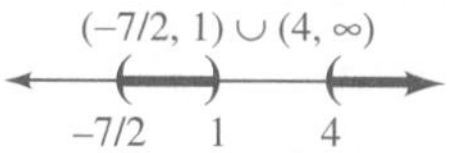

15. a.

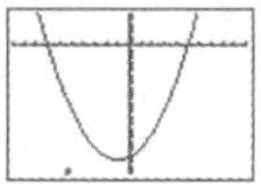

$x < -7$ or $x > 5$ **b.**

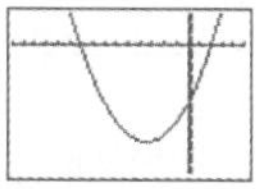

$-9 < x < 2$ **c.**

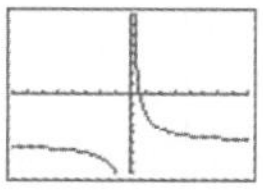

$x < 0$ or $x \geq \frac{3}{5}$

d.

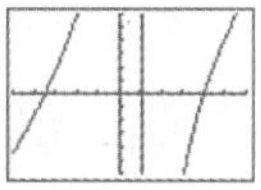

$-\frac{7}{2} < x < 1$ or $x > 4$ **16. a.**

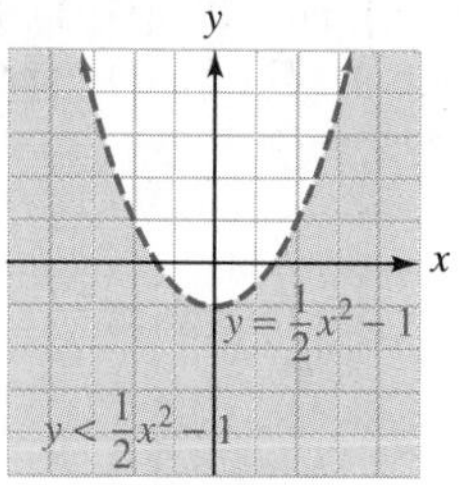

b.

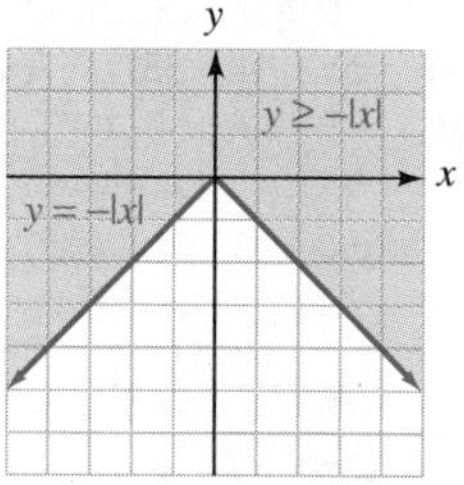

17. a. $(f + g)(x) = 3x + 1$
b. $(f - g)(x) = x - 1$
c. $(f \cdot g)(x) = 2x^2 + 2x$
d. $(f/g)(x) = \frac{2x}{x+1}$ **e.** 6
f. -1 **g.** $2(x + 1)$
h. $2x + 1$

18. a. yes

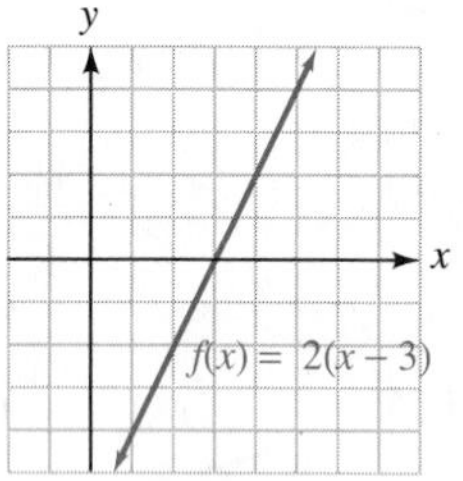

b. no

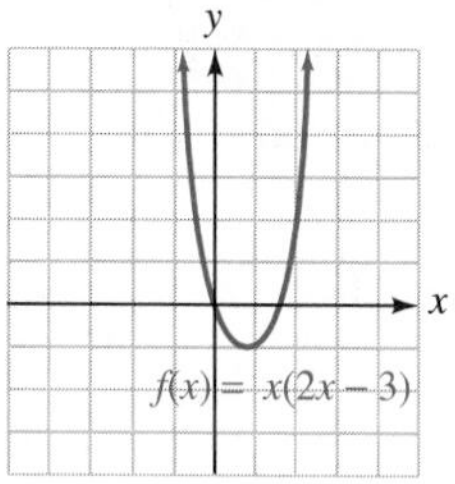

c. no

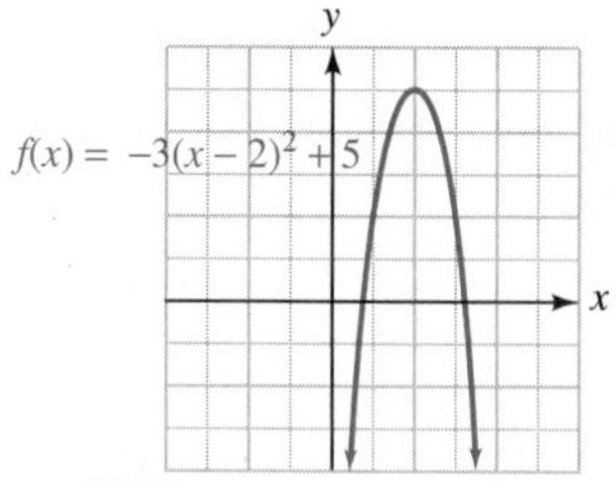

d. no

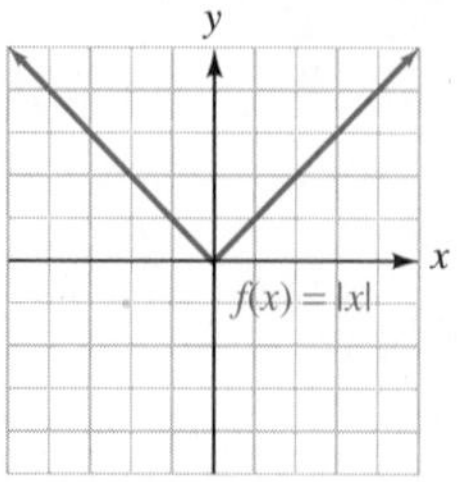

19. a. $f^{-1}(x) = \frac{x+3}{6}$ **b.** $f^{-1}(x) = \frac{x-5}{4}$ **c.** $y = \sqrt{\frac{x+1}{2}}$ **d.** $x = |y|$

Chapter 8 Test (page 652)

1. 3, −6 **2.** $-\frac{3}{2}, -\frac{5}{3}$ **3.** 144 **4.** 625 **5.** $-2 \pm \sqrt{3}$ **6.** $\frac{5 \pm \sqrt{37}}{2}$ **7.** $-1 + 11i$ **8.** $4 - 7i$ **9.** $8 + 6i$ **10.** $-10 - 11i$ **11.** $0 - \dfrac{\sqrt{2}}{2}i$ **12.** $\frac{1}{2} + \frac{1}{2}i$ **13.** nonreal **14.** 2 **15.** 10 in. **16.** 1, $\frac{1}{4}$

17.

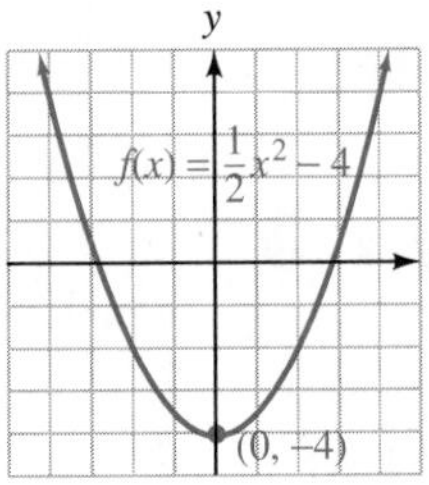

18.

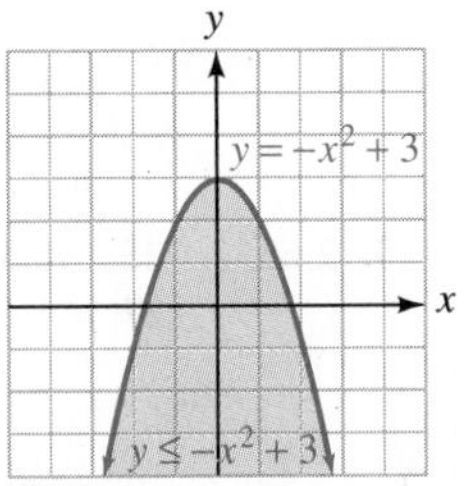

19.

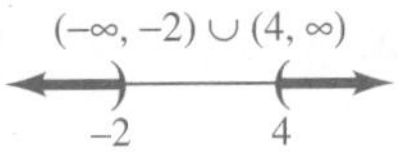

20.

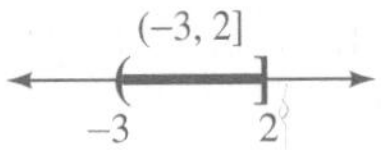

21. $(g + f)(x) = 5x - 1$ **22.** $(f - g)(x) = 3x + 1$ **23.** $(g \cdot f)(x) = 4x^2 - 4x$ **24.** $(g/f)(x) = \frac{x-1}{4x}$ **25.** 3 **26.** −4 **27.** −8 **28.** −9 **29.** $4(x - 1)$ **30.** $4x - 1$ **31.** $y = \frac{12 - 2x}{3}$ **32.** $y = -\sqrt{\frac{x-4}{3}}$

Cumulative Review Exercises (page 654)

1. D = $(-\infty, \infty)$; R = $[-3, \infty)$ **2.** D = $(-\infty, \infty)$; R = $(-\infty, 0]$ **3.** $y = 3x + 2$ **4.** $y = -\frac{2}{3}x - 2$ **5.** $-4a^2 + 12a - 7$ **6.** $6x^2 - 5x - 6$ **7.** $(x^2 + 4y^2)(x + 2y)(x - 2y)$ **8.** $(3x + 2)(5x - 4)$ **9.** 6, −1 **10.** 0, $\frac{2}{3}$, $-\frac{1}{2}$ **11.** $5x^2$ **12.** $4t\sqrt{3t}$ **13.** $-3x$ **14.** $4x$ **15.** $\frac{1}{2}$ **16.** 16 **17.** y^2 **18.** $x^{17/12}$ **19.**

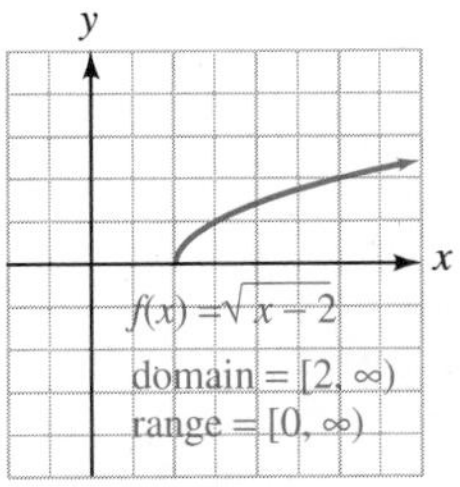

20.

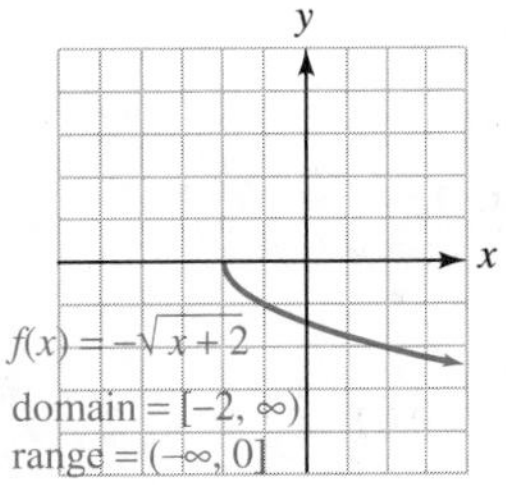

21. $x^{4/3} - x^{2/3}$

22. $\frac{1}{x} + 2 + x$ **23.** $7\sqrt{2}$ **24.** $-12\sqrt[4]{2} + 10\sqrt[4]{3}$ **25.** $-18\sqrt{6}$ **26.** $\dfrac{5\sqrt[3]{x^2}}{x}$ **27.** $\dfrac{x + 3\sqrt{x} + 2}{x - 1}$ **28.** $\sqrt{xy}$ **29.** 2, 7 **30.** $\frac{1}{4}$ **31.** $3\sqrt{2}$ in. **32.** $2\sqrt{3}$ in. **33.** 10 **34.** 9 **35.** 1, $-\frac{3}{2}$ **36.** $\dfrac{-2 \pm \sqrt{7}}{3}$

37.

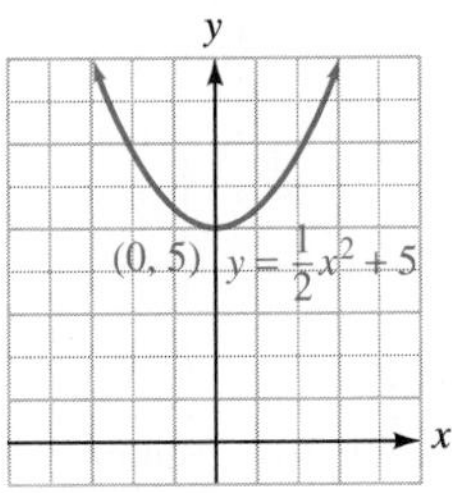

38.

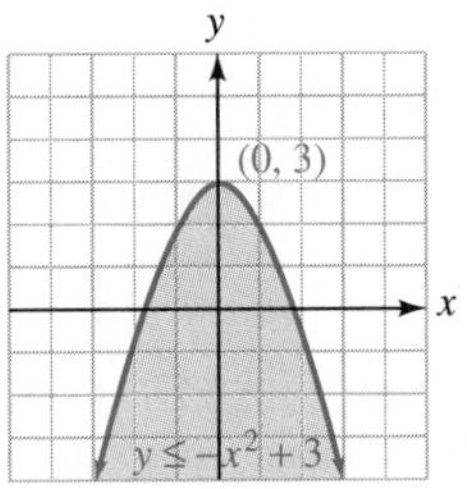

39. $7 + 2i$ **40.** $-5 - 7i$ **41.** 13 **42.** $12 - 6i$ **43.** $-12 - 10i$

44. $\frac{3}{2} + \frac{1}{2}i$ **45.** $\sqrt{13}$ **46.** $\sqrt{61}$ **47.** −2 **48.** 9, 16 **49.**

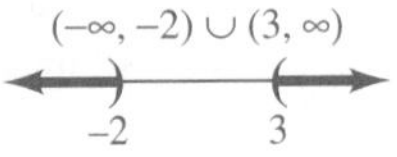

50.

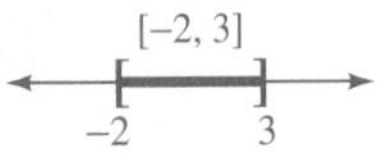

51. 5 **52.** 27

53. $12x^2 - 12x + 5$ **54.** $6x^2 + 3$ **55.** $f^{-1}(x) = \frac{x-2}{3}$ **56.** $f^{-1}(x) = \sqrt[3]{x - 4}$

Getting Ready (page 658)

1. 8 **2.** 5 **3.** $\frac{1}{25}$ **4.** $\frac{8}{27}$

Orals (page 669)

1. 4 **2.** 25 **3.** 18 **4.** 3 **5.** $\frac{1}{4}$ **6.** $\frac{1}{25}$ **7.** $\frac{2}{9}$ **8.** $\frac{1}{27}$

Exercise 9.1 (page 669)

1. 40 **3.** 120° **5.** exponential **7.** $(0, \infty)$ **9.** increasing **11.** $P\left(1 + \frac{r}{k}\right)^{kt}$ **13.** 2.6651 **15.** 36.5548 **17.** 8 **19.** $7^{3\sqrt{3}}$

21.

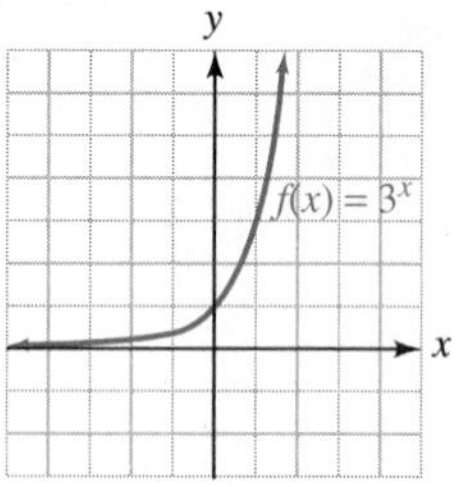

23.

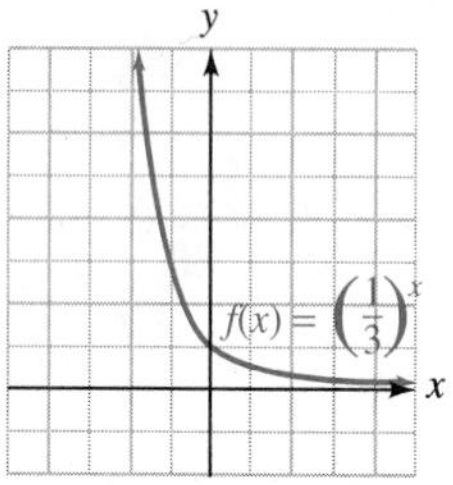

25.

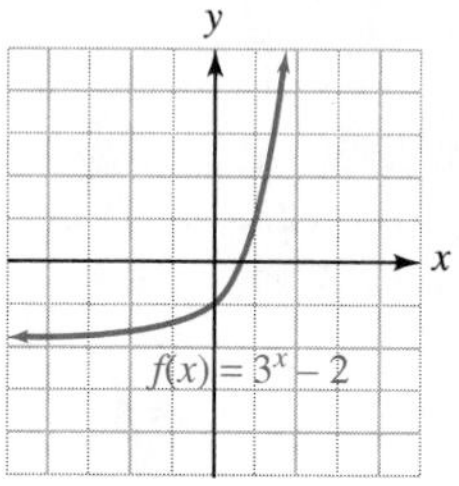

27.

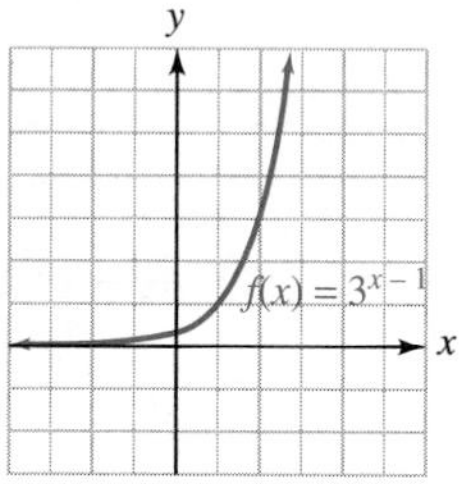

29. $b = \frac{1}{2}$

31. no value **33.** $b = 2$ **35.** $b = 3$ **37.**

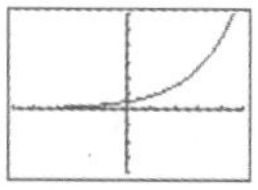

39.

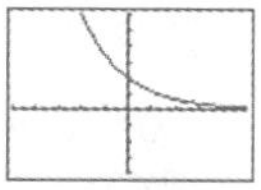

41. \$22,080.40 **43.** \$32.03 **45.** \$2,273,996.13

47. $\frac{32}{243}A_0$ **49.** 5.0421×10^{-5} coulombs **51.** \$1,115.33

Getting Ready (page 672)

1. 2 **2.** 2.25 **3.** 2.44 **4.** 2.59

Orals (page 677)

1. 1 **2.** 2.72 **3.** 7.39 **4.** 20.09 **5.** 2 **6.** 2

Exercise 9.2 (page 678)

1. $4x^2\sqrt{15x}$ **3.** $10y\sqrt{3y}$ **5.** 2.72 **7.** increasing **9.** $A = Pe^{rt}$ **11.**

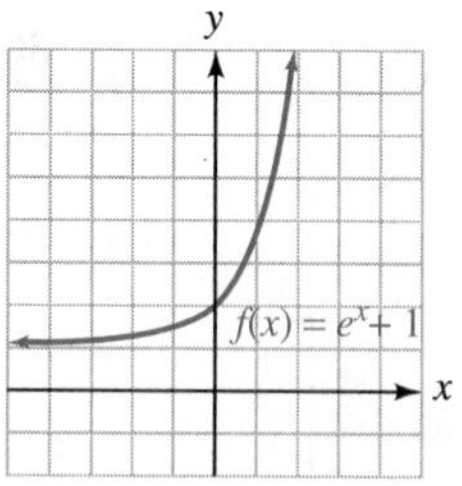

13.

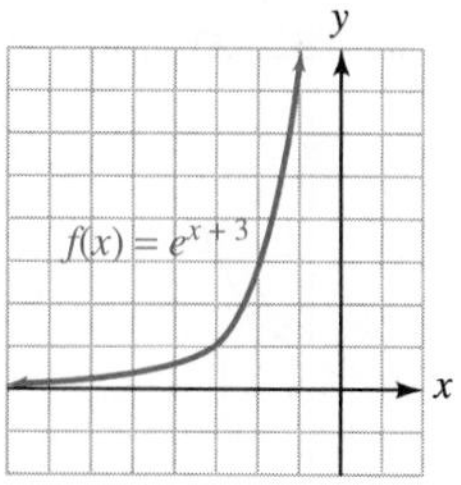

15.

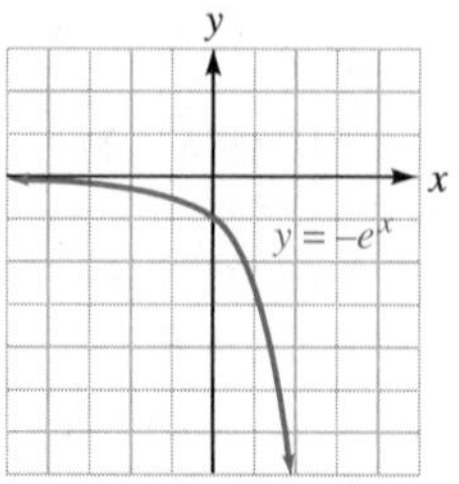

17.

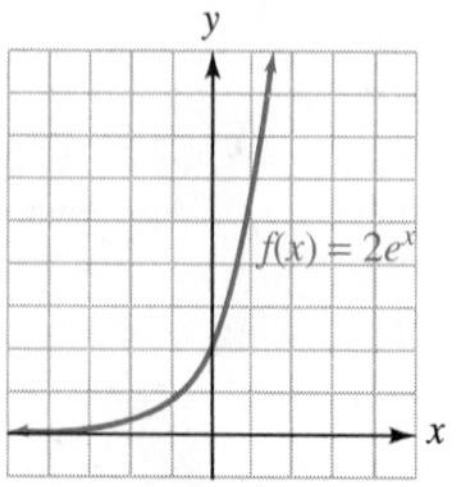

19. no **21.** no **23.** \$10,272.17 **25.** \$6,391.10 **27.** \$7,518.28 from annual compounding; \$7,647.95 from continuous compounding **29.** 10.6 billion **31.** 2.6 **33.** 6,817 **35.** 0.16 **37.** 0 **39.** 49 mps **41.** \$3,094.15 **43.** 72 yr **49.** $k = e^5$

Getting Ready (page 680)

1. 1 **2.** 25 **3.** $\frac{1}{25}$ **4.** 4

Orals (page 688)

1. 3 **2.** 2 **3.** 5 **4.** 2 **5.** 2 **6.** 2 **7.** $\frac{1}{4}$ **8.** $\frac{1}{2}$ **9.** 2

Exercise 9.3 (page 688)

1. 10 **3.** 0; $-\frac{5}{9}$ does not check **5.** $x = b^y$ **7.** range **9.** inverse **11.** exponent **13.** $(b, 1)$, $(1, 0)$ **15.** $20 \log \frac{E_O}{E_I}$ **17.** $3^3 = 27$ **19.** $\left(\frac{1}{2}\right)^2 = \frac{1}{4}$ **21.** $4^{-3} = \frac{1}{64}$ **23.** $\left(\frac{1}{2}\right)^3 = \frac{1}{8}$ **25.** $\log_6 36 = 2$ **27.** $\log_5 \frac{1}{25} = -2$ **29.** $\log_{1/2} 32 = -5$ **31.** $\log_x z = y$ **33.** 4 **35.** 2 **37.** 3 **39.** $\frac{1}{2}$ **41.** -3 **43.** 49 **45.** 6 **47.** 5 **49.** $\frac{1}{25}$ **51.** $\frac{1}{6}$ **53.** $-\frac{3}{2}$ **55.** $\frac{2}{3}$ **57.** 5 **59.** $\frac{3}{2}$ **61.** 4 **63.** 8 **65.** 4 **67.** 4 **69.** 3 **71.** 100 **73.** 0.9165 **75.** -2.0620 **77.** 25.25 **79.** 17,378.01 **81.** 0.00 **83.** 8

85. increasing

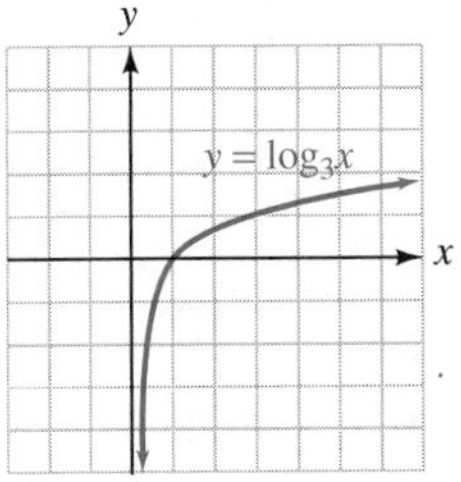

87. decreasing

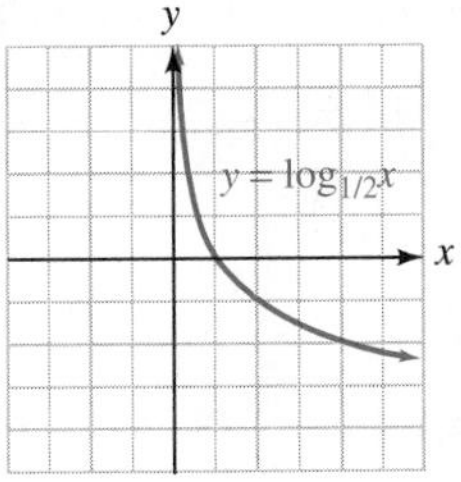

89.

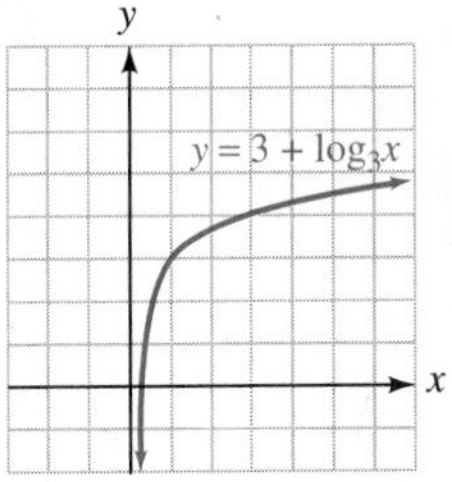

91.

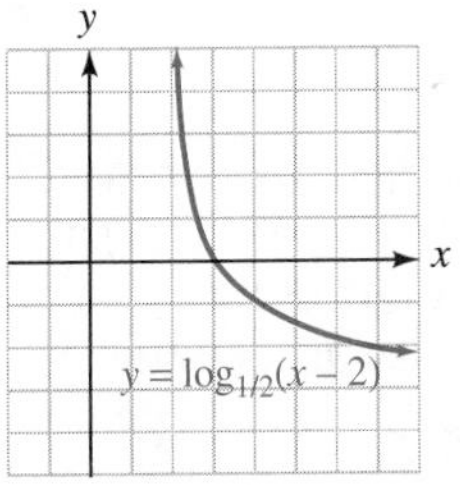

93.

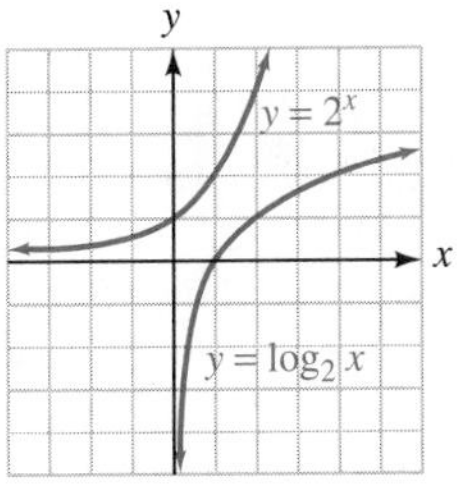

95.

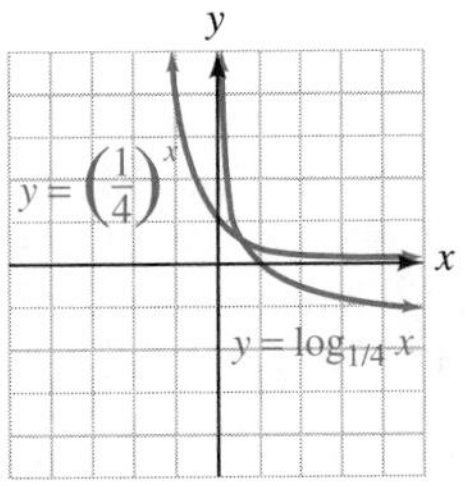

97. no value of b **99.** 3

101. 29.0 db **103.** 49.5 db **105.** 4.4 **107.** 4 **109.** 4.2 yr **111.** 10.8 yr

Getting Ready (page 692)

1. 2 **2.** -3 **3.** 1 **4.** 0

Orals (page 696)

1. $e^y = x$ **2.** $\ln b = a$ **3.** $t = \frac{\ln 2}{r}$

Exercise 9.4 (page 696)

1. $y = 5x$ **3.** $y = -\frac{3}{2}x + \frac{13}{2}$ **5.** $x = 5$ **7.** $\frac{1}{2x - 3}$ **9.** $\frac{x + 1}{3(x - 2)}$ **11.** $\log_e x$ **13.** $(-\infty, \infty)$ **15.** 10 **17.** $\frac{\ln 2}{r}$ **19.** 3.2288 **21.** 2.2915 **23.** -0.1592 **25.** none **27.** 9.9892 **29.** 23.8075 **31.** 0.0089 **33.** 61.9098 **35.** no **37.** no

39.

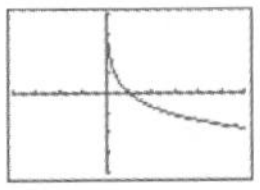

41.

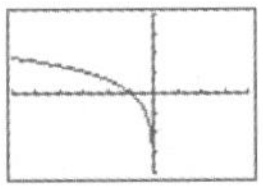

43. 5.8 yr **45.** 9.2 yr

Getting Ready (page 698)

1. x^{m+n} **2.** 1 **3.** x^{mn} **4.** x^{m-n}

Orals (page 707)

1. 2 **2.** 5 **3.** 343 **4.** $\frac{1}{4}$ **5.** 2 **6.** 2 **7.** $\frac{1}{4}$ **8.** $\frac{1}{2}$ **9.** 2

Exercise 9.5 (page 707)

1. $-\frac{7}{6}$ **3.** $\left(1, -\frac{1}{2}\right)$ **5.** 0 **7.** M, N **9.** x, y **11.** x **13.** $\neq$ **15.** 0 **17.** 7 **19.** 10 **21.** 1 **23.** 0 **25.** 7 **27.** 10 **29.** 1 **37.** $\log_b x + \log_b y + \log_b z$ **39.** $\log_b 2 + \log_b x - \log_b y$ **41.** $3 \log_b x + 2 \log_b y$ **43.** $\frac{1}{2}(\log_b x + \log_b y)$ **45.** $\log_b x + \frac{1}{2}\log_b z$ **47.** $\frac{1}{3} \log_b x - \frac{1}{4} \log_b y - \frac{1}{4} \log_b z$ **49.** $\log_b \frac{x+1}{x}$ **51.** $\log_b x^2y^{1/2}$ **53.** $\log_b \frac{z^{1/2}}{x^3y^2}$ **55.** $\log_b \frac{\frac{x}{z} + x}{\frac{y}{z} + y} = \log_b \frac{x}{y}$ **57.** false **59.** false **61.** true **63.** false **65.** true **67.** true **69.** 1.4472 **71.** 0.3521 **73.** 1.1972 **75.** 2.4014 **77.** 2.0493 **79.** 0.4682 **81.** 1.7712 **83.** -1.0000 **85.** 1.8928 **87.** 2.3219 **89.** 4.77 **91.** from 2.5119×10^{-8} to 1.585×10^{-7} **93.** It will increase by $k \ln 2$. **95.** The intensity must be cubed.

Getting Ready (page 709)

1. $2 \log x$ **2.** $\frac{1}{2} \log x$ **3.** 0 **4.** $2b \log a$

Orals (page 718)

1. $x = \frac{\log 5}{\log 3}$ **2.** $x = \frac{\log 3}{\log 5}$ **3.** $x = -\frac{\log 7}{\log 2}$ **4.** $x = -\frac{\log 1}{\log 6} = 0$ **5.** $x = 2$ **6.** $x = \frac{1}{2}$ **7.** $x = 10$ **8.** $x = 10$

Exercise 9.6 (page 718)

1. 0, 5 **3.** $\frac{2}{3}, -4$ **5.** exponential **7.** $A_0 2^{-t/h}$ **9.** 1.1610 **11.** 1.2702 **13.** 1.7095 **15.** 0 **17.** ± 1.0878 **19.** 0, 1.0566 **21.** 3, -1 **23.** $-2, -2$ **25.** 0 **27.** 0.2789 **29.** 1.8 **31.** 3, -1 **33.** 2 **35.** 3 **37.** -7 **39.** 4 **41.** 10, -10 **43.** 50 **45.** 20 **47.** 10 **49.** 10 **51.** 1, 100 **53.** no solution **55.** 6 **57.** 9 **59.** 4 **61.** 1, 7 **63.** 20 **65.** 8 **67.** 5.1 yr **69.** 42.7 days **71.** about 4,200 yr **73.** 5.6 yr **75.** 5.4 yr **77.** because $\ln 2 \approx 0.7$ **79.** 25.3 yr **81.** 2.828 times larger **83.** 13.3 **87.** $x \leq 3$

Chapter Summary (page 723)

1. a. $5^{2\sqrt{2}}$ **b.** $2^{\sqrt{10}}$ **2. a.**

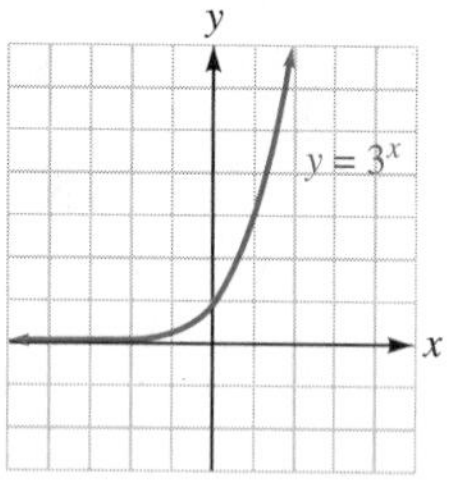

b.

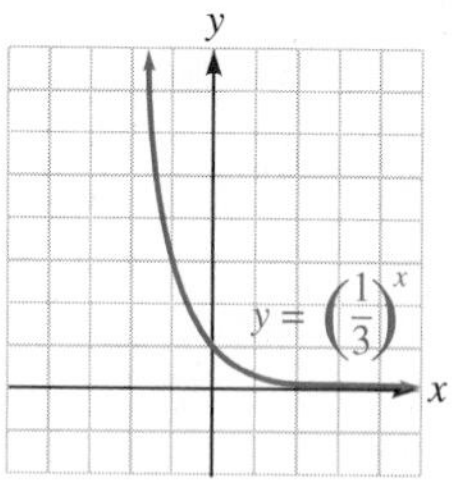

3. $x = 1, y = 6$ **4.** D: $(-\infty, \infty)$, R: $(0, \infty)$

5. a.

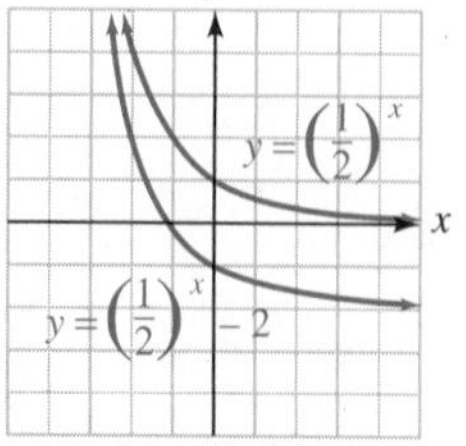

b.

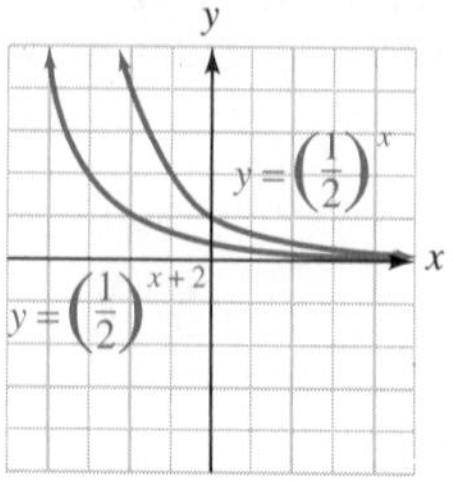

6. \$2,189,703.45 **7.** \$2,324,767.37

8. a.

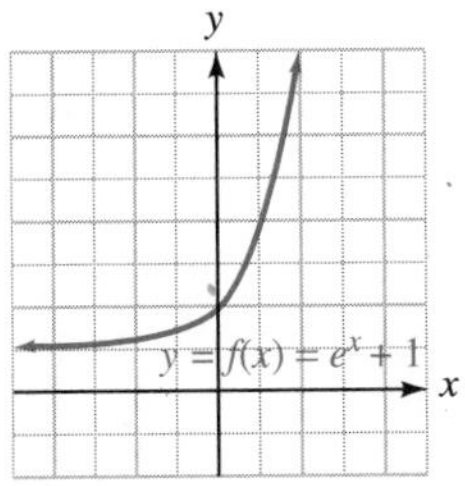

b.

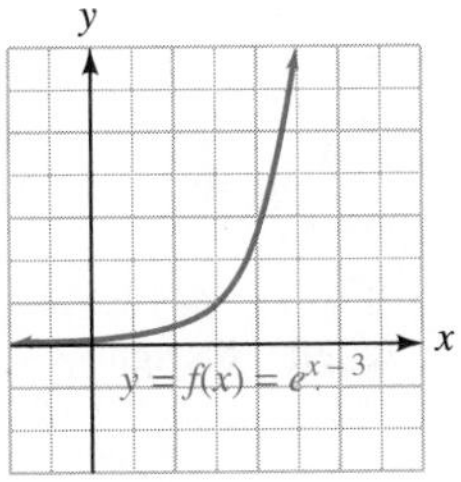

9. about 582,000,000 **10.** $(0, \infty)$, $(-\infty, \infty)$ **11. a.** 2 **b.** $-\frac{1}{2}$ **c.** 0 **d.** -2 **e.** $\frac{1}{2}$ **f.** $\frac{1}{3}$ **12. a.** 32 **b.** 9 **c.** 27 **d.** -1 **e.** $\frac{1}{8}$ **f.** 2 **g.** 4 **h.** 2 **i.** 10 **j.** $\frac{1}{25}$ **k.** 5 **l.** 3

13. a.

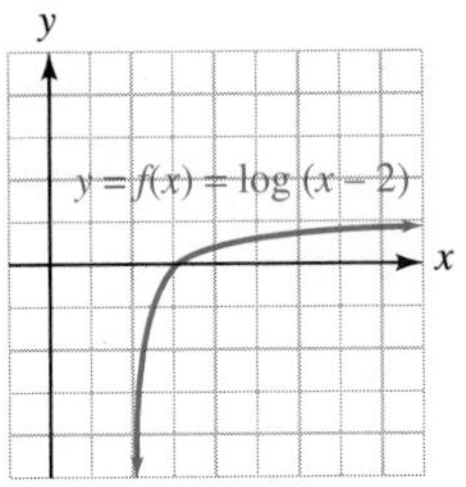

b.

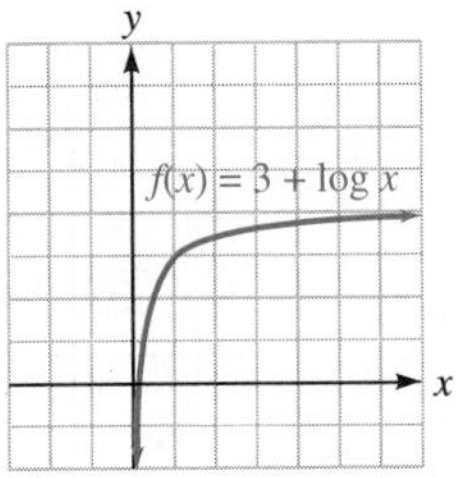

14. a.

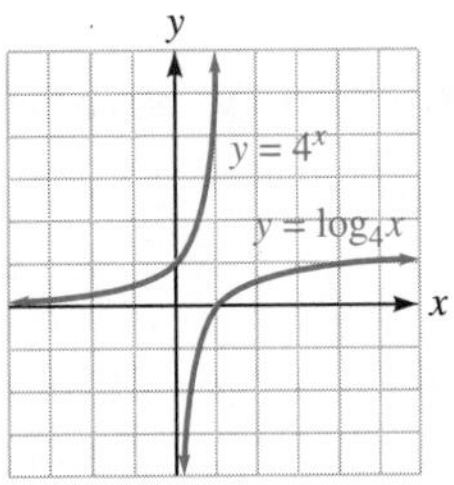

b.

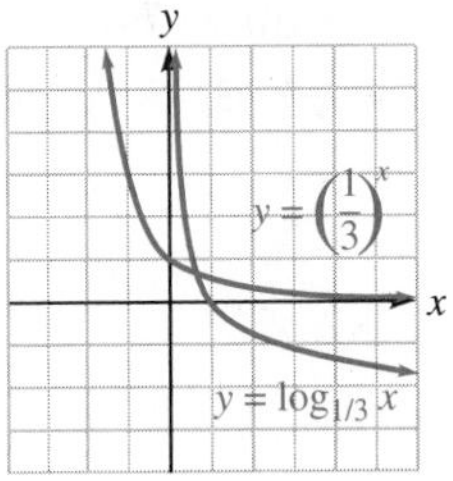

15. 53 db **16.** 4.4 **17. a.** 6.1137 **b.** -0.1111 **18. a.** 10.3398 **b.** 2.5715

19. a.

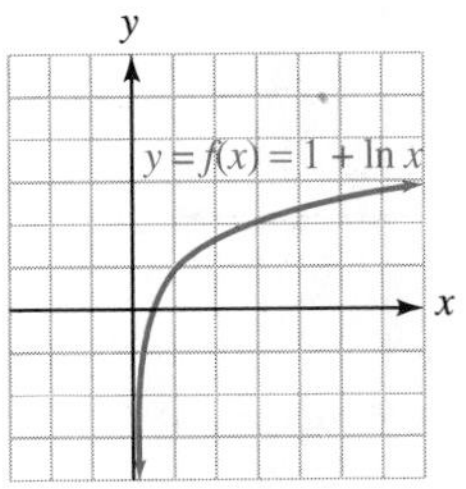

b.

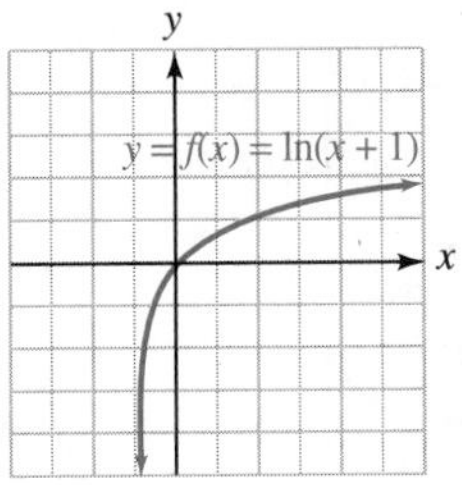

20. 23 yr **21. a.** 0 **b.** 1 **c.** 3 **d.** 4 **22. a.** 4 **b.** 0 **c.** 7 **d.** 3 **e.** 4 **f.** 9 **23. a.** $2 \log_b x + 3 \log_b y - 4 \log_b z$ **b.** $\frac{1}{2}(\log_b x - \log_b y - 2 \log_b z)$ **24. a.** $\log_b \frac{x^3 z^7}{y^5}$ **b.** $\log_b \frac{y^3\sqrt{x}}{z^7}$ **25. a.** 3.36 **b.** 1.56 **c.** 2.64 **d.** -6.72 **26.** 1.7604 **27.** about 7.94×10^{-4} gram-ions per liter **28.** $k \ln 2$ less **29. a.** $\frac{\log 7}{\log 3} \approx 1.7712$ **b.** 2 **c.** $\frac{\log 3}{\log 3 - \log 2} \approx 2.7095$ **d.** $-1, -3$ **30. a.** 25, 4 **b.** 4 **c.** 2 **d.** 4, 3 **e.** 6 **f.** 31 **g.** $\frac{\ln 9}{\ln 2} \approx 3.1699$ **h.** no solution **i.** $\frac{e}{e-1} \approx 1.5820$ **j.** 1 **31.** about 3,300 yr

Chapter 9 Test (page 727)

1.

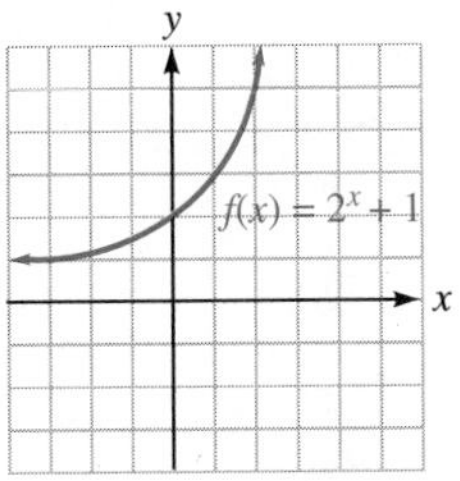

2.

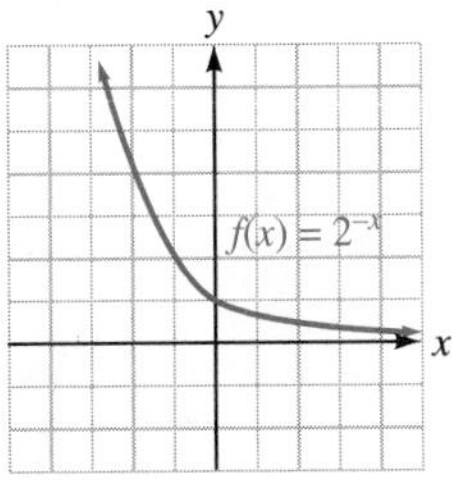

3. $\frac{3}{64}$ g **4.** \$1,060.90 **5.**

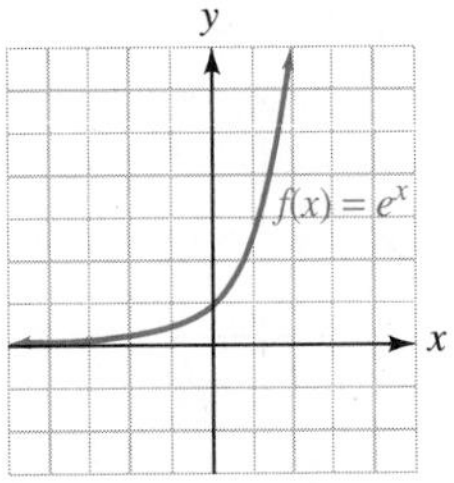

6. \$4,451.08 **7.** 2

8. 3 **9.** $\frac{1}{27}$ **10.** 10 **11.** 2 **12.** $\frac{27}{8}$ **13.**

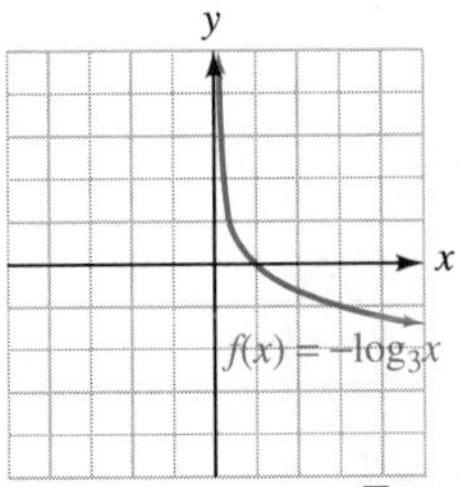

14.

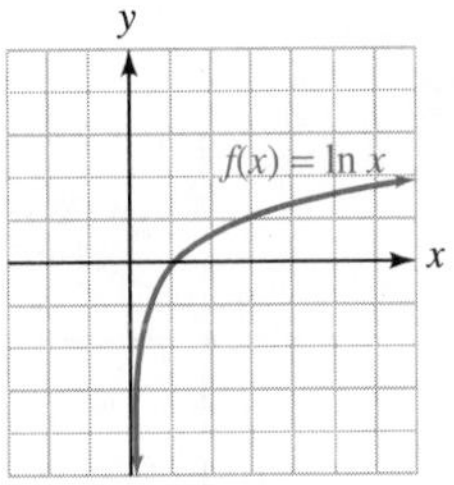

15. $2 \log a + \log b + 3 \log c$

16. $\frac{1}{2}(\ln a - 2 \ln b - \ln c)$ **17.** $\log \frac{b\sqrt{a+2}}{c^3}$ **18.** $\log \frac{\sqrt[3]{a}}{c\sqrt[3]{b^2}}$ **19.** 1.3801 **20.** 0.4259 **21.** $\frac{\log 3}{\log 7}$ or $\frac{\ln 3}{\ln 7}$ **22.** $\frac{\log e}{\log \pi}$ or $\frac{\ln e}{\ln \pi}$ **23.** true **24.** false **25.** false **26.** false **27.** 6.4 **28.** 46 **29.** $\frac{\log 3}{\log 5}$ **30.** $\frac{\log 3}{(\log 3) - 2}$ **31.** 1 **32.** 10

Getting Ready (page 731)

1. $x^2 - 4x + 4$ **2.** $x^2 + 8x + 16$ **3.** $\frac{81}{4}$ **4.** 36

Orals (page 741)

1. (0, 0), 12 **2.** (0, 0), 11 **3.** (2, 0), 4 **4.** (0, −1), 3 **5.** down **6.** up **7.** left **8.** right

Exercise 10.1 (page 741)

1. 5, $-\frac{7}{3}$ **3.** 3, $-\frac{1}{4}$ **5.** circle, plane **7.** $r^2 < 0$ **9.** parabola, (3, 2), right **11.**

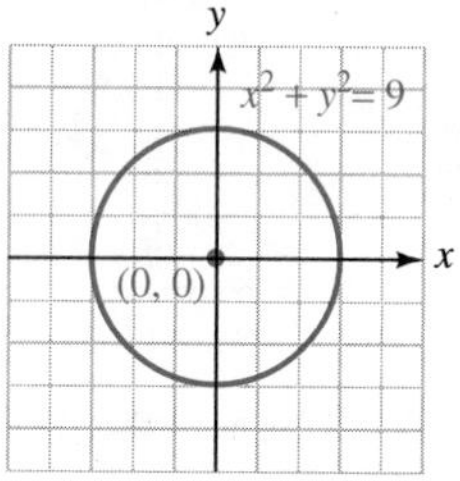

13.

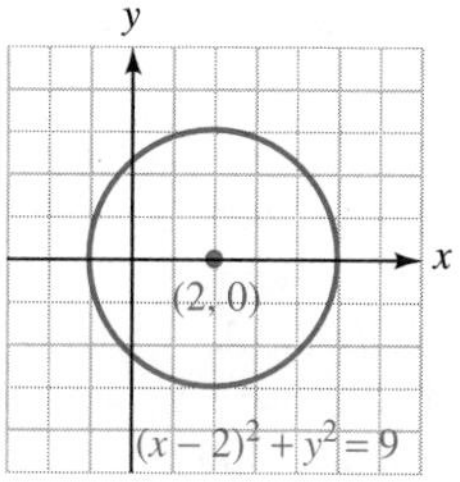

15.

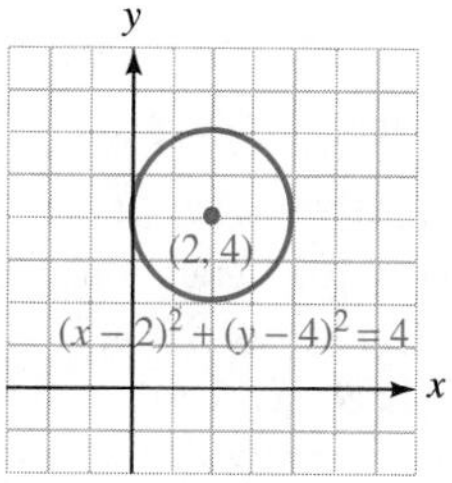

17.

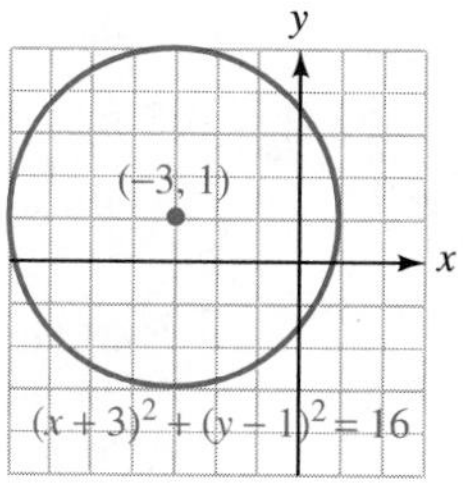

19.

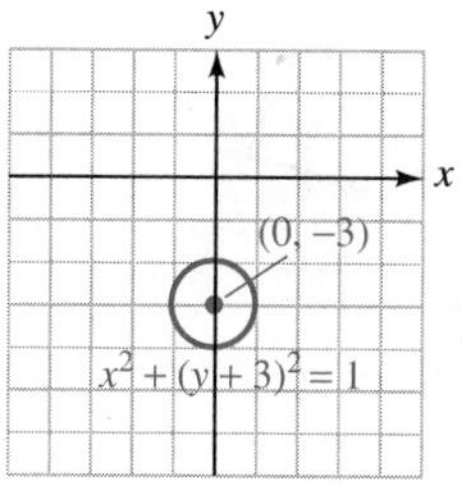

21.

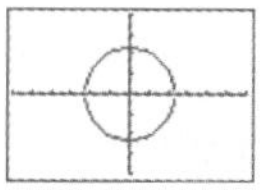

23.

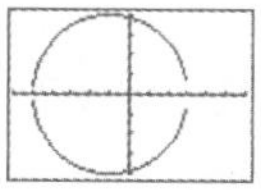

25. $x^2 + y^2 = 1$ **27.** $(x - 6)^2 + (y - 8)^2 = 25$ **29.** $(x + 2)^2 + (y - 6)^2 = 144$

31. $x^2 + y^2 = 2$ **33.**

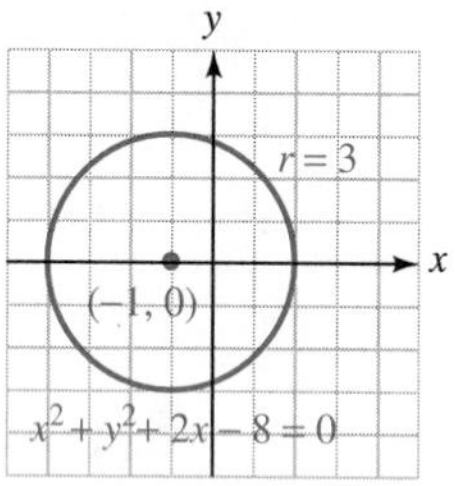

35.

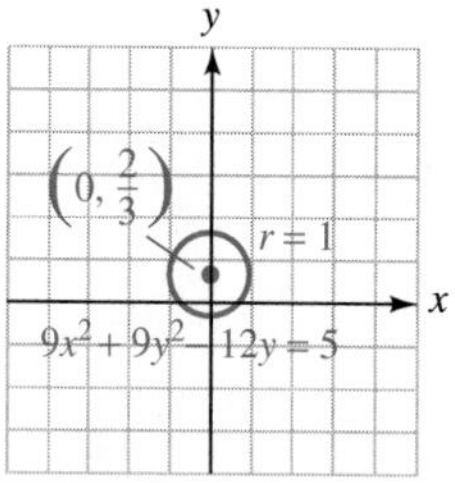

37.

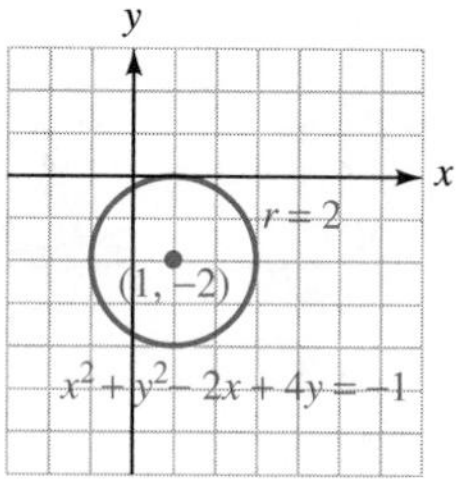

39.

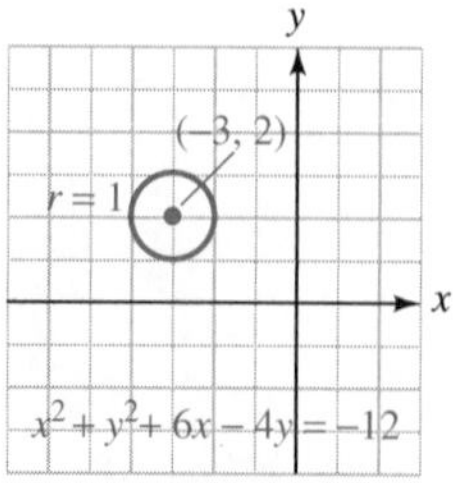

41.

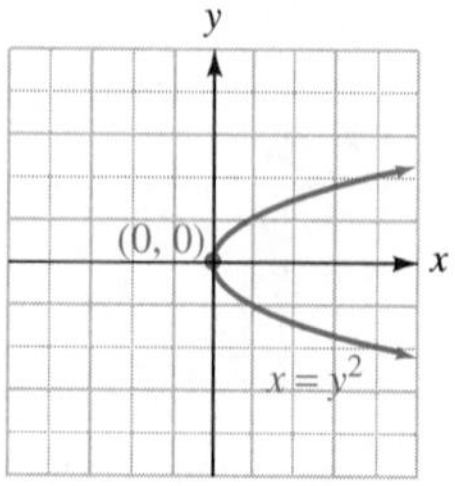

43.

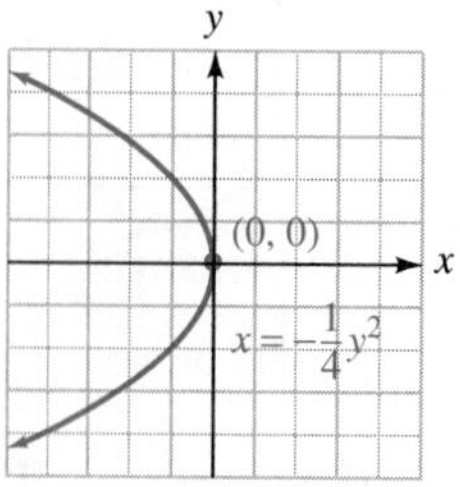

45.

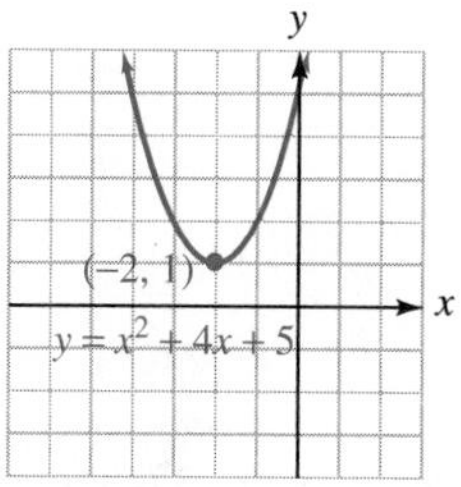

47. 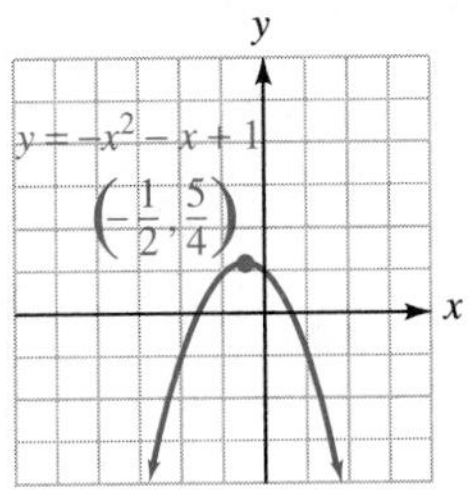

49. 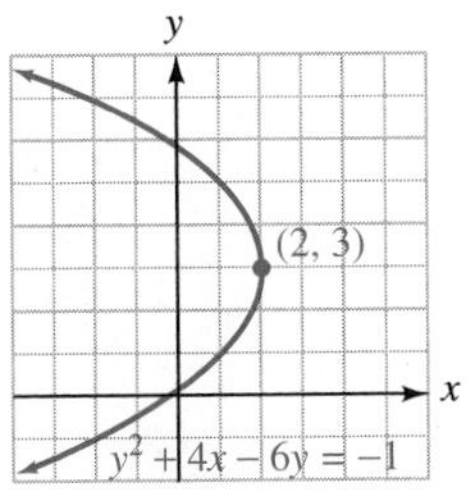

51. 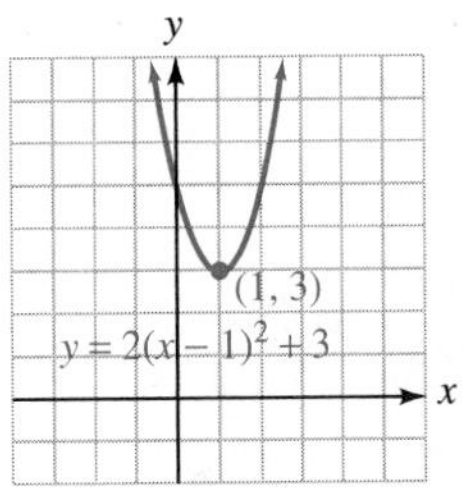

53. 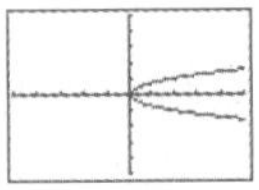

55. 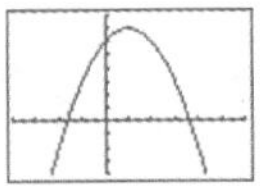

57. $(x - 7)^2 + y^2 = 9$ **59.** no **61.** 30 ft away **63.** 2 AU

Getting Ready (page 746)

1. $y = \pm b$ **2.** $x = \pm a$

Orals (page 754)

1. $(\pm 3, 0), (0, \pm 4)$ **2.** $(\pm 5, 0), (0, \pm 6)$ **3.** $(2, 0)$ **4.** $(0, -1)$

Exercise 10.2 (page 754)

1. $12y^2 + \dfrac{9}{x^2}$ **3.** $\dfrac{y^2 + x^2}{y^2 - x^2}$ **5.** ellipse, sum **7.** center **9.** $(0, 0)$ **11.** 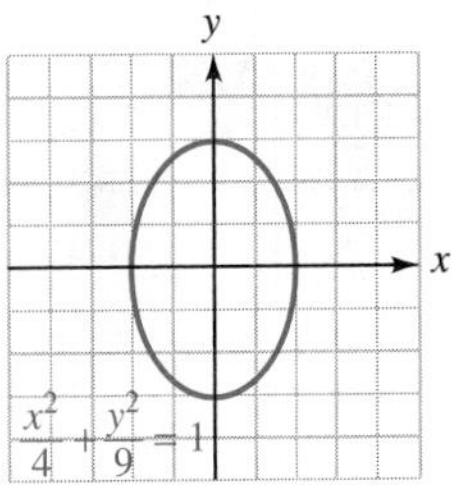

13. 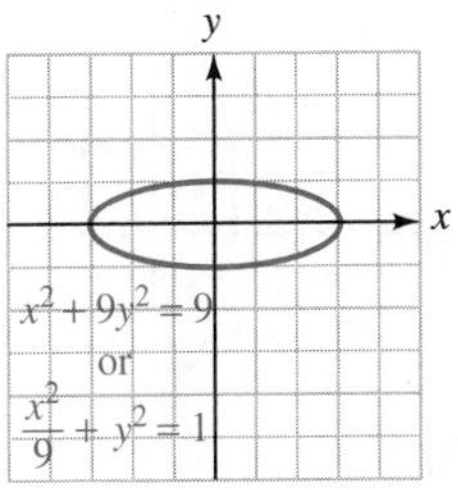

15. 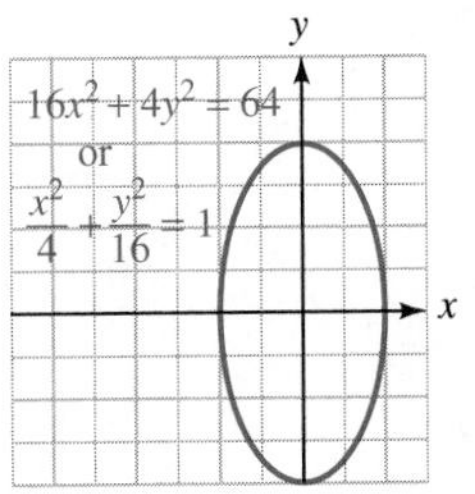

17. 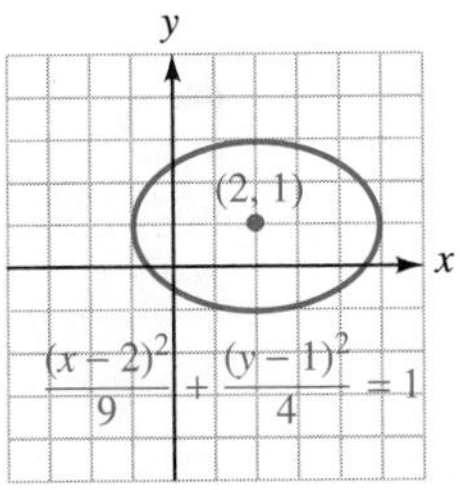

19. 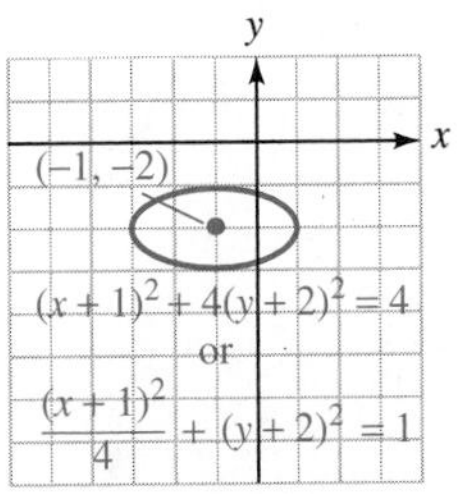

21. 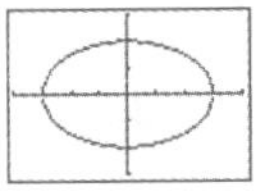

23. 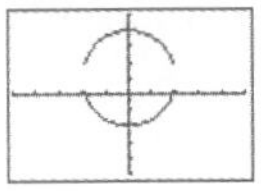

25. 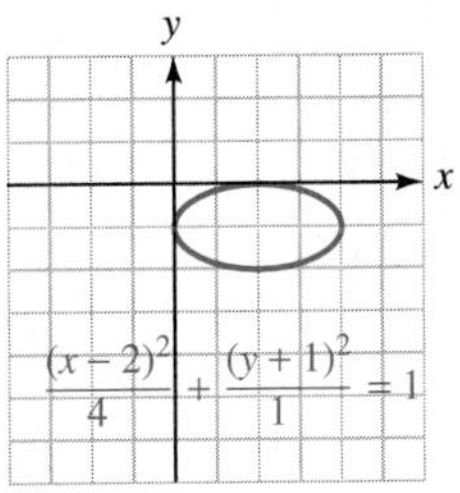

27. 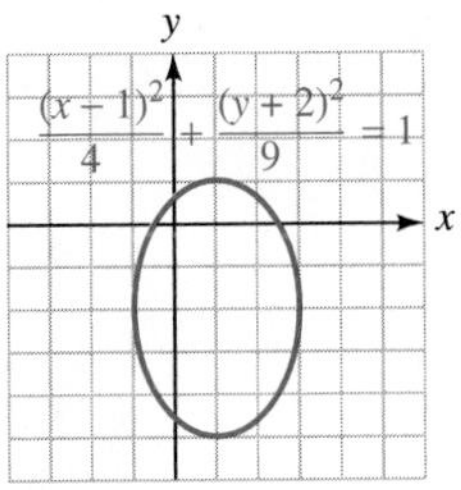

29. $y = \frac{1}{2}\sqrt{400 - x^2}$ **31.** 12π sq. units

Getting Ready (page 757)

1. $y = \pm 2.0$ **2.** $y = \pm 2.9$

Orals (page 765)

1. $(\pm 3, 0)$ **2.** $(0, \pm 5)$

Exercise 10.3 (page 765)

1. $-3x^2(2x^2 - 3x + 2)$ **3.** $(5a + 2b)(3a - 2b)$ **5.** hyperbola, difference **7.** center **9.** $(0, 0)$

11.

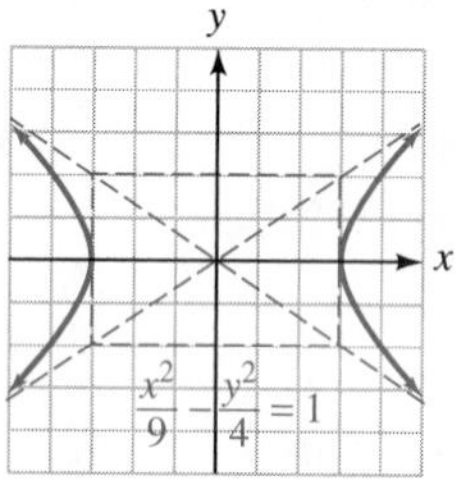

13.

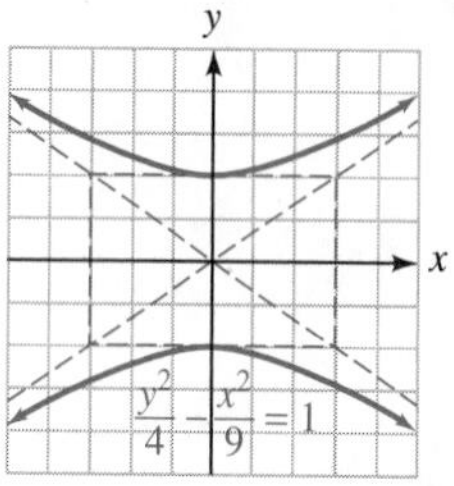

15.

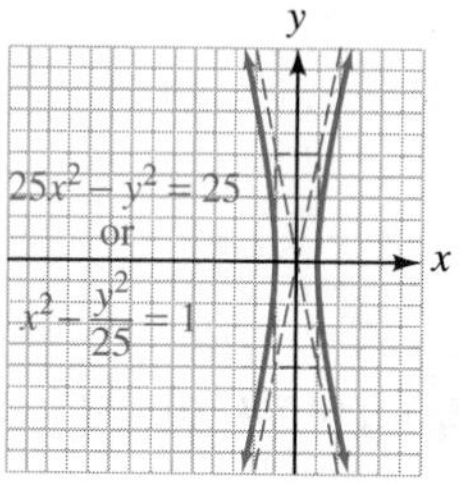

17.

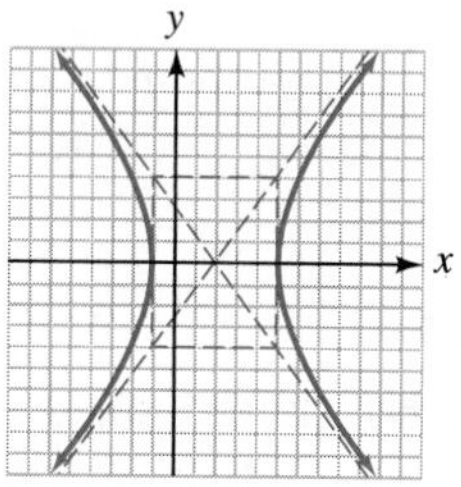

$\frac{(x-2)^2}{9} - \frac{y^2}{16} = 1$

19.

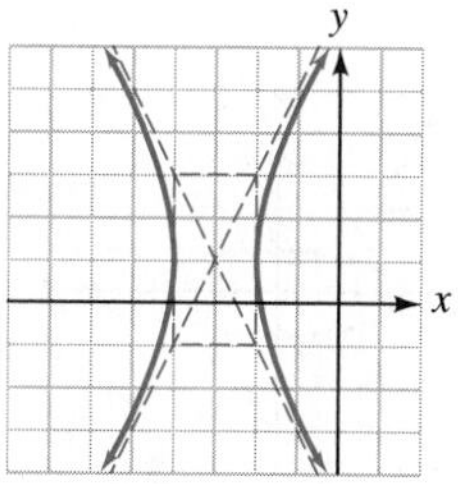

$4(x+3)^2 - (y-1)^2 = 4$

or $(x+3)^2 - \frac{(y-1)^2}{4} = 1$

21.

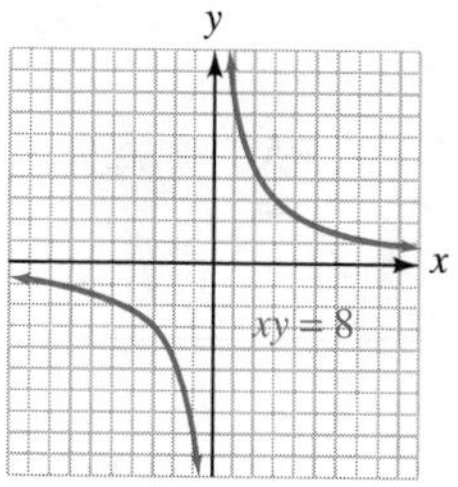

23.

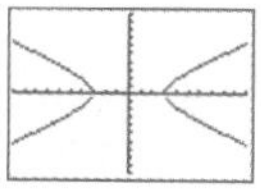

25.

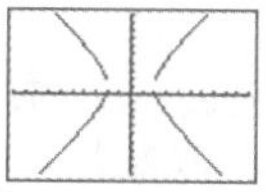

27.

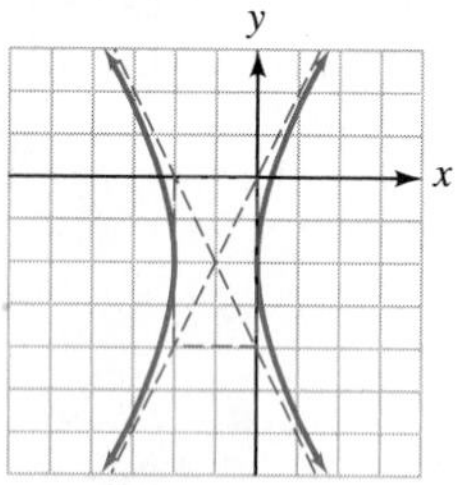

$\frac{(x+1)^2}{1} - \frac{(y+2)^2}{4} = 1$

29.

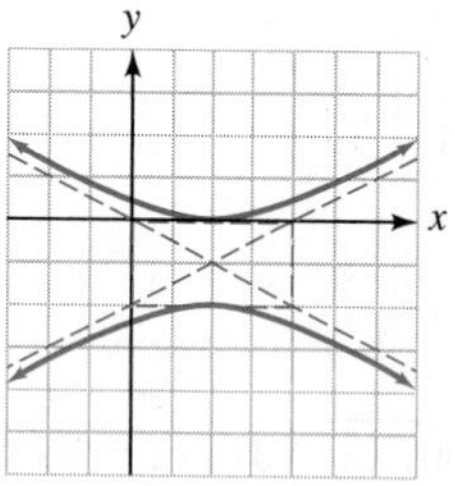

$\frac{(y+1)^2}{1} - \frac{(x-2)^2}{4} = 1$

31. 3 units **33.** $10\sqrt{3}$ units

Getting Ready (page 768)

1. $7x^2 - y^2 = 44$ **2.** $5x^2 - 3y^2 = 8$

Orals (page 772)

1. 0, 1, 2 **2.** 0, 1, 2 **3.** 0, 1, 2, 3, 4 **4.** 0, 1, 2, 3, 4

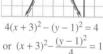

Exercise 10.4 (page 773)

1. $-11x\sqrt{2}$ **3.** $\frac{t}{2}$ **5.** graphing, substitution **7.**

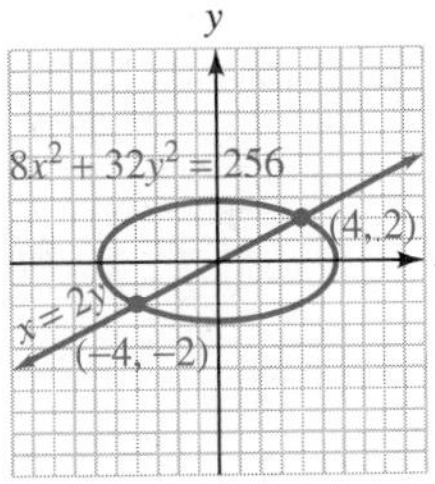

9.

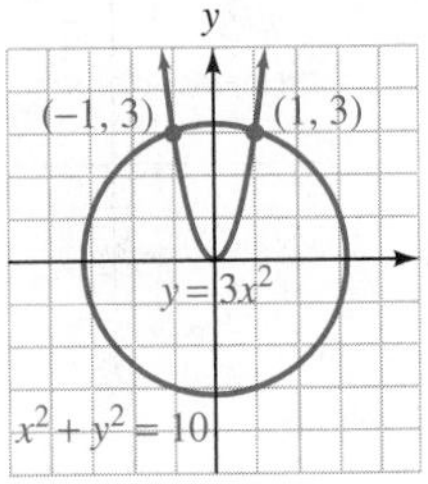

11.

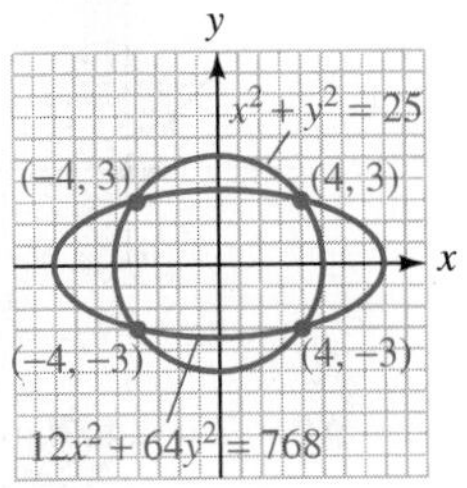

13.

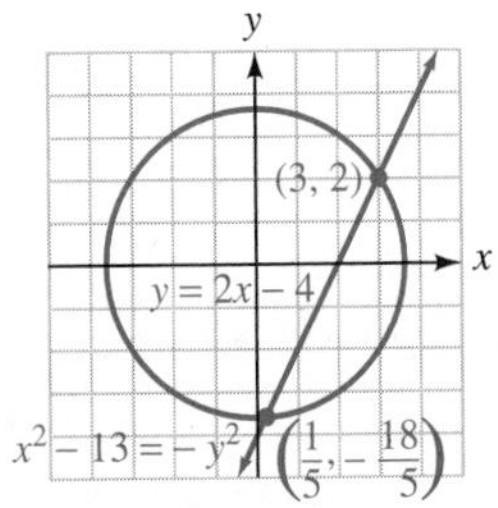

15. (1, 0), (5, 0)

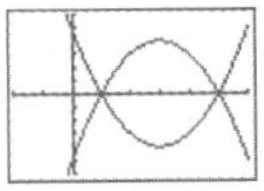

17. (3, 0), (0, 5) **19.** (1, 1) **21.** (1, 2), (2, 1) **23.** (−2, 3), (2, 3) **25.** $(\sqrt{5}, 5), (-\sqrt{5}, 5)$ **27.** (3, 2), (3, −2), (−3, 2), (−3, −2) **29.** (2, 4), (2, −4), (−2, 4), (−2, −4) **31.** $(-\sqrt{15}, 5), (\sqrt{15}, 5)$, (−2, −6), (2, −6) **33.** (0, −4), (−3, 5), (3, 5) **35.** (−2, 3), (2, 3), (−2, −3), (2, −3) **37.** (3, 3) **39.** (6, 2), (−6, −2), $(\sqrt{42}, 0), (-\sqrt{42}, 0)$ **41.** $\left(\frac{1}{2}, \frac{1}{3}\right), \left(\frac{1}{3}, \frac{1}{2}\right)$ **43.** 4 and 8 **45.** 7 cm by 9 cm **47.** either \$750 at 9% or \$900 at 7.5% **49.** 68 mph, 4.5 hr **53.** 0, 1, 2, 3, 4

Getting Ready (page 776)

1. positive **2.** negative **3.** 98 **4.** −3

Orals (page 780)

1. increasing **2.** decreasing **3.** constant **4.** increasing

Exercise 10.5 (page 780)

1. 20 **3.** domains **5.** constant, $f(x)$ **7.** step **9.** increasing on $(-\infty, 0)$, decreasing on $(0, \infty)$ **11.** decreasing on $(-\infty, 0)$, constant on (0, 2), increasing on $(2, \infty)$ **13.** constant on $(-\infty, 0)$, increasing on $(0, \infty)$

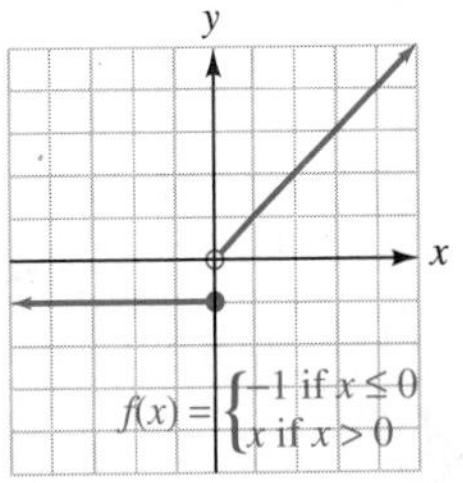

15. decreasing on $(-\infty, 0)$, increasing on (0, 2), decreasing on $(2, \infty)$

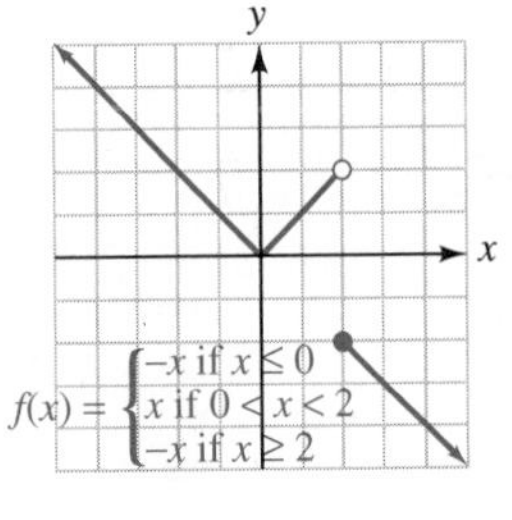

17.

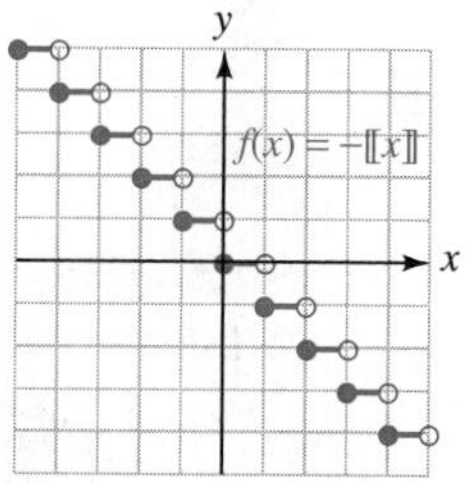

19.

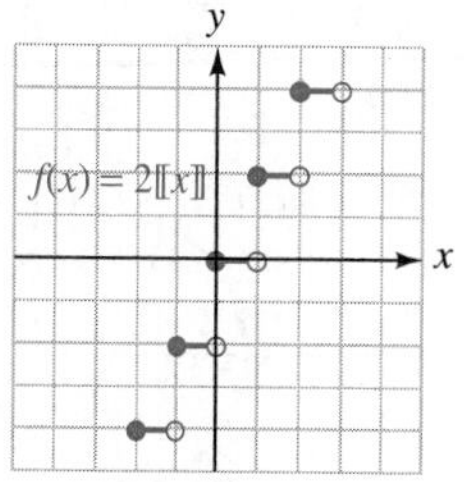

21.

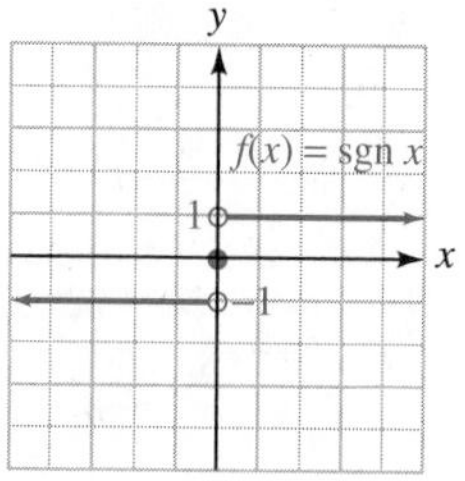

23. \$30

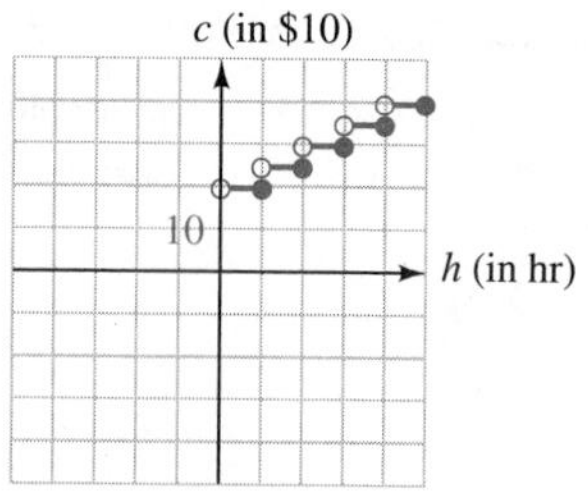

25. After 2 hours, network B is cheaper.

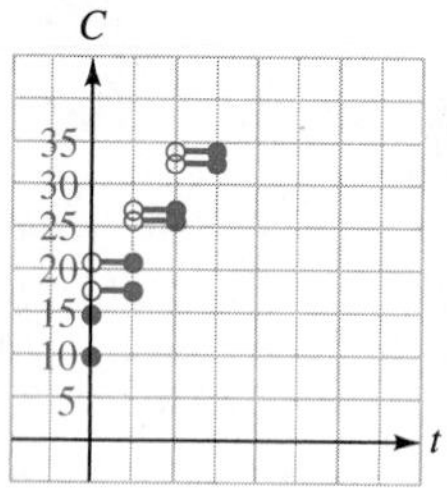

Chapter Summary (page 785)

1. a.

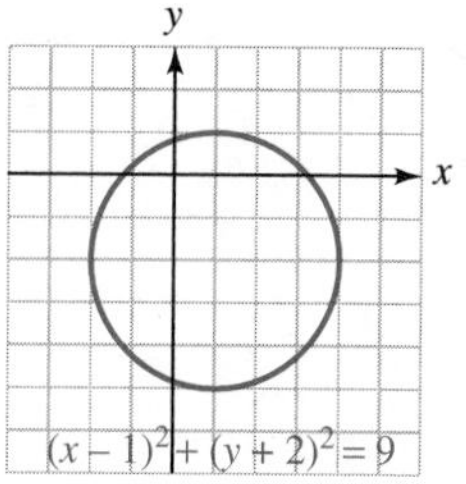

b.

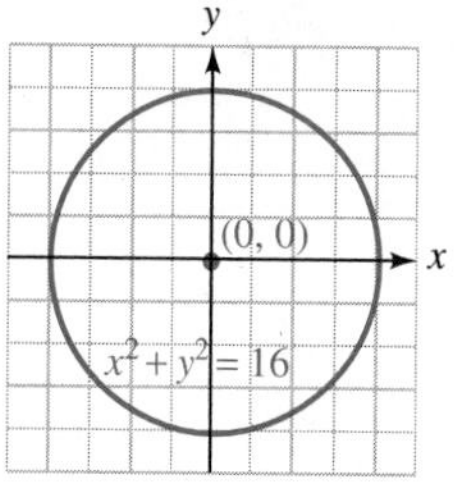

2.

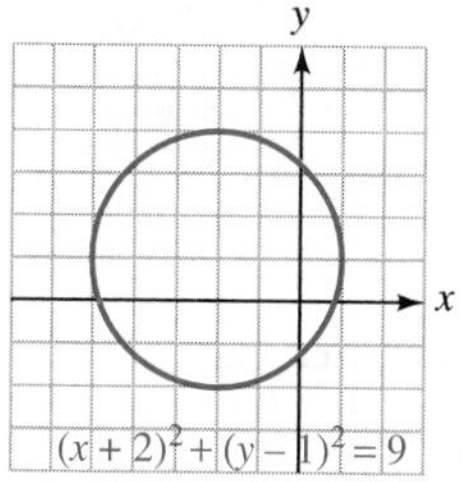

3. a.

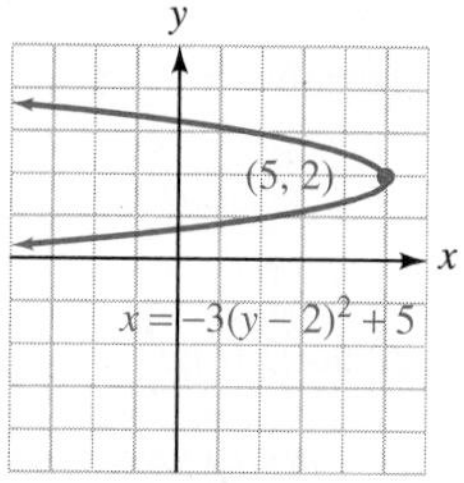

b.

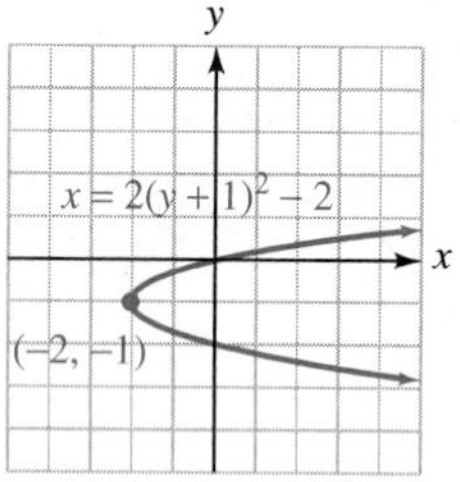

4. a.

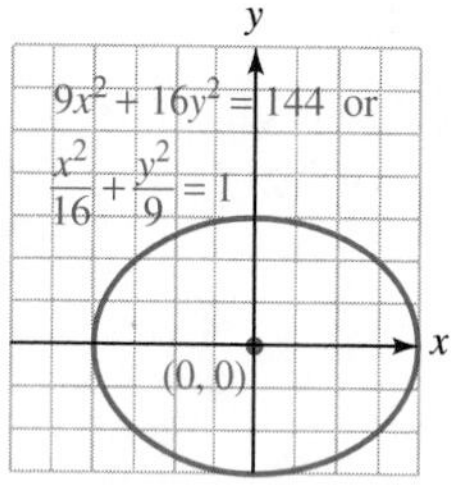

b.

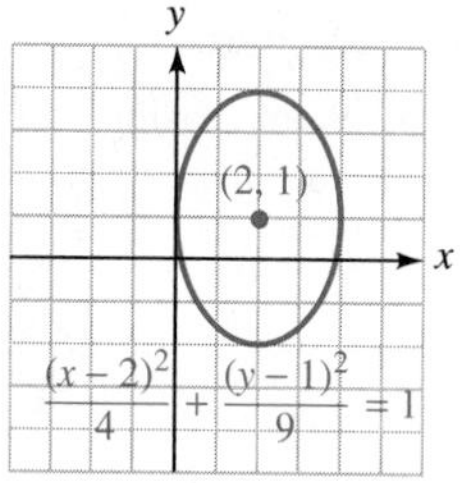

5.

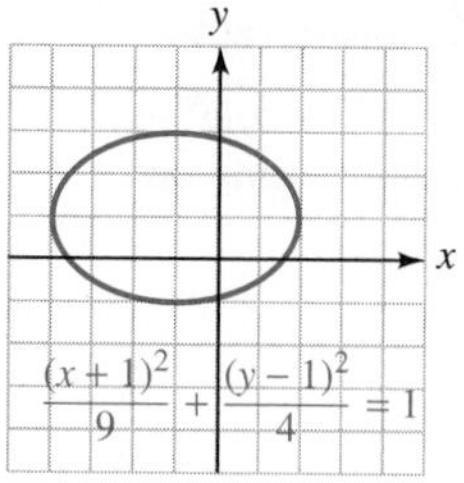

6. a.

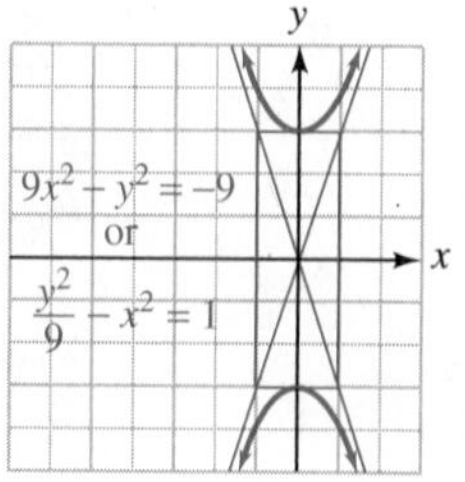

b.

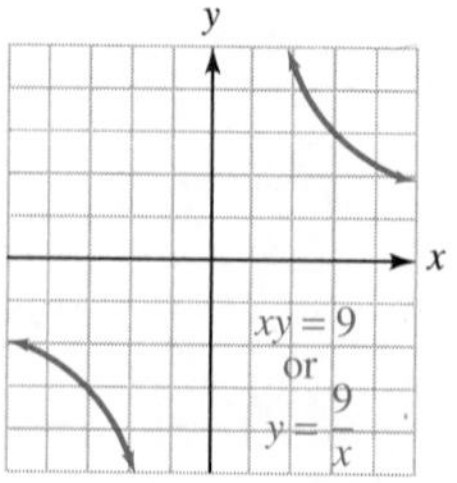

7. hyperbola

8.

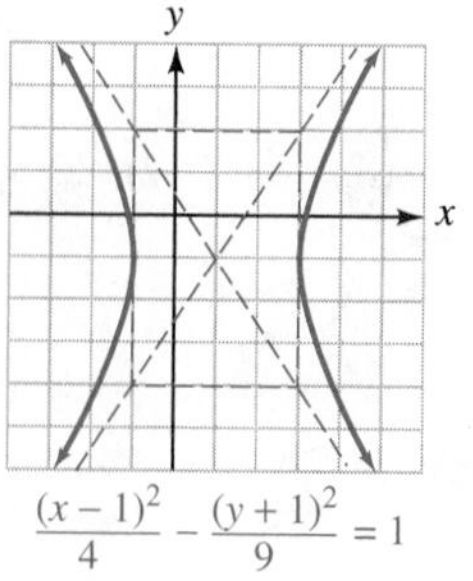

9. a. (4, 2), (4, −2), (−4, 2), (−4, −2) **b.** (2, 3), (2, −3), (−2, 3), (−2, −3) **10.** increasing on (−∞, −2), constant on (−2, 1), decreasing on (1, ∞) **11. a.**

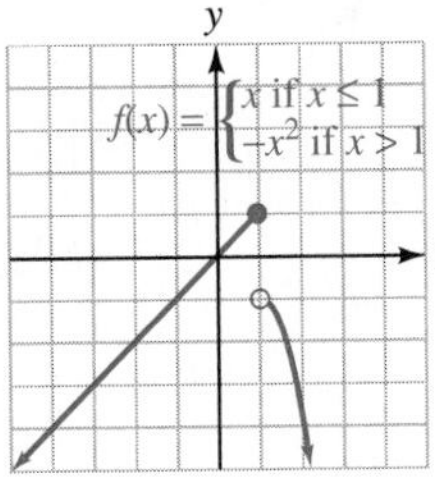

b.

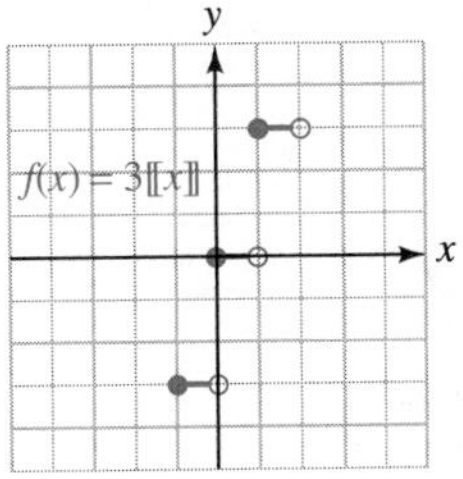

Chapter 10 Test (page 788)

1. (2, −3), 2 **2.** (−2, 3), 4 **3.**

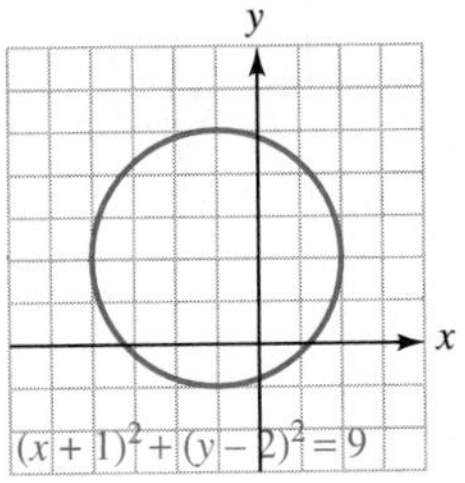

4.

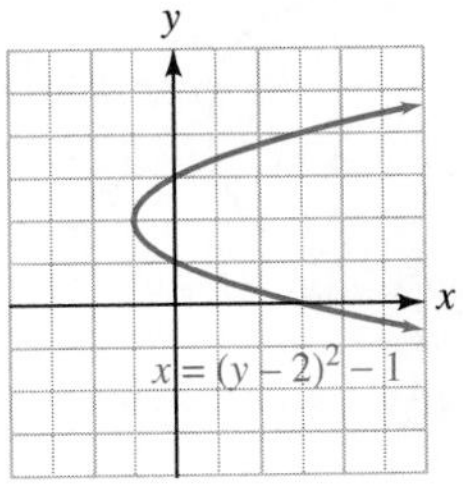

5.

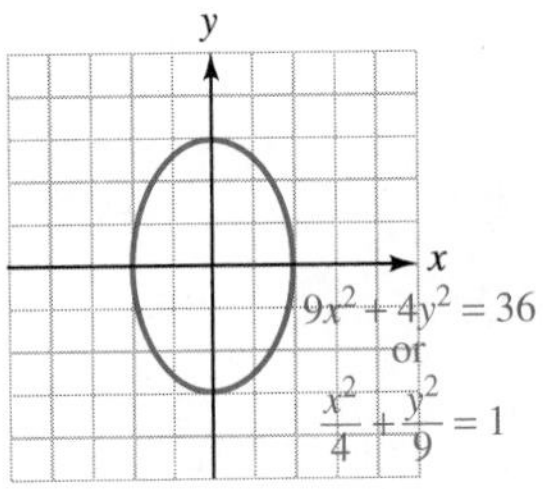

6.

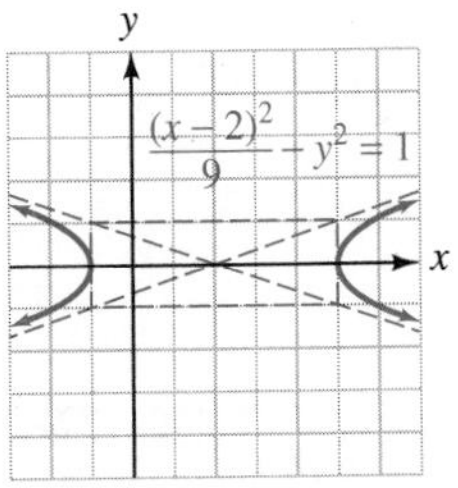

7.

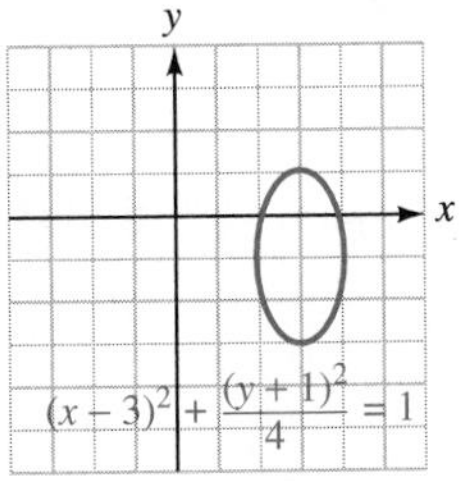

8.

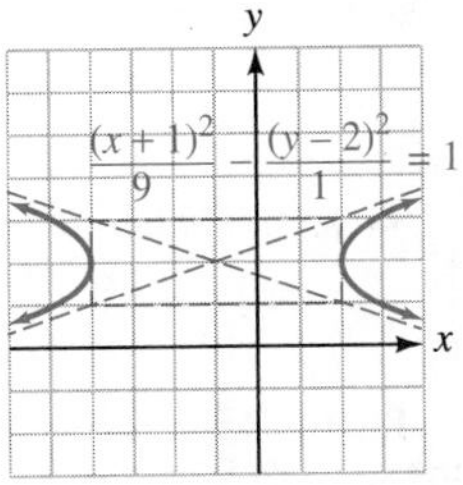

9. (2, 6), (−2, −2) **10.** (3, 4), (−3, 4)

11. increasing on (−3, 0), decreasing on (0, 3) **12.**

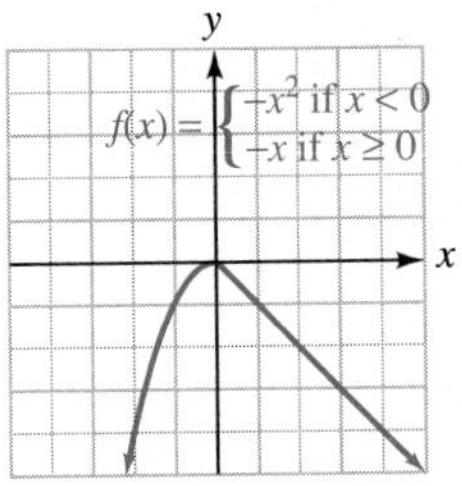

Cumulative Review Exercises (page 789)

1. $12x^2 - 5xy - 3y^2$ **2.** $a^{2n} - 2a^n - 3$ **3.** $\frac{5}{a-2}$ **4.** $a^2 - 3a + 2$ **5.** 1 **6.** $\frac{4a-1}{(a+2)(a-2)}$ **7.** parallel **8.** perpendicular **9.** $y = -2x + 5$ **10.** $y = -\frac{9}{13}x + \frac{7}{13}$ **11.**

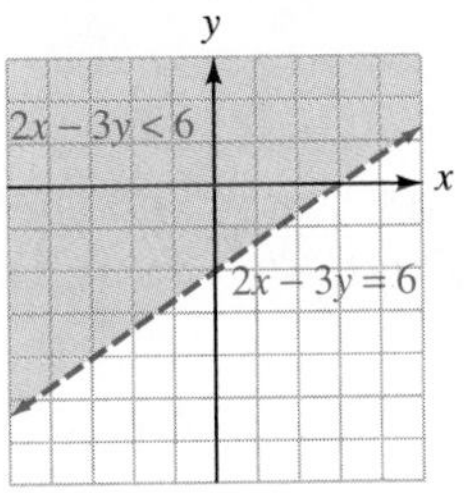

12.

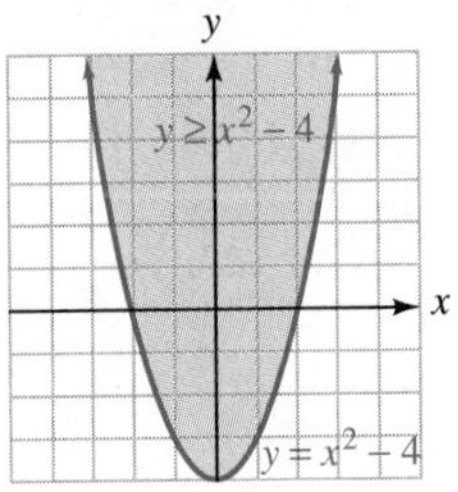

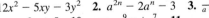

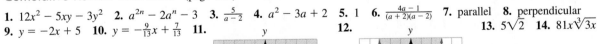

13. $5\sqrt{2}$ **14.** $81x\sqrt[3]{3x}$

15. 0, 5 **16.** 0 **17.** $\frac{2}{3}, -\frac{3}{2}$ **18.** $\dfrac{-4 \pm \sqrt{19}}{3}$ **19.** $4x^2 + 4x - 1$ **20.** $f^{-1}(x) = \sqrt[3]{\dfrac{x+1}{2}}$ **21.**

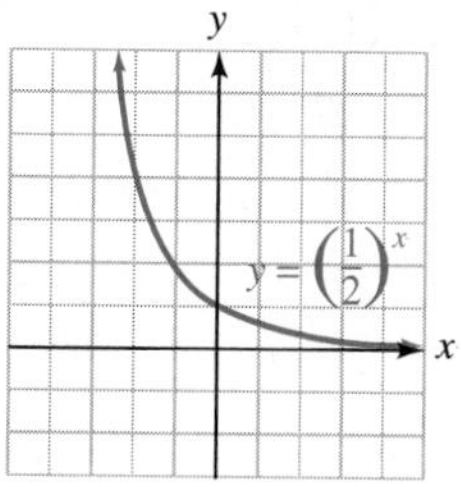

22. $2^y = x$ **23.** $\dfrac{2 \log 2}{\log 3 - \log 2}$ **24.** 16 **25.**

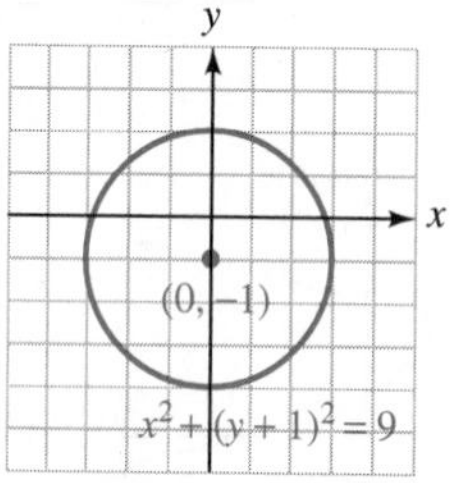

26.

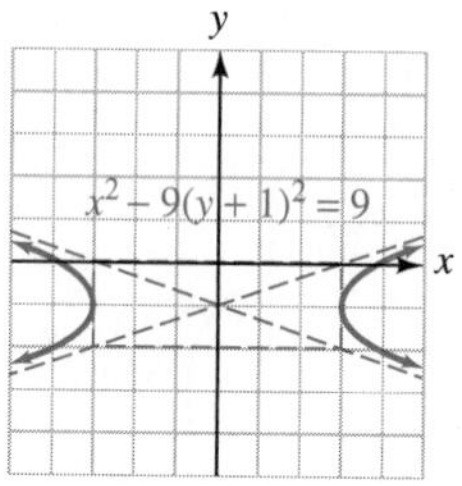

Getting Ready (page 793)

1. $x^2 + 4x + 4$ **2.** $x^2 - 6x + 9$ **3.** $x^3 + 3x^2 + 3x + 1$ **4.** $x^3 - 6x^2 + 12x - 8$

Orals (page 798)

1. 1 **2.** 24 **3.** 1 **4.** 120 **5.** $m^2 + 2mn + n^2$ **6.** $m^2 - 2mn + n^2$ **7.** $p^2 + 4pq + 4q^2$ **8.** $4p^2 - 4pq + q^2$

Exercise 11.1 (page 799)

1. 2 **3.** 5 **5.** one **7.** Pascal's **9.** 6! **11.** 1 **13.** 6 **15.** −120 **17.** 30 **19.** 144 **21.** 40,320 **23.** $\frac{1}{110}$ **25.** 2,352 **27.** 10 **29.** 21 **31.** $\frac{1}{168}$ **33.** $x^3 + 3x^2y + 3xy^2 + y^3$ **35.** $x^4 - 4x^3y + 6x^2y^2 - 4xy^3 + y^4$ **37.** $8x^3 + 12x^2y + 6xy^2 + y^3$ **39.** $x^3 - 6x^2y + 12xy^2 - 8y^3$ **41.** $8x^3 + 36x^2y + 54xy^2 + 27y^3$ **43.** $\dfrac{x^3}{8} - \dfrac{x^2y}{4} + \dfrac{xy^2}{6} - \dfrac{y^3}{27}$ **45.** $81 + 216y + 216y^2 + 96y^3 + 16y^4$ **47.** $\dfrac{x^4}{81} - \dfrac{2x^3y}{27} + \dfrac{x^2y^2}{6} - \dfrac{xy^3}{6} + \dfrac{y^4}{16}$ **51.** 1, 1, 2, 3, 5, 8, 13, . . . ; beginning with 2, each number is the sum of the previous two numbers.

Getting Ready (page 800)

1. 6 **2.** 10 **3.** −20 **4.** −4

Orals (page 802)

1. 2 **2.** 3 **3.** 6 **4.** 6 **5.** 5 **6.** 2 **7.** 1 **8.** 8

Exercise 11.2 (page 802)

1. (2, 3) **3.** 8 **5.** 3 **7.** 7 **9.** $3a^2b$ **11.** $-4xy^3$ **13.** $15x^2y^4$ **15.** $28x^6y^2$ **17.** $90x^3$ **19.** $640x^3y^2$ **21.** $-12x^3y$ **23.** $-70{,}000x^4$ **25.** $810xy^4$ **27.** $180x^4y^2$ **29.** $-\frac{1}{6}x^3y$ **31.** $\frac{n!}{3!(n-3)!}a^{n-3}b^3$ **33.** $\frac{n!}{4!(n-4)!}a^{n-4}b^4$ **35.** $\frac{n!}{(r-1)!(n-r+1)!}a^{n-r+1}b^{r-1}$ **39.** 252

Getting Ready (page 804)

1. 3, 5, 7, 9 **2.** 4, 7, 10, 13

Orals (page 809)

1. 14 **2.** 1 **3.** 5 **4.** −6 **5.** 3 **6.** 5

Exercise 11.3 (page 810)

1. $18x^2 + 8x - 3$ **3.** $\dfrac{6a^2 + 16}{(a + 2)(a - 2)}$ **5.** sequence **7.** arithmetic, difference **9.** arithmetic mean **11.** $1 + 2 + 3 + 4 + 5$ **13.** 3, 5, 7, 9, 11 **15.** −5, −8, −11, −14, −17 **17.** 5, 11, 17, 23, 29 **19.** −4, −11, −18, −25, −32 **21.** −118, −111, −104, −97, −90 **23.** 34, 31, 28, 25, 22 **25.** 5, 12, 19, 26, 33 **27.** 355 **29.** −179 **31.** −23 **33.** 12 **35.** $\frac{17}{4}, \frac{13}{2}, \frac{35}{4}$ **37.** 12, 14, 16, 18 **39.** $\frac{29}{2}$ **41.** $\frac{5}{4}$ **43.** 1,335 **45.** 459 **47.** 354 **49.** 255 **51.** 1,275 **53.** 2,500 **55.** 60 **57.** 31 **59.** 12 **61.** 60, 110, 160, 210, 260, 310; $6,060 **63.** 11,325 **65.** 368 ft **69.** $\frac{3}{2}$, 2, $\frac{5}{2}$, 3, $\frac{7}{2}$, 4

Getting Ready (page 813)

1. 10, 20, 40 **2.** 18, 54, 162

Orals (page 819)

1. 27 **2.** $\frac{1}{27}$ **3.** 2.5 **4.** $\sqrt{3}$ **5.** 6 **6.** 1

Exercise 11.4 (page 819)

1. [−1, 6] **3.** $(-\infty, -3) \cup (4, \infty)$ **5.** geometric **7.** common ratio **9.** $S_n = \dfrac{a - ar^n}{1 - r}$ **11.** 3, 6, 12, 24, 48 **13.** $-5, -1, -\frac{1}{5}, -\frac{1}{25}, -\frac{1}{125}$ **15.** 2, 8, 32, 128, 512 **17.** −3, −12, −48, −192, −768 **19.** −64, 32, −16, 8, −4 **21.** −64, −32, −16, −8, −4 **23.** 2, 10, 50, 250, 1,250 **25.** 3,584 **27.** $\frac{1}{27}$ **29.** 3 **31.** 6, 18, 54 **33.** −20, −100, −500, −2,500 **35.** −16 **37.** $10\sqrt{2}$ **39.** No geometric mean exists. **41.** 728 **43.** 122 **45.** −255 **47.** 381 **49.** $\frac{156}{25}$ **51.** $-\frac{21}{4}$ **53.** about 669 people **55.** $1,469.74 **57.** $140,853.75 **59.** $\left(\frac{1}{2}\right)^{11} \approx 0.0005$ **61.** $4,309.14 **67.** arithmetic mean

Getting Ready (page 822)

1. 4 **2.** 4.5 **3.** 3 **4.** 2.5

Orals (page 825)

1. 8 **2.** $\frac{1}{8}$ **3.** $\frac{1}{2}$ **4.** $\frac{1}{8}$ **5.** 27 **6.** 16

Exercise 11.5 (page 825)

1. yes **3.** no **5.** infinite **7.** $S = \frac{a}{1 - r}$ **9.** 16 **11.** 81 **13.** 8 **15.** $-\frac{135}{4}$ **17.** no sum **19.** $-\frac{81}{2}$ **21.** $\frac{1}{9}$ **23.** $-\frac{1}{3}$ **25.** $\frac{4}{33}$ **27.** $\frac{25}{33}$ **29.** 30 m **31.** 5,000 **35.** $\frac{4}{5}$ **39.** no; $0.999999 = \frac{999,999}{1,000,000} < 1$

Getting Ready (page 826)

1. 24 **2.** 120 **3.** 30 **4.** 168

Orals (page 835)

1. 15 **2.** 120 **3.** 3 **4.** 6 **5.** 1 **6.** 1

Exercise 11.6 (page 835)

1. 6, −3 **3.** 8 **5.** $p \cdot q$ **7.** $P(n, r)$ **9.** $n!$ **11.** $\binom{n}{r}$, combinations **13.** 1 **15.** 6 **17.** 60 **19.** 12 **21.** 5 **23.** 1,260 **25.** 10 **27.** 20 **29.** 50 **31.** 2 **33.** 1 **35.** $\frac{n!}{2!(n-2)!}$ **37.** $x^4 + 4x^3y + 6x^2y^2 + 4xy^3 + y^4$ **39.** $8x^3 + 12x^2y + 6xy^2 + y^3$

41. $81x^4 - 216x^3 + 216x^2 - 96x + 16$ **43.** $-1{,}250x^2y^3$ **45.** $-4x^6y^3$ **47.** 35 **49.** 1,000,000 **51.** 136,080 **53.** 8,000,000 **55.** 720 **57.** 2,880 **59.** 13,800 **61.** 720 **63.** 900 **65.** 364 **67.** 5 **69.** 1,192,052,400 **71.** 18 **73.** 7,920 **77.** 48

Chapter Summary (page 840)

1. a. 144 **b.** 20 **c.** 15 **d.** 220 **e.** 1 **f.** 8 **2. a.** $x^5 + 5x^4y + 10x^3y^2 + 10x^2y^3 + 5xy^4 + y^5$ **b.** $x^4 - 4x^3y + 6x^2y^2 - 4xy^3 + y^4$ **c.** $64x^3 - 48x^2y + 12xy^2 - y^3$ **d.** $x^3 + 12x^2y + 48xy^2 + 64y^3$ **3. a.** $6x^2y^2$ **b.** $-10x^2y^3$ **c.** $-108x^2y$ **d.** $864x^2y^2$ **4.** 42 **5.** 122, 137, 152, 167, 182 **6.** $\frac{41}{3}, \frac{58}{3}$ **7.** 1,550 **8.** $-\frac{45}{2}$ **9. a.** $\frac{15}{2}$ **b.** 378 **c.** 14 **d.** 360 **10.** 24, 12, 6, 3, $\frac{3}{2}$ **11.** 4 **12.** 24, −96 **13.** $\frac{2{,}186}{9}$ **14.** $-\frac{85}{8}$ **15.** 125 **16.** $\frac{5}{99}$ **17.** 136 **18. a.** 5,040 **b.** 1 **c.** 20,160 **d.** $\frac{1}{10}$ **19. a.** 1 **b.** 1 **c.** 28 **d.** 84 **e.** 700 **f.** $\frac{7}{4}$ **20.** \$1,638.40 **21.** \$134,509.57 **22.** 12 yr **23.** 1,600 ft **24.** 120 **25.** 720 **26.** 120 **27.** 150

Chapter 11 Test (page 844)

1. 210 **2.** 1 **3.** $-5x^4y$ **4.** $24x^2y^2$ **5.** 66 **6.** 306 **7.** 34, 66 **8.** 3 **9.** −81 **10.** $\frac{364}{27}$ **11.** 18, 108 **12.** $\frac{27}{2}$ **13.** 120 **14.** 40,320 **15.** 15 **16.** 56 **17.** 720 **18.** 322,560 **19.** 24 **20.** $\frac{7}{30}$ **21.** 35 **22.** 30

Cumulative Review Exercises (page 844)

1. (2, 1) **2.** (1, 1) **3.** (2, −2) **4.** (3, 1) **5.** −1 **6.** −1 **7.** (−1, −1, 3) **8.** 1

9.

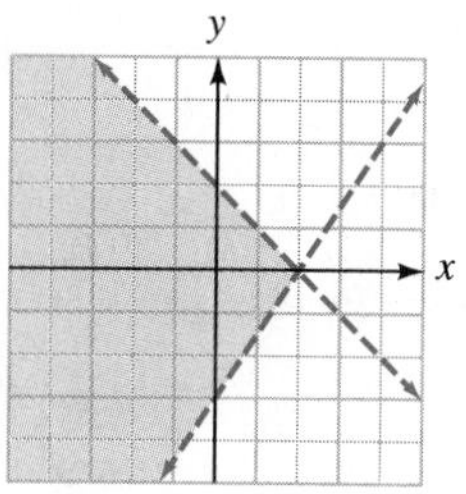

10.

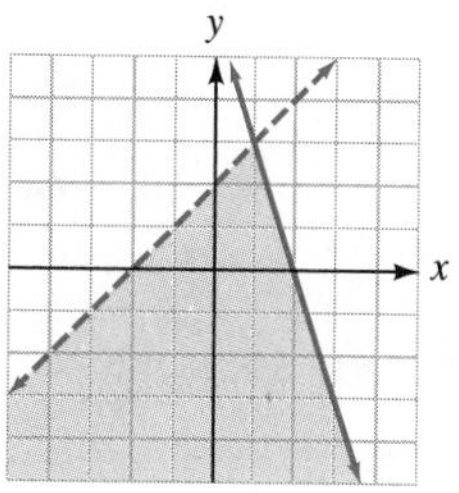

11.

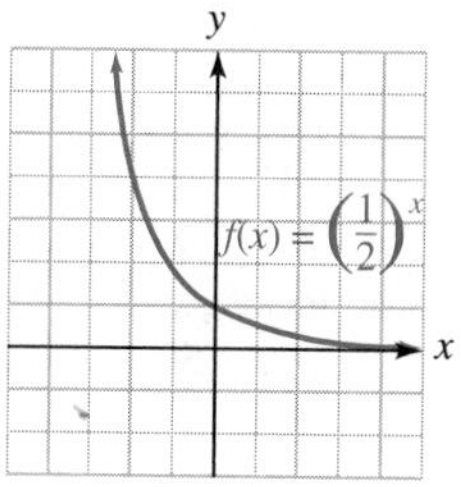

12. $2^y = x$ **13.** 5 **14.** 3 **15.** $\frac{1}{27}$ **16.** 1 **17.** $y = 2^x$ **18.** x **19.** 1.9912 **20.** 0.3010 **21.** 1.6902 **22.** 0.1461 **23.** $\frac{5 \log 2}{\log 3 - \log 2}$ **24.** 8 **25.** \$2,848.31 **26.** 1.16056 **27.** 30,240 **28.** $81a^4 - 108a^3b + 54a^2b^2 - 12ab^3 + b^4$ **29.** $112x^2y^6$ **30.** 103 **31.** 690 **32.** 8 and 19 **33.** 42 **34.** 27 **35.** 27 **36.** $\frac{1{,}023}{64}$ **37.** 12, −48 **38.** $\frac{27}{2}$ **39.** 504 **40.** 35 **41.** $\frac{7}{3}$ **42.** $C(n, n)$ **43.** 5,040 **44.** 84

Exercise I.1 (page A-4)

1. y-axis **3.** origin **5.** y-axis **7.** none **9.** none **11.** x-axis

13.

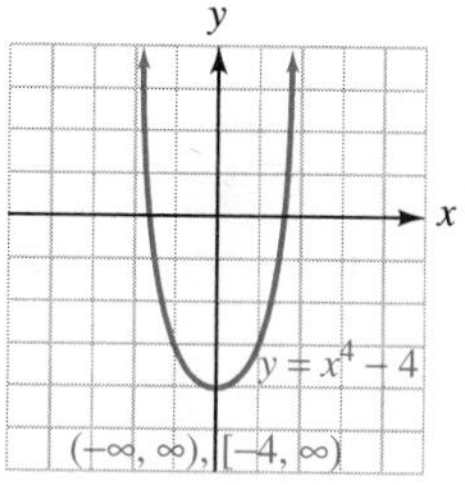

15.

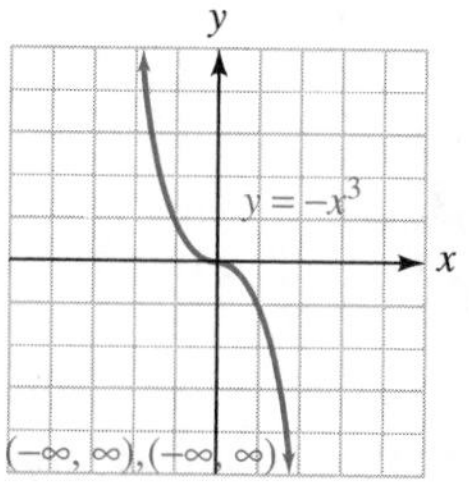

17.

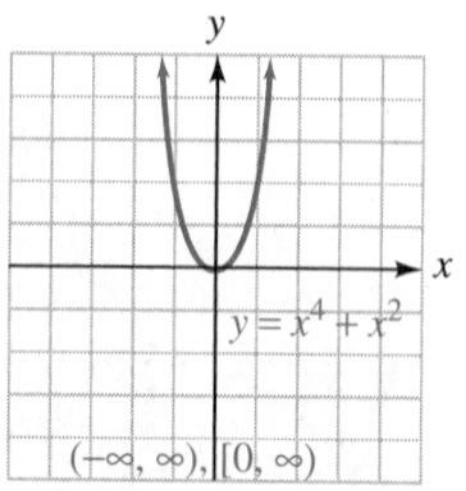

19.

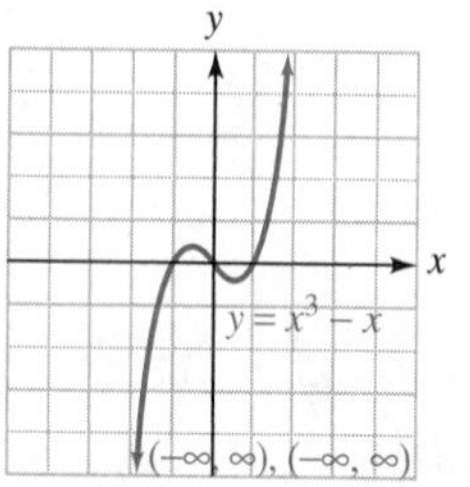

21.

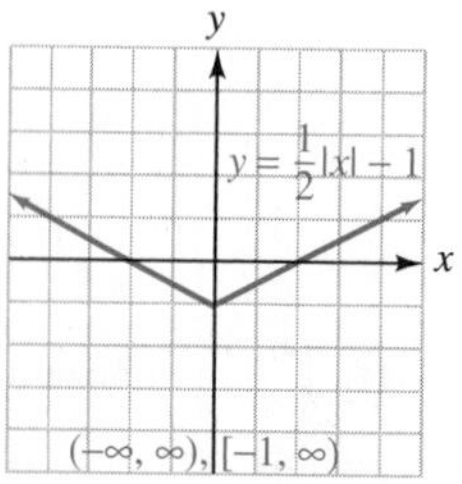

23.

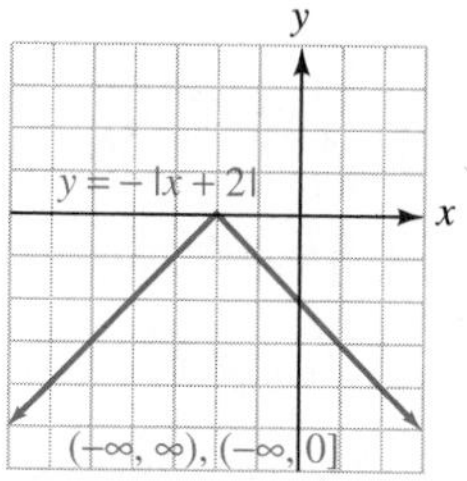

INDEX

CHAPTER 1 BASIC CONCEPTS

Properties of exponents:

If there are no divisions by 0, then

$$x^m x^n = x^{m+n} \qquad (x^m)^n = x^{mn}$$

$$(xy)^n = x^n y^n \qquad \left(\frac{x}{y}\right)^n = \frac{x^n}{y^n}$$

$$x^0 = 1 \qquad x^{-n} = \frac{1}{x^n}$$

$$\frac{x^m}{x^n} = x^{m-n} \qquad \left(\frac{x}{y}\right)^{-n} = \left(\frac{y}{x}\right)^n$$

CHAPTER 2 GRAPHS, EQUATIONS OF LINES, AND FUNCTIONS

Midpoint formula: If $P(x_1, y_1)$ and $Q(x_2, y_2)$, the midpoint of segment PQ is

$$M\left(\frac{x_1 + x_2}{2}, \frac{y_1 + y_2}{2}\right)$$

Slope of a nonvertical line: If $x_1 \neq x_2$, then

$$m = \frac{\Delta y}{\Delta x} = \frac{y_2 - y_1}{x_2 - x_1}$$

Equations of a line:

Point–slope form: $y - y_1 = m(x - x_1)$

Slope–intercept form: $y = mx + b$

General form: $Ax + By = C$

Horizontal line: $y = b$

Vertical line: $x = a$

CHAPTER 3 SYSTEMS OF EQUATIONS

$$\begin{vmatrix} a & b \\ c & d \end{vmatrix} = ad - bc$$

CHAPTER 4 INEQUALITIES

If $k > 0$, then

$|x| = k$ is equivalent to $x = k$ or $x = -k$.

$|a| = |b|$ is equivalent to $a = b$ or $a = -b$.

$|x| < k$ is equivalent to $-k < x < k$.

$|x| > k$ is equivalent to $x < -k$ or $x > k$.

CHAPTER 5 POLYNOMIALS AND POLYNOMIAL FUNCTIONS

Factoring formulas:

Difference of two squares:

$$x^2 - y^2 = (x + y)(x - y)$$

Sum and difference of two cubes:

$$x^3 + y^3 = (x + y)(x^2 - xy + y^2)$$

$$x^3 - y^3 = (x - y)(x^2 + xy + y^2)$$

CHAPTER 6 RATIONAL EXPRESSIONS

Variation: If k is a constant, then

$y = kx$ represents direct variation.

$y = \frac{k}{x}$ represents inverse variation.

$y = kxz$ represents joint variation.

$y = \frac{kx}{z}$ represents combined variation.

CHAPTER 7 RATIONAL EXPONENTS AND RADICALS

The distance formula:

$$d(PQ) = \sqrt{(x_2 - x_1)^2 + (y_2 - y_1)^2}$$

If x, y, and n are real numbers and $x = y$, then $x^n = y^n$.

Fractional exponents: If m and n are positive integers and $x > 0$, then

$$x^{m/n} = \sqrt[n]{x^m} = \left(\sqrt[n]{x}\right)^m \qquad x^{-m/n} = \frac{1}{x^{m/n}}$$

If at least one of a or b is positive, then

$$\sqrt[n]{ab} = \sqrt[n]{a}\sqrt[n]{b} \qquad \sqrt[n]{\frac{a}{b}} = \frac{\sqrt[n]{a}}{\sqrt[n]{b}} \quad (b \neq 0)$$

CHAPTER 8 QUADRATIC FUNCTIONS, INEQUALITIES, AND ALGEBRA OF FUNCTIONS

The quadratic formula:

$$x = \frac{-b \pm \sqrt{b^2 - 4ac}}{2a} \quad (a \neq 0)$$

Complex numbers: If a, b, c, and d are real numbers and $i^2 = -1$, then

$$(a + bi) + (c + di) = (a + c) + (b + d)i$$

$$(a + bi)(c + di) = (ac - bd) + (ad + bc)i$$

$$|a + bi| = \sqrt{a^2 + b^2}$$